Advances in Molecular Genetics and Breeding of Cattle, Sheep, and Goats

Advances in Molecular Genetics and Breeding of Cattle, Sheep, and Goats

Topic Editors

Xiukai Cao

Hui Li

Huitong Zhou

MDPI

Basel • Beijing • Wuhan • Barcelona • Belgrade • Novi Sad • Cluj • Manchester

Topic Editors
Xiukai Cao
Yangzhou University
Yangzhou
China

Hui Li
Guangxi University
Nanning
China

Huitong Zhou
Lincoln University
Lincoln
New Zealand

Editorial Office
MDPI AG
Grosspeteranlage 5
4052 Basel, Switzerland

This is a reprint of the Topic, published open access by the journals *Animals* (ISSN 2076-2615), *Current Issues in Molecular Biology* (ISSN 1467-3045), *Genes* (ISSN 2073-4425) and *International Journal of Molecular Sciences* (ISSN 1422-0067), freely accessible at: https://www.mdpi.com/topics/2VREF2667L.

For citation purposes, cite each article independently as indicated on the article page online and as indicated below:

Lastname, Firstname, Firstname Lastname, and Firstname Lastname. Article Title. *Journal Name* **Year**, *Volume Number*, Page Range.

ISBN 978-3-7258-8051-5 (Hbk)
ISBN 978-3-7258-8052-2 (PDF)
https://doi.org/10.3390/books978-3-7258-8052-2

Contents

Preface

Ruminant livestock—cattle, sheep, and goats—occupy a central place in global agriculture, supplying meat, milk, fiber, and other vital products to hundreds of millions of people across vastly different production environments. Yet the genetic improvement of these species has long been constrained by the complexity of economically important traits, lengthy generation intervals, and the practical limitations of conventional phenotypic selection. The rapid maturation of genomic technologies over the past two decades has opened transformative new possibilities: from high-density SNP arrays and whole-genome resequencing to CRISPR-based genome editing, machine learning-assisted genomic prediction, and integrated multi-omics analyses. This Reprint was assembled in recognition of that transformation and of the urgent need to consolidate its scientific fruits in a single, coherent reference.

The articles gathered here represent the collective efforts of research groups working across diverse breeds, production systems, and geographic contexts. The scope of this Reprint is deliberately broad, spanning genomic selection methodology, functional gene characterization, non-coding RNA biology, epigenomics, reproductive genetics, economic traits, disease resistance, and adaptation to extreme environments. In bringing these themes together, this Reprint aims to offer researchers, animal breeders, and graduate students a comprehensive and current overview of where the field stands and where it is headed.

The motivation for compiling this Reprint arose from the recognition that advances in molecular genetics are increasingly converging with practical breeding decisions. Genomic selection is moving beyond proof-of-concept in elite dairy populations and is now being applied to locally adapted and indigenous breeds whose genetic improvement has historically lagged. Multi-omics approaches are revealing the regulatory architecture that underlies complex traits, while genome-editing technologies are beginning to offer routes to genetic improvement that conventional selection cannot achieve. Documenting and synthesizing these developments in an accessible format serves both the scientific community and the broader livestock sector.

This Reprint is addressed to molecular geneticists, quantitative geneticists, animal scientists, and breeding program managers, as well as to policymakers and students seeking an authoritative entry point into precision ruminant breeding. It is our hope that the research presented here will stimulate further inquiry, foster interdisciplinary collaboration, and ultimately contribute to more productive, resilient, and sustainable ruminant production systems worldwide.

Xiukai Cao, Hui Li, and Huitong Zhou
Topic Editors

animals

MDPI

Editorial

Advances in Molecular Genetics and Breeding of Cattle, Sheep, and Goats

Xiukai Cao

Joint International Research Laboratory of Agriculture and Agri-Product Safety of Ministry of Education of China, Yangzhou University, Yangzhou 225009, China; cxkai0909@163.com

1. Introduction

Ruminant livestock—cattle, sheep, and goats—are cornerstones of global food security, collectively providing meat, milk, fiber, and other essential products that sustain the livelihoods of hundreds of millions of people across diverse agro-ecological zones [1]. As the global human population is expected to approach ten billion, the livestock sector faces an intensifying dual challenge: enhancing productivity and product quality while simultaneously improving resilience to climate change, emerging infectious diseases, and the societal imperative to reduce the environmental footprint of animal agriculture. Traditional breeding programs, which rely on phenotypic selection and pedigree information, have delivered steady but incremental genetic gains; however, they remain constrained by long generation intervals, the high cost of progeny testing, and limited ability to dissect the genetic architecture of complex, polygenic traits.

The genomic era has fundamentally transformed this landscape. The construction of reference genomes for cattle [2], sheep [3], and goats [4] has provided the essential framework for high-resolution genetic analyses, while the subsequent development of high-density single nucleotide polymorphism (SNP) arrays and whole-genome resequencing platforms has enabled genome-wide association studies (GWAS) and genomic selection (GS) at population scale. The landmark proposal by Meuwissen et al. of genomic selection using genome-wide dense marker maps [5] catalyzed a paradigm shift in dairy cattle breeding, which is now extending to sheep, goat, and beef cattle systems [6]. Concurrently, multi-omics technologies—integrating transcriptomics, epigenomics, proteomics, and metabolomics—have opened new windows into the molecular mechanisms connecting genotype to phenotype, highlighting the contributions of non-coding RNAs, epigenetic modifications, and post-translational regulation in shaping complex traits [7].

Genome-editing technologies, particularly CRISPR-Cas systems [8,9], have further expanded the toolkit of animal breeders, enabling the precise introduction or disruption of targeted alleles and offering the prospect of accelerating genetic improvement beyond what conventional selection can achieve. Alongside these biotechnological advances, the adoption of machine learning algorithms for genomic prediction has attracted growing attention, with ensemble methods demonstrating the potential to model non-linear marker effects and improve prediction accuracy for traits of economic importance [10].

Against this backdrop, the Topic "Advances in Molecular Genetics and Breeding of Cattle, Sheep, and Goats" was launched to consolidate cutting-edge research across these intersecting disciplines. The Topic attracted 35 contributions—33 original research articles and two comprehensive reviews—spanning genomic selection, functional gene characterization, non-coding RNA biology, multi-omics analyses, molecular mechanisms of economic traits, adaptive traits, and disease resistance. This editorial provides a thematic overview of these contributions and highlights their collective significance for advancing precision breeding and sustainable ruminant production.

2. An Overview of Published Articles

2.1. Genetic Adaptation to Extreme Environments

Several contributions examined how ruminants have evolved molecular strategies to cope with high-altitude hypoxia, cold stress, and other environmental extremes—challenges of particular relevance to plateau and mountain regions where indigenous breeds must thrive under conditions that impose strong natural selection.

A study on indigenous Yunnan goats distributed along an altitudinal gradient (500–3500 m) identified a non-synonymous SNP (g.86650A>T, p.Gln556Leu) in the *EPAS1* gene—a master regulator of the hypoxia-inducible factor (HIF) pathway—whose frequency increased significantly with elevation. The T allele was associated with elevated red blood cell counts and hemoglobin concentrations, while simultaneously reducing erythrocyte aggregation to maintain blood flow efficiency. Protein structural modeling confirmed the variant's functional significance, providing a molecular basis for high-altitude adaptation in goats analogous to mechanisms documented in Tibetan yak and high-altitude human populations [11]. In yak placenta, the adhesion molecules *CTNNB1* and *CDH1* were shown to regulate trophoblast cell adhesion and tight junction formation across gestational stages under hypoxic conditions, illuminating the maternal-fetal molecular interface essential for successful reproduction at altitude. Cold-stress resilience was addressed in bovine mammary epithelial cells, where ENO1—a glycolytic enzyme—was identified as a critical modulator of apoptosis and energy metabolism under acute cold exposure, offering a candidate marker for maintaining lactation performance in dairy systems subject to seasonal temperature extremes.

2.2. Wool, Fiber, and Coat Quality Genetics

Fiber quality is a paramount economic trait for the sheep and cashmere goat industries, and a substantial cluster of articles advanced the molecular genetics of wool and cashmere fiber production.

The keratin-associated protein (*KRTAP*) gene family, which encodes the matrix proteins that determine fiber ultrastructure and mechanical properties, was examined in multiple contexts. Polymorphisms in *KRTAP13-3* were associated with wool diameter, heterotypic fiber variability, and uniformity in Chinese Tan sheep. A newly identified gene, *KRTAP6-3*, was characterized for the first time in *Capra hircus*; its genetic analysis revealed significant associations with cashmere fiber diameter, identifying a new target for genetic improvement in the luxury fiber market. The *KRT71* keratin gene was examined in Gansu Alpine Fine-Wool Sheep, with KASP genotyping and immunofluorescence demonstrating that its polymorphisms influence wool length and that its expression is localized to the inner root sheath of the hair follicle.

A comprehensive review synthesized current knowledge of wool fiber properties, structure, and the genetic strategies available for sheep improvement, serving as an authoritative reference bridging molecular biology and field-applicable breeding. Complementing the keratin-gene studies, a screening study identified circRNAs associated with secondary wool follicle (SF) development in fine-wool sheep and constructed a competing endogenous RNA (ceRNA) network involving the Wnt and TGF-β pathways, establishing that non-coding RNA-mediated post-transcriptional regulation is an important dimension of SF morphogenesis. The genetics of coat color in local goats of Southwestern China was elucidated by identifying associations between polymorphisms in *ASIP*, *AHCY*, and *ITCH*

https://doi.org/10.3390/ani16081130

genes and diverse pigmentation patterns, providing tools for breed authentication and marker-assisted selection.

2.3. Growth, Carcass, and Meat Quality Traits

Growth performance, carcass composition, and intramuscular fat (IMF) deposition are economically critical polygenic traits, and multiple articles addressed their genetic and molecular basis through complementary approaches.

A GWAS study on Charolais cattle utilizing high-coverage whole-genome resequencing in a cohort of 240 animals identified 196 significant SNPs for growth traits and 29 for carcass traits, pinpointing *ACOX1* as a strong candidate gene influencing backfat thickness and body weight and demonstrating the resolution advantage of whole-genome sequencing compared with chip-based platforms [12]. In goats, missense mutations in the *COLQ* gene were found to affect neuromuscular junction stability with downstream consequences for carcass traits in Leizhou Black Goats. A histological and transcriptomic characterization of postnatal longissimus dorsi development in Ziwuling Black Goats mapped the transition from myofiber proliferation to hypertrophy, thereby providing a timeline for nutritional intervention windows. SNPs in the promoter region of the myogenin (*MyoG*) gene were shown to alter transcription factor binding sites and were significantly associated with body weight and length in Guizhou White Goats, illustrating how regulatory variants can exert quantitative effects on growth traits.

IMF deposition and fatty acid composition have received particular attention through multi-omics approaches. CIDEA was characterized as a key regulator of lipid deposition in goat intramuscular preadipocytes, promoting IMF accumulation via activation of the focal adhesion pathway while simultaneously suppressing cell proliferation. Transcriptome–metabolome integration in three-way crossbred sheep and in lambs identified coordinated regulation of lipid biosynthesis pathways; in lambs, the *MYH7* gene was found to drive a shift toward slow-twitch myofibers that enhances the accumulation of unsaturated fatty acids beneficial to meat nutritional quality. A study on Duroc pigs, included for its comparative insights into ruminant muscle biology, integrated bulk RNA sequencing and LC-MS proteomics across three postnatal muscle types and identified nine critical genes underlying muscle-type heterogeneity, with findings that informed parallel cattle and sheep research.

2.4. Reproductive Performance and Litter Size

Reproductive efficiency directly determines flock and herd profitability, and several articles expanded the catalog of functional genes and non-coding RNAs amenable to selection for prolificacy improvement.

The NR4A1 transcription factor was characterized in the Shaanbei White Cashmere Goat, where promoter polymorphisms affecting transcription factor binding were associated with litter size, implicating *NR4A1* in the regulatory network governing ovarian function. In Hetian sheep, the *ID2* gene was demonstrated to promote granulosa cell proliferation and modulate steroid hormone secretion, with specific SNPs directly linked to increased litter size, providing molecular targets for marker-assisted selection. Whole-genome sequencing of Dorper × Hu hybrid sheep, combined with selective sweep analysis and GWAS, identified candidate genomic regions harboring genes such as *SLC24A2* and *ABCA1* that were significantly associated with high litter size, establishing a genomic reference for improving prolificacy in this widely bred hybrid.

Non-coding RNA regulation of reproduction was addressed by two notable studies. In dairy goat endometrium, circRNA3890 was found to function as a molecular sponge for miR-26b-3p, de-repressing the MDM4/p53 signaling axis to establish endometrial

https://doi.org/10.3390/ani16081130

receptivity—a prerequisite for successful embryo implantation with direct implications for embryo-transfer programs. Proteomic profiling of semen from Duolang sheep rams carrying different FecB genotypes (wild-type, heterozygous BB+, and homozygous BB) via DIA quantitative proteomics revealed that differentially expressed seminal proteins are enriched in innate immune response and oxidative phosphorylation pathways, providing insight into how a major prolificacy gene affects male reproductive biology and sperm quality.

2.5. Milk Production, Genomic Diversity, and Breeding Strategy

Milk production and mammary gland biology were addressed in a small but impactful cluster of studies. An integrated analysis of whole-genome sequencing data from 78 water buffaloes, deploying multiple GWAS models (cBLUP, GMATs, and BayesR), identified *MAPK10* as a candidate gene for 305-day milk yield, offering a molecular entry point for improving dairy buffalo productivity. An integrated transcriptomic study combining data from 77 *Bos taurus* and 77 *Ovis aries* mammary gland samples, analyzed through WGCNA and protein–protein interaction networks, identified *RPS15* and *RPL7A* as robust hub genes regulating species-specific differences in milk quality and lactation performance, underscoring the value of cross-species comparative transcriptomics.

Genetic diversity assessment was addressed in two complementary studies. ROH analysis of the Kazakh Fat-Tailed Coarse-Wool sheep characterized genomic inbreeding levels, identified selection signatures associated with muscle formation (*MYF5*) and fat metabolism (*PRDM16*), and provided population-level insights relevant to conservation and selective breeding [13]. In crossbred cattle populations of the Colombian Caribbean, SNP-based PCA and ROH analyses demonstrated that despite the absence of formal breeding programs, these populations retain high genetic diversity and carry favorable allele frequencies for both milk and meat production—findings with immediate practical implications for sustainable breeding in tropical environments [14].

Genomic selection methodology was advanced by studies comparing ABLUP, GBLUP, and single-step GBLUP (ssGBLUP) for early growth traits in Inner Mongolian cashmere goats, demonstrating that ssGBLUP—which seamlessly blends pedigree and genomic data—outperformed traditional approaches [15]. Extending these methodological advances, machine learning algorithms including Random Forest, Gradient Boosting Decision Tree, XGBoost, and LightGBM were benchmarked against GBLUP for cashmere fiber traits in the same breed; hyperparameter-tuned LightGBM captured non-linear marker interactions and achieved superior prediction accuracy for fiber diameter and cashmere yield [10,16]. Genetic parameter estimation and model optimization for early growth traits in Yunnan semi-fine wool sheep provided a methodological foundation for national-level breeding program optimization in this breed.

2.6. Genome Editing, and Multi-Omics for Complex Traits

A forward-looking cluster of contributions addressed the emerging frontiers of epigenomics, non-coding RNA biology, CRISPR-based genome editing, and spatial multi-omics, collectively pointing toward the next generation of breeding tools [17].

The *MSTN* gene-edited cattle study combined whole-genome bisulfite sequencing (WGBS) and transcriptomics to reveal that loss of *MSTN* not only induces the expected skeletal muscle hypertrophy [18] (with improved shear force and cooking loss metrics) but also unexpectedly reshapes the transcriptomic and DNA methylation landscapes of cardiac and smooth muscle, with axon guidance signaling emerging as a pleiotropically affected pathway. This study highlighted both the transformative potential and the need for systematic phenotypic monitoring when applying gene-editing technologies in livestock. In a parallel disease-resistance application, base editing was used to introduce a

https://doi.org/10.3390/ani16081130

targeted mutation at the 2SP site of the bovine *NRAMP1* promoter; dual luciferase assays confirmed that this endogenous edit—achieved without synthetic selection markers—significantly enhanced *NRAMP1* expression upon Mycobacterium tuberculosis challenge while maintaining baseline levels in uninfected cells, offering a cleaner and safer route to tuberculosis-resistant cattle than transgenic approaches [8,9].

A comprehensive review on innovations in cattle breeding technology surveyed the current state of CRISPR-Cas applications in livestock, critically examining regulatory challenges and advocating for science-based policy frameworks to facilitate the translation of gene-editing discoveries into commercial breeding programs. For complex growth traits in cattle, a multi-omics study linking the blood transcriptome, fecal metabolome, and gut microbiome identified a cross-omics regulatory axis where host genes (*RADIL, ZBTB20*) interact with gut microbes (Oscillospira) to modulate bile acid metabolism and influence average daily gain, exemplifying how host–microbiome interactions can be integrated into genomic models [7]. Spatial multi-omics profiling of the Qianqiu goat gut—combining 16S rRNA sequencing, untargeted LC-MS metabolomics, and WGCNA—revealed a distinctive enrichment of methanogens in the small intestine that shapes host lipid metabolism in a region-specific manner. Metabolomic analysis of cauda epididymal fluid in yaks and cattle identified species-specific differences in antioxidant capacity and energy metabolites relevant to sperm maturation physiology in high-altitude environments.

3. Conclusions

The 35 articles assembled in this Topic collectively chart the frontiers of molecular genetics and breeding in cattle, sheep, and goats, and together they signal a clear paradigm shift: from the reductionist study of individual genes toward holistic, systems-biology approaches that integrate genomics, transcriptomics, epigenomics, proteomics, and metabolomics with advanced computational frameworks including machine learning and spatial multi-omics. GWAS, GS, and multi-omics integration are moving from proof-of-concept in major commercial breeds to practical implementation in locally adapted and indigenous populations, broadening the beneficiaries of genomic improvement.

Several cross-cutting themes deserve emphasis. First, non-coding RNA biology—particularly circRNA—has emerged as a productive research frontier in ruminant genetics, with roles demonstrated here in wool follicle development, endometrial receptivity, and beyond, confirming that the regulatory genome is as important a target as the protein-coding genome. Second, epigenomic approaches, exemplified by WGBS studies on *MSTN*-edited cattle and multi-omics analyses of growth traits, indicate that DNA methylation and chromatin state are tractable layers of biological information that should be incorporated into next-generation breeding models. Third, the application of CRISPR-Cas genome editing to both production traits (*MSTN*) and disease resistance (*NRAMP1*) demonstrates that animal scientists are no longer merely observing natural genetic variation but are becoming active architects of livestock genomes.

Looking ahead, four priorities stand out. First, scaling GS to smallholder systems in developing regions will require cost-effective low-density panels and robust imputation pipelines. Second, integrating microbiome and epigenomic data into breeding indices offers a promising avenue for capturing "missing heritability" and improving resilience to nutritional and environmental variation. Third, CRISPR-based functional validation of candidate genes identified through GWAS and multi-omics—particularly those affecting hypoxia tolerance, endometrial receptivity, and lipid metabolism—will accelerate the translation from discovery to commercial application. Fourth, ethical and regulatory frameworks for genome-edited livestock must evolve in parallel with technical advances to ensure societal acceptance and responsible deployment.

https://doi.org/10.3390/ani16081130

It is our hope that this collection will serve as both a valuable resource for researchers and a catalyst for continued innovation in the molecular genetics and precision breeding of ruminant livestock.

Acknowledgments: X.C. sincerely acknowledges the National Natural Science Foundation of China [32302694], Major Basic Research Project in Universities of Jiangsu Province [25KJA230001], and Yangzhou Municipal Science and Technology Plan Project [YZ2025052].

Conflicts of Interest: The author declares no conflicts of interest.

List of Contributions

1. Cui, J.; Li, L.; Su, Y.; Han, S.; Zhang, L.; Song, Y. Regulatory Effect of circRNA3890 on the Establishment of Receptive Endometrium in Dairy Goats. *Animals* **2026**, *16*, 696. https://doi.org/10.3390/ani16050696.
2. Song, X.; Long, H.; Meng, J.; Zhao, Y.; Wu, Z.; An, Q. Single Nucleotide Polymorphisms in the Promoter Region of *MyoG* Gene Affecting Growth Traits and Transcription Factor Binding Sites in Guizhou White Goat (*Capra hircus*). *Genes* **2026**, *17*, 14. https://doi.org/10.3390/genes17010014.
3. Zhang, Q.; Yang, L.; Li, Y.; Gu, P.; Si, R.; Li, S.; Zhu, L.; Zhang, W. Integrated Transcriptomic Analysis Identifies Core Hub Genes Regulating Mammary Gland Traits (Milk Quality/Lactation) in Dairy Livestock: *Bos taurus* and *Ovis aries*. *Genes* **2025**, *16*, 1483. https://doi.org/10.3390/genes16121483.
4. Guo, P.; Qin, W.; Song, W.; Chen, H. Spatial Multi-Omics Analysis of the Qianqiu Goat Gut Microbiome and Metabolome. *Int. J. Mol. Sci.* **2025**, *26*, 11815. https://doi.org/10.3390/ijms262411815.
5. Qiao, L.; Ma, K.; Yao, Q.; Zhang, S.; Pang, Z.; Wang, W.; Cai, K.; Liu, W. Whole-Genome Sequencing of Dorper × Hu Hybrid Sheep for Screening Selection Signatures Associated with Litter Size. *Animals* **2025**, *15*, 3505. https://doi.org/10.3390/ani15233505.
6. Zhang, F.; Wang, C.; Shangguan, A.; Suo, X.; Chen, M.; Tao, H.; Jiang, F.; Xu, T.; Zhang, N.; Hua, Z.; et al. Genome-Wide Association Studies of Growth and Carcass Traits in Charolais Cattle Based on High-Coverage Whole-Genome Resequencing. *Int. J. Mol. Sci.* **2025**, *26*, 11411. https://doi.org/10.3390/ijms262311411.
7. Sun, H.; Wang, X.; Zhu, L.; Shahbaz, G.; Gu, R.; Huang, Q.; Li, W.; Xing, F. Study on the Function of *ID2* Gene in Granulosa Cells of Ovaries of Hetian Sheep and Its Correlation Analysis with Lambing Traits. *Animals* **2025**, *15*, 3271. https://doi.org/10.3390/ani15223271.
8. Guo, H.; Shi, F.; Gu, L.; Wang, Y.; Yue, Y.; Huang, W.; Yang, Y.; Sun, P.; Xue, W.; Zhang, X.; et al. Study on the Development and Formation Specifics of Longissimus Dorsi Muscles in Ziwuling Black Goats. *Animals* **2025**, *15*, 3265. https://doi.org/10.3390/ani15223265.
9. Zhang, Y.; Niu, Z.; Yan, J.; Chen, Y.; Xue, Z.; Xu, J.; Ma, Y.; Shi, H. Analysis of Semen Proteomic Differences Among Three Genotypes of FecB Rams in Duolang Sheep. *Genes* **2025**, *16*, 1226. https://doi.org/10.3390/genes16101226.
10. Liu, J.; Yan, X.; Li, W.; Xue, S.; Wang, Z.; Su, R. Genomic Selection for Cashmere Traits in Inner Mongolian Cashmere Goats Using Random Forest, Gradient Boosting Decision Tree, Extreme Gradient Boosting and Light Gradient Boosting Machine Methods. *Animals* **2025**, *15*, 2940. https://doi.org/10.3390/ani15202940.
11. Yang, G.; Yang, X.; Liu, D.; Zhou, W.; Zhou, A.; Xiong, Y.; Xiong, X.; Fu, W.; Li, J.; Lan, D.; et al. Metabolomic Analysis of Cauda Epididymal Fluid in Yaks and Cattle. *Animals* **2025**, *15*, 2861. https://doi.org/10.3390/ani15192861.
12. Zhou, H.; Bai, L.; Li, S.; Wang, J.; Hickford, J. Wool: From Properties and Structure to Genetic Insights and Sheep Improvement Strategies. *Animals* **2025**, *15*, 2790. https://doi.org/10.3390/ani15192790.
13. Kozhakhmet, A.; Akhatayeva, Z.; Dossybayev, K.; Yermekova, M.; Kapassuly, T.; Yergali, K.; Torekhanov, A.; Bissenov, U.; Lan, X.; Kulataev, B. Genomic Characterization of the Kazakh Fat-Tailed Coarse-Wool Sheep Breed Using ROH Analysis. *Animals* **2025**, *15*, 2714. https://doi.org/10.3390/ani15182714.

https://doi.org/10.3390/ani16081130

14. Wang, K.; Li, X.; Liu, X.; Liufu, S.; Xiao, L.; Chen, B.; Chen, W.; Jiang, J.; Liu, Y.; Ma, H. Multi-Omics Insights into Postnatal Skeletal Muscle Development in Duroc Pigs. *Animals* **2025**, *15*, 2715. https://doi.org/10.3390/ani15182715.
15. Luo, Y.; Xi, B.; Song, Y.; Xiao, T.; Lu, Z.; Liu, J.; Yuan, C.; Guo, T. Screening of circRNAs Associated with Secondary Wool Follicle Development in Fine-Wool Sheep and Construction of Their ceRNA Network. *Animals* **2025**, *15*, 2629. https://doi.org/10.3390/ani15172629.
16. Huang, J.; Wang, K.; Zhang, Y.; Han, J.; Zhou, H.; Jiang, Q. Effects of *COLQ* Gene Missense Mutations on Growth and Meat Traits in Leizhou Black Goats. *Animals* **2025**, *15*, 2618. https://doi.org/10.3390/ani15172618.
17. Han, X.; Liu, X.; Fu, Y.; Chen, J.; Lai, C.; Yang, X.; Shan, X.; Chen, Y.; Jiang, H. Molecular Insights into Intramuscular Unsaturated Fatty Acid Deposition in Lambs Through Multi-Omics Profiling. *Animals* **2025**, *15*, 2617. https://doi.org/10.3390/ani15172617.
18. Cao, X.; Ru, W.; Cheng, J.; Sun, L.; Zhang, N.; Zhaxi, L.; Dunzhu, R.; Sun, F.; Yang, K.; Gao, Y.; et al. Transcriptome and Metabolome Profiles Reveal the Underlying Mechanism of Fat Deposition Changes in Three-Way Crossbred Yak for High-Quality Beef Production. *Animals* **2025**, *15*, 2599. https://doi.org/10.3390/ani15172599.
19. Li, W.; Chen, H.; Cao, D.; Yang, X. Identification of MAPK10 as a Candidate Gene for High Milk Production in Water Buffaloes Through a Genome-Wide Association Study. *Animals* **2025**, *15*, 2567. https://doi.org/10.3390/ani15172567.
20. Shen, N.; Wang, J.; Liao, J.; Yu, H.; Sun, W.; Jia, X.; Lai, S. ENO1 Regulates Apoptosis Induced by Acute Cold Stress in Bovine Mammary Epithelial Cells. *Animals* **2025**, *15*, 2559. https://doi.org/10.3390/ani15172559.
21. Shao, P.; Li, Q.; Liao, Y.; Wang, Y.; Lin, Y.; Xiang, H.; Du, Z.; Zhang, C.; Zhu, J.; Huang, L. Activation of Focal Adhesion Pathway by CIDEA as Key Regulatory Axis in Lipid Deposition in Goat Intramuscular Precursor Adipocytes. *Animals* **2025**, *15*, 2374. https://doi.org/10.3390/ani15162374.
22. Zhao, F.; He, Z.; Sun, H.; Wang, J.; Liu, X.; Hao, Z.; Li, M.; Li, S. Polymorphism of Keratin Gene *KRT71* and Its Relationship with Wool Properties in Gansu Alpine Fine-Wool Sheep. *Animals* **2025**, *15*, 2028. https://doi.org/10.3390/ani15142028.
23. Zhang, L.; Zhao, S.; Yu, H.; Duan, Y.; Zhang, J.; Xu, N.; Zhao, Y. ASIP, AHCY and ITCH Genes Are Associated with the Coat Color of Local Goats (*Capra hircus*) of Southwestern China. *Animals* **2025**, *15*, 1849. https://doi.org/10.3390/ani15131849.
24. Cao, J.; Chen, Z.; Zhang, J.; Cao, L.; Li, S. The Identification of a New Gene *KRTAP 6-3* in *Capra hircus* and Its Potential for the Diameter Improvement of Cashmere Fibers. *Genes* **2025**, *16*, 721. https://doi.org/10.3390/genes16060721.
25. Zhang, T.; Gao, L.; Zhou, B.; Xu, Q.; Liu, Y.; Li, J.; Lv, Q.; Zhang, Y.; Wang, R.; Su, R.; et al. Genomic Selection for Early Growth Traits in Inner Mongolian Cashmere Goats Using ABLUP, GBLUP, and ssGBLUP Methods. *Animals* **2025**, *15*, 1733. https://doi.org/10.3390/ani15121733.
26. Areb, E.; Bi, Y.; Bai, Y.; Zhu, Q.; Ma, L.; Pan, C.; Chen, X.; Lan, X. Functional and Genetic Insights into the Role of the *NR4A1* Gene in the Litter Size of the Shaanbei White Cashmere Goat. *Animals* **2025**, *15*, 1729. https://doi.org/10.3390/ani15121729.
27. Hai, C.; Liu, X.; Bai, C.; Su, G.; Yang, L.; Li, G. Loss of Myostatin Shapes the Transcriptomic and Epigenetic Landscapes Across Multiple Muscle Types in Cattle. *Curr. Issues Mol. Biol.* **2025**, *47*, 431. https://doi.org/10.3390/cimb47060431.
28. Rodríguez-Serrano, A.; Ahumada-Velasco, M.; Cárdenas Beltrán, J. Assessment of Genetic Diversity and Productive Traits in Crossbreed Cattle in the Caribbean Region, Colombia. *Genes* **2025**, *16*, 677. https://doi.org/10.3390/genes16060677.
29. Wang, Y.; Yang, H.; Zhao, X.; Ni, X.; Zhao, Y.; You, Z.; Lu, Q.; Tang, S.; Quan, G.; Fu, X. Genetic Evaluation of Early Growth Traits in Yunnan Semi-Fine Wool Sheep. *Animals* **2025**, *15*, 1512. https://doi.org/10.3390/ani15111512.
30. Wang, Y.; Cui, X.; Chen, Z. Innovations in Cattle Breeding Technology: Prospects in the Era of Gene Editing. *Animals* **2025**, *15*, 1364. https://doi.org/10.3390/ani15101364.

31. Wei, Y.; Yuan, M.; Zhang, Y.; Gao, Y. The 2SP Site Mutation in the Bovine Natural Resistance-Associated Macrophage 1 Promoter Exhibits Antituberculosis Potential. *Int. J. Mol. Sci.* **2025**, *26*, 4229. https://doi.org/10.3390/ijms26094229.
32. Bai, L.; Zhou, H.; Tao, J.; Hickford, J. Variation in Ovine KRTAP13-3 and Its Association with Wool Characteristics in Chinese Tan Sheep. *Animals* **2025**, *15*, 1069. https://doi.org/10.3390/ani15071069.
33. Zhang, B.; Song, C.; Zhou, B.; Zhang, J.; Dong, W.; Zhang, Y.; Zhao, X.; Zhang, Q. CTNNB1 and CDH1 Regulate Trophoblast Cell Adhesion and Junction Formation in Yak Placental Tissue at Different Gestational Stages. *Animals* **2025**, *15*, 876. https://doi.org/10.3390/ani15060876.
34. Gu, M.; Jiang, H.; Ma, F.; Li, S.; Guo, Y.; Zhu, L.; Shi, C.; Na, R.; Wang, Y.; Zhang, W. Multi-Omics Analysis Revealed the Molecular Mechanisms Affecting Average Daily Gain in Cattle. *Int. J. Mol. Sci.* **2025**, *26*, 2343. https://doi.org/10.3390/ijms26052343.
35. Zhu, L.; Tang, L.; Zhao, Y.; Li, S.; Gou, X.; Deng, W.; Kong, X. EPAS1 Variations and Hematological Adaptations to High-Altitude Hypoxia in Indigenous Goats in Yunnan Province, China. *Animals* **2025**, *15*, 695. https://doi.org/10.3390/ani15050695.

References

1. Scherf, B.; Pilling, D. *The Second Report on the State of the World's Animal Genetic Resources for Food and Agriculture*; FAO: Rome, Italy, 2015.
2. Sequencing, B.G.; Consortium, A.; Elsik, C.G.; Tellam, R.L.; Worley, K.C.; Gibbs, R.A.; Muzny, D.M.; Weinstock, G.M.; Adelson, D.L.; Eichler, E.E.; et al. The genome sequence of taurine cattle: A window to ruminant biology and evolution. *Science* **2009**, *324*, 522–528. [CrossRef]
3. Jiang, Y.; Xie, M.; Chen, W.; Talbot, R.; Maddox, J.F.; Faraut, T.; Wu, C.; Muzny, D.M.; Li, Y.; Zhang, W. The sheep genome illuminates biology of the rumen and lipid metabolism. *Science* **2014**, *344*, 1168–1173. [CrossRef]
4. Dong, Y.; Xie, M.; Jiang, Y.; Xiao, N.; Du, X.; Zhang, W.; Tosser-Klopp, G.; Wang, J.; Yang, S.; Liang, J. Sequencing and automated whole-genome optical mapping of the genome of a domestic goat (*Capra hircus*). *Nat. Biotechnol.* **2013**, *31*, 135–141. [PubMed]
5. Meuwissen, T.H.; Hayes, B.J.; Goddard, M. Prediction of total genetic value using genome-wide dense marker maps. *Genetics* **2001**, *157*, 1819–1829. [CrossRef] [PubMed]
6. Georges, M.; Charlier, C.; Hayes, B. Harnessing genomic information for livestock improvement. *Nat. Rev. Genet.* **2019**, *20*, 135–156. [CrossRef] [PubMed]
7. Clark, E.L.; Archibald, A.L.; Daetwyler, H.D.; Groenen, M.A.; Harrison, P.W.; Houston, R.D.; Kühn, C.; Lien, S.; Macqueen, D.J.; Reecy, J.M. From FAANG to fork: Application of highly annotated genomes to improve farmed animal production. *Genome Biol.* **2020**, *21*, 285. [CrossRef] [PubMed]
8. Cong, L.; Ran, F.A.; Cox, D.; Lin, S.; Barretto, R.; Habib, N.; Hsu, P.D.; Wu, X.; Jiang, W.; Marraffini, L.A. Multiplex genome engineering using CRISPR/Cas systems. *Science* **2013**, *339*, 819–823. [CrossRef] [PubMed]
9. Doudna, J.A.; Charpentier, E. The new frontier of genome engineering with CRISPR-Cas9. *Science* **2014**, *346*, 1258096. [CrossRef] [PubMed]
10. Montesinos-López, O.A.; Montesinos-López, A.; Pérez-Rodríguez, P.; Barrón-López, J.A.; Martini, J.W.; Fajardo-Flores, S.B.; Gaytan-Lugo, L.S.; Santana-Mancilla, P.C.; Crossa, J. A review of deep learning applications for genomic selection. *BMC Genom.* **2021**, *22*, 19. [CrossRef] [PubMed]
11. Yi, X.; Liang, Y.; Huerta-Sanchez, E.; Jin, X.; Cuo, Z.X.P.; Pool, J.E.; Xu, X.; Jiang, H.; Vinckenbosch, N.; Korneliussen, T.S. Sequencing of 50 human exomes reveals adaptation to high altitude. *Science* **2010**, *329*, 75–78. [CrossRef] [PubMed]
12. Daetwyler, H.D.; Capitan, A.; Pausch, H.; Stothard, P.; Van Binsbergen, R.; Brøndum, R.F.; Liao, X.; Djari, A.; Rodriguez, S.C.; Grohs, C. Whole-genome sequencing of 234 bulls facilitates mapping of monogenic and complex traits in cattle. *Nat. Genet.* **2014**, *46*, 858–865. [CrossRef] [PubMed]
13. Kijas, J.W.; Lenstra, J.A.; Hayes, B.; Boitard, S.; Porto Neto, L.R.; San Cristobal, M.; Servin, B.; McCulloch, R.; Whan, V.; Gietzen, K. Genome-wide analysis of the world's sheep breeds reveals high levels of historic mixture and strong recent selection. *PLoS Biol.* **2012**, *10*, e1001258. [CrossRef] [PubMed]
14. Burrow, H.M.; Mrode, R.; Mwai, A.O.; Coffey, M.P.; Hayes, B.J. Challenges and opportunities in applying genomic selection to ruminants owned by smallholder farmers. *Agriculture* **2021**, *11*, 1172. [CrossRef]
15. VanRaden, P.M. Efficient methods to compute genomic predictions. *J. Dairy Sci.* **2008**, *91*, 4414–4423. [CrossRef] [PubMed]
16. Hayes, B.J.; Lewin, H.A.; Goddard, M.E. The future of livestock breeding: Genomic selection for efficiency, reduced emissions intensity, and adaptation. *Trends Genet.* **2013**, *29*, 206–214. [CrossRef] [PubMed]

17. Rexroad, C.; Vallet, J.; Matukumalli, L.K.; Reecy, J.; Bickhart, D.; Blackburn, H.; Boggess, M.; Cheng, H.; Clutter, A.; Cockett, N. Genome to phenome: Improving animal health, production, and well-being—A new USDA blueprint for animal genome research 2018–2027. *Front. Genet.* **2019**, *10*, 327. [CrossRef] [PubMed]
18. Grobet, L.; Royo Martin, L.J.; Poncelet, D.; Pirottin, D.; Brouwers, B.; Riquet, J.; Schoeberlein, A.; Dunner, S.; Ménissier, F.; Massabanda, J. A deletion in the bovine myostatin gene causes the double–muscled phenotype in cattle. *Nat. Genet.* **1997**, *17*, 71–74. [CrossRef] [PubMed]

https://doi.org/10.3390/ani16081130

animals

MDPI

Article

EPAS1 Variations and Hematological Adaptations to High-Altitude Hypoxia in Indigenous Goats in Yunnan Province, China

Li Zhu [1,†], Lin Tang [1,†], Yunong Zhao [2], Shanshan Li [2], Xiao Gou [2], Weidong Deng [1,*] and Xiaoyan Kong [1,*]

[1] Faculty of Animal Science and Technology, Yunnan Agricultural University, Kunming 650201, China; zhuli18328815855@163.com (L.Z.); zero--xc@163.com (L.T.)
[2] School of Animal Science and Technology, Foshan University, Foshan 528231, China; lightrainooo@163.com (Y.Z.); 15036126218@163.com (S.L.); gouxiaosa@163.com (X.G.)
* Correspondence: dengwd@ynau.edu.cn (W.D.); 2020007@ynau.edu.cn (X.K.)
† These authors contributed equally to this work.

Simple Summary: Tibetan goats, native to high-altitude regions, have developed unique physiological and genetic adaptations to survive under low oxygen conditions. This study investigated the Endothelial PAS Domain Protein1 (*EPAS1*) gene, a critical regulator of hypoxia response, using blood samples from goats across elevations in Yunnan (500–3500 m). A key polymorphism (*g.86650 A>T*, p.Gln556Leu) was identified, with its frequency increasing at higher altitudes. This polymorphism was associated with enhanced oxygen transport (e.g., increased red blood cell count and hemoglobin concentration) while maintaining blood flow efficiency by reducing erythrocyte aggregation. Structural modeling confirmed its functional significance in hypoxia adaptation, offering valuable insights into genetic mechanisms underlying high-altitude survival.

Abstract: The *EPAS1* gene plays a central role in hypoxia adaptation in high-altitude animals. Using over 400 blood samples from goats across elevations in Yunnan (500–3500 m), this study examined hematological traits, genetic polymorphisms, and protein structure. Red blood cell (RBC) and hemoglobin (HGB) levels increased significantly with altitude ($p < 0.05$), reflecting improved oxygen transport. A non-synonymous SNP (*g.86650 A>T*, p.Gln556Leu) exhibited adaptive selection, with the *T* allele frequency rising at higher altitudes ($p < 0.05$). At 2500 m, *TT* genotype goats showed significantly higher RBC and HGB levels than AA genotypes ($p < 0.05$). Protein modeling revealed structural instability caused by the polymorphism, highlighting its role in enhancing hypoxia adaptation. These findings provide a foundation for improving high-altitude livestock genetics.

Keywords: Tibetan goat; *EPAS1* gene; genetic polymorphism; hematological traits; hypoxia adaptation

1. Introduction

High altitudes exceeding 2500 m are defined as plateau regions, representing the critical elevation at which arterial oxygen saturation (SaO_2) declines significantly [1]. These regions are characterized by extreme environmental factors, including low temperature, low humidity, intense ultraviolet radiation, low atmospheric pressure, and hypoxia, which present formidable challenges to the survival and reproduction of flora and fauna [2–4]. Among these factors, hypoxia is regarded as the primary environmental stressor, directly limiting physiological processes related to oxygen transport and utilization. For instance, at an altitude of 3700 m, oxygen partial pressure is only 63% of that at sea level [5], often

https://doi.org/10.3390/ani15050695

leading to hypoxemia [6]. Investigating the mechanisms underlying animal adaptation to such extreme environments not only unveils the principles of molecular evolution but also provides essential theoretical guidance for the development of plateau animal husbandry.

In recent years, studies on plateau adaptability have identified multiple candidate genes, with *EPAS1*, Egl Nine Homolog 1 (*EGLN1*), and Protein Kinase AMP-Activated Catalytic Subunit Alpha 1 (*PRKAA1*) playing pivotal roles in the hypoxia-inducible factor (HIF) pathway [7–14]. The HIF pathway serves as the central regulatory mechanism in hypoxic environments, mediating erythropoiesis, angiogenesis, and metabolic processes to help organisms maintain oxygen homeostasis [15,16]. In both high-altitude human populations [10,17–20] and animals (e.g., yak, Tibetan sheep, and Tibetan goats) [11,21–24], the *EPAS1* gene has been confirmed as a key gene exhibiting strong signals of selection for plateau adaptation. *EPAS1* encodes HIF-2α, which regulates hemoglobin synthesis (HGB), oxygen transport, and energy metabolism, playing a vital role in high-altitude adaptation [25,26]. Non-synonymous single nucleotide polymorphisms (SNPs), such as H194R, have been significantly associated with relatively lower hematocrit in Andean highlanders [19].

Tibetan goats, traditional livestock in the Tibetan Plateau [27], inhabit regions above 4000 m, displaying enhanced cardiopulmonary function, oxygen-carrying capacity, and metabolic efficiency [28]. For example, Song et al. [29] demonstrated that the Q579L variation of the *EPAS1* gene significantly influences mean corpuscular hemoglobin concentration (MCHC) in Tibetan cashmere goats. In Nepalese goats, the frequency of the T allele at the Q579L variation increases significantly with altitude [30], suggesting a strong association between this variation and high-altitude adaptation.

The aim of this study was to detect and analyze *EPAS1* gene polymorphism in Yunnan goats distributed along a continuous altitudinal gradient and to explore its potential association with blood physiological parameters, including hematocrit level, hemoglobin concentration, and oxygen saturation. Moreover, protein structure prediction was conducted, focusing on the potential impact of the analyzed variant on the EPAS1 protein. These comprehensive analyses provide new theoretical insights into the mechanisms of high-altitude adaptation in Yunnan goats and offer valuable scientific support for the genetic improvement and sustainable development of plateau animal husbandry.

2. Materials and Methods

2.1. Selection of Experimental Animals and Measurement of Hematological Parameters

A total of 407 blood samples from local indigenous Yunnan goats (healthy and adult) distributed along a continuous altitudinal gradient were collected for this study. All blood samples were collected from the jugular vein following standard veterinary procedures. This study adhered to animal welfare guidelines and was approved by the relevant ethical committee, ensuring that all animal handling and sample collection followed the principles of humane treatment and minimal distress. The samples were obtained from Yunnan Province, covering Honghe County (HH, 500 m, n = 61), Huize County (HZ, 1500 m, n = 109), Dali Prefecture (DL, 2500 m, n = 125), and Diqing Prefecture (DQ, 3800 m, n = 112). Detailed sample information is provided in Figure 1. Blood physiological parameters were measured using an automated veterinary hematology analyzer (Mindray, BC-30Vet, Shenzhen, China), which included red blood cell (RBC), hemoglobin concentration (HGB), hematocrit level (HCT), mean corpuscular volume (MCV), mean corpuscular hemoglobin concentration (MCHC), mean corpuscular hemoglobin (MCH), and red cell distribution width coefficient of variation (RDW). A total of seven parameters were assessed. Additionally, blood and plasma viscosity parameters were measured using a fully automated blood rheometer (Zhongchi, ZL9000, Beijing, China), combined with a cone-plate whole-blood testing system and a capillary plasma testing system. These included fibrinogen (FB), plasma viscosity (PV), whole-blood low-shear relative index (WLS), whole-blood medium-shear relative index (WMS), whole-blood high-shear relative index (WHS), erythrocyte aggregation index (EI), erythrocyte aggregation coefficient (EC), Casson viscosity (CV), red blood cell intraviscosity (RBCIV), low-shear flow resistance (LSFR), medium-shear flow resistance (MSFR), high-shear flow resistance (HSFR), and yield stress (YS), covering a total of 13 parameters.

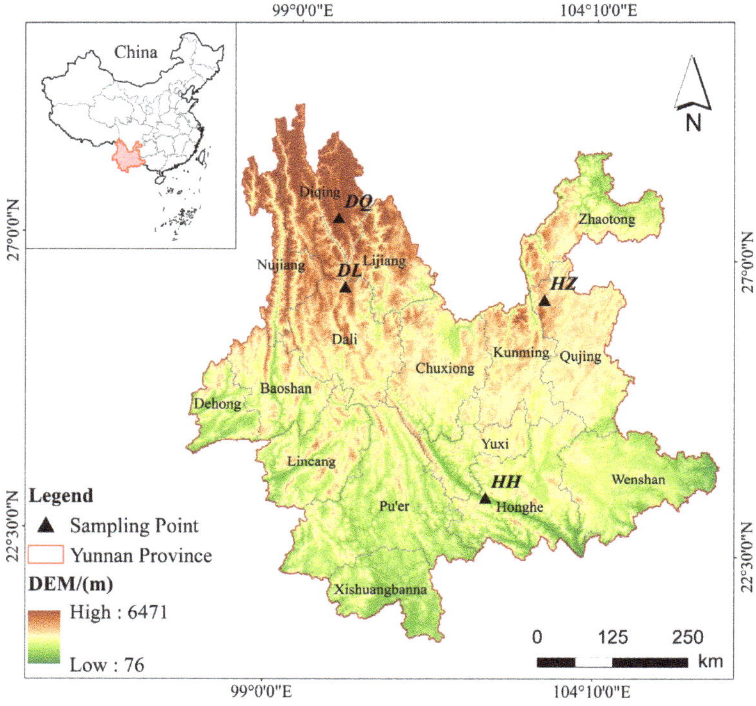

Figure 1. Geographic distribution and sample information of local goats at different altitudes in Yunnan Province.

To identify polymorphic variation in the goat genome, DNA was extracted from 10 individuals of both local Honghe goats (HHs) and Tibetan goats (DQs). The core region of the *EPAS1* gene (including all exonic regions) was amplified and sequenced. Subsequently, genomic DNA was extracted from an expanded set of 229 samples to further validate the identified core SNP variation. These included HHs (n = 58), HZs (n = 55), DLs (n = 35), and DQs (n = 81), ensuring a representative sampling across different altitudinal gradients. The amplified and genotyped SNP loci were then analyzed to explore their potential associations with blood physiological phenotypes. The samples used in this part of the experiment were all selected from the aforementioned population.

2.2. DNA Extraction from Blood Samples

DNA extraction from blood samples was performed strictly according to the instructions provided in the Blood Genomic DNA Extraction Kit (Tiangen, Beijing, China). Approximately 50 µL of high-concentration DNA solution was extracted from each sample. DNA concentration was measured using a NanoDrop 2000 spectrophotometer (Thermo Fisher Scientific, Waltham, MA, USA), and DNA integrity was assessed by 1% agarose gel electrophoresis to ensure it met the requirements for subsequent gene amplification and sequencing analysis.

2.3. Primer Design and PCR Amplification for the EPAS1 Gene

Goat gene sequences for *EPAS1* (Locus: NC_022303.1) were downloaded from the GenBank database of NCBI, and primers for the entire gene region were designed using Premier 5.0 software. All primers were synthesized by TsingKe Biological Technology (Beijing, China); their features are shown in Figure 2.

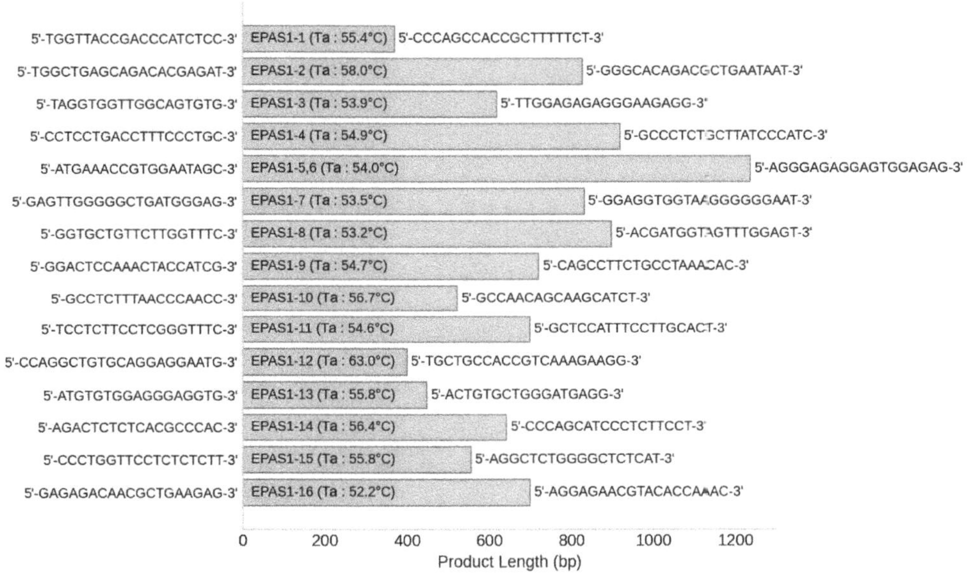

Figure 2. Primer information for the core region of the *EPAS1* gene.

Initially, 20 individuals (10 from Honghe County [HH] and 10 from Diqing Prefecture [DQ]) were sequenced to identify potential SNPs within the *EPAS1* gene. Based on these results, one SNP was selected for further genotyping in the remaining samples. Genotyping was performed by sequencing, using specific primers (including EPAS1-12) for PCR amplification.

The PCR reactions were carried out in a total volume of 25 μL, consisting of 2 μL genomic DNA, 1 μL of each forward and reverse primer, 12.5 μL of 2 × Taq Master Mix (TsingKe Biological Technology, Beijin, China), and 8.5 μL ddH_2O. The PCR thermal cycling program used is as follows: initial denaturation at 95 °C for 6 min, followed by 35 cycles, each consisting of denaturation at 94 °C for 50 s, annealing at the temperature specified in Figure 2 for 30 s, and extension at 72 °C for 50 s; the final extension step was performed at 72 °C for 8 min. PCR products were sequenced using the ABI 3730 DNA sequencer. The sequences were edited using Chromas (version 2.6.6) and aligned with MEGA 5.0 software to identify SNPs.

2.4. Statistical Analysis

Allele and genotype frequencies were compared using the chi-square test.

To evaluate the physiological and rheological indices of adult goats at different altitudes at the Yunnan–Tibet line and to perform genotype–phenotype association analysis, a two-factor least square analysis without interaction effects was conducted with the following model:

$$Y_{ijk} = \mu + H_i/G_i + S_j + e_{ijk}$$

where Y_{ikj} is the observed trait value, μ is the population mean, H_i is the altitude effect, G_i is the genotype effect, S_j is the sex effect, and e_{ijk} is the random error, compliant with the $N(0, \sigma_e^2)$ distribution.

The GLM (general linear models) procedure in SAS (version 9.0) was used to calculate least squares means, with results expressed as "mean ± SE." Statistical significance was set at $p < 0.05$ and $p < 0.01$.

2.5. 3D Structure Analysis

The functional impact of the Gln 556-Leu (Q556L) polymorphism was predicted using PolyPhen-3 [31]. Five independent predictions were performed for each polymorphism to ensure consistency and reliability. The structure with the highest pTM score was selected for further analysis. Polymorphism site interaction analysis was conducted using PyMOL (Schrödinger, LLC, New York, NY, USA, 2020, version 2.5.6). Hydrogen bonds and electrostatic interactions were examined to assess potential structural and functional alterations. Figures were generated in PyMOL and further annotated and rendered using Adobe Illustrator (Adobe Inc., San Jose, CA, USA, 2021, version 25.1).

3. Results

3.1. Physiological Adaptation of Yunnan Goats at Different Altitudes

To investigate the high-altitude physiological adaptation phenotypes of Yunnan goats, this study systematically measured blood physiological and rheological parameters of goats distributed across altitudes ranging from 500 m to 3500 m in Yunnan Province (Table 1). The results indicated significant changes in blood parameters with increasing altitude as atmospheric oxygen levels decreased ($p < 0.01$).

Table 1. Hematological and hemorheological indicators of goats at different altitude gradients.

	DQ (n = 81) 3500 m	**DL (n = 35) 2500 m**	**HZ (n = 55) 1500 m**	**HH (n = 58) 500 m**
Red Blood Cell (RBC, 10^{12}/L)	17.40 ± 2.25^{A}	15.23 ± 2.34^{B}	13.98 ± 2.48^{C}	11.97 ± 2.12^{D}
Hemoglobin Concentration (HGB, g/L)	101.01 ± 9.47^{A}	93.18 ± 14.90^{B}	76.87 ± 13.38^{C}	72.10 ± 12.49^{D}
Hematocrit (HCT, %)	30.02 ± 3.11^{A}	27.33 ± 3.76^{B}	25.14 ± 4.19^{C}	19.58 ± 3.23^{D}
Mean Corpuscular Volume (MCV, fL)	17.34 ± 1.85^{C}	18.14 ± 1.71^{A}	18.07 ± 1.79^{AB}	16.50 ± 1.51^{D}
Mean Corpuscular Hemoglobin (MCH, pg)	5.77 ± 0.54^{C}	6.13 ± 0.46^{A}	5.44 ± 0.41^{D}	5.99 ± 0.43^{B}
Mean Corpuscular Hemoglobin Concentration (MCHC, g/L)	335.44 ± 17.80^{B}	340.85 ± 22.78^{B}	305.00 ± 14.62^{C}	366.63 ± 12.51^{A}
Red Cell Distribution Width (RDW, %)	20.57 ± 1.84^{A}	20.42 ± 1.58^{A}	20.23 ± 1.71^{A}	20.04 ± 1.28^{A}
Plasma Viscosity (PV, mPa·s)	1.64 ± 0.24^{C}	2.24 ± 0.45^{A}	1.83 ± 0.30^{B}	1.52 ± 0.29^{C}
Fibrinogen (FB, g/L)	3.66 ± 0.39^{C}	4.92 ± 1.00^{A}	4.03 ± 0.65^{B}	3.34 ± 0.63^{D}
Whole-Blood Low-Shear Relative Index (WLS)	6.88 ± 1.21^{A}	6.07 ± 1.59^{B}	5.78 ± 1.25^{B}	6.59 ± 1.58^{A}
Whole-Blood Medium-Shear Relative Index (WMS)	1.83 ± 0.24^{B}	1.61 ± 0.43^{C}	1.69 ± 0.33^{C}	2.14 ± 0.43^{A}
Whole-Blood High-Shear Relative Index (WHS)	1.39 ± 0.19^{B}	1.23 ± 0.38^{C}	1.32 ± 0.28^{BC}	1.70 ± 0.31^{A}
Erythrocyte Aggregation Index (EI)	4.97 ± 0.69^{B}	5.36 ± 2.07^{A}	4.42 ± 0.78^{C}	3.83 ± 0.42^{D}
Erythrocyte Aggregation Coefficient (EC)	3.41 ± 0.47^{B}	3.67 ± 1.42^{A}	3.04 ± 0.54^{C}	2.63 ± 0.29^{D}
Casson Viscosity (CV, mPa·s)	1.88 ± 0.17^{B}	2.19 ± 0.67^{A}	2.02 ± 0.43^{A}	2.20 ± 0.34^{A}
Erythrocyte Intracellular Viscosity (RBCIV, mPa·s)	0.67 ± 0.07^{A}	0.90 ± 0.18^{A}	0.79 ± 0.12^{A}	0.61 ± 0.11^{A}
Low-Shear Flow Resistance (LSFR, 10^9 SI)	33.85 ± 4.95^{B}	39.83 ± 10.71^{A}	30.98 ± 5.09^{C}	28.73 ± 4.17^{C}
Medium-Shear Flow Resistance (MSFR, 10^9 SI)	21.16 ± 1.92^{B}	24.37 ± 5.15^{A}	21.34 ± 3.62^{B}	22.23 ± 3.37^{B}
High-Shear Flow Resistance (HSFR, 10^9 SI)	16.00 ± 1.33^{B}	18.52 ± 4.79^{A}	16.65 ± 3.06^{B}	18.03 ± 2.97^{AB}
Yield Stress (YS, mPa)	3.99 ± 0.99^{B}	4.97 ± 2.55^{A}	3.31 ± 0.91^{C}	2.64 ± 0.54^{D}

The data in the same line were compared; means with different capital letters are significantly different ($p < 0.01$).

RBC and HGB increased significantly with altitude, from 11.97 ± 2.12 × 10^{12}/L and 72.10 ± 12.49 g/L in the local goats at 500 m (HHs) to 17.40 ± 2.25 × 10^{12}/L and 101.01 ± 9.47 g/L in the DQs at 3500 m ($p < 0.01$). HCT followed a similar trend, rising from 19.58 ± 3.23% to 30.02 ± 3.11% ($p < 0.01$). PV peaked at 2.24 ± 0.45 mPa·s in the DLs at 2500 m ($p < 0.01$) but decreased to 1.64 ± 0.24 mPa·s in the DQs at 3500 m. FB and EI also showed significant changes, with fibrinogen at 4.92 ± 1.00 g/L in the DLs at 2500 m, significantly higher than 3.66 ± 0.39 g/L in the DQs at 3500 m ($p < 0.01$). Furthermore, with increasing altitude, both EI and LSFR declined, while YS peaked at 4.97 ± 2.55 mPa in the DLs at 2500 m ($p < 0.01$).

3.2. *Identification of EPAS1 Gene SNPs and Genotyping Results*

The *EPAS1* gene is a critical regulatory gene for hypoxia adaptation in high-altitude mammals. In this study, Sanger sequencing was used to analyze the core regions of the *EPAS1* gene in high-altitude DQs and low-altitude HHs. A total of six SNPs were identified (Figure 3). Among them, *g.86650 A>T* (Q556L) is a missense mutation that results in an amino acid substitution from glutamine (Q) to leucine (L) at position 556, suggesting a potential functional impact on protein structure or activity. Significant differences in genotype and allele frequencies at this position were observed between the two populations ($p < 0.01$). In the DQs (high-altitude, 3500 m), the *T* allele frequency was 0.85, with the *TT* genotype accounting for 0.70. In contrast, in the HHs (low-altitude, 500 m), the *A* allele frequency was 0.70, and the *AA* genotype accounted for 0.60. Similarly, for *g.41114C>T*, genotype distributions differed significantly between high- and low-altitude populations ($p < 0.01$), with the *TT* genotype frequency being 0.70 in DQs but only 0.20 in HHs, suggesting a potential role in hypoxia adaptation. *g.91117A>G* showed significant differentiation between high- and low-altitude populations ($p < 0.01$), with the *GG* genotype at 0.57 in DQs and fixed (1.00) in HHs. However, as a synonymous SNP with no amino acid change, its functional relevance remains unclear, and it was not included in the association analysis. In contrast, the remaining four SNPs (*g.83908 C>T*, *g.85046 C>A*, *g.88664 G>C*, and *g.91117 A>G*) did not show significant differences in genotype or allele frequencies between high- and low-altitude populations ($p > 0.05$).

To further validate the role of *g.86650 A>T* in altitude adaptation, we expanded genotyping analysis to include goat populations across different altitudes in Yunnan Province (DQ: 3500 m, DL: 2500 m, HZ: 1500 m, and HH: 500 m). The results showed that the *T* allele frequency increased with altitude. In the DQs, the *T* allele frequency was 0.78, with the *TT* genotype being dominant (0.62), whereas in the HHs, the *T* allele frequency decreased to 0.41, with a value of 0.36 for the *AA* genotype ($p < 0.01$) (Table 2).

Table 2. Genotype and allele frequencies of the *g.86650 A>T* position in goat populations across different altitudes.

Population	Genotype Frequencies			Allele Frequencies		*p* Value
	AA	AT	TT	A	T	
DQ	0.06	0.32	0.62	0.20	0.80	<0.01
DL	0.51	0.31	0.17	0.70	0.30	
HZ	0.47	0.27	0.25	0.61	0.39	
HH	0.40	0.40	0.20	0.60	0.40	

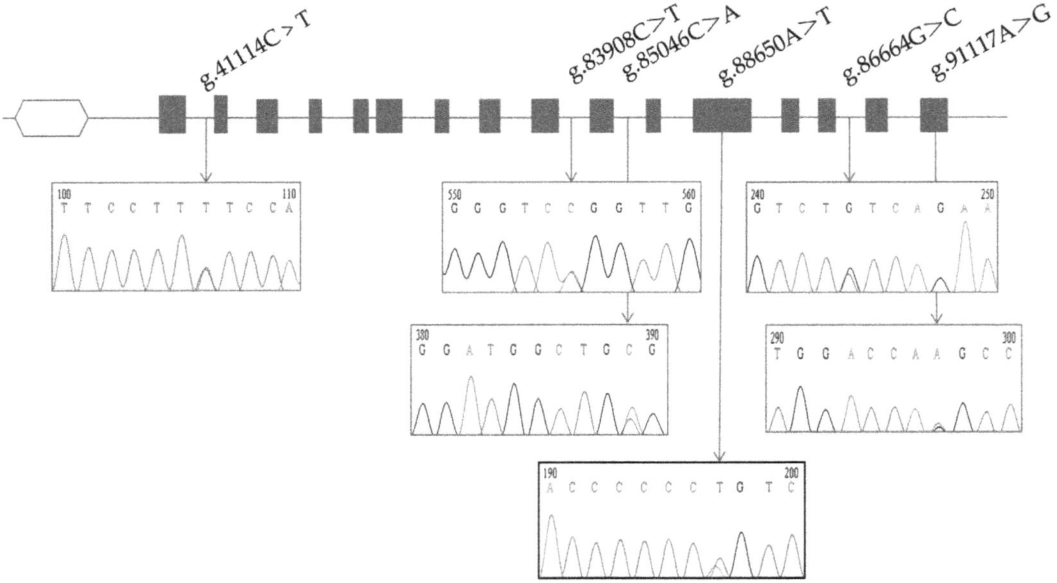

Figure 3. SNP variant information of the *EPAS1* gene.

3.3. Impact of Gln 556-Leu (Q556L) Polymorphism on Hematological and Hemorheological Traits Across Altitudes

To evaluate the effects of the Q556L missense polymorphism on hematological and hemorheological indices, a genotype–phenotype association analysis was conducted in goat populations across different altitudes in Yunnan Province (Tables 3 and 4). In the DQ (3500 m), HZ (1500 m), and HH (500 m) populations, no significant associations were observed between genotypes and either hematological or hemorheological indices ($p > 0.05$). However, in the DL (2500 m) population, individuals with the *TT* genotype (plateau type) exhibited significantly higher red blood cell (RBC) and hemoglobin (HGB) concentrations compared to those with the *AA* genotype (lowland type) (RBC: 16.02 ± 2.57 vs. 12.37 ± 1.88 × 10^{12}/L, HGB: 101.00 ± 16.09 vs. 79.83 ± 12.99 g/L, $p < 0.05$).

Table 3. Association analysis of hematological traits with genotypes.

Group	Genotype	RBC (10^{12}/L)	HGB (g/L)	HCT (%)	MCV (fL)	MCH (pg)	MCHC (g/L)	RDW (%)
DQ	*TT* (50)	18.81 ± 3.06	113.14 ± 18.19	31.72 ± 5.15	16.91 ± 1.17	5.95 ± 0.33	355.90 ± 13.63	20.53 ± 1.52
	AT (26)	19.16 ± 1.30	113.77 ± 6.43	31.89 ± 2.23	16.71 ± 1.09	5.90 ± 0.29	356.65 ± 11.11	20.97 ± 1.33
	AA (5)	18.92 ± 2.36	112.60 ± 8.88	31.56 ± 2.06	16.84 ± 1.55	5.94 ± 0.38	356.00 ± 11.66	20.02 ± 1.21
DL	*TT* (6)	16.02 ± 2.57 [a]	101.00 ± 16.09 [a]	29.08 ± 4.19 [a]	18.22 ± 0.39	6.25 ± 0.08	346.17 ± 8.75	20.62 ± 1.40
	AT (11)	14.50 ± 2.12 [a]	90.91 ± 14.24 [ab]	26.29 ± 3.90 [ab]	18.21 ± 1.21	6.22 ± 0.29	345.00 ± 9.61	19.70 ± 1.42
	AA (18)	12.37 ± 1.88 [b]	79.83 ± 12.99 [b]	23.54 ± 3.73 [b]	19.11 ± 1.29	6.41 ± 0.35	338.61 ± 12.23	19.46 ± 1.43
HZ	*TT* (14)	12.88 ± 2.46	69.21 ± 15.11	22.77 ± 4.67	17.76 ± 1.54	5.31 ± 0.47	303.07 ± 12.20	21.52 ± 1.47
	AT (15)	12.79 ± 3.04	69.47 ± 18.99	22.96 ± 5.72	18.02 ± 1.40	5.35 ± 0.39	300.13 ± 12.82	21.66 ± 1.29
	AA (26)	13.39 ± 2.31	73.04 ± 13.07	24.24 ± 4.17	18.27 ± 2.11	5.41 ± 0.37	300.92 ± 16.03	22.27 ± 1.84
HH	*TT* (11)	14.95 ± 2.44	94.55 ± 10.74	26.89 ± 2.73	18.34 ± 2.67	6.34 ± 0.61	351.18 ± 18.36	19.95 ± 1.81
	AT (26)	14.88 ± 2.24	93.69 ± 12.77	26.93 ± 3.28	18.39 ± 2.61	6.31 ± 0.68	348.12 ± 29.90	21.58 ± 2.06
	AA (21)	14.35 ± 2.79	95.38 ± 14.24	25.97 ± 5.75	18.07 ± 2.10	6.74 ± 1.18	352. 58 ± 13.75	21.60 ± 3.49

Values in the same row with different lowercase letters indicate significant differences ($p < 0.05$), while values without letters are not significantly different ($p > 0.05$).

Table 4. Association analysis of hemorheological traits with genotypes.

Group	Genotype	PV (mPa·s)	FB (g/L)	WLS	WMS	WHS	EI	EC	CV (mPa·s)	RBCIV (mPa·s)	LSFR (10^9 SI)	MSFR (10^9 SI)	HSFR (10^9 SI)	YS (mPa)
	TT (50)	1.16 ± 0.75	2.55 ± 1.64	5.73 ± 3.91	1.73 ± 1.18	1.36 ± 0.93	2.95 ± 1.96	2.02 ± 1.35	1.84 ± 1.24	0.45 ± 0.30	27.23 ± 18.18	19.28 ± 12.92	15.19 ± 10.20	2.77 ± 1.86
DQ	*AT* (26)	1.28 ± 0.73	2.83 ± 1.61	5.85 ± 3.35	1.78 ± 1.02	1.41 ± 0.80	3.20 ± 1.80	2.20 ± 1.24	1.98 ± 1.12	0.51 ± 0.29	28.89 ± 16.22	20.61 ± 11.61	16.27 ± 9.19	2.91 ± 1.64
	AA (5)	1.40 ± 0.80	3.08 ± 1.77	5.94 ± 3.44	1.84 ± 1.08	1.46 ± 0.86	3.26 ± 1.84	2.24 ± 1.26	2.15 ± 1.22	0.56 ± 0.32	30.66 ± 17.17	22.25 ± 12.51	17.65 ± 9.95	3.02 ± 1.71
	TT (6)	2.35 ± 0.34	5.16 ± 0.76 ^{a}	8.64 ± 1.90 ^{a}	1.82 ± 0.50 ^{a}	1.28 ± 0.39	6.93 ± 0.92 ^{b}	4.75 ± 0.63 ^{b}	2.34 ± 0.86 ^{a}	0.94 ± 0.14	60.21 ± 14.87 ^{a}	29.91 ± 9.49 ^{a}	21.08 ± 7.26 ^{a}	8.81 ± 1.77 ^{ab}
DL	*AT* (11)	2.51 ± 0.45	5.53 ± 1.00 ^{a}	8.06 ± 1.97 ^{ab}	1.57 ± 0.54 ^{ab}	1.08 ± 0.43	8.13 ± 2.28 ^{a}	5.58 ± 1.57 ^{a}	1.99 ± 0.81 ^{ab}	1.01 ± 0.18	59.05 ± 11.04 ^{a}	26.80 ± 7.94 ^{ab}	18.34 ± 6.53 ^{ab}	9.39 ± 2.13 ^{a}
	AA (18)	2.46 ± 0.29	3.76 ± 1.54 ^{b}	6.55 ± 1.76 ^{b}	1.24 ± 0.39 ^{b}	1.14 ± 0.30	8.14 ± 2.00 ^{a}	5.59 ± 1.37 ^{a}	1.53 ± 0.59 ^{b}	0.98 ± 0.12	47.44 ± 11.30 ^{b}	21.07 ± 6.32 ^{b}	14.29 ± 4.88 ^{b}	7.64 ± 2.05 ^{b}
	TT (14)	1.84 ± 0.71	4.05 ± 155	6.05 ± 1.74	1.81 ± 0.54	1.42 ± 0.44	4.32 ± 0.56	3.96 ± 0.38	2.02 ± 0.41	0.74 ± 0.28	30.47 ± 5.09	21.28 ± 3.67	16.71 ± 3.12	3.17 ± 0.75
HZ	*AT* (15)	1.83 ± 0.20	4.04 ± 0.44	5.82 ± 0.99	1.72 ± 0.33	1.35 ± 0.28	4.36 ± 0.65	3.00 ± 0.45	2.08 ± 0.49	0.73 ± 0.08	31.56 ± 4.26	21.97 ± 4.15	17.24 ± 3.68	3.30 ± 0.60
	AA (26)	1.76 ± 0.48	3.86 ± 1.06	5.97 ± 1.62	1.74 ± 0.44	1.36 ± 0.35	4.42 ± 0.64	3.03 ± 0.44	2.12 ± 0.30	0.70 ± 0.19	33.11 ± 6.04	22.58 ± 3.07	17.61 ± 2.43	3.53 ± 1.03
	TT (11)	1.39 ± 0.33	3.05 ± 0.73	8.68 ± 2.00	2.13 ± 0.45	1.58 ± 0.33	5.49 ± 0.46	3.77 ± 0.32	1.70 ± 0.15	0.55 ± 0.13	34.59 ± 5.03	19.87 ± 2.01	14.71 ± 1.37	4.42 ± 0.90
HH	*AT* (26)	1.36 ± 0.39	2.98 ± 0.85	8.81 ± 1.72	2.10 ± 0.37	1.54 ± 0.28	5.74 ± 0.71	3.94 ± 0.49	1.63 ± 0.38	0.54 ± 0.16	34.47 ± 6.19	19.35 ± 3.90	14.22 ± 3.10	4.52 ± 0.92
	AA (21)	1.26 ± 0.23	2.77 ± 0.50	10.12 ± 3.61	2.45 ± 0.90	1.80 ± 0.68	5.66 ± 0.61	3.88 ± 0.42	1.72 ± 0.30	0.50 ± 0.09	36.01 ± 6.29	20.33 ± 3.36	14.97 ± 2.55	4.69 ± 1.00

Values in the same row with different lowercase letters indicate significant differences ($p < 0.05$), while values without letters are not significantly different ($p > 0.05$).

In addition, in the DL population, individuals with the *TT* genotype had significantly higher hematocrit (HCT) values compared to those with the *AA* genotype (29.08 ± 4.19% vs. 23.54 ± 3.73%, $p < 0.05$), accompanied by elevated blood viscosity indices such as whole-blood low-shear relative index (WLS). However, *TT* genotype individuals exhibited significantly lower red blood cell aggregation (EI and EC) compared to the *AA* genotype ($p < 0.05$), suggesting that the reduced aggregation capacity of red blood cells may effectively mitigate the adverse effects of blood flow resistance.

3.4. Structural Analysis of the Gln 556-Leu (Q556L) Polymorphism

The Gln 556-Leu (Q556L) polymorphism's functional impact was assessed using structural modeling and interaction analysis Figure 4. The comparison between the wild-type and mutant forms revealed significant alterations in local structural stability and hydrogen bonding.

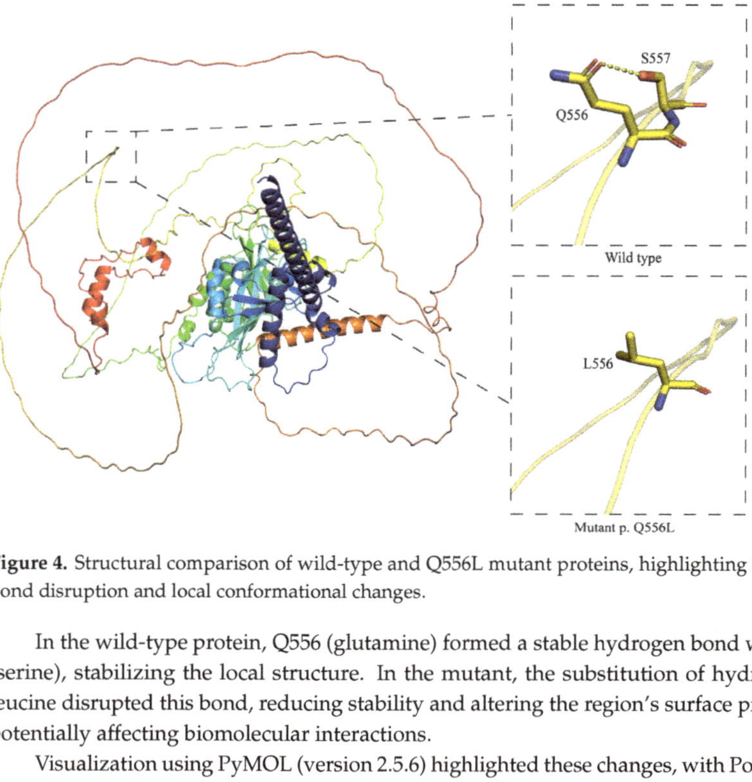

Figure 4. Structural comparison of wild-type and Q556L mutant proteins, highlighting hydrogen bond disruption and local conformational changes.

In the wild-type protein, Q556 (glutamine) formed a stable hydrogen bond with S557 (serine), stabilizing the local structure. In the mutant, the substitution of hydrophobic leucine disrupted this bond, reducing stability and altering the region's surface properties, potentially affecting biomolecular interactions.

Visualization using PyMOL (version 2.5.6) highlighted these changes, with PolyPhen-3 predicting a deleterious impact on functionality. The polymorphism's structural effects suggest impaired adaptation to hypoxic conditions, emphasizing its potential role in protein dysfunction. The figures generated in this study illustrate the structural impact of the polymorphism, offering deeper insights into its biological implications.

4. Discussion

High-altitude hypoxia poses significant challenges to the survival of flora, fauna, and humans. Populations indigenous to high-altitude regions exhibit distinct adaptive characteristics. For example, Tibetans and Sherpas maintain relatively low hemoglobin concentrations [32], which significantly reduces the risk of polycythemia [33,34] and chronic mountain sickness [35] compared to lowland populations. Similarly, this study found that high-altitude Tibetan goats significantly enhance oxygen transport capacity by increasing red blood cell count (RBC), hemoglobin concentration (HGB), and hematocrit level (HCT). This adaptive strategy is consistent with that observed in other high-altitude mammals, such as Tibetan donkeys [36], Tibetan pigs [37], and Tibetan horses [38]. These findings provide further evidence supporting the existence of convergent mechanisms of high-altitude hypoxia adaptation across species.

While increased hemoglobin concentration is an effective compensatory mechanism in hypoxia, excessive erythrocytosis may elevate blood viscosity, impairing tissue perfusion and oxygen delivery [37,39]. This study revealed a unique mechanism for regulating blood rheology in Tibetan goats. Despite slightly higher plasma viscosity and erythrocyte aggregation compared to low-altitude goats, their whole-blood high-shear relative index (WHS) and Casson viscosity (CV) were significantly lower, indicating maintained erythrocyte deformability. This trait, also observed in high-altitude species like yaks and Tibetan sheep, mitigates the adverse effects of increased blood viscosity on microcirculation by optimizing erythrocyte deformability.

At the molecular level, hypoxia-sensing gene pathways play central roles in regulating erythropoiesis, angiogenesis, cardiopulmonary adjustment, and energy metabolism. *EPAS1*, encoding HIF-2α, is a key member of the hypoxia signaling pathway. By regulating downstream genes such as erythropoietin (*EPO*) and vascular endothelial growth factor (*VEGF*), *EPAS1* enhances hypoxic adaptation [40,41]. Under normoxia, HIF-2α is hydroxylated by PHD and degraded via VHL-mediated ubiquitination. However, under hypoxia, HIF-2α translocates to the nucleus, dimerizes with HIF-1β, and activates downstream transcription [40]. In humans, SNP variations in *EPAS1* are strongly associated with the lower hemoglobin levels observed in Tibetans [42–44]. Similarly, specific variations in *EPAS1* are significantly correlated with hemoglobin concentration in yaks [45], indicating the conserved role of this gene in high-altitude adaptation.

The structural analysis further demonstrated the potential impact of the missense polymorphism (*g.86650 A>T*, p.Gln556Leu) in exon 12 of *EPAS1*. Protein modeling revealed that Q556 (glutamine) in the wild-type protein formed a stable hydrogen bond with S557 (serine), maintaining local structural stability. In contrast, the substitution to L556 (leucine) disrupted this hydrogen bond due to the change from a polar to a hydrophobic side chain, reducing structural stability. Such alterations may impair HIF-2α functionality, affecting downstream gene regulation and reducing hypoxic adaptation.

The sequencing results indicate a significant increase in the frequency of the T allele at this position with elevation ($p < 0.05$), with the high-altitude genotype (*TT*) predominant in Tibetan goats. Association analysis showed that in the DL (2500 m) population, individuals with the *TT* genotype exhibited significantly higher RBC, HGB, and HCT levels than those with the *AA* genotype ($p < 0.05$), suggesting that this polymorphism enhances oxygen transport capacity to promote high-altitude adaptation. Although the *TT* genotype was associated with increased blood viscosity (e.g., WLS), it also showed significantly reduced erythrocyte aggregation (EI and EC) ($p < 0.05$), likely mitigating the adverse effects of increased viscosity on blood circulation by maintaining erythrocyte deformability.

Notably, no significant differences in hematological or hemorheological traits were observed between the *TT* and *AA* genotypes in the DQ (3500 m), HZ (1500 m), and HH (500 m) populations. This may be attributed to the smaller sample size of the *AA* genotype in low-altitude populations, limiting statistical power. Additionally, Song et al. [29] reported a significant association between the p.Gln556Leu polymorphism and mean corpuscular hemoglobin concentration, a result not replicated in this study, suggesting that the functional effects of *EPAS1* may vary with genetic background and environmental factors.

In conclusion, this study further validates the critical role of *EPAS1* in high-altitude adaptation and provides new insights into the molecular mechanisms underlying adaptive evolution in high-altitude animals. Future studies with larger sample sizes and functional validation experiments are warranted to investigate the adaptive effects and regulatory mechanisms of this gene under varying environmental conditions.

5. Conclusions

This study highlights the critical role of the *EPAS1 g.86650 A>T* (p.Gln556Leu) polymorphism in enhancing oxygen transport capacity in Tibetan goats under hypoxic conditions. The polymorphism was associated with increased RBC, HGB, and HCT levels, particularly in the DL population, indicating its significance in high-altitude adaptation. Despite elevated blood viscosity, reduced erythrocyte aggregation indices (EI and EC) in the *TT* genotype suggest a compensatory mechanism to maintain microcirculation.

The structural analysis showed that the Q556L substitution disrupted hydrogen bonding, potentially impairing *EPAS1* functionality. These findings provide new insights into hypoxia adaptation mechanisms and offer a foundation for improving livestock breeding strategies in high-altitude regions.

Author Contributions: Conceptualization, X.G. and X.K.; methodology, L.Z. and Y.Z.; software, S.L.; validation, L.Z., L.T. and Y.Z.; formal analysis, L.Z. and S.L.; investigation, L.Z. and L.T.; resources, X.K.; data curation, S.L. and L.Z.; writing—original draft preparation, L.Z.; writing—review and editing, X.G. and W.D.; visualization, S.L.; supervision, X.K.; project administration, X.G.; funding acquisition, X.K. and X.G. All authors have read and agreed to the published version of the manuscript.

Funding: This research was funded by the Yunnan Basic Research Plan project—surface project, grant number 202101AT070711.

Institutional Review Board Statement: All research involving animals was conducted according to the ethical approval guidelines for animal studies provided by the animal ethics committee of Yunnan Agricultural University. Approval was given on 16 March 2021 (Approval No. 202103035).

Informed Consent Statement: Not applicable.

Data Availability Statement: The data presented in this study are available upon request from the corresponding author.

Acknowledgments: We thank all participants for their advice and support for this study.

Conflicts of Interest: All authors declare no conflicts of interest.

References

1. Moore, L.G. Human Genetic Adaptation to High Altitudes: Current Status and Future Prospects. *Quat. Int.* **2017**, *461*, 4–13. [CrossRef] [PubMed]
2. Wang, B.; Zhang, Y.B.; Zhang, F.; Lin, H.; Wang, X.; Wan, N.; Ye, Z.; Weng, H.; Zhang, L.; Li, X.; et al. On the Origin of Tibetans and Their Genetic Basis in Adapting High-Altitude Environments. *PLoS ONE* **2011**, *6*, e7002. [CrossRef] [PubMed]
3. Udpa, N.; Ronen, R.; Zhou, D.; Liang, J.; Stobdan, T.; Appenzeller, O.; Yin, Y.; Du, Y.; Guo, L.; Cao, R.; et al. Whole Genome Sequencing of Ethiopian Highlanders Reveals Conserved Hypoxia Tolerance Genes. *Genome Biol.* **2014**, *15*, R36. [CrossRef] [PubMed]
4. Zhao, Y.; Lu, X.; Cheng, Z.; Tian, M.; Qiangba, Q.Y.; Fu, Q.; Ren, Z. Comparative Proteomic Analysis of Tibetan Pig Spermatozoa at High and Low Altitudes. *BMC Genom.* **2019**, *20*, 569.
5. Lee, F.S. Hypoxia Inducible Factor Pathway Proteins in High-Altitude Mammals. *Trends Biochem. Sci.* **2024**, *49*, 79–92. [CrossRef]
6. Storz, J.F. Hemoglobin-Oxygen Affinity in High-Altitude Vertebrates: Is There Evidence for an Adaptive Trend? *J. Exp. Biol.* **2016**, *219*, 3190–3203. [CrossRef]
7. An, X.; Mao, L.; Wang, Y.; Xu, Q.; Liu, X.; Zhang, S.; Qiao, Z.; Li, B.; Li, F.; Kuang, Z.; et al. Genomic Structural Variation Is Associated with Hypoxia Adaptation in High-Altitude Zokors. *Nat. Ecol. Evol.* **2024**, *8*, 339–351. [CrossRef]

8. Wu, D.D.; Yang, C.P.; Wang, M.S.; Dong, K.Z.; Yan, D.W.; Hao, Z.Q.; Fan, S.Q.; Chu, S.Z.; Shen, Q.S.; Jiang, L.P.; et al. Convergent Genomic Signatures of High-Altitude Adaptation among Domestic Mammals. *Natl. Sci. Rev.* **2020**, *7*, 952–963. [CrossRef]
9. Liu, J.; Yuan, C.; Guo, T.; Wang, F.; Zeng, Y.; Ding, X.; Lu, Z.; Renqing, D.; Zhang, H.; Xu, X.; et al. Genetic Signatures of High-Altitude Adaptation and Geographic Distribution in Tibetan Sheep. *Sci. Rep.* **2020**, *10*, 18332. [CrossRef]
10. Bhandari, S.; Cavalleri, G.L. Population History and Altitude-Related Adaptation in the Sherpa. *Front. Physiol.* **2019**, *10*, 1116. [CrossRef]
11. Liu, X.; Zhang, Y.; Li, Y.; Pan, J.; Wang, D.; Chen, W.; Zheng, Z.; He, X.; Zhao, Q.; Pu, Y.; et al. EPAS1 Gain-of-Function Mutation Contributes to High-Altitude Adaptation in Tibetan Horses. *Mol. Biol. Evol.* **2019**, *36*, 2591–2603. [CrossRef] [PubMed]
12. Zhang, Z.; Du, H.; Bai, L.; Yang, C.; Li, Q.; Li, X.; Qiu, M.; Yu, C.; Jiang, Z.; Jiang, X.; et al. Whole Genome Bisulfite Sequencing Reveals Unique Adaptations to High-Altitude Environments in Tibetan Chickens. *PLoS ONE* **2018**, *13*, e0193596. [CrossRef] [PubMed]
13. Guo, J.; Tao, H.; Li, P.; Li, L.; Zhong, T.; Wang, L.; Ma, J.; Chen, X.; Song, T.; Zhang, H. Whole-Genome Sequencing Reveals Selection Signatures Associated with Important Traits in Six Goat Breeds. *Sci. Rep.* **2018**, *8*, 10405. [CrossRef] [PubMed]
14. Storz, J.F.; Cheviron, Z.A. Functional Genomic Insights into Regulatory Mechanisms of High-Altitude Adaptation. *Adv. Exp. Med. Biol.* **2016**, *903*, 113–128.
15. Storz, J.F.; Cheviron, Z.A. Physiological Genomics of Adaptation to High-Altitude Hypoxia. *Annu. Rev. Anim. Biosci.* **2021**, *9*, 149–171. [CrossRef]
16. Semenza, G.L. The Genomics and Genetics of Oxygen Homeostasis. *Annu. Rev. Genom. Hum. Genet.* **2020**, *21*, 183–204. [CrossRef]
17. Zhang, X.; Witt, K.E.; Bañuelos, M.M.; Ko, A.; Yuan, K.; Xu, S.; Nielsen, R.; Huerta-Sanchez, E. The History and Evolution of the Denisovan-EPAS1 Haplotype in Tibetans. *Proc. Natl. Acad. Sci. USA* **2021**, *118*, e2020803118. [CrossRef]
18. Li, C.; Li, X.; Xiao, J.; Liu, J.; Fan, X.; Fan, F.; Lei, H. Genetic Changes in the EPAS1 Gene between Tibetan and Han Ethnic Groups and Adaptation to the Plateau Hypoxic Environment. *PeerJ* **2019**, *7*, e7055. [CrossRef]
19. Lawrence, E.S.; Gu, W.; Bohlender, R.J.; Anza-Ramirez, C.; Cole, A.M.; Yu, J.J.; Hu, H.; Heinrich, E.C.; O'Brien, K.A.; Vasquez, C.A.; et al. Functional EPAS1/HIF2A Missense Variant Is Associated with Hematocrit in Andean Highlanders. *Sci. Adv.* **2024**, *10*, eabe7371. [CrossRef]
20. Xin, J.; Zhang, H.; He, Y.; Duren, Z.; Bai, C.; Chen, L.; Luo, X.; Yan, D.S.; Zhang, C.; Zhu, X.; et al. Chromatin Accessibility Landscape and Regulatory Network of High-Altitude Hypoxia Adaptation. *Nat. Commun.* **2020**, *11*, 4928. [CrossRef]
21. Schweizer, R.M.; Velotta, J.P.; Ivy, C.M.; Jones, M.R.; Muir, S.M.; Bradburd, G.S.; Storz, J.F.; Scott, G.R.; Cheviron, Z.A. Physiological and genomic evidence that selection on the transcription factor EPAS1 has altered cardiovascular function in high-altitude deer mice. *PLoS Genet.* **2019**, *15*, e1008371. [CrossRef] [PubMed]
22. Graham, A.M.; McCracken, K.G. Convergent evolution on the hypoxia-inducible factor (HIF) pathway genes EGLN1 and EPAS1 in high-altitude ducks. *Heredity* **2019**, *122*, 819–832. [CrossRef] [PubMed]
23. von Holdt, B.; Fan, Z.; Ortega-Del Vecchyo, D.; Wayne, R.K. EPAS1 variants in high-altitude Tibetan wolves were selectively introgressed into highland dogs. *PeerJ* **2017**, *5*, e3379.
24. Dong, K.Z.; Kang, Y.; Yao, N.; Shu, G.T.; Zuo, Q.Q.; Zhao, Q.J.; Ma, Y.H. Genetic variation of EPAS1 gene in Tibetan pigs and three low-altitude pig breeds in China. *J. Integr. Agric.* **2014**, *13*, 1990–1998. [CrossRef]
25. Erbel, P.J.A.; Card, P.B.; Karakuzu, O.; Bruick, R.K.; Gardner, K.H. Structural basis for PAS domain heterodimerization in the basic helix-loop-helix-PAS transcription factor hypoxia-inducible factor. *Proc. Natl. Acad. Sci. USA* **2003**, *100*, 15504–15509. [CrossRef]
26. Scortegagna, M.; Morris, M.A.; Oktay, Y.; Bennett, M.; Garcia, J.A. The HIF family member EPAS1/HIF-2α is required for normal hematopoiesis in mice. *Blood* **2003**, *102*, 1634–1640. [CrossRef]
27. Zeder, M.A.; Bradley, D.G.; Emshwiller, E.; Smith, B.D. Documenting Domestication: New Genetic and Archaeological Paradigms. *Science* **2006**, *311*, 1065–1068.
28. Deng, J.; Feng, J.; Li, L.; Zhong, T.; Wang, L.; Guo, J.; Ba, G.; Song, T.; Zhang, H. Polymorphisms, differentiation, and phylogeny of 10 Tibetan goat populations inferred from mitochondrial D-loop sequences. *Mitochondrial DNA Part A DNA Mapp. Seq. Anal.* **2018**, *29*, 439–445. [CrossRef]
29. Song, S.; Yao, N.; Yang, M.; Liu, X.; Dong, K.; Zhao, Q.; Pu, Y.; He, X.; Guan, W.; Yang, N.; et al. Exome sequencing reveals genetic differentiation due to high-altitude adaptation in the Tibetan cashmere goat (*Capra hircus*). *BMC Genom.* **2016**, *17*, 1–12. [CrossRef]
30. Sasazaki, S.; Tomita, K.; Nomura, Y.; Kawaguchi, F.; Kunieda, T.; Shah, M.K.; Mannen, H. FGF5 and EPAS1 gene polymorphisms are associated with high-altitude adaptation in Nepalese goat breeds. *Anim. Sci. J.* **2021**, *92*, e13640. [CrossRef]
31. Jumper, J.; Evans, R.; Pritzel, A.; Green, T.; Figurnov, M.; Ronneberger, O.; Tunyasuvunakool, K.; Bates, R.; Žídek, A.; Potapenko, A.; et al. Highly accurate protein structure prediction with AlphaFold. *Nature* **2021**, *596*, 583–589. [CrossRef] [PubMed]
32. Zhuang, J.; Droma, T.; Sutton, J.R.; Groves, B.M.; McCullough, R.E.; McCullough, R.G.; Sun, S.; Moore, L.G. Altitude, Hypoxia, and the Tibetan Plateau. *Respir. Physiol.* **1996**, *103*, 75–82. [CrossRef] [PubMed]
33. Garruto, R.M.; Chin, C.T.; Weitz, C.A.; Liu, J.C.; Liu, R.L.; He, X. Hematological differences during growth among Tibetans and Han Chinese born and raised at high altitude in Qinghai, China. *Am. J. Phys. Anthropol.* **2003**, *122*, 171–183. [CrossRef] [PubMed]

34. Beall, C.M.; Brittenham, G.M.; Strohl, K.P.; Blangero, J.; Williams-Blangero, S.; Goldstein, M.C.; Decker, M.J.; Vargas, E.; Villena, M.; Soria, R.; et al. Hemoglobin concentration of high-altitude Tibetans and Bolivian Aymara. *Am. J. Phys. Anthropol.* **1998**, *106*, 385–400. [CrossRef]
35. West, J.B. Hypoxia, polycythemia, and chronic mountain sickness. *Chest* **1988**, *94*, A-22–A-23. [CrossRef]
36. Han, X.; Wan, B.; Kong, X.; Li, J.; Liu, X.; Wang, X.; Jiang, Y.; Yang, S.; Yang, J.-F.; Yan, D.-W. Blood Physiology Characteristics of Tibet Donkey. *J. Econ. Anim.* **2015**, *19*, 47–50.
37. Kong, X.; Dong, X.; Yang, S.; Qian, J.; Yang, J.; Jiang, Q.; Li, X.; Wang, B.; Yan, D.; Lu, S.; et al. Natural selection on TMPRSS6 associated with the blunted erythropoiesis and improved blood viscosity in Tibetan pigs. *Comp. Biochem. Physiol. Part B Biochem. Mol. Biol.* **2019**, *233*, 11–22. [CrossRef]
38. Zhao, X.; Wei, X.; Lian, L.; Lu, S.; Yan, D.; Gou, X. Study on Blood Physiological Indicators of Adaptation to Hypoxia in Tibet Horse. *J. Yunnan Agric. Univ.* **2014**, *29*, 684–688.
39. Lee, S.R.; Jung, J.M.; Jung, L.Y.; Lee, J.H.; Lee, S.H.; Rhee, K.S.; Chae, J.K.; Kim, W.H.; Ko, J.K.; Lee, D.H.; et al. Elevated coronary whole blood viscosity in acute coronary syndrome patients. *Clin. Hemorheol. Microcirc.* **2013**, *55*, 85–94. [CrossRef]
40. Franke, K.; Gassmann, M.; Wielockx, B. Erythrocytosis: The HIF pathway in control. *Blood* **2013**, *122*, 1122–1128. [CrossRef]
41. Vriend, J.; Reiter, R.J. Melatonin and the von Hippel-Lindau/HIF-1 oxygen sensing mechanism: A review. *Biochim. Biophys. Acta -Rev. Cancer* **2016**, *1865*, 176–183. [CrossRef] [PubMed]
42. Beall, C.M.; Cavalleri, G.L.; Deng, L.; Elston, R.C.; Gao, Y.; Knight, J.; Li, C.; Li, J.C.; Liang, Y.; McCormack, M.; et al. Natural selection on EPAS1 (HIF2α) associated with low hemoglobin concentration in Tibetan highlanders. *Proc. Natl. Acad. Sci. USA* **2010**, *107*, 11459–11464. [CrossRef]
43. Peng, Y.; Cui, C.; He, Y.; Ouzhuluobu; Zhang, H.; Yang, D.; Zhang, Q.; Bianbazhuoma; Yang, L.; He, Y.; et al. Down-regulation of EPAS1 transcription and genetic adaptation of Tibetans to high-altitude hypoxia. *Mol. Biol. Evol.* **2017**, *34*, 818–830. [PubMed]
44. Tashi, T.; Reading, N.S.; Wuren, T.; Zhang, X.; Moore, L.G.; Hu, H.; Tang, F.; Shestakova, A.; Lorenzo, F.; Burjanivova, T.; et al. Gain-of-function EGLN1 prolyl hydroxylase (PHD2 D4E:C127S) in combination with EPAS1 (HIF-2α) polymorphism lowers hemoglobin concentration in Tibetan highlanders. *J. Mol. Med.* **2017**, *95*, 665–670. [CrossRef] [PubMed]
45. Wu, X.; Ding, X.; Chu, M.; Guo, X.; Bao, P.; Liang, C.; Yan, P. Novel SNP of EPAS1 gene associated with higher hemoglobin concentration revealed the hypoxia adaptation of yak (*Bos grunniens*). *J. Integr. Agric.* **2015**, *14*, 741–748. [CrossRef]

International Journal of
Molecular Sciences

MDPI

Article

Multi-Omics Analysis Revealed the Molecular Mechanisms Affecting Average Daily Gain in Cattle

Mingjuan Gu [1], Hongyu Jiang [1], Fengying Ma [1], Shuai Li [1], Yaqiang Guo [1], Lin Zhu [1], Caixia Shi [1], Risu Na [1], Yu Wang [2,*] and Wenguang Zhang [1,3,*]

1 College of Animal Science, Inner Mongolia Agricultural University, Hohhot 010010, China; gmj0119@yeah.net (M.G.); 15894949495@163.com (H.J.); fengyingma1997@163.com (F.M.); lishuai@emails.imau.edu.cn (S.L.); gggyaqiang@163.com (Y.G.); zhulinynacxhs@163.com (L.Z.); shicx98@163.com (C.S.); narisu@swu.edu.cn (R.N.)
2 College of Veterinary Medicine, Inner Mongolia Agricultural University, Hohhot 010010, China
3 College of Life Science, Inner Mongolia Agricultural University, Hohhot 010010, China
* Correspondence: wangyu@imau.edu.cn (Y.W.); wenguangzhang@imau.edu.cn (W.Z.)

Abstract: The average daily gain (ADG) is a critical index for evaluating growth rates in cattle and is closely linked to the economic benefits of the cattle industry. Heredity is one of the factors affecting the daily gain of cattle. However, the molecular mechanisms regulating ADG remain incompletely understood. This study aimed to systematically unravel the molecular mechanisms underlying the divergence in ADG between high average daily gain (HADG) and low average daily gain (LADG) Angus cattle through integrated multi-omics analyses (microbiome, metabolome, and transcriptome), hypothesizing that the gut microbiota–host gene–metabolism axis is a key regulatory network driving ADG divergence. Thirty Angus cattle were classified according to their HADG and LADG. Fecal and serum samples were collected for 16S, fecal metabolome, and blood transcriptome analysis. The results showed that compared with the LADG group, the abundance of Firmicutes increased in the HADG group, while the abundance of Bacteroidetes and Proteobacteria decreased. Metabolomics and transcriptomic analysis revealed that KEGG pathways associated with differentially expressed genes (DEGs) and differentially abundant metabolites (DAMs) were enriched in bile acid metabolism. Spearman correlation analysis showed that *Oscillospira* was positively correlated with *ZBTB20* and negatively correlated with *RADIL*. *ZBTB20* was negatively correlated with *dgA-11_gut_group*. This study analyzed the regulatory mechanism of average daily gain of beef cattle from genetic, metabolic, and microbial levels, providing a theoretical basis for analyzing the mechanism of differential daily gain of beef cattle, and has important significance for improving the production performance of beef cattle. The multi-omics network provides biomarker foundations for machine learning-based ADG prediction models, offering potential applications in precision breeding. While these biomarkers show promise for precision breeding, their causal roles require further validation. The conclusions are derived from a single breed (Angus) and gender (castrated males). Future studies should include females and diverse breeds to assess generalizability.

Keywords: average daily gain; Angus cattle; 16S rRNA; metabolome; transcriptome

1. Introduction

Beef, renowned for its rich nutritional value, is popular among consumers worldwide, with its consumption rising, particularly in developing nations [1]. The average daily gain

(ADG), as a key trait of beef cattle breeding, is closely related to economic benefits. An increase in ADG is helpful to improve the production efficiency and economic benefit of the beef cattle industry. Achieving the target market weight in a shorter time is possible for cattle with higher ADG compared to those with lower ADG, thus significantly cutting down on feeding costs. Many factors can affect cattle ADG, including nutrition, genetics, environment, management, health, and other factors [2]. For example, the daily gain of calves can be significantly improved by feeding the calves with astragalus root extract [3]. In addition, cattle temperament affects ADG, with cattle that are quieter and calmer during handling having greater average daily gains than cattle that become agitated during routine handling [4]. Daily gain is closely related to the digestion and absorption capacity of the small intestine. Studies have shown that jejunum gene expression profiles are significantly different in cattle of high and low daily gain [5]. Moreover, Sunirmal Sheet et al. conducted transcriptome analysis on three distinct skeletal muscles of Hanwoo cattle, including the longissimus dorsi (LD), semimembranosus (SB), and psoas major (PM), to explore genetic influences on ADG variation [6]. However, the molecular mechanisms regulating ADG remain incompletely understood, partly due to limitations in sample size, technical approaches, and data integration.

With the advancement of high-throughput sequencing and omics technologies (e.g., RNA sequencing, microbiome analysis, and metabolomics), these approaches have become common methodologies in growth and development of livestock research. Gut microbiota plays an important role in the average daily gain of livestock [7]. The intestinal microbiome influences gut and systemic health through its metabolites [8]. The fecal metabolome is a "functional reading" that links microbial activity to phenotypes [9]. The blood transcriptome reflects systemic host response [10]. Most published studies have focused on single-omics approaches. While individual omics analyses have significantly contributed to the mechanisms regulating growth and development in domestic animals, they are somewhat limited because they focus on a single domain. The use of multi-omics measurements can yield more information than single omics, resulting in insightful conclusions that cannot be derived from any single omics technique [11]. For this reason, multi-omics approaches are gaining popularity. In previous studies exploring the mechanism of ADG in ruminants, key gaps still exist, such as small sample size ($n < 20$) limiting the universality, single-omics perspective failing to reveal cross-layer regulatory networks, and insufficient integration of multi-omics data hindering the verification of critical pathways [7,12]. The physiological mechanism of regulating the average daily gain of beef cattle is still a complex scientific problem, and no single omics can fully uncover the mechanism. Integrating multiple omics strategies is beneficial to further understand the regulatory mechanism of differences in average daily gain of beef cattle from a systematic and holistic perspective. This study combines an expanded cohort (n = 30) with integrative microbiome–metabolome–transcriptome analyses and Spearman/Mantel tests to systematically decode ADG divergence, bridging technical and integrative gaps in current research. Therefore, this study aimed to answer how gut microbiota composition, host gene expression, and metabolite profiles differ between high- and low-ADG cattle. How do these differences synergistically regulate ADG through cross-omics interactions?

In this study, the microbiome, metabolome, and transcriptome of Angus cattle with high and low daily gain were analyzed under the same feeding management and dietary conditions, and the molecular mechanism of the difference in daily gain was investigated. Fecal and serum samples from 30 Angus steers (15 high-/low-ADG each) were subjected to 16S rRNA sequencing (microbiome), untargeted metabolomics (UHPLC-MS), and whole-blood transcriptomics (RNA-seq). Cross-omics regulatory networks were constructed via Spearman correlation analysis. This study provides a theoretical basis for analyzing the mechanism of differential daily gain of beef cattle and has important significance for improving the production performance of beef cattle.

2. Results

2.1. Comparison of the Fecal Microbiotas of High- and Low-Daily-Gain Performance Groups

A total of 30 Angus cattle were tested, with 15 randomly selected from high- and low-daily-gain groups, respectively. To assess whether the amount of sequencing in this study was adequate to capture the diversity of the original microbial communities, the 16S rRNA sequencing data were analyzed for α-diversity. The Chao 1 index, which measures species richness, indicated no significant differences between the high- and low-daily-gain groups ($p = 0.6576$, Figure 1A). Conversely, the Shannon index, which also reflects species richness, revealed significant differences between the two groups ($p = 0.0012$, Figure 1B). Statistical analysis of α-diversity indices revealed that the high-daily-gain group had greater community and species richness compared to the low-daily-gain group (Figure 1A,B). Although there were similarities in microbial community diversity among the groups, differences in community richness were observed. Beta diversity, which involves comparing biodiversity across different samples, was assessed using principal coordinate analysis (PCoA) (Figure 1C). The distribution of fecal microbiota composition between the high-daily-gain group and the low-daily-gain group is distinctly separated, showing significant differences in bacterial communities between the two groups. Species composition analysis provides insight into the community structure of samples at various taxonomic levels. Figure 1D,E illustrate the results of species annotation at the phylum and genus levels for the two groups, as shown by 16S rRNA sequencing. At the phylum level, Firmicutes was the dominant member of the gut microbiota in all individuals, constituting between 52.6% and 77.6% of all phyla, followed by Bacteroidota and Proteobacteria. Compared with the low-daily-increase group, the abundance of Firmicutes in the high-daily-increase group increased, while the abundance of Bacteroidota and Proteobacteria decreased. We selected the top 10 genera at the genus level. The category Others contains the remaining species, while Unassigned is for those without classification or annotation. At the genus level, the dominant bacteria in the gut of high daily gain for cattle included 10 species, such as *Rikenellaceae_RC9_gut_group, UCG-005, Solibacillus, Lysinibacillus, Bacteroides,* and *Alistipes* (Figure 1E). At the genus level, a total of 152 genera were identified, of which 94 were shared between the two groups (Figure 1F). Biomarkers with significant differences between groups, from the phylum to genus level, were identified using LefSe analysis (LDA score > 4) (Figure 1G). A total of 19 different microflora strains were isolated from the intestinal contents, according to the results, including 2 phyla, 3 class, 4 orders, 6 families, and 4 genera. Clostridia was the most abundant marker in intestinal contents in the HADG. Bacillales was the most abundant market in the LADG. The predicted functions of gut bacteria were analyzed using PICRUSt2. A total of 32 of the most important functions were detected for gut bacteria (Figure 1H). The top 10 functions were amino acid metabolism, carbohydrate metabolism, metabolism of cofactors and vitamins, metabolism of terpenoids and polyketides, metabolism of other amino acids, lipid metabolism, energy metabolism, xenobiotics biodegradation and metabolism, glycan biosynthesis and metabolism, and biosynthesis of other secondary metabolites.

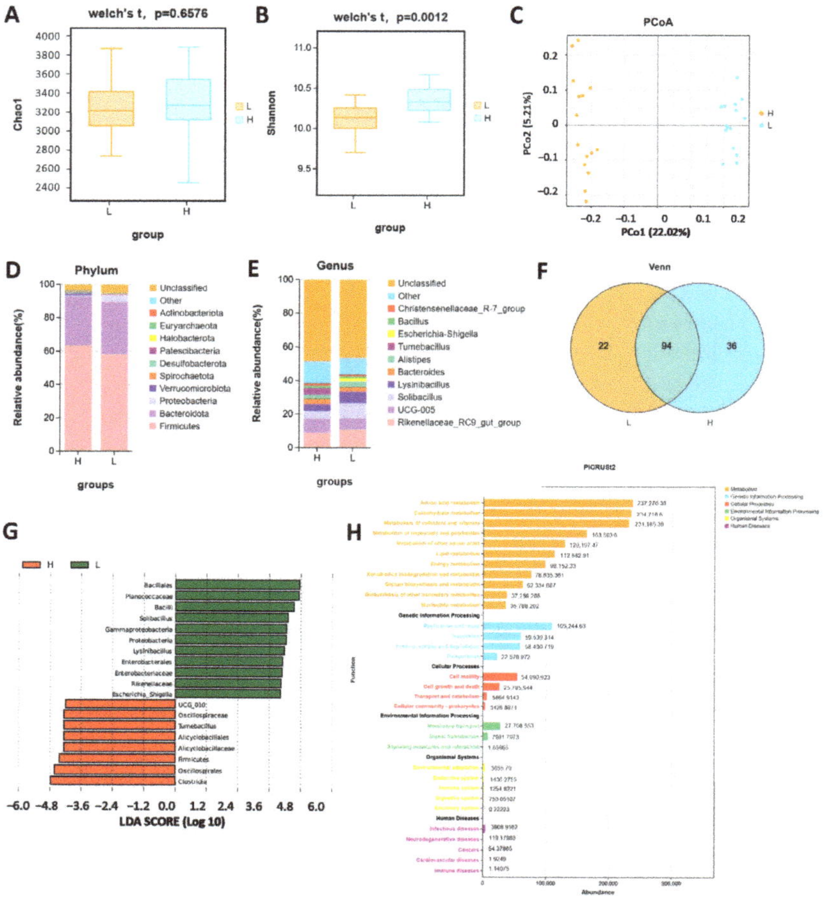

Figure 1. The 16S sequencing of fecal samples from HADG and LADG cattle. The α-diversity (alpha diversity) plots with chao1 index (**A**) and Shannon index (**B**). (**C**) Principal coordinate analysis (PCoA) of fecal microbial communities based on the Bray–Curtis distance. (**D**) The distribution of microbiota at the phylum level. (**E**) The distribution of microbiota at the genus level. (**F**) Venn diagram. (**G**) Histogram of LDA value distribution of LEfSe analysis. (**H**) PICRUSt2 analysis.

2.2. Comparison of the Fecal Metabolites of High- and Low-Daily-Gain Performance Groups

A non-targeted metabolomics analysis was performed on 30 cattle fecal samples. To compare the metabolomic analysis results of the two groups, Figure 2A,B shows the results of principal component analysis (PCA) and orthogonal partial least squares discriminant analysis (OPLS-DA) analyses. The two models showed significant clustering, indicating there was obvious separation between the two groups. In total, 566 differentially expressed metabolites were identified, with 123 metabolites upregulated and 443 downregulated in the HADG group compared with the levels in the LADG group ((VIP) ≥ 1, $p < 0.05$ and FC ≥ 2) (Figure 2C,D). In addition, we found through KEGG enrichment analysis that these metabolites were involved in purine metabolism, secondary bile acid biosynthesis, serotonergic synapse, and ABC transporters (Figure 2E).

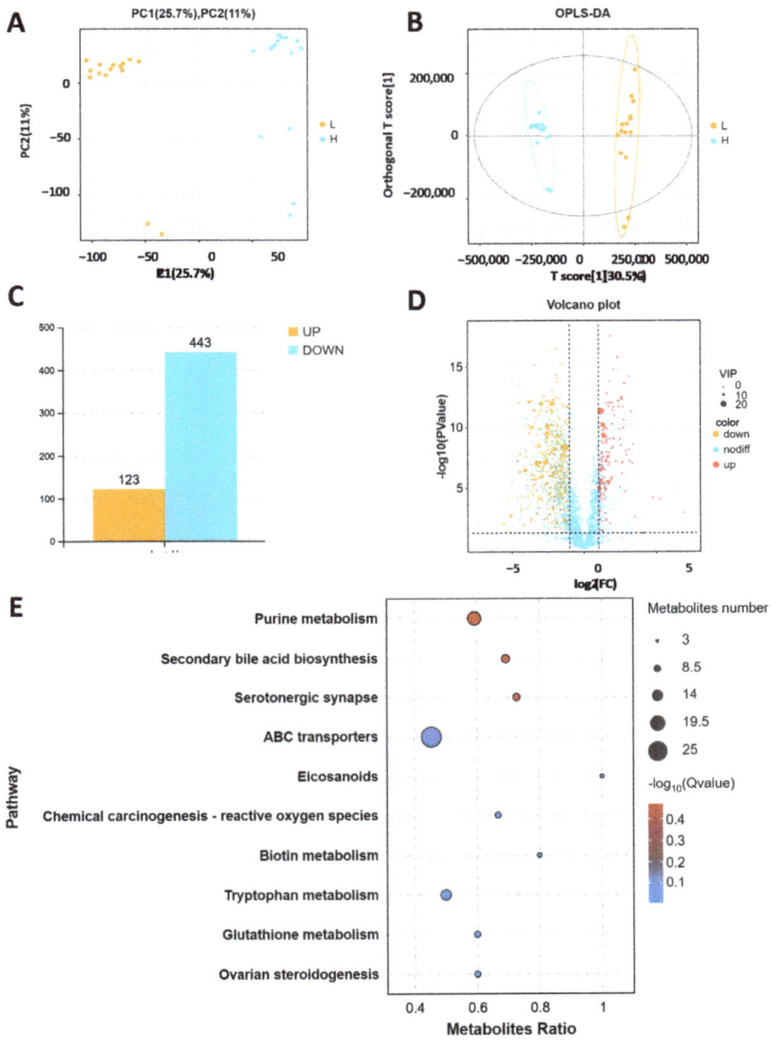

Figure 2. Metabolome analysis of fecal samples from HADG and LADG cattle. (**A**) PCA score plot of the metabolome. (**B**) OPLS-DA score plot of all the metabolite features. (**C**) Number of up- and down-regulated differentially abundant metabolites. (**D**) Volcano diagram of differentially abundant metabolites. (**E**) KEGG enrichment analysis of differentially abundant metabolites from HADG and LADG cattle.

2.3. Comparison of the Blood Transcriptome of High- and Low-Daily-Gain Performance Groups

To explore the molecular mechanism of the ADG of cattle, we used transcriptome sequencing technology to analyze the enrichment of DEGs and pathways in high and low daily weight gain in bovine blood. The results showed that there were 101 DEGs (FDR < 0.01; fold change > 2 or <−2), of which 45 were up-regulated and 56 were down-regulated (Figure 3A,B). Enrichment analysis showed that the top 10 signaling pathways were alcoholism, systemic lupus erythematosus, viral carcinogenesis, collecting duct acid secretion, rap1 signaling pathway, ECM–receptor interaction, PI3K-Akt signaling pathway, cell adhesion molecules, cGMP-PKG signaling pathway, and primary bile acid biosynthesis (Figure 3C). To anticipate how the identified DEGs interact, we utilized the online string platform to predict the protein–protein interaction network of differentially expressed genes (Figure 3D).

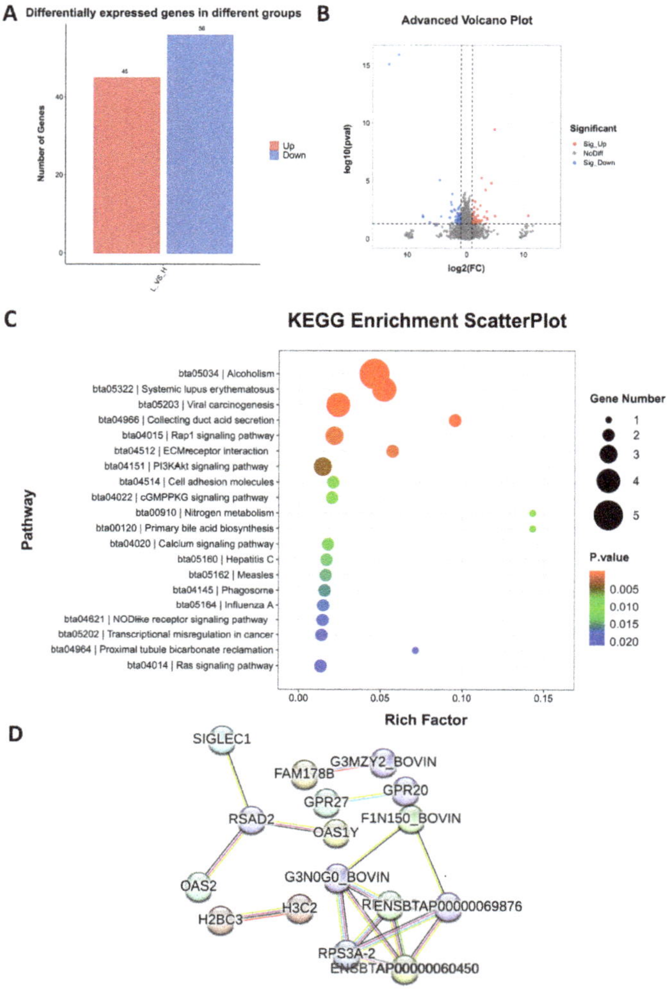

Figure 3. Transcriptome analysis of blood samples from HADG and LADG cattle. (**A**) The number of up- and down-regulated differentially expressed genes (DEGs). (**B**) Volcano diagram of the differentially expressed genes. (**C**) Differentially expressed gene KEGG enrichment. (**D**) PPI network of differentially expressed genes.

2.4. Integrative Crosstalk of Multi-Omics

To study the link between gene and microbiota and metabolites in high- and low-daily-weight-gain cattle, we determined the significant correlations between DEGs and various differential microorganisms and metabolites. The relationship between DEGs and gut microbiota and metabolites was examined using Spearman's correlation, which was visualized through a heatmap and a chord diagram (Figure 4). Significant gene–fecal microbiota and metabolite correlations were determined based on $|r| > 0.69$ and $p < 0.05$. The association analyses of differential gut microbiota and differential genes are presented in Figure 4A,B. We observed a positive correlation between *ZBTB20* and *Oscillospira* and a negative correlation between *RADIL* and *Oscillospira*. Additionally, we found a negative correlation between *ZBTB20* and *dgA-11_gut_group*. As for *RADIL*, there was no relationship between *RADIL* and *dgA-11_gut_group*. Association analyses of differentially expressed genes and differentially abundant metabolites demonstrated that *ZBTB20* and *RADIL* were associated with 26 fecal metabolites, and TFF2, bta-mir-6531, and GPR27 were associated with 20, 17, and 22 fecal metabolites.

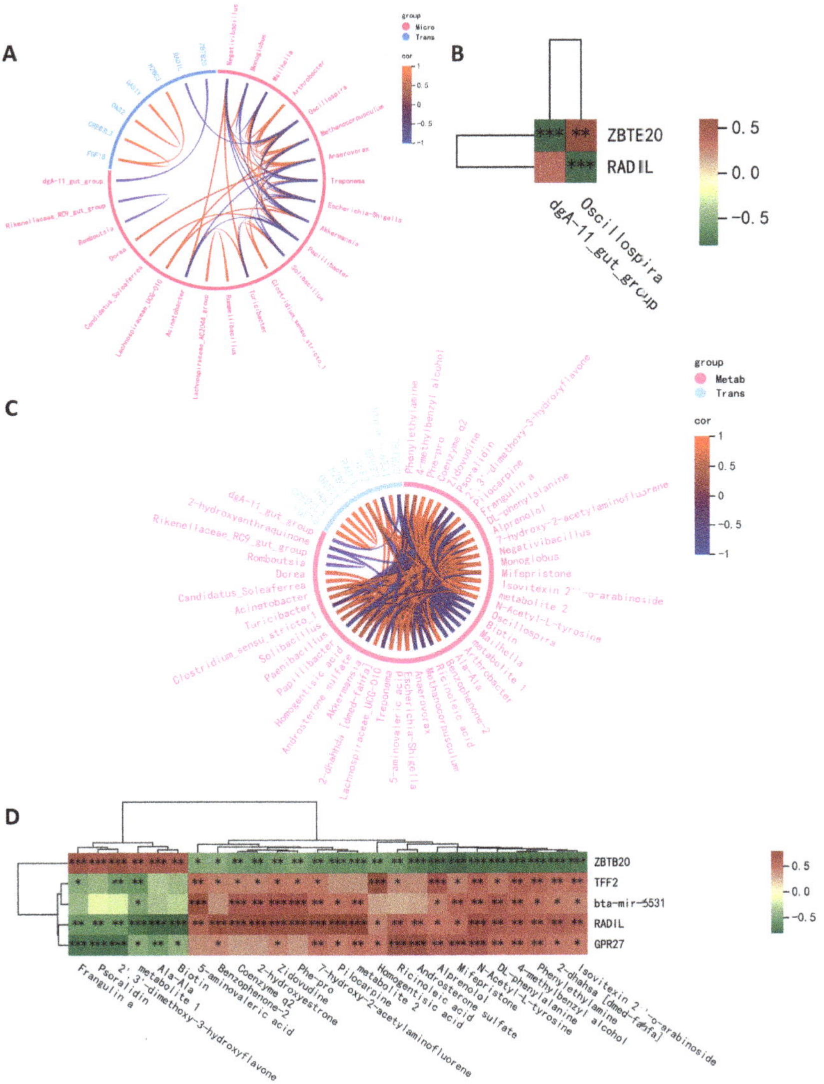

Figure 4. Multi-omics correlation analysis. (**A**) Spearman correlation chord plot of differential microbial and differentially expressed genes. (**B**) A heat map of the correlation analysis between differential microbial and differentially expressed genes. (**C**) Spearman correlation chord plot of differentially abundant metabolites and differentially expressed genes. (**D**) A heat map of the correlation analysis between differentially abundant metabolites and differentially expressed genes. metabolite 1 represents 6h-dibenzo[b,d]pyran, 3-(1,1-dimethylheptyl)-6a,7,10,10a-tetrahydro-1-methoxy-6,6,9-trimethyl-, (6ar,10ar)-; metabolite 2 represents 1h-pyrrole-3-propanoic acid, 5-[(1,2-dihydro-2-oxo-3h-indol-3-ylidene)methyl]-2,4-dimethyl-; * $p < 0.05$, ** $p < 0.01$, *** $p < 0.001$.

A comprehensive analysis of microbiome, metabolome, and transcriptome data can help elucidate the interactions between host gene expression, microbial composition, and the daily weight gain of cattle. It is generally believed that the high correlation may be caused by the relationship between different omics data, such as regulation and interaction, so the results based on the threshold sieve correlation can excavate the potential relationship. Therefore, we created a comprehensive multi-omics network using transcriptomics, metabolomics, and metagenomics data to demonstrate the functional connections between analytes within and across different omics datasets (Figure 5A). By constructing an interactive multi-omics correlation network, it was found that *RADIL* and *Oscillospira* played a central role in the daily weight gain of cattle (Figure 5B). The Sankey diagram further shows the linkages among the gut microbiome, fecal metabolites, and blood transcriptome (Figure 5C). These results suggest that *RADIL* may influence bovine daily weight gain by regulating metabolism through the regulation of intestinal enteric *Oscillospira* abundance.

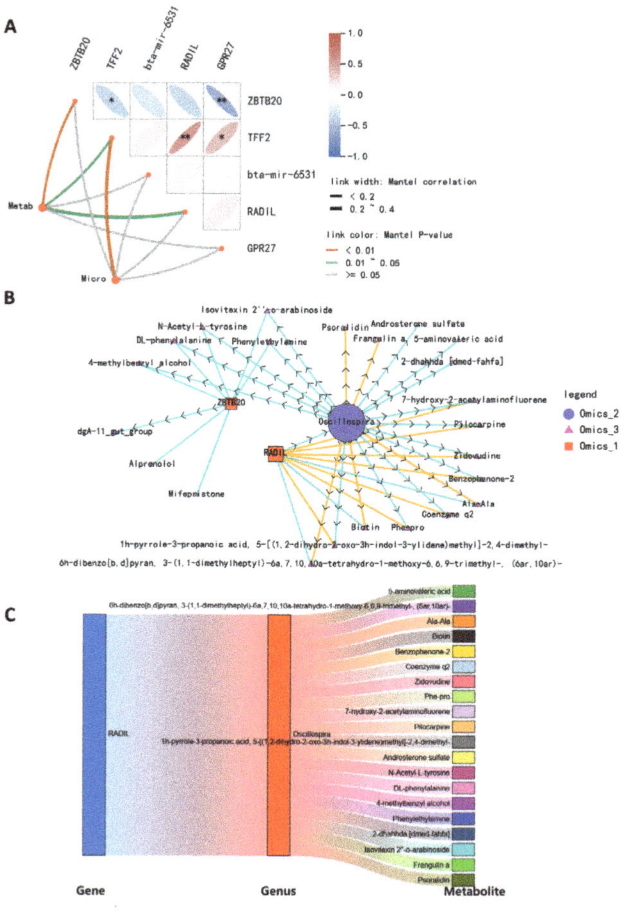

Figure 5. Integration of microbiome, transcriptome, and metabolome. (**A**) Mantel test correlation plot. The thickness of the line represents the size of the Mantel test correlation coefficient; the color of the line indicates the p-value of the Mantel test. * $p < 0.05$, ** $p < 0.01$. (**B**) A network diagram of microbiome, transcriptome, and metabolome. Different nodes in the diagram mark different microbiota, metabolites, or genes. The shape of the microbiota is circular, the shape of the metabolites is triangular, and the shape of the genes is square. (**C**) A Sankey diagram showing the correlation of genes–microorganisms–metabolites.

3. Discussion

The economic advantages for the beef cattle industry are linked to average daily gain (ADG), making it a vital metric of production performance in commercial beef cattle, but the specific genetic mechanism of regulating ADG is not completely clear. Gaining a complete understanding of the genetic mechanisms related to key growth and livestock traits facilitates the discovery of new genes and genetic pathways that can be beneficial for selecting livestock for breeding. In recent years, multi-omics combined analysis technology has been gradually applied to the analysis of key factors of complex biological pathways, which can be verified and interpreted from different levels, and is more conducive to revealing the complex regulatory mechanisms of animal growth and development [13,14]. Hence, microbiome, metabolome, and transcriptome analyses were performed on high- and low-daily-gain Angus cattle in this study to provide novel molecular functions that may influence average daily gain in Angus cattle.

Gut microbiota significantly influences host health, metabolism, growth, and development [15,16]. The gut microbiome is crucial for maintaining cattle health and enhancing their performance [17,18]. Previous studies have shown that there are significant differences in the intestinal microbes of high- and low-daily-gain livestock. Zhou et al. indicated that goats with higher ADG harbored more abundant *Ruminococcus* in the rumen after compound enzyme preparation supplementation [19]. Yin et al. investigated the rumen and rectum microbial community in lambs with high and low ADG and revealed significant differences in gut microbes with different daily gain [7]. In pigs, ADG positively correlates with abundance of *Erysipelothrix, Streptomyces, Dubosiella, Parolsenella,* and *Adlercreutzia,* and negatively correlates with abundance of *Lactobacillus* and *Prevotella* [20]. Fang et al. found that the abundance of *Ruminococcaceae* and *Bacteroidales_S24-7_group* in the HADG was significantly higher than in the LADG, while *Eubacterium_coprostanoligenes_group, Christensenellaceae_R-7_group,* and *opportunistic pathogens* in the HADG were significantly lower than in the LADG in meat rabbits [21]. Huang et al. found that the *Prevotellaceae_NK3B31_group* and *Alistipes* were significantly associated with an increase in ADG in broilers [22]. In addition, *Lactobacillus* was positively correlated with ADG, while *Clostridium_sensu_stricto_1* was negatively correlated with ADG in post-hatching chicks [23]. This study found that *Clostridia* was the most abundant in intestinal contents in the HADG, and *Bacillales* was the most abundant in the LADG. *Clostridium* is strictly anaerobic breathing [24]. Interestingly, clostridium produces short-chain fatty acids as metabolic end products, such as propionic acid and butyric acid, which promote intestinal development and health in producing animals and are essential for improving their growth performance [25–27]. Studies have shown that butyrate supplementation can increase average daily gain in cattle [28]. This may be one of the reasons for the high daily weight gain of cattle.

Many studies have shown that changes in metabolite profiles are related to livestock performance [29–31]. This includes an analysis of the metabolome of livestock with differential daily gain. Feng and colleagues examined the metabolite profiles in sheep serum to identify the metabolic traits of sheep with varying average daily gains under identical management conditions [32]. Jiang et al. integrated analyzed the microbiome and serum metabolome of Yorkshire pigs to reveal the molecular regulation mechanism of average daily gain, which provided a valuable reference for the identification of ADG-related molecular markers in pig farming [20]. In cattle, the mechanisms of different daily weight gain were investigated by assessing their plasma amine/phenol- and carbonyl-metabolome and whole-blood immune gene expression profiles [12]. In addition, Ogunade et al. also used the plasma carboxyl-metabolome to analyze the mechanism of differences in the average daily gain of beef cattle [33]. However, it was reported that 90 percent of gut bacterial species were associated with 82 percent of fecal metabolites, while only 34 percent of bacterial species were associated with 24 percent of blood metabolites [34]. In contrast to serum-based omics methods, the fecal metabolome shows direct interactions among genetic, environmental, and dietary influences [35]. As a result, employing fecal samples in metabolomics research could enhance the identification of biomarkers. In the study, fecal metabolome analysis revealed significant alterations in the metabolites between the LADG and HADG. We identified 566 differentially abundant metabolites, including 123 upregulated and 443 downregulated metabolites, those enriched in purine metabolism, and secondary bile acid biosynthesis metabolic and other pathways. These metabolites may be more indicative of the metabolite profile of differences in average daily gain when combined with fecal microorganisms. Xu et al. analyzed the metabolites of average daily gain differences in the fecal metabolome of pre-weaning Holstein heifers and found 35 fecal metabolites with significant differences, of which 17 were up-regulated and 18 were down-regulated [36]. Nine metabolites were found to be significantly associated with ADG, including pretyrosine, 3-keto-b-D-galactose, 4-hydroxyglucobrassicin, venlafaxine, convalloside, D-fructose, (3R)-beta-leucine, nicoboxil, and dinoterb. The differentially abundant metabolites found in this study did not contain these nine metabolites, which may be related to the differences in breed and months of age of cattle.

Daily gain is a complex carcass trait, and a large part of the variation in average daily gain can be attributed to genetic factors, as average daily gain heritability in cattle has been reported to be moderate to high, between 0.30 and 0.6 [37]. The use of high-throughput RNA sequencing has been demonstrated to be a successful approach for gene expression studies [38]. Numerous transcriptomics studies related to productivity traits have been conducted in cattle, including skeletal muscle profiling among breeds [39], gene expression profiling among distinct muscle cuts [40], and heat stress response in blood [41]. In this research, we utilized RNA-seq technology to analyze the entire blood transcriptome of the LADG and HADG to identify their genetic distinctions. We identified 101 differentially expressed genes, including 45 upregulated and 56 downregulated genes. Enrichment analysis showed that the top 10 signaling pathways were alcoholism, systemic lupus erythematosus, viral carcinogenesis, collecting duct acid secretion, Rap1 signaling pathway, ECM–receptor interaction, PI3K-Akt signaling pathway, cell adhesion molecules, cGMP-PKG signaling pathway, and primary bile acid biosynthesis. One of the metabolic pathways is primary bile acid biosynthesis. Bile acids are cholesterol derivatives that play an important role in fat metabolism [42]. Primary bile acids are synthesized in the liver and converted to secondary and tertiary bile acids by intestinal bacterial flora [43]. Studies have shown that the ADG of production traits in livestock and poultry is related to bile acid metabolism, such as broilers [44,45]. These results suggest that the regulation of bile acid metabolism during beef cattle production may affect the ADG of cattle. *DPP6*, *CDKN1A*, and *FZD5* may be candidate genes for ADG variation in the transcriptomic analysis of high- and low-daily-gain Hanwoo cattle [6]. There was no difference in these genes in this study, which may be caused by different samples and varieties.

We performed joint analyses of the transcriptome, metabolome, and microbiome to systematically assess interdependencies at multiple levels, rather than looking at the omics individually, to identify the transcriptome–microbiome–metabolome relationships that are important for average daily gain. Differentially expressed genes were correlated with differentially abundant metabolites and indicator species. The genes that were significantly correlated with both differentially abundant metabolites and indicator species were *RADIL* and *ZBTB20*, and the microorganisms that were significantly correlated with *RADIL* and *ZBTB20* were *Oscillospira*. Fecal metabolomics is a functional reading of the gut microbiome that can reflect changes in the gut microbiome [9]. There are 20 metabolites related to *Oscillospira*. These metabolites were enriched and analyzed. The metabolic pathways involved included oxidative phosphorylation, arginine and proline metabolism, phenylalanine metabolism, and tyrosine metabolism. We then mapped the gene–microbial–metabolite network, the *RADIL–Oscillospira*–metabolite, by integrating transcriptomic, microbiome, and metabolomic data. *RADIL* is a downstream effector of Rap1, a universally expressed small GTPase that plays a key role in regulating cell adhesion, angiogenesis, and endothelial barrier function [46]. *Oscillospira* belongs to the family Ruminococcaceae, order Clostridiales, and class Clostridia in the phylum Firmicutes [47]. *Oscillospira* may be the next generation of candidate probiotics with great potential for health applications [48]. However, this study found that the abundance of *Oscillospira* at high daily gain was lower than that at low daily gain. Several studies have shown that there is a significant positive correlation between *Oscillospira* and low fat, leanness, and human health with weight loss, lipid reduction, and slimming [49]. This may explain that although *Oscillospira* is a probiotic, its abundance decreases at high daily growth. These results indicate that *RADIL* and *Oscillospira* could have potential application to estimate cattle with different ADG. But further larger-size studies with more various herds of cattle are desired to validate our finding. It is crucial to emphasize the variation in ADG from the collective influence of various factors working in unison. Although *RADIL* and *Oscillospira* were obtained by correlation analysis, it may be a key candidate gene and microbe, but not the only one. Host heredity is a key factor in determining the co-occurring microbiota in the host gut by host–microbial interaction [50]. This study revealed the possible potential role of genes–microbes–metabolites in high ADG cattle through Spearman correlation networks: *RADIL* activates bile acid metabolism by regulating *Oscillospira* and ultimately achieves higher daily gain, with precise mechanisms awaiting functional validation. This pattern of cross-level regulation explains the complexity that cannot be captured by single omics and provides a multi-dimensional intervention target for the targeted regulation of ADG.

This study also has limitations. Although our cohort (n = 30) exceeds typical single-omics studies, it remains insufficient to capture low-abundance microbes or rare metabolites, limiting generalizability. Furthermore, the exclusive focus on Angus steers precludes insights into breed or gender-specific ADG mechanisms. Future studies should include larger cohorts across breed, genders, and age groups to dissect genetic, developmental, and gender-specific mechanisms. Although Spearman networks revealed significant correlations, statistical associations do not imply causation and require further experimental verification.

4. Materials and Methods

4.1. Animals and Sample Collection

Fecal and serum samples were taken rectally from 30 Angus cattle from Inner Mongolia Muniuhai Livestock Development Co., Ltd. A total of 200 Angus cattle (all castrated males, aged 24.7 ± 1.44 months) were initially screened under standardized feeding conditions. Basal diet: 60% corn silage (DM basis), 30% concentrate (containing 16% crude protein, 5% fat, 2.5% calcium), 10% alfalfa hay. Cattle were ranked by ADG, and the top 7.5% (HADG, n = 15) and bottom 7.5% (LADG: n = 15) were selected to represent extreme phenotypes. The study involved cattle with an initial body weight (BW) of 210 ± 15 kg at weaning. The final BW was recorded as 550 ± 25 kg (HADG) and 480 ± 20 kg (LADG). All cattle were confirmed to be disease-free through veterinary inspection, and no antibiotics were administered during the trial. The study included only castrated male cattle to eliminate variations in growth related to gender. Following collection, fecal samples were collected in labeled containers, frozen in liquid nitrogen, and stored at −80 °C. Serum was isolated by centrifugation at 2000 rpm for 15 min at 4 °C and stored at −80 °C.

4.2. Amplicon Sequencing 16S rRNA and Subsequent Analysis

Microbial DNA was extracted by Guangzhou Magen Biotechnology, targeting the V3–V4 region of the 16S rRNA gene, with amplicons purified using AMPure XP Beads (Beckman Coulter, Brea, CA, USA). The 16S rRNA V3–V4-specific primers were 16S_F 5′-(CCTACGGGNGGCWGCAG)-3′ and 16S_R 5′-(GGACTACHVGGGTATCTAAT)-3′. Subsequently, the sequencing libraries were generated, and library quality was assessed. DNA library sequencing was performed on the Illumina NovaSeq 6000/PacBio Revivo platform by Gene Denovo Biotechnology Co., Ltd. (Guangzhou, China). Raw data containing adapters or low-quality reads would affect the following analysis. Thus, to obtain high-quality clean reads, raw reads were further filtered according to the following rules using FASTP (version 0.18.0): (1) removing reads containing more than 10% of unknown nucleotides (N), (2) removing reads containing more than 50% of bases with quality (Q-value) < 20, and (3) removing adapter contamination. The clean tags were clustered into operational taxonomic units (OTUs) of ≥97% similarity using the UPARSE (version 9.2.64) pipeline. All chimeric tags were removed using the UCHIME algorithm and finally effective tags were obtained for further analysis. The tag sequence with the highest abundance was selected as a representative sequence within each cluster. The representative OTU sequences or ASV sequences were classified into organisms by a naive Bayesian model using the RDP classifier (version 2.2) based on the SILVA database (version 138.1) or the UNITE database (version 8.3), with a confidence threshold value of 0.8. The bioinformatics analysis was performed at www.omicshare.com.

4.3. Untargeted Metabolomics Study and Analysis

Fecal samples were subjected to untargeted metabolomics analyses. Fecal samples (100 mg) were dried overnight under a low-flow nitrogen stream to remove water. A total of 300 µL precooled acetonitrile and 100 µL precooled distilled water that acted as extraction solvent were added to each sample. Then, the mixture was vigorously vortexed for 3 min and sonicated for 5 min. After blending and centrifuging at 15,000 rpm for 15 min at 4 °C, the supernatant was obtained, and 10 µL of supernatant from each sample were mixed to obtain a quality control (QC) sample. Samples were filtered through a 0.22 µm microporous membrane for UHPLC-MS analysis. Analysis was performed using an UHPLC (1290 Infinity LC, Agilent Technologies, Waldbronn, Germany) coupled to a quadrupole time-of-flight (AB Sciex TripleTOF 6600, Shanghai, China). For HILIC separation, samples were analyzed using a 2.1 mm × 100 mm ACQUIY UPLC BEH Amide 1.7 µm column (waters, Wexford, Ireland). Solvent A was 0.1% formic acid in water and solvent B was 0.1% formic acid in acetonitrile. Metabolites were eluted using the following gradient at a flow of 0.4 mL/min: 0–2 min, 0–1% B; 2–6.5 min, 1–20% B; 6.5–11.5 min, 20–95% B; 11.5–13.5 min, 95–99% B; 13.5–16.5 min, 99–1% B; 16.5–20 min, 1% B; 20–21 min, 1–0% B; 21–22 min, 0% B. Injection volumes were set to 5 µL. Samples from all the groups were assayed in mixed and random order. Raw data underwent a conversion to the mzML format using Proteo Wizard software (version 3.0), followed by the extraction of peak data values using the XCMS program (version 3.18.0). The SIMCA-P program (version 14.1) was used to perform the analysis. A significant threshold of p-value ≤ 0.05 was set depending on the univariate T-test analysis. VIP > 1 and $p < 0.05$ were used to identify significantly changed metabolites between groups. The ultimate data in the current study were presented using www.omicshare.com.

4.4. RNA-Seq and Transcriptome Data Analysis

Total RNA was extracted from the blood samples following the instructions for TRlzol Reagent (Life Technologies, Carlsbad, CA, USA). The quality of RNA was assessed using a

NanoDrop 2000 (Thermo Fisher Scientific, Wilmington, DE, USA), while the integrity of the RNA was evaluated using an Agilent Bioanalyzer 2100 system (Agilent Technologies, Santa Clara, CA, USA) with an RNA Nano 6000 detection kit. Biomarker Technologies Co., Ltd. (Beijing, China) carried out the construction of gene libraries and RNA-seq using qualified RNA samples. After the libraries passed quality control, we proceeded with PE150 pattern sequencing by utilizing the Illumina NovaSeq6000 sequencing platform. To obtain clean data, we employed perl scripting to remove reads that contained connectors and low-quality reads (reads with a proportion of N > 10% or a proportion of bases with Q $\leq$ 10 exceeding 50%) from the raw data. The clean data were compared with the reference genome (Bos_taurus_UMD_3.1.1) using Hisat2 version 2.0.4. Gene expression levels were quantified by fragments per kilobase per million reads (FPKM), and differential gene expression analysis of the two groups of samples was conducted through the utilization of DESeq2 version 1.30.1. Genes with a |log2 (fold change, FC)| > 1 and a false discovery rate (FDR) < 0.01 were deemed DEGs. The function of DEGs was annotated through databases such as Gene Ontology (GO) and the KEGG Ortholog database (KO). The DEGs underwent GO enrichment analysis and Kyoto Encyclopedia of Genes and Genomes (KEGG) pathway enrichment analysis using clusterProfiler version 4.4.4. Protein–protein interaction (PPI) detection was constructed for predicting present study-related multiple proteins and the visualizing based on the integrated data requesting from the STRING database (https://string-db.org/, v11.5, accessed on 1 December 2024).

4.5. Correlation Analysis of Microbiome, Metabolome, and Transcriptom

Using the Metware Cloud 1.0, a free online platform for data analysis (https://cloud.metware.cn, accessed on 12 December 2024), correlations between high- and low-daily-gain bovine differentially expressed genes and levels of various microorganisms and metabolites were analyzed and visualized. Correlation analysis was based on Spearman's rank correlation coefficient |r| > 0.69 and p <0.05.

5. Conclusions

In summary, the microbiome, metabolome, and transcriptome of HADG and LADG Angus cattle were analyzed in this study. Different microorganisms, metabolites, and genes of the two groups were investigated to explore potential biomolecular markers affecting the average daily gain of Angus cattle. Compared with the LADG group, the abundance of Firmicutes increased in the HADG group, and the abundance of Bacteroidetes and Proteobacteria decreased. Metabolome and transcriptome analysis showed that the KEGG pathway involved in differentially expressed genes and differentially abundant metabolites was related to bile acid metabolism. Spearman's correlation analysis showed that *Oscillospira* was positively correlated with *ZBTB20* and negatively correlated with *RADIL*. *ZBTB20* was negatively correlated with *dgA-11_gut_group*. This suggests a potential axis where Oscillospira modulates host gene expression to suppress fat deposition and enhance nutrient partitioning toward muscle growth, thereby improving ADG. The correlation network between gut microbiota (*Oscillospira*), host genes (*ZBTB20, RADIL*), and bile acid metabolites provides a foundation for developing machine learning models to predict ADG. In addition, *ZBTB20* and *RADIL* could serve as candidate genes for marker-assisted selection, accelerating genetic gains in beef cattle breeding programs. In summary, this multi-omics study reveals that the interplay between gut microbiota (*Oscillospira*), host genes (*ZBTB20, RADIL*), and bile acid metabolism underpins ADG divergence in Angus cattle. While these biomarkers show promise for precision breeding, their causal roles require further validation. The conclusions are derived from a single breed (Angus) and gender (castrated males). Future studies should include females and diverse breeds to assess generalizability.

Author Contributions: Conceptualization, M.G. and W.Z.; Methodology, M.G. and H.J.; Software, F.M. and S.L.; Validation, H.J.; Formal analysis, F.M.; Investigation, R.N.; Resources, W.Z. Data curation, Y.G. and C.S.; Writing—original draft, M.G.; Writing—review & editing, L.Z. and R.N.; Visualization, M.G. and S.L.; Supervision, C.S.; Project administration, R.N. and Y.W.; Funding acquisition, Y.W. and W.Z. All authors have read and agreed to the published version of the manuscript.

Funding: This research was supported by the Technology Plan Project in Inner Mongolia Autonomous Region (2021GG0025, 2023YFHH0058), the Natural Science Foundation of Inner Mongolia (2024QN03051), the Inner Mongolia Open Competition Projects (2022JBGS0025), the Tongliao City Open Competition Projects (TL2024TW002), and the College of Animal Science, Inner Mongolia Agricultural University special project of high-level achievement cultivation (QF202202).

Institutional Review Board Statement: This study was approved by the Animal Welfare Committee of the Inner Mongolia Agricultural University of China. All experiments were carried out in strict accordance with the recommendations in the Guide for the Care and Use of Animals of Inner Mongolia Agricultural University and approved by the Animal Welfare Committee of Inner Mongolia Agricultural University.

Informed Consent Statement: Not applicable.

Data Availability Statement: The original contributions presented in the study are included in the article; further inquiries can be directed to the corresponding authors.

Acknowledgments: We thank Bang Guo, Senior Livestock Breeder at Inner Mongolia Muniuhai Livestock Development Co., Ltd., and his colleagues for their support in this research. Thank you https://www.home-for-researchers.com/ (accessed on 1 January 2025) for your help in writing this article.

Conflicts of Interest: The authors declare no conflicts of interest.

References

1. Magalhaes, D.R.; Çakmakçı, C.; Campo, M.D.M.; Çakmakçı, Y.; Makishi, F.; Silva, V.; Trindade, M.A. Changes in the Current Patterns of Beef Consumption and Consumer Behavior Trends-Cross-Cultural Study Brazil-Spain-Turkey. *Foods* **2023**, *12*, 475. [CrossRef] [PubMed]
2. Place, N.T.; Heinrichs, A.J.; Erb, H.N. The effects of disease, management, and nutrition on average daily gain of dairy heifers from birth to four months. *J. Dairy Sci.* **1998**, *81*, 1004–1009. [CrossRef] [PubMed]
3. Wei, H.; Ding, L.; Wang, X.; Yan, Q.; Jiang, C.; Hu, C.; Wang, G.; Zhou, Y.; Henkin, Z.; Degen, A.A. Astragalus root extract improved average daily gain, immunity, antioxidant status and ruminal microbiota of early weaned yak calves. *J. Sci. Food Agric.* **2021**, *101*, 82–90. [CrossRef] [PubMed]
4. Voisinet, B.D.; Grandin, T.; Tatum, J.D.; O'Connor, S.F.; Struthers, J.J. Feedlot cattle with calm temperaments have higher average daily gains than cattle with excitable temperaments. *J. Anim. Sci.* **1997**, *75*, 892–896. [CrossRef]
5. Foote, A.P.; Keel, B.N.; Zarek, C.M.; Lindholm-Perry, A.K. Beef steers with average dry matter intake and divergent average daily gain have altered gene expression in the jejunum. *J. Anim. Sci.* **2017**, *95*, 4430–4439. [CrossRef]
6. Sheet, S.; Jang, S.S.; Kim, J.H.; Park, W.; Kim, D. A transcriptomic analysis of skeletal muscle tissues reveals promising candidate genes and pathways accountable for different daily weight gain in Hanwoo cattle. *Sci. Rep.* **2024**, *14*, 315. [CrossRef]
7. Yin, X.; Duan, C.; Ji, S.; Tian, P.; Ju, S.; Yan, H.; Zhang, Y.; Liu, Y. Average Daily Gain in Lambs Weaned at 60 Days of Age Is Correlated with Rumen and Rectum Microbiota. *Microorganisms* **2023**, *11*, 348. [CrossRef]
8. Weng, Y.; Xu, T.; Wang, C.; Jin, Y. Oral Exposure to Epoxiconazole Disturbed the Gut Micro-Environment and Metabolic Profiling in Male Mice. *Metabolites* **2023**, *13*, 522. [CrossRef]
9. Zierer, J.; Jackson, M.A.; Kastenmüller, G.; Mangino, M.; Long, T.; Telenti, A.; Mohney, R.P.; Small, K.S.; Bell, J.T.; Steves, C.J.; et al. The fecal metabolome as a functional readout of the gut microbiome. *Nat. Genet.* **2018**, *50*, 790–795. [CrossRef]
10. Song, F.; Qian, Y.; Peng, X.; Li, X.; Xing, P.; Ye, D.; Lei, H. The frontline of immune response in peripheral blood. *PLoS ONE* **2017**, *12*, e0182294. [CrossRef]

1. Miao, Z.; Humphreys, B.D.; McMahon, A.P.; Kim, J. Multi-omics integration in the age of million single-cell data. *Nat. Rev. Nephrol.* **2021**, *17*, 710–724. [CrossRef] [PubMed]
2. Ogunade, I.M.; McCoun, M. Average daily gain divergence in beef steers is associated with altered plasma metabolome and whole blood immune-related gene expression. *Transl. Anim. Sci.* **2020**, *4*, txaa074. [CrossRef]
3. Du, L.; Li, K.; Chang, T.; An, B.; Liang, M.; Deng, T.; Cao, S.; Du, Y.; Cai, W.; Gao, X.; et al. Integrating genomics and transcriptomics to identify candidate genes for subcutaneous fat deposition in beef cattle. *Genomics* **2022**, *114*, 110406. [CrossRef]
4. Yu, H.; Wang, J.; Zhang, K.; Cheng, G.; Mei, C.; Zan, L. Integrated multi-omics analysis reveals variation in intramuscular fat among muscle locations of Qinchuan cattle. *BMC Genom.* **2023**, *24*, 367. [CrossRef] [PubMed]
5. Valdes, A.M.; Walter, J.; Segal, E.; Spector, T.D. Role of the gut microbiota in nutrition and health. *BMJ* **2018**, *361*, k2179. [CrossRef]
6. Perlman, D.; Martínez-Álvaro, M.; Moraïs, S.; Altshuler, I.; Hagen, L.H.; Jami, E.; Roehe, R.; Pope, P.B.; Mizrahi, I. Concepts and Consequences of a Core Gut Microbiota for Animal Growth and Development. *Annu. Rev. Anim. Biosci.* **2022**, *10*, 177–201. [CrossRef]
7. Lourenco, J.M.; Welch, C.B.; Krause, T.R.; Wieczorek, M.A.; Fluharty, F.L.; Rothrock, M.J.; Pringle, T.D.; Callaway, T.R. Fecal Microbiome Differences in Angus Steers with Differing Feed Efficiencies during the Feedlot-Finishing Phase. *Microorganisms* **2022**, *10*, 1128. [CrossRef] [PubMed]
8. Liu, J.H.; Zhang, M.L.; Zhang, R.Y.; Zhu, W.Y.; Mao, S.Y. Comparative studies of the composition of bacterial microbiota associated with the ruminal content, ruminal epithelium and in the faeces of lactating dairy cows. *Microb. Biotechnol.* **2016**, *9*, 257–268. [CrossRef]
9. Zhou, G.; Liang, X.; He, X.; Li, J.; Tian, G.; Liu, Y.; Wang, X.; Chen, Y.; Yang, Y. Compound enzyme preparation supplementation improves the production performance of goats by regulating rumen microbiota. *Appl. Microbiol. Biotechnol.* **2023**, *107*, 7287–7299. [CrossRef]
10. Jiang, S.; Si, J.; Mo, J.; Zhang, S.; Chen, K.; Gao, J.; Xu, D.; Bai, L.; Lan, G.; Liang, J. Integrated Microbiome and Serum Metabolome Analysis Reveals Molecular Regulatory Mechanisms of the Average Daily Weight Gain of Yorkshire Pigs. *Animals* **2024**, *14*, 278. [CrossRef]
11. Fang, S.; Chen, X.; Pan, J.; Chen, Q.; Zhou, L.; Wang, C.; Xiao, T.; Gan, Q.F. Dynamic distribution of gut microbiota in meat rabbits at different growth stages and relationship with average daily gain (ADG). *BMC Microbiol.* **2020**, *20*, 116. [CrossRef] [PubMed]
12. Zhao, G.; Niu, Y.; Wang, H.; Qin, S.; Zhang, R.; Wu, Y.; Xiao, X.; Xu, Y.; Yang, C. Effects of three different plant-derived polysaccharides on growth performance, immunity, antioxidant function, and cecal microbiota of broilers. *J. Sci. Food Agric.* **2024**, *104*, 1020–1029. [CrossRef]
13. Huang, T.; Han, J.; Liu, Y.; Fei, M.; Du, X.; He, K.; Zhao, A. Dynamic distribution of gut microbiota in posthatching chicks and its relationship with average daily gain. *Poult. Sci.* **2023**, *102*, 103008. [CrossRef] [PubMed]
14. Meurens, F.; Carlin, F.; Federighi, M.; Filippitzi, M.E.; Fournier, M.; Fravalo, P.; Ganière, J.P.; Grisot, L.; Guillier, L.; Hilaire, D.; et al. Clostridium botulinum type C, D, C/D, and D/C: An update. *Front. Microbiol.* **2022**, *13*, 1099184. [CrossRef]
15. Alonazi, M.; Ben Bacha, A.; Al Suhaibani, A.; Almnaizel, A.T.; Aloudah, H.S.; El-Ansary, A. Psychobiotics improve propionic acid-induced neuroinflammation in juvenile rats, rodent model of autism. *Transl. Neurosci.* **2022**, *13*, 292–300. [CrossRef] [PubMed]
16. Kearney, S.M.; Gibbons, S.M.; Poyet, M.; Gurry, T.; Bullock, K.; Allegretti, J.R.; Clish, C.B.; Alm, E.J. Endospores and other lysis-resistant bacteria comprise a widely shared core community within the human microbiota. *ISME J.* **2018**, *12*, 2403–2416. [CrossRef]
17. Yang, W.; Yu, T.; Huang, X.; Bilotta, A.J.; Xu, L.; Lu, Y.; Sun, J.; Pan, F.; Zhou, J.; Zhang, W.; et al. Intestinal microbiota-derived short-chain fatty acids regulation of immune cell IL-22 production and gut immunity. *Nat. Commun.* **2020**, *11*, 4457. [CrossRef] [PubMed]
18. Liu, W.; La, A.T.Z.; Evans, A.; Gao, S.; Yu, Z.; Bu, D.; Ma, L. Supplementation with sodium butyrate improves growth and antioxidant function in dairy calves before weaning. *J. Anim. Sci. Biotechnol.* **2021**, *12*, 2. [CrossRef]
19. Yu, H.; Yu, S.; Guo, J.; Wang, J.; Mei, C.; Abbas Raza, S.H.; Cheng, G.; Zan, L. Comprehensive Analysis of Transcriptome and Metabolome Reveals Regulatory Mechanism of Intramuscular Fat Content in Beef Cattle. *J. Agric. Food Chem.* **2024**, *72*, 2911–2924. [CrossRef]
20. Goldansaz, S.A.; Guo, A.C.; Sajed, T.; Steele, M.A.; Plastow, G.S.; Wishart, D.S. Livestock metabolomics and the livestock metabolome: A systematic review. *PLoS ONE* **2017**, *12*, e0177675. [CrossRef]
21. Xue, M.Y.; Sun, H.Z.; Wu, X.H.; Liu, J.X.; Guan, L.L. Multi-omics reveals that the rumen microbiome and its metabolome together with the host metabolome contribute to individualized dairy cow performance. *Microbiome* **2020**, *8*, 64. [CrossRef] [PubMed]
22. Feng, T.; Ding, H.; Wang, J.; Xu, W.; Liu, Y.; Kenéz, Á. Metabolite Profile of Sheep Serum With High or Low Average Daily Gain. *Front. Vet. Sci.* **2021**, *8*, 662536. [CrossRef]
23. Ogunade, I.; Oyebade, A.; Osa-Andrews, B.; Peters, S. Plasma Carboxyl-Metabolome Is Associated with Average Daily Gain Divergence in Beef Steers. *Animals* **2021**, *11*, 67. [CrossRef]

34. Visconti, A.; Le Roy, C.I.; Rosa, F.; Rossi, N.; Martin, T.C.; Mohney, R.P.; Li, W.; de Rinaldis, E.; Bell, J.T.; Venter, J.C.; et al. Interplay between the human gut microbiome and host metabolism. *Nat. Commun.* **2019**, *10*, 4505. [CrossRef] [PubMed]
35. Goedert, J.J.; Sampson, J.N.; Moore, S.C.; Xiao, Q.; Xiong, X.; Hayes, R.B.; Ahn, J.; Shi, J.; Sinha, R. Fecal metabolomics: Assay performance and association with colorectal cancer. *Carcinogenesis* **2014**, *35*, 2089–2096. [CrossRef]
36. Xu, S.Y.; Feng, X.R.; Zhao, W.; Bi, Y.L.; Diao, Q.Y.; Tu, Y. Rumen and hindgut microbiome regulate average daily gain of preweaning Holstein heifer calves in different ways. *Microbiome* **2024**, *12*, 131. [CrossRef] [PubMed]
37. Mehrban, H.; Naserkheil, M.; Lee, D.H.; Ibáñez-Escriche, N. Genetic parameters and correlations of related feed efficiency, growth, and carcass traits in Hanwoo beef cattle. *Anim. Biosci.* **2021**, *34*, 824–832. [CrossRef]
38. Wang, Z.; Gerstein, M.; Snyder, M. RNA-Seq: A revolutionary tool for transcriptomics. *Nat. Rev. Genet.* **2009**, *10*, 57–63. [CrossRef]
39. Guo, B.; Greenwood, P.L.; Cafe, L.M.; Zhou, G.; Zhang, W.; Dalrymple, B.P. Transcriptome analysis of cattle muscle identifies potential markers for skeletal muscle growth rate and major cell types. *BMC Genom.* **2015**, *16*, 177. [CrossRef]
40. Sheet, S.; Jang, S.S.; Lim, J.A.; Park, W.; Kim, D. Molecular signatures diversity unveiled through a comparative transcriptome analysis of longissimus dorsi and psoas major muscles in Hanwoo cattle. *Anim. Biotechnol.* **2024**, *35*, 2379883. [CrossRef]
41. Garner, J.B.; Chamberlain, A.J.; Vander Jagt, C.; Nguyen, T.T.T.; Mason, B.A.; Marett, L.C.; Leury, B.J.; Wales, W.J.; Hayes, B.J. Gene expression of the heat stress response in bovine peripheral white blood cells and milk somatic cells in vivo. *Sci. Rep.* **2020**, *10*, 19181. [CrossRef] [PubMed]
42. Yin, Y.; Gong, S.; Han, M.; Wang, J.; Shi, H.; Jiang, X.; Guo, L.; Duan, Y.; Guo, Q.; Chen, Q.; et al. Leucine regulates lipid metabolism in adipose tissue through adipokine-mTOR-SIRT1 signaling pathway and bile acid-microbe axis in a finishing pig model. *Anim. Nutr. (Zhongguo Xu Mu Shou Yi Xue Hui)* **2024**, *16*, 158–173. [CrossRef]
43. Russell, D.W. The enzymes, regulation, and genetics of bile acid synthesis. *Annu. Rev. Biochem.* **2003**, *72*, 137–174. [CrossRef] [PubMed]
44. Cao, G.; Yu, Y.; Wang, H.; Yang, H.; Tao, F.; Yang, S.; Liu, J.; Li, Z.; Yang, C. Dietary Clostridium butyricum and 25-Hydroxyvitamin D(3) modulate bone metabolism of broilers through the gut-brain axis. *Poult. Sci.* **2024**, *103*, 103966. [CrossRef] [PubMed]
45. Chen, L.; Shi, Y.; Li, J.; Shao, C.; Ma, S.; Shen, C.; Zhao, R. Dietary bile acids improve breast muscle growth in chickens through FXR/IGF2 pathway. *Poult. Sci.* **2024**, *103*, 103346. [CrossRef] [PubMed]
46. Choi, B.H.; Kou, Z.; Colon, T.M.; Chen, C.H.; Chen, Y.; Dai, W. Identification of Radil as a Ras binding partner and putative activator. *J. Biol. Chem.* **2021**, *296*, 100314. [CrossRef]
47. Yanagita, K.; Manome, A.; Meng, X.Y.; Hanada, S.; Kanagawa, T.; Tsuchida, T.; Mackie, R.I.; Kamagata, Y. Flow cytometric sorting, phylogenetic analysis and in situ detection of Oscillospira guillermondii, a large, morphologically conspicuous but uncultured ruminal bacterium. *Int. J. Syst. Evol. Microbiol.* **2003**, *53*, 1609–1614. [CrossRef]
48. Yang, J.; Li, Y.; Wen, Z.; Liu, W.; Meng, L.; Huang, H. Oscillospira—A candidate for the next-generation probiotics. *Gut Microbes* **2021**, *13*, 1987783. [CrossRef]
49. Konikoff, T.; Gophna, U. Oscillospira: A Central, Enigmatic Component of the Human Gut Microbiota. *Trends Microbiol.* **2016**, *24*, 523–524. [CrossRef]
50. Zhang, Z.; Li, D. Thermal processing of food reduces gut microbiota diversity of the host and triggers adaptation of the microbiota: Evidence from two vertebrates. *Microbiome* **2018**, *6*, 99. [CrossRef]

animals

MDPI

Article

CTNNB1 and CDH1 Regulate Trophoblast Cell Adhesion and Junction Formation in Yak Placental Tissue at Different Gestational Stages

Bohao Zhang [1,2,3], Chen Song [1,3], Bin Zhou [2,3], Junjun Zhang [2,3], Weitao Dong [1,3], Yong Zhang [1,2,3], Xingxu Zhao [1,2,3] and Quanwei Zhang [1,2,3,*]

1 College of Veterinary Medicine, Gansu Agricultural University, Lanzhou 730070, China; zhangbhgs@163.com (B.Z.); scott1217163@163.com (C.S.); d.wt2008@163.com (W.D.); zhangy@gsau.edu.cn (Y.Z.); zhaoxx@gsau.edu.cn (X.Z.)

2 College of Life Science and Biotechnology, Gansu Agricultural University, Lanzhou 730070, China; zhoub@st.gsau.edu.cn (B.Z.); 107332202298@st.gsau.edu.cn (J.Z.)

3 Gansu Key Laboratory of Animal Generational Physiology and Reproductive Regulation, Lanzhou 730070, China

* Correspondence: zhangqw@gsau.edu.cn; Tel.: +86-9317632509

Simple Summary: As a major cell type in the placenta, trophoblasts (TCs) play a crucial role in supporting placental development and fetal growth. Investigating the molecular mechanisms underlying TC adhesion and junction formation during different gestational stages is essential for improving the reproductive health of yaks. In this study, candidate differentially expressed proteins associated with cell adhesion and junction formation were identified using data-independent acquisition proteomics analysis. Bioinformatic analysis of six Gene Ontology terms and two Kyoto Encyclopedia of Genes pathways revealed that β-catenin (CTNNB1) and E-cadherin (CDH1) are involved in cell adhesion and junction formation, influencing both placental development and the pregnancy process. High expression of CTNNB1 and CDH1 in the TCs of placental tissues at five months of gestation confirmed their critical role in maintaining pregnancy homeostasis. This study provides new insights into the molecular mechanisms underlying TC adhesion and junction formation during yak placental development.

Abstract: The molecular mechanisms underlying trophoblast (TC) adhesion and junction formation, essential processes for mammal placental development and pregnancy maintenance, are unclear in yaks. Here, we investigated adhesion- and junction-related phenotypes in placental tissue and identified differentially expressed proteins (DEPs) related to TC adhesion and junction formation through data-independent acquisition proteomics analysis, then validated DEP functions and expression patterns through morphological and molecular biology techniques. Yak placental tissues at five months of gestation showed enhanced TC proliferation and differentiation compared to those at two months of gestation, and upregulated intercellular adhesion molecule-1 and occludin, highlighting the importance of cell adhesion and junction formation in TC differentiation and placental development. We identified 54 downregulated and 49 upregulated DEPs enriched in six Gene Ontology terms related to cell adhesion and junction formation, with β-catenin (CTNNB1) and E-cadherin (CDH1) playing pivotal roles. Bioinformatic analysis of Kyoto Encyclopedia of Genes pathways suggested that CTNNB1 and CDH1 are involved in cell adhesion and cancer, with implications for placental development. Immunohistochemistry and immunofluorescence staining revealed the strong expression of CTNNB1 and CDH1 in TC cytoplasm, particularly in trophoblast giant cells and uninucleate TCs at five months of gestation. qPCR and Western blot analyses revealed upregulated *CTNNB1* and *CDH1* mRNA and protein levels at five months of gestation, implicating their critical role in maintaining placental homeostasis and TC adhesion and junction formation. These findings elucidate the molecular mechanisms by which cell adhesion and junction formation regulate TC function and placental development during yak gestation.

Keywords: yak; placenta; DIA proteomics; β-catenin (CTNNB1); E-cadherin (CDH1); bioinformatic analysis

1. Introduction

Yaks (*Bos grunniens*), which are distributed across the Tibetan Plateau and other high-altitude regions, are vital livestock that provide essential resources for local herders and have significant economic and ecological value [1,2]. However, reproductive diseases, such as miscarriage, intrauterine growth restriction, and preeclampsia are prevalent during yak pregnancy. Pregnancy-related diseases severely affect yak reproductive performance, causing severe losses to the yak industry and local economies [2]. The placenta plays a critical role in material exchange, hormone secretion, immune regulation, and blood flow modulation during pregnancy [3,4]; the coordinated functioning of these processes is crucial for fetal development, successful pregnancy maintenance, and pregnancy-related disease avoidance [4,5]. Therefore, investigating the structure and function of yak placenta, particularly the molecular regulatory mechanisms involved in different gestational stages, is essential for a comprehensive understanding of yak reproductive physiology and for preventing and treating pregnancy-related diseases in yaks.

Trophoblast cells (TCs), a major cell population in the placenta, help develop a stable placental structure through coordinated invasion, differentiation, and secretory functions, providing critical support for fetal growth and development [6]. Defects in TC adhesion and junction formation lead to structural abnormalities in the placenta and contribute to various pregnancy-related diseases [7]. For example, preeclampsia, a hypertensive pregnancy disorder caused by reduced TC invasion and impaired vascular remodeling, often leads to postpartum hemorrhage, intrauterine fetal growth restriction, or even fetal death [8]; intrauterine growth restriction increases the risk of fetal hypoxia, acidosis, and hypoglycemia [9]; and placental abruption, typically resulting from weak TC adhesion to the uterine wall, can lead to maternal renal failure, uterine bleeding, and fetal asphyxia [10]. Recurrent miscarriages are also a common consequence of impaired TC adhesion and junction formation [7,11]. Research on TC adhesion and junctional defects in pregnancy-related disorders is well established in humans but remains limited in yaks. Thus, exploring the molecular mechanisms underlying TC adhesion and junction formation is crucial for understanding normal placental development and its role in pregnancy in yaks.

Intercellular adhesion and junction formation, which occur through interactions between cell adhesion molecules (e.g., integrins and cadherins) and the extracellular matrix (ECM), regulate cell migration, differentiation, and proliferation [12]. Disruption of these adhesive interactions leads to reduced cell migration, limited differentiation, and an unstable tissue structure [13,14], which is also common in TCs and can impede placental growth and destabilize the maternal–fetal interface. In humans and cattle, several signaling pathways regulate TC adhesion and junction formation, for example, the integrin/focal adhesion kinase signaling pathway promotes TC migration and differentiation by modulating interactions between the cytoskeleton and ECM [13]; the mitogen-activated protein kinase signaling pathway, activated by external growth factors such as transforming growth factor beta and epidermal growth factor, enhances TC proliferation during mid-pregnancy, maintaining the stability of the maternal–fetal interface and ensuring fetal development [15]; and the wingless/integrated (Wnt) signaling pathway plays a vital role in TC adhesion and junction formation, helping maintain placental function and structural stability [16,17]. In summary, TC adhesion and junction formation are indispensable for normal placental development and pregnancy progression. Despite these advances in other species, little is known about the molecular regulation of TC adhesion and junction formation in yaks, particularly across different gestational stages. Accordingly, the aim of this study was to explore the dynamic differentiation process and potential molecular mechanisms of TCs in yak placental tissues at two and five months of gestation. Differentially expressed proteins (DEPs) associated with cell adhesion and junction formation were identified using data-independent acquisition (DIA) proteomics and bioinformatics analysis. The functional roles and expression patterns of key DEPs were systematically verified using morphological assessments and molecular biology techniques. This research provides a novel and comprehensive understanding of TC adhesion and junction formation during yak pregnancy, contributing to improved strategies for preventing pregnancy-related disorders and enhancing yak reproductive performance.

2. Materials and Methods

2.1. Sample Preparation and Collection

Samples were obtained from a commercial slaughterhouse in Tianzhu Tibetan Autonomous County (Wuwei City, Gansu Province, China). Healthy pregnant yaks (six years old) with a normal metabolism and no evidence of disease were selected. The pregnancy stage of the fetuses was determined and classified based on the crown–rump length [18] and information provided by the herders. Uterine tissues were immediately collected after slaughter and rinsed three to five times with physiological saline. The uteri were then dissected using sterile surgical scissors to expose the fetal membranes, ensuring the integrity of the amnion, allantois, and chorion. Placental tissues corresponding to early pregnancy (two-month gestation group) and middle pregnancy (five-month gestation group) stages were collected (n = 3/group). A portion of the tissues was fixed in 4% paraformaldehyde, and the remaining tissues were stored at −80 °C for subsequent analyses. This study was approved by the Ethics Committee of Gansu Agricultural University (code GSAU-Eth-LST-2021-003).

2.2. Hematoxylin–Eosin (H&E) Staining

Fixed placental tissues were processed following standard paraffin-embedding protocols and sectioned into 5 μm thick slices using a microtome (Leica, Wetzlar, Germany) [19]. Sections were deparaffinized in xylene and rehydrated using a graded ethanol series. H&E staining was performed as described previously [19,20]. Briefly, nuclei were stained with hematoxylin, followed by differentiation and bluing treatment. The cytoplasm was then counterstained with eosin, followed by gradient dehydration. Finally, the sections were sealed with neutral balsam (Servicebio, Wuhan, China). All images were captured using an optical microscope and image processing system (Nikon, Tokyo, Japan). All staining assays were performed at least in triplicate.

2.3. Immunofluorescence (IF) Staining

IF staining assays were performed for the co-localization analysis of occludin, intercellular adhesion molecule-1 (ICAM1), β-catenin (CTNNB1), E-cadherin (CDH1), and cytokeratin (CK)-7 proteins, as previously described [20,21]. Specifically, the sections were incubated with sodium citrate for antigen retrieval and with H_2O_2 to inactivate endogenous peroxidase. After blocking, the sections were incubated with primary antibodies at different dilutions (Table S1), then incubated with appropriate secondary antibodies (Alexa Fluor® 488 for CK-7, Alexa Fluor® 647 for CTNNB1 and occludin, and Alexa Fluor® 594 for CDH1 and ICAM1; Abcam, Cambridge, UK), as described previously [20–22]. Nuclei were stained with 4′,6-diamidino-2-phenylindole (DAPI) solution (Solarbio, Beijing, China). Fluorescence signals and images were captured using a fluorescence microscope (ECHO; Broomfield, CO, USA). All staining assays were performed at least in triplicate.

2.4. DIA Sequence and Bioinformatics Analysis

Raw DIA proteomics data were acquired using an Orbitrap Exploris 480 system (Thermo Fisher Scientific, Waltham, MA, USA) coupled with nano-LC, where peptides are separated on a C18 column and analyzed using the DIA method with multiple isolation windows covering a broad m/z range, as previously described [23]. The acquired spectra were then processed using SpectronAut v19.7 (Biognosys AG, Schlieren, Switzerland) to identify and quantify peptides based on a reference protein database with raw DIA data mapped to the reference genome (NCBI-GCA_005887515.3) and analyzed using default parameters to obtain the protein quantification results. A 1% Q value (false discovery rate) cutoff was applied at both the precursor and protein levels. Proteins were quantified using the top three peptides that passed the 1% Q value cutoff. DEPs were identified based on an absolute fold change > 1.50 and a p value < 0.05 for functional annotation and analysis. Gene Ontology (GO) and Kyoto Encyclopedia of Genes and Genomes (KEGG) pathway analyses were conducted to annotate the functions of DEPs, as described previously [20], with terms and pathways considered significant at $p < 0.05$. DIA proteomic sequence data from the placental tissues of two-month and five-month gestation groups of yaks, with accession numbers IPX0006098000/PXD040910 in the ProteomeXchange database (https://www.iprox.cn/page/project.html, accessed on 12 July 2022), were used for GO and KEGG annotation and analysis. We focused on DEPs and GO terms ($p < 0.01$) related to cell adhesion and junction formation. Enrichment analysis of key DEPs was performed using the KEGG pathway database ($p < 0.05$) to further elucidate DEP functions. Enrichment circle plots, volcano plots, heatmaps, and UpSet–Venn diagrams were created using the OmicShare online platform (http://www.omicshare.com/tools, accessed on 20 November 2024) [19,21]. Protein–protein interaction networks of candidate DEPs were generated using STRING v12.0 (EMBL, Heidelberg, Germany) and Cytoscape 2.8.1 (Cytoscape Consortium, La Jolla, CA, USA) using the ClueGO plugin (Cytoscape Consortium, La Jolla, CA, USA) [24,25]. The potential mechanisms of the candidate DEPs were inferred through KEGG analysis and a literature review. Illustrative mechanistic diagrams were generated using Adobe Illustrator 2022 (San Jose, CA, USA).

2.5. Immunohistochemistry (IHC) Staining

After deparaffinization in xylene and rehydration in graded ethanol, sections were treated with sodium citrate (Beyotime, Shanghai, China) for antigen retrieval. Subsequently, the sections were incubated sequentially with 3% H_2O_2 and bovine serum albumin (Solarbio, Beijing, China) to block non-specific binding, followed by overnight incubation at 4 °C with the primary antibodies at different dilutions. IHC staining was performed using a Streptavidin-Biotin Complex (Bioss, Beijing, China) staining system according to the manufacturer's instructions [20,22]. Positive signals were visualized using a 3,3′-diaminobenzidine detection kit (Service Bio, Wuhan, China). All images were captured and analyzed using an optical microscope and image processing system (Nikon, Tokyo, Japan). All staining assays were performed at least in triplicate.

2.6. RNA Isolation, cDNA Synthesis, and Quantitative PCR Assays

Total RNA was extracted from yak placental tissues in the two groups using a FastP.RNA isolation kit (Vazyme, Nanjing, China) then used for cDNA synthesis according to the manufacturer's instructions [19,20]. The RNA concentration was quantified using a NanoDrop-8000 (Thermo Fisher Scientific, Waltham, MA, USA). with RNA integrity evaluated using 1% denaturing formaldehyde agarose gel electrophoresis. Subsequently, 1 μg of total RNA was reverse-transcribed into cDNA using an Evo M-MLV RT Kit (AG, Wuhan, China). The relative expression levels of *CTNNB1* and *CDH1* mRNA in the placental tissues were quantified using quantitative PCR (qPCR), as described previously [19,21]. The expression of *β-actin* was used as an endogenous control. qPCR primers (Table S2) were designed using Primer 3.0 (https://primer3.ut.ee/, accessed on 10 August 2024) and synthesized by Qingke Biotech (Xi'an, China). The qPCR reactions were performed on a LightCycler 96 real-time PCR system (Roche, Basel, Switzerland), and relative expression levels were calculated using the $2^{-\Delta\Delta CT}$ method [26]. All reactions were performed in triplicate to ensure reproducibility.

2.7. Western Blotting

The relative expression patterns of CTNNB1 and CDH1 proteins in placental tissues at different pregnancy stages were examined by Western blot. Total protein was extracted from 100 mg of each tissue sample using pre-chilled radioimmunoprecipitation assay buffer (Solarbio, Beijing, China) with 1 mM phenylmethylsulfonyl fluoride (Solarbio, Beijing, China). Protein concentrations were measured using a bicinchoninic acid protein assay kit (Boster, Wuhan, China). A total of 30 μg protein was used for Western blot analysis, as previously described [21,27]. The proteins were then transferred to polyvinylidene fluoride membranes for co-incubation with the primary antibodies overnight at 4 °C. The membranes were then incubated with the secondary antibody, visualized using enhanced chemiluminescent reagent (New Cell & Molecular Biotech, Jiangsu, China), and imaged with a chemiluminescence imaging system (General Electric, Boston, MA, USA). The integrated optical density (IOD) values of the bands were visualized and quantified using ImageJ software v1.44p (National Institutes of Health, Bethesda, MD, USA). β-actin was used as an endogenous control, and the protein expression levels of CDH1 or CTNNB1 in the two-month gestation group were used as the reference for normalization. All immunoblot assays and statistical analyses were performed in triplicate.

2.8. Statistical Analysis

All data were presented as the mean ± SD. qPCR and Western blot data were analyzed using Student's *t*-test (between two groups) or one-way analysis of variance (within multiple groups). Statistical analyses and graph construction were performed using Prism software (version 9.0; GraphPad, San Diego, CA, USA) and Adobe Illustrator (Adobe, San Jose, CA, USA). Statistical significance was set to $p < 0.05$.

3. Results

3.1. Morphology of TCs in Yak Placenta at Different Gestation Stages

H&E staining results at two months of gestation revealed that the chorion extended villi while forming folds, with primary villi progressively branching into shorter secondary villi with sparse blood vessels in the villous stroma (Figure 1(A1)). At this stage, the trophoblast was structured into inner cytotrophoblast (CTB) and outer syncytiotrophoblast (STB) layers, with abundant trophoblast giant cells (TGCs) and uninucleate trophoblast cells (UTCs) distributed throughout the tissue (Figure 1(A2)). At five months of gestation, the placental STB and CTB showed clear demarcation and vacuolated structures. The volume of the fetal cotyledonary villus (FCV) was significantly increased through branching into secondary/tertiary villi, enhancing the maternal–fetal interface, and a dense capillary network was observed in the villous stroma (Figure 1(B1)). The nuclei of UTCs appeared round or oval upon light staining, whereas those of multinucleated TGCs were larger and more intensely stained (Figure 1(B2)). The IF results revealed clearly defined DAPI-labeled nuclei in the placental tissues of both groups (Figure 1C). Compared with the two-month gestation group, the five-month gestation group exhibited significantly increased fluorescence intensity of the TC marker CK-7, tight junction protein occludin, and adhesion molecule marker ICAM1 (Figure 1D–F). The co-localization results showed that CK-7, occludin, and ICAM1 were widely expressed in FCV and co-localized in the cytoplasm of both TGCs and UTCs (Figure 1G). These findings highlight the dynamic structural changes that occur in the placenta during early and middle pregnancy, particularly during TC invasion and differentiation. The upregulation of occludin and ICAM1 underscores the critical role of cell adhesion and junction formation in TC invasion and differentiation.

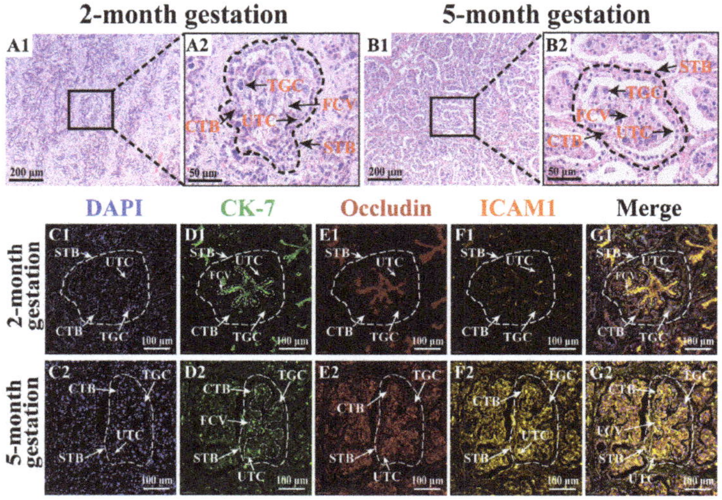

Figure 1. Morphological observation of yak placental tissues at two and five months of gestation. (**A**,**B**) H&E staining results. (**C**–**G**) Co-localization analysis of occludin and ICAM1: nuclei (blue, (**C1**,**C2**)), CK-7 (green, (**D1**,**D2**)), occludin (red, (**E1**,**E2**)), ICAM1 (orange, (**F1**,**F2**)), and merged CK-7, occludin, and ICAM1 staining (**G1**,**G2**). CTB, cytotrophoblast; STB, syncytiotrophoblast; UTC, uninucleate trophoblast cell; TGC, trophoblast giant cell; FCV, fetal cotyledonary villi. Scale bars: 200 μm (100× magnification), 100 μm (200× magnification), 50 μm (400× magnification).

3.2. Proteome Analysis and Protein Identification and Annotation Results

Compared with the two-month gestation group, 1622 DEPs, including 836 upregulated and 786 downregulated DEPs, were screened in the five-month gestation group (Figure 2A, Table S3). These DEPs were subjected to GO term ($p < 0.01$, Table S4) and KEGG pathway ($p < 0.05$, Table S5) annotation, which identified 53 biological processes (BP), 28 molecular functions (MF), 34 cellular components (CC), and 52 pathways (Figure 2B). Most BPs were associated with cell adhesion and migration, immune responses, and developmental processes (Figure 2C). Most MFs were related to their binding activities (ions, proteins, and lipids), receptor activities, and enzymatic functions (Figure 2D). Most CCs were associated with membrane structure and function, extracellular and adhesion processes, and cell junctions (Figure 2E). KEGG enrichment analysis revealed that most pathways were related to the immune response and inflammation, cancer and cell growth, and hormone/steroid regulation and metabolism (Figure 2F). Proteome analysis identified 1622 DEPs enriched in GO terms and pathways, suggesting their relevance to the molecular mechanisms underlying placental development.

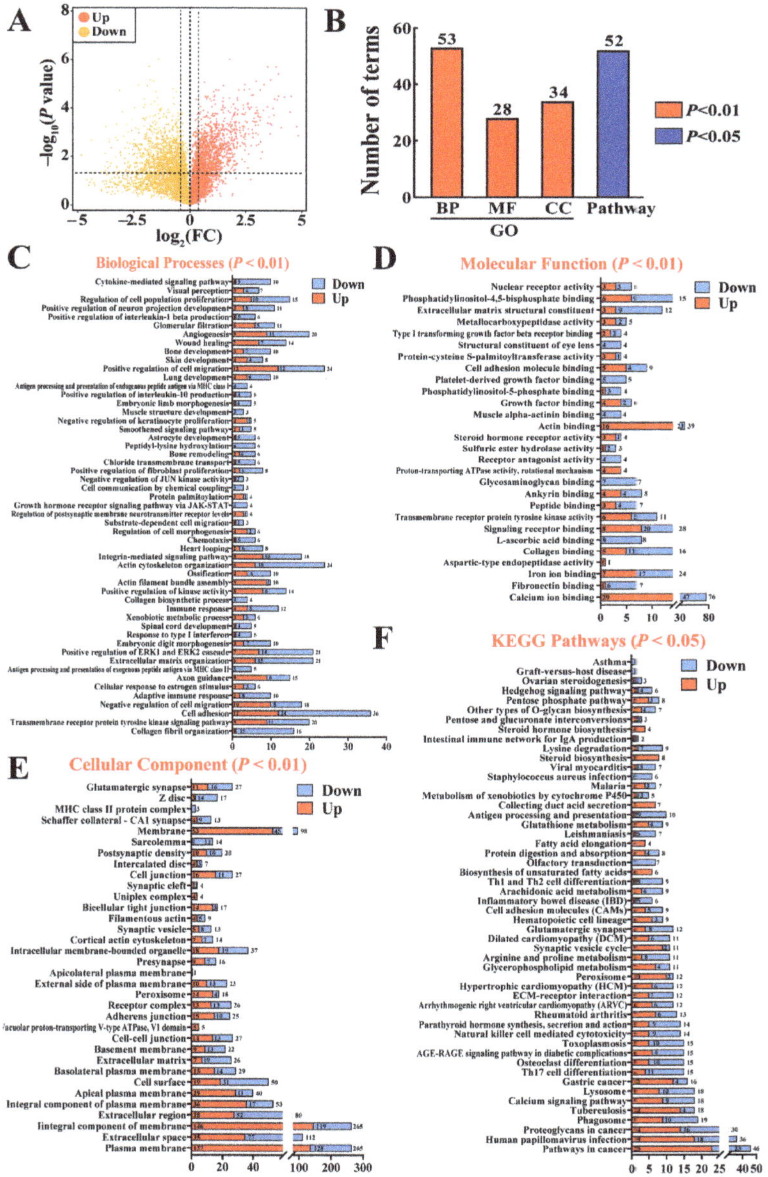

Figure 2. Summary of proteome analysis and protein identification and annotation results from DIA analysis. (**A**) Volcano plot of 1622 DEPs, including 836 upregulated and 786 downregulated DEPs. (**B**) Summary of GO terms for biological processes (BP), molecular functions (MF), cellular components (CC) ($p < 0.01$), and pathway annotation ($p < 0.05$). (**C–F**) GO annotation results for BPs, MFs, and CCs, along with pathway annotation.

3.3. Identification of Candidate DEPs Related to Cell Adhesion and Junctions from GO Terms

GO terms related to cell adhesion and junction formation, along with associated candidate DEPs, were screened based on the 1622 DEPs identified from DIA proteomics (Figure 3). Six GO terms ($p < 0.01$), including cell junctions, adherens junctions, cell–cell junctions, cell adhesion, bicellular tight junctions, and cell adhesion molecule binding, and 103 candidate DEPs were identified (Figure 3A, Table S6). The UpSet–Venn diagram indicated that CTNNB1 was shared among five GO terms, whereas CDH1 was shared among four GO terms (Figure 3B). The volcano plot and heatmap further showed that the candidate proteins, which included 54 downregulated and 49 upregulated proteins, were differentially expressed between the two-month and five-month gestation groups. Notably, CTNNB1 and CDH1 were upregulated in the placental tissue at five months of gestation (Figure 3C,D). ClueGO analysis revealed that 15 of the 103 DEPs were directly related to 23 GO terms, and that the important candidate DEPs CTNNB1 and CDH1 were associated with 18 GO terms related to cell adhesion and seven GO terms related to cell junctions, respectively (Figure 3E). These findings indicate that CTNNB1 and CDH1 play pivotal roles in yak pregnancy, likely by regulating cell adhesion and junction formation.

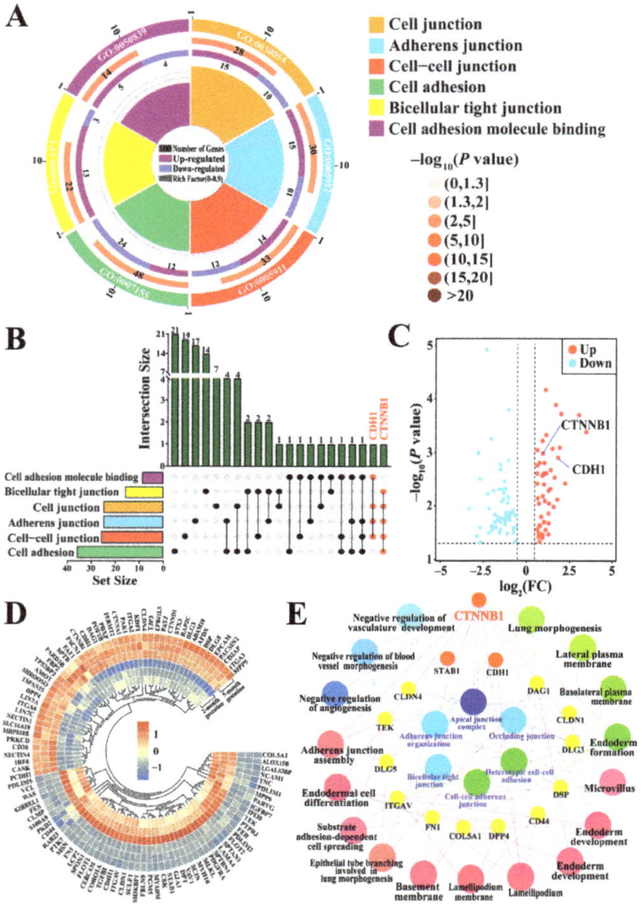

Figure 3. Identification of candidate DEPs related to cell adhesion and junction formation based on GO enrichment data. (A) Candidate DEPs and GO terms. (B) UpSet–Venn diagram of six GO terms

and 103 DEPs. (**C**) Volcano plot of 103 DEPs, including 54 downregulated and 49 upregulated DEPs. (**D**) Heatmap of 103 candidate DEPs. (**E**) Interactive relationship analysis of candidate DEPs and GO terms performed using ClueGO.

3.4. Identification of Pathways and Candidate DEPs Related to CTNNB1 and CDH1

Pathways involving CTNNB1 and CDH1 were selected to identify their potential roles in cell adhesion and junction formation in yak placental tissues at two and five months of gestation (Figure 4). Six pathways, including cell adhesion molecules, arrhythmogenic right ventricular cardiomyopathy, proteoglycans in cancer, pathways in cancer, human papillomavirus infection, and gastric cancer, were identified by DIA proteomic analysis. Among these, gastric cancer and pathways in cancer were the two key pathways associated with CTNNB1 and CDH1 (Figure 4A, Table S7). The interaction network of these two pathways suggests that CTNNB1 is involved in four pathways, whereas CDH1 is involved in three. The DEPs associated with the two key pathways were significantly differentially expressed between the two-month and five-month gestation groups (Figure 4B). A total of 48 candidate DEPs, including 23 downregulated and 25 upregulated DEPs, were identified within these two pathways (Figure 4C). ClueGO analysis indicated that 16 of the 48 DEPs were directly related to 17 pathways, with CTNNB1 and CDH1 involved in eight pathways related to cell adhesion and junction formation and seven cancer-related KEGG pathways, respectively (Figure 4D). These findings suggest that CTNNB1 and CDH1 may regulate placental development during pregnancy, likely through their involvement in cell adhesion and junction formation, as well as cancer-related pathways.

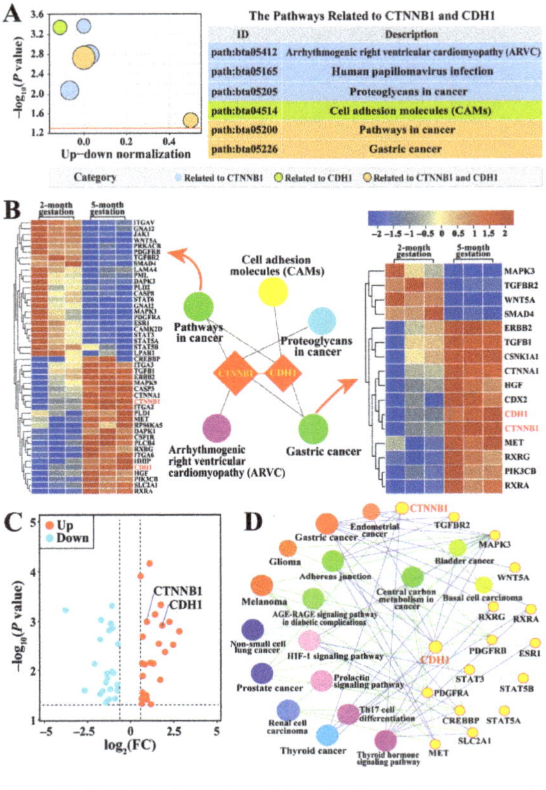

Figure 4. Identification of candidate DEPs and pathways related to cell adhesion and junction formation involving CTNNB1/CDH1. (**A**) Candidate DEPs and pathways. (**B**) Protein–protein

interaction network and heatmap analysis of pathways. (**C**) Volcano plot of 48 DEPs, including 23 downregulated and 25 upregulated DEPs. (**D**) Interactive relationship analysis of candidate DEPs and pathways performed using ClueGO.

3.5. Distribution and Co-Localization Analysis of CTNNB1 and CDH1 Proteins in Yak Placental Tissues

Positive expression of CTNNB1 and CDH1 was observed in the FCV and cytoplasm of TCs at both two and five months of gestation (Figure 5A,B). In the two-month gestation group, CTNNB1 and CDH1 were detected in UTCs and TGCs and sparsely detected in FCV (Figure 5(A1,B1)). The staining intensity of CTNNB1 and CDH1 was enhanced at five months of gestation in UTCs, TGCs, and FCV (Figure 5(A2,B2)). No positive staining for CTNNB1 or CDH1 was observed in the negative control group (Figure 5(C1,C2)). IOD analysis revealed significantly upregulated expression of CTNNB1 and CDH1 in the five-month gestation group compared to that in the two-month gestation group ($p < 0.01$) (Figure 5D,E). The IF results revealed clearly defined nuclei labeled with DAPI in placental tissues at both two and five months of gestation (Figure 5F). The fluorescence intensity of CK-7 was significantly higher at five months of gestation than at two months of gestation (Figure 5G). Positive fluorescence signals for CTNNB1 and CDH1 were present in the cytoplasm of TGCs and UTCs in both two-month and five-month gestation groups, with stronger signal intensity in the latter group (Figure 5H,I). The co-localization results indicated that CK-7, CTNNB1, and CDH1 were co-localized in the cytoplasm of TGCs and UTCs (Figure 5J). These findings highlight the upregulated expression of CTNNB1 and CDH1 and their cooperative roles in TC adhesion and junction formation during pregnancy.

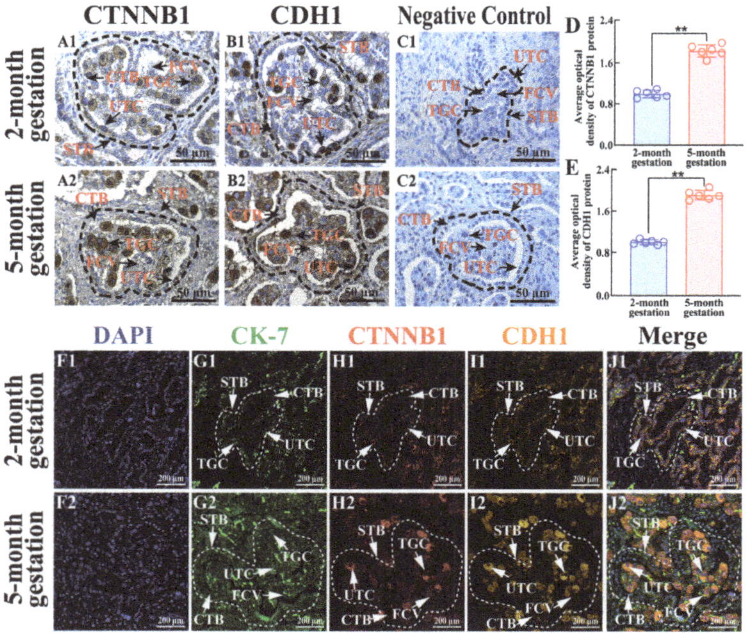

Figure 5. Distribution and co-localization analysis of CTNNB1/CDH1 in yak placental tissues at two and five months of gestation. (**A**,**B**) Intracellular localization analysis. (**C**) Negative control. (**D**,**E**) Gray values of positive expression of CTNNB1 and CDH1 proteins quantified using ImageJ software. (**F**–**J**) Co-localization analysis: nuclei (blue, (**F1**,**F2**)), CK-7 (green, (**G1**,**G2**)), CTNNB1 (red, (**H1**,**H2**)), CDH1 (orange, (**I1**,**I2**)), merged with CK7, CTNNB1, and CDH1 (**J1**,**J2**) staining. CTB, cytotrophoblast;

STB, syncytiotrophoblast; UTC, uninucleate trophoblast cell; TGC, trophoblast giant cell; FCV, fetal cotyledonary villi. Scale bars: 200 μm (100× magnification), 50 μm (400× magnification). ** represents $p < 0.01$.

3.6. *CTNNB1* and *CDH1* mRNA and Protein Expression Patterns in Yak Placental Tissues

To further investigate the impact of placental TC adhesion and junction formation on pregnancy, the relative expression levels of *CTNNB1* and *CDH1* mRNA and protein were evaluated in yak placental tissues using qPCR and Western blot (Figure 6). Compared to the two-month gestation group, the five-month gestation group showed significantly upregulated relative mRNA expression levels of *CTNNB1* and *CDH1* ($p < 0.01$) (Figure 6A,B). Moreover, CTNNB1 and CDH1 protein expressions were present but were different in each sample of the two groups (Figure 6C). IOD values indicated that CTNNB1 and CDH1 expressions were significantly higher in the five-month gestation group than in the two-month gestation group ($p < 0.01$) (Figure 6D,E). Overall, these findings suggest that CTNNB1 and CDH1 may play critical roles in maintaining placental structure and function, and that their upregulation is strongly correlated with TC adhesion and junctional integrity in the placenta.

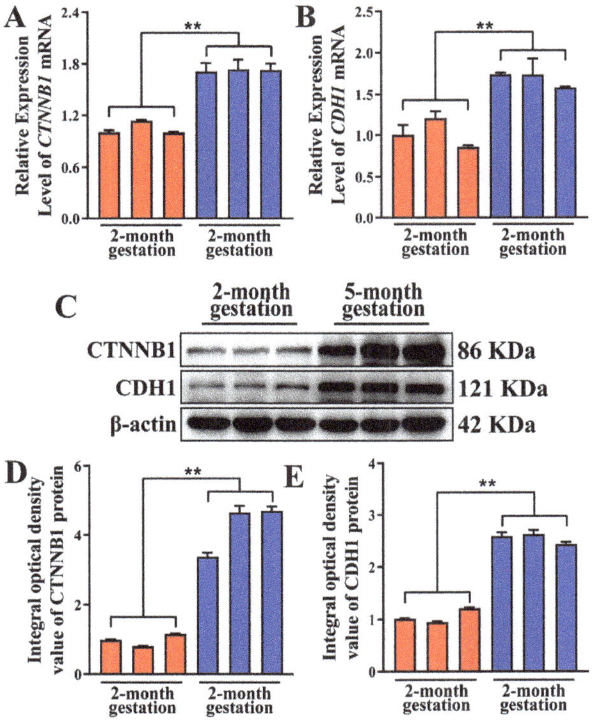

Figure 6. Pattern analysis of *CTNNB1* and *CDH1* mRNA and protein expression in yak placental tissues. (**A**,**B**) Expression levels of *CTNNB1* and *CDH1* mRNA monitored by qPCR. (**C**) Protein bands of CTNNB1, CDH1, and β-actin in placental tissues. (**D**,**E**) Integral optical density values of CTNNB1 and CDH1 proteins in yak placental tissues, respectively. Data are presented as means ± SEM. ** represents $p < 0.01$.

4. Discussion

Cell adhesion and junction formation are critical to maintaining TC structural stability and functional integrity, which directly influence placental development and pregnancy maintenance [12,28]. Abnormalities in these processes can disrupt placental growth, material transport, immune regulation, maternal–fetal interface stability, and placental tissue remodeling [7,11]. Thus, investigating TC morphological changes and identifying key functional proteins involved in cell adhesion and junction formation can offer valuable insights into the mechanisms underlying yak placental development.

Morphological changes in TCs during early- and mid-pregnancy are essential for normal placental development [29,30]. During early pregnancy, TCs migrate and invade the surrounding decidua and maternal blood vessels, forming an intervillous space serving as the foundation for placental growth and function [31]. In middle pregnancy, TC invasiveness decreases, whereas adhesion, junction formation, and differentiation capacities increase, leading to clearer boundaries between STBs and CTBs, thus ensuring maternal–fetal nutrient exchange and pregnancy stability [32]. Likewise, our findings suggest that TC invasion, migration, and differentiation are closely associated with cell adhesion and junction formation. TC invasiveness and migration are finely regulated by the uterine microenvironment, with various proteases and adhesion molecules being directly or indirectly involved in the infiltration and adhesion processes [8,12]. For example, increased ICAM1 expression in TCs promotes the adhesion between trophoblasts and endometrial cells, facilitating proper differentiation and syncytiotrophoblast formation [33]. Additionally, occludin, a key tight junction protein, is positively expressed in both STBs and CTBs, sealing intercellular gaps and preventing the non-selective permeability of ions, water, and other molecules, thus maintaining placental barrier function and preventing immune rejection [12,34]. Compared to early pregnancy TCs, those in mid-pregnancy showed significantly upregulated expression of occludin and ICAM1, highlighting the importance of cell adhesion and junction formation processes in placental homeostasis. This phenomenon may result from the increasing demand for enhanced placental barrier function as pregnancy progresses, which increases cell junction protein expression. During mid-pregnancy, the increased need for maternal–fetal material exchange promotes stable cell adhesion and tight junction formation, maintaining placental structure and function [34,35]. However, the relationship between these processes and placental development, and the underlying molecular mechanisms, requires further investigation in yaks.

Based on DIA proteomics data, we identified six key GO terms and 103 DEPs, particularly CTNNB1 and CDH1, involved in cell junctions, adhesion junctions, and cell adhesion GO terms and are likely critical to regulating placental development in yaks. CTNNB1 and CDH1 form cell adhesion complexes that promote stable intercellular adhesion and cell polarity [35,36]. This adhesion mechanism is essential for placental barrier function. Specifically, CDH1 expression level regulates trophoblast cell invasiveness, with lower expression in early pregnancy facilitating trophoblast migration, while upregulation in mid-pregnancy contributes to establishing and stabilizing the placental barrier [37,38]. CTNNB1 regulates cell proliferation and differentiation through the Wnt signaling pathway, further influencing placental maturation [17,35,39]. Additionally, CTNNB1 and CDH1 are involved in endoderm formation during early embryonic development; however, their precise mechanisms require further exploration [40]. Pathway analysis revealed that they participate in cancer-related pathways, including gastric cancer and pathways in cancer. CTNNB1 regulates cell proliferation and differentiation through the Wnt signaling pathway, and its mutations can lead to uncontrolled cell proliferation, suppressed apoptosis, and activation of tumor invasion and metastasis [35,36]. Downregulation of CDH1 disrupts intercellular adhesion, promotes epithelial–mesenchymal transition in tumor cells, and enhances their invasive and metastatic capabilities [41]. Similarly to the regulation of cancer cell proliferation, TC adhesion and junction functions during placental development may be regulated by CTNNB1 and CDH1. Abnormal expression of CTNNB1 and CDH1 was linked to pregnancy-related placental disorders. For example, in patients with preeclamptic, dysfunction and abnormal differentiation of the CTB in the placenta are linked to the dysregulated expression of CTNNB1 and CDH1 [42]. Additionally, gestational diabetes mellitus may be linked to excessive CTNNB1 activation, which interferes with placental metabolic function and downregulates CDH1, thus disrupting placental barrier function [42]. ClueGO analysis suggested that other potential key DEPs in related GO processes and cancer-related pathways may also regulate placental development and pregnancy processes, warranting further investigation. For example, cluster of differentiation 44 [43] in endoderm development, stabilin-1 [44] in the negative regulation of vasculature development, and fibronectin [45] and integrin subunit alpha V [46] in substrate adhesion-dependent cell spreading promote TC migration and invasion, maintain placental stability, and facilitate embryo implantation. Importantly, interactions between CTNNB1/CDH1 and these key DEPs highlight their critical role in placental development.

CTNNB1 and CDH1 form an adhesion complex that maintains tight intercellular connections during cell adhesion [35,36]. Mutations in CTNNB1/CDH1 may disrupt Wnt signaling, hinder TC adhesion during embryonic development, impair embryo implantation, and affect normal fetal development [17,39,40]. CTNNB1/CDH1 are critical to inhibiting cell migration and maintaining normal embryonic development during pregnancy, suggesting that they may regulate pregnancy through cell adhesion and junction processes. Our results are consistent with such findings, but studies on the role of CTNNB1/CDH1 in yak placental development during pregnancy are limited. Our DIA proteomics results showed that CTNNB1/CDH1 expression was higher in mid-pregnancy, suggesting its involvement in TC adhesion and junction formation in yak placental tissues, which is consistent with the mRNA and protein expression results. CTNNB1, a cytoskeletal component, forms complexes with CDH1 on the inner side of the epithelial cell membrane, promoting cell adhesion, morphological stability, and signal transduction via cytoskeletal interactions [47,48]. CTNNB1 is localized in the cytoplasm and involved in cytoskeletal connections [48], whereas CDH1 is localized in the Golgi apparatus and participates in intracellular transport and degradation processes [49]. CTNNB1/CDH1 is primarily localized in the cell membrane and cytoplasm, maintaining intercellular adhesion and tissue structure integrity [47–49]. We confirmed that CTNNB1/CDH1 was predominantly expressed in the cytoplasm of FCV, UTCs, and TGCs, suggesting that high CTNNB1/CDH1 expression in mid-pregnancy supports TC adhesion, junction formation, and differentiation, thereby contributing to the gradual development and stability of the placental structure and ensuring pregnancy progression.

Based on these results and those of previous studies, we propose a potential mechanism underlying the roles of CTNNB1 and CDH1 in pregnancy (Figure 7). During pregnancy, TCs in the placenta undergo key adhesion changes, transitioning from weak to strong intercellular adhesions to appropriate invasion and subsequent stabilization, which are essential for placental development [12,34]. Specifically, early in pregnancy, limited CDH1 expression results in weaker adhesion between TCs, promoting their migration into and invasion of the uterine lining [12,50]. CTNNB1 was internalized and degraded, further reducing adhesion strength [51]. As pregnancy progresses to mid-pregnancy, the increased CDH1 and CTNNB1 expression enhance adhesion by forming stable CTNNB1–CDH1 complexes aided by calcium ions (Ca^{2+}) [37]. Additionally, recycling of CTNNB1 to the cell membrane further strengthens adhesion, providing stability to the developing placenta [38]. Free CTNNB1 can also activate the Wnt signaling pathway, promoting TC growth and differentiation [39]. Therefore, the dynamic regulation of CTNNB1/CDH1 is crucial for maintaining the balance between TC invasion and stable adhesion during placental development.

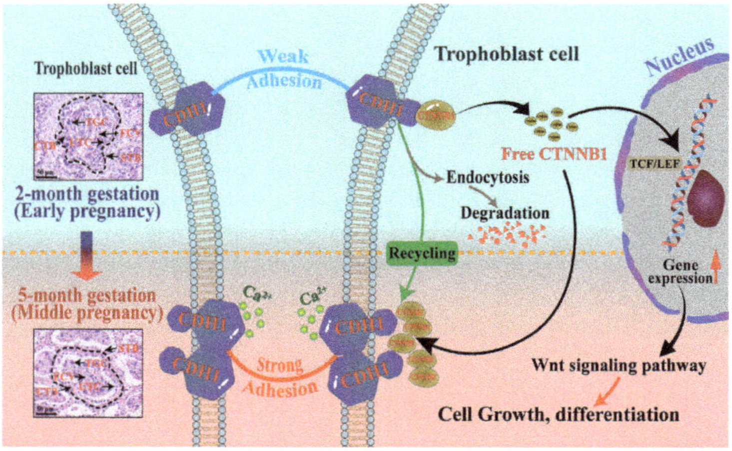

Figure 7. Potential molecular mechanisms by which CTNNB1 and CDH1 mediate adhesion and junction formation in placental trophoblast cells. Scale bar: 50 μm (400× magnification).

However, our study has limitations. Specifically, further validation using TC experiments and pregnancy animal models is required to verify the mechanisms of cell adhesion and junction formation and improve the reliability of our findings. Nevertheless, we have systematically identified candidate DEPs associated with cell adhesion and junction formation, providing new insights into the molecular mechanisms underlying yak placental development during pregnancy.

5. Conclusions

This study revealed the crucial role of CTNNB1/CDH1 in maintaining placental homeostasis and regulating trophoblast adhesion and junctions in yaks during pregnancy. Based on DIA proteomics data, we identified 103 DEPs associated with six GO terms and 48 DEPs associated with two pathways. Bioinformatics analysis indicated that CTNNB1 and CDH1 are involved in placental development and pregnancy maintenance via cell adhesion and junction processes as well as cancer-related pathways. Furthermore, IHC and IF staining localized CTNNB1 and CDH1 predominantly in the cytoplasm of TCs, especially UTCs and TGCs. The mRNA and protein expression levels of *CTNNB1* and *CDH1* were significantly upregulated at five months of gestation compared to that at two months of gestation. In conclusion, this study provides new insights into the molecular mechanisms underlying TC adhesion and junction formation during yak placental development, as well as a theoretical foundation for improving yak reproductive performance.

Supplementary Materials: The following supporting information can be downloaded at: https://www.mdpi.com/xxx/s1. Table S1: The used antibodies and dilution ratio in the present study. Table S2: qPCR primer sequences. Table S3: The 1622 DEPs generated from the DIA proteomics sequence database. Table S4: Data of KEGG pathway with $p < 0.05$. Table S5: Data for GO terms with $p < 0.05$. Table S6: The six significant GO terms and 103 DEPs identified from DIA proteomics. Table S7: The six significant KEGG pathways and 91 DEPs identified from DIA proteomics. Figures S1–S3: Whole Western blot of CTNNB1 (86 kDa), CDH1 (121 kDa), and β-actin (42 kDa).

Author Contributions: Conceptualization, B.Z. (Bohao Zhang) and Q.Z.; methodology, B.Z. (Bohao Zhang), C.S. and J.Z.; software, B.Z. (Bohao Zhang) and B.Z. (Bin Zhou); validation, B.Z. (Bohao Zhang) and J.Z.; formal analysis, C.S. and J.Z.; investigation, W.D. and Q.Z.; resources, Y.Z., X.Z. and Q.Z.; data curation, B.Z. (Bohao Zhang), W.D. and Q.Z.; writing—original draft preparation, B.Z. (Bohao Zhang) and W.D.; writing—review and editing, Y.Z. and Q.Z.; visualization, B.Z. (Bohao Zhang) and Q.Z.; supervision, Y.Z. and X.Z.; project administration, X.Z. and Q.Z.; funding acquisition, Y.Z. and Q.Z. All authors have read and agreed to the published version of the manuscript.

Funding: This study was supported by the National Natural Science Foundation of China (32360836), the Key Project of Natural Science Foundation of Gansu Province (23YFNA0012), and the National Natural Science Foundation of China (32060752).

Institutional Review Board Statement: The animal study protocol was approved by the Ethics Committee of Gansu Agricultural University (approval number: GSAU-Eth-LST-2021-003).

Informed Consent Statement: Not applicable.

Data Availability Statement: The data that support the findings of this study are available from the corresponding author upon reasonable request.

Conflicts of Interest: The authors declare that no conflicts of interest exist.

References

1. Zhang, Q.; Wang, Q.; Gong, J.; Du, J.; Zhang, Y.; Zhao, X.J.G. Yak igf2 promotes fibroblast proliferation via suppression of igf1r and pi3kcg expression. *Genes* **2018**, *9*, 169. [CrossRef] [PubMed]
2. Pei, J.; Zhao, S.; Yin, M.; Wu, F.; Li, J.; Zhang, G.; Wu, X.; Bao, P.; Xiong, L.; Song, W.; et al. Differential proteomics of placentas reveals metabolic disturbance and oxidative damage participate yak spontaneous miscarriage during late pregnancy. *BMC Vet. Res.* **2022**, *18*, 248. [CrossRef] [PubMed]
3. Ner-Kluza, J.; Wawrzykowski, J.; Franczyk, M.; Siberring, J.; Kankofer, M. Identification of protein patterns in bovine placenta at early-mid pregnancy–Pilot studies. *Rapid Commun. Mass Spectrom.* **2019**, *33*, 1084–1090. [CrossRef] [PubMed]
4. Ortega, M.A.; Fraile-Martínez, O.; García-Montero, C.; Sáez, M.A.; Álvarez-Mon, M.A.; Torres-Carranza, D.; Álvarez-Mon, M.; Bujan, J.; García-Honduvilla, N.; Bravo, C.J.C. The pivotal role of the placenta in normal and pathological pregnancies: A focus on preeclampsia, fetal growth restriction, and maternal chronic venous disease. *Cells* **2022**, *11*, 568. [CrossRef]
5. Tanner, A.R.; Kennedy, V.C.; Lynch, C.S.; Hord, T.K.; Winger, Q.A.; Rozance, P.J.; Anthony, R.V. In vivo investigation of ruminant placenta function and physiology—A review. *J. Anim. Sci.* **2022**, *100*, skac045. [CrossRef]
6. Igwebuike, U. Trophoblast cells of ruminant placentas—A minireview. *Anim. Reprod. Sci.* **2006**, *93*, 185–198. [CrossRef]
7. Watson, E.D.; Hughes, M.; Simmons, D.G.; Natale, D.R.; Sutherland, A.E.; Cross, J.C. Cell–cell adhesion defects in Mrj mutant trophoblast cells are associated with failure to pattern the chorion during early placental development. *Dev. Dyn.* **2011**, *240*, 2505–2519. [CrossRef]
8. Cao, C.; Dai, Y.; Wang, Z.; Zhao, G.; Duan, H.; Zhu, X.; Wang, J.; Zheng, M.; Weng, Q.; Wang, L. The role of junctional adhesion molecule-C in trophoblast differentiation and function during normal pregnancy and preeclampsia. *Placenta* **2022**, *118*, 55–65. [CrossRef]
9. Zhou, H.; Zhao, C.; Wang, P.; Yang, W.; Zhu, H.; Zhang, S. Regulators involved in trophoblast syncytialization in the placenta of intrauterine growth restriction. *Front. Endocrinol.* **2023**, *14*, 1107182. [CrossRef]
10. Gumina, D.L.; Su, E.J. Mechanistic insights into the development of severe fetal growth restriction. *Clin. Sci.* **2023**, *137*, 679–695. [CrossRef]
11. Luan, X.; Zhai, J.; Li, S.; Du, Y. Downregulation of FHL2 suppressed trophoblast migration, invasion and epithelial–mesenchymal transition in recurrent miscarriage. *Reprod. BioMedicine Online* **2024**, *48*, 103342. [CrossRef] [PubMed]
12. Adu-Gyamfi, E.A.; Czika, A.; Gorleku, P.N.; Ullah, A.; Panhwar, Z.; Ruan, L.-L.; Ding, Y.-B.; Wang, Y.-X. The involvement of cell adhesion molecules, tight junctions, and gap junctions in human placentation. *Reprod. Sci.* **2021**, *28*, 305–320. [CrossRef] [PubMed]
13. Bridger, P.S.; Haupt, S.; Leiser, R.; Johnson, G.A.; Burghardt, R.C.; Tinneberg, H.-R.; Pfarrer, C. Integrin activation in bovine placentomes and in caruncular epithelial cells isolated from pregnant cows. *Biol. Reprod.* **2008**, *79*, 274–282. [CrossRef] [PubMed]
14. Johnson, G.A.; Burghardt, R.C.; Bazer, F.W.; Seo, H.; Cain, J.W. Integrins and their potential roles in mammalian pregnancy. *J. Anim. Sci. Biotechnol.* **2023**, *14*, 115. [CrossRef]

15. Xie, M.; McCoski, S.R.; Johnson, S.E.; Rhoads, M.L.; Ealy, A.D. Combinatorial effects of epidermal growth factor, fibroblast growth factor 2 and insulin-like growth factor 1 on trophoblast cell proliferation and embryogenesis in cattle. *Reprod. Fertil. Dev.* **2017**, *29*, 419–430. [CrossRef]
16. Lu, W.; Tu, Z.; Wang, S.; Lu, J.; Wang, Q.; Wang, W.; Wang, B.; Wang, H.; Ni, H.; Guo, Y. Spatiotemporal expression of Wnt signaling pathway components during bovine placental development. *Theriogenology* **2013**, *80*, 893–902. [CrossRef]
17. Dietrich, B.; Haider, S.; Meinhardt, G.; Pollheimer, J.; Knöfler, M.J.C.; Sciences, M.L. WNT and NOTCH signaling in human trophoblast development and differentiation. *Cell. Mol. Life Sci.* **2022**, *79*, 292. [CrossRef]
18. Assis Neto, A.C.; Pereira, F.; Santos, T.C.d.; Ambrosio, C.; Leiser, R.; Miglino, M. Morpho-physical recording of bovine conceptus (Bos indicus) and placenta from days 20 to 70 of pregnancy. *Reprod. Domest. Anim.* **2010**, *45*, 760–772. [CrossRef]
19. Zhang, Q.; Bai, X.; Lin, T.; Wang, X.; Zhang, B.; Dai, L.; Shi, J.; Zhang, Y.; Zhao, X. HMOX1 promotes ferroptosis in mammary epithelial cells via FTH1 and is involved in the development of clinical mastitis in dairy cows. *Antioxidants* **2022**, *11*, 2221. [CrossRef]
20. Zhang, Q.; Wang, Q.; Zhang, Y.; Cheng, S.; Hu, J.; Ma, Y.; Zhao, X. Comprehensive analysis of microrna–messenger rna from white yak testis reveals the differentially expressed molecules involved in development and reproduction. *Int. J. Mol. Sci.* **2018**, *19*, 3083. [CrossRef]
21. Zhang, Q.; Bai, X.; Shi, J.; Wang, X.; Zhang, B.; Dai, L.; Lin, T.; Gao, Y.; Zhang, Y.; Zhao, X. DIA proteomics identified the potential targets associated with angiogenesis in the mammary glands of dairy cows with hemorrhagic mastitis. *Front. Vet. Sci.* **2022**, *9*, 980963. [CrossRef] [PubMed]
22. Dai, L.; Zhang, Q.; Shi, J.; Bai, X.; An, X.; Zhang, B.; Zhang, Y.; Zhao, X. The distribution, expression patterns and functional analysis of NR1D1 and NR4A2 in the reproductive axis tissues of the male Tianzhu White Yak. *Animals* **2021**, *11*, 3117. [CrossRef] [PubMed]
23. Quan, J.; Kang, Y.; Li, L.; Zhao, G.; Sun, J.; Liu, Z. Proteome analysis of rainbow trout (Oncorhynchus mykiss) liver responses to chronic heat stress using DIA/SWATH. *J. Proteom.* **2021**, *233*, 104079. [CrossRef]
24. Bindea, G.; Galon, J.; Mlecnik, B. CluePedia Cytoscape plugin: Pathway insights using integrated experimental and in silico data. *Bioinformatics* **2013**, *29*, 661–663. [CrossRef]
25. Bathla, S.; Sindhu, A.; Kumar, S.; Dubey, S.K.; Pattnaik, S.; Rawat, P.; Chopra, A.; Mohanty, A.K. Quantitative proteomics revealed the putative biomarker for detection of early-stage intra-mammary gland infection in cow. *J. Proteins Proteom.* **2020**, *11*, 173–181. [CrossRef]
26. Livak, K.J.; Schmittgen, T.D. Analysis of relative gene expression data using real-time quantitative PCR and the 2−ΔΔCT method. *Methods* **2001**, *25*, 402–408. [CrossRef]
27. Zhang, B.; Lin, T.; Bai, X.; An, X.; Dai, L.; Shi, J.; Zhang, Y.; Zhao, X.; Zhang, Q. Sulfur Amino Acid Metabolism and the Role of Endogenous Cystathionine-γ-lyase/H2S in Holstein Cows with Clinical Mastitis. *Animals* **2022**, *12*, 1451. [CrossRef]
28. Sağsöz, H.; Liman, N.; Akbalık, M.E.; Alan, E.; Saruhan, B.G.; Ketani, M.A.; Erdoğan, S. Expression of cadherins and some connective tissue components in cow uterus and placenta during pregnancy. *Res. Vet. Sci.* **2022**, *151*, 64–79. [CrossRef]
29. Liu, B.; Cui, Y.; Yang, B.; Fan, J.; Zhao, Z.; Yu, S. Morphometric analysis of yak placentomes during gestation. *Anat. Rec. Adv. Integr. Anat. Evol. Biol.* **2010**, *293*, 1873–1879. [CrossRef]
30. Xu, T.; Liu, B.; Cui, Y.; He, J.; Fan, J.; Yu, S. A light and electron microscopy study of the yak placentome. *Folia Morphol.* **2019**, *78*, 818–826. [CrossRef]
31. Haeger, J.-D.; Hambruch, N.; Pfarrer, C. Placental development and its control in cattle. *Biosci. Proc.* **2019**, *8*, 10–1530. [CrossRef]
32. Huppertz, B. Traditional and new routes of trophoblast invasion and their implications for pregnancy diseases. *Int. J. Mol. Sci.* **2019**, *21*, 289. [CrossRef] [PubMed]
33. Boro, P.; Kumaresan, A.; Singh, A.; Gupta, D.; Kumar, S.; Manimaran, A.; Mohanty, A.; Mohanty, T.; Pathak, R.; Attupuram, N. Expression of short chain fatty acid receptors and pro-inflammatory cytokines in utero-placental tissues is altered in cows developing retention of fetal membranes. *Placenta* **2014**, *35*, 455–460. [CrossRef]
34. Aplin, J.; Jones, C.; Harris, L. Adhesion molecules in human trophoblast—A review. I. Villous trophoblast. *Placenta* **2009**, *30*, 293–298. [CrossRef]
35. Medrek, C.; Landberg, G.; Andersson, T.; Leandersson, K. Wnt-5a-CKIα signaling promotes β-catenin/E-cadherin complex formation and intercellular adhesion in human breast epithelial cells. *J. Biol. Chem.* **2009**, *284*, 10968–10979. [CrossRef]
36. Der, B.; Bugacov, H.; Briantseva, B.-M.; McMahon, A.P. Cadherin adhesion complexes direct cell aggregation in the epithelial transition of Wnt-induced nephron progenitor cells. *Development* **2024**, *151*, dev202303. [CrossRef]
37. Nelson, W.J. Regulation of cell–cell adhesion by the cadherin–catenin complex. *Biochem. Soc. Trans.* **2008**, *36*, 149–155. [CrossRef]
38. Nakano, H.; Shimada, A.; Imai, K.; Takahashi, T.; Hashizume, K. The cytoplasmic expression of E-cadherin and β-catenin in bovine trophoblasts during binucleate cell differentiation. *Placenta* **2005**, *26*, 393–401. [CrossRef]
39. Sonderegger, S.; Pollheimer, J.; Knöfler, M. Wnt signalling in implantation, decidualisation and placental differentiation–review. *Placenta* **2010**, *31*, 839–847. [CrossRef]

40. van der Wal, T.; Van Amerongen, R. Walking the tight wire between cell adhesion and WNT signalling: A balancing act for β-catenin. *Open Biol.* **2020**, *10*, 200267. [CrossRef]
41. Ma, F.; Li, W.; Liu, C.; Li, W.; Yu, H.; Lei, B.; Ren, Y.; Li, Z.; Pang, D.; Qian, C. MiR-23a promotes TGF-β1-induced EMT and tumor metastasis in breast cancer cells by directly targeting CDH1 and activating Wnt/β-catenin signaling. *Oncotarget* **2017**, *8*, 69538. [CrossRef] [PubMed]
42. Nonn, O.; Debnath, O.; Valdes, D.S.; Sallinger, K.; Secener, A.K.; Haider, S.; Fischer, C.; Tiesmeyer, S.; Nimo, J.; Kuenzer, T. Disturbed trophoblast transition links preeclampsia progression from placenta to the maternal syndrome. *bioRxiv* **2022**. [CrossRef]
43. Takahashi, H.; Takizawa, T.; Matsubara, S.; Ohkuchi, A.; Kuwata, T.; Usui, R.; Matsumoto, H.; Sato, Y.; Fujiwara, H.; Okamoto, A. Extravillous trophoblast cell invasion is promoted by the CD44–hyaluronic acid interaction. *Placenta* **2014**, *35*, 163–170. [CrossRef] [PubMed]
44. Kim, S.-Y.; Lee, E.-H.; Kim, E.N.; Son, W.-C.; Kim, Y.H.; Park, S.-Y.; Kim, I.-S.; Kim, J.-E. Identifying stabilin-1 and stabilin-2 double knockouts in reproduction and placentation: A descriptive study. *Int. J. Mol. Sci.* **2020**, *21*, 7235. [CrossRef]
45. Hua, Q.; Li, Z.; Zhou, Y.; Wang, Y.; Yu, Y.; Sun, L.; Ye, J.; Li, L. Single-cell RNA sequencing reveals association of aberrant placental trophoblasts and FN1 reduction in late-onset fetal growth restriction. *Placenta* **2024**, *146*, 30–41. [CrossRef]
46. Frank, J.W.; Seo, H.; Burghardt, R.C.; Bayless, K.J.; Johnson, G.A. ITGAV (alpha v integrins) bind SPP1 (osteopontin) to support trophoblast cell adhesion. *Reproduction* **2017**, *153*, 695–706. [CrossRef]
47. Ishiguro, H.; Wakasugi, T.; Terashita, Y.; Sakamoto, N.; Tanaka, T.; Mizoguchi, K.; Sagawa, H.; Okubo, T.; Takeyama, H. Decreased expression of CDH1 or CTNNB1 affects poor prognosis of patients with esophageal cancer. *World J. Surg. Oncol.* **2016**, *14*, 240. [CrossRef]
48. Tanabe, S.; Kawabata, T.; Aoyagi, K.; Yokozaki, H.; Sasaki, H. Gene expression and pathway analysis of CTNNB1 in cancer and stem cells. *World J. Stem Cells* **2016**, *8*, 384. [CrossRef]
49. Kusafuka, K.; Yamada, H.; Ishino, K.; Maeda, M.; Yamanegi, K.; Baba, S.; Ohuchi, T.; Inagaki, H.; Yamamoto, H.; Iwasaki, T. Salivary Duct Carcinoma with Rhabdoid Features—No or Aberrant Expression of E-cadherin and Genetic Changes in CDH1: Immunohistochemical and Genetic Analyses of 17 Cases. *Am. J. Surg. Pathol.* **2021**, *45*, 439–449. [CrossRef]
50. Jalali, B.M.; Lukasik, K.; Witek, K.; Baclawska, A.; Skarzynski, D.J. Changes in the expression and distribution of junction and polarity proteins in the porcine endometrium during early pregnancy period. *Theriogenology* **2020**, *142*, 196–206. [CrossRef]
51. Herington, J.L.; Bi, J.; Martin, J.D.; Bany, B.M. β-Catenin (CTNNB1) in the mouse uterus during decidualization and the potential role of two pathways in regulating its degradation. *J. Histochem. Cytochem.* **2007**, *55*, 963–974. [CrossRef]

animals

MDPI

Article

Variation in Ovine *KRTAP13-3* and Its Association with Wool Characteristics in Chinese Tan Sheep

Lingrong Bai [1,2], Huitong Zhou [1,2], Jinzhong Tao [1,3] and Jon G. H. Hickford [1,2,*]

1 International Wool Research Institute, Faculty of Animal Science and Technology, Gansu Agricultural University, Lanzhou 730070, China; lingrong.bai@lincolnuni.ac.nz (L.B.); huitong.zhou@lincoln.ac.nz (H.Z.); tao_jz@nxu.edu.cn (J.T.)

2 Gene-Marker Laboratory, Faculty of Agriculture and Life Sciences, Lincoln University, Lincoln 7647, New Zealand

3 College of Animal Science and Technology, Ningxia University, Yinchuan 750021, China

* Correspondence: jon.hickford@lincoln.ac.nz

Simple Summary: Wool is a valuable natural material, but its quality can vary. This research explored how a gene that encodes a specific wool protein affects some wool characteristics in Chinese Tan sheep. This is a breed known for producing fibre with a unique crimp pattern. Five different variants of the gene *KRTAP13-3* were identified, and this genetic variation was found to be associated with two traits that describe variability in the heterotypic hair fibres of the sheep. This gene may have potential value as a gene marker for assisting with the genetic improvement of wool traits.

Abstract: Understanding the genetic factors that influence wool quality is essential for enhancing wool quality and uniformity. This study investigated the *KRTAP13-3* gene in Chinese Tan sheep, a breed known for its unique wool characteristics. We analysed 232 sheep and revealed five previously identified sequence variants of *KRTAP13-3* and then modelled to ascertain whether there was any association between the nucleotide sequence variation and variation in mean fibre diameter (MFD), fibre diameter standard deviation (FDSD), coefficient of variation of fibre diameter (CVFD), and mean fibre curvature (MFC). Twelve genotypes were observed, with the five variants having frequencies that ranged from 64.0% to 1.1%. Among the four variants with frequencies above 5%, nucleotide sequence variation was associated with heterotypic hair fibre diameter variation. The most common variant (*A*) was linked to increased FDSD and CVFD, while two other variants (*B* and *D*) revealed trends towards being associated with decreased CVFD. No associations were found with variation in the fine wool fibres from the Tan sheep. This suggests that *KRTAP13-3* plays a role in regulating heterotypic hair fibre diameter variability and that it could possibly be a gene marker for improving wool traits.

Keywords: keratin-associated protein 13-3 gene; polymorphism; FDSD; CVFD; heterotypic hair; Chinese Tan sheep

1. Introduction

Mammalian genomes contain large gene families, including olfactory receptor genes, immunoglobulin genes, zinc finger protein genes, solute carrier protein genes, T-cell receptor genes, ATPase genes, keratin genes, and keratin-associated protein (KAP) genes. The latter two families play a key role in defining the structure and function of the sheep pelage called wool. Wool has unique properties, yet its use is restricted because of variation in its physical and chemical characteristics. This can affect its value. The heritability of many valued wool traits is moderate to high [1,2], and this presents opportunities for genetic improvement. However, achieving this improvement would require a comprehensive understanding of the genetic factors that regulate these traits.

Wool primarily consists of proteins, with the wool keratins forming structures called intermediate filaments. The KAPs create a matrix that embeds these filaments through various interactions [3]. They are classified into the high sulphur (HS) group, the ultra-high sulphur (UHS) group, and the high glycine–tyrosine (HGT) group [3]. While the interactions between the HGT-KAPs and the intermediate filaments are not fully understood, the HS- and UHS-KAPs cross-link with the intermediate filaments via disulfide bond formation [3].

The full set of KAP genes (*KRTAPs*) has yet to be identified in sheep [4]; however, in humans, 89 functional *KRTAPs* across 25 KAP families (KAP1 to KAP13, KAP15 to KAP17, and KAP19 to KAP27) have been identified [5–7]. Genes from two of the HS-KAP families, KAP11 and KAP13, are the first to be expressed in hair follicles from humans, with their expression occurring in tandem with the expression of keratins and prior to expression of the first HGT-KAP [5,7–9].

The human KAP13 family comprises four members [9]. Recently, three members of the KAP13 family were described in sheep, including how they affected selected wool traits [10]. Despite their chromosomal proximity, these three KAP13 genes exhibited different levels of sequence variation, and they had different associations with variation in wool traits. Specifically, *KRTAP13-1* was less polymorphic compared with *KRTAP13-2* and *KRTAP13-4*. In Chinese Tan sheep, an indigenous breed that produces 'spring-like' crimped wool in early life [11], *KRTAP13-1* appeared to be monomorphic, whereas variation in *KRTAP13-2* was associated with fibre diameter standard deviation (FDSD) and the coefficient of variation of fibre diameter (CVFD) for the heterotypic hair fibres (fibres that can change seasonally at times, tending towards being more hair-like and potentially medullated in the summer, but reverting to being wool-like in the winter for warmth) [10]. In contrast, no association was found between *KRTAP13-4* variation and the four fibre diameter-related traits investigated.

Although *KRTAP13-3* has been identified and is polymorphic [12], its effect on wool traits remains unexplored. This study aimed to investigate the variation of *KRTAP13-3* in Chinese Tan sheep, assess its association with four fibre diameter-related traits, and compare these findings with other *KRTAP13-n* family members.

2. Materials and Methods

2.1. Chinese Tan Sheep Studied and the Measurement of Key Wool Traits

Two hundred and thirty-eight lambs of the Tan breed were studied. These were the offspring of ten unrelated sires, with all the lambs born in the same year, at the same farm. They were all exposed to similar environmental conditions and feeding regimes. All but six lambs (three sets of twins) were singletons at birth, and to avoid possibly confounding subsequent analyses, these six lambs were removed from the study. This left 232 lambs for genetic analysis.

Samples of fibre were collected from each lamb. These were collected from the mid-side region of the lambs near the middle of the Er-mao period (at day 35 postpartum). Next, the fine wool fibres were manually separated from the heterotypic hair fibres in the sample, capitalising on their differences in fibre diameter and length. This was undertaken by spreading each fibre sample on a flannel-coated panel. A stiff plastic card was then used to press the bottom of all the fibres onto the panel, which allowed the longer heterotypic hair fibres to be pulled out of the sample. The process was repeated several times to ensure that the heterotypic hair fibres and the fine wool fibres were fully separated.

The mean fibre diameter (MFD), FDSD, and mean fibre curvature (MFC) of the heterotypic hair fibres and finer wool fibres were then measured. The fine wool samples were evaluated by Pastoral Measurements Limited (Timaru, New Zealand), while the hair samples were measured by the New Zealand Wool Testing Authority (NZWTA, Ahuriri, Napier, New Zealand) using the International Wool Textile Organisation (IWTO) standardised Laserscan method IWTO 12 [13]. The CVFD was calculated from MFD and FDSD by dividing the FDSD by the MFD and multiplying by 100 to express it as a percentage.

Venous blood samples from each lamb were collected onto blotting paper (TFN paper; Munktell Filter AB, Falun, Sweden). These samples were allowed to dry in the air and were then kept in the dark at room temperature until further use. With this blood collection approach, the genomic DNA in the leukocytes becomes bound to the paper, and the DNA was then purified for use in PCR amplification as follows.

A 1.2 mm punch was taken from the blood spot on the TFN blotting paper and placed in a 0.7 mL PCR tube. Each tube contained a 200 µL volume of 20 mM NaOH. These tubes were then incubated at 60 °C for 30 min. Next, the NaOH solution was carefully removed by aspiration, and the disks with attached DNA were equilibrated in 200 µL of TE^{-1} buffer (10 mM Tris-HCl, 0.1 mM EDTA, pH 8.0). This buffer was then removed by aspiration, and the disks were air-dried until used to prime Polymerase Chain Reaction (PCR) amplifications.

2.2. PCR Amplification and SSCP Analysis of *KRTAP13-3*

Two PCR primers were created to amplify a 310 bp region of ovine *KRTAP13-3* that was known to be polymorphic based on the sequences of five ovine *KRTAP13-3* variants in GenBank (Accession numbers JN377429-JN377433). These primers were 5′-GCTCTTCTCTCTACAGTCGA-3′ (forward primer) and 5′-CAGGGAAAATTCATTGGG TGC-3′ (reverse primer), and they were synthesized by Integrated DNA Technologies (Coralville, IA, USA).

The PCR amplification reactions were undertaken in a 15-µL reaction that contained 0.5 U Taq DNA polymerase (Qiagen, Hilden, Germany), 2.5 mM Mg^{2+}, 0.25 µM each primer, 150 µM each dNTP, 1× reaction buffer supplied with the enzyme, and the purified genomic DNA on a 1.2-mm TFN card disk, made up to volume with double-deionized water. The amplifications were performed in thermal cyclers (Model S1000; Bio-Rad, Hercules, CA, USA) with an initiating denaturation of 2 min at 94 °C, followed by 35 cycles of 94 °C for 30 s, 60 °C for 30 s, 72 °C for 30 s, and a final extension step at 72 °C for 5 min.

Single-stranded conformation polymorphism (SSCP) analyses were used to examine the PCR amplicons. With this analytical method, which can resolve DNA sequence variation to the single nucleotide level, a 0.7-µL aliquot of each PCR amplification reaction was mixed with 7 µL of gel loading dye (0.025% bromophenol blue, 0.025% xylene cyanol, 10 mM EDTA, and 98% formamide). These solutions were denatured (90 °C for 5 min), and the samples were placed on wet ice prior to being loaded immediately onto 16 × 18 cm, 12% acrylamide:bisacrylamide (37.5:1; Bio-Rad) gels. Protean II xi cells (Bio-Rad) were used for electrophoresis with coolant circulating at 15 °C. The samples were electrophoresed at 320 V in 0.5× TBE buffer for 18 h. Ovine blood samples that produced the DNA sequence variants identified previously by Gong et al. [12] were obtained, and PCR products from these samples were run on each gel as a reference to determine the variants present in each Tan sheep sample.

The polyacrylamide gels were silver-stained in a solution of 10% ethanol, 0.5% acetic acid, and 0.2% silver nitrate for 10 min and then rinsed in distilled water and developed with a solution of 3% NaOH and 0.1% HCOH until dark-staining bands appeared against the yellow background of the gel.

2.3. Statistical Analyses

The genetic diversity of *KRTAP13-3* variation was assessed using two genetic parameters: genetic heterogeneity and polymorphism information content (PIC). These parameters were determined using an online calculator (https://www.genecalculators.net/pq-chwe-polypicker.html, accessed on 5 February 2025).

General Linear Models (GLMs) were developed using Minitab version 16 (Minitab Inc., State College, PA, USA) and used to assess the influence of the presence or absence of the *KRTAP13-3* variants detected on the fibre traits, but only for variants with frequencies greater than 5% in the Tan lambs. This was because variants present in only a small number of lambs can bias the analyses if these lambs have extreme phenotypes. Gender and sire were included as fixed and random factors, respectively, in the models.

The model used to examine the effect of variant presence/absence was $Y_{ijkl} = \mu + V_i + Gender_j + Sire_k + e_{ijkl}$, where Y_{ijkl} is a phenotype of the ijkl-th sheep (for either MFD, CVFD, FDSD, or MFC), μ = is the arithmetic mean for the characteristic, V_i is the effect of the i-th variant (presence and absence), $Gender_j$ is the fixed effect of j-th gender, $Sire_k$ is the random effect of k-th sire, and e_{ijkl} is the random residual effect. When $p < 0.05$, the associations were considered significant.

3. Results

All five variants identified previously [12] were detected in the Chinese Tan sheep using PCR-SSCP (Figure 1). Twelve different genotypes were revealed: *AA* (n = 92), *AB* (n = 51), *AC* (n = 40), *AD* (n = 18), *AE* (n = 4), *BB* (n = 5), *BC* (n = 11), *BD* (n = 5), *BE* (n = 1), *CC* (n = 2), *CD* (n = 2), and *DD* (n = 1). The frequencies of variants *A* through *E* were 64.0%, 16.8%, 12.3%, 5.8%, and 1.1%, respectively. The heterogeneity and PIC of this gene were 54.4% and 50.4%, respectively.

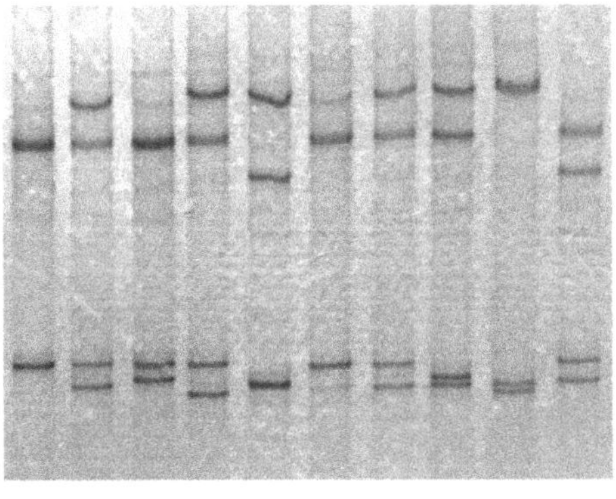

Figure 1. PCR-SSCP patterns of ovine *KRTAP13-3* variants. Five different banding patterns (*A* to *E*), corresponding to five different DNA haplotypes were observed. These occurred in either homozygous or heterozygous forms, with different nucleotide sequences expected to produce different banding patterns on the gels. Each DNA sequence produces two bands that correspond to the two strands of the DNA for any given variant, with heterozygous sheep usually having four bands.

As variant *E* occurred at a frequency under 5%, five sheep carrying this variant were excluded from the association analyses, resulting in a final dataset for modelling analysis of 227 sheep. Among these sheep, variation in *KRTAP13-3* was associated with FDSD and

CVFD in heterotypic hair fibres (Table 1). The presence of variant *A* was linked to increased FDSD and CVFD in heterotypic hair fibres, while trends of association were also observed for variants *B* and *D*, which were associated with decreased CVFD. No associations were found with variation in fine wool fibres.

Table 1. Association of *KRTAP13-3* variants with wool traits in Chinese Tan sheep.

Fibre Type	Fibre Trait [1]	Variant	Present		Absent		*p*-Value [2]
			Mean ± SE	*n*	Mean ± SE	*n*	
Fine wool	MFD (μm)	*A*	16.7 ± 0.17	201	16.7 ± 0.33	26	0.429
		B	16.9 ± 0.23	72	16.6 ± 0.18	155	0.133
		C	16.8 ± 0.25	55	16.6 ± 0.17	172	0.404
		D	16.4 ± 0.31	26	16.7 ± 0.18	201	0.349
	FDSD (μm)	*A*	4.2 ± 0.12	201	4.0 ± 0.24	26	0.429
		B	4.3 ± 0.17	72	4.2 ± 0.13	155	0.558
		C	4.2 ± 0.18	55	4.2 ± 0.13	172	0.799
		D	4.0 ± 0.23	26	4.2 ± 0.13	201	0.371
	CVFD (%)	*A*	25.0 ± 0.58	201	23.9 ± 1.12	26	0.285
		B	25.1 ± 0.78	72	24.9 ± 0.61	155	0.809
		C	24.8 ± 0.86	55	25.0 ± 0.59	172	0.852
		D	24.3 ± 1.05	26	25.1 ± 0.62	201	0.473
	MFC (°/mm)	*A*	63.9 ± 1.14	201	64.0 ± 2.21	26	0.961
		B	62.7 ± 1.54	72	64.4 ± 1.20	155	0.245
		C	65.2 ± 1.68	55	63.6 ± 1.16	172	0.303
		D	65.9 ± 2.07	26	63.3 ± 1.22	201	0.237
Heterotypic hair	MFD (μm)	*A*	29.7 ± 0.34	201	30.0 ± 0.67	26	0.564
		B	29.7 ± 0.46	72	29.7 ± 0.35	155	0.999
		C	29.7 ± 0.51	55	29.7 ± 0.34	172	0.877
		D	30.1 ± 0.62	26	29.6 ± 0.37	201	0.432
	FDSD (μm)	*A*	8.3 ± 0.15	201	7.7 ± 0.31	26	**0.033**
		B	8.1 ± 0.21	72	8.4 ± 0.16	155	0.105
		C	8.3 ± 0.23	55	8.3 ± 0.16	172	0.906
		D	8.0 ± 0.28	26	8.4 ± 0.17	201	0.261
	CVFD (%)	*A*	28.1 ± 0.45	201	25.7 ± 0.90	26	**0.005**
		B	27.2 ± 0.63	72	28.2 ± 0.48	155	*0.094*
		C	27.8 ± 0.69	55	28.0 ± 0.47	172	0.789
		D	26.7 ± 0.84	26	28.3 ± 0.50	201	*0.082*
	MFC (°/mm)	*A*	46.6 ± 0.72	201	45.3 ± 1.43	26	0.309
		B	46.2 ± 1.00	72	46.6 ± 0.76	155	0.609
		C	46.8 ± 1.09	55	46.5 ± 0.74	172	0.742
		D	45.8 ± 1.33	26	46.7 ± 0.79	201	0.524

[1] MFD—mean fibre diameter; FDSD—fibre diameter standard deviation; CVFD—coefficient of variation of fibre diameter; MFC—mean fibre curvature. [2] Values of $p < 0.05$ are shown in bold; values of $p < 0.10$ are shown in italics.

4. Discussion

This study identified multiple variants of *KRTAP13-3* in Chinese Tan sheep, but with differences in their frequency distribution compared with the New Zealand (NZ) Romney sheep described by Gong et al. [12]. In the NZ Romney sheep, the common variants (*A*, *B*, and C) had similar frequencies, while the rare variants (*D* and *E*) were much less common. In contrast, within the Chinese Tan sheep population, variant *A* was the most common, with variants *B* and *C* occurring at lower frequencies. Variant *D* was less common again, while variant *E* was the rarest.

Within the Chinese Tan sheep population, *KRTAP13-3* exhibited a greater genetic heterogeneity and PIC when compared with *KRTAP13-1* and *KRTAP13-4* [10], but it was comparable to *KRTAP13-2*. This might suggest that the increased diversity of *KRTAP13-3* and *KRTAP13-2* has greater functional effect, with this supported by the observed associations revealed in this study and by Bai et al. [10], respectively. Equally, Bai et al. [10] suggested that *KRTAP13-1* was monomorphic; hence, no effects on wool traits could be investigated. Care is needed in drawing these conclusions though, as both studies are limited in size and involved only a single breed or group of wool sheep. This work needs to be expanded to the more common wool breeds, such as the Merino, and include other important wool traits too.

The associations detected for *KRTAP13-3* in this study are like those found with *KRTAP13-2*, which was also associated with the FDSD and CVFD of heterotypic hair fibres [10]. *KRTAP13-3* is located approximately 11.2 kb from *KRTAP13-2*, suggesting that the proximity of the two genes underpins a common effect. However, *KRTAP15-1*, a member of a different gene family is also co-located, being approximately 15.0 kb from *KRTAP13-3*. In Merino × Southdown-cross sheep, variation in *KRTAP15-1* was strongly associated with wool yield, although an association with FDSD was also observed [14]. Since these studies involved different sheep breeds and measured different wool traits, comparisons should again be made with caution, and it remains a challenge to determine whether the associations detected are due to the genes themselves, linkage with nearby genes, or the effect of being part of a gene family that may have coordinated activity; albeit, the differences between the four *KRTAP13* genes would suggest that the latter is unlikely. Further investigation into the associations of *KRTAP15-1* and *KRTAP13-n* and within the same populations may provide additional insight.

The association between *KRTAP13-3* and FDSD and CVFD, but not with MFD, suggests that variation in the gene principally affects the uniformity of the measured fibre diameters, rather than the average diameter of the fibres sampled. The exact mechanism underlying these associations remains unclear, but one possible explanation is that *KRTAP13-3* may influence wool fibre ellipticity. Wool fibres typically have an elliptical rather than a circular cross-section, and studies have shown that higher FDSD and CVFD are linked to increased ellipticity [15]. Alternatively, the variation in FDSD and CVFD may reflect along fibre variation in diameter, with this possibly reflecting changes in follicle activity over time; albeit, this too needs further study.

Interestingly, the associations are only observed for the heterotypic hair fibres and not the fine wool fibres. One possible explanation is the differential expression of *KRTAP13-3* between these two fibre types. Heterotypic hair fibres, which are coarser and exhibit greater variability in diameter, may express *KRTAP13-3* differently compared with fine wool fibres, potentially leading to a more pronounced impact on fibre diameter uniformity. The structural differences between coarse and fine fibres could affect how KAP13-3 protein influences fibre ellipticity and/or uniformity. Coarser fibres, with their large surface area, are generally more susceptible to external forces, which might further amplify the gene's effect on fibre structure. In contrast, fine wool fibres, with their more consistent diameter and reduced susceptibility to external forces, may show a diminished impact of the gene on their structure and, hence, not reveal the same associations. This suggests that more research is needed to validate these findings and to further explore the differential expression of *KRTAP13-3* in various fibre types. Investigating how *KRTAP13-3* expression varies among different fibre types and under different conditions will provide a better understanding of its role in fibre traits and confirm the observed associations.

The key sequence difference between *KRTAP13-3* variants *A* and *B* or *D* is the c.242G/A substitution, which results in the amino acid change p.Arg81Gln [12]. The substitution of a positively charged arginine with a neutral glutamine could alter the protein's structure and its interactions with other proteins, potentially leading to effects on increased FDSD and CVFD in heterotypic hair. This too needs further investigation at the functional level.

5. Conclusions

Five previously identified variants of *KRTAP13-3* were detected in the Chinese Tan sheep studied using PCR-SSCP. The most common variant (*A*) was linked to increased FDSD and CVFD. No associations were found with variation in the fine wool fibres from the Chinese Tan sheep. These findings suggest that *KRTAP13-3* plays a role in regulating heterotypic hair fibre diameter variability.

Author Contributions: Conceptualization, H.Z., J.T. and J.G.H.H.; formal analysis, L.B. and H.Z.; investigation, L.B. and J.T.; methodology, L.B., H.Z. and J.T.; supervision, H.Z., J.T. and J.G.H.H.; writing—original draft, L.B. and H.Z.; writing—review and editing, H.Z., J.T. and J.G.H.H. All authors have read and agreed to the published version of the manuscript.

Funding: This research was funded by the National Natural Science Foundation of China (32460818).

Institutional Review Board Statement: The collection of sheep blood by the nicking of their ears is covered by Section 7.5 Animal Identification, in: Code of Welfare: Sheep and Beef Cattle (2016); a code of welfare issued under the Animal Welfare Act 1999 (New Zealand Government) and is deemed to be an acceptable practice in commercial sheep farming.

Informed Consent Statement: Informed consent was obtained from sheep farm owners.

Data Availability Statement: The raw data supporting the conclusions of this article will be made available by the authors on request.

Acknowledgments: The authors thank Freeman Fang for providing technique support and assistance with the PCR-SSCP typing and analysis.

Conflicts of Interest: The authors declare no conflict of interest.

References

1. Pickering, N.; Blair, H.; Hickson, R.; Dodds, K.; Johnson, P.; McEwan, J. Genetic relationships between dagginess, breech bareness, and wool traits in New Zealand dual-purpose sheep. *J. Anim. Sci.* **2013**, *91*, 4578–4588. [PubMed]
2. Wuliji, T.; Dodds, K.; Andrews, R.; Turner, P. Selection response to fleece weight, wool characteristics, and heritability estimates in yearling Romney sheep. *Livest. Sci.* **2011**, *135*, 26–31.
3. Gong, H.; Zhou, H.; Forrest, R.H.; Li, S.; Wang, J.; Dyer, J.M.; Luo, Y.; Hickford, J.G. Wool keratin-associated protein genes in sheep—A review. *Genes* **2016**, *7*, 24. [CrossRef] [PubMed]
4. Zhou, H.; Gong, H.; Wang, J.; Luo, Y.; Li, S.; Tao, J.; Hickford, J.G. The complexity of the ovine and caprine keratin-associated protein genes. *Int. J. Mol. Sci.* **2021**, *22*, 12838. [CrossRef] [PubMed]
5. Rogers, M.A.; Langbein, L.; Praetzel Wunder, S.; Giehl, K. Characterization and expression analysis of the hair keratin associated protein KAP26.1. *Br. J. Dermatol.* **2008**, *159*, 725–729. [PubMed]
6. Rogers, M.A.; Schweizer, J. Human KAP genes, only the half of it? Extensive size polymorphisms in hair keratin-associated protein genes. *J. Investig. Dermatol.* **2005**, *124*, vii–ix. [PubMed]
7. Rogers, M.A.; Winter, H.; Langbein, L.; Wollschläger, A.; Praetzel-Wunder, S.; Jave-Suarez, L.F.; Schweizer, J. Characterization of human KAP24.1, a cuticular hair keratin-associated protein with unusual amino-acid composition and repeat structure. *J. Investig. Dermatol.* **2007**, *127*, 1197–1204. [CrossRef] [PubMed]
8. Rogers, M.A.; Langbein, L.; Winter, H.; Ehmann, C.; Praetzel, S.; Korn, B.; Schweizer, J. Characterization of a cluster of human high/ultrahigh sulfur keratin-associated protein genes embedded in the type I keratin gene domain on chromosome 17q12-21. *J. Biol. Chem.* **2001**, *276*, 19440–19451. [PubMed]
9. Rogers, M.A.; Langbein, L.; Winter, H.; Ehmann, C.; Praetzel, S.; Schweizer, J. Characterization of a first domain of human high glycine-tyrosine and high sulfur keratin-associated protein (KAP) genes on chromosome 21q22.1. *J. Biol. Chem.* **2002**, *277*, 48993–49002. [PubMed]
10. Bai, L.; Zhou, H.; Li, W.; Tao, J.; Hickford, J.G. Characterisation of three ovine *KRTAP13* family genes and their association with wool traits in Chinese Tan sheep. *Animals* **2024**, *14*, 2862. [CrossRef] [PubMed]

1. Tao, J.; Zhou, H.; Yang, Z.; Gong, H.; Ma, Q.; Ding, W.; Li, Y.; Hickford, J.G. Variation in the KAP8-2 gene affects wool crimp and growth in Chinese Tan sheep. *Small Rumin. Res.* **2017**, *149*, 77–80. [CrossRef]
2. Gong, H.; Zhou, H.; Dyer, J.M.; Plowman, J.E.; Hickford, J.G.H. Identification of the keratin-associated protein 13-3 (KAP13-3) gene in sheep. *Open J. Genet.* **2011**, *1*, 60–64.
3. Available online: https://www.member.iwto.org/store/viewproduct.aspx?id=12408762 (accessed on 20 February 2025).
4. Li, W.; Gong, H.; Zhou, H.; Wang, J.; Liu, X.; Li, S.; Luo, Y.; Hickford, J. Variation in the ovine keratin-associated protein 15-1 gene affects wool yield. *J. Agric. Sci.* **2018**, *156*, 922–928.
5. McGregor, B. *Properties, Processing and Performance of Rare and Natural Fibres: A Review and Interpretation of Existing Research Results*; Rural Industries Research and Development Corporation (RIRDC): Canberra, Australia, 2012.

International Journal of
Molecular Sciences

MDPI

Article

The 2SP Site Mutation in the Bovine Natural Resistance-Associated Macrophage 1 Promoter Exhibits Antituberculosis Potential

Yongke Wei [1,2,†], Mengke Yuan [1,2,†], Yong Zhang [1,2,*] and Yuanpeng Gao [1,2,*]

1 College of Veterinary Medicine, Northwest A&F University, Yangling 712100, China; weitom2012@163.com (Y.W.); m15291999052@163.com (M.Y.)
2 Key Laboratory of Animal Biotechnology, Northwest A&F University, Yangling 712100, China
* Correspondence: zhangyong1956@nwsuaf.edu.cn (Y.Z.); gaoyuanpeng1990@163.com (Y.G.)
† These authors contributed equally to this work.

Abstract: Gene-edited cattle overexpressing natural resistance-associated macrophage 1 (NRAMP1) have demonstrated enhanced resistance to tuberculosis (TB). However, introducing synthetic sequences and selection markers may pose potential risks. The endogenous editing of target gene promoters could effectively mitigate these risks. To date, no available mutation sites in the bovine *NRAMP1* promoter have been identified to enhance host resistance to TB. In this study, we identified a unique mutation editing site, designated as 2SP, within the bovine *NRAMP1* promoter, using bioinformatics analysis and dual luciferase assays. The mutation at the 2SP site specifically increased *NRAMP1* promoter activity by 2.3-fold after *Mycobacterium tuberculosis* H37Ra infection, without modifying promoter activity in non-infected groups. By using base editing techniques, an endogenously edited THP-1 cell line with a mutation at the homologous region of the 2SP site was generated, without introducing screening markers. In H37Ra infection experiments, the edited THP-1 cells specifically upregulated NRAMP1 expression and significantly inhibited H37Ra proliferation, while maintaining baseline NRAMP1 expression levels in the absence of infection. In this research, we identified a novel mutation site and provided a fundamental reference for the development of gene-edited cattle with enhanced resistance to TB.

Keywords: base editing; NRAMP1; promoter analysis; tuberculosis

1. Introduction

Tuberculosis (TB), mainly caused by *Mycobacterium tuberculosis* (Mtb), remains one of the most prevalent infectious diseases worldwide. Approximately 25% of the global human population exhibits immunological evidence of prior Mtb infection, and approximately 10% of infected individuals develop clinical disease during their lifetime [1].

The fact that only a small number of individuals infected with Mtb progress to TB indicates that genetic factors play a crucial role in resistance to TB, as affirmed by numerous studies [2–9]. *Mycobacterium bovis* (*M. bovis*) is the most prevalent pathogen among mycobacteria [10]. As a zoonotic disease, bovine TB not only has serious implications for animal production but also poses substantial risks to public health [11].

The natural resistance-associated macrophage 1 (*NRAMP1*) gene, also known as the solute carrier family 11 member 1 gene, is one of the most extensively investigated genes related to TB resistance [12]. *NRAMP1* was initially identified through positional cloning as a candidate gene for resistance against *M. bovis*, *Leishmania donovani*, and *lethal Salmonella typhimurium* [13]. These intracellular pathogens share survival strategies and replicate within mildly acidic macrophage vacuoles [14]. *NRAMP1* is specifically expressed in the membrane of late endosomes/lysosomes of professional phagocytes, functioning as a divalent cation transporter [15–17].

Although the related mechanisms remain to be further explored, genetic variation in *NRAMP1* is associated with susceptibility/resistance to TB [18–22]. and it is considered to play a crucial role in resisting TB development [23]. The structure of *NRAMP1* and its function in preventing intracellular pathogen infections are both ancient and conserved [24], as evidenced by their analogous roles in natural resistance to intracellular pathogens, ranging from amoeba *D. discoideum* [25] to mouse macrophages [13].

TB resistance has been correlated with *NRAMP1* gene polymorphisms. The minor allele frequency (MAF) of genetic polymorphisms occurring in the exon of *NRAMP1* is relatively low, and these polymorphisms are frequently identified as risk factors [18,26–29]. Conversely, the MAF of polymorphisms located in regulatory regions of *NRAMP1* is comparatively higher and may confer positive effects [18,30–32]. Hence, gene editing targeting the regulatory regions to enhance a host's resistance to TB may represent a more rational strategy.

Our previous research successfully produced gene-edited cattle with enhanced TB resistance by overexpressing NRAMP1 [33,34]. However, the introduction of synthetic sequences and marker genes poses potential risks. In this study, we report a novel mutation site available for endogenous base editing, designated as 2SP, and offer preliminary evidence of its distinctive effect on the regulation of bovine *NRAMP1* expression. As an unconventional variant recognition site for the *NRAMP1* transcription factors (TFs) SP1/SP3, the 2SP site exhibits a weakened affinity for SP1/SP3. We discovered that SP1/SP3 are TFs that activate *NRAMP1* expression. Nevertheless, increasing the affinity for SP1/SP3 at the 2SP site results in a significant downregulation of NRAMP1 expression. Interestingly, mutating the 2SP site to eliminate its affinity significantly enhances the promoter activity after H37Ra infection while leaving the promoter activity unchanged in the non-infected group.

Owing to the deficiency of bovine macrophage cell lines and based on the similarity in promoter sequences between bovine and humans, we established a 2SP site homologous region-edited THP-1 cell line for testing. The edited THP-1 cell line exhibited increased NRAMP1 protein expression after H37Ra infection and more effective inhibition of H37Ra proliferation.

2. Results

2.1. Detection of the Core Promoter of Bovine NRAMP1 and Comparative Analysis Between Mouse and Cattle

By integrating and analyzing RNA sequencing data from multiple independent laboratories deposited in the NCBI Gene Expression Omnibus profiles (https://www.ncbi.nlm.nih.gov/geoprofiles/, accessed on 19 June 2022), we observed that Mtb infection induced a substantial increase in Nramp1 expression in mice, whereas the induction in cattle was comparatively limited (Supplementary Figure S1A–D). In this study, the mouse macrophage cell line RAW264.7 and Mtb strain H37Ra were used to verify the results and for further experiments. In the subsequent experiments, we utilized a multiplicity of infection (MOI) of 10 to ensure the comparability and consistency of our experimental results with the analyzed datasets. This MOI was chosen to match the experimental conditions predominantly reported in the analyzed transcriptomic datasets. In RAW264.7 cells, *Nramp1* expression was significantly upregulated at 4 h post-infection (hpi), peaked at 12 hpi, and returned to baseline levels by 36 hpi (Supplementary Figure S1E).

To detect and compare the core promoter regulating bovine *NRAMP1* expression after H37Ra infection, we progressively truncated the promoter fragment, analyzed transcriptional changes, and compared the relative transcriptional activities of the truncated promoters between the H37Ra-infected and non-infected groups. We generated pGL4.10 containing the sequence from −1611 to +73 bp (ARS-UCD 1.2 bovine assembly, relative to the TSS) and then progressively deleted the 5′ sequence of the *NRAMP1* promoter to generate eight truncated promoter reporter constructs, namely, pGL4.1-T1(−1242/+73), pGL4.1-T2(−792/+73), pGL4.1-T3(−278/+73), pGL4.1-T4(−264/+73), pGL4.1-T5(−214/+73), pGL4.1-T6(−162/+73), pGL4.1-T7(−104/+73), and pGL4.1-T8(−60/+73). pGL4.1-T3(−278/+73) exhibited sufficient activity compared to the other truncated promoters and demonstrated significantly higher activity after infection, indicating that this region is crucial for *NRAMP1* expression and may contain cis-regulatory elements responsive to H37Ra infection (Figure 1A).

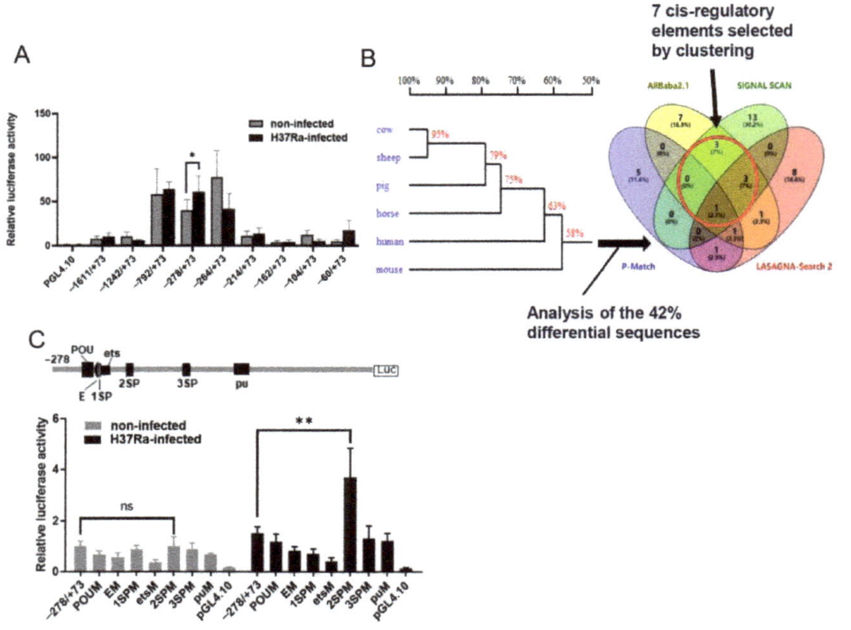

Figure 1. Analysis of the regulatory region of the *NRAMP1* promoter in cattle and screening of candidate mutation sites. (**A**) Luciferase assays were performed to test the transcriptional activity of different truncated promoters after H37Ra infection or no infection. Bovine *NRAMP1* promoters with different lengths and internal reference vectors were transfected into RAW264.7 cells for 48 h. The infected group was infected with H37Ra (MOI = 10) at 24 h after transfection. Values were normalized to the mean of the non-infected PGL4.10 group (±SD). Student's *t*-test was used to evaluate the differences (*, $p < 0.05$). (**B**) A homology analysis of the *NRAMP1* promoter −278/+73 region across different species was conducted using DNAMAN. Cis-regulatory elements within this region were predicted by online software and those with JASPAR scores above the 75% threshold were identified. Elements within the 42% differential sequences of the *NRAMP1* promoters in mice and cattle were screened and classified using the online tool Venny 2.1.0. (bioinfogp.cnb.csic.es/tools/venny, accessed on 15 July 2022), and seven candidate sites were obtained, as indicated by the red circles that highlight the clustered candidate sites. (**C**) The effects of mutations at the seven candidate sites on promoter activity. The relative transcriptional activities of the promoters were determined by luciferase assays. The pGL4.1 vectors that contained the different promoter fragments were co-transfected with pRL-TK in RAW264.7. The H37Rainfected groups were infected with H37Ra (MOI = 10) 24 h after transfection and then incubated for 24 h. Values were normalized to the mean of the non-infected pGL4.1-T3(−278/+73) group (±SD). Student's *t*-test was used to evaluate the differences (**, $p < 0.01$).

Homology analysis revealed that the bovine *NRAMP1* promoter −278/+73 region is highly homologous to sheep, pigs, horses, and humans, while it diverges widely from the mouse *Nramp1* promoter (Figure 1B). We wanted to investigate whether sequence differences between cattle and mouse *Nramp1* promoters lead to variations in the recruitment of TB-associated TFs. To screen candidate TFs related to TB stimulation and their binding sites, cis-regulatory elements were predicted by P-Match, AliBaba2.1, SIGNAL SCAN (http://gene-regulation.com, accessed on 11 July 2022), and LASAGNA-Search 2 (https://biogrid-lasagna.engr.uconn.edu/, accessed on 11 July 2022, the website is currently unavailable), using the bovine *NRAMP1* promoter −278/+73 region or the homologous region in the mouse *Nramp1* promoter (−224/+176), respectively. Cis-regulatory elements for immune- and inflammatory-associated TFs were selected, and their sequences were scored using JASPAR (http://jaspar2016.genereg.net/, accessed on 11 July 2022). Those scoring above the 75% threshold were retained. Among the predicted cis-regulatory elements, those located within the differential sequence regions of bovine and mouse promoters were analyzed and clustered, leading to the identification of seven candidate sites for further investigation (Figure 1B).

2.2. *Mutation at the 2SP Site Specifically Increases NRAMP1 Promoter Activity After H37Ra Infection*

To determine the effects of the seven cis-regulatory elements in H37Ra infection, pGL4.1-T3(−278/+73) plasmids with mutations in each core motif were created, named POUM, EM,1SPM,2SPM,3SPM, etsM, and puM, respectively (Table S2). These seven recombinant plasmids were co-transfected with the internal reference vector pRL-TK into RAW264.7 cells, and the relative transcriptional activities were compared between the H37Ra-infected and non-infected groups. As shown in Figure 1C, 1SPM showed a significant decrease in *NRAMP1* promoter activity after H37Ra infection, whereas it only presented a faint decrease with no significant difference in the absence of H37Ra infection. Interestingly, after H37Ra infection, 2SPM led to a significant increase (2.3-fold) in *NRAMP1* promoter activity, but no difference was manifested in the non-infected group. On the other hand, 3SPM did not exhibit significant differences in *NRAMP1* expression in either the infected or the non-infected groups (Supplementary Figure S2A). POUM, EM, etsM, and puM all significantly reduced *NRAMP1* promoter activity, exhibiting similar magnitudes of reduction in both H37Ra-infected and non-infected groups (Supplementary Figure S2B–D).

2.3. *Altering the Affinity of the 2SP Site for SP1/SP3 Affects NRAMP1 Promoter Activity After H37Ra Infection*

We analyzed the homology of the bovine *NRAMP1* promoter with 108 species of ruminants and 32 other species using the Animal Omics Database (http://animal.omics.pro/, accessed on 5 August 2022). Bovine *NRAMP1* promoter was highly homologous to 130 of the total 140 species (Supplementary Figure S3). Further analysis of the homologous region of the 2SP site in 130 species revealed two primary conserved core motifs: 5′-GGGAGAGG-3′ (in cattle, sheep, and related species) and 5′-GGGAGG-3′ (in humans, pigs, and related species). These sequences were identified as variants of the consensus SP1/SP3 recognition sites. The TF-predicted scores from the JASPAR database suggest that the two variants exhibit weakened affinity for SP1/SP3 (7.18 and 4.83). However, six species lost affinity for SP1/SP3 due to mutations or deletions in the homologous region of 2SP site: they are gerenuks, horses, bats, hedgehogs, mice, and manatees.

Next, we aimed to determine whether the altered SP1/SP3 binding affinity at the 2SP site was a contributing factor to the changes in bovine *NRAMP1* promoter activity after H37Ra infection. We first investigated the effect of eliminating or weakening SP1/SP3 binding at the 2SP site on *NRAMP1* expression. Sequence analysis revealed that the region corresponding to the 2SP site in the mouse *Nramp1* promoter lacks SP1/SP3 binding affinity, while equivalent regions in human and pig promoters exhibit relatively weaker affinity compared to cattle. We introduced mutations into pGL4.1-T3(−278/+73) using the sequences corresponding to the 2SP site from mice, humans, and pigs, designated as m2SPM, h2SPM, and p2SPM, respectively. Promoter activities were assessed under both H37Ra-infected and uninfected conditions. In H37Ra-infected groups, substituting the 2SP site with the corresponding mouse sequence resulted in a substantial increase in promoter activity (1.7-fold), whereas substitution with the human and pig sequences also significantly increased promoter activity, but to a lesser extent (1.28-fold and 1.24-fold). On the other hand, these three mutated promoters showed no significantly different activities to pGL4.1-T3(−278/+73) in the non-infected group (Figure 2A). To further evaluate the reliability of the differences in SP1/SP3 binding affinity for the 2SP site across different species, 3D structural models of bovine SP1 and SP3 were constructed, and molecular docking simulations were performed. Molecular docking provides a computational method for predicting the binding modes and affinities of protein–ligand interactions, which can help elucidate the structural basis of species-specific differences in binding behavior. The 3D model structures of SP1 and SP3 proteins, along with their predicted binding pockets, showed a high degree of similarity (Table 1). Additionally, both proteins exhibited high average conservation scores, indicating their evolutionary conservation. These findings align with previous reports suggesting that SP1 and SP3 recognize identical binding sites and are conserved across species [35–37]. Using bovine SP1 as an example, we observed that the High Ambiguity Driven DOCKing (HADDOCK) scores of the 2SP site sequences from cattle, humans, pigs, and mice increased progressively, indicating a corresponding decrease in SP1 binding affinity. Consistently, replacing the bovine 2SP site sequence with those from humans, pigs, and mice resulted in a corresponding increase in promoter activity following H37Ra infection. This finding supports the relationship between 2SP site sequence variation and its functional impact on promoter activity.

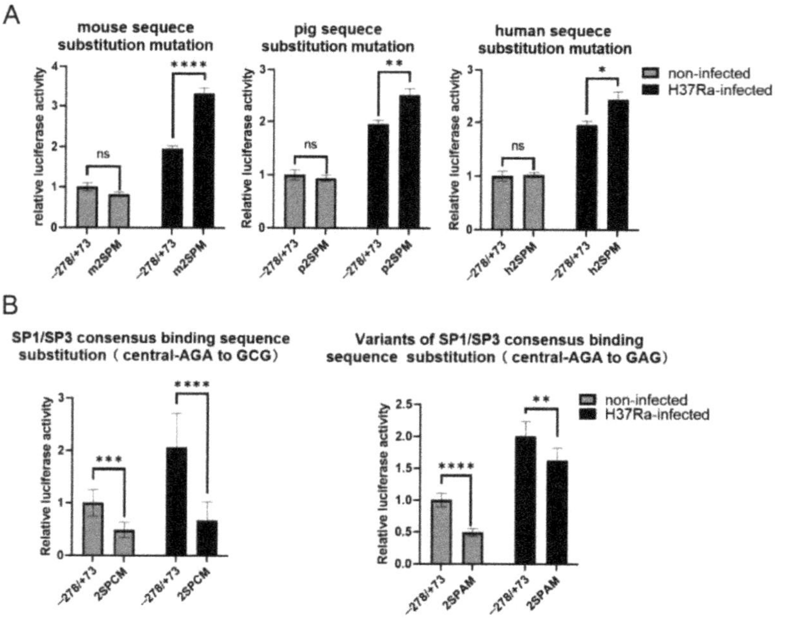

Figure 2. Altering the 2SP site affinity for SP1/SP3 affects *NRAMP1* promoter activity. (**A**) Mouse sequence substitution (no affinity for SP1/SP3) resulted in a substantial increase (1.7-fold) in *NRAMP1*

promoter activity after H37Ra infection, while pig sequence substitution and human sequence substitution (weaker affinity for SP1/SP3) resulted in a moderate increase (1.28-fold and 1.24-fold, respectively) in *NRAMP1* promoter activity after H37Ra infection. The relative transcriptional activities of the promoters were determined by luciferase assays. Values were normalized to the mean of the non-infected pGL4.1-T3(−278/+73) group (±SD). Student's *t*-test was used to evaluate the differences (ns, $p \geq 0.05$; *, $p < 0.05$; **, $p < 0.01$; ****, $p < 0.0001$). Different mutated promoter activities in B were detected in the same way. (**B**) SP1/SP3 consensus binding sequence substitution resulted in a substantial decrease in *NRAMP1* promoter activity in both the H37Ra infection group (0.52-fold) and the non-infection group (0.28-fold), and repeated experiments using the THP-1 cell line yielded consistent results (0.48-fold and 0.32-fold, respectively). The variants of SP1/SP3 consensus binding sequence substitution resulted in a moderate decrease in *NRAMP1* promoter activity in both the H37Ra infection group (0.85-fold) and the non-infection group (0.49-fold) (**, $p < 0.01$; ***, $p < 0.001$; ****, $p < 0.0001$).

Table 1. Prediction of binding pockets in bovine SP1/SP3-modeled protein and corresponding scoring and parameters of the predicted binding pockets (top). The energy of interaction between bovine SP1-modeled protein and ligands: dsDNA sequences of the 2SP site in different species and the mutant sequence used for gene editing (bottom).

Binding Pocket Prediction								
	Bovine SP1				Bovine SP3			
P2Rank	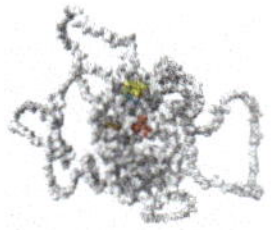				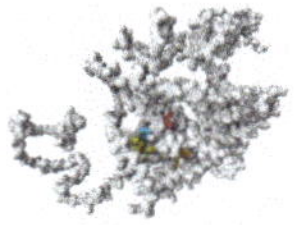			
Rank *	1	2	3	4	1	2	3	4
Score	5.72	4.11	2.33	2.00	7.14	5.23	3.65	1.09
Probability	0.245	0.146	0.049	0.036	0.341	0.215	0.064	0.007
Number of residues	9	8	5	5	10	12	6	6
Average conservation score	78.637	84.98	80.862	88.962	77.02	80.285	81.635	65.845

Energy of interaction (docking with bovine SP1)	2SP (cattle)	h2SPM (humans)	p2SPM (pigs)	m2SPM (mice)	2SPM2 (mutation)
HADDOCK score **	−88.9 ± 14.3	−72.0 ± 15.8	−65.5 ± 10.4	−64.8 ± 13.5	−37.5 ± 7.9
Van der Waals energy	−91.4 ± 6.2	−72.5 ± 12.0	−80.1 ± 9.1	−74.7 ± 16.7	−70.4 ± 3.8
Electrostatic energy	−379.3± 35.3	−335.1 ± 32.6	−198.2 ± 46.8	−215.1 ± 28.9	−272.3 ± 4.3
Desolvation energy	32.9 ± 2.5	28.0 ± 2.3	17.5 ± 1.7	18.9 ± 1.4	27.5 ± 1.3
Sum of energies	−526.7	−451.6	−326.3	−335.6	−352.7
Buried surface area	2487.8 ± 121.2	2114.5 ± 303.0	1881.0 ± 238.7	1930.4 ± 319.8	2019.9 ± 96.7

* Ranks of different colors correspond to pockets of the same color in the 3D structure. ** A composite energy metric combining van der Waals, electrostatic, and desolvation contributions; lower scores indicate stronger binding.

Conversely, we investigated the impact of enhanced SP1/SP3 binding affinity at the 2SP site on *NRAMP1* promoter activity. Previous studies have reported that the consensus SP1-binding sequence is GGGGCGGGG, and SP3 shares structural similarities with SP1, resulting in similar affinities for the SP1-binding site [38]. The central A-substituted variant (GGGGAGGG) represents a weaker affinity variant of the consensus SP1/SP3 recognition site [39]. To test whether a stronger SP1/SP3 binding affinity at the 2SP site would alter bovine *NRAMP1* promoter activity, we mutated the 2SP site to 5′-GGGGAGGG-3′ (2SPAM) and 5′-GGGGCGGG-3′ (2SPCM), which exhibited a stronger affinity and the strongest affinity for SP1/SP3, respectively. Both 2SPCM and 2SPAM mutations significantly decreased bovine *NRAMP1* promoter activity in both H37Ra-infected and non-infected groups, with 2SPCM showing a more pronounced reduction (0.32-fold) compared to 2SPAM (0.82-fold) in the H37Ra-infected group (Figure 2B).

We also generated pGL4.10 containing the mouse *Nramp1* promoter sequence from −1611 bp to +156 bp. In accordance with the sequences of bovine and humans, the mouse homologous region sequence of the 2SP site was substituted to 5′-GGGAGAGG-3′ and 5′-GGGAGG-3′, namely, MbM and MhM, respectively. Luciferase assays revealed that the two substitutions significantly reduced promoter activity in the H37Ra-infected group; however, no significant difference was manifested in the non-infected group (Supplementary Figure S4).

2.4. SP1 and SP3 Play Different Roles in Regulating the Activity of the Nramp1 Promoter

SP1 is a key regulator of myeloid-specific genes including *NRAMP1* [40]. SP1 and SP3 are members of the SP (specificity protein) TF family; they are structurally similar and have similar affinities for the same binding site [38]. The highly conserved protein structure of SP1 and SP3 in mammals suggests a similar role in regulating *NRAMP1* expression in different mammals.

To elucidate how the altered binding affinity of SP1/SP3 at the 2SP site specifically enhances *NRAMP1* promoter activity after H37Ra infection, we aimed to investigate the roles of SP1 and SP3 in this process. We first detected the expression levels of Sp1 and Sp3 in RAW264.7 cells following H37Ra infection. As shown in Figure 3A, *Sp1* expression peaked at 24 hpi and subsequently declined, returning to near baseline levels by 48 hpi. *Sp3* expression peaked at 12 hpi and returned to baseline by 36 hpi.

Although the mechanism is not fully elucidated, both SP1 and SP3 are ubiquitously expressed factors but can regulate specific gene expression. One explanation is that SP3 is a weaker transcriptional activator than SP1, and they compete for binding to the same binding site; therefore, the ratio of SP1:SP3 proteins may affect target gene expression [41,42]. Another study found that both SP1 and SP3 can regulate the basal expression of target genes, but only SP3 participates in the IFN-inducible complex formation with other TFs, including STAT1, STAT2, and IRF9, at the target gene promoter to regulate specific gene expression [43]. To further explore whether SP1/SP3 and several other potential candidate TFs (STAT1, STAT3, IRF1, IRF2, IRF3, and IRF9) are involved in the specific gene expression of *Nramp1* induced by H37Ra, the overexpression recombinant plasmids pCMV-SP1, pCMV-SP3, pCMV-STAT1, pCMV-STAT3, pCMV-IRF1, pCMV-IRF2, pCMV-IRF3, or pCMV-IRF9 were transfected into RAW264.7 cells and infected with H37Ra or not. The results showed that in the non-infected group, SP1, SP3, IRF1, IRF3, and STAT1 significantly enhanced *Nramp1* expression, while in the H37Ra-infected group, SP1, SP3, IRF2, and IRF3 significantly increased *Nramp1* expression (Supplementary Figure S5). It is worth noting that in the non-infected group, SP3 overexpression slightly upregulated *Nramp1* expression (1.3-fold), whereas after H37Ra infection, it markedly enhanced the activation of *Nramp1* expression (1.9-fold). However, SP1 overexpression exhibited similar activation ability (1.8-fold and 2.0-fold) regardless of infection status (Figure 3B).

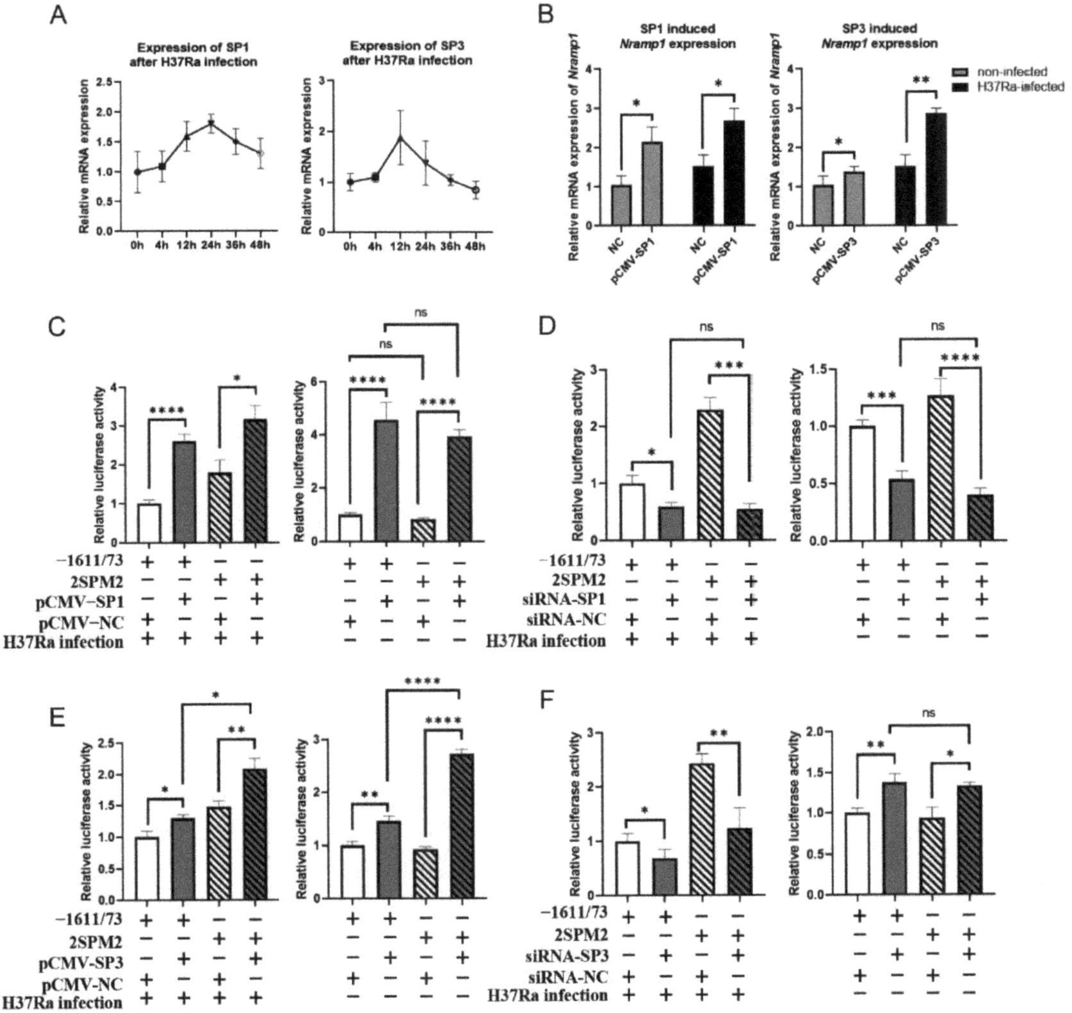

Figure 3. The roles of SP1 and SP3 in regulating *NRAMP1* expression and the impact of the 2SP site mutations on SP1 or SP3's regulatory effects. (**A**) The expression levels of *Sp1* and *Sp3* in response to H37Ra infection. RAW264.7 cells were infected with H37Ra (MOI = 10) for 0 h (uninfected), 4 h, 12 h, 24 h, 36 h, and 48 h. The expressions of *Sp1* and *Sp3* were detected by qPCR. (**B**) The regulation of SP1 or SP3 overexpression on *Nramp1* expression in H37Ra-infected/non-infected cells, respectively. pCMV-SP1, pCMV-SP3, and NC groups were transfected into RAW264.7 cells after H37Ra infection or no infection. The infected group cells were infected with H37Ra (MOI = 10) at 24 h after transfection. The expressions of *Nramp1* were detected by qPCR (*, $p < 0.05$; **, $p < 0.01$). (**C**,**D**) The impact of the 2SPM2 mutation on SP1 overexpression or knockdown in H37Ra-infected and non-infected groups, respectively. On the left side is the H37Ra-infected group, and on the right side is the non-infected group. The white and dark columns represent the blank control and the overexpression/knockout treatment, respectively. The diagonally hatched columns indicate that the promoter used in this group are the 2SPM2 mutant promoter. In (**E**,**F**), the representations are consistent. Values were normalized to the mean of the −1611/+73 + pCMV group (ns, $p \geq 0.05$; *, $p < 0.05$; ***, $p < 0.001$; ****, $p < 0.0001$). (**E**,**F**) The impact of the 2SPM2 mutation on SP3 overexpression or knockdown in H37Ra-infected and non-infected groups, respectively. Values were normalized to the mean of the −1611/+73 + pCMV group (ns, $p \geq 0.05$; *, $p < 0.05$; **, $p < 0.01$; ****, $p < 0.0001$).

2.5. Validating the Impact of the 2SP Site Mutations on the SP1/SP3-Mediated Regulation of NRAMP1 Expression

Previous studies have demonstrated that for promoters containing multiple Sp-binding sites, Sp1 may bind at both proximal and distal transcriptional sites to exert its transcriptional synergism [38,44,45]. According to the prediction, the bovine *NRAMP1* promoter contains multiple SP-binding sites. To more accurately assess the impact of 2SP site mutations, we utilized the pGL4.10 vector (pGL4.10−1611/+73) containing the bovine *NRAMP1* promoter sequence from −1611 to +73 bp for the functional validation of the 2SP site mutations in subsequent experiments. To minimize potential adverse effects and enhance safety, we aimed to reduce the number of mutated bases at the 2SP site. Two editing sequences that reduced the number of mutant bases were designed, along with a complete deletion of the 2SP site as a comparison, namely, PGL4.10-2SPM1, PGL4.10-2SPM2, and PGL4.10-2SPM3, respectively. Both 2SPM1 and 2SPM2 mutations significantly increased promoter activity after H37Ra infection. However, the 2SPM2 mutant promoter exhibited a more similar activity to the −1611/+73 promoter in the uninfected group (Supplementary Figure S6). Consequently, the 2SPM2 mutation (5′-GGGAGAGGTG-3′ to 5′-AAAAGAGGTG-3′) was selected for further investigation.

To investigate the impact of the 2SPM2 mutation on the SP1/SP3-mediated regulation of bovine *NRAMP1* expression, the overexpression recombinant plasmid of SP1/SP3 or siRNA against SP1/SP3 was co-transfected with PGL4.10-2SPM2 or pGL4.10−1611/+73 in RAW264.7 cells. The results indicated that the overexpression of SP1 significantly increased the activity of the −1611/+73 promoter and the 2SPM2 mutant promoter in both H37Ra-infected and non-infected cells. In non-infected cells, there was no significant difference in promoter activity between the −1611/+73 promoter and the 2SPM2 mutant promoter, regardless of SP1 overexpression (Figure 3C). The knockdown of SP1 significantly decreased −1611/+73 promoter and 2SPM2 mutant promoter activities in both H37Ra-infected and non-infected cells. In both the infected and non-infected cells, there was no significant difference in the activity of the 2SPM2 mutant promoter and the −1611/+73 promoter following SP1 knockdown (Figure 3D). Based on the previous results, after H37Ra infection, SP1 expression was upregulated, leading to the activation of *NRAMP1* expression. In molecular docking, the HADDOCK score of 2SPM2 increased from −88.9 to −37.5 upon mutation, which suggests a marked reduction in its binding affinity. Collectively, these findings suggest that the 2SPM2 mutation does not impair the ability of SP1 to activate *NRAMP1* expression.

The overexpression of SP3 significantly increased the activities of the −1611/+73 promoter and the 2SPM2 mutant promoter in both H37Ra-infected and non-infected cells. Following SP3 overexpression, the upregulation was greater in the 2SPM2 mutant promoter (1.42-fold in infected and 2.95-fold in non-infected cells) compared to the −1611/+73 promoter (1.29-fold in infected and 1.45-fold in non-infected cells) (Figure 3E). The knockdown of SP3 significantly decreased the activities of both promoters in H37Ra-infected cells (to 0.68-fold for −1611/+73 and 0.5-fold for 2SPM2), while in non-infected cells, SP3 knockdown resulted in increased promoter activities (Figure 3F). Based on the previous results, after H37Ra infection, SP3 expression is upregulated, and its ability to activate *NRAMP1* expression is also enhanced. These results indicate that following H37Ra infection, the 2SPM2 mutant promoter exhibits increased sensitivity to SP3 regulation. On the other hand, in non-infected cells, although SP3 overexpression slightly increases *NRAMP1* expression, its activating capacity is weaker compared to SP1. SP3 competes with SP1 for binding to the same regulatory site, potentially inhibiting SP1′s activation. Consequently, SP3 knockdown results in enhanced *NRAMP1* expression.

2.6. Base Editing of the Homologous Region of the 2SP Site in THP-1 Cells and Phenotypic Verification

To verify the endogenous expression of *NRAMP1* after editing, we managed to generate macrophage cell lines with the 2SP site edited. Due to the lack of bovine macrophage cell lines and the high homology of bovine and human *NRAMP1* promoters, a base editing system for the 2SP mutation of the THP-1 cell line was designed and constructed. A previous study proved that a central G-substitution at the Sp1/Sp3 recognition site would reduce binding affinity by at least 30-fold; thus, the guide was designed to make a central A-to-G mutation in the homologous region of the 2SP site. The epi-ABEmax-NG-h2SPM vector, containing the designed sgRNA, was constructed based on epi-ABE4max-NG (Addgene plasmid #135976) for THP-1 cell editing, with the aim of introducing a central A-to-G mutation (5′-GGGAGG-3′ to 5′-GGGGGG-3′) (Figure 4A). The frequencies of precisely edited cells were detected by sequencing the mixed cell population. Under several different experimental systems we tested, the frequency of precisely edited cells reached a maximum of 0.33% (Supplementary Figure S7). Out of twenty-five 96-well plates, a total of 272 clones survived the sorting process, and 1 clone was confirmed to be precisely edited as a heterozygote with the expected single nucleotide mutation (A to G) at the target site. The sequence is shown in Figure 4B.

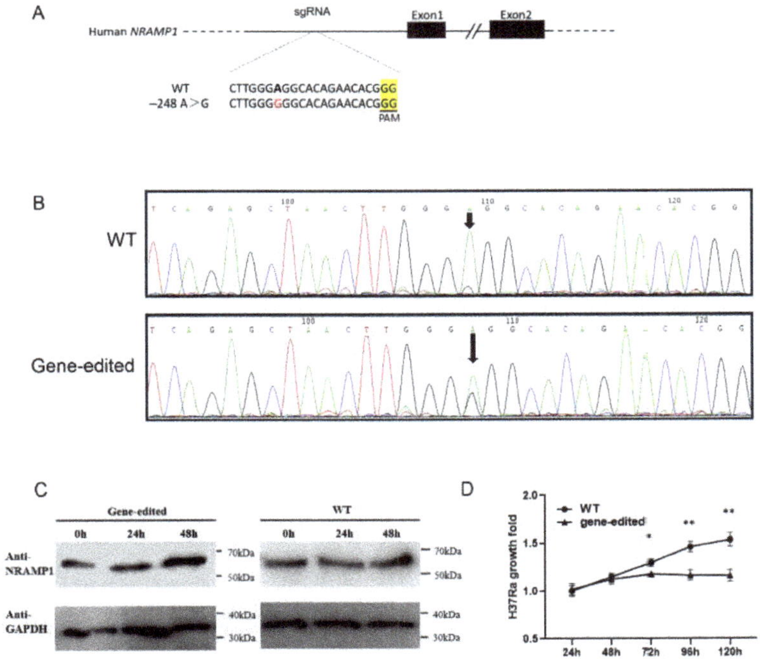

Figure 4. Base editing of the homologous region of the 2SP site in THP-1 cells and phenotypic verification. (**A**) The sgRNA designed for editing the homologous region of the 2SP site in THP-1 cells. The bolded A denotes the target site for editing in the wild-type sequence, whereas the red-highlighted G signifies the mutated base being G, and the yellow highlighting indicates the location of the PAM site. (**B**) The sequence of the *NRAMP1* promoter in the edited clone. The edited site is indicated by the arrow. The nucleotide sequence before and after the edited site was identical to the wild type. (**C**) Western blot analyses to detect NRAMP1 expression in gene-edited cells and wild-type THP-1 cells after H37Ra infection. (**D**) Multiplication of H37Ra in wild-type THP-1 cells or gene-edited cells. The values of the gene-edited groups and the wild-type group at 48 h, 72 h, 96 h, and 120 h were normalized to the mean of the wild-type group at 24 h (±SD). Student's *t*-test was used to evaluate the differences (*, $p < 0.05$; **, $p < 0.01$).

The production of the NRAMP1 protein in the gene-edited clones and wild-type THP-1 cells before/after H37Ra infection was assessed by Western blotting. Before H37Ra infection, the gene-edited group exhibited comparable NRAMP1 expression levels to the wild-type group; however, as the duration of infection progressed, NRAMP1 expression in the gene-edited group increased more remarkably compared to the wild-type group (Figure 4C). To further evaluate the potential of gene-edited THP-1 cells in enhancing the inhibition of Mtb proliferation, CFU counts were enumerated at 24, 48, 72, 96, and 120 h post-H37Ra infection. The gene-editing group exhibited a significantly enhanced inhibition of H37Ra multiplication compared to those in the wild-type group after 72h (Figure 4D).

3. Discussion

In this paper, we describe the screening of bovine *NRAMP1* promoter mutation site gene editing to enhance TB resistance. A novel mutation site designated as 2SP was identified. The mutation of this site specifically upregulates the activity of the bovine *NRAMP1* promoter after H37Ra infection while it does not affect basal expression in uninfected groups. Subsequently, THP-1 cell lines with the homologous region of the 2SP site edited were generated using the CRISPR-Cas9 base editing technique. In fact, we also established a screening system for bovine fetal fibroblasts (BFFs) with 2SP site editing and obtained correct edited BFFs. These BFFs are used to produce edited-gene cattle.

NRAMP1 has been shown to be critical for host resistance to TB and a previous study in our laboratory demonstrated that gene-edited *NRAMP1* overexpression cattle increased TB resistance [33,34]. However, the insertion of synthetic sequences still poses potential security risks. Thus, the aim of this study was to specifically upregulate bovine *NRAMP1* expression after TB infection through endogenous mechanisms, while minimizing its impact on basal transcription in the absence of infection.

Studies have shown that *NRAMP1* is evolutionarily conserved, indicating that its structure and function have remained stable and sufficient throughout evolution. Mutations in its structure are more likely to be harmful. However, mutations in non-coding regions that alter the expression rather than the protein structure may be beneficial for TB resistance [18,30]. On the other hand, despite the conserved structure of *NRAMP1* being shared among different species, there are variations in their TB resistance. For example, certain strains of mice are considered to be resistant to TB [46,47], while most hoofed stock are considered to be susceptible [48]. This variation could be associated with different expression levels of *NRAMP1*. To test this hypothesis, we compared the levels of TB-induced *NRAMP1* gene expression between cows and mice using the data uploaded from different labs. As expected, mice showed stronger TB-induced *Nramp1* upregulation than cows, both in vivo and in vitro.

In eukaryotes, gene transcription is controlled by fundamental transcriptional mechanisms consisting of TFs and various sequence motifs that interact with basal TFs in the core promoter regions [49]. Recently, with the development of gene editing technologies, base editing has been used to generate or revert mutations in core promoter regulatory regions, aiming to regulate the expression of target genes [50–52]. These edits typically occur at frequently mutated sites within core promoter regions, and the efficient editing of key sites is critical for the success of this technique. However, studies on natural mutations in the bovine *NRAMP1* promoter associated with TB resistance remain insufficient, and no mutation editing site within the *NRAMP1* promoter has been reported to enhance TB resistance. In this study, we aimed to identify and screen potential mutation editing sites using interspecies sequence differences in the *NRAMP1* promoter and predicted TF binding sites as references.

Among the screened candidate mutation sites, the 2SP site showed great potential for mutation editing. The homologous region of the 2SP site exhibited two conserved sequences in 124 out of the total 130 species with *NRAMP1* gene promoter homology, which were variants of the consensus SP1/SP3 recognition site. Six species lost affinity for SP1/SP3 due to mutations or deletions in the homologous region of the 2SP site. These include mice and horses, with some strains exhibiting relative resistance to TB [47,53], as well as bats, which are thought to have strong immune system [54,55], implying that the loss of affinity for SP1/SP3 at the 2SP site may benefit to TB resistance. By substituting the 2SP site with corresponding sequences from humans, pigs, and mice and predicting their respective affinities for SP1/SP3, we observed that a lower affinity of the 2SP site correlates with higher promoter activity following H37Ra infection. In contrast, replacing the 2SP site with conserved SP1/SP3 binding sequences significantly reduced promoter activity, with a more pronounced reduction observed post-infection. In a study investigating the SNP within the SP1/SP3 binding site of the CD14 promoter, it was found that this SNP reduces the affinity of the binding site while enhancing promoter activity. The authors further suggested that this effect might be attributed to the decreased affinity of the binding site for SP3 [56].

In this study, we found that the overexpression of both SP1 and SP3 activates bovine *NRAMP1* basal transcription. SP1 demonstrated a greater ability to activate basal transcription, while only SP3, not SP1, specifically enhanced its ability to activate *NRAMP1* expression after H37Ra infection. We speculate that only SP3 forms an Mtb-induced complex with other TFs to activate *NRAMP1* expression during infection specifically. This is consistent with a previous study that suggested that SP1 plays a role in basal transcription, while SP3 is involved in inducible transcription [43]. However, further experiments are still needed to verify this assumption.

Except for the 2SP site, mutations at the 1SP site (−262/−253bp), an adjacent upstream SP1/SP3 recognition site relative to the 2SP site, also showed a specific response to H37Ra infection, but with the opposite effect. Specifically, the 1SP site mutation did not alter promoter activity in uninfected conditions but significantly reduced promoter activity after infection. These imply that the 1SP site may be a "correct" recognition site for the specific regulation of SP3; however, the 2SP site may not be a "correct" recognition site, and its weak affinity makes it compete with the 1SP site to bind SP3.

Therefore, the mechanism through which the 2SP site mutation can specifically upregulate *NRAMP1* expression after infection but not affect its expression in uninfected conditions may be as follows: in uninfected conditions, SP1 plays a dominant role in activating *NRAMP1* expression, and its activation is independent of binding to the 1SP site or the 2SP site. Consequently, mutation at the 2SP site does not affect the basal expression of *NRAMP1*. During H37Ra infection, SP3 specifically activates *NRAMP1* expression by binding to the 1SP site. However, the adjacent 2SP site competes for SP3 binding, thereby inhibiting this activation. Therefore, mutations at the 2SP site reduce competition for SP3 binding, enhance the binding efficiency of SP3 at the 1SP site, and thus specifically upregulate *NRAMP1* expression during infection.

Finally, we established a base-editing screening system targeting the homologous region of the 2SP site in THP-1 cells and successfully generated THP-1 cells with an edited 2SP site. Upon validation, the edited cells exhibited significantly elevated NRAMP1 expression levels following H37Ra infection and effectively suppressed bacterial proliferation.

In summary, through interspecific comparison and sequence analysis, we identified for the first time a potential editing site in the bovine *NRAMP1* promoter that may specifically upregulate *NRAMP1* expression in response to Mtb infection. Subsequently, we generated THP-1 cell lines with the homologous region of the 2SP site edited, which exhibited enhanced TB resistance. Moreover, BFFs with the 2SP site edited were also generated and used for nuclear transfer. We will further optimize the experimental system to obtain a sufficient number of precisely edited BFFs for screening high-quality clones suitable for nuclear transfer. Consequently, a limitation of this study is that the findings were not verified in vivo, which will be addressed in subsequent experiments.

4. Materials and Methods

4.1. qPCR

Total RNA was isolated using TransZol reagent (TransGen, Beijing, China) and subsequently transcribed into cDNA using EasyScript One-Step RT-PCR SuperMix (TransGen, Beijing, China). The quantitative primers of mouse *Nramp1*, mouse *Sp1*, mouse *Sp3*, and human *NRAMP1* were designed, and mouse glyceraldehyde 3-phosphate dehydrogenase (*Gapdh*) and human *Gapdh* were selected as normalization controls, respectively (Table S1). The PCR cycling conditions were performed according to the manufacturer's protocols for TransStart FastPfu DNA Polymerase (TransGen, Beijing, China).

4.2. Construction of Vectors

The bovine *NRAMP1* promoter reporter vector pGL4.10 −1611/+73 and mouse *Nramp1* promoter reporter vector pGL4.10 m−1611/+156 were generated by inserting the bovine promoter fragment and mouse promoter fragment into the luciferase reporter vector pGL4.10, respectively. The truncated bovine *NRAMP1* promoter reporter vectors pGL4.1-T1(−1242/+73), pGL4.1-T2(−792/+73), pGL4.1-T3(−278/+73), pGL4.1-T4(−264/+73), pGL4.1-T5(−214/+73), pGL4.1-T6(−162/+73), pGL4.1-T7(−104/+73), and pGL4.1-T8(−60/+73) were constructed based on the pGL4.10 −1611/+73. The primers are shown in Table S2.

The 7 potential TF binding site mutations for the bovine *NRAMP1* promoter fragment (POUM, EM,1SPM,2SPM,3SPM, etsM, 2SPM1, 2SPM2, and 2SPM3) and the homologous regions of 2SP site substitution/deletion for the mouse *Nramp1* promoter fragment (MbM and MhM) were generated by overlap PCR or directly using the primers contain the mutation. The primers are shown in Table S3. Then, the fragments were inserted into the luciferase reporter vector pGL4.10.

To verify the effects of optimized mutations at the 2SP locus, PGL4.10 vectors harboring the 2SPM1, 2SPM2, and 2SPM3 mutations were constructed using an overlap extension method (Table S4).

In the overexpression experiments, cDNA extracted from RAW264.7 cells or BFF was used as a template to amplify the coding region sequence of the target gene. The amplified sequences were then subcloned into the pCMV vector (Takara, Beijing, China). An empty pCMV vector served as the negative control. In the knockdown experiments, specific siRNAs against targets SP1 and SP3 and negative control siRNA were synthesized from Genepharma.

epi-ABEmax-NG-h2SPM was generated based on epi-ABE4max-NG (Addgene plasmid #135976) [57]. The guide was designed to target the homologous region of the 2SP site in the human *NRAMP1* promoter (5′-CTTGGGAGGCACAGAACACG-3′), aiming to produce a central A to G mutation (5′-GGGAGG-3′ to 5′-GGGGGG-3′).

4.3. Cell Culture

RAW264.7 cells, obtained from the American Type Culture Collection, were cultured in Dulbecco's modified Eagle's medium (Gibco, Waltham, MA, USA) containing 10% fetal bovine serum (Gibco, Waltham, MA, USA) at 37 °C in a 5% CO_2 environment.

THP-1 cells, obtained from the American Type Culture Collection, were cultured in RPMI1640 medium (Gibco, Waltham, MA, USA) containing 10% fetal bovine serum (Gibco, Waltham, MA, USA) at 37 °C in a 5% CO_2 environment.

4.4. H37Ra Infection

H37Ra strains were cultured in 7H9 media (Becton Dickinson, Franklin Lakes, NJ, USA) supplemented with 10% OADC at 37 °C in a 5% CO_2 environment to logarithmic phase. Before infection, the final concentration of bacteria was calculated. At 24 h post-transfection, the appropriate number of bacteria were separated from the stock culture, centrifuged at 12,000× *g* for 1 min, and resuspended in RAW264.7 or THP-1 media for MOI = 10. Then, the bacteria were added to RAW264.7 cells or THP-1 cells for 4 h at 37 °C, 5% CO_2, before being washed twice with PBS to remove extracellular bacteria, and were then cultured in fresh media containing 10% fetal bovine serum.

4.5. Luciferase Assays

RAW264.7 cells were seeded in 24-well plates one day before transfection and grown to a confluence of 60%. Then, 0.5 μg of pGL4.10-promoter or empty vector pGL4.10 and 0.1 μg of pRL-TK normalizing vector were co-transfected according to the protocol of the PEI Transfection Reagent (MedChemExpress, Monmouth Junction, NJ, USA).

In total, 1×10^6 THP-1 cells per sample were centrifuged at 1050 rpm for 5 min, resuspended in RPMI1640 without fetal bovine serum, and seeded in a 24-well plate one day before transfection. Then, 0.5 μg of pGL4.10-promoter or empty vector pGL4.10 and 0.1 μg of pRL-TK normalizing vector were co-transfected according to the protocol of the PEI Transfection Reagent (MedChemExpress, Monmouth Junction, NJ, USA).

At 48h post-transfection, RAW264.7 or THP-1 cell lysates were collected for luciferase activity analysis by using the Double-Luciferase Reporter Assay Kit (TransGen, Beijing, China).

4.6. Editing of NRAMP1 in THP-1 Cells

In this study, we performed a central A to G mutation in the homologous region of the 2SP site in the *NRAMP1* promoter (−248A to G) with THP-1 cells, to mimic the editing at the 2SP site in the bovine *NRAMP1* promoter and validate the expression level. The human *NRAMP1* promoter sequences containing the homologous region of the 2SP site from −30 bp upstream to +30 bp downstream were entered into the base editing online design software BE-Designer (http://www.rgenome.net/be-designer/, accessed on 4 October 2022) [58], selecting '5′-NG-3′' for PAM type, 'Homo sapiens (GRCh38/hg38) -Human' for target genome, 'ABE (A to G)' for base editing type, and '14–17' for window, and outputting the results. SgRNA selection was performed using an evaluation model as described in Reference [34]. Three sgRNAs were obtained and input into Cas-OFFinder (http://www.rgenome.net/cas-offinder/, accessed on 4 October 2022) [59] to select the sgRNA with fewer potential off-target sites (the option "Mismatch Number (equal to or less than)" was, respectively, set to 3 and 5). The detailed data are provided in Table S5. The selected sequence was 5′-CTTGGGAGGCACAGAACACG-3′.

The epi-ABEmax-NG-h2SPM vector containing the designed sgRNA was generated for THP-1 cell editing. The transfection of epi-ABEmax-NG-h2SPM into THP-1 cells was performed using the same method and proportion of plasmid as that used in the luciferase assays. Five days after transfection, 20 μg/mL of blasticidin was used for screening. The screening lasted 2 days before diluting into 10 cells/mL and seeding in 96-well plates. The number of cells in each well was examined under a microscope and the wells with a single cell were marked and left to grow for 2 weeks or more. When the number was sufficient, half of the cells from each marked well were used for positive identification, and the correct clones were kept for further experimentation.

4.7. Western Blot Analyses

Cells were lysed in ice-cold RIPA cell buffer containing protease inhibitors (Thermo Scientific, Waltham, MA, USA). The proteins were separated with 12% acrylamide gels and subsequently transferred to PVDF membranes (Millipore, Burlington, MA, USA). The primary antibodies used to detect NRAMP1 (Rockville, Dubai, United Arab Emirates, Catalogue No. ab59696) and GAPDH (Cell Signaling Technology, Danvers, MA, USA, Cat# 2118L) were diluted 1:500. Goat anti-rabbit IgG (zhuangzhibio, Xi'an, China, EK020) was used as secondary antibody for all primary antibodies, diluted 1:5000.

4.8. SP1/SP3 3D Structure Modeling and Ligand Docking

The SP1 and SP3 protein structure modeling was carried out in AlphaFold3 [60]. The dsDNA secondary structures of the 2SP site sequences from different species and the mutated sequences were predicted using the Sequence to Structure web tool, which is accessible at the following URL (https://scfbio-iitd.res.in/software/drugdesign/bdna.jsp, accessed on 27 March 2025). Functional domain was identified using the InterPro tool (https://www.ebi.ac.uk/interpro/, accessed on 27 March 2025) [61], and P2Rank (https://prankweb.cz/, accessed on 27 March 2025) [62] was used for predicting, visualizing, and scoring the binding pocket. The docking of ligand molecules (DNA) was performed and scored in Haddock [63,64]

Supplementary Materials: The following supporting information can be downloaded at: https://www.mdpi.com/article/10.3390/ijms26094229/s1. The Supplementary Materials have cited the following references [65–69].

Author Contributions: Conceptualization, Y.W. and M.Y.; methodology, Y.W.; validation, Y.W. and M.Y.; formal analysis, Y.W. and M.Y.; investigation, Y.W. and M.Y.; resources, Y.G.; data curation, Y.W.; writing—original draft preparation, Y.W.; writing—review and editing, Y.G.; visualization, Y.W.; supervision, Y.G.; project administration, Y.Z.; funding acquisition, Y.G. and Y.Z. All authors have read and agreed to the published version of the manuscript.

Funding: This work was supported by the National Natural Science Foundation of China (number: 32172706), the National Dairy Industry Technology Innovation Center Project Topics (number: 2022-Scientific Research Attack-3), and the Shaanxi Provincial Key Project (number: 2022GD-TSLD-46-0101).

Institutional Review Board Statement: Not applicable.

Informed Consent Statement: Not applicable.

Data Availability Statement: The data supporting the findings of this study are available from the corresponding author upon reasonable request. These data encompass all relevant experimental results and analyses and will be provided to researchers who wish to verify or extend this work.

Conflicts of Interest: The authors declare no conflict of interest.

Abbreviations

The following abbreviations are used in this manuscript:

TB	tuberculosis
Mtb	*Mycobacterium tuberculosis*
M. bovis	*Mycobacterium bovis*
NRAMP1	natural resistance-associated macrophage 1
MAF	minor allele frequency
TF	transcription factor
Gapdh	glyceraldehyde 3-phosphate dehydrogenase
MOI	multiplicity of infection
hpi	hours post-infection

HADDOCK	High Ambiguity Driven DOCKing
SP	specificity protein
BFFs	bovine fetal fibroblasts

References

1. Zumla, A.; Raviglione, M.; Hafner, R.; von Reyn, C.F. Tuberculosis. *N. Engl. J. Med.* **2013**, *368*, 745–755. [CrossRef]
2. Comstock, G.W. Tuberculosis in twins: A re-analysis of the Prophit survey. *Am. Rev. Respir. Dis.* **1978**, *117*, 621–624. [CrossRef]
3. Wiart, A.; Jepson, A.; Banya, W.; Bennett, S.; Whittle, H.; Martin, N.G.; Hill, A.V. Quantitative association tests of immune responses to antigens of Mycobacterium tuberculosis: A study of twins in West Africa. *Twin. Res.* **2004**, *7*, 578–588. [CrossRef] [PubMed]
4. Abel, L.; El-Baghdadi, J.; Bousfiha, A.A.; Casanova, J.L.; Schurr, E. Human genetics of tuberculosis: A long and winding road. *Philos. Trans. R. Soc. Lond. B Biol. Sci.* **2014**, *369*, 20130428. [CrossRef] [PubMed]
5. Bellamy, R.; Beyers, N.; McAdam, K.P.; Ruwende, C.; Gie, R.; Samaai, P.; Bester, D.; Meyer, M.; Corrah, T.; Collin, M.; et al. Genetic susceptibility to tuberculosis in Africans: A genome-wide scan. *Proc. Natl. Acad. Sci. USA* **2000**, *97*, 8005–8009. [CrossRef] [PubMed]
6. Cooke, G.S.; Campbell, S.J.; Bennett, S.; Lienhardt, C.; McAdam, K.P.; Sirugo, G.; Sow, O.; Gustafson, P.; Mwangulu, F.; van Helden, P.; et al. Mapping of a novel susceptibility locus suggests a role for MC3R and CTSZ in human tuberculosis. *Am. J. Respir. Crit. Care Med.* **2008**, *178*, 203–207. [CrossRef]
7. Ghanavi, J.; Farnia, P.; Farnia, P.; Velayati, A.A. Human genetic background in susceptibility to tuberculosis. *Int. J. Mycobacteriol.* **2020**, *9*, 239–247. [CrossRef]
8. Hoal, E.G.; Lewis, L.A.; Jamieson, S.E.; Tanzer, F.; Rossouw, M.; Victor, T.; Hillerman, R.; Beyers, N.; Blackwell, J.M.; Van Helden, P.D. SLC11A1 (NRAMP1) but not SLC11A2 (NRAMP2) polymorphisms are associated with susceptibility to tuberculosis in a high-incidence community in South Africa. *Int. J. Tuberc. Lung Dis.* **2004**, *8*, 1464–1471.
9. Meyer, C.G.; Thye, T. Host genetic studies in adult pulmonary tuberculosis. *Semin. Immunol.* **2014**, *26*, 445–453. [CrossRef]
10. Cosivi, O.; Grange, J.M.; Daborn, C.J.; Raviglione, M.C.; Fujikura, T.; Cousins, D.; Robinson, R.A.; Huchzermeyer, H.F.; de Kantor, I.; Meslin, F.X. Zoonotic tuberculosis due to Mycobacterium bovis in developing countries. *Emerg. Infect. Dis.* **1998**, *4*, 59–70. [CrossRef]
11. O'Reilly, L.M.; Daborn, C.J. The epidemiology of Mycobacterium bovis infections in animals and man: A review. *Tuber. Lung Dis.* **1995**, *76* (Suppl. 1), 1–46. [CrossRef] [PubMed]
12. Jin, J.; Sun, L.; Jiao, W.; Zhao, S.; Li, H.; Guan, X.; Jiao, A.; Jiang, Z.; Shen, A. SLC11A1 (Formerly NRAMP1) gene polymorphisms associated with pediatric tuberculosis in China. *Clin. Infect. Dis.* **2009**, *48*, 733–738. [CrossRef]
13. Vidal, S.M.; Malo, D.; Vogan, K.; Skamene, E.; Gros, P. Natural resistance to infection with intracellular parasites: Isolation of a candidate for Bcg. *Cell* **1993**, *73*, 469–485. [CrossRef] [PubMed]
14. Amer, A.O.; Swanson, M.S. A phagosome of one's own: A microbial guide to life in the macrophage. *Curr. Opin. Microbiol.* **2002**, *5*, 56–61. [CrossRef]
15. Wessling-Resnick, M. Nramp1 and Other Transporters Involved in Metal Withholding during Infection. *J. Biol. Chem.* **2015**, *290*, 18984–18990. [CrossRef]
16. Forbes, J.R.; Gros, P. Iron, manganese, and cobalt transport by Nramp1 (Slc11a1) and Nramp2 (Slc11a2) expressed at the plasma membrane. *Blood* **2003**, *102*, 1884–1892. [CrossRef]
17. Gruenheid, S.; Pinner, E.; Desjardins, M.; Gros, P. Natural resistance to infection with intracellular pathogens: The Nramp1 protein is recruited to the membrane of the phagosome. *J. Exp. Med.* **1997**, *185*, 717–730 [CrossRef]
18. Archer, N.S.; Nassif, N.T.; O'Brien, B.A. Genetic variants of SLC11A1 are associated with both autoimmune and infectious diseases: Systematic review and meta-analysis. *Genes Immun.* **2015**, *16*, 275–283. [CrossRef] [PubMed]
19. Azad, A.K.; Sadee, W.; Schlesinger, L.S. Innate immune gene polymorphisms in tuberculosis. *Infect. Immun.* **2012**, *80*, 3343–3359. [CrossRef]
20. Brites, D.; Gagneux, S. Old and new selective pressures on Mycobacterium tuberculosis. *Infect. Genet. Evol.* **2012**, *12*, 678–685. [CrossRef]
21. Moller, M.; Hoal, E.G. Current findings, challenges and novel approaches in human genetic susceptibility to tuberculosis. *Tuberculosis* **2010**, *90*, 71–83. [CrossRef]
22. Shahzad, F.; Bashir, N.; Ali, A.; Nadeem, A.; Ammar, A.; Kashif, M.; Javaid, K.; Jahan, S.; Tahir, R.; Rizwan, M.; et al. SLC11A1 genetic variation and low expression may cause immune response impairment in TB patients. *Genes Immun* **2022**, *23*, 85–92. [CrossRef]
23. Cellier, M.F. Nutritional immunity: Homology modeling of Nramp metal import. *Adv. Exp. Med. Biol.* **2012**, *946*, 335–351. [CrossRef] [PubMed]
24. Cellier, M.F. Cell-Type Specific Determinants of NRAMP1 Expression in Professional Phagocytes. *Biology* **2013**, *2*, 233–283. [CrossRef] [PubMed]

25. Peracino, B.; Wagner, C.; Balest, A.; Balbo, A.; Pergolizzi, B.; Noegel, A.A.; Steinert, M.; Bozzaro, S. Function and mechanism of action of Dictyostelium Nramp1 (Slc11a1) in bacterial infection. *Traffic* **2006**, *7*, 22–38. [CrossRef] [PubMed]
26. Hu, B.; Wang, Y.; Wang, Z.; He, X.; Wang, L.; Yuan, D.; He, Y.; Jin, T.; He, S. Association of SLC11A1 Polymorphisms with Tuberculosis Susceptibility in the Chinese Han Population. *Front. Genet.* **2022**, *13*, 899124. [CrossRef]
27. Vidal, S.M.; Pinner, E.; Lepage, P.; Gauthier, S.; Gros, P. Natural resistance to intracellular infections: Nramp1 encodes a membrane phosphoglycoprotein absent in macrophages from susceptible (Nramp1 D169) mouse strains. *J. Immunol.* **1996**, *157*, 3559–3568. [CrossRef]
28. Liu, K.; Zhang, B.; Teng, Z.; Wang, Y.; Dong, G.; Xu, C.; Qin, B.; Song, C.; Chai, J.; Li, Y.; et al. Association between SLC11A1 (NRAMP1) polymorphisms and susceptibility to tuberculosis in Chinese Holstein cattle. *Tuberculosis* **2017**, *103*, 10–15. [CrossRef]
29. Malik, S.; Abel, L.; Tooker, H.; Poon, A.; Simkin, L.; Girard, M.; Adams, G.J.; Starke, J.R.; Smith, K.C.; Graviss, E.A.; et al. Alleles of the NRAMP1 gene are risk factors for pediatric tuberculosis disease. *Proc. Natl. Acad. Sci. USA* **2005**, *102*, 12183–12188. [CrossRef]
30. Barnes, I.; Duda, A.; Pybus, O.G.; Thomas, M.G. Ancient urbanization predicts genetic resistance to tuberculosis. *Evolution* **2011**, *65*, 842–848. [CrossRef]
31. Zaahl, M.G.; Robson, K.J.; Warnich, L.; Kotze, M.J. Expression of the SLC11A1 (NRAMP1) 5′-(GT)n repeat: Opposite effect in the presence of −237C→T. *Blood Cells Mol. Dis.* **2004**, *33*, 45–50. [CrossRef] [PubMed]
32. Searle, S.; Blackwell, J.M. Evidence for a functional repeat polymorphism in the promoter of the human NRAMP1 gene that correlates with autoimmune versus infectious disease susceptibility. *J. Med. Genet.* **1999**, *36*, 295–299. [CrossRef]
33. Yuan, M.; Zhang, J.; Gao, Y.; Yuan, Z.; Zhu, Z.; Wei, Y.; Wu, T.; Han, J.; Zhang, Y. HMEJ-based safe-harbor genome editing enables efficient generation of cattle with increased resistance to tuberculosis. *J. Biol. Chem.* **2021**, *296*, 100497. [CrossRef]
34. Gao, Y.; Wu, H.; Wang, Y.; Liu, X.; Chen, L.; Li, Q.; Cui, C.; Liu, X.; Zhang, J.; Zhang, Y. Single Cas9 nickase induced generation of NRAMP1 knockin cattle with reduced off-target effects. *Genome. Biol.* **2017**, *18*, 13. [CrossRef] [PubMed]
35. Gilmour, J.; O'Connor, L.; Middleton, C.P.; Keane, P.; Gillemans, N.; Cazier, J.B.; Philipsen, S.; Bonifer, C. Robust hematopoietic specification requires the ubiquitous Sp1 and Sp3 transcription factors. *Epigenetics Chromatin* **2019**, *12*, 33. [CrossRef] [PubMed]
36. Zhou, W.; Fang, J.; Jia, Q.; Meng, H.; Liu, F.; Mao, J. Transcription factor specificity protein (SP) family in renal physiology and diseases. *PeerJ* **2025**, *13*, e18820. [CrossRef]
37. Safe, S. Specificity Proteins (Sp) and Cancer. *Int. J. Mol. Sci.* **2023**, *24*, 5164. [CrossRef]
38. Li, L.; He, S.; Sun, J.M.; Davie, J.R. Gene regulation by Sp1 and Sp3. *Biochem. Cell Biol.* **2004**, *82*, 460–471. [CrossRef]
39. Letovsky, J.; Dynan, W.S. Measurement of the binding of transcription factor Sp1 to a single GC box recognition sequence. *Nucleic Acids Res.* **1989**, *17*, 2639–2653. [CrossRef]
40. Richer, E.; Campion, C.G.; Dabbas, B.; White, J.H.; Cellier, M.F. Transcription factors Sp1 and C/EBP regulate NRAMP1 gene expression. *FEBS J.* **2008**, *275*, 5074–5089. [CrossRef]
41. Resendes, K.K.; Rosmarin, A.G. Sp1 control of gene expression in myeloid cells. *Crit. Rev. Eukaryot. Gene. Expr.* **2004**, *14*, 171–181. [CrossRef] [PubMed]
42. Hata, Y.; Duh, E.; Zhang, K.; Robinson, G.S.; Aiello, L.P. Transcription factors Sp1 and Sp3 alter vascular endothelial growth factor receptor expression through a novel recognition sequence. *J. Biol. Chem.* **1998**, *273*, 19294–19303. [CrossRef] [PubMed]
43. Ward, S.V.; Samuel, C.E. The PKR kinase promoter binds both Sp1 and Sp3, but only Sp3 functions as part of the interferon-inducible complex with ISGF-3 proteins. *Virology* **2003**, *313*, 553–566. [CrossRef]
44. Su, W.; Jackson, S.; Tjian, R.; Echols, H. DNA looping between sites for transcriptional activation: Self-association of DNA-bound Sp1. *Genes Dev.* **1991**, *5*, 820–826. [CrossRef] [PubMed]
45. Mastrangelo, I.A.; Courey, A.J.; Wall, J.S.; Jackson, S.P.; Hough, P.V. DNA looping and Sp1 multimer links: A mechanism for transcriptional synergism and enhancement. *Proc. Natl. Acad. Sci. USA* **1991**, *88*, 5670–5674. [CrossRef]
46. Chackerian, A.A.; Behar, S.M. Susceptibility to Mycobacterium tuberculosis: Lessons from inbred strains of mice. *Tuberculosis* **2003**, *83*, 279–285. [CrossRef]
47. Shepelkova, G.; Pommerenke, C.; Alberts, R.; Geffers, R.; Evstifeev, V.; Apt, A.; Schughart, K.; Wilk, E. Analysis of the lung transcriptome in Mycobacterium tuberculosis-infected mice reveals major differences in immune response pathways between TB-susceptible and resistant hosts. *Tuberculosis* **2013**, *93*, 263–269. [CrossRef]
48. LoBue, P.A.; Enarson, D.A.; Thoen, C.O. Tuberculosis in humans and animals: An overview. *Int. J. Tuberc. Lung Dis.* **2010**, *14*, 1075–1078.
49. Sato, M.P.; Makino, T.; Kawata, M. Natural selection in a population of Drosophila melanogaster explained by changes in gene expression caused by sequence variation in core promoter regions. *BMC Evol. Biol.* **2016**, *16*, 35. [CrossRef]
50. Li, X.; Qian, X.; Wang, B.; Xia, Y.; Zheng, Y.; Du, L.; Xu, D.; Xing, D.; DePinho, R.A.; Lu, Z. Programmable base editing of mutated TERT promoter inhibits brain tumour growth. *Nat. Cell Biol.* **2020**, *22*, 282–288. [CrossRef]
51. Liao, J.; Chen, S.; Hsiao, S.; Jiang, Y.; Yang, Y.; Zhang, Y.; Wang, X.; Lai, Y.; Bauer, D.E.; Wu, Y. Therapeutic adenine base editing of human hematopoietic stem cells. *Nat. Commun.* **2023**, *14*, 207. [CrossRef]

2. Ravi, N.S.; Wienert, B.; Wyman, S.K.; Bell, H.W.; George, A.; Mahalingam, G.; Vu, J.T.; Prasad, K.; Bandlamudi, B.P.; Devaraju, N.; et al. Identification of novel HPFH-like mutations by CRISPR base editing that elevate the expression of fetal hemoglobin. *eLife* **2022**, *11*, e65421. [CrossRef] [PubMed]
3. Chen, Q.; Tong, Q.; Ge, H.; Li, W.; Liu, J.; Wang, Y.; Guo, Z.; Quan, F.; Zhang, Y. Identification of SP110 in horse (Equus caballus): Isolation of novel splice variants and evidence of activation effects on macrophages. *Tuberculosis* **2016**, *101*, 85–94. [CrossRef] [PubMed]
4. Wang, L.F.; Gamage, A.M.; Chan, W.O.Y.; Hiller, M.; Teeling, E.C. Decoding bat immunity: The need for a coordinated research approach. *Nat. Rev. Immunol.* **2021**, *21*, 269–271. [CrossRef]
5. Friedrichs, V.; Toussaint, C.; Schafer, A.; Rissmann, M.; Dietrich, O.; Mettenleiter, T.C.; Pei, G.; Balkema-Buschmann, A.; Saliba, A.E.; Dorhoi, A. Landscape and age dynamics of immune cells in the Egyptian rousette bat. *Cell Rep.* **2022**, *40*, 111305. [CrossRef]
6. LeVan, T.D.; Bloom, J.W.; Bailey, T.J.; Karp, C.L.; Halonen, M.; Martinez, F.D.; Vercelli, D. A common single nucleotide polymorphism in the CD14 promoter decreases the affinity of Sp protein binding and enhances transcriptional activity. *J. Immunol.* **2001**, *167*, 5838–5844. [CrossRef]
7. Qi, T.; Wu, F.; Xie, Y.; Gao, S.; Li, M.; Pu, J.; Li, D.; Lan, F.; Wang, Y. Base Editing Mediated Generation of Point Mutations Into Human Pluripotent Stem Cells for Modeling Disease. *Front. Cell Dev. Biol.* **2020**, *8*, 590581. [CrossRef]
8. Hwang, G.H.; Park, J.; Lim, K.; Kim, S.; Yu, J.; Yu, E.; Kim, S.T.; Eils, R.; Kim, J.S.; Bae, S. Web-based design and analysis tools for CRISPR base editing. *BMC Bioinform.* **2018**, *19*, 542. [CrossRef]
9. Bae, S.; Park, J.; Kim, J.S. Cas-OFFinder: A fast and versatile algorithm that searches for potential off-target sites of Cas9 RNA-guided endonucleases. *Bioinformatics* **2014**, *30*, 1473–1475. [CrossRef]
0. Abramson, J.; Adler, J.; Dunger, J.; Evans, R.; Green, T.; Pritzel, A.; Ronneberger, O.; Willmore, L.; Ballard, A.J.; Bambrick, J.; et al. Accurate structure prediction of biomolecular interactions with AlphaFold 3. *Nature* **2024**, *630*, 493–500. [CrossRef]
1. Paysan-Lafosse, T.; Blum, M.; Chuguransky, S.; Grego, T.; Pinto, B.L.; Salazar, G.A.; Bileschi, M.L.; Bork, P.; Bridge, A.; Colwell, L.; et al. InterPro in 2022. *Nucleic Acids Res.* **2023**, *51*, D418–D427. [CrossRef]
2. Krivak, R.; Hoksza, D. P2Rank: Machine learning based tool for rapid and accurate prediction of ligand binding sites from protein structure. *J. Cheminform.* **2018**, *10*, 39. [CrossRef] [PubMed]
3. Honorato, R.V.; Trellet, M.E.; Jimenez-Garcia, B.; Schaarschmidt, J.J.; Giulini, M.; Reys, V.; Koukos, P.I.; Rodrigues, J.; Karaca, E.; van Zundert, G.C.P.; et al. The HADDOCK2.4 web server for integrative modeling of biomolecular complexes. *Nat. Protoc.* **2024**, *19*, 3219–3241. [CrossRef] [PubMed]
4. Honorato, R.V.; Koukos, P.I.; Jimenez-Garcia, B.; Tsaregorodtsev, A.; Verlato, M.; Giachetti, A.; Rosato, A.; Bonvin, A. Structural Biology in the Clouds: The WeNMR-EOSC Ecosystem. *Front. Mol. Biosci.* **2021**, *8*, 729513. [CrossRef] [PubMed]
5. Abdelaal, H.F.M.; Thacker, T.C.; Wadie, B.; Palmer, M.V.; Talaat, A.M. Transcriptional Profiling of Early and Late Phases of Bovine Tuberculosis. *Infect. Immun.* **2022**, *90*, e0031321. [CrossRef]
6. Nalpas, N.C.; Magee, D.A.; Conlon, K.M.; Browne, J.A.; Healy, C.; McLoughlin, K.E.; Rue-Albrecht, K.; McGettigan, P.A.; Killick, K.E.; Gormley, E.; et al. RNA sequencing provides exquisite insight into the manipulation of the alveolar macrophage by tubercle bacilli. *Sci. Rep.* **2015**, *5*, 13629. [CrossRef]
7. Nalpas, N.C.; Park, S.D.; Magee, D.A.; Taraktsoglou, M.; Browne, J.A.; Conlon, K.M.; Rue-Albrecht, K.; Killick, K.E.; Hokamp, K.; Lohan, A.J.; et al. Whole-transcriptome, high-throughput RNA sequence analysis of the bovine macrophage response to Mycobacterium bovis infection in vitro. *BMC Genom.* **2013**, *14*, 230. [CrossRef]
8. Olson, G.S.; Murray, T.A.; Jahn, A.N.; Mai, D.; Diercks, A.H.; Gold, E.S.; Aderem, A. Type I interferon decreases macrophage energy metabolism during mycobacterial infection. *Cell Rep.* **2021**, *35*, 109195. [CrossRef]
9. Rothchild, A.C.; Olson, G.S.; Nemeth, J.; Amon, L.M.; Mai, D.; Gold, E.S.; Diercks, A.H.; Aderem, A. Alveolar macrophages generate a noncanonical NRF2-driven transcriptional response to Mycobacterium tuberculosis in vivo. *Sci. Immunol.* **2019**, *4*, eaaw6693. [CrossRef]

animals

MDPI

Review

Innovations in Cattle Breeding Technology: Prospects in the Era of Gene Editing

Yu Wang [1], Xiangshun Cui [2] and Zhi Chen [1,*]

[1] College of Animal Science and Technology, Yangzhou University, Yangzhou 225009, China; 221902118@stu.yzu.edu.cn
[2] Laboratory of Animal Developmental Biology, Department of Animal Science, Chungbuk National University, Cheongju 28644, Republic of Korea; xscui@cbnu.ac.kr
* Correspondence: zhichen@yzu.edu.cn

Simple Summary: With the rapid advancement of modern society and the continuous growth of the global population, consumers' demand for cattle and their products, including high-quality beef and dairy products, has been steadily increasing. However, the existing cattle breeds currently known to people are confronted with a multitude of production-related challenges. These issues range from low meat and milk production efficiency and susceptibility to diseases to suboptimal adaptation to diverse environmental conditions, all of which necessitate continuous efforts to enhance their traits in real-world production scenarios. To address these challenges, scholars are actively exploring new ways to breed novel cattle breeds. By introducing cutting-edge gene editing technology and in-depth analysis of candidate genes that impact various important traits of cattle, such as growth rate, disease resistance, and milk quality, we strive to provide practical and innovative ideas and methodologies for the successful cultivation of new cattle breeds.

Abstract: As a core species in the global livestock industry, cattle play an irreplaceable role in human food security and economic development. Beef cattle and dairy cattle meet the dietary needs of billions of people around the world by providing high-quality protein and dairy products, respectively. With the growth in population and the intensification of the pressure of climate change, traditional breeding techniques may be unsuitable to meet the increasingly growing demands for sustainable and highly adaptable processes. In recent years, the rapid development of genomics, bioinformatics, and gene-editing technologies has provided unprecedented tools and perspectives for the genetic improvement of cattle, driving the precise design and efficient development of new cattle breeds. However, the development of new cattle breeds still faces multiple bottlenecks pertaining to scientific, ethical, and industrialization aspects, which can be addressed through interdisciplinary collaboration. In this review, we will systematically assess the technological progress in the genetic breeding of beef cattle and dairy cattle, analyze the integration path of traditional breeding and modern biotechnology, and explore the future directions of cattle breeding research under the sustainable development goals, with the aim of providing theoretical support for cattle breeding.

Keywords: cattle; breeding; gene editing; biotechnology

1. Introduction

The demand for beef and dairy products continues to increase, driven by the global population growth and rising living standards [1]. Cattle, as one of the most important domestic animals, hold an irreplaceable position in human society. Globally, several breeds of dairy and beef cattle are indispensable in the livestock industry due to their distinct genetic traits, superior production performance, and substantial economic value. Dairy breeds are renowned for their high milk production or superior milk quality. Among them, the *Holstein breed* stands out as the most globally widespread breed, easily recognizable by its distinctive black-and-white patches and renowned for its high average annual milk yield. In contrast, the *Jersey breed*, though smaller in size, is prized for its exceptionally high milk fat content [2]. The *Ayrshire breed* thrives in hilly pasture environments, and the *Guernsey breed* is renowned for the high carotenoid content of its milk, which contributes to its distinctive richness. Beef breeds are distinguished by their core strengths in terms of growth rate, muscle development, and meat quality. The *Angus breed* is globally celebrated for its hornless black coat, exceptional marbling, and tender meat. The *Charolais cattle* from France are known for their large size and rapid weight gain, making them ideal for intensive fattening programs. The *Aubrac cattle*, prized for their juiciness and tenderness, are another notable breed. Meanwhile, the *Japanese Wagyu cattle* stand at the pinnacle of beef production, renowned for their extreme marbling and unique, rich flavor profile. In addition, multipurpose breeds, such as the *Swiss Simmental* and *Danish Red*, integrate both milk and meat production capabilities, offering versatile options for diversified farming systems. Through continuous selective breeding and crossbreeding efforts, these breeds have significantly contributed to the advancement and efficiency of modern animal husbandry practices (Figure 1).

Figure 1. Common breeds of cattle in the world (including dairy cattle, beef cattle, and dual-purpose cattle).

The development of new cattle breeds is crucial for driving the progress of the livestock industry, meeting evolving market demands, and enhancing the industry's overall competitiveness [3]. High-quality cattle breeds are the key to improving farming efficiency and production levels; however, dairy cattle, beef cattle, and dual-purpose cattle breeds face their own diverse set of problems. Traditional cattle breeds often face limitations in terms of their growth rate, meat and milk production performance, and disease resistance, which hinder their ability to meet the demands of modern animal husbandry for scale, intensification, and high efficiency. As a critical component of modern animal husbandry, dairy cattle breeding encounters numerous challenges while striving to enhance milk production efficiency and economic returns. The declining genetic diversity is a core problem, and overreliance on a few high-yielding sires leads to increased inbreeding rates. The accumulation of recessive harmful genes may trigger risks such as reduced fertility in offspring [4], weakened immunity, and a prevalence of metabolic diseases. Secondly, achieving a balance between high production traits and health traits remains a significant challenge. An overemphasis on milk production may compromise a cow's disease resistance, limb health, or longevity, ultimately undermining the long-term sustainability of the herd [5]. Ethical and biosafety controversies in the application of reproduction technologies such as embryo transfer and gene editing techniques, which may lead to the spread of genetic defects, compromise animal welfare, or produce unforeseen ecosystem chain reactions [6]. In addition, the tension between environmental adaptability and intensive farming practices is becoming increasingly pronounced. Many high-yielding dairy breeds exhibit significant sensitivity to specific climatic or feeding conditions, which limits their ability to adapt to climate change or meet the production demands of resource-constrained regions. The technological and germplasm barriers between different countries are one of the important factors contributing to the slow development of the beef cattle industry. For example, China has long relied on imported breeds (such as Simmental and Angus). However, due to insufficient local adaptive breeding efforts post-introduction, these breeds often experience performance degradation, trapping the industry in a cycle of 'introduction–degradation–re-introduction'. Meanwhile, indigenous specialty breeds, such as Qinchuan cattle and Luxi cattle, are increasingly marginalized due to the lack of systematic selection and breeding programs. Additionally, the whole-genome databases for breeding cattle have insufficient coverage, and the reliance on imported performance measurement equipment further limits the accuracy of genetic evaluations. On the breeding front, small-scale free-range operations dominate, often lacking performance measurement systems and genealogical records. This results in a fragmented breeding data chain, compounded by an inadequate female stock, which collectively hinders the sustainability of continuous breeding efforts [7].

Gene editing is significant in the field of breeding new cattle lines [8]. Gene editing technology enables precise gene localization and modification, overcoming the limitations of traditional breeding methods, such as a long generation interval. This breakthrough significantly reduces the breeding time and accelerates the development of superior varieties. Additionally, gene editing unlocks and utilizes previously inaccessible genetic resources, facilitating targeted innovations that lead to higher yields and improved quality in new varieties. In addition, cattle breeding has fully entered the era of genomic selection [9]. GS can be applied to cattle breeding for predicting the precocious puberty of cows and selecting some traits of young bulls that are difficult to measure [10]. In addition, GS can also be applied in fields such as the paternity testing of cattle herds and the prediction of embryo genomes [11]. Furthermore, high-density genotyping is the basis for the genome-wide linkage analyses and genetic evaluation of cattle. The use of gene chips allows for easier, faster, and more accurate large-scale typing. A large number of studies have shown that the inclusion of functional loci that are significantly associated with target traits in gene chips can significantly improve the accuracy of genomic selection [12,13]. Gene chips act as a pivotal bridge connecting various technologies in biological breeding. In the realm of gene editing, they provide essential technological support. For screening and identifying gene targets, gene chips enable high-throughput genome-wide scanning, pinpointing the precise entry points for gene editing. When assessing gene editing efficiency, gene microarrays can detect changes in gene expression before and after editing, offering a comprehensive view of the editing outcomes. They also verify editing specificity and identify potential off-target risks. Furthermore, in studying gene editing mechanisms, gene microarrays facilitate a deeper understanding of intracellular signaling pathway remodeling and gene regulatory network dynamics by monitoring the behavior of relevant genes throughout the gene editing process.

New cattle lines can be developed by introducing or enhancing superior traits, such as improving the growth rates and meat quality of beef cattle, increasing the milk yield and quality of dairy cattle, and boosting the disease and environmental resilience of herds. These advancements contribute to the overall productivity and economic efficiency of the livestock industry, fostering its sustainable development. Therefore, this review will provide a brief overview of gene editing techniques commonly used in breeding and assess the potential candidate genes for developing new dairy and beef cattle lines. The aim of this review is to offer a theoretical foundation and practical insights for breeding innovative cattle varieties.

2. From CRISPR/Cas9 to Multi-Gene Editing: New Pathways for Genetic Improvement of Cattle

Gene editing is a genetic engineering activity that involves altering an organism's specific genomic DNA by means of DNA sequence knockouts, insertions, and base editing to change its properties and functions [14]. Major gene editing technologies include zinc finger nucleases (ZFNs), transcription activator-like effector nucleases (TALENs), and clusters of regularly interspaced short palindromic repeats (CRISPR)/Cas9 gene editing [15,16].

Since the beginning of the 21st century, the most popular gene editing technology has been the CRISPR/Cas9 system. Some scholars have shown that the CRISPR/Cas9 system can be successfully applied in breeding to improve reproductive traits and production performance and introduce strong disease resistance traits in a variety of livestock. This can be achieved due to the CRISPR/Cas9 system's advantages of a low application cost, a wide editing scope, a high targeting efficiency, simple operation, and support of multi-site operations [17]. Cattle are monozygotic animals, their reproduction cycle and growth cycle are longer than those of poultry, and the associated cost of experimentation is high. The development of gene editing is relatively lagging behind, and through the learning and application of CRISPR/Cas9 technology, we will be able to promote the rapid development of research in this field. However, despite the rapid development of gene editing technology, this technology is still limited to editing a single gene and improving a single trait, which no longer meets the current stage of breeding needs. Moreover, during the breeding process, some traits may be controlled by multiple genes [18]. In such circumstances, it is rather challenging to edit a single gene to change these traits. In order to resolve this situation, a simple and efficient multi-gene editing system is needed. The use of gene editing technology for the simultaneous editing or transcriptional regulation of multiple loci has become a major trend, which is of great significance for the improvement in livestock germplasm.

In multiple gene editing technology, we need to introduce two or more sgRNAs at the same time so that the Cas9 protein can be guided by multiple sgRNAs to effectively recognize multiple target sites [19]. Different scholars have given different answers to the question of how to introduce multiple sgRNAs at the same time [20]. The more common methods include the establishment of sgRNA expression boxes containing multiple monocistronic sgRNA expression boxes and the establishment of multicistronic sgRNA expression boxes. For multiple individual sgRNA expression cassettes, it is generally possible to tandemly connect multiple repetitive Pol III-type promoter-initiated sgRNA expression cassettes, each of which is transcribed by a separate RNA polymerase III (Pol III) promoter [21]. The common promoters in this system are U3 and U6, and these Pol III promoters start transcription at specific nucleotides. For example, the U3 promoter transcribes from 'A', and the U6 promoter transcribes from 'G' [22]. Although this method allows for the editing of multiple genes, the design of the system necessitates the separate assembly of promoters, terminators, and sgRNA backbones, which is not only cumbersome to execute and requires repetitive work but also affects each of the transcriptional activities of neighboring promoters [23]. Thus, multi-cis sgRNA expression cassettes are preferred today. For such systems, there are three common constructs: nuclease Csy4-mediated multi-cis sgRNA processing, tRNA-based multi-cis sgRNA processing, and self-cleaving nuclease-mediated multi-cis sgRNA processing. Next, we will briefly describe three ways of building multi-cis sgRNA expression cassettes. Csy4, a crRNA-processing protein from Pseudomonas aeruginosa, is a specific ribonucleic acid endonuclease. By artificially installing Csy4 excision sites on both sides of sgRNAs to construct a system of Cys4-processed-sgRNAs with a tandem structure in which Csy4 can cleave each of the repetitive sequences, the Csy4-mediated arrays of gRNAs can efficiently release a large number of sgRNAs at the same time [24–26]. As for the multi-cis-transon sgRNA processing of tRNAs, the 5′ and 3′ end sequences of tRNAs, which are present in every organism, can be specifically recognized and cleaved by the cellular endogenous RNaseP and RNaseZ enzymes, releasing independent tRNAs and mature sgRNAs that direct the Cas9 protein to cleave the target sites. By constructing an artificial multiple cis-tRNA-gRNA (PTG) gene tandem expression system with the tRNA-sgRNA-sgRNA backbone as the minimal repeating unit, it is possible to enable the efficient expression of multiple sgRNAs from a single Pol III promoter [27]. Since multiple gene editing using self-cleaving nucleases has been developed on a large scale in microorganisms and because there are very few reports of multiple sgRNA expression in animals using self-cleaving systems, we will not cover them in this review. Different multi-gene editing systems have their own specific constraints, yet they all face the same pressing problem: too many sgRNA species compete for the limited number of nucleases in a cell, a phenomenon that makes knockdown less efficient and causes more sgRNAs to be in tandem. However, we cannot deny that the simultaneous editing of multiple genes is a key breakthrough in the use of CRISPR gene editing technology to genetically breed new lines of cattle, which is of great significance to the field of livestock production (Figure 2).

Single-gene and multi-gene editing technologies collectively provide precise and efficient pathways for advancing cattle breeding. Single-gene editing enables the rapid introduction of specific traits by targeting key functional genes, while multi-gene editing addresses the genetic antagonism between traits through the synergistic regulation of complex genetic networks. This approach overcomes the bottleneck of optimizing multiple traits in traditional breeding methods. Some cattle breeds have already been modified through single-gene editing to confer heat-tolerance mutations, thereby enhancing their resistance to heat stress [28]. Additionally, there are cattle that have had their reproductive capabilities improved through single-gene editing [29]. Although there are not many reports showing the application of multi-gene editing in cattle breeding as of today, i.e., up to 2025, it has already achieved an adequate level of success in the poultry industry [30]. This bolsters our confidence in the application of this technology in cattle breeding. The combination of the two strategies not only accelerates the breeding of 'customized' cattle but also promotes the upgradation of the livestock sector towards efficiency and sustainability.

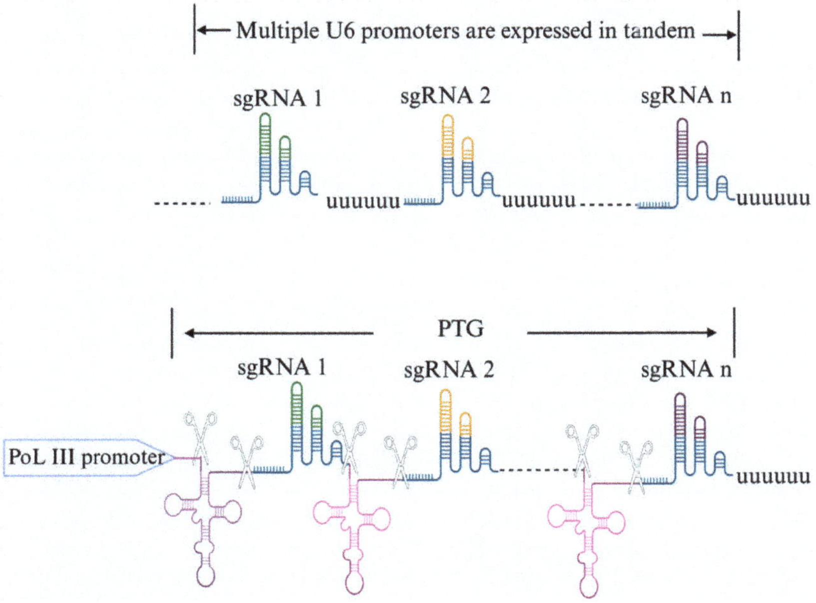

Figure 2. Schematic representation of multiple single cis–trans sgRNA expression cassettes versus multiple cis–trans sgRNA expression cassettes.

3. Breeding New Lines of Dairy Cattle

3.1. Selection of Key Genes in Dairy Cattle Breeding

If people want to breed new high-quality cows, they must pay attention to the expression of relevant genes that can affect the characteristics of the progenitor cows. The number of genes that can influence lactation and reproduction traits in dairy cows is very large (Table 1) and cannot be exhaustively reviewed in this review. Therefore, this review will focus on several gene families widely recognized by the academic community for their roles in lactation initiation, milk fat synthesis, and milk fat transport in dairy cows. This review will summarize their known functions and explore their potential applications.

Table 1. Classification of genes affecting traits in dairy cattle.

Gene Names	Main Function	References
FABP3, PPARG, AGPAT6, SCD1, DGAT1, SREBP1	Regulation of milk fat metabolism	[31–35]
CSN2, CSN3, LGB	Regulation of milk protein metabolism	[36,37]
IQGAP2	Regulation of lactose metabolism	[38]
FSHR, PRLR, BMP15, KISS1	Influence reproductive function	[28,39–41]
CD14, CXCR1, MBL1	Modulation of immune function	[42–44]
HSP70, NR3C1	Improvement in stress capacity	[45,46]
TERT	Regulation of rate of aging	[47]

3.1.1. *DGAT*

Triglycerides are neutral lipids and are the main energy storage lipids in mammals and are essential substances for increasing the amount of milk fat in dairy cows. *DGAT1* and *DGAT2* have long been found to be ubiquitously expressed in mammalian tissues, with the highest levels of expression in tissues with active triglyceride synthesis, such as adipose tissue, the small intestine, liver, and mammary gland. Both have been shown to catalyze the final step of triglyceride synthesis and are involved in the formation of triglycerides [48]. Although both *DGAT1* and *DGAT2* have the acyl coenzyme A–glycerol diester acyltransferase activity, they have no sequence homology, belonging to different gene families and exhibiting different biochemical, cellular, and physiological functions [49]; thus, we need to treat both of them differently when exploring their role in cattle breeding.

We focus on the *DGAT1* gene, located on bovine chromosome 14, which is a key gene affecting milk production traits in a wide range of cattle breeds [50]. It has been shown that the exonic region of the *DGAT1* gene causes different dairy breeds to exhibit different milk production traits [51]. For example, the g.8525C>T SNP genotype of the *DGAT1* gene showed a highly significant association with milk proteins in Bangladeshi river buffaloes [52], whereas the same SNP genotype (g.8525C>T) showed a significant correlation with milk fat in the Riverside buffalo population [53]; thus, its role in Holstein cows needs to be further explored. In addition, mutations of only *DGAT1* play a very important role in milk production in dairy cows. The non-synonymous mutation K232A in exon 8 of *DGAT1* leads to a change in the encoded amino acid from lysine to alanine, which alters the yield and composition of milk [54] and is also associated with a higher fat and protein content [55]. In addition, the K232A mutation has been reported to have a positive effect on the clotting ability of the milk of *Holstein cows* from some locations [56], which has led us to focus on the utilization of this mutation when breeding new Holstein cows. Moreover, differences in the genotype of *DGAT1* may also alter the lipid composition of the milk fat globule membrane, changing its PL/TAG ratio and, thus, affecting the amount of milk fat [57].

Focusing on the role of *DGAT1* in cow lactation, researchers cannot ignore this key gene, as it is considered an immune modulator [58]. Another powerful function of *DGAT1* is to protect cells from the pro-inflammatory effects and lipotoxicity of lipids while promoting lipid accumulation with immunoactivity regulation [48]. Therefore, *DGAT1* can effectively reduce the probability of mastitis in new *Holstein cows*, which is highly favorable for the breeding and promotion of new breeds.

3.1.2. *GHR*

Previous studies have shown that growth hormone (GH) plays a significant role in lactogenesis, and the growth hormone receptor (GHR), which can mediate its action, is considered a functional candidate gene for milk performance in dairy cows [59]. Through some of its functions, we know that *GHR* is abundantly expressed on the membrane and in the cytoplasm of bovine mammary epithelial cells (BMECs) [60]. The bovine *GHR* gene was also found to be localized on chromosome 20 and consists of 10 exons (E), of which E1 is very small and has a non-coding sequence [61,62].

Interestingly, similar to *DGAT*, genetic polymorphisms in *GHR* also influence lactation yield, milk fat, milk protein, and other relevant contents of cow's milk [63,64]. When a single-nucleotide substitution from T to A occurs in exon 8 of the *GHR* gene, it results in the substitution of a neutral phenylalanine (279F) at position 279 of the amino acid sequence with an uncharged but polar tyrosine (279Y), a change that is located in the transmembrane structural domain of the *GHR* gene and has an impact on the milk yield of some dairy cow populations [65]. Based on this variation in milk yield, it has been found that the *GHR F279Y* genotype has a significant effect on the milk yield, protein and fat contents, casein production and content, and lactose production and content of the *German Holstein cow* population [66], and this genotype also has a positive effect on the milk production traits of *Chinese Holstein cows* [67]. The *GHR* gene also regulates the uptake and utilization of nutrients by mammary cells, increasing the supply of raw materials for milk synthesis, thus increasing milk production. From these studies, it is evident that the *GHR F279Y* genotype plays a significant role in dairy cattle populations from diverse places; thus, it is necessary to pay attention to this gene when attempting to breed new types of dairy cattle.

In addition, *GHR* can activate various signaling pathways, including JAK2/STAT, MAPK, and PI3K [68]. By activating the downstream signaling pathways related to milk synthesis, it can more effectively promote the uptake of nutrients by mammary cells for milk synthesis, thus increasing milk production. For example, the JAK-STAT5 pathway is the main pathway by which growth affects the expression of genes encoding milk proteins (e.g., β-casein); thus, *GHR* not only acts alone but also undertakes important tasks in different pathways involved in milk production [69], making it a more versatile gene. These pathways promote bone growth, muscle formation, and overall body development in young cows, helping them to achieve a suitable adult body shape and laying the foundation for subsequent performance.

3.1.3. *PRL*

It is well known that the prolactin (*PRL*) gene is on chromosome 23 in cattle, spans 9.4 kb, and contains five exons and four introns, and it is not contained in any chain-group quantitative trait loci (QTLs).

PRL is not only a key regulator of mammalian reproductive processes but also is involved in breast proliferation and differentiation during pregnancy [70] and is a key gene in breast development [71]. Specifically, it is similar to the *GHR* gene, and the important pathway induced is the Jak-STAT signaling pathway [72]. *PRL* can also stimulate the STAT5 pathway to influence cell proliferation and differentiation [73], thus playing a role in the mammary gland development and milk production in animals. In the *PRL* gene studies of cattle conducted by other scholars, a variety of polymorphisms have been found in the bovine *PRL* gene [74]. To date, 45 SNPs located within exons have been recorded in the Ensembl Variation database. By exploring these SNPs, researchers have demonstrated that a common polymorphism within exon 4 of the bovine *PRL* gene (rs211032652 SNP, c.396G>A) can have a definite effect on the milk yield, milk protein, and milk fat of some dairy cattle populations [75].

Mastitis is a serious, unavoidable problem of dairy production, with implications for both public health and milk production [76]. The study of the effect of *PRL* on bovine mastitis resistance is a very interesting topic. For example, it has been found that *PRL* promotes the inflammatory response of bovine mammary epithelial cells through NF-kB activity [77,78]. However, it is crucial to assess whether polymorphisms in the *PRL* gene are related to the immune functions of dairy cows. Fortunately, some researchers have found a high correlation between genetic polymorphisms associated with prolactin and mastitis resistance and current dairy breeds' susceptibility to mastitis [79]. This bolsters the importance of the *PRL* gene in dairy cattle breeding.

3.2. *Outlook on Dairy Breeding*

For animals with long generation intervals, such as dairy cows, genome selection using gene chip typing can greatly shorten the generation interval and accelerate the process of genetic improvement. The numerous studies mentioned earlier have identified genetic markers associated with economically important traits such as milk production, reproduction, growth, and health [31,36,38], which can provide technical and theoretical support for breeding new lines of dairy cattle. For example, Chinese scholars have previously reported GWASs for milk production traits in *Chinese Holstein cows* and identified 105 significantly associated SNP loci using the Illumina Bovine SNP50 Microarray [80]. By using gene chips to mine and tag genetic information related to milk production traits, we can select dairy breeds with a higher milk yield and a better milk composition, which can directly improve the economic benefits of dairy farming. Moreover, association analyses have been carried out using multiple SNP microarrays with adult Holstein heifers to assess their conception and pregnancy rates and with young Holstein heifers to assess their conception rate, identifying a number of validated genes, including GN-RHR, and new regions in the cluster of pregnancy-associated glycoprotein genes [81], which provide a reference for unraveling the genetic variation and genomic regions of the fertility traits of dairy cows. Genetic markers related to reproduction can help us to select cows with higher fertility and shorter reproduction cycles and to improve the reproductive efficiency of new lines of cows. In addition, the genes affecting the body size and the limbs and hooves of current cow breeds (e.g., *ADIPOR2, INPP4A, DNMT3A*), genes causing mastitis in current cow breeds (e.g., *ABCC9, ACHE*), and genes causing ketosis in current cow breeds (e.g., *NRXN3, ACOXL, BCL2L11*) have been investigated to a certain extent [82–85], which facilitates the selection and editing of genes for future studies.

Scientists can use CRISPR/Cas9 to knock out the β-lactoglobulin gene (*BLG*) when breeding cows. β-lactoglobulin is an allergen found in milk, and knocking out *BLG* would make the milk free of this allergen. Editing *SCD1,* a key regulator of the fatty acid metabolic pathway, to produce low-fat milk or editing *ASMT* and *AANAT* to increase the amount of melatonin in milk are ideas to further improve the milk quality of new dairy cow breeds. For the breeding of new lines of dairy cattle in China, we can exploit the function of indigenous Sanhe cattle rumen microbial interactions (e.g., BOLA-DRA and *MUC1*) that significantly enhance the efficiency of roughage degradation and use the CRISPR-Cas12f system to achieve the simultaneous editing of multiple genes in order to breed feed-saving dairy cattle lines with a high efficiency of feed conversion. For the optimization of milk composition, we can first construct a model of the milk protein anabolic network and precisely regulate the κ-casein phosphorylation site (Ser149→Glu) using the base editing technology so as to reduce the particle sizes of casein micelles and significantly improve the processing characteristics of dairy products. In the field of disease resistance breeding, a real-time monitoring system for mastitis resistance based on SNP gene chips has been developed, which, in combination with genome in situ synthesis technology, integrates the lysostaphin gene into the mammary gland-specific expression locus in a targeted manner, reducing the incidence of clinical mastitis. The aforementioned methods can be used to establish programs for global dairy cattle breeding (Figure 3).

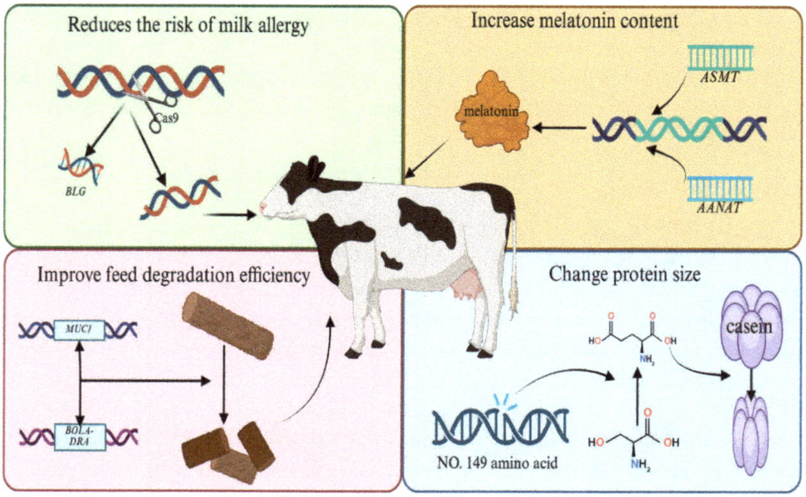

Figure 3. Schematic representation of breeding new lines of dairy cattle.

4. Breeding New Lines of Beef Cattle

4.1. Selection of Key Genes in Beef Cattle Breeding

Consumer perceptions of beef quality are shaped by both intrinsic attributes, such as nutritional value, and extrinsic attributes, including appearance and sensory characteristics. These factors play a pivotal role in influencing consumer preferences. Therefore, similar to dairy cattle breeding, it is essential to identify and understand the diverse range of genes that impact meat quality, tenderness, and other traits when developing new beef cattle lines (Table 2). Therefore, we summarize the candidate genes involved in the important traits of beef cattle breeding. This provides a basis for the development of new lines of beef cattle.

Table 2. Functions of genes affecting traits in beef cattle.

Gene Name	Trait	Reference
ADIPOQ	Marble pattern	[86]
PRKAG3	Muscle pH	[87]
	Cartilage length	[88]
LEP	Fatty acid composition	[89]
	Eye muscle area	[90]
	Muscle pH	[91]
	Marble pattern	[92]
CAST	Muscle iron content	[93]
	Muscle pH	[94]
	Juiciness	[95]
FASN	MUFA	[96]
	SFA	[97]
	C14:0	[98]
	C16:0	[99]
	C18:0	[100]
	Carcass weight	[101]
MC4R	Muscle pH	[87]
	Carcass weight	[102]
	Marble pattern	[103]

4.1.1. *MSTN*

The genetic variation in the *MSTN* gene (also known as the *GDF8* gene), located on bovine chromosome 2, has been found to be associated with an increase in the number of muscle fibers in cattle, leading to more muscular animals or double-muscled animals [104]. When animals exhibit dual-muscle genotypes, they have low water-holding capacity (WHC), fat coverage, and fat content in the LMA (muscle cross-sectional area). Furthermore, double-muscled animals have high carcass yield and muscle weight and area. Although the reduced fat content and low WHC negatively affect the organoleptic attributes of meat compared to other genotypes with a higher fat content, because of their higher CW and body size, higher efficiency, and lower feeding costs (due to a better feed conversion), these animals provide higher economic returns in intensive production systems [105]. Moreover, a lower IMF and the different proportions of FAs are considered positive for human health. Compared to other animals with normal genotypes, herds with the dual-muscle genotype have lower total FAs, and they have a higher proportion of polyunsaturated fats (PUFAs) in their bodies, resulting in a higher polyunsaturated/saturated fat ratio and a lower n-6/n-3 ratio [106]. This double-muscled trait has been associated with reduced fertility [107]; thus, it is important to focus on the muscle growth rate of cattle in tandem with their reduced fertility.

4.1.2. *CAPN* and *CAST*

CAST and *CAPN*, two of the most important genes for meat production and composition, are located in cattle on chromosomes 7 and 29, respectively, and they are associated with performances related to meat production traits. Meat tenderness is a prominent characteristic among the quality characteristics of meat that are most desired by consumers. Beef tenderness is associated with higher CSET concentrations. This is attributed to the inhibition of *CAPN* by *CAST* during post-mortem muscle protein hydrolysis [108], as the *CAST* gene leads to more calpain inhibitor proteins and, therefore, a higher rate of calpain inhibition. This suggests that *CAPN* is the main reason for the increase in meat tenderness during the post-slaughter process [109]. Another important reason for selecting *CAST* as a candidate gene is that *CAST* activity has a very high heritability and is closely related to meat shear, which means that meat tenderness can be improved in animals selected for *CAST*-based modifications; given the high heritability of *CAST*, a rapid genetic response can be achieved by selecting animals for CAST-based modifications [110].

4.1.3. *LEP*

The *LEP* gene is located on chromosome 4 and encodes the gene for the leptin hormone, which acts on receptors in the hypothalamus controlled by the leptin receptor (*LEPR*) gene to suppress appetite. As such, it affects the regulation of feed intake, so it is related to body composition, energy balance, and body weight [91]. Furthermore, the serum leptin concentration is positively correlated with the marbling score, SF, and USDA quality grade, but it is negatively correlated with LMA [111]. These findings imply that leptin can be used as an indicator for assessing carcass composition in live animals and is useful in breeding programs.

4.2. Beef Cattle Breeding Expectations

Genetic technology innovation is the core driving force to overcome the bottleneck of traditional breeding in the breeding of new lines of beef cattle. Firstly, whole-genome sequencing technology should be used to systematically analyze the genetic characteristics of local cattle breeds (e.g., *Qinchuan* and *Nanyang cattle*) to explore the functional loci related to meat quality (e.g., the *FABP4* gene), growth rate (e.g., the *MSTN* gene), and resistance, and to set up a gene database. The early and precise selection of bulls through genomic selection (GS) technology shortens the selection cycle from the traditional 5–6 generations to 2–3 generations, dramatically advancing genetic progress. Secondly, CRISPR/Cas9 gene editing can be used to target key traits, such as knocking out the myostatin gene to promote muscle growth or introducing natural resistance genes (e.g., the *BoLA-DRB3* variant) to increase resistance to Mycoplasma bovis pneumonia. At the same time, SNP molecular marker-assisted breeding chips have been developed to implement polygenic selection for complex traits such as intramuscular fat deposition (e.g., the *SCD* gene) and feed conversion rate (e.g., the *GH* gene). In addition, it is necessary to strengthen the research on epigenetic regulatory mechanisms and to enhance the efficacy of gene expression by means of methylation modifications and miRNA interventions. Genomic selection technology can be used to screen breeders with the high feed conversion rate and high-quality marbling deposition to establish an efficient breeding system. Moreover, we can analyze gene–nutrient interactions through metabolomics and transcriptomics, develop customized feeding programs, and use single-cell sequencing to reveal the synergistic regulatory network between rumen microbes and host genes so as to select ecologically adapted lines for extreme environments, such as plateaus and hot and humid environments. Moreover, we should simultaneously improve the ethical norms and biosafety assessment systems of gene editing and create a whole chain breeding mode of 'gene design–precision selection–

ecological adaptation–industry empowerment' based on industrial applications, eventually creating a modern beef cattle breeding system based on gene technology, germplasm innovation, precision selection, and resistance enhancement, thus promoting the leapfrog development of the beef cattle breeding industry (Figure 4).

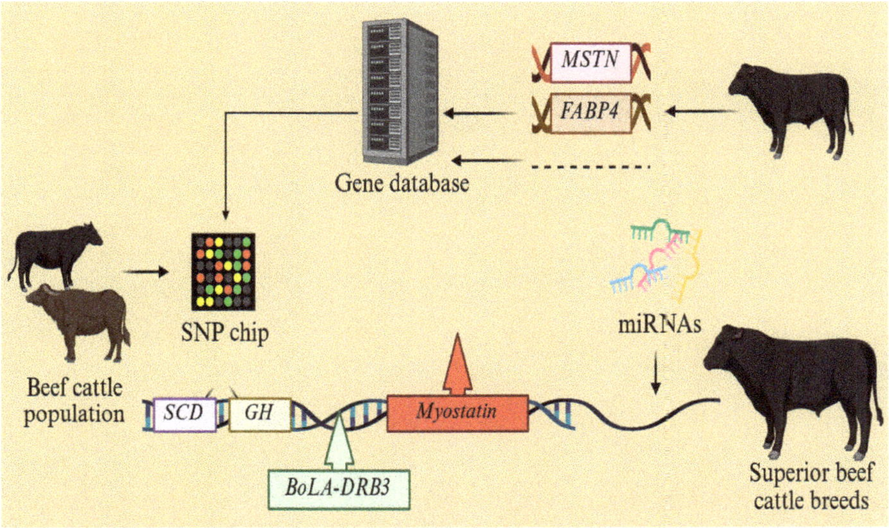

Figure 4. Schematic of breeding new lines of beef cattle.

5. Breeding New Lines of Dual-Purpose Cattle

All genes related to lactation and meat production mentioned earlier have similar applications in dual-purpose cattle. Therefore, we will next discuss feasible ideas for the future breeding of dual-purpose cattle. The breeding of new lines of dual-purpose cattle requires a combination of genomics and gene editing to ensure a balance of traits based on the existing breeds. Scientists should first screen cattle with a high milk protein content (e.g., a β-casein B allele phenotype) and high-quality intramuscular fat deposition (e.g., a high expression of the *FABP4* gene) as core parents. The rapid growth characteristics (*GH* gene variation) of Charolais cattle or the antioxidant components in milk fat (regulated by the *SCD* gene) of Guernsey cattle should be introduced through hybridization. At the same time, tick-resistant genes (such as the TICAM1 pathway) of African Sanga cattle or cold-resistant genes (such as the allelic variation in *UCP2* [112]) of Nordic cattle breeds should be integrated into the genes of dual-purpose cattle to enhance their environmental adaptability. Single-cell transcriptome technology should be utilized to decipher the key regulatory networks of mammary epithelial cells and skeletal muscle stem cells. The targeted editing of milk production-related genes (for example, knocking out the *LGB* gene to reduce the allergenicity of β-lactoglobulin and inserting the *CSN3* promoter to enhance the expression of κ-casein) and meat production-related genes (such as activating the imprinted regulatory region of *IGF2* to improve muscle differentiation efficiency and inhibiting the *ADRB3* gene to reduce excessive fat deposition) should be carried out. Simultaneously, disease resistance modules (such as using the *MBL2* gene to enhance innate immunity and the *SLC11A1* mutant to improve anti-tuberculosis ability) and stress response pathways (such as the heat-resistant allele of *HSF1* [113] and its optimization [114]) should be integrated. The off-target risk of CRISPR/Cas12a and base editing technologies should be reduced. Embryo chimerism technology should be used to rapidly obtain gene-chimeric populations. Nanopore real-time sequencing should be utilized to monitor epigenetic modifications (such as the influence of DNA methylation on milk production traits [115]) and to simulate extreme climates in a controlled-environment chamber to verify the adaptive phenotypes. Finally, gut microbiota transplantation should be combined with dynamic metabolomic analysis to screen new cattle lines with synergistically improved milk and meat performance, stress resistance, and genetic stability [116] (Figure 5).

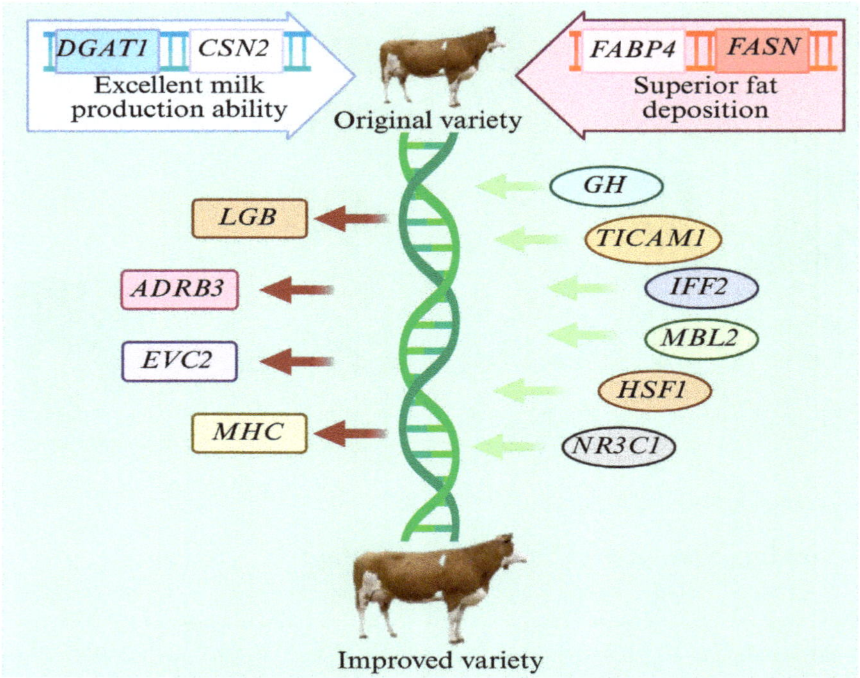

Figure 5. Schematic diagram of the breeding of a new dual-purpose cattle line (green arrows represent insertion, and red arrows represent deletion).

6. Discussion and Conclusions

With the continuous progress of science and technology and the increasing diversification of market demands, the breeding of new lines of cattle has shown a series of new development trends [117]. The in-depth integration of molecular and traditional breeding has become an important future direction for the livestock industry. Traditional breeding methods first involve the selection of appearance or quality characteristics and then the use of techniques such as crossbreeding improvement and variety selection to improve cattle breeds [118]. However, traditional breeding methods have limitations, such as being time-consuming and having poor predictability and low accuracy. With the rapid development of modern biotechnology and genomics, molecular breeding techniques have emerged. Molecular breeding enables the precise selection and utilization of superior genes by locating and determining genes for economically important traits in cattle. The use of genome-wide selection techniques can improve the efficiency and accuracy of gene selection by analyzing genome-wide genetic markers to accurately assess the genetic value of individuals during the early growth stages of cattle [10]. Combining molecular breeding techniques with traditional breeding methods can greatly accelerate breeding progress and improve breeding results.

Mining the good genes of local breeds is also an important trend in the breeding of new lines of cattle. Many countries and regions have rich resources of local cattle breeds, and these local breeds are undergoing long-term natural selection as well as artificial selection and are forming many excellent traits, such as high disease resistances, high roughage-degradation efficiencies, and excellent meat qualities. *China's Xiangxi yellow cattle* are adapted to the grazing environment of the grassy hills and slopes in the Western Hunan region of China and are characterized by their tolerance towards rough fodder and excellent meat quality [119]. Through molecular biology technology, promising genes of local breeds can be explored and studied, and these genes can be introduced into the breeding of new lines of cattle to enrich their genetic diversity and improve their adaptability and production performance; at the same time, this can also help to protect and utilize the resources of local cattle breeds.

As the public awareness of food safety and health continues to rise, there is an urgent need to develop cattle breeds with enhanced disease resistance and superior meat and milk quality. In terms of disease resistance, technologies such as gene editing and molecular marker-assisted selection enable the identification and breeding of cattle with genetic resistance to common diseases, including foot-and-mouth disease, bovine viral diarrhea, and brucellosis. This approach reduces the risk of disease outbreaks, minimizes drug use, and safeguards both herd health and food safety [120]. Regarding meat and milk quality, greater emphasis is being placed on breeding cattle with desirable traits such as vibrant meat color, rich marbling, balanced fat content, and high milk fat and protein contents to meet the consumer demand for premium products. Furthermore, intelligent breeding techniques are expected to play an increasingly vital role in the development of new cattle breeds. With the rapid advancement of the Internet of Things (IoT), big data, and artificial intelligence (AI), intelligent breeding technologies now enable the real-time monitoring and precise regulation of cattle growth environments, feeding management, and health status, offering significant support for breeding efforts [121]. For instance, sensors can be used to monitor barn conditions such as temperature, humidity, and air quality in real time, while automated systems can adjust ventilation, heating, and cooling equipment to create an optimal environment for cattle growth. Additionally, big data analytics can process growth, reproduction, and health data to provide a scientific basis for farming decisions, enhancing both operational efficiency and management standards [122]. Furthermore, AI technologies, including image and voice recognition, can facilitate disease diagnosis and early warning systems by promptly detecting abnormalities in the herd, enabling proactive preventive and curative measures [123].

Despite the numerous issues associated with gene editing, such as the challenges faced by animals in their survival after mutagenesis, the insufficient accuracy of gene editing, consumers' resistance towards animal products resulting from gene editing, and the complex ethical and regulatory environment, we still firmly believe that gene editing will offer substantial assistance in the creation of new cattle breeds [124,125]. The development of new cattle breeds is not only a significant leap in livestock farming technology but also an innovative response to global challenges. By integrating gene editing, whole-genome selection, and intelligent biological breeding technologies, not only can the efficiency and quality of meat and dairy production be significantly improved, but global carbon emission reduction targets can also be achieved by optimization of the feed conversion rate and reductions in methane emissions. Under the multiple pressures of population growth, climate change, and resource constraints, if the required cattle breeds can be efficiently created through gene editing, global food security will be enhanced, and the development of sustainable agriculture will be promoted. Furthermore, this will create a new avenue for the livestock industry to contribute to climate governance. In addition, the disease resistance and environmental adaptability of these new cattle breeds will also provide solutions for livestock production in ecologically vulnerable areas such as tropical and arid regions, helping to overcome resource constraints. Ultimately, this 'genetic revolution' in the field of cattle breeding represents a major effort made by humanity to ensure a sustainable future.

Author Contributions: Writing—original draft preparation, Y.W. and X.C.; writing—review and editing, Y.W. and Z.C. All authors have read and agreed to the published version of the manuscript.

Funding: This review was supported by the Taizhou Science and Technology Support Plan (Agriculture) Project (TN202313), the National Natural Science Foundation of China (Grant Nos. 32472851 and 32272825), the Independent Innovation in Jiangsu Province of China (Grant No. CX(24)3080) and Innovation and Entrepreneurship Training Program for College Students (202411117095Z).

Institutional Review Board Statement: The ethical review was not required for this study, as only data were analyzed and no research was performed on animals.

Informed Consent Statement: Not applicable.

Data Availability Statement: No new data were created or analyzed in this study.

Acknowledgments: The authors extend their sincere gratitude to the following individuals for their invaluable contributions to the refinement of this review. We are deeply indebted to Xiaofen Li of the School of Animal Science and Technology, Jiangsu Agri-animal Husbandry Vocational College, Taizhou 225300, China (lxf@jsahvc.edu.cn); Yu Hou from the Institute of Nutrition, Zhejiang NHU Co., Ltd., Xinchang 312500, China (houyu23@outlook.com); and Ying Jiang of the College of Animal Science and Technology, Yangzhou University, Yangzhou 225009, China (MZ220230522@stu.yzu.edu.cn), for their rigorous revisions and critical feedback during the manuscript preparation. Their expertise in content refinement, attention to detail, and constructive suggestions significantly enhanced the scientific accuracy, clarity, and organization of this work.

Conflicts of Interest: The authors declare no conflicts of interest.

References

1. Thornton, P.K. Livestock production: Recent trends, future prospects. *Philos. Trans. R. Soc. London. Ser. B Biol. Sci.* **2010**, *365*, 2853–2867. [CrossRef] [PubMed]
2. Oleinik, S.A.; Maltsev, A.E.; Filatov, D.A. Comparative assessment of the productive qualities of Holstein and Jersey dairy cattle. *BIO Web. Conf.* **2024**, *139*, 11010. [CrossRef]
3. Pascottini, O.B.; Crowe, A.D.; Ramil, U.Y.; Hostens, M.; Opsomer, G.; Crowe, M.A. Perspectives in cattle reproduction for the next 20 years—A European context. *Theriogenology* **2025**, *233*, 8–23. [CrossRef] [PubMed]
4. Scott, B.A.; Mariam, M.H.; Tiezzi, F.; Berg, I.v.d.; Maltecca, C.; Pryce, J.E. Optimizing genetic diversity in Australian Holsteins and Jerseys: A comparative analysis of whole-genome and regional inbreeding depression effects. *J. Dairy Sci.* **2024**, *108*, 2658–2668. [CrossRef] [PubMed]
5. Berry, D.P.; Wall, E.; Pryce, J.E. Genetics and genomics of reproductive performance in dairy and beef cattle. *Animal* **2014**, *8*, 105–121. [CrossRef]
6. Ahmad, S.; Agung, I.; Hendra, H.; Jasmadi; Ainsyar, H.M.; Angger, S.A.; Erma, S.A.; Hilda, N.; Taufik, K.; Guna, D.I.N.; et al. Effects of various macroalgae species on methane production, rumen fermentation, and ruminant production: A meta-analysis from in vitro and in vivo experiments. *Anim. Feed Sci. Technol.* **2022**, *294*, 115503.
7. Mikkola, M.; Desmet, K.L.J.; Kommisrud, E.; Riegler, M.A. Recent advancements to increase success in assisted reproductive technologies in cattle. *Anim. Reprod.* **2024**, *21*, e20240031. [CrossRef] [PubMed]
8. Mueller, M.L.; Van Eenennaam, A.L. Synergistic power of genomic selection, assisted reproductive technologies, and gene editing to drive genetic improvement of cattle. *CABI Agric. Biosci.* **2022**, *3*, 13. [CrossRef]
9. Xu, L.; Bickhart, D.M.; Cole, J.B.; Schroeder, S.G.; Song, J.; Tassell, C.P.V.; Sonstegard, T.S.; Liu, G.E. Genomic signatures reveal new evidences for selection of important traits in domestic cattle. *Mol. Biol. Evol.* **2015**, *32*, 711–725. [CrossRef]
10. Alves, F.J.G.; Alves, S.D.; Macedo, M.L.F.; Pinto, d.M.T.; Simielli, F.L.F.; Santos, S.D.B.d.; Roberto, C.; Galvão, A.L. Sustainable Intensification of Beef Production in the Tropics: The Role of Genetically Improving Sexual Precocity of Heifers. *Animals* **2022**, *12*, 174. [CrossRef]

1. Alves, F.J.G.; Elisa, P.; Iana, S.P.; Soares, C.G.; Macedo, M.L.F.; Zerlotti, M.M.E.; Fernando, B.; Roberto, C.; Galvão, d.A.L. Current applications and perspectives of genomic selection in Bos indicus (Nellore) cattle. *Livest. Sci.* **2022**, *263*, 105001.
2. Zhang, Z.; Erbe, M.; He, J.; Ober, U.; Gao, N.; Zhang, H.; Simianer, H.; Li, J. Accuracy of whole-genome prediction using a genetic architecture-enhanced variance-covariance matrix. *G3* **2015**, *5*, 615–627. [CrossRef] [PubMed]
3. Brøndum, R.F.; Su, G.; Janss, L.; Sahana, G.; Guldbrandtsen, B.; Boichard, D.; Lund, M.S. Quantitative trait loci markers derived from whole genome sequence data increases the reliability of genomic prediction. *J. Dairy Sci.* **2015**, *98*, 4107–4116. [CrossRef] [PubMed]
4. John, v.d.O.; Constantinos, P. The genome editing revolution. *Trends Biotechnol.* **2023**, *41*, 396–409.
5. Christian, M.; Cermak, T.; Doyle, E.L.; Schmidt, C.; Zhang, F.; Hummel, A.; Bogdanove, A.J.; Voytas, D.F. Targeting DNA double-strand breaks with TAL effector nucleases. *Genetics* **2010**, *186*, 757–761. [CrossRef]
6. Cong, L.; Ran, F.A.; Cox, D.; Lin, S.; Barretto, R.; Habib, N.; Hsu, P.D.; Wu, X.; Jiang, W.; Marraffini, L.A.; et al. Multiplex Genome Engineering Using CRISPR/Cas Systems. *Science* **2013**, *339*, 819–823. [CrossRef]
7. Magdalena, H.; Daniel, L.; Joanna, Z.; Ryszard, S. CRISPR/Cas9 Immune System as a Tool for Genome Engineering. *Arch. Immunol. Ther. Exp.* **2017**, *65*, 233–240.
8. Kuznetsov, V.; Revina, G.; Astashenkova, L. Additive-Polygenic Inheritance of Reproductive System Diseases in Holstein Cows in Subpopulations. In *International Scientific Conference the Fifth Technological Order: Prospects for the Development and Modernization of the Russian Agro-Industrial Sector (TFTS 2019)*; Atlantis Press: Dordrecht, The Netherlands, 2020.
9. Dong, F.; Xie, K.; Chen, Y.; Yang, Y.; Mao, Y. Polycistronic tRNA and CRISPR guide-RNA enables highly efficient multiplexed genome engineering in human cells. *Biochem. Biophys. Res. Commun.* **2017**, *482*, 889–895. [CrossRef] [PubMed]
10. Feng, X.; Zhao, D.; Zhang, X.; Ding, X.; Bi, C. CRISPR/Cas9 assisted Multiplex Genome Editing Technique in *Escherichia coli*. *Biotechnol. J.* **2018**, *13*, e1700604. [CrossRef]
11. Tetsushi, S.; Ayami, N.; Satoshi, K.; Kazuaki, C.; Takashi, Y. Multiplex genome engineering in human cells using all-in-one CRISPR/Cas9 vector system. *Sci. Rep.* **2014**, *4*, 5400.
12. Kor, S.D.; Chowdhury, N.; Keot, A.K.; Yogendra, K.; Chikkaputtaiah, C.; Sudhakar Reddy, P. RNA Pol III promoters-key players in precisely targeted plant genome editing. *Front. Genet.* **2022**, *13*, 989199. [CrossRef] [PubMed]
13. Nie, L.; Thakur, M.D.; Wang, Y.; Su, Q.; Zhao, Y.; Feng, Y. Regulation of U6 Promoter Activity by Transcriptional Interference in Viral Vector-Based RNAi. *Genom. Proteom. Bioinform.* **2010**, *8*, 170–179. [CrossRef] [PubMed]
14. Muntazir, M.; Aejaz, A.D.; Milan, S.; Anshika, T.; Nancy, B.; Umer, B.; Ahmad, B.B.; Abbu, Z.; Sajad, A.; TanvirUlHassan, D.; et al. CRISPR-Based Genome Editing Tools: Insights into Technological Breakthroughs and Future Challenges. *Genes* **2021**, *12*, 797. [CrossRef]
15. Kishimoto, T.; Nishimura, K.; Morishita, K.; Fukuda, A.; Miyamae, Y.; Kumagai, Y.; Sumaru, K.; Nakanishi, M.; Hisatake, K.; Sano, M. An engineered ligand-responsive Csy4 endoribonuclease controls transgene expression from Sendai virus vectors. *J. Biol. Eng.* **2024**, *18*, 9. [CrossRef]
16. Haurwitz, R.E.; Sternberg, S.H.; Doudna, J.A. Csy4 relies on an unusual catalytic dyad to position and cleave CRISPR RNA. *EMBO J.* **2012**, *31*, 2824–2832. [CrossRef] [PubMed]
17. Zalatan, J.G.; Lee, M.E.; Almeida, R.; Gilbert, L.A.; Whitehead, E.H.; Russa, M.L.; Tsai, J.C.; Weissman, J.S.; Dueber, J.E.; Qi, L.S.; et al. Engineering Complex Synthetic Transcriptional Programs with CRISPR RNA Scaffolds. *Cell* **2015**, *160*, 339–350. [CrossRef]
18. Cuellar, C.J.; Amaral, T.F.; Rodriguez-Villamil, P.; Ongaratto, F.; Martinez, D.O.; Labrecque, R.; Losano, J.D.D.A.; Estrada-Cortés, E.; Bostrom, J.R.; Martins, K.; et al. Consequences of gene editing of PRLR on thermotolerance, growth, and male reproduction in cattle. *FASEB BioAdvances* **2024**, *6*, 223–234. [CrossRef] [PubMed]
19. Zhao, Y.; Yang, L.; Su, G.; Wei, Z.; Liu, X.; Song, L.; Hai, C.; Wu, D.; Hao, Z.; Wu, Y.; et al. Growth Traits and Sperm Proteomics Analyses of Myostatin Gene-Edited Chinese Yellow Cattle. *Life* **2022**, *12*, 627. [CrossRef]
20. Wang, Y.; Bi, D.; Qin, G.; Song, R.; Yao, J.; Cao, C.; Zheng, Q.; Hou, N.; Wang, Y.; Zhao, J. Cytosine Base Editor (hA3A-BE3-NG)-Mediated Multiple Gene Editing for Pyramid Breeding in Pigs. *Front. Genet.* **2020**, *11*, 592623. [CrossRef]
21. Kumar, D.P.; Shubham, G.; Kumar, M.S.; Reena, A.; Manishi, M.; Kumar, N.S.; Periasamy, K.; Singh, K.R. Identification of polymorphism in fatty acid binding protein 3 (FABP3) gene and its association with milk fat traits in riverine buffalo (*Bubalus bubalis*). *Trop. Anim. Health Prod.* **2016**, *48*, 849–853.
22. Wang, C.; Zhao, J.; Feng, X.; Zhao, W.; Ma, R.; Yu, B.; Xue, L.; Wang, H.; Chen, Y.; Zhang, J.; et al. bta-miR-224 regulates milk fat metabolism by targeting FABP4 in bovine mammary epithelial cells. *Genomics* **2024**, *116*, 110955. [CrossRef] [PubMed]
23. Zhou, F.; Xue, J.; Shan, X.; Qiu, L.; Miao, Y. Functional roles for AGPAT6 in milk fat synthesis of buffalo mammary epithelial cells. *Anim. Biotechnol.* **2022**, *34*, 11–12. [CrossRef]
24. Ramos, M.C.G.; Souza, F.P.A.d.; Diniz, P.M.G.C.; Cruz, R.I.; Ferraz, L.F.C.; Thalia, Z.; Vercesi, F.A.E.; Tomita, B.F.Â.; Santos, C.M.R.; Sundfeld, G.M.A. Phenotypic variation in milk fatty acid composition and its association with stearoyl-CoA desaturase 1 (SCD1) gene polymorphisms in Gir cows. *J. Anim. Breed. Genet. = Z. Fur Tierz. Und Zucht.* **2023**, *140*, 532–548.

35. Li, N.; Zhao, F.; Wei, C.; Liang, M.; Zhang, N.; Wang, C.; Li, Q.-Z.; Gao, X.-J. Function of SREBP1 in the Milk Fat Synthesis of Dairy Cow Mammary Epithelial Cells. *Int. J. Mol. Sci.* **2014**, *15*, 16998–17013. [CrossRef]
36. Amalfitano, N.; Mota, L.F.M.; Rosa, G.M.; Cecchinato, A.; Bittante, G. Role of CSN2, CSN3, and BLG genes and the polygenic background in the cattle milk protein profile. *J. Dairy Sci.* **2022**, *105*, 6001–6020. [CrossRef]
37. Molee, A.; Poompramun, C.; Mernkrathoke, P. Effect of casein genes—Beta-LGB, DGAT1, GH, and LHR—On milk production and milk composition traits in crossbred Holsteins. *Genet. Mol. Res. GMR* **2015**, *14*, 2561–2571. [CrossRef]
38. Zhang, J.; Yang, G.; Zha, X.; Ma, X.; La, Y.; Wu, X.; Guo, X.; Chu, M.; Bao, P.; Yan, P.; et al. Polymorphisms Within the IQGAP2 and CRTAC1 Genes of Gannan Yaks and Their Association with Milk Quality Characteristics. *Foods* **2024**, *13*, 3720. [CrossRef]
39. Pambuko, G.; Vanessa, R.; Widyastuti, R.; Prastowo, S. FSHR gene polymorphism and its association to reproductive traits in Friesian Holstein cattle. *IOP Conf. Ser. Earth Environ. Sci.* **2024**, *1341*, 012023. [CrossRef]
40. Yang, X.; Wang, Z.; Chen, Y.; Ding, H.; Fang, Y.; Ma, X.; Liu, H.; Guo, J.; Zhao, J.; Wang, J.; et al. ALKBH5 Reduces BMP15 mRNA Stability and Regulates Bovine Puberty Initiation Through an m6A-Dependent Pathway. *Int. J. Mol. Sci.* **2024**, *25*, 11605. [CrossRef]
41. Umesh, S.; Rani, A.; Sushil, K.; Rajib, D.; Thiruvothur, V.R.; Shaily, S.; Singh, S.G.; Bharat, S.R.; Kotikalapudi, S.M.; Nitin, V.R.P. Association of bovine KISS1 single nucleotide polymorphisms with reproductive traits in Indian Cattle. *Reprod. Domest. Anim. = Zuchthyg.* **2020**, *55*, 922–930.
42. Gupta, J.P.; Bhushan, B.; Asaf, V.N.M.; Kumar, A.; Ranjan, S.; Panigrahi, M.; Kumar, A.; Kumar, P. Association and expression analysis of single nucleotide polymorphisms of CD14 gene with somatic cell score in crossbred cattle. *Gene Rep.* **2018**, *12*, 255–260. [CrossRef]
43. Muslimova, Z.; Abdualiyeva, A.; Shaugimbayeva, N.; Orynkhanov, K.; Ussenbekov, Y. Genotyping of Holstein Cows by SELL, MX1 and CXCR1 Gene Loci Associated With Mastitis Resistance. *Reprod. Domest. Anim. = Zuchthyg.* **2024**, *59*, e14713. [CrossRef] [PubMed]
44. Kamaldeep; Magotra, A.; Pander, B.L.; Dalal, D.S.; Malik, B.S.; Garg, A.R.; Malik, A. Evaluation of candidate genotype of immune gene MBL1 associated with udder health and performance traits in dairy cattle and buffalo of India. *Trop. Anim. Health Prod.* **2021**, *53*, 429. [CrossRef]
45. Lim, J.W.; Lee, J.H.; Nejad, J.G.; Lee, H.G. Effects of L-leucine and sodium acetate on milk protein synthesis under heat stress conditions in bovine mammary epithelial cells in vitro. *J. Therm. Biol.* **2024**, *126*, 103975.
46. Vilela, P.B.; Bonvino, S.N.; Afonso, d.F.L.; Zerlotti, M.M.E.; Silveira, R.E.; Paro, P.C.C. Expression of candidate genes for residual feed intake in tropically adapted Bos taurus and Bos indicus bulls under thermoneutral and heat stress environmental conditions. *J. Therm. Biol.* **2021**, *99*, 102998.
47. Xu, L.; Idrees, M.; Joo, M.D.; Sidrat, T.; Wei, Y.; Song, S.H.; Li, K.L.; Kong, I.K. Constitutive Expression of TERT Enhances β-Klotho Expression and Improves Age-Related Deterioration in Early Bovine Embryos. *Int. J. Mol. Sci.* **2021**, *22*, 5327. [CrossRef]
48. Oleszycka, E.; Kwiecień, K.; Grygier, B.; Cichy, J.; Kwiecińska, P. The many faces of DGAT1. *Life Sci.* **2024**, *362*, 123322. [CrossRef]
49. Anonymous. Thematic Review Series: Glycerolipids. DGAT enzymes and triacylglycerol biosynthesis. *Int. News Fats Oils Relat. Mater. Inform* **2009**, *20*, 175.
50. Elzaki, S.; Korkuć, P.; Arends, D.; Reissmann, M.; Brockmann, G.A. Effects of DGAT1 on milk performance in Sudanese Butana × Holstein crossbred cattle. *Trop. Anim. Health Prod.* **2022**, *54*, 142. [CrossRef]
51. Zahoor, K.M.; Yulin, M.; Jiaying, M.; Jianxin, X.; Yue, L.; Shuai, L.; Adnan, K.; Muhammad, K.I.; Zhijun, C. Association of DGAT1 With Cattle, Buffalo, Goat, and Sheep Milk and Meat Production Traits. *Front. Vet. Sci.* **2021**, *8*, 712470.
52. Mou, M.A.; Deb, G.K.; Hridoy, M.F.A.; Alam, M.A.; Barai, H.R.; Haque, M.A.; Bhuiyan, M.S.A. Detection of Polymorphisms in FASN, DGAT1, and PPARGC1A Genes and Their Association with Milk Yield and Composition Traits in River Buffalo of Bangladesh. *Animals* **2024**, *14*, 1945. [CrossRef] [PubMed]
53. Li, J.; Liu, S.; Li, Z.; Zhang, S.; Hua, G.; Salzano, A.; Campanile, G.; Gasparrini, B.; Liang, A.; Yang, L. DGAT1 polymorphism in Riverine buffalo, Swamp buffalo and crossbred buffalo. *J. Dairy Res.* **2018**, *85*, 412–415. [CrossRef]
54. Spelman, R.J.; Ford, C.A.; McElhinney, P.; Gregory, G.C.; Snell, R.G. Characterization of the DGAT1 Gene in the New Zealand Dairy Population. *J. Dairy Sci.* **2002**, *85*, 3514–3517. [CrossRef]
55. Bovenhuis, H.; Visker, M.H.P.W.; Poulsen, N.A.; Sehested, J.; Valenberg, H.J.F.v.; Arendonk, J.A.M.v.; Larsen, L.B.; Buitenhuis, A.J. Effects of the diacylglycerol o-acyltransferase 1 (DGAT1) K232A polymorphism on fatty acid, protein, and mineral composition of dairy cattle milk. *J. Dairy Sci.* **2016**, *99*, 3113–3123. [CrossRef]
56. Bobbo, T.; Tiezzi, F.; Penasa, M.; Marchi, M.D.; Cassandro, M. Short communication: Association analysis of diacylglycerol acyltransferase (DGAT1) mutation on chromosome 14 for milk yield and composition traits, somatic cell score, and coagulation properties in Holstein bulls. *J. Dairy Sci.* **2018**, *101*, 8087–8091. [CrossRef] [PubMed]
57. Argov-Argaman, N.; Mida, K.; Cohen, B.-C.; Visker, M.; Hettinga, K. Milk fat content and DGAT1 genotype determine lipid composition of the milk fat globule membrane. *PLoS ONE* **2017**, *8*, e68707. [CrossRef] [PubMed]

58. DGAT1-dependent triacylglycerol storage by macrophages protects mice from diet-induced insulin resistance and inflammation. *J. Clin. Investig.* **2011**, *121*, 1667. [CrossRef]
59. El-Komy, S.M.; Saleh, A.A.; Abdel-Hamid, T.M.; El-Magd, M.A. Association of GHR Polymorphisms with Milk Production in Buffaloes. *Animals* **2020**, *10*, 1203. [CrossRef]
60. Sakamoto, K.; Komatsu, T.; Kobayashi, T.; Rose, M.T.; Aso, H.; Hagino, A.; Obara, Y. Growth hormone acts on the synthesis and secretion of α-casein in bovine mammary epithelial cells. *J. Dairy Res.* **2005**, *72*, 264–270. [CrossRef]
61. Maj, A.; Oprzadek, J.; Oprzadek, A.; Dymnicki, E.; Zwierzchowski, L. Polymorphism in the 5′-noncoding region of the bovine growth hormone receptor gene and its association with meat production traits in cattle *Anim. Res.* **2004**, *53*, 503–514. [CrossRef]
62. Andrzej, M.; Lech, Z. Molecular evolution of coding and non-coding sequences of the growth hormone receptor (GHR) gene in the family Bovidae. *Folia Biol.* **2006**, *54*, 31–36.
63. Waters, S.M.; McCabe, M.S.; Howard, D.J.; Giblin, L.; Magee, D.A.; MacHugh, D.E.; Berry, D.P. Associations between newly discovered polymorphisms in the Bos taurus growth hormone receptor gene and performance traits in Holstein-Friesian dairy cattle. *Anim. Genet.* **2011**, *42*, 39–49. [CrossRef] [PubMed]
64. Sanchez, M.P.; Govignon-Gion, A.; Ferrand, M.; Gelé, M.; Pourchet, D.; Amigues, Y.; Fritz, S.; Boussaha, M.; Capitan, A.; Rocha, D.; et al. Whole-genome scan to detect quantitative trait loci associated with milk protein composition in 3 French dairy cattle breeds. *J. Dairy Sci.* **2016**, *99*, 8203–8215. [CrossRef]
65. Sarah, B.; Jong-Joo, K.; Sirja, M.; Anne, S.-K.; Anne, C.; Paulette, B.; Nadine, C.; Christine, F.; Bernard, G.; Dave, J.; et al. Molecular dissection of a quantitative trait locus: A phenylalanine-to-tyrosine substitution in the transmembrane domain of the bovine growth hormone receptor is associated with a major effect on milk yield and composition. *Genetics* **2003**, *163*, 253–266.
66. Rahmatalla, S.A.; Müller, U.; Strucken, E.M.; Reissmann, M.; Brockmann, G.A. The F279Y polymorphism of the GHR gene and its relation to milk production and somatic cell score in German Holstein dairy cattle. *J. Appl. Genet.* **2011**, *52*, 459–465. [CrossRef] [PubMed]
67. Ma, Y.N.; He, P.J.; Zhu, J.; Lei, Z.M.; Liu, Z.; Wu, J.P. The effect of polymorphism F279Y of GHR gene on milk production trait in Chinese Holstein cattle. *Zhongguo Ying Yong Sheng Li Xue Za Zhi = Zhongguo Yingyong Shenglixue Zazhi = Chin. J. Appl. Physiol.* **2013**, *29*, 400–404.
68. Perry, J.K.; Mohankumar, K.M.; Emerald, B.S.; Mertani, H.C.; Lobie, P.E. The contribution of growth hormone to mammary neoplasia. *J. Mammary Gland Biol. Neoplasia* **2008**, *13*, 131–145. [CrossRef]
69. Cui, Y.; Sun, X.; Jin, L.; Yu, G.; Li, Q.; Gao, X.; Ao, J.; Wang, C. MiR-139 suppresses β-casein synthesis and proliferation in bovine mammary epithelial cells by targeting the GHR and IGF1R signaling pathways. *BMC Vet. Res.* **2017**, *13*, 350. [CrossRef]
70. Cordero, A.; Pellegrini, P.; Sanz-Moreno, A.; Trinidad, E.M.; Serra-Musach, J.; Deshpande, C.; Dougall, W.C.; Pujana, M.A.; González-Suárez, E. Rankl Impairs Lactogenic Differentiation Through Inhibition of the Prolactin/Stat5 Pathway at Midgestation. *Stem Cells* **2016**, *34*, 1027–1039. [CrossRef]
71. Oakes, S.R.; Rogers, R.L.; Naylor, M.J.; Ormandy, C.J. Prolactin regulation of mammary gland development. *J. Mammary Gland Biol. Neoplasia* **2008**, *13*, 13–28. [CrossRef]
72. Hennighausen, L.; Robinson, G.W.; Wagner, K.-U.; Liu, X. Prolactin Signaling in Mammary Gland Development. *J. Biol. Chem.* **1997**, *272*, 7567–7569. [CrossRef] [PubMed]
73. Mapes, J.; Li, Q.; Kannan, A.; Anandan, L.; Laws, M.; Lydon, J.P.; Bagchi, I.C.; Bagchi, M.K. CUZD1 is a critical mediator of the JAK/STAT5 signaling pathway that controls mammary gland development during pregnancy. *PLoS Genet.* **2017**, *13*, e1006654. [CrossRef]
74. Mehmannavaz, Y.; Amirinia, C.; Bonyadi, M.; Torshizi, R. Effects of bovine prolactin gene polymorphism within exon 4 on milk related traits and genetic trends in Iranian Holstein bulls. *Afr. J. Biotechnol.* **2009**, *8*, 19.
75. Elena, I.D.; Eugeniu, M.A.; Valentin, M.C.; Ionel, N.R.; Toma, C.L.; Mihai, C.; Cristian, G.A. Polymorphism of the Prolactin (PRL) Gene and Its Effect on Milk Production Traits in Romanian Cattle Breeds. *Vet. Sci.* **2023**, *10*, 275. [CrossRef]
76. Corina, P.; Viorel, H.; Ionica, I.; Luminita, C. Etiology of Mastitis and Antimicrobial Resistance in Dairy Cattle Farms in the Western Part of Romania. *Antibiotics* **2022**, *11*, 5. [CrossRef]
77. Salgado-Lora, M.G.; Medina-Estrada, I.; López-Meza, J.E.; Ochoa-Zarzosa, A. Prolactin and Estradiol are Epigenetic Modulators in Bovine Mammary Epithelial Cells during Staphylococcus aureus Infection. *Pathogens* **2020**, *9*, 520. [CrossRef]
78. Antonio, B.M.M.; Guadalupe, S.L.M.; Edmundo, L.M.J.; Alejandra, O.Z. Prolactin regulates H3K9ac and H3K9me2 epigenetic marks and miRNAs expression in bovine mammary epithelial cells challenged with *Staphylococcus aureus*. *Front. Microbiol.* **2022**, *13*, 990478.
79. Beishova, I.; Belaya, A.; Kuzhebayeva, U.; Ulyanova, T.; Ulyanov, V.; Beishov, R.; Ginayatov, N.; Kovalchuk, A.; Kharzhau, A.; Sidarova, A. Association of polymorphic variants of prolactin (PRL) and beta-lactoglobulin (BLG) genes with resistance/susceptibility to mastitis in holstein cows. *Braz. J. Biol. = Rev. Brasleira De Biol.* **2024**, *84*, e284961. [CrossRef] [PubMed]
80. Jiang, L.; Liu, J.; Sun, D.; Ma, P.; Ding, X.; Yu, Y.; Zhang, Q. Genome wide association studies for milk production traits in Chinese Holstein population. *PLoS ONE* **2010**, *5*, e13661. [CrossRef]

81. Liang, Z.; Prakapenka, D.; VanRaden, P.M.; Jiang, J.; Ma, L.; Da, Y. A Million-Cow Genome-Wide Association Study of Three Fertility Traits in U.S. Holstein Cows. *Int. J. Mol. Sci.* **2023**, *24*, 10496. [CrossRef]
82. Mohamed, A.I.; Xubin, L.; Mudasir, N.; Idriss, A.A.A.; Tianle, X.; Husien, Y.M.; Yongjiang, M.; Zhangping, Y. Genome-Wide Association Study Identifies Candidate Genes Associated with Feet and Leg Conformation Traits in Chinese Holstein Cattle. *Animals* **2021**, *11*, 2259. [CrossRef]
83. Klein, S.L.; Scheper, C.; May, K.; König, S. Genetic and nongenetic profiling of milk β-hydroxybutyrate and acetone and their associations with ketosis in Holstein cows. *J. Dairy Sci.* **2020**, *103*, 10332–10346. [CrossRef]
84. Freebern, E.; Santos, D.J.; Fang, L.; Jiang, J.; Parker Gaddis, K.L.; Liu, G.E.; VanRaden, P.M.; Maltecca, C.; Cole, J.B.; Ma, L. GWAS and fine-mapping of livability and six disease traits in Holstein cattle. *BMC Genom.* **2020**, *21*, 41. [CrossRef] [PubMed]
85. Nayeri, S.; Schenkel, F.; Fleming, A.; Kroezen, V.; Sargolzaei, M.; Baes, C.; Cánovas, A.; Squires, J.; Miglior, F. Genome-wide association analysis for β-hydroxybutyrate concentration in Milk in Holstein dairy cattle. *BMC Genet.* **2019**, *20*, 58. [CrossRef]
86. Choi, Y.; Davis, M.E.; Chung, H. Effects of genetic variants in the promoter region of the bovine adiponectin (ADIPOQ) gene on marbling of Hanwoo beef cattle. *Meat Sci.* **2015**, *105*, 57–62. [CrossRef]
87. Ribeca, C.; Bonfatti, V.; Cecchinato, A.; Albera, A.; Gallo, L.; Carnier, P. Effect of polymorphisms in candidate genes on carcass and meat quality traits in double muscled Piemontese cattle. *Meat Sci.* **2014**, *96*, 1376–1383. [CrossRef] [PubMed]
88. Gill, J.L.; Bishop, S.C.; McCorquodale, C.; Williams, J.L.; Wiener, P. Associations between single nucleotide polymorphisms in multiple candidate genes and carcass and meat quality traits in a commercial Angus-cross population. *Meat Sci.* **2010**, *86*, 985–993. [CrossRef] [PubMed]
89. Orrù, L.; Cifuni, G.F.; Piasentier, E.; Corazzin, M.; Bovolenta, S.; Moioli, B. Association analyses of single nucleotide polymorphisms in the LEP and SCD1 genes on the fatty acid profile of muscle fat in Simmental bulls. *Meat Sci.* **2010**, *87*, 344–348. [CrossRef]
90. Da Silva, R.C.G.; Ferraz, J.B.S.; Meirelles, F.V.; Eler, J.P.; Balieiro, J.C.D.C.; Cucco, D.D.C.; Mattos, E.C.; Rezende, F.M.D.; Silva, S.D.L. Association of single nucleotide polymorphisms in the bovine leptin and leptin receptor genes with growth and ultrasound carcass traits in Nellore cattle. *Genet. Mol. Res. GMR* **2012**, *11*, 3721–3728. [CrossRef]
91. Tian, J.; Zhao, Z.; Zhang, L.; Zhang, Q.; Yu, Z.; Li, J.; Yang, R. Association of the leptin gene E2-169T>C and E3-299T>A mutations with carcass and meat quality traits of the Chinese Simmental-cross steers. *Gene* **2013**, *518*, 443–448. [CrossRef]
92. Melucci, L.M.; Panarace, M.; Feula, P.; Villarreal, E.L.; Grigioni, G.; Carduza, F.; Soria, L.A.; Mezzadra, C.A.; Arceo, M.E.; Mazzucco, J.P.; et al. Genetic and management factors affecting beef quality in grazing Hereford steers. *Meat Sci.* **2012**, *92*, 768–774. [CrossRef] [PubMed]
93. Casas, E.; Duan, Q.; Schneider, M.J.; Shackelford, S.D.; Wheeler, T.L.; Cundiff, L.V.; Reecy, J.M. Polymorphisms in calpastatin and mu-calpain genes are associated with beef iron content. *Anim. Genet.* **2014**, *45*, 283–284. [CrossRef] [PubMed]
94. Cinzia, R.; Valentina, B.; Alessio, C.; Andrea, A.; Fabio, M.; Luigi, G.; Paolo, C. Association of polymorphisms in calpain 1, (mu/I) large subunit, calpastatin, and cathepsin D genes with meat quality traits in double-muscled Piemontese cattle. *Anim. Genet.* **2013**, *44*, 193–196.
95. SeungHwan, L.; SeungChang, K.; HanHa, C.; SooHyun, C.; HyeongCheol, K.; DaJeong, L.; BongHwan, C.; ChangGwan, D.; Aditi, S.; Gondro, C.; et al. Mutations in calpastatin and micro -Calpain are associated with meat tenderness, flavor and juiciness in Hanwoo (Korean cattle): Molecular modeling of the effects of substitutions in the calpastatin/ micro -Calpain complex. *Meat Sci.* **2014**, *96*, 1501–1508.
96. Joo, K.H.; Aditi, S.; Hyun, L.S.; Ho, L.D.; Jeong, L.D.; Min, C.Y.; Suk, Y.B.; Hwan, L.S. Genetic association of PLAG1, SCD, CYP7B1 and FASN SNPs and their effects on carcass weight, intramuscular fat and fatty acid composition in Hanwoo steers (Korean cattle). *Anim. Genet.* **2017**, *48*, 251–252.
97. Bartoň, L.; Bureš, D.; Kott, T.; Řehák, D. Associations of polymorphisms in bovine DGAT1, FABP4, FASN, and PPARGC1A genes with intramuscular fat content and the fatty acid composition of muscle and subcutaneous fat in Fleckvieh bulls. *Meat Sci.* **2016**, *114*, 18–23. [CrossRef]
98. Bhuiyan, M.S.A.; Kim, Y.K.; Kim, H.J.; Lee, D.H.; Lee, S.H.; Yoon, H.B.; Lee, S.H. Genome-wide association study and prediction of genomic breeding values for fatty-acid composition in Korean Hanwoo cattle using a high-density single-nucleotide polymorphism array 1. *J. Anim. Sci.* **2018**, *96*, 4063–4075. [CrossRef]
99. Mazzucco, J.P.; Goszczynski, D.E.; Ripoli, M.V.; Melucci, L.M.; Pardo, A.M.; Colatto, E.; Rogberg-Muñoz, A.; Mezzadra, C.A.; Depetris, G.J.; Giovambattista, G.; et al. Growth, carcass and meat quality traits in beef from Angus, Hereford and cross-breed grazing steers, and their association with SNPs in genes related to fat deposition metabolism. *Meat Sci.* **2016**, *114*, 121–129. [CrossRef]
100. Dongyep, O.; Yoonseok, L.; Boomi, L.; Jungsou, Y.; Euiryong, C.; Younyoung, K.; Chaeyoung, L. Fatty acid composition of beef is associated with exonic nucleotide variants of the gene encoding FASN. *Mol. Biol. Rep.* **2012**, *39*, 4083–4090.
101. Rempel, L.A.; Casas, E.; Shackelford, S.D.; Wheeler, T.L. Relationship of polymorphisms within metabolic genes and carcass traits in crossbred beef cattle. *J. Anim. Sci.* **2012**, *90*, 1311–1316. [CrossRef]

02. Liu, H.; Tian, W.; Zan, L.; Wang, H.; Cui, H. Mutations of MC4R gene and its association with economic traits in Qinchuan cattle. *Mol. Biol. Rep.* **2010**, *37*, 535–540. [CrossRef] [PubMed]
03. Jiyeon, S.; Sang, S.D.; Do, P.K.; Kyo, L.H.; Sik, K.H. Identification and analysis of MC4R polymorphisms and their association with economic traits of Korean cattle (Hanwoo). *Mol. Biol. Rep.* **2012**, *39*, 3597–3601.
04. Haruna, I.L.; Ekegbu, U.J.; Ullah, F.; Amirpour-Najafabadi, H.; Zhou, H.; Hickford, J.G.H. Genetic variations and haplotypic diversity in the Myostatin gene of New Zealand cattle breeds. *Gene* **2020**, *740*, 144400. [CrossRef]
05. Martínez, A.; Aldai, N.; Celaya, R.; Osoro, K. Effect of breed body size and the muscular hypertrophy gene in the production and carcass traits of concentrate-finished yearling bulls. *J. Anim. Sci.* **2010**, *88*, 1229–1239. [CrossRef]
06. Raes, K.; De Smet, S.; Demeyer, D. Effect of double-muscling in Belgian Blue young bulls on the intramuscular fatty acid composition with emphasis on conjugated linoleic acid and polyunsaturated fatty acids *Anim. Sci.* **2001**, *73*, 253–260. [CrossRef]
07. Fiems, L.O. Double Muscling in Cattle: Genes, Husbandry, Carcasses and Meat. *Animals* **2012**, *2*, 472–506. [CrossRef]
08. Robinson, D.L.; Cafe, L.M.; McIntyre, B.L.; Geesink, G.H.; Barendse, W.; Pethick, D.W.; Thompson, J.M.; Polkinghorne, R.; Greenwood, P.L. Production and processing studies on calpain-system gene markers for beef tenderness: Consumer assessments of eating quality. *J. Anim. Sci.* **2012**, *90*, 2850–2860. [CrossRef] [PubMed]
09. Alves, D.D.; Goes, R.H.d.T.e.B.d.; Mancio, A.B. Maciez da carne bovina. *Ciência Anim. Bras.* **2006**, *6*, 135.
10. Shackelford, S.D.; Koohmaraie, M.; Cundiff, L.V.; Gregory, K.E.; Rohrer, G.A.; Savell, J.W. Heritabilities and phenotypic and genetic correlations for bovine postrigor calpastatin activity, intramuscular fat content, Warner-Bratzler shear force, retail product yield, and growth rate1. *J. Anim. Sci.* **1994**, *72*, 857–863. [CrossRef]
11. Geary, T.W.; McFadin, E.L.; MacNeil, M.D.; Grings, E.E.; Short, R.E.; Funston, R.N.; Keisler, D.H. Leptin as a predictor of carcass composition in beef cattle. *J. Anim. Sci.* **2003**, *81*, 1–8. [CrossRef]
12. Wang, Y.; Yang, W.; Gui, L.; Wang, H.; Zan, L. Association and expression analyses of the Ucp2 and Ucp3 gene polymorphisms with body measurement and meat quality traits in Qinchuan cattle. *J. Genet.* **2016**, *95*, 939–946. [CrossRef]
13. Molinari, P.C.; Bromfield, J.J. Inflammatory responses of bovine endometrial epithelial cells are increased under in vitro heat stress conditions. *J. Therm. Biol.* **2023**, *114*, 103564. [CrossRef]
14. Poleti, M.D.; DeRijk, R.H.; Rosa, A.F.; Moncau, C.T.; Oliveira, P.S.; Coutinho, L.L.; Eler, J.P.; Balieiro, J.C.C. Genetic variants in glucocorticoid and mineralocorticoid receptors are associated with concentrations of plasma cortisol, muscle glycogen content, and meat quality traits in male Nellore cattle. *Domest. Anim. Endocrinol.* **2015**, *51*, 105–113. [CrossRef]
15. Wang, L.; Sun, H.Z.; Guan, L.L.; Liu, J.X. Short communication: Relationship of blood DNA methylation rate and milk performance in dairy cows. *J. Dairy Sci.* **2019**, *102*, 5208–5211. [CrossRef]
16. Geng, Q.; Lin, W.; Yang, L.; Hu, X.; Qiu, X. Rumen-protected guanidinoacetic acid improves growth performance in beef cattle under chronic heat stress by reshaping gut microbiota and modulating serum metabolism. *Front. Microbiol.* **2025**, *16*, 1529596. [CrossRef]
17. Tian, R.; Mahmoodi, M.; Tian, J.; Koshkoiyeh, S.E.; Zhao, M.; Saminzadeh, M.; Li, H.; Wang, X.; Li, Y.; Esmailizadeh, A. Leveraging Functional Genomics for Understanding Beef Quality Complexities and Breeding Beef Cattle for Improved Meat Quality. *Genes* **2024**, *15*, 1104. [CrossRef]
18. Cole, J.B.; VanRaden, P.M. Symposium review: Possibilities in an age of genomics: The future of selection indices 1. *J. Dairy Sci.* **2018**, *101*, 3686–3701. [CrossRef]
19. Chen, D.; Wang, X.; Guo, Q.; Deng, H.; Luo, J.; Yi, K.; Sun, A.; Chen, K.; Shen, Q. Muscle Fatty Acids, Meat Flavor Compounds and Sensory Characteristics of Xiangxi Yellow Cattle in Comparison to Aberdeen Angus. *Animals* **2022**, *12*, 1161. [CrossRef]
20. Kasimanickam, R.; Ferreira, J.C.P.; Kastelic, J.; Kasimanickam, V. Application of Genomic Selection in Beef Cattle Disease Prevention. *Animals* **2025**, *15*, 277. [CrossRef]
21. Kaur, D.; Virk, A.K. Smart neck collar: IoT-based disease detection and health monitoring for dairy cows. *Discov. Internet Things* **2025**, *5*, 12. [CrossRef]
22. Schnidrig, G.A.; Struchen, R.; Schärrer, S.; Heim, D.; Hadorn, D.; Regula, G.S.; Paternoster, G. Improved cattle farm classification: Leveraging machine learning and linked national datasets. *Front. Vet. Sci.* **2025**, *12*, 1517173. [CrossRef]
23. Kittichai, V.; Kaewthamasorn, M.; Arnuphaprasert, A.; Jomtarak, R.; Naing, K.M.; Tongloy, T.; Chuwongin, S.; Boonsang, S. A deep contrastive learning-based image retrieval system for automatic detection of infectious cattle diseases. *J. Big Data* **2025**, *12*, 2. [CrossRef]
24. Dilger, A.C. Gene editing in livestock: Advancements, opportunities and challenges. *J. Anim. Sci.* **2019**, *97*, 60. [CrossRef]
25. Villamil, P.R.; Beaton, B.P.; Krisher, R.L. Gene editing in livestock: Innovations and applications. *Anim. Reprod.* **2024**, *21*, e20240054. [CrossRef]

animals

MDPI

Article

Genetic Evaluation of Early Growth Traits in Yunnan Semi-Fine Wool Sheep

Yaqian Wang [1,2], Hongyuan Yang [3], Xiaoqi Zhao [3], Xiaojun Ni [3], Yuanchong Zhao [4], Zhengrong You [5], Qingwei Lu [1,2], Sen Tang [2], Guobo Quan [3,*] and Xuefeng Fu [2,*]

1 College of Animal Science, Xinjiang Agricultural University, Urumqi 830052, China
2 Xinjiang Key Laboratory of Animal Biotechnology, Key Laboratory of Herbivorous Animal Genetics, Breeding and Reproduction, Ministry of Agriculture and Rural Affairs, Institute of Biotechnology, Xinjiang Academy of Animal Sciences, Urumqi 830011, China
3 Yunnan Animal Science and Veterinary Institute, Kunming 650224, China
4 Qiaojia Agriculture and Rural Bureau, Qiaojia 654600, China
5 Zhaotong Agriculture and Rural Bureau, Zhaotong 657000, China
* Correspondence: waltq20020109@163.com (G.Q.); fuxuefengfeng@126.com (X.F.)

Simple Summary: The Yunnan semi-fine wool sheep is the first semi-fine wool sheep breed developed in China. This study investigates the non-genetic factors influencing early growth traits and estimates their genetic parameters, providing a scientific basis for the selective breeding of this breed and other semi-fine wool sheep populations. The findings indicate that birth year, dam age, sex, flock and litter size significantly affect Birth Weight (BWT) and Birth Weight (BWT) ($p < 0.01$). Additionally, birth month was found to significantly influence Birth Weight (BWT) ($p < 0.01$), while weaning month had a statistically significant effect on Weaning Weight (WWT) ($p < 0.05$). No significant effects of farm location were observed on either trait ($p > 0.05$). The most accurate genetic evaluation model determined the heritability of the Birth Weight (BWT) and Weaning Weight (WWT) as 0.3123 and 0.3471.

Abstract: With economic development and improved living standards, the demand for mutton and wool continues to grow, and improving the production performance and genetic potential of sheep breeds has become the key to promoting the high-quality development of the sheep industry. Thus, this study analyzes the influencing factors of the early production traits of Yunnan semi-fine wool sheep, optimizes the genetic evaluation model, and relies on accurate genetic parameter estimation to provide a theoretical basis for formulating a scientific and efficient breeding strategy for this breed. Data were collected from the Laishishan and Xiaohai breeding farms in Qiaojia, Yunnan, covering production records of the core flock from 2018 to 2022. Using the GLM procedure in SAS 9.4 software, this study analyzed the non-genetic influences on early production traits in Yunnan semi-fine wool sheep. Concurrently, Danish Milk Unit 5 (DMU 5) software estimated the variance components across various animal models for each trait. Employing the Akaike Information Criterion (AIC) and likelihood ratio test (LRT), six models were tested, incorporating or excluding maternal inheritance and environmental impacts, to identify the optimal model for deriving the genetic parameters. The results show that the birth year, dam age, sex, flock and litter size significantly affect both the Birth Weight (BWT) and Weaning Weight (WWT) ($p < 0.01$). Additionally, the birth month was found to exert a significant effect on Birth Weight (BWT) ($p < 0.01$), the weaning month has a significant effect on the Weaning Weight (WWT) ($p < 0.05$). No significant effects of farm location were observed on either trait ($p > 0.05$). The most accurate genetic evaluation model determined the heritability of the Birth Weight (BWT) and Weaning Weight (WWT) as 0.3123 and 0.3471. From a production perspective, improving lamb birth, Weaning Weight (WWT), feed composition,

 https://doi.org/10.3390/ani15111512

and maternal nutrition during gestation is vital for breeding efficiency. This study not only identified the optimal animal models for early growth traits in Yunnan semi-fine wool sheep, offering a precise basis for estimating genetic parameters but also provides theoretical guidance for genetic selection and breed improvement in this population.

Keywords: Yunnan semi-fine wool sheep; genetic parameters; early growth traits

1. Introduction

Economic growth and rising living standards have led to an increased demand for mutton and wool, fostering the expansion of the sheep industry [1]. Amid this growing demand, developing high-yield and well-adapted breeds is crucial for ensuring the sustainability and productivity of the sheep industry [2]. The Yunnan semi-fine wool sheep, China's first semi-fine wool breed, is a dual-purpose breed for both meat and wool production, primarily distributed in Zhaotong, Yunnan Province [3]. This breed was developed in the late 1960s by the Yunnan Academy of Animal Science and Veterinary Medicine through successive crossbreeding of long-wool semi-fine sheep (e.g., Romney and Lincoln) with local coarse-wool sheep. This breeding strategy enabled the Yunnan semi-fine wool sheep to adapt to cold mountainous regions and southern grasslands [4,5]. Due to its strong adaptability and favorable production performance, this breed has become an important genetic resource in the region. As of 2023, the original breeding farm in Qiaojia County maintains a core flock of 1425 breeding sheep (including 1143 at Laishishan Breeding Farm and 282 at Xiaohai Breeding Farm) [6]. The Yunnan semi-fine wool sheep breeding farm primarily focuses on conservation, but the selective breeding of the core flock faces numerous challenges, including limited population size. Otherwise, most of the breeding sheep produced by the original farm are sold as commercial sheep, resulting in low coverage of improved breeds and difficulty in realizing the benefits of the new variety [6,7].

Early growth traits, particularly the Birth Weight and Weaning Weight, are critical indicators in selective breeding, as they directly influence lamb survival, growth rate, and overall production efficiency [8,9]. Selecting for these traits can significantly accelerate genetic progress and improve economic returns. However, achieving genetic improvement requires precise genetic parameter estimation, which serves as the foundation for effective breeding programs [10,11].

Accurate genetic parameter estimation is essential for assessing population characteristics and refining breeding strategies [12]. This requires reliable statistical models with individual additive genetic effects, maternal genetic effects, maternal permanent environmental effects, and a variety of non-genetic factors influencing trait expression [13]. Therefore, selecting an optimal animal model tailored to the specific breeding conditions is crucial for ensuring reliable genetic evaluations and driving sustainable genetic improvement.

Currently, there are limited studies on genetic parameter estimates of Yunnan semi-fine wool sheep; this study analyzes early growth traits using data collected from the Laishishan and Xiaohai Breeding Farms between 2018 and 2022. SAS software was employed to assess non-genetic factors, while variance components were estimated using the Danish Milk Unit 5 (DMU 5) software under a single-trait animal model. The likelihood ratio test (LRT) and Akaike Information Criterion (AIC) were used to compare different models and identify the optimal model for estimating genetic parameters of the Birth Weight and Weaning Weight. This study provides a theoretical foundation for selective breeding, offering key insights into the genetic improvement of Yunnan semi-fine wool sheep.

2. Materials and Methods

2.1. Data Sources and Processing

The Yunnan semi-fine wool sheep are raised under a grazing-based husbandry system. This study focused on newborn lambs of Yunnan semi-fine wool sheep from the Laishishan and Xiaohai Breeding Farms in Qiaojia, Yunnan, and collected lambing records, mating records, and pedigree data from these two farms during 2018–2022. Microsoft Excel software 2019 was used for data cleaning and preprocessing. Missing values in the raw data were identified and excluded, and extreme or abnormal data were removed based on practical production conditions. After these steps, a total of 3252 complete datasets of early growth traits were obtained. The early growth traits included the Birth Weight and Weaning Weight. Descriptive statistical analysis results for each trait are presented in Table 1.

Table 1. Descriptive statistics of valid records for early growth traits and reproductive traits.

Traits	Sample Size	Mean ± SD	Min (KG)	Max (KG)	Coefficient of Variation, %
BWT (kg)	3252	4.39 ± 0.62	2.10	7	14.27
WWT (kg)	3252	21.67 ± 4.28	6.50	46	19.78

2.2. Stratification of Environmental Fixed Effects

The classification of different factor levels is detailed in Table 2. Based on the actual production conditions and data structure of the sheep farm, the effects of seven factors—the lamb birth year, dam age, birth month, birth type, birth sex, weaning month, and weaning age—on the Birth Weight and Weaning Weight of Yunnan semi-fine wool sheep were analyzed.

Table 2. Level division of non-genetic factors for each trait.

Factor								Level
Birth Year	Farm	Dam Age	Birth Month	Birth Type	Sex	Weaning Month	Weaning Age	
2022	1	1	February	Singleton	Male	June	79–110 days	1
2021	2	2	March	Twins	Female	July	110–130 days	2
2020	-	3	April	-	-	-	130–140 days	3
2019	-	4	-	-	-	-	>140 days	4
2018	-	-	-	-	-	-	-	5

"-" indicates "none".

2.3. Statistical Analysis Methods

The general linear model (GLM) procedure in SAS 9.4 software was used for the least squares analysis of the primary traits [14].

The statistical model for the Birth Weight is as follows:

$$Y_{ijklmcrn} = u + a_i + b_j + d_k + t_l + h_m + f_c + O_r + e_{ijklmcrn} \quad (1)$$

where $Y_{ijklmcrn}$ is the observed trait value; u is the overall mean; a_i is the effect of the i level of the birth year; b_j is the effect of the j level of the birth month; d_k is the effect of the k level of the birth sex; t_l is the effect of the l level of the birth type; h_m is the effect of the m level of the maternal age; f_c is the effect of the c level of the farm; O_r is the effect of the r level of the herd group; $e_{ijklmcrn}$ is the random residual effect.

The statistical model for the Weaning Weight is as follows:

$$Y_{ijklmcrvn} = u + a_i + b_j + d_k + t_l + h_m + f_c + O_r + P_v + e_{ijklmcrvn} \quad (2)$$

where $Y_{ijklmcrvn}$ is the observed trait value; u is the overall mean; a_i is the effect of the i level of the birth year; b_j is the effect of the j level of the weaning month; d_k is the effect of the k level of the birth sex; t_l is the effect of the l level of the birth type; h_m is the effect of the m level of the maternal age; f_c is the effect of the c level of the farm; O_r is the effect of the r level of the herd group; P_v is the effect of the v weaning age days; $e_{ijklmcrvn}$ is the random residual effect.

2.4. Genetic Parameter Estimation Model

Six distinct animal models were used for each trait by including or excluding the following effects: individual additive genetic effects, maternal additive genetic effects, maternal permanent environmental effects, and the covariance between direct and maternal additive genetic effects [15]. The models were structured as follows:

$$\text{Model 1} \;\; Y = X_b + Z_{1a} + e \quad (3)$$

$$\text{Model 2} \;\; Y = X_b + Z_{1a} + Z_{3mp} + e \quad (4)$$

$$\text{Model 3} \;\; Y = X_b + Z_{1a} + Z_{2mg} + e \;\; \mathrm{COV(a,m)} = 0 \quad (5)$$

$$\text{Model 4} \;\; Y = X_b + Z_{1a} + Z_{2mg} + e \;\; \mathrm{COV(a,m)} = \mathrm{A}\sigma_{am} \quad (6)$$

$$\text{Model 5} \;\; Y = X_b + Z_{1a} + Z_{2mg} + Z_{3mp} + e \;\; \mathrm{COV(a,m)} = 0 \quad (7)$$

$$\text{Model 6} \;\; Y = X_b + Z_{1a} + Z_{2mg} + Z_{3mp} + e \;\; \mathrm{COV(a,m)} = \mathrm{A}\sigma_{am} \quad (8)$$

In the model, Y is the observation vector of the traits; b represents the fixed effects; a denotes the individual additive genetic effects; mg refers to the maternal additive genetic effects; mp indicates the maternal permanent environmental effects; e is the residual effect. X, Z_1, Z_2, and Z_3 are the incidence matrices corresponding to the fixed effects, individual additive genetic effects, maternal additive genetic effects, and maternal permanent environmental effects, respectively. A represents the additive genetic relationship matrix, which accounts for the covariance (σ_{am}) between the individual additive genetic effects (a) and maternal additive genetic effects (mg).

2.5. Comparison of Different Models

To accurately estimate the genetic parameters for different traits of the Yunnan semi-fine wool sheep, this study designed six different animal models for fitting and performance evaluation, in order to identify the optimal animal model for each trait.

The Akaike Information Criterion (AIC) was used to evaluate the variance components estimated by the different models. The formula for calculating AIC is as follows [16]:

$$\mathrm{AIC} = 2\mathrm{k} - 2\mathrm{LogL} \quad (9)$$

In the formula, L represents the log-likelihood function value, and K denotes the number of parameters to be estimated [17]. The Akaike Information Criterion (AIC) reflects the impact of the number of parameters to be estimated on the estimation accuracy. While increasing the number of parameters in a model can improve the goodness of fit, an excessive number of parameters can lead to overfitting. Therefore, AIC rewards goodness of fit while penalizing the increase in the number of estimated parameters. When selecting the best model from a set of models, the model with the smallest AIC value is preferred [18], as it provides the best estimation of variance components [19].

The likelihood ratio test was used to compare the relative fit of the different models. The test formula is as follows:

$$\mathrm{LR} = -2\log\frac{L1}{L2} = [-2\log(L_1)] - [-2\log(L_2)] \quad (10)$$

In the formula, LR represents the likelihood ratio, and L_1 and L_2 represent the log-likelihood values of Model 1 and Model 2, respectively, where Model 1 is a sub-model of Model 2. Additionally, LR follows a chi-square distribution (χ^2) with degrees of freedom equal to the difference in the number of estimated parameters between Model 2 and Model 1. If the test result is significant, it indicates that the added parameters have a significant impact on the trait; otherwise, the parameters do not have a significant effect [20–22].

2.6. Genetic Parameter Calculation

The variance components estimated by the Danish Milk Unit 5 (DMU 5) software were substituted into the following formula to calculate the genetic parameters:

Individual heritability:

$$h_\alpha^2 = \sigma_\alpha^2/\sigma_p^2 \quad (11)$$

Maternal heritability:

$$h_m^2 = \sigma_m^2/\sigma_p^2 \quad (12)$$

Maternal permanent environmental heritability, respectively:

$$h_c^2 = \sigma_c^2/\sigma_p^2 \quad (13)$$

Phenotypic variance:

$$\sigma_p^2 = \sigma_a^2 + \sigma_m^2 + \sigma_e^2 \quad (14)$$

where $\sigma_\alpha^2, \sigma_m^2, \sigma_c^2$, σ_e^2 and σ_p^2 represent the direct genetic variance, mother genetic variance, mother permanence environmental variance, residual variance, and direct and mother genetic covariance, σ_{am} represents the covariance between individual additive genetic effects and maternal additive genetic effects. h_α^2, h_m^2, and h_c^2 represent individual heritability, maternal heritability, and maternal permanent environmental heritability, respectively.

3. Results

3.1. Least Squares Analysis of Birth Weight (BWT) and Weaning Weight (WWT) in Yunnan Semi-Fine Wool Sheep

3.1.1. Least Squares Analysis of Birth Weight in Yunnan Semi-Fine Wool Sheep

As shown in Table 3, the birth year, sex, lambing date, flock, litter size, and dam age all had highly significant effects on the Birth Weight ($p < 0.01$). However, there were no significant effects of the farm on the Birth Weight of Yunnan semi-fine wool sheep ($p > 0.05$). The fixed effects of Birth Weight for each trait are shown in Table 4.

Table 3. Fixed effects tests (F-values) for Birth Weight (BWT) in Yunnan semi-fine wool sheep.

Factor	Degrees of Freedom	Type III Sum of Squares	Mean Square	F-Value
Year	4	39.63 **	9.90	25.17
Sex	1	20.43 **	20.43	51.90
Farm	1	0.00 NS	0.00	0.00
Herd Group	9	47.02 **	5.22	13.27
Litter Size	1	339.70 **	339.70	862.90
Dam Age	3	79.06 **	26.35	66.95
Lambing Month	2	9.68 **	4.84	12.31

** $p < 0.01$. NS: non-significant ($p > 0.05$).

Table 4. Least squares analysis of variance for Birth Weight (BWT) of Yunnan semi-fine wool sheep.

Birth Year	*n*	LSM ± SE	Dam Age	*n*	LSM ± SE
2022	819	4.05 ± 0.05 ^{bc}	1	786	3.86 ± 0.05 ^{b}
2021	598	4.32 ± 0.05 ^{a}	2	1283	4.25 ± 0.05 ^{a}
2020	557	4.03 ± 0.05 ^{c}	3	903	4.26 ± 0.05 ^{a}
2019	709	4.10 ± 0.05 ^{bc}	4	280	4.24 ± 0.06 ^{a}
2018	569	4.26 ± 0.05 ^{ab}	Herd Group	*n*	LSM ± SE
Sex	*n*	LSM ± SE	1	444	4.24 ± 0.05 ^{b}
Male	1645	4.07 ± 0.05	2	346	4.21 ± 0.06 ^{b}
Female	1607	4.23 ± 0.05	3	293	3.9 ± 0.05 ^{d}
Farm	*n*	LSM ± SE	4	314	3.98 ± 0.05 ^{c}
Laishishan	2643	4.15 ± 0.04	5	246	3.92 ± 0.06 ^{c}
Xiaohai	609	4.15 ± 0.05	6	388	4.23 ± 0.05 ^{b}
Birth Type	*n*	LSM ± SE	7	430	4.22 ± 0.05 ^{b}
Singleton	2720	4.61 ± 0.04	8	374	4.13 ± 0.05 ^{b}
Twins	532	3.70 ± 0.05	9	366	4.21 ± 0.05 ^{b}
Birth Month	*n*	LSM ± SE	10	51	4.47 ± 0.10 ^{a}
February	2075	4.05 ± 0.02 ^{b}	-	-	-
March	1156	4.19 ± 0.02 ^{a}	-	-	-
April	21	4.22 ± 0.13 ^{a}	-	-	-

The data in the same column with different lowercase letters indicate significance ($p < 0.05$), while identical letters suggest no significance ($p > 0.05$), "-" indicates "none".

3.1.2. Least Squares Analysis of Weaning Weight (WWT) in Yunnan Semi-Fine Wool Sheep

The fixed effects of the Weaning Weight traits are presented in Table 5. As shown, the birth year, sex, weaning age, flock, litter size, and dam age all exhibited highly significant effects on the Weaning Weight of Yunnan semi-fine wool sheep ($p < 0.01$). Additionally, the weaning month had a significant effect on the Weaning Weight ($p < 0.05$), while the farm location had no significant impact on the Weaning Weight ($p > 0.05$). The fixed effects of the Birth Weight for each trait are shown in Table 6.

Table 5. Fixed effects tests (F-values) for Weaning Weight (WWT) in Yunnan semi-fine wool sheep.

Factor	Degrees of Freedom	Type III Sum of Squares	Mean Square	F-Value
Year	4	17,381.09 **	4345.27	236.41
Sex	1	1569.98 **	1569.98	85.42
Farm	1	4.23 NS	4.23	0.23
Herd Group	9	5256.81 **	584.09	251.55
Litter Size	1	5700.75 **	5700.75	246.09
Dam Age	3	2712.35 **	904.11	49.19
Weaning Month	1	180.84 *	180.84	9.84
Weaning Age Days	3	2712.35 **	904.11	49.19

** $p < 0.01$, * $p < 0.05$. NS: non-significant ($p > 0.05$).

Table 6. Least squares analysis of variance for Weaning Weight (WWT) of Yunnan semi-fine wool sheep.

Birth Year	*n*	LSM ± SE	Weaning Month	*n*	LSM ± SE
2022	819	18.48 ± 0.18^{c}	June	2869	20.60 ± 0.20
2021	598	25.89 ± 0.20^{a}	July	383	22.33 ± 0.30
2020	557	20.50 ± 0.22^{b}	Weaning age days	*n*	LSM ± SE
2019	709	21.16 ± 0.22^{b}	79–110 days	640	19.33 ± 0.23^{c}
2018	569	21.28 ± 0.25^{b}	110–130 days	1580	20.94 ± 0.17^{b}
Sex	*n*	LSM ± SE	130–140 days	794	22.51 ± 0.20^{a}
Male	1645	20.77 ± 0.16^{a}	>140 days	238	23.08 ± 0.29^{a}
Female	1607	22.16 ± 0.16^{b}	Herd Group	*n*	LSM ± SE
Farm	*n*	LSM ± SE	1	444	21.90 ± 0.24^{bc}
Laishishan	2643	21.65 ± 0.24	2	346	23.43 ± 0.27^{a}
Xiaohai	609	21.27 ± 0.21	3	293	20.44 ± 0.27^{e}
Birth Type	*n*	LSM ± SE	4	314	20.56 ± 0.27^{de}
Singleton	2720	23.33 ± 0.14	5	246	21.42 ± 0.30^{cd}
Twins	532	19.60 ± 0.20	6	388	22.42 ± 0.25^{ab}
Dam Age	*n*	LSM ± SE	7	430	19.36 ± 0.25^{e}
1	786	19.41 ± 0.20^{b}	8	374	22.37 ± 0.24^{ab}
2	1283	22.27 ± 0.16^{a}	9	366	20.38 ± 0.26^{de}
3	903	22.23 ± 0.18^{a}	10	51	22.33 ± 0.61^{ab}
4	280	21.94 ± 0.27^{a}	-	-	-

The data in the same column with different lowercase letters indicate significance ($p < 0.05$), while identical letters suggest no significance ($p > 0.05$), "-" indicates "none".

3.2. *Estimation of Variance Components for Early Growth Traits in Yunnan Semi-Fine Wool Sheep Using Different Animal Models*

3.2.1. Estimation of Variance Components for Birth Weight (BWT) Using Different Animal Models

As shown in Table 7, the heritability estimates for Birth Weight varied across models, with the lowest value observed in Model 5 (0.2430) and the highest in Model 6 (0.3828). Maternal genetic effects also exhibited considerable variation, ranging from 0.1205 in Model 3 to 0.2011 in Model 5. Similarly, maternal permanent environmental effects showed notable differences across the models, ranging from 0.0831 to 0.1295.

Table 7. Variance components for Birth Weight (BWT) estimated by different animal models.

Model	σ^2_α	σ^2_m	σ^2_c	σ^2_e	σ^2_p	h^2_α	h^2_m	h^2_c	σ_{am}
1	0.1289	-	-	0.2620	0.3909	0.3298	-	-	-
2	0.1034	-	0.0504	0.2354	0.3892	0.2657	-	0.1295	-
3	0.1015	0.0470	-	0.2413	0.3898	0.2605	0.1205	-	-
4	0.1162	0.0643	-	0.2280	0.3722	0.3123	0.1726	-	-3.63×10^{-2}
5	0.1249	0.1034	0.0504	0.2354	0.5142	0.2430	0.2011	0.0980	-
6	0.2322	0.1034	0.0504	0.2355	0.6065	0.3828	0.1704	0.0831	-1.49×10^{-2}

σ^2_α: direct genetic variance; σ^2_m: mother genetic variance; σ^2_e: residual variance; σ^2_c: mother permanence environmental variance; σ_{am}: direct and mother genetic covariance; σ^2_p: phenotypic variance; h^2_α: direct genetic effect; h^2_m: mother genetic effect; h^2_c: mother permanence environmental effect; "-": without this effect in the model.

3.2.2. Estimation of Variance Components for Weaning Weight (WWT) Using Different Animal Models

As presented in Table 8, the heritability estimates for Weaning Weight varied across the models, with the lowest value observed in Model 5 (0.1429) and the highest in Model 4 (0.6635). Maternal genetic effects also exhibited substantial variation, ranging from 0.0967 in Model 3 to 0.2908 in Model 5. Similarly, maternal permanent environmental effects showed notable differences across the models, ranging from 0.0879 to 0.1298.

Table 8. Estimation of variance components for Weaning Weight (WWT) using different animal models.

Model	σ^2_α	σ^2_m	σ^2_c	σ^2_e	σ^2_p	h^2_α	h^2_m	h^2_c	σ_{am}
1	5.8876	-	-	11.9501	17.8377	0.3301	-	-	-
2	6.0876	-	2.2762	9.1744	17.5382	0.3471	-	0.1298	-
3	5.5776	1.6255	-	9.5990	16.8021	0.3320	0.0967	-	-
4	6.0876	2.2762	-	3.5083	9.1744	0.6635	0.2481	-	1.52×10^{-2}
5	2.7988	5.6967	1.7224	9.3686	19.5865	0.1429	0.2908	0.0879	-
6	2.8152	5.6934	1.7236	9.3704	19.6062	0.1436	0.2904	0.0879	3.57×10^{-3}

σ^2_α: direct genetic variance; σ^2_m: mother genetic variance; σ^2_e: residual variance; σ^2_c: mother permanence environmental variance; σ_{am}: direct and mother genetic covariance; σ^2_p: phenotypic variance; h^2_α: direct genetic effect; h^2_m: mother genetic effect; h^2_c: mother permanence environmental effect; "-": without this effect in the model.

3.3. *Comparison of Different Animal Models*

3.3.1. Comparison of Different Models Using Akaike Information Criterion (AIC)

The −2LogL and AIC values for different animal models are listed in Table 9. Based on the AIC analysis, Model 4 provided the best genetic parameter estimates for Birth Weight, indicating that individual additive genetic effects, maternal effects, and the interaction between individual additive genetic effects and maternal additive genetic effects have a significant impact on the Birth Weight. Model 2 provided the best genetic parameter estimates for Weaning Weight, suggesting that individual additive genetic effects and maternal permanent environmental effects significantly influence the Weaning Weight.

Table 9. Standard values of AIC information for traits in different animal models.

Model	BWT		WWT	
	−2logL	AIC	−2logL	AIC
1	201.50	205.50	12,543.80	12,547.80
2	158.27	164.27	12,291.67	12,297.67
3	162.66	168.66	12,298.77	12,304.77
4	151.98	159.98	12,291.67	12,299.67
5	153.82	161.82	12,303.05	12,311.05
6	155.23	165.23	12,303.04	12,313.04

3.3.2. Comparison of Different Models Using Likelihood Ratio Test (LRT)

The likelihood ratio and chi-square test results for the comparison of the six models are presented in Table 10. The results show that for Birth Weight in early traits of Yunnan semi-fine wool sheep, no significant differences were observed between Model 2 and Model 6, as well as Model 5 and Model 6 ($p > 0.05$). However, significant differences were found between the other models ($p < 0.05$). For Weaning Weight, no significant differences were found between Model 5 and Model 6 ($p > 0.05$), while significant differences were observed between the other models ($p < 0.05$).

Table 10. Likelihood ratios and χ2 test results for different model comparisons of each trait.

Model Comparison	df	BWT	WWT
1 vs. 2	1	43.2314 **	252.1310 **
1 vs. 3	1	38.8369 **	245.0297 **
1 vs. 4	2	49.5147 **	252.1310 **
1 vs. 5	2	47.6751 **	240.7535 **
1 vs. 6	3	46.2666 **	240.7673 **
2 vs. 5	1	4.4437 *	11.3776 **
2 vs. 6	2	3.0352 NS	11.3638 **
3 vs. 4	1	10.6779 **	7.1014 **
3 vs. 5	1	8.8383 **	4.2762 *
3 vs. 6	2	7.4298 *	4.2624 *
4 vs. 6	1	3.2481 *	11.3638 **
5 vs. 6	1	1.4085 NS	0.0138 NS

** $p < 0.01$, * $p < 0.05$; NS means no significance.

4. Discussion

4.1. Effects of Different Fixed Effects on Early Growth Traits

This study considered seven fixed effects, including the lamb's birth year, maternal age, birth month, birth type, birth sex, weaning month, and weaning age. For the Birth Weight of Yunnan semi-fine wool sheep, five factors were primarily considered: birth year, maternal age, birth month, birth type, and birth sex. For the Weaning Weight, the above five factors were included along with the weaning month and weaning age. This study found that the average Weaning Weight of Yunnan semi-fine wool sheep showed an increasing trend over time, with longer weaning periods resulting in a higher Weaning Weight. The birth year, sex, lambing date, herd, litter size, and maternal age influenced the Birth Weight of Yunnan semi-fine wool sheep, while the birth year, sex, weaning age, herd, litter size, and maternal age influenced the Weaning Weight. Bradford [23] suggested that factors such as the breed, nutritional level, litter size, maternal age, and season affect lambs through the ewe. Sabri Gul's study on Kilis goats found that the sex, birth type, herd, and year influenced early growth traits, which aligns with the findings of this study. Similar results were reported by Yang Jie et al. [24] for the Birth Weight of Chinese Merino (Xinjiang type) lambs, Li Xiuli et al. [25] for the Weaning Weight of South African Meat Merino sheep, and Masala et al. [26] for the average daily gain before weaning in Avikalin sheep.

Yunnan semi-fine wool sheep are raised under grazing conditions, and variations in the birth year may affect traits due to differences in feeding management, feed supply, climate changes, and personnel adjustments on the farm. The influence of the birth season on the Birth Weight, Weaning Weight, and average daily gain before weaning may be attributed to differences in environmental conditions across months. Since Yunnan semi-fine wool sheep are primarily pasture-raised, temperature variations across months could also affect forage moisture and nutritional quality. To improve early growth traits in Yunnan semi-fine wool sheep, ensuring adequate feed supply during the ewe's gestation period is essential to meet the nutritional needs of both the ewe and the fetus. Additionally, supplementary feeding before and after birth may enhance the lambs' Birth Weight and Weaning Weight.

4.2. Comparison of Variance Component Estimation Across Different Animal Models

The coefficient of variation for the Birth Weight of Yunnan semi-fine wool sheep was 14.27%, a result similar to the 17% coefficient of variation for the Birth Weight reported by Tian Haining [27] of Qinghai Plateau semi-fine wool sheep. In the model by Behrem et al. [28], which considered both individual additive genetic effects and maternal additive genetic effects, the heritability estimates for Birth Weight were 0.13 for Sardi sheep and

0.22 for Central Anatolian Merino sheep. Mohammadi et al. [29] included random effects such as individual additive genetic effects, maternal additive genetic effects, and maternal temporary environmental effects in their model, estimating the heritability of the Birth Weight of Shal sheep at 0.13. Li et al. [30] and Mandal et al. [31] estimated the heritability of the Birth Weight of Australian Merino sheep and Muzaffarnagari sheep at 0.15, which aligns with the findings of this study. However, the heritability estimate in this study (0.3123) is higher than those reported by Behrem et al. (0.13 and 0.22) and Mohammadi et al. (0.13), as well as Li et al. and Mandal et al. (0.15).

The heritability of the Weaning Weight of Yunnan semi-fine wool sheep was 0.3471. Hanford et al. [32] estimated the heritability of the Weaning Weight of Columbia sheep at 0.07, with the model including individual additive genetic effects, maternal additive genetic effects, and maternal permanent environmental effects, all of which indicated low heritability. Jalil-saghale et al. [33] estimated the heritability of the Weaning Weight of Baluchi sheep at 0.12, with the model incorporating individual additive genetic effects, maternal additive genetic effects, and maternal permanent environmental effects. Safari et al. [34] estimated the heritability of the Weaning Weight of Australian Merino sheep at 0.29, with the model including individual additive genetic effects, litter effects, maternal permanent environmental effects, and the interaction between individual direct genetic effects and maternal genetic effects, a result similar to that of this study. Di et al. [35] studied Chinese super-fine Merino sheep and estimated the heritability of the Weaning Weight at 0.16, with the model considering individual additive genetic effects and maternal permanent environmental effects.

In mammals, early growth traits are influenced by both direct additive genetic effects and maternal effects due to the prolonged maternal dependence of offspring [36,37]. Maternal effects, including maternal genetic contributions and maternal environmental influences, can significantly impact lamb growth and survival. Studies on Iranian Baluchi and Munjal sheep emphasize the necessity of incorporating maternal effects into statistical models to enhance the accuracy of genetic parameter estimation [36,38]. If maternal effects are ignored, models may overestimate additive genetic variance, leading to biased heritability estimates and reduced selection efficiency [39]. Accurately partitioning maternal genetic and maternal permanent environmental effects ensures unbiased variance estimation, thereby optimizing breeding strategies and enhancing genetic gain [40,41]. Lambs receive supplemental feeding after birth, and their Weaning Weight may be influenced by feeding management practices, resulting in lower heritability. Variations in growth performance and genetic structure among different sheep breeds can also lead to differences in Weaning Weight, thereby increasing the heritability estimates. Additionally, factors such as feeding management and environmental conditions, including different nutritional levels and growth environments, may contribute to varying degrees of phenotypic variation.

5. Conclusions

This study identified optimal animal models for estimating the genetic parameters of early growth traits in Yunnan semi-fine wool sheep, with Model 4 best suited for the Birth Weight and Model 2 for the Weaning Weight. The heritability estimates for these traits were moderate (0.3123) for Birth Weight; (0.3471) for Weaning Weight, indicating substantial genetic potential for selection. The selection for higher birth and Weaning Weight, combined with the strategic culling of older or low-performing ewes based on reproductive performance and maternal traits, can accelerate genetic gain. Implementing these strategies in breeding programs is expected to improve flock productivity, reduce feeding costs, and promote sustainable sheep production.

Author Contributions: Conceptualization and methodology, Y.W. and H.Y.; software, X.Z. and X.N.; validation, Q.L. and S.T.; formal analysis, Y.W. and Q.L.; investigation, Y.W., Q.L. and G.Q.; resources, Y.Z. and Z.Y.; data curation, Y.Z. and Z.Y.; writing—original draft preparation, Y.W.; writing—review and editing, G.Q.; visualization and supervision, G.Q.; project administration and funding acquisition, X.F. and G.Q. All authors have read and agreed to the published version of the manuscript.

Funding: This work was supported by the earmarked fund for CARS (Grant No. CARS-39), and Xinjiang Uygur Autonomous Region "Tianshan Talents" Training Program, Youth Top Talent Project (Grant No. 2023TSYCCX0031).

Institutional Review Board Statement: Not applicable.

Informed Consent Statement: Not applicable.

Data Availability Statement: The data presented in this study are available from the corresponding authors upon reasonable request. The data are not publicly available due to privacy or ethical restrictions.

Acknowledgments: We gratefully acknowledge the support and assistance from the Innovation Team for Conservation and Utilization of Sheep Genetic Resources at the Yunnan Academy of Animal Science and Veterinary Medicine, the Zhaotong Municipal Bureau of Agriculture and Rural Affairs, and the Qiaojia County Bureau of Agriculture and Rural Affairs for this research. All authors have read and agreed to the published version of the manuscript.

Conflicts of Interest: The authors declare no conflict of interest.

Abbreviations

The following abbreviations are used in this manuscript:

BWT	Birth Weight
WWT	Weaning Weight

References

1. Pu, X.; Ma, S.; Zhao, B.; Tang, S.; Lu, Q.; Liu, W.; Wang, Y.; Cen, Y.; Wu, C.; Fu, X. Transcriptome Meta-Analysis Reveals the Hair Genetic Rules in Six Animal Breeds and Genes Associated with Wool Fineness. *Front. Genet.* **2024**, *15*, 1401369. [CrossRef]
2. Alberto, F.J.; Boyer, F.; Orozco-terWengel, P.; Streeter, I.; Servin, B.; de Villemereuil, P.; Benjelloun, B.; Librado, P.; Biscarini, F.; Colli, L.; et al. Convergent Genomic Signatures of Domestication in Sheep and Goats. *Nat. Commun.* **2018**, *9*, 813. [CrossRef] [PubMed]
3. Zhao, Y.C.; Yang, H.Y.; Zhou, D.Y.; Hu, Z.R.; Xiao, M.F.; Liu, W.B. Genetic Resource Investigation and Analysis of Yunnan Semi-fine Wool Sheep. *Livest. Poult. Breed.* **2009**, *5*, 56–57. (In Chinese)
4. Wang, W.H. Conservation and Utilization of Yunnan Semi-fine Wool Sheep. *Yunnan J. Anim. Sci. Vet.* **2009**, *S1*, 42–43. (In Chinese)
5. Zhao, Y.G.; Hong, Q.H.; Xie, P.; Chen, G.P.; Wang, W.H. Study on Yunnan Semi-wool Sheep Production Performance. *J. Domest. Anim. Ecol.* **2011**, *32*, 51–56. (In Chinese)
6. Zhao, Y.C.; Long, H.Y.; Liang, K.P.; Liao, X.H.; You, Z.R. Current Status and Development Strategies of Yunnan Semi-fine Wool Sheep Industry. *Contemp. Anim. Husb.* **2023**, *10*, 137–140. (In Chinese)
7. Zhao, Y.C.; You, Z.R.; Ni, X.J.; Zhao, X.Q.; Zhou, D.Y.; Quan, G.B. Breeding and Industrialization Promotion of Yunnan Semi-fine Wool Sheep. *China Livest. Poult. Breed.* **2024**, *20*, 151–157. (In Chinese)
8. Liu, B.D. Study on Genetic Parameter Estimation of Early Growth Traits in Dumeng Sheep. Master's Thesis, Inner Mongolia Agricultural University, Hohhot, China, 2024. (In Chinese). [CrossRef]
9. Zhang, W.S.; Fang, Z.H. Analysis of Fixed Effects on Birth Weight and Weaning Weight in Cashmere Goats. *Chin. J. Anim. Sci.* **1995**, *1*, 40–41. (In Chinese)
10. Boujenane, I.; Hazzab, A.E. Genetic Parameters for Direct and Maternal Effects on Body Weights of Draa Goats. *Small Rumin. Res.* **2008**, *80*, 16–21. [CrossRef]
11. Zhang, C.; Yang, L.; Shen, Z. Variance Components and Genetic Parameters for Weight and Size at Birth in the Boer Goat. *Livest. Sci.* **2008**, *115*, 73–79. [CrossRef]
12. Bangar, Y.C.; Magotra, A.; Yadav, A.S. Variance Components and Genetic Parameter Estimates for Pre-Weaning and Post-Weaning Growth Traits in Jakhrana Goat. *Small Rumin. Res.* **2020**, *193*, 106278. [CrossRef]

3. Wei, Y.L. Study on Genetic Evaluation Models and Parameter Estimation for Growth Traits in White Cashmere Goats. Master's Thesis, Inner Mongolia Agricultural University, Hohhot, China, 2014. (In Chinese).
4. Wang, R.; Wang, X.; Liu, B.; Zhang, L.; Li, J.; Chen, D.; Ma, Y.; He, H.; Liu, J.; Liu, Y.; et al. Estimation of Genetic Parameters of Early Growth Traits in Dumeng Sheep. *Animals* **2024**, *14*, 2298. [CrossRef] [PubMed]
5. Liu, W.; Lu, Q.; Tang, S.; Pu, X.; Wang, Y.; Wu, C.; Hu, X.; Hong, W.; Fu, X. Comparison of Different Animal Models for Estimating Genetic Parameters for Early Growth Traits and Reproductive Traits in Tianmu Sainuo Sheep. *Front. Vet. Sci.* **2024**, *11*, 1349790. [CrossRef]
6. Sharif, N.; Ali, A.; Dawood, M.; Khan, M.I.U.R.; Do, D.N. Environmental Effects and Genetic Parameters for Growth Traits of Lohi Sheep. *Animals* **2022**, *12*, 3590. [CrossRef]
7. Akaike, H. A New Look at the Statistical Model Identification. *IEEE Trans. Autom. Control* **1974**, *19*, 716–723. [CrossRef]
8. Gowane, G.R.; Chopra, A.; Prince, L.L.L.; Paswan, C.; Arora, A.L. Estimates of (Co)Variance Components and Genetic Parameters for Body Weights and First Greasy Fleece Weight in Bharat Merino Sheep. *Animal* **2010**, *4*, 425–431. [CrossRef] [PubMed]
9. Tao, C.Q.; Yang, H.W. Spatial Econometric Model Selection and Simulation Analysis. *Stat. Res.* **2014**, *31*, 88–96. (In Chinese) [CrossRef]
0. Li, L.J.; Yang, L.; Shen, X.Y. Comparison of Different Models for Genetic Parameter Estimation in Guizhou Semi-fine Wool Sheep. *Anim. Husb. Vet. Med.* **2014**, *46*, 37–40. (In Chinese)
1. Alves, K.; Schenkel, F.S.; Brito, L.F.; Robinson, A. Estimation of Direct and Maternal Genetic Parameters for Individual Birth Weight, Weaning Weight, and Probe Weight in Yorkshire and Landrace Pigs. *J. Anim. Sci.* **2018**, *96*, 2567–2578. [CrossRef]
2. Buse, A. The Likelihood Ratio, Wald, and Lagrange Multiplier Tests: An Expository Note. *Am. Stat.* **1982**, *36*, 153–157.
3. Bradford, G.E. The Role of Maternal Effects in Animal Breeding. VII. Maternal Effects in Sheep. *J. Anim. Sci.* **1972**, *35*, 1324–1334. [CrossRef] [PubMed]
4. Yang, J.; Tian, K.C.; Huang, X.X.; Fu, X.F.; Gou, X.X.; Zhao, G.P.; Wubuli, N.; Tian, Y.Z.; Maimaiti, A. Birth Weight and Its Influencing Factors in Chinese Merino Sheep (Xinjiang Type). *Xinjiang Agric. Sci.* **2014**, *51*, 735–740. (In Chinese)
5. Li, X.L. NonGenetic Factors Affecting Growth Traits of South African Mutton Merino Sheep. *China Herb. Sci.* **2014**, *34*, 13–15+18. (In Chinese)
6. Mahala, S.; Saini, S.; Kumar, A.; Prince, L.L.L.; Gowane, G.R. Effect of Non-Genetic Factors on Growth Traits of Avikalin Sheep. *Small Rumin. Res.* **2019**, *174*, 47–52. [CrossRef]
7. Tian, H.N. Comparison of Different Models for Genetic Parameter Estimation in Qinghai Plateau Semi-Fine Wool Sheep. *Heilongjiang Anim. Sci. Vet. Med.* **2016**, *3*, 82–84. (In Chinese) [CrossRef]
8. Behrem, S. Estimation of Genetic Parameters for Pre-Weaning Growth Traits in Central Anatolian Merino Sheep. *Small Rumin. Res.* **2021**, *197*, 106319. [CrossRef]
9. Mohammadi, H.; Shahrebabak, M.M.; Shahrebabak, H.M.; Bahrami, A.; Dorostkar, M. Model Comparisons and Genetic Parameter Estimates of Growth and the Kleiber Ratio in Shal Sheep. *Arch. Anim. Breed.* **2013**, *56*, 264–275. [CrossRef]
0. Li, L.; Brown, D.J. Estimation of Genetic Parameters for Lambing Ease, Birthweight and Gestation Length in Australian Sheep. *Anim. Prod. Sci.* **2016**, *56*, 934. [CrossRef]
1. Mandal, A.; Karunakaran, M.; Sharma, D.K.; Baneh, H.; Rout, P.K. Variance Components and Genetic Parameters of Growth Traits and Kleiber Ratio in Muzaffarnagari Sheep. *Small Rumin. Res.* **2015**, *132*, 79–85. [CrossRef]
2. Hanford, K.J.; Van Vleck, L.D.; Snowder, G.D. Estimates of Genetic Parameters and Genetic Change for Reproduction, Weight, and Wool Characteristics of Columbia Sheep. *J. Anim. Sci.* **2002**, *80*, 3086–3098. [CrossRef]
3. Jalil-Sarghale, A.; Kholghi, M.; Shahrebabak, M.M.; Shahrebabak, H.M.; Mohammadi, H.; Abdollahi-Arpanahi, R. Model Comparisons and Genetic Parameter Estimates of Growth Traits in Baluchi Sheep. *Slovck J. Anim. Sci.* **2014**, *47*, 12–18.
4. Safari, E.; Fogarty, N.M.; Gilmour, A.R.; Atkins, K. Across Population Genetic Parameters for Wool, Growth, and Reproduction Traits in Australian Merino Sheep. 2. Estimates of Heritability and Variance Components. *Aust. J. Agric. Res.* **2007**, *58*, 177–184. [CrossRef]
5. Di, J.; Zhang, Y.; Tian, K.C.; Lazate; Liu, J.F.; Xu, X.M.; Zhang, Y.J.; Zhang, T.H. Estimation of (Co)Variance Components and Genetic Parameters for Growth and Wool Traits of Chinese Superfine Merino Sheep with the Use of a Multi-Trait Animal Model. *Livest. Sci.* **2011**, *138*, 278–288. [CrossRef]
6. Xu, X.M.; Li, Y.F.; Fu, X.F.; Yu, L.J.; Zhang, Y.H.; Huang, X.X.; Tian, K.C. Heritability Estimates and Non-Genetic Factors Analysis of Birth Weight of Chinese Merino Sheep (Xinjiang Type). *China Anim. Husb. Vet. Med.* **2014**, *41*, 168–171. (In Chinese)
7. Bai, J.Y.; Li, J.Q.; Jia, X.P.; Zhang, Q.; Daoerji. Influence of Maternal Genetic Effect on Genetic Parameter Estimates of Production Traits of Cashmere Goat. *Hereditas* **2006**, *9*, 1083–1086. [CrossRef]
8. Liu, Y. The Genetic Parameter Estimation for Meat and WoolTraits in LiangShan Semi-Wool Sheep. Master's Thesis, Sichuan Agricultural University, Ya'an, China, 2007. (In Chinese).

39. Zhang, J.B.; Yuan, C.; Yue, Y.J.; Guo, J.; Niu, C.E.; Wang, X.J.; Wang, L.J.; Lü, H.Q.; Yang, B.-H. Comparison and Analysis of Genetic Parameters Estimation of Early Growth Traits of Alpine Merino Sheep by Different Animal Models. *Sci. Agric. Sin.* **2018**, *51*, 1202–1212.
40. Maghsoudi, A.; Torshizi, R.V.; Jahanshahi, A.S. Estimates of (Co)variance Components for Productive and Composite Reproductive Traits in Iranian Cashmere Goats. *Livest. Sci.* **2009**, *126*, 162–167. [CrossRef]
41. Abbasi, M.A.; Abdollahi-Arpanahi, R.; Maghsoudi, A.; Torshizi, R.V.; Nejati-Javaremi, A. Evaluation of Models for Estimation of Genetic Parameters and Maternal Effects for Early Growth Traits of Iranian Baluchi Sheep. *Small Rumin. Res.* **2012**, *104*, 62–69. [CrossRef]

genes

MDPI

Article

Assessment of Genetic Diversity and Productive Traits in Crossbreed Cattle in the Caribbean Region, Colombia

Andrés Rodríguez-Serrano, Marcos Ahumada-Velasco * and Jesús María Cárdenas Beltrán

Profesional de Investigación, Universidad de Cartagena, Cartagena de Indias 130001, Colombia; camilo.rodriguez.serrano@gmail.com (A.R.-S.); jcardenas@unisalle.edu.co (J.M.C.B.)
* Correspondence: ganaderiaguajira@unicartagena.edu.co

Abstract: Objectives: Evaluate the genetic diversity and productive traits of crossbred cattle in the Caribbean region of Colombia, through analyses derived from the assessment of the genome-wide single-nucleotide polymorphism (SNP). Methods: A total of 590 individuals and 66,098 SNPs were analyzed by principal components analysis (PCA) and detection of runs of homozygosity (ROH). The population was composed of 531 heifers marked as crossbreed and a group of 59 heifers marked as purebred Gyr. Additionally, allele frequencies were calculated for commercially important traits (*CSN2*, *CSN3*, *LGB*, *DGAT1*, *GH1*, *CAPN1_316*, *CAPN1_350*, *CAPN1_4751*, *CAST_282*, *CAST_2870*, and *CAST_2959*). Results: Global differences in PCA were 7.35%, and principal components explained 1.94% and 5.41% of the variation. Five ROH islands were identified in crossbred animals on chromosomes 2, 5, 7, 8, and 12. The majority of observed ROH classes were shorter than 2 Mb, 54% in crossbreed cattle and 47% in Gyr cattle. Individual inbreeding was 5.2% in crossbreed and 12% in Gyr cattle. Both groups had similar allelic and genotypic frequencies for most of the evaluated commercial traits. Only a wide variation was observed in the genes related to growth hormone (*GH1*) and Calpastatin (*CAST_2870* and *CAST_22959*). Crossbreed heifers had desired allele frequencies for better milk production and quality in the genes *CSN2*, *LGB*, *DGAT1*, and *GH1*, as well as in the genes *CAST_2870* and *CAST_2959*. Conclusions: Crossbreed cattle in the Colombian Caribbean region possess high genetic diversity and desirable allele frequencies to implement breeding and intense selection programs aimed at improving production yields.

Keywords: dual-purpose systems; genomic analysis; Gyr cattle; SNPs; homozygosity; inbreeding

1. Introduction

In biological terms, genetic diversity reflects the degree to which organisms can adapt to changing environmental conditions [1]. In livestock, it is the extent of selection processes that a population has undergone and the capacity to implement genetic improvement programs with the aim of increasing the frequency of desirable economic traits in animals [2]. Domestication and further selection of cattle have allowed the creation of a large number of breeds through factors such as mutations, migrations, and isolation [3]. These breeds have well-defined characteristics of interest to breeders, primarily dairy or meat production [4].

Due to this selection, and other practices such as intensive selective breeding, for a limited number of traits, intensification of production systems, controlled environmental conditions, and the globalization of breeding programs using common bulls, pure breeds have increasing inbreeding levels [2]. This increase in inbreeding could reduce the effective population size and result in inbreeding depression, probably leading to a decline in the expression of economically important traits [5]. For this reason, breeding programs and selection schemes must consider strategies to maximize genetic gain with minimal impact of inbreeding levels [6].

In contrast, crossbreed cattle in Colombia, due to the production system in which they develop, which are characterized by dual-purpose systems managed by low-income livestock farmers and a lack of technification, are often raised with minimal or no breeding programs, where production level is not the primary focus. Dairy production yield and average daily gains are about 3.5 L cow^{-1} and 420 g day^{-1}, respectively. These animals compose 35% of the Colombian cattle herd [7], and despite their production levels, they represent nearly 50% of the national milk supply [8] and 40.5% of the total beef production chain [9].

Crossbreed cattle in dual-purpose systems of the Caribbean region of Colombia are primarily a product of crosses between different breeds, predominantly of the *Bos indicus* type; however, due to the lack of animal and productive records, their genealogical information is often partial or entirely unknown. It is common to have the indiscriminate use of breeders [10] and genetic diversity, inbreeding levels, and important economic traits are either unknown or under-evaluated. Consequently, implementing effective selection programs is difficult [11].

Some producers in the region incorporate pure breeds in their herds to increase productive performance and achieve heterosis. However, given the aforementioned challenges and lack of knowledge, not all crossbreeding efforts yield the expected advantages [10]. Faced with this situation, genomic analysis using single-nucleotide polymorphism (SNP) could be a tool to overcome the deficient genealogical information [12] and measure genetic diversity through methodologies like the detection of runs of homozygosity (ROH), which can identify identical homozygote genomic regions by descendance across the entire autosome genome and measure the coefficient of genomic inbreeding [13]. Additionally, SNP detection allows the identification of genomic regions associated with specific phenotypical parameters known as quantitative trait loci (QTL), among which are CSN2, CSN3, GH1, and LGB for milk production, and CAPN, CSN, and CAST for meat quality [14,15]. Therefore, the objective of this study was to evaluate the genetic diversity and productive traits of crossbred cattle in the Caribbean region of Colombia, through analyses derived from the assessment of the genome-wide single-nucleotide polymorphism (SNP), and through this analysis, test the hypothesis that the management of crossbred cattle in dual-purpose systems in the region leads to reduced genetic diversity and a lower prevalence of commercially productive traits in the genome.

2. Materials and Methods

The study was conducted in the Caribbean Region of Colombia, and hair follicle samples were collected from 590 heifers from Bolivar n = 75, Cordoba n = 506, and Sucre n = 6. Heifers were selected in order to make a descriptive and exploratory analysis of the genetic diversity and prevalence of commercially productive traits in the genome, which works as a tool for reproduction and genetic improvement programs. Heifers were selected from small-scale producers, body weights ranging between 300–350 kg and body condition scores between 5–8 (on a 1–10 scale). Sampling was conducted during the mandatory cattle vaccination campaigns organized by the Colombian government, during which each producer confined the herd on their own properties. The confinement was conducted only once and lasted no more than 20 min. Tail hair removal was performed with a quick and instantaneous upward movement, extracting at least 20 hairs with intact follicles stored in paper envelopes. Initially, the plan was to sample crossbred heifers, but the cattle population of the producers who participated in the sampling included a significant proportion of pure Gyr breed animals, which are often used as genetic improvement material. For this reason, it was decided to differentiate the two groups—crossbred and pure Gyr—to conduct the analyses.

DNA extraction was performed using the Termofisher, MagMAX™ CORE Nucleic Acid Purification kit, Thermo Fisher Scientific, Waltham, MA, USA. Genotyping and commercially important genes identifications were performed using the Termofisher's software, Axiom™ V. 4.0, Bovine Genotyping 100 K. Data was pruned by quality control using software PLINK 1.9 [16] with a call rate (number of markers successfully identified) $\geq$ 95%, minor allele frequency (MAF) < 0.05, Hardy–Weinberg equilibrium > 0.000001, and missing genotype < 10%. After quality control, 66,098 SNPs were preserved, and no heifers were removed. The population was composed of 531 heifers marked as crossbreed and a group of 59 heifers marked as pure breed Gyr.

2.1. Genomic Analysis

A principal component analysis (PCA) was performed, using the genetic distances between individuals, calculated with the command "--distance-matrix" from PLINK 1.9 software. Additionally, to measure the inbreeding coefficient, a runs of homozygosity analysis (FROH) was performed using the RSdtudio V.4.4.1 package "DetectRuns" [17] with the "SlidingRuns" method, using the following parameters: (a) Window size 50 SNP; (b) minimum ROH length 1000 bps; (c) Maximum gap per ROH 10^6; and (d) Minimum SNP 20. The FROH was estimated following the method described by [18].

$$F_{ROH}\frac{L_{ROH}}{L_{AUT}}$$

where LROH is the total length of all ROH detected in the individual's autosomes, and LAUT is the entire length of the autosomal genome. Additionally, the population was categorized into five length classes (1–2 Mb, 2–4 Mb, 4–8 Mb, 8–16 Mb, and >16 Mb) to analyze the FROH distribution. The percentage of SNPs existing within a ROH was calculated, and potential hotspots or ROH in the genome were identified when the percentage exceeded a 50% threshold [19]. SNPs found within a ROH were explored for annotated genes on the *Bos taurus* assembly ASR-UCD1.2 in the genomic data viewer available at https://www.ensembl.org/Bos_taurus/Info/Index, accessed on 15 February 2025.

2.2. Commercially Important Traits

Genotypic frequencies for commercially important traits (CSN2, CSN3, LGB, DGAT1, GH1, CAPN1_316, CAPN1_350, CAPN1_4751, CAST_282, CAST_2870, and CAST_2959) were calculated using the Hard–Weinberg equation $(p^2 + 2pq + q^2)$, where p^2 is the frequency of the dominant homozygous dominant genotype, $2pq$ is the frequency of the heterozygous genotype, and q^2 is the frequency of the homozygous recessive genotype. Allele frequencies were calculated from genotypic frequencies, and the proportion of individuals that have important genes for milk and beef production was established.

3. Results

According to the PCA, the two evaluated groups (Crossbreed and Gyr cattle) exhibit molecular differences. However, they are closely related, primarily because crossbreed cattle in the region consist of *Bos indicus* breeds, and Gyr cattle have an important presence in the Caribbean region of Colombia. Notably, some of the farmer's heifers registered ass crossbred show a closer relationship to the Gyr population; this observation underscores the lack of accurate records, which is a common situation in small livestock farmers.

Global differences in PCA were 7.35%, and principal components explained 1.94 % and 5.41% of the variation (Figure 1). When the origin of the crossbred heifers (farm) was included in the PCA analysis to assess whether small-scale producers utilized different breeds in their farms according to specific production goals, no differences were detected, probably because the crossbreed cattle across the evaluated population were composed of similar breeds.

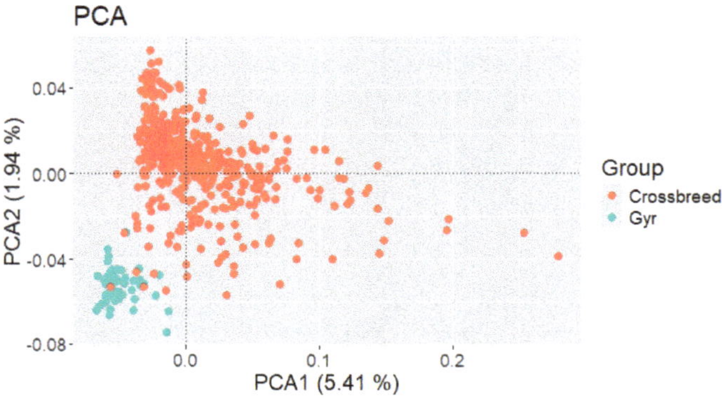

Figure 1. Population groups according to principal component analysis (PCA).

Consequently, with a widely dispersed genomic population, crossbred cattle had fewer ROH islands compared to the Gyr population, as illustrated in the Manhattan plot (Figure 2). Using the 50% threshold, five ROH islands were identified in crossbred heifers on BTA 2, 5, 7, 8, and 12, whereas the Gyr cattle showed 31 ROH islands distributed across BTA 2, 3, 5, 7, 8, 10, 11, 12, 19, 21, 23, 24, 26, and 29.

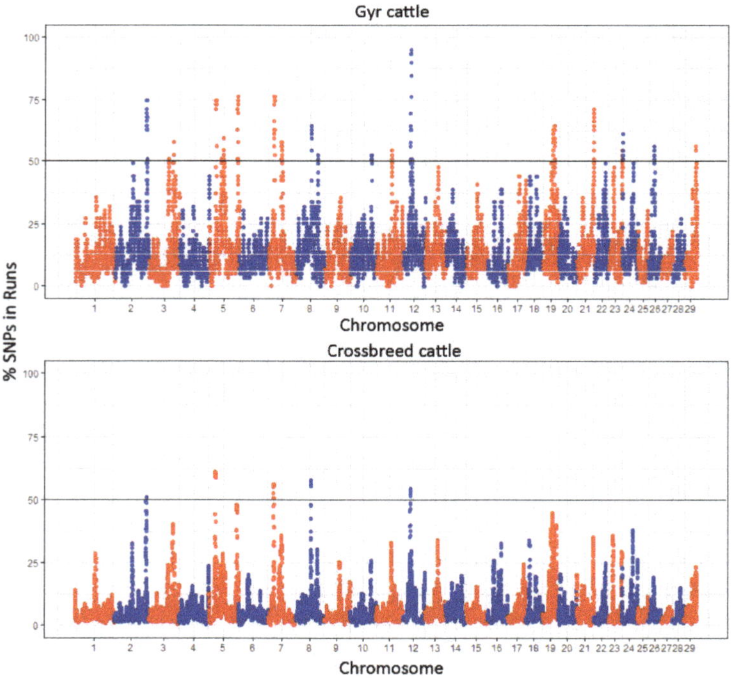

Figure 2. SNP frequency within run of homozygosity in the autosome of crossbreed and Gyr cattle in the Caribbean region of Colombia.

ROH islands were found because the SNPs are shared by more than 50% of crossbred heifers, serving as evidence of selection. These islands revealed 60 annotated genes (Table 1).

Table 1. Description of the ROH islands detected in the autosomal genome of the crossbreed cattle population in the Colombian Caribbean Region.

BTA	nSNP	From	To	Associated Gene
2	22	125,691,492	126,383,381	*FGR, AHDC1, WASF2, MAP3K6, TMEM222, SYYL1, GPN2, KDF1, TENT5B*
5	52	25,259,925	26,964,045	*GLYCAM1, PDE1B, GTSF1, COPZ1, MAP3K12, TF7, CALCOCO1, HOXC4-8-10-11-13, SMUG1, ATP5MC2*
7	51	19,764,739	21,612,020	*FSD1, NMRK2, ATCAY, DAPK3, TEKTIP1, GIPC3, SHF ANKRD24, SHD, TMIGD2, MATK, CELF5, NCLN, TLE6, MOB3A, S1PR4, APBA3*
8	51	58,010,694	59,902,102	*RHF32, SPATA31G1, FIGO, STOML2, ATOSB, RUSC2, CIMIP2B, TESK1, CD72, SIT1*
12	48	27,933,503	29,563,846	*STARD13, KL, PDS5B, N4BP2L2, ZAR1L, FRY, RXFP2*

The majority of observed ROH classes were shorter than 2 Mb, 54% in crossbreed cattle and 47% in Gyr cattle (Table 2). The maximum number of ROH observed per cow was 105 in crossbreed cattle and 114 in Gyr cattle, while the minimum ROH for the same groups was 3 and 58, respectively.

Table 2. Number (nROH) and mean length (in Mb) of the ROH islands in different length classes.

Group		<2	2–4	4–8	8–16	>16
Crossbreed	n ROH	14,290	9399	2028	663	181
	Mean length	1.45	2.64	5.4	10.85	23.5
Gyr	n ROH	2493	1860	529	236	142
	Mean length	1.44	2.68	5.42	11.19	26.11

Mean inbreeding values are low in both groups; individual inbreeding rates were 5.2% in crossbreed and 12% in Gyr cattle. Mean FROHCrh across the autosome was low, especially in crossbreed cattle, compared with the Gyr population, 7% and 12%, respectively; the highest mean values were found for BTA 5, 12, and 19, in both groups (Figure 3). These indicate that even in the pure breed, genetic diversity is high; these values are lower in the crossbreed group and similar in the pure breed group.

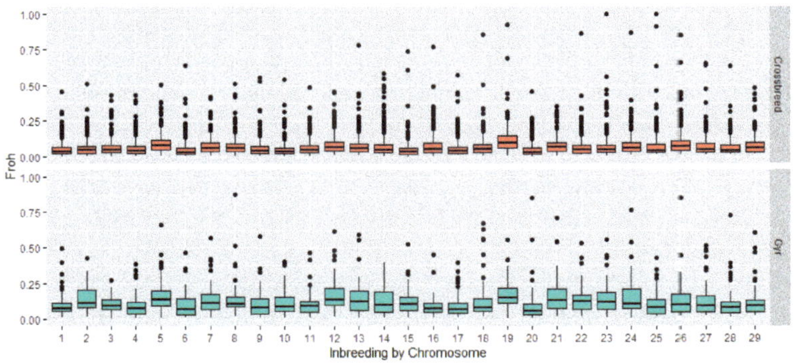

Figure 3. Inbreeding based on ROH among the autosome of crossbreed and Gyr cattle from the Colombian Caribbean region.

Both groups have similar allelic and genotypic frequencies for most of the evaluated commercial traits; only a wide variation was observed in the genes related to growth hormone (*GH1*) and Calpastatin (*CAST_2870* and *CAST_22959*) (Table 3). Homozygote variants and desired alleles were less frequent in the evaluated population for genes of commercial importance in meat production, probably because crossbreed cattle in dual-purpose systems in the Colombian Caribbean region are not selected and raised for high average daily gains, and breeding with *Bos taurus* breeds specially for meat production is uncommon.

Table 3. Genomic and allelic frequencies for commercial traits in crossbreed and Gyr cattle from the Colombian Caribbean region.

GEN	SNP ID	Description	Genotypic Frequencies						Allelic Frequencies			
			Gyr	Cross	Gyr	Cross	Gyr	Cross	Gyr	Cross	Gyr	Cross
			A1A1		**A1A2**		**A2A2**		**A1**		**A2**	
CSN2	rs43703013	Beta Casein	0.00	0.00	0.17	0.08	0.83	0.92	0.08	0.04	0.92	0.96
	rs43703011		0.02	0.01	0.19	0.17	0.8	0.82	0.11	0.09	0.89	0.91
			AA		**AB**		**BB**		**A**		**B**	
CSN3	rs43706475	Kappa Casein	1.00	1.00	0.00	0.00	0.00	0.00	1.00	1.00	0.00	0.00
	rs110014544		0.78	0.80	0.22	0.19	0.00	0.01	0.89	0.11	0.89	0.11
LGB	rs110180463	Beta lactoglobulin	0.00	0.01	0.02	0.17	0.98	0.9	0.01	0.10	0.99	0.82
	rs110066229		0.19	0.11	0.46	0.48	0.36	0.41	0.42	0.35	0.58	0.65
	rs110641366		0.00	0.06	0.10	0.42	0.90	0.51	0.05	0.27	0.95	0.73
	rs109625649		0.17	0.10	0.47	0.48	0.36	0.42	0.41	0.34	0.59	0.66
			AA		**AG**		**GG**		**A**		**G**	
DGAT1	rs109234250	CoA:diacylglycerol acyltransferase 1	0.84	0.65	0.14	0.31	0.02	0.04	0.91	0.81	0.09	0.19
			TT		**TG**		**GG**		**T**		**G**	
GH1	rs109191047	Growth hormone	0.00	0.50	0.20	0.35	0.98	0.60	0.01	0.22	0.99	0.78
			CC		**CG**		**GG**		**C**		**G**	
CAPN1_316	rs17872000	Calpain	0.00	0.00	0.00	0.10	1.00	0.90	0.00	0.05	1.00	0.95
			AA		**AG**		**GG**		**A**		**G**	
CAPN1_530	rs17871051		0.00	0.00	0.07	0.12	0.93	0.88	0.03	0.06	0.97	0.94
			TT		**TC**		**CC**		**T**		**G**	
CAPN1_4571	rs17872050		0.73	0.69	0.27	0.27	0.00	0.03	0.86	0.83	0.14	0.17
			CC		**CG**		**GG**		**C**		**G**	
CAST_282	rs110955059	Calpastatin	0.29	0.26	0.42	0.53	0.29	0.22	0.50	0.52	0.50	0.48
			AA		**AG**		**GG**		**A**		**G**	
CAST_22870	rs41255587		0.05	0.19	0.27	0.51	0.58	0.30	0.19	0.45	0.81	0.55
CAST_22959	rs109221039		0.69	0.39	0.29	0.48	0.02	0.13	0.84	0.63	0.16	0.37

4. Discussion

The PCA results showed a clear differentiation between the two evaluated groups. This is evidence that small-scale producers try to maintain their purebred population, either as a personal preference or because Gyr cattle are a locally adapted breed that significantly improves farm milk yields. When only crossbred heifers were analyzed, no genetic differences were observed. These results are consistent with those reported by Toro-Ospina et al. [20], who found no genetic differentiation among Caqueteño Cattle (a local breed from Colombia) from different farms. Rosero et al. [10] registered that crossbreed cattle in dual-purpose systems in Colombia did not exhibit genetic separation compared with crossbreed individuals with a clear phenotype of *Bos indicus* or *Bos taurus* animals, which leads to the use of control breeds to define better the groups that constitute multiracial individuals; when these authors incorporated pure breeds in their genomic analysis, they observed that crossbreed cattle of Piedemonte Llanero-Colombia have predominance of Brahman, Gyr, and Holstein breeds.

The PCA results in the present study showed greater dispersion among crossbred individuals compared to the Gyr population. Indicating higher genetic variability, this outcome is expected because, historically, small-scale producers in dual-purpose systems of the Colombian Caribbean region, among their herds that have a high presence of various breeds of *Bos indicus* animals, can eventually include animals with greater *Bos taurus* influence, either as crossbred or purebred individuals. These results are similar to those reported by Naranjo et al. [21] in crossbreed cattle for milk production from various regions of Colombia, which exhibited greater genetic dispersion compared with pure breeds. In addition, the study also compared the genomic information of Brahman, Gyr, Brown Swiss,

Holstein, Romosinuano, Jersey, and Senepol breeds, highlighting that crossbreed animals are in the center of these groups, revealing a multiplicity of lineages used for crossbreeding, with an important component of Brahman and Gyr breeds.

The wide genomic dispersion observed in crossbred cattle within dual-purpose systems in Colombia, through different breeds of *Bos indicus* and *Bos taurus* type, suggests a historically indiscriminate use of breeds in the absence of a breeding program [22].

The genes with an associated QTL were as follows: *AHDC1*, which could be related to body weight, energy metabolism, and clinical and subclinical ketosis [23,24]; *TENT5B*, involved in innate immune response [25]; *GLYCAM1*, which have an important role in regulating milk protein [26]; *PEDE1B*, identified as important to growth and carcass quality [27]; *GTSF1*, involved in reproduction processes through its role in the Argonaute proteins activity [28]; *CALCOCO1*, a gene candidate for meat quality in Angus cattle through cellular phosphate and glucose metabolism, as well as protein synthesis and degradation [29,30]; *ATP5MC2*, related with oxidative phosphorylation [31]; *MOB3A* and *PIGO* genes, which could be associated with weight gain and meat quality [32,33]; *STARD13*, a candidate gene for average daily gain (ADG) in cattle [34]; and *PDS5B*, a candidate gene for hematological parameters in Holstein [35]. No commercially important traits were observed in the ROH islands founded for the crossbreed heifer population evaluated, probably because very high production yields are not the principal selection criteria for small-scale producers of the region. Normally for them, it is more important to have resistant and adapted animals that can live and produce in an environment that does not always provide the resources for higher production levels or maintain specialized cattle.

In creole breeds raised in similar environmental conditions, it had been reported similar proportions of ROH length classes shorter than 2 Mb, 62.82% in the Caqueteño breed, and 76.0% in the Tropical milking Criollo breed [20,36]. In Gyr populations, Peripolli et al. [37] registered a proportion of 59.42% in the length class <2 Mb, indicating ancient inbreeding in both groups despite the lack of breeding programs.

Inbreeding values were similar to those registered by Martinez-Rocha et al. [36], in the Tropical milking Criollo, who reported a mean FROHCrh of 11.26%, and mean values across the autosome between 0 and 20%. FROH inbreeding can be higher in specialized breeds due to the high rate of selection to which they are subjected. Lozada-Soto et al. [38] reported a FROH between 15% and 17% in the Brown Swiss and Jersey Cattle, respectively, and Betancur-Zambrano et al. [39] reported a FROH of 28% in the Colombian Holstein cattle.

In dual-purpose cattle, the presence of desired commercial traits for both dairy and meat production is important. Among the genes of commercial interest for dairy production is Kappa casein (*CSN3*), where the BB allele is reported to improve cheese yield [40]. In the evaluated heifers, this allele is the least frequent, particularly in its homozygote form.

For Beta casein (*CSN2*), the A2A2 allele has been associated with enhanced production of essential amino acids and health benefits for humans [41], whereas the A1A1 allele is linked to the best milk quality for cheese production [42]. In the present study, the evaluated individuals showed high frequencies of the A2A2 allele, which is important in small-scale producers, where part of the milk produced is designated for self-consumption.

Regarding the Beta lactoglobulin (*LGB*), the presence of the A allele is associated with the best milk protein concentration (ideal for cheese making), whereas genotype BB is associated with better fats and casein content as well as high milk production [43]. In the evaluated population, the B allele was present at a high frequency in both groups.

For the CoA:diacylglycerol acyltransferase 1 (*DGAT1*), the AA variant codes for K lysine, which is associated with improved production of milk fat and protein [44]. The Gyr population showed a higher frequency of the AA variant compared to the crossbreed group. A similar trend was observed for the Growth hormone (*GH1*) gene, where the G allele has been reported as beneficial to milk composition and production [45], and Gyr group had a higher frequency of the GG variant compared to the crossbreed individuals.

In Calpain Genes (*CAPN1_316*, *CAPN1_530*, and *CAPN1_4751*), the desired alleles due to their relationship with meat tenderness are C, A, and C, respectively [46,47]. In the evaluated population, both crossbreed and Gyr cattle showed minimal frequencies of these alleles or, in some cases, their complete absence. Regarding the Calpastatin genes (*CAST_282*, *CAST_2870*, and *CAST_2959*), the desired alleles are C, G, and A, respectively [48], which in the evaluated population have frequencies close to 50%.

5. Conclusions

The evaluated population of crossbreed cattle in dual-purpose systems of the Colombian Caribbean Region showed a wide genomic dispersion, indicating high genetic diversity. No genomic differentiation was observed among individuals within the crossbreed group, probably because the evaluated crossbreed heifers are composed of similar breeds with a strong *Bos indicus* genetic component.

Despite a lack of accurate records and the absence of a breeding program, homozygous regions are short, and the inbreeding coefficient based on ROH is low even in heifers recorded as pure Gyr. A total of five ROH islands were detected; however, no genes with associated common QTLs were identified, indicating that probably the population has not been subjected to intense selection processes aimed at achieving high production yields in either milk or weight gains.

Nevertheless, crossbreed heifers have the desired allele frequencies for better milk production and quality in the genes *CSN2*, *LGB*, *DGAT1*, and *GH1*, as well as in the genes *CAST_2870* and *CAST_2959*, which are important for meat production and tenderness.

Crossbreed cattle in the Colombian Caribbean region possess high genetic diversity and desirable allele frequencies to implement breeding and intense selection programs aimed at improving production yields. However, this should be in parallel with proper records management, improved feed and water supply, and the maintenance of optimal sanitary conditions to enhance gene expression for either milk or meat production.

Author Contributions: All authors made and validated the study and contributed to writing, methods, and the paper. Data curation and writing, A.R.-S. Review, supervision, and editing. M.A.-V. and J.M.C.B. All authors have read and agreed to the published version of the manuscript.

Funding: This research was funded by the SGR "Sistema Nacional de Regalías" of the Colombian Government, under BPIN Code 021000100255.

Institutional Review Board Statement: Not applicable, the principles of replacement, reduction, and refinement from the NSW Government and the scientific procedures of the United Kingdom do not apply to the study conducted. However, the procedures carried out were designed to minimize stress in the animals. The animals used in the study were not experimental; they were animals native to the region, grazing under natural conditions and belonging to different producers. Neither the handling activities nor the living conditions of the individuals were altered for the collection of samples. The hair sample collection was conducted only once. The animals were confined for just 20 min to obtain the sample. The removal of follicles (taken from the tail) was harmless, painless, and instantaneous. After this, the individuals returned to their pastures and were not confined again.

Informed Consent Statement: Not applicable.

Data Availability Statement: The data presented in this study are available on request from the corresponding author due to confidentiality.

Acknowledgments: The authors thank the CTIE "Centro de Transferencia, Innovación y Emprendimiento" of the Universidad de Cartagena for the producers call for participation and research administration, and the SGR "Sistema general de regalías" of the Colombian government for funding this study.

Conflicts of Interest: The authors declare no conflicts of interest.

References

1. Ellegren, H.; Galtier, N. Determinants of Genetic Diversity. *Nat. Rev. Genet.* **2016**, *17*, 422–433. [CrossRef] [PubMed]
2. Brito, L.F.; Bédère, N.; Douhard, F.; Oliveira, H.R.; Arnal, M.; Peñagaricano, F.; Schinckel, A.P.; Baes, C.F.; Miglior, F. Review: Genetic Selection of High-Yielding Dairy Cattle Toward Sustainable Farming Systems in a Rapidly Changing World. *Animal* **2021**, *15*, 100292. [CrossRef] [PubMed]
3. Herron, J.C.; Freeman, S. Mendelian Genetics in Populations II: Migration, Drift and Nonrandom Mating. In *Evolutionary Analysis*, 5th ed.; Pearson Education: London, UK, 2014; pp. 259–264.
4. Groeneveld, L.F.; Lenstra, J.A.; Eding, H.; Toro, M.A.; Scherf, B.; Pilling, D.; Negrini, R.; Finlay, E.K.; Groeneveld, E.; Weigend, S. Genetic Diversity in Farm Animals—A Review. *Anim. Genet.* **2010**, *41* (Suppl. S1), 6–31. [CrossRef]
5. Makanjoula, B.O.; Maltecca, C.; Miglior, F.; Schenkel, F.S.; Baes, C.F. Effect of Recent and Ancient Inbreeding on Production and Fertility Traits in Canadian Holsteins. *BMC Genom.* **2020**, *21*, 605. [CrossRef]
6. Meuwissen, T.H.E. Maximizing the Response of Selection with a Predefined Rate of Inbreeding. *J. Anim. Sci.* **1997**, *75*, 934–940. [CrossRef]
7. González-Quintero, R.; Barahona-Rosales, R.; Bolívar-Vergara, D.M.; Chirinda, N.; Arango, J.; Pantévez, H.A.; Correa-Londoño, G.; Sánchez-Pinzón, M.S. Technical and Environmental Characterization of Dual-Purpose Cattle Farms and Ways of Improving Production: A Case Study in Colombia. *Pastoralism* **2020**, *10*, 19. [CrossRef]
8. Cortés Mora, J.A.; Cotes Torres, A.; Cotes Torres, J.M. Structural Features of Dual-Purpose Cattle Production System in the Colombian Humid Tropic. *Rev. Colomb. Cienc. Pecu.* **2012**, *25*, 229–239. [CrossRef]
9. González-Quintero, R.; Sánchez-Pinzón, M.S.; Bolívar-Vergara, D.M.; Chirinda, N.; Arango, J.; Pantévez, H.A.; Correa-Londoño, G.; Barahona-Rosales, R. Technical and Environmental Characterization of Very Small, Small, Medium, and Large Cow-Calf Operations in Colombia. *Rev. Mex. Cienc. Pecu.* **2020**, *11*, 183–200. [CrossRef]
10. Rosero Alpala, J.A.; Rangel Garcia, W.D.; Rojas Barreto, A.; Burgos-Paz, W.O. Contribution of Genomic Data in Defining the Breed Composition of Dual-Purpose Cattle. *Rev. Mex. Cienc. Pecu.* **2021**, *12*, 1008–1024. [CrossRef]
11. Schwarz, L.; Križanac, A.M.; Schneider, H.; Falker-Gieske, C.; Heise, J.; Liu, Z.; Bennewitz, J.; Thaller, G.; Tetens, J. Genetic and Genomic Analysis of Reproduction Traits in Holstein Cattle Using SNP Chip Data and Imputed Sequence Level Genotypes. *BMC Genom.* **2024**, *25*, 880. [CrossRef]
12. Berry, D.P.; Spangler, M.L. Animal Board Invited Review: Practical Applications of Genomic Information in Livestock. *Animal* **2023**, *17*, 100996. [CrossRef] [PubMed]
13. Bernini, F.; Punturiero, C.; Vevey, M.; Blanchet, V.; Milanesi, R.; Delledonne, A.; Bagnato, A.; Strillacci, M.G. Assessing Major Genes Allele Frequencies and the Genetic Diversity of the Native Aosta Cattle Female Population. *Ital. J. Anim. Sci.* **2023**, *22*, 1008–1022. [CrossRef]
14. Igoshin, A.V.; Romashov, G.A.; Chernyaeva, E.N.; Elatkin, N.P.; Yudin, N.S.; Larkin, D.M. Comparative Analysis of Allele Frequencies for DNA Polymorphisms Associated with Disease and Economically Important Traits in the Genomes of Russian and Foreign Cattle Breeds. *Russ. J. Genet.* **2022**, *26*, 298–307. [CrossRef]
15. Lopdell, T.J. Using QTL to Identify Genes and Pathways Underlying the Regulation and Production of Milk Components in Cattle. *Animals* **2023**, *13*, 911. [CrossRef]
16. Purcell, S.; Neale, B.; Todd-Brown, K.; Thomas, L.; Ferreira, M.A.R.; Bender, D.; Maller, J.; Sklar, P.; de Bakker, P.I.W.; Daly, M.J.; et al. PLINK: A Tool Set for Whole-Genome Association and Population-Based Linkage Analyses. *Am. J. Hum. Genet.* **2007**, *81*, 559–575. [CrossRef]
17. Biscarini, F.; Cozzi, P.; Gaspa, G.; Marras, G. *detectRUNS: Detect Runs of Homozygosity and Runs of Heterozygosity in Diploid Genomes. R Package 2019*; Institute for Statistics and Mathematics, Vienna University of Economics and Business: Vienna, Austria, 2019.
18. McQuillan, R.; Leutenegger, A.-L.; Abdel-Rahman, R.; Franklin, C.S.; Pericic, M.; Barac-Lauc, L.; Smolej-Narancic, N.; Janicijevic, B.; Polasek, O.; Tenesa, A.; et al. Runs of Homozygosity in European Populations. *Am. J. Hum. Genet.* **2008**, *83*, 359–372. [CrossRef]
19. Cesarani, A.; Sorbolini, S.; Criscione, A.; Bordonaro, S.; Pulina, G.; Battacone, G.; Marletta, D.; Gaspa, G.; Macciotta, N.P.P. Genome-wide Variability and Selection Signatures in Italian Island Cattle Breeds. *Anim. Genet.* **2018**, *49*, 371–383. [CrossRef]
20. Toro-Ospina, A.M.; Herrera Rios, A.C.; Pimenta Schettini, G.; Vallejo Aristizabal, V.H.; Bizarria dos Santos, W.; Zapata, C.A.; Ortiz Morea, E.G. Identification of Runs of Homozygosity Islands and Genomic Estimated Inbreeding Values in Caqueteño Creole Cattle (Colombia). *Genes* **2022**, *13*, 1232. [CrossRef]

21. Naranjo Guerrero, L.F.; Rogberg-Muñoz, A.; Rodríguez, N.; González Herrera, L.G. Genomic Diversity Study of Highly Crossbred Cattle Population in a Low and Middle Tropical Environment. *Trop. Anim. Health Prod.* **2024**, *56*, 258. [CrossRef]
22. Galina, C.S.; Geffroy, M. Dual-Purpose Cattle Raised in Tropical Conditions: What Are Their Shortcomings in Sound Productive and Reproductive Function? *Animals* **2023**, *13*, 2224. [CrossRef]
23. Soares, B.D.; Vargas, G.; Duffield, T.; Schenkel, F.; Squires, E.J. Genome-wide Association Study and Functional Analyses for Clinical and Subclinical Ketosis in Holstein Cattle. *J. Dairy Sci.* **2021**, *104*, 6789–6802. [CrossRef] [PubMed]
24. Li, L.; Shao, S.; Wang, Y.; Du, Z.; Yu, H.; Li, F.; Qin, Y. Ahdc1 is a Potent Regulator of Obesity and Energy Metabolism. *Am. J. Physiol. Endocrinol. Metab.* **2023**, *325*, E638–E648. [CrossRef] [PubMed]
25. Liudkovska, V.; Krawczyk, P.S.; Brouze, A.; Gumińska, N.; Wegierski, T.; Cysewski, D.; Mackiewicz, Z.; Ewbank, J.J.; Drabikowski, K.; Mroczek, S.; et al. TENT5 Cytoplasmic Noncanonical Poly(A) Polymerases Regulate the Innate Immune Response in Animals. *Sci. Adv.* **2023**, *9*, eadd9468. [CrossRef]
26. Li, J.; Huang, X.; Chen, Q.; Lei, X. Non-Additive Association of the Rs41257413 SNP in the GLYCAM1 Gene with Milk Protein Percentage in Chinese Holstein Cattle. *SSRN* **2025**, *20*, 5129267. [CrossRef]
27. Shin, S.; Heo, J.; Yeo, J.; Lee, C.; Chung, E. Genetic Association of Phosphodiesterase 1B (PDE1B) with Carcass Traits in Korean Cattle. *Mol. Biol. Rep.* **2012**, *39*, 4869–4874. [CrossRef]
28. Arif, A.; Bailey, S.; Izumi, N.; Anzelon, T.A.; Ozata, D.M.; Andersson, C.; Gainetdinov, I.; MacRae, I.J.; Tomari, Y.; Zamore, P.D. GTSF1 Accelerates Target RNA Cleavage by PIWI-Clade Argonaute Proteins. *Nature* **2022**, *608*, 618–625. [CrossRef]
29. Mateescu, R.G.; Garrick, D.J.; Reecy, J.M. Network Analysis Reveals Putative Genes Affecting Meat Quality in Angus Cattle. *Front. Genet.* **2017**, *8*, 171. [CrossRef]
30. Wu, J.; Wu, T.; Xie, X.; Niu, Q.; Zhao, Z.; Zhu, B.; Chen, Y.; Zhang, L.; Gao, X.; Niu, X.; et al. Genetic Association Analysis of Copy Number Variations for Meat Quality in Beef Cattle. *Foods* **2023**, *12*, 3986. [CrossRef]
31. Jiang, W.; Mooney, M.H.; Shirali, M. Unveiling the Genetic Landscape of Feed Efficiency in Holstein Dairy Heifers: Insights into Heritability, Genetic Markers, and Pathways via Meta-Analysis. *J. Anim. Sci.* **2024**, *102*, skae040. [CrossRef]
32. Carvalheiro, R.; Costilla, R.; Neves, H.H.R.; Albuquerque, L.G.; Moore, S.; Hayes, B.J. Unraveling Genetic Sensitivity of Beef Cattle to Environmental Variation Under Tropical Conditions. *Genet. Sel. Evol.* **2019**, *51*, 29. [CrossRef]
33. Li, Y.; Yang, M.; Lou, A.; Yun, J.; Ren, C.; Li, X.; Xia, G.; Nam, K.; Yoon, D.; Jin, H.; et al. Integrated Analysis of Expression Profiles with Meat Quality Traits in Cattle. *Sci. Rep.* **2022**, *12*, 5926. [CrossRef] [PubMed]
34. Zhang, W.; Li, J.; Guo, Y.; Zhang, L.; Xu, L.; Gao, X.; Zhu, B.; Gao, H.; Ni, H.; Chen, Y. Multi-Strategy Genome-Wide Association Studies Identify the DCAF16-NCAPG Region as a Susceptibility Locus for Average Daily Gain in Cattle. *Sci. Rep.* **2016**, *6*, 38073. [CrossRef]
35. Yang, T.; Luo, H.; Lou, W.; Chang, Y.; Brito, L.F.; Zhang, H.; Ma, L.; Hu, L.; Wang, A.; Li, S. Genetic Background of Hematological Parameters in Holstein Cattle Based on Genome-Wide Association and RNA Sequencing Analyses. *J. Dairy Sci.* **2024**, *107*, 24345. [CrossRef]
36. Martínez-Rocha, R.; Hidalgo, J.; Cesarani, A.; Ramírez-Valverde, R.; Núñez-Domínguez, R.; García-Muñiz, J.G.; Domínguez-Viveros, J. Assessment of Genetic Diversity, Runs of Homozygosity, and Signatures of Selection in Tropical Milking Criollo Cattle Using Pedigree and Genomic Data. *Genes* **2022**, *13*, 1896. [CrossRef]
37. Peripolli, E.; Stafuzza, N.B.; Munari, D.P.; Lima, A.L.F.; Irgang, R.; Machado, M.A.; Panetto, J.C.D.C.; Ventura, R.V.; Baldi, F.; da Silva, M.V.G.B. Assessment of Runs of Homozygosity Islands and Estimates of Genomic Inbreeding in Gyr (*Bos indicus*) Dairy Cattle. *BMC Genom.* **2018**, *19*, 34. [CrossRef]
38. Lozada-Soto, E.A.; Tiezzi, F.; Jiang, J.; Cole, J.B.; VanRaden, P.M.; Maltecca, C. Genomic Characterization of Autozygosity and Recent Inbreeding Trends in All Major Breeds of US Dairy Cattle. *J. Dairy Sci.* **2022**, *105*, 8956–8971. [CrossRef]
39. Betancur Zambrano, M.F.; Rincón Flórez, J.C.; Herrera Rios, A.C.; Solarte Portilla, C.E.; Bedoya Berrio, G. de J. Evaluation of Runs of Homozygosity and Genomic Inbreeding in Holstein Cattle from Colombia. *Semin. Ciências Agrárias* **2020**, *41* (Suppl. S2), 3397. [CrossRef]
40. Rosero, J.A.; Álvarez, L.A.; Muñoz, J.E.; Durán, C.V.; Rodas, Á.G. Allelic Frequency of the Kappa-Casein Gene in Colombian and Creole Cattle Breeds. *Rev. Colomb. Cienc. Pecu.* **2012**, *25*, 97–108. [CrossRef]
41. de Vitte, K.; Kerziene, S.; Klementavičiūtė, J.; de Vitte, M.; Mišeikienė, R.; Kudlinskienė, I.; Čepaitė, J.; Dilbiene, V.; Stankevičius, R. Relationship of β-casein Genotypes (A1A1, A1A2, and A2A2) to the Physicochemical Composition and Sensory Characteristics of Heifers' Milk. *J. Appl. Anim. Res.* **2022**, *50*, 161–166. [CrossRef]
42. Bisutti, V.; Pegolo, S.; Giannuzzi, D.; Mota, L.F.M.; Vanzin, A.; Toscano, A.; Trevisi, E.; Ajmone Marsan, P.; Brasca, M.; Cecchinato, A. The β-casein (CSN2) A2 Allelic Variant Alters Milk Protein Profile and Slightly Worsens Coagulation Properties in Holstein Heifers. *J. Dairy Sci.* **2022**, *105*, 3794–3809. [CrossRef]
43. Andrade Ortega, S.G.; López Flores, J.O.; Sánchez Gómez, I.E. Allelic and Genotypic Frequencies of the Kappa Casein and βeta Lactoglobulin Gene in Brown Swiss Cattle. *Calera* **2021**, *21*, 42–48. [CrossRef]

4. Bobbo, T.; Cipolat-Gotet, C.; Bittante, G.; Cecchinato, A. Short Communication: Association Analysis of Diacylglycerol Acyltransferase (DGAT1) Mutation on Chromosome 14 for Milk Yield and Composition Traits, Somatic Cell Score, and Coagulation Properties in Holstein Bulls. *J. Dairy Sci.* **2018**, *101*, 7075–7081. [CrossRef] [PubMed]
5. Echeverri Zuluaga, J.J.; Vázquez Araque, N.A.; Gallo Garcia, Y.M. Polimorfismo del gen de la Somatotropina Bovina y su asociación con características de importancia en la producción lechera. *Rev. Lasallista Investig.* **2010**, *7*, 67–76.
6. Bonilla, C.A.; Rubio, M.S.; Sifuentes, A.M.; Parra-Bracamonte, G.M.; Arellano, V.W.; Méndez, M.R.D.; Berruecos, J.M.; Ortiz, R. Association of CAPN1 316, CAPN1 4751 and TG5 Markers with Bovine Meat Quality Traits in Mexico. *Genet. Mol. Res.* **2010**, *9*, 2395–2405. [CrossRef]
7. Desgarennes Alcalá, C.M.; Del Moral Ventura, S.; Meza Villalvazo, V.M.; Peña Castro, J.M.; Zárate Martínez, J.P.; Zavaleta, J.A. Allelic and Genotypic Frequencies of CAPN1 and CAST, Two Genes Associated to Meat Quality, in Cattle from the Papaloapan Basin. *Nova Sci.* **2017**, *9*, 211–228. [CrossRef]
8. Casas, E.; White, S.N.; Wheeler, T.L.; Shackelford, S.D.; Koohmaraie, M.; Riley, D.G.; Chase, C.C., Jr.; Johnson, D.D.; Smith, T.P.L. Effects of Calpastatin and μ-Calpain Markers in Beef Cattle on Tenderness Traits. *J. Anim. Sci.* **2006**, *84*, 520–525. [CrossRef]

current issues in molecular biology

MDPI

Article

Loss of *Myostatin* Shapes the Transcriptomic and Epigenetic Landscapes Across Multiple Muscle Types in Cattle

Chao Hai, Xuefei Liu, Chunling Bai, Guanghua Su, Lei Yang * and Guangpeng Li *

State Key Laboratory of Reproductive Regulation and Breeding of Grassland Livestock, College of Life Science, Inner Mongolia University, Hohhot 010070, China; h15248037201@163.com (C.H.); liuxuefei1006@126.com (X.L.); chunling1980_0@163.com (C.B.); suguanghua0707@163.com (G.S.)
* Correspondence: leiyang@imu.edu.cn (L.Y.); gpengli@imu.edu.cn (G.L.)

Abstract: *Myostatin* (*MSTN*) is a critical regulator of muscle development. This study aimed to investigate the transcriptional and epigenetic mechanisms by which *MSTN* gene editing affects skeletal, cardiac, and smooth muscle function in cattle. The results showed that the *MSTN* gene-edited (MT) cattle skeletal muscle exhibited significantly larger myofiber cross-sectional areas ($p = 0.049$), accompanied by reduced shear force ($p = 0.044$), cooking loss rate ($p = 0.0029$), and pH ($p = 0.014$). Transcriptomic and whole-genome bisulfite sequencing (WGBS) revealed distinct expression and methylation patterns across muscle types. Notably, axon guidance signaling was identified as a shared enriched pathway in both transcriptional and CG/CHG/CHH methylation profiles of the gluteus. Further, 102 differentially expressed genes (DEGs) were commonly identified across all three muscle types; their KEGG enrichment included immune-related and cellular interaction pathways (e.g., antigen processing and presentation, and cell adhesion molecules), many of which intersect with axon guidance functions. Core regulators such as *SEMA3A*, *PLXNA1*, and *NTN1* were epigenetically modulated in MT gluteus and heart. These findings suggest that *MSTN* knockout remodels neuromuscular signaling through muscle-type-specific transcriptional and epigenetic reprogramming.

Keywords: *Myostatin* gene editing; meat quality traits; DNA methylation; axon guidance pathway

1. Introduction

Muscles play a fundamental role in animal physiology, facilitating movement, circulation, and various metabolic functions. Based on their structural and functional properties, animal muscles are categorized into three main types: skeletal muscle, heart muscle, and smooth muscle [1,2]. Among these, skeletal muscle accounts for approximately 40–50% of total body mass in livestock and directly determines meat yield and quality, making it a central target for genetic and nutritional interventions [3–5]. Skeletal muscle, responsible for voluntary movements, consists of multinucleated striated fibers controlled by the somatic nervous system. This muscle type exhibits high plasticity in response to genetic and environmental factors. In contrast, heart muscle is an involuntary, striated muscle exclusive to the heart, regulated by intrinsic pacemaker cells and the autonomic nervous system, ensuring continuous rhythmic contractions for blood circulation [6–8]. Smooth muscle, found in internal organs such as the digestive tract and blood vessels, is also under involuntary control and plays crucial roles in visceral functions, including peristalsis and vascular regulation [9,10].

Among the key regulators of muscle development and metabolism, myostatin (MSTN), a member of the transforming growth factor-beta (TGF-β) superfamily, acts as a negative regulator of muscle growth [11,12]. *MSTN* gene knockout (*MSTN*-KO) has been shown to promote increased muscle mass in multiple species, including cattle, pigs, and mice [13–15]. Mechanistically, *MSTN* inhibition leads to enhanced myoblast proliferation and reduced differentiation during the prenatal period, resulting in increased muscle fiber number (hyperplasia) [16]. Postnatally, *MSTN* suppression promotes satellite cell activation and fusion to existing fibers, contributing to muscle fiber enlargement (hypertrophy) [17]. Recent studies suggest that *MSTN* may be involved in heart muscle function and smooth muscle homeostasis through its regulatory effects on muscle-specific transcription factors and epigenetic modifications [18–20]. In addition to its well-characterized role in regulating skeletal muscle mass, *MSTN* also participates in the functional regulation of cardiac and smooth muscle tissues. This study systematically investigates these effects across multiple muscle types using integrative transcriptomic and epigenetic analyses, providing a comprehensive view of *MSTN*'s broader physiological roles. By uncovering distinct gene expression and DNA methylation patterns in heart and smooth muscle following *MSTN* knockout, our findings expand the understanding of *MSTN* as a systemic modulator of muscle function. These findings will enhance our understanding of *MSTN*'s role beyond skeletal muscle growth and contribute to the development of genetically optimized livestock with improved production traits.

2. Materials and Methods

2.1. Animals

All animal procedures were conducted in accordance with the ethical guidelines of the Committee on the Ethics of Animal Experiments at Inner Mongolia University (Approval number: IMU-CATTLE-2023-059).

The *MSTN* gene-edited (MT) cattle used in this study were generated in our laboratory through precise gene editing based on CRISPR/Cas9 technology, which targeted exon 1 of the bovine *MSTN* gene. Single guide RNAs (sgRNAs) were designed and transfected into bovine fetal fibroblasts (BFFs), and gene-edited positive clones were identified through sequencing. These edited cells were subsequently used as nuclear donors in somatic cell nuclear transfer (SCNT) to produce cloned animals carrying a biallelic knockout of the *MSTN* gene [21]. Compared to naturally double-muscled breeds such as Belgian Blue or Piedmontese, the use of gene-edited cattle enabled us to control for breed-specific genetic backgrounds and isolate the effects of *MSTN* disruption. Moreover, these gene-edited animals were more accessible within our institutional resources and have been previously shown by our group to exhibit muscle phenotypes similar to those observed in naturally occurring *MSTN* mutant cattle, thus representing a reliable and tractable model for studying *MSTN* function.

To assess the impact of *MSTN* knockout on muscle gene expression and epigenetic modifications, three MT cattle and three age-matched wild-type (WT) cattle from the same breed were selected for analysis. All animals, including MT and WT controls, were Mongolian cattle raised under uniform environmental and management conditions, including consistent feeding and housing practices, to ensure comparability between groups (Supplementary Tables S1 and S2). At approximately 24 months of age, cattle were humanely slaughtered, and samples of three distinct muscle tissues—skeletal muscle from the gluteus medius, cardiac muscle from the left ventricle of the heart, and smooth muscle from the middle segment of the esophagus—were collected. For each tissue type, three replicates (approximately 10 g per sample) were obtained and immediately preserved for subsequent meat quality analysis, as well as transcriptome (RNA-seq) and DNA methylation (whole-genome bisulfite sequencing) analyses (Figure 1). The gluteus muscle was selected as a representative skeletal muscle due to its anatomical consistency and suitability for transcriptomic and epigenetic analyses, offering reliable tissue characteristics. The esophagus was chosen to represent smooth muscle because of its relatively large size and accessibility during dissection.

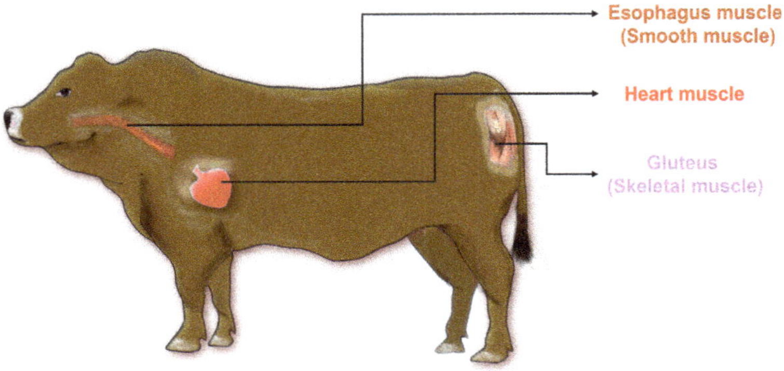

Figure 1. Muscle samples representing all three major muscle types in cattle were collected from distinct anatomical sites. Skeletal muscle samples were obtained from the gluteus muscle, while smooth muscle samples were collected from the esophageal smooth muscle.

2.2. Hematoxylin and Eosin (HE) Staining of Muscle Tissues

Three types of muscle tissues were collected from each animal within 2 h post-slaughter to ensure tissue integrity. Samples were cut into approximately 0.5 cm × 0.5 cm blocks using sterile scalpels and immediately fixed in 4% paraformaldehyde for 24 h at room temperature. Fixed tissues were then dehydrated through a graded ethanol series (70%, 80%, 95%, and 100%), cleared in xylene, and embedded in paraffin using an automatic tissue processor (Leica TP1020, Leica Biosystems, Wetzlar, Germany). The tissues were sectioned at a thickness of 2–4 µm using a Leica SM 2000 R microtome (Leica Microsystems, Wetzlar, Germany). Sections were then dried overnight at 37 °C.

For histological analysis, paraffin sections were deparaffinized in xylene and rehydrated through a graded ethanol series, and then they were stained with hematoxylin for 5 min, followed by eosin for 2 min, as previously described [22]. After staining, slides were dehydrated through graded alcohols, cleared in xylene, and coverslipped using mounting medium. Images were captured using a light microscope (Nikon Eclipse E100, Tokyo, Japan) equipped with a digital camera.

For histological analysis, transverse sections of gluteus, heart, and esophagus muscles were stained with HE and imaged under a light microscope (Nikon Eclipse E100, Japan). Digital images were analyzed using ImageJ software (v1.53t, NIH, Bethesda, MD, USA). The cross-sectional area of individual muscle fibers was delineated manually using the polygon selection tool and calculated in square micrometers (μm^2). At least 10 muscle fibers were measured per sample to ensure statistical reliability.

2.3. Measurement of Meat Tenderness-Related Traits

Meat quality evaluations, including physicochemical measurements (pH, shear force, cooking loss, and pressing loss), were limited to the gluteus muscle, which represents a commercially relevant skeletal muscle [23]. Cardiac and esophageal muscles were not subjected to these tests, as they are not conventional meat products. For each analysis, approximately 5 g of tissue was used, and all tests were performed in triplicate.

Muscle pH was measured 24 h postmortem after rigor resolution at 4 °C. A portable pH meter (PH5, Tianxiang Feiyu Technology Co., Ltd., Beijing, China) was inserted directly into the center of the muscle sample to obtain readings without homogenization or dilution [24].

Muscle shear force was determined using a texture analyzer (C-LM3B, Beijing Tianxiang Feiyu Technology Co., Ltd., Beijing, China), following the *Warner–Bratzler* shear force protocol recommended by the *American Meat Science Association* (AMSA, 2016). After aging for 7 days at 0–4 °C, cylindrical muscle samples (~1.5 cm diameter, 2.0 cm height, and ~5 g weight) were prepared. Prior to testing, all samples were equilibrated to room temperature (~20 °C). A Warner–Bratzler V-shaped blade was used to shear the samples at a constant crosshead speed of 200 mm/min. The shear force was applied perpendicularly to the fiber orientation. The shear force values are reported in kilograms-force (kg·f) [25].

Pressing loss was measured using a meat water-holding capacity analyzer (TenovoMeat-1, Tianxiang Feiyu Technology Co., Ltd., Beijing, China). Pressure was applied perpendicularly to the muscle fibers. The samples were sampled 24 h postmortem and weighed (W1). The sample was placed between 16 layers of filter paper on both sides and compressed between two hard plastic plates under a pressure of 68.66 kPa for 5 min. After compression, the sample was wiped dry and weighed again (W2). Pressing loss was calculated using the formula: pressing loss (%) = $[(W1 - W2)/W1] \times 100\%$ [24].

Cooking loss was assessed by sealing samples in plastic cooking pouches and immersing them in a 75 °C water bath (GD120, Grant, Shepreth, UK) for 45 min until the internal temperature reached 70 ± 2 °C. After cooling to room temperature, samples were wiped and re-weighed. Cooking loss was calculated as the percentage of weight loss during cooking [26].

2.4. Whole-Genome Bisulfite Sequencing and Data Analysis

Muscle tissue samples were dissected within 30 min postmortem and immediately frozen in liquid nitrogen, followed by long-term storage at −80 °C until genomic DNA extraction. No additional stabilizing reagents were used, as the downstream analysis focused on DNA methylation. Whole-genome bisulfite sequencing (WGBS) libraries were constructed following the protocol recommended by the sequencing service provider (BGI-Shenzhen, China). Sequencing was performed on the MGISEQ-2000 platform, generating 150 bp paired-end (PE150) reads.

The raw sequencing data were subjected to quality control to remove reads containing adapter contamination, poly-N, or low-quality bases. Specifically, reads were filtered based on the following criteria: (i) reads with more than 10% unknown bases (N); (ii) reads containing more than 50% bases with Phred quality score ≤ 5; and (iii) adapter-contaminated reads. After filtering, high-quality clean reads were retained for downstream analysis.

The clean reads were aligned to the Bos taurus reference genome (GCF_000003205.7_Btau_5.0.1, downloaded from NCBI) using Bismark (version 0.22.3) [27], a specialized bisulfite aligner. Bismark was run in default mode, and only uniquely mapped reads were retained. Methylation calls for cytosines in the CG, CHG, and CHH contexts were extracted using Bismark's methylation extractor module. The methylation level of each cytosine was calculated as the number of reads supporting methylated cytosines divided by the total number of reads covering the site.

Differentially methylated regions (DMRs) between MT and WT cattle were identified using the DSS (Dispersion Shrinkage for Sequencing data) R package. Regions were considered differentially methylated if they satisfied the criteria of a p-value < 0.05 and a minimum methylation fold change of ≥ 1.5. DMRs were annotated to nearby genes based on the Bos taurus reference genome annotation. KEGG pathway enrichment analysis of DMR-associated genes was performed using the clusterProfiler package (v4.6.2) in R [28].

2.5. Transcriptome Sequencing and Differential Expression Analysis of Muscle Tissues

Total RNA was isolated from skeletal muscle, heart muscle, and smooth muscle tissues using standard protocols, followed by integrity assessment to ensure high-quality RNA for library construction. Transcriptome libraries were generated and sequenced using the MGISEQ-2000 platform, producing paired-end reads of 150 bp in length.

Raw RNA-seq reads were aligned to the *Bos taurus* reference genome (GCF_000003205.7_Btau_5.0.1) using HISAT2 (v2.2.1) [29]. Transcript assembly and quantification were performed using StringTie (v2.2.1), and raw read counts were extracted using the prepDE.py script provided by the StringTie package. Differentially expressed genes (DEGs) between MT and WT groups were identified using DESeq2 (v1.38.3) with default parameters [30]. Genes with an adjusted p-value < 0.05 and $|\log_2 \text{fold change}| \geq 1$ were considered significantly differentially expressed. Ballgown (v2.30.0) was used only for transcript visualization and descriptive statistics of normalized expression values [31].

Principal component analysis (PCA) was performed based on $\log_2$-transformed fragments per kilobase of transcript per million mapped reads (FPKM) values. Genes with zero expression across all samples were removed. The PCA was conducted using the prcomp() function in R (v4.6.2), with centering and scaling enabled. The proportion of variance (PV) and standard deviation (SD) for each principal component were computed from the eigenvalues. PCA visualization was carried out using the ggplot2 packages to assess overall expression divergence among muscle types and genotypes. KEGG pathway enrichment analysis of the DEGs was performed using the *clusterProfiler* R package (v4.6.2) to explore the signaling pathways involved in gene expression differences across skeletal, cardiac, and smooth muscle tissues.

2.6. Data Statistical Analysis

Statistical analyses of muscle tenderness indicators (including pH, shear force, pressing loss rate, and cooking loss rate) and muscle area measurements were conducted using GraphPad Prism software (version 8.3.0, GraphPad Software, San Diego, CA, USA). DEGs identified by DESeq2 were visualized using GraphPad Prism. Differences between two muscle groups (MT vs. WT) were analyzed using unpaired two-tailed Student's t-tests. Results are presented as mean ± standard deviation (SD). Each group consisted of at least three biological replicates. $p < 0.05$ was considered statistically significant.

3. Results

3.1. Differences in Morphology Across Three Muscle Types Between MT and WT Cattle

The three selected muscles tissue—heart, esophagus (smooth muscle), and gluteus (skeletal muscle)—exhibited distinct morphological structures. Histologically, the heart muscle displayed densely packed and uniformly arranged fibers with narrow intercellular spaces, characteristic of highly oxidative tissue. The esophagus muscle featured a looser fiber arrangement, abundant connective tissue, and a relatively coarse fiber structure. In contrast, the gluteus muscle showed a more typical skeletal muscle architecture with relatively thick and aligned fibers, indicating its primary role in locomotion. Compared to WT cattle, MT cattle exhibited enlarged muscle fiber size in the gluteus muscle. However, no significant differences were observed in the heart and esophageal muscles between the two groups (Figure 2A). Quantification of muscle fiber area (Figure 2B) revealed a significant increase in the gluteus muscle of MT cattle compared to WT cattle (3263.06 ± 327.30 vs. 2512.56 ± 332.52 μm^2, $p = 0.049$). In the esophageal smooth muscle, where fibers are spindle-shaped and variably oriented, we specifically assessed the cross-sectional area of smooth muscle cells. No significant differences were detected in muscle fiber area in the heart (382.30 ± 24.36 vs. 363.94 ± 38.29, $p > 0.05$) and esophagus muscles (65.69 ± 13.58 vs. 65.54 ± 8.78, $p > 0.05$) between the two groups.

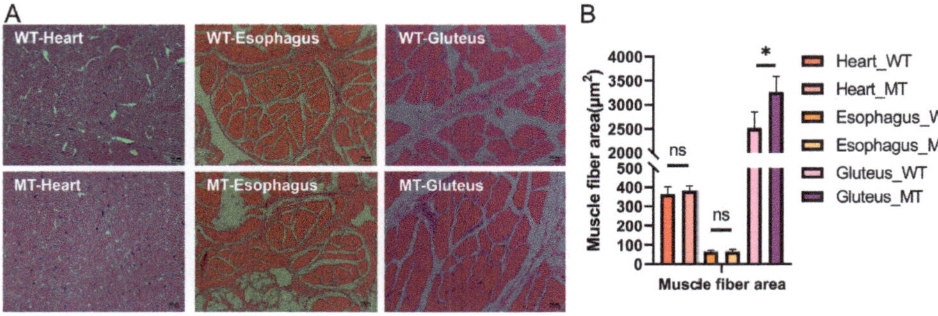

Figure 2. Histological morphology and muscle fiber area analysis of heart muscle, esophageal muscle, and gluteus muscle in MT and WT cattle. (**A**) Representative HE-stained sections showing tissue morphology of the three muscle types. (**B**) Quantification of muscle fiber area for each muscle type. $p < 0.05$ was considered statistically significant. * $p < 0.05$; ns, not significant.

These morphological differences were reflected in the physical and chemical properties of the gluteus muscle (Table 1). In terms of tenderness, measured by shear force, the gluteus muscle exhibited significantly lower shear force in MT cattle compared to WT (MT: 1.743 ± 0.460 kg·f; WT: 2.424 ± 0.443 kg·f; $p = 0.044$), indicating improved tenderness. The gluteus muscle also showed a lower cooking loss rate in MT cattle (MT: 30.99 ± 1.296%; WT: 35.66 ± 2.106, $p = 0.0029$), along with a significant decrease in pH in the MT group ($p = 0.0143$).

Table 1. Comparison of gluteus muscle tenderness parameters in MT and WT cattle.

Indicator	MT	WT	*p*-Value
Shear force, SF (kg·f)	1.743 ± 0.46	2.424 ± 0.443	0.044118
Extrusion loss rate (%)	5.406 ± 1.028	4.886 ± 0.765	0.390872
Cooking loss rate (%)	30.99 ± 1.296	35.66 ± 2.106	0.002877
pH	5.316 ± 0.09	5.634 ± 0.21	0.014338

Taken together, these data demonstrate that muscle-specific morphological traits are closely associated with differences in meat quality parameters, and that *MSTN* knockout may exert muscle-type-dependent effects on postmortem pH, water retention, and tenderness.

3.2. Comparative Analysis of DNA Methylation and Gene Expression in Different Muscle Types of Cattle

To further explore the relationship between methylation changes and transcriptional regulation, genome-wide DNA methylation profiles and RNA-seq data were analyzed in the heart, esophagus, and gluteus tissues from MT and WT cattle.

The cytosine coverage statistics show that the majority of cytosines in all tissues and groups (MT and WT) had relatively low sequencing depth (0–10×), with over 75% of sites falling into this range, while only a small proportion of sites had high coverage (>30×), indicating an overall moderate sequencing depth across samples (Supplementary Table S3). The methylation landscape of CG, CHG, and CHH contexts exhibited distinct distribution patterns across the three muscle types (heart, esophagus, and gluteus). CG methylation showed the highest levels (98.06–99.00%, Supplementary Table S4) among the three contexts, with a sharp decrease near the transcription start site (TSS), whereas CHG and CHH methylation levels remained relatively low and stable across genomic regions (Figure 3A).

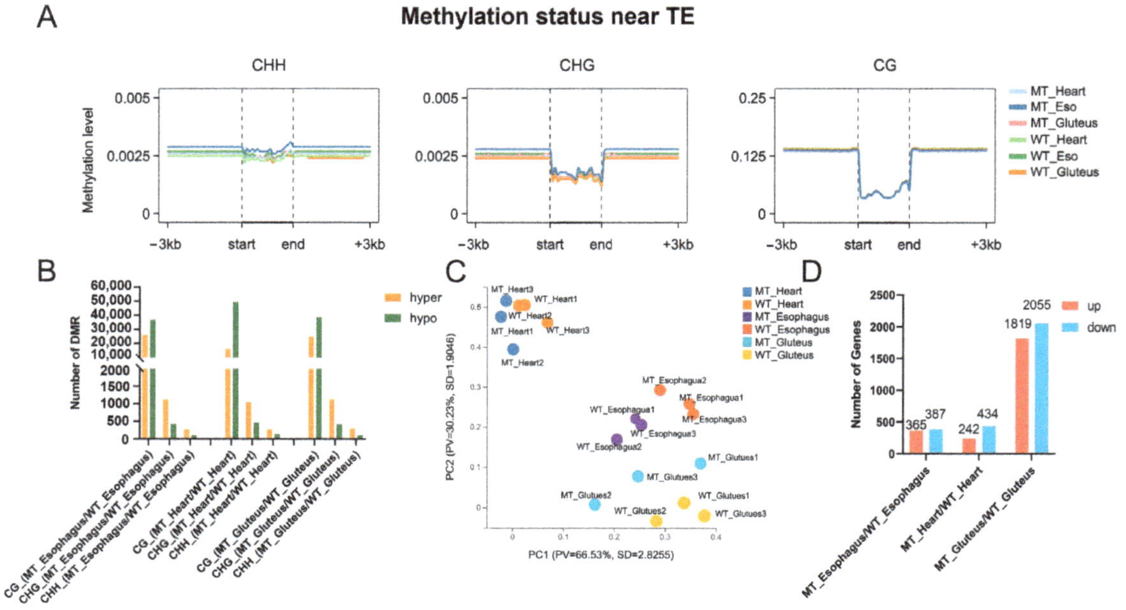

Figure 3. DNA methylation status and gene expression profiles in three muscle types of cattle. (**A**) DNA methylation levels near transposable elements (TEs) across skeletal muscle (gluteus), heart muscle, and smooth muscle (esophagus). (**B**) Number of differentially methylated regions (DMRs) identified among the three muscle types. (**C**) Principal component analysis (PCA) of transcriptomic profiles showing clustering patterns of skeletal muscle, heart, and smooth muscle. (**D**) Expression patterns of differentially expressed genes (DEGs) among the three muscle types.

Differentially methylated regions (DMRs) were identified between MT and WT groups, with CG context displaying the highest number of DMRs across all muscle types (Figure 3B). The gluteus muscle exhibited the greatest number of DMRs in all three muscles, followed by the heart and esophagus (Supplementary Table S4). In all cases, hypermethylation events were more prevalent than hypomethylation, suggesting a potential regulatory role in gene expression. This distribution also mirrors the morphological and meat quality results, where the gluteus muscle showed the most pronounced differences between MT and WT cattle, further highlighting its sensitivity to *MSTN* gene editing.

Transcriptome profiling via principal component analysis (PCA) revealed distinct clustering of tissue types, with MT and WT gluteus samples exhibiting clear separation, suggesting significant transcriptional reprogramming in muscle tissue. In contrast, heart and esophagus samples showed smaller genotype-driven variation. Notably, the heart samples clustered farther from the gluteus and esophagus tissues, indicating greater divergence in gene regulatory programs and expression profiles between cardiac muscle and the other two muscle types (Figure 3C). Consistently, differential expression analysis revealed the greatest number of differentially expressed genes (DEGs) in gluteus tissue, with 2055 genes downregulated and 1819 upregulated in MT compared to WT. Fewer DEGs were identified in the esophagus (365 upregulated and 387 downregulated) and heart (242 upregulated and 434 downregulated, Figure 3D), suggesting that *MSTN* deletion exerts a more pronounced transcriptional impact on skeletal muscle than on other muscle-rich tissues.

3.3. Comparative Functional Transcriptomic Analysis Across Muscle Types

KEGG pathway enrichment analysis revealed distinct functional differences among the three muscle types between MT and WT cattle. In the heart muscle, differentially expressed genes (DEGs) were significantly enriched in pathways related to complement and coagulation cascades, cholesterol metabolism, cell adhesion molecules, MAPK signaling pathway, and antigen processing and presentation (Figure 4A). In the esophageal muscle, significant enrichment was observed in pathways associated with phagosome formation, fatty acid degradation, AMPK signaling, butanoate metabolism, arginine biosynthesis, and circadian rhythm. Additionally, immune-related pathways, such as allograft rejection, and antigen processing and presentation, were also enriched (Figure 4B). In the gluteus muscle, DEGs were predominantly enriched in pathways related to oxidative phosphorylation, carbon metabolism, axon guidance, and muscle contraction. Notably, several pathways associated with neurodegenerative diseases, including Alzheimer's disease, Parkinson's disease, and Huntington's disease, were significantly enriched (Figure 4C).

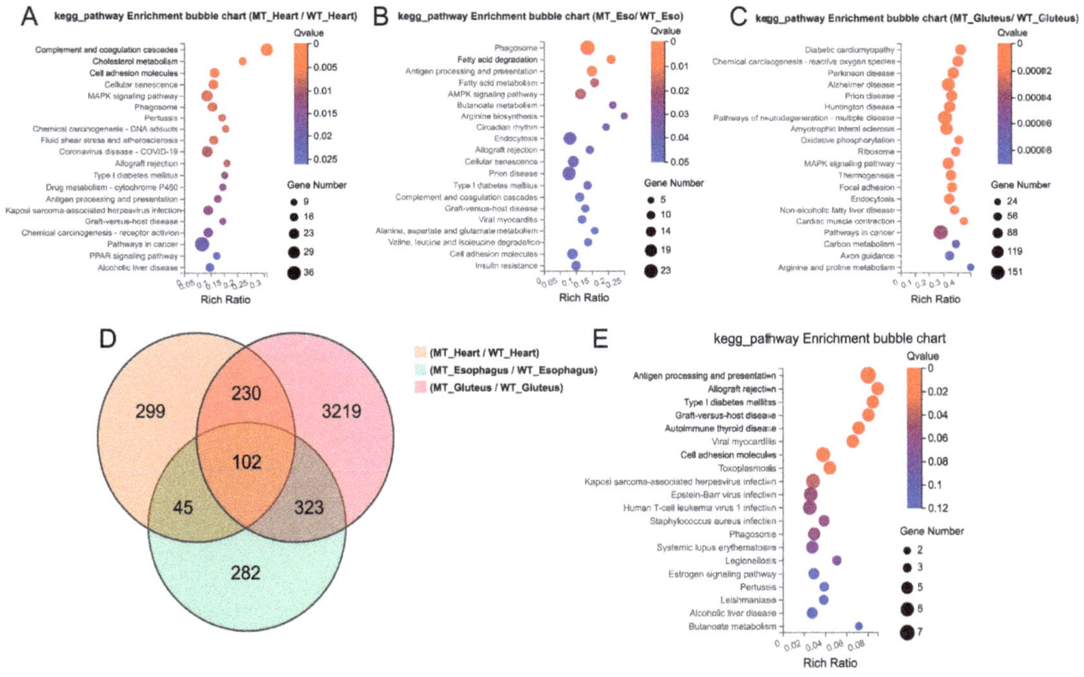

Figure 4. Functional enrichment analysis of differentially expressed genes (DEGs) among the three muscle types. (**A–C**) KEGG pathway bubble plots of DEGs in heart, smooth muscle, and skeletal muscle, respectively. (**D**) Venn diagram illustrating the overlap of DEGs among the three muscle types. (**E**) KEGG enrichment bubble plot of 102 common DEGs shared across all three muscle types.

A Venn diagram analysis (Figure 4D) revealed a total of 102 DEGs commonly shared across the three muscle types. KEGG analysis of these shared genes (Figure 4E) demonstrated significant enrichment in antigen processing and presentation, type I diabetes mellitus, cell adhesion molecules, and various viral infection and autoimmune-related pathways, indicating that *MSTN* gene-editing consistently affects immune and stress-response mechanisms in muscle tissues. Taken together, these results demonstrate that *MSTN* gene editing leads to muscle hypertrophy and alters key physiological and molecular features in a muscle type-specific manner, with shared transcriptional reprogramming involving immune regulation and metabolic remodeling.

3.4. Comparative Functional Analysis of DNA Methylation Patterns Among Muscle Types

In the gluteus muscle, differentially methylated sites (DMSs) in all three methylation contexts (CHH, CHG, and CG) were significantly enriched in axon guidance and axon regeneration, suggesting epigenetic regulation of neuronal signaling and muscle remodeling. Other enriched pathways included dopaminergic synapse, cholinergic synapse, insulin signaling, and PI3K-Akt signaling, indicating potential influences on neuromuscular function and metabolic regulation (Figure 5A–C). Consistent with previous gene expression analysis, KEGG enrichment of gluteus DEGs revealed pathways related to axon guidance, muscle contraction, and neurodegenerative diseases. These transcriptomic changes, aligned with methylation patterns, suggest coordinated regulation of neuronal and metabolic processes in MT cattle muscle.

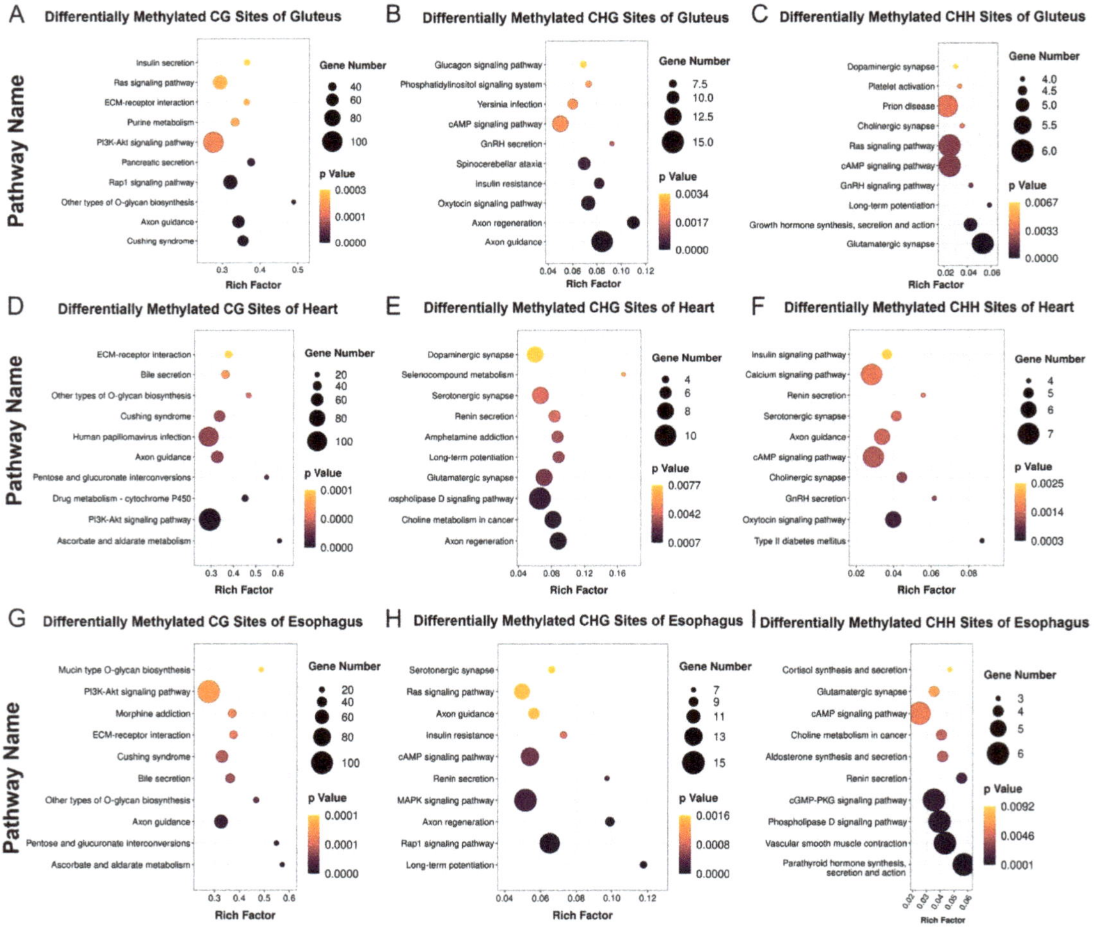

Figure 5. KEGG enrichment analysis of differentially methylated CH, CHG, and CHH sites between MT and WT cattle. (**A–C**) KEGG pathway enrichment of differentially methylated cytosines (mCH, mCHG, and mCHH) in skeletal muscle. (**D–F**) KEGG pathway enrichment of mCH, mCHG, and mCHH in heart muscle. (**G–I**) KEGG pathway enrichment of mCH, mCHG, and mCHH in smooth muscle.

In the heart muscle, DMSs were significantly enriched in axon guidance, as well as cAMP signaling, serotonergic synapse, and calcium signaling pathways (Figure 5D–F). These findings suggest that DNA methylation may epigenetically regulate cardiac contractility and intracellular signal transduction. Correspondingly, DEGs in the heart were enriched in pathways such as cell adhesion molecules and MAPK signaling, indicating potential transcriptional modulation.

In the esophagus (smooth muscle), DMSs across all three methylation contexts were consistently enriched in the axon guidance pathway, as well as in the PI3K-Akt signaling pathway, ECM–receptor interactions, and glycosaminoglycan biosynthesis. These findings suggest that *MSTN* gene editing may influence extracellular matrix remodeling and neuromuscular regulation in smooth muscle through DNA methylation. In line with this, the previously identified DEGs in smooth muscle were significantly enriched in pathways related to phagosome formation, fatty acid degradation, and AMPK signaling—metabolic and immune-associated processes (Figure 5G–I).

3.5. Differential Expression of Axon Guidance-Related Genes Across Muscle Types

To further explore *MSTN*-dependent transcriptional regulation of the axon guidance pathway across different muscle types, we utilized RNA-Seq data obtained from gluteus, heart, and esophageal muscles to quantify the expression levels of four representative axon guidance-related key genes, *SEMA3A*, *PLXNA1*, *ROBO1*, and *NTN1*, that were both significantly differentially expressed and functionally enriched in the axon guidance pathway (Figure 6).

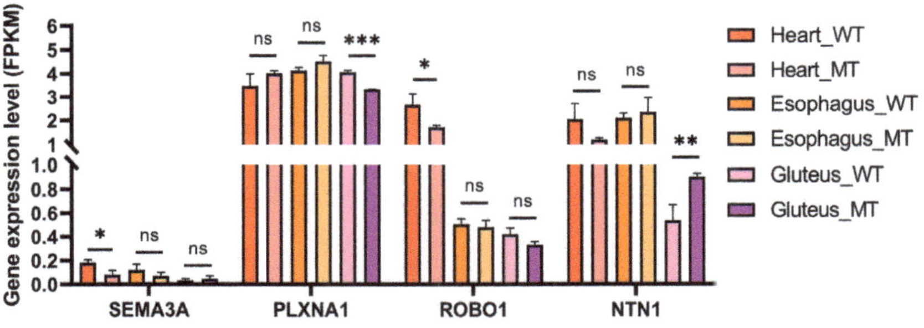

Figure 6. Expression levels of axon guidance-related genes in three muscle types of WT and MT cattle. FPKM values of *SEMA3A*, *PLXNA1*, *ROBO1*, and *NTN1* were compared between WT and MT cattle in heart, esophageal, and gluteus muscles. Data are presented as mean ± SD ($n = 3$ per group). * $p < 0.05$, ** $p < 0.01$, and *** $p < 0.001$; ns, not significant.

The expression of *SEMA3A* was significantly decreased in the heart muscle of MT cattle compared with WT ($p < 0.05$), while no significant changes were observed in esophageal or gluteus muscles. *PLXNA1* exhibited a significant decrease only in gluteus muscle ($p < 0.001$), whereas its expression remained stable in the heart and esophagus across genotypes. *ROBO1* showed a downward trend in MT cattle in all muscle types, but without statistical significance.

Interestingly, *NTN1* expression was significantly upregulated in gluteus muscle of MT cattle ($p < 0.01$), while no changes were detected in heart and esophagus. These results suggest muscle-specific regulatory effects of myostatin deficiency on axon guidance gene expression, particularly in skeletal muscle.

4. Discussion

4.1. Morphology and Meat Quality of MSTN-Edited Cattle Muscles

Histological examination using HE staining revealed distinct fiber morphology among skeletal, cardiac, and smooth muscles. In particular, the cross-sectional area of muscle fibers was significantly increased in MT cattle, most prominently in skeletal muscle. This finding is in line with *MSTN*'s established role as a negative regulator of myogenesis. *MSTN* signals through activin type IIB receptors, activating SMAD2/3 transcription factors, which suppress myoblast proliferation and differentiation [13,32]. Disruption of *MSTN* removes this inhibitory pathway, thereby promoting muscle fiber hypertrophy through enhanced satellite cell activity and protein synthesis [33].

Compared to WT cattle, MT cattle demonstrated significantly improved tenderness and reduced cooking loss in the gluteus muscle. These changes are likely associated with hypertrophic muscle fiber growth and altered extracellular matrix remodeling following *MSTN* disruption: Myostatin normally inhibits satellite cell proliferation and regulates connective tissue deposition [34]. Thus, its deletion may enhance muscle fiber diameter and affect collagen content, potentially contributing to improved tenderness and water-holding capacity. Although we did not directly quantify intramuscular collagen or connective tissue density in this study, previous studies have reported reduced perimysial collagen content and lower collagen cross-linking in *MSTN*-deficient animals [35]. These findings support the hypothesis that lower shear force values observed in *MSTN*-KO cattle may be partially attributed to such structural changes. Further investigation using histological or biochemical collagen assays is warranted to validate this speculation.

In summary, these findings suggest that *MSTN* editing exerts measurable effects on muscle morphology and skeletal muscle meat-quality parameters.

4.2. DNA Methylation and Transcriptional Modulation of Muscle Function

In addition to phenotypic changes in muscle morphology, it is important to understand the underlying molecular mechanisms. DNA methylation is a key epigenetic modification regulating gene transcription and is known to affect muscle development, metabolism, and differentiation [36,37].

The axon guidance pathway, although traditionally associated with neural development, has been increasingly recognized for its role in muscle fiber formation, neuromuscular junction establishment, and cell–cell communication within muscle tissue. These changes were especially prominent in skeletal muscle, consistent with its higher plasticity and relevance to meat production [38,39]. Mechanistically, this may be explained by the overall lower DNA methylation level observed in skeletal muscle, particularly in promoter regions. Such hypomethylation is known to support a more open chromatin state and enhanced transcriptional activity [40]. These regulatory layers may contribute to the activation of growth-related pathways, increased protein synthesis, and improved muscle quality traits in *MSTN*-KO animals. Although post-translational mechanisms were not directly assessed, their role in fine-tuning these processes cannot be excluded and warrants future study.

Interestingly, recent studies have shown that *MSTN* also plays important roles in axon guidance signaling. MSTN and its homolog GDF11 are expressed in neurons and have been reported to affect synapse formation and neuronal morphology [41]. For example, in Drosophila, the MSTN homolog Myoglianin (MYO) regulates synaptic strength and composition at the neuromuscular junction via a Smad2-dependent mechanism. Moreover, MSTN and GDF11 were found to inhibit neurite outgrowth and synaptogenesis in cultured rat cortical neurons, suggesting a conserved role in neurodevelopmental regulation [42]. These findings support the hypothesis that epigenetic and transcriptional modulation of axon guidance genes may influence muscle development, potentially contributing to observed differences in muscle structure. Several axon guidance molecules, such as *SEMA3A*, *NTN1*, and *EPH/ephrin* families, have been shown to regulate myoblast adhesion, differentiation, and muscle innervation [43–45]. However, the direct impact of these molecular changes on meat quality traits such as tenderness or cooking loss remains to be further investigated.

The enrichment of cAMP and MAPK-related pathways in cardiac tissue suggests the presence of a signaling cascade analogous to the cAMP/PKA–MAPK axis, known for its role in synaptic plasticity in neurons. PKA may inhibit Ras activity, indicating competitive regulation between PKA- and Ras-mediated MAPK activation [46,47]. This convergence may influence cardiac plasticity and contractile function. In the esophagus, coordinated changes in methylation and expression implicate *MSTN* in metabolic adaptation and immune regulation, potentially via AMPK/NOX4/PI3K/AKT signaling, which has established links to glucose homeostasis and insulin resistance [48,49].

To further investigate this connection, we analyzed the expression profiles of four representative axon guidance-related genes—*SEMA3A*, *PLXNA1*, *ROBO1*, and *NTN1*—across heart, esophageal, and gluteus muscles. Notably, *SEMA3A*, a well-characterized inhibitory cue for axonal pathfinding and muscle fiber outgrowth [50], was significantly downregulated in the heart of MT cattle. This downregulation may reflect a disruption of *MSTN*-dependent regulatory pathways. MSTN, as a member of the TGF-β superfamily, signals primarily through SMAD2/3-mediated transcriptional mechanisms. Previous studies have demonstrated that *SEMA3A* transcription is positively regulated by TGF-β signaling via SMAD binding elements in its promoter region [51,52]. Therefore, the loss of *MSTN* could lead to decreased SMAD2/3 activation, reducing transcriptional activation of *SEMA3A*. This reduced expression of *SEMA3A* could relieve its inhibitory effect on cardiomyocyte outgrowth and cytoskeletal remodeling, thereby contributing to tissue-specific adaptations in muscle structure or performance. Interestingly, both *ROBO1* and *SEMA3A* were upregulated in the cardiac muscle of MT cattle, suggesting a coordinated regulatory mechanism. This pattern is consistent with previous findings that mutations in axon guidance-related genes such as *ROBO1* and *SEMA3A* are often associated with congenital heart defects and diaphragm malformations, implying that these genes play critical roles in cardiac morphogenesis and structural integrity [53,54]. *PLXNA1*, a receptor for *SEMA3A*, exhibited relatively stable expression across most tissues but was significantly downregulated in the gluteus muscle of MT cattle. Although the precise significance of *PLXNA1* and *SEMA3A* expression patterns across different muscle types remains unclear, the observed gluteus-specific suppression of *PLXNA1* may attenuate *SEMA3A*-mediated repulsive signaling, potentially reducing inhibitory cues and thereby creating a more permissive environment for axonal extension or myofiber growth [55]. Importantly, *NTN1* (netrin-1), a guidance molecule known for its dual role in mediating both attractive and repulsive signaling, is significantly upregulated in the gluteal muscle of MT cattle. This suggests a potential role for *NTN1* in promoting muscle innervation and maintaining fiber integrity. Notably, this alteration has been associated with intramuscular fat deposition in pigs, indicating that *NTN1* may synergize with *MSTN* to regulate both adipogenesis and muscle development [56,57]. These expression changes support the hypothesis that *MSTN* deletion induces a coordinated regulation of axon guidance genes, contributing to improved muscle innervation, fiber integrity, and enhanced muscle structure and function.

We acknowledge that the current study is limited by the relatively small sample size, which is mainly due to the challenges in producing and maintaining *MSTN* gene-edited cattle with consistent genetic backgrounds. In addition, the study lacks direct functional validation of these genes in muscle development. The relationships between altered axon guidance gene expression and muscle phenotype remain correlative, and the precise mechanisms through which these molecular changes influence myofiber formation, fiber type composition, or innervation remain to be elucidated. Additionally, while the integrative transcriptomic and epigenetic analyses provide strong indirect support, experimental validation using loss- or gain-of-function assays in cellular or animal models would be essential to confirm the causal roles of these genes and pathways.

5. Conclusions

Our findings reveal that *MSTN* gene editing in cattle enhances skeletal muscle structure and tenderness, as evidenced by physicochemical analyses performed on the gluteus muscle. These results indicate improvements in specific physical characteristics of skeletal muscle but do not encompass sensory, commercial, or consumer-based measures of meat quality. Additionally, *MSTN* disruption induced distinct transcriptional and epigenetic alterations in cardiac and smooth muscles. The identification of axon guidance as a shared regulatory axis underscores the broader role of *MSTN* in muscle system regulation beyond skeletal muscle growth.

Supplementary Materials: The following supporting information can be downloaded at https://www.mdpi.com/article/10.3390/cimb47060431/s1.

Author Contributions: Conceptualization, L.Y. and G.L.; methodology, C.H.; software, C.H.; investigation, C.H. and X.L.; resources, L.Y. and G.L.; data curation, C.H., C.B. and G.S.; writing—original draft preparation, C.H.; writing—review and editing, L.Y. and G.L.; visualization, C.H.; supervision, X.L., C.B. and G.S.; project administration, L.Y. and G.L.; funding acquisition, L.Y. and G.L. All authors have read and agreed to the published version of the manuscript.

Funding: This study was supported by the Scientific and Technological Innovation 2030 (2023ZD0404803); National Natural Science Foundation of China (32360837, 32341052); Inner Mongolia Open Competition Projects (2022JBGS0025); Inner Mongolia Science and Technology Leading Team (2022LJRC0006); Inner Mongolia New technology for Grassland Livestock Biological Breeding (2025KYPT0101); Mongolia Science and Technology Major Projects (2021ZD0009, 2021ZD0008, 2022ZD0008, and 2023KJHZ0028); Inner Mongolia Young Talents Projects (NJYT23138); Inner Mongolia Natural Science Foundation (2023MS03004); Central Government Guides Development (2022ZY0212); National Agricultural Science and Technology Project (NK2022130203); Collaborative Innovation among Universities in Hohhot (XTCX202306); Ministry of Education Engineering Centre Project (JYBGC-SYS2022); and Xinjiang Uygur Science and Technology Major Project (2023A0201116); TongLiao Open Competition Projects (TL2024TW0020103).

Institutional Review Board Statement: All protocols were approved by the Institutional Animal Care and Use Committee at Inner Mongolia University (IMU-CATTLE-2023-059).

Informed Consent Statement: Not applicable.

Data Availability Statement: The original contributions presented in the study are included in the article/Supplementary Materials; further inquiries can be directed to the corresponding author.

Conflicts of Interest: The authors declare no conflicts of interest.

References

1. Katzemich, A.; Liao, K.A.; Czerniecki, S.; Schöck, F. Alp/Enigma family proteins cooperate in Z-disc formation and myofibril assembly. *PLoS Genet.* **2013**, *9*, e1003342. [CrossRef] [PubMed]
2. Katter, K.; Geurts, A.M.; Hoffmann, O.; Mátés, L.; Landa, V.; Hiripi, L.; Moreno, C.; Lazar, J.; Bashir, S.; Zidek, V.; et al. Transposon mediated transgenesis, transgenic rescue, and tissue-specific gene expression in rodents and rabbits. *FASEB J.* **2013**, *27*, 930–941. [CrossRef] [PubMed]
3. Reza, M.M.; Subramaniyam, N.; Sim, C.M.; Ge, X.; Sathiakumar, D.; McFarlane, C.; Sharma, M.; Kambadur, R. Irisin is a pro-myogenic factor that induces skeletal muscle hypertrophy and rescues denervation-induced atrophy. *Nat. Commun.* **2017**, *8*, 1104. [CrossRef]

4. Bellmann, J.; Goswami, R.Y.; Girardo, S.; Rein, N.; Hosseinzadeh, Z.; Hicks, M.R.; Busskamp, V.; Pyle, A.D.; Werner, C.; Sterneckert, J. A customizable microfluidic platform for medium-throughput modeling of neuromuscular circuits. *Biomaterials* **2019**, *225*, 119537. [CrossRef]
5. Frontera, W.R.; Ochala, J. Skeletal muscle: A brief review of structure and function. *Calcif. Tissue Int.* **2015**, *96*, 183–195. [CrossRef]
6. Pieperhoff, S. Gene Mutations Resulting in the Development of ARVC/D Could Affect Cells of the Cardiac Conduction System. *Front. Physiol.* **2012**, *3*, 22. [CrossRef]
7. Chang, A.N.; Kamm, K.E.; Stull, J.T. Role of myosin light chain phosphatase in cardiac physiology and pathophysiology. *J. Mol. Cell. Cardiol.* **2016**, *101*, 35–43. [CrossRef] [PubMed]
8. Günther, A.; Baumann, A. Distinct expression patterns of HCN channels in HL-1 cardiomyocytes. *BMC Cell Biol.* **2015**, *16*, 18. [CrossRef]
9. Hemming, M.L.; Fan, C.; Raut, C.P.; Demetri, G.D.; Armstrong, S.A.; Sicinska, E.; George, S. Oncogenic Gene-Expression Programs in Leiomyosarcoma and Characterization of Conventional, Inflammatory, and Uterogenic Subtypes. *Mol. Cancer Res.* **2020**, *18*, 1302–1314. [CrossRef]
10. Webb, R.C. Smooth muscle contraction and relaxation. *Adv. Physiol. Educ.* **2003**, *27*, 201–206. [CrossRef] [PubMed]
11. Tang, R.; Harasymowicz, N.S.; Wu, C.L.; Collins, K.H.; Choi, Y.R.; Oswald, S.J.; Guilak, F. Gene therapy for follistatin mitigates systemic metabolic inflammation and post-traumatic arthritis in high-fat diet-induced obesity. *Sci. Adv.* **2020**, *6*, eaaz7492. [CrossRef] [PubMed]
12. Liu, Z.; Chen, M.; Chen, S.; Deng, J.; Song, Y.; Lai, L.; Li, Z. Highly efficient RNA-guided base editing in rabbit. *Nat. Commun.* **2018**, *9*, 2717. [CrossRef] [PubMed]
13. McPherron, A.C.; Lee, S.J. Double muscling in cattle due to mutations in the myostatin gene. *Proc. Natl. Acad. Sci. USA* **1997**, *94*, 12457–12461. [CrossRef]
14. McPherron, A.C.; Lawler, A.M.; Lee, S.J. Regulation of skeletal muscle mass in mice by a new TGF-beta superfamily member. *Nature* **1997**, *387*, 83–90. [CrossRef] [PubMed]
15. Cui, W.T.; Xiao, G.J.; Jiang, S.W.; Qian, L.L.; Cai, C.B.; Li, B.; Xie, S.S.; Gao, T.; Li, K. Effect of ZFN-edited myostatin loss-of-function mutation on gut microbiota in Meishan pigs. *PLoS ONE* **2019**, *14*, e0210619. [CrossRef]
16. Wang, X.H.; Mitch, W.E. Mechanisms of muscle wasting in chronic kidney disease. *Nat. Rev. Nephrol.* **2014**, *10*, 504–516. [CrossRef]
17. Nielsen, C.; Potter, R.M.; Borowy, C.; Jacinto, K.; Kumar, R.; Carlson, C.G. Postnatal Hyperplasic Effects of ActRIIB Blockade in a Severely Dystrophic Muscle. *J. Cell. Physiol.* **2017**, *232*, 1774–1793. [CrossRef]
18. Gu, M.; Wang, S.; Di, A.; Wu, D.; Hai, C.; Liu, X.; Bai, C.; Su, G.; Yang, L.; Li, G. Combined Transcriptome and Metabolome Analysis of Smooth Muscle of Myostatin Knockout Cattle. *Int. J. Mol. Sci.* **2023**, *24*, 8120. [CrossRef]
19. Luo, Z.B.; Luo, Q.R.; Xuan, M.F.; Han, S.Z.; Wang, J.X.; Guo, Q.; Choe, Y.G.; Jin, S.S.; Kang, J.D.; Yin, X.J. Comparison of internal organs between myostatin mutant and wild-type piglets. *J. Sci. Food Agric.* **2019**, *99*, 6788–6795. [CrossRef]
20. Gu, M.; Zhou, X.; Zhu, L.; Gao, Y.; Gao, L.; Bai, C.; Yang, L.; Li, G. Myostatin Mutation Promotes Glycolysis by Increasing Phosphorylation of Phosphofructokinase via Activation of PDE5A-cGMP-PKG in Cattle Heart. *Front. Cell Dev. Biol.* **2021**, *9*, 774185. [CrossRef]
21. Zhao, Y.; Yang, L.; Su, G.; Wei, Z.; Liu, X.; Song, L.; Hai, C.; Wu, D.; Hao, Z.; Wu, Y.; et al. Growth Traits and Sperm Proteomics Analyses of Myostatin Gene-Edited Chinese Yellow Cattle. *Life* **2022**, *12*, 627. [CrossRef] [PubMed]
22. Feldman, A.T.; Wolfe, D. Tissue processing and hematoxylin and eosin staining. *Methods Mol. Biol.* **2014**, *1180*, 31–43. [CrossRef] [PubMed]
23. Luo, S.; Liu, Y.; Bu, L.; Wang, D.; Wen, Z.; Yang, Y.; Xu, Y.; Wu, D.; Li, G.; Yang, L. The Impact of MSTN Gene Editing on Meat Quality and Metabolomics: A Comparative Study Among Three Breeds of MSTN-Edited and Non-Edited Cattle. *Animals* **2025**, *15*, 47. [CrossRef]
24. Weng, K.; Huo, W.; Li, Y.; Zhang, Y.; Zhang, Y.; Chen, G.; Xu, Q. Fiber characteristics and meat quality of different muscular tissues from slow- and fast-growing broilers. *Poult. Sci.* **2022**, *101*, 101537. [CrossRef]
25. Van Oeckel, M.J.; Warnants, N.; Boucqué, C.V. Pork tenderness estimation by taste panel, Warner-Bratzler shear force and on-line methods. *Meat Sci.* **1999**, *53*, 259–267. [CrossRef]
26. Wang, Y.; Tian, X.; Liu, X.; Zhang, Y.; Zhao, K.; Zhang, K.; Wang, W. Effects of different cooking methods on physicochemical, textural properties of yak meat and its changes with intramuscular connective tissue during in vitro digestion. *Food Chem.* **2023**, *422*, 136188. [CrossRef]
27. Krueger, F.; Andrews, S.R. Bismark: A flexible aligner and methylation caller for Bisulfite-Seq applications. *Bioinformatics* **2011**, *27*, 1571–1572. [CrossRef] [PubMed]
28. Yu, G.; Wang, L.G.; Han, Y.; He, Q.Y. clusterProfiler: An R package for comparing biological themes among gene clusters. *Omics* **2012**, *16*, 284–287. [CrossRef]
29. Thakur, V. RNA-Seq Data Analysis for Differential Gene Expression Using HISAT2-StringTie-Ballgown Pipeline. *Methods Mol. Biol.* **2024**, *2812*, 101–113. [CrossRef]

30. Love, M.I.; Huber, W.; Anders, S. Moderated estimation of fold change and dispersion for RNA-seq data with DESeq2. *Genome Biol.* **2014**, *15*, 550. [CrossRef]
31. Pertea, M.; Kim, D.; Pertea, G.M.; Leek, J.T.; Salzberg, S.L. Transcript-level expression analysis of RNA-seq experiments with HISAT, StringTie and Ballgown. *Nat. Protoc.* **2016**, *11*, 1650–1667. [CrossRef] [PubMed]
32. Wang, H.; Guo, S.; Gao, H.; Ding, J.; Li, H.; Kong, X.; Zhang, S.; He, M.; Feng, Y.; Wu, W.; et al. Myostatin regulates energy homeostasis through autocrine- and paracrine-mediated microenvironment communication. *J. Clin. Investig.* **2024**, *134*, e178303. [CrossRef] [PubMed]
33. McFarlane, C.; Vajjala, A.; Arigela, H.; Lokireddy, S.; Ge, X.; Bonala, S.; Manickam, R.; Kambadur, R.; Sharma, M. Negative auto-regulation of myostatin expression is mediated by Smad3 and microRNA-27. *PLoS ONE* **2014**, *9*, e87687. [CrossRef]
34. Paek, H.J.; Quan, B.H.; Choe, H.M.; Li, Z.Y.; Yin, X.J. Myostatin deficiency decreases cardiac extracellular matrix in pigs. *Transgenic Res.* **2022**, *31*, 553–565. [CrossRef] [PubMed]
35. Paek, H.J.; Luo, Z.B.; Choe, H.M.; Quan, B.H.; Gao, K.; Han, S.Z.; Li, Z.Y.; Kang, J.D.; Yin, X.J. Association of myostatin deficiency with collagen related disease-umbilical hernia and tippy toe standing in pigs. *Transgenic Res.* **2021**, *30*, 663–674. [CrossRef]
36. Kottakis, F.; Nicolay, B.N.; Roumane, A.; Karnik, R.; Gu, H.; Nagle, J.M.; Boukhali, M.; Hayward, M.C.; Li, Y.Y.; Chen, T.; et al. LKB1 loss links serine metabolism to DNA methylation and tumorigenesis. *Nature* **2016**, *539*, 390–395. [CrossRef]
37. Fan, Y.; Liang, Y.; Deng, K.; Zhang, Z.; Zhang, G.; Zhang, Y.; Wang, F. Analysis of DNA methylation profiles during sheep skeletal muscle development using whole-genome bisulfite sequencing. *BMC Genom.* **2020**, *21*, 327. [CrossRef]
38. Whitman, M.C.; Engle, E.C. Ocular congenital cranial dysinnervation disorders (CCDDs): Insights into axon growth and guidance. *Hum. Mol. Genet.* **2017**, *26*, R37–R44. [CrossRef]
39. Martins, A.F.; Xavier Neto, J.; Azambuja, A.; Sereno, M.L.; Figueira, A.; Campos-Junior, P.H.; Rosário, M.F.; Toledo, C.B.; Silva, G.A.; Kitten, G.T.; et al. Repulsive Guidance Molecules a, b and c Are Skeletal Muscle Proteins, and Repulsive Guidance Molecule a Promotes Cellular Hypertrophy and Is Necessary for Myotube Fusion. *Cells Tissues Organs* **2015**, *200*, 326–338. [CrossRef]
40. Riebler, A.; Menigatti, M.; Song, J.Z.; Statham, A.L.; Stirzaker, C.; Mahmud, N.; Mein, C.A.; Clark, S.J.; Robinson, M.D. BayMeth: Improved DNA methylation quantification for affinity capture sequencing data using a flexible Bayesian approach. *Genome Biol.* **2014**, *15*, R35. [CrossRef]
41. Abati, E.; Manini, A.; Comi, G.P.; Corti, S. Inhibition of myostatin and related signaling pathways for the treatment of muscle atrophy in motor neuron diseases. *Cell. Mol. Life Sci.* **2022**, *79*, 374. [CrossRef] [PubMed]
42. Augustin, H.; McGourty, K.; Steinert, J.R.; Cochemé, H.M.; Adcott, J.; Cabecinha, M.; Vincent, A.; Halff, E.F.; Kittler, J.T.; Boucrot, E.; et al. Myostatin-like proteins regulate synaptic function and neuronal morphology. *Development* **2017**, *144*, 2445–2455. [CrossRef] [PubMed]
43. McLoon, L.K. A new role for satellite cells: Control of reinnervation after muscle injury by semaphorin 3A. Focus on "Possible implication of satellite cells in regenerative motoneuritogenesis: HGF upregulates neural chemorepellent Sema3A during myogenic differentiation". *Am. J. Physiol. Cell Physiol.* **2009**, *297*, C227–C230. [CrossRef]
44. Li, S.; Karri, D.; Sanchez-Ortiz, E.; Jaichander, P.; Bassel-Duby, R.; Liu, N.; Olson, E.N. Sema3a-Nrp1 Signaling Mediates Fast-Twitch Myofiber Specificity of Tw2^{+} Cells. *Dev. Cell* **2019**, *51*, 89–98.e84. [CrossRef]
45. Stark, D.A.; Karvas, R.M.; Siegel, A.L.; Cornelison, D.D. Eph/ephrin interactions modulate muscle satellite cell motility and patterning. *Development* **2011**, *138*, 5279–5289. [CrossRef]
46. Waltereit, R.; Weller, M. Signaling from cAMP/PKA to MAPK and synaptic plasticity. *Mol. Neurobiol.* **2003**, *27*, 99–106. [CrossRef]
47. Vogt Weisenhorn, D.M.; Roback, L.J.; Kwon, J.H.; Wainer, B.H. Coupling of cAMP/PKA and MAPK signaling in neuronal cells is dependent on developmental stage. *Exp. Neurol.* **2001**, *169*, 44–55. [CrossRef] [PubMed]
48. Schultze, S.M.; Hemmings, B.A.; Niessen, M.; Tschopp, O. PI3K/AKT, MAPK and AMPK signalling: Protein kinases in glucose homeostasis. *Expert Rev. Mol. Med.* **2012**, *14*, e1. [CrossRef]
49. Yan, J.; Wang, C.; Jin, Y.; Meng, Q.; Liu, Q.; Liu, Z.; Liu, K.; Sun, H. Catalpol ameliorates hepatic insulin resistance in type 2 diabetes through acting on AMPK/NOX4/PI3K/AKT pathway. *Pharmacol. Res.* **2018**, *130*, 466–480. [CrossRef]
50. Daneshvar, N.; Anderson, J.E. Preliminary Study of S100B and Sema3A Expression Patterns in Regenerating Muscle Implicates P75-Expressing Terminal Schwann Cells and Muscle Satellite Cells in Neuromuscular Junction Restoration. *Front. Cell Dev. Biol.* **2022**, *10*, 874756. [CrossRef]
51. Nakao, A.; Imamura, T.; Souchelnytskyi, S.; Kawabata, M.; Ishisaki, A.; Oeda, E.; Tamaki, K.; Hanai, J.; Heldin, C.H.; Miyazono, K.; et al. TGF-beta receptor-mediated signalling through Smad2, Smad3 and Smad4. *EMBO J.* **1997**, *16*, 5353–5362. [CrossRef] [PubMed]
52. Lai, Y.J.; Chang, S.H.; Tung, Y.C.; Chang, G.J.; Almeida, C.; Chen, W.J.; Yeh, Y.H.; Tsai, F.C. Naringin activates semaphorin 3A to ameliorate TGF-β-induced endothelial-to-mesenchymal transition related to atrial fibrillation. *J. Cell. Physiol.* **2024**, *239*, e31248. [CrossRef] [PubMed]
53. Kardon, G.; Ackerman, K.G.; McCulley, D.J.; Shen, Y.; Wynn, J.; Shang, L.; Bogenschutz, E.; Sun, X.; Chung, W.K. Congenital diaphragmatic hernias: From genes to mechanisms to therapies. *Dis. Model. Mech.* **2017**, *10*, 955–970. [CrossRef]

4. Johnson, V.; Junge, H.J.; Chen, Z. Temporal regulation of axonal repulsion by alternative splicing of a conserved microexon in mammalian Robo1 and Robo2. *eLife* **2019**, *8*, e46042. [CrossRef] [PubMed]
5. Oleari, R.; Caramello, A.; Campinoti, S.; Lettieri, A.; Ioannou, E.; Paganoni, A.; Fantin, A.; Cariboni, A.; Ruhrberg, C. PLXNA1 and PLXNA3 cooperate to pattern the nasal axons that guide gonadotropin-releasing hormone neurons. *Development* **2019**, *146*, dev176461. [CrossRef]
6. Wang, L.; Zhao, L.; Zhang, L.; Liu, X.; Hou, X.; Gao, H.; Yan, H.; Zhao, F.; Wang, L. NTN1 Affects Porcine Intramuscular Fat Content by Affecting the Expression of Myogenic Regulatory Factors. *Animals* **2019**, *9*, 609. [CrossRef]
7. Nakayama, H.; Ohnuki, H.; Nakahara, M.; Nishida-Fukuda, H.; Sakaue, T.; Fukuda, S.; Higashiyama, S.; Doi, Y.; Mitsuyoshi, M.; Okimoto, T.; et al. Inactivation of axon guidance molecule netrin-1 in human colorectal cancer by an epigenetic mechanism. *Biochem. Biophys. Res. Commun.* **2022**, *611*, 146–150. [CrossRef]

animals

MDPI

Article

Functional and Genetic Insights into the Role of the *NR4A1* Gene in the Litter Size of the Shaanbei White Cashmere Goat

Ebadu Areb [1,2,3,†], Yutian Bi [1,†], Yangyang Bai [1], Qihui Zhu [1], Lingyuan Ma [1], Chuanying Pan [1], Xiaolei Chen [2,*] and Xianyong Lan [1,*]

[1] Key Laboratory of Animal Genetics, Breeding and Reproduction of Shaanxi Province, College of Animal Science and Technology, Northwest A & F University, Xianyang 712100, China; ebaduareb3@nwafu.edu.cn (E.A.); biyutian0927@163.com (Y.B.); bai345@126.com (Y.B.); zqhzqhzqh2002@163.com (Q.Z.); malingyuan15@163.com (L.M.); panyu1980@126.com (C.P.)

[2] College of Science, Northwest A & F University, Xianyang 712100, China

[3] Central Ethiopia Agricultural Research Institute at Worabe Agricultural Research Center, Worabe P.O. Box 021, Ethiopia

* Correspondence: chenchen100@nwsuaf.edu.cn (X.C.); lanxianyong79@126.com (X.L.)

† These authors contributed equally to this work.

Simple Summary: Nuclear receptor subfamily 4 group A member 1 (*NR4A1*) is a transcription factor that plays a significant role in various physiological activities, including cell proliferation and differentiation, as well as reproductive signaling pathways. In this study, we identified genetic variations in the *NR4A1* gene, both insertion/deletion (InDel) and single nucleotide polymorphism (SNP), that are associated with litter size in goats. These findings offer new insights into the role of *NR4A1* in goat reproduction. Therefore, *NR4A1* could serve as a potential molecular marker for improving reproductive performance in goat breeding programs.

Abstract: Nuclear receptor subfamily 4 group A member 1 (*NR4A1*) plays a crucial role in regulating various physiological processes. As gene mining for reproductive traits is essential, this study aimed to investigate the mRNA expression, genetic variation, and association of the *NR4A1* gene with goat litter size. We examined the mRNA expression levels of the *NR4A1* gene in eight different tissues of female Shaanbei White Cashmere (SBWC) goats ($n = 6$). Then, a novel 11-bp insertion/deletion (InDel) variant was genotyped in 1136 SBWC goats, 87 SNPs were identified through resequencing ($n = 120$), and selection signal analysis was undertaken. The *NR4A1* gene was expressed in all examined tissues, including the ovary and the oviduct, suggesting its role in goat reproduction. Both the 11-bp InDel and 13 SNP variants showed significant association with litter size. Additionally, four potential transcription factor binding sites were predicted within the insertion allele, which may contribute to increased litter size. Selection signal analysis revealed strong pressure on the *NR4A1* gene region in the Cashmere goat population. These findings suggest that *NR4A1* is a promising candidate gene for improving litter size in goats and could be utilized as a genetic marker in breeding programs.

Keywords: goat; gene expression; genetic variation; litter size; *NR4A1 gene*; association analysis

1. Introduction

The litter size trait has a significant economic value, which is regulated by genetic and non-genetic factors [1]. Hereditary genetic factors that control litter size include a combination of major genes and polygenes [1]. Several biological processes, such as folliculogenesis and ovulation [2], signal transduction pathways [2], hormonal regulation [3], neuroendocrine regulation [2], and genetic factors, such as the *BMPR1P* gene, play a role in reproductive processes and influence litter size [4]. The low-heritability nature of litter size and maternal effect results in prolificacy traits' improvement being challenging [5]. To alleviate this challenge, a candidate gene approach in goat genomics is increasingly utilized to uncover the molecular mechanisms underlying variations in traits [6]. The Shaanbei White Cashmere (SBWC) goat is a widely recognized goat breed and valued for its high-quality cashmere and superior meat quality [7] but characterized by low litter size performance [8]. Various studies have reported differing litter size performance values for this breed at different times, including 1.5 [9], 1.7 [10], 1.5 [11], and 1.38 [12]. Therefore, further exploration of candidate genes is required for the breed improvement program.

While increasing litter size has long been a desirable trait in genetic improvement programs for enhanced productivity, this trend raises significant concerns regarding animal welfare and sustainability [13]. Large litters, especially in pigs and small ruminants, are often associated with higher pre-weaning mortality, lower average birth weights, and increased variability among littermates, which can compromise neonatal survival and long-term health [13,14]. Moreover, the metabolic demands placed on dams may elevate risks of exhaustion, insufficient maternal care, and impaired reproductive longevity [13,15]. Therefore, breeding strategies should emphasize optimal litter size rather than maximal, integrating phenotypic traits like birth weight uniformity, maternal behavior, and survival rates to align with animal welfare standards and sustainability [14,16,17].

Nuclear receptor subfamily 4 group A member 1 (*NR4A1*), also referred to as *Nur 77* or *TR3*, belongs to the orphan nuclear receptor transcription factor family. The *Capra hircus NR4A1* gene is localized on chromosome 5 with ten exons. The *NR4A1-NR4A3* subfamily comprises immediate early genes that are activated by various stimuli, including peptide hormones, growth factors, cytokines, inflammatory and physiological signals, and cellular stress [18]. It was one of the first genes identified as an early response gene, rapidly activated by nerve growth factor in pheochromocytoma cells [19]. *NR4A1* is involved in regulating metabolism, cardiovascular and neurological function, vascular stability, and immune cell homeostasis during inflammation and cancer [20]. It is expressed in different tissues and plays a role in cell proliferation and apoptosis [21]. Overexpressing *NR4A1* accelerated the development of goat intramuscular preadipocytes through the PI3K/AKT pathway [22]. The *NR4A1* gene is not only associated with reproduction and intramuscular preadipocytes but also used as a marker for feed restriction in dairy goats [23]. On the other hand, the reduced methylation level of the *NR4A1* gene affects gene expression in adult sheep longissimus dorsi muscle [24], and it is a glucose indicator, as its expression was related to the insulin sensitivity of skeletal muscle [25]. The expression of *NR4A1* was down-regulated in diabetic mice [26]. Compared with wild mice, *NR4A1* knockout mice were more likely to be obese [26]. This research output indicates that the *NR4A1* gene might have an association with growth traits.

The *NR4A1* gene was one of the up-regulated genes in the ovaries of Xiang pigs with large litter sizes [27]. It is suggested that this gene might play important roles in promoting litter size by increasing the level of steroid and peptide hormone supply through the ovary and facilitating oocyte ovulation and in vivo fertilization [27]. Additionally, this gene plays a role in regulating progesterone synthesis in goat luteal cells [28], indicating its participation in the reproductive process. Furthermore, the adenosine monophosphate-activated protein kinase/*NR4A1* signaling axis is involved in ovarian function changes in premature ovarian insufficiency rats following umbilical cord mesenchymal stem cell transplantation [29].

Similarly, a significant association between the single nucleotide polymorphism (SNP) exploration of *NR4A1* g.3952A>G and the total number of piglets born and weaned for commercial sows was reported [30]. According to the report, Landrace sows with the AG genotype variant exhibited significantly higher body weight and reproductive performance than the GG genotype. Specifically, they produced 2.98 kg more piglets in total litter weight at birth, 2.41 more piglets were born alive, 2.52 more piglets were weaned alive, and 15.68 kg more in total weaning litter weight was achieved [30]. Also, this gene is involved in the cell cycle and differentiation of granulosa cells, theca cells, and oocytes [31]. Similarly, the *NR4A1* gene regulates the expression of key genes involved in ovarian steroidogenesis and enhances ovarian reserve in aged mice [32]. These studies have implicated that the *NR4A1* gene has a role in the regulation of ovulation and luteal function. However, the function of this gene in goat reproductive performance remains unclear.

It is necessary to use marker-assisted selection (MAS) through the gene mining method in goat genomics to find out how genetic differences are caused at the molecular level, with a focus on traits of interest [6]. The previous reports from our group have identified a variety of potential genetic marker candidate genes that could aid goat breeding programs. Some examples are *PPP6C* [9], *AKAP12* [33], *IGF2BP2* [34], *CLSTN2* [12], *SMAD2* [10], *KMT2A* [35], *GATA4* [36], and *KDM6A* [37].

According to the research findings described above, we hypothesized that genetic variants within the *NR4A1* gene might have an effect on litter size by affecting mRNA expression profiles. Among genetic variants, insertion/deletion (InDel) and SNP offer high accuracy and stability for directly detecting gene polymorphisms [9,38]. Therefore, the aim of this study was to investigate the mRNA expression of the *NR4A1* gene in multiple tissues, as well as the association between genetic variants (InDels and SNPs) and goat litter size. This approach aims to facilitate further gene exploration and provide a potential marker for breeding programs.

2. Materials and Methods

2.1. Bioinformatics Analysis of Regulation and Conservation of the NR4A1 Gene

The networks of physical interaction, pathways, co-localization, co-expression, and other interactions were examined by using the online bioinformatics tool GeneMANIA (https://genemania.org/ accessed on 20 January 2025), for the *NR4A1* gene with other genes. The conservation analysis of this gene across different taxonomic groups was also examined. To do this, the sequence of the *NR4A1* gene for six species was retrieved from the NCBI; these include *Capra hircus* (NC 030812.1), *Bos taurus* (NC 037332.1), *Ovis aries* (NC 056056.1), *Homo sapiens* (NC 000012.12), *Mus musculus* (NC 000081.7), and *Sus scrofa* (NC 010447.5). MEGA11 software (version 11.0.13) was used to perform different species sequence muscle alignment [9]. The evolutionary tree was constructed using the neighbor-joining method in MEGA11 [39], and similarity analysis was performed using the Clustal Omega Multiple Sequence Alignment (MSA) (Clustal Omega, EMBL-EBI, https://www.ebi.ac.uk/jdispatcher/msa/clustalo accessed on 20 January 2025) bioinformatics tool.

2.2. Tissue Sample Collection, Total RNA Extraction, and cDNA Synthesis

Tissue samples were collected from healthy SBWC goats from Yulin, Shaanxi Province, China. A total of eight types of tissues (liver, heart, lung, spleen, oviduct, fat, kidney, and ovary) were collected from the selected firstborn female goats (n = 6). The age of the selected animals was between 2 and 3 years and in the estrous cycle stage. All goats were maintained under uniform environmental, management, and feeding conditions. To determine the mRNA expression levels of the *NR4A1* gene, total RNA was extracted from the aforementioned tissues by using TRIzol reagent (TaKaRa Biotech Co., Ltd., Dalian, China).

The extracted total RNA was stored at −80 °C following the manufacturer's protocol. RNA purity and quality were evaluated using a NanoDrop ND-1000 spectrophotometer (Thermo Scientific, Waltham, MA, USA) to ensure the A_{260}/A_{280} ratio falls within the acceptable range, which is 1.8 to 2.1 [11]. RNA integrity was assessed via 1% agarose gel electrophoresis [9]. First-strand complementary DNA (cDNA) was synthesized using the Prime-Script™ RT reagent kit (TaKaRa Biotech Co., Ltd.) based on the manufacturer's protocol. The synthesized cDNA was kept at a temperature of −20 °C [40].

2.3. Reverse Transcriptase Quantitative Real-Time PCR (RT-qPCR)

The sequences for the *NR4A1* gene and reference gene (*GAPDH*) primers were designed using NCBI's primer designing tool (https://blast.ncbi.nlm.nih.gov/Blast.cgi accessed on 12 March 2024). The coding sequence of transcript variant X3 (XM_018047732.1) of the *NR4A1* gene and *Capra hircus GAPDH* as reference gene sequences were used to design primers for RT-qPCR (Table 1). Primer pairs were designed from the exon–exon junction to amplify a specific product without nonspecific amplification or genomic DNA contamination. The primers were produced by Sangon Biotech Ltd., Xi'an, China. The cDNA of the tissues of female goats, including the ovary, spleen, liver, oviduct, heart, kidney, fat, and lung, was used to analyze the mRNA expression. The RT-qPCR mixture (20 μL) comprised 2 μL of cDNA (1:100 dilution), 10 μL of 2× SYBR Premix Ex Taq (Takara Biotech), 7.2 μL of distilled water, and 0.4 μL of each forward and reverse primer. To improve the accuracy, three replications were performed for each sample, and relative gene expression was normalized by the fold change of the $2^{-\Delta\Delta Ct}$ method using the reference gene [9]. The RT-qPCR was conducted in real time using the CFX96 Real-Time PCR Detection System (Bio-Rad, Hercules, CA, USA). The thermal cycling protocol was 40 cycles of pre-cycling at 95 °C for 30 s, denaturation at 95 °C for 10 s, and annealing and extension at 60 °C for 30 s.

Table 1. Primer pair for polymorphism detection, genotyping, and determining mRNA expression of the *NR4A1* gene.

Primers Name	Primer Sequence (5′->3′)	Sizes (bp)	Tm (°C)	Region
P1	F: TCTGCCTTTGGGACAGCAAG R: GCCTTGAGCCCTATTCACCC	235	60.54 60.47	upstream
P2	F: GGGTGAGATGTGGAGAGCAG R: TAGGGGGATTTGCTCTGTGG	199	59.82 59.08	upstream
P3	F: AGCCCCATCTCTGGACATACT R: AATGGGAGCGTTGTCTGGG	200	60.06 60	intron
P4	F: CACCAGGAAGAGGTCCCAAC R: GGGTCAACAGGAGAGGAGGA	368	59.96 60.25	intron
P5	F: CCTCGCCACACCTTGCATTT R: CACATTCCTCTCCCCACCTC	233	61.53 59.46	upstream
P6	F: AGGCTGTGTGTTTGTCCCAG R: GGGTTCGGCCATATCCTCAG	197	60.47 59.96	intron
P7	F: GGTTGTAAGAAGGCGCAGAG R: GCCCTTCCAACCAATAGCAC	486	58.92 59.18	upstream
P8	F: ACACCTTTACCCGAGAGAGC R: ACTCTCTGGACCCTGAACAC	223	59.1 58.66	intron

Table 1. *Cont.*

Primers Name	Primer Sequence (5′->3′)	Sizes (bp)	Tm (°C)	Region
NR4A1-qPCR	F: ACAGACAGCCTGAAAGGACG R: GACCAGGGAGGTGAGTAGGT	93	59.97 59.96	exon 1–2
GAPDH-qPCR	F: AAAGTGGACATCGTTGCCAT R: CCGTTCTCTGCCTTGACTGT	116	58.09 59.97	exon 2

Abbreviations: P = primer; Tm = melting temperature; F = forward; R = reverse; bp = base pair.

2.4. Animal Selection, Ear Tissue Sample Collection, and Genomic DNA Extraction

Firstborn litter size data were collected from 1136 randomly selected goats from the SBWC goat in Yulin, Shaanxi Province, China. All sampled goats were raised at the Shaanbei Cashmere Goat Breeding Farm in Hengshan County. The Agricultural Technical Station of Yulin City, Shaanxi Province, provided the recorded data. According to the pedigree, there is no genetic relationship between goat individuals. The selected goats were clustered into two based on their birth type (mother with single birth and mother with multiple births). All goats were maintained under suitable and uniform environmental, feeding, and management conditions [9,41]. The sampled goats were bred through natural mating without controlled reproductive techniques. The does were mated with bucks at a ratio of 1:18–24. Does could kid up to 3 times for 2 years, and kids were grown with their does until a weaning age of 3 months [42].

Ear samples were excised and placed in centrifuge tubes containing 70% alcohol for preservation [9]. Genomic DNA was extracted from the ear tissue using a high-salt extraction method. This is a reliable and cost-effective approach for obtaining DNA [43]. DNA quality and purity were assessed by measuring the A_{260}/A_{280} ratio using a NanoDrop™ 1000 Spectrophotometer (Thermo Scientific, Waltham, MA, USA) [11]. Then, the extracted DNA was diluted to a concentration of 20 ng/μL and stored at −20 °C for further use [44].

2.5. Insertion–Deletion Identification and Primer Design

InDels for the goat *NR4A1* gene were identified by using bioinformatics tools of Ensembl (https://asia.ensembl.org/index.html accessed on 23 March 2024) and GGVD (Goat VariationDB (GGVD) of the Animal Omics database (omicsDB) [45], both database were accessed on 12 March 2024. Additionally, the GGVD database was used to examine minor allele frequencies (MAF), geographical origin group frequencies, and genetic component group frequencies. InDels with >10 bp and MAF > 0.05 were selected to design primers. Primers were designed for identified variants of InDels for polymorphism detection and genotyping by using the NCBI primer designing tool (https://www.ncbi.nlm.nih.gov/tools/primer-blast/ accessed on 23 March 2024). Then, the primers were synthesized by Sangon Biotech Ltd., Xi'an, China (Table 1).

2.6. PCR Amplification, Polymorphism Detection, and Genotyping

Polymorphism was detected for the identified InDels by using the DNA mixed pool technique of 24 to 48 randomly selected individuals [9,46]. For the touchdown PCR program, a reaction for mixed pool PCR with a volume of 13 μL was employed. The setup mixture included 0.3 μL of DNA from each individual DNA sample, 6.5 μL of PCR mix, 5 μL of double-distilled water (ddH_2O), and 0.5 μL of each forward and reverse primer. To enhance detection efficiency, each primer was replicated three times. The mixed pool PCR products were detected on a 3.5% agarose gel and then stained with ethidium bromide. The presence of two distinct bands or differences in band location across the three replicates for a given primer indicated the availability of polymorphism.

PCR amplification for genotyping used a reaction mixture with a total volume of 13 μL with 5 μL of ddH_2O, 0.5 μL of each forward and reverse primer, 0.5 μL of DNA, and 6.5 μL of PCR mix. The temperature adjustment was an initial denaturation step at 95 °C for 5 min, followed by 35 cycles of denaturation at 94 °C for 30 s, annealing at 65 °C for 30 s, and extension at 72 °C for 20 s with a final extension at 72 °C for 5 min. Additionally, at the final stage, the temperature was maintained at 12 °C for 20 min to preserve and maintain the integrity of the PCR product. Lastly, the PCR product was detected using 3.5% agarose gel electrophoresis, and the gel was stained with ethidium bromide. Subsequently, from each genotype, two PCR products were randomly selected and sent for Sanger sequencing. Sequence synthesization was performed by Sangon Biotech Ltd., Xi'an, China. Then, sequence alignment was performed by using FinchTV version 1.4.0 and BioXM software version 2.7.1.

2.7. SNP Loci Identification for the NR4A1 Gene Through Resequencing

By resequencing the 120 SBWC goats, SNP loci were found from the *NR4A1* gene. First, the original FASTQ file was filtered using fastp v0.20.0, and then the whole genome data were aligned with the Oar_rambouillet_v2.0 reference genome using BWA-MEM v0.7.13-r1126 to obtain a binary alignment bam file [47]. SNPs were detected using GATK v3.6-0-g89b7209, and SNP calling was performed using the "HaplotypeCaller" and "GenotypeGVCFs" modules. Finally, bcftools-1.13 was used to remove sequencing and alignment errors, and the parameters were "QD < 2.0, | | QUAL < 30.0, | | SOR > 3.0, | | FS > 60.0, | | MQ < 40.0, | | MQRankSum < −12.5, | | ReadPosRankSum < −8.0", with further filtered SNPs using the parameters "-m2 -m2-I", MAF ≥ 0.05, and F_MISSING < 0.2.

2.8. Population Genetic Parameter Estimation

Nei's method was used to measure genetic parameters of the populations, including the effective allele number (Ne), polymorphism information content (PIC), heterozygosity (He), and homozygosity (Ho) (http://www.msrcall.com/Gdicall.aspx [11], accessed on 2 January 2025). The polymorphism information content value was classified as PIC < 0.25, 0.25 < PIC < 0.5, and PIC > 0.5 for low, moderate, and high diversity, respectively [11]. Similarly, Hardy–Weinberg equilibrium (HWE), Weir and Cockerham (W&C), and allele frequency were calculated by using the Genepop on the Web online tool.

2.9. Analysis of Selection Signal at the Region of the NR4A1 Gene

The *NR4A1* gene's 0.5 Mb flanking region was examined for the strength of selection signals. The Cashmere goat population was compared with goat ancestral groups, such as Bezoar (BEZ), Southwest Asian goat (SWA), African dairy goat (AFD), Europe (EUR), and East Asian goat (EAS). The statistical methods, such as nucleotide diversity (−log10(Pi)), Tajima's D, heterozygosity (Z(Hp)), integrated haplotype score (|iHS| > 2.0 ratio), cross-population extended haplotype homozygosity (XPEHH), Z-transformed Fst (Z(Fst)), difference in nucleotide diversity (ln(Pi ratio)), and composite likelihood ratio (CLR), were used to determine the selection sweep. A sliding window of 30 kb of the Animal Omics of the GGVD database was used to retrieve the annotation data.

2.10. Identification of Transcription Factor Binding Site

An online bioinformatics transcription factor binding site (TFBS) prediction tool, AImodules (https://www.biozentrum.uni-wuerzburg.de/bioinfo/computing/aimodules, accessed on 29 January 2025), using predefined matrices from the JASPAR database, was used. AIModules recognizes conserved motifs and combinations of motifs (modules), allowing for a number of interesting biological applications, such as the analysis of promoter

TFBS [48]. For both insertion and deletion sequences of the flanking region, it was used to predict TFBS for InDel mutation.

2.11. Data Analysis

The mRNA relative expression levels for various tissues of the *NR4A1* gene were determined by using the $2^{-\Delta\Delta Ct}$ method [9,11]. Nei's online tool was used to analyze population genetic parameters, such as Ho, He, Ne, and PIC values. Additionally, the Genepop online tool was used to analyze HWE, allele frequency, and W&C parameters. Similarly, to detect the linkage disequilibrium (LD) between SNPs of the *NR4A1* gene, D′ (normalized measure of linkage) and r^2 (squared correlation coefficient) were estimated and visualized by using the SHEsis online program (http://analysis.bio-x.cn [49], accessed on 2 January 2025). The linkage disequilibrium threshold was categorized as D′ or $r^2 > 0.8$, $0.5 < r^2 < 0.8$, and $r^2 < 0.5$, which were high, moderate, and low, respectively [50]. GraphPad Prism version 10.4.1 was used to determine the chi-square test (χ^2) of the relative mRNA expression profile across tissues and allele and genotype frequency for the sampled population. The association between litter size (single-born mother vs. multiple-born mother) (odds of multiple kids) and the 11-bp InDel genotype (II, ID, and DD) was analyzed by using a binomial logistic regression model in R software version 2024.12.1 [51].

$$\log\left(\frac{P(multiple\ kid)}{1 - P(multiple\ kid)}\right) = \beta 0 + \beta 1\ x\ Genotype\ ID + \beta 2\ x\ Genotype\ II \quad (1)$$

where *P* represents the probability of multiple kids, with the DD genotype used as the reference category. The reference genotype was taken alphabetically by default. On the other hand, SPSS version 20 was used to analyze the association between litter size and SNP variants. The model used was $Y_{jk} = \mu + G_j + e_{jk}$, where Y_{jk} = litter size, μ = population mean, G_j = effect of genotype (each SNP have either 2 or 3 levels), and e_{jk} = random error.

The gene regulatory network was visualized by using the platform https://genemania.org accessed on 2 January 2025). GeneMania allows for visualization of up to 100 genes linked to a query gene; however, in this study, only 30 genes were selected to construct the network diagram in order to enhance clarity and simplify the visualization. For the graphical network visualization of the *NR4A1* gene, multiple interaction parameters were used, including physical interaction, co-expression, co-localization, genetic interaction, pathway involvement, and shared protein domains. A maximum of 30 resultant genes were allowed for automatically selected weighting method. A network with thicker lines indicates strong evidence of linkage [52]. Selection signal statistical methods, such as -log10(Pi), Tajima's D, Z(Hp), ln(Pi ratio), |iHS| > 2.0 ratio, XPEHH, Z(Fst), and CLR, were determined by using annotation data of GGVD under the Animal Omics database.

3. Results

3.1. NR4A1 Gene Interaction Network

According to GeneMania, (https://genemania.org/ accessed on 2 January 2025), the *NR4A1* gene physically and genetically interacted with, co-expressed, co-localized, and shared protein domains with other genes and transcription factors (Figure 1). Among many genes, only the top 30 genes were filtered out to construct a network diagram. Among the listed genes, the majority of them have a physical interaction with the *NR4A1* gene (77.64%) and are co-expressed (8.01%) (Figure 1). Genes like *NR4A2*, *EGR1*, *ATF3*, *ZFP36*, and others were co-expressed with *NR4A1*. On the other hand, genes like *IFI27*, *AKT1*, *SLC25A4*, *BNIP3L*, and others have a physical interaction with the *NR4A1* gene. Similarly, *RXRA*, *RXRG*, *IFI27*, and others have a direct genetic interaction with the *NR4A1* gene.

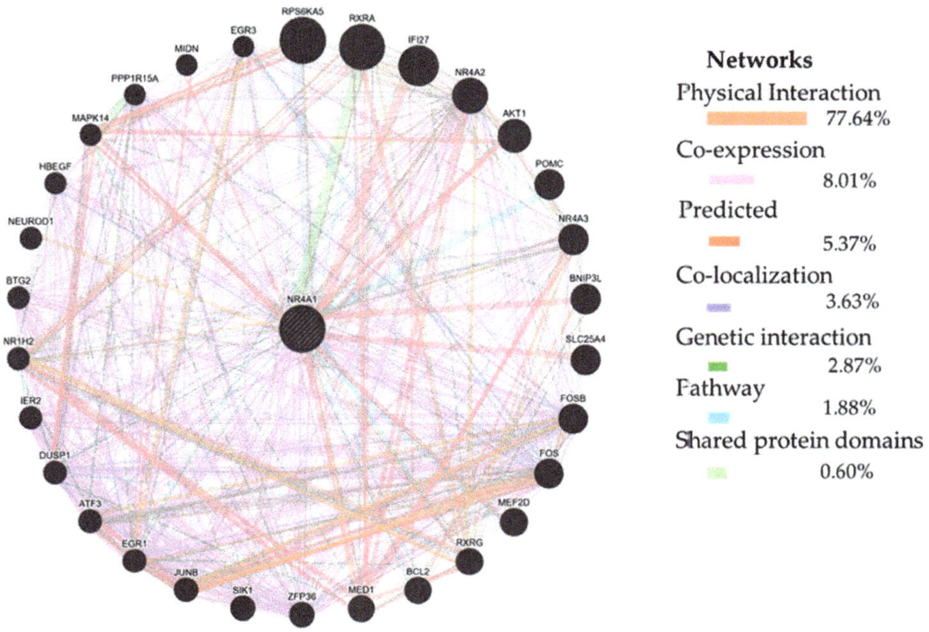

Figure 1. Prediction of regulatory network for *NR4A1* gene. Thirty gene interaction networks, including genes, transcription factors, and pathways that have been directly or indirectly associated with the *NR4A1* gene, were identified. The network's legend indicates the proportions of physical interaction, co-expression, predicted co-localization, genetic interaction, pathways, and shared protein domains with regulatory networks.

3.2. *NR4A1* Gene Conservation Analysis Across Taxonomic Order

Five other species' nucleotide sequences were compared to the goat *NR4A1* gene. The taxonomic order of the species, including artiodactyl (*Ovis aries*, *Bos taurus*, and *Sus scrofa*), primate (*Homo sapiens*), and rodent (*Mus musculus*), was used for comparison. Artiodactyls, or even-toed ungulates, exhibit a high degree of similarities among taxonomic orders. In particular, the sequences of *Ovis aries* (NC 056056.1), *Bos taurus* (NC 037332.1), *Sus scrofa* (NC 010447.5), *Homo sapiens* (NC 000012.12), and *Mus musculus* (NC 000081.7) were found to have homological similarity with *Capra hircus* of 99%, 97%, 91%, 88%, and 79%, respectively. The resulting phylogenetic tree shows a more distant relationship with primates and rodents but a close evolutionary relationship among goats, sheep, and cattle (Figure 2).

3.3. *NR4A1* Gene mRNA Transcriptional Profiles

The relative mRNA expression profile of the *NR4A1* gene (NC 030812.1 chromosome 5) across various tissues of the firstborn female SBWC goat was evaluated. For both *NR4A1* and *GAPDH*, a single sharp peak of the melting curve was observed at approximately 87 °C and 83 °C, respectively, confirming the specificity of the amplified product. This gene was expressed in all tissues examined, including the lung, ovary, spleen, oviduct, fat, heart, kidney, and liver. The lung, ovary, and spleen had relatively high expression of the *NR4A1* gene compared to others ($p < 0.05$). Moderate expression in the ovary could suggest a role in reproductive functions (Figure 3). However, minimal expression includes the liver, heart, kidney, and others, indicating that the gene may have less relevance for the metabolic or circular functions.

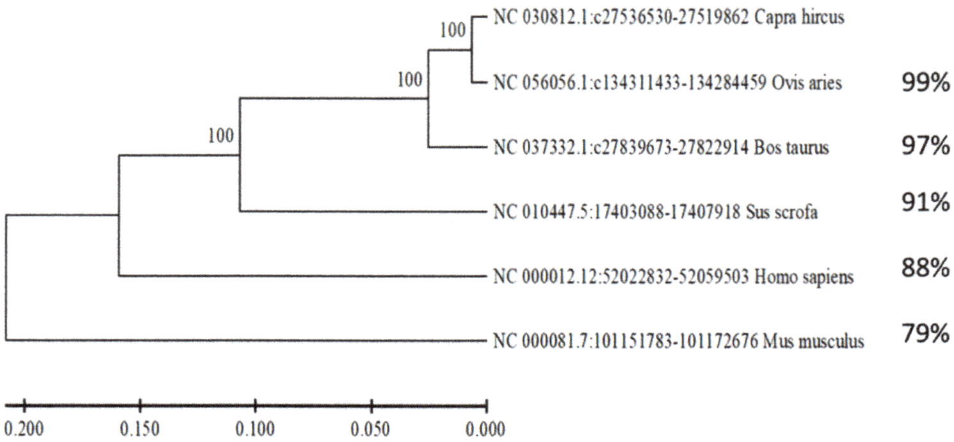

Figure 2. Phylogeny tree for the *NR4A1* gene across different species. The evolutionary conservation of the *NR4A1* gene for *Capra hircus*, *Ovis aries*, *Bos taurus*, *Sus scrofa*, *Homo sapiens*, and *Mus musculus* with their similarity percentage was determined.

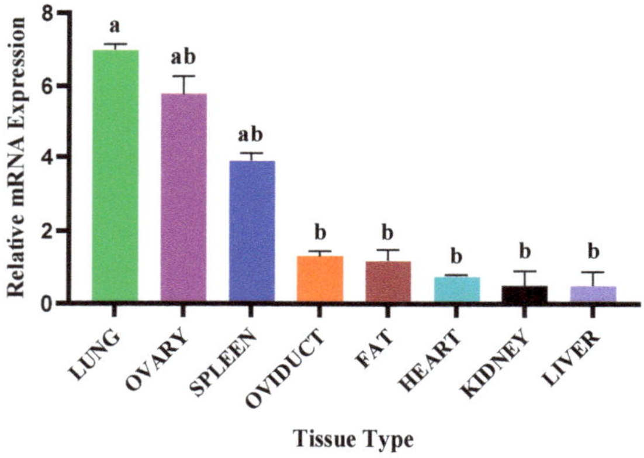

Figure 3. Relative mRNA expression profile of the *NR4A1* gene in tissues of the SBWC goat. Tissues from the lung, ovary, spleen, oviduct, fat, heart, kidney, and liver of six firstborn female goats were evaluated. Data represent means ± SEM; letters on the bar (a, b) represent significant difference of means at $p < 0.05$.

3.4. Genotyping and Population Genetic Parameter Analysis

To detect polymorphism, eight primer pairs were designed from both upstream and intron regions of the *NR4A1* gene. After mixed pool detection, a novel variant of primer 3 (NC_030812.1 g.2900-2911del CGACTAGGGGG), exhibited polymorphism (Table 1). Three genotyping variants were identified from 1136 firstborn female goats through PCR amplification, gel electrophoresis (Figure 4a), and Sanger sequencing (Figure 4b). The genotypes include insertion/insertion (II: $n = 245$), insertion/deletion (ID: $n = 637$), and deletion/deletion (DD: $n = 254$) (Table 2). According to the sampled population, the genotype frequency of a population was estimated as 21.6%, 56.1%, and 22.4% for II, ID, and DD, respectively. Additionally, homozygosity and heterozygosity population parameters were nearly equal in proportion with the sampled population (Table 2). The detailed analysis of genetic variation within a population based on genotypes and allele frequencies, as well as metrics that measure genetic diversity within a population, is illustrated in Table 2.

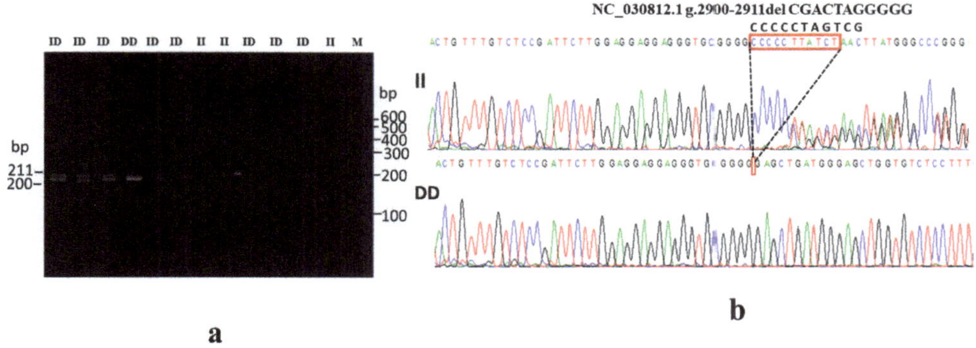

Figure 4. Identification of the 11 bp-del of nucleotide sequence variants within the goat *NR4A1* gene. (**a**) Gel electrophoresis diagram with variant genotypes of II, homozygote insertion/insertion; ID, heterozygote insertion/deletion; DD, homozygote deletion/deletion; M, marker; and bp, base pair. (**b**) Sanger sequencing diagram of PCR product. The upper panel shows the sequence chromatograms of the II genotype, whereas the lower one shows the DD genotype, and the dashed line highlighted triangle indicates the inserted and deleted sequence.

Table 2. Genetic parameters of the 11 bp InDel locus mutation within the *NR4A1* gene in the SBWC goat.

Genotype Frequencies			Allele Frequencies		Genetic Variation Metrics					
II	ID	DD	I	D	Ho	He	Ne	PIC	HWE	W&C
0.216 $n = 245$	0.561 $n = 637$	0.224 $n = 254$	0.496	0.504	0.500	0.499	1.999	0.374	$p < 0.05$	−0.1211
					$N = 1136$					

Where: II = insertion/insertion; ID = insertion/deletion; DD = deletion/deletion; I = insertion; D = deletion; Ho = homozygosity; He = heterozygosity; Ne = effective allele number; PIC = polymorphic information content; HWE = Hardy–Weinberg equilibrium; W&C-Weir and Cockerham.

The proportionality of the nearly equal chance of both alleles (I and D) was confirmed by the effective allele number (Ne), which was 1.999 (Table 2). The PIC value (0.374) indicates that the population has a moderate diversity. On the other hand, according to the chi-square test (χ^2) of the sampled population, the novel variant NC_030812.1 g.2900-2911del CGACTAGGGGG of the *NR4A1* gene in the SBWC goat population significantly deviates from HWE ($p < 0.05$) (Table 2). The HWE deviation was also confirmed by the W&C statistical parameter. The W&C value of −0.1211 indicated that the population at the target locus shows moderate heterozygote excess (12.11%) compared to expectations under HWE. This could suggest heterozygote advantage or gene flow in the population. This might be due to the selection program of SBWC goats for the breed improvement program, including growth and reproduction improvement, which results in non-random mating, and, simultaneously, the genomic region of the *NR4A1* gene might be under selection pressure.

On the other hand, according to the SNP analysis results, the population was in HWE, except for SNP 8. The genetic parameters, such as Ho, He, Ne, and PIC, for each SNP were different. Similarly, there is moderate diversity for the population because the PIC values for SNP2 and SNP8 were 0.27 (Table 3). According to the Ne value, all of the SNPs had different allele proportionality. This indicates that the population's diversity is decreasing due to the selective breeding program for the target goat population. In addition, the values for most of the SNPs, such as SNP1, SNP4, SNP5, SNP6, and others, verified that the population's genetic diversity is low (Table 3).

Table 3. Genetic parameters of resequencing locus SNPs within the *NR4A1* gene in the SBWC goat.

SNPs	Genotypes	Genomic Location	Variant ID	Allele Frequencies	Ho	He	Ne	PIC	HWE
SNP1	TT (115) TC (5)	NC 030812.1: g. 27520435T>C	Novel	0.96 (T) 0.04 (C)	0.96	0.04	1.04	0.03	$p > 0.05$
SNP2	GG (76) GC (39) CC (5)	NC 030812.1: g. 27521198G>C	rs668853994	0.80 (G) 0.20 (C)	0.68	0.32	1.48	0.27	$p > 0.05$
SNP3	CC (101) CG (18) GG (1)	NC 030812.1: g. 27521415C>G	rs657635772	0.92 (C) 0.08 (G)	0.85	0.15	1.18	0.14	$p > 0.05$
SNP4	CC (108) CT (12)	NC 030812.1: g. 27522352C>T	Novel	0.95 (C) 0.05 (T)	0.91	0.09	1.10	0.09	$p > 0.05$
SNP5	CC (116) CT (4)	NC 030812.1: g. 27526206C>T	Novel	0.98 (C) 0.02 (T)	0.97	0.03	1.03	0.03	$p > 0.05$
SNP6	AA (115) AG (5)	NC 030812.1: g. 27527050A>G	rs683100351	0.98 (A) 0.02 (G)	0.96	0.04	1.04	0.04	$p > 0.05$
SNP7	CC (115) CT (5)	NC 030812.1: g. 27527625C>T	rs677241829	0.98 (C) 0.02 (T)	0.96	0.04	1.04	0.04	$p > 0.05$
SNP8	CC (81) CA (27) AA (10)	NC 030812.1: g. 27528181C>A	Novel	0.79 (C) 0.21 (A)	0.68	0.32	1.47	0.27	$p < 0.05$
SNP9	GG (96) GC (23) CC (1)	NC 030812.1: g. 27531784G>C	rs661365536	0.90 (G) 0.10 (C)	0.81	0.19	1.23	0.17	$p > 0.05$
SNP10	GG (103) GA (17)	NC 030812.1: g. 27532725G>A	Novel	0.93 (G) 0.07 (A)	0.87	0.13	1.15	0.12	$p > 0.05$
SNP11	CC (3) CG (34) GG (83)	NC 030812.1: g. 27533799C>G	rs669578807	0.17 (C) 0.83 (G)	0.72	0.27	1.38	0.24	$p > 0.05$
SNP12	AA (3) AG (34) GG (83)	NC 030812.1: g. 27533803A>G	rs658171703	0.17 (A) 0.83 (G)	0.72	0.27	1.38	0.24	$p > 0.05$
SNP13	GG (104) GA (16)	NC 030812.1: g. 27534820G>A	Novel	0.93 (G) 0.07 (A)	0.88	0.12	1.14	0.12	$p > 0.05$

SNP = single nucleotide polymorphism; ID = identification; Ho = homozygosity; He = heterozygosity; Ne = effective allele number; PIC = polymorphic information content; HWE = Hardy–Weinberg equilibrium; numbers in the bracket under the genotype column indicate the number of goats; g. = genomic.

Additionally, for both the ancestral and worldwide goat groups, the aforementioned InDel loci had different alternate allele frequencies (Tables 4 and S1 and supplementary file Figures S1 and S2). Data for six ancestral goat groups and worldwide goats, including both reference and alternate allele frequencies, were obtained from the Animal Omics database. The alternate allele frequency is an important parameter to study variability at a given locus. Populations with higher alternate allele frequency tend to exhibit greater genetic diversity, which can enhance their adaptability and resilience.

Table 4. Allele frequency distribution of six ancestral goat groups.

Ancestral Goat Group	Number of Samples	Reference Allele Frequency	Alternate Allele Frequency
Bezoar	24	0.771	0.229
Africa	54	0.713	0.287
Africa Dairy	15	0.767	0.233
Europe	28	0.786	0.214
Southwest Asia	34	0.735	0.265
East Asia	57	0.868	0.132

3.5. Association of NR4A1 Gene InDel Variants with Goat Litter Size

The main effect of the 11bp-del mutation of the *NR4A1* gene significantly impacted litter size ($p = 1.62 \times 10^{-9}$) (Table 5). Firstborn mothers with the II genotype exhibited approximately four times higher odds of producing multiple kids compared to those with the DD genotype of 11-bp deletion (odds ratio = 3.975, $p = 1.58 \times 10^{-6}$) (Table 5). Consequently, firstborn SBWC goats carrying the DD genotype for the aforementioned InDel loci had a significantly lower probability of producing multiple kids.

Table 5. Coefficient table of association analysis of 11-bp-del loci in the *NR4A1* gene with litter size (multiple kids born) in SBWC goats.

Coefficient Table						Goodness of Fit		
Predictor	Estimate (β)	SE	z-Value	*p*-Value	Odd Ratio	χ^2 Test	AIC	*p*-Value
Intercept	−2.5735	0.2445	−10.524	$<2 \times 10^{-16}$	0.076			
Genotype ID	+0.1320	0.2848	+0.463	0.643	1.141	40.484	751.14	1.62×10^{-9}
Genotype II	+1.3801	0.2875	+4.800	1.58×10^{-6}	3.975			

SE = standard error; AIC = Akaike information criteria.

Mothers with II genotypes have significantly higher litter size performance than ID and DD genotypes (Figure 5a). However, there is no statistically significant difference between mothers with ID and DD genotypes. The distribution of 11 bp-del InDel variations varied among different litter size types according to the χ^2 test of the entire sampled population of the intra-group result ($p < 0.0001$) (Figure 5b). On the other hand, the χ^2 test distribution analysis of the mother with single kids and mother with multiple kid groups of SBWC goats showed no statistically significant difference between allotypes at the 11 bp-del loci of the *NR4A1* gene (Figure 5c).

3.6. Association of the NR4A1 Gene SNP Loci with Goat Litter Size

SNPs were identified by resequencing the sampled goat population ($n = 120$) for the *NR4A1* gene. The resequencing approach helps to genotype individuals, determine genetic variation, and perform association analysis. A total of 87 SNPs were identified, and 49 SNPs were evaluated for the association analysis. The remaining SNPs did not have enough variability within the sampled population. Among the analyzed SNPs, 13 SNPs were statistically significant with goat litter size (Table 6). SNPs with non-significant associations are available in the supplementary file, Table S2. Among SNPs, SNP1, SNP4, SNP5, SNP6, SNP7, SNP11, and SNP12 were highly significantly associated with litter size ($p < 0.0001$) (Table 6). Certain genotypes within the SNPs, such as TC (SNP1), CC (SNP2), GG (SNP3), CT (SNP5), AG (SNP6), CT (SNP7), CC (SNP9), CC (SNP10), and AA (SNP11), are represented by a low number of individuals (Table 6). This indicates that these variants are rare in the studied samples of individuals. Due to the limited number of individuals with these genotypes, further validation with a larger sample size may be necessary.

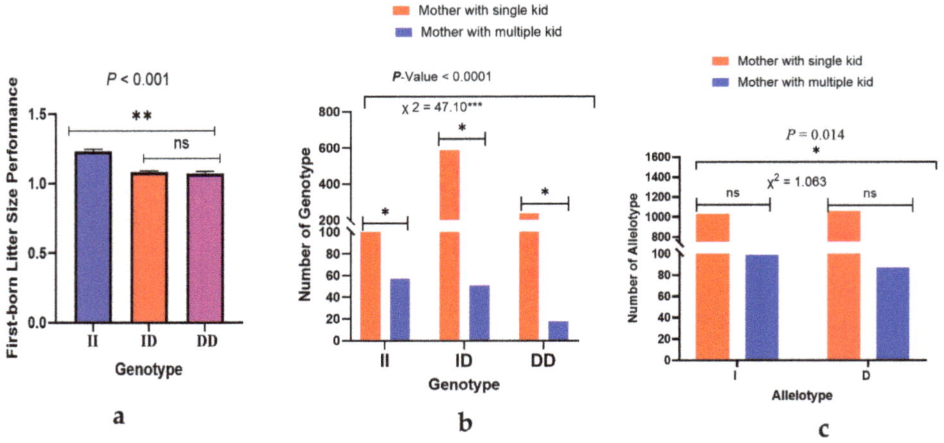

Figure 5. Litter size performance and distribution of the genotypes of the 11-bp-del loci in the first-born mother with single and multiple kid groups of the *NR4A1* gene. (**a**) Litter size performance across different genotypes of firstborn mothers in SBWC goats; data represent the mean ± SEM, and mean values are considered different: ** = $p < 0.001$. (**b**) The intra-group of genotype χ^2 tests of the 11-bp loci within the *NR4A1* gene in SBWC goats; number of genotype distributions considered different: *** = $p < 0.0001$. (**c**) The intra-group allele type χ^2 test of the 11-bp locus within the *NR4A1* gene in SBWC goats; data represent the number of allele type distributions for the sampled population. Distribution of alleles considered different: * = $p < 0.05$, ns-not significant.

Table 6. Significantly associated resequencing SNPs of the *NR4A1* gene with goat litter size.

SNP	Genotypes	N	Mean ± SEM	*p*-Value
SNP1	TT	115	1.65 ᵃ ± 0.07	8.24×10^{-17}
	TC	5	1.00 ᵇ ± 0.00	
SNP2	GG	76	1.75 ᵃ ± 0.08	0.032
	GC	39	1.38 ᶜ ± 0.11	
	CC	5	1.60 ᵇ ± 0.24	
SNP3	CC	101	1.70 ᵃ ± 0.07	0.003
	CG	18	1.22 ᵇ ± 0.13	
	GG	1	1.00	
SNP4	CC	108	1.69 ᵃ ± 0.07	5×10^{-6}
	CT	12	1.08 ᵇ ± 0.08	
SNP5	CC	116	1.65 ᵃ ± 0.07	9.20×10^{-17}
	CT	4	1.00 ᵇ ± 0.00	
SNP6	AA	115	1.65 ᵃ ± 0.07	8.24×10^{-17}
	AG	5	1.00 ᵇ ± 0.00	
SNP7	CC	115	1.65 ᵃ ± 0.07	8.24×10^{-17}
	CT	5	1.00 ᵇ ± 0.00	
SNP8	CC	81	1.52 ᶜ ± 0.08	0.034
	CA	27	1.70 ᵇ ± 0.12	
	AA	10	2.10 ᵃ ± 0.23	

Table 6. *Cont.*

SNP	Genotypes	N	Mean ± SEM	*p*-Value
SNP9	GG	96	$1.70^{a} \pm 0.07$	0.033
	GC	23	$1.35^{b} \pm 0.13$	
	CC	1	1.00	
SNP10	GG	103	$1.69^{a} \pm 0.07$	0.001
	GA	17	$1.24^{b} \pm 0.11$	
SNP11	CC	3	$1.00^{c} \pm 0.00$	1.95×10^{-11}
	CG	34	$1.56^{b} \pm 0.10$	
	GG	83	$1.67^{a} \pm 0.03$	
SNP12	AA	3	$1.00^{c} \pm 0.00$	1.95×10^{-11}
	AG	34	$1.56^{b} \pm 0.10$	
	GG	83	$1.67^{a} \pm 0.03$	
SNP13	GG	104	$1.68^{a} \pm 0.07$	0.003
	GA	16	$1.25^{b} \pm 0.11$	

SNP = single nucleotide polymorphism; N = number of genotyped animals; letters across columns indicate significant differences within SNPs.

3.7. Linkage Disequilibrium Estimation Among SNPs

To determine linkage disequilibrium among 13 SNPs of the *NR4A1* gene, D′ and r^2 were estimated for SBWC goats. For both graphs, red boxes represent strong linkage disequilibrium, light red represents moderate LD, and white boxes represent weak LD (Figure 6). Most of the SNPs had strong linkage disequilibrium (D′). For example, SNP 1 was highly linked (complete LD, D′ = 1) with SNP 5, SNP 6, and SNP 7, which means they could have been inherited together (Figure 6a). On the other hand, the linkage analysis (correlation coefficient) expressed by r^2 varied across SNPs (from strong r^2 = 1 to no association r^2 = 0) (Figure 6b).

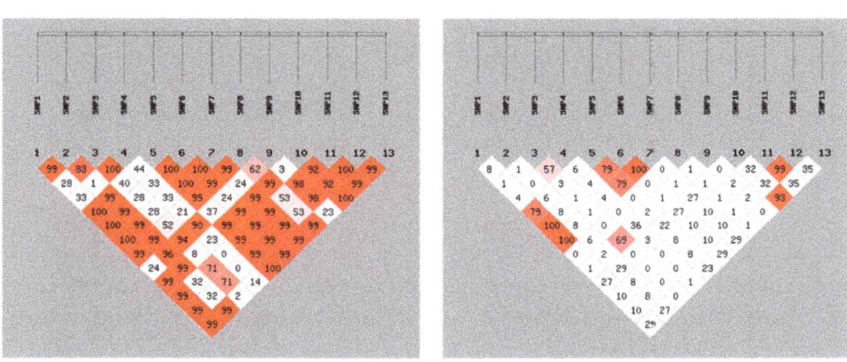

a b

Figure 6. Linkage disequilibrium (LD) analysis between 13 SNPs of the *NR4A1* gene. (**a**) D′ test LD (**b**) r^2 correlation among SNPs. For both plots, red boxes indicate strong LD, light red boxes indicate moderate LD, and white boxes indicate low or no LD among SNPs.

3.8. Analysis of Selection Signals on the NR4A1 Gene's Genomic Region

When compared to the ancestral group, the *NR4A1* gene's 0.5 Mb flanking region (shown by a red dashed box) showed stronger selection signals, indicating a higher level of selection pressure (Figure 7). The statistical methods used resulted in observations in Cashmere goats that the ancestral group along chromosome 5 ranged from 27.05 Mb to 28.0 Mb. A positive and peak value of the −log10(Pi) at the genomic region of 27.52 Mb indicated that the region was under selection pressure (Figure 7). Similarly, Tajima's D and the | iHS | > 2.0 ratio confirmed that the region was under selection. On the other hand, Z (Fst) evidenced that there is a differentiation of Bezoar with domestic goats. Furthermore, other statistics, such as Z(Hp), XPEHH, ln(Pi ratio), and CLR, were used to determine selection sweep and confirm selection pressure of the specified region (Figure S3).

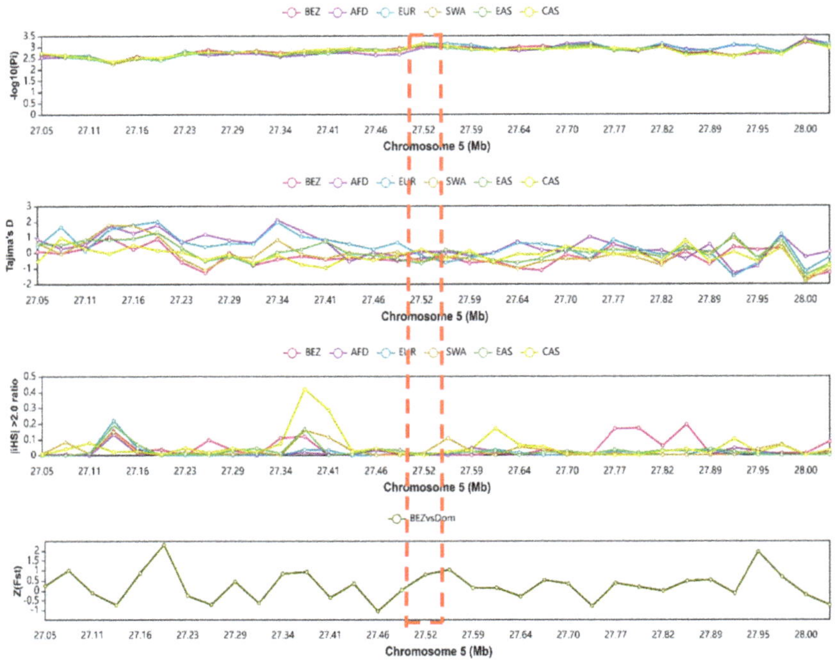

Figure 7. Selective sweep analysis across different statistical methods for the 0.5 Mb flanking region of the *NR4A1* gene: nucleotide diversity (−log10(Pi)); Tajima's D, iHS (>2.0 ratio), and Z-transformed Fst. The red dashed box refers to the genomic location of the *NR4A1* gene. Ancestral goat groups: Bez = Bezoar, AFD = Africa dairy, EUR = Europe, SWA = Southwest Asia, EAS = East Asia, CAS = Cashmere, and Dom = domestic goat. Mb = megabyte.

3.9. Prediction of Transcription Factor Binding Site

The distribution of predicted transcription factor binding sites was determined by using insertion and deletion sequences (Figure 8). Many transcriptions factor binding sites were identified by using the AImodule of the online bioinformatics tool. A straight black line indicates the sequence of a gene. Among the detected transcription factor binding sites highlighted in red rectangles (on the graph and the legend), PLAG2 like zinc finger 2 (PLAGL2) (GGGCCCCTA), CCTC-binding factor-like (CTCFL) (GGGCCCCCTAGT), insulin gene enhancer protein 1 (Isl1) (AGCTGATGGCC), and nuclear receptor subfamily 5 group A member 1 (NR5A1) (ATGGCCCTGGT) were only found in the insertion sequence (Figure 8). These transcription factor binding sites (motifs) might have a role in litter size because II genotypes exhibit significantly higher litter size than the other genotypes (Table 5 and Figure 5a).

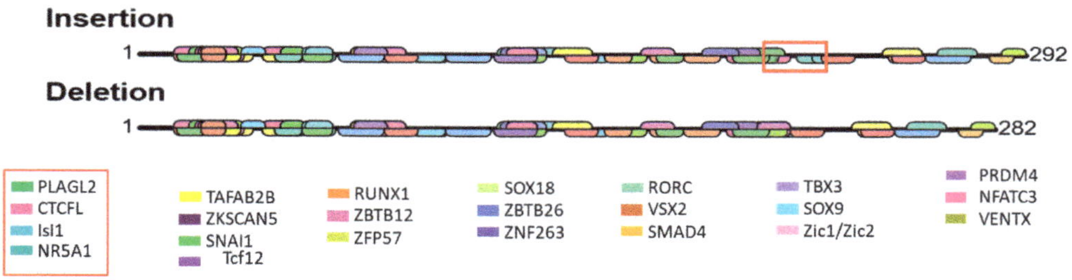

Figure 8. Prediction of transcription factor binding sites for insertion and deletion variants of 11 bp in the goat *NR4A1* gene by using the AImodules online bioinformatics tool. The black line represents the insertion and deletion sequence for the flanking region of InDel. Above the line is the transcription factor binding site for the forward strand (+), and below the line is the transcription factor binding site for the reverse strand (−). Red rectangles highlight binding sites that were found only from insertion sequences.

4. Discussion

4.1. Function, Evolutionary Conservation, and Expression of Goat NR4A1

In goats, litter size is governed by numerous polygenes with minor effects. Among those polygenes, *NR4A1* was identified as a candidate gene that affects the litter size performance of goats. This gene has been directly or indirectly associated with many other genes (Figure 1). Some of these linked genes directly participate in the reproductive process. For example, *RXRA* and *RXRG* mRNA expression was significantly higher in the endometrial tissue of normal cows compared to repeat breeder cows and repeat breeder cows with subclinical endometritis [53]. The authors explained that altered expression of *RXRA*, *RXRG*, and related genes could potentially impair embryo elongation and implantation, promoting embryonic loss in repeat breeder cows [53]. Similarly, *EGR1*-ablated (knockout) mice are infertile, highlighting its essential role in reproduction [54]. The study provided mechanistic insights by demonstrating that *ERG1* mediates VEGF A and FDF 2 signaling in buffalo luteal cells, which affects critical reproductive processes, including angiogenesis, cell proliferation, and steroidogenesis in the corpus luteum [54]. Furthermore, genes like *AKT*, which is responsible for placental development and embryo development [55], *NR4A2*, and *NR4A3* are involved in immune response and inflammation, which are essential for successful reproduction [56]. The crucial significance of the *NR4A1* gene was demonstrated by the conservation study, which revealed that it was evolutionarily conserved in *Capra hircus*, *Ovis aries*, *Bos taurus*, *Sus scrofa*, *Homo sapiens*, and *Mus musculus* (Figure 2).

NR4A1 gene expression was evaluated through RT-qPCR in several tissues. The lung had a comparatively high level of *NR4A1* mRNA expression, followed by the ovary and the spleen. The findings suggest that *NR4A1* may play a significant role in mammalian reproduction, specifically in the maturation of oocytes and follicles. As previously explained, this gene is involved in the cell cycle and differentiation of granulosa cells, theca cells, and oocytes [31]. Moreover, lipid droplets linked to the *NR4A1* contribute to the regulation of progesterone synthesis in the luteal cells of goats [28]. Similarly, *NR4A1–AKR1C1* interaction mediated miR-99b, leading to sustainable progesterone production by the luteal tissue of pigs [57]. It has been shown to regulate the expression of key genes involved in ovarian steroidogenesis and enhance ovarian reserve in aged mouse ovaries [32]. Similarly, several outputs of previous studies have shown that gene expression levels and genetic variations, including InDels and SNPs, significantly influence litter size in goats. For example, studies on genes like *SNX29* and *AKAP12* have identified specific InDel variants strongly associated with increased litter sizes in goat populations. Hence, based on expression results and previously reported evidence, we speculated that the *NR4A1* gene may influence the litter size of goats. To confirm our speculation, the InDel experiment was performed.

4.2. *Effect of NR4A1 Gene Genetic Variability and Selection Signal*

The SBWC goat breed was generally categorized as a low prolific performance breed [8]. Heritability of goat litter size was reported as 0.09 ± 0.02 [58], indicating low heritability and difficulty to improve through direct selection. Therefore, exploration of gene mining for marker-assisted selection is required. As a result, a total of 1136 firstborn does of SBWC goat genomic DNA were utilized to examine the effect of the 11-bp InDel variant in the *NR4A1* gene on goat litter size. According to the association study, goats with the II genotypes had more multiple kids than the ID and DD genotypes (Table 5). Similarly, out of 49 identified SNPs, 13 showed a significant association with litter size. In agreement with the current findings, SNP exploration of *NR4A1* g.3952A>G showed a significant association with litter size at birth and litter size at weaning for commercial sows [30]. Structural variations in a gene can affect its function by disrupting the coding sequence or modifying regulatory regions, leading to changes in protein expression, structure, or activity. [59]. There are various genes that were identified as molecular markers through gene mining for litter size. For example, *DNAH1* has shown increased expression levels in goats with higher litter sizes, emphasizing the necessity of precise genetic research in increasing reproductive performance [60]. Various additional genes were excavated through gene mining techniques to obtain candidate genes for goat litter size. These include *BMP15* and *GDF9* [61], *GNRHR* [62], *SMAD1* [11], *PPP3CA* [63], and *Runx2* [64].

Also, the analysis results of SNPs indicated that most of the SNPs had strong linkage disequilibrium and weak correlations (nearly independent) (Figure 6a,b). These SNPs' variability may have an effect on individual litter size performance. Through linkage disequilibrium, variation in a single gene can affect phenotypic performance [65]. In agreement with the current findings, two SNPs (P27R and A85G loci) for the *GDF9* gene of SBWC goats were strongly linked [66].

Furthermore, all of the analyzed selective signals suggested that the 11-bp InDel was linked to reproductive traits, as the *NR4A1* gene was under higher selection than in ancestral goat groups (Figure 7). The reason for selecting the Cashmere goat for comparison was due to its genetic relation to the SBWC goat. The SBWC goat is a crossbred breed developed using the Liaoning Cashmere goat as the male line and the Shaanbei Black goat as the female line [67]. Its development involved a straightforward hybridization process and progressed through three key stages: hybrid improvement, crossbreeding stabilization, and selection improvement to upgrade [67]. Also, the reason for selecting Bezoar and other ancestral goat groups was because of their higher alternative allele frequency than others (Table 4 and Supplementary files Figures S1 and S2).

Our investigation used a prediction technique to find transcription factor binding sites because transcription factors can attach to particular DNA sequences. Among the identified transcription factor binding sites, PLAGL2, CTCFL, Isl1, and NR5A1 are capable of binding the transcription factors for the insertion sequence, which results in overexpression of the *NR4A1* gene and has an effect on litter size. Gene regulatory activity is significantly impacted by the orientation and arrangement of transcription factor binding sites [68]. Gene transcription is precisely and specifically regulated by the binding of transcription factors, which also affects the enrollment of the transcriptional mechanism [69]. Our results highlighted the potential relevance of the 11-bp-del and SNP loci of the *NR4A1* gene as a possible marker for marker-assisted selection in goats. In the future, more investigation is needed to uncover the complex molecular pathways that underlie the *NR4A1* gene's impact on goat reproductive traits.

5. Conclusions

This study suggests that the *NR4A1* gene plays a role in goat reproductive performance, particularly litter size. Bioinformatics analysis revealed that *NR4A1* may interact functionally and physically with other reproductive genes, such as *RXRA*, *RXRG*, and *EGR1*. Relative mRNA expression profiling in multiple tissues, including the ovary and the oviduct, supports its potential involvement in fertility-related traits. The association analysis using InDel and SNP markers further indicated significant relationships between *NR4A1* variants and litter size, highlighting its possible utility as a candidate gene for marker-assisted selection in goats. However, the sample size for some variants of SNPs may still be insufficient for a robust conclusion regarding genotypes with low frequency. Therefore, further investigation in a large population is required to ensure consistency. Generally, *NR4A1* appears to be a promising candidate gene for goat litter size, but further research is essential to verify its biological function and potential as a marker in breeding programs.

Supplementary Materials: The following supporting information can be downloaded at: https://www.mdpi.com/article/10.3390/ani15121729/s1, Figure S1. Allele frequencies distribution of 11 bp-del locus of *NR4A1* gene for eight ancestral goat groups. The alternate allele frequencies of Africa, Southwest Asia, Africa dairy, and Bezoar ancestral goat groups were relatively higher, which indicates differentiation of the breed through time and results in diversity. Frequencies data was retrieved from the Animal Omics database. Ref = reference allele, Alt = alternate allele; Figure S2. Allele frequencies distribution of 11 bp-del locus of *NR4A1* gene for world-wide goat groups. Relatively higher alternate allele frequencies were available for Morocco, Tanzania, Cashmere, Switzerland, and Saanen goat. These high frequencies indicate that there is higher diversity, which enables adaptation and resilience. Data was generated from the Animal Omics data base. Ref = reference allele, Alt = alternate allele; Figure S3. Selective sweep analysis across different statistical methods for the 0.5 Mb flanking region of the *NR4A1* gene: heterozygosity (Z(Hp)), composite likelihood ratio (CLR), cross-population extended haplotype homozygosity (XPEHH), and difference in nucleotide diversity (ln(Pi ratio)). The red dashed box referred to the genomic location of the *NR4A1* gene. Ancestral goat groups Bez = Bezoar, AFD = Africa dairy, EUR = Europe, SWA = Southwest Asia, EAS = East Asia, CAS = Cashmere, Dom = Domestic goat. Mb = megabyte; Table S1. Allele frequencies distribution of worldwide goat groups for the 11 bp-del locus of the *NR4A1 gene*; Table S2. Selected non-significantly associated resequencing loci SNPs of the *NR4A1* gene with goat litter size.

Author Contributions: Conceptualization, E.A., Y.B. (Yutian Bi), X.C. and X.L.; writing original draft, E.A.; writing—review and editing, E.A., Y.B. (Yutian Bi), Y.B. (Yangyang Bai), Q.Z., L.M., C.P., X.C. and X.L.; methodology, E.A., Y.B. (Yutian Bi), Y.B., Y.B. (Yangyang Bai), Q.Z. and L.M.; data curation, E.A., Y.B. (Yutian Bi), Y.B. (Yangyang Bai) and Q.Z.; software, E.A., Y.B. (Yutian Bi), Y.B. (Yangyang Bai) and Q.Z.; formal analysis, E.A., visualization, E.A.; supervision, C.P., X.C. and X.L.; project administration, C.P., X.C. and X.L.; investigation, C.P., X.C. and X.L.; resources, C.P., X.C. and X.L.; funding acquisition, C.P., X.C. and X.L. All authors have read and agreed to the published version of the manuscript.

Funding: This work was supported by the National Natural Science Foundation of China (No. 32272850), the National Key R&D Program of China (2022YFF1000100), the National Key R&D Program of China (2022YFD1300200), and the Project of Science and Technology Innovation of Undergraduates of Northwest A&F University (2024004609F).

Institutional Review Board Statement: The animal study protocol was approved by the Institutional Review Board (or Ethics Committee) of the Animal Policy and Welfare Committee of Northwest A&F University (file number No. NWAFAC1008 in June 2023).

Informed Consent Statement: Not applicable.

Data Availability Statement: The raw data supporting the conclusions of this article will be made available by the authors upon request.

Acknowledgments: We are grateful to the staff of the SBWC Goat Breeding Farm, Shaanxi province, Yulin City, China.

Conflicts of Interest: The authors declare no conflicts of interest.

Abbreviations

The following abbreviations are used in this manuscript:

SBWC	Shaanbei White Cashmere goat
NR4A1	Nuclear receptor subfamily 4 group A member 1
InDel	Insertion–deletion
SNP	Single nucleotide polymorphism
MAS	Marker-assisted selection
MSA	Multiple Sequence Alignment
GGVD	Goat Genomic Variation Database
MAF	Minor allele frequencies
Ne	Effective allele number
PIC	Polymorphism information content
He	Heterozygosity
Ho	Homozygosity
HWE	Hardy–Weinberg equilibrium
W&C	Weir and Cockerham
BEZ	Bezoar goat
SWA	Southwest Asian goat
AFD	African dairy goat
EUR	European goat
EAS	East Asian goat
iHS	Integrated haplotype score
XPEHH	Cross-population extended haplotype homozygosity
Fst	Z-transformed
CLR	Composite likelihood ratio
TFBS	Transcription factor binding site
LD	Linkage disequilibrium
AIC	Akakie Information Criteria

References

1. Zhong, T.; Hou, D.; Zhao, Q.; Zhan, S.; Wang, L.; Li, L.; Zhang, H.; Zhao, W.; Yang, S.; Niu, L. Comparative Whole-Genome Resequencing to Uncover Selection Signatures Linked to Litter Size in Hu Sheep and Five Other Breeds. *BMC Genom.* **2024**, *25*, 480. [CrossRef] [PubMed]
2. Chen, W.; Han, Y.; Chen, Y.; Liu, X.; Liang, H.; Wang, C.; Khan, M.Z. Potential Candidate Genes Associated with Litter Size in Goats: A Review. *Animals* **2025**, *15*, 82. [CrossRef] [PubMed]
3. Guo, C.; Ye, J.; Liu, J.; Li, Z.; Deng, M.; Guo, Y.; Liu, G.; Sun, B.; Li, Y.; Liu, D. Whole-Genome Sequencing Identified Candidate Genes Associated with High and Low Litter Size in Chuanzhong Black Goats. *Front. Vet. Sci.* **2024**, *11*, 1420164. [CrossRef] [PubMed]
4. El-Shorbagy, H.M.; Abdel-Aal, E.S.; Mohamed, S.A.; El-Ghor, A.A. Association of PRLR, IGF1, and LEP Genes Polymorphism with Milk Production and Litter Size in Egyptian Zaraibi Goat. *Trop. Anim. Health Prod.* **2022**, *54*, 321. [CrossRef]
5. Zhao, Y.J.; Xu, H.Z.; Zhao, Z.Q.; Narisu; Mao, J.W.; Guan, D.L.; Xie, C. Polymorphisms of Osteopontin Gene and Their Association with Placental Efficiency and Prolificacy in Goats. *J. Appl. Anim. Res.* **2015**, *43*, 272–278. [CrossRef]
6. Zonaed Siddiki, A.M.A.M.; Miah, G.; Islam, M.S.; Kumkum, M.; Rumi, M.H.; Baten, A.; Hossain, M.A. Goat Genomic Resources: The Search for Genes Associated with Its Economic Traits. *Int. J. Genomics* **2020**, 5940205. [CrossRef]
7. Hui, Y.; Zhang, Y.; Wang, K.; Pan, C.; Chen, H.; Qu, L.; Song, X.; Lan, X. Goat DNMT3B: An Indel Mutation Detection, Association Analysis with Litter Size and mRNA Expression in Gonads. *Theriogenology* **2020**, *147*, 108–115. [CrossRef]

. Wang, K.; Liu, X.; Qi, T.; Hui, Y.; Yan, H.; Qu, L.; Lan, X.; Pan, C. Whole-Genome Sequencing to Identify Candidate Genes for Litter Size and to Uncover the Variant Function in Goats (*Capra hircus*). *Genomics* **2021**, *113*, 142–150. [CrossRef]

. Bai, Y.; Bo, D.; Bi, Y.; Areb, E.; Zhu, H.; Pan, C.; Lan, X. Analysis of Goat PPP6C mRNA Profile, Detection of Genetic Variations, and Their Associations with Litter Size. *Anim. Reprod. Sci.* **2024**, *268*, 107544. [CrossRef]

0. Wijayanti, D.; Zhang, S.; Bai, Y.; Pan, C.; Chen, H.; Qu, L.; Guo, Z.; Lan, X. Investigation on mRNA Expression and Genetic Variation within Goat SMAD2 Gene and Its Association with Litter Size. *Anim. Biotechnol.* **2023**, *34*, 2111–2119. [CrossRef]

1. Wijayanti, D.; Zhang, S.; Yang, Y.; Bai, Y.; Akhatayeva, Z.; Pan, C.; Zhu, H.; Qu, L; Lan, X. Goat SMAD Family Member 1 (SMAD1): mRNA Expression, Genetic Variants, and Their Associations with Litter Size. *Theriogenology* **2022**, *193*, 11–19. [CrossRef] [PubMed]

2. Wijayanti, D.; Bai, Y.; Hanif, Q.; Chen, H.; Zhu, H.; Qu, L.; Guo, Z.; Lan, X. Goat CLSTN2 Gene: Tissue Expression Profile, Genetic Variation, and Its Associations with Litter Size. *Anim. Biotechnol.* **2023**, *34*, 2674–2683. [CrossRef] [PubMed]

3. Buthelezi, N.L.; Mtileni, B.; Nephawe, K.A.; Idowu, P.A.; Modiba, M.C.; Mpedi, H.; Mpofu, T.J. The Impact of Parity, Litter Size and Birth Weight Variations within a Litter on Piglet Pre-Weaning Performance. *Czech J. Anim. Sci.* **2024**, *69*, 255–268. [CrossRef]

4. Rutherford, K.; Baxter, E.; D'Eath, R.; Turner, S.; Arnott, G.; Roehe, R.; Ask, B.; SandØe, P.; Moustsen, V.; Thorup, F.; et al. The Welfare Implications of Large Litter Size in the Domestic Pig I: Biological Factors. *Anim. Welf.* **2013**, *22*, 199–218. [CrossRef]

5. Ju, M.; Wang, X.; Li, X.; Zhang, M.; Shi, L.; Hu, P.; Zhang, B.; Han, X.; Wang, K.; Li, X.; et al. Effects of Litter Size and Parity on Farrowing Duration of Landrace × Yorkshire Sows. *Animals* **2021**, *12*, 94. [CrossRef]

6. Boonkum, W.; Permthongchoochai, S.; Chankitisakul, V.; Duangjinda, M. Genetic Strategies for Enhancing Litter Size and Birth Weight Uniformity in Piglets. *Front. Vet. Sci.* **2025**, *12*, 1512701. [CrossRef]

7. Parra-Vargas, M.; Ramon-Krauel, M.; Lerin, C.; Jimenez-Chillaron, J.C. Size Does Matter: Litter Size Strongly Determines Adult Metabolism in Rodents. *Cell Metab.* **2020**, *32*, 334–340. [CrossRef]

8. Crean, D.; Murphy, E.P. Targeting NR4A Nuclear Receptors to Control Stromal Cell Inflammation, Metabolism, Angiogenesis, and Tumorigenesis. *Front. Cell Dev. Biol.* **2021**, *9*, 1–10. [CrossRef]

9. Pearen, M.A.; Muscat, G.E.O. Orphan Nuclear Receptors and the Regulation of Nutrient Metabolism: Understanding Obesity. *Physiology* **2012**, *27*, 156–166. [CrossRef]

0. Cruz-Nieves, R.P.; Ramírez-Rosales, G.E.; González-Ramírez, J.; Sánchez-Muñoz, F.; Ruiz-Hernández, A. Analysing the Gene Expression Profiles of the Orphan Nuclear Receptors NR4A1, NR4A2 and NR4A3 in Premalignant Lesions of the Cervix and Cervicitis. *Eur. J. Obstet. Gynecol. Reprod. Biol. X* **2024**, *24*, 100355. [CrossRef]

1. Safe, S.; Jin, U.-H.; Morpurgo, B.; Abdayyeh, A.; Singh, M.; Tjalkens, R.B.; Steroid, J.; Mol, B.; Author, B. Nuclear Receptor 4A (NR4A) Family-Orphans No More. *J. Steroid. Biochem. Mol. Biol.* **2016**, *157*, 48–60. [CrossRef] [PubMed]

2. Xing, J.; Zheng, J.; Cui, S.; Wang, J.; Wang, Y.; Li, Y.; Zhu, J.; Lin, Y. Nuclear Receptor Subfamily 4 Group A Member 1 (NR4A1) Promotes the Adipogenesis of Intramuscular Preadipocytes through PI3K/AKT Pathway in Goats. *Animals* **2024**, *14*, 2051. [CrossRef] [PubMed]

3. Parreira, J.R.; Hernández-Castellano, L.E.; Argüello, A.; Capote, J.; Castro, N.; De Sousa Araújo, S.; De Almeida, A.M. Understanding Seasonal Weight Loss Tolerance in Dairy Goats: A Transcriptomics Approach. *BMC Genom.* **2020**, *21*, 629. [CrossRef] [PubMed]

4. Zhang, Z.-K.; Li, J.; Liu, J.; Guo, B.; Leung, A.; Zhang, G.; Zhang, B.-T. Icaritin Requires Phosphatidylinositol 3 Kinase (PI3K)/Akt Signaling to Counteract Skeletal Muscle Atrophy Following Mechanical Unloading. *Sci. Rep.* **2016**, *6*, 20300. [CrossRef]

5. Kasch, J.; Kanzleiter, I.; Saussenthaler, S.; Schürmann, A.; Keijer, J.; van Schothorst, E.; Klaus, S.; Schumann, S. Insulin Sensitivity Linked Skeletal Muscle Nr4a1 DNA Methylation Is Programmed by the Maternal Diet and Modulated by Voluntary Exercise in Mice. *J. Nutr. Biochem.* **2018**, *57*, 86–92. [CrossRef]

6. Chao, L.C.; Wroblewski, K.; Zhang, Z.; Pei, L.; Vergnes, L.; Ilkayeva, O.R.; Ding, S.Y.; Reue, K.; Watt, M.J.; Newgard, C.B.; et al. Insulin Resistance and Altered Systemic Glucose Metabolism in Mice Lacking Nur77. *Diabetes* **2009**, *58*, 2788–2796. [CrossRef]

7. Ran, X.; Hu, F.; Mao, N.; Ruan, Y.; Yi, F.; Niu, X.; Huang, S.; Li, S.; You, L.; Zhang, F.; et al. Differences in Gene Expression and Variable Splicing Events of Ovaries between Large and Small Litter Size in Chinese Xiang Pigs. *Porc. Health Manag.* **2021**, *7*, 52. [CrossRef]

8. Yu, H.; Li, X.; Zhao, J.; Wang, W.; Wei, Q.; Mao, D. NR4A1-Mediated Regulation of Lipid Droplets in Progesterone Synthesis in Goat Luteal Cells. *Biol. Reprod.* **2024**, *111*, 640–654. [CrossRef]

9. Cui, L.; Bao, H.; Zhu, W.; Tang, Y.; Luo, Q.; Si, Y.; Fu, Q.; Jiang, Z. hUMSCs Transplantation Regulates AMPK/NR4A1 Signaling Axis to Inhibit Ovarian Fibrosis in POI Rats. *Stem Cell Rev. Rep.* **2023**, *19*, 1449–1465. [CrossRef]

0. Kumchoo, T.; Mekchay, S. Association of NR4A1 and GNB2L1 Genes with Reproductive Traits in Commercial Pig Breeds. *Genet. Mol. Res.* **2015**, *14*, 16276–16284. [CrossRef]

1. Yu, H.; Zhao, J.; Pei, X.; Xia, Y.; Li, H.; Wang, W.; Mao, D. Dual Role of NR4A1 in Porcine Ovarian Granulosa Cell Differentiation and Granulosa-Lutein Cell Regression in Vitro. *Theriogenology* **2023**, *198*, 292–304. [CrossRef] [PubMed]

32. Yao, Y.; Wang, B.; Yu, K.; Song, J.; Wang, L.; Zhang, X.; Li, Y. Nur77 Improves Ovarian Function in Reproductive Aging Mice by Activating Mitophagy and Inhibiting Apoptosis. *Reprod. Biol. Endocrinol.* **2024**, *22*, 86. [CrossRef]
33. Kang, Z.; Bai, Y.; Lan, X.; Zhao, H. Goat *AKAP12*: Indel Mutation Detection, Association Analysis with Litter Size and Alternative Splicing Variant Expression. *Front. Genet.* **2021**, *12*, 648256. [CrossRef] [PubMed]
34. Xin, D.; Bai, Y.; Bi, Y.; He, L.; Kang, Y.; Pan, C.; Zhu, H.; Chen, H.; Qu, L.; Lan, X. Insertion/Deletion Variants within the IGF2BP2 Gene Identified in Reported Genome-Wide Selective Sweep Analysis Reveal a Correlation with Goat Litter Size. *J. Zhejiang Univ. Sci. B* **2021**, *22*, 757–766. [CrossRef]
35. Tang, Q.; Zhang, Y.; Yang, Y.; Hu, H.; Lan, X.; Pan, C. The KMT2A Gene: mRNA Differential Expression in the Ovary and a Novel 13-Nt Nucleotide Sequence Variant Associated with Litter Size in Cashmere Goats. *Domest. Anim. Endocrinol.* **2021**, *74*, 106538. [CrossRef]
36. Wang, Z.; Wang, X.; Lan, X.; Zhu, H.; Qu, L.; Pan, C. Polymorphism within the GATA Binding Protein 4 Gene Is Significantly Associated with Goat Litter Size. *Anim. Biotechnol.* **2023**, *34*, 4291–4300. [CrossRef]
37. Cui, Y.; Yan, H.; Wang, K.; Xu, H.; Zhang, X.; Zhu, H.; Liu, J.; Qu, L.; Lan, X.; Pan, C. Insertion/Deletion within the KDM6A Gene Is Significantly Associated with Litter Size in Goat. *Front. Genet.* **2018**, *9*, 91. [CrossRef]
38. Lin, G.; He, Q.; Cai, J.; Yang, X.; Wang, Y. Detection of Single Nucleotide Polymorphisms Based on Triple-Helix Molecular Switch Combined with Invader Assay. *Microchem. J.* **2024**, *199*, 109954. [CrossRef]
39. Tamura, K.; Stecher, G.; Kumar, S. MEGA11: Molecular Evolutionary Genetics Analysis Version 11. *Mol. Biol. Evol.* **2021**, *38*, 3022–3027. [CrossRef]
40. Zhang, S.; Jiang, E.; Kang, Z.; Bi, Y.; Liu, H.; Xu, H.; Wang, Z.; Lei, C.; Chen, H.; Lan, X. CircRNA Profiling Reveals an Abundant circBDP1 That Regulates Bovine Fat Development by Sponging miR-181b/miR-204 Targeting Sirt1/TRARG1. *J. Agric. Food Chem.* **2022**, *70*, 14312–14328. [CrossRef]
41. Bi, Y.; Zhang, S.; Li, J.; He, L.; Kang, Y.; Chen, H.; Lan, X.; Pan, C. The mRNA Expression Profile of the Goat Prion Protein Testis-Specific (PRNT) Gene and Its Associations with Litter Size. *Theriogenology* **2021**, *165*, 69–75. [CrossRef] [PubMed]
42. Wang, K.; Yan, H.; Xu, H.; Yang, Q.; Zhang, S.; Pan, C.; Chen, H.; Zhu, H.; Liu, J.; Qu, L.; et al. A Novel Indel within Goat Casein Alpha S1 Gene Is Significantly Associated with Litter Size. *Gene* **2018**, *671*, 161–169. [CrossRef] [PubMed]
43. Aljanabi, S.M.; Martinez, I. Universal and Rapid Salt-Extraction of High Quality Genomic DNA for PCR-Based Techniques. *Nucleic Acids Res.* **1997**, *25*, 4692–4693.
44. Yue, C.; Bai, W.L.; Zheng, Y.Y.; Hui, T.Y.; Sun, J.M.; Guo, D.; Guo, S.L.; Wang, Z.Y. Correlation Analysis of Candidate Gene SNP for High-Yield in Liaoning Cashmere Goats with Litter Size and Cashmere Performance. *Anim. Biotechnol.* **2021**, *32*, 43–50. [CrossRef]
45. Fu, W.; Wang, R.; Yu, J.; Hu, D.; Cai, Y.; Shao, J.; Jiang, Y. GGVD: A Goat Genome Variation Database for Tracking the Dynamic Evolutionary Process of Selective Signatures and Ancient Introgressions. *J. Genet. Genom.* **2021**, *48*, 248–256. [CrossRef]
46. Akhmet, N.; Zhu, L.; Song, J.; Akhatayeva, Z.; Zhang, Q.; Su, P.; Li, R.; Pan, C.; Lan, X. Exploring the Sheep MAST4 Gene Variants and Their Associations with Litter Size. *Animals* **2024**, *14*, 591. [CrossRef]
47. Li, H.; Durbin, R. Fast and Accurate Short Read Alignment with Burrows-Wheeler Transform. *Bioinformatics* **2009**, *25*, 1754–1760. [CrossRef]
48. Aydinli, M.; Liang, C.; Dandekar, T. Motif and Conserved Module Analysis in DNA (Promoters, Enhancers) and RNA (lncRNA, mRNA) Using AlModules. *Sci. Rep.* **2022**, *12*, 17588. [CrossRef]
49. Shi, Y.Y.; He, L. SHEsis, a Powerful Software Platform for Analyses of Linkage Disequilibrium, Haplotype Construction, and Genetic Association at Polymorphism Loci. *Cell Res.* **2005**, *15*, 97–98, Erratum in *Cell Res.* **2006**, *16*, 851. [CrossRef]
50. Myers, A.J.; Gibbs, J.R.; Webster, J.A.; Rohrer, K.; Zhao, A.; Marlowe, L.; Kaleem, M.; Leung, D.; Bryden, L.; Nath, P.; et al. A Survey of Genetic Human Cortical Gene Expression. *Nat. Genet.* **2007**, *39*, 1494–1499. [CrossRef]
51. R Core Team. *R: A Language and Environment for Statistical Computing*; R Foundation for Statistical Computing: Vienna, Austria, 2024.
52. Warde-Farley, D.; Donaldson, S.L.; Comes, O.; Zuberi, K.; Badrawi, R.; Chao, P.; Franz, M.; Grouios, C.; Kazi, F.; Lopes, C.T.; et al. The GeneMANIA Prediction Server: Biological Network Integration for Gene Prioritization and Predicting Gene Function. *Nucleic Acids Res.* **2010**, *38*, W214–W220. [CrossRef] [PubMed]
53. Kasimanickam, R.; Kasimanickam, V.; Grende, K. Endometrial Expression of Various Genes (ISGs, PPARs, RXRs and MUC1) on Day 16 Post-Ovulation in Repeat Breeder Cows, with or without Subclinical Endometritis. *Theriogenology* **2020**, *142*, 251–259. [CrossRef] [PubMed]
54. Punetha, M.; Chouhan, V.S.; Sonwane, A.; Singh, G.; Bag, S.; Green, J.A.; Whitworth, K.; Sarkar, M. Early Growth Response Gene Mediates in VEGF and FGF Signaling as Dissected by CRISPR in Corpus Luteum of Water Buffalo. *Sci. Rep.* **2020**, *10*, 6849. [CrossRef] [PubMed]
55. Kalous, J.; Aleshkina, D.; Anger, M. A Role of PI3K/Akt Signaling in Oocyte Maturation and Early Embryo Development. *Cells* **2023**, *12*, 1830. [CrossRef]

56. Prince, L.R.; Prosseda, S.D.; Higgins, K.; Carlring, J.; Prestwich, E.C.; Ogryzko, N.V.; Rahman, A.; Basran, A.; Falciani, F.; Taylor, P.; et al. NR4A Orphan Nuclear Receptor Family Members, NR4A2 and NR4A3, Regulate Neutrophil Number and Survival. *Blood* **2017**, *130*, 1014–1025. [CrossRef]
57. Przygrodzka, E.; Sokołowska, G.; Myszczynski, K.; Krawczynski, K.; Kaczmarek, M.M. Clustered microRNAs: The Molecular Mechanism Supporting the Maintenance of Luteal Function during Early Pregnancy. *FASEB J.* **2020**, *34*, 6582–6597. [CrossRef]
58. Mucha, S.; Tortereau, F.; Doeschl-Wilson, A.; Rupp, R.; Conington, J. Animal Board Invited Review: Meta-Analysis of Genetic Parameters for Resilience and Efficiency Traits in Goats and Sheep. *Animal* **2022**, *16*, 100456. [CrossRef]
59. Li, Y.; Liu, Q.; Pan, C.-Y.; Lan, X.-Y. The Free Fatty Acid Receptor 2 (FFA2): Mechanisms of Action, Biased Signaling, and Clinical Prospects. *Pharmacol. Ther.* **2025**, *272*, 108878. [CrossRef]
60. Wang, Z.; Wang, R.; Pan, C.; Chen, H.; Qu, L.; Wu, L.; Guo, Z.; Zhu, H.; Lan, X. Genetic Variations and mRNA Expression of Goat DNAH1 and Their Associations with Litter Size. *Cells* **2022**, *11*, 1371. [CrossRef]
61. Abuzahra, M.; Abu Eid, L.; Effendi, M.H.; Mustofa, I.; Lamid, M.; Rehman, S. Polymorphism Studies and Candidate Genes Associated with Litter Size Traits in Indonesian Goats. a Systematic Review. *F1000Research* **2023**, *12*, 61. [CrossRef]
62. Saranya, S.K.; Thomas, M.; Aravindakshan, T.V.; Venkatachalapathy, R.T.; Sukumaran, I.; Kanakkaparambil, R. Diplotype-Based Assessment of GNRHR Gene Polymorphisms and Their Role in Determining Litter Size at Birth in the Malabari Goat Population of South India. *Small Rumin. Res.* **2024**, *233*, 107229. [CrossRef]
63. Bai, Y.; Li, J.; Zhu, H.; Liu, J.; Dong, S.; Li, L.; Qu, L.; Chen, H.; Song, X.; Lan, X. Deletion Mutation within the Goat PPP3CA Gene Identified by GWAS Significantly Affects Litter Size. *Reprod. Fertil. Dev.* **2021**, *33*, 476–483. [CrossRef] [PubMed]
64. Jiang, E.; Kang, Z.; Wang, X.; Liu, Y.; Liu, X.; Wang, Z.; Li, X.; Lan, X. Detection of Insertions/Deletions (InDels) within the Goat Runx2 Gene and Their Association with Litter Size and Growth Traits. *Anim. Biotechnol.* **2021**, *32*, 169–177. [CrossRef] [PubMed]
65. Huang, D.W.; Wang, J.X.; Liu, Q.Y.; Chu, M.X.; Di, R.; He, J.N.; Cao, G.L.; Fang, L.; Feng, T.; Li, N. Analysis on DNA Sequence of TSHB Gene and Its Association with Reproductive Seasonality in Goats. *Mol. Biol. Rep.* **2013**, *40*, 1893–1904. [CrossRef]
66. Bi, Y.; Li, J.; Wang, X.; He, L.; Lan, K.; Qu, L.; Lan, X.; Song, X.; Pan, C. Two Novel Rare Strongly Linked Missense SNPs (P27R and A85G) Within the GDF9 Gene Were Significantly Associated With Litter Size in Shaanbei White Cashmere (SBWC) Goats. *Front. Vet. Sci.* **2020**, *7*, 406. [CrossRef]
67. Yuan, C.; Guo, T.T.; Liu, J.B.; Yue, Y.J.; Yang, B.H. Conservation and Utilization of Indigenous Goats and Breeding of New Breeds in China. In *Sustainable Goat Production in Adverse Environments*; Springer: Cham, Switzerland, 2017; Volume 1, pp. 457–472. [CrossRef]
68. Georgakopoulos-Soares, I.; Deng, C.; Agarwal, V.; Chan, C.S.Y.; Zhao, J.; Inoue, F.; Ahituv, N. Transcription Factor Binding Site Orientation and Order Are Major Drivers of Gene Regulatory Activity. *Nat. Commun.* **2023**, *14*, 2333. [CrossRef]
69. Muley, V.Y. Prediction and Analysis of Transcription Factor Binding Sites: Practical Examples and Case Studies Using R Programming. In *Reverse Engineering of Regulatory Networks*; Springer: Berlin/Heidelberg, Germany, 2024; pp. 199–225.

animals
MDPI

Article

Genomic Selection for Early Growth Traits in Inner Mongolian Cashmere Goats Using ABLUP, GBLUP, and ssGBLUP Methods

Tao Zhang [1,2], Linyu Gao [1,3,4], Bohan Zhou [1,3,4], Qi Xu [1,3,4], Yifan Liu [1,3,4], Jinquan Li [3,4], Qi Lv [1,3,4], Yanjun Zhang [1,3,4], Ruijun Wang [1,3,4], Rui Su [1,3,4,*] and Zhiying Wang [1,3,4,*]

[1] College of Animal Science, Inner Mongolia Agricultural University, Hohhot 010018, China
[2] Grassland Research Institute, Chinese Academy of Agricultural Sciences, Hohhot 010018, China
[3] Inner Mongolia Key Laboratory of Sheep & Goat Genetics Breeding and Reproduction, Hohhot 010018, China
[4] Key Laboratory of Mutton Sheep & Goat Genetics and Breeding, Ministry of Agriculture and Rural Affairs, Hohhot 010018, China
* Correspondence: suruiyu@126.com (R.S.); wzhy0321@126.com (Z.W.)

Simple Summary: This study focused on improving the growth rates of Inner Mongolian cashmere goats (IMCGs) by identifying the most effective genomic selection methods for early growth traits. Faster growth rates are important for farmers because they can lead to increased meat production, ultimately enhancing the economic value of these goats. We analyzed data from 2256 cashmere goats, looking at traits such as birth weight, weaning weight, daily weight gain, and yearling weight. By using advanced statistical models, we aimed to determine which method would provide the most accurate predictions for breeding values based on genetic information. Our research found that specific factors like birth year, herd, sex, birth type, and the age of dams play important roles in growth traits. Among the models we tested, the one that included additive genetic effects from both the individual and its mother, as well as environmental effects, was best. Notably, the method known as ssGBLUP offered the highest accuracy in predicting breeding values, making it the preferred choice for breeding programs aimed at enhancing growth rates in IMCGs. In conclusion, we recommend using the ssGBLUP method for improving breeding efficiency in the growth traits of IMCGs.

Abstract: This study aimed to identify the best model and method for the genomic selection of early growth traits in Inner Mongolian cashmere goats (IMCGs). Using data from 50,728 SNPs, the phenotypes (birth weight, BW; weaning weight, WW; daily weight gain, DWG; and yearling weight, YW) of 2256 individuals, and pedigree information from 14,165 individuals, fixed effects were analyzed using a generalized linear model. Four single-trait animal models with varying combinations of individual and maternal effects were evaluated using the ABLUP, GBLUP, and ssGBLUP methods. The best model was selected based on a likelihood ratio test. Five-fold cross-validation was used to assess the accuracy and reliability of the genomic estimated breeding values (GEBVs). Birth year and herd significantly affected BW ($p < 0.05$) and WW, DWG, and YW ($p < 0.01$), while sex, birth type, and dam age had highly significant effects on all traits ($p < 0.01$). Model 4, incorporating direct and maternal additive genetic effects, maternal environmental effects, and their covariance, was optimal. Additionally, ssGBLUP achieved the highest GEBV accuracy (0.61–0.70), outperforming the GBLUP and ABLUP methods. Thus, ssGBLUP is recommended for enhancing the genetic progress in IMCGs. Under the best method, the heritability estimates for BW, WW, DGW, and YW were 0.11, 0.25, 0.15, and 0.23, respectively.

Keywords: genomic selection; accuracy; GEBV; Inner Mongolian cashmere goats

1. Introduction

The Inner Mongolian cashmere goat (IMCG), a highly regarded breed known worldwide, has been developed over a long period through natural selection and artificial breeding for both cashmere and meat production [1]. This breed has the advantages of a high cashmere production, good cashmere quality, and stable genetic architecture. It is used as the paternal parent for the cultivation of many cashmere goat breeds in China and was listed as one of the first breeds for the protection of genetic resources of livestock and poultry in China.

Early growth traits have strong implications for the reproductive and production performance of animals [2–4]. The growth and development of young lambs are directly linked to the economic benefits of livestock production, serving as key indicators of growth rate, health status, and overall productivity. Gätjens [5] reported that individuals born small had a higher risk for cardio-metabolic disease. Early growth traits, such as birth weight and weaning weight, are the basis for selection in genetic improvement programs for meat production due to their strong association with each other [6]. Therefore, the assessment of individual growth performance is imperative for determining the genetic potential of breeds and subsequently devising effective genetic improvement programs.

Genomic selection was described by Meuwissen et al. [7]. It is widely used in the selection of superior animals and plants [8–10]. There have been numerous reports on the genetic evaluation of early growth traits in livestock. Lavvaf and Noshary [11] used a single-trait animal model to perform a genetic evaluation of early growth traits in the Lori breed of sheep. Early growth traits were significantly affected by direct additive genetic effects, direct maternal genetic effects, and environmental effects. Balasundaram et al. [12] evaluated the genetic parameters of the growth traits and quantitative genetic metrics of Mecheri sheep in Tamil Nadu. The best model included direct and maternal genetic effects as random effects with no covariance.

Due to their economic importance, early growth traits in most species have been subjected to genomic selection to improve breeding efficiency. Fei Ge et al. [13] evaluated the accuracy of genomic prediction for growth traits at weaning and yearling ages in yak. The result indicated that the prediction accuracy for early growth traits in yak ranged from 0.147 to 0.391. Terakado et al. [14] compared methods for predicting genomic breeding values for growth traits in Nellore cattle. Bayesian methods provided slightly more accurate predictions of genomic breeding values than the GBLUP. Song et al. [15] performed genomic prediction for growth traits in pig using an admixed reference population. It was demonstrated that ssGBLUP was the best method for genomic selection in pigs. Tianfei Liu et al. [16] evaluated the accuracy of genomic prediction for growth traits in Chinese triple-yellow chickens. The accuracy of the genomic prediction of growth traits ranged from 0.448 to 0.468. However, there are few reports on the genomic selection of early growth traits in goats. The purpose of this study is to determine the best model and method for the genomic prediction of early growth traits in IMCGs. This is of great significance for performing the genome selection of early growth traits in Inner Mongolian cashmere goats.

2. Materials and Methods

2.1. Data and Population Sources

The data were collected from the Inner Mongolia Yi-Wei White Cashmere Goat Limited Liability Company, located in the southwestern part of Inner Mongolia (Ordos, Inner

Mongolia, China). Birth weight (BW), weaning weight (WW), daily weight gain before weaning (DWG), and yearling weight (YW) were considered in this study. The phenotypes of the above traits were recorded from 2014 to 2019. The total number of individuals with pedigree records was 14,165 from 1990 to 2019, from which 2256 were genotyped using the 70K SNP chip (Illumina Inc., San Diego, CA, USA).

2.2. Data Preprocessing

For the phenotype data of the early growth traits collected from this population, an Excel spreadsheet was used for preprocessing. Firstly, the individuals with recorded genotypes were extracted from a database of breeding information, and the outliers for each trait were further excluded. Values outside the range of the mean plus or minus 2.58 times the standard deviation were defined as outliers. Then, quality control of the genotype data was performed using PLINKv1.90 software [17]. The quality control criteria were as follows: (1) exclude individuals with SNP missing rate of >10%; (2) exclude SNPs with individual missing rate of >10%; (3) exclude SNPs with minor allele frequency (MAF) of <5%; and (4) exclude SNPs violating the Hardy–Weinberg equilibrium (HWE) at $p < 1 \times 10^{-5}$. The missing genotypes were imputed using Beagle [18]. Finally, a total of 50,728 SNPs in 2256 individuals remained for subsequent analyses.

2.3. Determination of Fixed and Random Effects

The determination of fixed and random effects is essential for genetic evaluation. The fixed effects considered in this study include sex (male or female), year of birth (2014–2021), herds (1–12, including 6 adult dam herds, 3 yearling dam herds, 2 yearling ram herds, and 1 adult ram herd), age of dam (2–8 years old), and birth type (single, twins, and triplets). The environmental factors affecting early growth traits in IMCGs were further determined by constructing generalized linear models of the phenotype and environmental effects using the GLM procedure of SAS 9.2 [19]. The generalized linear model is as follows.

$$y_{ijklmn} = u + S_i + Y_j + H_k + D_l + B_m + e_{ijklmn},$$

where y is the observations of individuals in each trait, u is the overall mean, S_i is the effect of the ith sex, Y_j is the effect of the jth year of birth, H_k is the effect of the kth herds, D_l is the effect of the lth dam's age, B_m is the effect of the mth birth type, and e_{ijklmn} is the random error.

After determining the fixed effects, four single-trait animal models with various combinations of individual and maternal effects were constructed. The formula of each model is as follows:

$$\begin{aligned}
&\text{Model 1}: \ y = Xb + Z_1\,a + e;\\
&\text{Model 2}: \ y = Xb + Z_1\,a + Z_2\,m + e;\\
&\text{Model 3}: \ y = Xb + Z_1\,a + Z_2\,m + Z_3\,c + e\ \text{cov}(a, m) = 0;\\
&\text{Model 4}: \ y = Xb + Z_1\,a + Z_2\,m + Z_3\,c + e\ \text{cov}(a, m) \neq 0;
\end{aligned}$$

where y is the vector of observations for each trait, b is the vector of fixed effects, a is the vector of individual additive genetic effects, m is the vector of maternal genetic effects, c is the vector of maternal permanent environmental effects, and X, Z_1, Z_2, and Z_3 are the incidence matrices for the effects in b, a, m, and c, respectively. e is the vector of residual effects. The difference between Model 3 and Model 4 is the presence or absence of covariance between the direct additive and maternal genetic effects. The total heritability was calculated by the following formula.

$$h_T^2 = \left(\sigma_a^2 + 0.5\sigma_m^2 + 1.5\sigma_{am}\right)/\sigma_p^2,$$

where h_T^2 is the total heritability. σ_a^2 is the individual additive genetic variance. σ_m^2 is the maternal genetic variance. σ_{am} is the covariance between the individual direct additive genetic effect and maternal genetic effect.

The most appropriate model was obtained using the likelihood ratio test [20]. The formula used to calculate the likelihood ratio value was as follows:

$$LR = -2Log\left(\frac{L_1}{L_2}\right),$$

where LR is the value of the likelihood ratio, and $LogL_1$ and $LogL_2$ are the maximum likelihood function values for Model 1 and Model 2, respectively. Model 1 was a submodel of Model 2 [21]. The LR statistic follows a chi-square distribution ($\aleph^2$), with degrees of freedom equal to the difference in the number of parameters between the two models. The best model can be determined by comparing the LR value to the critical value of the chi-square distribution.

2.4. Estimates of Variance Components and Genetic Parameters

The variance components and genetic parameters were estimated for each early growth trait with three methods (ABLUP, GBLUP, and ssGBLUP) using the ASRgenomics package of ASREML4.2 software [22]. For each method, the mixed animal model equations were as follows.

ABLUP: The A matrix was constructed using the pedigree.

$$\begin{bmatrix} X'Z & X'Z \\ X'Z & Z'Z + \frac{\sigma_e^2}{\sigma_a^2}A^{-1} \end{bmatrix} \begin{bmatrix} \hat{b} \\ \hat{a} \end{bmatrix} = \begin{bmatrix} X'y \\ Z'y \end{bmatrix},$$

where X' is the transpose matrix of the incidence matrix of fixed effects. Z is the incidence matrix for the effects in individual additive genetic effects. Z' is the transpose matrix of Z. σ_a^2 is the variance of the individual additive genetic effects. σ_e^2 is the residual variance. A^{-1} is the inverse matrix of A.

GBLUP: The G matrix was constructed using genotype data.

$$G = \frac{ZZ'}{2\sum P_i\,(1 - P_i)}$$

In the G matrix above, P_i refers to the allele frequency of the jth marker.

$$\begin{bmatrix} X'X & X'Z \\ Z'X & Z'Z + G^{-1}\frac{\sigma_e^2}{\sigma_a^2} \end{bmatrix} \begin{bmatrix} b \\ \mu \end{bmatrix} = \begin{bmatrix} X'y \\ Z'y \end{bmatrix},$$

where G^{-1} is the inverse matrix of G.

ssGBLUP: The H matrix was constructed by combining the pedigree and genotype.

$$H = \begin{bmatrix} H_{11} & H_{12} \\ H_{21} & H_{22} \end{bmatrix},$$

$$= \begin{bmatrix} A_{11} + A_{12}A_{22}^{-1}(G - A_{22})A_{22}^{-1}A_{21} & GA_{12}A_{22}^{-1} \\ GA_{22}^{-1}A_{21} & G \end{bmatrix},$$

$$H^{-1} = A^{-1} + \begin{bmatrix} 0 & 0 \\ 0 & G^{-1} - A_{22}^{-1} \end{bmatrix},$$

where A_{11} is the relationship matrix among known pedigree individuals, A_{12} and A_{21} are the relationship matrices between known pedigree individuals and individuals with molecular marker data, respectively, and A_{22} is the pedigree-based relationship matrix among individuals with molecular marker data. G is the genomic relationship matrix among individuals with molecular marker data. A^{-1} is the inverse of the submatrix A_{11}, and A_{22}^{-1} is the inverse of the submatrix A_{22}. The meanings of the other letters are the same as in the ABLUP mixed model equations.

2.5. Assessment of the Genomic Prediction Accuracy

The research population was divided into five groups, four of which were combined as the reference group and the remaining as the validation group. Each group was sequentially used as the validation group, and this was repeated five times; finally, the average of the correlation coefficients between the corrected phenotype values and the breeding values obtained from the five repetitions was used as a criterion for evaluating the genomic prediction accuracy. When the result is closer to 1, its prediction ability is better [23]. The formula was as follows:

$$r = \frac{cov(a,p)}{\sqrt{var(a)var(p)}},$$

where cov(a, p) is the covariance of the estimated breeding and phenotype values in each trait, respectively. var(a) and var(p) are the variances of the estimated breeding values and phenotype values in each trait, respectively.

The formula for the reliability of the genomic estimated breeding value (GEBV) is as follows:

$$r_i^2 = 1 - \frac{PEV_i}{\sigma_a^2},$$

where PEV_i is the prediction error variance of the ith individual, which was obtained by inverting the mixed model equations. It reflects the uncertainty of the estimated breeding value. σ_a^2 is the additive genetic variance.

3. Results

3.1. Descriptive Statistics of Early Growth Traits in IMCGs

The basic descriptive statistics of the phenotype data of early growth traits in Inner Mongolian cashmere goats are shown in Table 1. The mean values of BW, WW, DWG, and YW were 2.76 kg, 21.97 kg, 0.17 kg, and 37.20 kg, respectively, and the coefficients of variation ranged from 15.42% to 21.04%. It was found that the differences in each trait among individuals were small.

Table 1. Basic descriptive statistics of early growth traits in IMCGs.

Traits	No.	Mean	C.V (%)	Max	Min
BW (kg)	1811	2.76	16.96	4.53	1.00
WW (kg)	1809	21.97	15.42	34.77	13.60
DWG (kg/d)	967	0.17	17.79	0.27	0.09
YW (kg)	1953	37.20	21.04	80.50	20.50

Note: BW: birth weight, WW: weaning weight, DWG: daily weight gain before weaning, YW: yearling weight, No.: number of individuals, C.V: coefficient of variation, Max: maximum, Min: minimum.

3.2. Determination of Fixed Effects

The analysis of variance of non-genetic factors, including sex, year of birth, herd, age of dam, and birth type, on the early growth traits of IMCGs is shown in Table 2. It is clear that year of birth and herds had a significant effect on BW ($p < 0.05$). The effect of sex on the four traits was highly significant ($p < 0.001$). Year of birth had a highly significant effect on DWG and WW ($p < 0.001$). Herd had a highly significant impact on WW and DWG ($p < 0.01$) and on YW ($p < 0.001$). The age of the dam had a highly significant effect on all the traits in this study ($p < 0.01$). The impact of birth type was highly significant for BW, WW, and DWG ($p < 0.001$) and YW ($p < 0.01$). Therefore, all the factors were included as fixed effects in the animal models used for genetic evaluation in this study.

Table 2. Effects of non-genetic factors on early growth traits in IMCGs.

Traits / Factors	BW		WW		DWG		YW	
	F-Value	*p*-Value	F-Value	*p*-Value	F-Value	*p*-Value	F-Value	*p*-Value
Sex	167.02	2.0×10^{-16} ***	1272.77	2.0×10^{-16} ***	838.74	2.0×10^{-16} ***	436.31	2.0×10^{-16} ***
Year of birth	6.18	0.0131 *	6.51	0.0022 **	29.24	8.1×10^{-8} ***	255.27	2.0×10^{-16} ***
Herds	6.04	0.0142 *	7.21	0.0073 **	9.01	0.0093 **	491.17	2.0×10^{-16} ***
Age of dam	10.08	0.0029 **	14.27	0.0002 **	10.48	0.0013 **	7.03	0.0086 **
Birth type	229.33	2.0×10^{-16} ***	548.10	2.0×10^{-16} ***	195.22	2.0×10^{-16} ***	7.52	0.0062 **

Note: BW: birth weight, WW: weaning weight, DWG: daily weight gain before weaning, YW: yearling weight, ***: indicates highly significant difference ($p < 0.001$); **: indicates highly significant difference ($p < 0.01$); *: indicates significant difference ($p < 0.05$).

3.3. Model Comparison

The likelihood ratio test values of each model for the early growth traits of IMCGs are shown in Table 3. The value of −2lnL for the early growth traits of IMCGs in Model 4 under each method was lower than that in the other models, indicating that Model 4 is the best model for estimating the genomic breeding values of the early growth traits of IMCGs. In addition, the chi-square test was performed to evaluate the reliability of the models. The results are shown in Table 4. For each trait with three methods, significant or highly significant differences between Model 4 and the other models were observed. No significant differences were observed between Model 1 and Model 2 for YW using either the ABLUP or GBLUP methods. Similarly, Model 1 and Model 3 showed no significant differences for YW under the ABLUP method. For BW and YW, comparisons between Model 2 and Model 3 using the ABLUP method also revealed no significant differences. Additionally, Model 2 and Model 3 did not differ significantly for DWG when analyzed with the GBLUP method.

Table 3. The values of $-2lnL$ in each model for early growth traits with the three methods.

Traits	Model	$-2lnL$		
		ABLUP	GBLUP	ssGBLUP
BW	1	−811.4	−816.1	−810.9
	2	−837.8	−841.4	−815.6
	3	−842.0	−841.4	−843.5
	4	**−849.5**	**−847.4**	**−849.9**
WW	1	4735.2	4725.0	4861.3
	2	4718.8	4721.0	4855.2
	3	4692.1	4717.1	4728.3
	4	**4561.6**	**4683.8**	**4712.9**
DWG	1	−6426.9	−6549.3	−6575.8
	2	−6543.3	−6602.7	−6591.4
	3	−6566.5	−6603.5	−6600.2
	4	**−6597.0**	**−6610.6**	**−6615.1**

Table 3. *Cont.*

Traits	Model	−2lnL		
		ABLUP	GBLUP	ssGBLUP
YW	1	7610.2	7610.1	7743.2
	2	7608.8	7608.8	7734.2
	3	7605.3	7603.8	7602.9
	4	**7583.0**	**7517.1**	**7575.9**

Note: Bold indicates the best model.

Table 4. The chi-square test of the likelihood ratio values among the different animal models.

Model	df	BW			WW			DGW			YW		
		ABLUP	GBLUP	ssGBLUP	ABLUP	GBLUP	ssGBLUP	ABLUP	GBLUP	ssGBLUP	ABLUP	GBLUP	ssGBLUP
1:2	1	26.46 **	25.28 **	4.70 *	16.38 **	4.00 *	6.09 *	116.37 **	53.41 **	15.68 **	1.37 ^{ns}	1.26 ^{ns}	8.96 **
1:3	2	30.65 **	25.30 **	32.54 **	43.07 **	7.85 *	133.03 **	139.56 **	54.18 **	24.46 **	4.82 ^{ns}	6.30 *	140.27 **
1:4	3	38.11 **	31.31 **	39.00 **	173.57 **	41.20 **	148.41 **	170.08 **	61.30 **	39.30 **	27.18 **	92.98 **	167.32 **
2:3	1	4.19 *	0.01 ^{ns}	27.84 **	26.69 **	3.85 *	126.94 **	23.19 **	0.77 ^{ns}	8.78 **	3.45 ^{ns}	5.04 **	131.31 **
2:4	2	11.65 **	6.02 *	34.29 **	157.19 **	37.20 **	142.32 **	53.70 **	7.89 *	23.62 **	25.81 **	91.73 **	158.36 **
3:4	1	7.46 **	6.01 *	6.45 *	130.52 **	33.35 **	15.38 **	30.52 **	7.12 **	14.84 **	22.35 **	86.68 **	27.05 **

Note: ** means the difference is highly significant ($p < 0.01$); * means the difference is significant ($p < 0.05$); ^{ns} means the difference is not significant ($p > 0.05$). BW: birth weight, WW: weaning weight, DWG: daily weight gain before weaning, YW: yearling weight.

3.4. Estimation of Variance Components and Genetic Parameters

It was confirmed that Model 4 is the best model for the genetic evaluation of early growth traits in IMCGs. The variance components and genetic parameters obtained by the ABLUP, GBLUP, and ssGBLUP methods for each trait in this study under the optimal model are shown in Tables 5–8. For BW, the range of direct additive heritability with the three methods was 0.09–0.11, the maternal additive heritability was 0.06–0.19, and the contribution of the maternal permanent environmental effect to the phenotypic variance was 2.45×10^{-8}–0.12. The covariance between the direct additive genetic effects and maternal additive genetic effects was negative in the ABLUP and GBLUP analyses. The correlation coefficients were −0.41 and −0.35, respectively (Table 5). For WW, the range of direct additive heritability with the three methods was 0.17–0.43, the maternal additive heritability was 0.14–0.22, and the contribution of the maternal permanent environmental effect to the phenotypic variance was 1.90×10^{-8} to 1.36×10^{-5}. The covariance between the direct additive genetic effects and the maternal additive genetic effects was negative. The correlation coefficients varied from −0.54 to −0.91 (Table 6). For DWG, the range of direct additive heritability with the three methods was 0.12–0.15, the maternal additive heritability was 0.08–0.11, and the contribution of the maternal permanent environmental effect to the phenotypic variance was 5.84×10^{-3} to 7.49×10^{-3}. The covariance between the direct and maternal additive genetic effects was negative. The correlation coefficients ranged from −0.41 to −0.32 (Table 7). For YW, the range of direct additive heritability with the three methods was 0.23–0.32, the maternal additive heritability was 0.12–0.26, and the contribution of the maternal permanent environmental effect to the phenotypic variance was 1.6×10^{-5} to 2.0×10^{-5}. The covariance between the direct additive and maternal additive genetic effects was negative. The correlation coefficients ranged from −0.84 to −0.56 (Table 8).

Overall, the contribution of the maternal additive genetic effects to the phenotype was relatively large, while the contribution of the maternal permanent environmental effects to the phenotype was relatively small. The heritability of the weaning weight and yearling weight of IMCGs was medium to high, while the heritability of the other two traits was moderate to low.

Table 5. Estimation of the variance components and genetic parameters of birth weight using different methods under the optimal model.

Methods	σ^2_a	σ^2_m	$\sigma_{a,m}$	σ^2_c	σ^2_e	σ^2_p	$h^2_a \pm SE$	h^2_m	C^2	h^2_T	$r_{a,m}$
ABLUP	0.02	0.03	−0.01	7.91×10^{-6}	0.12	0.16	0.10 ± 0.07	0.19	4.94×10^{-5}	0.13	−0.41
GBLUP	0.01	0.01	7.93×10^{-4}	3.67×10^{-9}	0.13	0.15	0.09 ± 0.05	0.07	2.45×10^{-8}	0.11	0.08
ssGBLUP	0.02	0.01	-4.93×10^{-3}	0.02	0.12	0.17	0.11 ± 0.06	0.06	0.12	0.10	−0.35

Note: σ^2_a: direct additive genetic variance; σ^2_m: maternal additive genetic variance; $\sigma_{a,m}$: covariance of the individual and maternal genetic effects; σ^2_c: maternal permanent environment variance; σ^2_e: residual variance; σ^2_p: phenotypic variance; h^2_a: direct heritability; h^2_m: maternal heritability; C^2: the ratio of the maternal environmental effects variance to the phenotypic variance; h^2_T: total heritability; $r_{a,m}$ is the correlation coefficient between the additive and maternal genetic effects.

Table 6. Estimation of the variance components and genetic parameters of the weaning weight using different methods under the optimal model.

Methods	σ^2_a	σ^2_m	$\sigma_{a,m}$	σ^2_c	σ^2_e	σ^2_p	$h^2_a \pm SE$	h^2_m	C^2	h^2_T	$r_{a,m}$
ABLUP	2.25	1.13	−1.44	3.24×10^{-5}	3.25	5.19	0.43 ± 0.13	0.22	6.24×10^{-6}	0.13	−0.91
GBLUP	0.90	0.70	−0.43	9.79×10^{-8}	3.98	5.15	0.17 ± 0.10	0.14	1.90×10^{-8}	0.12	−0.54
ssGBLUP	1.23	0.77	−0.74	6.82×10^{-5}	3.75	5.01	0.25 ± 0.05	0.15	1.36×10^{-5}	0.10	−0.76

Note: σ^2_a: direct additive genetic variance; σ^2_m: maternal additive genetic variance; $\sigma_{a,m}$: covariance of the individual and maternal genetic effects; σ^2_c: maternal permanent environment variance; σ^2_e: residual variance; σ^2_p: phenotypic variance; h^2_a direct heritability; h^2_m: maternal heritability; C^2: the ratio of the maternal environmental effects variance to the phenotypic variance; h^2_T: total heritability; $r_{a,m}$ is the correlation coefficient between the additive and maternal genetic effects.

Table 7. Estimation of the variance components and genetic parameters of the daily weight gain using different methods under the optimal model.

Methods	σ^2_a	σ^2_m	$\sigma_{a,m}$	σ^2_c	σ^2_e	σ^2_p	$h^2_a \pm SE$	h^2_m	C^2	h^2_T	$r_{a,m}$
ABLUP	3.38×10^{-5}	2.97×10^{-5}	-1.15×10^{-5}	2.08×10^{-6}	2.24×10^{-4}	2.78×10^{-4}	0.12 ± 0.05	0.11	7.49×10^{-3}	0.11	−0.37
GBLUP	4.05×10^{-5}	2.68×10^{-5}	-1.44×10^{-5}	2.37×10^{-6}	2.64×10^{-4}	3.23×10^{-4}	0.14 ± 0.04	0.08	7.34×10^{-3}	0.11	−0.41
ssGBLUP	5.78×10^{-5}	3.06×10^{-5}	-1.34×10^{-5}	2.21×10^{-6}	3.02×10^{-4}	3.79×10^{-4}	0.15 ± 0.05	0.08	5.84×10^{-3}	0.14	−0.32

Note: σ^2_a: direct additive genetic variance; σ^2_m: maternal additive genetic variance; $\sigma_{a,m}$: covariance of the individual and maternal genetic effects;σ^2_c: maternal permanent environment variance; σ^2_e: residual variance; σ^2_p: phenotypic variance; h^2_a: direct heritability; h^2_m: maternal heritability; C^2: the ratio of the maternal environmental effects variance to the phenotypic variance; h^2_T: total heritability; $r_{a,m}$ is the correlation coefficient between the additive and maternal genetic effects.

Table 8. Estimation of the variance components and genetic parameters of the yearling weight using different methods under the optimal model.

Methods	σ^2_a	σ^2_m	$\sigma_{a,m}$	σ^2_c	σ^2_e	σ^2_p	$h^2_a \pm SE$	h^2_m	C^2	h^2_T	$r_{a,m}$
ABLUP	6.01	4.94	−4.58	3.68×10^{-4}	12.30	18.67	0.32 ± 0.09	0.26	1.97×10^{-5}	0.09	−0.84
GBLUP	4.33	2.57	−2.13	3.67×10^{-4}	16.50	21.27	0.20 ± 0.07	0.12	1.73×10^{-5}	0.11	−0.64

Table 8. *Cont.*

Methods	σ_a^2	σ_m^2	$\sigma_{a,m}$	σ_c^2	σ_e^2	σ_p^2	$h_a^2 \pm SE$	h_m^2	C^2	h_T^2	$r_{a,m}$
ssGBLUP	5.31	2.98	−2.24	3.64×10^{-4}	17.10	23.16	0.23 ± 0.03	0.13	1.57×10^{-5}	0.15	−0.56

Note: σ_a^2: direct additive genetic variance; σ_m^2: maternal additive genetic variance; $\sigma_{a,m}$: covariance of the individual and maternal genetic effects; σ_c^2: maternal permanent environment variance; σ_e^2: residual variance; σ_p^2: phenotypic variance; h_a^2: direct heritability; h_m^2: maternal heritability; C^2: the ratio of the maternal environmental effects variance to the phenotypic variance; h_T^2: total heritability; $r_{a,m}$ is the correlation coefficient between the additive and maternal genetic effects.

3.5. Evaluation of the Accuracy and Reliability of the Estimated Genomic Breeding Values

The accuracy and reliability of the genomic estimated breeding values (GEBVs) with the three methods for early growth traits in IMCGs are shown in Table 9. The accuracy of the GEBV with the three methods for BW, WW, DWG, and YW was 0.52–0.61, 0.68–0.70, 0.59–0.70, and 0.60–0.67, respectively. It was found that the accuracy and reliability of each trait with the ssGBLUP method were the highest, which were significantly higher than with the other two methods. The lowest accuracy of the GEBV was observed with the ABLUP methods.

Table 9. Evaluation of the accuracy and reliability of the estimated genomic breeding values for early growth traits in the three methods.

Traits	Methods	Accuracy	Accuracy Error	Reliability
BW	ABLUP	0.53 [c]	0.128	0.40
	GBLUP	0.56 [b]	0.008	0.41
	ssGBLUP	0.61 [a]	0.006	0.43
WW	ABLUP	0.68 [b]	0.165	0.50
	GBLUP	0.68 [b]	0.011	0.51
	ssGBLUP	0.70 [a]	0.004	0.52
DWG	ABLUP	0.59 [c]	0.160	0.48
	GBLUP	0.66	0.020	0.53
	ssGBLUP	0.70 [a]	0.008	0.58
YW	ABLUP	0.60 [c]	0.149	0.41
	GBLUP	0.64 [b]	0.021	0.52
	ssGBLUP	0.67 [a]	0.007	0.59

Note: the same letters indicate no significant difference, while different letters denote significant differences.

4. Discussion

Early growth traits are not only an important index for assessing the growth and development of livestock, but also play an important role in a series of production and economic benefits of livestock. There are many factors affecting the early growth traits of cashmere goats, such as sex, birth type, age of the dam, and herd. In addition, other environmental factors such as nutrition and disease also have some effects on the early growth traits of livestock [24]. In this study, all factors, including sex, age of the dam, birth type, herd, and years of measurement, had significant effects on the early traits of IMCGs. Wang et al. [25] showed that year of birth, herd, birth type, sex, and mother's age had an effect on the preweaning traits of Inner Mongolian Arbas cashmere goats, which is in agreement with the results of this study. Mahala [26] indicated that sex had significant effect on the early growth traits of Avikalin sheep. Lalit et al. [27] showed that year of birth and sex had significant effects on the birth weight, weaning weight, daily weight gain, and weekly weight of Harnali sheep. Jalil-Sarghale [28] showed that year of birth, sex, and birth type had significant effects on birth weight, weaning weight, and weekly weight, which is consistent with the results of our study. Mohammadi et al. [29] reported that year of birth,

birth type, sex, and age of the dam had significant effects on the birth and weaning weights of Zandi lambs. Therefore, it can be seen that early growth traits for goats and sheep are greatly influenced by environmental factors. The observed variations could potentially be attributed to interannual differences in temperature, humidity, and pasture conditions, as well as divergent feeding management practices adopted by different herders [30]. In our study, the early growth weights in males were higher than those in females. This may be attributed to the fact that male lambs have a stronger growth intensity than ewes. The influence of age of the dam on early growth traits could be associated with the physiological status of the ewe. Additionally, the competition for nutrients between twin lambs results in a reduced average absorption of nutrients per lamb during the embryonic stage of development, leading to single lambs having higher early body weights compared to twin lambs [31].

As important economic traits, it is necessary to perform accurate genetic assessments of early growth traits. In this study, four animal models were constructed to assess the maternal genetic effect, maternal environmental effect, and covariance between the direct additive effect and maternal genetic effect. The model including all four effects estimated the genetic parameters of the early growth traits in IMCGs best. Illa et al. [32] reported that the direct additive effect, maternal genetic effect, and covariance between the direct additive effect and maternal genetic effect had significant effects on the average daily body weight gain in Nellore sheep. Since the conception rate, lactation ability, and litter protection ability of ewes are affected by both genetics and environment, maternal effects can be divided into maternal genetic effects and maternal environmental effects [33]. Hoque et al. [34] reported that the accuracy of the genetic evaluation of early body weight traits in Japanese Black cattle will decrease if the maternal effects are ignored in the animal model. Gowane et al. [35] and Hanford et al. [36] compared different models for estimating the genetic parameters for the body weight of Bharat Merino sheep and Columbian sheep. The results indicated that maternal additive genetic effects had significant effects on the early growth traits of sheep. Dige et al. [37] reported that the direct additive genetic effect, maternal genetic effects, maternal environmental effects, and covariance between individual additive genetic effects and maternal additive genetic effects had significant effects on growth and feed efficiency traits in Jamunapari goats, which is consistent with the results in our study. However, Ulutas [38] suggested that animal models with only the direct additive genetic effect and maternal genetic effect should be used for the genetic evaluation of preweaning in Suckler cattle. The differences among the results of the above studies may be due to differences in the genetic basis of breeds or population sizes.

In this study, the range of heritability for birth weight was calculated to be 0.09–0.11. Cloete et al. [39] reported that the heritability for birth weight of Merino lambs was 0.16. Di et al. [40] indicated that the heritability for birth weight of ultrafine Chinese Merinos was 0.15. It can be seen that the birth weight of lambs is a trait with a moderate to low heritability. The range of heritability for weaning weight of IMCGs in this study was 0.17–0.43. Menezes et al. [41] demonstrated that the heritability for weaning weight in Boer goats was 0.28, which is higher than that in our study. This may be due to the differences between breeds. The heritability of daily weight gain of IMCGs was consistent with the results in Boer goats from Zhang et al. [42]. The heritability of yearling weight of IMCGs ranged from 0.20 to 0.32, which is similar to the results from Gowane et al. [35] and Abdalla et al. [43]. However, some studies illustrated that the heritability of yearling traits in South African Angora goats and Black Bengal goats was high [44,45]. The differences among the results between studies may be due to various geographic environments, management conditions, and sample size.

As a breeding technology, genomic selection is used to derive genomic breeding values using genetic markers. It has been performed in dairy cattle, beef cattle, pigs, fish, and so on. Genomic selection can effectively improve the genetic progress of livestock and poultry, reduce the cost of progeny testing, and shorten the generation interval, which has been a hot research topic in livestock and poultry breeding in recent years [46]. In this study, three methods, including ABLUP, GBLUP, and ssGBLUP, were used to achieve estimations of the genomic breeding values for early growth traits in IMCGs. It was found that the accuracy and reliability of the GEBV of early growth traits in IMCGs was highest using the ssGBLUP method. Hayes et al. [47] reported that the accuracy of the GEBV of milk production traits with the GBLUP method was almost as high as that with Bayes. Siavash et al. [48] compared the prediction accuracy of the genomic selection of slaughter traits in Canadian regional pig breeds using the BLUP, GBLUP, ssGBLUP, and Bayes methods. The results showed that the ssGBLUP method had a higher prediction accuracy than the other methods. Mancisidor et al. [49] analyzed the accuracy of the genomic selection of important economic traits in Huacaya alpacas with BLUP, ssGBLUP, and other methods. The ssGBLUP method was better for medium- and high-heritability traits when genomic prediction was carried out on small samples. Abdalla et al. [43] reported that the accuracy of the GEBV using ssGBLUP (0.51) was higher than that with ABLUP (0.35). Thus, it can be seen that genomic selection can improve the accuracy of the estimated breeding value. The ssGBLUP method fully utilizes pedigree and genotype information to estimate breeding values, which may be the reason for its high accuracy in genetic evaluation.

5. Conclusions

Animal models for the genetic evaluation of early growth traits in IMCGs should include direct additive genetic effects, maternal genetic effects, maternal environmental effects, and the covariance between the direct additive and maternal genetic effects. The maternal effect has an impact on the early growth traits of IMCGs. The direct additive heritability of birth weight and daily weight gain were 0.09–0.11 and 0.12–0.15, respectively. Both were low heritable traits. Therefore, selection for BW and DWG in IMCGs based on the GEBV will result in slow genetic progress. The direct heritability estimates for WW and YW were moderate (0.17–0.43 and 0.20–0.32, respectively). Thus, rapid genetic improvement can be obtained. Furthermore, when appropriate selection intensity, observed phenotypic variation, and potential generation interval reduction through genomic selection are collectively considered, more rapid genetic progress can be anticipated. Additionally, we determined that the accuracy of the GEBV could be improved by performing genomic selection with the ssGBLUP method.

Author Contributions: Conceptualization, Z.W. and R.S.; methodology, J.L.; software, Q.L., Y.Z., R.W. and Y.L.; formal analysis, L.G.; data curation, B.Z. and Q.X.; writing—original draft preparation, T.Z.; writing—review and editing, T.Z. and Z.W.; visualization, L.G.; supervision, J.L. and Z.W.; project administration, Z.W.; funding acquisition, Z.W. All authors have read and agreed to the published version of the manuscript.

Funding: This study was supported by the Project for Enhancing Young Teacher's Research Ability (BR220112), the Program for Young Talents of Science and Technology in Universities of Inner Mongolia Autonomous Region (NJYT22038), the Central Government Guides Local Science and Technology Development Fund Projects (2022ZY0211), the Project for 2030 Science and Technology Innovation Major (2023ZD0405102), the Program for Inner Mongolia Autonomous Region Higher Education Innovation Team Development (NMGIRT2322), and the China Agriculture Research System of MOF and MARA (CARS-39).

Institutional Review Board Statement: The phenotype data of cashmere goats in the study were collected from the Inner Mongolia Yiwei White Cashmere Goats Company. The genotype data came from the previous research results of the group. There is no ethical statement involved, because the study did not include animal experiments.

Informed Consent Statement: Not applicable.

Data Availability Statement: The data that support the findings of this study are available from the corresponding authors upon reasonable request.

Acknowledgments: We would like to thank Su R and Wang ZY for their excellent technical assistance in preparing experimental data and constructing models. We also greatly appreciate the assistance with sample collection and phenotype determination provided by the herdsmen and workers in the Inner Mongolia Yiwei White Cashmere Goats Company. The authors also sincerely thank the editors and anonymous reviewers for their constructive criticism and helpful comments, which greatly improved the manuscript.

Conflicts of Interest: The authors declare no competing interests.

References

1. Su, R.; Gong, G.; Zhang, L.; Yan, X.; Wang, F.; Zhang, L.; Qiao, X.; Li, X.; Li, J. Screening the key genes of hair follicle growth cycle in Inner Mongolian Cashmere goat based on RNA sequencing. *Arch. Anim. Breed.* **2020**, *63*, 155–164. [CrossRef] [PubMed]
2. Balasundaram, B.; Gupta, A.K.; Dongre, V.; Mohanty, T.K.; Sharma, P.C.; Khate, K.; Singh, R.K. Influence of genetic and non-genetic factors on incidence of calving abnormalities in Karan fries cows. *Indian J. Anim. Res.* **2011**, *45*, 26–31.
3. Meng, G.; La, Y.; Bao, Q.; Wu, X.; Ma, X.; Huang, C.; Chu, M.; Liang, C.; Yan, P. Early growth and development and nonlinear model fitting analysis of Ashidan Yak. *Animals* **2023**, *13*, 1545. [CrossRef]
4. Fogarty, N.M. The correction for age and type of birth effects on weaning weight in sheep. *Proc. Aust. Soc. Anim. Prod.* **2022**, *11*, 425–429.
5. Gätjens, I.; Fedde, S.; Schmidt, S.C.E.; Hasler, M.; Plachta-Danielzik, S.; Müller, M.J.; Bosy-Westphal, A. Relationship between birth weight, early growth rate, and body composition in 5- to 7-year-old children. *Obes. Facts* **2022**, *15*, 519–527. [CrossRef]
6. Pereira, L.S.; Brunes, L.C.; Baldi, F.; do Carmo, A.S.; Soares, B.B.; Magnabosco, V.; da Costa Eifert, E.; Magnabosco, C.U. Genetic association between feed efficiency, growth, scrotal circumference, and carcass traits in Guzerat cattle. *Trop. Anim. Health Prod.* **2023**, *55*, 132–143. [CrossRef]
7. Meuwissen, T.H.; Hayes, B.J.; Goddard, M.E. Prediction of total genetic value using genome-wide dense marker maps. *Genetics* **2001**, *157*, 1819–1829. [CrossRef] [PubMed]
8. Berry, D.P.; Garcia, J.F.; Garrick, D.J. Development and implementation of genomic predictions in beef cattle. *Anim. Front.* **2016**, *6*, 32–38. [CrossRef]
9. Badke, Y.M.; Bates, R.O.; Ernst, C.W.; Fix, J.; Steibel, J.P. Accuracy of estimation of genomic breeding values in pigs using low-density genotypes and imputation. *G3 (Bethesda)* **2014**, *4*, 623–631. [CrossRef]
10. Liu, T.; Luo, C.; Ma, J.; Wang, Y.; Shu, D.; Qu, H.; Su, G. Including dominance effects in the prediction model through locus-specific weights on heterozygous genotypes can greatly improve genomic predictive abilities. *Heredity (Edinb)* **2022**, *128*, 154–158. [CrossRef]
11. Lavvaf, A.; Noshary, A. Estimation of genetic parameters and environmental factors on early growth traits for Lori breed sheep using single trait animal model. *Pak. J. Biol. Sci.* **2008**, *11*, 74–79. [CrossRef] [PubMed]
12. Balasundaram, B.; Thiruvenkadan, A.K.; Murali, N.; Muralidharan, J.; Cauveri, D.; Peters, S.O. Genetic parameters of growth traits and quantitative genetic metrics for selection and conservation of mecheri sheep of Tamil Nadu. *Animals* **2023**, *13*, 454. [CrossRef]
13. Ge, F.; Jia, C.; Bao, P.; Wu, X.; Liang, C.; Yan, P. Accuracies of genomic prediction for growth traits at weaning and yearling ages in Yak. *Animals* **2020**, *10*, 1793. [CrossRef] [PubMed]
14. Terakado, A.P.N.; Costa, R.B.; Irano, N.; Bresolin, T.; de Oliveira, H.N.; Carvalheiro, R.; Baldi, F.; Del Pilar Solar Diaz, I.; de Albuquerque, L.G. Comparison of methods for predicting genomic breeding values for growth traits in Nellore cattle. *Trop. Anim. Health Prod.* **2021**, *53*, 349–356. [CrossRef]
15. Song, H.; Zhang, J.; Jiang, Y.; Gao, H.; Tang, S.; Mi, S.; Yu, F.; Meng, Q.; Xiao, W.; Zhang, Q.; et al. Genomic prediction for growth and reproduction traits in pig using an admixed reference population. *J. Anim. Sci.* **2017**, *95*, 3415–3424. [CrossRef]
16. Liu, T.; Qu, H.; Luo, C.; Shu, D.; Wang, J.; Lund, M.S.; Su, G. Accuracy of genomic prediction for growth and carcass traits in Chinese triple-yellow chickens. *BMC Genet.* **2014**, *15*, 110–122. [CrossRef]

17. Chang, C.C.; Chow, C.C.; Tellier, L.C.; Vattikuti, S.; Purcell, S.M.; Lee, J.J. Second-generation PLINK: Rising to the challenge of larger and richer datasets. *Gigascience* **2015**, *4*, 7. [CrossRef] [PubMed]
18. Samad, H.; Coll, F.; Preston, M.D.; Ocholla, H.; Fairhurst, R.M.; Clark, T.G. Imputation-based population genetics analysis of Plasmodium falciparum malaria parasites. *PLoS Genet.* **2015**, *12*, e1005131. [CrossRef] [PubMed]
19. Publishing, S. *SAS/STAT(R) 9.22 User's Guide*; SAS: Cary, NC, USA, 2010.
20. Singh, M.K.; Dige, M.S.; Pourouchottamane, R.; Kumar, A.; Gowane, G.R. Influences of maternal factors on the estimate of genetic parameters for goat feed efficiency traits. *Trop. Anim. Health Prod.* **2022**, *54*, 376–385. [CrossRef]
21. Chernoff, H. On the distribution of the likelihood Ratio. *Ann. Math. Stat.* **1954**, *25*, 573–578. [CrossRef]
22. Butler, D.; Cullis, B.; Gilmour, A.; Gogel, B. *ASReml-R Reference Manual*; VSN International Ltd.: Hemel Hempstead, UK, 2009.
23. Wiggans, G.R.; Cooper, T.A.; Vanraden, P.M.; Cole, J.B. Technical note: Adjustment of traditional cow evaluations to improve accuracy of genomic predictions. *J. Dairy Sci.* **2011**, *94*, 6188–6193. [CrossRef] [PubMed]
24. Val, J.P.; Ramos, R.; Horcas, E. Analysis of factors affecting the rearing of early-weaned lambs of dairy breeds for the meat market. *Agronomy* **2019**, *9*, 694. [CrossRef]
25. Wang, Z.; Wang, R.J.; Zhang, W.G.; Wang, Z.X.; Wang, P.Y.; Liu, H.; Gao, L.X.; Bai, K.; Meng, R.Q.; Zhou, J.J.; et al. Estimation of genetic parameters for fleece traits in yearling Inner Mongolia Cashmere goat. *Small Rumin. Res.* **2013**, *109*, 15–21. [CrossRef]
26. Mahala, S.; Saini, S.; Kumar, A.; Prince, L.; Gowane, G.R. Effect of Non-Genetic factors on greasy fleece yield in Avikalin sheep under semi-arid conditions of Rajasthan. *Indian J. Anim. Sci.* **2006**, *76*, 557–559.
27. Lalit; Malik, Z.S.; Dalal, D.S.; Dahiya, S.P.; Patil, C.S.; Dahiya, R. Genetic analysis of growth traits in Harnali sheep. *Vet. World* **2016**, *9*, 128–132. [CrossRef]
28. Jalil-Sarghale, A.; Kholghi, M.; Moradi, S.M.; Moradi, S.H.; Arpanahi, R. Model comparisons and genetic parameter estimates of growth and the Kleiber ratio in Shal sheep. *Slovak J. Anim. Sci.* **2014**, *47*, 12–18.
29. Mohammadi, K.; Nass, M.T.B.; Fayazi, J.; Roshanfekr, H. Effects of environmental factors on pre-weaning growth traits in Zandi Lambs. *J. Anim. Vet. Adv.* **2010**, *9*, 837–840.
30. Lupi, T.M.; Nogales, S.; León, J.M.; Barba, C.; Delgado, J.V. Analysis of the non-genetic factors affecting the growth of Segureño sheep. *Ital. J. Anim. Sci.* **2016**, *14*, 3683. [CrossRef]
31. Shelton, M.; Willingham, T.; Thompson, P.; Roberts, E.M. Influence of docking and castration on growth and carcass traits of fat-tail Karakul, Rambouillet and crossbred lambs. *Small Rumin. Res.* **1991**, *4*, 235–243. [CrossRef]
32. Illa, S.K.; Gollamoori, G.; Nath, S. Direct and maternal variance components and genetic parameters for average daily body weight gain and Kleiber ratios in Nellore sheep. *Trop. Anim. Health Prod.* **2019**, *51*, 155–163. [CrossRef]
33. Vrtílek, M.; Chuard, P.J.C.; Iglesias-Carrasco, M.; Zhang, Z.; Jennions, M.D.; Head, M.L. The role of maternal effects on offspring performance in familiar and novel environments. *Cold Spring Harb. Lab.* **2021**, *127*, 52–65. [CrossRef]
34. Hoque, M.A.; Arthur, P.F.; Hiramoto, K.; Gilmour, A.R.; Oikawa, T. Variance components due to direct genetic, maternal genetic and permanent environmental effect for growth and feed-efficiency traits in young male Japanese Black cattle. *J. Anim. Breed. Genet.* **2007**, *124*, 102–107. [CrossRef]
35. Gowane, G.R.; Chopra, A.; Prince, L.L.; Paswan, C.; Arora, A.L. Estimates of (co)variance components and genetic parameters for body weights and first greasy fleece weight in Bharat Merino sheep. *Animal* **2010**, *4*, 425–431. [CrossRef]
36. Hanford, K.J.; Snowder, G.D.; Van Vleck, L.D. Models with nuclear, cytoplasmic, and environmental effects for production traits of Columbia sheep. *J. Anim. Sci.* **2003**, *81*, 1926–1932. [CrossRef]
37. Dige, M.S.; Rout, P.K.; Singh, M.K.; Dass, G.; Kaushik, R.; Gowane, G.R. Estimation of co (variance) components and genetic parameters for growth and feed efficiency traits in Jamunapari goat. *Small Rumin. Res.* **2021**, *196*, 106317. [CrossRef]
38. Ulutas, Z.; Dewi, I.A.; Saatci, M. Six different models for estimation of genetic parameters and breeding values for preweaning and post-weaning in suckler cattle. *Turk. J. Vet. Anim. Sci.* **2001**, *25*, 179–184.
39. Cloete, S.W.; Misztal, I.; Olivier, J.J. Genetic parameters and trends for lamb survival and birth weight in a Merino flock divergently selected for multiple rearing ability. *J. Anim. Sci.* **2009**, *87*, 2196–2208. [CrossRef]
40. Di, J.; Zhang, Y.; Tian, K.C.; Lazate Liu, J.F.; Xu, X.M.; Zhang, Y.J.; Zhang, T.H. Estimation of (co) variance components and genetic parameters for growth and wool traits of Chinese superfine merino sheep with the use of a multi-trait animal model. *Livest. Sci.* **2011**, *138*, 278–288. [CrossRef]
41. Menezes, L.M.; Sousa, W.H.; Cavalcanti, E.P.; Gama, L.T. Genetic parameters for reproduction and growth traits in Boer goats in Brazil. *Small Rumin. Res.* **2016**, *136*, 247–256. [CrossRef]
42. Zhang, C.Y.; Ying, Z.; Xu, D.Q.; Li, X.; Jie, S.; Yang, L.G. Genetic and phenotypic parameter estimates for growth traits in Boer goat. *Livest. Sci.* **2009**, *124*, 66–71. [CrossRef]
43. Abdalla, E.A.; Makanjuola, B.O.; van Staaveren, N.; Wood, B.J.; Baes, C.F. Accuracy of genomic selection for reducing susceptibility to pendulous crop in turkey (*Meleagris gallopavo*). *Poult. Sci.* **2022**, *101*, 101601. [CrossRef] [PubMed]
44. Snyman, G. Genetic analysis of body weight in South African Angora kids and young goats. *S. Afr. J. Anim. Sci.* **2012**, *42*, 146–155. [CrossRef]

5. Mia, M.M.; Khandoker, M.A.M.Y.; Husain, S.S.; Faruque, M.O.; Notter, D.R.; Haque, M.N. Genetic evaluation of growth Traits of Black Bengal Goat. *Iran. J. Appl. Anim. Sci.* **2013**, *3*, 845–852.
6. Mapholi, N.O.; Marufu, M.C.; Maiwashe, A.; Banga, C.B.; Muchenje, V.; MacNeil, M.D.; Chimonyo, M.; Dzama, K. Towards a genomics approach to tick (Acari: Ixodidae) control in cattle: A review. *Ticks Tick-Borne Dis.* **2014**, *5*, 475–483. [CrossRef]
7. Hayes, B.J.; Bowman, P.J.; Chamberlain, A.J.; Goddard, M.E. Invited review: Genomic selection in dairy cattle: Progress and challenges. *J. Dairy Sci.* **2009**, *92*, 433–443. [CrossRef]
8. Ardestani, S.S.; Jafarikia, M.; Sullivan, B.; Sargolzaei, M.; Miar, Y. A comparison of different genomic prediction methods in the Canadian Landrace swine population. *J. Anim. Sci.* **2020**, *98*, 10–16. [CrossRef]
9. Mancisidor, B.; Cruz, A.; Gutiérrez, G.; Burgos, A.; Morón, J.A.; Wurzinger, M.; Gutiérrez, J.P. Ssgblup method improves the accuracy of breeding value prediction in huacaya alpaca. *Animals* **2021**, *11*, 3052. [CrossRef]

genes

MDPI

Article

The Identification of a New Gene *KRTAP 6-3* in *Capra hircus* and Its Potential for the Diameter Improvement of Cashmere Fibers

Jian Cao [1,†], Zhanzhao Chen [2,†], Jianmin Zhang [1], Liang Cao [1] and Shaobin Li [2,*]

[1] College of Bioengineering, Jiuquan Vocational and Technical University, Jiuquan 735000, China; caojian@st.gsau.edu.cn (J.C.)
[2] Gansu Key Laboratory of Herbivorous Animal Biotechnology, Faculty of Animal Science and Technology, Gansu Agricultural University, Lanzhou 730070, China
* Correspondence: lisb@gsau.edu.cn
† These authors contributed equally to this work.

Abstract: Background: Cashmere is one of the important economic products of goats, and the KRTAP gene family, as an important family of regulatory genes in the growth process of cashmere fiber, largely affects the quality of cashmere. Methods: In this study, the *KRTAP6-3* gene was identified and located on goat chromosome 1 using a goat genome homology search combined with a phylogenetic tree approach. The Longdong cashmere goat *KRTAP6-3* gene variation and its effect on cashmere quality were explored by using the polymerase chain reaction single-stranded conformation polymorphism (PCR-SSCP) technique, in situ hybridization, and the allele presence/absence model. Results: The results identified a total of six SNPs in *KRTAP6-3*, three of which were located in the coding region and two of which were synonymous mutations, in addition to 45- bp deletion sequences detected in alleles C and F. Moreover, the *KRTAP6-3* mRNA showed a strong expression signal in the cortical layer of the primary and secondary follicles in the inner root sheaths, as well as in the cells of the hair papillae and the matrices during the anagen phase, and signaling at the sites described above is attenuated during the telogen phase. The presence of allele C was associated with increased MFD (mean fiber diameter) ($p < 0.01$). The MFD of goats with allele C genotype (genotype AC) was significantly higher ($p < 0.05$) than that of goats without allele C genotype (genotypes AA and AB). Conclusions: This indicates that genetic variation in the *KRTAP6-3* gene in goats is significantly associated with cashmere traits and can serve as a candidate gene for molecular markers of cashmere traits.

Keywords: *Capra hircus*; *KRTAP6-3*; cashmere quality; mRNA expression localization

1. Introduction

Cashmere, produced from the secondary hair follicles of the cashmere goat, is one of the finest natural fibers in the world [1]. The reason why cashmere is very precious is not only because of the scarcity of production, but, more importantly, because of its excellent quality and characteristics, it is considered to be 'fiber gem' or 'fiber queen'. Its characteristics involve it being an all-textile raw material that human beings can use and are incomparable, and, therefore, it is also known as 'soft gold' [2]. As one of the most important sources of economic value for goats, differences in the traits of cashmere can have an impact on the economic benefits of cashmere [3] and are of increasing interest due to the fact that cashmere is a natural and sustainable raw material. Over the past two decades, China's cashmere production has shown a trend of first rising and then falling.

The low economic returns from cashmere production may be one of the main reasons for the decline in cashmere production. The economic efficiency of cashmere depends on its own quality. Under equal cost conditions, higher-quality cashmere often leads to higher incomes for farmers, so researchers expect to use modern biotechnology to actively and effectively improve the quality of cashmere. The composition of keratin-related proteins and the genetic diversity of genes in cashmere are closely related to the quality of cashmere fiber [4], which provides important ideas and a basis for the selection and breeding research of cashmere goats.

Keratin-associated proteins (KAPs) are important protein components of cashmere fibers that form the matrix of cashmere fibers and, together with keratin, constitute the basic structure of cashmere fibers. The KRTAP gene family, as genes encoding KAPs, plays an important role in maintaining normal wool growth. It has been shown that variations in the KRTAP gene family members, sheep *KRTAP20-1* [5] and *KRTAP28-1* [6] genes, were significantly associated with MFD, suggesting that different genes may significantly affect the same wool fiber trait. Shaobin Li et al. [7] found significant associations between variation in *KRTAP26-1* and wool fiber yield, mean fiber length (MFL), MFD, and other traits, which, in turn, suggests that variation in the same gene may also affect multiple wool traits simultaneously. This is not only seen in sheep, but also in goats. The seasonal growth of cashmere fibers is a typical feature of cashmere goats [8], where the cashmere fleece undergoes cyclic renewal, and its growth process roughly consists of three phases: the anagen phase, the catagen phase, and the telogen phase [9]. Hair follicles generally consist of primary hair follicles (PHFs) and secondary hair follicles (SHFs), and breeding for cashmere goats is often focused on increasing the number of SHFs [10]. Researchers have found that the fineness of cashmere is the main factor affecting the quality of cashmere [11], so the mechanism of regulating the fineness of cashmere has received more and more attention from researchers. The average GC content of all sheep KRTAPs identified to date is relatively high, the processing capacity of the DNA polymerase is reduced by the high GC content, and the high GC content also increases the slippage of the DNA polymerase, leading to a higher rate of gene sequence mutations [12]. The correlation between the occurrence of nucleotide insertions and deletions and substitutions is universal across all prokaryotic and eukaryotic genomes tested [13–15]. KAPs generally contain high levels of cysteine, or glycine and tyrosine [16,17]. Cysteine plays an important role in the synthesis of cashmere fibers. It is the first limiting amino acid and forms the disulfide bond between KAPs and KIFs that are cross-linked to each other. A disulfide bond is a chemical bond that connects two cysteines in different peptide chains or in the same peptide chain, but no cysteine is present in sheep KAP36-1 [18], suggesting that there may be other forms of cross-linking or interactions between HGT-KAPs and KIFs. The presence of tyrosine in KAPs may play a crucial role, and the benzene ring in the tyrosine side chain allows tyrosine residues in HGT-KAPs to interact with other tyrosine residues or other aromatic amino acids [19]. In addition, in HGT-KAPs, tyrosine residues are usually surrounded by glycine residues, and if the glycine is close to the tyrosine, then the tyrosine residue can have greater freedom to move its benzene ring into the appropriate spatial orientation, (i.e., conformational freedom), which facilitates amino acid interactions. Tyrosine also has a reactive hydroxyl group at the end of its side chain, which can act as a hydrogen donor and, therefore, may form a hydrogen bond with another aromatic amino acid at the center of the benzene ring [20], which would make the ring layer interactions stronger and may lead to the further strengthening of the wool fibers while giving a degree of flexibility [21].

In recent years, the research on cashmere goats has become increasingly scarce. However, with the gradual improvement of people's living standards, cashmere market demand is growing rapidly [22,23]; but, cashmere production is low and the contradiction between the market and the demand has been formed, so the progress of the improvement of goat breeds need to be further strengthened. With technological innovation, i.e., the use of genetic breeding to improve the yield and quality of cashmere [24], as a sustainable production of raw materials, the cashmere industry will have a broad space for development.

2. Materials and Methods

All experiments were approved by Gansu Agricultural University (Approval No. 2006-398) and complied with the Guidelines for the Keeping and Use of Laboratory Animals developed by the Ministry of Science and Technology of the People's Republic of China.

2.1. Animal Species and Sources

Animals were sourced from the Qingyang City Breeding Company in Gansu Province, and 356 Longdong cashmere goats of about one year of age were selected and marked on the ears to differentiate them from each other, and to ensure that the animals were in similar physical condition, healthy, and kept under the same conditions.

2.2. Collection of Cashmere and Blood Samples

In mid-April, cashmere samples were collected from the central region of the posterior margin of the left scapula of each goat, and the corresponding cashmere traits were determined. At the time of cashmere collection, a partial blood sample was collected from the ear of each goat on an FTA card (Whatman, Maidstone, UK). The blood samples were naturally air-dried at room temperature and stored in a sample chamber for the subsequent extraction and analysis of goat genomic DNA.

2.3. Collection of Skin Tissues

Six Longdong cashmere goat rams at about three years of age, kept under the same conditions, were selected for the collection of skin tissues in March and October, i.e., during the anagen and telogen phases of wool development. The collection method was as follows: fix the goat so that the left side of the body is upward, remove the downy hair on the left side of the posterior edge of the scapula so that the skin is exposed, sterilize the skin with iodine volts and alcohol, and anaesthetize it with a needle locally; then, use a scalpel and tweezers to obtain the skin tissues of the Longdong cashmere goat of about 2 cm^2. Completely spread out the collected skin tissue, place it between two glass slides to fix it, and then store it in the in situ hybridization solution.

2.4. Identification of the *KRTAP6-3* Gene

The coding region sequence of sheep *KRTAP6-3* gene (GenBank accession number: KT725833) was used as a template, and BLAST (version: 2.14.0) was applied to search for homology in the goat genome (NC_030808.1). The homologous sequences were identified as the target sequences by constructing phylogenetic tree analysis.

2.5. DNA Extraction and Amplification of Target Fragments

Genomic DNA was obtained using the method described by Zhou et al. [25]. The *KRTAP6-3* sequence of sheep (GenBank accession number: KT725833) was used as a template, and the BLAST function of GenBank was used to search for Caprine Genome Assembly NC_030808.1. The sequence with the highest homology to the *KRTAP6-3* sequence of sheep was assumed to be the *KRTAP6-3* sequence of Longdong cashmere goat, and the specific primers were designed using primer 5.0 to amplify the sequence of goat *KRTAP6-3* gene. The primer sequence information was F: 5′-TCTACCCGAGAACAACCTC-3′, R: 5′-CTTCCATGATGCAGCCTAAC-3′. The annealing temperature was 57 °C; these primers were synthesized by Sangon Biotech Co., Ltd. (Shanghai, China).

2.6. Characterization of Genetic Variation in the Goat KRTAP6-3 Gene

Amplification product variation was analyzed using polymerase chain reaction single-stranded conformational polymorphism (PCR-SSCP) technology with silver staining for color development and gene sequencing. For specific experimental procedures, refer to the research conducted by Hua Gong et al. [26]. The amplification products were first subjected to polyacrylamide gel electrophoresis and, at the end of electrophoresis, the polyacrylamide gel was stained to show the bands according to the method described by Byun et al. [27]. Based on the staining results, the sequencing method was selected by determining the type of genotype of the sample. Genotypes included both pure and heterozygous types, and were sequenced by means of the direct sequencing of PCR amplification products and cut-gel sequencing, respectively. Sequencing was performed in both directions by Shanghai Sangong Bioengineering Co. (Shanghai, China). The optimal conditions for the SSCP electrophoresis of primed PCR amplification products were 180 v, 25 °C, and 16 h, and the concentration of the gel was 12%.

2.7. In Situ Hybridization and RT-qPCR

Specific primers were designed according to the sequence of the genes submitted in GeneBank (KT725833), which were used to detect the relative expression level of goat *KRTAP6-3* gene in Longdong cashmere goats at different times. RT-qPCR primers and the internal reference information are shown in Table 1. The RT-qPCR reaction was carried out with β-actin as the internal reference gene using the Applied Biosystems QuantStudio® 6 Flex qPCR instrument, and the relative expression amount of *KRTAP6-3* gene was calculated by using the $2^{-\Delta\Delta Ct}$ method [28]. The morphologically intact body side skin of Longdong cashmere goats was selected, embedded in dipping wax, and sectioned, and then unfolded in DEPC water and baked overnight at 37 °C for spare use. The *KRTAP6-3* gene RT-PCR product was cloned and sequenced and then used to make the template for digoxin-labelled probe after repeated purification and DNA template denaturation. The steps of Roche digoxin labeling kit instructions were used.

Table 1. RT-qPCR primer information and amplification condition.

Gene	Primer Sequences	Length of Products (bp)	Annealing Temperature (°C)
KRTAP6-3	F: 5′-AAGCTATGGCTATGGAGGC-3′ R: 5′-CCAGAGCCGTAGCAGGAG-3′	99	60
β-actin	F: 5′-ACTACAGCAACTACTACGGTG-3′ R: 5′-GGACAGCACCGTGTTGGCGTAGA-3′	112	60

2.8. Data Collation and Analysis

The collated data were analyzed for association between individuals with genotype frequencies and allele frequencies greater than 5% and cashmere traits using General Linear Mixed-Effect Models (GLMMs) using SPSS V24.0 (IBM, Armonk, NY, USA) software. ANOVA results showed that both gender and sire had an effect ($p < 0.05$) on cashmere traits in the model, so gender was fitted as a fixed factor and sire was fitted as a random factor in the model.

3. Results

3.1. The Result of KRTAP6-3 Gene Identification

The coding region sequence of the sheep *KRTAP6-3* gene (GenBank accession number: KT725833) was used as a template, and the BLAST program was applied to conduct a homology search in the goat genome (NC_030808.1). The results showed that an open reading frame (ORF) with a size of 255-bp and 89% homology to the coding region sequence of the sheep *KRTAP6-3* gene was searched. The fragment is located on goat chromosome 1. Since sheep *KRTAP6-n* has high sequence similarity in the coding region, in order to confirm that this 255-bp ORF is the goat *KRTAP6-3* gene, the upstream and downstream 300-bp gene sequences of this ORF were further analyzed in this experiment, respectively, and, through the construction of the phylogenetic tree analysis, the amplified ORF was found to be the closest to the sheep *KRTAP6-3* gene sequence and more distant from other *KRTAP6-n* gene sequences as compared to other *KRTAP6-n* sequences in sheep. The upstream and downstream sequences had the closest genetic distance to the sheep *KRTAP6-3* gene sequence, whereas they were more distant from the other *KRTAP6-n* gene sequences, indicating that this sequence was the goat *KRTAP6-3* gene (Figure 1).

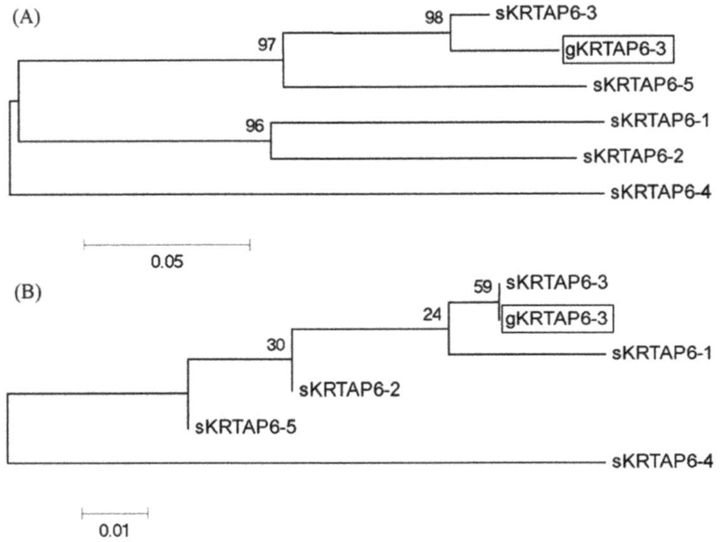

Figure 1. Phylogenetic trees constructed for the 300-bp sequences (**A**) upstream and (**B**) downstream of the coding region of the putative goat *KRTAP6-3* and sheep *KRTAP6-n*. The goat and sheep KRTAPs are indicated with a prefix "g" and "s", respectively; the caprine *KRTAP6-3* sequences identified in the study are shown in a box, the same as below.

The amino acid sequences of all HGT-KAPs already identified in humans, sheep, and goats were downloaded from the NCBI database, and phylogenetic trees were constructed with the six goat *KRTAP6-3* sequences obtained in this experiment. The results showed that the newly identified goat *KRTAP6-3* sequences clustered together with sheep *KRTAP6-3* with the closest genetic distance and clustered more distantly with other known HGT-KAP sequence genes (Figure 2). Combined with the results of 300-bp gene sequence analysis upstream and downstream of the newly identified ORF of the goat *KRTAP6-3* gene, this suggests that the six newly identified sequences are for the goat *KRTAP6-3* gene.

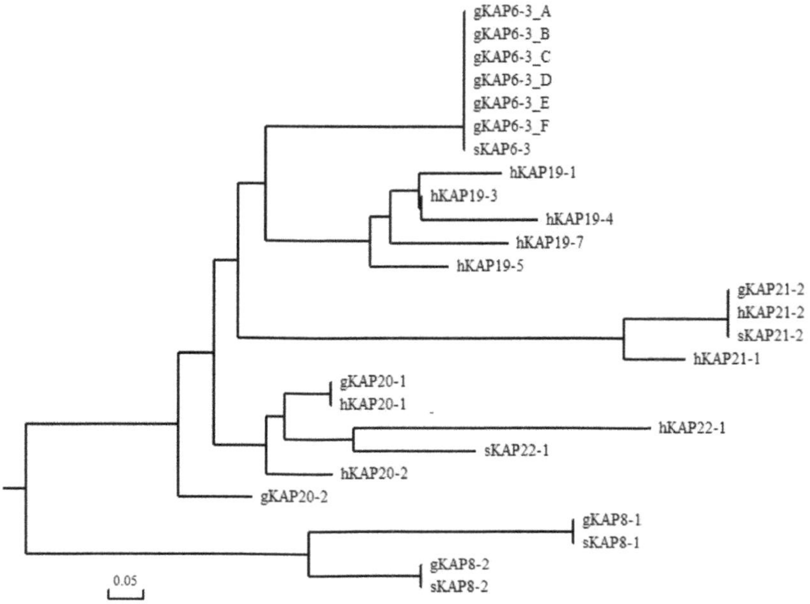

Figure 2. Phylogenetic tree constructed with newly discovered goat *KRTAP6-3* sequences and other HGT-KAPs found in goats, sheep, and humans.

3.2. Genetic Characterization of the KRTAP6-3 Gene

After the polyacrylamide gel electrophoresis analysis of 337 Longdong cashmere goats, a total of 12 PCR-SSCP band types were identified (Figure 3), and six specific bands were identified after discrimination, which appeared individually or in combination in the tested individual goats.

The subsequent DNA sequencing of the PCR amplicons showed that the six SSCP bands represented six different nucleotide sequences, all of which differed from the DNA sequences of the goat genome (NC_030808.1) but showed 89% similarity, indicating that these six sequences represented the alleles of goat *KRTAP6-3*, and were named as alleles A, B, C, D, E and F, with allele frequencies A-F of 62.8%, 18.2%, 15.3%, 1.5%, 1.8% and 0.4%, respectively, and allele A being the advantageous allele. These six alleles in the form of pure and heterozygous individuals comprised a total of 12 genotypes AA, BB, CC, AB, AC, AD, AE, AF, BC, BE, CD, and DE with genotype frequencies of 39.8%, 2.7%, 2.7%, 24.3%, 17.8%, 0.9%, 2.1%, 0.9%, 6.2%, 0.6%, 1.2%, and 0.9%, respectively (Table 2). The AA genotype was the advantageous genotype.

Sequencing results showed that six SNPs were identified in six nucleotide sequences (Figure 4), three of them were located in the coding region, c.103C/T and c.138C/A were synonymous mutations, c.149G/A was a non-synonymous mutation, which resulted in the change of arginine to histidine at position 50 (p.Arg50His), and the other three SNPs were located in the 3′UTR region, which were c.*16G/C, c.*22C/T, and c.*82T/C. In addition, sequences C and F have a deletion of 45 bp each (c.82_c.126delTGTGGCTACGGCTCTGGCTTCTGCAGGCTGGGCTGTGGCTATGGC and c.91_c.135delGGCTCTGGCTTCCGCAGGCTGGGCTGTGGCTATGGCTCCTGCTAC). The six allele sequences of goat *KRTAP6-3* encode polypeptide chains consisting of 84 or 69 amino acid residues (Figure 5). Among these three polypeptide chains, glycine and tyrosine had the highest content of 39.13–39.29 mol% and 21.19–23.19 mol%, respectively, cysteine had a moderate content (13.04–13.10 mol%), and other amino acids had a low content, such as serine 8.33–8.70 mol%, leucine 5.80–5.95 mol%, arginine 4.35–5.95 mol%, phenylalanine 2.90–3.57 mol%, methionine, and aspartic acid 1.19–1.45 mol%. Alleles C and F both have deletions of 15 amino acid residues (p.28_c.42delCGYGSGFRRLGCGYG and p.31_c.45delGSGFRRLGCGYGSCY) compared to alleles A, B, D, and E. In addition, four amino acid or seven amino acid sequence repeats (CGYG or SGFRRLG) were found in the amino acid sequence encoded by goat *KRTAP6-3*.

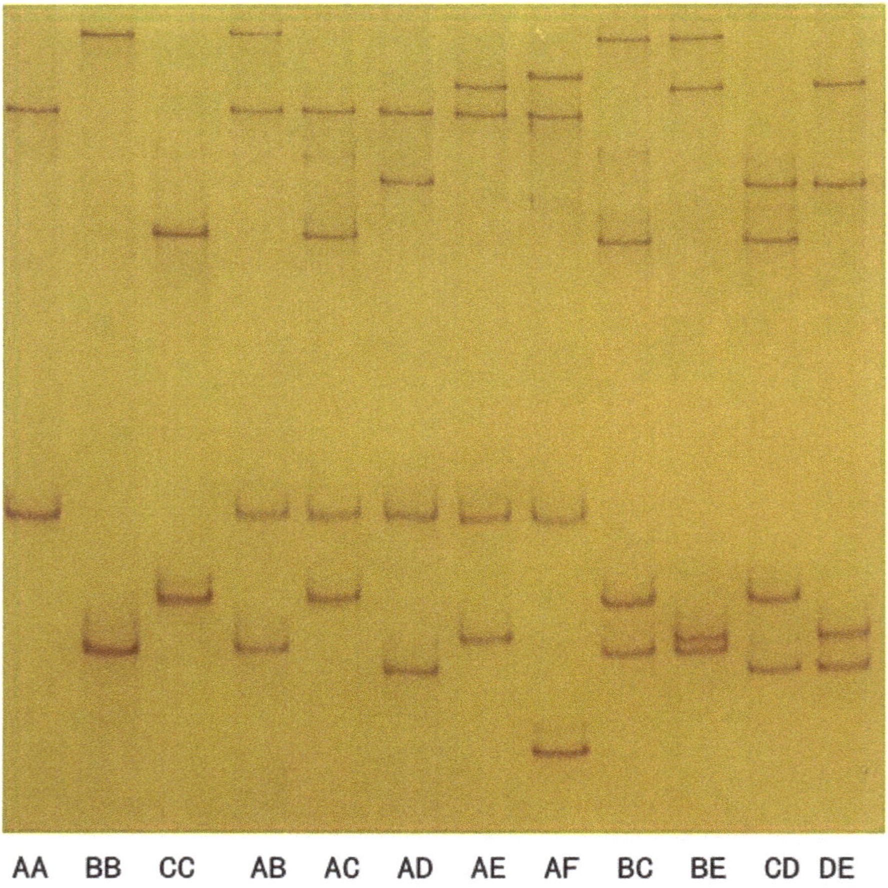

Figure 3. PCR-SSCP results of *KRTAP6-3* in Longdong cashmere goats.

Table 2. Genotypes and their corresponding gene frequencies.

Genotype (*n*)	Genotype Frequency (%)	Genotype (*n*)	Genotype Frequency (%)
AA (134)	39.8%	AE (7)	2.1%
AB (82)	24.3%	CD (4)	1.2%
AC (60)	17.8%	AF (3)	0.9%
BC (21)	6.2%	AD (3)	0.9%
BB (9)	2.7%	DE (3)	0.9%
CC (9)	2.7%	BE (2)	0.6%

```
gKAP6-3*A tctacccgagaacaacctcaacaaccaacaccATGTGTGGCTACTACGGAAACTACTACG 60
gKAP6-3*B ------------------------------------------------------------ 60
gKAP6-3*C ------------------------------------------------------------ 60
gKAP6-3*D ------------------------------------------------------------ 60
gKAP6-3*E ------------------------------------------------------------ 60
gKAP6-3*F ------------------------------------------------------------ 60

gKAP6-3*A GCGGCCTCGGCTGTGGGAGCTATGGCTATGGAGGCCTGGGCTGTGGCTATGGCTGTGGCT 120
gKAP6-3*B ------------------------------------------------------------ 120
gKAP6-3*C -----------------------------------------------------....... 120
gKAP6-3*D ------------------------------------------------------------ 120
gKAP6-3*E ------------------------------------------------------------ 120
gKAP6-3*F ------------------------------------------------------------ 120

gKAP6-3*A ACGGCTCTGGCTTCCGCAGGCTGGGCTGTGGCTATGGCTCCTGCTACGGCTCTGGCTTCC 180
gKAP6-3*B --------------T--------------------------------------------- 180
gKAP6-3*C ......................................---------------------- 169
gKAP6-3*D -------------------------------------------------A---------- 180
gKAP6-3*E ------------------------------------------------------------ 180
gKAP6-3*F --.............................................------------- 169

gKAP6-3*A GCAGGCTGGGCTGTGGCTATGGCTGTGGCTACGGCTATGGCTCCCGCTCTCTCTGTGGCT 240
gKAP6-3*B A----------------------------------------------------------- 240
gKAP6-3*C ------------------------------------------------------------ 195
gKAP6-3*D ------------------------------------------------------------ 240
gKAP6-3*E A----------------------------------------------------------- 240
gKAP6-3*F ------------------------------------------------------------ 195

gKAP6-3*A GTGGCTATGGATGCGGCTATGGCTCTGGCTTTGGCTACTACTATTGAagacgccaagaaa 300
gKAP6-3*B ------------------------------------------------------------ 300
gKAP6-3*C ------------------------------------------------------------ 255
gKAP6-3*D ------------------------------------------------------------ 300
gKAP6-3*E ------------------------------------------------------------ 300
gKAP6-3*F ------------------------------------------------------------ 255

gKAP6-3*A gagtctcaccctctatatctggacaccaggattcaccagttctgaagccctcatgttctg 360
gKAP6-3*B ------------------------------------------------------------ 360
gKAP6-3*C --c-----t--------------------------------------------------- 315
gKAP6-3*D ------------------------------------------------------------ 360
gKAP6-3*E ------------------------------------------------------------ 360
gKAP6-3*F ------------------------------------------------------------ 315

gKAP6-3*A tgtctttctcctgggtgatggttaggctgcatcatggaag 400
gKAP6-3*B ---------------------------------------- 400
gKAP6-3*C --------c------------------------------- 355
gKAP6-3*D ---------------------------------------- 400
gKAP6-3*E ---------------------------------------- 400
gKAP6-3*F ---------------------------------------- 355
```

Figure 4. Sequence alignment of goat *KRTAP6-3*. Note: nucleotides in coding and non-coding regions are represented by upper- and lowercase letters, respectively. Dashes indicate nucleotide sequences that are identical to the sequence above, and nucleotide deletions are indicated by dots. Transcription start and stop codons are indicated by red letters.

```
sKAP6-3     MCGYYGNYYGGLGCGSYGYGGLGYGYGCGYGSGFRRLGCGYGSCYGSGFRRLGCGYGCGY..........GYGCGYGSGFGYYY*
gKAP6-3*A   -----------------------C------------------------------------GYGSRSLCGC--------------*
gKAP6-3*B   -----------------------C-------------------------H---------------------------------*
gKAP6-3*C   -----------------------C---...............---------------------------------------*
gKAP6-3*D   -----------------------C-----------------------------------------------------------*
gKAP6-3*E   -----------------------C-------------------------H---------------------------------*
gKAP6-3*F   -----------------------C------..............----------------------------------------*
```

Figure 5. Alignment of the predicted amino acid sequences of the *KRTAP6-3* gene identified in goats and sheep. Dots represent amino acid deletions. Goats and sheep are labelled with the prefix "g" and the prefix "s", respectively. Repeat sequences are indicated by boxes.

3.3. Localization and Relative Expression Levels of KRTAP6-3 mRNA at Different Periods of Time

The results of the study on the relative expression levels of the newly discovered gene *KRTAP6-3* mRNA in different stages of goats showed that the gene was expressed in both the anagen and telogen, and the expression level was significantly higher in the anagen than in the telogen, with a multiplicity of difference of 14.42 times. A1, A2, A3, and A4 are the expression results of the *KRTAP6-3* mRNA of Longdong cashmere goats in anagen and telogen skin, respectively. The results showed that the hair follicle structure was more complete and clear and the cellular arrangement was more tightly ordered when the hair follicle was in the anagen phase compared to the telogen phase; the *KRTAP6-3* mRNA was strongly expressed in the cortical layer of the hair shaft of the primary and secondary hair follicles, the inner root sheath, and the hair papilla cells and hair matrices of Longdong cashmere goats during the anagen phase; and the expression of the signals in the above-described sites was weakened during the telogen phase (Figure 6).

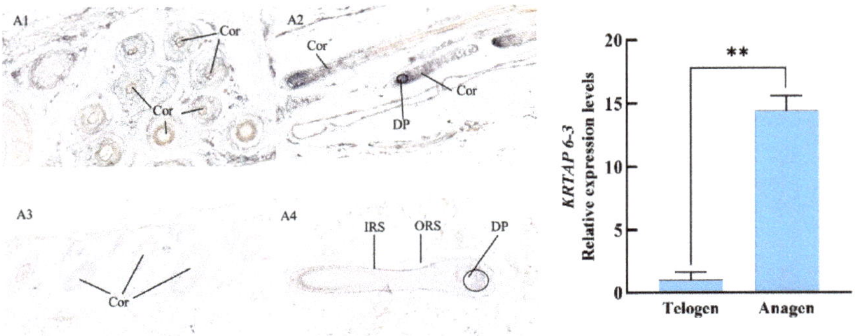

Figure 6. Expression localization and expression level of *KRTAP6-3* mRNA in the skin of Longdong goats at different periods of time. Note: **A1** and **A2** are anagen-phase-transverse and longitudinal cuts (30×), and **A3** and **A4** are telogen-phase-transverse and longitudinal cuts (30×), respectively. IRS, ORS, Cor, DP, and Mx are the inner root sheath, the outer root sheath, the cortical layer, the dermal papilla cells, and the hairy parent cells, respectively, same as below. **: Indicates the existence of extremely significant differences.

3.4. Association Analysis of KRTAP6-3 Gene with Cashmere Traits

In the allele presence/absence model, among 337 Longdong cashmere goats, only three of the six alleles detected had frequencies greater than 5%, namely A (62.8%), B (18.2%), and C (15.3%). The other alleles, D, E, and F, had frequencies lower than 5% and were, therefore, excluded from the association analysis, leaving only the remaining 315 for statistical analysis. The results showed that the presence of allele C was associated with an increase in the mean diameter of cashmere (13.5 ± 0.04 µm in its absence; 13.9 ± 0.12 µm in its presence; $p = 0.003$), i.e., the presence of the allele C led to an increase in cashmere diameter. The remaining two traits measured were not associated with the presence (or absence) of alleles A, B, and C ($p > 0.05$) (Table 3).

Genotypes with less than 5% frequency were excluded from the association analysis with cashmere traits, including genotypes BB, CC, AD, AE, AF, BE, BC, CD, and DE. The association results showed that the mean fiber diameter was significantly higher in the AC genotype than in the AA and AB genotypes ($p = 0.010$), which suggests that allele C is associated with an increase in mean fiber diameter (Table 4).

Table 3. Association between alleles in *KRTAP6-3* and variation in cashmere traits in Longdong cashmere goats.

Traits	Variant	Absent		Present		p Value
		Mean ± SE	n	Mean ± SE	n	
Cashmere yield (g)	A	415 ± 8.5	39	413 ± 5.1	276	0.805
	B	419 ± 4.5	203	406 ± 8.3	112	0.252
	C	412 ± 4.7	225	428 ± 13.1	90	0.313
Mean fiber diameter (μm)	A	13.5 ± 0.06	39	13.5 ± 0.05	276	0.629
	B	13.6 ± 0.04	203	13.5 ± 0.08	112	0.219
	C	**13.5 ± 0.04**	**225**	**13.9 ± 0.12**	**90**	**0.003**
Crimped fiber length (μm)	A	4.3 ± 0.10	39	4.2 ± 0.05	276	0.864
	B	4.2 ± 0.05	203	4.3 ± 0.09	112	0.556
	C	4.3 ± 0.05	225	4.2 ± 0.14	90	0.528

Note: Estimated marginal means and standard errors derived from general linear mixed-effects models that included "gender" as a fixed factor and "sire" as a random factor, respectively. $p < 0.01$ and $p < 0.05$ are shown in bold, same as below.

Table 4. Association analysis between *KRTAP6-3* genotype and cashmere traits in Longdong cashmere goats.

Traits	Mean ± SE			p Value
	AA (n = 134)	AB (n = 82)	AC (n = 60)	
Cashmere yield (g)	416 ± 8.6	407 ± 8.7	423 ± 15.1	0.314
Mean fiber diameter (μm)	**13.4 ± 0.08** [b]	**13.6 ± 0.08** [b]	**14.0 ± 0.14** [a]	**0.010**
Crimped fiber length (μm)	4.2 ± 0.09	4.3 ± 0.09	4.2 ± 0.16	1.000

a,b: The same letters indicate no significant difference, while different letters indicate significant difference.

4. Discussion

The keratin-associated protein gene family has been extensively studied, and it has been found that multiple members of this gene family have an impact on the economic traits of cashmere. A new gene, *KRTAP6-3*, was identified in goats in this study, adding a new member to the HGT-KAPs gene family, which is located on chromosome 1. In phylogenetic analysis, the goat *KRTAP6-3* gene was closest to the sheep *KRTAP6-3* gene in genetic distance and clustered together in the phylogenetic tree, suggesting that the gene is *KRTAP6-3* for goats. It is noteworthy that the goat KAP6-3 protein possesses a conserved "MCGYYGNY" and "GSGFGYYY" sequence at the amino-terminal and carboxy-terminal ends, respectively, which are unique to KAP6 proteins in mammals. Up to now, the same structure has been identified in the sheep *KRTAP6* family genes [29,30].

The sheep *KRTAP6-3* [31] gene showed abundant polymorphism in Merino sheep. Similarly, *KRTAP6-3* was found to be equally abundant in Longdong cashmere goats. A total of six SNPs were detected in the 399-bp fragments amplified in the Longdong cashmere goat population, and there was a 45-bp deletion sequence in both allele C and allele F with different deletion sequences. Sheep and goats are different genera in the same family, and although the *KRTAP6-3* gene is polymorphic in both, the sequence variants detected in both are different. The difference is that the SNPs identified in the sheep *KRTAP6-3* gene were all located in the coding region, and most of them (80%) resulted in amino acid changes, with non-synonymous mutations dominating. In contrast, in goat, three of the six SNPs were located in the coding region, only one was a nonsynonymous mutation, and the remaining three SNPs were located in the noncoding region, and the polymorphisms observed in both may be derived from different mechanisms. Similarly, a 45-bp insertion/deletion was detected in the central region of the coding region of both goat and sheep *KRTAP6-3* genes, and this new insertion/deletion in goat *KRTAP6-3* further supports the idea that sequence length variation is also a feature of the KRTAPs gene family [30]. To date, length variants are present in all *KAP6* genes except *KRTAP6-4*, but the number of insertions/deletions found in the KAP6 genes varies, with length variants detected in higher numbers in the *KRTAP6-1* and *KRTAP6-3* genes, 57-bp and 45-bp, respectively, and in lower numbers in the *KRTAP6-5* gene, 18-bp [29]. A 45-bp deletion was detected in goat *KRTAP6-3* alleles C and F. The effect of allele F on cashmere fiber traits was not analyzed because the genotype frequency of allele F was less than 5%, and only the effect of C on cashmere fiber traits was analyzed using the presence of deletion model, and a significant correlation was found between allele C and cashmere mean fiber diameter. The presence of allele C increases the diameter of cashmere fibers. The association analysis of the different genotypes of the *KRTAP6-3* gene with cashmere fiber traits showed that the average fiber diameter of the AB genotype, which did not contain allele C, was finer than the MFD ratio of the genotype AC, which contained allele C. There was no association of the different genotypes with cashmere yield and fiber curl length, which is consistent with the results of the presence/absence model. In the presence/absence model, the MFD of cashmere was smaller when allele C was absent than when C was present, suggesting that the deletion of allele C was associated with a reduction in MFD.

The results of this study show that there are significant differences in the localization of *KRTAP6-3* mRNA expression in telogen and anagen hair follicles, manifested as obvious differences in positive signal intensity and expression levels. During the anagen phase, strong mRNA expression signals were observed in the cortex and inner root sheath as well as papilla and hair matrix cells, while weaker signals were observed in the telogen phase, suggesting that the expression of this gene is spatio-temporally specific and exerts its main regulatory function mostly during the anagen phase. Furthermore, studies on *KRTAP6-3* in sheep have revealed that it also exhibits strong expression signals in the inner root sheath [32]; this indicates that the expression of the protein encoded by this gene is similar in goats and sheep. Compared with the telogen phase, the mRNA expression level of the gene was significantly higher in the anagen phase with a 14.42-fold difference, which was consistent with the results of previous studies [33]; expression levels of KRTAP gene family members were significantly differentiated between anagen and telogen phases. The newly identified gene had expression signals in the cortical layer, whereas no positive expression signals were detected in the medullary layer of the hair follicle, and it is hypothesized that the expression of this gene is more likely to benefit the improvement of cashmere traits. It has been shown that members of the KRTAP gene family show a typical co-evolutionary pattern, i.e., there is synergy between members of the same gene family, further confirming the similarity of expression patterns between members, and it has been suggested that the KRATP gene could play a more active role in improving the structure of cashmere fibers in domesticated mammals such as goats [34].

In the sheep *KRTAP6* family, both the 57- bp sequence deletion detected in allele C of *KRTAP6-1* [30] and the 45-bp sequence deletion detected in allele G of sheep *KRTAP6-3* [31] were significantly associated with wool fiber diameter. This suggests that sequence length variation may have an effect on cashmere wool fiber diameter. Sequence length variation in goat *KRTAP6-3* may affect cashmere fiber traits in several ways. First, the deletion sequences found in allele C and allele F may result in the deletion of 15 amino acid residues containing repetitive sequences in the central region of the protein, and these amino acid changes may affect fiber properties by affecting the structure, properties, or interactions with the IFs of the KAP6-3 protein. Secondly, deletion of the allele C amino acid sequence would result in an 18% reduction in the occurrence of the aromatic amino acids tyrosine (two deletions) and phenylalanine (one deletion) and a 50% reduction in the occurrence of the basic amino acid arginine (two deletions) in the protein. Aromatic and basic amino acids regulate the arrangement of intermediate filaments (IFs) by interacting with cation- π [35], and a reduction in the number of these residues may affect the strength of the interaction between KAP6-3 and keratin intermediate filament proteins (KIFs). Thirdly, 45-bp deletion also resulted in the absence of cysteine (two deletions) and glycine (six deletions). Cysteine is usually the first limiting amino acid for wool fiber synthesis and is required for the formation of disulfide bonds between KAPs and IFs, which play a very important role in wool growth. Although the exact role of glycine in HGT-KAPs has not been determined, the fact that glycine is the smallest amino acid and has no side chains may make HGT-KAPs proteins more flexible and, thus, better able to form tight structures with KIF proteins [31]. Finally, the deletion of the goat KAP6-3 protein also leads to a reduction in the number of residues that may be phosphorylated, which may result in changes in keratin solubility, the organization of keratin filaments, and interactions with other proteins [36,37], thus affecting the traits of cashmere fibers. The effect of the *KRTAP6-3* gene on cashmere fiber traits may also be caused by the independent action of the *KRTAP6-3* gene or by a cascade effect with upstream and downstream neighboring genes. However, as far as the association between the goat *KRTAP6-3* gene and the sheep *KRTAP6* subfamily of genes with MFD has been identified, it has the potential to be used as a molecular marker for cashmere fiber traits in goats. In subsequent studies, the association between genes and cashmere traits can be further verified through methods such as gene editing validation and cell experiments. The functional conservation of these genes in hair follicle development can be analyzed across different species. Based on modern genotyping technology, this research can be applied to molecular-assisted breeding in goats.

5. Conclusions

KAPs are the main protein components of goat cashmere and play a key role in the economic characteristics of cashmere. The newly identified gene *KRTAP6-3* in Longdong cashmere goats is significantly associated with MFD. Therefore, the newly discovered gene can be used as a candidate gene for molecular markers of cashmere traits for the improvement of the cashmere quality of breeds.

Author Contributions: Conceptualization, J.C.; Data curation, J.Z.; Funding acquisition, J.C. and S.L.; Methodology, Z.C., J.Z. and L.C.; Resources, S.L.; Visualization, L.C.; Writing—original draft, J.C.; Writing—review and editing, Z.C. All authors have read and agreed to the published version of the manuscript.

Funding: This work was financially supported by the Gansu Provincial Science and Technology Programme Funding Gansu Provincial Department of Science and Technology Special Programme for Science and Technology Specialists (25CXNF006), Science and technology program of Gansu Province (23JDKA0010), Jiuquan Vocational Technical College Hu Yang Science and Technology Innovation Team (Herbivorous Animals Efficiency Production Innovation Team) Project (2024XJTXM03) and Jiuquan Science and Technology Support Programme (2024MA1012).

Institutional Review Board Statement: The animal study protocol adhered to the guidelines for the care and use of experimental animals issued by the Ministry of Science and Technology of the People's Republic of China (Approval Number 2006-398) and was also approved by Gansu Agricultural University, Lanzhou, China.

Informed Consent Statement: Informed consent was obtained from all subjects involved in the study.

Data Availability Statement: The authors affirm that all data necessary for confirming the conclusions of the article are present within the article, figures, and tables.

Conflicts of Interest: The authors declare no conflicts of interest.

References

1. Wu, C.; Qin, C.; Fu, X.; Huang, X.; Tian, K. Integrated analysis of lncRNAs and mRNAs by RNA-Seq in secondary hair follicle development and cycling (anagen, catagen and telogen) of Jiangnan cashmere goat (*Capra hircus*). *BMC Vet. Res.* **2022**, *18*, 167. [CrossRef] [PubMed]
2. Ge, W.; Zhang, W.; Zhang, Y.; Zheng, Y.; Li, F.; Wang, S.; Liu, J.; Tan, S.; Yan, Z.; Wang, L.; et al. A Single-cell Transcriptome Atlas of Cashmere Goat Hair Follicle Morphogenesis. *Genom. Proteom. Bioinform.* **2021**, *19*, 437–451. [CrossRef]
3. Wu, C.; Li, J.; Xu, X.; Xu, Q.; Qin, C.; Liu, G.; Wei, C.; Zhang, G.; Tian, K.; Fu, X. Effect of the FA2H Gene on cashmere fineness of Jiangnan cashmere goats based on transcriptome sequencing. *BMC Genom.* **2022**, *23*, 527. [CrossRef] [PubMed]
4. Jin, M.; Wang, L.; Li, S.; Xing, M.X.; Zhang, X. Characterization and expression analysis of KAP7.1, KAP8.2 gene in Liaoning new-breeding cashmere goat hair follicle. *Mol. Biol. Rep.* **2011**, *38*, 3023–3028. [CrossRef]
5. Gong, H.; Zhou, H.; Bai, L.; Li, W.; Li, S.; Wang, J.; Luo, Y.; Hickford, J.G.H. Associations between variation in the ovine high glycine-tyrosine keratin-associated protein gene KRTAP20-1 and wool traits. *J. Anim. Sci.* **2019**, *97*, 587–595. [CrossRef]
6. Bai, L.; Wang, J.; Zhou, H.; Gong, H.; Tao, J.; Hickford, J.G.H. Identification of Ovine KRTAP28-1 and Its Association with Wool Fibre Diameter. *Animals* **2019**, *9*, 142. [CrossRef]
7. Li, S.; Zhou, H.; Gong, H.; Zhao, F.; Hu, J.; Luo, Y.; Hickford, J.G.H. Identification of the Ovine Keratin-Associated Protein 26-1 Gene and Its Association with Variation in Wool Traits. *Genes* **2017**, *8*, 225. [CrossRef]
8. Liu, B.; Zhao, R.; Wu, T.; Ma, Y.; Gao, Y.; Wu, Y.; Hao, B.; Yin, J.; Li, Y. Transcriptomes reveal microRNAs and mRNAs in different photoperiods influencing cashmere growth in goat. *PLoS ONE* **2023**, *18*, e0282772. [CrossRef] [PubMed]
9. Rile, N.; Liu, Z.; Gao, L.; Qi, J.; Zhao, M.; Xie, Y.; Su, R.; Zhang, Y.; Wang, R.; Li, J.; et al. Expression of Vimentin in hair follicle growth cycle of inner Mongolian Cashmere goats. *BMC Genom.* **2018**, *19*, 38. [CrossRef]
10. Dai, B.; Liang, H.; Guo, D.D.; Bi, Z.W.; Yuan, J.L.; Jin, Y.; Huan, L.; Guo, X.D.; Cang, M.; Liu, D.J. The Overexpression of Tβ4 in the Hair Follicle Tissue of Alpas Cashmere Goats Increases Cashmere Yield and Promotes Hair Follicle Development. *Animals* **2019**, *10*, 75. [CrossRef]
11. Hui, T.; Zheng, Y.; Yue, C.; Wang, Y.; Bai, Z.; Sun, J.; Cai, W.; Zhang, X.; Bai, W.; Wang, Z. Screening of cashmere fineness-related genes and their ceRNA network construction in cashmere goats. *Sci. Rep.* **2021**, *11*, 21977. [CrossRef] [PubMed]
12. Kiktev, D.A.; Sheng, Z.; Lobachev, K.S.; Petes, T.D. GC content elevates mutation and recombination rates in the yeast Saccharomyces cerevisiae. *Proc. Natl. Acad. Sci. USA* **2018**, *115*, E7109–E7118. [CrossRef]
13. Tian, D.; Wang, Q.; Zhang, P.; Araki, H.; Yang, S.; Kreitman, M.; Nagylaki, T.; Hudson, R.; Bergelson, J.; Chen, J.Q. Single-nucleotide mutation rate increases close to insertions/deletions in eukaryotes. *Nature* **2008**, *455*, 105–108. [CrossRef] [PubMed]
14. Chen, J.Q.; Wu, Y.; Yang, H.; Bergelson, J.; Kreitman, M.; Tian, D. Variation in the ratio of nucleotide substitution and indel rates across genomes in mammals and bacteria. *Mol. Biol. Evol.* **2009**, *26*, 1523–1531. [CrossRef]
15. McDonald, M.J.; Wang, W.C.; Huang, H.D.; Leu, J.Y. Clusters of nucleotide substitutions and insertion/deletion mutations are associated with repeat sequences. *PLoS Biol.* **2011**, *9*, e1000622. [CrossRef]
16. Powell, B.C.; Rogers, G.E. The role of keratin proteins and their genes in the growth, structure and properties of hair. *Exs* **1997**, *78*, 59–148. [CrossRef] [PubMed]
17. Rogers, M.A.; Langbein, L.; Praetzel-Wunder, S.; Winter, H.; Schweizer, J. Human hair keratin-associated proteins (KAPs). *Int. Rev. Cytol.* **2006**, *251*, 209–263. [CrossRef]
18. Gong, H.; Zhou, H.; Wang, J.; Li, S.; Luo, Y.; Hickford, J.G.H. Characterisation of an Ovine Keratin Associated Protein (KAP) Gene, Which Would Produce a Protein Rich in Glycine and Tyrosine, but Lacking in Cysteine. *Genes* **2019**, *10*, 848. [CrossRef]
19. McGaughey, G.B.; Gagné, M.; Rappé, A.K. pi-Stacking interactions. Alive and well in proteins. *J. Biol. Chem.* **1998**, *273*, 15458–15463. [CrossRef]
20. Levitt, M.; Perutz, M.F. Aromatic rings act as hydrogen bond acceptors. *J. Mol. Biol.* **1988**, *201*, 751–754. [CrossRef]
21. Fraser, R.D.; Parry, D.A. Filamentous Structure of Hard β-Keratins in the Epidermal Appendages of Birds and Reptiles. *Sub-Cell. Biochem.* **2017**, *82*, 231–252. [CrossRef]
22. Dressler, E.A.; Bormann, J.M.; Weaber, R.L.; Merkel, R.C.; Rolf, M.M. A review of cashmere fiber phenotypes: Production, heritabilities, and genetic correlations. *Small Rumin. Res.* **2024**, *240*, 107369. [CrossRef]
23. Gao, Y.; Duo, L.; Zhe, X.; Hao, L.; Song, W.; Gao, L.; Cai, J.; Liu, D. Developmental Mapping of Hair Follicles in the Embryonic Stages of Cashmere Goats Using Proteomic and Metabolomic Construction. *Animals* **2023**, *13*, 3076. [CrossRef] [PubMed]
24. Toigonbaev, S.; Kerven, C.; Mueller, J.P.; Khaliapov, T.; Abdikarimov, A.; McGregor, B.A. Characterization of the growth, productivity and physical fiber properties of indigenous cashmere goats in southern Kyrgyzstan. *Small Rumin. Res.* **2023**, *226*, 107057. [CrossRef]
25. Zhou, H.; Hickford, J.G.; Fang, Q. A two-step procedure for extracting genomic DNA from dried blood spots on filter paper for polymerase chain reaction amplification. *Anal. Biochem.* **2006**, *354*, 159–161. [CrossRef]
26. Gong, H.; Zhou, H.; Yu, Z.; Dyer, J.; Plowman, J.E.; Hickford, J. Identification of the ovine keratin-associated protein KAP1-2 gene (KRTAP1-2). *Exp. Dermatol.* **2011**, *20*, 815–819. [CrossRef] [PubMed]

27. Byun, S.O.; Fang, Q.; Zhou, H.; Hickford, J.G. An effective method for silver-staining DNA in large numbers of polyacrylamide gels. *Anal. Biochem.* **2009**, *385*, 174–175. [CrossRef]
28. Livak, K.J.; Schmittgen, T.D. Analysis of relative gene expression data using real-time quantitative PCR and the 2(-Delta Delta C(T)) Method. *Methods* **2001**, *25*, 402–408. [CrossRef]
29. Zhou, H.; Gong, H.; Wang, J.; Dyer, J.M.; Luo, Y.; Hickford, J.G. Identification of four new gene members of the KAP6 gene family in sheep. *Sci. Rep.* **2016**, *6*, 24074. [CrossRef]
30. Zhou, H.; Gong, H.; Li, S.; Luo, Y.; Hickford, J.G. A 57-bp deletion in the ovine KAP6-1 gene affects wool fibre diameter. Journal of animal breeding and genetics. *J. Anim. Breed. Genet.* **2015**, *132*, 301–307. [CrossRef]
31. Li, S.; Zhou, H.; Gong, H.; Zhao, F.; Wang, J.; Luo, Y.; Hickford, J.G.H. Variation in the Ovine KAP6-3 Gene (KRTAP6-3) Is Associated with Variation in Mean Fibre Diameter-Associated Wool Traits. *Genes* **2017**, *8*, 204. [CrossRef] [PubMed]
32. Sun, H.; He, Z.; Zhao, F.; Hu, J.; Wang, J.; Liu, X.; Zhao, Z.; Li, M.; Luo, Y.; Li, S. Spatiotemporal Expression Characterization of KRTAP6 Family Genes and Its Effect on Wool Traits. *Genes* **2024**, *15*, 95. [CrossRef] [PubMed]
33. Bhat, N.A.; Iqbal, Z.; Mir, S.S.; Bhat, B.A.; Ayaz, A.; Bhat, Z.F.; Shah, R.A.; Ganai, N.A.; Andrabi, S.M.; Bhat, H.F. Characterization and Expression Profiling of KRTAP9.2 at Different Developmental Stages of Pashmina Fiber in Changthangi Goat. *Indian. J. Anim. Res.* **2022**, *56*, 1214–1220. [CrossRef]
34. Wu, D.D.; Irwin, D.M. Evolution of Trichocyte Keratin Associated Proteins. *Adv. Exp. Med. Biol.* **2018**, *1054*, 47–56. [CrossRef] [PubMed]
35. Liao, J.; Lowthert, L.A.; Ku, N.O.; Fernandez, R.; Omary, M.B. Dynamics of human keratin 18 phosphorylation: Polarized distribution of phosphorylated keratins in simple epithelial tissues. *J. Cell Biol.* **1995**, *131*, 1291–1301. [CrossRef]
36. Strnad, P.; Windoffer, R.; Leube, R.E. Induction of rapid and reversible cytokeratin filament network remodeling by inhibition of tyrosine phosphatases. *J. Cell Sci.* **2002**, *115*, 4133–4148. [CrossRef]
37. Gallivan, J.P.; Dougherty, D.A. Cation-pi interactions in structural biology. *Proc. Natl. Acad. Sci. USA* **1999**, *96*, 9459–9464. [CrossRef]

animals

MDPI

Article

ASIP, *AHCY* and *ITCH* Genes Are Associated with the Coat Color of Local Goats (*Capra hircus*) of Southwestern China

Linyun Zhang, Shengnan Zhao, Houmo Yu, Yixin Duan, Jipan Zhang, Naiyi Xu * and Yongju Zhao *

College of Animal Science and Technology, Southwest University, Chongqing Key Laboratory of Herbivore Science, Chongqing Key Laboratory of Forage & Herbivore, Chongqing Engineering Research Center for Herbivores Resource Protection and Utilization, Chongqing Herbivore Engineering Research Center, Chongqing 400715, China; lyzhang023@163.com (L.Z.); zsnyolo@163.com (S.Z.); yhm7791@163.com (H.Y.); duanyx315@163.com (Y.D.); jpanzhang@live.com (J.Z.)

* Correspondence: naiyixu@126.com (N.X.); zyongju@163.com (Y.Z.); Tel.: +86-15136462019 (N.X.); +86-13883399148 (Y.Z.)

Simple Summary

Coat color is an important economic trait in goats, influencing breed identification and market value. Chongqing goats, representative local breeds in Southwestern China, exhibit diverse coat colors; however, the genetic basis underlying this variation remains poorly understood. In this study, we identified genes associated with the coat color of Chongqing goats via a genome-wide association study (GWAS). The results revealed that the *ASIP*, *AHCY*, and *ITCH* genes on chromosome 13 play a significant role in coat color variation in Chongqing goats. Further haplotype and functional analyses supported the association of these genes with coat color, with *ASIP* and *AHCY* related to black coat color and the *ITCH* gene potentially influencing white coat color expression. These findings enhance our understanding of the genetic mechanisms regulating coat color in goats and provide valuable genetic resources for breed conservation and improvement.

Abstract

The local goat breeds of Chongqing represent the typical goat populations of Southwestern China and play a significant role in global goat research. However, studies on these goats remain limited. Chongqing goats exhibit diverse coat colors, a crucial economic trait essential for individual and breed identification. In this study, we performed genome-wide association analysis (GWAS) of Chongqing goats to identify candidate genes associated with coat color. The results of the GWAS revealed that the *ASIP*, *AHCY*, and *ITCH* genes on chromosome 13 are significantly associated with coat color variation in Chongqing goats. Furthermore, haplotype analysis, gene function annotation, and pathway analysis enrichment further confirmed that a significant region on chromosome 13 is associated with coat color in local Chongqing goats, and that *ASIP* and *AHCY* are related to black coat color, whereas *ITCH* is potentially associated with white coat expression. Additionally, we also identified *TPK1* as being associated with coat color in goats, which has hardly been reported. The newly identified genes expand our understanding of the complex genetic architecture underlying pigmentation traits in livestock. Our study identifies the genes related to coat color in local goats, which enriches the genetic resource pool of goats and contributes to a deeper understanding of the genetic mechanisms underlying coat color in these breeds.

Keywords: goat; coat color; GWAS; haplotype

1. Introduction

Goats, as highly adaptable and economically sustainable livestock, exhibit coat coloration that not only determines the market value of their pelts and fibers but also correlates with multiple significant traits, including growth performance, disease resistance, and environmental adaptability. Additionally, coat color serves as a critical identifier for individual recognition, breed (strain) classification, and taxonomic differentiation. The expression and regulation of coat coloration represent a genetically intricate process, governed by the polygenic interactions of multiple loci that collectively orchestrate pigment synthesis, distribution, and phenotypic manifestation [1]. The diverse coat color observed in goats represents a fascinating biological phenomenon. This diversity is closely linked to significant variations in melanin content among different breeds, which is reflected in the wide spectrum of pigmentation patterns across goat populations [2]. In recent years, substantial progress has been made in elucidating the genetic mechanisms responsible for coat color variation in goats, largely driven by advancements in high-throughput sequencing technologies. Key genes such as *ASIP* and *MC1R* have been functionally characterized, with their regulatory mechanisms becoming increasingly well understood [3–5]. These genes play pivotal roles in modulating the biosynthesis of eumelanin and pheomelanin, thereby governing both the intensity and spatial distribution of coat pigmentation patterns.

China possesses abundant indigenous goat genetic resources. Chongqing goats represent typical local breeds in Southwestern China, with six local varieties recognized as national-level genetic resources: Dazu black goat (DZ), Yudong black goat (YD), Banjiao goat (BJ), Hechuan white goat (HC), Chuandong white goat (CD), and Youzhou black goat (YZ). Due to long-term adaptation to unique natural environment and traditional farming practices, these goats have developed distinct regional coat color patterns. In addition to the black and white coat colors commonly found in these breeds, YZ exhibits a relatively unique phenotype characterized by a predominantly white coat with a black mid-dorsal stripe. The distinct phenotypic characteristics of all six local breeds of Chongqing are illustrated in Figure S1. Research on the genetic mechanisms underlying coat color in Chongqing indigenous goats remains relatively limited, and the regulatory networks controlling coat color in these breeds have yet to be fully elucidated. With the advancement of genomic technologies, GWASs have become a powerful tool for identifying genetic variations influencing these traits [6]. Compared to genotyping arrays, whole-genome resequencing provides a significantly larger amount of variant information, enabling more precise analysis results [7]. Therefore, in this study, we used whole-genome resequencing data that our lab previously generated to perform a GWAS.

Given the crucial role of coat color in breed improvement and the conservation of local genetic resources, this study aims to identify candidate genomic regions associated with coat color in indigenous Chongqing goats using whole-genome resequencing. This study of Chongqing goats with different coat colors utilizes a GWAS to identify candidate functional genes related to coat color. In addition, haplotype analysis, candidate gene protein interactions, functional annotation, and pathway enrichment analysis are performed to further explore the potential regulatory roles of these genes in coat color determination.

2. Materials and Methods

A complete list of abbreviations used in this manuscript is provided in the abbreviations section (Table S1) of the Supplementary Materials.

2.1. Data and Quality Control

We used 776 goat whole-genome resequencing data determined by our previously generated data (NCBI accession numbers: PRJNA1212315) [8] using the DNBSEQ-T7 platform and aligned the clean reads to the goat reference genome (ARS1.2) with BWA. Quality control was performed with PLINK [9], removing SNPs with missing rates > 10%, minor allele frequency < 1%, and Hardy–Weinberg equilibrium p-values $< 1 \times 10^{-6}$.

2.2. Genome-Wide Association Study

In this study, a GWAS between single-nucleotide polymorphisms (SNPs) and coat color was conducted using a mixed linear model (MLM) in the GEMMA [10] software. The model is specified as follows:

$$y = X\beta + Zg + Qk + e \quad (1)$$

In this equation, y represents the phenotypic data, β is the vector of fixed effects excluding SNPs and kinship (such as PCA, sex, etc.), g is the vector of SNP effects, k represents the vector of random effects (kinship), and e is the vector of residual effects. X, Z, and Q are the corresponding design matrices.

The GWAS results were visualized using a Manhattan plot and QQ plot generated with the "qqman" [11] and "CMplot" packages [12] in R.

To determine the significance thresholds, Bonferroni correction was applied. The genome-wide significance threshold was set at 1/n, while an extremely significant threshold was established at 0.05/n: genes existing within ±200 kb of the SNPs noted.

2.3. Haplotype Analysis

The haplotype analysis was explored using Beagle [13] and the "pegas" [14] and "ape" [15] packages in R; the network diagram was drawn using Network 10.02 [16].

2.4. Gene Function Annotation and Pathway Analysis

The protein–protein interaction (PPI) network diagram was constructed by using the STRING database [17]. The gene ontology (GO) database classifies gene functions into three main categories: cellular component (CC), molecular function (MF), and biological process (BP). Kyoto Encyclopedia of Genes and Genomes (KEGG) is a database for metabolic pathways. Candidate genes were identified based on their positions in the ARS1.2, using the upstream and downstream 200 kb distances of significant loci as the interval. SNP loci were annotated to corresponding genes using bedtools software version 2.18. GO and KEGG functional enrichment analyses were performed using the DAVID online tool (https://davidbioinformatics.nih.gov/ accessed on 20 January 2025), and the results were visualized using the "ggplot2" package [18] in R.

3. Results

3.1. GWAS of Coat Color in Chongqing Local Goats

A GWAS was performed for the six local goat breeds from Chongqing, encompassing three coat color types: black (DZ, YD), white (BJ, HC, CD), and mainly white with a black mid-dorsal stripe (YZ), abbreviated as black-white. The results revealed a highly significant region on chromosome 13 associated with the *ASIP* gene, as well as significant regions on chromosomes 4 and 26 associated with the *TPK1* and *PCDH15* genes, respectively, and the *KIT* gene was identified within the regulatory region spanning ±200 kb around a highly significant SNP on chromosome 6 (Figure 1, Table S2).

In addition to black and white coat colors, Chongqing indigenous goats also include the black-white YZ breed. To investigate the genetic basis of coat color variation, we conducted a three-way comparison among these coat color groups, as well as pairwise comparisons between the two groups in combination. This strategy aimed to identify genetic variants primarily responsible for black or white coat color formation and to determine whether a unique genetic mechanism underlies the development of black-white patterned coats. First, we performed a GWAS by comparing the black coat color group (DZ, YD) with white coat color group (BJ, HC, CD). The significant SNPs on chromosome 13 were mapped to *ASIP* and *AHCY*. Additionally, within the ±200 kb region surrounding the significant SNPs on chromosome 13, several genes were identified, including *ITCH*, *EIF2S2*, *RALY*, and *LOC10219053* and several genes which were mapped on chromosomes 2, 10, 15, and 18, such as *PDE1A*, *RYR3* and *CMIP*, were also detected (Figure 2A, Table S3).

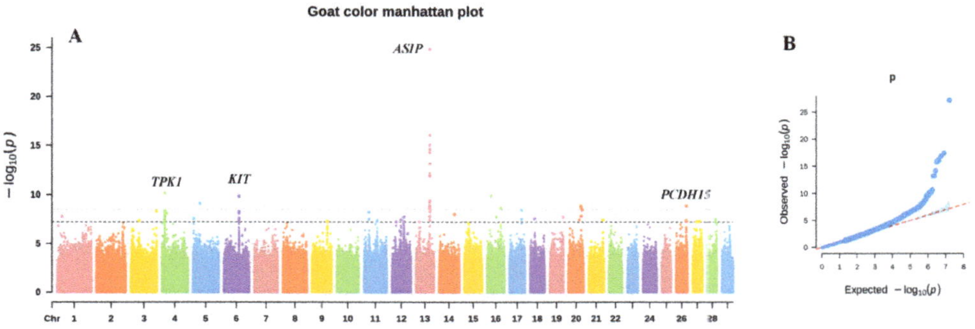

Figure 1. GWAS results for black, white, and black-white coat colors of Chongqing local goats. Note: (**A**) is a Manhattan plot, the gray horizontal line is the very significant threshold line (0.05/n), and the black is the significant threshold line (1/n), the different colors of dots represent SNPs on different chromosomes; (**B**) is a QQ plot, with the red dotted line indicating the expected distribution under the null hypothesis and the blue dots showing the observed *p*-values.

Goats with a black coat color (DZ, YD) were then compared with those with a black-white coat color (YZ), revealing a highly significant SNP region on chromosome 13. The significant SNPs were mapped to the *ASIP*, *BPIFB4*, and *DNMT3B* genes, while the ±200 kb flanking region contained *AHCY*, *ITCH*, and *EIF2S2*, as well as significant regions on chromosomes 2, 5, 7, 8, 9, 10, 14, 17, 20, and 27 associated with genes such as *IRS1*, *DNM2*, and *ESR1*, which may be related to the expression of coat color in goats (Figure 2B, Table S4). Finally, comparing goats with a white coat color (BJ, HC, CD) with those with a black-white coat color (YZ) revealed the most significant SNP region located on chromosome 13. The significant SNPs were mapped to the *ASIP* and *ITCH*, with the ±200 kb flanking region containing genes such as *AHCY*, *EIF2S2*, *RALY*, and *LOC102190531*. In addition, significant regions on chromosomes 1, 4, 11, 16, 25, and 26, containing genes such as *CTNNA2*, *SLC29A4*, and *LOC102185739*, were also detected and may be associated with coat color variation in goats (Figure 2C, Table S5).

3.2. Haplotype Analysis of Coat Color Significant SNP Region

Haplotypes were constructed using the significant regions identified through the GWAS across all the above four combinations (Figures 1 and 2) to further validate the association between candidate genes and coat color variation. The *ASIP* gene exhibited strong associations across multiple statistical models, and a highly significant region within the *ASIP* gene (chr13: 63,232,000–63,263,000 bp) was selected for haplotype construction (Figure 3A). The results reveal that individuals in the black coat color group (red) possessed distinct haplotypes within this region, which were closely related to each other and clearly differentiated from those observed in the white coat color (blue) and black-white coat color (purple) groups.

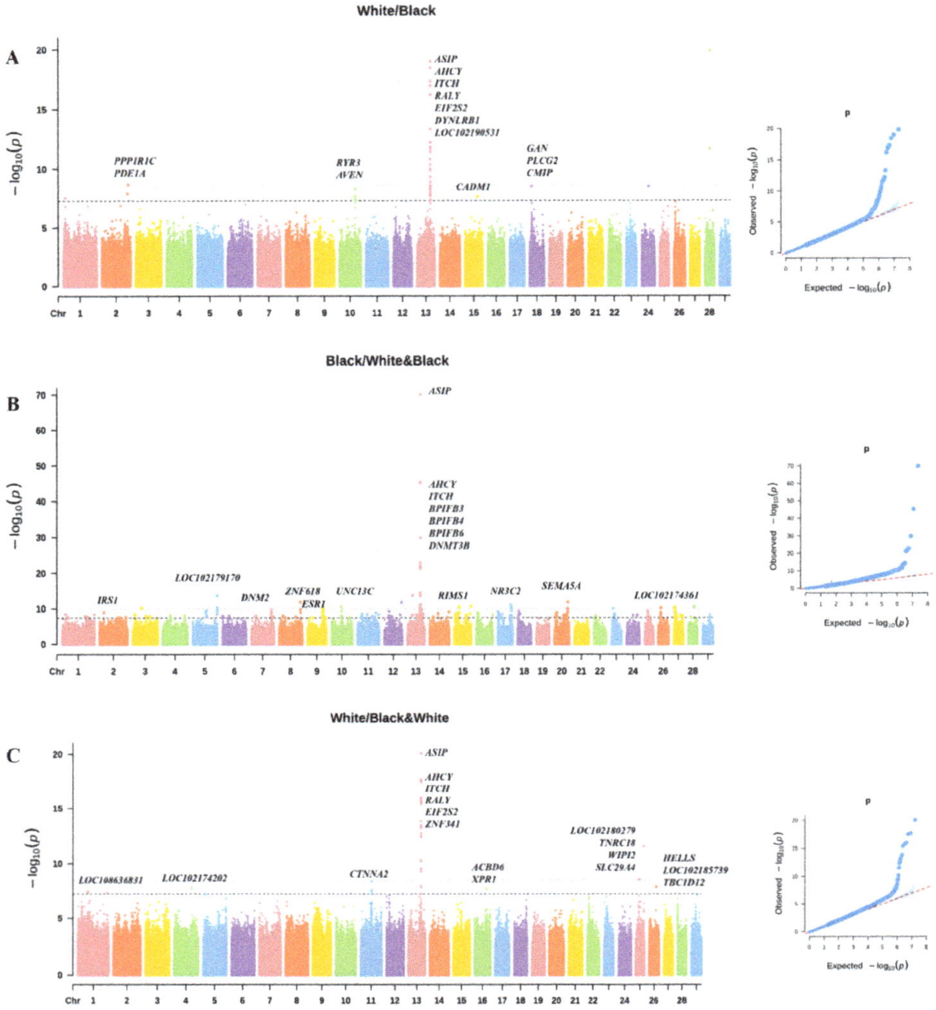

Figure 2. GWAS results for different combinations of coat color. Note: (**A**) is the Manhattan plot and QQ plot of the black and white coat colors; the gray horizontal line is the very significant threshold line (0.05/n) and the black line is the significant threshold line (1/n). (**B**) is the Manhattan plot and QQ plot of the black and black-white coat colors. (**C**) is the Manhattan plot and QQ plot of the white and black-white coat colors. In QQ plots, the red dotted line represents the expected distribution under the null hypothesis, and blue dots represent observed *p*-values.

In the analysis comparing black with white and black-white coat colors, significant SNPs were consistently identified within or near the *AHCY* gene across groups. Therefore, haplotypes were constructed, and a haplotype network was generated for the highly significant region of the *AHCY* gene (chr13: 63,257,000–63,276,000 bp) (Figure 3B). Within this region, the black coat color group (red) possessed a unique haplotype, with most individuals and haplotypes clustering closely together in the network, suggesting a strong association with this pigmentation phenotype. Additionally, the white coat color group (blue) and the black-white coat color group (purple) exhibited distinct but less tightly clustered haplotypes in this region.

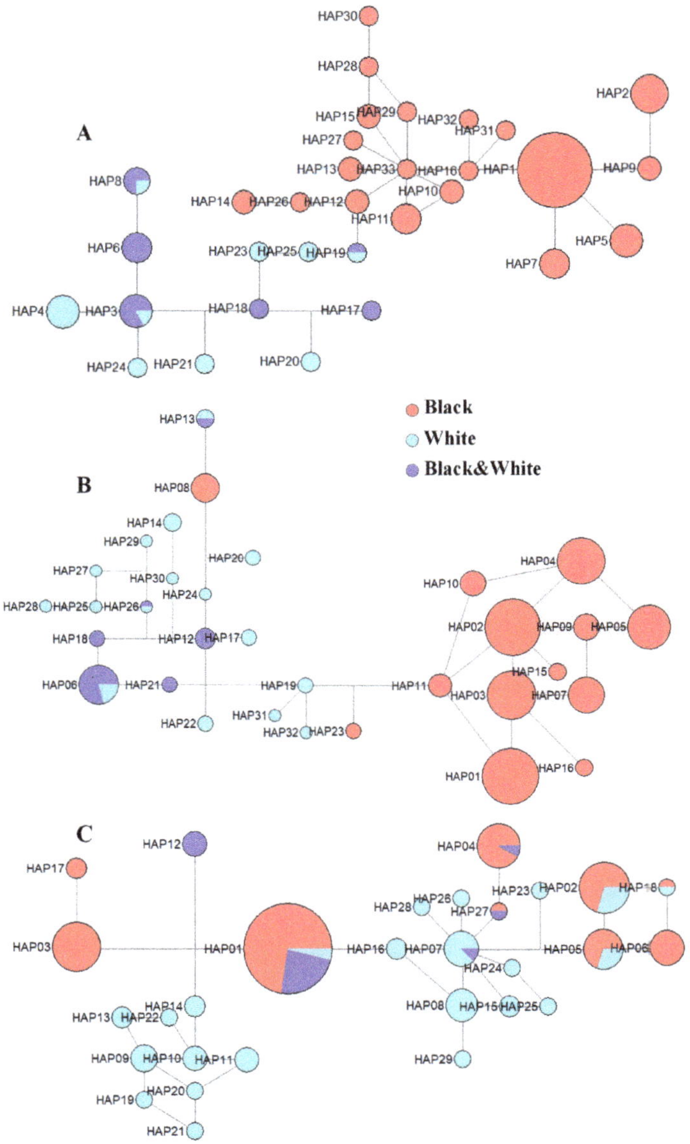

Figure 3. Haplotype analysis results for the significant genes related to coat color in Chongqing local goats. Note: (**A**) represents the haplotype network of the *ASIP* gene across different coat colors; (**B**) represents the haplotype network of the *AHCY* gene across different coat colors; (**C**) represents the haplotype network of the *ITCH* gene across different coat colors. The red color in the network represents the group with a black coat color; the purple color represents the group with a white coat color; and the blue color represents the group with a black-white coat color.

In the analysis comparing white with black-white coat colors, significant SNPs were identified within the *ITCH* gene. Additionally, *ITCH* was also found to be near significant SNPs in the comparisons between black and white coat color groups, as well as between black coat color and black-white coat color groups. Haplotypes were constructed, and a haplotype network was generated for the *ITCH* gene (chr13: 63,363,104–63,462,630) (Figure 3C). In the haplotypes of the *ITCH* gene, the black coat color group (red) possessed unique haplotypes, though most individuals showed relatively minor differences from other coat colors. However, the white coat color group (blue) exhibited relatively distinct and clustered haplotypes, suggesting that the *ITCH* gene may influence coat color by affecting the expression of genes associated with white coat color in goats.

3.3. Results of Gene Function Annotation and Pathway Analysis

To investigate the functional relevance of the genes prioritized in coat color regulation, the PPI network was constructed using significant genes (*ASIP*, *KIT*, *AHCY*, *ITCH*, and *RALY*) identified via the GWAS (Figure 4A). Notably, *ASIP* and *KIT* occupy hub positions within the network topology, exhibiting direct or indirect interactions with core regulators of melanogenesis and pigment deposition, including *MC1R*, *TYR*, and *TYRP1*. This connectivity pattern underscores their pivotal regulatory roles in modulating coat color phenotypes. Furthermore, *AHCY*, *ITCH*, and *RALY* demonstrate network connectivity with *ASIP* or *MC1R*, suggesting their potential involvement in melanogenic signaling cascades via indirect regulatory mechanisms.

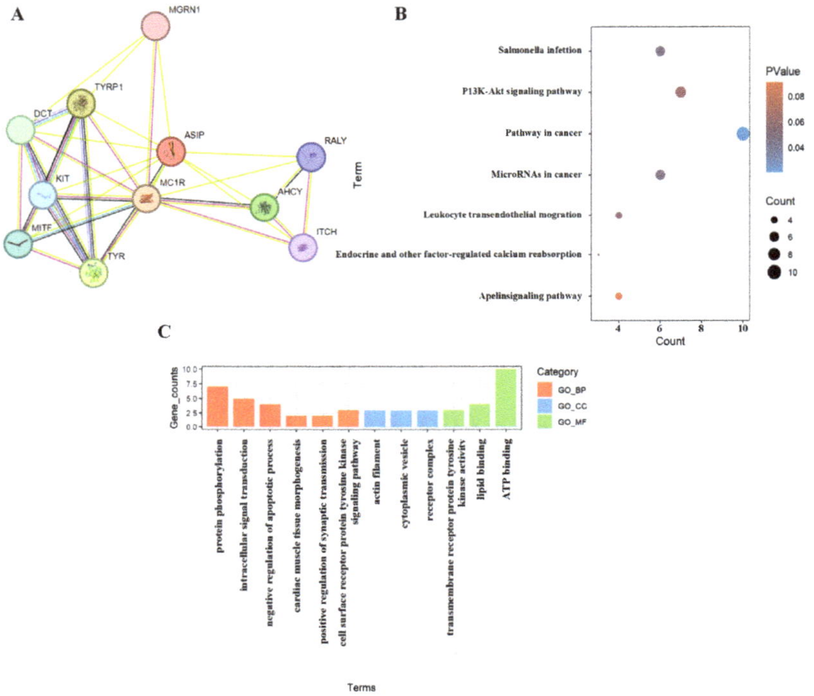

Figure 4. PPI (**A**), KEGG (**B**), and GO (**C**) results for significant genes ($p < 1/n$).

KEGG pathway enrichment was performed on the differentially enriched genes of coat color in Chongqing local goats (Figure 4B). Among the identified pathways, several pathways are potentially involved in pigmentation regulation. Notably, the PI3K-Akt signaling pathway and pathways in cancer were significantly enriched, both of which are known to regulate cell proliferation, differentiation, and processes key to survival in melanocyte development and function [19]. Additionally, pathways such as salmonella infection and leukocyte transendothelial migration may suggest a potential link between immune regulation and coat color determination [20]. These findings imply that coat color variation may be influenced by multiple signaling pathways, particularly those involved in melanogenesis, cellular signaling, and immune-related processes.

The GO enrichment analysis identified functional categories associated with genes potentially involved in coat color variation. The GO terms are classified into three categories (Figure 4C): biological processes (GO_BP, red), cellular components (GO_CC, blue), and molecular functions (GO_MF, green). The most enriched biological processes included protein phosphorylation and intracellular signal transduction, both of which are closely associated with melanogenesis and pigment cell signaling [21]. Additionally, the enrichment of GO terms related to receptor activity, lipid binding, and kinase signaling pathways suggests a potential role in regulating pigment cell differentiation and melanin synthesis [22]. These functional categories collectively indicate that the genes associated with coat color may influence pigmentation in local Chongqing goats through mechanisms involving signal transduction, phosphorylation, and metabolic regulation.

4. Discussion

Coat color is a critical economic trait in goats, and recent advances in molecular biology have significantly enhanced our understanding of the genetic basis underlying this trait. Our study identified key genetic regions associated with coat color in local goat breeds from Chongqing, including the genes *ASIP*, *AHCY*, and *ITCH*. *ASIP*, located on chromosome 13, plays a crucial role in pigmentation regulation across various species, including goats, with studies highlighting its key involvement in coat color patterns [3]. During the association analysis of the three different coat colors, a significant regulatory SNP region on chromosome 6 was mapped to the *KIT* gene, which is significantly associated with coat color. We identified the genes (*AHCY*, *ITCH*, and *KIT*) related to coat color, which were also observed in breeds such as the Angora, Chongming white goat, cashmere goat, Pakistani Angora, and Barbari goats [23–27]. However, many investigations still rely on SNP arrays due to their low cost and high throughput. And SNP chips are limited to predefined loci and may fail to capture rare or novel variants. Compared to previous research, our study employed whole-genome resequencing, which offers broader genomic coverage and higher resolution, enabling more comprehensive variant detection. Additionally, many previous studies have been limited by small sample sizes, which reduce statistical power for detecting minor allele effects and increase the risk of false negatives. In this study, we utilized a large dataset comprising 776 individuals, thereby enhancing the robustness and reliability of association signals related to coat color in goats.

In this study, we found that the *AHCY* and *ITCH* genes were significantly associated with coat color variation. In the *AHCY* gene haplotypes, the black coat color group was almost entirely distinct from other coat color groups; the haplotype distribution in different groups of *AHCY* is similar to that of the *ASIP* gene, suggesting that the expression in this region may affect the black coat color of goats. In the *ITCH* gene haplotypes, two independent regions were observed in the white coat color group, one of which was almost entirely found in BJ, while the other showed a high frequency in HC, indicating potential breed-specific differences in white coat expression. In particular, there were multiple highly significant SNPs of *ITCH* in the comparison between white coat color and black-white coat color goats. These SNPs showed significant differences from the other groups. Combined with its specificity regarding the white population in haplotypes, it may have a certain relationship with the degree of expression of the white coat phenotype in goats. More importantly, the association between the *ITCH* gene and white coat color in goats has not been previously explored, so the current study provides novel insights into the genetic basis of white coat color expression in goats. In addition, several significant SNPs were identified within the *TPK1* gene on chromosome 1, which has not been previously linked to pigmentation. This finding suggests that *TPK1* is a potential novel candidate gene involved in coat color regulation in goats. In addition, multiple genes located within 200 kb upstream and downstream of significant SNPs were detected, indicating that a broader regulatory network may contribute to coat color variation. These genes (e.g., *TPK1*, *EIF2S2*, and *LOC102190531*), which have hardly been reported, provide valuable targets for future studies on the genetic architecture and molecular mechanisms underlying pigmentation traits in goats. Further research and verification can be conducted in the future. Nevertheless, all significant SNPs located within the ASIP gene region were annotated as intron variants or 5′ untranslated region (5′UTR) variants (Table S6). None of them resulted in protein-coding changes, suggesting that these loci may influence coat color through transcriptional regulation or splicing-related mechanisms rather than direct amino acid alterations. The precise regulatory roles of these intronic and UTR variants remain to be elucidated and warrant further functional investigations. Our study fills the gap in research on the coat color of Chongqing goats and provides a comprehensive understanding of goat coat color. These findings may be applied to the genetic improvement of local breeds through marker-assisted selection, particularly in regions where a black coat color is favored for its market value or adaptive benefits. Incorporating these loci into breeding programs could enhance population consistency and economic performance.

However, due to the complexity of coat color formation and regulation, the specific mechanisms determining animal coat color are not yet fully elucidated. With the advancement of genomic sequencing technologies and molecular testing technology, there is potential for deeper exploration and validation of these mechanisms.

5. Conclusions

This study identifies *ASIP*, *AHCY*, and *ITCH* as key coat color candidate genes through a GWAS of Chongqing goats from Southwestern China, which provides novel insights into the genetic basis of coat color variation in goats. By integrating haplotype analysis, gene function annotation, and pathway enrichment, we confirmed the involvement of this gene region in the black coat phenotype of local Chongqing goats and uncovered a potential link between *ITCH* and white coat color expression.

These findings have important implications for local goat populations, particularly in the selective breeding and maintenance of desirable coat color traits in indigenous breeds. Understanding the genetic basis of pigmentation can aid in maintaining the genetic diversity of local goat populations and optimizing breeding strategies for improved economic value.

Supplementary Materials: The following supporting information can be downloaded at: https://www.mdpi.com/article/10.3390/ani15131849/s1, Table S1: list of abbreviations; Table S2: The genes associated with significant SNPs for black, white, and black-white coat colors in Chongqing local goats; Table S3: The genes associated with significant SNPs for black and white coat colors in Chongqing local goats; Table S4: The genes associated with significant SNPs for black and black-white coat color in Chongqing local goats; Table S5: The genes associated with significant SNPs for white and black-white coat colors in Chongqing local goats; Table S6. Functional annotation of significant SNPs within the ASIP gene region associated with coat color in Chongqing goats. Figure S1: Pictures of six local goat breeds in Chongqing.

Author Contributions: Conceptualization, N.X. and Y.Z.; methodology, L.Z., J.Z. and N.X.; software, L.Z., S.Z. and H.Y.; validation, L.Z., S.Z. and Y.D.; formal analysis, L.Z., S.Z., H.Y. and Y.D.; investigation, S.Z., H.Y. and Y.D.; resources, N.X. and Y.Z.; data curation, S.Z., H.Y. and Y.D.; writing—original draft preparation, L.Z. and S.Z.; writing—review and editing, J.Z., N.X. and Y.Z.; visualization, L.Z., S.Z. and Y.D.; supervision, N.X. and Y.Z.; project administration, N.X. and Y.Z. All authors have read and agreed to the published version of the manuscript.

Funding: This work was funded by the Chongqing Science and Technology Innovation Key Project of China (No. CSTB2022TIAD-CUX0013), the Collection, Utilization and Innovation of Animal Resources by Research Institutes and Enterprises of Chongqing (No.Cqnyncw-kqlhtxm), Chongqing Modern Agricultural Industry Technology System (CQMAITS202313), the National Key Research and Development Program of China (No.2022YFD1300202) and the National Training Program of Innovation and Entrepreneurship for Undergraduates (No. S202310635040).

Institutional Review Board Statement: All animal management and experimental procedures followed the animal care protocols approved by the Ethics Committee of Southwest University (No. IACUC-20230417-02).

Informed Consent Statement: Informed consent was obtained from all subjects involved in the study.

Data Availability Statement: Data sharing is not applicable to this article as no datasets were generated or analyzed during the current study.

Acknowledgments: We thank the Hefei Advanced Computing Center for providing numerical computation resources for this study.

Conflicts of Interest: The authors declare that they have no conflicts of interest to this work.

References

1. Cieslak, M.; Reissmann, M.; Hofreiter, M.; Ludwig, A. Colours of domestication. *Biol. Rev. Camb. Philos. Soc.* **2011**, *86*, 885–899. [CrossRef] [PubMed]
2. Bhat, B.; Singh, A.; Iqbal, Z.; Kaushik, J.; Rao, A.; Ahmad, S.; Bhat, H.; Ayaz, A.; Sheikh, F.; Kalra, S.; et al. Comparative transcriptome analysis reveals the genetic basis of coat color variation in Pashmina goat. *Sci. Rep.* **2019**, *9*, 6361. [CrossRef] [PubMed]
3. Henkel, J.; Saif, R.; Jagannathan, V.; Schmocker, C.; Zeindler, F.; Bangerter, E.; Herren, U.; Posantzis, D.; Bulut, Z.; Ammann, P.; et al. Selection signatures in goats reveal copy number variants underlying breed-defining coat color phenotypes. *PLoS Genet.* **2019**, *15*, e1008536. [CrossRef]
4. Arenas-Báez, P.; Torres-Hernández, G.; Castillo-Hernández, G.; Hernández-Rodríguez, M.; Sánchez-Gutiérrez, R.; Vargas-López, S.; González-Maldonado, J.; Domínguez-Martínez, P.; Granados-Rivera, L.; Maldonado-Jáquez, J. Coat Color in Local Goats: Influence on Environmental Adaptation and Productivity, and Use as a Selection Criterion. *Biology* **2023**, *12*, 929. [CrossRef] [PubMed]
5. Ma, L.; Zhao, W.; Ma, Q.; Wang, J.; Zhao, Z.; Zhang, J.; Gu, Y. Genome-Wide Association Study of Birth Wool Length, Birth Weight, and Head Color in Chinese Tan Sheep Through Whole-Genome Re-Sequencing. *Animals* **2024**, *14*, 3495. [CrossRef] [PubMed]
6. Bush, W.; Moore, J. Chapter 11: Genome-wide association studies. *PLoS Comput. Biol.* **2012**, *8*, e1002822. [CrossRef]
7. Pavan, S.; Delvento, C.; Ricciardi, L.; Lotti, C.; Ciani, E.; D'Agostino, N. Recommendations for Choosing the Genotyping Method and Best Practices for Quality Control in Crop Genome-Wide Association Studies. *Front. Genet.* **2020**, *11*, 447. [CrossRef]
8. Zhang, L.; Duan, Y.; Zhao, S.; Yu, H.; Zhang, J.; Xu, N.; Zhao, Y. Genome-wide association studies reveal novel loci associated with carcass and body measures in goats (*Capra hircus*). *bioRxiv* **2025**. [CrossRef]
9. Chang, C.; Chow, C.; Tellier, L.; Vattikuti, S.; Purcell, S.; Lee, J. Second-generation PLINK: Rising to the challenge of larger and richer datasets. *GigaScience* **2015**, *4*, 7. [CrossRef]
10. Zhou, X.; Matthew, S. Efficient multivariate linear mixed model algorithms for genome-wide association studies. *Nat. Methods* **2014**, *11*, 407–409. [CrossRef]
11. Stephen, D.T. qqman: An R package for visualizing GWAS results using Q-Q and manhattan plots. *J. Open Source Softw.* **2018**, *3*, 731. [CrossRef]
12. Yin, L. CMplot: Circle Manhattan Plot. R Package Version 4.5.0. 2023. Available online: https://CRAN.R-project.org/package=CMplot (accessed on 12 September 2024).
13. Browning, B.L.; Tian, X.; Zhou, Y.; Browning, S.R. Fast two-stage phasing of large-scale sequence data. *Am. J. Hum. Genet.* **2021**, *108*, 1880–1890. [CrossRef]
14. Paradis, E. pegas: An R package for population genetics with an integrated-modular approach. *Bioinformatics* **2010**, *26*, 419–420. [CrossRef] [PubMed]
15. Paradis, E.; Schliep, K. ape 5.0: An environment for modern phylogenetics and evolutionary analyses in R. *Bioinformatics* **2019**, *35*, 526–528. [CrossRef]
16. Bandelt, H.; Forster, P.; Röhl, A. Median-joining networks for inferring intraspecific phylogenies. *Mol. Biol. Evol.* **1999**, *16*, 37–48. [CrossRef] [PubMed]

17. Szklarczyk, D.; Kirsch, R.; Koutrouli, M.; Nastou, K.; Mehryary, F.; Hachilif, R.; Gable, A.L.; Fang, T.; Doncheva, N.T.; Pyysalo, S.; et al. The STRING database in 2023: Protein-protein association networks and functional enrichment analyses for any sequenced genome of interest. *Nucleic Acids Res.* **2023**, *51*, D638–D646. [CrossRef] [PubMed] [PubMed Central]
18. Wickham, H. *ggplot2: Elegant Graphics for Data Analysis*; Springer: Berlin/Heidelberg, Germany, 2016.
19. Peng, Y.; Wang, Y.; Zhou, C.; Mei, W.; Zeng, C. PI3K/Akt/mTOR Pathway and Its Role in Cancer Therapeutics: Are We Making Headway? *Front. Oncol.* **2022**, *12*, 819128. [CrossRef]
20. Videira, I.F.; Moura, D.F.; Magina, S. Mechanisms regulating melanogenesis. *An. Bras. Dermatol.* **2013**, *88*, 76–83. [CrossRef]
21. Slominski, A.; Tobin, D.J.; Shibahara, S.; Wortsman, J. Melanin pigmentation in mammalian skin and its hormonal regulation. *Physiol. Rev.* **2004**, *84*, 1155–1228. [CrossRef] [PubMed]
22. Hearing, V.J. Determination of melanin synthetic pathways. *J. Investig. Dermatol.* **2011**, *131*, E8–E11. [CrossRef]
23. Tarricone, S.; Schlosserová, N.; Bruno, S.; Sardina, M.T.; Landi, V.; Giannico, F.; Colonna, M.A.; Sarti, F.M.; Lasagna, E.; Ceccobelli, S.; et al. Selection Signatures in Italian Goat Populations Sharing the "facciuto" Phenotype. *Genes* **2025**, *16*, 390. [CrossRef] [PubMed]
24. Mendoza, M.N.; Raudsepp, T.; Alshanbari, F.; Gutiérrez, G.; Ponce, F.A. Chromosomal Localization of Candidate Genes for Fiber Growth and Color in Alpaca (*Vicugna pacos*). *Front. Genet.* **2019**, *10*, 583. [CrossRef] [PubMed]
25. Gao, J.; Lyu, Y.; Zhang, D.; Reddi, K.K.; Sun, F.; Yi, J.; Liu, C.; Li, H.; Yao, H.; Dai, J.; et al. Genomic Characteristics and Selection Signatures in Indigenous Chongming White Goat (*Capra hircus*). *Front. Genet.* **2020**, *11*, 901. [CrossRef] [PubMed]
26. Wan, X.; Jing, J.N.; Wang, D.F.; Lv, F.H. Whole-genome selective scans detect genes associated with important phenotypic traits in goat (*Capra hircus*). *Front. Genet.* **2023**, *14*, 1173017. [CrossRef]
27. Wu, D.; Fan, J.; Pang, Y.; Wen, B.; Li, W.; Yang, G.; Cheng, H.; Shi, J.; Wang, T.; Hu, S.; et al. Identification and Expression Patterns of Critical Genes Related to Coat Color in Cashmere Goats. *Genes* **2025**, *16*, 222. [CrossRef]

animals

Article

Polymorphism of Keratin Gene *KRT71* and Its Relationship with Wool Properties in Gansu Alpine Fine-Wool Sheep

Fangfang Zhao [1,*], Zhaohua He [1,2], Hongxian Sun [1], Jiqing Wang [1], Xiu Liu [1], Zhiyun Hao [1], Mingna Li [1] and Shaobin Li [1]

1 Gansu Key Laboratory of Herbivorous Animal Biotechnology, Faculty of Animal Science and Technology, International Wool Research Institute, Gansu Agricultural University, Lanzhou 730070, China; hezh1250148718@163.com (Z.H.); sunhx@st.gsau.edu.cn (H.S.); wangjq@gsau.edu.cn (J.W.); liux@gsau.edu.cn (X.L.); haozy@gsau.edu.cn (Z.H.); limn@gsau.edu.cn (M.L.); lisb@gsau.edu.cn (S.L.)
2 College of Biology and Food, Shangqiu Normal University, Shangqiu 476000, China
* Correspondence: zhaofangfang@gsau.edu.cn

Simple Summary

As the structural proteins of wool, the keratin family's structural and functional changes are likely to cause changes in wool traits. Previous studies have found that KRT71 is a differentially expressed protein in different types of wool, but there is still a lack of research on its expression localization and gene variation. This study analyzed the expression localization and gene polymorphism of *KRT71* using protein immunofluorescence and kompetitive allele-specific PCR methods. The results showed that KRT71 had strong positive expression in the inner root sheath, and the variation in *KRT71* showed a notable association with the wool length.

Abstract

Keratin plays a crucial role in wool formation. Conducting polymorphism studies on key keratin genes is helpful in identifying key SNP sites that might influence wool traits. In this research, kompetitive allele-specific PCR (KASP) genotyping and protein immunofluorescence techniques were used to explore the polymorphisms of the *KRT71* gene in Gansu alpine fine-wool sheep, analyze the relationship between the gene polymorphisms and wool production traits, and examine the expression and localization of the KRT71 protein in the hair follicles of fine-wool sheep. The results indicated that there were two single-nucleotide polymorphisms (SNPs) in the 5′ UTR and exon 9 of the *KRT71* gene, named SNP1 (C.-7G/C) and SNP2 (C.1500G/A), respectively. Regarding SNP1, the mean stable length (MSL) of *GG* genotype individuals was significantly longer than that of *GC* genotype individuals ($p < 0.05$). Similarly, for SNP2, the MSL of *GG* genotype individuals was significantly greater than that of *GA* genotype individuals ($p < 0.05$). Moreover, the KRT71 protein showed moderate positive expression in the cuticle, outer root sheath, and sebaceous gland. It had strong positive expression in the inner root sheath, while no positive expression was detected in the hair medulla and hair papilla. In summary, the sheep *KRT71* gene could be an important candidate gene for improvements in wool length.

Keywords: *KRT71*; wool characteristics; hair follicle development; Gansu alpine fine-wool sheep

1. Introduction

Wool, a raw textile material with a long-standing history, is well-liked by consumers due to its softness, warmth, flame retardancy, elasticity, and wearing comfort [1]. In the consumer market, wool textiles and other wool-based products, like wool sound insulators in cosmetics, wool antimicrobial bandages, and lanolin in medical supplies, have become indispensable in daily life [2,3]. The quality and yield of wool serve as crucial factors in enhancing the economic efficiency of the wool industry [4,5]. Relevant research has indicated that wool quality traits (such as fineness, crimp, length, etc.) have moderate–high heritability, which implies that variations in wool traits might be significantly affected by genetic factors [6,7]. Hence, when the nutritional levels and environmental factors remain constant, genetic factors are significant determinants of wool quality. At the same time, protein-coding genes play vital regulatory roles in the development of various traits in both plants and animals [8].

Keratin and keratin-associated protein genes play crucial roles in hair formation in both humans and animals [9–11]. Research has revealed that keratins not only offer structural support to cells but also participate in intracellular signaling pathways. These pathways are responsible for regulating metabolic processes and promoting the growth of epithelial cells [12]. The *KRT71* gene, which is part of the keratin family, belongs to the type II keratins. In humans, it is located on chromosome 12, while, in sheep, it is situated on chromosome 3. Moreover, the *KRT71* gene is primarily expressed in the root sheath of the hair follicle. Mutations in this gene might serve as the genetic foundation for curly hair and oligohemorrhagia disease [13]. It has been discovered that base mutations in the *KRT71* gene could interact with the counterparts of this gene in type I keratin. This interaction subsequently results in curly hair in animals or humans [14]. Additionally, Duan et al. demonstrated that the mRNA expression levels of genes like *KRT1*, *KRT6A*, and *KRT71* in the skin of curly-haired Yanshan cashmere goats were significantly higher compared to those of non-curly-haired goats. Furthermore, cashmere goats with curly-fiber cashmere could produce cashmere with a finer diameter [15]. The aforementioned studies imply that the *KRT71* gene may have a significant impact on hair formation and hair follicle development. Nevertheless, the influence of mutations in the gene base sequence of the *KRT71* gene on hair follicle development and wool traits in fine-haired goats requires further exploration.

Gansu alpine fine-wool sheep hold significant economic value and ecological status in the alpine pastoral areas of the Qinghai–Tibet Plateau and Qilian Mountains of China. Nevertheless, when compared with fine-wool sheep breeds like Australian Merino and Chinese Merino, there remains a disparity in the wool quality of Gansu alpine fine-wool sheep [16]. Traits including the fineness of wool and the number of hair follicles all have an impact on the wool price, and these traits are co-regulated by multiple genes [17]. Consequently, it is essential to thoroughly explore the functional genes associated with the regulation of wool traits in Gansu alpine fine-wool sheep from a genetic perspective and subsequently uncover the gene functions.

In this research, based on our pre-sequencing results [18], we further selected the candidate gene *KRT71* for gene polymorphism detection. We employed kompetitive allele-specific PCR (KASP) genotyping technology and protein immunofluorescence to examine the polymorphisms of the *KRT71* gene in Gansu alpine fine-wool sheep and the relationship between gene polymorphisms and wool production traits. Additionally, we investigated the expression localization of the KRT71 protein in the hair follicles of fine-wool sheep. This study enriches the molecular genetic data on economic traits in fine-wool sheep.

2. Materials and Methods

2.1. Ethics Statement

The animals used in the study were all from a farm in Tianzhu Zangzu Autonomous County, China. Informed consent was obtained from the animal owners. The animal experiments were approved by the Gansu Agricultural University Ethics Committee under the code GSAU-ETH-AST-2021-028.

2.2. Experimental Animals and Sample Collection

In this research, blood and wool samples were obtained from 315 one-year-old Gansu alpine fine-wool sheep ewes in Songshan Township, Tianzhu Tibetan Autonomous County to identify single-nucleotide polymorphisms (SNPs) in the *KRT71* gene. All these sheep were kept under the same nutritional levels and feeding environment. The collected blood samples were stored at −20 °C for the subsequent extraction of blood DNA. Wool samples were taken from the left scapula area of the sheep to measure various wool economic traits. Additionally, a one-year-old Gansu alpine fine-wool sheep ewe was chosen, and a skin sample was collected from its left scapula. This skin sample was then fixed in a 4% paraformaldehyde solution at room temperature for protein immunofluorescence identification. When the samples were collected, the wool fibers were in the growth stage of the hair follicle.

2.3. Determination of Wool Production Traits

Wool samples were used for the determination of various wool economic traits, such as the mean fiber diameter (MFD), coefficient of variation of fiber diameter (CVFD), fiber diameter standard deviation (FDSD), mean fiber curvature (MFC), comfort factor (CF), MSL, and mean stable strength (MSS), etc. The wool analysis was carried out with reference to the National Standard of the People's Republic of China for Sheep Wool (GB1523-2013) [19].

2.4. Protein Immunofluorescence Localization Analysis

Sheep skin tissue was immersed in 4% paraformaldehyde for fixation over a period of 48 h. Subsequently, an immunofluorescence assay was carried out to detect the KRT71 protein, following the method described by Diao et al. [20]. For the immunofluorescence detection of the KRT71 protein, the primary antibody employed was K71 rabbit anti-rabbit (Product Item No.: GTX107343, GeneTex, Irvine, CA, USA), and the secondary antibody was FITC-labeled goat anti-rabbit (Product Item No.: GB22303, Servicebio, Wuhan, China). The microscopic images were observed and analyzed using the CaseViewer 2.4 (3DHISTECH, Budapest, Hungary) image analysis software. Positive expression scores were assigned to the cuticle, dermal papilla, sebaceous gland, inner root sheath, outer root sheath, and hair medulla sites within three fields of view of the sections.

2.5. Blood Genomic DNA Extraction

Whole blood samples from Gansu alpine fine-wool sheep were thawed and genomic DNA was extracted according to the steps of the EasyPure® Blood Genomic DNA Kit (Beijing All Style Gold Biotechnology Co., Ltd., Beijing, China). Agarose gel electrophoresis was employed to examine the quality of the DNA, while the NanoDropTM Lite Ultra-Micro Spectrophotometer (Thermo Fisher Scientific Inc., Waltham, MA, USA) was utilized to measure the concentration and purity of the genomic DNA. The extracted genomic DNA was stored at −20 °C for preservation.

2.6. Detection of SNPs in the KRT71 Gene

The ENSEMBL database was employed to forecast the polymorphisms of the *KRT71* gene. Based on the prediction of SNPs in this gene, the target fragments of the *KRT71* gene to be amplified were chosen. The gene sequence of sheep *KRT71* (accession number: NC_056056.1) was retrieved from the NCBI database. Subsequently, specific primers were designed using the Premier 5.0 software (Version 5.0, Premier Biosoft International, Palo Alto, CA, USA), and the details of these primers are presented in Table 1. These primers were synthesized by Sangon Biotech in Shanghai, China.

Table 1. Primer information of *KRT71* gene.

Gene	Position	Forward Primer (5′ → 3′)	Reverse Primer (5′ → 3′)	Amplification Product Length (bp)
KRT71	5′ UTR	TTCATGTTGTTTCAGAGGTGGGAG	CTCAGGGAAGACTATACGGGAAACCTT	835
	E9	TCTCTGCCATATTGTCCCCAGTGG	TCAGGAAGACACAGTCTGGGAGAAGTG	535

In this research, genomic DNA from 20 Gansu alpine fine-wool sheep was selected. The target fragments of the *KRT71* gene were individually amplified. Agarose gel electrophoresis was utilized to assess the integrity of the amplified products. After this, the amplified products were sent to Sangon Biotech in Shanghai, China for one-generation Sanger sequencing. Once the sequencing was completed, the SNPs of the target fragments were screened by integrating the sequencing peak maps with the base sequence information.

2.7. Genotypic Detection of SNP Sites

The number of SNP sites within the target fragment of the *KRT71* gene was determined. Subsequently, the genomic DNA of all Gansu alpine fine-wool sheep, along with the details of the identified SNP sites, was submitted to Gentides Biotechnology Co., Ltd. (Gentides, Wuhan, China) for KASP genotyping. A microplate reader equipped with a fluorescence resonance energy transfer (FRET) probe was employed to gather fluorescence data. Online software (http://www.snpway.com/snpdecoder/, accessed on 7 July 2025) was utilized to generate genotyping maps.

2.8. Statistical Analysis

Microsoft Excel 2021 was employed for the collation and statistical analysis of the SNP genotype data. Based on the method reported by Botstein et al. [21], the allele frequencies (AF), genotype frequencies (GF), chi-squared values ($\chi 2$), gene heterozygosity (He), homozygosity (Ho), polymorphism information content (PIC), and effective number of alleles (Ne) were calculated.

The detected SNPs were subjected to the chi-squared ($\chi 2$) test to assess their Hardy–Weinberg equilibrium status. The online software SHEsis (http://shesisplus.bio-x.cn/SHEsis.html, accessed on 8 July 2025) was utilized to analyze the linkage disequilibrium (LD) at the gene SNP site.

A general linear model ($Yijk = \mu + Gi + \varepsilon ijk$) in SPSS 26.0 (version 26.0, IBM Corp., Armonk, NY, USA) was used to analyze the relationships between genotypes and wool production traits. Y represents the phenotypic value of the trait; μ is the overall mean; Gi is the genotype effect; and εijk is the random error. The results are presented as the mean $\pm$ standard error, and a p-value < 0.05 indicates significant differences.

3. Results

3.1. Localization of KRT71 Protein in Sheep Skin Tissue

The immunofluorescence results showed that there were some differences in the intensity of KRT71 positivity in different parts of the skin and hair follicles of Gansu alpine fine-wool sheep during the wool growth period. Among them, the KRT71 protein was moderately positively expressed at the cuticle, outer root sheath, and sebaceous gland of the skin; strongly positively expressed in the inner root sheath; and not positively expressed in the hair medulla and dermal papilla (Figure 1).

A

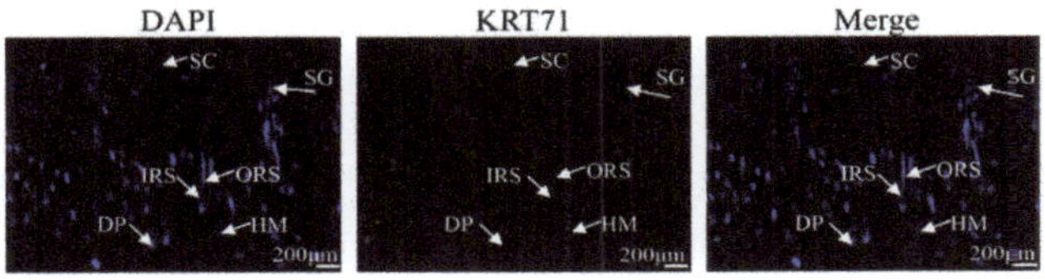

B

Protein	Stratum corneum	INS	ORS	Hair medulla	Sebaceous gland	DP
KRT71	++	+++	++	−	++	−

Figure 1. Localization of KRT71 in the skin tissue of Gansu alpine fine-wool sheep. (**A**) KRT71 protein immunofluorescence; (**B**) positive intensity of KRT71 protein in skin tissue. KRT71 fluorescence staining is shown in green, while DAPI-labeled cell nucleus fluorescence staining is shown in blue. DP: dermal papilla; SC: stratum corneum; SG: sebaceous gland; IRS: inner root sheath; ORS: outer root sheath; HM: hair medulla. –: no positive expression; ++: moderate positive expression; +++: strong positive expression.

3.2. Results of SNP Site Detection of KRT71 Gene in Gansu Alpine Fine-Wool Sheep

The results showed that there were two SNPs in the two target fragments of the *KRT71* gene in Gansu alpine fine-wool sheep (Figure 2). Among them, the SNPs in the 5′ UTR region and exon 9 region of the *KRT71* gene were named SNP1 (C.-7G/C) and SNP2 (C.1500G/A), respectively (Table 2). The SNP at exon 9 was a synonymous mutation.

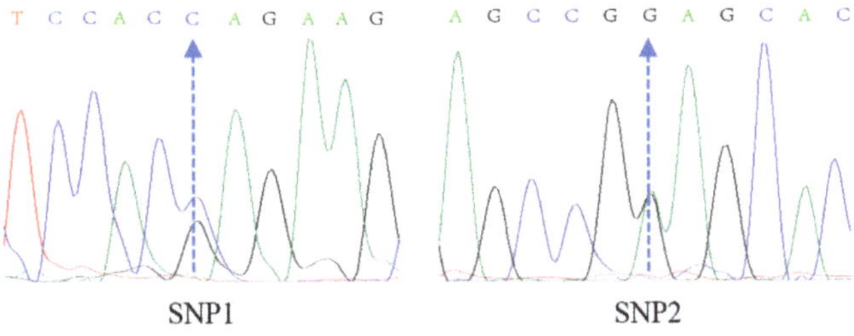

Figure 2. Sequencing peak map of SNP screening for *KRT71* gene in Gansu alpine fine-wool sheep.

Table 2. SNP-related information for *KRT71* gene.

Gene	SNP	Position	Physical Position/bp	Variation Type	Amino Acid Change
KRT71	SNP1	5′ UTR	g.133805316	G → C	No change
	SNP2	Exon 9	g.133812850	G → A	No change

3.3. Genotyping Results of SNPs in KRT71 Gene

KASP technology was used to type the SNPs of the *KRT71* gene in Gansu alpine fine-wool sheep. As shown in Figure 3, the two SNPs were both polymorphic and showed three genotypes. Meanwhile, the gene and genotype frequency statistics showed that the dominant genotypes of SNP1 and SNP2 were GG (92.36%) and GG (91.95%), and the dominant alleles were G (0.960) and G (0.958), respectively (Table 3).

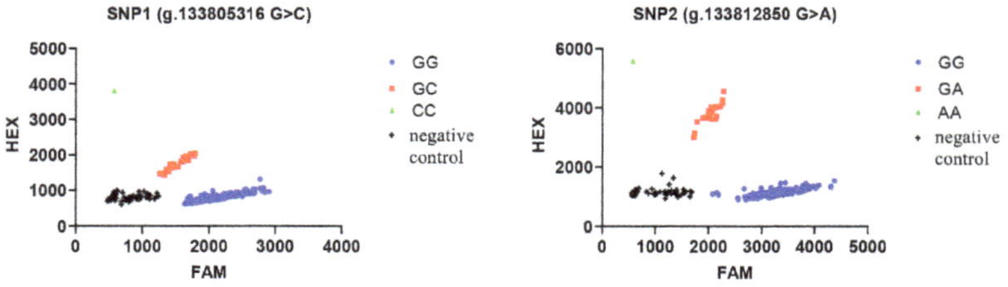

Figure 3. Genotyping results of *KRT71* gene in Gansu alpine fine-wool sheep. The green and blue groups clustered near the horizontal and vertical axes represent pure heterozygotes, respectively. The red groups clustered separately between the two are heterozygotes, and the black groups clustered near the intersection of the axes are negative controls.

Table 3. Genotype frequency and allele frequency of *KRT71* gene.

Genotype	Genotype Frequency(n)			Allele Frequency		χ^2	*p*-Value
SNP1	*GG* 0.924 (278)	*GC* 0.073 (22)	*CC* 0.003 (1)	G 0.960	C 0.040	0.617	0.432
SNP2	*GG* 0.920 (274)	*GA* 0.077 (23)	*AA* 0.003 (1)	G 0.958	A 0.042	0.470	0.493

3.4. Population Genetic Polymorphism Analysis of the KRT71 Gene

The population genetic polymorphism results showed that the two SNPs in the *KRT71* gene had high purity (Ho > 0.5), and the SNP with the highest number of effective alleles was SNP2 (Ne = 1.087). In addition, the SNP loci all showed low polymorphism (PIC < 0.25). The heterozygosity (He) results showed that the heterozygote frequencies at the SNP1 and SNP2 loci were 0.076 and 0.080, respectively (Table 4).

Table 4. Analysis of genetic parameters of *KRT71* gene population.

SNP	Ho	He	PIC	Ne
SNP1	0.923	0.076	0.074	1.083
SNP2	0.919	0.080	0.077	1.087

3.5. Association Analysis of KRT71 Genotype with Wool Production Traits

As shown in Table 5, the two SNPs of the *KRT71* gene all showed a significant association with the MSL ($p < 0.05$). Specifically, in SNP1, the MSL of GG genotype individuals were significantly longer than that of GC genotype individuals ($p < 0.05$). Similarly, in SNP2, the MSL of GG genotype individuals was significantly longer compared to that of GA genotype individuals ($p < 0.05$).

Table 5. The associations between gene variations in *KRT71* and wool traits.

SNP	Genotype	MFD (μm)	FDSD	CVFD (%)	MSL (mm)	MFC (Number/mm)	CF (%)	MSS (cN/dt)
SNP1	*GG*	21.508 ± 0.156	5.624 ± 0.073	25.142 ± 0.213	80.047 ± 0.768 [a]	104.998 ± 0.815	87.482 ± 0.805	13.982 ± 0.303
	GC	21.726 ± 0.628	5.395 ± 0.204	24.766 ± 0.551	75.454 ± 1.841 [b]	106.138 ± 2.989	89.285 ± 3.034	15.848 ± 1.051
SNP2	*GG*	21.516 ± 0.158	5.619 ± 0.074	25.139 ± 0.214	79.963 ± 0.768 [a]	104.913 ± 0.818	87.566 ± 0.805	13.986 ± 0.305
	GA	21.733 ± 0.601	5.368 ± 0.196	24.645 ± 0.539	75.304 ± 1.766 [b]	106.386 ± 2.861	89.500 ± 2.900	15.806 ± 1.005

Note: The shoulder labels of data in the same column with different lowercase letters indicate significant differences ($p < 0.05$).

3.6. Linkage Disequilibrium Analysis of SNP Loci in the KRT71 Gene

In this study, two indicators (D′ and R^2 values) were combined to determine the linkage disequilibrium between the SNP loci of the *KRT71* gene. The results showed that there was strong linkage disequilibrium between the two SNP loci of the *KRT71* gene (D′ = 1, R^2 = 0.96) (Figure 4).

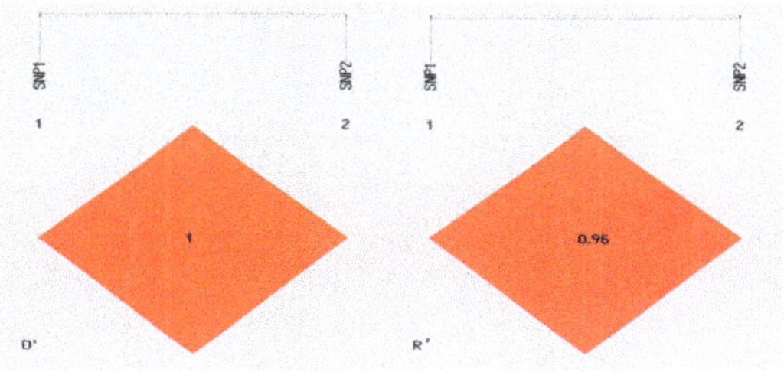

Figure 4. Linkage disequilibrium analysis of SNPs in *KRT71* gene of Gansu alpine fine-wool sheep. D′ = 0, R^2 = 0, indicates that the loci of the two SNPs are in a state of complete linkage equilibrium; D′ = 1, R^2 = 1, indicates that the loci of the two SNPs are in a state of complete linkage disequilibrium.

4. Discussion

Detecting SNPs of functional genes is an effective approach to identifying the key SNP sites of genes. This method has been extensively utilized in analyzing the polymorphisms of functional genes associated with various animal production traits, including wool, meat, milk, and fat deposition. Chai et al. discovered that a base mutation (c.-364A/G) in the promoter fragment of the *KRT34* gene might influence the wool economic traits of Merino sheep, such as the mean fiber diameter (MFD), fiber diameter standard deviation (FDSD), and mean staple length (MSL) [22]. Zhao et al. demonstrated that five SNP loci (g. 105914953 G > A, g. 105922232 T > C, g. 105922244 A > G, g. 1059223 A > T, and g. 105922340 G > T) in the FGF5 gene were significantly correlated with traits like the wool length, wool weight, and growth performance in fine-wool sheep [23]. Base mutations (c.600 G/A and c.652 + 93 C/A) in the exon 3–4 region of the *KRT85* gene were related to wool production in Merino sheep [24]. There was a strong correlation between polymorphisms (G > A) in the FASN gene and the wool fiber diameter [25]. Moreover, mutations in SLIT3 and ZNF280B were found to be correlated with the wool fiber diameter of alpine merino sheep [26]. In summary, the research method of detecting gene SNPs and conducting an association analysis between these SNPs and various animal production traits is feasible. It is more beneficial for application in production practice, thereby facilitating improvements in various production traits of farm animals.

The efficient and convenient screening of candidate genes is of great significance for subsequent gene validation. At present, there are numerous methods for the screening of functional genes, including whole-genome resequencing, transcriptome analysis, proteome analysis, DNA methylation (WGBS), RNA methylation (such as m6G, m5C, m7G, etc.), and other epigenetic modifications, along with other emerging technological tools [27].

In this research, after integrating the pre-sequencing results and referring to the relevant literature [18,28], we carried out further screening and selected the *KRT71* gene for in-depth investigation. We explored the potential relationships between these functional genes and wool economic traits through gene and protein expression localization and polymorphism detection.

Genes and proteins exhibit tissue-specific expression patterns, which subsequently lead to diverse biological functions [29,30]. Immunofluorescence, a potent tool in biological research and medical diagnosis, boasts high sensitivity in antigen detection. It can precisely localize and detect specific proteins or antigens and has been extensively applied in various biological and medical research fields [31,32]. Moreover, immunofluorescence techniques enable the direct visualization of target antigens within cells or tissue, offering intuitive visual data that aid in comprehending antigen distribution and localization. Consequently, to explore the expression and localization of KRT71 in skin hair follicles, this study was carried out. The results indicated that the KRT71 protein showed strong positive expression in the inner root sheaths of skin hair follicles. Relevant research has demonstrated that the length and curvature of the inner root sheath significantly influence the shape of wool [33]. This finding further implies that the *KRT71* gene might play a crucial role in hair follicle development as well as wool growth. Therefore, it is essential to conduct research on its genetic polymorphisms and explore the relationship between SNPs of the *KRT71* gene and wool economic traits.

SNPs could induce changes in wool traits through multiple molecular and biological mechanisms. These primarily encompass the following: SNPs within coding regions can modify the structure and function of proteins, influence gene expression levels, and cause chain imbalances and are related to quantitative trait loci (QTL); key gene variants can impact signaling pathways; there are alterations in epigenetic regulation (such as SNPs in methylation-sensitive loci and variants in histone modification loci), as well as indirect physiological effects [34,35]. Moreover, apart from being present in coding regions, SNPs are highly distributed in non-coding regions. They can also be detected in non-coding RNAs, introns, or the 5′ and 3′ untranslated regions (UTRs) [36]. Even though these changes do not lead to the production of modified proteins associated with specific diseases, they can still affect RNA–protein interactions by altering the secondary structure of RNA [37,38]. Although the change in the *KRT71* gene SNP identified in this study is a synonymous mutation and does not result in any amino acid substitutions, it may affect protein expression or structure. It has been reported that synonymous mutations may affect the rate of mRNA translation, thereby potentially altering the way in which proteins fold [39]. The observed effect of the *KRT71* gene may also be due to its linkage to other KRTs on the same chromosome. KRT71 belongs to the keratin family and its biological significance needs to be further investigated.

In the polymorphism analysis of genes, conducting an association analysis between genotypes and production traits is beneficial in identifying crucial loci of SNPs, which might play a role in enhancing the economic traits of farm animals [40]. In this research, we discovered that the variation in the *KRT71* gene was related to the MSL. The MSL is one of most important wool economic traits and is closely linked to the wool processing performance. Wool with a long MSL is more commercially desirable as it tends to be easier to spin, leads to fewer stoppages, and ultimately can form stronger and more even yarns when compared to shorter wool [41]. This suggests that, if the wool length is used as the improvement target, individuals with the GG genotype will be more popular in the wool industry.

Even though the association between variations in the *KRT71* gene and wool curvature did not show a significant difference, in both SNP1 and SNP2, there was still a trend in which the wool curvature of GC genotype and GA genotype individuals was higher than that of GG genotype individuals. Moreover, numerous previous studies have also demonstrated a strong correlation between *KRT71* gene variants and the curliness of animal hair [13,42,43]. In subsequent studies, researchers could consider increasing the sample size to further examine the associations between this gene variation and wool production traits.

5. Conclusions

This research indicates that variations in the *KRT71* gene are related to the mean stable length of wool in Gansu alpine fine-wool sheep, while the KRT71 protein is strongly positively expressed mainly at the root sheath within the hair follicle in fine-wooled sheep.

Author Contributions: Formal Analysis, Z.H. (Zhiyun Hao); Investigation, J.W.; Methodology, H.S. and M.L.; Software, Z.H. (Zhaohua He) and X.L.; Writing—Original Draft. F.Z.; Writing—Review and Editing, F.Z. and S.L. All authors contributed to the article. All authors have read and agreed to the published version of the manuscript.

Funding: This study was financially supported by the Science and Technology Innovation Talent Plan of Gansu Province (25JR6KA011; 23JDKA0010), the Discipline Team Project of Gansu Agricultural University (GAU-XKTD-2022-21), and the Fuxi Young Talents Fund of Gansu Agricultural University (Gaufx-03Y04).

Institutional Review Board Statement: The animal experiments were approved by the Gansu Agricultural University Ethics Committee with the approved code GSAU-ETH-AST-2021-028.

Informed Consent Statement: Not applicable.

Data Availability Statement: The authors affirm that all data necessary to confirm the conclusions of the article are present within the article, figures, and tables.

Acknowledgments: The authors would like to thank all the reviewers who participated in the review.

Conflicts of Interest: There are no conflicts of interest to declare.

References

1. Ye, Z.; Wells, C.M.; Carrington, C.G.; Hewitt, N.J. Thermal conductivity of wool and wool–hemp insulation. *Int. J. Energy Res.* **2006**, *30*, 37–49. [CrossRef]
2. Patnaik, A.; Mvubu, M.; Muniyasamy, S.; Botha, A.; Anandjiwala, R.D. Thermal and sound insulation materials from waste wool and recycled polyester fibers and their biodegradation studies. *Energy Build.* **2015**, *92*, 161–169. [CrossRef]
3. Pei, H.; Ma, X.; Pan, Y.; Han, T.; Lu, Z.; Wu, R.; Cao, X.; Zheng, J. Separation and purification of lanosterol, dihydrolanosterol, and cholesterol from lanolin by high-performance counter-current chromatography dual-mode elution method. *J. Sep. Sci.* **2019**, *42*, 2171–2178. [CrossRef] [PubMed]
4. Kicinska-Jakubowska, A.; Morales Villavicencio, A.; Zimniewska, M.; Przybylska, P.; Kwiatkowska, E. Evaluation of Wool Quality Parameters of Polish Sheep Breeds. *J. Nat. Fibers* **2022**, *19*, 5880–5887. [CrossRef]
5. Zenda, M.; Malan, P.J.; Geyer, A.C. An analysis of the contribution of wool characteristics on price determination of Merino Wool and White Wool all combined in South Africa. *Small Rumin. Res.* **2023**, *219*, 106890. [CrossRef]
6. Bromley, C.M.; Snowder, G.D.; Van Vleck, L.D. Genetic parameters among weight, prolificacy, and wool traits of Columbia, Polypay, Rambouillet, and Targhee sheep. *J. Anim. Sci.* **2000**, *78*, 846–858. [CrossRef]
7. Burton, D.J.; Ludden, P.A.; Stobart, R.H.; Alexander, B.M. 50 years of the Wyoming ram test: How sheep have changed. *J. Anim. Sci.* **2015**, *93*, 1327–1331. [CrossRef]
8. Cipriano, A.; Ballarino, M. The Ever-Evolving Concept of the Gene: The Use of RNA/Protein Experimental Techniques to Understand Genome Functions. *Front. Mol. Biosci.* **2018**, *5*, 20. [CrossRef]
9. Gong, H.; Zhou, H.; Forrest, R.H.J.; Li, S.; Wang, J.; Dyer, J.M.; Luo, Y.; Hickford, J.G.H. Wool Keratin-Associated Protein Genes in Sheep—A Review. *Genes* **2016**, *7*, 24. [CrossRef]
10. Zernov, N.V.; Skoblov, M.Y.; Marakhonov, A.V.; Shimomura, Y.; Vasilyeva, T.A.; Konovalov, F.A.; Abrukova, A.V.; Zinchenko, R.A. Autosomal Recessive Hypotrichosis with Woolly Hair Caused by a Mutation in the Keratin 25 Gene Expressed in Hair Follicles. *J. Investig. Dermatol.* **2016**, *136*, 1097–1105. [CrossRef]

11. Coulombe, P.A.; Omary, M.B. 'Hard' and 'soft' principles defining the structure, function and regulation of keratin intermediate filaments. *Curr. Opin. Cell Biol.* **2002**, *14*, 110–122. [CrossRef] [PubMed]
12. Pan, X.; Hobbs, R.P.; Coulombe, P.A. The expanding significance of keratin intermediate filaments in normal and diseased epithelia. *Curr. Opin. Cell Biol.* **2013**, *25*, 47–56. [CrossRef] [PubMed]
13. Fujimoto, A.; Farooq, M.; Fujikawa, H.; Inoue, A.; Ohyama, M.; Ehama, R.; Nakanishi, J.; Hagihara, M.; Iwabuchi, T.; Aoki, J.; et al. A Missense Mutation within the Helix Initiation Motif of the Keratin K71 Gene Underlies Autosomal Dominant Woolly Hair/Hypotrichosis. *J. Investig. Dermatol.* **2012**, *132*, 2342–2349. [CrossRef] [PubMed]
14. Kuramoto, T.; Hirano, R.; Kuwamura, M.; Serikawa, T. Identification of the Rat Rex Mutation as a 7-bp Deletion at Splicing Acceptor Site of the Krt71 Gene. *J. Vet. Med. Sci.* **2010**, *72*, 909–912. [CrossRef]
15. Duan, C.; Zhang, L.; Gao, K.; Guo, Y.; Liu, Y.; Zhang, Y. Cashmere production, skin characteristics, and mutated genes in crimped cashmere fibre goats. *Animal* **2022**, *16*, 100565. [CrossRef]
16. Li, W.; Guo, J.; Li, F.; Niu, C. Evaluation of Crossbreeding of Australian Superfine Merinos with Gansu Alpine Finewool Sheep to Improve Wool Characteristics. *PLoS ONE* **2016**, *11*, e0166374. [CrossRef]
17. Bolormaa, S.; Swan, A.A.; Brown, D.J.; Hatcher, S.; Moghaddar, N.; van der Werf, J.H.; Goddard, M.E.; Daetwyler, H.D. Multiple-trait QTL mapping and genomic prediction for wool traits in sheep. *Genet. Sel. Evol.* **2017**, *49*, 62. [CrossRef]
18. He, Z.; Sun, H.; Zhao, F.; Ma, L.; Wang, J.; Liu, X.; Li, M.; Hao, Z.; Li, S. MicroRNA expression profiles reveal wool development and fineness regulation in Gansu alpine fine-wool sheep. *Genomics* **2024**, *116*, 110922. [CrossRef]
19. *GB1523-2013;* Sheep Wool. Standardization Administration of the People's Republic of China (SAC): Beijing, China, 2013.
20. Diao, X.; Yao, L.; Wang, X.; Li, S.; Qin, J.; Yang, L.; He, L.; Zhang, W. Hair Follicle Development and Cashmere Traits in Albas Goat Kids. *Animals* **2023**, *13*, 617. [CrossRef]
21. Botstein, D.; White, R.L.; Skolnick, M.; Davis, R.W. Construction of a genetic linkage map in man using restriction fragment length polymorphisms. *Am. J. Hum. Genet.* **1980**, *32*, 314–331.
22. Chai, W.; Zhou, H.; Gong, H.; Hickford, J.G.H. Variation in the ovine KRT34 promoter region affects wool traits. *Small Rumin. Res.* **2022**, *206*, 106586. [CrossRef]
23. Zhao, H.; Hu, R.; Li, F.; Yue, X. Five SNPs Within the FGF5 Gene Significantly Affect Both Wool Traits and Growth Performance in Fine-Wool Sheep (*Ovis aries*). *Front. Genet.* **2021**, *12*, 732097. [CrossRef] [PubMed]
24. Chai, W.; Zhou, H.; Gong, H.; Wang, C.; Hickford, J.G.H. Variation in the Exon 3–4 Region of Ovine KRT85 and Its Effect on Wool Traits. *Animals* **2024**, *14*, 2272. [CrossRef] [PubMed]
25. Almaamory, Y.A.; Al-Anbari, N.N. Relationship of FASN Gene Polymorphism in Growth Performance and Wool Traits of Awassi Sheep. *IOP Conf. Ser. Earth Environ. Sci.* **2023**, *1214*, 012032. [CrossRef]
26. Yue, L.; Lu, Z.; Guo, T.; Liu, J.; Yuan, C.; Yang, B. Association of SLIT3 and ZNF280B Gene Polymorphisms with Wool Fiber Diameter. *Animals* **2023**, *13*, 3552. [CrossRef]
27. Xue, Q.; Huang, Y.; Cheng, C.; Wang, Y.; Liao, F.; Duan, Q.; Wang, X.; Miao, C. Progress in epigenetic regulation of milk synthesis, with particular emphasis on mRNA regulation and DNA methylation. *Cell Cycle* **2023**, *22*, 1675–1693. [CrossRef]
28. He, Z.; Zhao, F.; Sun, H.; Hu, J.; Wang, J.; Liu, X.; Li, M.; Hao, Z.; Zhao, Z.; Shi, B.; et al. Screened of long non-coding RNA related to wool development and fineness in Gansu alpine fine-wool sheep. *BMC Genom.* **2025**, *26*, 8. [CrossRef]
29. Kryuchkova-Mostacci, N.; Robinson-Rechavi, M. A benchmark of gene expression tissue-specificity metrics. *Brief. Bioinform.* **2016**, *18*, 205–214. [CrossRef]
30. Grover, M.; Gang, S.S.; Troemel, E.R.; Barkoulas, M. Proteasome inhibition triggers tissue-specific immune responses against different pathogens in *C. elegans*. *PLoS Biol.* **2024**, *22*, e3002543. [CrossRef]
31. Chen, R.; Zhou, H.; Li, A.; Cheng, X.; Liu, X.; Huang, F.; Wang, Y.; Liu, Y.; Gong, H.; Liu, X.; et al. Chemical Sectioning for Immunofluorescence Imaging. *Anal. Chem.* **2021**, *93*, 8698–8703. [CrossRef]
32. Cohen, R.; Lee-Pullen, T.; Miller, T.J.; Meehan, K.; Fuller, K.; McCoy, M.J. Optimising multiplex immunofluorescence staining for characterising the tumour immune micro-environment. *Methods* **2023**, *219*, 48–57. [CrossRef] [PubMed]
33. Paus, R.; Cotsarelis, G. The Biology of Hair Follicles. *N. Engl. J. Med.* **1999**, *341*, 491–497. [CrossRef] [PubMed]
34. Teama, S. DNA polymorphisms: DNA-based molecular markers and their application in medicine. In *Genetic Diversity and Disease Susceptibility*; Intechopen: London, UK, 2018; Volume 25.
35. Gebert, M.; Jaśkiewicz, M.; Moszyńska, A.; Collawn, J.F.; Bartoszewski, R. The Effects of Single Nucleotide Polymorphisms in Cancer RNAi Therapies. *Cancers* **2020**, *12*, 3119. [CrossRef] [PubMed]
36. Shatoff, E.; Bundschuh, R. Single nucleotide polymorphisms affect RNA-protein interactions at a distance through modulation of RNA secondary structures. *PLoS Comput. Biol.* **2020**, *16*, e1007852. [CrossRef]
37. Van Kouwenhove, M.; Kedde, M.; Agami, R. MicroRNA regulation by RNA-binding proteins and its implications for cancer. *Nat. Rev. Cancer* **2011**, *11*, 644–656. [CrossRef]
38. Burd, C.G.; Dreyfuss, G. Conserved structures and diversity of functions of RNA-binding proteins. *Science* **1994**, *265*, 615–621. [CrossRef]

9. Hurst, L.D. Molecular genetics: The sound of silence. *Nature* **2011**, *471*, 582–583. [CrossRef]
0. Zhao, J.; Sauvage, C.; Bitton, F.; Causse, M. Multiple haplotype-based analyses provide genetic and evolutionary insights into tomato fruit weight and composition. *Hortic. Res.* **2022**, *9*, uhab009. [CrossRef]
1. Holman, B.; Malau-Aduli, A. A review of sheep wool quality traits. *Annu. Rev. Res. Biol.* **2012**, *2*, 1–14.
2. Runkel, F.; Klaften, M.; Koch, K.; Böhnert, V.; Büssow, H.; Fuchs, H.; Franz, T.; de Angelis, M.H. Morphologic and molecular characterization of two novel Krt71 (Krt2-6g) mutations: Krt71rco12 and Krt71rco13. *Mamm. Genome* **2006**, *17*, 1172–1182. [CrossRef]
3. Shimomura, Y.; Wajid, M.; Petukhova, L.; Kurban, M.; Christiano, A.M. Autosomal-dominant woolly hair resulting from disruption of keratin 74 (KRT74), a potential determinant of human hair texture. *Am. J. Hum. Genet.* **2010**, *86*, 632–638. [CrossRef]

animals

MDPI

Article

Activation of Focal Adhesion Pathway by *CIDEA* as Key Regulatory Axis in Lipid Deposition in Goat Intramuscular Precursor Adipocytes

Peng Shao [1], Qi Li [1], Yu Liao [1], Yong Wang [1], Yaqiu Lin [1], Hua Xiang [1], Zhanyu Du [1], Changhui Zhang [1], Jiangjiang Zhu [1,2] and Lian Huang [1,*]

1 Qinghai-Tibetan Plateau Animal Genetic Resource Reservation and Utilization Key Laboratory of Sichuan Province, Southwest Minzu University, Chengdu 610225, China; s935246946@163.com (P.S.); liqiligexiao@outlook.com (Q.L.); 19828626311@163.com (Y.L.); wangyong010101@hotmail.com (Y.W.); linyq1999@163.com (Y.L.); xianghua2008411@163.com (H.X.); yuzhan.du@outlook.com (Z.D.); zhangchanghui0074@163.com (C.Z.); zhujiang4656@hotmail.com (J.Z.)

2 Key Laboratory of Qinghai-Tibetan Plateau Animal Genetic Resource Reservation and Utilization, Southwest Minzu University, Ministry of Education, Chengdu 610041, China

* Correspondence: kin8248806@163.com

Simple Summary

CIDEA, a protein associated with lipid droplets, plays a crucial role in maintaining lipid homeostasis. Our research demonstrates that *CIDEA* increases the levels of lipid droplets and triglycerides while concurrently repressing cell proliferation. Furthermore, we discovered that *CIDEA* facilitates lipid accumulation in goat intramuscular adipocytes through the activation of the focal adhesion pathway. This study enhances our understanding of the molecular processes governing intramuscular fat deposition and provides a theoretical framework that could inform molecular breeding efforts aimed at improving the quality of goat meat.

Abstract

Intramuscular fat (IMF) content determines the quality of goat meat and is regulated by the comprehensive effect of the proliferation and adipogenesis of intramuscular preadipocytes. Our previous RNA-seq data revealed that cell death-inducing DNA fragmentation factor alpha (DFFA)-like effector (CIDE) A was upregulated during the development of intramuscular fat in the longissimus dorsi muscle tissue, implying an important role in lipid homeostasis. However, the mechanism by which *CIDEA*, a member of the CIDE family, regulates intramuscular fat deposition in goat muscle is unknown, so we explored the function and underlying mechanism of *CIDEA* in goat intramuscular preadipocytes. To address this, we altered *CIDEA* in intramuscular preadipocytes and resolved the effect and mechanism of *CIDEA* in adipogenesis through RT-PCR, Western blot, triglyceride and LD determinations, CCK-8, and RNA-seq. It was found that *CIDEA* increased lipid droplets (LDs) and triglyceride contents and inhibited cell proliferation. Meanwhile, the lipid metabolism-related genes *PPARγ*, *C/EBPα*, *SREBP1c*, *PLIN1*, *TIP47*, *ADFP*, *DGAT1*, *ACC*, *FASN*, *ACSL1*, and *FABP3* were upregulated, while the lipolysis and β-oxidation genes *HSL*, *ACOX1*, and *CPT1B*, as well as the proliferation marker gene *CDK1*, were all downregulated upon *CIDEA* overexpression. Differentially expressed genes in *CIDEA* dysregulation groups through RNA-seq were selected and were enriched in the apelin and focal adhesion signaling pathways. Specifically, the Western blot and rescue assays found that focal adhesion, but not apelin, was the key signaling pathway in *CIDEA* regulating lipid deposition in goat intramuscular preadipocytes. In summary, this study reveals that *CIDEA* promotes lipid deposition in intramuscular preadipocytes through the focal adhesion pathway and inhibits cell proliferation. This work clarifies the functional role and downstream signaling pathway of *CIDEA* in intramuscular fat deposition and provides theoretical support for improving meat quality by targeting key phenotype-related genes.

Keywords: *CIDEA*; intramuscular fat; RNA-seq; focal adhesion pathway; fat deposition

1. Introduction

Goat meat is favored by consumers as a high-quality meat food with high protein and low cholesterol [1]. Intramuscular fat (IMF), also known as marbling, plays a crucial role in the tenderness, juiciness, and flavor of meat products, determining the quality of meat [2,3]. The IMF content is controlled by adipocytes' number and volume, which is regulated by the comprehensive effect of proliferation and adipogenesis processes of intramuscular preadipocytes. Thus, understanding the underlying molecular mechanism of preadipocytes' proliferation and adipogenesis is essential to understand IMF deposition and, thereby, improve the quality of goat meat.

Cell death-inducing DNA fragmentation factor alpha-like effector A (*CIDEA*), a member of the CIDE gene family, is a lipid droplet-associated protein [4,5], mainly localized on the surface of lipid droplets and the endoplasmic reticulum [6,7]. *CIDEA* plays an important role in lipid droplet expansion and TAG accumulation [8,9]. In mice, *CIDEA*, induced by a high-fat diet (HFD), was highly expressed in brown adipocytes [10,11] and correlated with the development of hepatic steatosis [12,13]. High expression of *CIDEA* showed TAG accumulation in hepatocytes and maintained a healthy obese phenotype in the adipose tissue of transgenic mice [14,15]. Its deficiency resulted in the accumulation of smaller LDs and improved insulin sensitivity in brown adipocytes [6,10] and HFD-induced fatty liver [13], as well as increased whole-body energy expenditure in HFD-fed mice. However, the role and underlying molecular mechanism of *CIDEA* in regulating goat IMF deposition remain to be studied.

Apelin is a peptide hormone identified as the only endogenous ligand for the previously orphaned G protein-coupled APJ receptor [16]. In a previous study, apelin, as an adipokine, inhibited adipogenesis via the mitogen-activated protein kinase (MAPK) pathway [17]. In endothelial cells, apelin promotes cell proliferation by activating the downstream phosphoinositide-3 kinase (PI3K)/Akt signaling pathway [18]. Focal adhesion kinase (FAK) is a key component of the membrane proximal signaling layer in focal adhesion complexes and regulates important cellular processes, including cell migration, proliferation, and survival [19]. By controlling the supply of precursors and the enzyme activity of proteins involved in lipid synthesis, FAK can affect lipid metabolism [20].

This study investigates the effect of *CIDEA* on the adipogenesis of goat intramuscular precursor adipocytes and explores the molecular mechanism of *CIDEA* inducing lipid deposition through the FAK signaling pathway. These data reveal the crucial role of *CIDEA* in the lipid deposition of goat intramuscular preadipocytes, which can help to thoroughly interpret the molecular mechanism of intramuscular fat deposition, thus providing a theoretical basis for the improvement of goat meat quality by molecular breeding.

2. Materials and Methods

2.1. Ethics Statement

All experimental exercises were isolated and approved by the Institutional Animal Care and Use Committee, Southwest Minzu University (Chengdu, China) (permit number:

S2020-013). Tissue-sampling procedures were conducted in strict accordance with the state and institutional animal care and ARRIVE guidelines.

2.2. Isolation and Culture of Goat Intramuscular Preadipocytes

Purchased from Sichuan Jianyang Dageda Animal Husbandry Co., Ltd. (Jianyang, China), the 2-day-old Jianzhou big-eared goats were rapidly euthanized via carotid artery bleeding in the laboratory. The longissimus dorsi muscle tissues were collected from 3 male Jianzhou big-eared goats, 2 days old, in a sterile atmosphere, followed by washing with PBS containing 5% penicillin–streptomycin (Boster, PYG0016, Pleasanton, CA, USA) three times. The tissues were minced and then transferred into 50 mL centrifuge tubes. Then, type II collagenase (Sigma-Aldrich, C2-BIOC, St. Louis, MO, USA) was added for digestion at 37 °C for 1.5 h (shaking every 5 min), and digestion was terminated with complete culture medium. The digested tissue was filtered to remove the large undigested tissue. Then, the cell suspension was filtered with a 200-mesh (75 μm) sieve and centrifuged at 2000 rpm for 5 min to obtain the mixed cells. The red blood cell lysate (Boster, AR1118, CA, USA) was used to resuspend the precipitate, followed by standing for 5 min to remove the red cells. Subsequently, the cell suspension underwent centrifugation at 2000 revolutions per minute for a duration of 5 min, followed by resuspension in complete culture medium sourced from Gibco (C11330500BT, Beijing, China). The cells were seeded in a 25 cm^2 culture plate and cultured at 37 °C in 5% CO_2. After 2 h, the medium was replaced with fresh complete medium, and the attached cells were considered goat primary intramuscular preadipocytes. All subsequent experiments were performed in three biological replicates and three technical replicates.

The goat primary intramuscular preadipocytes were cultured with the complete culture medium containing 10% fetal bovine serum (12483012, Gibco, Carlsbad, CA, USA) and 90% DMEM/F12 (SH30023−01, Hyclone, Logan, UT, USA) supplemented with 100 U/mL penicillin–streptomycin (080092569, Harbin Pharmaceutical Group, Harbin, China). The medium was changed every two days during this process, and there were no significant changes in the adipocyte morphology. To promote adipogenesis, cells were cultured with an adipogenic medium consisting of complete medium and 50 μM oleic acid (112-80-1, Sigma, MO, USA) for 48 h.

In the FAK pathway inhibition assay, the FAK inhibitor was added at a concentration of 100 nM along with the oleic acid induction solution. The control group was an equal volume of DMSO, and the phenotyping examinations were conducted after adipogenic induction for 48 h.

2.3. Goat CIDEA Overexpression Vector Construction and siRNA Synthesis

Based on the CDS sequence of goat *CIDEA* in the NCBI database, the CDS region of the *CIDEA* gene was amplified and inserted into the pcDNA3.1 (+) plasmid, which was double-digested by EcoRI and HindIII, and named the CIDEA-OVER vector. Next, the recombinant plasmid was confirmed through enzyme digestion and DNA sequencing, with the empty pcDNA3.1 (+) plasmid serving as the negative control, designated as pcDNA3.1.

The siRNAs for goat *CIDEA* mRNA were designed and synthesized by Shanghai GenePharma company. siRNA-NC S: UUCUCCGAACGUGUCACGUTT; A: ACGUGACACGUUCGGAGAATT. siRNA-CIDEA393 S: CCACCAUGUACGAGAUGUATT; A: UACAUCUCGUACAUGGUGGTT.

To achieve the overexpression and interference of *CIDEA*, plasmids and siRNAs were transfected into goat primary intramuscular precursor adipocytes in a 6-well plate once the cells reached 80% confluence. These transfections were carried out using Lipofectamine™ 3000 transfection reagent (L3000015, Invitrogen, Carlsbad, CA, USA), following the manufacturer's guidelines. In addition, 1 μg of plasmids or 120 μM of siRNAs was used for the transfection in each well of the 6-well plate in our research.

2.4. Oil Red O Staining and Triglyceride Determination

Two days post-transfection, the relative lipid droplet content was assessed using Oil Red O staining (Solarbio, G1262, Beijing, China). Briefly, cells were fixed with formaldehyde for a 30-min duration. Subsequently, the lipid droplets within the cells were stained with a filtered Oil Red O solution for 20 min following a PBS wash. After three additional PBS washes to remove excess stain, the lipid droplets were visualized and imaged under a microscope. The stained lipid droplets were then dissolved in isopropanol, and their absorbance was quantified at a wavelength of 510 nm.

For the detection of triglyceride content, the cells were rinsed three times with PBS and then incubated with 200 μL of lysis buffer for 10 min at room temperature. The supernatant was heated at 70 °C for 10 min, followed by centrifugation at 2000 rpm for 5 min. The resulting supernatant was then used for triglyceride quantification (Applygen, E1014-105, Beijing, China). The optical density (OD) value of triglycerides was measured at a wavelength of 550 nm. To normalize the triglyceride content, the protein concentration of each sample was determined using the BCA method (Thermo Fisher Scientific, 23225, Beijing, China).

2.5. Cell-Counting Kit-8 (CCK-8) Assay

Cells were seeded into a 96-well plate and transfected with *CIDEA* overexpression vectors or siRNAs and their controls individually. After 0, 24, 36, and 48 h of transfection, 10 μL of the CCK-8 reagent (AC11L054, Life-iLab, Shanghai, China) was added to each well. Then, the cells were incubated for 0.5 h at 37 °C. At last, the absorbance was measured using a microplate reader at a 450 nm wavelength.

2.6. Western Blot

Cellular proteins were extracted using RIPA buffer (Solarbio Tech Inc., Beijing, China) containing a protease inhibitor (04693132001, Roche, Mannheim, Germany) and a phosphatase inhibitor. The total proteins were then separated via SDS-PAGE electrophoresis and transferred onto PVDF membranes for abundance analysis. This was followed by occlusion using 5% BSA (4240GR100, Biofroxx, Einhausen, Germany), incubation of the primary antibody overnight at 4 °C, TBST washing of the PVDF membrane three times, and 1 h of room-temperature incubation of the secondary antibody. The primary antibodies utilized were anti-β-actin (1:6000, BM0627, BOSTER, Wuhan, China), anti-p-p38-MAPK (1:1000, 3285S, Cell Signaling Technology, Danvers, MA, USA), anti-p38-MAPK (1:1000, 8690S, Cell Signaling Technology), anti-p-FAK (1:1000, ab81298, Abcam, Cambridge, UK), anti-FAK (1:1000, #3285, Cell Signaling Technology), anti-p-AKT (1:2000, 4060, Cell Signaling Technology), and anti-AKT (1:1000, ab32505, Abcam). Goat anti-mouse IgG-HRP (1:6000, BA1050, Boster) and goat anti-rabbit IgG-HRP (1:6000, BA1054, Boster) served as the secondary antibodies. The target proteins were visualized using an enhanced chemiluminescence (ECL) detection system from Thermo (Waltham, MA, USA).

2.7. Reverse Transcription–Quantitative PCR (RT-qPCR)

Total RNA was extracted using RNAiso Plus (Takara, Cat. No. 9109), and its concentration and purity were measured. The RNA was then reverse-transcribed using a reverse transcription kit from Vazyme (Cat. No. R32301). RT-qPCR was performed with a Bio-Rad CFX96 PCR System, utilizing the Taq Pro Universal SYBR qPCR Master Mix from Vazyme (Cat. No. Q712-02, Nanjing, China) and gene-specific primers (Table S1). UXT served as the internal reference gene, and the relative expression levels were calculated according to the

$2^{-\Delta\Delta CT}$ method. To clarify the expression pattern of *CIDEA* in goat intramuscular precursor adipocytes, cellular RNA was extracted during the adipogenic induction period (days 0, 2, 4, 6, and 8), and the abundance of *CIDEA* was subsequently detected by RT-qPCR.

2.8. RNA Sequencing (RNA-seq)

After 48 h of oleic acid induction, total RNA was extracted using TRIzol (Takara, 9109, Beijing, China) reagent for RNA-seq analysis. The samples were named as the NO group (negative control for overexpression), the AO group (*CIDEA* overexpression), the AS group (*CIDEA* knockdown), and the NS group (*CIDEA* knockdown control). Samples with three replicates were used for high-throughput transcriptome sequencing (Shanghai OE Biotechnology Co, Ltd. Shanghai, China). The Illumina Novaseq 6000 sequencing platform was used to sequence the library and generate 150 bp paired-end reads with 100-300 cycles. DEsq2 was used to screen differentially expressed genes with $p < 0.05$. DEGs were subjected to GO and KEGG analysis.

2.9. Statistical Analysis

All experiments were carried out in three biological replicates and repeated three times. GraphPad Prism 9.0 was used for statistical analysis and plotting. Student's *t*-test and one-way ANOVA were used to calculate the difference. $p < 0.05$ was considered significant, and $p < 0.01$ was highly significant.

3. Results

3.1. CIDEA Is Associated with Intramuscular Fat Deposition

Our previous RNA-seq data [21] and RNA abundance detected by RT-qPCR showed that the expression of *CIDEA* was upregulated in the longissimus dorsi muscle tissue of 24-month-old goats compared with that of 2-month-old goats (Figure 1A). To further investigate the influence of *CIDEA* on intramuscular adipogenesis, we examined the expression of *CIDEA* during the differentiation of intramuscular preadipocytes in goats. The findings indicated a gradual increase in the expression of *CIDEA* from day 0 to day 8 (Figure 1B), implying that *CIDEA* may play a pivotal role in the adipogenesis of intramuscular preadipocytes.

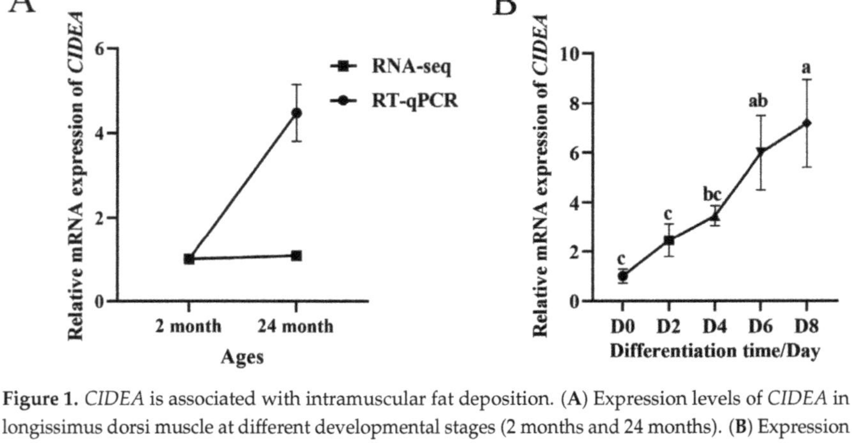

Figure 1. *CIDEA* is associated with intramuscular fat deposition. (**A**) Expression levels of *CIDEA* in longissimus dorsi muscle at different developmental stages (2 months and 24 months). (**B**) Expression pattern of *CIDEA* during differentiation of goat preadipocytes. Distinct lowercase letters signify statistically significant differences ($p < 0.05$).

3.2. Overexpression of CIDEA Promotes Lipid Deposition in Goat Primary Intramuscular Preadipocytes

To elucidate the role of *CIDEA* in lipid deposition in goat intramuscular preadipocytes, we overexpressed *CIDEA* by transfecting the overexpression vector (CIDEA OVER). The results showed that the expression level of *CIDEA* was increased by about 23-fold ($p < 0.01$; Figure 2A). Lipid droplets and TAG contents were both increased after *CIDEA* overexpression ($p < 0.01$; Figure 2B–D). The CCK-8 assay revealed that cell viability was reduced in the CIDEA OVER group ($p < 0.05$; Figure 2E). Correspondingly, the mRNA abundances of transcription factors (*PPARγ*, *C/EBPα*, and *SREBP1c*), lipid droplet accumulation genes (*PLIN1* and *ADFP*), triglyceride synthesis genes (*DGAT1* and *DGAT2*), and fatty acid synthesis and transport genes (*ACC*, *FASN*, *ACSL1*, and *FABP3*) were increased after overexpressing *CIDEA* in intramuscular preadipocytes (Figure 2F–I). However, the mRNA expressions of lipolysis and β-oxidation genes (*HSL*, *ACOX1*, and *CPT1B*) and proliferation genes (*CCND2* and *CDK1*) were both downregulated (Figure 2J,K).

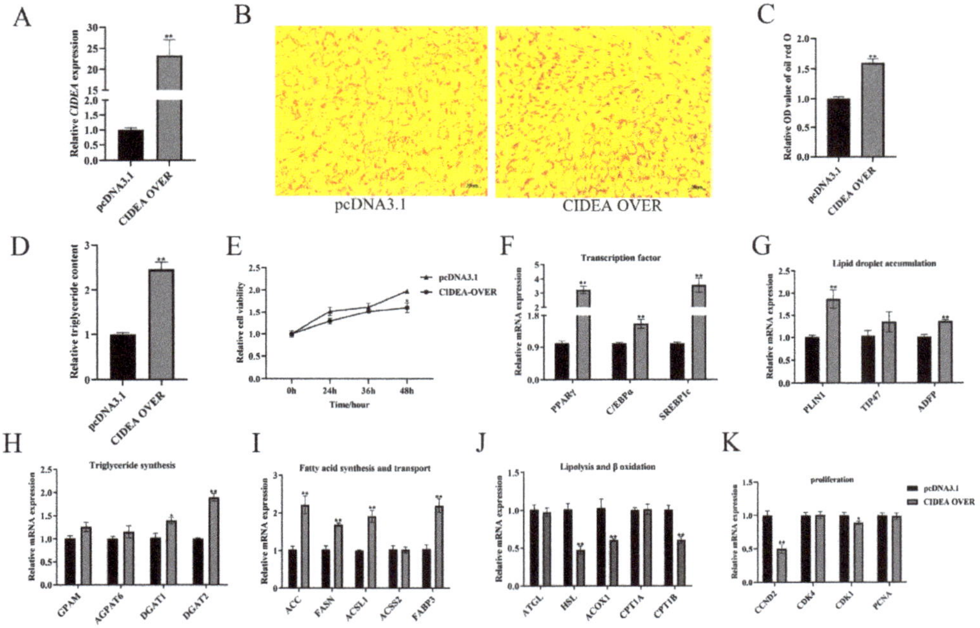

Figure 2. Overexpression of *CIDEA* promotes lipid deposition in goat intramuscular preadipocytes. (**A**) Overexpression efficiency detection. With UXT as the internal reference gene and the negative control as reference. (**B**,**C**) Oil red O staining and quantification of lipid droplets after *CIDEA* overexpression. (**D**) Relative triglyceride content in *CIDEA*-overexpressing cells. (**E**) Cell viability detection after overexpression of *CIDEA*. (**F**–**K**) Effects of *CIDEA* overexpression on the expression levels of genes related to lipid metabolism and cell proliferation. * $p < 0.05$; ** $p < 0.01$.

3.3. Knockdown of CIDEA Inhibits Adipogenesis in Intramuscular Preadipocytes

Then, we performed siRNA knockdown of *CIDEA* in goat intramuscular preadipocytes, which resulted in a reduction of up to 82% in transcript levels (Figure 3A). Lipid droplets and TAG contents were both decreased in the *CIDEA* knockdown group ($p < 0.01$; Figure 3B–D). The CCK-8 assay showed that siRNA-mediated suppression increased the viability of intramuscular preadipocytes ($p < 0.05$; Figure 3E). Correspondingly, mRNA expressions of transcription factors (*PPARγ*, *C/EBPα*, and *SREBP1c*), lipid droplet accumulation genes (*PLIN1*, *TIP47*, and *ADFP*), triglyceride synthesis genes (*GPAM*, *AGPAT6*, and

DGAT1), and fatty acid synthesis and transport genes (*ACC*, *FASN*, *ACSL1*, *ACSS2*, and *FABP3*) were all downregulated after knockdown of *CIDEA* in intramuscular preadipocytes (Figure 3F–I). Moreover, the mRNA abundances of lipolysis and β-oxidation genes (*ATGL*, *HSL*, *ACOX1*, *CPT1A*, and *CPT1B*) and proliferation genes (*CDK1* and *PCNA*) were both upregulated (Figure 3J,K).

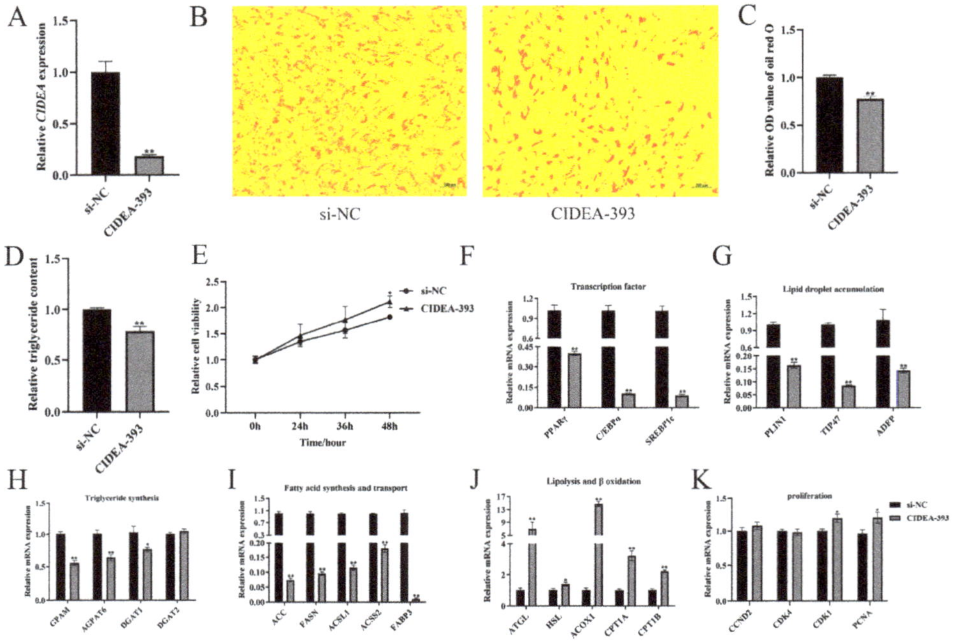

Figure 3. Knockdown of *CIDEA* inhibits adipogenesis of intramuscular preadipocytes. (**A**) Knockout efficiency detection after transfection of CIDEA-393 (siCIDEA) in preadipocytes. (**B**,**C**) Oil red O staining and quantification of lipid droplets content after knockdown with *CIDEA*. (**D**) Relative triglyceride content detection in *CIDEA* knockdown cells. (**E**) Cell viability detection by CCK-8 assay kit after *CIDEA* knockdown. (**F**–**K**) Effect of *CIDEA* knockdown on expression levels of genes related to lipid metabolism and cell proliferation. * $p < 0.05$; ** $p < 0.01$.

3.4. Screening and Analysis of Differentially Expressed Genes (DEGs) in Intramuscular Preadipocytes with Dysregulated CIDEA Expression

To elucidate the molecular mechanism by which *CIDEA* regulates lipid deposition in intramuscular preadipocytes, the transcriptional profiles of CIDEA OVER CIDEA-393 groups, together with their controls, were identified by RNA-seq. The RNA-sequencing data and their analysis are listed in Table S2. After *CIDEA* overexpression, we identified 134 differentially expressed genes (DEGs) ($p < 0.05$), of which 56 genes were upregulated and 78 genes were downregulated (Figure 4A). After *CIDEA* expression was reduced in goat intramuscular adipocytes, 1493 DEGs were identified ($p < 0.05$), of which 977 were upregulated and 516 were downregulated (Figure 4C). The heat map revealed that, despite significant differences between groups, the expression patterns were comparable within groups among samples, suggesting minimal variations between individual samples (Figure S2B,D). GO enrichment analysis showed that dysregulated expression of *CIDEA* DEGs was enriched in biological processes related to lipid metabolism, such as C-terminal protein lipidation, fatty acid omega-oxidation, lipid binding, cellular response to lipids, lipid transport involved in lipid storage, and other biological processes associated with lipid metabolism (Figure S2A,C). KEGG pathway analysis indicated that differential mRNAs were involved in apelin and focal adhesion signaling pathways (Figure 4B,D).

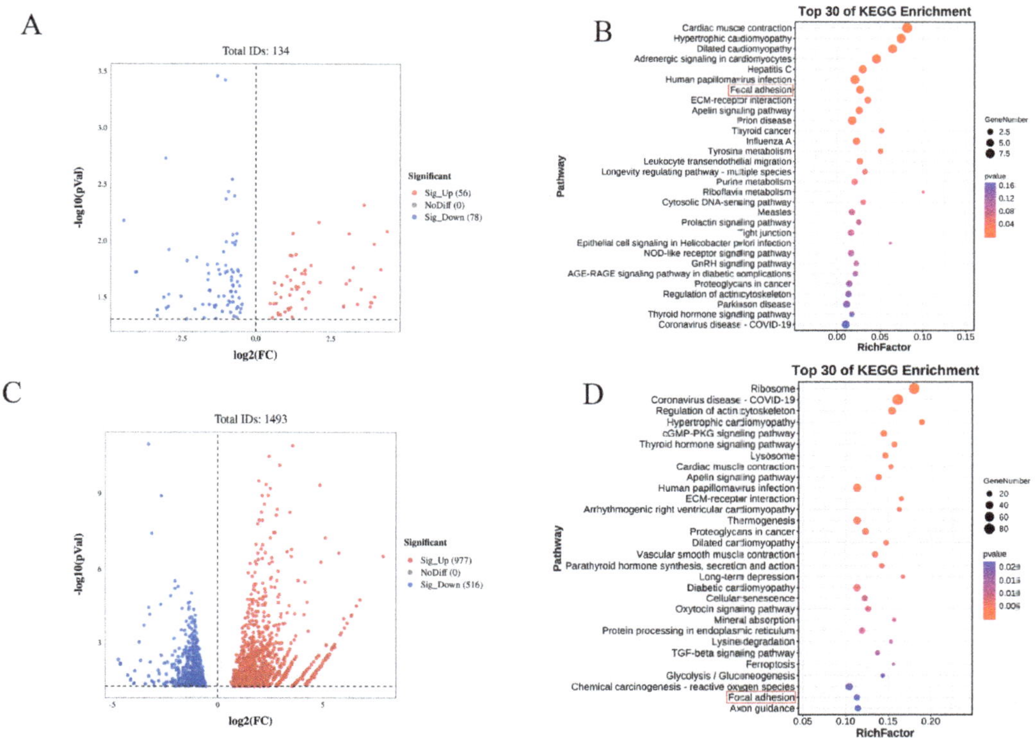

Figure 4. Screening and analysis of differentially expressed genes (DEGs) with dysregulated *CIDEA* expression. (**A**) Volcano plot of DEGs in *CIDEA*-overexpressing precursor adipocytes. Red dots indicate significant upregulation of genes; blue dots indicate significant downregulation of genes. (**B**) KEGG pathway analysis of DEGs in *CIDEA* overexpression group. (**C**) Volcano plot of DEGs in *CIDEA* knockdown precursor adipocytes. Red dots indicate significant upregulation of genes; blue dots indicate significant downregulation of genes. (**D**) KEGG pathway analysis of DEGs in *CIDEA* knockdown group. The red border indicates that the focal adhesion signaling pathway was enriched in the differential genes analysis of transcriptome sequencing results following both overexpression and knockdown of CIDEA. Subsequent experiments confirmed the activation of this pathway.

3.5. CIDEA Regulates Lipid Deposition in Goat Intramuscular Preadipocytes via Focal Adhesion Pathway

RNA-seq analysis showed that the apelin and focal adhesion pathways were enriched in both *CIDEA* overexpression and interference cells. This suggests that the effect of *CIDEA* on lipid deposition in goat preadipocytes may be mediated via these two pathways [22]. Therefore, we examined the abundances of the signaling proteins and their phosphorylated forms in these two pathways. The results showed that the abundance of p-FAK and the ratio of p-FAK/FAK were both elevated, while p-38, one key downstream signaling protein of the apelin pathway, exhibited no change (Figure 5A,B). In contrast, knockdown of *CIDEA* decreased the abundances of p-FAK and p-AKT, together with the ratios of p-FAK/FAK and p-AKT/AKT (Figure S4A,B). On this basis, we detected the mRNA expression of FAK and found that overexpression of *CIDEA* promoted FAK mRNA expression, while knockdown of *CIDEA* suppressed its expression (Figure S1A,B). We subsequently employed FAK inhibitors to investigate whether *CIDEA* regulates intramuscular adipogenesis via the focal adhesion pathway [23]. Our findings demonstrated that FAK inhibitors effectively suppressed the focal adhesion pathway (Figure S3A,B). The Oil red O staining results displayed that inhibition of FAK signaling could rescue the lipid droplet content increase induced by *CIDEA* overexpression (Figure 5C,D). The cellular triglyceride content change was consistent with the finding for the lipid droplet content (Figure 5E). Moreover, we found that inhibiting FAK activity further decreased the lipid droplet and TAG contents induced by *CIDEA* interference (Figure S4C,D).

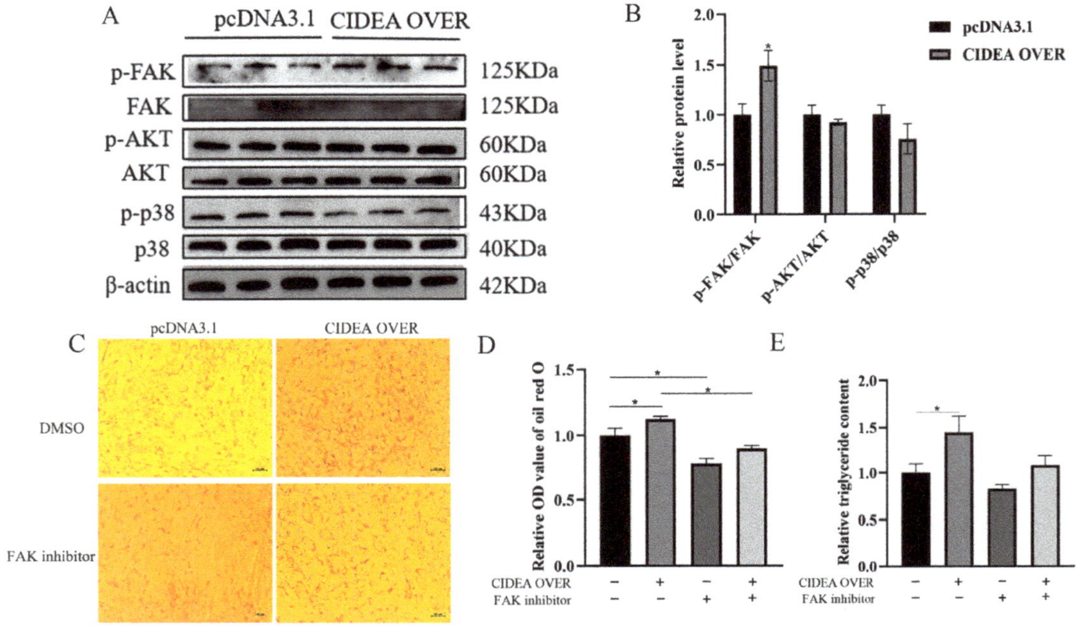

Figure 5. *CIDEA* regulates lipid deposition in goat intramuscular preadipocytes via focal adhesion pathway. (**A**) Detecting protein levels of p-p38, p38, p-AKT, AKT, p-FAK, and FAK after overexpression of *CIDEA* using Western blot. (**B**) Determining ratios of p-FAK/FAK, p-AKT/AKT, and p-p38/p38 upon overexpressing *CIDEA*. (**C**) Lipid droplet content detection after co-transfection of FAK inhibitor or DMSO and CIDEA OVER or pcDNA3.1 by Oil red O staining. (**D**) Determination of relative OD value of lipid droplets extracted after Oil red O staining. (**E**) Intracellular triglyceride content detection after co-transfection of FAK inhibitor or DMSO and CIDEA OVER or pcDNA3.1. * $p < 0.05$.

4. Discussion

Previous research indicates that *CIDEA* is a pivotal player in both disease pathogenesis and lipid metabolism. Nonetheless, the precise function and regulatory mechanisms of *CIDEA* with regard to intramuscular fat accumulation in goats have yet to be fully understood. Based on our preliminary RNA-seq data, it was observed that the expression level of *CIDEA* was significantly elevated in the longissimus dorsi muscle tissue of 24-month-old goats compared with their 2-month-old counterparts. Consequently, the objective of this study was to explore the impact of *CIDEA* on lipid deposition within goat intramuscular preadipocytes.

Intramuscular fat is mainly determined by adipocyte number and adipocyte volume [24]. Mature adipocytes are unable to divide and differentiate, so IMF deposition is due to the comprehensive effect of progenitor adipocyte cell proliferation and differentiation. Our research shows that the expression of *CIDEA* is upregulated in the longissimus dorsi muscle of 24-month-old goats compared with 2-month-old goats. Since 2-month-old goats are in the juvenile stage, the accumulation of IMF has not yet begun. In contrast, 24-month-old goats have reached adulthood, during which IMF gradually accumulates, leading to a significant increase in its content. This finding aligns with the increase in IMF content reported in our previous study [21]. Despite the different detection methods, the accuracy and depth of the results vary accordingly. Moreover, our study observed a gradual upregulation of *CIDEA* expression from day 0 to day 8 during adipogenic differentiation in goat intramuscular tissue. Analogously, the expression level of *CIDEC*, a homolog of *CIDEA*, was also increased during the differentiation process of human preadipocytes [25]. These findings suggest that *CIDEA* may play a beneficial role in the adipogenesis of intramuscular preadipocytes in goats.

In our study, we observed that the expression levels of several lipid metabolism-related genes were closely associated with *CIDEA*. *DGAT1* and *DGAT2* are rate-limiting enzymes required for triglyceride synthesis [26,27], and our research showed that *CIDEA* promotes the expression of *DGAT1* and *DGAT2* in goat intramuscular adipocytes. *ATGL* and *HSL* are enzymes that facilitate the breakdown of triglycerides into free fatty acids [28,29]. Subsequently, the free fatty acids are transported to mitochondria by CPT1s for β-oxidation [30], or they undergo oxidation by *ACOX1* within peroxisomes [31]. Our research also supports the finding that expression of *CIDEA* leads to the downregulation of a series of lipolysis (*ATGL*, *HSL*, and *ACOX1*) and β-oxidation (*CPT1A* and *CPT1B*) related genes. *FASN* and *ACC*, as the key rate-limiting enzymes in de novo fatty acid synthesis, play a crucial role in lipid production. The upregulation of *CIDEA* expression enhances the expression of these enzymes. Consequently, *CIDEA* promotes lipid deposition by facilitating fatty acid uptake and triglyceride synthesis.

Transcriptional regulatory factors play a significant role in the synthesis of lipids. Previous research has indicated that *CIDEA* acts as an activator of CCAAT/enhancer-binding protein (C/EBP) in the mammary glands of lactating mice [32]. Additionally, the promoter region of *CIDEA* contains a sterol regulatory element (SRE) and peroxisome proliferator-activated receptor (PPAR) elements that can be bound by *SREBP1c* and *PPARγ* [33]. Interestingly, we observed that the expressions of *SREBP1c* and *PPARγ* were affected by *CIDEA* in goat intramuscular preadipocytes ($p < 0.01$), suggesting a potential interaction between *SREBP1c*, *PPARγ*, and *CIDEA*. In dairy goats, *C/EBPα* enhances triacylglycerol synthesis by modulating the activity of the *PPARG* promoter [34]. Our observations are in good agreement with the known effects of *C/EBPα* on *PPARγ*. Other studies have shown that *CIDEA* expression inhibits AMP-activated protein kinase (AMPK) activity, which enhances *PPARγ* expression, thereby increasing triglyceride content, and that *CIDEA* expression promotes the nuclear translocation of *SREBP1c* [35]. Furthermore, expression of *PPARγ* and *SREBP1c* can directly activate the transcription of *FASN* and *ACC*, thereby promoting the formation of lipids [36,37]. On the basis of these results, we hypothesized that *CIDEA* regulated lipid metabolism through *PPARγ* and *SREBP1c*, thereby regulating their downstream genes (*ACC* and *FASN*) expression.

To further unravel the potential molecular mechanism by which *CIDEA* affects IMF deposition in goats, we performed RNA-seq on intramuscular preadipocytes after overexpression and knockdown of *CIDEA*. Interestingly, KEGG pathway enrichment analysis revealed obvious enrichments of apelin and focal adhesion pathways after *CIDEA* dysregulation. It is known that the apelin pathway enhances insulin sensitivity, promotes glucose uptake and utilization, inhibits fatty acid synthesis, and stimulates fatty acid oxidation by activating downstream signaling pathways such as PI3K/Akt and MAPK [38–40]. The PI3K-Akt signaling pathway also serves as the downstream signaling of the focal adhesion pathway [23]. In our study, we found that *CIDEA* activated the FAK and AKT signaling proteins, two key signaling proteins in the focal adhesion pathway, but not the p38 signaling protein, a downstream signaling protein of the MAPK pathway, which is consistent with previous findings in mice [41,42]. Intriguingly, our data revealed that interfering with *CIDEA* inhibited the activation of AKT, while overexpression of *CIDEA* did not influence the signaling protein, which needed further exploration. On this basis, we inquired into whether *CIDEA* regulates lipid deposition through the focal adhesion pathway. The FAK inhibitor was used to inhibit FAK signaling. The results suggested that the inhibition of FAK rescued the lipid droplet content increase induced by *CIDEA* overexpression and further decreased the lipid droplet content in *CIDEA*-interfering cells. Unfortunately, due to species limitations, we did not find the antibody matching the goat *CIDEA* protein to characterize *CIDEA* expression at the protein level.

In summary, this study reveals that *CIDEA* promotes lipid deposition in intramuscular preadipocytes through the focal adhesion pathway and inhibits cell proliferation. These works clarify the functional role and downstream signaling pathway of CIDEA in intramuscular fat deposition and provide theoretical support for improving meat quality through manipulating phenotype-related key genes.

5. Conclusions

Our study provides evidence for the role of *CIDEA* in intramuscular fat deposition in goats. *CIDEA* promotes adipogenesis through the focal adhesion pathway and inhibits cell proliferation in goat intramuscular preadipocytes. These findings contribute to our understanding of the comprehensive effect of *CIDEA* on intramuscular fat deposition and lay the theoretical foundation for the development of goat molecular breeding technology.

Supplementary Materials: The following supporting information can be downloaded at https://www.mdpi.com/article/10.3390/ani15162374/s1. Figure S1. mRNA abundance detection of FAK in CIDEA-dysregulated intramuscular preadipocytes. Figure S2. Screening and analysis of differentially expressed genes (DEGs) with dysregulated CIDEA expression. Figure S3. Inhibition of focal adhesion pathway by FAK inhibitors. Figure S4. CIDEA regulates lipid deposition in goat intramuscular preadipocytes via focal adhesion pathway. Table S1. Primers for quantitative real-time PCR (RT-qPCR). Table S2. Sequencing data quality control. In addition, the original Western Blot images are available in the Supplementary Materials.

Author Contributions: Conceptualization, P.S. and L.H.; methodology, J.Z.; software, Q.L.; validation, P.S. and Y.L. (Yu Liao); formal analysis, Y.L. (Yaqiu Lin); investigation, Y.W.; resources, L.H.; data curation, Y.L. (Yaqiu Lin); writing—original draft preparation, P.S.; writing—review and editing, H.X.; visualization, Z.D.; supervision, C.Z.; project administration, L.H.; funding acquisition, L.H. All authors have read and agreed to the published version of the manuscript.

Funding: This work was supported by the Sichuan Science and Technology Program (Chengdu, China; 2021YFYZ0003 and 2024NSFSC0029), the National Natural Science Foundation of China (Beijing, China; 32302702), the Zhejiang Science and Technology Program (Hangzhou, China; 2022C04017), the Southwest Minzu University Research Startup Funds (RQD2023018), and the Scientific and Technological Innovation Team for Qinghai–Tibetan Plateau Research in Southwest Minzu University (2024CXTD13).

Institutional Review Board Statement: All experimental procedures were approved by the Institutional Animal Care and Use Committee, Southwest Minzu University (Chengdu, China). Permit number: S2020-013.

Informed Consent Statement: Not applicable.

Data Availability Statement: These RNA-seq data are deposited in the NCBI Sequence Read Archive (SRA) under bio project number PRJNA995405.

Acknowledgments: We thank Shanghai OE Biotechnology Co, Ltd. for providing the sequencing service.

Conflicts of Interest: The authors declare no conflicts of interest.

References

1. Kumar, P.; Sharma, N.; Sharma, S.; Mehta, N.; Verma, A.K.; Chemmalar, S.; Sazili, A.Q. In-vitro meat: A promising solution for sustainability of meat sector. *J. Anim. Sci. Technol.* **2021**, *63*, 693–724. [CrossRef]
2. Baik, M.; Kang, H.J.; Park, S.J.; Na, S.W.; Piao, M.; Kim, S.Y.; Fassah, D.M.; Moon, Y.S. TRIENNIAL GROWTH AND DEVELOPMENT SYMPOSIUM: Molecular mechanisms related to bovine intramuscular fat deposition in the longissimus muscle. *J. Anim. Sci.* **2017**, *95*, 2284–2303. [CrossRef]
3. Tan, Z.; Jiang, H. Molecular and Cellular Mechanisms of Intramuscular Fat Development and Growth in Cattle. *Int. J. Mol. Sci.* **2024**, *25*, 2520. [CrossRef]
4. Chen, F.J.; Yin, Y.; Chua, B.T.; Li, P. CIDE family proteins control lipid homeostasis and the development of metabolic diseases. *Traffic* **2020**, *21*, 94–105. [CrossRef]
5. Xu, L.; Zhou, L.K.; Li, P. CIDE Proteins and Lipid Metabolism. *Arterioscler. Thromb. Vasc.* **2012**, *32*, 1094–1098. [CrossRef] [PubMed]
6. Qi, J.; Gong, J.; Zhao, T.; Zhao, J.; Lam, P.; Ye, J.; Li, J.Z.; Wu, J.; Zhou, H.M.; Li, P. Downregulation of AMP-activated protein kinase by Cidea-mediated ubiquitination and degradation in brown adipose tissue. *EMBO J.* **2008**, *27*, 1537–1548. [CrossRef]
7. Christianson, J.L.; Boutet, E.; Puri, V.; Chawla, A.; Czech, M.P. Identification of the lipid droplet targeting domain of the Cidea protein. *J. Lipid Res.* **2010**, *51*, 3455–3462. [CrossRef]
8. Nishimoto, Y.; Nakajima, S.; Tateya, S.; Saito, M.; Ogawa, W.; Tamori, Y. Cell death-inducing DNA fragmentation factor A-like effector A and fat-specific protein 27β coordinately control lipid droplet size in brown adipocytes. *J. Biol. Chem.* **2017**, *292*, 10824–10834. [CrossRef] [PubMed]
9. Zhang, S.; Shui, G.; Wang, G.; Wang, C.; Sun, S.; Zouboulis, C.C.; Xiao, R.; Ye, J.; Li, W.; Li, P. Cidea control of lipid storage and secretion in mouse and human sebaceous glands. *Mol. Cell Biol.* **2014**, *34*, 1827–1838. [CrossRef] [PubMed]
10. Zhou, Z.; Yon Toh, S.; Chen, Z.; Guo, K.; Ng, C.P.; Ponniah, S.; Lin, S.C.; Hong, W.; Li, P. Cidea-deficient mice have lean phenotype and are resistant to obesity. *Nat. Genet.* **2003**, *35*, 49–56. [CrossRef]
11. Barbatelli, G.; Murano, I.; Madsen, L.; Hao, Q.; Jimenez, M.; Kristiansen, K.; Giacobino, J.P.; De Matteis, R.; Cinti, S. The emergence of cold-induced brown adipocytes in mouse white fat depots is determined predominantly by white to brown adipocyte transdifferentiation. *Am. J. Physiol. Endocrinol. Metab.* **2010**, *298*, E1244–E1253. [CrossRef]
12. Do, G.M.; Oh, H.Y.; Kwon, E.Y.; Cho, Y.Y.; Shin, S.K.; Park, H.J.; Jeon, S.M.; Kim, E.; Hur, C.G.; Park, T.S.; et al. Long-term adaptation of global transcription and metabolism in the liver of high-fat diet-fed C57BL/6J mice. *Mol. Nutr. Food Res.* **2011**, *55* (Suppl. 2), S173–S185. [CrossRef]
13. Zhou, L.; Xu, L.; Ye, J.; Li, D.; Wang, W.; Li, X.; Wu, L.; Wang, H.; Guan, F.; Li, P. Cidea promotes hepatic steatosis by sensing dietary fatty acids. *Hepatology* **2012**, *56*, 95–107. [CrossRef]
14. Abreu-Vieira, G.; Fischer, A.W.; Mattsson, C.; de Jong, J.M.; Shabalina, I.G.; Rydén, M.; Laurencikiene, J.; Arner, P.; Cannon, B.; Nedergaard, J.; et al. Cidea improves the metabolic profile through expansion of adipose tissue. *Nat. Commun.* **2015**, *6*, 7433. [CrossRef] [PubMed]
15. Puri, V.; Ranjit, S.; Konda, S.; Nicoloro, S.M.; Straubhaar, J.; Chawla, A.; Chouinard, M.; Lin, C.; Burkart, A.; Corvera, S.; et al. Cidea is associated with lipid droplets and insulin sensitivity in humans. *Proc. Natl. Acad. Sci. USA* **2008**, *105*, 7833–7838. [CrossRef] [PubMed]
16. Zhang, W.; Liao, Y.; Shao, P.; Yang, Y.; Huang, L.; Du, Z.; Zhang, C.; Wang, Y.; Lin, Y.; Zhu, J. Integrated analysis of differently expressed microRNAs and mRNAs at different postnatal stages reveals intramuscular fat deposition regulation in goats (*Capra hircus*). *Anim. Genet.* **2024**, *55*, 238–248. [CrossRef] [PubMed]
17. Than, A.; Cheng, Y.; Foh, L.C.; Leow, M.K.; Lim, S.C.; Chuah, Y.J.; Kang, Y.; Chen, P. Apelin inhibits adipogenesis and lipolysis through distinct molecular pathways. *Mol. Cell Endocrinol.* **2012**, *362*, 227–241. [CrossRef] [PubMed]
18. Zhang, J.; Liu, Q.; Hu, X.; Fang, Z.; Huang, F.; Tang, L.; Zhou, S. Apelin/APJ signaling promotes hypoxia-induced proliferation of endothelial progenitor cells via phosphoinositide-3 kinase/Akt signaling. *Mol. Med. Rep.* **2015**, *12*, 3829–3834. [CrossRef]
19. Acebrón, I.; Righetto, R.D.; Schoenherr, C.; de Buhr, S.; Redondo, P.; Culley, J.; Rodríguez, C.F.; Daday, C.; Biyani, N.; Llorca, O.; et al. Structural basis of Focal Adhesion Kinase activation on lipid membranes. *EMBO J.* **2020**, *39*, e104743. [CrossRef]
20. Zhou, J.; Yi, Q.; Tang, L. The roles of nuclear focal adhesion kinase (FAK) on Cancer: A focused review. *J. Exp. Clin. Cancer Res.* **2019**, *38*, 250. [CrossRef]
21. Lin, Y.; Zhu, J.; Wang, Y.; Li, Q.; Lin, S. Identification of differentially expressed genes through RNA sequencing in goats (*Capra hircus*) at different postnatal stages. *PLoS ONE* **2017**, *12*, e0182602. [CrossRef] [PubMed]

22. Yuan, C.; Gou, X.; Deng, J.; Dong, Z.; Ye, P.; Hu, Z. FAK and BMP-9 synergistically trigger osteogenic differentiation and bone formation of adipose derived stem cells through enhancing Wnt-β-catenin signaling. *Biomed. Pharmacother.* **2018**, *105*, 753–757. [CrossRef]
23. Xiong, Y.; Wang, Y.; Xu, Q.; Li, A.; Yue, Y.; Ma, Y.; Lin, Y. LKB1 Regulates Goat Intramuscular Adipogenesis Through Focal Adhesion Pathway. *Front. Physiol.* **2021**, *12*, 755598. [CrossRef]
24. Tang, Y.; Zhang, W.; Wang, Y.; Li, H.; Zhang, C.; Wang, Y.; Lin, Y.; Shi, H.; Xiang, H.; Huang, L.; et al. Expression Variation of *CPT1A* Induces Lipid Reconstruction in Goat Intramuscular Precursor Adipocytes. *Int. J. Mol. Sci.* **2023**, *24*, 13415. [CrossRef]
25. Li, F.; Gu, Y.; Dong, W.; Li, H.; Zhang, L.; Li, N.; Li, W.; Zhang, L.; Song, Y.; Jiang, L.; et al. Cell death-inducing DFF45-like effector, a lipid droplet-associated protein, might be involved in the differentiation of human adipocytes. *FEBS J.* **2010**, *277*, 4173–4183. [CrossRef] [PubMed]
26. Wang, L.; Xu, S.; Zhou, M.; Hu, H.; Li, J. The role of DGAT1 and DGAT2 in tumor progression via fatty acid metabolism: A comprehensive review. *Int. J. Biol. Macromol.* **2024**, *278 Pt 3*, 134835. [CrossRef]
27. Yang, C.; Li, Q.; Lin, Y.; Wang, Y.; Shi, H.; Xiang, H.; Zhu, J. Diacylglycerol acyltransferase 2 promotes the adipogenesis of intramuscular preadipocytes in goat. *Anim. Biotechnol.* **2022**, *34*, 2376–2383. [CrossRef]
28. Brejchova, K.; Radner, F.P.W.; Balas, L.; Paluchova, V.; Cajka, T.; Chodounska, H.; Kudova, E.; Schratter, M.; Schreiber, R.; Durand, T.; et al. Distinct roles of adipose triglyceride lipase and hormone-sensitive lipase in the catabolism of triacylglycerol estolides. *Proc. Natl. Acad. Sci. USA* **2021**, *118*, e2020999118. [CrossRef]
29. Zheng, Y.; Chen, J.; Macwan, V.; Dixon, C.L.; Li, X.; Liu, S.; Yu, Y.; Xu, P.; Sun, Q.; Hu, Q.; et al. S-acylation of ATGL is required for lipid droplet homoeostasis in hepatocytes. *Nat. Metab.* **2024**, *6*, 1549–1565. [CrossRef]
30. Liu, Y.C.; Wei, G.; Liao, Z.Q.; Wang, F.X.; Zong, C.; Qiu, J.; Le, Y.; Yu, Z.L.; Yang, S.Y.; Wang, H.S.; et al. Design and Synthesis of Novel Indole Ethylamine Derivatives as a Lipid Metabolism Regulator Targeting PPARα/CPT1 in AML12 Cells. *Molecules* **2023**, *29*, 12. [CrossRef] [PubMed]
31. Zhang, F.; Xiong, Q.; Tao, H.; Liu, Y.; Zhang, N.; Li, X.F.; Suo, X.J.; Yang, Q.P.; Chen, M.X. ACOX1, regulated by C/EBPα and miR-25-3p, promotes bovine preadipocyte adipogenesis. *J. Mol. Endocrinol.* **2021**, *66*, 195–205. [CrossRef]
32. Wang, W.; Lv, N.; Zhang, S.; Shui, G.; Qian, H.; Zhang, J.; Chen, Y.; Ye, J.; Xie, Y.; Shen, Y.; et al. Cidea is an essential transcriptional coactivator regulating mammary gland secretion of milk lipids. *Nat. Med.* **2012**, *18*, 235–243. [CrossRef]
33. Li, Q.; Wang, W.; Duan, F.; Wang, Y.; Chen, S.; Shi, K.; Xia, Y.; Li, X.; Gao, Y.; Liu, G. DNMT3B Alleviates Liver Steatosis Induced by Chronic Low-grade LPS via Inhibiting CIDEA Expression. *Cell Mol. Gastroenterol. Hepatol.* **2024**, *17*, 59–77. [CrossRef]
34. Tian, H.; Luo, J.; Guo, P.; Li, C.; Zhang, X. C/EBPα promotes triacylglycerol synthesis via regulating PPARG promoter activity in goat mammary epithelial cells. *J. Anim. Sci.* **2023**, *101*, skac412. [CrossRef]
35. Cheng, J.; Xu, D.; Chen, L.; Guo, W.; Hu, G.; Liu, J.; Fu, S. CIDEA Regulates De Novo Fatty Acid Synthesis in Bovine Mammary Epithelial Cells by Targeting the AMPK/PPARγ Axis and Regulating SREBP1. *J. Agric. Food Chem.* **2022**, *70*, 11324–11335. [CrossRef]
36. Li, J.; Huang, Q.; Long, X.; Zhang, J.; Huang, X.; Aa, J.; Yang, H.; Chen, Z.; Xing, J. CD147 reprograms fatty acid metabolism in hepatocellular carcinoma cells through Akt/mTOR/SREBP1c and P38/PPARα pathways. *J. Hepatol.* **2015**, *63*, 1378–1389. [CrossRef]
37. Ho, T.C.; Wan, H.T.; Lee, W.K.; Lam, T.K.Y.; Lin, X.; Chan, T.F.; Lai, K.P.; Wong, C.K.C. Effects of In Utero PFOS Exposure on Epigenetics and Metabolism in Mouse Fetal Livers. *Environ. Sci. Technol.* **2023**, *57*, 14892–14903. [CrossRef] [PubMed]
38. Lee, D.K.; George, S.R.; O'Dowd, B.F. Unravelling the roles of the apelin system: Prospective therapeutic applications in heart failure and obesity. *Trends Pharmacol. Sci.* **2006**, *27*, 190–194. [CrossRef] [PubMed]
39. Yao, F.; Lv, Y.-C.; Zhang, M.; Xie, W.; Tan, Y.-L.; Gong, D.; Cheng, H.-P.; Liu, D.; Li, L.; Liu, X.-Y.; et al. Apelin-13 impedes foam cell formation by activating Class III PI3K/Beclin-1-mediated autophagic pathway. *Biochem. Biophys. Res. Commun.* **2015**, *466*, 637–643. [CrossRef] [PubMed]
40. Xie, F.; Liu, W.; Feng, F.; Li, X.; He, L.; Lv, D.; Qin, X.; Li, L.; Li, L.; Chen, L. Apelin-13 promotes cardiomyocyte hypertrophy via PI3K-Akt-ERK1/2-p70S6K and PI3K-induced autophagy. *Acta Biochim. Biophys. Sin.* **2015**, *47*, 969–980. [CrossRef]
41. Chen, H.J.; Yan, X.Y.; Sun, A.; Zhang, L.; Zhang, J.; Yan, Y.E. High-Fat-Diet-Induced Extracellular Matrix Deposition Regulates Integrin-FAK Signals in Adipose Tissue to Promote Obesity. *Mol. Nutr. Food Res.* **2022**, *66*, e2101088. [CrossRef] [PubMed]
42. Lin, X.; Zhuang, S.; Chen, X.; Du, J.; Zhong, L.; Ding, J.; Wang, L.; Yi, J.; Hu, G.; Tang, G.; et al. lncRNA ITGB8-AS1 functions as a ceRNA to promote colorectal cancer growth and migration through integrin-mediated focal adhesion signaling. *Mol. Ther.* **2022**, *30*, 688–702. [CrossRef] [PubMed]

animals

Article

ENO1 Regulates Apoptosis Induced by Acute Cold Stress in Bovine Mammary Epithelial Cells

Na Shen, Jie Wang, Jiayu Liao, Hengwei Yu, Wenqiang Sun, Xianbo Jia and Songjia Lai *

College of Animal Science and Technology, Sichuan Agricultural University, Chengdu 611130, China; 13289700731@163.com (N.S.); m18483220592@163.com (J.W.); ljiayu0121@163.com (J.L.); 18792427097@163.com (H.Y.); wqsun2021@163.com (W.S.); jaxb369@sicau.edu.cn (X.J.)
* Correspondence: laisj5794@163.com

Simple Summary

Acute cold stress adversely affects *dairy cow* health and lactation, but the specific mechanism by which damages bovine mammary epithelial cells (BMECs) remains unclear. Our study shows that acute cold stress harms BMECs by increasing cell apoptosis and disrupting the balance of apoptosis-related genes. We further identified that alpha-enolase (ENO1), a key glycolytic enzyme with multiple non-glycolytic functions in pathophysiological processes including signal transduction and apoptosis, plays a critical protective role, since its downregulation exacerbates apoptosis. These findings explain how acute cold stress impairs the normal function of BMECs and suggest a new direction for protecting *dairy cows* from cold stress by targeting *ENO1* as a potential target. This research could lead to better strategies for maintaining cow mammary health, ensuring stable milk production, and reducing economic losses for farmers in cold environments.

Abstract

Low-temperature environments in cold regions pose a significant threat to cattle farming. Bovine mammary epithelial cells (BMECs) are highly sensitive to cold stress, and acute cold stress can induce apoptosis, adversely affecting lactation performance and health. To explore the mechanism of acute cold stress-induced apoptosis in BMECs, we established an in vitro acute cold stress model. Results showed that mRNA levels of HSP90 increased significantly in a time-dependent manner after 2 h of cold stress, confirming successful model establishment. Following 4 h of cold stress, pro-apoptotic genes (Caspase-3, Bax) exhibited significantly elevated mRNA levels, while the anti-apoptotic gene (BCL-2) showed significantly reduced mRNA levels. Concurrently, the apoptosis rate increased significantly, indicating that acute cold stress induces apoptosis and suggesting the 4 h mark may represent a critical transition point. Integrated transcriptomic and functional analyses identified ENO1 as a core metabolic regulator counteracting acute cold stress-induced apoptosis in BMECs. As a multifunctional protein, ENO1 (alpha-enolase) acts as a central enzyme in glycolysis while exerting additional roles in cellular signaling and apoptotic processes, thereby participating in various pathophysiological regulations. Both mRNA and protein levels of *ENO1* were significantly elevated in cold-stressed cells compared to untreated controls. Importantly, interference with *ENO1* expression aggravated the extent of cold stress-induced apoptosis, demonstrating the regulatory role of *ENO1* in this process. To our knowledge, this is the first report elucidating the core regulatory function of *ENO1* in acute cold stress-induced apoptosis in BMECs. These findings provide a theoretical basis for understanding apoptotic mechanisms under stress.

Keywords: acute cold stress; bovine mammary epithelial cells (BMECs); apoptosis; enolase 1 (ENO1)

1. Introduction

Against the backdrop of intensifying global climate change and frequent extreme cold events, the adverse effects of cold environments on the cattle industry have become increasingly prominent. In major dairy-producing regions such as Russia and Ontario, Canada, extreme winter cold has emerged as a critical constraint on dairy industry development. Studies indicate that winter cold waves can induce cold stress responses in *dairy cows* [1]. This response exerts multifaceted detrimental effects on production performance. For one thing, it causes significant declines in milk yield; research by Thompson et al. has confirmed that reduced mammary blood flow under cold conditions may be a key mechanism inhibiting milk secretion [2]. For another, it increases the incidence of diseases such as mastitis, elevating morbidity and mortality rates in calves, and ultimately reducing economic returns [3,4].

Cold stress refers to a series of defensive responses and functional adjustments initiated by the body when animals are exposed to temperatures below their temperature tolerance threshold [5,6]. As an important type of environmental stress parallel to heat stress, both can activate the neuro-endocrine-immune network, trigger oxidative stress and metabolic disorders, and affect the physiological health of animals [7–9]. However, there are significant differences in physiological mechanisms between cold stress and heat stress. Heat stress is centered on heat dissipation disorders, where cell damage stems from high-temperature-induced protein denaturation, and cell homeostasis is mainly maintained by activating heat shock proteins (HSPs) [10,11]. In contrast, cold stress focuses on the surge in heat production demand, and copes with low temperatures by enhancing metabolic heat production and activating cold resistance-related genes [12–14].

Under normal circumstances, animals maintain stable core body temperature by enhancing metabolic heat production. However, when the ambient temperature is excessively low, it leads to an imbalance between heat production and heat dissipation, disrupting the homeostasis of the body's internal environment and subsequently triggering physiological dysfunction [15]. Studies have revealed that cold stimulation directly reduces blood flow in the mammary tissues of ruminants. On one hand, this is because low temperatures activate the sympathoadrenal medullary axis to release norepinephrine, and simultaneously activate oxidative stress-related pathways, which further exacerbate norepinephrine-mediated vasoconstriction. On the other hand, to maintain core body temperature, the body prioritizes blood distribution to core organs, thereby reducing mammary blood flow [16–19]. Reduced mammary blood flow results in insufficient nutrient supply and impaired excretion of metabolic waste, placing *bovine* mammary epithelial cells (BMECs) in an adverse metabolic environment. This leads to mitochondrial dysfunction, insufficient ATP production, and triggers the mitochondrial apoptotic pathway. Additionally, direct damage caused by low temperatures further exacerbates BMEC apoptosis [20–22]. The phenomenon of cold stress-induced cell apoptosis has been confirmed in various cell types, including mouse lymphocytes and bovine Sertoli cells [23–26]. As a genetically regulated programmed death process, cell apoptosis is of great significance in maintaining tissue homeostasis and responding to environmental stress [27,28]. However, excessive apoptosis leading to large-scale cell death can have adverse effects on the organism [29,30].

To counteract cold stress-induced damage, cells activate intrinsic cold resistance mechanisms. Alpha-enolase (ENO1), a key enzyme in the glycolytic pathway, also possesses diverse non-glycolytic functions [31,32]. It is involved in various pathophysiological processes such as cellular signal transduction and apoptosis [33,34]. Existing studies have suggested that under environmental stress conditions, the expression level and functional status of *ENO1* may undergo adaptive changes, thereby affecting the process of cell apoptosis. For *dairy cows*, the number of mammary epithelial cells and their lactation capacity directly determine milk yield, and reducing their apoptosis rate is an important prerequisite for maintaining lactation performance [35].

However, research on the mechanisms underlying cold stress-induced apoptosis in bovine mammary epithelial cells (BMECs) remains limited, and the role of *ENO1* in this process is yet to be defined. Therefore, this study employed BMECs to establish an in vitro model simulating acute cold stress. Integrating transcriptomic analysis and functional validation, we systematically investigated the signaling pathways of acute cold stress-induced apoptosis and the core regulatory function of *ENO1*. Our findings will provide a theoretical basis for deciphering the apoptotic mechanisms in bovine mammary cells under acute cold stress, while also offering foundational insights for breeding cold-resistant dairy cattle lines and informing the development of cold-wave prevention strategies for dairy farms.

2. Materials and Methods

2.1. Cell Culture and Treatment

The bovine mammary epithelial cell line (BMECs, MAC-T) was cultured in DMEM/F12 medium supplemented with 10% fetal bovine serum (Gibco, Carlsbad, CA, USA) and 1% penicillin-streptomycin (Corning, Corning, NY, USA) at 37 °C under 5% CO_2. To investigate the effects of acute cold stress on BMECs, cells at 85–90% confluence were seeded into 6-well or 12-well plates (Corning, Corning, NY, USA) and incubated overnight at 37 °C according to established protocols. The following day, cells were transferred to a 4 °C environment and subjected to cold stress treatment for 0, 2, 4, 6, or 8 h in medium containing 0.1 M HEPES (pH 7.4; Thermo Fisher Scientific, Waltham, MA, USA). Cells were harvested post-treatment for subsequent experiments. Based on combined expression analysis of HSP90 and apoptosis-related genes, the optimal condition for establishing the acute cold stress model was determined to be 4 h of exposure at 4 °C.

2.2. Preparation of Sequencing Samples

RNA sequencing (RNA-seq) analysis was performed on bovine mammary epithelial cells. This included three biological control replicates (CON1, CON2, CON3; cultured at 37 °C for 4 h) and three biological replicates of the cold stress-treated group (LT1, LT2, LT3; treated at 4 °C for 4 h). All sequencing services were provided by LC-Bio Technology Co., Ltd. (Hangzhou, China).

2.3. Apoptosis Assay

Cell apoptosis rates were determined using an Annexin V-FITC Apoptosis Detection Kit (Beyotime, Nantong, China). Following the manufacturer's protocol, cells were collected after cold stress treatment, stained with Annexin V-FITC and propidium iodide (PI), and incubated at room temperature for 15 min. DAPI (Beyotime, Nantong, China) was added to stain nuclei. Cells were observed under an inverted fluorescence microscope (Olympus, Tokyo, Japan): nuclei appeared blue (DAPI), early apoptotic cells fluoresced green (Annexin V-FITC positive only), and late apoptotic cells fluoresced red (double positive for Annexin V-FITC and PI). At least three random fields per group were captured. Quantification using ImageJ2 software (National Institutes of Health, Bethesda, MD, USA) involved: (1) converting images to 8-bit grayscale; (2) applying a uniform threshold to all images to distinguish signal from background; (3) counting positive cells using the "Analyze Particles" tool, with size and circularity parameters set to exclude debris (0.5–50 μm^2, circularity 0.3–1.0). Results are expressed as the percentage of PI- and FITC-positive cells relative to the total number of DAPI-stained nuclei.

2.4. Total RNA Extraction and RT-qPCR

Total RNA was extracted using TRIzol reagent, and its concentration and purity were measured. RNA samples were deemed qualified when the A260/A230 and A260/A280 ratios both exceeded 1.80. RNA integrity was assessed via 1.5% agarose gel electrophoresis; qualified samples exhibited clear, bright 28S and 18S rRNA bands, with the 28S band intensity approximately twice that of the 18S band. Qualified RNA was stored at −80 °C. Reverse transcription was performed using an mRNA Reverse Transcription Kit (Vazyme, Nanjing, China) according to the manufacturer's instructions. Quantitative real-time PCR (RT-qPCR) was conducted using SYBR® Premix Ex Taq™ (Foregene, Chengdu, China) on a ForeQuant F4 Sequence Detection System. The thermal cycling protocol was: 95 °C for 10 min (initial denaturation); 40 cycles of 95 °C for 10 s (denaturation) and 60 °C for 20 s (annealing/extension); followed by melt curve analysis to verify primer specificity. All primer sequences used in this study are listed in Table 1. The relative expression level of each gene was calculated using the $2^{(-\Delta\Delta Ct)}$ method, with β-actin as the internal reference gene.

Table 1. Primer sequences.

Primer Name	Sequences
β-actin-F	GCCCATCTATGAGGGGTACGC
β-actin-R	CTCCTTGATGTCACGGACGATTTC
BCL-2-F	GCCTTTGTGGAGCTGTATGG
BCL-2-R	CTGTGGGCTTCACTTATGGC
NFIA-F	GAAAGGATCCCACTTCCGGT
NFIA-R	GTCTCCCCACAGCCATCAC
KITLG-F	CCTCTCGTCCACACTCAAGG
KITLG-R	AGTTGTTCCACCATCTCGCT
ALDH1A1-F	GGACCTGTGCAGCAAATCA
ALDH1A1-R	ATAGCAGTTCACCCACACGG
HMGN1-F	CGGGAGGCTTTGATTGTCTTG
HMGN1-R	GCACAACTGACTCCAAACTGC
ANXA1-F	AGTGAGCCCCTATCCTACGTT
ANXA1-R	ACTTCATCCAGGGGCTTTCC
ENO1-F	TGCCACTTATCGGTCATCCTT
ENO1-R	GCGCGTCTTATCATTGTCCC
BAX-F	CCCGAGTTGATCAGGACCAT
BAX-R	GTGGGTGTCCCAAAGTAGGA
CASPASE3-F	TTGAGACAGACAGTGGTGCT
CASPASE3-R	TCTTTGCATTTCGCCAGGAA
ZC3H10-F	GATCTGCTTCTGGTGGGACT
ZC3H10-R	AAAAGGGAACTGACCTGCTGA
HSP90-F	TGCTTGGGAGTCTTCTGCTG
HSP90-R	CACTTCTTTGACCCGCCTCT
HGF-F	TGGCATCAAATGTCAGCCCT
HGF-R	CACCAAGGTCCCCCTTCTTC

Note: Primer name: The identifier for each primer, indicating the gene target and whether it is the forward (F) or reverse (R) strand. Sequence: The nucleotide sequence of each primer.

2.5. siRNA Synthesis and Transfection

Small interfering RNAs (siRNAs) and negative control (NC) siRNA were designed and synthesized by GenePharma (Shanghai, China). Their sequences are listed in Table 2. For transfection, cells were seeded into 6-well plates. Upon reaching 50–60% confluence, siRNA (final concentration: 50 nM) and Lipofectamine 3000 (Invitrogen, Carlsbad, CA, USA) were mixed in Opti-MEM medium (Gibco, Grand Island, NY, USA) according to the manufacturer's protocol. The mixture was incubated at room temperature for 15 min and then added to the cells. Cells were subsequently incubated at 37 °C for 48 h. Subsequent experiments were performed following verification of transfection efficiency by RT-qPCR or Western blotting.

Table 2. Sequences of forward and reverse siRNA oligonucleotides targeting *ENO1*.

Name	Primer Sequences
Si-*ENO1*-F	GAGAAGAUCGACAAGCUGA
Si-*ENO1*-R	UCAGCUUGUCGAUCUUCUC

Note: siRNA sequences (5′→3′) targeting Bos taurus ENO1 (NCBI NC__037343). Name: The identifier for each siRNA oligonucleotide, indicating whether it is the forward (F) or reverse (R) strand. Primer Sequences: The nucleotide sequences of the siRNA oligonucleotides.

2.6. Protein Extraction and Western Blotting

Cells were lysed using radioimmunoprecipitation assay (RIPA) buffer (Beyotime, Nantong, China). Protein concentration was determined with a BCA protein assay kit (Beyotime, Nantong, China). Protein expression levels were detected by Western blotting, with β-actin serving as the internal reference for normalization. Protein samples were separated using SDS-polyacrylamide gel electrophoresis (SDS-PAGE). Subsequently, separated proteins were transferred onto polyvinylidene difluoride (PVDF) membranes (Beyotime, Nantong, China), followed by a 2 h blocking step. After blocking, membranes were incubated with appropriate primary antibodies at 4 °C overnight (antibody dilution concentrations are listed in Table 3). Following secondary antibody incubation, protein bands were visualized using Sparkjade ECLsuper (ED0015-A; Sparkjade Biotechnology Co., Jinan, China).

Table 3. Antibody dilution concentration.

Antibody	Company	Accession Number	Dilution Ratio
anti-BCL-2	HUABIO (Hangzhou, China)	HA721235	1:5000
anti-BAX	HUABIO (Hangzhou, China)	ET1603-34	1:20,000
anti-HSP90	HUABIO (Hangzhou, China)	HA722689	1:10,000
anti-ENO1	HUABIO (Hangzhou, China)	ET1705-56	1:2000
anti-CYT-C	HUABIO (Hangzhou, China)	ET1701-65	1:2000
anti--β-Actin	GenuIN (Hefei, China)	#2885	1:5000

Note: Antibody: The name of the specific antibody used in the experiments. Company: The supplier of the antibody. Accession Number: The unique identifier provided by the supplier for the specific antibody product. Dilution Ratio: The recommended dilution ratio for the antibody as used in the experiments.

2.7. Statistical Analysis

All data are presented as mean ± standard error of the mean (SEM). Experiments were independently repeated at least three times. Comparisons between two groups were analyzed using Student's *t*-test, while one-way ANOVA was employed for multiple comparisons. Statistical analyses were performed using GraphPad Prism 9. Significance levels are denoted as follows: * $p < 0.05$, ** $p < 0.01$.

3. Results

3.1. Establishment of Acute Cold Stress Model and Apoptosis Assessment

After acute cold stress treatment, apoptosis rates, cold adaptation responses, and expression levels of apoptosis-related genes in BMECs were analyzed, with the following results: Annexin V-FITC assays revealed a highly significant increase in apoptosis rates at 4, 6, and 8 h post-cold stress compared to controls ($p < 0.01$, Figure 1A,B), with apoptosis severity escalating over time. This indicates a positive correlation between cold stress duration and apoptotic progression. RT-qPCR analysis of relevant genes (Figure 1C–G) revealed: Heat shock protein 90 (HSP90, a canonical stress-response marker involved in cytoprotection and protein repair) and the pro-apoptotic gene *Bax* were upregulated in a time-dependent manner. *Caspase3* mRNA levels increased significantly at 4 h and remained elevated thereafter, driving apoptotic progression. The anti-apoptotic gene *Bcl-2* exhibited a contrasting trend: its expression decreased in a time-dependent manner starting from 4 h, opposite to the trend of pro-apoptotic genes. In addition, the mRNA expression of *ZC3H10* (involved in the regulation of cellular cold acclimatization) peaked at 4 h.

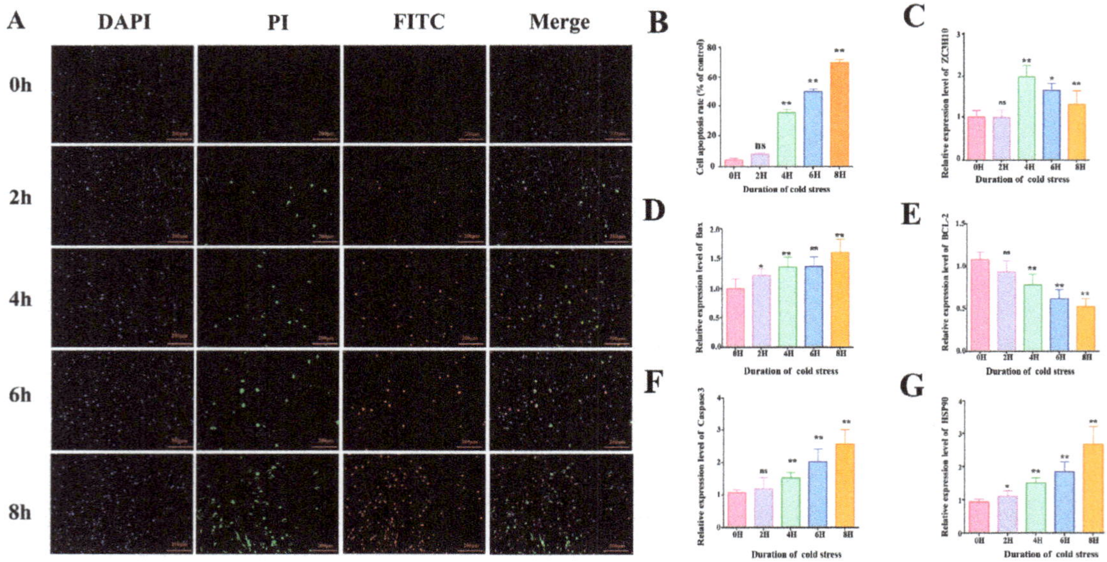

Figure 1. Acute cold stress induces apoptosis. (**A**) Annexin V-FITC fluorescence apoptosis assay of BMECs treated at 4 °C (0 h, 2 h, 4 h, 6 h, 8 h); blue fluorescence represents DAPI-stained cells, green fluorescence represents PI-staining-positive early apoptotic cells, and red fluorescence represents FITC-staining-positive late apoptotic cells. (**B**) Percentage of apoptotic cells. Apoptosis rate = positive apoptotic cells/number of DAPI-stained cells × 100% ($n = 3$) (**C–G**) Relative mRNA expression levels of *ZC3H10*, *Bax*, *BCL-2*, *Caspase-3*, and *HSP90* in 4–treated (0 h, 2 h, 4 h, 6 h, 8 h) BMECs ($n = 6$). Data are expressed as mean ± standard error. * $p < 0.05$; ** $p < 0.01$, ns (not significantly different).

Based on the above findings, subsequent experiments used a 4 h acute cold stress treatment to further investigate the molecular mechanisms underlying apoptosis induced by acute cold stress and the functions of related genes.

3.2. Analysis of Transcriptome Sequencing Results

Transcriptomic analysis identified 1205 differentially expressed genes (DEGs) between the control group (CON) and acute cold stress-treated group (LT), comprising 368 upregulated and 837 downregulated genes (Figure 2A). This provides a critical gene set for investigating cold stress effects on cells. To validate sample reliability, hierarchical clustering analysis of DEGs (Figure 2D) revealed pronounced distinctions between CON and LT groups. Principal component analysis (PCA) further demonstrated tight clustering of CON samples, while LT samples formed a separate cluster distinct from CON (Figure 2B). Validation of 10 randomly selected DEGs by RT-qPCR showed high concordance with sequencing data (Figure 2C), further confirming the accuracy of transcriptome sequencing.

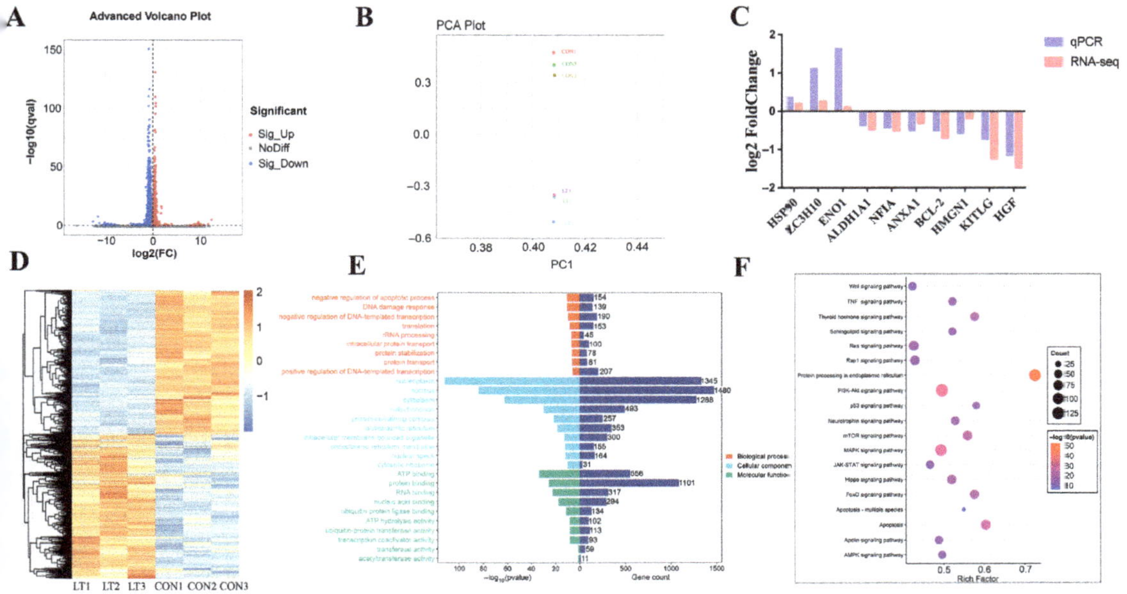

Figure 2. Analysis of differentially expressed genes. (**A**) Volcano plot of differentially expressed genes (DEGs) between the control group (CON) and the 4 h acute cold stress-treated group (LT). (**B**) Principal component analysis (PCA) of the CON group and the 4 h acute cold stress-treated group (LT). (**C**) Validation of 10 DEGs by RT-qPCR. (**D**) Heatmap of DEGs between the CON group and the 4 h acute cold stress-treated group (LT). (**E**) Gene Ontology (GO) enrichment bar plot of DEGs in the CON group versus the 4 h acute cold stress-treated group (LT). (**F**) Kyoto Encyclopedia of Genes and Genomes (KEGG) enrichment bubble plot of DEGs in the CON group versus the 4 h acute cold stress-treated group (LT).

Functionally, Gene Ontology (GO) enrichment analysis (Figure 2E) elucidated cold stress impacts across three domains: In the Biological process dimension, genes were significantly enriched in entries such as "negative regulation of apoptotic process". In the Cellular component dimension, differentially expressed genes were significantly enriched in nucleoplasm and other subcellular structures. The differential distribution of genes involved in processes such as rRNA processing, intracellular protein transport, and protein stabilization was observed in different subcellular structures. In the Molecular function dimension, genes related to functions such as "ATP binding" and "protein binding" were significantly enriched.

KEGG enrichment analysis (Figure 2F) showed that "Apoptosis" and "Apoptosis—multiple species" pathways were significantly enriched, which directly confirmed that the apoptosis pathway was activated in BMECs under cold stress. apoptosis pathway was activated under cold stress, and a large number of differentially expressed genes were involved in it, which was direct evidence of the cellular initiation of apoptosis program. At the same time, PI3K-Akt, MAPK, p53, and other signaling pathways indirectly related to apoptosis were also activated, which intertwined with each other to constitute the apoptosis regulatory network of BMECs under acute cold stress. Through multi-faceted analysis, functional annotation, and literature research, the key differential gene ENO1 was identified, and the multi-level apoptotic regulatory network of bovine mammary epithelial cells (BMECs) under acute cold stress was clarified.

3.3. ENO1 Regulates Cold Stress-Induced Apoptosis in BMECs

To investigate the role of the *ENO1* gene in cold stress-induced apoptosis, this study detected the mRNA and protein levels of *ENO1* using RT-qPCR and Western blot. The results showed that compared with the non-interfered control group, the mRNA and protein expression levels of *ENO1* in both the control group (37 °C) and cold stress-treated group (4 °C) were significantly decreased in BMECs with *ENO1* expression interference ($p < 0.05$, Figure 3A–C,H,J), indicating successful *ENO1* interference.

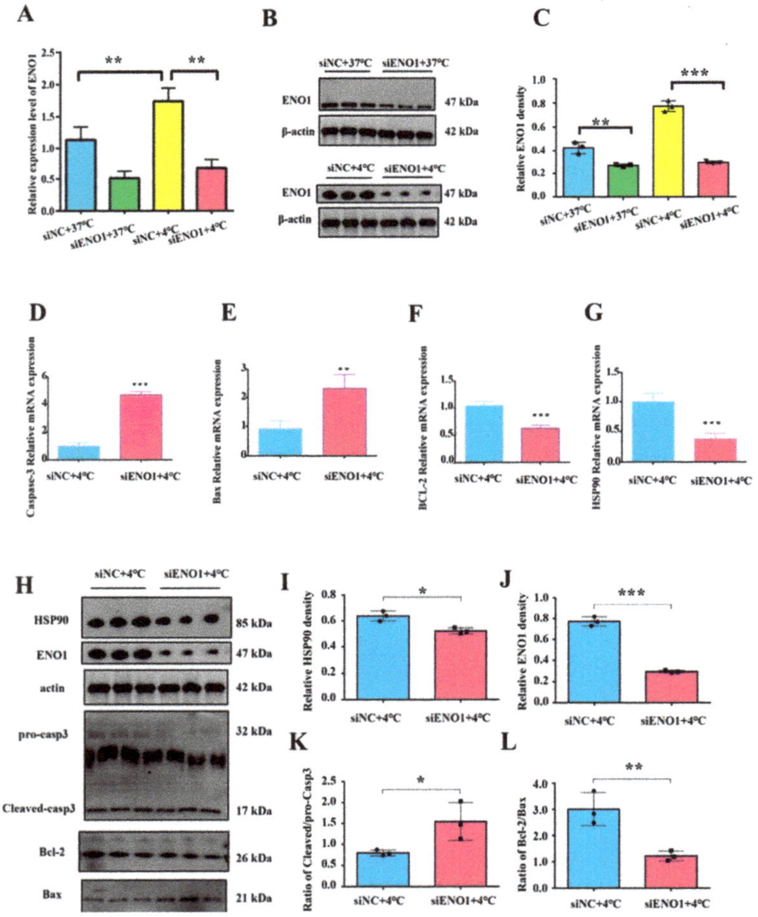

Figure 3. Interference with *ENO1* and acute cold stress co-treatment promotes apoptosis in BMECs. (**A–C**) Transfection efficiency of *ENO1* in BMECs. (**D–G**) Relative mRNA expression levels of *Caspase3*,

Bax, *BCL-2*, and *HSP90* in BMECs. (**H–L**) Relative protein expression levels of HSP90, ENO1 and apoptosis-related genes in BMECs. The original Western blot (WB) images are available in the Supplementary Materials. ($n = 3$). Data are expressed as mean ± standard error. * $p < 0.05$; ** $p < 0.01$, *** $p < 0.001$, ns (not significantly different).

To clarify the correlation between *ENO1* and apoptosis in BMECs under acute cold stress conditions, BMECs were co-treated with acute cold stress (4 °C) and *ENO1* interference. results showed that after co-treatment, the mRNA expression of the pro-apoptotic genes *Bax* and *Caspase3* was significantly increased (Figure 3D,E), and the mRNA expression of anti-apoptotic gene *Bcl-2* and cold stress-related gene *HSP90* was significantly decreased ($p < 0.05$, Figure 3F,G). The results of protein level detection showed that the expression level of *HSP90* was significantly reduced (Figure 3I); the ratios of Cleaved/procasp3 and Bax/Bcl-2 were significantly elevated (Figure 3K,L).

4. Discussion

In this study, a reliable in vitro model of acute cold stress in bovine mammary epithelial cells (BMECs) was established to investigate the molecular mechanism of apoptosis induced by acute cold stress. mRNA expression of *HSP90*, a classical stress biomarker [36,37], was upregulated with the increase in the cold stress treatment time, which indicated that the model was successfully constructed. Based on the successfully constructed model, in order to further investigate the effect of acute cold stress on apoptosis, we further detected apoptosis-related indexes. Given the significant dysregulation of apoptosis-related genes and the elevated apoptosis rate, we identified 4 h as the critical time point for apoptosis initiation, which provides a key reference for exploring the early apoptotic events triggered by cold stress.

Under acute cold stress, excessive accumulation of intracellular reactive oxygen species (ROS) causes oxidative damage and mitochondrial dysfunction [38]. Mitochondria are the main source of intracellular ROS and play a key role in initiating and executing cell apoptosis. When mitochondrial dysfunction occurs, it triggers the mitochondria-dependent apoptotic pathway, leading to the release of cytochrome c, formation of apoptosomes, and subsequent activation of downstream caspase-3, thereby initiating cell apoptosis [39]. Early studies have demonstrated that acute cold stress can induce apoptosis in various mammalian cells [23–26], and excessive apoptosis of BMECs exerts negative effects on milk production efficiency and mammary gland health [40]. Consistent with these early findings, sequencing results showed that acute cold stress treatment significantly increased the activation of apoptotic pathways. KEGG functional analysis revealed that the "Apoptosis" and "Apoptosis-multiple species" pathways were significantly enriched, which directly confirms that the apoptotic pathways of BMECs are activated under cold stress. A large number of differentially expressed genes are involved in these pathways, providing direct evidence that cells initiate apoptotic programs. Meanwhile, signaling pathways indirectly related to apoptosis, such as PI3K–Akt, MAPK, and p53, are also activated. These pathways interact with each other, collectively forming a regulatory network for BMEC apoptosis under acute cold stress.

Studies by Ma J et al. have indicated that *ENO1* can reduce intracellular ROS production and cell apoptosis by regulating mitochondrial homeostasis [41]. As a multifunctional protein, *ENO1* participates in the regulation of apoptotic pathways through multiple mechanisms, and its action mechanisms exhibit significant differences in different cell types and pathological conditions. A study on pancreatic ductal adenocarcinoma (PDAC) found that ENO1 mediates cell apoptosis by regulating ERK activation [42]. In a study on hypoxic pulmonary hypertension models, it was observed that *ENO1* improves mitochondrial function in endothelial cells and reduces apoptosis through the PI3K-Akt-mTOR pathway [43].

It is thus hypothesized that *ENO1* may play an important role in coping with mitochondrial damage caused by acute cold stress by affecting mitochondrial function or related apoptotic pathways.

Based on previous reports and sequencing results, we identified *ENO1* as a key regulatory factor. After treating BMECs at 4 °C for 4 h, the expression level of *ENO1* was significantly upregulated, indicating that cells have an active compensatory mechanism. GO annotation revealed that ENO1 has both core glycolytic enzyme activity and apoptosis regulatory potential. From the perspective of cellular metabolism, ATP depletion caused by cold stress further exacerbates the apoptotic process. As a key glycolytic enzyme, the upregulation of *ENO1* may enhance the glycolytic process and maintain ATP production, thereby alleviating the energy crisis caused by cold stress. This suggests its important role in linking energy metabolism and cell survival. In addition, previous studies have confirmed that *ENO1* not only promotes cell survival by regulating glycolysis in breast cancer but also may interact with *HSP70* to protect hepatocytes from heat stress [44,45]. Meanwhile, our experimental results confirmed that interfering with *ENO1* expression exacerbated acute cold stress-induced apoptosis of BMECs, and these results clarified the core regulatory role of *ENO1* in the survival of BMECs under acute cold stress conditions.

Although our research is insightful, there are limitations, including the singularity of the cell model, the failure to consider the interaction between cells in mammary tissues in vivo, the lack of involvement in differences in cellular stress responses under different physiological states, and the failure to explore the role of *ENO1* in other types of stress (such as oxidative stress). Subsequent studies need to verify the results in in vivo models (such as cold-stressed lactating *dairy cows*). Meanwhile, the precise molecular interactions (such as binding partners and pathway crosstalk) through which *ENO1* inhibits apoptosis have not been clarified.

In conclusion, we confirmed that acute cold stress induces apoptosis of BMECs and identified *ENO1* as a key metabolic factor against this damage. This work not only deepens the understanding of the cold stress response mechanism in bovine mammary epithelial cells but also pioneeringly proposes *ENO1* as a potential target for improving the cold resistance of *dairy cows*.

5. Conclusions

In this study, an in vitro cold stress model of bovine mammary epithelial cells (BMECs) was established (treated at 4 °C), confirming that 4 h is the key time point for apoptosis activation. This is the first study to reveal that α-enolase (ENO1), as a key regulatory factor, plays a central role in resisting apoptosis of BMECs induced by acute cold stress. This discovery provides an important theoretical basis for in-depth understanding of the apoptotic mechanism of BMECs under acute cold stress, clarifies the value of *ENO1* as a potential target in improving cellular cold resistance, and lays a foundation for subsequent research on cold resistance breeding of *dairy cows*. Subsequent animal experiments are required to further verify the regulatory role of *ENO1* in vivo and explore its specific molecular mechanism. However, this study has provided a new direction for the improvement strategy of cold resistance traits in *dairy cow* gene regulation. The findings are expected to offer theoretical support for pastures to formulate more scientific early-warning and intervention plans against cold waves, thereby helping reduce production losses in winter and further facilitating the sustainable development of animal husbandry.

Supplementary Materials: The following supporting information can be downloaded at: https://www.mdpi.com/article/10.3390/ani15172559/s1, Figure S1: Original Western blot images.

Author Contributions: N.S.: writing—review and editing. J.W.: methodology. J.L.: methodology. H.Y.: data curation. X.J.: data curation. W.S.: investigation. S.L.: funding acquisition. All authors have read and agreed to the published version of the manuscript.

Funding: This research was funded by the Innovation in High-quality Cattle Breeding Materials and Methods, and New Breed Selection, Key R&D Program of Sichuan Province, 2021YFYZ0001; Molecular Mechanism of p73 in Regulating Ferroptosis of Bovine Ovarian Granulosa Cells, National Natural Science Foundation of China, 32202670; Exploration of Germplasm Resources and Key Technologies for Efficient Breeding of Yingjing Yellow Cattle, kczx2023-2025-02.

Institutional Review Board Statement: All experiments in the present study involving animals were performed under the direction of the Institutional Animal Care and Use Committee from the College of Animal Science and Technology, Sichuan Agricultural University, China (Certification No. SYXK2019-187; Approval time: 29 January 2019).

Informed Consent Statement: Not applicable.

Data Availability Statement: The original contributions presented in this study are included in the article. Further inquiries can be directed to the corresponding author.

Conflicts of Interest: The authors declare that the research was conducted in the absence of any commercial or financial relationships that could be construed as a potential conflict of interest.

Abbreviations

BMECs	Bovine Mammary Epithelial Cells
ENO1	α-enolase 1
β-actin	Beta-Actin
LT	Low Temperature
HSP90	Heat Shock Protein 90
Caspase3	Cysteine Aspartyl Protease 3
Bax	BCL-2-associated X protein (BAX)
BCL-2	B-cell lymphoma 2
ZC3H10	zinc finger CCCH—type containing 10
ALDH1A1	Aldehyde Dehydrogenase 1 Family Member A1
HGF	Hepatocyte Growth Factor
NFIA	Nuclear Factor I A
ANXA1	Annexin A1
HMGN1	High Mobility Group Nucleosome-Binding Protein 1
KITLG	Kit Ligand

References

1. Angrecka, S.; Herbut, P. Conditions for Cold Stress Development in Dairy Cattle Kept in Free Stall Barn During Severe Frosts. *Czech J. Anim. Sci.* **2015**, *60*, 81–87. [CrossRef]
2. Thompson, G.E.; Thomson, E.M. Effect of Cold Exposure on Mammary Circulation Oxygen Consumption and Milk Secretion in the Goat. *J. Physiol.* **1977**, *272*, 187–196. [CrossRef]
3. Vander Zaag, A.; Le Riche, E.; Baldé, H.; Kallil, S.; Ouellet, V.; Charbonneau, É.; Coates, T.; Wright, T.; Luimes, P.; Gordon, R. Comparing Thermal Conditions inside and Outside Lactating Dairy Cattle Barns in Canada. *J. Dairy. Sci.* **2023**, *106*, 4738–4758. [CrossRef]
4. Lim, D.H.; Mayakrishnan, V.; Ki, K.S.; Kim, Y.; Kim, T.I. The Effect of Seasonal Thermal Stress on Milk Production and Milk Compositions of Korean Holstein and Jersey Cows. *Anim. Biosci.* **2021**, *34*, 567–574. [CrossRef]
5. Zhang, H.W.; Wang, Y.N.; Qiu, D.R. Key Points of Feeding and Management Techniques for Dairy Cows Under Cold Stress Conditions. *Breed. Feed.* **2022**, *21*, 47–49.
6. Su, S.H.; Hu, Y.C.; Wang, Y.; An, X.P.; Qi, J.W. Research Progress on Cold Stress and Intelligent Moni-toring in Ruminants. *Chin. J. Anim. Sci.* **2024**, *60*, 47–49.
7. Bai, H.; Kawahara, M.; Kusama, K.; Sakurai, T.; Pfarrer, C.; Takahashi, M. Heat Stress Induces Oxidative Stress and Activates the KEAP1-NFE2L2-ARE Pathway in Reproduction-Related Cells. *Anim. Sci. J.* **2025**, *96*, e70023. [CrossRef] [PubMed]

8. Zhao, F.Q.; Zhang, Z.W.; Qu, J.P.; Yao, H.D.; Li, M.; Li, S.; Xu, S.W. Cold Stress Induces Antioxidants and Hsps in Chicken Immune Organs. *Cell Stress. Chaperones* **2014**, *19*, 635–648. [CrossRef]
9. Gujar, G.; Tiwari, M.; Yadav, N.; Monika, D. Heat Stress Adaptation in Cows—Physiological Responses and Underlying Molecular Mechanisms. *J. Therm. Biol.* **2023**, *118*, 103740. [CrossRef]
10. Becker, C.A.; Stone, A.E. Graduate Student Literature Review: Heat Abatement Strategies Used to Reduce Negative Effects of Heat Stress in Dairy Cows. *J. Dairy Sci.* **2020**, *103*, 9667–9675. [CrossRef]
11. Bettaieb, A.; Averill-Bates, D.A. Thermotolerance Induced at a Mild Temperature of 40 °C Alleviates Heat Shock-Induced ER Stress and Apoptosis in HeLa Cells. *Biochim. Biophys. Acta* **2015**, *1853*, 52–62. [CrossRef]
12. Bornstein, M.R.; Neinast, M.D.; Zeng, X.; Chu, Q.; Axsom, J.; Thorsheim, C.; Li, K.; Blair, M.C.; Rabinowitz, J.D.; Arany, Z. Comprehensive Quantification of Metabolic Flux During Acute Cold Stress in Mice. *Cell Metab.* **2023**, *35*, 2077–2092.e6. [CrossRef] [PubMed]
13. Zhang, X.; Xiao, J.; Jiang, M.; Phillips, C.J.C.; Shi, B. Thermogenesis and Energy Metabolism in Brown Adipose Tissue in Animals Experiencing Cold Stress. *Int. J. Mol. Sci.* **2025**, *26*, 3233. [CrossRef]
14. Yudin, N.S.; Larkin, D.M. Candidate Genes for Domestication and Resistance to Cold Climate According to Whole Genome Sequencing Data of Russian Cattle and Sheep Breeds. *Vavilovskii Zhurnal Genet. Sel.* **2023**, *27*, 463–470. [CrossRef]
15. Ma, H.J.; Wang, P.C.; Yu, Y.; Ren, C.H.; Zhang, Z.J.; Wang, Q.J. Mechanisms and intervention measures of cold stress-induced thermogenesis in brown adipose tissue of lambs. *Chin. J. Anim. Sci.* **2025**, *61*, 61–67.
16. Broucek, J.; Letkovicová, M.; Kovalcuj, K. Estimation of Cold Stress Effect on Dairy Cows. *Int. J. Biometeorol.* **1991**, *35*, 29–32. [CrossRef]
17. Ameka, M.; Markan, K.R.; Morgan, D.A.; BonDurant, L.D.; Idiga, S.O.; Naber, M.C.; Zhu, Z.; Zingman, L.V.; Grobe, J.L.; Rahmouni, K.; et al. Liver-Derived FGF21 Maintains Core Body Temperature During Acute Cold Exposure. *Sci. Rep.* **2019**, *9*, 630. [CrossRef]
18. Zha, S.; Ao, R.G. Effects of cold stress on antioxidant function and blood indices of grazing beef cattle. *Contemp. Livest. Poult. Breed. Ind.* **2021**, *6*, 12–14.
19. Fu, K.; Li, Z.Y.; Cui, M.J. Research progress on heat and cold stress in dairy cow production. *Today's Anim. Husb. Vet. Med.* **2020**, *36*, 65.
20. Davis, S.R.; Collier, R.J. Mammary Blood Flow and Regulation of Substrate Supply for Milk Synthesis. *J. Dairy. Sci.* **1985**, *68*, 1041–1058. [CrossRef] [PubMed]
21. Kong, X.; Liu, H.; He, X.; Sun, Y.; Ge, W. Unraveling the Mystery of Cold Stress-Induced Myocardial Injury. *Front. Physiol.* **2020**, *11*, 580811. [CrossRef]
22. Li, T.; Bai, H.; Yang, L.; Hao, W.; Wei, S.; Yan, P. Low Temperature Exposure Inhibits Proliferation and Induces Apoptosis of Bovine Subcutaneous Preadipocytes via P38 MAPK/JNK Activation. *Comp. Biochem. Physiol. B Biochem. Mol. Biol.* **2023**, *264*, 110813. [CrossRef]
23. Cheng, C.H.; Guo, Z.X.; Wang, A.L. The Protective Effects of Taurine on Oxidative Stress, Cytoplasmic Free-Ca(2+) and Apoptosis of Pufferfish (*Takifugu obscurus*) under Low Temperature Stress. *Fish. Shellfish. Immunol.* **2018**, *77*, 457–464. [CrossRef]
24. Chen, L.; Chen, H. Effect of Mahuang Gancao Ganjiang Decoction on Fusion and Fission of Mitochondria and Apoptosis of Lymphocytes in Mice under Cold Stress. *Evid. Based Complement Altern. Med.* **2017**, *2017*, 5132963. [CrossRef]
25. Zhang, H.; Huang, H.; Zheng, P.; Feng, R.; Wang, X.; Huang, F.; Ma, M.; Tian, Y.; Zhang, G. The Alleviative Effect of Thyroid Hormone on Cold Stress-Induced Apoptosis Via Hsp70 and Mitochondrial Apoptosis Signal Pathway in Bovine Sertoli Cells. *Cryobiology* **2022**, *105*, 63–70. [CrossRef]
26. Doeppner, T.R.; Grune, T.; de Groot, H.; Rauen, U. Cold-Induced Apoptosis of Rat Liver Endothelial Cells: Involvement of the Proteasome. *Transplantation* **2003**, *75*, 1946–1953. [CrossRef] [PubMed]
27. Kawamoto, Y.; Nakajima, Y.I.; Kuranaga, E. Apoptosis in Cellular Society: Communication between Apoptotic Cells and Their Neighbors. *Int. J. Mol. Sci.* **2016**, *17*, 2144. [CrossRef] [PubMed]
28. Taabazuing, C.Y.; Okondo, M.C.; Bachovchin, D.A. Pyroptosis and Apoptosis Pathways Engage in Bidirectional Crosstalk in Monocytes and Macrophages. *Cell Chem. Biol.* **2017**, *24*, 507–514.e4. [CrossRef] [PubMed]
29. Wang, L.; Zhang, X.; Zhang, H. Excessive Apoptosis of RIP1-Deficient T Cells Leads to Premature Aging. *EMBO Rep.* **2023**, *24*, e57925. [CrossRef]
30. Green, K.A.; Streuli, C.H. Apoptosis Regulation in the Mammary Gland. *Cell Mol. Life Sci.* **2004**, *61*, 1867–1883. [CrossRef]
31. Li, Y.; Liu, L.; Li, B. Role of ENO1 and Its Targeted Therapy in Tumors. *J. Transl. Med.* **2024**, *22*, 1025. [CrossRef]
32. Díaz-Ramos, A.; Roig-Borrellas, A.; García-Melero, A.; López-Alemany, R. A-Enolase, a Multifunctional Protein: Its Role on Pathophysiological Situations. *J. Biomed. Biotechnol.* **2012**, *2012*, 156795. [CrossRef] [PubMed]
33. Wang, N.; Qiao, H.; Hao, J.; Deng, C.; Zhou, N.; Yang, L.; Zeng, M.; Guan, Q. RNA-Binding Protein ENO1 Promotes the Tumor Progression of Gastric Cancer by Binding to and Regulating Gastric Cancer-Related Genes. *J. Gastrointest. Oncol.* **2023**, *14*, 585–598. [CrossRef] [PubMed]

34. Aziguli, T.; Xiao, S.Y.; Yang, Y.; Musitaba, M. ENO1 Promotes PDAC Progression by Inhibiting CD8(+) T Cell Infiltration through Upregulating PD-L1 Expression Via HIF-1α Signaling. *Transl. Oncol.* **2025**, *52*, 102261. [CrossRef]
35. Herve, L.; Quesnel, H.; Lollivier, V.; Boutinaud, M. Regulation of Cell Number in the Mammary Gland by Controlling the Exfoliation Process in Milk in Ruminants. *J. Dairy Sci.* **2016**, *99*, 854–863. [CrossRef]
36. Wang, Y.; Qu, P. Cold stress promotes HSP90 expression and disease progression in PCOS rats. *Chin. J. Birth Health Hered.* **2022**, *30*, 597–601.
37. Sajad, S.; Jiang, S.; Anwar, M.; Dai, Q.; Luo, Y.; Hassan, M.A.; Tetteh, C.; Song, J. Genome-Wide Study of HSP90 Gene Family in Cabbage (*Brassica oleracea* Var. Capitata, L.) and Their Imperative Roles in Response to Cold Stress. *Front. Plant Sci.* **2022**, *13*, 908511. [CrossRef]
38. Blagojevic, D.P.; Grubor-Lajsic, G.N.; Spasic, M.B. Cold Defence Responses: The Role of Oxidative Stress. *Front. Biosci.* **2011**, *3*, 416–427. [CrossRef]
39. Lee, J.H.; Won, Y.S.; Park, K.H.; Lee, M.K.; Tachibana, H.; Yamada, K.; Seo, K.I. Celastrol Inhibits Growth and Induces Apoptotic Cell Death in Melanoma Cells Via the Activation ROS-Dependent Mitochondrial Pathway and the Suppression of PI3K/Akt Signaling. *Apoptosis* **2012**, *17*, 1275–1286. [CrossRef]
40. Liu, L.; Lu, O.; Li, D.; Tian, Y.; Liu, Z.; Wen, Y.; Peng, T.; Song, Y.; Du, X.; Wang, Z.; et al. Sirtuin 3 Mitigates Oxidative-Stress-Induced Apoptosis in Bovine Mammary Epithelial Cells. *J. Dairy Sci.* **2023**, *106*, 7266–7280. [CrossRef] [PubMed]
41. Ma, J.; Zhu, J.; Li, J.; Liu, J.; Kang, X.; Yu, J. Enhanced E6AP-Mediated Ubiquitination of ENO1 Via Linc00663 Contributes to Radiosensitivity of Breast Cancer by Regulating Mitochondrial Homeostasis. *Cancer Lett.* **2023**, *560*, 216118. [CrossRef]
42. Sun, H.; Mo, J.; Cheng, R.; Li, F.; Li, Y.; Guo, Y.; Li, Y.; Zhang, Y.; Bai, X.; Wang, Y.; et al. ENO1 Expression and ERK Phosphorylation in PDAC and Their Effects on Tumor Cell Apoptosis in a Hypoxic Microenvironment. *Cancer Biol. Med.* **2022**, *19*, 1598–1616. [CrossRef]
43. Shi, Y.; Liu, J.; Zhang, R.; Zhang, M.; Cui, H.; Wang, L.; Cui, Y.; Wang, W.; Sun, Y.; Wang, C. Targeting Endothelial ENO1 (Alpha-Enolase) PI3K-Akt-mTOR Axis Alleviates Hypoxic Pulmonary Hypertension. *Hypertension* **2023**, *80*, 1035–1047. [CrossRef] [PubMed]
44. Zhang, J.; Li, H.; Miao, L.; Ding, J. Silencing of ENO1 Inhibits the Proliferation, Migration and Invasion of Human Breast Cancer Cells. *J. BUON* **2020**, *25*, 696–701. [PubMed]
45. Zeng, T.; Cao, Y.; Gu, T.; Chen, L.; Tian, Y.; Li, G.; Shen, J.; Tao, Z.; Lu, L. Alpha-Enolase Protects Hepatocyte against Heat Stress through Focal Adhesion Kinase-Mediated Phosphatidylinositol 3-Kinase/Akt Pathway. *Front. Genet.* **2012**, *12*, 693780. [CrossRef] [PubMed]

animals

MDPI

Article

Identification of MAPK10 as a Candidate Gene for High Milk Production in Water Buffaloes Through a Genome-Wide Association Study

Wangchang Li [1,2,†], Huan Chen [1,†], Duming Cao [1] and Xiaogan Yang [1,*]

[1] Guangxi Key Laboratory of Animal Breeding, Disease Control and Prevention, College of Animal Science & Technology, Guangxi University, Nanning 530004, China; liwangchang@st.gxu.edu.cn (W.L.)
[2] Hunan Provincial Key Laboratory of Animal Intestinal Function and Regulation, Hunan International Joint Laboratory of Animal Intestinal Ecology and Health, Laboratory of Animal Nutrition and Human Health, College of Life Sciences, Hunan Normal University, Changsha 410081, China
* Correspondence: xgyang@gxu.edu.cn
† These authors contributed equally to this work.

Simple Summary

This study identifies key genetic markers and candidate genes, such as MAPK10, offering new insights into enhancing milk production in water buffaloes through targeted breeding strategies. These findings pave the way for more productive dairy buffalo populations, benefiting farmers and global dairy industries.

Abstract

Buffaloes are a vital genetic resource for dairy production, yet advancements in improving milk production have been somewhat limited. In this study, we performed an integrated analysis of genomic sequencing data from 78 water buffaloes and their milk production traits, with a focus on 305-day milk yield (MY). Leveraging advancements in sequencing technology alongside genome-wide association study (GWAS) methods such as cBLUP, GMATs, and BayesR, we aimed to identify genetic factors that could facilitate the breeding of high-quality buffaloes. Our analysis revealed two significant SNPs associated with milk production traits. Based on these markers, four candidate genes were identified within the surrounding genomic regions. These genes showed significant enrichment in lactation-related pathways, including the prolactin signaling pathway (mitogen-activated protein kinase 10, MAPK10), IL-17 signaling pathway (MAPK10), MAPK signaling pathway (MAPK10), and adipocytokine signaling pathway (MAPK10). The identification of these candidate genes, particularly MAPK10, provides a robust theoretical basis for molecular breeding strategies aimed at enhancing milk production in buffaloes. This work paves the way for more targeted and effective breeding programs in the future.

Keywords: buffalo; genome-wide association study; milk production traits

1. Introduction

China possesses abundant genetic resources of indigenous cattle breeds, characterized by high biodiversity and wide geographical distribution, making them a crucial component of global bovine (Bos spp.) genetic diversity [1]. The country currently maintains 26 water buffalo (Bubalus bubalis) breeds, including 24 native breeds and 2 introduced breeds. While water buffalo are occasionally used for meat production, their primary economic value lies in dairy output. Water buffalo milk, renowned for its rich trace element content, diverse fatty acid profiles, and enhanced digestibility, has garnered significant research interest and emerged as a focal point in China's dairy industry [2].

Water buffalo milk exhibits superior economic value compared to conventional cow milk, owing to its unique nutritional profile (e.g., higher fat content, calcium density, and digestibility), positioning it as a premium product in the dairy market. However, the average lactation productivity of water buffalo (5–8 kg/day) remains substantially lower than that of Holstein dairy cows (25–30 kg/day) [3]. This disparity underscores the urgency of implementing scientific interventions, such as genomic selection and gene-editing technologies, to enhance milk yield in water buffalo. Addressing this challenge through targeted strategies could not only drive industrial-scale production but also reshape the supply dynamics of China's dairy market [4,5].

The revolutionary application of molecular genetic technologies has transformed livestock breeding, with Single Nucleotide Polymorphisms (SNPs) emerging as the cornerstone of genomic research [6–8]. Whole-genome resequencing, powered by high-throughput platforms like Illumina, enables comprehensive profiling of genetic variation, offering precise molecular tools for livestock improvement. Building on this foundation, Genome-Wide Association Studies (GWAS) employ rigorous statistical frameworks—such as using three statistical models (cBLUP [9], GMATs [10], and BayesR [11]) and false discovery rate (FDR < 0.05) corrections—to significantly enhance the detection efficiency of candidate genes (e.g., MAK1P0) associated with lactation traits. This methodological advancement provides a critical basis for elucidating the molecular mechanisms underlying milk production in water buffalo and advancing targeted genetic breeding strategies.

2. Materials and Methods

2.1. Animal Welfare Statement

The protocol for this study was approved by the Attitude of the Animal Care and Welfare Committee of Guangxi University (GXU2019-021), Nanning, China.

2.2. Phenotypes and Animal Resources

This study collected 305-day milk yield (MY) data from 78 water buffaloes, including purebred Murrah buffaloes (imported from Pakistan) n = 21; purebred Nili-Ravi buffaloes (imported from India) n = 21; and Murrah × Nili-Ravi crossbred buffaloes (F1 or backcross generations) n = 36. All farmed buffalo samples mentioned above were obtained from the research farm of the Institute of Buffalo, Chinese Academy of Agricultural Sciences.

All individuals were multiparous cows (2–5 lactations) with complete lactation records standardized to a 305-day lactation period. The target trait was the 305-day milk yield (MY), used to evaluate inter-breed productivity differences and heterosis effects.

2.3. Sample Collection, Sequencing and Data Storage

Genomic DNA was extracted from water buffalo blood samples collected via caudal vein puncture using a vacuum blood collection system (Vacutainer, Shanghai, China). Blood was drawn from the caudal vein at the junction of the 4th and 5th caudal vertebrae (10 cm from the tail base) after disinfection with 75% ethanol. A 5 mL sterile syringe (24 G needle) was vertically inserted 0.5 cm into the skin, followed by blood collection using vacuum tubes. Genomic DNA extraction was performed using the phenol-chloroform method: anticoagulated blood was lysed with CTAB buffer (100 mM Tris-HCl, 1.4 M NaCl, 20 mM EDTA, 2% CTAB) supplemented with β-mercaptoethanol (0.2%) at 65 °C for 1 h. Proteins were removed by phenol/chloroform/isoamyl alcohol (25:24:1) extraction, followed by chloroform/isoamyl alcohol (24:1) purification. DNA was precipitated with isopropanol, resuspended in TE buffer, and quantified using NanoDrop (OD260/OD280 ratio 1.7–2.0). DNA integrity was assessed by 0.8% agarose gel electrophoresis (TAE buffer, 120 V, 30 min). Qualified DNA was fragmented and prepared for Illumina library construction, followed by paired-end 150 bp sequencing on the Illumina HiSeq 2000 platform at Genedenovo Biotechnology (Guangzhou, China) [10]. The sequencing data have been uploaded to (https://doi.org/10.6084/m9.figshare.28629623.v1, accessed on 1 April 2025).

2.4. Alignments and Variant Identification

Clean reads were aligned to the reference genome (UOA_WB_1) using BWA-MEM (v0.7.17) [12], followed by SNP detection with standardized pipelines including Samtools (v1.9), Picard (v3.1.1), and GATK (v4.0) [13,14]. The detected SNPs were subjected to hard filtering via GATK's "VariantFiltration" module with the following criteria: QD (QualByDepth) < 2.0; FS (FisherStrand) > 60.0; MQRankSum (MappingQualityRankSumTest) < −12.5; ReadPosRankSum (ReadPosRankSumTest) < −8.0; MQ (RMSMappingQuality) < 40.0; SOR (StrandOddsRatio) > 3.0. This filtering strategy follows GATK's recommended hard-filtering thresholds, ensuring the retention of statistically reliable SNPs for downstream analyses.

2.5. Variation Filtering

In genome-wide association studies (GWAS), rare alleles (minor allele frequency [MAF] < 0.05), high missingness rates (>30% genotype data missing), and multi-allelic variants may introduce biases. To mitigate these errors, we employed PLINK (v1.9) to filter detected SNPs according to the following criteria: exclusion of multi-allelic sites (retaining only bi-allelic SNPs); removal of SNPs with MAF < 0.05; and filtering SNPs with >30% missing genotype data (parameter: --geno 0.3) [15]. This quality control pipeline ensures data reliability for downstream association analyses. Following stringent filtering criteria, a refined set of 448,470 markers was retained, comprising 381,569 SNPs and 66,901 Indels.

2.6. Population Structure Analysis

GCTA software (v1.92.2) was employed to construct a genetic relationship matrix (GRM) based on genome-wide SNP data, followed by principal component analysis (PCA) to explore population structure and genetic relationships among samples [16]. Concurrently, SNP markers were pruned for linkage disequilibrium (LD) using Admixture software (v1.3): A PLINK-based pipeline applied a 50 kb sliding window step size and a 10 SNP window size, removing one marker from each pair with an LD coefficient (r^2) > 0.2. This filtering retained 448,470 independent SNPs for subsequent population structure analysis. Finally, PopHelper software (v2.2.7) [17] was utilized to generate bar plots illustrating the ancestral composition proportions and genetic architecture of individuals within each subpopulation.

2.7. Genome-Wide Association Mapping

To effectively control for potential confounding factors (e.g., population structure and kinship) in the analysis, principal component analysis (PCA) and a genetic relationship matrix (GM) were incorporated as random effects [18,19]. Subsequently, genome-wide association studies (GWAS) were conducted using three models, cBLUP, GMATs, and BayesR, to identify associations between molecular markers and target traits in a mixed population. Candidate loci were identified based on a significance threshold adjusted for multiple testing (0.05/n, where n represents the total number of markers). Candidate genes were identified based on their proximity to candidate SNP loci within approximately 50 kb upstream and downstream regions (total ~100 kb genomic span). The NCBI-maintained dbSNP database [20] (Database of Single Nucleotide Polymorphisms) was utilized to determine whether these candidate loci corresponded to previously reported genetic variants (i.e., rsID-annotated sites in dbSNP) or represented novel SNPs. This approach integrates genomic positional association with database annotation to facilitate the functional characterization of candidate genes.

2.8. Candidate-Associated Gene Pathway Enrichment

The study performed systematic functional enrichment analysis on the identified candidate genes using the Kyoto Encyclopedia of Genes and Genomes (KEGG) and Gene Ontology (GO) databases integrated within the KOBAS (v3.0) platform [21], aiming to elucidate their potential involvement in biological pathways and molecular functions. Pathway enrichment analysis was performed to evaluate the enrichment of candidate genes in specific functions or pathways, with a significance threshold of p-value ≤ 0.05 used to determine statistically significant functional enrichment.

3. Results

3.1. Phenotypic Value Statistics of the Traits

Based on the histogram analysis of 305-day milk yield (MY) phenotypic distribution, the average milk production in water buffalo was 2321.3 kg (Figure 1), with approximately 15% of individuals classified as high yielders (MY > 3000 kg). The genetic and molecular characteristics of these high-yielding individuals constituted a core focus of this study.

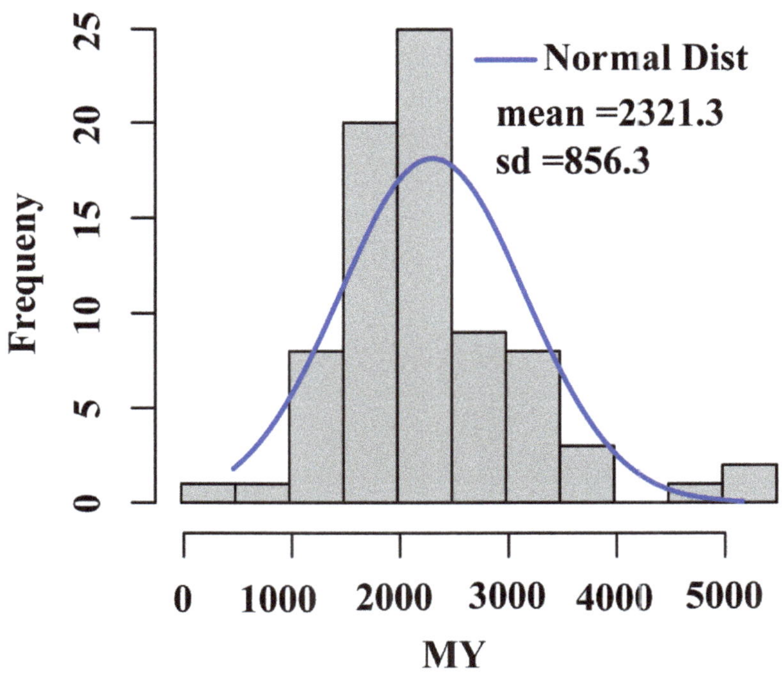

Figure 1. MY histogram. MY, 305-day milk yield.

3.2. Population Structure

To systematically evaluate genetic differentiation among water buffalo populations, unsupervised principal component analysis (PCA) was conducted to resolve population structure (Figure 2). Key findings include the following: Distinct Population Divergence (Figure 2A): Nili-Ravi (NB) and Murrah (MB) populations exhibited significant separation, indicating inherent genetic divergence between these taurine-derived breeds. Subpopulation Discrimination (Figure 2B,C): NB and MB populations achieved ~90% classification accuracy in PC1-PC2 space, whereas hybrid water buffalo (ZB) individuals showed marked clustering overlap with NB and MB, suggesting closer genetic relatedness between ZB and the two riverine populations. Principal Component Efficacy (Figure 2D): The first three principal components (PCs) captured 11.3% of total genetic variance, significantly outperforming subsequent components. This demonstrates that the top three PCs adequately represent lower-order genetic information.

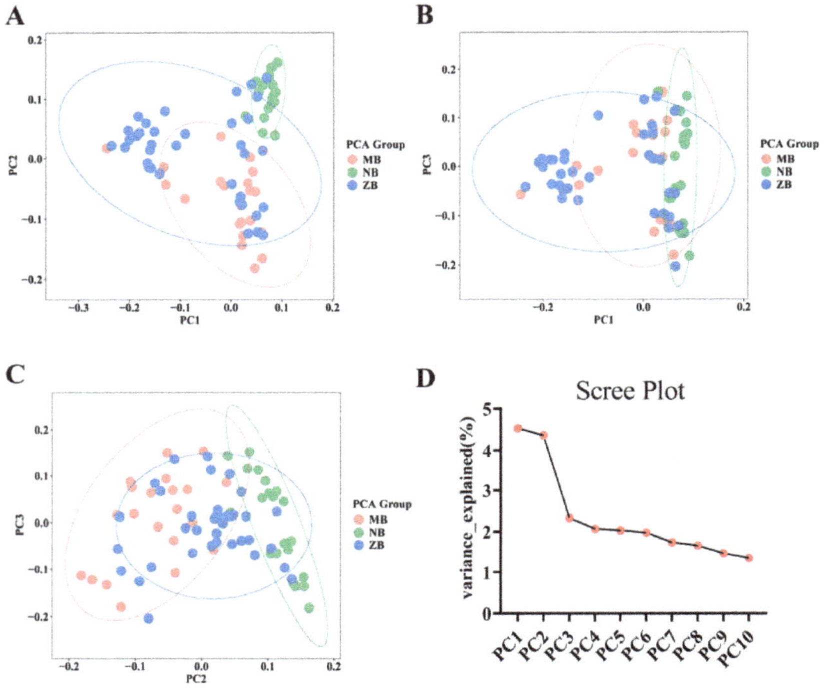

Figure 2. The population structure was visualized through three 2D PCA scatter plots (**A**–**C**) and one 3D PCA scatter plot (**D**), with the percentage of variance explained by each principal component (PC) indicated in parentheses (PC1: 28.5%, PC2: 19.3%, PC3: 14.7%). In the 2D plots, Panel A: PC1 vs. PC2 clustering; Panel B: PC1 vs. PC3 clustering; Panel C: PC2 vs. PC3 clustering. Each data point represents an individual buffalo, color-coded as follows: Murrah buffalo (MB): 21 individuals (pink); Nili-Ravi buffalo (NB): 21 individuals (green); hybrid water buffaloes (ZB): 36 individuals (green).

To dissect the genetic structure and ancestral composition of water buffalo populations, Bayesian clustering analysis was performed across K = 2–9 subpopulations, with genetic ancestry proportions visualized via bar plots (Figure 3). Each color represents a genetic cluster (K value), with the y-axis showing individual ancestry proportions (0–1). At K = 3, Nili-Ravi (NB), Murrah (MB), and hybrid water buffalo (ZB) exhibit distinct population-specific signatures. Cross-validation errors (CV) identified K = 3 as optimal, with minimum CV error (0.801) (Figure 4). This aligns with PCA-identified structure (Figure 2), confirming three biologically meaningful genetic clusters.

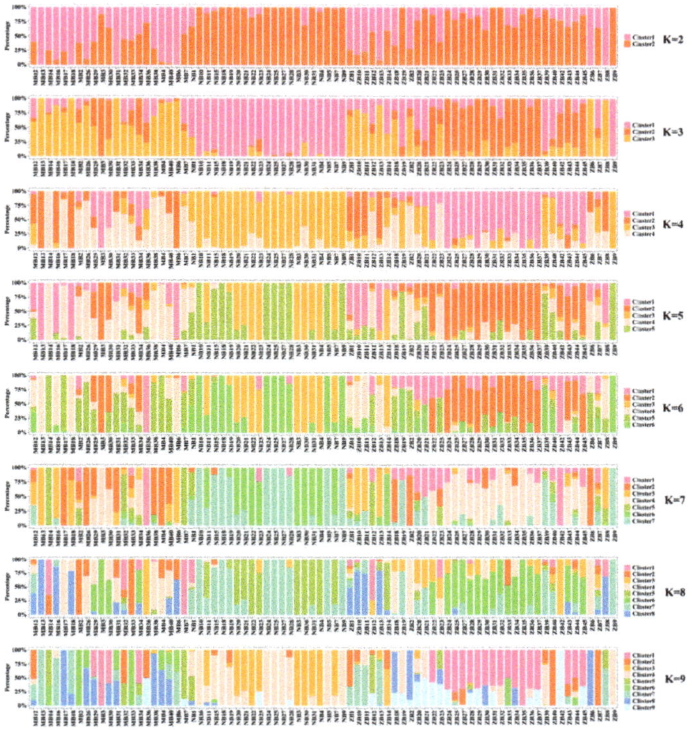

Figure 3. Genetic bar chart illustration for K-means clustering with varying numbers of clusters (K = 2 to 9).

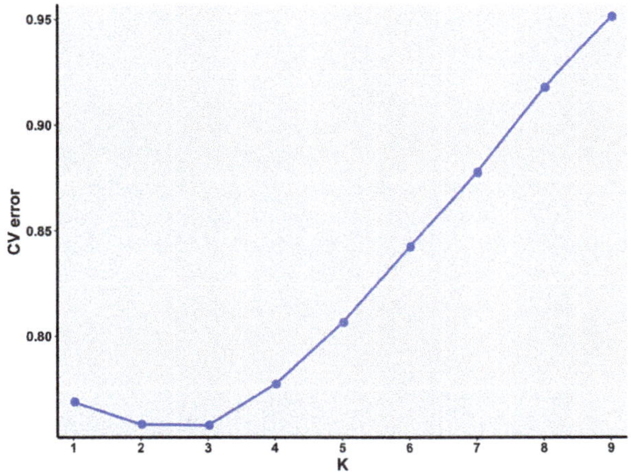

Figure 4. Line plot of cross-validation error rates across varying cluster numbers (K) in population structure analysis. The X-axis represents the number of clusters (K), and the Y-axis indicates the corresponding cross-validation error rate (CV error).

3.3. Results of the Genome-Wide Associations

In our GWAS, we have identified a large number of SNPs associated with MY (Figure 5). We identified two statistically significant SNPs and four candidate genes within a 50 Kb range surrounding these loci that were associated with the trait MY (Table 1).

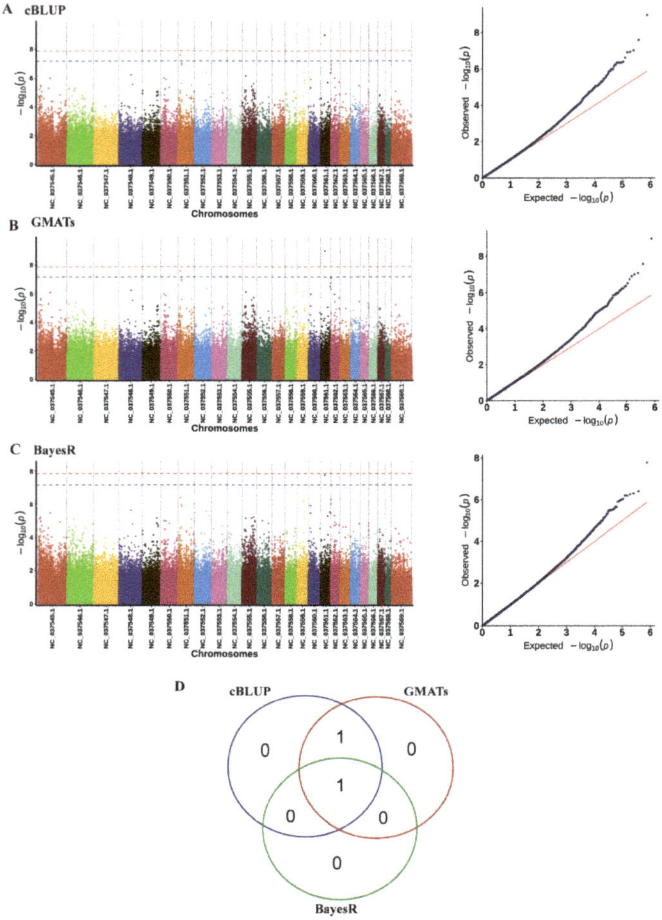

Figure 5. Association analysis of MY traits in water buffalo using three GWAS methods. (**A–C**) Comparative visualization of genome-wide association results from cBLUP, GMATs, and BayesR methods: Left panel (Manhattan plot) displays −log10(*p*) values for 30,145 SNPs across 25 chromosomes (24 autosomes + X). The blue horizontal line indicates a significance threshold of 0.05/n, The red horizontal line indicates a significance threshold of 0.01/n. SNPs exceeding this threshold are considered significant. Right panel (Q-Q plot) compares observed (−log10 *p*) values (x-axis) against expected uniform distribution values (y-axis). (**D**) Venn diagram summarizing overlapping significant SNPs identified by all three methods.

Table 1. The SNPs identified via genome-wide association analysis encompass detailed information *.

Methods	SNP	Chr	Pos	*p*	Candidate Genes
cBLUP	1	NC_037551.1	16,156,790	3.8×10^{-8}	*MAPK10*
cBLUP	2	NC_037561.1	28,948,029	1.8×10^{-9}	*ZNF84, ZNF26, ZNF605*
BayesR	1	NC_037551.1	28,948,029	2.8×10^{-8}	*ZNF84, ZNF26, ZNF605*
GMATs	1	NC_037561.1	16,156,790	1.4×10^{-9}	*MAPK10*
GMATs	2	NC_037561.1	28,948,029	1.5×10^{-8}	*ZNF84, ZNF26, ZNF605*

* Mitogen-activated protein kinase10, MAPK10; zinc finger protein 84, ZNF84; zinc finger protein 26, ZNF26; zinc finger protein 605, ZNF605.

3.4. Kyoto Encyclopedia of Genes and Genome Pathway Analysis of Candidate Genes

To elucidate the biological roles of candidate genes, we performed Gene Ontology (GO) and Kyoto Encyclopedia of Genes and Genomes (KEGG) pathway enrichment analyses. GO results revealed 20 significantly enriched terms (FDR < 0.05) across three domains: biological process (BP, n = 9), cellular component (CC, n = 5), and molecular function (MF, n = 3). Notable BP terms included regulation of biological processes (GO:0048518), immune responses (GO:0002376), metabolism (GO:0008152), and cellular activities (GO:0009987) (Figure 6A). KEGG analysis highlighted 20 enriched pathways (Figure 6B, Supplementary Table S1), grouped into five functional categories: genetic/environmental information processing, cellular processes, organismal systems, and human diseases. Key pathways included the following: Environmental sensing: Hippo (ko04391), MAPK (ko04013), and ErbB (ko04012) signaling pathways. Cellular homeostasis: Apoptosis (ko04215) and mitophagy (ko04137). Immune regulation: Toll/IMD (ko04624), IL-17 (ko04657), Th1/Th2 differentiation (ko04658), and prolactin signaling (ko04917).

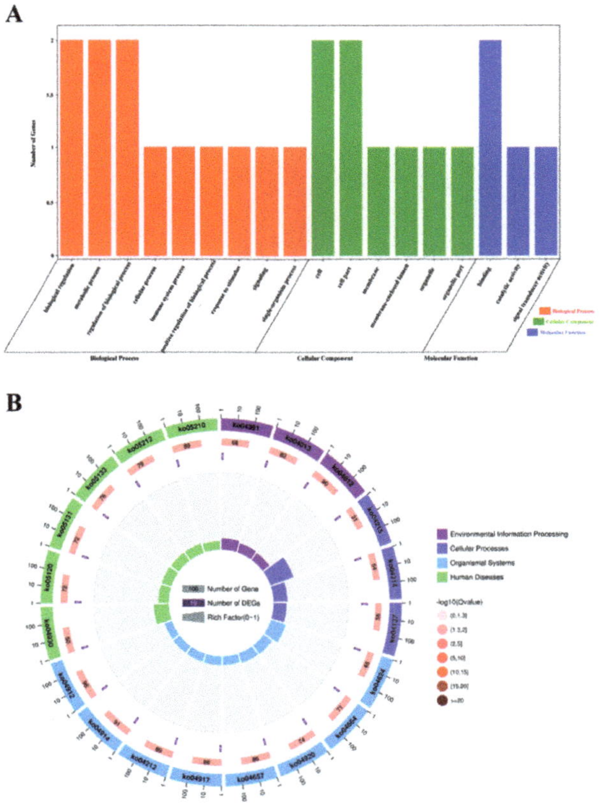

Figure 6. GO and KEGG enrichment analysis of candidate genes. (**A**) Top 20 significantly enriched Gene Ontology (GO) terms across three domains, cellular component, biological process, and molecular function, visualized as a horizontal bar chart. (**B**) Four-layer KEGG pathway enrichment plot: Outermost layer: Functional classification of enriched pathways, with scale bars indicating gene counts. Second layer: Gene counts and adjusted p-values (FDR) for each background category. Longer bars represent higher gene counts; red intensity correlates with statistical significance (smaller p-values). Third layer: Total candidate gene counts per pathway. Innermost layer: Enrichment factor (candidate gene count in category/background gene count per category), with grid lines spaced at 0.1 intervals. Color gradients reflect enrichment significance (darker shades = higher significance).

Our findings demonstrate that MAPK10, a central candidate gene, potentially modulates lactation performance through interactions with prolactin, IL-17, mitophagy, and MAPK signaling pathways. These insights uncover molecular mechanisms underlying dairy production and offer actionable targets for genetic improvement in buffalo breeding programs.

3.5. Significant Association of Milk Protein Content with SNP Validation

To validate the association between SNP NC_037551.1:16156790 (MAPK gene) and 305-day milk yield (MY) in water buffalo, genotyping-based histogram analysis was performed (Figure 7). Key findings include the T/T homozygous mutation exhibiting a significantly higher average MY (4008.8 ± 312.5 kg) compared to C/T heterozygotes (2712.3 ± 289.1 kg) and C/C wild type (2076.8 ± 265.4 kg) (One-way ANOVA, $p = 0.0017$).

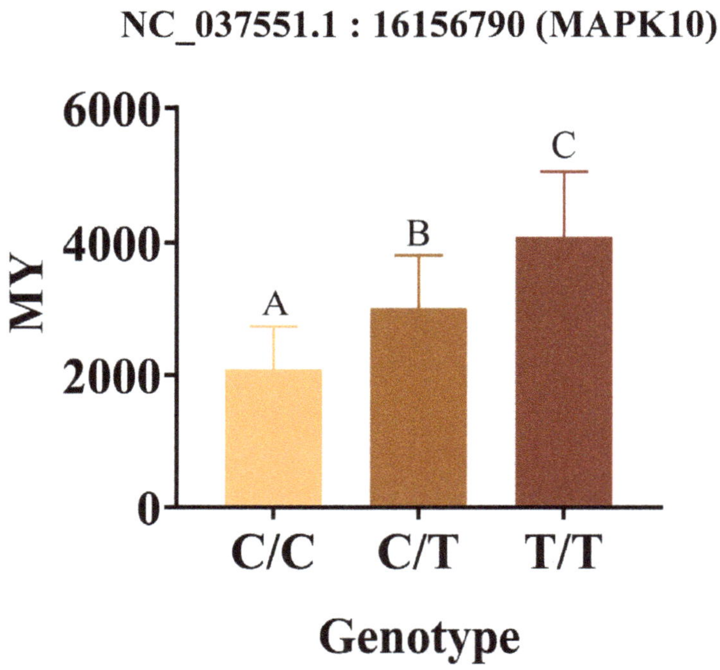

Figure 7. The phenotypic values of milk protein content. Different letters in the same column indicate significant differences ($p < 0.05$); T/T (homozygous mutation) type; C/T (heterozygous mutation); C/C (reference genotype) type.

4. Discussion

4.1. Molecular Genetic Structure

In genome-wide association studies (GWAS), population ancestry differences represent critical confounding factors that may lead to spurious associations by altering allele frequency patterns [22–24]. To address this challenge, we integrated principal component analysis (PCA), kinship matrix, and cross-validation error (CV error) methods to systematically resolve genetic structure. Although ZB shares partial genetic overlap with MB/NB, kinship, CV error, and PCA collectively support dividing the 78 buffalo cohort into three genetic clusters corresponding to Murrah (MB), hybrid water buffalo (ZB), and Nili-Ravi (NB) populations. These results establish a robust population structure framework for subsequent GWAS, effectively mitigating ancestry-related confounding effects.

4.2. Genome-Wide Association Analysis of Reproductive-Related Traits

In the association analysis of MY, we identified a total of four candidate genes (mitogen-activated protein kinase10, MAPK10; Zinc Finger Protein 84, ZNF84; Zinc Finger Protein 26, ZNF26; Zinc Finger Protein 605, ZNF605). The high milk yield in water buffaloes is a highly complex process that involves the coordinated action of multiple signaling pathways. Among the identified genes, some are associated with key pathways, including the prolactin signaling pathway, IL-17 signaling pathway, mitophagy pathway, and MAPK signaling pathway.

Previous studies have indicated that the prolactin signaling pathway and MAPK signaling pathway plays a crucial role in regulating cellular growth, development, cell cycle progression, and other functions, all of which contribute to MY efficiency [25–28]. We have identified the candidate gene MAPK10 as playing a significant biological role within the prolactin signaling pathway and MAPK signaling pathway. Specifically, MAPK10 (Mitogen-Activated Protein Kinase 10) is involved in regulating various intracellular signal cascades, including those related to prolactin secretion and cell proliferation. These processes are crucial for enhancing milk production in water buffaloes. For instance, MAPK10 can interact with key components of the MAPK signaling pathway, such as ERK and JNK transcription factors, to regulate cellular processes [29]. This interaction helps maintain cellular homeostasis and supports efficient lactation function. By modulating the activity of JNK, MAPK10 further influences the state of ERK proteins, controlling their nuclear localization and transcriptional activity [30]. This regulatory mechanism enables cells to respond appropriately to environmental signals, producing a variety of physiological effects, including promoting growth, differentiation, survival, and apoptosis.

4.3. The Role of MAPK10 in Enhancing Buffalo Milk Production: Current Findings and Future Directions

This study identified a regulatory locus within the MAPK10 gene through complementary GWAS methodologies, providing critical insights for enhancing water buffalo milk production. However, the findings primarily serve as a preliminary molecular genetic marker hypothesis for high-yield traits, largely constrained by the limited sample size (n = 78) which precluded precise and robust inferences. This study systematically interrogated multiple QTL databases (e.g., AnimalQTLdb and QTLbase) but found no previously reported QTLs associated with the identified locus (NC_037551.1:16156790). Although the analysis adopted stringent quality control measures—including permutation tests, Bonferroni corrections, and false discovery rate adjustments—to eliminate non-significant associations, the conclusions remain exploratory in nature [31,32]. Future investigations aim to systematically integrate these genetic determinants with functional genomics and multi-omics approaches to fully elucidate the mechanistic role of MAPK10 in buffalo lactation physiology.

5. Conclusions

This study conducted comprehensive whole-genome sequencing and genome-wide association analysis (GWAS) to identify genetic markers associated with high milk yield in water buffalo, ultimately detecting two significant SNP loci. Functional annotation revealed that NC_037551.1:16156790 (MAPK gene) could serve as a candidate molecular marker for high-yield traits. However, due to the limited sample size (n = 78), the precise functional mechanisms of this gene remain inadequately characterized. Future research will prioritize expanding the cohort size and incorporating functional genomics approaches—such as transcriptomic profiling—to validate the regulatory role of MAPK in buffalo lactation physiology and provide robust genetic evidence for breeding applications.

Supplementary Materials: The following supporting information can be downloaded at https://www.mdpi.com/article/10.3390/ani15172567/s1, Table S1: KEGG pathways of candidate genes.

Author Contributions: W.L., H.C. and D.C. participated in writing the initial draft of the paper. All authors contributed to the writing, review, and editing of the manuscript. X.Y. supervised this project. All authors have read and agreed to the published version of the manuscript.

Funding: This work was financially supported by grants from Guangxi Key R&D Program (GuiKe AB25069485), the Central Government Guided Local Science and Technology Development Fund (GuangXi) (Guike ZY24212019), the Guangxi Science and Technology Major Project (GuiKe AA24263009), and the Guangxi Science and Technology Major Project (GuiKe AA22068099-3).

Institutional Review Board Statement: The protocol for this study was approved by the Attitude of the Animal Care and Welfare Committee of the Guangxi university (GXU2019-021), Nanning, China.

Informed Consent Statement: Not applicable.

Data Availability Statement: https://doi.org/10.6084/m9.figshare.28629623.v1, accessed on 1 February 2025.

Conflicts of Interest: The authors declare that the research was conducted in the absence of any commercial or financial relationships that could be construed as a potential conflict of interest.

References

1. Lin, Y.X.; Sun, H.; Shaukat, A.; Deng, T.X.; Abdel-Shafy, H.; Che, Z.X.; Zhou, Y.; Hu, C.M.; Li, H.Z.; Wu, Q.P.; et al. Novel Insight Into the Role of ACSL1 Gene in Milk Production Traits in Buffalo. *Front. Genet.* **2022**, *13*, 896910. [CrossRef]
2. Barłowska, J.; Szwajkowska, M.; Litwińczuk, Z.; Król, J. Nutritional Value and Technological Suitability of Milk from Various Animal Species Used for Dairy Production. *Compr. Rev. Food Sci. Food Saf.* **2011**, *10*, 291–302. [CrossRef]
3. Pisanu, S.; Cacciotto, C.; Pagnozzi, D.; Puggioni, G.M.G.; Uzzau, S.; Ciaramella, P.; Guccione, J.; Penati, M.; Pollera, C.; Moroni, P.; et al. Proteomic changes in the milk of water buffaloes (water buffalo) with subclinical mastitis due to intramammary infection by Staphylococcus aureus and by non-aureus staphylococci. *Sci. Rep.* **2019**, *9*, 15850. [CrossRef]
4. Deng, T.; Liang, A.; Liang, S.; Ma, X.; Lu, X.; Duan, A.; Pang, C.; Hua, G.; Liu, S.; Campanile, G.; et al. Integrative Analysis of Transcriptome and GWAS Data to Identify the Hub Genes Associated With Milk Yield Trait in Buffalo. *Front. Genet.* **2019**, *10*, 36. [CrossRef] [PubMed]
5. Srirattana, K.; Hufana-Duran, D.; Atabay, E.P.; Duran, P.G.; Atabay, E.C.; Lu, K.; Liang, Y.; Chaikhun-Marcou, T.; Theerakittayakorn, K.; Parnpai, R. Current status of assisted reproductive technologies in buffaloes. *Anim. Sci. J.* **2022**, *93*, e13767. [CrossRef] [PubMed]
6. Zhang, S.; Yao, Z.; Li, X.; Zhang, Z.; Liu, X.; Yang, P.; Chen, N.; Xia, X.; Lyu, S.; Shi, Q.; et al. Assessing genomic diversity and signatures of selection in Pinan cattle using whole-genome sequencing data. *BMC Genom.* **2022**, *23*, 460. [CrossRef]
7. Bush, W.S.; Moore, J.H. Chapter 11: Genome-wide association studies. *PLoS Comput. Biol.* **2012**, *8*, e1002822. [CrossRef] [PubMed]
8. Wang, J.; Zhang, Z. GAPIT version 3: Boosting power and accuracy for genomic association and prediction. *Genom. Proteom. Bioinform.* **2021**, *19*, 629–640. [CrossRef]
9. Wang, J.; Zhou, Z.; Zhang, Z.; Li, H.; Liu, D.; Zhang, Q.; Bradbury, P.J.; Buckler, E.S.; Zhang, Z. Expanding the BLUP alphabet for genomic prediction adaptable to the genetic architectures of complex traits. *Heredity* **2018**, *121*, 648–662. [CrossRef]
10. Ning, C.; Wang, D.; Kang, H.; Mrode, R.; Zhou, L.; Xu, S.; Liu, J.F. A rapid epistatic mixed-model association analysis by linear retransformations of genomic estimated values. *Bioinformatics* **2018**, *34*, 1817–1825. [CrossRef]
11. Breen, E.J.; MacLeod, I.M.; Ho, P.N.; Haile-Mariam, M.; Pryce, J.E.; Thomas, C.D.; Daetwyler, H.D.; Goddard, M.E. BayesR3 enables fast MCMC blocked processing for largescale multi-trait genomic prediction and QTN mapping analysis. *Commun. Biol.* **2022**, *5*, 661. [CrossRef] [PubMed]
12. Pham, M.; Tu, Y.; Lv, X. Accelerating BWA-MEM Read Mapping on GPUs. In Proceedings of the ICS '23: Proceedings of the 37th International Conference on Supercomputing, Orlando, FL, USA, 21–23 June 2023; pp. 155–166.
13. Danecek, P.; Bonfield, J.K.; Liddle, J.; Marshall, J.; Ohan, V.; Pollard, M.O.; Whitwham, A.; Keane, T.; McCarthy, S.A.; Davies, R.M.; et al. Twelve years of SAMtools and BCFtools. *GigaScience* **2021**, *10*, giab008. [CrossRef]
14. Bathke, J.; Lühken, G. OVarFlow: A resource optimized GATK 4 based Open source Variant calling workFlow. *BMC Bioinform.* **2021**, *22*, 402. [CrossRef]
15. Chang, C.C.; Chow, C.C.; Tellier, L.C.; Vattikuti, S.; Purcell, S.M.; Lee, J.J. Second-generation PLINK: Rising to the challenge of larger and richer datasets. *Gigascience* **2015**, *4*, 7. [CrossRef]

16. Yang, J.; Lee, S.H.; Goddard, M.E.; Visscher, P.M. GCTA: A tool for genome-wide complex trait analysis. *Am. J. Hum. Genet.* **2011**, *88*, 76–82. [CrossRef] [PubMed]
17. Francis, R.M. pophelper: An R package and web app to analyse and visualize population structure. *Mol. Ecol. Resour.* **2017**, *17*, 27–32. [CrossRef]
18. Bradbury, P.J.; Zhang, Z.; Kroon, D.E.; Casstevens, T.M.; Ramdoss, Y.; Buckler, E.S. TASSEL: Software for association mapping of complex traits in diverse samples. *Bioinformatics* **2007**, *23*, 2633–2635. [CrossRef]
19. Hosseini, S.M.; Tingzhu, Y.; Pasandideh, M.; Liang, A.; Hua, G.; Farmanullah; Schreurs, N.M.; Raza, S.H.A.; Salzano, A.; Campanile, G.; et al. Genetic Association of PPARGC1A Gene Single Nucleotide Polymorphism with Milk Production Traits in Italian Mediterranean Buffalo. *BioMed Res. Int.* **2021**, *2021*, 1–8. [CrossRef] [PubMed]
20. Sherry, S.T.; Ward, M.H.; Kholodov, M.; Baker, J.; Phan, L.; Smigielski, E.M.; Sirotkin, K. dbSNP: The NCBI database of genetic variation. *Nucleic Acids Res.* **2001**, *29*, 308–311. [CrossRef]
21. Kanehisa, M.; Furumichi, M.; Sato, Y.; Kawashima, M.; Ishiguro-Watanabe, M. KEGG for taxonomy-based analysis of pathways and genomes. *Nucleic Acids Res.* **2023**, *51*, D587–D592. [CrossRef]
22. Haldar, T.; Ghosh, S. Effect of population stratification on false positive rates of population-based association analyses of quantitative traits. *Ann. Hum. Genet.* **2012**, *76*, 237–245. [CrossRef]
23. Persyn, E.; Redon, R.; Bellanger, L.; Dina, C. The impact of a fine-scale population stratification on rare variant association test results. *PLoS ONE* **2018**, *13*, e0207677. [CrossRef]
24. Elhaik, E. Principal Component Analyses (PCA)-based findings in population genetic studies are highly biased and must be reevaluated. *Sci. Rep.* **2022**, *12*, 14683. [CrossRef] [PubMed]
25. Liu, Y.; Yao, S.; Meng, Q.; Liu, X.; Han, H.; Kan, C.; Liu, G. A novel signaling transduction pathway of melatonin on lactose synthesis in cows via melatonin receptor 1 (MT1) and prolactin receptor (PRLR). *PeerJ* **2023**, *11*, e15932. [CrossRef]
26. Qian, L.; Lopez, V.; Seo, Y.A.; Kelleher, S.L. Prolactin regulates ZNT2 expression through the JAK2/STAT5 signaling pathway in mammary cells. *Am. J. Physiol.-Cell Physiol.* **2009**, *297*, C369–C377. [CrossRef] [PubMed]
27. Guo, Y.J.; Pan, W.W.; Liu, S.B.; Shen, Z.F.; Xu, Y.; Hu, L.L. ERK/MAPK signalling pathway and tumorigenesis. *Exp. Ther. Med.* **2020**, *19*, 1997–2007. [CrossRef] [PubMed]
28. Gu, Y.; Zhang, N.; Zhu, S.; Lu, S.; Jiang, H.; Zhou, H. Estradiol reduced 5-HT reuptake by downregulating the gene expression of Plasma Membrane Monoamine Transporter (PMAT, Slc29a4) through estrogen receptor β and the MAPK/ERK signaling pathway. *Eur. J. Pharmacol.* **2022**, *924*, 174939. [CrossRef]
29. Maya-Bernal, J.L.; Ramírez-Santiago, G. Spatio-temporal dynamics of a cell signal pathway with negative feedbacks: The MAPK/ERK pathway. *Eur. Phys. J. E* **2016**, *39*, 28. [CrossRef]
30. Haftcheshmeh, S.M.; Abedi, M.; Mashayekhi, K.; Mousavi, M.J.; Navashenaq, J.G.; Mohammadi, A.; Momtazi-Borojeni, A.A. Berberine as a natural modulator of inflammatory signaling pathways in the immune system: Focus on NF-κB, JAK/STAT, and MAPK signaling pathways. *Phytother. Res.* **2022**, *36*, 1216–1230. [CrossRef]
31. Lange, M.; Begolli, R.; Giakountis, A. Non-Coding Variants in Cancer: Mechanistic Insights and Clinical Potential for Personalized Medicine. *Non-Coding RNA* **2021**, *7*, 47. [CrossRef] [PubMed]
32. Hynes, A.C.; Sreenan, J.; MKane, M.T. The phosphatidylinositol signalling system in elongating bovine blastocysts; formation of phosphoinositides, inositol phosphates and stimulation by growth factors. *Reprod. Fertil. Dev.* **2002**, *14*, 515–523. [CrossRef] [PubMed]

animals

MDPI

Article

Transcriptome and Metabolome Profiles Reveal the Underlying Mechanism of Fat Deposition Changes in Three-Way Crossbred Yak for High-Quality Beef Production

Xiukai Cao [1,2], Wenxiu Ru [3], Jie Cheng [4], Le Sun [3], Nan Zhang [5], Lawang Zhaxi [5], Renzeng Dunzhu [5], Fengbo Sun [5], Kai Yang [6], Yue'e Gao [6], Xixia Huang [1], Bizhi Huang [6,*] and Hong Chen [1,*]

1 College of Animal Science, Xinjiang Agricultural University, Urumqi 830052, China; cxkai0909@163.com (X.C.)
2 Joint International Research Laboratory of Agriculture and Agri-Product Safety of Ministry of Education of China, Yangzhou University, Yangzhou 225009, China
3 Key Laboratory of Animal Genetics, Breeding and Reproduction of Shaanxi Province, College of Animal Science and Technology, Northwest A&F University, Yangling 712100, China
4 Jiangsu Key Laboratory of Sericultural and Animal Biotechnology, Jiangsu University of Science and Technology, Zhenjiang 212100, China
5 Tibet Autonomous Region Center for Animal Disease Control and Prevention, Lhasa 850032, China
6 Yunnan Academy of Grassland and Animal Science, Kunming 650212, China
* Correspondence: ynbphbz@163.com (B.H.); chenhong1212@263.net (H.C.)

Simple Summary

Yaks are important animals that live in the high mountains of Tibet, providing food and income for local people. However, yak meat quality could be improved to meet growing consumer demands. We created a new type of cattle called Yajiangxue by carefully breeding yaks with two other cattle breeds over multiple generations. This breeding approach combined the best traits from each animal: the ability to survive in harsh mountain conditions from yaks, fewer calving difficulties from Tibetan yellow cattle, and better meat quality from Angus cattle. We studied the fat tissue from these animals using advanced laboratory techniques to understand what makes their meat better. We found that Yajiangxue cattle had higher levels of healthy fats and beneficial compounds compared to their parent breeds. These cattle also showed better resistance to cellular damage and had more favorable fat composition that could improve meat flavor and nutritional value. The fat tissue contained more of certain good fatty acids that are beneficial for human health, as well as compounds that act as natural antioxidants. Our research helps explain why this new crossbred cattle produces higher quality meat and provides scientific evidence for developing better livestock breeding programs. This work could help farmers in Tibet and similar regions produce more nutritious and flavorful beef while maintaining animals well-adapted to challenging mountain environments.

Abstract

Yajiangxue cattle (XF) is three-way crossbred cattle developed specifically for producing high-quality beef in the Tibetan Plateau by introducing the bloods of Tibetan yellow cattle (HF) and Angus cattle into Tibetan yak (MF). In the present study, we mainly focused on fat deposition and metabolism changes and used RNA-seq and LC-MS/MS-based metabolomics to partially explain the meat quality improvement in Yajiangxue cattle. Differential expression analysis revealed 1762, 2949, and 2931 different expression genes in XF vs. HF, XF vs. MF, and XF vs. cattle–yak (PF), respectively, such as *BMP2*, *WISP2*, *FGF1*, *IL1B*, *IL6*, and *WNT5B*. Immune response, oxidation–reduction processes, and fatty acid metabolism were markedly enriched. Furthermore, an initial identification revealed 319 metabolites using positive ion mode and 289 metabolites using negative ion mode

 https://doi.org/10.3390/ani15172599

in bovine adipose tissue across four breeds/populations. Of these, 143 were differential metabolites in positive ion mode, while 166 were in negative ion mode. The main pathways of metabolism affected by breed/population were unsaturated fatty acid biosynthesis, tryptophan and tyrosine biosynthesis, primary bile acid biosynthesis, cholesterol metabolism, beta-alanine metabolism, etc. Similarly, both the transcriptome and the metabolome results highlighted fatty acid metabolism. These results could help elucidate the biological mechanisms involved in fat deposition and identify valuable biomarkers for specific metabolite accumulation.

Keywords: transcriptomics; metabolomics; LC-MS/MS; fat deposition; yak

1. Introduction

The Tibetan Plateau, a vast, elevated region often referred to as the 'Roof of the World', is the highest and most expansive plateau globally. Its average altitude of 5 km and area of roughly 3 million square kilometers have earned it the nickname the 'Third Pole' [1,2]. Yaks (Bos grunniens), a sacred religious symbol, evolved special physiological features for the harsh climatic environments of the Tibetan Plateau characterized by extreme cold, low oxygen levels, and intense ultraviolet radiation. It was essential for sustaining life on the plateau, providing resources like transportation, food, shelter, and fuel and intricately linked to the Tibetan nationality. Nowadays, yak husbandry is thriving and has greatly improved Tibetans' socio-cultural life.

Due to the rising demand for beef, approximately 14 million domestic yaks in China have become a valuable alternative beef source, providing a significant boost to the Tibetan economy. Improving yak meat production and quality is one of major goals of the yak industry for the next few decades, given customers' growing concerns about the health and nutritional benefits of their food [3]. Despite the success of pure-breeding in increasing meat production, improving meat quality traits is challenging due to their low-to-moderate heritability (0.14–0.33), including fat composition and fat content [4]. Crossbreeding accompanied with heterosis is an efficient strategy for generating desirable improvements in both of the two traits [5]. Heterosis means the performance of crossbred descendants is better than the expected average of their parents for a particular trait. Studies have shown that the specific breeds used in crossbreeding systems significantly impact the quality of yak meat [6]. To address the limitations of two-way crossbreeding systems, such as calving difficulties caused by large size disparities between breeds like Simmental or Holstein and yaks, which often lead to dystocia and increased mortality of dams and calves [7], we developed a novel three-way crossbreeding system to produce Yajiangxue cattle (XF). By using Tibetan yellow cattle (HF) as an intermediate parental breed, this system reduces dystocia risks due to HF's medium-sized stature, which facilitates easier calving compared to direct Angus–yak crosses. Furthermore, Angus was used as a terminal sire to enhance heterosis and integrate superior meat quality and rapid growth by breeding with half-yak (F1) females called cattle–yak (PF) from ♂Tibetan yellow cattle × ♀Tibetan yak (MF). Yajiangxue cattle exhibited rapid growth, high productivity, excellent meat quality, and exceptional adaptability to the plateau environment, giving them a comprehensive advantage over MF, HF, PF, and Angus cattle. Here, we mainly focused on fat deposition changes and used omics analyses of transcription and metabolization to partially explain the meat quality improvement in XF.

It is widely believed that fat deposition (composition and content) is a crucial indicator of meat quality. Meatier cattle were those that were small and blocky, representing selection for fat and not muscle. Japanese Wagyu has good quality, obvious marbling, and strong fat deposition capacity [8]. Fatty acids, the metabolites of fat, are important factors affecting the flavor and nutrition of beef [9]. Fatty acids are composed of monounsaturated fatty acids (MUFAs), saturated fatty acids (SFAs), polyunsaturated fatty acids (PUFAs), nutraceutical fatty acids (n-3, n-6 PUFAs), and functional unsaturated fatty acids (FUFAs). Yak kidney fat is rich in functional fatty acids and has a valuable composition. Specifically, fifteen fatty acids were identified in yak kidney fat, comprising 49.68% SFAs and 48.58% UFAs. This composition indicates the potential for developing yak fat into commercial products for the food industry [10]. A comparison of lipid profiles revealed that plateau PF and MF muscle contains higher levels of phospholipids enriched with long-chain polyunsaturated fatty acids than plain cattle muscle, suggesting superior lipid nutritional quality [11]. The lipid composition of yak shanks and flanks may also differ significantly, with the metabolic pathways of sphingolipids and glycerophospholipids playing key roles in driving these differences in lipid components [12]. The differences in fatty acid composition in yak meat under grazing and stall feeding regimes, as well as the regulatory mechanisms of the feeding system on fatty acid content in yak meat, have been explored [13,14]. Therefore, analyzing meat quality traits by comparing fat tissue across different cattle breeds holds great significance.

However, most of existing studies have focused on production traits, and the systematic comparison of fat deposition among yak and its crossbred descendants of different generations has yet to be conducted. Integrative omics analysis of transcription and metabolization has been demonstrated as an efficient method for uncovering the functions of unknown genes and establish gene–metabolite relationships, which outperform individual omics and the candidate gene method [15]. Transcriptome analysis estimates the gene mRNA expression profile through RNA-Seq. Metabolomics is a comprehensive approach that involves both the quantitative and qualitative analysis of low-molecular-weight metabolites. These metabolites, which are the final products of cellular processes, have the potential to directly affect the phenotype. A range of advanced metabolomic technologies is employed for the determination and analysis of small molecules, including 1H nuclear magnetic resonance (1H NMR), liquid chromatography–mass spectrometry (LC-MS), and gas chromatography–mass spectrometry (GC-MS) [16–20]. The combination of transcriptomics and non-targeted metabolomics has unveiled the mechanisms underlying follicular atresia in Chinese buffalo [21]. Using a similar method, another integrated analysis provided insights into the intricate molecular regulatory network underlying meat quality in Enshi black pigs [22]. Additionally, a metabolomic and transcriptomic study was conducted to investigate the alterations in metabolic and immune responses of steers exposed to heat stress [23]. Here, we integrated RNA-Seq and LC-MS/MS of fat tissues from yak and its crossbred descendants of different generations to find potential valuable biomarkers of beef quality, which may facilitate further molecular breeding of yak.

2. Materials and Methods

2.1. Ethics Statement

The protocols and animals used in this study were approved by the Faculty Animal Policy and Welfare Committee of Northwest A&F University, NWAFAC1008.

2.2. Animals and Data Collection

MF (Niangya yak breed) were crossbred with HF to produce PF hybrids (1/2 yak). Subsequently, Angus cattle were used as terminal sires to artificially inseminate PF, resulting in the creation of XF cattle (1/4 yak). During the summer and autumn, these animals were grazed on alpine pastures in Tibet. In spring and winter, they were housed in barns with ad libitum access to feed and water, following the NY/T 815-2004 standard [24]. The diet was formulated with a concentrate-to-roughage ratio of 35:65. Roughages included alfalfa hay (50%) and wheat straw (15%), while concentrates consisted of corn flour (14%), wheat bran (6.83%), soybean meal (4.9%), cottonseed meal (3.85%), fermented distiller's grains (3.5%), NaCl (0.17%), and a premix (1.75%). The diet provided a net energy for maintenance and fattening of 7.04 MJ/kg, with 11.42% crude protein, 0.68% total calcium, 0.39% total phosphorus, 27.96% neutral detergent fiber, and 15.02% acid detergent fiber, ensuring adequate nutrition for growth, fat deposition, and adaptation to the Tibetan Plateau's high-altitude environment. Animals were randomly selected at 25 months of age, housed in pens, and serially slaughtered at 36 months of age for experimental purposes and subsequent meat quality analysis. A total of 23 fasted animals were randomly selected for quality analysis, including MF (n = 6, 3 males and 3 females), HF (n = 5, 2 males and 3 females), PF (n = 6, all females), and XF (n = 6, 3 males and 3 females). The animals were fasted for 24 h prior to slaughter, and the carcasses were subsequently chilled for 24 h at 4 °C. After slaughter, data on head weight, front hoof weight, back hoof weight, tare weight, liver weight, heart weight, lung weight, spleen weight, kidney weight, stomach weight, visceral fat, oxtail, bullwhip, small intestine, large intestine, intestinal fat, marbling score, and fat color were collected for analysis. Subcutaneous fat samples were collected from the backfat over the longissimus dorsi muscle (between the 12th and 13th ribs), a standard site for assessing fat deposition in beef cattle. These samples were collected immediately after slaughter, within 10 min, and were then rapidly snap-frozen in liquid nitrogen to preserve RNA integrity. Both the marbling score and fat color were evaluated according to the Chinese national standard GB/T 29392-2022 [25]. Marbling score was assessed using a 5-point scale, where 1 indicates scarce marbling and 5 indicates abundant marbling. Fat color was evaluated using an 8-point scale, where 1 represents white and 8 represents yellow.

2.3. RNA Extraction and Transcriptome Analysis

Total RNA was extracted from subcutaneous fat samples using Trizol reagent (Invitrogen, Carlsbad, CA, USA) according to the manufacturer's guidelines. RNA integrity was assessed using an Agilent 2100 Bioanalyzer (Agilent Technologies, Santa Clara, CA, USA) and agarose gel electrophoresis, while RNA purity was measured with a NanoPhotometer spectrophotometer (Thermo Scientific, Waltham, MA, USA). Trimmomatic v0.32 was used to remove sequence adapters and low-quality reads, followed by FastQC v0.10.1 quality assessment. The sequencing reads were then aligned to the ARS-UCD1.2 cattle reference genome using HISAT2 v2.2.4. Differential gene expression analysis was conducted with DESeq2 v1.24.0, and genes with a false discovery rate (FDR) value less than 0.05 were considered differentially expressed genes (DEGs). To perform gene set enrichment analysis (GSEA), all the genes were ranked based on fold-change and significance derived from the differential analysis [26]. GSEA was conducted using mSigDB C2 gene sets and curated gene sets.

2.4. Non-Targeted LC-MS/MS Metabolomics Analysis

The pretreatment of LC-MS/MS samples was carried out according to a previous report [27]. LC-MS/MS analysis (Novogene Co., Ltd., Beijing, China) was performed using a UHPLC system equipped with a Hypesil Gold column (2.1 mm × 100 mm, 1.9 μm) coupled to a Q Exactive HF-X mass spectrometer. The mobile phase consisted of water containing 0.1% formic acid (for positive ion mode) and 5 mmol/L ammonium acetate (for negative ion mode) (A) and methanol (B), with a flow rate of 0.2 mL/min. The elution gradient was 2% B for 1.5 min; 2% to 100% B over 12.0 min; 100% B for 14.0 min; 100% to 2% B over 14.1 min; and 2% B for 17 min. The Exactive HF-X mass spectrometer was operated in both positive and negative polarity modes with parameters set as 10 arb (auxiliary gas), 40 arb (sheath gas), 3.2 kV (spray voltage), and 320 °C (capillary temperature).

Compound Discoverer 3.1 (Thermo Fisher Scientific, Waltham, MA, USA) was used for the alignment and picking of peak and metabolite quantitation with raw data to obtain qualitative and quantitative results of metabolites. Based on additive ions, fragment ions, and molecular ion peaks, the total spectral intensity obtained by normalizing the peak intensities was employed to calculate the molecular formula. Peaks were subsequently compared with the MassList, mzVault, and mzCloud databases to achieve precise qualitative and relative quantitative results. Following identification, metabolites were functionally annotated using the online databases HMDB, KEGG, and Lipid Maps. Multivariate statistical analyses, including principal component analysis (PCA) and partial least squares discriminant analysis (PLS-DA), were performed with MetaX software (version 1.0.4) to evaluate the processed data [28]. With R2 and Q2 values used to assess the goodness of fit, we validated PLS-DA models through permutation tests. To investigate the changes in metabolite composition of bovine fat caused by different breeds/populations, various comparisons between XF and HF, XF and MF, XF and PF were conducted, respectively. Metabolites were classified as differential if they had a variable importance in projection (VIP) > 1, a t-test p-value < 0.05, and a fold change (FC) ≥ 1.2 or FC ≤ 0.833. The relative metabolite levels were determined using Z-score plots, and the trends in metabolite changes across the four breeds/populations were analyzed. The KEGG database was adopted to predict the functions of these metabolites.

2.5. Integrated Analysis of the Transcriptome and Metabolome

To assess the relationship between DEGs and differential metabolites (DMs), we calculated pairwise Pearson correlation coefficients using R (Version 4.3.2). All identified DEGs and DMs were mapped to the KEGG pathway database (https://www.genome.jp/kegg/pathway.html, accessed on 20 June 2022) to find the shared pathways information using Microsoft Excel, and the main biochemical pathways and signal transduction pathways in which the DEGs and DMs were determined.

2.6. Statistical Analyses

One-way ANOVA was used to compare the performances of economic traits of yak and its crossbred descendants of different generations. After conducting the one-way ANOVA, we used least significant difference (LSD) post hoc test for multiple comparisons among the four groups. * $p < 0.05$, ** $p < 0.01$, *** $p < 0.001$.

3. Results

3.1. XF Possess Enhanced Growth Characteristics and Meat Quality

We analyzed the growth traits and meat quality of Niangya yak and its crossbred descendants of different generations (Figure 1). Notably, a direct comparison with Angus cattle was not feasible due to the use of artificial insemination with Angus semen, which prevented Angus samples under identical environmental and management conditions from being obtained. The breeds significantly affected head weight, front hoof weight, back hoof weight, tare weight, liver weight, heart weight, lung weight, spleen weight, kidney weight, stomach weight, visceral fat, oxtail, bullwhip, small intestine, large intestine, and intestinal fat ($p < 0.01$). These growth traits were much higher in XF than in HF, MF, and PF ($p < 0.01$) (Table 1). Additionally, we compared the marbling score and fat color among the four breeds, and the marbling score of XF was found to be higher than other breeds on average (Table 1). These data suggest that XF developed through a three-way cross system on the Tibetan Plateau have high yield, rapid growth, and superior meat quality.

Figure 1. Three-way cross system for producing Yajiangxue cattle with high meat production and quality.

Table 1. Growth traits and meat quality of yak and its crossbred descendants.

Economic Traits	Breeds/Populations (LSM ± SE)				p
	MF (n = 6)	HF (n = 5)	XF (n = 6)	PF (n = 6)	
Head weight	27.91 ± 1.10 ^{Bc}	19.74 ± 1.32 ^{Cd}	44.44 ± 1.23 ^{Aa}	32.51 ± 2.13 ^{Bb}	0.000
Front hoof weight	4.53 ± 0.26 ^{Bb}	2.76 ± 0.31 ^{Cc}	8.07 ± 0.29 ^{Aa}	5.24 ± 0.50 ^{Bb}	0.000
Back hoof weight	4.53 ± 0.28 ^{Bb}	2.82 ± 0.34 ^{Cc}	8.06 ± 0.32 ^{Aa}	5.00 ± 0.55 ^{BCb}	0.000
Tare weight	31.46 ± 1.85 ^{Bc}	20.22 ± 2.22 ^{Cd}	56.27 ± 2.06 ^{Aa}	38.84 ± 3.58 ^{Bb}	0.000
Heart weight	1.74 ± 0.13 ^{Bb}	1.06 ± 0.15 ^{Cc}	4.93 ± 0.14 ^{Aa}	2.27 ± 0.25 ^{Bb}	0.000
Liver weight	5.74 ± 0.50 ^{Bb}	3.92 ± 0.60 ^{Bc}	11.57 ± 0.56 ^{Aa}	7.22 ± 0.97 ^{Bb}	0.000
Lung weight	4.89 ± 0.48 ^{Bb}	2.45 ± 0.58 ^{Cc}	7.53 ± 0.54 ^{Aa}	5.13 ± 0.94 ^{ABCb}	0.000
Spleen weight	0.58 ± 0.08 ^{Bc}	0.36 ± 0.10 ^{Bc}	1.69 ± 0.09 ^{Aa}	1.20 ± 0.16 ^{Ab}	0.000
Kidney weight	0.86 ± 0.07 ^{Bb}	0.61 ± 0.09 ^{Bc}	1.52 ± 0.08 ^{Aa}	1.01 ± 0.14 ^{ABbc}	0.000
Stomach weight	15.34 ± 1.14 ^{Bb}	12.37 ± 1.37 ^{Bb}	32.00 ± 1.27 ^{Aa}	17.39 ± 2.21 ^{Bb}	0.000
Visceral fat	na	na	53.94 ± 4.27 ^{Aa}	11.82 ± 4.57 ^{Bb}	0.001
Oxtail	0.85 ± 0.43 ^{Bb}	0.69 ± 0.52 ^{Bb}	4.31 ± 0.48 ^{Aa}	1.46 ± 0.84 ^{ABb}	0.000
Bullwhip	0.99 ± 2.03 ^{b}	1.82 ± 2.32 ^{b}	10.67 ± 1.64 ^{a}	na	0.027
Small intestine	5.68 ± 0.57 ^{Bb}	3.59 ± 0.68 ^{Bc}	8.27 ± 0.64 ^{Aa}	5.39 ± 1.10 ^{ABabc}	0.000
Large intestine	6.61 ± 1.66 ^{Bb}	4.61 ± 1.99 ^{Bb}	15.96 ± 1.85 ^{Aa}	5.60 ± 3.21 ^{ABb}	0.000
Intestinal fat	3.76 ± 1.97 ^{Bb}	3.14 ± 2.37 ^{Bb}	27.42 ± 2.20 ^{Aa}	13.34 ± 3.82 ^{ABb}	0.000
Marbling score	0.00 ± 0.00 ^{Bb}	1.00 ± 0.00 ^{Bb}	3.17 ± 0.60 ^{Aa}	1.78 ± 0.31 ^{Aa}	0.000
Fat color	7.00 ± 0.00 ^{Aa}	6.00 ± 0.00 ^{Bb}	2.00 ± 0.00 ^{Cc}	6.33 ± 0.21 ^{Bb}	0.000

Note: Different letters in the same row indicate significant differences among groups (uppercase letters for $p < 0.01$, lowercase letters for $p < 0.05$). LSM, least square means; SE, standard error; MF, yak; HF, Tibetan yellow cattle; XF, Yajiangxue cattle; PF, cattle–yak; na, not available.

3.2. Identification of Differentially Expressed Genes (DEGs)

RNA-Seq was performed to assess the subcutaneous fat of the four breeds. The raw reads per sample ranged from 38,050,738 to 48,497,648, with an average of 45,116,929 reads. The average clean reads per sample was 44,341,967. The average Q20 and Q30 percentages were 96.88% and 91.85%, respectively, indicating high sequencing quality. The average unique mapping rate was 82.76%, indicating a high quality of alignment. Detailed information is given in Table S1. The expression abundance of all genes is listed in Table S2. We found 1762, 2949, and 2931 DEGs in XF vs. HF, XF vs. MF, and XF vs. PF, respectively, as shown in volcano plots (Figure S1). More specifically, there were 937 upregulated and

825 downregulated genes in XF vs. HF, 1594 upregulated and 1355 downregulated genes in XF vs. MF, and 1587 upregulated genes and 1344 downregulated genes in XF vs. PF. These DEGs had clearly different expression patterns, as visualized in the heatmap shown in Figure S1. Among these DEGs, there were 507 exclusively expressed genes in XF, 397 genes in HF, 480 genes in MF, and 350 genes in PF (Figure 2A). The lipid-related DEGs in XF are significantly higher than in the HF, MF, and PF groups, such as *HADH*, and XF is significantly higher than HF and PF for genes such as *ESR1* and *APOL3*, which can promote growth and fat deposition.

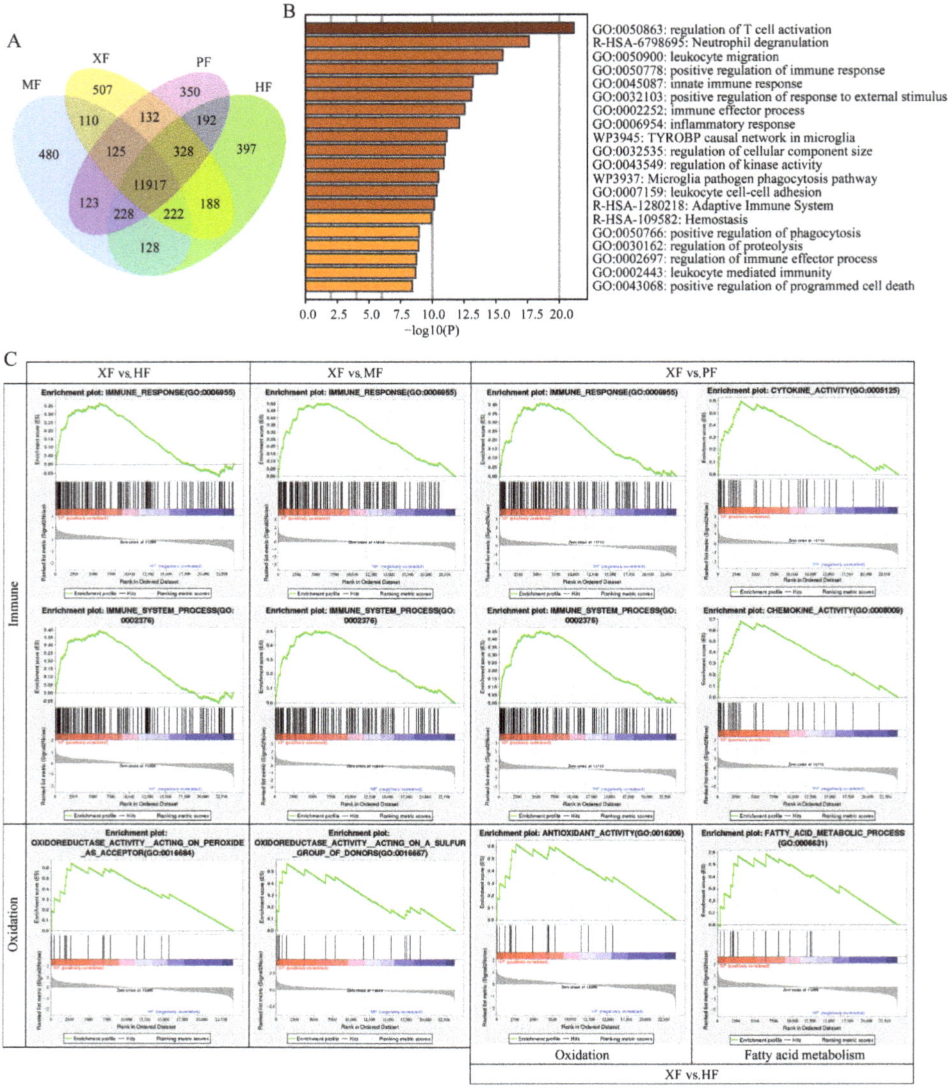

Figure 2. Gene annotation analyses. (**A**) Venn diagram of expressed genes between all breed samples. (**B**) GO and KEGG of 360 XF-specific highly expressed genes by Metascape. (**C**) GSEA analysis.

3.3. DEGs Are Enriched in Immune, Oxidation, and Fat Metabolism

First, gene ontology (GO) analysis was performed for functional annotation of DEGs, including molecular function (MF), cellular component (CC), and biological process (BP).

For the XF vs. HF group ($p < 0.05$; Table S3), the DEGs were grouped into 55 significantly different subcategories, encompassing 21 MF terms, 6 CC terms, and 28 BP terms. For the MF category, 2.5% of the genes (20 of 788 genes) were located in enzyme binding (GO:0005488). For the CC category, 11.6% of the genes (31 of 268 genes) were annotated to the extracellular region (GO:0044421). For the BP category, 4.5% (22/489), 4.3% (21/489 genes), and 8.6% (42/489) of the genes were involved in the immune response (GO:0006955), immune system process (GO:0002376), and oxidation–reduction process (GO:0055114), respectively.

For the XF vs. MF group ($p < 0.05$; Table S4), the DEGs were grouped into 37 significantly different subcategories, including 14 MF terms, 5 CC terms, and 18 BP terms. For the MF category, 6.5% of the genes (86 of 1320 genes) were located in oxidoreductase activity (GO:0016491). For the CC category, 11.16% of the genes (52 of 466 genes) were annotated into the extracellular region (GO:0005576). For the BP category, 3.9% (32/823) and 3.8% (31/823) of the genes were involved in the immune system process (GO:0002376) and immune response (GO:0006955), respectively. For the XF vs. PF group ($p < 0.05$; Table S5), the DEGs were grouped into 79 significantly different subcategories, including 33 MF terms, 21 CC terms, and 25 BP terms. For the MF category, 4.2% (55/1310) of the genes were located in the structural constituent of the ribosome (GO:0003735). For the CC category, 21.4% (98/459) of the genes were annotated into the mitochondrion (GO:0005739). For the BP category, 7.7% (63/813) and 7.9% (64/813) of the genes were involved in the peptide biosynthetic process (GO:0043043) and peptide metabolic process (GO:0006518), respectively.

To further explore the potential biological significance of phenotype divergence, the identified DEGs were mapped to various pathways using KEGG bioinformatics databases. For XF vs. HF group ($p < 0.01$; Table S6), the enrichment analysis revealed that the top enriched terms were primarily related to immune-related pathways, such as phagosome, Th1/Th2/Th17 cell differentiation, NF-kappa B signaling pathway, human T-cell leukemia virus 1 infection, and hematopoietic cell lineage, as well as fat-related metabolism, such as type I diabetes mellitus and the thyroid hormone signaling pathway. For XF vs. MF group ($p < 0.01$; Table S7), enrichment analysis revealed that the top enriched terms were primarily related to immune-related pathways, such as phagosome, hematopoietic cell lineage, NF-kappa B signaling pathway, leukocyte transendothelial migration, and B cell receptor signaling pathway, as well as fat-related metabolism, such as the PPAR signaling pathway. For XF vs. PF group ($p < 0.01$; Table S8), enrichment analysis revealed that the top enriched terms were primarily related to immune-related pathways, such as phagosome and Th1/Th2/Th17 cell differentiation, as well as fat-related metabolism, such as FA metabolism, UFA biosynthesis, FA elongation, FA degradation, and non-alcoholic fatty liver disease. Interestingly, 360 genes in XF were expressed at significantly higher levels than in HF, MF, and PF. These genes, including *BMP2*, *WISP2*, *FGF1*, *IL1B*, *IL6*, *WNT5B*, *TNFSF14*, *FST*, *CCBE1*, and *HSPB1*, were mainly involved in immune-related pathways, such as regulation of T cell activation, neutrophil degranulation, and leukocyte migration ($p < 0.01$; Table S9, Figure 2B).

GSEA suggested that, compared with HF, XF showed upregulated genes enriched in oxidoreductase activity acting on peroxide as an acceptor (GO:0016684), antioxidant activity (GO:0016209), immune system process (GO:0002376), immune response (GO:0006955), and FA metabolic process (GO:0006631). Compared with MF, XF showed upregulated genes enriched in oxidoreductase activity acting on a sulfur group of donors (GO:0016667), purine nucleoside binding (GO:0001883), purine ribonucleoside binding (GO:0032550), immune system process (GO:0002376), and immune response (GO:0006955). Compared with PF, XF showed upregulated genes enriched in cytokine activity (GO:0005125), chemokine activity (GO:0008009), G protein coupled receptor binding (GO:0001664), immune response

(GO:0006955), and immune system process (GO:0002376) ($p < 0.05$; Figure 2C). Overall, the immune response, oxidation–reduction, and fatty acid metabolism were markedly enriched in XF compared with HF, MF, and PF, showing that XF have good meat quality, resistance to oxidation, and strong plateau adaptability.

3.4. Identification of Differential Metabolites

To ensure the stability and quality of our metabolomics data, we used three quality control samples, which were analyzed as biological replicates. The stability of the entire detection process was assessed by calculating the Pearson correlation coefficient between the relative quantitative values of the metabolites in the quality control samples. Highly repeatable and reliable data were produced, given the observation that the Pearson correlation coefficient among quality control samples ranged from 0.991 to 0.995. Subsequently, PLS-DA was employed as a supervised pattern recognition method for multivariate statistical analysis. The negative ion mode clearly differentiates between any two breeds/populations (Figure 3), and the positive ion mode is presented in Figure S2. For the cross-validation, Q2Y indicates the predictive ability of models, while R2Y indicates the interpretation ability of the models. The R2Y values exceeded the Q2Y values, indicating that the models exhibited good cumulative interpretation and predictive abilities. To prevent model overfitting, subsequent permutation tests were conducted (Figures 3 and S2). The R2 values (blue line) consistently exceeded the Q2 values (red line), suggesting a low risk of overfitting in these models.

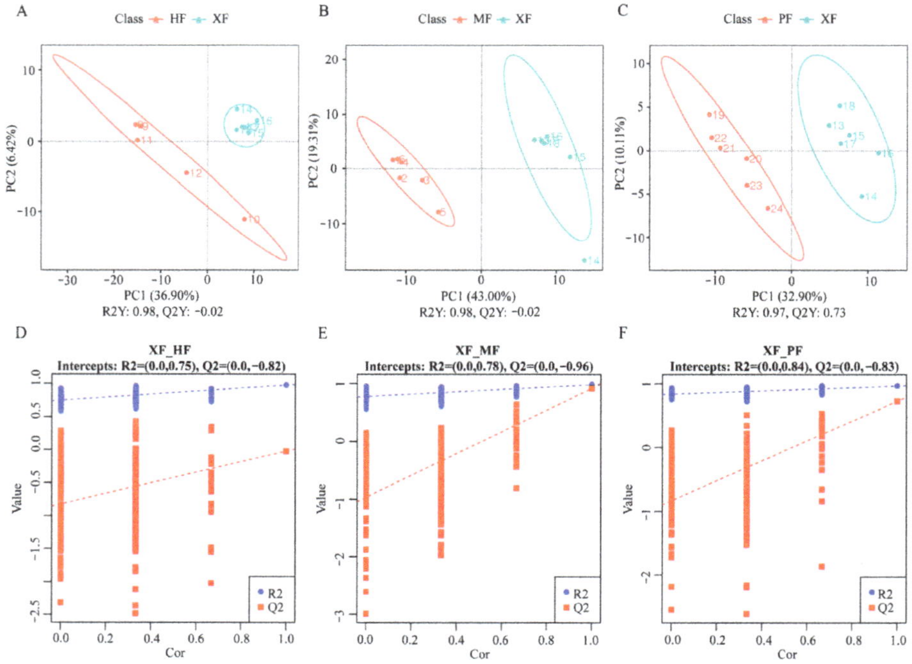

Figure 3. The PLS-DA (**A–C**) and its models validated by permutation tests (**D–F**) between different breed groups in negative ion mode. R2Y = 0.98, Q2Y = −0.02 in XF vs. HF; R2Y = 0.98, Q2Y = 0.92 in XF vs. MF; R2Y = 0.97, Q2Y = 0.73 in XF vs. PF.

A total of 319 and 289 metabolites were initially identified in positive and negative ion modes, respectively. After filtering out metabolites with no significant changes, 143 and 166 DMs remained in positive and negative ion modes, respectively. Results showed that

8, 20, and 9 metabolites were significantly upregulated and 109, 175, and 145 metabolites were downregulated in the comparison of XF vs. HF, XF vs. MF, XF vs. PF, respectively (Figure 4, Table S10).

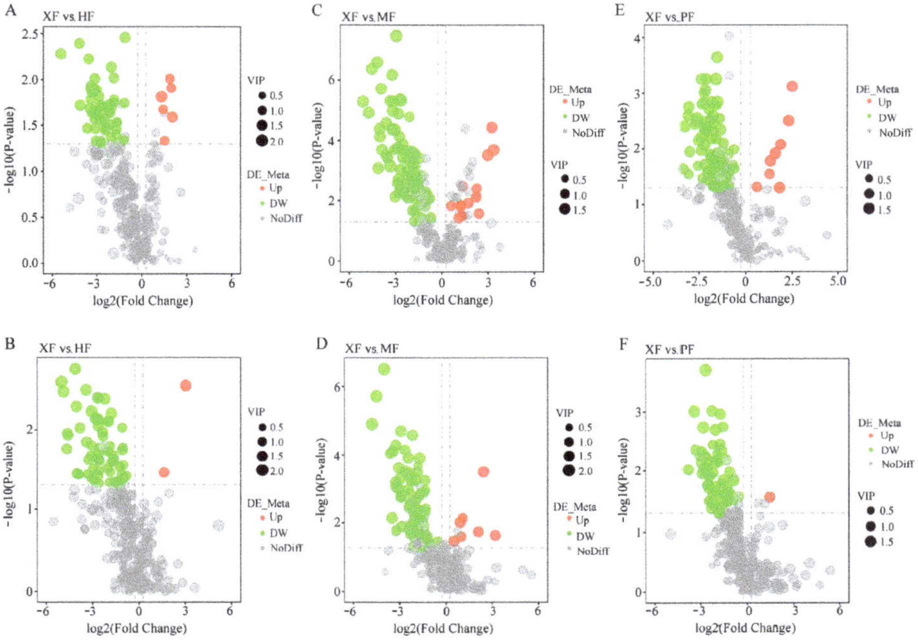

Figure 4. Analysis of metabolome profile diversity between different breeds. (**A**) XF vs. HF in neg model; (**B**) XF vs. HF in pos model; (**C**) XF vs. MF in neg model; (**D**) XF vs. MF in pos model; (**E**) XF vs. PF in neg model; (**F**) XF vs. PF in neg model. Each dot corresponds to a metabolite, with red dots indicating significantly upregulated metabolites and green dots indicating significantly downregulated metabolites. Gray dots represent metabolites with no significant differential expression. The size of each dot reflects the VIP numeric value.

Lipid-related metabolites play an essential role in meat quality. In XF vs. HF (Table S11), two differential fatty acids and conjugate metabolites including 16-hydroxyhexadecanoic acid (HMDB0006294) and trans-10-heptadecenoic acid (HMDB0244268) experienced an uptrend. Both positive and negative ions also detected oxidized glutathione (HMDB0003337,). Vitamin B2 (HMDB0000244), taurine (HMDB0000251), and DL-lanthionine (HMDB0251521) were also detected. In XF vs. MF (Table S11), two differential fatty acids and conjugate metabolites including 16-hydroxyhexadecanoic acid (HMDB0006294) and palmitic acid (HMDB0000220) were detected. Seven organic acids and derivative metabolites including taurine (HMDB0000251), oxidized glutathione (HMDB0003337), Asp-Glu (HMDB0028752), cyclohexaneacetic acid (HMDB0031403), O-phosphorylethanolamine (HMDB0000224), N-acetyl-L-leucine (HMDB0011756), and DL-lanthionine (HMDB0251521pos) were identified. Two glycerophospholipids, LPC 20:3 (LMGP01050139) and PE (18:0/13:1) (LMGP02010301), were also detected. In XF vs. PF (Table S11), the differential fatty acid and conjugate metabolite is trans-10-heptadecenoic acid (HMDB0244268), and the differential pregnane steroid metabolite is pregnenolone (HMDB0000253,). Two differential glycerophosphocholines, LPC 20:3 (LMGP01050139) and PC (18:0/18:1) (LMGP01010761), were identified. 13(S)-HOTrE (cpd:C16316, map00592, alpha-Linolenic acid metabolism) was detected in alpha-linolenic acid metabolism.

Notably, several metabolites of fatty acid and umami amino acids influencing beef quality and flavor were significantly altered in XF (Figure 5). 16-hydroxyhexadecanoic acid in XF was more significantly increased than in HF and MF, trans-10-heptadecenoic acid in XF was more significantly increased than in HF and PF, and palmitic acid in XF was more significantly increased than in MF. The differential steroid and steroid derivative metabolite pregnenolone experienced an upward trend in the breed comparison of XF vs. PF, while glycolithocholic acid, glycocholic acid, and palmitoyl sphingomyelin decreased in XF. The above findings suggest that XF have good meat quality and are rich in nutrition and flavor.

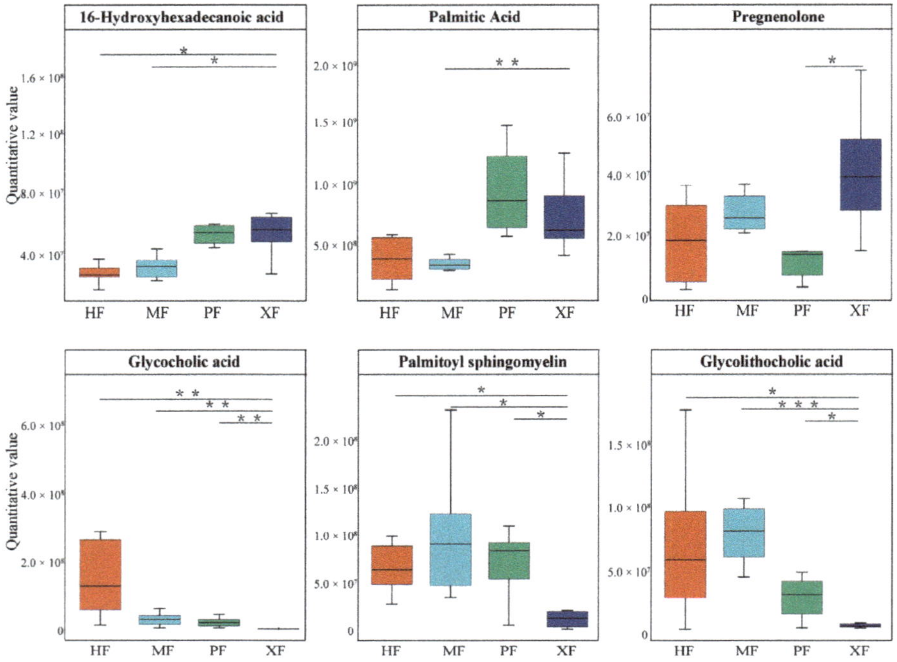

Figure 5. The relative peak intensities of pivotal metabolites affecting meat flavor and quality across four breeds/populations. * $p < 0.05$, ** $p < 0.01$, *** $p < 0.001$.

3.5. DMs Are Enriched in Fatty Acid Biosynthesis and Hormone Metabolism

The top pathways of differential metabolites of XF vs. HF, XF vs. MF, and XF vs. PF are outlined in Figure S3. DMs identified in the XF and HF were most significantly enriched in the oxytocin signaling pathway and caffeine metabolism (Figure S3A,B). UFA biosynthesis was the most significant pathway detected in the breed/population comparison of XF and MF (Figure S3C,D). Antifolate resistance, the AMPK signaling pathway, the longevity regulating pathway, pyrimidine metabolism, and the sphingolipid signaling pathway were significant pathways detected in the comparison of XF and PF (Figure S3E,F). These findings support the roles of FAs and hormones in yak growth and meat quality.

3.6. Integrated Analysis Reveals Promising Candidates Associated with Meat Quality

To elucidate the correlation between DEGs and DMs, we conducted an integrated analysis of metabolome and transcriptome profiles. The complete results are presented in Tables S11–S17, and the top integration KEGG pathways are visualized in Figure 6. We extracted integrated KEGG pathways associated with meat quality or flavor, and the significant correlations are summarized in Table 2. In detail, in the comparison between XF and HF ($p < 0.05$; Figure 7A, Tables S12 and S13), four differential metabolites—oxidized glutathione, arachidonic acid, prostaglandin E2, and 5′AMP—involved in FA metabolism, were significantly associated with several key genes, including *LAP3*, *GPX1*, *PGD*, *SAT1*, *ACSL4*, and *IDH2*. Interestingly, these metabolites and genes are all involved in glutamate metabolism, biosynthesis of UFAs, arachidonic acid metabolism, taurine and hypotaurine metabolism, cysteine and methionine metabolism, and linoleic acid metabolism, indicating their critical role in fat formation during the breeding process.

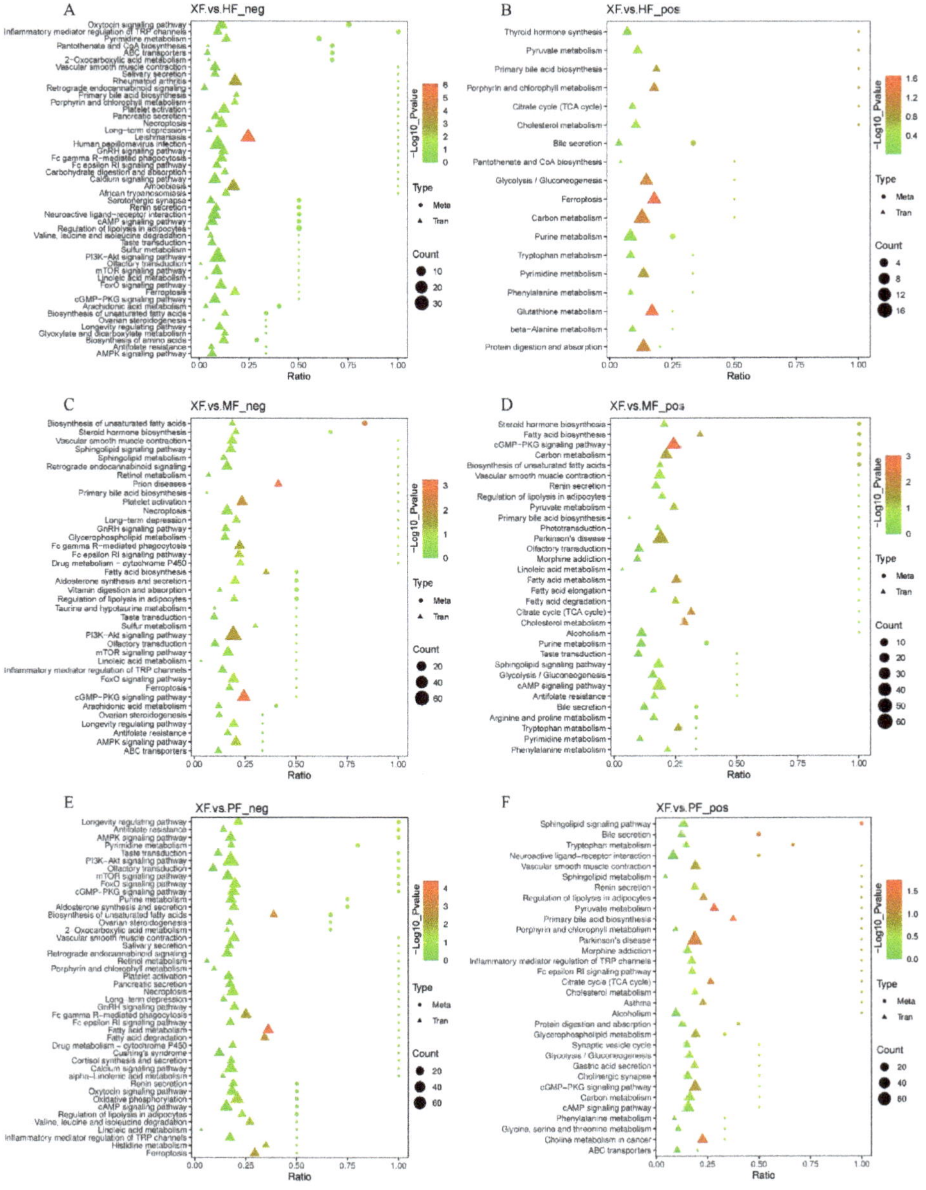

Figure 6. Integrated KEGG analysis of metabolome and transcriptome profiles across different breed groups. (**A**) XF vs. HF_neg; (**B**) XF vs. HF_pos; (**C**) XF vs. MF_neg; (**D**) XF vs. MF_pos; (**E**) XF vs. PF_neg; (**F**) XF vs. PF_pos.

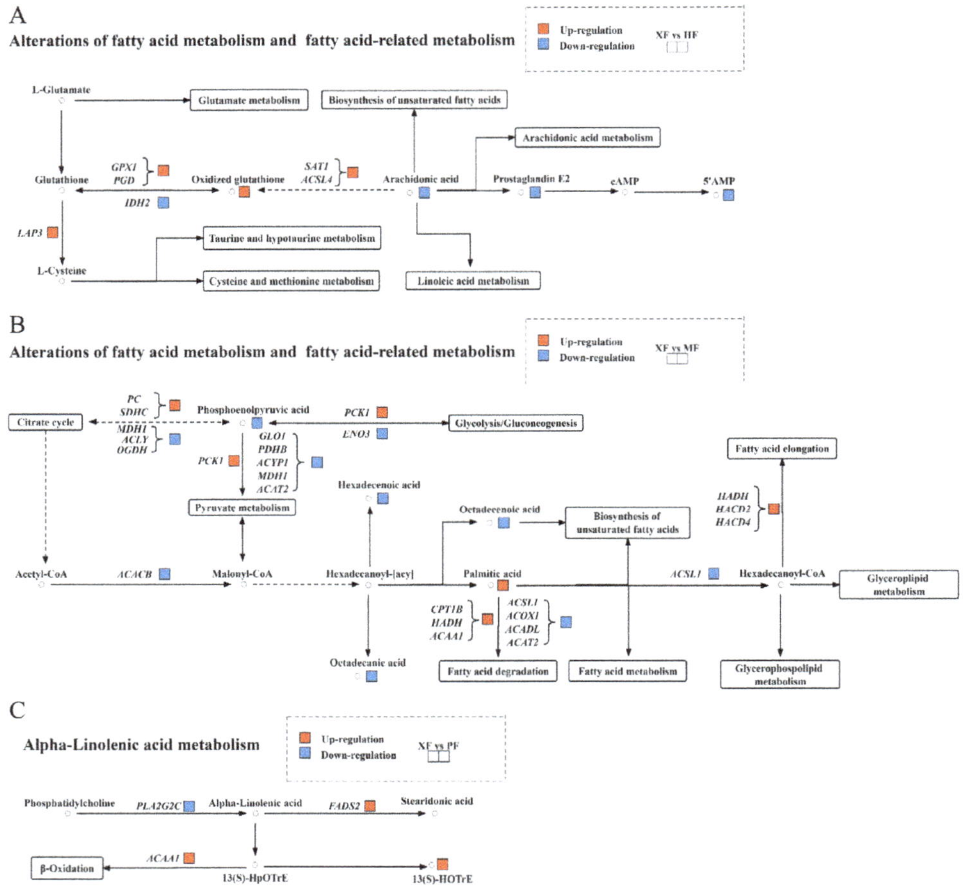

Figure 7. Alterations in fatty acid metabolism in different breed groups. (**A**) Schematic representation of the genes and metabolites involved in fatty acid metabolism (XF vs. HF). (**B**) Schematic representation of the genes and metabolites involved in fatty acid metabolism (XF vs. MF). (**C**) Schematic representation of the genes and metabolites involved in alpha-linolenic acid metabolism (XF vs. PF). Blue indicates significant downregulation. Red indicates significant upregulation. Genes are italicized to differentiate them from metabolites.

Table 2. Significant correlations between metabolite profiles and gene expressions potentially associated with meat quality.

Metabolites	Groups	Trend	KEGG Pathway	Significantly Related Genes ($p < 0.05$)
oxidized glutathione	XF vs. HF	up	ferroptosis	*CP* (gene id 514194), *FTL* (286861), *FTH1* (281173), *ACSL4* (536628), *CYBB* (281112), *LOC788801* (788801), *TF* (280705), *novel.1303*, *ATG* (5532686), *SAT1* (508861)
oxidized glutathione	XF vs. HF	up	glutathione metabolism	*RRM1* (505537), *GPX1* (281209), *LAP3* (781648), *GSTA1* (777644), *PGD* (514939), *CHAC1* (505991), *IDH2* (327669), *LOC100295687* (100295687), *RRM2B* (528960), *ODC1* (281365), and *GSTA3* (768055)
palmitic acid	XF vs. MF	up	fatty acid biosynthesis	*HSD17B8* (532422), *ACACB* (515338), *CBR4* (533020), *RPP14* (515208), *ACSF3* (509209), and *ACSL1* (537161)

Table 2. *Cont.*

Metabolites	Groups	Trend	KEGG Pathway	Significantly Related Genes ($p < 0.05$)
palmitic acid	XF vs. MF	up	fatty acid metabolism	*FADS2* (521822), *ACOX1* (513996), *HACD4* (618814), *ACAT2* (512044), *HSD17B8* (532422), *ACADL* (614508), *CBR4* (533020), *HADH* (532785), *CPT1B* (509459), *RPP14* (515208), *HACD2* (613886), *LOC613570* (613570), *ACSF3* (509209), *ELOVL1* (540348), *ACSL1* (537161), and *ACAA1* (508324)
sedoheptulose 1,7-bisphosphate	XF vs. MF	up	carbon metabolism	*LOC101902656* (101902656), *HK3* (510616), *PSPH* (533630), *PDHB* (613610), *ACOX1* (513996), *ENO3* (540303), *ACAT2* (512044), *PRPS2* (537688), *HIBCH* (535883), *MDH2* (281306), *HK2* (788926), *PSAT1* (533044), *ALDH6A1* (327692), *IDH3B* (613338), *MDH1* (535182), *LOC788293* (788293), *PCCA* (614302), *GPI* (280808), *GOT2* (286886), *GLYCTK* (507949), *OGDH* (534599), *LOC614208* (614208), *FBP1* (513483), *PHGDH* (505103), *IDH3G* (614145), *SDHC* (327696), *LOC616200* (616200), *GCSH* (317723), and *PC* (338471)
13(S)-HOTrE	XF vs. PF	up	alpha-linolenic acid metabolism	*ACAA1* (508324), *PLA2G2C* (504978), and *FADS2* (521822)

In the comparison between XF and MF ($p < 0.05$; Figure 7B, Tables S14 and S15), five differential metabolites responsible for fatty acid metabolism—phosphoenolpyruvic acid, hexadecenoic acid, octadecanic acid, octadecenoic acid, and palmitic acid—were significantly correlated with several key genes, including *PC, SDHC, MDH1, ACLY, OGDH, PCK1, ENO3, GLO1, PDHB, ACYP1, ACAT2, ACACB, CPT1B, HADH, ACAA1, ACSL1, ACOX1, ACADL, HADH, HACD2*, and *HACD4*. Interestingly, these metabolites and genes are all involved in fatty acid metabolism, indicating their critical role in fat formation during the breeding process. Meanwhile, these results imply that citrate cycle, glycolysis/gluconeogenesis, pyruvate metabolism, FA degradation, FA metabolism, biosynthesis of UFAs, FA elongation, glycerophospolipid metabolism, and glyceroplipid metabolism might affect fat formation in XF compared with MF, XF, and PF.

In the comparison between XM and PM ($p < 0.05$; Figure 7C, Tables S16 and S17), one differential metabolite responsible for alpha-linolenic acid metabolism, 13(S)-HOTrE, was significantly correlated with several key genes, including *PLA2G2C, FADS2*, and *ACAA1*. This differential metabolite was responsible for β-oxidation. These results imply that alpha-linolenic acid metabolism might affect fat formation in XF compared with PF.

4. Discussion

As efficient utilizers of highland herbage resources, yaks are vital to the Tibetan Plateau's livestock industry. The yak industry has flourished in the past decade, significantly boosting the regional economy. However, ensuring consumer satisfaction by evaluating the quality of yaks and their hybrids is a pressing concern. Given the relatively low heritability of fat deposition, introducing exogenous blood with favorable traits may be a promising approach from a genetic and breeding perspective.

In this study, a three-way cross system was created with Angus cattle, a globally renowned beef breed known for its exceptional marbled appearance. By introducing HF as the parental F0, this three-way cross system aims to mitigate the calving difficulties associated with two-way crossbreeding between Angus and yak. The native breed, with a medium-sized stature compared to Angus and yak, provides an advantage for the crossbred offspring in handling the challenging foraging conditions of the Qinghai–Tibetan

Plateau [29]. All in all, the excellent reproductive performance of Angus and HF makes them well-suited for yak hybridization on the Tibetan Plateau. This is the first three-way cross system designed for high-quality beef production on the Tibetan Plateau. Our results demonstrate that this system effectively combines optimal fat deposition with alpine adaptability, rapid growth, high yield, and excellent meat quality. Our results showed that XF individuals generally grow faster and reach larger sizes than yaks, particularly in later stages of development. The introduction of Angus into the breeding program prevented inbreeding depression or heterosis loss commonly observed in backcrosses to yaks or local cattle breeds. As a result, XF populations exhibited enhanced growth performance from birth compared to PF individuals.

Among these desirable traits, meat quality is crucial for economic success in the beef cattle industry. Numerous factors influence meat quality, including tenderness, hardness, color, flavor, muscle fiber characteristics, oxidative stability, and fat composition and content [4]. Meat quality is a complex trait influenced by various physicochemical characteristics, including tenderness, pH, color, FAs composition, intramuscular fat content, and sensory attributes [30]. It is widely believed that fat deposition (composition and content) is a crucial indicator of meat quality. Adipose-derived secretory factors, such as lipokines and adipokines, play a role in regulating skeletal muscle development and maintaining homeostasis and may have potential in human health and husbandry production [31]. Therefore, analyzing fat tissue from different cattle breeds is crucial for understanding and improving meat quality traits.

Extensive research has demonstrated that animal fat deposition is a complex biological process influenced by various transcriptional factors, including SREBP-1c, STAT5, PPARγ, KLF family, and C/EBP family, which have been shown to play significant roles in promoting adipogenesis [32]. Differences in gene expression related to fat deposition have been extensively characterized using RNA-Seq. In this research, lipid-related DEGs such as *HADH* were found to be significantly higher in XF than in HF, MF, and PF, with *ESR1* and *APOL3* being significantly higher in XF than in HF and PF, which can promote growth and fat deposition. HADH protein affects lipid deposition in pigs, as revealed through iTRAQ-based proteomic analysis [33]. *ESR1*, related to fat deposition, was significant higher in fat tail than small tail tissues [34]. *APOL3*, which plays a role in lipid transport and metabolism, was identified as a highly duplicate gene in the beef breeds. It was found to have a higher copy number in Angus compared to Holstein, Hereford, and Nelore [35]. The *APOL3* gene is also suggested to be a candidate gene for IMF deposition [36]. Notably, key genes such as *BMP2* and *FGF1* were significantly upregulated in XF compared to HF, MF, and PF. *BMP2* is known to promote adipogenesis by inducing preadipocyte differentiation through the activation of the PPARγ signaling pathway, as evidenced by studies showing reduced fat accumulation in *BMP2* knockout models [37]. Similarly, *FGF1* enhances lipid accumulation and insulin sensitivity in adipose tissue, with overexpression experiments demonstrating increased fat deposition in transgenic mice [38]. These findings suggest that the upregulation of *BMP2* and *FGF1* in XF contributes to enhanced fat deposition, supporting the observed improvements in marbling score and meat quality.

A total of 360 genes were found to be significantly higher in XF than in HF, MF, and PF. Enrichment analysis indicated that the top enriched terms were primarily associated with immune-related pathways, showing that XF have strong immune capacity and environmental suitability. The enrichment of immune pathways in XF, such as NF-kappa B signaling and cytokine activity, suggests a link to fat deposition, as inflammation can modulate adipogenesis. For instance, the inflammatory factor IL6, upregulated in XF, has a controversial role in adipose tissue. It may promote chronic inflammation, leading to impaired adipogenesis and insulin resistance in obesity [39], yet it also potentiates BMP2-induced adipogenesis by activating pathways like p38 MAPK, enhancing lipid accumulation in mesenchymal stem cells [40]. This dual role implies that IL6 contributes to the observed fat deposition balancing immune responses and adipogenic processes, supporting improved meat quality traits like marbling. In the XF vs. HF and XF vs. MF groups, oxidoreductase activity (GO:0016491) and oxidation–reduction (GO:0055114) were enriched, showing that the oxidoreductase response plays an important role in the change in F0 and F2. In the XF vs. HF group comparison, XF were found to have higher antioxidant activity compared to HF. For example, for *GPX1* and *CAT*, their expression levels in the two populations were both XF > HF. *GPX1* belongs to the glutathione peroxidase family, which reduces organic hydroperoxides and hydrogen peroxide (H_2O_2) by glutathione, thereby protecting cells from oxidative damage. Similarly, *CAT* encodes catalase, a crucial antioxidant enzyme involved in the body's defense against oxidative stress [41]. As well as the immune response and oxidation–reduction being markedly enriched, fatty acid metabolism was also important. Fatty acid metabolism and purine-related metabolism were associated with umami flavor [18]. Compared with HF and MF, the beef umami in XF has a better taste. Collectively, these findings provided a novel molecular understanding of the observed improvement in XF from the perspective of environmental suitability, oxidation–reduction, and good fat deposition.

The metabolic profile of chicken meat was analyzed through 1H NMR spectroscopy [17,18]. A comparative metabolomics analysis of milk components was conducted using LC-MS/MS technology, focusing on Chinese Holstein cows versus Italian Mediterranean buffaloes [19]. A combined LC-MS and GC-MS approach was employed to analyze the metabolite characteristics of thigh meat from fast- and slow-growing broilers at market age [20]. Since LC-MS/MS detected more metabolites than GC-MS/MS, the effects of breed need to be explored across a wider metabolite spectrum. Here, non-targeted LC-MS/MS technology was used to determine the metabolite composition of bovine fat in XF, HF, MF, and PF.

The composition of dietary FAs is also closely linked to human health. Trans-10-heptadecenoic acid, 16-hydroxyhexadecanoic acid, and palmitic acid are fatty acids and conjugates, among which trans-10-heptadecenoic acid is an unsaturated fatty acid and essential fatty acid, which cannot be synthesized by the body and therefore must be obtained through the diet [42]. 16-hydroxyhexadecanoic acid (also known as 16-hydroxypalmitic acid, ω-hydroxypalmitic acid, lanopalmitic acid, or juniperic acid; $C_{16}H_{32}O_3$), a hydroxylated derivative of palmitic acid, was significantly increased in XF compared to HF and MF. While our study highlights the elevated levels of this metabolite in XF, we acknowledge that no direct association with meat quality phenotypes, such as marbling score, has been established in the current literature or our data. However, hydroxy fatty acids, including 16-hydroxyhexadecanoic acid, are known to influence fat deposition [43], contribute to the formation of flavor compounds like lactones in meat [44], and improve insulin sensitivity [45], suggesting potential indirect contributions to meat quality and nutritional value in XF. Bovine fat is an important source of unsaturated fatty acid. Dietary supplementation of food with 16-hydroxyhexadecanoic acid, trans-10-heptadecenoic acid, and palmitic acid has been used as a potential method to supplement saturated and monounsaturated FAs in foods. Intramuscular fat refers to the lipids stored within the muscle, primarily consisting of cholesterol, phospholipids, and triglycerides [46,47]. Lipid oxidation significantly contributes to the development of meat flavor, with early studies indicating that phospholipids are more likely to generate flavor compared to triglycerides [48]. Three differential glycerophospholipid metabolites—PE (18:0/18:1), LPC (20:3), and PC (18:0/18:1)—experienced a significant uptrend in the breed comparison of XF vs. MF, XF vs. MF, and XF vs. PF, respectively, showing that XF has better lipid deposition compared to MF and PF, inheriting from HF or Angus.

Regarding umami amino acids, Asp-Glu was more significantly increased in XF than in MF [49]. Amino acids not only serve as important nutritional components for humans but also act as precursors to the flavor of cooked meat and contribute to its umami taste. In total, 37 different amino acids, peptides, and analogues were detected in bovine fat through metabonomic. DL-lanthionine was more significantly increased in XF than in HF and MF. The biological value of fat largely depends on the presence of digestible proteins that contain essential amino acids, which the human body cannot synthesize. In the comparison of amino acid pathways between XF and HF, MF, and PF, the identified differential metabolites were enriched in pathways related to isoleucine, leucine, and valine biosynthesis; tryptophan, tyrosine, and phenylalanine biosynthesis; and tryptophan, phenylalanine, and beta-alanine metabolism. Among these, isoleucine, leucine, valine, tryptophan, tyrosine, and phenylalanine are essential amino acids in the related pathways. We recognize that umami amino acids are more directly associated with meat flavor when evaluated in muscle tissue rather than subcutaneous fat. However, our analysis of umami amino acids such as Asp-Glu in subcutaneous fat provides valuable insights into the systemic metabolic differences among breeds/populations. Subcutaneous adipose tissue serves as an active metabolic organ that influences overall carcass composition and indirectly affects meat quality through several mechanisms. Adipose tissue acts as a reservoir for amino acids and their metabolites, which can be mobilized and transported to muscle tissue during meat aging and processing [50]. The metabolic activity of subcutaneous fat influences the overall amino acid profile in circulation, potentially affecting muscle metabolism [51]. Subcutaneous fat metabolites can serve as biomarkers reflecting the breed's overall metabolic capacity for flavor compound production. While our subcutaneous fat results cannot directly predict muscle umami intensity, they indicate the breed's inherent metabolic potential for amino acid synthesis and metabolism, suggesting that XF may have superior capacity for developing desirable flavor compounds. Future studies should indeed analyze muscle tissue directly to validate these metabolic differences in terms of actual eating quality and sensory attributes.

Metabolic pathway analysis is a valuable approach for examining the direct internal relationships between metabolites, allowing the reconstruction of biochemical reaction networks [52]. KEGG pathway analysis found that biosynthesis of UFAs; biosynthesis of phenylalanine, tryptophan, and tyrosine; metabolism of tryptophan; biosynthesis of primary bile acid; citrate cycle (TCA cycle); and metabolism of cholesterol, beta-alanine, linoleic acid, alpha-linoleic acid, and purine were the significant pathways of bovine fat metabolism affected by breed/population. Alanine metabolism and purine metabolism have been identified as key pathways influencing the meat flavor of Wuding chicken, suggesting that breed differences may account for the variation in metabolic pathways [18].

Integrating multi-omics data can aid in revealing potential gene functions related to specific metabolite accumulation. Using RNA-seq data, we performed the analysis of Pearson correlation between metabolite profiles and gene expression during the cattle breeding process. In this study, we paid attention to the differential metabolites enriched in the most significant pathways across the four breeds/populations and investigated the corresponding key genes in each breed. These genes were considered as the most promising potential markers for explaining the influence of breed on the metabolic profiles of cattle. In the XF vs. HF and XF vs. MF groups, oxidoreductase activity (GO:0016491) and the oxidation–reduction process (GO:0055114) were enriched, showing that oxidoreductase response played an important role in the change in F0 and F2. Raw meat is prone to oxidative deterioration during processing and storage due to chemical reactions. Nevertheless, the inclusion of antioxidants can mitigate the rate and extent of this oxidative deterioration, thereby prolonging the shelf life of meat products. Meat is a source of various endogenous antioxidants and bioactive compounds, such as creatine, carnosine, carnitine, glutathione, taurine, and ubiquinone [53]. High levels of taurine, oxidized glutathione, and vitamin B2 were detected in the XF group, and these compounds are known to have antioxidant properties. Taurine is recognized for its significant potential in the development of functional foods. It helps prevent liver steatosis through the inhibition of lipogenesis and promotion of energy expenditure, and it also mitigates oxidative damage by reducing ROS levels and stabilizing the mitochondrial membrane [54]. Glutathione reductase (GR) reduces oxidized glutathione (GSSG) to glutathione (GSH) by utilizing NADPH, as previously described [55]. The oxidized form of glutathione, GSSG, is known for its antioxidant properties and detoxification capabilities, and has potential therapeutic applications for numerous diseases, including cancer and chronic diseases [56]. In a previous study, it was observed that an antioxidant-rich diet led to a decrease ($p < 0.0001$) in GSH and GSH/GSSG values, and an increase ($p < 0.0001$) in GSSG levels in the liver of animals compared to those not receiving such a diet. Furthermore, the activity of the glutathione peroxidase enzyme was found to be higher ($p < 0.01$) in animals on an antioxidant-rich diet [57]. These findings suggest that the modulation of glutathione levels through dietary interventions may have significant implications for the prevention and treatment of various diseases. The antioxidant nature of vitamin B2 can help safeguard the body from oxidative stress, particularly by mitigating lipid peroxidation and oxidative damage resulting from reperfusion [58]. Our study observed increased taurine, oxidized glutathione, and vitamin B2 levels in the fat of XF than that in HF and MF, suggesting that different breeds may have distinct metabolic mechanisms. These findings demonstrated that the XF group exhibited optimal oxidation resistance, suggesting that this breed/population may contribute to improved meat quality and extended shelf life.

Taurine, oxidized glutathione, and vitamin B2 also have reducibility and were higher in XF than in HF and MF. Taurine is a particularly noteworthy amino acid found in fat. It is a crucial component of meat quality, contributing to taste and also playing essential physiological roles in humans [59]. Studies in animal models have shown that taurine supplementation can help alleviate metabolic disorders such as diabetes, hypertension, hyperlipidemia, and obesity [60]. Oxidized glutathione was detected in both positive and negative ion mode and contributes to certain flavor characteristics, especially in meat [61]. It is commonly assumed that vitamin B2 intake is overestimated based on dietary intakes, as compared to the levels measured in biochemical status indexes. In fact, studies have found that B2 intakes are often more than 5% below the recommended intake, with young women and seniors being particularly affected. To address this issue, increasing the B2 content in beef would not only help meet recommended intake levels but would also provide labeling advantages [62]. Moreover, research has suggested that B2 may indirectly impact cancer development through its influence on folate, highlighting the potential importance of increasing B2 content in beef [63]. In conclusion, XF cattle meat could be introduced to the market.

5. Conclusions

In summary, XF is a three-way crossbreed with rapid growth, high yield, superior meat quality, and robust adaptability to plateau conditions, giving them a comprehensive advantage over HF, MF, PF, and Angus cattle. In the present study, we mainly focused on fat deposition and fat metabolism changes and used transcriptome sequencing and LC-MS/MS-based metabolomics to partially explain the meat quality improvement in XF. Overall, differential expression analysis revealed 1762, 2949, and 2931 DEGs in XF vs. HF,

XF vs. MF, and XF vs. PF, respectively. Lipid-related DEGs such as *HADH* are significantly higher in XF than in HF, MF, and PF, while *ESR1* and *APOL3* are significantly higher in XF than in HF and PF, which can promote growing and fat deposition. The DEGs were markedly enriched in immune response, oxidation–reduction, and fatty acid metabolism, including *BMP2, WISP2, FGF1, IL1B, IL6,* and *WNT5B*. Additionally, 319 metabolites were initially identified in bovine adipose tissue using positive ion mode and 289 metabolites using negative ion mode, including 143 differential metabolites in positive ion mode and 166 differential metabolites in negative ion mode across four breeds/populations. Overall, fat in XF has the highest level of trans-10-heptadecenoic acid, indicating a lower risk of cardiovascular disease. Asp-Glu, which contributes to fat flavor, presented an upward trend in XF. Vitamin B2, taurine, and oxidized glutathione, which are involved in antioxidant processes, presented an upward trend in XF. The main pathways of XF fat metabolism influenced by breed were UFA biosynthesis; phenylalanine, tyrosine, and tryptophan biosynthesis; tryptophan metabolism; primary bile acid biosynthesis; citrate cycle (TCA cycle); cholesterol metabolism; beta-alanine metabolism; linoleic acid metabolism; alpha-linoleic acid metabolism; and purine metabolism. Integrated analysis of the RNA-Seq and metabolome data uncovered potentially functional genes that impact chemical compositions and metabolic pathways. These findings provide insights into the biological processes driving fat deposition and help identify important biomarkers of specific metabolite accumulation. Additionally, future studies should prioritize the quantification of the differential metabolites identified in this research to accurately reflect the nutritional composition of bovine fat. It is also essential to investigate potential gene markers and their roles in the accumulation of specific metabolites at the cellular level. In summation, XF fat offers higher energy, a stronger umami flavor, and optimal oxidation resistance. Similarly, the transcriptome results and the metabolome results also focus on fatty acid metabolism. Integrated analysis of the metabolome and transcriptome can provide valuable insights into the biological mechanisms underlying fat deposition and facilitate the identification of biomarkers associated with specific metabolite accumulation.

Supplementary Materials: The following supporting information can be downloaded at: https://www.mdpi.com/article/10.3390/ani15172599/s1, Figure S1: Preliminary analysis of transcriptome profiles of XF, HF, MF and PF. Figure S2: The PLS-DA and its models validated by permutation tests between different breed groups in positive ion mode. Figure S3: The KEGG pathways of differential metabolites between the four breed groups. Table S1: RNA-Seq_statistic. Table S2: RNA-Seq_expression. Table S3: GO analysis for functional annotation of DEGs of XF vs. HF. Table S4: GO analysis for functional annotation of DEGs of XF vs. MF. Table S5: GO analysis for functional annotation of DEGs of XF vs. PF. Table S6: KEGG analysis for DEGs of XF vs. HF. Table S7: KEGG analysis for DEGs of XF vs. MF. Table S8: KEGG analysis for DEGs of XF vs. PF. Table S9: GO annotation for XF specific genes. Table S10: Statistic of different metabolites. Table S11: Different lipid metabolites. Table S12: Integrated KEGG analysis of XF vs. HF in negative ions. Table S13: Integrated KEGG analysis of XF vs. HF in positive ions. Table S14: Integrated KEGG analysis of XF vs. MF in negative ions. Table S15: Integrated KEGG analysis of XF vs. MF in positive ions. Table S16: Integrated KEGG analysis of XF vs. PF in negative ions. Table S17: Integrated KEGG analysis of XF vs. PF in positive ions.

Author Contributions: X.C.: writing—review and editing, writing—original draft, visualization, validation, software, methodology, data curation. W.R.: software, data curation. J.C.: methodology, data curation. L.S.: software. N.Z.: resources. L.Z.: resources. R.D.: resources. F.S.: resources. K.Y.: writing—review and editing. Y.G.: writing—review and editing. X.H.: resources, writing—review and editing. B.H.: writing—review and editing, funding acquisition. H.C.: resources, investigation, writing—review and editing, supervision, project administration, funding acquisition, conceptualization. All authors have read and agreed to the published version of the manuscript.

Funding: This work was supported by the Identification of Key Genes of High-quality Beef of Multi-cross Tibetan yak of Animal Husbandry Station of Tibet Autonomous Region (2022-6-1) and Breeding Project of Yunling Cattle of Yunnan Joint Laboratory of Seed and Seed Industry (202205AR070001-18).

Institutional Review Board Statement: The animal study protocol was approved by the Faculty Animal Policy and Welfare Committee of Northwest A&F University (protocol code NWAFAC1008 and date of approval 10 May 2021).

Informed Consent Statement: Not applicable.

Data Availability Statement: The raw data supporting the conclusions of this article will be made available by the authors on request.

Conflicts of Interest: The authors declare no conflicts of interest.

Abbreviations

The following abbreviations are used in this manuscript:

DEG	differentially expressed gene
DM	differential metabolite
FC	fold change
FUFA	functional unsaturated fatty acid
GC-MS	gas chromatography–mass spectrometer
GO	gene ontology
GSEA	gene set enrichment analysis
HF	Tibetan yellow cattle
LC-MS	liquid chromatography–mass spectrometry
LC-MS/MS	liquid chromatography with tandem mass spectrometry
MF	Tibetan yak
MUFA	monounsaturated fatty acid
PCA	principal component analysis
PF	cattle–yak
PLS-DA	partial least squares discriminant analysis
PUFA	polyunsaturated fatty acid
SFA	saturated fatty acid
VIP	variable importance in projection
XF	Yajiangxue cattle

References

1. Royden, L.H.; Burchfiel, B.C.; van der Hilst, R.D. The geological evolution of the Tibetan Plateau. *Science* **2008**, *321*, 1054–1058. [CrossRef]
2. Kang, S.; Xu, Y.; You, Q.; Flügel, W.-A.; Pepin, N.; Yao, T. Review of climate and cryospheric change in the Tibetan Plateau. *Environ. Res. Lett.* **2010**, *5*, 015101. [CrossRef]
3. Cao, X.-K.; Cheng, J.; Huang, Y.-Z.; Wang, X.-G.; Ma, Y.-L.; Peng, S.-J.; Chaogetu, B.; Zhuoma, Z.; Chen, H. Growth performance and meat quality evaluations in three-way cross cattle developed for the Tibetan Plateau and their molecular understanding by integrative omics analysis. *J. Agric. Food Chem.* **2018**, *67*, 541–550. [CrossRef] [PubMed]
4. Wood, J.; Enser, M.; Fisher, A.; Nute, G.; Sheard, P.; Richardson, R.; Hughes, S.; Whittington, F. Fat deposition, fatty acid composition and meat quality: A review. *Meat Sci.* **2008**, *78*, 343–358. [CrossRef] [PubMed]
5. Pitchford, W.; Deland, M.; Siebert, B.; Malau-Aduliand, A.; Bottema, C. Genetic variation in fatness and fatty acid composition of crossbred cattle. *J. Anim. Sci.* **2002**, *80*, 2825–2832. [CrossRef] [PubMed]
6. Song, C.; Huang, Y.; Yang, Z.; Ma, Y.; Chaogetu, B.; Zhuoma, Z.; Chen, H. RNA-Seq analysis identifies differentially expressed genes in subcutaneous adipose tissue in qaidaford cattle, cattle-yak, and angus cattle. *Animals* **2019**, *9*, 1077. [CrossRef]
7. You, Z.; Jin, Z.; Mingliang, Z. Research progress on interspecific hybridization and utilization of yak. *Cao Xue* **2025**, *1*, 67–70+74.
8. Motoyama, M.; Sasaki, K.; Watanabe, A. Wagyu and the factors contributing to its beef quality: A Japanese industry overview. *Meat Sci.* **2016**, *120*, 10–18. [CrossRef]

9. Wood, J.; Richardson, R.; Nute, G.; Fisher, A.; Campo, M.; Kasapidou, E.; Sheard, P.; Enser, M. Effects of fatty acids on meat quality: A review. *Meat Sci.* **2004**, *66*, 21–32. [CrossRef]
10. Wang, Q.; Wu, J.; Zhang, S.; Zhang, Y.; Zhang, H.; Fan, E. GC analysis of the fatty acid composition of yak kidney. *Chromatographia* **2009**, *69*, 139–143. [CrossRef]
11. Gu, X.; Sun, W.; Yi, K.; Yang, L.; Chi, F.; Luo, Z.; Wang, J.; Zhang, J.; Wang, W.; Yang, T. Comparison of muscle lipidomes between cattle-yak, yak, and cattle using UPLC–MS/MS. *J. Food Compos. Anal.* **2021**, *103*, 104113. [CrossRef]
12. Li, W.; Wang, J.; Zhang, C.; Wang, N.; Zhang, C.; Chen, W.; Wu, T. Using an integrated feature-based molecular network and lipidomics approach to reveal the differential lipids in yak shanks and flanks. *Food Chem.* **2023**, *403*, 134352. [CrossRef]
13. Xiong, L.; Pei, J.; Wang, X.; Guo, S.; Guo, X.; Yan, P. Lipidomics and transcriptome reveal the effects of feeding systems on fatty acids in yak's meat. *Foods* **2022**, *11*, 2582. [CrossRef]
14. Xiong, L.; Pei, J.; Wang, X.; Guo, S.; Guo, X.; Yan, P. Effect of lipids in yak muscle under different feeding systems on meat quality based on untargeted lipidomics. *Animals* **2022**, *12*, 2814. [CrossRef]
15. Li, J.; Zhang, D.; Yin, L.; Li, Z.; Yu, C.; Du, H.; Jiang, X.; Yang, C.; Liu, Y. Integration analysis of metabolome and transcriptome profiles revealed the age-dependent dynamic change in chicken meat. *Food Res. Int.* **2022**, *156*, 111171. [CrossRef]
16. Bischof, G.; Witte, F.; Terjung, N.; Januschewski, E.; Heinz, V.; Juadjur, A.; Gibis, M. Effect of sampling position in fresh, dry-aged and wet-aged beef from M. Longissimus dorsi of Simmental cattle analyzed by 1H NMR spectroscopy. *Food Res. Int.* **2022**, *156*, 111334. [CrossRef]
17. de Oliveira, E.S.; Lião, L.M.; Silva, A.K.; Prado, C.S.; Sena, M.M.; Oliveira, G.d.A.R. A t-test ranking-based discriminant analysis for classification of free-range and barn-raised broiler chickens by 1H NMR spectroscopy. *Food Chem.* **2023**, *399*, 134004. [CrossRef]
18. Xiao, Z.; Ge, C.; Zhou, G.; Zhang, W.; Liao, G. 1H NMR-based metabolic characterization of Chinese Wuding chicken meat. *Food Chem.* **2019**, *274*, 574–582. [CrossRef]
19. Yuan, X.; Shi, W.; Jiang, J.; Li, Z.; Fu, P.; Yang, C.; Rehman, S.U.; Pauciullo, A.; Liu, Q.; Shi, D. Comparative metabolomics analysis of milk components between Italian Mediterranean buffaloes and Chinese Holstein cows based on LC-MS/MS technology. *PLoS ONE* **2022**, *17*, e0262878. [CrossRef]
20. Chen, X.; Cao, J.; Geng, A.; Zhang, X.; Wang, H.; Chu, Q.; Yan, Z.; Zhang, Y.; Liu, H.; Zhang, J. Integration of GC-MS and LC-MS for metabolite characteristics of thigh meat between fast-and slow-growing broilers at marketable age. *Food Chem.* **2023**, *403*, 134362. [CrossRef]
21. Cheng, J.; Pan, Y.; Yang, S.; Wei, Y.; Lv, Q.; Xing, Q.; Zhang, R.; Sun, L.; Qin, G.; Shi, D. Integration of transcriptomics and non-targeted metabolomics reveals the underlying mechanism of follicular atresia in Chinese buffalo. *J. Steroid Biochem. Mol. Biol.* **2021**, *212*, 105944. [CrossRef]
22. Zhan, H.; Xiong, Y.; Wang, Z.; Dong, W.; Zhou, Q.; Xie, S.; Li, X.; Zhao, S.; Ma, Y. Integrative analysis of transcriptomic and metabolomic profiles reveal the complex molecular regulatory network of meat quality in Enshi black pigs. *Meat Sci.* **2022**, *183*, 108642. [CrossRef]
23. Eom, J.S.; Lee, S.J.; Gu, B.-H.; Lee, S.J.; Lee, S.-S.; Kim, S.-H.; Kim, B.-W.; Lee, S.S.; Kim, M. Metabolomic and transcriptomic study to understand changes in metabolic and immune responses in steers under heat stress. *Anim. Nutr.* **2022**, *11*, 87–101. [CrossRef]
24. *NY/T 815-2004*; Feeding Standard of Beef Cattle. MAPRC: Beijing, China, 2004.
25. *GB/T 29392-2022*; Quality Grading for Livestock and Poultry Meat-Beef. Standards Press of China: Beijing, China, 2022.
26. Subramanian, A.; Tamayo, P.; Mootha, V.K.; Mukherjee, S.; Ebert, B.L.; Gillette, M.A.; Paulovich, A.; Pomeroy, S.L.; Golub, T.R.; Lander, E.S. Gene set enrichment analysis: A knowledge-based approach for interpreting genome-wide expression profiles. *Proc. Natl. Acad. Sci. USA* **2005**, *102*, 15545–15550. [CrossRef]
27. Want, E.J.; Masson, P.; Michopoulos, F.; Wilson, I.D.; Theodoridis, G.; Plumb, R.S.; Shockcor, J.; Loftus, N.; Holmes, E.; Nicholson, J.K. Global metabolic profiling of animal and human tissues via UPLC-MS. *Nat. Protoc.* **2013**, *8*, 17–32. [CrossRef]
28. Wen, B.; Mei, Z.; Zeng, C.; Liu, S. metaX: A flexible and comprehensive software for processing metabolomics data. *BMC Bioinform.* **2017**, *18*, 183. [CrossRef]
29. Zhou, J.; Liu, H.; Zhong, C.; Degen, A.; Yang, G.; Zhang, Y.; Qian, J.; Wang, W.; Hao, L.; Qiu, Q. Apparent digestibility, rumen fermentation, digestive enzymes and urinary purine derivatives in yaks and Qaidam cattle offered forage-concentrate diets differing in nitrogen concentration. *Livest. Sci.* **2018**, *208*, 14–21. [CrossRef]
30. Sierra, V.; Guerrero, L.; Fernández-Suárez, V.; Martínez, A.; Castro, P.; Osoro, K.; Rodríguez-Colunga, M.; Coto-Montes, A.; Oliván, M. Eating quality of beef from biotypes included in the PGI "Ternera Asturiana" showing distinct physicochemical characteristics and tenderization pattern. *Meat Sci.* **2010**, *86*, 343–351. [CrossRef] [PubMed]
31. Gu, X.; Wang, L.; Liu, S.; Shan, T. Adipose tissue adipokines and lipokines: Functions and regulatory mechanism in skeletal muscle development and homeostasis. *Metabolism* **2023**, *139*, 155379. [CrossRef] [PubMed]
32. White, U.A.; Stephens, J.M. Transcriptional factors that promote formation of white adipose tissue. *Mol. Cell. Endocrinol.* **2010**, *318*, 10–14. [CrossRef] [PubMed]

33. Wang, Z.; Shang, P.; Li, Q.; Wang, L.; Chamba, Y.; Zhang, B.; Zhang, H.; Wu, C. iTRAQ-based proteomic analysis reveals key proteins affecting muscle growth and lipid deposition in pigs. *Sci. Rep.* **2017**, *7*, 46717. [CrossRef]
34. Wang, X.; Fang, C.; He, H.; Cao, H.; Liu, L.; Jiang, L.; Ma, Y.; Liu, W. Identification of key genes in sheep fat tail evolution Based on RNA-seq. *Gene* **2021**, *781*, 145492. [CrossRef]
35. Bickhart, D.M.; Hou, Y.; Schroeder, S.G.; Alkan, C.; Cardone, M.F.; Matukumalli, L.K.; Song, J.; Schnabel, R.D.; Ventura, M.; Taylor, J.F. Copy number variation of individual cattle genomes using next-generation sequencing. *Genome Res.* **2012**, *22*, 778–790. [CrossRef]
36. Poleti, M.D.; Regitano, L.C.; Souza, G.H.; Cesar, A.S.; Simas, R.C.; Silva-Vignato, B.; Oliveira, G.B.; Andrade, S.C.; Cameron, L.C.; Coutinho, L.L. Longissimus dorsi muscle label-free quantitative proteomic reveals biological mechanisms associated with intramuscular fat deposition. *J. Proteom.* **2018**, *179*, 30–41. [CrossRef] [PubMed]
37. Huang, H.; Song, T.-J.; Li, X.; Hu, L.; He, Q.; Liu, M.; Lane, M.D.; Tang, Q.-Q. BMP signaling pathway is required for commitment of C3H10T1/2 pluripotent stem cells to the adipocyte lineage. *Proc. Natl. Acad. Sci. USA* **2009**, *106*, 12670–12675.
38. Jonker, J.W.; Suh, J.M.; Atkins, A.R.; Ahmadian, M.; Li, P.; Whyte, J.; He, M.; Juguilon, H.; Yin, Y.-Q.; Phillips, C.T. A PPARγ–FGF1 axis is required for adaptive adipose remodelling and metabolic homeostasis. *Nature* **2012**, *485*, 391–394. [CrossRef]
39. Han, M.S.; White, A.; Perry, R.J.; Camporez, J.-P.; Hidalgo, J.; Shulman, G.I.; Davis, R.J. Regulation of adipose tissue inflammation by interleukin 6. *Proc. Natl. Acad. Sci. USA* **2020**, *117*, 2751–2760. [CrossRef]
40. Huang, R.-L.; Sun, Y.; Ho, C.-K.; Liu, K.; Tang, Q.-Q.; Xie, Y.; Li, Q. IL-6 potentiates BMP-2-induced osteogenesis and adipogenesis via two different BMPR1A-mediated pathways. *Cell Death Dis.* **2018**, *9*, 144. [CrossRef]
41. Ighodaro, O.; Akinloye, O. First line defence antioxidants-superoxide dismutase (SOD), catalase (CAT) and glutathione peroxidase (GPX): Their fundamental role in the entire antioxidant defence grid. *Alex. J. Med.* **2018**, *54*, 287–293. [CrossRef]
42. Di Pasquale, M.G. The essentials of essential fatty acids. *J. Diet. Suppl.* **2009**, *6*, 143–161. [CrossRef] [PubMed]
43. Hammarstedt, A.; Syed, I.; Vijayakumar, A.; Eliasson, B.; Gogg, S.; Kahn, B.B.; Smith, U. Adipose tissue dysfunction is associated with low levels of the novel Palmitic Acid Hydroxystearic Acids. *Sci. Rep.* **2018**, *8*, 15757. [CrossRef]
44. Domínguez, R.; Pateiro, M.; Gagaoua, M.; Barba, F.J.; Zhang, W.; Lorenzo, J.M. A comprehensive review on lipid oxidation in meat and meat products. *Antioxidants* **2019**, *8*, 429. [CrossRef] [PubMed]
45. Zhou, P.; Santoro, A.; Peroni, O.D.; Nelson, A.T.; Saghatelian, A.; Siegel, D.; Kahn, B.B. PAHSAs enhance hepatic and systemic insulin sensitivity through direct and indirect mechanisms. *J. Clin. Investig.* **2019**, *129*, 4138–4150. [CrossRef]
46. Liu, L.; Liu, X.; Cui, H.; Liu, R.; Zhao, G.; Wen, J. Transcriptional insights into key genes and pathways controlling muscle lipid metabolism in broiler chickens. *BMC Genom.* **2019**, *20*, 863. [CrossRef]
47. Hocquette, J.-F.; Gondret, F.; Baéza, E.; Médale, F.; Jurie, C.; Pethick, D. Intramuscular fat content in meat-producing animals: Development, genetic and nutritional control, and identification of putative markers. *Animal* **2010**, *4*, 303–319. [CrossRef] [PubMed]
48. Huang, Y.-C.; Li, H.-J.; He, Z.-F.; Wang, T.; Qin, G. Study on the flavor contribution of phospholipids and triglycerides to pork. *Food Sci. Biotechnol.* **2010**, *19*, 1267–1276. [CrossRef]
49. Wang, W.; Zhou, X.; Liu, Y. Characterization and evaluation of umami taste: A review. *TrAC Trends Anal. Chem.* **2020**, *127*, 115876. [CrossRef]
50. Moya, V.-J.; Flores, M.; Aristoy, M.; Toldrá, F. Pork meat quality affects peptide and amino acid profiles during the ageing process. *Meat Sci.* **2001**, *58*, 197–206. [CrossRef]
51. Listrat, A.; Lebret, B.; Louveau, I.; Astruc, T.; Bonnet, M.; Lefaucheur, L.; Picard, B.; Bugeon, J. How muscle structure and composition influence meat and flesh quality. *Sci. World J.* **2016**, *2016*, 3182746. [CrossRef]
52. Klamt, S.; Stelling, J. Two approaches for metabolic pathway analysis? *Trends Biotechnol.* **2003**, *21*, 64–69. [CrossRef] [PubMed]
53. Williams, P. Nutritional composition of red meat. *Nutr. Diet.* **2007**, *64*, S113–S119. [CrossRef]
54. Song, Q.; Guo, J.; Zhang, Y.; Chen, W. The beneficial effects of taurine in alleviating fatty liver disease. *J. Funct. Foods* **2021**, *77*, 104351. [CrossRef]
55. Sordillo, L.M.; Aitken, S.L. Impact of oxidative stress on the health and immune function of dairy cattle. *Vet. Immunol. Immunopathol.* **2009**, *128*, 104–109. [CrossRef]
56. Gaucher, C.; Boudier, A.; Bonetti, J.; Clarot, I.; Leroy, P.; Parent, M. Glutathione: Antioxidant properties dedicated to nanotechnologies. *Antioxidants* **2018**, *7*, 62. [CrossRef] [PubMed]
57. Correa, L.B.; Netto, A.S.; da Silva, J.S.; Consolo, N.R.B.; Pugine, S.M.P.; de Melo, M.P.; de Souza Santana, R.S.; Zanetti, M.A. Changes on meat fatty acid profile, cholesterol and hepatic metabolism associated with antioxidants and canola oil supplementation for Nellore cattle. *Livest. Sci.* **2022**, *257*, 104850. [CrossRef]
58. Lonergan, E.H.; Zhang, W.; Lonergan, S.M. Biochemistry of postmortem muscle—Lessons on mechanisms of meat tenderization. *Meat Sci.* **2010**, *86*, 184–195. [CrossRef] [PubMed]
59. Imanari, M.; Higuchi, M.; Shiba, N.; Watanabe, A. Accurate analysis of taurine, anserine, carnosine and free amino acids in a cattle muscle biopsy sample. *Anim. Sci. J.* **2010**, *81*, 369–376. [CrossRef] [PubMed]

60. Murakami, S. Role of taurine in the pathogenesis of obesity. *Mol. Nutr. Food Res.* **2015**, *59*, 1353–1363. [CrossRef] [PubMed]
61. Ueda, Y.; Yonemitsu, M.; Tsubuku, T.; Sakaguchi, M.; Miyajima, R. Flavor characteristics of glutathione in raw and cooked foodstuffs. *Biosci. Biotechnol. Biochem.* **1997**, *61*, 1977–1980. [CrossRef] [PubMed]
62. Egelandsdal, B.; Oostindjer, M.; Hovland, E.-M.; Okholm, B.; Saarem, K.; Bjerke, F.; Ruud, L.; Grabež, V.; Haug, A. Identifying labelling and marketing advantages of nutrients in minced beef meat: A case study. *Meat Sci.* **2020**, *159*, 107920. [CrossRef] [PubMed]
63. Thakur, K.; Tomar, S.K.; Singh, A.K.; Mandal, S.; Arora, S. Riboflavin and health: A review of recent human research. *Crit. Rev. Food Sci. Nutr.* **2017**, *57*, 3650–3660. [CrossRef] [PubMed]

animals

MDPI

Article

Molecular Insights into Intramuscular Unsaturated Fatty Acid Deposition in Lambs Through Multi-Omics Profiling

Xuewen Han †, Xudong Liu †, Yu Fu, Jinlin Chen, Cuiyu Lai, Xiaofan Yang, Xuesong Shan *, Yang Chen * and Huaizhi Jiang

College of Animal Science and Technology, Jilin Agricultural University, Changchun 130118, China; hanxuewen1123@163.com (X.H.); 19845486668@163.com (X.L.); 15164460626@163.com (Y.F.); chenjinlin7314@163.com (J.C.); 19843927673@163.com (C.L.); jianghz6806@126.com (H.J.)

* Correspondence: shanxuesong@jlau.edu.cn (X.S.); chenyang7419@163.com (Y.C.)

† These authors contributed equally to this work.

Simple Summary

Fat in meat plays an important role in its flavor and nutritional value. Among different types of fat, unsaturated fatty acids are considered healthier for humans. Improving the content of these healthy fats in lamb could benefit both consumers and the livestock industry. In this study, we selected two groups of lambs with high and low levels of unsaturated fatty acids in their muscles. We examined their gene activity, chemical compounds in the muscle, and the types of muscle fibers they had. We found that lambs with more unsaturated fat had different patterns of gene activity related to fat metabolism, energy production in cells, and immune function. We also detected changes in chemical compounds linked to fat and amino acid processing. A key finding was that the *MYH7* gene, which is related to the development of slow muscle fibers, was more active in lambs with higher unsaturated fat. These lambs also had more slow-twitch muscle fibers. Our findings provide new insights that could help farmers produce lamb with healthier fat, improving the quality of meat for consumers.

Abstract

Unsaturated fatty acids are key contributors to the nutritional and sensory quality of lamb meat. To investigate the molecular basis of intramuscular unsaturated fatty acid variation, we selected lambs with divergent fatty acid profiles and performed integrated transcriptomic and untargeted metabolomic analyses of the longissimus dorsi muscle. The high unsaturated fatty acid group exhibited distinct gene expression patterns in pathways related to lipid metabolism, mitochondrial function, and immune responses. Metabolomic profiling revealed significant enrichment of metabolites involved in both the biosynthesis and degradation of fatty acids. Among the differentially expressed genes, *MYH7* was markedly upregulated in lambs with higher unsaturated fatty acid content, suggesting a potential regulatory role in energy metabolism or lipid homeostasis. These findings provide new molecular insights into the mechanisms underlying unsaturated fatty acid deposition in lamb and identify *MYH7* and other candidates as potential targets for improving meat quality through breeding or nutritional strategies.

Keywords: unsaturated fatty acids; lamb; transcriptome; metabolome; muscle lipid metabolism; muscle fiber type

1. Introduction

Mutton, a traditional dietary staple in many cultures, is prized for its distinctive flavor and rich nutritional profile, making it an important component of the human diet. In addition to its palatability, mutton is an excellent source of high-quality protein, vitamins, and trace elements, while containing relatively low levels of fat and cholesterol, thus offering significant health benefits compared with pork [1]. Mutton provides 25 to 30 g of complete protein per 100 g, which is essential for tissue growth, repair, and immune function [2,3]. Furthermore, mutton is abundant in B vitamins, including B12, B6, riboflavin, and niacin, as well as important minerals such as iron, zinc, and phosphorus. These nutrients play a crucial role in maintaining overall health and preventing anemia [4–6]. Consequently, global demand for mutton has grown rapidly in recent years [7].

Evaluations of mutton quality usually include physicochemical parameters such as pH, water-holding capacity, marbling, tenderness, cooking loss, and drip loss, together with nutritional attributes [8,9]. Among these parameters, the composition of intramuscular fat—especially the content and profile of unsaturated fatty acids (UFAs)—has garnered increasing attention. Mutton fat is primarily composed of monounsaturated fatty acids (MUFAs) and polyunsaturated fatty acids (PUFAs) [10]. These fatty acids not only contribute to the flavor of the meat but also offer positive health benefits. Diets high in UFAs are linked to lower levels of serum low-density lipoprotein (LDL) cholesterol and reduced cardiovascular risk. In particular, MUFAs such as oleic acid have been shown to improve lipid profiles and decrease atherosclerosis risk [11–13]. Thus, UFAs are critical determinants of both the nutritional value and sensory quality of meat, and in mutton specifically, their abundance and composition directly influence health outcomes and consumer ac-acceptance [14,15].

In response to increasing consumer demand for healthier meat products, enhancing UFA deposition in muscle has become a major objective in both sheep breeding and nutritional strategies. However, the regulatory mechanisms underlying UFA accumulation in skeletal muscle remain poorly defined and appear to involve complex networks of lipid synthesis, transport, oxidation, and storage [16,17]. Some well-known lipid metabolism enzymes, such as FASN, SCD, and ELOVLs, are considered important regulatory factors [18,19], and our previous research suggests that muscle fiber composition may also play a significant role [20]. Myosin heavy chain 7 (MYH7), a structural gene encoding the β-myosin heavy chain specific to slow-twitch fibers, is essential for determining fiber type and regulating metabolic characteristics in muscle tissue [21]. While MYH7 is classically associated with contractile function, its role in lipid metabolism has not been extensively investigated.

Currently, multi-omics approaches integrating transcriptomic, proteomic, and metabo lomic data have provided powerful tools for dissecting the molecular basis of meat quality traits in livestock. In cattle and pigs, such strategies have revealed key regulatory genes and metabolites associated with fat deposition, muscle development, and flavor compound formation [22,23]. For instance, integrated transcriptomic and metabolomic profiling has linked fatty acid metabolism to flavor differences in beef and identified lipid–amino acid interactions critical for pork quality [24,25].

By contrast, studies applying multi-omics profiling to sheep remain relatively limited. Existing reports mainly focus on growth performance, wool traits, or general meat quality parameters, with fewer efforts directed at elucidating the molecular mechanisms underlying intramuscular fatty acid deposition [26]. Although some transcriptomic analyses have identified candidate genes related to lipid metabolism in sheep muscle, they often lack validation through metabolomics or integration with biochemical data [27]. Consequently, a comprehensive understanding of the regulatory networks that drive UFA accumulation in lamb meat is still missing.

In this study, we aimed to fill this gap by conducting an integrative transcriptomic and metabolomic analysis of longissimus dorsi tissues from lambs with divergent UFA contents. By combining differential expression, functional enrichment, and metabolic profiling, we particularly focused on the potential regulatory role of *MYH7*, given its established link to muscle metabolism. Our findings provide novel insights into the molecular basis of UFA deposition in lambs and highlight *MYH7* as a potential genetic marker for improving mutton quality.

2. Materials and Methods

2.1. Animals and Sample Collection

A total of 60 six-month-old crossbred female lambs (Small-tailed Han × Australian White), reared under standardized management conditions at Guofeng Livestock Farm in Changling County, were randomly selected for this study. The animals were housed indoors with ad libitum access to a nutritionally balanced diet, which included Leymus chinensis (50.0%), corn (20.0%), soybean meal (5.4%), DDGS (7.0%), soybean oil (15.0%), $CaHPO_4$ (0.5%), limestone (1.4%), NaCl (0.5%), and vitamin and mineral additives (0.2%). Nutrient levels were formulated to meet the requirements for growing lambs, with metabolic energy at 11.99 MJ/kg, crude protein at 10.1%, calcium at 0.65%, and total phosphorus at 0.33%. Pre-slaughter body weights ranged from 30 to 50 kg, showing a normal distribution without outliers. All lambs were fasted for 24 h with ad libitum access to water before slaughter. Following exsanguination via the jugular vein, samples of the longissimus dorsi muscle were immediately collected. Samples for meat quality assessment were stored at 4 °C, while those for transcriptomic, metabolomic, and quantitative PCR analyses were snap-frozen in liquid nitrogen and stored at −80 °C. Samples intended for immunohistochemistry and immunofluorescence were fixed in 4% paraformaldehyde.

2.2. Meat Quality Traits and Fatty Acid Composition

Meat quality parameters were assessed as previously described by Fu et al. [20]. Briefly, longissimus dorsi muscle samples were collected to measure intramuscular fat content (IMF), pH, meat color (L*, a*, b*), shear force (SF), and cooking loss. IMF was determined using the Soxhlet extraction method. Ph 24 was measured within 2 h postmortem at the thoracolumbar junction after homogenization in 0.1 mol/L KCl. Meat color was evaluated after 20 min of blooming using a CR-400 chromameter (Konica Minolta, Tokyo, Japan) with an 8 mm aperture under D65 illumination. Shear force was quantified using a TA-XT Plus texture analyzer (Stable Micro Systems, Godalming, UK) on cooked cylindrical samples aligned parallel to muscle fibers. Cooking loss was calculated by weighing standardized meat cubes (2 × 2 × 2 cm, ~30 g) before and after heating to 70 °C in a water bath. All measurements were performed in triplicate and averaged. Fatty acid composition was analyzed by liquid chromatography following the Chinese National Food Safety Standard GB 5009.168-2016 [28].

2.3. Transcriptome Sequencing and Analysis

Total RNA was extracted from longissimus dorsi muscle tissues of 6 lambs (n = 6 per group, high UFA vs. low UFA) using TRIzol reagent (Invitrogen, Carlsbad, CA, USA), following the manufacturer's protocol. RNA purity and concentration were assessed using a NanoDrop 2000 (Thermo Scientific, Waltham, MA, USA), and integrity was evaluated using an Agilent 2100 Bioanalyzer (RIN > 7.0). mRNA sequencing (mRNA-seq) libraries were prepared using the NEBNext®® Ultra™ RNA Library Prep Kit (NEB, Ipswich, MA,

USA) and sequenced on the Illumina NovaSeq 6000 platform with 150 bp paired-end reads. Clean reads were aligned to the sheep reference genome Oar_rambouillet_v1.0 (GenBank assembly accession: GCA_002742125.1) [29] using HISAT2 v2.2.1 (Johns Hopkins University, Baltimore, MD, USA) [30], and gene expression levels were quantified as fragments per kilobase of transcript per million mapped reads (FPKM). Differentially expressed genes (DEGs) were identified using DESeq2 v1.30.1 (Bioconductor, Fred Hutchinson Cancer Research Center, Seattle, WA, USA) [31], with FoldChange > 1.5 and $p < 0.05$ as thresholds. Functional enrichment analyses including Gene Ontology (GO) and KEGG pathway enrichment were conducted using the ClusterProfiler package v4.0.5 (Bioconductor, Yu Lab, Southern Medical University, Guangzhou, China) [32] in R (R Foundation for Statistical Computing, Vienna, Austria), with Ovis aries Oar_rambouillet_v1.0 as the reference genome background.

2.4. Metabolomics Analysis

Untargeted metabolomics profiling was conducted on longissimus dorsi tissues (n = 6 per group) using ultra-performance liquid chromatography coupled to tandem mass spectrometry (UPLC–MS/MS). Tissues were homogenized in 80% methanol, followed by centrifugation and filtration. Chromatographic separation was performed using a Waters ACQUITY UPLC system equipped with a BEH Amide column, and metabolites were detected using a QTRAP 6500+ mass spectrometer (AB Sciex). Quality control samples were included throughout the run. Peak extraction, alignment, and quantification were conducted using Analyst software v1.7.1. Metabolites were identified and annotated based on m/z values and MS/MS fragmentation patterns using the HMDB databases. Metabolites with variable importance in projection (VIP) > 1 and $p < 0.05$ (t-test) were considered significantly different. Multivariate statistical analyses, including principal component analysis (PCA), were conducted using MetaboAnalyst v6.0 (Xia Lab, McGill University, Montreal, QC, Canada) [33].

2.5. Quantitative Real-Time PCR (qPCR)

Total RNA (1 μg) was reverse-transcribed into cDNA using the PrimeScript RT Reagent Kit (Takara, Tokyo, Japan). Quantitative PCR was performed with TB Green Premix Ex Taq II (Takara) on a CFX96 Real-Time PCR System (Bio-Rad, Hercules, CA, USA). Primer sequences were designed using Primer-BLAST (NCBI) and are listed in Supplementary Table S1. PCR conditions were as follows: initial denaturation at 95 °C for 30 s, followed by 40 cycles of 95 °C for 5 s and 60 °C for 30 s, with a melt-curve analysis performed at the end of amplification. Gene expression was normalized to β-actin, and relative quantification was calculated using the $2^{-\Delta\Delta Ct}$ method. Each reaction was run in three technical replicates and six biological replicates.

2.6. Western Blot Analysis (WB)

Total protein was extracted from muscle tissues using RIPA lysis buffer with protease inhibitors (Beyotime, Shanghai, China). Protein concentrations were measured using a BCA Protein Assay Kit (Thermo Scientific). Equal amounts of protein (20 μg) were separated by 10% SDS-PAGE and transferred to PVDF membranes (Millipore, Billerica, MA, USA). After blocking with 5% non-fat milk in TBST, membranes were incubated overnight at 4 °C with primary antibodies against FADD, MYH7, TMEM74, CSNK2A2, CTSB, and β-actin (loading control), followed by HRP-conjugated secondary antibodies. Immunoreactive bands were visualized using enhanced chemiluminescence (ECL, Thermo), and band intensities were quantified using ImageJ v1.54g. Each group included three biological replicates. Antibody details are listed in Supplementary Table S2.

2.7. Gene Set Enrichment Analysis (GSEA)

During the differential gene analysis, *MYH7* attracted our attention. To highlight the mechanism of *MYH7*-mediated UFA deposition, we performed a single-gene GSEA using *MYH7* as the identifier. Gene expression matrices were input into GSEA software v4.1.0 (Broad Institute, Cambridge, MA, USA), and enrichment was assessed using KEGG and Hallmark gene sets from MSigDB. Samples were divided into high- and low-expression groups based on the median expression of the target gene. Normalized enrichment score (NES), nominal *p*-value, and false discovery rate (FDR q-value) were calculated, with significance defined as FDR < 0.25 and $p < 0.05$.

2.8. Statistical Analysis

Data are presented as mean ± SEM. Differences between high and low slow-twitch groups were analyzed using unpaired *t*-tests with SPSS Statistics v27.0 (IBM Corp., Armonk, NY, USA). Statistical significance was set at $p < 0.05$, and highly significant differences at $p < 0.01$. Graphs were generated using GraphPad Prism 9.5 (GraphPad Software, San Diego, CA, USA).

3. Results

3.1. Comparison of Meat Quality Traits and Fatty Acid Composition Between High- and Low-UFA Groups

To investigate the potential mechanisms underlying intramuscular unsaturated fatty acid (UFA) variation, six lambs with the highest UFA content (H) and six with the lowest (L) were selected from a cohort of sixty individuals based on fatty acid profiles of the longissimus dorsi muscle. As shown in Figure 1, the H group exhibited significantly higher UFA levels than the L group ($p = 0.0001$), while body weight showed no significant difference between groups ($p = 0.8915$), suggesting that UFA variation was independent of growth traits and that the selected samples were suitable for subsequent comparative analyses.

Meat quality traits were further assessed (Figure 1). Among these, the H group showed a significantly lower intramuscular fat content compared to the L group ($p = 0.0117$) and a significant increase in muscle moisture ($p = 0.0459$). Other parameters, including pH24, shear force, cooking loss, meat color indices (L*, a*, b*), protein content, and muscle fiber diameter, did not differ significantly between groups ($p > 0.05$), indicating relatively consistent overall meat quality characteristics aside from lipid content.

Fatty acid profiling revealed several key differences (Table 1). The H group had significantly higher proportions of oleic acid (C18:1n9, $p = 0.0016$), arachidonic acid (C20:4n6, $p = 0.0023$), and total UFAs (57.31% vs. 52.08%, $p = 0.0001$), including both monounsaturated (MUFAs, $p = 0.0064$) and polyunsaturated fatty acids (PUFAs, though not statistically significant, $p = 0.1094$). Conversely, saturated fatty acids (SFAs) were significantly lower in the H group (42.43% vs. 47.94%, $p = 0.0001$), particularly palmitic acid (C16:0) and stearic acid (C18:0).

These findings indicate that higher UFA content in lamb muscle is associated with distinct alterations in fatty acid composition—particularly enrichment of oleic acid and depletion of SFAs—without compromising most meat quality parameters. The reduced fat content and increased moisture in the H group may also reflect a shift in lipid deposition patterns.

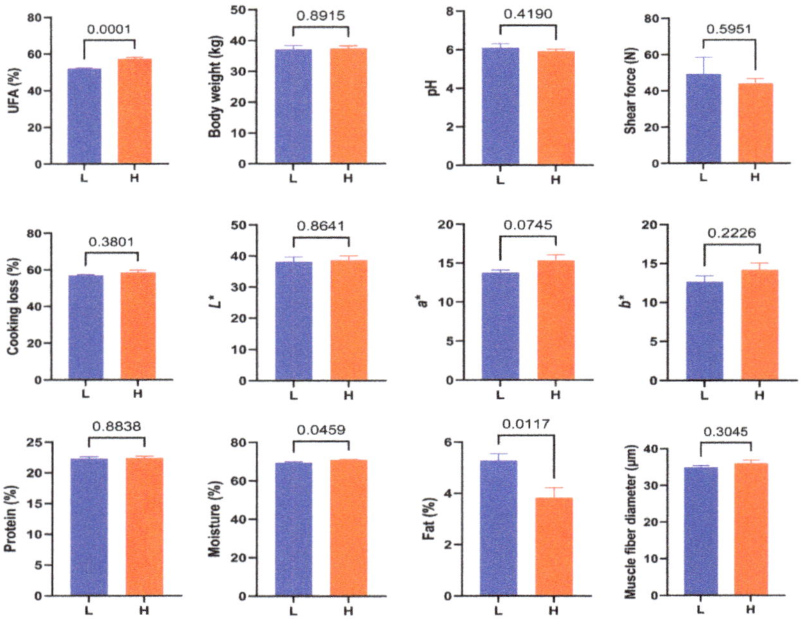

Figure 1. Comparison of unsaturated fatty acid (UFA) content, body weight, and meat quality traits between lambs with high UFA content (H group) and low UFA content (L group). Meat quality traits include ultimate pH (pH value of muscle measured 24 h postmortem), tenderness (shear force, N), cooking loss (%), meat color parameters (L*, lightness; a*, redness; b*, yellowness), protein content (%), moisture content (%), intramuscular fat content (%), and muscle fiber diameter (μm). Data are presented as mean ± SEM. $p < 0.05$ was considered statistically significant.

Table 1. Comparison of fatty acid composition (% of total fatty acids) between different groups.

Fatty Acid	L	H	*p*-Value	Cohen's d
Oleic acid (C18:1n9)	44.17 ± 0.57	47.73 ± 0.50	0.0016	2.17
Arachidonic acid (C20:4n6)	1.18 ± 0.05	1.93 ± 0.19	0.0023	1.73
Palmitic acid (C16:0)	25.27 ± 0.47	23.05 ± 0.54	0.0128	1.4
Stearic acid (C18:0)	19.55 ± 0.59	17.27 ± 0.58	0.0572	1.29
Myristic acid (C14:0)	2.08 ± 0.09	1.63 ± 0.13	0.0522	1.33
Heptadecanoic acid (C17:0)	0.89 ± 0.07	0.69 ± 0.04	0.0797	1.17
Linoleic acid (C18:2n6)	4.78 ± 0.32	5.56 ± 0.52	0.1801	0.63
Palmitoleic acid (C16:1)	1.61 ± 0.10	1.68 ± 0.11	0.738	0.21
Pentadecanoic acid (C15:0)	0.19 ± 0.01	0.11 ± 0.03	0.0601	1.22
cis-10-Heptadecenoic acid (C17:1)	0.61 ± 0.03	0.55 ± 0.02	0.1732	0.83
cis-10-Pentadecenoic acid (C15:1)	0.07 ± 0.03	0.03 ± 0.02	0.4143	0.57
Saturated Fatty Acids (SFAs)	47.94 ± 0.35	42.43 ± 0.80	0.0001	3.44
Unsaturated Fatty Acids (UFAs)	52.08 ± 0.36	57.31 ± 0.79	0.0001	3.48
Monounsaturated Fatty Acids (MUFAs)	46.46 ± 0.74	49.99± 0.71	0.0064	1.98
Polyunsaturated Fatty Acids (PUFAs)	5.62 ± 0.40	7.31 ± 0.87	0.1094	1.01

H: high UFA content group, L: low UFA content group; data are presented as mean ± SEM. $p < 0.05$ was considered statistically significant.

3.2. Transcriptomic Analysis Between High- and Low-UFA Groups

To investigate the transcriptional differences in muscle tissue between high- and low-unsaturated fatty acid (UFA) groups, we performed differential gene expression and functional enrichment analyses. Hierarchical clustering revealed a clear separation in expression profiles between the two groups (Figure 2A), which was further supported by principal component analysis (PCA), where the first principal component (PC1) explained 49% of the total variance (Figure 2B).

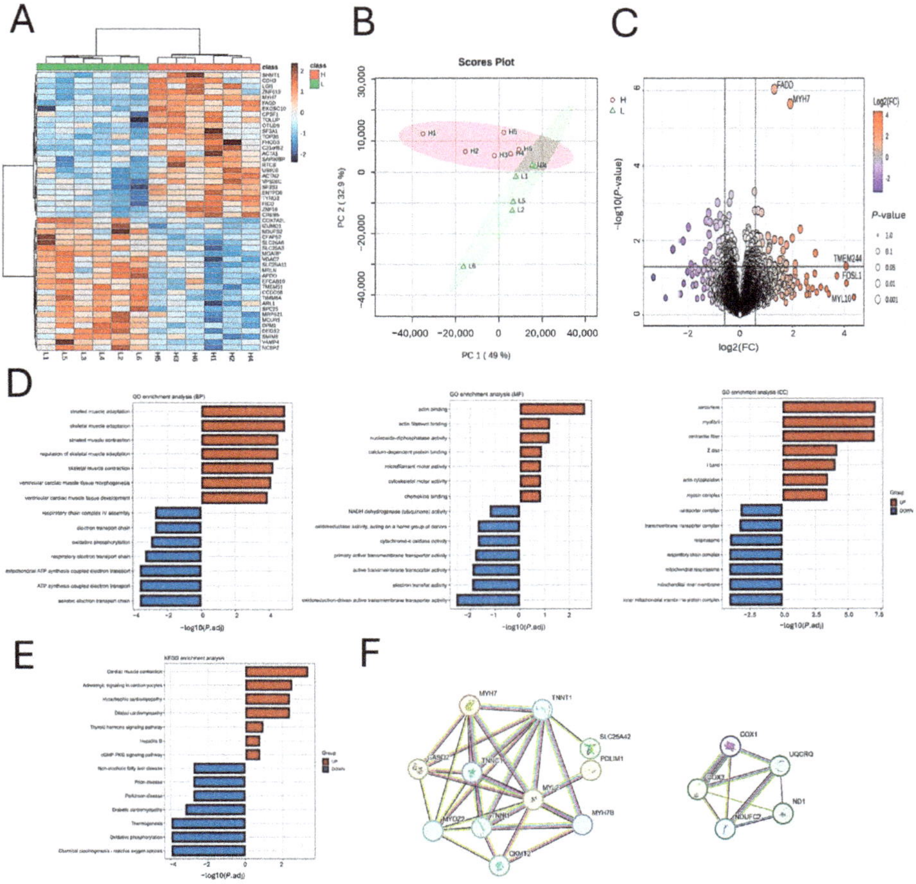

Figure 2. Transcriptomic analysis of longissimus dorsi muscle tissues from lambs with high (H) and low (L) UFA content. (**A**) Heatmap showing hierarchical clustering of differentially expressed genes (DEGs). Each column represents one sample, and each row represents one gene. Red indicates high expression and blue indicates low expression. (**B**) Principal component analysis (PCA) plot showing separation between H and L groups based on transcriptomic profiles (PC1 and PC2 represent the first and second principal components). (**C**) Volcano plot of DEGs. The x-axis shows $\log_2$(fold change, FC), and the y-axis shows $-\log_{10}$(p value). Red dots represent significantly upregulated genes, and blue dots represent significantly downregulated genes (FC > 1.5, $p < 0.05$). (**D**) Gene Ontology (GO) enrichment analysis of DEGs. The top enriched terms are shown in the biological process (BP), molecular function (MF), and cellular component (CC) categories. (**E**) Kyoto Encyclopedia of Genes and Genomes (KEGG) pathway enrichment analysis of DEGs. (**F**) Protein–protein interaction (PPI) network of DEGs. Each node represents a gene, and the edges indicate predicted interactions.

A total of 111 differentially expressed genes (DEGs) were identified based on a threshold of fold change > 1.5 and p < 0.05. Compared to the low-UFA (L) group, 64 genes were significantly upregulated, and 47 genes were downregulated in the high-UFA (H) group (Figure 2C). Notably, genes such as FADD, MYH7, TMEM74, CSNK2A2, and CTSB were markedly upregulated in the high-UFA group, suggesting potential links to fatty acid composition.

Gene Ontology (GO) enrichment analysis of DEGs (Figure 2D) revealed distinct functional patterns. Upregulated genes were enriched in biological processes (BPs) associated with muscle development and contraction, including striated muscle tissue development, muscle tissue development, and cardiac muscle development. In contrast, downregulated genes were associated with metabolic processes such as mitochondrial respiratory chain, ATP synthesis coupled electron transport, and fatty acid metabolism. Regarding molecular functions (MFs), upregulated genes were involved in actin binding, myosin binding, and calcium-dependent protein binding, while downregulated genes were enriched in NADH dehydrogenase activity and mitochondrial proton transport. Cellular component (CC) analysis showed that upregulated genes were primarily localized to myofibrils, contractile fibers, and Z discs, whereas downregulated genes were associated with mitochondrial inner membrane protein complexes.

KEGG pathway enrichment (Figure 2E) further indicated that upregulated genes were significantly involved in pathways such as cardiac muscle contraction, calcium signaling, and PI3K-Akt signaling. Conversely, downregulated genes were enriched in fatty acid degradation, peroxisome, and oxidative phosphorylation pathways. These findings suggest that the high-UFA group exhibits enhanced muscle contraction and signaling activities, whereas the low-UFA group features more active fatty acid metabolism and mitochondrial functions.

Protein–protein interaction (PPI) network analysis (Figure 2F) revealed that upregulated genes formed a muscle structure and contraction-related module centered on MYH7, TNNT1, and TNNI2, while downregulated genes clustered around COX1, ND1, and UQCRC1, key components of the mitochondrial respiratory chain complex.

Collectively, the transcriptomic data suggest that muscles in the high-UFA group exhibit a more oxidative fiber phenotype, with enhanced structural remodeling and signal transduction, whereas those in the low-UFA group show more pronounced fatty acid and energy metabolism. These molecular mechanisms may underline the observed differences in intramuscular UFA content.

3.3. Metabolomic Differences in Muscle Tissues Between High- and Low-UFA Groups

To investigate the impact of varying levels of unsaturated fatty acids (UFAs) on muscle metabolite profiles, we conducted untargeted metabolomic profiling of muscle tissues from high-UFA (H) and low-UFA (L) groups. The results revealed significant differences in metabolite composition between the two groups.

As shown in Figure 3A, hierarchical clustering of differential metabolites demonstrated clear separation between H and L groups, with good within-group consistency. Several metabolites, including N-acetylneuraminic acid, nicotinamide, and 18-hydroxy-5Z,8Z,11Z,14Z-eicosatetraenoic acid, were significantly differentially expressed between the groups.

Principal component analysis (PCA) further confirmed the metabolic divergence (Figure 3B). The first two principal components (PC1 and PC2) accounted for 69.6% of the total variance (PC1: 53.3%, PC2: 16.3%). A clear separation of the H and L groups was observed, indicating substantial differences in muscle metabolite profiles.

Volcano plot analysis identified 25 significantly different metabolites (Figure 3C, Table 2), including compounds upregulated in the H group such as 2(1H)-pyridinone and N-acetylneuraminic acid, and metabolites downregulated in the L group such as NADH and 18-hydroxyeicosatetraenoic acid. Differential metabolites were screened based on fold change ($FC > 1.5$) and statistical significance ($p < 0.05$).

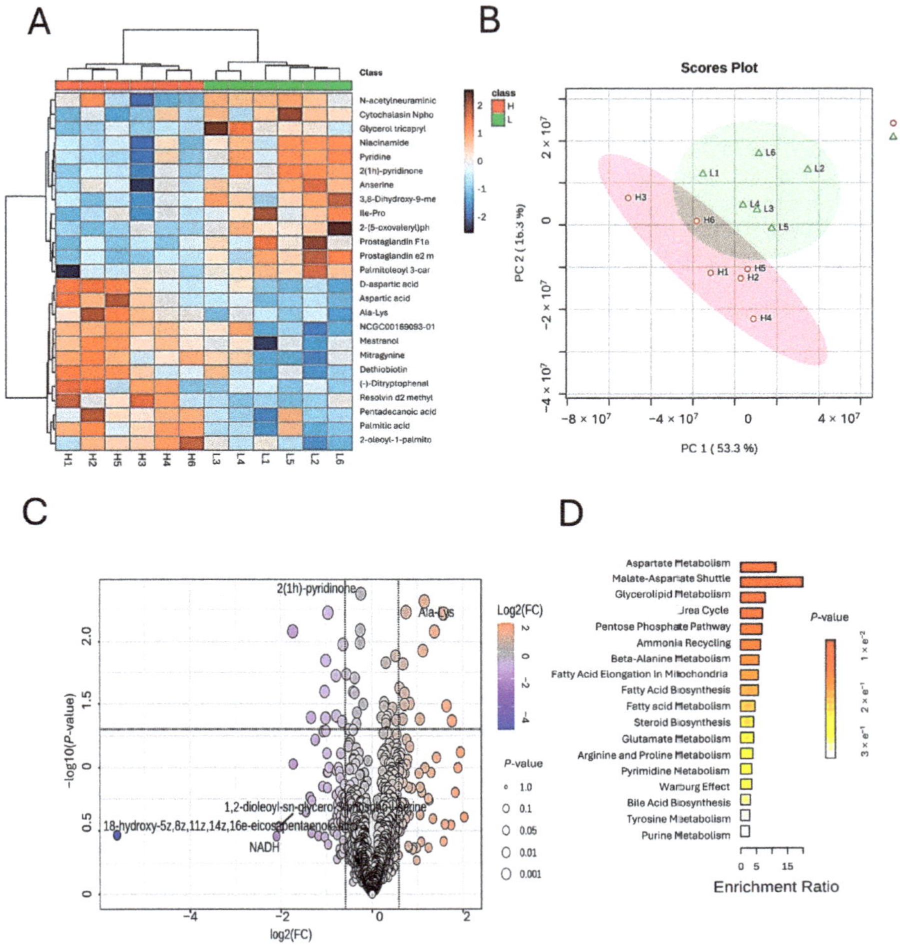

Figure 3. Metabolomic profiling of muscle tissues from lambs with high (H) and low (L) UFA content. (**A**) Heatmap of differentially expressed metabolites (DEMs). Each column represents a sample, and each row represents a metabolite. Red indicates high abundance, and blue indicates low abundance. (**B**) PCA plot showing clustering of samples (H vs. L) based on metabolic profiles (PC1 and PC2 represent the first and second principal components). (**C**) Volcano plot of DEMs. The x-axis shows $\log_2$(FC), and the y-axis shows −log10(p value). Red dots indicate significantly upregulated metabolites, and blue dots indicate significantly downregulated metabolites (FC > 1, $p < 0.05$). (**D**) KEGG pathway enrichment analysis of DEMs.

These differential metabolites span amino acids, fatty acids and their derivatives, and small molecule regulators, suggesting a metabolic reprogramming of lipid and amino acid metabolism in the H group. For example, glutamyllysine, Lys-Trp-Arg, and D-aspartic acid were significantly upregulated in the H group, indicative of enhanced amino acid metabolism. In contrast, metabolites such as 6-phospho-D-gluconate and 2′-deoxycytidine were enriched in the L group, potentially reflecting altered carbohydrate and nucleotide metabolism. Notably, lipid-derived signaling molecules such as Resolvin D2 methyl ester and Prostaglandin E2 methyl ester were differentially expressed in the H group, suggesting roles in modulating lipid accumulation and immune microenvironment in muscle tissue.

Table 2. Differential metabolites between high- and low-UFA muscle groups.

	FC	$\log_2$(FC)	p-Value
Ala-Lys	2.235	1.1603	0.0048582
Palmitic acid	1.6827	0.75082	0.0058927
Prostaglandin e2 methyl ester	0.51287	−0.96334	0.0059498
Aspartic acid	2.9882	1.5793	0.0060464
2-(5-oxovaleryl)phosphatidylcholine	0.30046	−1.7348	0.008306
Resolvin d2 methyl ester	2.6036	1.3805	0.0083364
3,8-Dihydroxy-9-methoxypterocarpan	0.64406	−0.63474	0.010545
D-aspartic acid	2.1818	1.1255	0.011801
Prostaglandin F1a	0.4956	−1.0127	0.014185
NCGC00169093-01	1.6099	0.68698	0.015147
Glycerol tricaprylate	0.57869	−0.78914	0.018894
Pentadecanoic acid	1.524	0.60791	0.023797
Cytochalasin Npho	0.48678	−1.0387	0.025344
Dethiobiotin	1.6431	0.7164	0.026165
(-)-Ditryptophenaline	1.7832	0.83444	0.031486
Isoanhydroicaritin	1.6412	0.71477	0.031982
Lys-Trp-Arg	3.1238	1.6433	0.032825
2′-deoxycytidine	0.58106	−0.78323	0.039292
N6-2-(4-aminophenyl)ethyladenosine	0.3957	−1.3375	0.040449
6-Phospho-D-gluconate	0.49474	−1.0153	0.040617
Prostaglandin d1	0.4935	−1.0189	0.040847
Ciprofloxacin piperazinyl-n4-sulfate	2.0831	1.0587	0.0416
3-Hydroxytetradecanoic acid	1.5959	0.67441	0.041679
Glutamyllysine	3.391	1.7617	0.042961
Idebenone	0.63487	−0.65546	0.043434

Differential metabolites were screened based on fold change (FC > 1.5) and statistical significance ($p < 0.05$).

Regarding fatty acid-related metabolites, palmitic acid, pentadecanoic acid, and 3-hydroxytetradecanoic acid were elevated in the H group, indicating increased fatty acid synthesis and associated enzymatic activity. In contrast, prostaglandins D1 and F1α were enriched in the L group, suggesting differential activation of the cyclooxygenase/prostaglandin pathway in lipid metabolism.

Pathway enrichment analysis revealed that the differential metabolites were predominantly involved in aspartate metabolism, the malate–aspartate shuttle, glycerolipid metabolism, and the urea cycle (Figure 3D). Among these, aspartate metabolism and the malate–aspartate shuttle were most significantly enriched, implying potential roles in regulating intramuscular UFA content.

3.4. Validation by Quantitative PCR and Western Blotting

To verify the reliability of the transcriptomic screening results, five differentially expressed candidate genes were randomly selected for quantitative real-time PCR (qRT-PCR) analysis in muscle tissues from the high-UFA (n = 6) and low-UFA (n = 6) groups. The qRT-PCR results showed expression trends consistent with those observed in the RNA-seq data (Figure 4A), confirming the robustness and reliability of the transcriptomic analysis.

To further validate protein-level expression changes of key genes, we performed Western blotting for two genes closely associated with fatty acid metabolism. As shown in Figure 4B, the protein levels of these targets were significantly higher in the high-UFA group compared to the low-UFA group (n = 3), in agreement with the transcriptomic trends. These results further support the potential regulatory roles of these genes in intramuscular unsaturated fatty acid deposition.

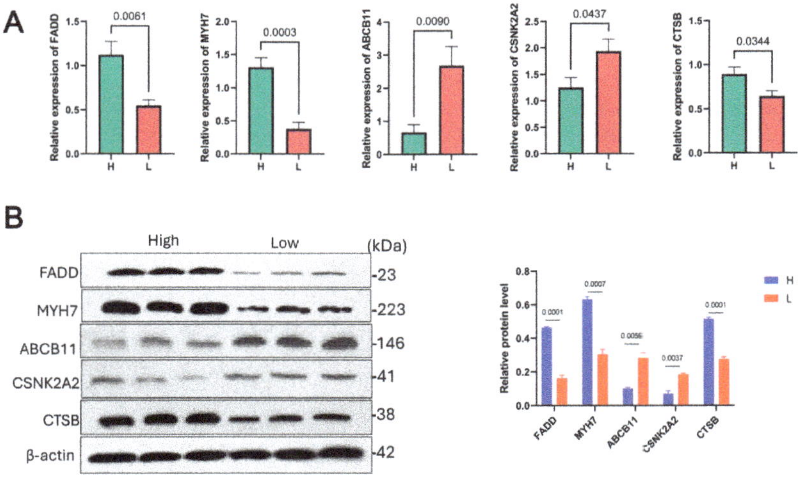

Figure 4. Validation of differentially expressed genes between high (H) and low (L) UFA content groups using qRT-PCR and Western blot analysis. (**A**) Quantitative real-time PCR (qRT-PCR) analysis of five randomly selected DEGs (n = 6 per group). Relative mRNA levels were normalized to β-actin expression. (**B**) Western blot analysis of protein expression levels (n = 3 per group). Band intensities were quantified and expressed as mean ± SEM. $p < 0.05$ was considered statistically significant.

3.5. Comparison of Muscle Fiber Composition Between Groups

Among the genes identified by transcriptome analysis, MYH7 attracted particular attention, as it is a well-known marker gene involved in slow-twitch muscle fiber formation. To determine whether muscle fiber type composition differed between the high- and low-UFA groups, we performed immunofluorescence staining on longissimus dorsi muscle sections to quantify the proportions of slow and fast fibers (Figure 5). The results showed that the proportion of slow-twitch fibers was significantly higher in the high-UFA group compared to the low-UFA group ($p = 0.0101$), consistent with the transcriptomic findings. This suggests that muscles with a higher content of slow fibers tend to have a higher concentration of unsaturated fatty acids.

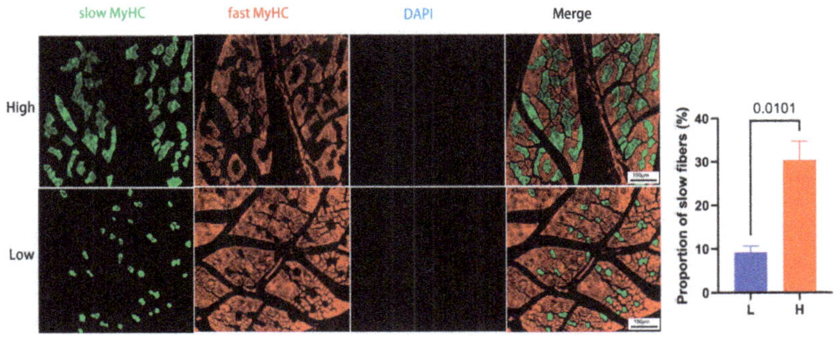

Figure 5. Immunofluorescence staining of muscle fiber types in longissimus dorsi from high (H) and low (L) UFA content groups. Muscle sections were stained with antibodies against slow myosin heavy chain (slow MyHC, green) and fast myosin heavy chain (fast MyHC, red). Cell nuclei were counterstained with DAPI (blue). Representative images are shown (scale bar = 100 μm). Quantitative results are expressed as mean ± SEM (n = 3 per group).

3.6. GSEA Analysis of the MYH7 Gene

To further explore the potential mechanisms by which MYH7 influences the intramuscular unsaturated fatty acid (UFA) content, a gene set enrichment analysis (GSEA) was performed using MYH7 expression levels as the phenotype. The results revealed several significantly enriched pathways associated with differential expression of MYH7 (Figure 6).

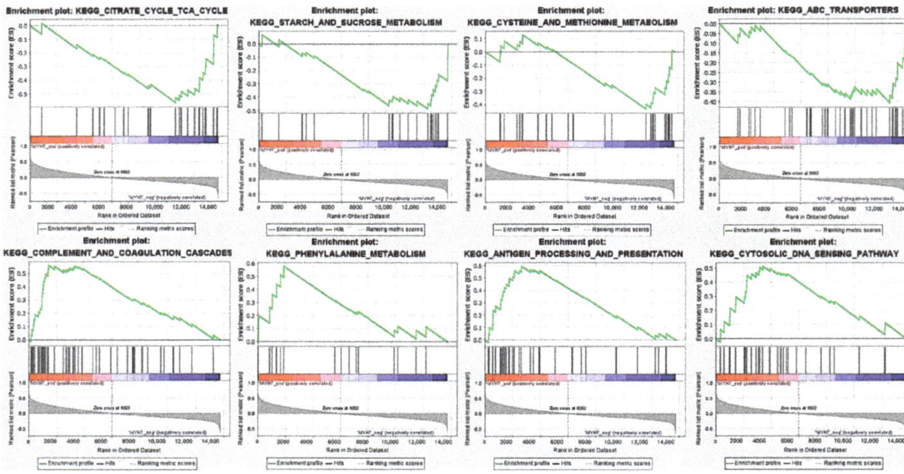

Figure 6. Gene Set Enrichment Analysis (GSEA) of MYH7 in skeletal muscle. Enrichment plots display the distribution of genes ranked by expression level, with curves representing the enrichment score (ES) for significantly associated biological processes and signaling pathways. $p < 0.05$ was considered statistically significant.

Notably, pathways related to energy and substrate metabolism, including the citrate cycle (TCA cycle), starch and sucrose metabolism, cysteine and methionine metabolism, and ABC transporters, were significantly enriched in the MYH7 low-expression group. This suggests that downregulation of MYH7 may enhance fundamental metabolic processes in skeletal muscle, potentially contributing to lipid metabolism and UFA deposition.

Conversely, pathways involved in immune regulation, such as complement and coagulation cascades, antigen processing and presentation, and the cytosolic DNA-sensing pathway, were significantly enriched in the MYH7 high-expression group, indicating a possible link between MYH7 expression and immune response. Additionally, the phenylalanine metabolism pathway was upregulated in the MYH7 high-expression group, reflecting shifts in amino acid metabolism.

Taken together, these findings suggest that MYH7 may influence UFA content in skeletal muscle by modulating metabolic and immune-related pathways. The crosstalk between these biological processes could represent an important mechanism underlying intramuscular fatty acid composition.

4. Discussion

Unsaturated fatty acid (UFA) content is a key determinant of meat nutritional value, flavor attributes, and its impact on consumer health [34]. In this study, we conducted a comprehensive analysis of transcriptomic and metabolomic data from muscle tissues with high versus low UFA levels, which revealed candidate genes, signaling pathways, and metabolic features associated with UFA deposition. These findings provide new molecular insights into lipid metabolism regulation in ruminant skeletal muscle.

4.1. Potential Regulatory Networks Revealed by Differentially Expressed Genes

Differential expression analysis showed significant upregulation of genes involved in energy metabolism, mitochondrial function, and lipid homeostasis in the high-UFA group, suggesting their possible role in intramuscular lipid deposition. Among them, FADD (Fas-associated death domain protein) displayed the highest statistical significance. Traditionally recognized as a key adaptor in apoptosis, recent studies have highlighted its "non-canonical" roles in metabolic regulation. FADD has been shown to modulate adipocyte metabolic activity via the NF-κB and mTOR signaling pathways, thereby influencing lipid droplet formation, lipid storage, and tissue homeostasis [35,36]. Consistently, FADD-deficient mice exhibit reduced respiratory quotient, diminished adipose mass, and decreased expression of lipogenic genes [37], suggesting its potential regulatory role in muscle lipid metabolism.

Other DEGs such as MCUR1, DEGS2, APOO, and ADIG may also participate in lipid metabolic regulation. MCUR1 regulates mitochondrial Ca^{2+} uptake, which has been reported to affect fatty acid oxidation and energy balance [38]. DEGS2, a key enzyme in sphingolipid synthesis, contributes to membrane stability and lipid signaling [39]. APOO is implicated in mitochondrial lipid transport [40], whereas ADIG (adipogenin) has been reported as an adipocyte-enriched transmembrane protein whose expression is strongly induced during adipogenesis; ADIG knockdown impairs adipocyte differentiation in vitro, and Adig knockout mice exhibit reduced fat mass, altered lipid droplet formation and impaired leptin secretion, indicating a functional role in adipogenesis and lipid droplet biology [41]. Thus, their upregulation in the high-UFA group may contribute to enhanced lipid storage capacity in muscle tissue.

Bidirectional GO enrichment analysis showed that upregulated genes were mainly enriched in processes related to muscle development, myofibril assembly, and actin/myosin binding, consistent with previous reports linking structural muscle genes to lipid deposition [42]. Conversely, downregulated genes were enriched in fatty acid metabolism, mitochondrial respiratory chain function, NADH dehydrogenase activity, and ATP synthesis, suggesting a shift toward reduced oxidative metabolism. Similar metabolic remodeling favoring lipid accumulation over catabolism has been reported in livestock muscle with higher intramuscular fat content [43]. Together, these functional enrichments suggest that muscle tissues in the high-UFA group favor lipid storage over oxidative metabolism, creating a metabolic environment conducive to UFA accumulation and flavor compound formation.

4.2. MYH7 as a Key Link Between Slow-Twitch Fibers and Lipid Deposition

MYH7 was significantly upregulated in the high-UFA group, with a fold change even higher than FADD (3.76 vs. 2.46), drawing particular attention. MYH7 encodes β-myosin heavy chain, a molecular marker of slow-twitch muscle fibers, whose elevated expression is typically associated with enhanced oxidative metabolism [21]. Slow-twitch fibers are known to be rich in mitochondria and possess high fatty acid oxidation capacity [44]. Previous studies in cattle and pigs have demonstrated that muscle with higher slow fiber content tends to accumulate more intramuscular lipids, improving meat quality traits [45]. Increased MYH7 expression reflects a higher proportion of slow fibers, which are enriched in lipid droplets and lipid-metabolizing enzymes [46], favoring intramuscular UFA storage. Notably, both FADD and MYH7 are closely linked to mitochondrial function, suggesting they may co-regulate lipid droplet formation and the surrounding metabolic milieu during lipid accumulation.

To further explore potential mechanisms by which MYH7 influences UFA deposition, we conducted GSEA to characterize the broader metabolic context associated with MYH7 expression. Genes negatively correlated with MYH7 were significantly enriched in the TCA cycle, carbohydrate metabolism, and amino acid metabolism, pathways known to provide energy and precursors for lipid synthesis [47]. Conversely, genes positively correlated with MYH7 were enriched in immune-related pathways such as complement and coagulation cascades and antigen processing. Similar immune–metabolic crosstalk has been reported in skeletal muscle, where immune pathways influence lipid turnover and deposition [48,49]. This dual metabolic–immune regulation offers a novel perspective on how MYH7 may influence intramuscular fatty acid composition.

4.3. Metabolomic Characteristics Reveal the Biochemical Basis of Lipid Deposition

Metabolomic analysis revealed significant upregulation of various lipid and amino acid metabolites in the high-UFA group. Elevated levels of N-acetylneuraminic acid and nicotinamide are consistent with enhanced NAD^+-dependent energy metabolism and mitochondrial activity [50], which may promote lipid synthesis or inhibit lipolysis. Upregulated lipid metabolites, including stearic acid, pentadecanoic acid, and 3-hydroxytetradecanoic acid, indicate increased activity of fatty acid biosynthesis pathways [51].

Pathway enrichment analysis further indicated involvement of aspartate metabolism, the malate–aspartate shuttle, triglyceride metabolism, and the urea cycle. These pathways have been reported to provide carbon skeletons and reducing equivalents for fatty acid synthesis in ruminant muscle [52].

Moreover, the upregulation of lipid signaling molecules such as resolvin D2 methyl ester and prostaglandin E2 methyl ester in the high-UFA group may be involved in modulating lipid accumulation and local immune status. These findings align with the immune pathway enrichment observed in MYH7-high samples, suggesting that muscle immune environments may influence lipid homeostasis via metabolic signaling.

4.4. Limitations and Future Perspectives

This study, based on RNA-seq and untargeted metabolomics, systematically identified candidate genes and pathways associated with differential UFA content and proposed a potential regulatory mechanism involving the interplay between muscle structure, metabolism, and immunity. However, several limitations remain: the findings are correlative and require functional validation (e.g., gene knockout/overexpression, lipid droplet staining, or metabolic flux analysis); the use of sheep as the model species raises questions about the translatability of the results to other livestock such as pigs, cattle, or poultry; and whether shifts in muscle fiber type are causal to UFA deposition or merely associative needs to be addressed in future longitudinal or interventional studies.

5. Conclusions

By integrating transcriptomic, metabolomic, and muscle phenotypic data, this study systematically elucidates the molecular basis underlying differences in unsaturated fatty acid (UFA) content in the longissimus dorsi muscle of sheep. Individuals with higher UFA content exhibited distinct transcriptional signatures associated with lipid metabolism, mitochondrial function, and immune responses. Metabolomic profiling further confirmed the accumulation of metabolites related to both fatty acid synthesis and degradation. Notably, MYH7, a hallmark gene of slow-twitch muscle fibers, was upregulated in the high-UFA group, suggesting that muscle fiber-type transition may contribute to UFA deposition. As a potential regulatory nexus linking muscle fiber composition and fatty acid metabolism, MYH7 may play a key role in promoting UFA accumulation in muscle. These findings offer molecular insights and potential targets for improving the fatty acid profile and nutritional quality of lamb meat.

Supplementary Materials: The following supporting information can be downloaded at: https://www.mdpi.com/article/10.3390/ani15172617/s1, Supplementary Table S1: Primer sequences used for RT-qPCR; Supplementary Table S2: Information of primary antibodies used in this study.

Author Contributions: Conceptualization, X.S. and Y.C.; methodology, Y.F., J.C. and X.L.; software, X.H. and X.Y.; validation, Y.F. and J.C.; formal analysis, X.H. and C.L.; writing—original draft preparation, X.H. and X.L.; writing—review and editing, X.S. and Y.C.; visualization, X.Y.; supervision, H.J.; project administration, C.L.; funding acquisition, H.J. All authors have read and agreed to the published version of the manuscript.

Funding: This research was funded by the National Key Research and Development Program of China, grant number 2021YFF1000702, and the Jilin Provincial Key Research and Development Plan Project, grant number 20220202038NC.

Institutional Review Board Statement: This study and included experimental procedures were approved by the Ethics Committee of Jilin Agricultural University (Approval No. 2023-12-1-057) and were conducted in accordance with the guidelines of the Animal Experimentation Center of Jilin Agricultural University (No. 2021 10 20 001).

Informed Consent Statement: Not applicable.

Data Availability Statement: Primer sequences used for RT-qPCR experiments are provided in Supplementary Table S1. RNA-seq data are included in NCBI Sequence Read Archive (SRA) PRJNA1139683. Other data will be shared on reasonable request to the corresponding author.

Acknowledgments: The authors acknowledge the technical support provided by Guofeng Zhang during animal husbandry and sampling.

Conflicts of Interest: The authors declare no conflicts of interest.

Abbreviations

The following abbreviations are used in this manuscript:

UFAs	Unsaturated Fatty Acids
SFAs	Saturated Fatty Acids
DEGs	Differentially Expressed Genes
RT-qPCR	Real-Time Quantitative Polymerase Chain Reaction
WB	Western Blot
GO	Gene Ontology
KEGG	Kyoto Encyclopedia of Genes and Genomes
PCA	Principal Component Analysis
TCA cycle	Tricarboxylic Acid Cycle
LC–MS	Liquid Chromatography–Mass Spectrometry
FASN	Fatty Acid Synthase
FABP3	Fatty Acid-Binding Protein 3
MYH7	Myosin Heavy Chain 7
IMF	Intramuscular Fat
TAG	Triacylglycerol
RNA-Seq	RNA Sequencing
FC	Fold Change
ATP	Adenosine Triphosphate
NADH	Nicotinamide Adenine Dinucleotide (Reduced form)

References

1. Ding, W.; Lu, Y.; Xu, B.; Chen, P.; Li, A.; Jian, F.; Yu, G.; Huang, S. Meat of Sheep: Insights into Mutton Evaluation, Nutritive Value, Influential Factors, and Interventions. *Agriculture* **2024**, *14*, 1060. [CrossRef]
2. Trbović, D.; Lukić, M.; Vranić, D.; Dordevic, V.; Stanisic, N.; Parunović, N.; Ćirić, J. Comprehensive Levels of Fatty Acids in Meat: Implications for Human Health. *Baltica* **2023**, *36*, 45–63. [CrossRef]
3. Fowler, S.; Morris, S.; Hopkins, D. Nutritional Composition of Lamb Retail Cuts from the Carcases of Extensively Finished Lambs. *Meat Sci.* **2019**, *154*, 126–132. [CrossRef] [PubMed]
4. Mahmud, I.; Karadaş, F. Effects of Melatonin and Vitamin E (Alpha-Tocopherol Acetate) as Antioxidants on Biochemical Blood Parameters, Lipid Profile, and Muscle Vitamin E Concentration in Awassi Lambs Fed a High-Energy Diet and Normal Diet. *Arch. Razi Inst.* **2023**, *78*, 249–259.
5. Keady, T.; Hanrahan, J.; Fagan, S. Cobalt Supplementation, Alone or in Combination with Vitamin B and Selenium: Effects on Lamb Performance and Mineral Status. *J. Anim. Sci.* **2017**, *95*, 379–386. [PubMed]
6. Sainsbury, J.; Schönfeldt, H.; Heerden, S. The nutrient composition of South African mutton. *J. Food Compos. Anal.* **2011**, *24*, 720–726. [CrossRef]
7. Li, S.; Li, X.; Ma, Q.; Wang, Z.; Fang, F.; Zhang, D. Consumer Preference, Behaviour and Perception about Lamb Meat in China. *Meat Sci.* **2022**, *192*, 108878. [CrossRef] [PubMed]
8. Lu, Z.; Li, J.; Yuan, C.; Xi, B.; Yang, B.; Meng, X.; Guo, T.; Yue, Y.; Gao, Y.; Liu, J.; et al. Evaluation of Mutton Quality Characteristics of Dongxiang Tribute Sheep Based on Membership Function and Gas Chromatography and Ion Mobility Spectrometry. *Front. Nutr.* **2022**, *9*, 852399. [CrossRef]
9. Xu, X.; Liu, H.; Wang, X.; Zhang, Q.; Guo, T.; Hu, L.; Xu, S. Evaluation of the Longissimus Thoracis et Lumborum Muscle Quality of Chaka and Tibetan Sheep and the Analysis of Possible Mechanisms Regulating Meat Quality. *Animals* **2023**, *13*, 2494. [CrossRef]
10. Xu, X.; Guo, T.; Zhang, Q.; Liu, H.; Wang, X.; Li, N.; Wang, Y.; Wei, L.; Hu, L.; Xu, S. Comparative Evaluation of the Nutrient Composition and Lipidomic Profile of Different Parts of Muscle in the Chaka Sheep. *Food Sci. Anim. Resour.* **2024**, *44*, 1305–1326. [CrossRef]
11. McPherson, R.; Spiller, G. Effects of Dietary Fatty Acids and Cholesterol on Cardiovascular Disease Risk Factors in Man. In *Handbook of Lipids in Human Nutrition*; CRC Press: Boca Raton, FL, USA, 2020; pp. 41–50.
12. Borges, M.C.; Haycock, P.C.; Zheng, J.; Hemani, G.; Holmes, M.V.; Smith, G.D.; Hingorani, A.D.; Lawlor, D.A. Role of Circulating Polyunsaturated Fatty Acids on Cardiovascular Diseases Risk: Analysis Using Mendelian Randomization and Fatty Acid Genetic Association Data from Over 114,000 UK Biobank Participants. *BMC Med.* **2022**, *20*, 210. [CrossRef]
13. De Souza, R.; Mente, A.; Maroleanu, A.; Cozma, A.; Ha, V.; Kishibe, T.; Uleryk, E.; Budylowski, P.; Schünemann, H.; Beyene, J.; et al. Intake of Saturated and Trans Unsaturated Fatty Acids and Risk of All Cause Mortality, Cardiovascular Disease, and Type 2 Diabetes: Systematic Review and Meta-Analysis of Observational Studies. *BMJ* **2015**, *351*, h3978. [CrossRef]
14. Mykolaichuk, L.P.; Lesnovskaya, O.V.; Sanzhara, R.A.; Pokhyl, O.M.; Gutyj, B.V.; Dochkin, D.O.; Mylostyvyi, R.V. The Fatty Acid Composition of Lamb Makes It a Valuable Raw Material for the Meat Processing Industry. *Sci. Messenger LNU Vet. Med. Biotechnol.* **2024**, *26*, 109–112. [CrossRef]
15. Yao, Y.; Wang, H.; Lu, Z.; Nian, F.; Zheng, C.; Li, F.; Tang, D. Improving Shelf Life and Content of Unsaturated Fatty Acids in Meat of Lambs Fed a Diet Supplemented with Grape Dregs. *Foods* **2023**, *12*, 4204. [CrossRef]
16. Liu, Y.; Li, R.; Ying, Y.; Zhang, Y.; Huang, Y.; Wu, H.; Lin, K. Non-genetic Factors Affecting the Meat Quality and Flavor of Inner Mongolian Lambs: A Review. *Front. Vet. Sci.* **2022**, *9*, 1067880. [CrossRef]
17. Sacchetti, M.; Saltin, B.; Osada, T.; Hall, G. Intramuscular Fatty Acid Metabolism in Contracting and Non-contracting Human Skeletal Muscle. *J. Physiol.* **2002**, *540*, 387–395. [CrossRef]
18. Schultz, K.; Marmorstein, R. Molecular Basis for Regulation of Mammalian Fatty Acid Synthase. *FASEB J.* **2022**, *36*, L8108. [CrossRef]
19. Zheng, Q.; Liu, L.; Guo, X.; Zhu, F.; Huang, Y.; Qin, Q.; Huang, X. Fish ELOVL7a Is Involved in Virus Replication via Lipid Metabolic Reprogramming. *Fish Shellfish Immunol.* **2024**, *149*, 109530. [CrossRef] [PubMed]
20. Fu, Y.; Chen, Y.; Han, X.; Tan, D.; Chen, J.; Lai, C.; Yang, X.; Shan, X.; Silva, L.H.P.; Jiang, H. Effects of Muscle Fiber Composition on Meat Quality, Flavor Characteristics, and Nutritional Traits in Lamb. *Foods* **2025**, *14*, 2309. [CrossRef] [PubMed]
21. Lee, L.A.; Barrick, S.K.; Meller, A.; Walklate, J.; Lotthammer, J.M.; Tay, J.W.; Stump, W.T.; Bowman, G.; Geeves, M.A.; Greenberg, M.J.; et al. Functional divergence of the sarcomeric myosin, MYH7b, supports species-specific biological roles. *J. Biol. Chem.* **2023**, *299*, 1102657. [CrossRef] [PubMed]
22. Carrillo, J.A.; He, Y.; Li, Y.; Liu, J.; Erdman, R.A.; Sonstegard, T.S.; Song, J. Integrated metabolomic and transcriptome analyses reveal finishing forage affects metabolic pathways related to beef quality and animal welfare. *Sci. Rep.* **2016**, *6*, 25948. [CrossRef] [PubMed]
23. Diniz, W.; Ward, A. 282 Multi-omics approaches to improve animal production. *J. Anim. Sci.* **2021**, *99*, 20–21. [CrossRef]

24. Kumar, S.T.; Zheng, Y.; Xu, J.; Zhao, Z.; Zhang, Q.; Zhang, Y.; Li, M.; Zou, H.; Azeem, R.M.; Sun, W.S.; et al. Transcriptome and Metabolome Insights into Key Genes Regulating Fat Deposition and Meat Quality in Pig Breeds. *Animals* **2024**, *14*, 3560. [CrossRef] [PubMed]
25. Frezarim, G.B.; Mota, L.F.M.; Fonseca, L.F.S.; Salatta, B.M.; Arikawa, L.M.; Schmidt, P.I.; Silva, D.B.S.; Albuquerque, L.G. Multi-omics integration identifies molecular markers and biological pathways for carcass and meat quality traits in Nellore cattle. *Sci. Rep.* **2025**, *15*, 10467. [CrossRef] [PubMed]
26. Zhao, B.; Luo, H.; He, J.; Huang, X.; Chen, S.; Fu, X.; Zeng, W.; Tian, Y.; Liu, S.; Li, C.; et al. Comprehensive transcriptome and methylome analysis delineates the biological basis of hair follicle development and wool-related traits in Merino sheep. *BMC Biol.* **2021**, *19*, 197. [CrossRef] [PubMed]
27. Wen, W.-X.; Mei, Z.-Q.; Wang, X.-Y.; Wang, J.-X.; Zhang, J.-H. Lipid metabolism in longissimus thoracis muscle of Tibetan sheep at different ages: Transcriptomic insights. *Front. Nutr.* **2022**, *9*, 847077.
28. *GB 5009.168–2016*; The National Health and Family Planning Commission of the People's Republic of China. National Food Safety Standard: Determination of Fatty Acid in Food. China Standards Press: Beijing, China, 2016; pp. 1–6.
29. Salavati, M.; Caulton, A.; Clark, R.; Gazova, I.; Smith, T.P.L.; Worley, K.C.; Cockett, N.E.; Archibald, A.L.; Clarke, S.; Murdoch, B.M.; et al. Global analysis of transcription start sites in the new ovine reference genome (Oar rambouillet v1.0). *Front. Genet.* **2020**, *11*, 580580. [CrossRef]
30. Kim, D.; Paggi, J.M.; Park, C.; Bennett, C.; Salzberg, S.L. Graph-based genome alignment and genotyping with HISAT2 and HISAT-genotype. *Nat. Biotechnol.* **2019**, *37*, 907–915. [CrossRef]
31. Love, M.I.; Huber, W.; Anders, S. Moderated estimation of fold change and dispersion for RNA-seq data with DESeq2. *Genome Biol.* **2014**, *15*, 550. [CrossRef]
32. Yu, G.; Wang, L.G.; Han, Y.; He, Q.Y. clusterProfiler: An R package for comparing biological themes among gene clusters. *OMICS A J. Integr. Biol.* **2012**, *16*, 284–287. [CrossRef]
33. Pang, Z.; Lu, Y.; Zhou, G.; Hui, F.; Xu, L.; Viau, C.; Spigelman, A.F.; MacDonald, P.E.; Wishart, D.S.; Li, S. MetaboAnalyst 6.0: Towards a unified platform for metabolomics data processing, analysis and interpretation. *Nucleic Acids Res.* **2024**, *52*, W398–W406. [CrossRef] [PubMed]
34. Hou, M.; Sun, W.; Ye, H.; Zhai, X.; Xue, Y.; Tang, R.; Teng, S.; Wu, R.; Luo, H. Comparative Analysis for Nutrients, Flavor Compounds, and Lipidome Revealed the Edible Value of Pond-Cultured Male Pelodiscus sinensis with Different Ages. *Food Chem.* **2024**, *454*, 139795. [CrossRef] [PubMed]
35. Wang, T.; Lin, P.; Wang, Y.; Chen, Y.; Zhang, Z.; Li, F.; Feng, J. FADD Cooperates with Caspase-8 to Positively Regulate the Innate Immune Response and Promote Apoptosis Following Bacterial Infection in Japanese Eel. *Fish Shellfish Immunol.* **2025**, *157*, 110110. [CrossRef]
36. Ranjan, K.; Pathak, C. Cellular Dynamics of Fas-Associated Death Domain in the Regulation of Cancer and Inflammation. *Int. J. Mol. Sci.* **2024**, *25*, 3228. [CrossRef]
37. Tang, J.; Li, M.; Liu, X.; Wang, Y.; Zhang, J.; Shu, H.; Liu, Z.; Zhang, C.; Fu, L.; Hu, J.; et al. FADD Regulates Adipose Inflammation, Adipogenesis, and Adipocyte Survival. *Cell Death Discov.* **2024**, *10*, 323. [CrossRef] [PubMed]
38. Mallilankaraman, K.; Doonan, P.; Cardenas, C.; Chandramoorthy, H.C.; Muller, M.; Miller, R.; Hoffman, N.E.; Gandhirajan, R.K.; Molgo, J.; Birnbaum, M.J.; et al. MICU1 is an essential gatekeeper for MCU-mediated mitochondrial Ca2+ uptake that regulates cell survival. *Cell* **2012**, *151*, 630–644. [CrossRef]
39. Tidhar, R.; Futerman, A.H. The complexity of sphingolipid biosynthesis in the endoplasmic reticulum. *Biochim. Et Biophys. Acta* **2013**, *1833*, 2511–2518. [CrossRef]
40. Koob, S.; Reichert, A.S. Novel intracellular functions of apolipoproteins: The ApoO protein family as constituents of the Mitofilin/MINOS complex determines cristae morphology in mitochondria. *Biol. Chem.* **2014**, *395*, 285–296. [CrossRef]
41. Álvarez-Guaita, A.; Patel, S.; Lim, K.; Haider, A.; Dong, L.; Conway, O.J.; Ma, M.K.; Chiarugi, D.; Saudek, V.; O'rAhilly, S.; et al. Phenotypic characterization of Adig null mice suggests roles for adipogenin in the regulation of fat mass accrual and leptin secretion. *Cell Rep.* **2021**, *34*, 108810. [CrossRef]
42. Picard, B.; Lefaucheur, L.; Berri, C.; Duclos, M.J. Muscle fibre ontogenesis in farm animal species. *Reprod. Nutr. Dev.* **2002**, *42*, 415–431. [CrossRef]
43. Wood, J.D.; Richardson, R.I.; Nute, G.R.; Fisher, A.V.; Campo, M.M.; Kasapidou, E.; Sheard, P.R.; Enser, M. Effects of fatty acids on meat quality: A review. *Meat Sci.* **2004**, *66*, 21–32. [CrossRef]
44. Komiya, Y.; Sakazaki, Y.; Goto, T.; Kawabata, F.; Suzuki, T.; Sato, Y.; Sawano, S.; Nakamura, M.; Tatsumi, R.; Ikeuchi, Y.; et al. Eicosapentaenoic Acid Increases Proportion of Type 1 Muscle Fibers Through PPARδ and AMPK Pathways in Rats. *iScience* **2024**, *27*, 109816. [CrossRef]
45. Ryu, Y.C.; Kim, B.C. The relationship between muscle fiber characteristics, postmortem metabolic rate, and meat quality of pig longissimus dorsi muscle. *Meat Sci.* **2005**, *71*, 351–357. [CrossRef] [PubMed]
46. Saedi, A.; Debruin, D.; Hayes, A.; Hamrick, M. Lipid Metabolism in Sarcopenia. *Bone* **2022**, *164*, 116539. [CrossRef]

47. Arnold, P.; Finley, L. Regulation and function of the mammalian tricarboxylic acid cycle. *J. Biol. Chem.* **2023**, *299*, 102838. [CrossRef] [PubMed]
48. Frisard, M.I.; McMillan, R.P.; Marchand, J.; Wahlberg, K.A.; Wu, Y.; Voelker, K.A.; Heilbronn, L.; Haynie, K.; Muoio, B.; Li, L.; et al. Toll-like receptor 4 modulates skeletal muscle substrate metabolism. *Am. J. Physiol.-Endocrinol. Metab.* **2010**, *298*, E988–E998. [CrossRef]
49. Hoene, M.; Weigert, C. The role of interleukin-6 in insulin resistance, body fat distribution and energy balance. *Obes. Rev.* **2008**, *9*, 20–29. [CrossRef]
50. Yoshino, J.; Mills, K.F.; Yoon, M.J.; Imai, S.I. Nicotinamide mononucleotide, a key NAD+ intermediate, treats the pathophysiology of diet- and age-induced diabetes in mice. *Cell Metab.* **2011**, *14*, 528–536. [CrossRef]
51. Wakil, S.J.; Stoops, J.K.; Joshi, V.C. Fatty acid synthesis and its regulation. *Annu. Rev. Biochem.* **1983**, *52*, 537–579. [CrossRef] [PubMed]
52. Wu, G. Functional amino acids in nutrition and health. *Amino Acids* **2013**, *45*, 407–411. [CrossRef]

animals

MDPI

Article

Effects of *COLQ* Gene Missense Mutations on Growth and Meat Traits in Leizhou Black Goats

ing Huang [1], Ke Wang [2,3], Yuelang Zhang [4], Jiancheng Han [2], Hanlin Zhou [2,3,*] and Qinyang Jiang [1,*]

1 Guangxi Key Laboratory of Animal Breeding, Disease Control and Prevention, College of Animal Science and Technology, Guangxi University, Nanning 530004, China; huangjing07182024@163.com
2 Zhanjiang Experimental Station, Chinese Academy of Tropical Agricultural Sciences, Zhanjiang 524013, China; lp_wangke@163.com (K.W.); hanjiancheng810@163.com (J.H.)
3 Sanya Research Institute, Chinese Academy of Tropical Agricultural Sciences, Sanya 572024, China
4 Hainan Institute of Zhejiang University, Sanya 572024, China; zhangyuelang@zju.edu.cn
* Correspondence: zhouhanlin8@163.com (H.Z.); jiangqinyang2013@gxu.edu.cn (Q.J.); Tel./Fax: +86-0771-3235635 (H.Z. & Q.J.)

Simple Summary

The Leizhou Black Goat is a prized local breed from southern China, known for growing quickly and providing high-quality meat. However, many young goats suffer from poor growth and weak muscles, which is a problem for farmers. We suspected that this might be caused by tiny, natural changes in their genes. Our study focused on one specific gene called *COLQ*, which is important for muscle function. We discovered several natural variations within this gene. Two of these variations were found to be directly linked to the goats' size and meat quality. One variation, in particular, significantly reduces the activity of the *COLQ* gene. This discovery is very useful because it allows breeders to use a simple genetic test to select parent goats that are more likely to produce healthy, strong offspring with better growth. This will help improve the sustainability and productivity of goat farming.

Abstract

As an indigenous goat breed unique to southern China, Leizhou Black Goats (LZBGs) are highly valued for their rapid growth, high reproductive performance, and superior meat quality. However, their offspring frequently exhibit symptoms of muscle atrophy and malnutrition, suggesting potential genetic defects underlying these adverse phenotypes. As a unique extracellular matrix component, collagen Q (*COLQ*) is specifically enriched within the synaptic basal lamina at vertebrate neuromuscular junctions (NMJs), where it anchors acetylcholinesterase (AChE) to facilitate efficient acetylcholine hydrolysis, ensuring precise neuromuscular signaling. The current investigation sought to characterize the spectrum of genetic polymorphisms within the *COLQ* gene and assess their correlation with key production traits, including growth performance and meat quality parameters, in the LZBG population. Previously, through whole-genome sequencing and transcriptome sequencing analyses of an LZBG population, we identified four SNPs in the *COLQ* gene, namely, two missense mutations (SNP1: p.238A/S and SNP3: p.47G/S), one intronic variant (SNP2), and one synonymous mutation (SNP4: p.101P/P). Population genetic analysis revealed strong linkage disequilibrium between SNP1 and SNP2. Computational modeling of protein structures predicted that the identified missense mutations may lead to alterations in protein conformation. Association analyses demonstrated significant correlations of SNP1 and SNP3 with growth and meat quality traits ($p < 0.05$), where SNP3 reduced *COLQ* expression by 0.64-fold in homozygotes. Association analysis revealed that both SNP1 and SNP3 showed significant correlations with growth and meat quality traits in LZBGs

($p < 0.05$). Notably, SNP3 (p.47G/S) was found to regulate *COLQ* gene expression, reducing its levels by 0.64-fold in homozygous individuals, suggesting its potential as a genetic marker for selecting goats with superior growth performance and muscular development characteristics. The identified genetic variants establish a foundation for marker-assisted selection in LZBG breeding programs with particular relevance to growth performance enhancement, while also advancing the understanding of *COLQ*'s functional mechanisms in muscle development.

Keywords: *COLQ*; genetic variants; goat; productivity performance; association analysis

1. Introduction

Goat husbandry, concentrated in tropical and semi-arid regions of developing countries, serves as a key income source for smallholders within mixed farming systems [1]. Compared to cattle and monogastric livestock, goats offer higher economic viability due to lower costs, faster reproduction, diversified products (meat, milk, and fiber), efficient use of byproducts, and strong resilience to drought and heat stress [2]. Furthermore, goat meat provides a nutritious red meat option with low cholesterol and fat content [3].

The Leizhou Black Goat (LZBG), an indigenous ruminant breed native to tropical China, demonstrates remarkable thermo-hygric adaptability, coupled with superior production traits, including accelerated growth kinetics, enhanced reproductive performance, and premium meat quality characteristics [4,5]. However, this breed frequently exhibits juvenile myopathic manifestations and nutritional deficiencies, even under conditions of adequate milk supply and proper nutrition, suggesting an underlying hereditary myopathy. These congenital conditions, often attributed to genetic mutations affecting neuromuscular signal transmission [6], muscle protein structure [7], or cellular metabolism [8], ultimately compromise production efficiency and commercial viability [9]. Therefore, screening for candidate genes and genetic variations related to growth and muscle development is of great significance for the breeding of LZBGs.

Collagen Q (COLQ), a specialized non-fibrillar collagen, plays an essential role in organizing the vertebrate neuromuscular junction by anchoring synaptic acetylcholinesterase (AChE) and participates in MuSK-dependent signaling for synaptic (CMS) regulation [10]. CMS are a group of inherited disorders caused by genetic mutations in the neuromuscular junction (NMJ), characterized by activity-dependent muscle weakness and deficient AChE function. Mutations in COLQ are known to cause congenital myasthenic syndromes across species [11]. For instance, Rinz et al. identified a pathogenic *COLQ* variant (I337T) in Labrador Retrievers leading to generalized myopathy [12], while Abitbol et al. reported a homozygous missense mutation in Sphinx and Devon Rex cats resulting in progressive muscle weakness due to disrupted AChE aggregation [13]. Laforgia et al. documented a case involving a large homozygous deletion in the *COLQ* C-terminal domain [14]. More recently, Ncube et al. associated the *COLQ* gene with carcass fat content and muscle content, suggesting a broader influence on muscle traits [15]. Once the CMS caused by such genetic mutations is inherited and accumulates within the population, it will lead to a sharp decline in both the quality and yield of livestock products. Fortunately, this genetic disorder can be efficiently eliminated through early screening. However, as of now, there are fewer reports related to *COLQ* gene in goats. Similar to these studies, our previous analysis in goats screened out *COLQ* too [16]. Given this context, we hypothesize that genetic variations in the *COLQ* gene regulate muscle development in goats, thereby influencing growth and meat quality traits.

This study sought to characterize *COLQ* gene expression profiles and identify potential SNPs within *COLQ* in LZBGs, investigating their associations with growth and muscle development. Additionally, correlations between *COLQ* genotypes and mRNA expression levels were assessed. These findings may serve as a valuable reference for goat breeding programs, advance functional studies of the *COLQ* gene, and uncover potential molecular markers for MAS applications.

2. Materials and Methods

2.1. Feeding Environment and Management

This study utilized experimental animals housed at the Leizhou Black Goat Farm in Danzhou, Hainan Province. All animals were maintained on an identical feeding regimen and raised under standardized conditions [17]. All experimental protocols received approval from the Animal Care and Use Committee of the Chinese Academy of Tropical Agricultural Sciences (CATAS-2025002ZES), with all procedures, including sample collection, conducted in strict compliance with institutional ethical guidelines (GB/T 35892-2018) [18].

2.2. Sample and Phenotypic Data Collection

Ear tissue samples were randomly collected from healthy, multiparous (having given birth multiple times) female Leizhou Black Goats, all approximately 2 years of age. Growth-related traits, including CC (chest circumference), CW (chest width), BL (body length), CD (Chest depth), BH (body height), WH (withers height), HW (hip width), and BW (body weight), were systematically measured and documented in all 1010 experimental animals. A subset of 98 two-year-old animals was randomly selected from the population for carcass evaluation, including measurements of water loss rate (WLR), shear force (SF), carcass weight (CW), cross-section area of longissimus dorsi lumbois muscle (CALM), and water holding capacity (WHC). The specific collection method strictly follows the Chinese agricultural industry standard NY/T 630-2002 [19]. For comprehensive gene expression analysis, multiple tissue types (including skin, heart, gluteofemoral biceps, longissimus dorsi muscle, rumen, liver, kidney, abdominal adipose, cerebrum, and cerebellum) were harvested from 18 randomly selected adult female animals. Longissimus dorsi muscle specimens ($N = 5 \times 6 = 30$) were obtained from LZBGs at defined developmental timepoints, namely, 0 days (neonatal), 6 months, and 1, 2, 3, and 4 years of age, enabling developmental expression profiling. Concurrently, longissimus dorsi muscle samples ($N = 32$) from 2-year-old females were analyzed to determine genotype-dependent effects of missense mutations on *COLQ* transcript levels. After the slaughtering and cutting of all the tissue samples, except for the fresh samples required for meat quality testing, they were immediately snap-frozen in liquid nitrogen and subsequently stored at −80 °C for long-term preservation.

2.3. Genomic DNA and Total RNA Extraction

Genomic DNA extraction from ear tissue samples was performed with the Animal Tissues/Cells Genomic DNA Extraction Kit (Solarbio, Beijing, China), followed by purity and concentration assessment using a NanoDrop™ 1000 Spectrophotometer (Thermo Fisher Scientific, Waltham, MA, USA). The DNA samples were diluted to 20 ng/μL using ddH_2O and stored at −20 °C until further analysis. Total RNA was extracted using the TRIzol method (Solarbio, Beijing, China), and the first-strand cDNA was synthesized by PrimeScript™ RT Reagent Kit (Takara, Tokyo, Japan) following the manufacturer's protocol.

2.4. Primer Synthesis, Amplifications, and Genotyping

Based on the GGVD (Goat Genome Variation Database, http://animal.omics.pro/code/index.php/GoatVar, accessed on 20 July 2024), 18 LZBGs' whole genome sequencing (WGS) and RNA-seq results (Supplementary Materials, File S1), 2 SNPs (chr1:152339819C>A; chr1:152348368C>T) on the LZBG *COLQ* gene were identified [16]. Subsequently, using the caprine *COLQ* gene reference sequence (NC_030808.1) as a template, four specific primer pairs were designed using the NCBI Primer-BLAST tool (last accessed: 1 August 2024) to amplify target regions. Exon-spanning primers were designed to quantify *COLQ* mRNA expression levels, using *GAPDH* as the endogenous control (Table S1). Conventional and quantitative PCR amplifications were performed following established protocols (Chen et al., 2015) [20]. *COLQ* variants were subsequently identified through Sanger sequencing. SNPs were detected via sequence alignment against the reference genome using BioXM software (v2.7.1; Nanjing Agricultural University, China).

2.5. Statistical Analysis

The analysis was performed using the Wei Sheng Xin program (https://www.bioinformatics.com.cn/, accessed on 8 May 2025) and GenAlEx (accessed on 12 May 2025) to calculate the linkage disequilibrium (LD) structure, population genetic parameters, Hardy–Weinberg equilibrium (HWE), and polymorphic information content (PIC) of SNP loci in the goat *COLQ* gene. Phylogenetic conservation and evolutionary relationships of *COLQ* across species were analyzed using Ensembl's gene tree and ortholog prediction tools. The COLQ protein structure was predicted via homology modeling using Swiss-Model (accessed 29 May 2025). Associations between *COLQ* SNPs and growth/muscle traits in Leizhou Goats were evaluated via the general linear model [9,21] (SPSS v23.0; IBM, Armonk, NY, USA), with the following model:

$$\mathrm{Y_{sj}} = \mu + \mathrm{G_s} + \mathrm{e_j}$$

where $\mathrm{Y_{sj}}$ represents the phenotypic value, μ denotes the population mean, $\mathrm{G_s}$ indicates the genotype effect, and $\mathrm{e_j}$ signifies random error. Relative *COLQ* expression levels were calculated using the $2^{-\Delta\Delta Ct}$ method [22], and the data derived from qPCR were analyzed using one-way analysis of variance followed by a post-hoc test. All of the data were presented as the mean $\pm$ standard error (S.E.), and $p < 0.05$ was considered to be significant.

3. Results

3.1. Identification of Missense Mutations in Goat COLQ Gene

Sanger sequencing revealed four genetic variants (Figure 1A); two were identified as missense mutations (chr1:152339819 C>A, COLQ p.238A/S; chr1:152348368 C>T, COLQ p.47G/S). A mutation was identified in an intronic region (chr1:152339884 T>A, COLQ), while the other mutation (chr1:152348390 A>G, COLQ p.101P/P) was classified as synonymous mutation based on CDS alignment using UniProt and Ensembl databases. Genotype frequency distributions for the four identified SNPs are presented in Table 1. All variants displayed intermediate polymorphic information content (PIC values: 0.25–0.50), indicating moderate genetic diversity. Furthermore, strong linkage disequilibrium was observed between SNP1 (p.238A/S) and SNP2 (chr1:152339884 T>A) (Figure S1).

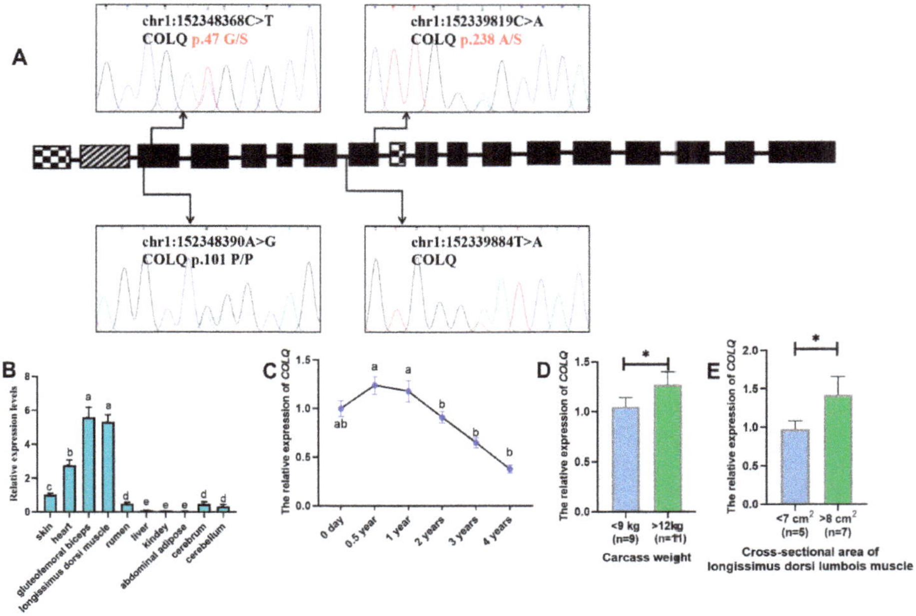

Figure 1. Identification of SNPs and expression of the *COLQ* gene in goats. (**A**) Localization and identification of SNPs in the *COLQ* gene by Sanger sequencing. The black module represents the exon regions shared by all three transcript variants. The mosaic module on the far left indicates the exon region unique to the COLQ-201 transcript (ENSCHIT00000016986.1). The central mosaic module denotes the exon regions common to both COLQ-201 and COLQ-202 transcripts (ENSCHIT00000016986.1; ENSCHIT00000017002.1). The hatched module shows the exon regions shared between COLQ-202 and COLQ-203 transcripts (ENSCHIT00000017002.1; ENSCHIT00000017011.1). The two missense mutations are marked in red. (**B**) Tissue expression profile of the *COLQ* gene in adult female goats. Letters (a–e) indicate significant differences ($p < 0.05$) in expression levels among tissues. (**C**) Temporal expression profile of the *COLQ* gene in longissimus dorsi muscle. Letters (a, b) indicate significant differences ($p < 0.05$) in expression levels across time points. (**D**) Expression of the *COLQ* gene in individuals with different extreme carcass weights. The * indicates significant differences ($p < 0.05$) between groups. (**E**) Expression of the *COLQ* gene in individuals with different extreme cross-sectional areas of longissimus dorsi muscle. The * indicates significant differences ($p < 0.05$) between groups.

Table 1. Genotypic frequencies and population parameters in the *COLQ* gene.

Loci	Size	Genotypic Frequencies			HWE	Population Parameters			
	N	Ref	Ref/Mut	Mut	*p*-Value	Ho	He	Ne	PIC
SNP1 p.238A/S	996	0.089	0.422	0.489	$p > 0.05$	0.422	0.420	1.720	0.332
SNP2 g.152339884T>A	1002	0.076	0.423	0.501	$p > 0.05$	0.423	0.409	1.690	0.325
SNP3 p.47G/S	997	0.540	0.378	0.082	$p > 0.05$	0.378	0.395	1.650	0.317
SNP4 p.101P/P	1005	0.191	0.598	0.211	$p > 0.05$	0.598	0.500	2.000	0.375

Note: HWE, Hardy–Weinberg equilibrium. Ho, observed homozygosity. He, heterozygosity. Ne, effective allele numbers. PIC, polymorphism information content.

3.2. The Impact of Missense Mutations on the Structure of the COLQ Protein

The gene homology analysis revealed that the goat *COLQ* gene exhibited the closest phylogenetic relationships with those of sheep, Bovinae, Siberian musk deer, and Yarkand deer (Figure S2), which is consistent with their evolutionary divergence patterns. Structural predictions of wild-type and mutant COLQ protein tertiary structures (Figures S3 and S4) revealed conformational differences between the two variants. The magnified views highlight regions of structural divergence (denoted by black dashed circles) that may result from amino acid substitutions, insertions, or deletions. These structural variations could potentially affect protein stability, enzymatic activity, or molecular interaction properties.

3.3. *The mRNA Expression of COLQ in LZBGs*

We systematically analyzed the mRNA expression profile of *COLQ* in multiple tissues from adult LZBGs. Ubiquitous expression of *COLQ* was detected in all examined tissues (Figure 1B), with the highest expression levels observed in cardiac and skeletal muscles, particularly in the heart, gluteofemoral biceps, and longissimus dorsi. Notably, the longissimus dorsi muscle exhibited significant differential *COLQ* mRNA expression between juvenile (0.5–1 year) and sexually/physically mature (2–4 years) developmental stages (Figure 1C). Furthermore, individuals with extreme phenotypic values for the cross-sectional area of longissimus dorsi (<7 cm^2 vs. >8 cm^2) and carcass weight (<9 kg vs. >12 kg) demonstrated significant differences in *COLQ* expression levels (Figure 1D).

3.4. *Association Analysis of COLQ Genetic Variants with Phenotypic Traits*

Table 1 summarizes the genotype frequency distributions and population genetic parameters of the four identified genetic variants. With the exception of SNP4 (p.101P/P), all three mutations were in Hardy–Weinberg equilibrium. The polymorphism information content (PIC) analysis demonstrated that all four loci exhibited intermediate polymorphism levels, rendering them suitable as genetic markers for population genetic or association studies. Notably, SNP4 displayed the highest PIC value (0.375), indicating the greatest informativeness; however, its deviation from Hardy–Weinberg equilibrium ($p < 0.05$) suggests potential population stratification or selection pressure, warranting cautious interpretation in subsequent analyses. Genetic association analysis of goat *COLQ* polymorphisms with growth and carcass traits identified SNP1 (p.Ala238Ser) as significantly associated ($p < 0.01$) with multiple zootechnical parameters: body height, chest width, chest circumference, withers height, hip width, body weight, carcass weight, and longissimus dorsi muscle cross-sectional area (Table 2). Significant phenotypic advantages were observed in wild-type homozygotes over mutant homozygotes for seven traits (excluding body height) and over heterozygotes for six traits (excluding body weight and longissimus dorsi area). SNP2 (g.152339884T>A) showed significant correlations with body height, withers height, and body weight (Table S2). SNP3 (p.47G/S) demonstrated significant associations with body height, chest circumference, body weight, carcass weight, cross-section area of longissimus dorsi lumbois muscle, and shear force (Table 3). Significant phenotypic advantages were observed in wild-type homozygotes over mutant homozygotes for six traits and over heterozygotes for five traits (excluding body weight and longissimus dorsi area). Of particular significance, SNP4 (p.101P/P) was exclusively associated with shear force (Table S3).

Table 2. The association analysis between the traits and SNP1 p.238A/S in the goat *COLQ* gene.

Traits	Genotypes (Mean ± SE)			*p* Values
	Ref	Ref/Mut	Mut	
body height (cm)	$54.21^{a} \pm 0.42$	$53.02^{b} \pm 0.36$	$53.44^{ab} \pm 0.33$	0.031
chest depth (cm)	26.47 ± 0.30	26.41 ± 0.19	26.30 ± 0.23	0.341
chest width (cm)	$15.83^{a} \pm 0.17$	$15.68^{b} \pm 0.23$	$15.64^{b} \pm 0.19$	0.026
body length (cm)	63.77 ± 0.32	63.69 ± 0.35	63.74 ± 0.41	0.842

Table 2. *Cont.*

Traits	Genotypes (Mean ± SE) Ref	Ref/Mut	Mut	p Values
chest circumference (cm)	71.33 a ± 0.43	69.69 b ± 0.46	70.15 b ± 0.38	0.041
withers height (cm)	56.17 a ± 0.43	54.62 b ± 0.35	54.73 b ± 0.28	0.019
hip width (cm)	17.77 ± 0.28	17.39 ± 0.19	17.51 ± 0.24	0.042
body weight (kg)	28.85 a ± 0.32	27.33 b ± 0.38	27.41 b ± 0.47	0.037
carcass weight (kg)	9.65 a ± 0.23	9.44 a ± 0.11	9.20 b ± 0.17	0.047
cross-section area of longissimus dorsi muscle (cm^2)	7.71 a ± 0.14	7.53 a ± 0.26	7.16 b ± 0.21	0.022
water loss rate (%)	4.51 ± 0.23	4.73 ± 0.11	4.65 ± 0.14	0.181
water holding capacity (%)	4.76 ± 0.14	4.53 ± 0.17	4.58 ± 0.08	0.407
shear force (N)	48.88 ± 0.24	49.01 ± 0.28	48.32 ± 0.31	0.083

Note: letters (a, b) indicate significant differences ($p < 0.05$) between different genotypes and traits.

Table 3. The association analysis between the traits and SNP3 p.47G/S in the goat *COLQ* gene.

Traits	Genotypes (Mean ± SE) Ref	Ref/Mut	Mut	p Values
body height (cm)	52.91 a ± 0.19	52.92 a ± 0.28	51.24 b ± 0.49	0.047
chest depth (cm)	26.33 ± 0.18	26.21 ± 0.19	26.31 ± 0.27	0.341
chest width (cm)	15.54 ± 0.15	15.56 ± 0.27	15.23 ± 0.33	0.268
body length (cm)	63.91 ± 0.37	62.80 ± 0.42	62.66 ± 0.42	0.427
chest circumference (cm)	70.82 a ± 0.24	69.35 a ± 0.27	67.37 b ± 0.41	0.024
withers height (cm)	54.97 ± 0.36	54.85 ± 0.24	53.42 ± 0.47	0.062
hip width (cm)	17.52 ± 0.21	17.09 ± 0.17	16.92 ± 0.11	0.207
body weight (kg)	27.88 a ± 0.25	27.07 a ± 0.32	25.96 b ± 0.21	0.021
carcass weight (kg)	9.71 a ± 0.19	9.33 b ± 0.16	9.29 b ± 0.14	0.039
cross-section area of longissimus dorsi muscle (cm^2)	7.74 a ± 0.12	7.31 b ± 0.20	7.39 b ± 0.28	0.044
water loss rate (%)	4.61 ± 0.15	4.52 ± 0.18	4.83 ± 0.21	0.506
water holding capacity (%)	4.68 ± 0.10	4.64 ± 0.13	4.59 ± 0.09	0.072
shear force (N)	49.03 a ± 0.22	48.77 a ± 0.38	47.38 b ± 0.25	0.023

Note: letters (a, b) indicate significant differences ($p < 0.05$) between different genotypes and traits.

3.5. Functional Impact of Missense Variants on COLQ Transcriptional Regulation

We investigated the differences in the corresponding mRNA expression levels in the longissimus dorsi muscle of female goats (Figure 2) among different SNP genotype groups. The genotype did not affect ($p > 0.05$) the *COLQ* expression in SNP1 (p.238A/S), while the *COLQ* expression in SNP3 (p.47G/S) was lower ($p < 0.01$) in the Mut genotype than in the Ref and Ref/Mut genotypes, being similar ($p > 0.05$) between the latter. Notably, the mutant homozygote of SNP3 (p.47G/S) exhibited a 0.64-fold reduction in *COLQ* expression compared to the wild-type.

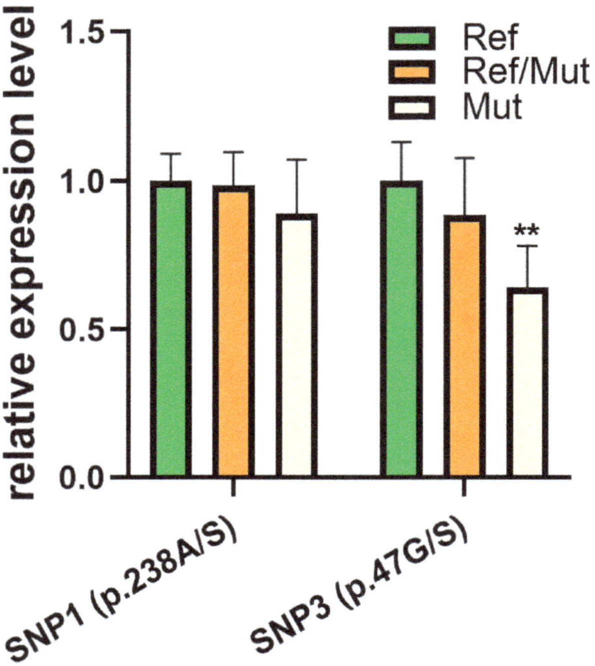

Figure 2. Missense mutations affect the expression of the COLQ gene in goat longissimus dorsi muscle. Comparison with normal genotype; ** indicates $p < 0.01$.

4. Discussion

Functioning as a specialized collagen, *COLQ* serves a crucial structural role by mediating the stable attachment of acetylcholinesterase (AChE) to the synaptic basal lamina within neuromuscular junctions (NMJs) [23]. Moreover, *COLQ* participates in the long-term maintenance of neuromuscular junction (NMJ) stability through its interaction with the MuSK-mediated signaling pathway [24]. MuSK is a receptor tyrosine kinase specifically expressed on the muscle cell membrane. As a key regulator of neuromuscular junction (NMJ) formation, its activation recruits downstream proteins (such as Dok-7 and Rapsyn), mediates the clustering and stabilization of acetylcholine receptors (AChRs), and modulates the expression profile of synapse-related genes [25]. The function of *COLQ*-anchored acetylcholinesterase (AChE) depends on the integrity of the MuSK signaling pathway. Mutations in *COLQ* may disrupt this pathway, leading to impaired neuromuscular transmission and subsequently affecting muscle development and meat traits in goats [26]. Previous studies have confirmed that pathogenic mutations in the *COLQ* gene are a known cause of congenital myasthenic syndrome related to acetylcholinesterase deficiency (CMS-EAD) [27]. The phenotypic manifestations are likely linked to mechanisms of MuSK signaling deficiency that cause structural abnormalities at the NMJ and myasthenic symptoms, thereby providing a theoretical basis for explaining the phenotypes induced by *COLQ* mutations in Leizhou Black Goats [28]. In the present study, *COLQ* exhibited predominant expression in muscular tissues. Notably, significant differential expression of *COLQ* was observed in the longissimus dorsi muscle between juvenile and sexually/physically mature stages in goats (*Capra hircus*), suggesting its potential involvement in myogenesis and muscle maturation regulatory processes. Furthermore, individuals exhibiting extreme phenotypic values for carcass weight and longissimus dorsi cross-sectional area demonstrated significant variations in *COLQ* expression levels, indicating a putative dose-dependent relationship between its expression and muscular development.

Herein, we identified four SNPs, namely, two missense mutations (SNP1: p.238A/S; SNP3: p.47G/S), one synonymous mutation (SNP4: p.101P/P), and one intronic variant (SNP2: chr1:152339884 T>A). Both missense mutations were located within functional domains of the COLQ protein. Homology modeling revealed that SNP1 and SNP3 induced localized conformational changes, potentially impairing acetylcholine esterase binding capacity and thereby disrupting neuromuscular signal transduction [29,30]. Notably, SNP3, situated in a conserved N-terminal region, significantly reduced *COLQ* mRNA expression and exhibited associations with both growth and meat quality traits. This functional impact may be mediated through diminished AChE stability or via interference with *COLQ* transcriptional activity [10], suggesting its role as a key causative mutation. Although SNP1 did not markedly alter expression levels, it correlated with multiple growth traits (e.g., body height and cross-sectional area of longissimus dorsi lumbois muscle), possibly exerting effects through post-translational modifications such as protein folding or enzymatic activity [31]. The intronic mutation (SNP2) showed strong linkage disequilibrium with SNP1, implying potential synergistic regulation of *COLQ* function and highlighting haplotype effects that may surpass individual SNP contributions. The synonymous SNP4, while not altering the amino acid sequence, demonstrated an association with shear force, possibly attributable to effects on mRNA stability [32], splicing efficiency [33], or linkage disequilibrium with functional SNPs via epigenetic mechanisms [34]. In the association analysis, the wild-type homozygotes consistently exhibited optimal phenotypic values across all measured traits for each SNP, indicating that these mutations represent loss-of-function variants. In practical breeding applications, early screening and elimination of these deleterious alleles could effectively reduce their frequency within the population, thereby improving the overall phenotypic mean [35,36].

These data indicate that the SNP3 (p.47G/S) variant modulates COLQ expression and shows potential as a molecular marker for breeding goats with improved growth and muscling phenotypes. However, its functional effects require validation in larger populations. Further in vitro studies, including *COLQ* knockdown/overexpression in myocytes, are needed to elucidate its direct impact on myocyte differentiation, along with enzymatic activity assays of the mutant protein, to facilitate its application in molecular breeding programs.

5. Conclusions

This study identified two missense mutations (SNP1 and SNP3) in the *COLQ* gene that are significantly associated with growth and carcass traits in Leizhou Black Goats. SNP3 reduces *COLQ* expression, making it a promising marker for breeding and providing new insights into its role in muscle development.

Supplementary Materials: The following supporting information can be downloaded at: https://www.mdpi.com/article/10.3390/ani15172618/s1. File S1: List of SNPs obtained by transcriptome sequencing annotation of LZBG. Table S1: Primer information; Table S2: Association analysis between traits and SNP2 g.152339884T>A in the goat *COLQ* gene; Table S3: Association analysis between traits and SNP4 p.101P/P in the goat *COLQ* gene; Figure S1: Linkage disequilibrium (LD) heatmap of SNPs in the goat *COLQ* gene; Figure S2: Ensembl gene trees of the *COLQ* gene; Figures S3 and S4: Predicted tertiary structures of wild-type and mutant COLQ proteins.

Author Contributions: Conceptualization, H.Z. and Q.J.; Formal Analysis, J.H. (Jing Huang) and K.W.; Project Administration, J.H. (Jiancheng Han) and H.Z.; Resources, Y.Z.; Validation, J.H. (Jing Huang); Visualization, K.W. and J.H. (Jiancheng Han); Writing—Original Draft, J.H. (Jing Huang) and K.W. All authors have read and agreed to the published version of the manuscript.

Funding: This research was supported with funding from Guangxi Key Laboratory of Animal Breeding, Disease Control and Prevention (ABDC-b202402), the Joint Funds of the Chinese Central Public-interest Scientific Institution Basal Research Fund (1630012025802, 1630012025807), and Hainan Provincial Natural Science Foundation of China Youth Fund Project (SQ2024QNJJ0594).

Institutional Review Board Statement: Sample collection followed China's national standard for Laboratory Animal Welfare and Ethical Review (GB/T 35892-2018) [18]. The experimental procedures were authorized by the Review Committee for the Use of Animal Subjects at the Chinese Academy of Tropical Agricultural Sciences and conducted in compliance with the ethics commission guidelines (CATAS-2025002ZES).

Informed Consent Statement: Not applicable.

Data Availability Statement: The original contributions presented in this study are included in the article/Supplementary Materials. Further inquiries can be directed to the corresponding author.

Acknowledgments: We thank the staff from the Demonstration Base of Tropical Herbage and Livestock Circular Agriculture and the Danzhou Hainan Goat Breeding Farm for help with sampling and measuring goat traits.

Conflicts of Interest: The authors declare no conflicts of interest.

References

1. Escareño, L.; Salinas-Gonzalez, H.; Wurzinger, M.; Iñiguez, L.; Sölkner, J.; Meza-Herrera, C. Dairy goat production systems: Status quo, perspectives and challenges. *Trop. Anim. Health Prod.* **2013**, *45*, 17–34. [CrossRef]
2. Akinmoladun, O.F.; Muchenje, V.; Fon, F.N.; Mpendulo, C.T. Small ruminants: Farmers' hope in a world threatened by water scarcity. *Animals* **2019**, *9*, 456. [CrossRef]
3. Aung, S.H.; Abeyrathne, E.D.N.S.; Hossain, M.A.; Jung, D.Y.; Kim, H.C.; Jo, C.; Nam, K.C. Comparative quality traits, flavor compounds, and metabolite profile of Korean native black goat meat. *Food Sci. Anim. Resour.* **2023**, *43*, 639–658. [CrossRef]
4. Li, Y.; Xu, X.; Deng, M.; Zou, X.; Zhao, Z.; Huang, S.; Liu, D.; Liu, G. Identification and comparative analysis of long non-coding RNAs in high- and low-fecundity goat ovaries during estrus. *Front. Genet.* **2021**, *12*, 648158. [CrossRef]
5. Yu, X.; Wang, H.; Li, Y.; Mu, X.; Yuan, K.; Wu, A.; Guo, J.; Hong, Y.; Zhang, H. Occurrence and genotypic identification of *Blastocystis* spp., *Enterocytozoon bieneusi*, and *Giardia duodenalis* in Leizhou Black goats in Zhanjiang City, Guangdong Province, China. *Animals* **2023**, *13*, 2777. [CrossRef] [PubMed]
6. Belaya, K.; Rodríguez Cruz, P.M.; Liu, W.W.; Maxwell, S.; McGowan, S.; Farrugia, M.E.; Petty, R.; Walls, T.J.; Sedghi, M.; Basiri, K.; et al. Mutations in GMPPB cause congenital myasthenic syndrome and bridge myasthenic disorders with dystroglycanopathies. *Brain* **2015**, *138*, 2493–2504. [CrossRef]
7. Johari, M.; Topf, A.; Folland, C.; Duff, J.; Dofash, L.; Marti, P.; Robertson, T.; Vilchez, J.; Cairns, A.; Harris, E.; et al. Loss-of-function variants in JPH1 cause congenital myopathy with prominent facial and ocular involvement. *J. Med. Genet.* **2024**, *61*, 992–998. [CrossRef]
8. De Palma, S.; Leone, R.; Grumati, P.; Vasso, M.; Polishchuk, R.; Capitanio, D.; Braghetta, P.; Bernardi, P.; Bonaldo, P.; Gelfi, C. Changes in muscle cell metabolism and mechanotransduction are associated with myopathic phenotype in a mouse model of collagen VI deficiency. *PLoS ONE* **2013**, *8*, e56716. [CrossRef]
9. Wu, Q.; Han, X.; Zhang, Y.; Liu, H.; Zhou, H.; Wang, K.; Han, J. One copy number variation within the angiopoietin-1 gene is associated with Leizhou Black goat meat quality. *Animals* **2024**, *14*, 2682. [CrossRef]
10. Uyen Dao, T.M.; Barbeau, S.; Messéant, J.; Della-Gaspera, B.; Bouceba, T.; Semprez, F.; Legay, C.; Dobbertin, A. The collagen ColQ binds to LRP4 and regulates the activation of the Muscle-Specific Kinase-LRP4 receptor complex by agrin at the neuromuscular junction. *J. Biol. Chem.* **2023**, *299*, 104962. [CrossRef]

11. Legay, C. Congenital myasthenic syndromes with acetylcholinesterase deficiency, the pathophysiological mechanisms. *Ann. N. Y. Acad. Sci.* **2018**, *1413*, 104–110. [CrossRef]
12. Rinz, C.J.; Levine, J.; Minor, K.M.; Humphries, H.D.; Lara, R.; Starr-Moss, A.N.; Guo, L.T.; Williams, D.C.; Shelton, G.D.; Clark, L.A. A COLQ missense mutation in Labrador Retrievers having congenital myasthenic syndrome. *PLoS ONE* **2014**, *9*, e106425. [CrossRef]
13. Abitbol, M.; Hitte, C.; Bossé, P.; Blanchard-Gutton, N.; Thomas, A.; Martignat, L.; Blot, S.; Tiret, L. A COLQ missense mutation in Sphynx and Devon Rex cats with congenital myasthenic syndrome. *PLoS ONE* **2015**, *10*, e0137019. [CrossRef]
14. Laforgia, N.; De Cosmo, L.; Palumbo, O.; Ranieri, C.; Sesta, M.; Capodiferro, D.; Pantaleo, A.; Iapicca, P.; Lastella, P.; Capozza, M.; et al. The first case of congenital myasthenic syndrome caused by a large homozygous deletion in the C-terminal region of COLQ (collagen like tail subunit of asymmetric acetylcholinesterase) protein. *Genes* **2020**, *11*, 1519. [CrossRef]
15. Ncube, K.T.; Dzomba, E.F.; Hadebe, K.; Soma, P.; Frylinck, L.; Muchadeyi, F.C. Carcass quality profiles and associated genomic regions of South African goat populations investigated using goat SNP50K genotypes. *Animals* **2022**, *12*, 364. [CrossRef]
16. Wang, K.; Xu, M.; Han, X.; Liu, H.; Han, J.; Sun, W.; Zhou, H. Transcriptome analysis of muscle atrophy in Leizhou black goats: Identification of key genes and insights into limb-girdle muscular dystrophy. *BMC Genom.* **2025**, *26*, 80. [CrossRef]
17. Huang, J.; Xu, M.; Zhang, Y.; Han, J.; Zhou, H.; Wang, K. Missense Mutations in FDNC5 Associated with Morphometric Traits and Meat Quality in Hainan Black Goats. *Animals* **2025**, *15*, 565. [CrossRef] [PubMed]
18. *GB/T 35892-2018*; Laboratory Animal—Guideline for Ethical Review of Animal Welfare. National Laboratory Animal Standardization Technical Committee: Beijing, China, 2018.
19. *NY/T 630-2002*; Mutton Quality Grading. Ministry of Agriculture of the People's Republic of China: Beijing, China, 2002.
20. Chen, Y.; Yang, L.; Lin, X.; Peng, P.; Shen, W.; Tang, S.; Lan, X.; Wan, F.; Yin, Y.; Liu, M. Effects of genetic variation of the sorting nexin 29 (SNX29) gene on growth traits of Xiangdong Black goat. *Animals* **2022**, *12*, 3461. [CrossRef] [PubMed]
21. Wang, K.; Yan, H.; Xu, H.; Yang, Q.; Zhang, S.; Pan, C.; Chen, H.; Zhu, H.; Liu, J.; Qu, L.; et al. A novel indel within goat casein alpha S1 gene is significantly associated with litter size. *Gene* **2018**, *671*, 161–169. [CrossRef] [PubMed]
22. Wang, K.; Kang, Z.; Jiang, E.; Yan, H.; Zhu, H.; Liu, J.; Qu, L.; Lan, X.; Pan, C. Genetic effects of DSCAML1 identified in genome-wide association study revealing strong associations with litter size and semen quality in goat (*Capra hircus*). *Theriogenology* **2020**, *146*, 20–25. [CrossRef]
23. Krejci, E.; Legay, C.; Thomine, S.; Sketelj, J.; Massoulié, J. Differences in expression of acetylcholinesterase and collagen Q control the distribution and oligomerization of the collagen-tailed forms in fast and slow muscles. *J. Neurosci.* **1999**, *19*, 10672–10679. [CrossRef]
24. Karmouch, J.; Dobbertin, A.; Sigoillot, S.; Legay, C. Developmental consequences of the ColQ/MuSK interactions. *Chem. Biol. Interact.* **2013**, *203*, 287–291. [CrossRef]
25. Wargon, I.; Richard, P.; Kuntzer, T.; Sternberg, D.; Nafissi, S.; Gaudon, K.; Lebail, A.; Bauche, S.; Hantaï, D.; Fournier, E.; et al. Long-term follow-up of patients with congenital myasthenic syndrome caused by COLQ mutations. *Neuromuscul Disord.* **2012**, *22*, 318–324. [CrossRef]
26. Cole, R.N.; Reddel, S.W.; Gervásio, O.L.; Phillips, W.D. Anti-MuSK patient antibodies disrupt the mouse neuromuscular junction. *Ann. Neurol.* **2008**, *63*, 782–789. [CrossRef]
27. Ohno, K.; Otsuka, K.; Ito, M. Roles of collagen Q in MuSK antibody-positive myasthenia gravis. *Chem. Biol. Interact.* **2016**, *259 Pt B*, 266–270. [CrossRef]
28. Shigemoto, K.; Konishi, T.; Ohta, M. Myasthenia gravis induced by autoantibodies against MuSK. *Nihon Rinsho* **2008**, *66*, 1149–1154. [PubMed]
29. Liu, Y.; Li, Z.; Shi, Y.; Xu, Y.; Wang, Z.; Wang, N.; Yang, K. Characterization of Novel Splicing Mutations and a Recurrent Deletion in COLQ Congenital Myasthenic Syndrome. *FASEB J.* **2025**, *39*, e70865. [CrossRef]
30. Dingová, D.; Kučera, M.; Hodbod, T.; Fischmeister, R.; Krejci, E.; Hrabovská, A. Cardiac acetylcholinesterase and butyrylcholinesterase have distinct localization and function. *Am. J. Physiol. Heart Circ. Physiol.* **2025**, *328*, H526–H542. [CrossRef]
31. Deepak Shyl, E.S.; Malgija, B.; Iniyan, A.M.; Vincent, S.G.P. Mutation in MCL1 predicted loop to helix structural transition stabilizes MCL1-Bax binding interaction favoring cancer cell survival. *Proteins* **2022**, *90*, 1699–1713. [CrossRef] [PubMed]
32. Hunt, R.C.; Simhadri, V.L.; Iandoli, M.; Sauna, Z.E.; Kimchi-Sarfaty, C. Exposing synonymous mutations. *Trends Genet.* **2014**, *30*, 308–321. [CrossRef] [PubMed]
33. Khabou, B.; Siala-Sahnoun, O.; Gargouri, L.; Mkaouar-Rebai, E.; Keskes, L.; Hachicha, M.; Fakhfakh, F. In silico investigation of the impact of synonymous variants in ABCB4 gene on mRNA stability/structure, splicing accuracy and codon usage: Potential contribution to PFIC3 disease. *Comput. Biol. Chem.* **2016**, *65*, 103–109. [CrossRef]
34. Garcia, J.A.; Lohmueller, K.E. Negative linkage disequilibrium between amino acid changing variants reveals interference among deleterious mutations in the human genome. *PLoS Genet.* **2021**, *17*, e1009676. [CrossRef] [PubMed]

35. Boshove, A.; Derks, M.F.L.; Sevillano, C.A.; Lopes, M.S.; van Son, M.; Knol, E.F.; Dibbits, B.; Harlizius, B. Large scale sequence-based screen for recessive variants allows for identification and monitoring of rare deleterious variants in pigs. *PLoS Genet.* **2024**, *20*, e1011034. [CrossRef] [PubMed]
36. Zhang, Y.; Guo, G.; Huang, H.; Lu, L.; Wang, L.; Fang, L.; Liu, L.; Wang, Y.; Zhang, S. Screening for JH1 genetic defect carriers in Jersey cattle by a polymerase chain reaction and restriction fragment length polymorphism assay. *J. Vet. Diagn. Investig.* **2015**, *27*, 596–599. [CrossRef] [PubMed]

animals

MDPI

Article

Screening of circRNAs Associated with Secondary Wool Follicle Development in Fine-Wool Sheep and Construction of Their ceRNA Network

Yu Luo [1,2], Binpeng Xi [1,2], Yufang Song [1,2], Tong Xiao [1,2], Zengkui Lu [1,2], Jianbin Liu [1,2], Chao Yuan [1,2] and Tingting Guo [1,2,*]

[1] Key Laboratory of Animal Genetics and Breeding on the Tibetan Plateau, Ministry of Agriculture and Rural Affairs, Lanzhou Institute of Husbandry and Pharmaceutical Sciences, Chinese Academy of Agricultural Sciences, Lanzhou 730050, China; 17393144478@163.com (Y.L.); w5562080w@163.com (B.X.); songyf805@163.com (Y.S.); xiaotong3110134994@163.com (T.X.); luzengkui@caas.cn (Z.L.); liujianbin@caas.cn (J.L.); yuanchao@caas.cn (C.Y.)

[2] Sheep Breeding Engineering Technology Research Center of Chinese Academy of Agricultural Sciences, Lanzhou 730050, China

* Correspondence: guotingting@caas.cn

Simple Summary

Wool is an important raw material of animal origin that is irreplaceable in national economies. Improving the yield of high-quality fine-wool sheep has been the core goal of fine-wool-sheep-breeding research. Wool growth is a complex physiological and biochemical process regulated by a variety of factors. Secondary follicle (SF) development directly determines wool yield and quality. In this study, we investigated the role of circular RNA (circRNA) in regulating SF development at ten key embryonic stages in Alpine Merino sheep. The expression profiles of circRNAs were constructed by RNA sequencing, and 173 differentially expressed circRNAs were found to be significantly enriched in core pathways of hair follicle development, such as Wnt/β-catenin, transforming growth factor-β/Smad, Notch, and mitogen-activated protein kinase. Further construction of 88 circRNA–miRNA–mRNA regulatory networks showed that 16 key circRNAs interacted with 44 miRNAs and 65 mRNAs through a competing endogenous RNA mechanism. Among these, six candidate molecules, including circRNA_07722 and circRNA_08249, provided important targets for analysing the molecular regulatory network of autologous fat development. These findings not only deepen our understanding of the regulatory mechanism of wool growth but also provide a new theoretical basis and technical route for the molecular breeding of fine-wool sheep.

Abstract

Various signalling molecules and pathways critical for wool production and quality regulate wool secondary follicle (SF) development. Circular RNAs (circRNAs) regulate SF morphogenesis through a competing endogenous RNA (ceRNA) mechanism; these novel cyclic non-coding RNAs are not known to regulate the development of SFs in the foetal period of fine-wool sheep. Here, we analysed circRNA expression profiles in the foetal skin of Gansu Alpine fine-wool sheep at 10 developmental stages (E87, E90, E93, E96, E99, E102, E105, E108, E111, and E138) using RNA sequencing. Among the differentially expressed circRNAs (DE circRNAs), 173 were significantly enriched in signalling pathways related to hair follicle (HF) development, such as Wnt/β-catenin, transforming growth factor-β/Smad, Notch, and mitogen-activated protein kinase. Six HF-development-related circRNAs were expressed at different stages and potentially regulated SF development through the ceRNA network. In total, 16 DE circRNAs and their targets, 44 miRNAs, and 65 mRNAs were screened, and 88 ceRNA regulatory network pathways related to SF development were constructed. These key DE circRNAs could be candidate genes for further exploration of the molecular HF development mechanism, providing an important theoretical basis for unravelling the regulatory network of SF development in fine-wool sheep and genetic wool trait improvement.

Keywords: fine-wool sheep; secondary wool follicle; development; circRNA; circRNA–miRNA–mRNA

1. Introduction

Wool, an important animal-derived textile raw material, plays a significant role in national economies. Increasing the yield of high-quality fine wool is an important direction in the current research on fine-wool-sheep breeding. Wool growth is a complex physiological and biochemical process influenced by multiple factors, such as genetics, the environment, and nutrition. The hair follicle (HF) is the central structure for wool growth and, although small, has a complex structure consisting of a population of cells from the ectoderm and mesoderm [1]. HFs are mainly composed of the hair bulb (HB), the inner root sheath (IRS), the hair shaft (HS), the outer root sheath (ORS), the connective tissue sheath (CTS), etc. [2]. HFs are divided into primary HFs (PFs) and secondary HFs (SFs), of which PFs develop earlier and SFs later [3,4]. The growth and development status of SFs directly impact the yield and quality of wool [5]. The redifferentiation process is a key factor in regulating the density of wool follicles, which, in turn, effectively improves its fineness. HF morphogenesis involves the synergistic effects of a series of signalling pathways between the epidermis and dermis. HF development is regulated by multiple signalling pathways, including the Wnt, sonic hedgehog (SHH) conduction, Notch conduction, transforming growth factor (TGF) family, bone morphogenetic protein (BMP), and fibroblast growth factor (FGF) pathways [6–11].

Previous studies focused primarily on the functional level of genes. However, in recent years, the role of non-coding RNAs (ncRNAs) (such as long circular RNAs (circRNAs), non-coding RNAs (ncRNAs), and microRNAs (miRNAs)) at different stages of HF development has attracted considerable attention. Numerous studies have shown that ncRNAs are important post-transcriptional regulators of gene expression during HF development. circRNAs are novel endogenous ncRNA molecules widely present in mammalian cells, formed by the reverse-splicing of the 3′ and 5′ ends of mRNA precursors to create single-stranded closed circular structures. CircRNAs can be classified into four types based on their source and composition: intron circRNAs (ciRNAs), exon circRNAs (ecircRNAs), exon–intron circRNAs (elciRNAs), and intergene circRNAs (icircRNAs) [12]. Unlike linear RNA, circRNA, due to the lack of a 5′ cap structure and 3′ poly(A) tail in its circular structure, can resist exonuclease degradation and exhibit high stability and conservation [12,13]. CircRNAs are rich in miRNA binding sites and can act as miRNA molecular sponges to regulate the expression of host genes [14,15], interact with RNA-binding proteins [16], participate in protein translation [17], or act as competitive endogenous RNAs (ceRNAs) to regulate gene expression [18,19]. CircRNAs that regulate gene expression and participate in the regulation of life activities through competitive binding of miRNAs have become a research hotspot in related fields. ceRNA molecules competitively bind to the same miRNA through miRNA response elements (MREs), thereby regulating their expression [20]. Recent studies have shown that circRNAs are involved in the regulation of HF growth via a ceRNA mechanism. For example, circFTO and circCSPP1 competitively bind to miR-148a and miR-10a, enhancing the expression of *BMP7* to promote the proliferation of lakeal papillae cells [21]; circRNA-1967 upregulates the expression of *LEF1* by adsorbing miR-93-3p and regulates the proliferation and differentiation of HF stem cells in cashmere goats [22]. However, studies on the regulatory mechanism of circRNAs during embryonic wool follicle development in fine-wool sheep remain relatively scarce.

The development of wool follicles in fine-wool sheep shows clear temporal nodes and phase characteristics. Specifically, when the gestational age reaches 87 days, SFs begin to occur, and at 102 days, the primary SFs are redifferentiated at the cervical protrusion site to form SFs. At a gestational age of 108 days, a large number of redifferentiation phenomena occur in primary SFs. At a gestational age of 138 days, most primary wool follicles and some SFs have matured [23]. Therefore, based on these previous findings, in this study, ten foetal skin tissue samples from fine-wool sheep with gestational ages of E87, E90, E93, E96, E99, E102, E105, E108, E111, and E138 were selected, and RNA sequencing (RNA-seq) technology combined with bioinformatics analysis was used to resolve the expression profiles of circRNAs at different developmental stages, identify DE circRNAs, reveal the biological functions and signalling pathway characteristics, and construct a circRNA–miRNA–mRNA regulatory network. The ultimate objective was to provide a new perspective for clarifying the functional mechanism of circRNAs in wool follicle development in fine-wool sheep.

2. Materials and Methods

2.1. Moral Declaration

All animals were handled in strict accordance with the Animal Ethics Procedures and Guidelines of the People's Republic of China. This study was approved by the Animal Ethics Committee of the Lanzhou Institute of Animal Husbandry and Veterinary Medicine, Chinese Academy of Agricultural Sciences (licence no.: SYXK-2019-010).

2.2. Sample Collection and Animal Processing

All fine-wool sheep used in this experiment were obtained from the Gansu Province Sheep Breeding Technology Extension Station (Huangcheng Town, Sunan County, Zhangye, China). Thirty pregnant ewes aged 2–3 years old with the same feeding conditions were selected (average fibre diameter of 19.0–21.5 μm), and these ewes were artificially inseminated with the sperm of rams of the same strain (wool fibre diameter of 19.20 μm). The day of fertilisation was designated as the 0th day of embryo development. Thirty foetuses were delivered by caesarean section from thirty different female animals at ten stages (E87, E90, E93, E96, E99, E102, E108, E111, and E138) of embryonic development, with three replicates at each time point. A section of the skin sample, collected 10 cm behind the foetal scapula, was obtained using skin biopsy punches (10 mm) (Biosharp, rxz509, Hefei, China). The skin samples were rinsed with phosphate-buffered saline, quickly loaded into freezing tubes before freezing in a liquid-nitrogen tank, and stored at −80 °C for total RNA extraction and used for transcriptome sequencing and qPCR analysis.

2.3. circRNA Sequencing

2.3.1. Extraction of Total RNA from the Skin and RNA Inversion

Total RNA was extracted from the skin tissue of 30 fine-wool sheep foetuses by sampling 50–100 mg of sheep skin in liquid nitrogen using TRIzol reagent (Takara, Shiga, Japan) according to the manufacturer's instructions. A NanoPhotometer® spectrophotometer (IMPLEN, Palo Alto, CA, USA) and an Agilent 2100 Bioanalyzer (Agilent Technologies, Palo Alto, CA, USA) were used to detect the concentration and quality of extracted total

RNA. The datasets used in the experiments were derived from high-throughput sequencing performed by Ouyi Biologicals (Shanghai, China) under Project No. HT2020-16026.

2.3.2. Library Preparation and Sequencing

The extracted total RNA samples were prepared to build a library using TruSeq Stranded Total RNA with the Ribo-Zero Gold Kit (Illumina, San Diego, CA, USA), in accordance with the manufacturer's instructions. The constructed RNA library was inspected and qualified using an Agilent 2100 Bioanalyzer (Agilent Technologies, Palo Alto, CA, USA) and sequenced using an Illumina sequencer.

2.3.3. Identification and Characteristic Analysis of circRNA

After obtaining a large amount of double-ended sequencing data, the whole transcription library was constructed using Trimmomatic software (v0.39) [24] to remove adapters and low-quality base reads and obtain high-quality data. Subsequently, Hisat2 (v2.2.1.0) [25] was used to align the clean reads with the reference genome Oar_v4.0 to obtain the position and sample-specific sequence information, and the RscqQC suite (v2.6.4) was used to statistically analyse the proportions of alignment types and evaluate the results. After obtaining the BAM file, the reads of the genes were assembled and aligned using StringTie software (v1.3.3b) [26], and the transcripts of each sample were fused and spliced into a complete transcript. Based on the structural characteristics of circRNAs and the characteristics of splicing sequences, circRNAs were predicted using CIRI [27] software (v2.0.3), classified, and annotated, and the chromosome distribution and length were statistically analysed.

2.4. DE circRNA Screening and Functional Enrichment Analysis of Source Genes

After standardising the counts of each sample circRNA using DESeq software (v1.18.0) [28], differential expression analysis of circRNAs was performed using the edgeR package (v3.12.1) [29]. Differences between comparison groups (E90 vs. E87, E93 vs. E87, E96 vs. E87, E99 vs. E87, E102 vs. E87, E105 vs. E87, E108 vs. E87, E111 vs. E87, and E138 vs. E87) and the analysis involved three main steps: ① normalisation of the read count values of circRNAs; ② calculation of the hypothesis test probability (p-value) under the model; and ③ multiple hypothesis testing correction to obtain the false discovery rate (FDR) < 0.05. Based on the differential analysis results, the circRNAs exhibiting $p < 0.05$ and $|\log2FC| > 1$ were chosen as DE circRNAs. The source genes of the DE circRNAs were analysed using Gene Ontology (GO, https://www.geneontology.org/, Date of visit: 3 July 2025) and the Kyoto Encyclopedia of Genes and Genomes (KEGG, http://www.genome.jp/kegg/, Date of visit: 5 July 2025).

2.5. Validation of DE circRNAs

To validate the RNA-seq data, nine DE circRNAs were randomly selected for the quantitative fluorescence assay by quantitative real-time polymerase chain reaction (qRT-PCR), using β-actin as the internal reference gene. All primers were designed and synthesised by Shenggong Bioengineering Co., Ltd. (Shanghai, China) (Table 1). DE circRNAs were amplified by PCR using the cDNA of Alpine Merino sheep embryonic skin as a template and detected by Sanger sequencing. The amplified product sequences were compared with the circRNA sequences obtained by RNA-Seq using MEGA 7.0. Then, qRT-PCR was performed using the reagent SYBR Premix Ex Taq II (Takara, Japan) on the Roche LightCycler® 96 real-time fluorescence quantitative PCR instrument with the following program: 95 °C for 30 s, 40 cycles at 95 °C for 5 s, and 60 °C for 30 s. β-Actin was used as the reference gene, and the equivalent expression level was calculated via the $2^{-\Delta\Delta CT}$ method [30].

Table 1. Information on qRT-PCR primer sequences.

circRNAs	Forward Primer Sequence	Reverse Primer Sequence
circRNA_07722	AGCCTTCCAATAGTGAACCTCA	GCAACCAAGTGTAGTGCAGG
circRNA_02253	TGAAACACAGTTGGCAGAGTCT	ACGGCAAGGCACCAAACT
circRNA_06688	TGGTGAAGACATTGCAGACGA	AGCTCATTCACTTGTTCTCTCCA
circRNA_04873	GCGGATGAAAGGAATGGGGA	TGCTCTGTTGACTCCCTGAAA
circRNA_06881	GAGAACCAGTGGCTGCGG	GCTTCTCCAACTTGTCCTCCT
circRNA_04670	TCGAGTTTGAATGGCTGAGACA	AGATGTATTCCAAGGTCCCCG
circRNA_04804	GTTGGAGCAGGAAGAGGAGC	TGAGATAGCAGGAGTTTGGAAGAC
circRNA_07666	TTTCTCGATTGGACCTGCGA	GGGTGAATGATCCTCTGGTGG
circRNA_05413	GGAGATCCCGCAAGCAGAG	AGCTTGCTATCCGAGTCTTTCT
circRNA_03494	ATTGTTGCCTACGCCCACTT	GCCATTCGTGAACAGCATCG
circRNA_08249	AAGGGTGATTCTGGTGCTCC	CAGACCAGGTGTACCAGCAG
β-actin	CAGTCGGTTGGATGGAGCAT	AGGCAGGGACTTCCTGTAAC

2.6. Construction of circRNA-miRNA-mRNA Interaction Network

CircRNA-targeted miRNAs were predicted using the miRanda software (v3.3a) and TargetScan (v7.0) [31]. miRTarBase (v6.1) was used to predict the mRNA interactions between circRNAs and miRNAs [32]. The circRNA–miRNA–mRNA interaction network was constructed using the Cytoscape software (v3.10.3) [33].

2.7. Statistical Analysis—Quantification of Gene Expression and Analysis

Statistical analysis was performed using SPSS software (version 22.0; SPSS Inc., Chicago, IL, USA), and graphs were plotted using Origin 2021 (OriginLab Corp., Northampton, MA, USA). Data were expressed as the mean ± standard error (SEM), and the RT-PCR results were analysed for relative quantification by the $2^{-\Delta\Delta Ct}$ method. Differences between groups were compared using the independent-samples *t*-test or one-way ANOVA, and the significance levels were set at $p < 0.05$ and <0.01 for significant and highly significant differences, respectively.

Statistical analysis was performed using SPSS software (version 22.0; SPSS Inc., Chicago, IL, USA), and graphs were plotted using Origin 2021 (OriginLab Corp., Northampton, MA, USA). First, the normality of the data distribution in each treatment group was evaluated using the Shapiro–Wilk test. For the independent-samples *t*-test, the F-test was applied to check the homogeneity of variance between the two groups before comparison. When conducting one-way ANOVA, Levene's test was used to assess the equality of variances across multiple groups. Data were expressed as the mean ± standard error (SEM), and the RT-PCR results were analysed for relative quantification by the $2^{-\Delta\Delta Ct}$ method. Differences between groups were compared using the independent-samples *t*-test or one-way ANOVA, and the significance levels were set at $p < 0.05$ and < 0.01 for significant and highly significant differences, respectively.

3. Results

3.1. High-Throughput Sequencing Quality Assessment

To explore the expression pattern of circRNAs in the skin HFs of alpine fine-wool sheep during the developmental period, high-throughput sequencing was performed on skin tissue samples from fine-wool sheep foetuses at 10 different developmental stages (E87, E90, E93, E96, E99, E102, E105, E108, E111, and E138). Thirty RNA libraries of deribosomal HF development at different stages were constructed. After RNA sequencing on the Illumina HiSeq 4000 platform, 3,050,120,000 original reads were obtained. After quality control, 2,960,630,000 clean reads were obtained. The Q30 values of all of the samples were >98.10%. The GC content ranged from 45.90 to 49.44%. After removing ribosomal RNA, 94.03–95.53%

of the clean reads were compared with the reference genome of fine-wool sheep. The overall assessment of the sequencing data is presented in Table S1. These clean data met both the quantity and quality requirements for subsequent circRNA analysis. In this study, 18232 circRNAs were identified, and their source genes were analysed. Novel circRNAs were categorised into five types: antisense (n = 398), exonic (n = 509), intergenic (n = 772), intronic (n = 303), and sense-overlapping (n = 16,250) circRNAs. The chromosome distribution showed that circRNAs were distributed on almost all chromosomes, with chromosomes 1–3 being the most enriched, and chromosomes 24 and 26 the least (Figure 1A). CircRNA types were predominantly sense-overlapping (89.13%), followed by intergenic (4.23%), exonic (2.79%), antisense (2.18%), and intronic (1.66%) in decreasing order (Figure 1B). Lengthwise, the highest number of circRNAs were over 2000 bp (3157), followed by 301–400 bp (2091) (Figure 1C). The GC content of the circRNA sequences was in the range of 35–50% (Figure 1D).

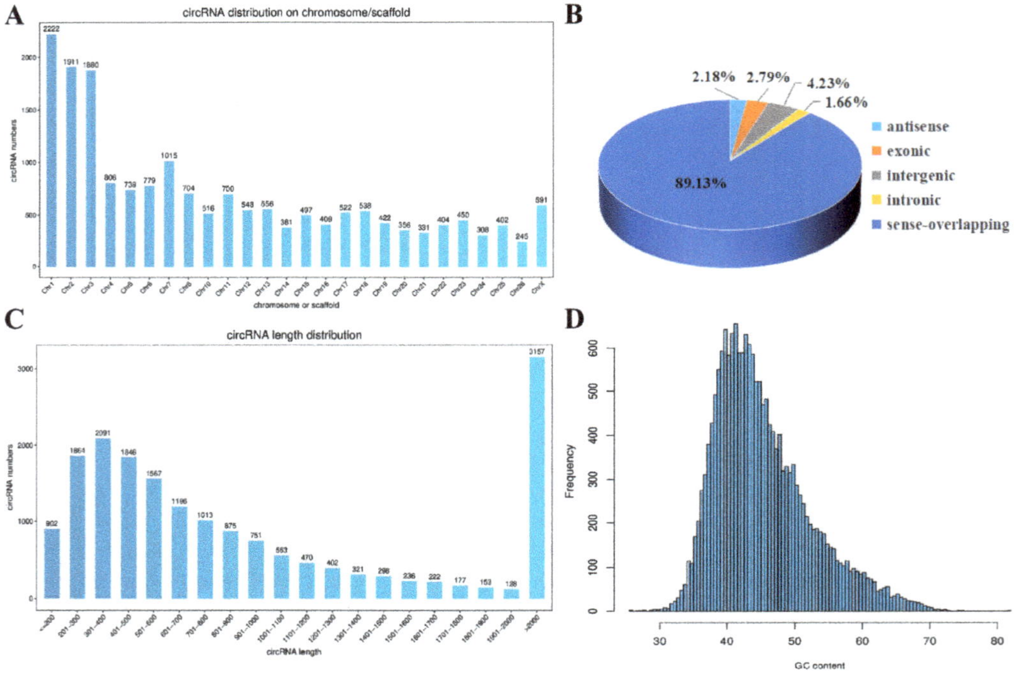

Figure 1. Analysis of the expression characteristics of circRNA in the skin tissue of alpine fine-wool sheep. (**A**) Distribution of circRNA numbers on each chromosome. (**B**) Classification of circRNAs in the wool follicles of fine-wool sheep. (**C**) Distribution map of circRNA sequence lengths. (**D**) Distribution map of circRNA GC contents.

3.2. circRNA Identification and Expression Analysis

Using p-value < 0.05 and $|\log_2 FC| \geq 1$ as screening conditions, a total of 173 DE circRNAs were identified. Differentially expressed circRNAs were screened in the comparison groups (E90 vs. E87, E93 vs. E87, E96 vs. E87, E99 vs. E87, E102 vs. E87, E105 vs. E87, E108 vs. E87, E111 vs. E87, and E138 vs. E87) (Table S2). The results showed that, in the E90 vs. E87, E93 vs. E87, E96 vs. E87, E99 vs. E87, E102 vs. E87, E105 vs. E87, E108 vs. E87, E111 vs. E87, and E138 vs. E87 groups, 18, 6, 7, 0, 11, 13, 7, 10, and 8 were upregulated, whereas 13, 13, 8, 0, 17, 10, 13, 8, and 10 were downregulated (Figure 2). This lays the foundation for the subsequent screening of circRNAs related to SF development in fine-wool sheep.

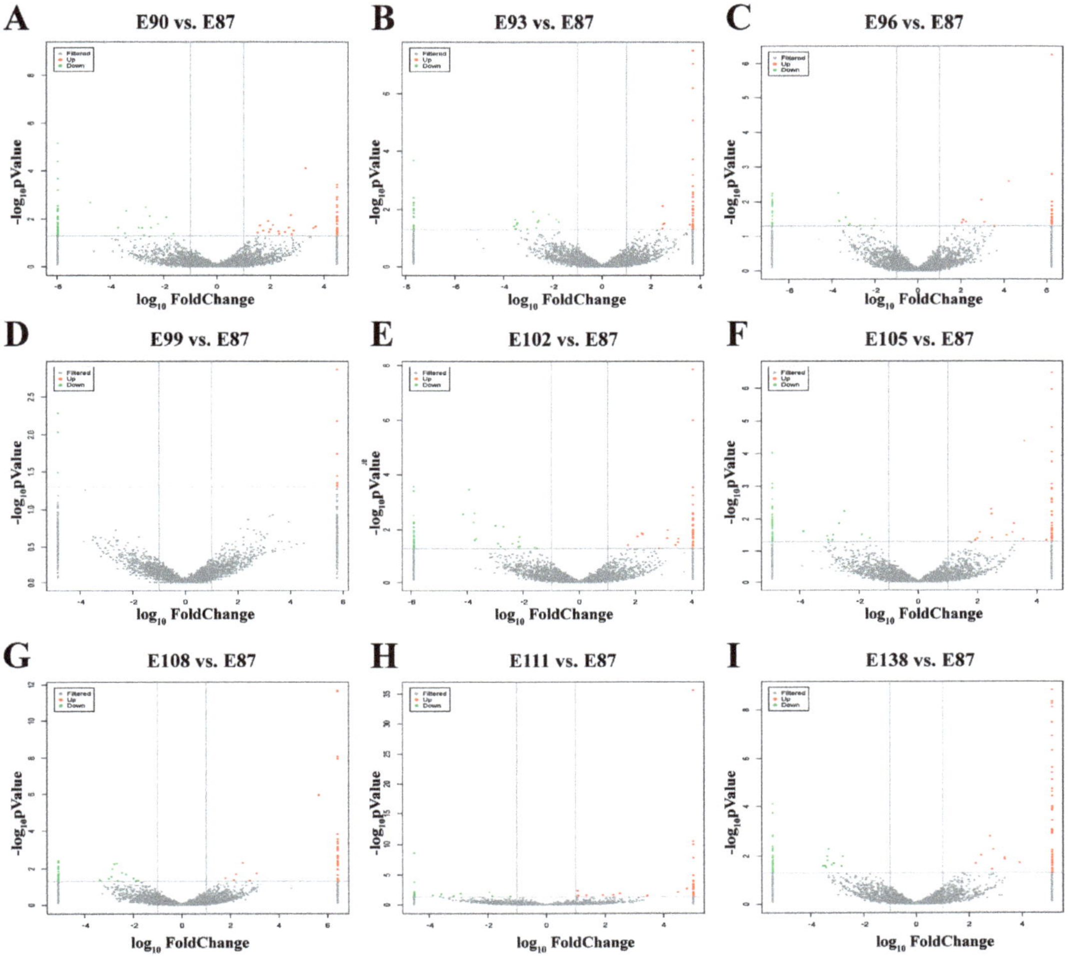

Figure 2. Volcano map of differentially expressed circRNAs. Grey represents circRNAs with non-significant differences, while red and green represent circRNAs with significant differences. (**A**) E90 vs. E87. (**B**) E93 vs. E87. (**C**) E96 vs. E87. (**D**) E99 vs. E87. (**E**) E102 vs. E87. (**F**) E105 vs. E87. (**G**) E108 vs. E87. (**H**) E111 vs. E87. (**I**) E138 vs. E87.

3.3. Functional Enrichment Analysis of Target Genes

To explore the regulatory role of the target genes of the DE circRNAs in SF development in fine-wool sheep, GO and KEGG pathway enrichment analyses were performed (Tables S3 and S4). The results of GO functional annotation showed that the regulation of epidermis development (GO:0045682), HF development (GO:0001942), HF morphogenesis (GO:0031069), HF maturation (GO:0048820), Wnt signalling pathway (GO:0016055), skin morphogenesis (GO:0043589), SMAD protein signal transduction (GO:0060395), skin development (GO:0043588), regulation of Notch signalling pathway (GO:0008593), activation of MAPK activity (GO:0000187), keratin filament (GO:0045095), and epidermal growth factor receptor (GO:0005154) were significantly enriched (Figure 3). We used the KEGG database to analyse the transcriptome results and identified eight enriched pathways associated with secondary HF development: the Wnt (ko04310), TGF-β (ko04350), Notch (ko04330), NF-κB (ko04064), PI3K-Akt (ko04151), P53 (ko04115), and AMPK (ko04152) and mTOR (ko04150) signalling pathways (Figure 4).

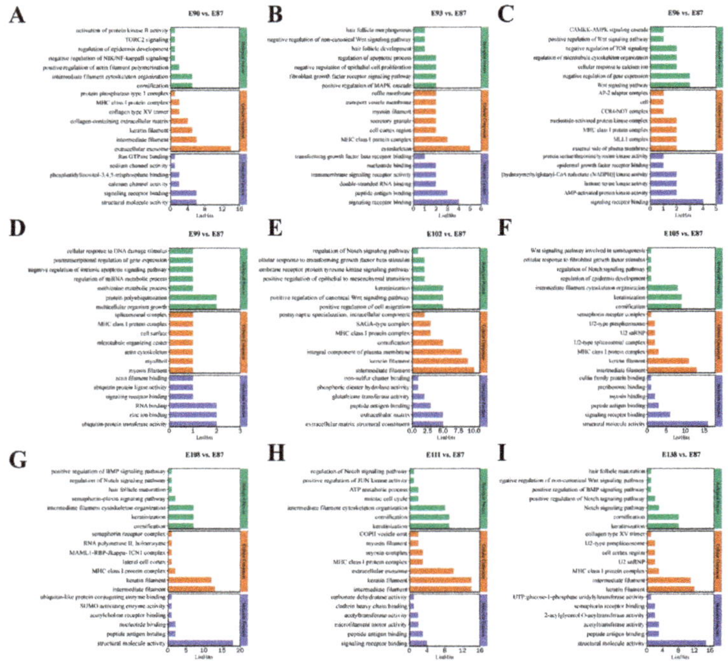

Figure 3. Functional annotation analysis of GO source genes. (**A**) E90 vs. E87. (**B**) E93 vs. E87. (**C**) E96 vs. E87. (**D**) E99 vs. E87. (**E**) E102 vs. E87. (**F**) E105 vs. E87. (**G**) E108 vs. E87. (**H**) E111 vs. E87. (**I**) E138 vs. E87.

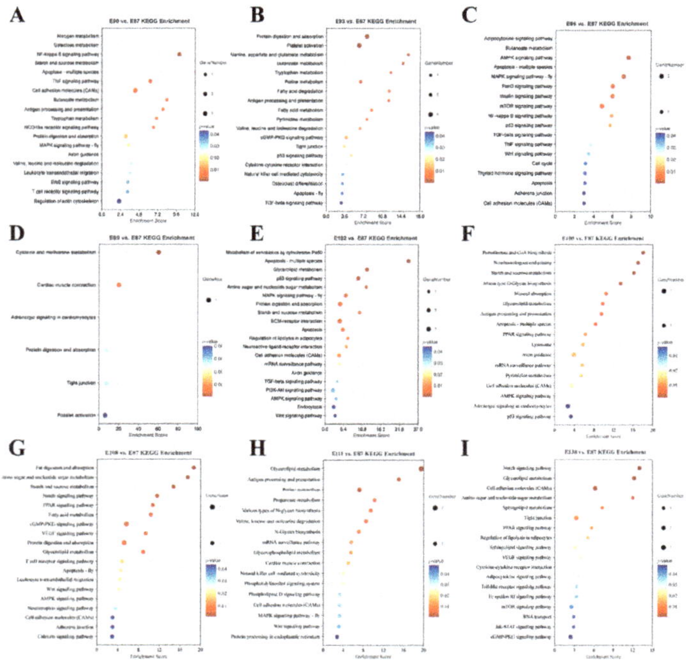

Figure 4. Analysis of the KEGG pathways of DE circRNAs' target genes in each comparison group. (**A**) E90 vs. E87. (**B**) E93 vs. E87. (**C**) E96 vs. E87. (**D**) E99 vs. E87. (**E**) E102 vs. E87. (**F**) E105 vs. E87. (**G**) E108 vs. E87. (**H**) E111 vs. E87. (**I**) E138 vs. E87.

3.4. Validation of the RNA-Seq Data

To verify the accuracy of the RNA-seq data, we randomly selected nine DE circRNAs for qRT-PCR analysis: circRNA_07722, circRNA_02253, circRNA_06688, circRNA_04873, circRNA_04670, circRNA_04804, circRNA_05413, circRNA_06881, and circRNA_07666 (Figures 5 and 6). The qRT-PCR-identified circRNA expression trends were consistent with the RNA-seq data (Figure 5B). The Sanger sequencing results were also consistent with the RNA-seq results (Figure 5C). PCR amplification confirmed the presence of the selected circRNAs (Figure 6). These results confirmed the authenticity of the circRNAs and reliability of the RNA-seq data.

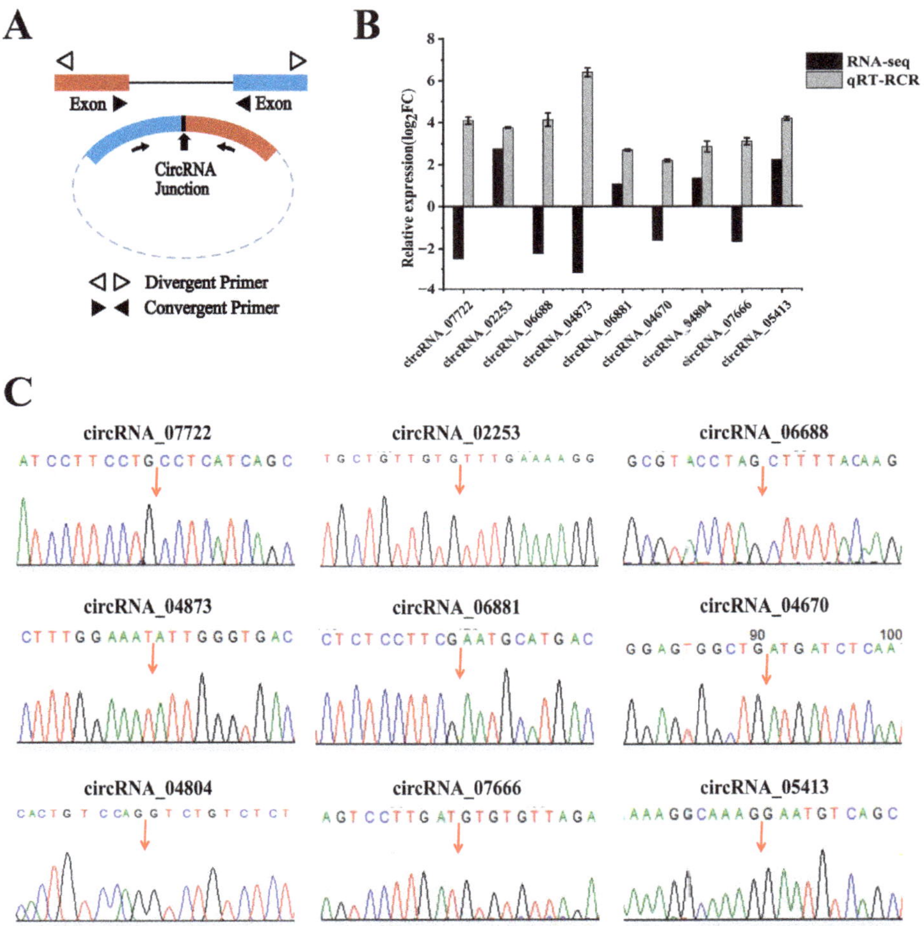

Figure 5. Verification of DE circRNAs from RNA-seq. (**A**) Schematic design of primers for DE circRNAs in qRT-PCR. (**B**) Comparison of 9 DE circRNA expression levels between RNA-seq and qRT-PCR. (**C**) Sanger sequencing of 9 DE circRNAs. β-Actin was used as the internal reference gene. Three replicates were set for qRT-PCR, and the results are expressed as the mean ± standard error (±SE). The cyclic site sequence of circRNAs in circbase are as follows: circRNA_07722 (ATCCTTCCT-GCCTCATCAGC); circRNA_02253 (TGCTGTTGTGTTTGAAAAGG); circRNA_06688 (GCGTACC-TAGCTTTTACAAG); circRNA_04873 (CTTTGGAAATATTGGGTGAC); circRNA_06881 (CTCTC-CTTCGAATGCATGAC); circRNA_04670 (GGAGTGGCTGATGATCTCAA); circRNA_04804 (CACT-GTCCAGGTCTGTCTCT); circRNA_07666 (AGTCCTTGATGTGTGTTAGA); and circRNA_05413 (AAAGGCAAAGGAATGTCAGC).

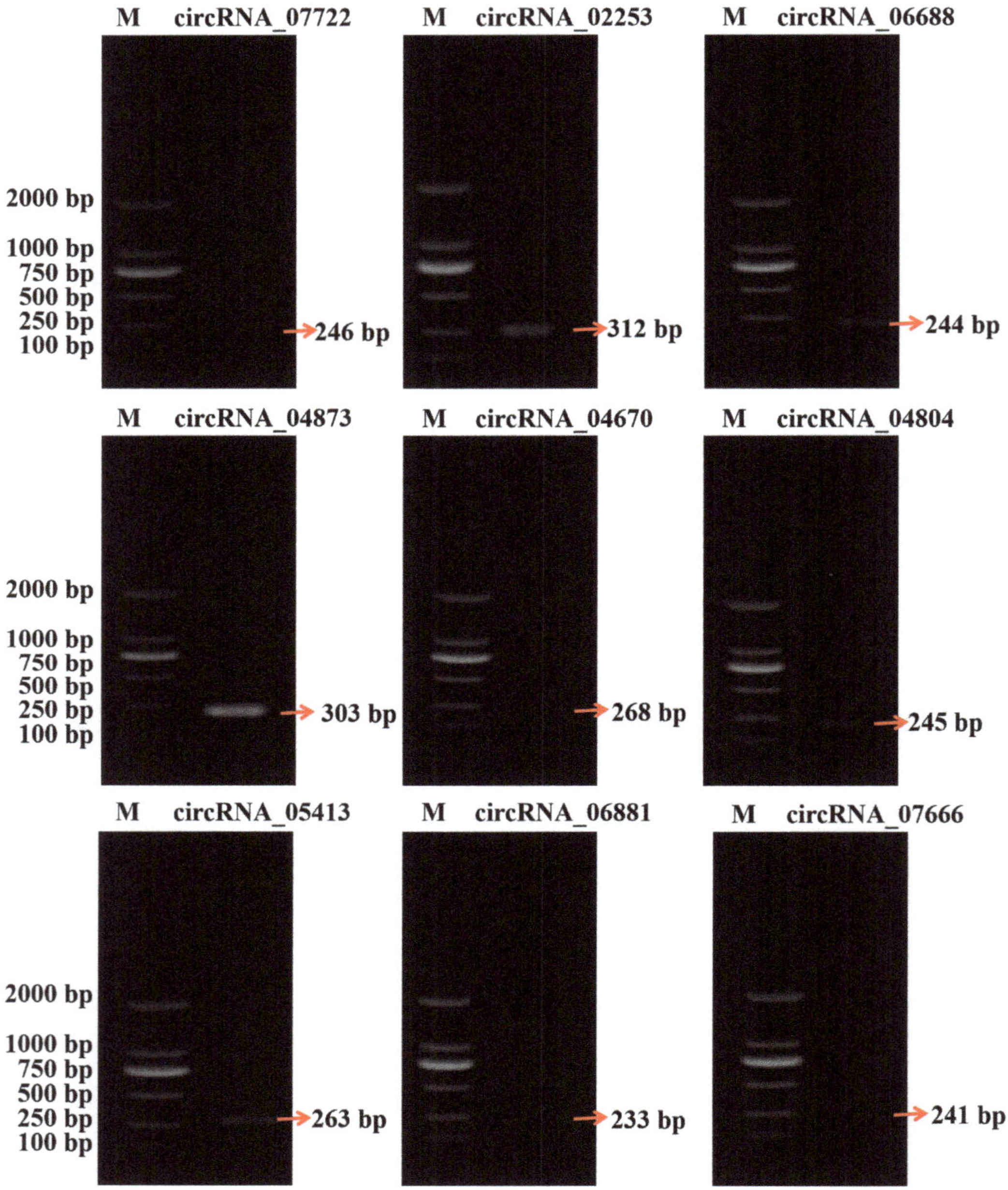

Figure 6. Agarose gel electrophoresis image of circRNAs amplified by PCR. Amplification bands of circRNA_07722, circRNA_02253, circRNA_06688, circRNA_04873, circRNA_04670, circRNA_04804, circRNA_ 05413, circRNA_06881, and circRNA_07666 on agarose gel electrophoresis.

3.5. Screening of DE circRNAs Related to Secondary Hair Follicle Development

We screened 16 DE circRNAs related to SF development to construct a circRNA–miRNA–mRNA interaction network. The 16 circRNAs were circRNA_07722, circRNA_00369, circRNA_01130, circRNA_15222, circRNA_08249, circRNA_00618, circRNA_04007, circRNA_01256, circRNA_02438, circRNA_12843, circRNA_13884, circRNA_05165, circRNA_05413, circRNA_04670, circRNA_04873, and circRNA_07666 (Table 2).

Table 2. Statistics of 16 circRNAs' information related to the development of secondary HFs.

circRNA ID	Source Gene	*p*-Value	Up/Down	circRNA Size	Isoform Name
circRNA_07722	COL15A1	2.29×10^{-2}	Down	362	XM_027964395.1
circRNA_00369	DOCK7	7.36×10^{-3}	Down	429	XM_015092012.2
circRNA_01130	ADAMTS5	2.03×10^{-3}	Down	301	XM_015092381.2
circRNA_15222	HADH	4.82×10^{-2}	Down	504	XM_004009637.4
circRNA_08249	COL3A1	7.73×10^{-3}	Up	315	XM_004004514.4
circRNA_00618	TGFBR3	4.99×10^{-2}	Up	871	XM_012176015.3
circRNA_04007	ISM1	2.15×10^{-2}	Up	739	XM_027976478.1
circRNA_01256	ARL13B	1.43×10^{-2}	Up	1077	XM_004002849.4
circRNA_02438	POSTN	4.29×10^{-2}	Down	570	XM_004012110.4
circRNA_07666	KCTD9	3.92×10^{-2}	Down	314	XM_027964318.1
circRNA_12843	KRT77	8.74×10^{-4}	Up	12,347	XM_027967277.1
circRNA_13884	SEMA3C	1.06×10^{-2}	Down	1027	XM_004007807.4
circRNA_05413	CTNND1	2.03×10^{-2}	Up	272	XM_027979807.1
circRNA_05165	DKK3	2.54×10^{-2}	Down	222	XM_027979387.1
circRNA_04873	ZNF114	4.36×10^{-2}	Down	1359	XM_027978554.1
circRNA_04670	FTO	4.35×10^{-2}	Down	344	XM_027977360.1

3.6. Analysis of circRNA-miRNA-mRNA Interaction Network

To investigate the regulation of targeting and post-transcriptional effects of circRNAs on miRNAs in the skin of alpine fine-wool sheep, we used miRanda (v3.3a) and TargetScan (v7.0) to predict their interactions (Table S5). Based on the 16 differentially expressed circRNAs, 44 targeted miRNAs, and 65 associated mRNAs, 88 circRNA–miRNA–mRNA regulatory networks related to SF development were constructed (Figure 7, Table S6). The analysis revealed that multiple circRNAs synergistically regulated the same miRNAs, such as circRNA_00369, circRNA_04670, and circRNA_01130, all of which targeted miR-30a-5p.

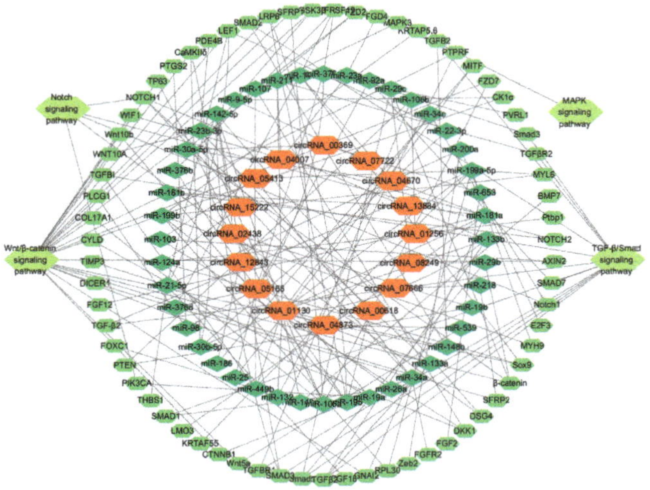

Figure 7. The circRNA–miRNA–mRNA regulatory network related to SF genesis and development. From the inner circle to the outer circle of circRNA–miRNA–mRNA are DE circRNAs, miRNAs, mRNAs, and signalling pathways.

3.7. *Analysis of the Expression Levels of circRNAs*

We randomly selected 6 circRNAs (circRNA_07722, circRNA_08249, circRNA_00618, circRNA_07666, circRNA_05413, and circRNA_04873) from 16 circRNAs associated with the occurrence of SFs in fine-wool sheep. Foetal skin was examined at 10 different developmental stages (E87, E90, E93, E96, E99, E102, E105, E108, E111, and E138) to understand the role of circRNAs in the foetal skin of fine-wool sheep (Figure 8). The results indicated that the overall trends of the expression levels of circRNA_07722, circRNA_07666, and circRNA_04873 gradually decreased as the number of foetal days increased (Figure 8A,D,F). However, the overall trends of the expression levels of circRNA_08249, circRNA_00618, and circRNA_05413 gradually increased with increasing foetal days and reached a maximum value at 138d (Figure 8B,C,E). These findings suggested that the above six circRNAs may have potential regulatory functions in SF development in the foetal skin of fine-wool sheep through the ceRNA network.

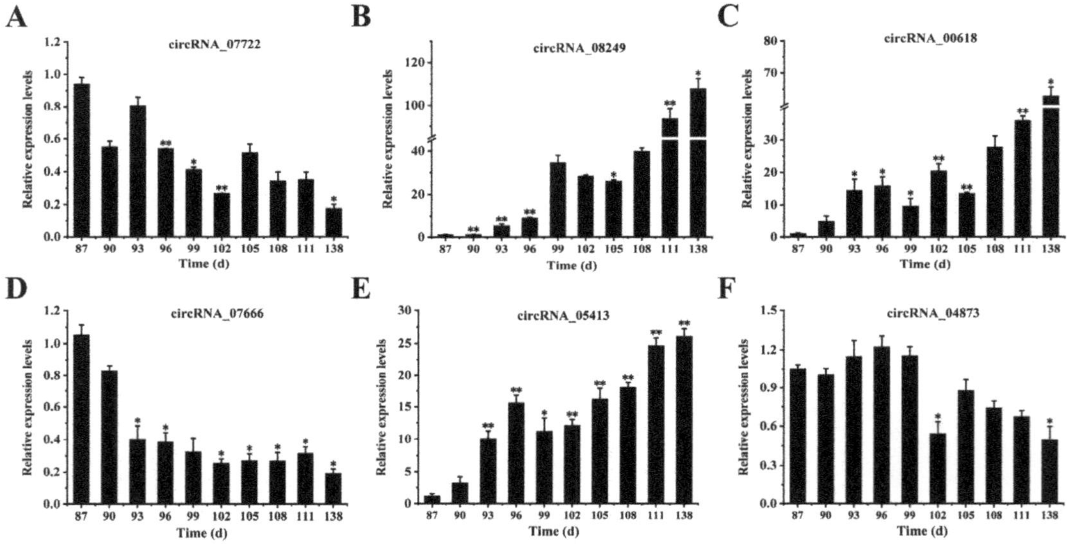

Figure 8. Expression patterns of circRNA. (**A–F**) Expression levels of six circRNAs (circRNA_07722, circRNA_08249, circRNA_00618, circRNA_07666, circRNA_05413, and circRNA_04873) related to SF development at different stages. * $p < 0.05$; ** $p < 0.01$.

4. Discussion

The growth and development of SFs in the embryonic stage as a regenerative system involves a variety of intricate molecular regulatory processes, including the genesis, proliferation, differentiation, and apoptosis of a wide range of cells [34]. Systematic research on the down production process and SF development mechanisms at different embryonic developmental stages in fine-wool sheep is lacking. Existing studies have demonstrated that the SFs of alpine fine-wool sheep originate at a gestational age of 87 days, the cervical protuberances of the primary SFs protrude and redifferentiate at 102 days, the primary SFs redifferentiate extensively at 108 days, and most of the primary SFs and some others redifferentiate at 138 days [23]. Based on this, we used RNA-seq technology to screen ceRNAs and constructed a circRNA–miRNA–mRNA regulatory network related to the occurrence and development of SFs in fine-wool sheep.

In this study, we analysed the expression characteristics of circRNAs in the HFs of fine-wool sheep at different developmental stages using RNA-seq and identified 173 DE circRNAs. Some circRNA-targeted genes have been reported to be involved in HF growth and may be important targets for research [35]. In addition, the expression profiles of circRNAs were consistent with those of their source genes [36,37]. KEGG is a pathway database for the systematic analysis of gene functions. In this study, KEGG pathway analysis revealed that Wnt, TGF-β, Notch, NF-κB, PI3K-Akt, P53, and AMPK were significantly involved in SFs ($p < 0.05$). Based on the significant KEGG pathways, 16 circRNAs were initially screened (circRNA_07722, circRNA_00369, circRNA_01130, circRNA_15222, circRNA_08249, circRNA_00618, circRNA_04007, circRNA_01256, circRNA_02438, circRNA_12843, circRNA_13884, circRNA_05165, circRNA_05413, circRNA_04670, circRNA_04873, and circRNA_07666). The genes from which these circRNAs originate (*COL15A1, DOCK7, ADAMTS5, HADH, COL3A1, TGFBR3, ISM1, ARL13B, POSTN, KCTD9, KRT77, SEMA3C, CTNND1, DKK3, ZNF114,* and *FTO*) are involved in the HF growth process [38–52]. A comparison of our results with those of previous studies suggests that the selected circRNAs may play important roles in signalling pathways at different stages of SF development in fine-wool sheep. Further studies are required to determine the exact mechanisms involved.

According to the ceRNA hypothesis, circRNAs regulate the expression of target genes by competitively binding to miRNAs [53]. In this study, 44 circRNA-targeting miRNAs associated with the development of SFs were predicted using miRanda (v3.3a) and TargetScan (v7.0) software. The Wnt/β–catenin [54], TGF-β/Smad [55], Notch [56], and MAPK [57] signalling pathways are essential in HF growth. The Wnt signalling pathway plays a central role in the regulation of follicle formation and cycle regeneration induced by HF papilla cells [58]. In the TGF-β signalling pathway, TGF-β1 knockdown delays the transition of the HF from the anagen to the anagen phase [59]. TGF-β2 inhibits the elongation of the hair shaft and induces changes in the morphology of the HF during the anagen phase of the HF [60]. The Notch signalling pathway is activated under the regulation of Tβ4, which in turn drives the proliferation, differentiation, and migration of HF cells [61]. The MAPK signalling pathway is essential for the regulation of mammalian cell proliferation [56].

Recent studies revealed that some circRNAs contain multiple miRNA binding sites [18,53]. As circRNAs cannot directly regulate their target genes, they act as 'miRNA sponges'. CircRNAs are involved in many biological processes by acting as miRNA sponges, thus eliminating the inhibitory effects of miRNAs on their target genes [53]. CircRNAs can regulate gene expression through the circRNA–miRNA–mRNA pathway [36]. However, the dynamic changes and functions of circRNAs at different developmental stages in fine-wool sheep SFs remain understudied. CircZNF423, a ceRNA that regulates the expression of oar-miR-541-3p and targets *CALM3*, inhibits the proliferation of sheep myoblasts [62]. In Hoop, circFTO and circCSPP1 competitively bind to miR-148a and miR-10a to deregulate the inhibitory effect of miRNAs on *BMP7*, which in turn promotes *BMP7* expression and drives the proliferation of sheep wool cells [63]. Furthermore, circRNA-1967 upregulates the expression of *LEF1* by adsorbing miR-93-3p, which is involved in HF-related regulation of cellular function [64]. In the hair papilla sheaths of velvet goat SFs, circRNA-1926 was shown to target and regulate cell cycle protein-dependent kinase 19 (*CDK19*) gene expression and mediate HF stem cell differentiation into SHFs via miR-148a/b-3p [64]. These findings suggest that circRNAs competitively bind miRNAs through the ceRNA mechanism to regulate the expression of key genes, which in turn affect HF growth and differentiation.

Based on the results of KEGG pathway analysis combined with DE circRNA data related to the occurrence of SFs in fine-wool sheep, we successfully constructed a circRNA–miRNA–mRNA interaction network containing 16 circRNAs, 44 miRNAs, and 65 mRNAs. This network provides important clues for the in-depth characterisation of the SF in sheep at different developmental stages. Previous studies have successfully predicted 102 pairs of circRNA–miRNA interactions and 126 pairs of miRNA–mRNA interactions [65], confirming that there is a targeted regulatory relationship between circRNAs and specific miRNAs, and that circRNAs can indirectly regulate target gene expression through miRNAs. In this study, 16 screened circRNAs were integrated and analysed with the predicted miRNA and mRNA sequence data, and 88 sets of circRNA–miRNA–mRNA interactions were identified. We hypothesised that the newly identified circRNAs might be involved in regulating the growth and development of HFs in fine-wool sheep through similar molecular mechanisms. In addition, the expression levels of six circRNAs (circRNA_07722, circRNA_08249, circRNA_00618, circRNA_07666, circRNA_05413, and circRNA_04873) were examined in this study at 10 different developmental stages of wool follicles in sheep embryos. The results showed that the expression of these circRNAs changed dynamically with the developmental stage. Future studies should focus on the specific mechanisms underlying the roles of the candidate circRNAs (circRNA_07722, circRNA_08249, circRNA_00618, circRNA_07666, circRNA_05413, and circRNA_04873) in HF development.

5. Conclusions

In this study, we constructed the expression profiles of circRNAs in the HFs of fine-wool sheep at different embryonic stages (E87, E90, E93, E96, E99, E102, E105, E108, E111, and E138), and a total of 173 DE circRNAs were identified. The results of GO and KEGG analyses suggest that there may be a relationship between circRNAs and their target genes. CircRNAs were found to play a role in HF growth and development in fine-wool sheep through signalling pathways such as Wnt/β-catenin, TGF-β/Smad, Notch, and MAPK. We examined the expression levels of six circRNAs associated with HF development at different embryonic stages that may regulate SF development in fine-wool sheep through the ceRNA network. A circRNA–miRNA–mRNA regulatory network was also constructed, and 88 interacting miRNA–mRNA pairs were investigated. The candidate circRNAs, including circRNA_07722, circRNA_08249, circRNA_00618, circRNA_07666, circRNA_05413, and circRNA_04873, provide targets for future studies on wool follicle regulation.

Supplementary Materials: The following supporting information can be downloaded from https://www.mdpi.com/article/10.3390/ani15172629/s1: Table S1: Summary table of sequencing data screening; Table S2: Differentially expressed circRNAs (DE circRNAs) among different groups; Table S3: KEGG enrichment results of DE circRNA source genes; Table S4: GO enrichment results of DE circRNA source genes; Table S5: All predicted target miRNAs of DE circRNAs; Table S6. Construction of DE circRNA–miRNA–mRNA interaction network.

Author Contributions: Conceptualisation, Y.L., T.X., and T.G.; methodology, Y.L., T.X., and T.G.; software, B.X., Y.S., and Y.L.; validation, T.X. and Y.L.; formal analysis, T.X. and Y.L.; investigation, T.G. and T.X.; resources, T.G. and Z.L.; data curation, Y.L., T.X., and T.G.; writing—original draft preparation, Y.L.; writing—review and editing, T.G.; visualisation, C.Y.; supervision, C.Y.; project administration, T.G. and, C.Y.; funding acquisition, J.L. and T.G. All authors have read and agreed to the published version of the manuscript.

Funding: This study was supported by grants from the Science and Technology Innovation 2030-Major Program (2023ZD0405101), the Natural Science Foundation of Gansu Province (21JR7RA028), the China Agriculture Research System (CARS-39-02), and the Chinese Academy of Agricultural Sciences of Technology Innovation Project(25-LZIHPS-07).

Institutional Review Board Statement: The animal study protocol was approved by the Institutional Ethics Committee of the Institute of Animal Sciences, Chinese Academy of Agricultural Sciences (IAS-CAAS) (permit no. SYXK-2019-010).

Informed Consent Statement: Informed Consent are obtained from the owner of Gansu Provincial Sheep Breeding Technology Extension Station.

Data Availability Statement: Data are available upon request due to privacy/ethical restrictions.

Conflicts of Interest: The authors declare no conflicts of interest.

References

1. Yang, H.Y.; Yang, H.; Shi, G.Q.; Shen, M.; Yang, J.Q.; Yang, Y.L.; Liu, X.J. Expression profile analysis of microRNAs during hair follicle development in the sheep foetus. *Biosci. Biotechnol. Biochem.* **2019**, *83*, 1045–1061. [CrossRef] [PubMed]
2. Sperling, L.C. Hair anatomy for the clinician. *J. Am. Acad. Dermatol.* **1991**, *25*, 1–17. [CrossRef] [PubMed]
3. Lyne, A. The development of the epidermis and hair canals in the merino sheep foetus. *Aust. J. Biol. Sci.* **1957**, *10*, 390. [CrossRef]
4. Parry, A.L.; Nixon, A.J.; Craven, A.J.; Pearson, A.J. The microanatomy, cell replication, and keratin gene expression of hair follicles during a photoperiod-induced growth cycle in sheep. *Cells Tissues Organs.* **1995**, *154*, 283–299. [CrossRef]
5. Gao, Y.; Wang, X.; Yan, H.; Zeng, J.; Ma, S.; Niu, Y.; Zhou, G.; Jiang, Y.; Chen, Y. Comparative transcriptome analysis of fetal skin reveals key genes related to hair follicle morphogenesis in cashmere goats. *PLoS ONE* **2016**, *11*, e0151118. [CrossRef]
6. Li, G.; Ji, Y.; Li, Y. Molecular mechanism of hair follicle morphogenesis. *Int. J. Dermatol Venereol.* **2004**, *30*, 38.
7. McMahon, A.P.; Ingham, P.W.; Tabin, C.J. Developmental roles and clinical significance of Hedgehog signaling. *Curr. Top. Dev. Biol.* **2003**, *53*, 1–114. [CrossRef]
8. Crowe, R.; Henrique, D.; Ish-Horowicz, D.; Niswander, L. A new role for Notch and Delta in cell fate decisions: Patterning the feather array. *Development* **1998**, *125*, 767–775. [CrossRef]
9. Schmidt-Ullrich, R.; Paus, R. Molecular principles of hair follicle induction and morphogenesis. *BioEssays* **2005**, *27*, 247–261. [CrossRef]
10. Thomadakis, G.; Ramoshebi, L.N.; Crooks, J.; Rueger, D.C.; Ripamonti, U. Immunolocalization of bone morphogenetic protein-2 and-3 and osteogenic protein-1 during murine tooth root morphogenesis and in other craniofacial structures. *Eur. J. Oral. Sci.* **1999**, *107*, 368–377. [CrossRef] [PubMed]
11. Millar, S.E. Molecular mechanisms regulating hair follicle development. *J. Invest. Dermatol.* **2002**, *118*, 216–225. [CrossRef]
12. Chen, L.-L.; Yang, L. Regulation of circRNA biogenesis. *RNA Biol.* **2015**, *12*, 381–388. [CrossRef]
13. Jeck, W.R.; Sharpless, N.E. Detecting and characterizing circular RNAs. *Nat. Biotechnol.* **2014**, *32*, 453–461. [CrossRef]
14. Zhang, Y.; Zhang, X.-O.; Chen, T.; Xiang, J.-F.; Yin, Q.-F.; Xing, Y.-H.; Zhu, S.; Yang, L.; Chen, L.-L. Circular intronic long noncoding RNAs. *Mol. Cell.* **2013**, *51*, 792–806. [CrossRef] [PubMed]
15. You, X.; Vlatkovic, I.; Babic, A.; Will, T.; Epstein, I.; Tushev, G.; Akbalik, G.; Wang, M.; Glock, C.; Quedenau, C.; et al. Neural circular RNAs are derived from synaptic genes and regulated by development and plasticity. *Nat. Neurosci.* **2015**, *18*, 603–610. [CrossRef]
16. Xu, Z.; Li, P.; Fan, L.; Wu, M. The potential role of circRNA in tumor immunity regulation and immunotherapy. *Front. Immunol.* **2018**, *9*, 9. [CrossRef]
17. Conn, S.J.; Pillman, K.A.; Toubia, J.; Conn, V.M.; Salmanidis, M.; Phillips, C.A.; Roslan, S.; Schreiber, A.W.; Gregory, P.A.; Goodall, G.J. The RNA binding protein quaking regulates formation of circRNAs. *Cell* **2015**, *160*, 1125–1134. [CrossRef]
18. Memczak, S.; Jens, M.; Elefsinioti, A.; Torti, F.; Krueger, J.; Rybak, A.; Maier, L.; Mackowiak, S.D.; Gregersen, L.H.; Munschauer, M.; et al. Circular RNAs are a large class of animal RNAs with regulatory potency. *Nature* **2013**, *495*, 333–338. [CrossRef]
19. Westholm, J.O.; Miura, P.; Olson, S.; Shenker, S.; Joseph, B.; Sanfilippo, P.; Celniker, S.E.; Graveley, B.R.; Lai, E.C. Genome-wide analysis of drosophila circular RNAs reveals their structural and sequence properties and age-dependent neural accumulation. *Cell Rep.* **2014**, *9*, 1966–1980. [CrossRef] [PubMed]
20. Salmena, L.; Poliseno, L.; Tay, Y.; Kats, L.; Pandolfi, P.P. A ceRNA hypothesis: The Rosetta Stone of a hidden RNA language? *Cell* **2011**, *146*, 353–358. [CrossRef] [PubMed]
21. Lv, X.; Wang, J.; Xu, Y.; Zhou, H.; Li, Y.; Sun, W. CircCSPP1 Competitively Binds miR-10a to Regulate BMP7 Expression and Affects the Proliferation of Dermal Papilla Cells. *Int. J. Mol. Sci.* **2024**, *25*, 11547. [CrossRef] [PubMed]
22. Zhu, Y.; Wang, Y.; Zhao, J.; Shen, J.; Wang, Z.; Bai, M.; Fan, Y.; Yin, R.; Mao, Y.; Bai, W. CircRNA-1967 participates in the differentiation of goat SHF-SCs into hair follicle lineage by sponging miR-93-3p to enhance LEF1 expression. *Anim. Biotechnol.* **2023**, *34*, 482–494. [CrossRef]
23. Wu, Y.Y.; Yue, Y.J.; Guo, T.T.; Wang, T.X.; Guo, J.; Li, G.Y.; Han, J.L.; Yang, M.; Liu, J.B.; Sun, X.P. Fetal skin hair follicle development and morphological structure of Chinese super-fine wool sheep (Gansu type). *Sci. Agric. Sin.* **2013**, *46*, 1923. [CrossRef]
24. Bolger, A.M.; Lohse, M.; Usadel, B. Trimmomatic: A flexible trimmer for Illumina sequence data. *Bioinformatics* **2014**, *30*, 2114–2120. [CrossRef]
25. Kim, D.; Langmead, B.; Salzberg, S.L. HISAT: A fast spliced aligner with low memory requirements. *Nat. Methods* **2015**, *12*, 357–360. [CrossRef]

26. Pertea, M.; Pertea, G.M.; Antonescu, C.M.; Chang, T.-C.; Mendell, J.T.; Salzberg, S.L. StringTie enables improved reconstruction of a transcriptome from RNA-seq reads. *Nat. Biotechnol.* **2015**, *33*, 290–295. [CrossRef]
27. Gao, Y.; Wang, J.; Zhao, F. CIRI: An efficient and unbiased algorithm for de novo circular RNA identification. *Genome Biol.* **2015**, *16*, 4. [CrossRef]
28. Frazee, A.C.; Sabunciyan, S.; Hansen, K.D.; Irizarry, R.A.; Leek, J.T. Differential expression analysis of RNA-seq data at single-base resolution. *Biostatistics* **2014**, *15*, 413–426. [CrossRef]
29. Robinson, M.D.; McCarthy, D.J.; Smyth, G.K. edgeR: A Bioconductor package for differential expression analysis of digital gene expression data. *Bioinformatics* **2010**, *26*, 139–140. [CrossRef] [PubMed]
30. Livak, K.J.; Schmittgen, T.D. Analysis of relative gene expression data using real-time quantitative PCR and the 2− ΔΔCT method. *Methods* **2001**, *25*, 402–408. [CrossRef] [PubMed]
31. Qiao, Y.; Mao, Y.; Wang, J.; Chen, R.; Libing, Z.; Su, Y.-Q.; Chen, J.; Zheng, W.-Q. Analysis of liver and gill miRNAs of Larimichthys crocea against Cryptocryon irritans challenge. *Fish. Shellfish Immunol.* **2016**, *59*, 484–491. [CrossRef] [PubMed]
32. Hsu, S.-D.; Lin, F.-M.; Wu, W.-Y.; Liang, C.; Huang, W.-C.; Chan, W.-L.; Tsai, W.-T.; Chen, G.-Z.; Lee, C.-J.; Chiu, C.-M.; et al. miRTarBase: A database curates experimentally validated microRNA–target interactions. *Nucleic Acids Res.* **2011**, *39* (Suppl. S1), D163–D169. [CrossRef]
33. Shannon, P.; Markiel, A.; Ozier, O.; Baliga, N.S.; Wang, J.T.; Ramage, D.; Amin, N.; Schwikowski, B.; Ideker, T. Cytoscape: A software environment for integrated models of biomolecular interaction networks. *Genome Res.* **2003**, *13*, 2498–2504. [CrossRef]
34. McDonald, B.; Hoey, W.; Hopkins, P.S. Cyclical fleece growth in cashmere goats. *Aust. J. Agric. Res.* **1987**, *38*, 597. [CrossRef]
35. Zhao, B.; Chen, Y.; Hu, S.; Yang, N.; Wang, M.; Liu, M.; Li, J.; Xiao, Y.; Wu, X. Systematic analysis of non-coding RNAs involved in the Angora rabbit (*Oryctolagus cuniculus*) hair follicle cycle by RNA sequencing. *Front. Genet.* **2019**, *10*, 407. [CrossRef]
36. Zhu, Z.; Li, Y.; Liu, W.; He, J.; Zhang, L.; Li, H.; Li, P.; Lv, L. Comprehensive circRNA expression profile and construction of circRNA-associated ceRNA network in fur skin. *Exp. Dermatol.* **2018**, *27*, 251–257. [CrossRef]
37. Tan, W.L.W.; Lim, B.T.S.; Anene-Nzelu, C.G.O.; Ackers-Johnson, M.; Dashi, A.; See, K.; Tiang, Z.; Lee, D.P.; Chua, W.W.; Luu, T.D.A.; et al. A landscape of circular RNA expression in the human heart. *Cardiovasc. Res.* **2017**, *113*, 298–309. [CrossRef]
38. Zhu, L.; Jiang, Q.; Meng, J.; Zhao, H.; Lin, J. Pan-cancer analysis of COL15A1: An immunological and prognostic biomarker. *Discov. Oncol.* **2024**, *15*, 325. [CrossRef]
39. Nakamuta, S.; Yang, Y.-T.; Wang, C.-L.; Gallo, N.B.; Yu, J.-R.; Tai, Y.; Van Aelst, L. Dual role for DOCK7 in tangential migration of interneuron precursors in the postnatal forebrain. *J. Cell Biol.* **2017**, *216*, 4313–4330. [CrossRef] [PubMed]
40. Verma, P.; Dalal, K. ADAMTS-4 and ADAMTS-5: Key enzymes in osteoarthritis. *J. Cell. Biochem.* **2011**, *112*, 3507–3514. [CrossRef] [PubMed]
41. Foomani, F.H.; Jarzembowski, J.A.; Mostaghimi, S.; Mehrvar, S.; Kumar, S.N.; Ranji, M. Optical metabolic imaging of mitochondrial dysfunction on HADH mutant newborn rat hearts. *IEEE J. Transl. Eng. Health Med.* **2021**, *9*, 1800407. [CrossRef]
42. Kuivaniemi, H.; Tromp, G. Type III collagen (COL3A1): Gene and protein structure, tissue distribution, and associated diseases. *Gene* **2019**, *707*, 151–171. [CrossRef] [PubMed]
43. Yeung, W.; Qi, X.; Pang, L.; Liu, H.; Ng, K.; Liu, J.; Lo, M.; Man, K. Type III TGF-β Receptor Down-Regulation Promoted Tumor Progression via Complement Component C5a Induction in Hepatocellular Carcinoma. *Cancers* **2021**, *13*, 1503. [CrossRef]
44. Gao, G.; Li, X.; Jiang, Z.; Osorio, L.; Tang, Y.L.; Yu, X.; Jin, G.; Zhou, Z. Isthmin-1 (Ism1) modulates renal branching morphogenesis and mesenchyme condensation during early kidney development. *Nat. Commun.* **2023**, *14*, 2378. [CrossRef]
45. Higginbotham, H.; Guo, J.; Yokota, Y.; Umberger, N.L.; Su, C.-Y.; Li, J.; Verma, N.; Hirt, J.; Ghukasyan, V.; Caspary, T.; et al. Arl13b-regulated cilia activities are essential for polarized radial glial scaffold formation. *Nat. Neurosci.* **2013**, *16*, 1000–1007. [CrossRef]
46. Ge, W.; Zhang, W.; Zhang, Y.; Zheng, Y.; Li, F.; Wang, S.; Liu, J.; Tan, S.; Yan, Z.; Wang, L.; et al. A Single-cell Transcriptome Atlas of Cashmere Goat Hair Follicle Morphogenesis. *Genom. Proteom. Bioinform.* **2021**, *19*, 437–451. [CrossRef]
47. Zhang, X.; Wang, P.; Chen, T.; Yan, W.; Guan, X.; Shen, G.; Luo, X.; Wan, X.; Ning, Q. Kctd9 deficiency impairs natural killer cell development and effector function. *Front. Immunol.* **2019**, *10*, 744. [CrossRef]
48. Gao, W.Z.; Xue, H.L.; Yang, J.C. Proteomics analysis of the secondary hair follicle cycle in Liaoning cashmere goat. *Small Rumin. Res.* **2021**, *201*, 106408. [CrossRef]
49. Mumblat, Y.; Kessler, O.; Ilan, N.; Neufeld, G. Full-length semaphorin-3C is an inhibitor of tumor lymphangiogenesis and metastasis. *Cancer Res.* **2015**, *75*, 2177–2186. [CrossRef]
50. McCrea, P.D.; Park, J.-I. Developmental functions of the P120-catenin sub-family. *Biochim. Biophys. Acta* **2007**, *1773*, 17–33. [CrossRef] [PubMed]
51. Sennett, R.; Wang, Z.; Rezza, A.; Grisanti, L.; Roitershtein, N.; Sicchio, C.; Mok, K.W.; Heitman, N.J.; Clavel, C.; Ma'ayan, A.; et al. An integrated transcriptome atlas of embryonic hair follicle progenitors, their niche, and the developing skin. *Dev. Cell* **2015**, *34*, 577–591. [CrossRef]

52. Merkestein, M.; Laber, S.; McMurray, F.; Andrew, D.; Sachse, G.; Sanderson, J.; Li, M.; Usher, S.; Sellayah, D.; Ashcroft, F.M.; et al. FTO influences adipogenesis by regulating mitotic clonal expansion. *Nat. Commun.* **2015**, *6*, 6792. [CrossRef]
53. Hansen, T.B.; Jensen, T.I.; Clausen, B.H.; Bramsen, J.B.; Finsen, B.; Damgaard, C.K.; Kjems, J. Natural RNA circles function as efficient microRNA sponges. *Nature* **2013**, *495*, 384–388. [CrossRef]
54. Romero-Carvajal, A.; Navajas Acedo, J.N.; Jiang, L.; Kozlovskaja-Gumbrienė, A.; Alexander, R.; Li, H.; Piotrowski, T. Regeneration of sensory hair cells requires localized interactions between the Notch and Wnt pathways. *Dev. Cell* **2015**, *34*, 267–282. [CrossRef]
55. Diike, P.; Heiden, C.H. *Smad Signal Transduction: Smads in Proliferation, Differentiation and Disease*; Springer: Berlin/Heidelberg, Germany, 2006. [CrossRef]
56. Zhang, W.; Liu, H.T. MAPK signal pathways in the regulation of cell proliferation in mammalian cells. *Cell Res.* **2002**, *12*, 9–18. [CrossRef]
57. Akilli Öztürk, Ö.A.; Pakula, H.; Chmielowiec, J.; Qi, J.; Stein, S.; Lan, L.; Sasaki, Y.; Rajewsky, K.; Birchmeier, W. Gab1 and Mapk signaling are essential in the hair cycle and hair follicle stem cell quiescence. *Cell Rep.* **2015**, *13*, 561–572. [CrossRef] [PubMed]
58. Wang, C.; Zang, K.; Tang, Z.; Yang, T.; Ye, X.; Dang, Y. Hordenine activated dermal papilla cells and promoted hair regrowth by activating Wnt signaling pathway. *Nutrients* **2023**, *15*, 694. [CrossRef]
59. Foitzik, K.; Lindner, G.; Mueller-Roever, S.; Maurer, M.; Botchkareva, N.; Botchkarev, V.; Handjiski, B.; Metz, M.; Hibino, T.; Soma, T.; et al. Control of murine hair follicle regression (catagen) by TGF-β1 in vivo. *FASEB J.* **2000**, *14*, 752–760. [CrossRef] [PubMed]
60. Foitzik, K.; Paus, R.; Doetschman, T.; Dotto, G.P. The TGF-β2 isoform is both a required and sufficient inducer of murine hair follicle morphogenesis. *Dev. Biol.* **1999**, *212*, 278–289. [CrossRef]
61. Dai, B.; Sha, R.N.; Yuan, J.L.; Liu, D.J. Multiple potential roles of thymosin β4 in the growth and development of hair follicles. *J. Cell Mol. Med.* **2021**, *25*, 1350–1358. [CrossRef] [PubMed]
62. RunQing, C.; HaiYin, H.; Peng, W.; KaiYang, L.; MingXing, C.; YuFang, L. Estrogen mediates CircZNF423 as a sponge for oar-miR-541-3p to target CALM3 for regulating myoblast proliferation in sheep. *Sci. Agric. Sin.* **2024**, *57*, 597. [CrossRef]
63. Yin, R.H.; Zhao, S.J.; Jiao, Q.; Wang, Z.Y.; Bai, M.; Fan, Y.X.; Zhu, Y.B.; Bai, W.L. CircRNA-1926 promotes the differentiation of goat SHF stem cells into hair follicle lineage by miR-148a/b-3p/CDK19 axis. *Animals* **2020**, *10*, 1552. [CrossRef]
64. Tian, D.; Pei, Q.; Jiang, H.; Guo, J.; Ma, X.; Han, B.; Li, X.; Zhao, K. Comprehensive analysis of the expression profiles of mRNA, lncRNA, circRNA, and miRNA in primary hair follicles of coarse sheep fetal skin. *BMC Genom.* **2024**, *25*, 574. [CrossRef]
65. Chen, F.; Feng, Z.; Zhu, J.; Liu, P.; Yang, C.; Huang, R.; Deng, Z. Emerging roles of circRNA_NEK6 targeting miR-370-3p in the proliferation and invasion of thyroid cancer via Wnt signaling pathway. *Cancer Biol. Ther.* **2018**, *19*, 1139–1152. [CrossRef] [PubMed]

animals

Article

Multi-Omics Insights into Postnatal Skeletal Muscle Development in Duroc Pigs

Kaiming Wang [1,†], Xin Li [1,†], Xibing Liu [1], Sui Liufu [1], Lanlin Xiao [1], Bohe Chen [1], Wenwu Chen [1], Jun Jiang [1], Yan Liu [2,*] and Haiming Ma [1,3,4,*]

1 College of Animal Science and Technology, Hunan Agricultural University, Changsha 410128, China; 15116529648@stu.hunau.edu.cn (K.W.); 13243663120@163.com (X.L.); hunheshuiguocha@163.com (X.L.); liufusui0816@163.com (S.L.); finafantacyla@163.com (L.X.); chenhe0213914@163.com (B.C.); cww1242646778@163.com (W.C.); jiangjun1121@126.com (J.J.)

2 Tangrenshen Group Co., Ltd., Zhuzhou 412007, China

3 Key Laboratory of Livestock and Poultry Resources (Pig) Evaluation and Utilization, Ministry of Agriculture and Rural Affairs, Changsha 410128, China

4 Yuelushan Laboratory, Changsha 410128, China

* Correspondence: ly9030@trsgroup.cn (Y.L.); mahaiming@hunau.edu.cn (H.M.)

† These authors contributed equally to this work.

Simple Summary

Skeletal muscles exhibit substantial heterogeneity in their developmental patterns and functions, depending on their location in the body. In this study, we identified six genes and eight metabolites that are potentially involved in regulating both skeletal muscle development and muscle fiber-type transformation. Notably, the divergence in SOL vs. GAS is minor at birth but increases gradually during development, whereas divergence in PMM vs. SOL and PMM vs. GAS is great from birth. Furthermore, we found that a developmental shift occurred from the MAPK signaling pathway (1–21 d) to the regulation of the actin cytoskeleton (21–120 d).

Abstract

Skeletal muscles, accounting for 40% of mammalian body mass, exhibit pronounced heterogeneity due to their distinct anatomical locations. Animal husbandry has focused excessively on *longissimus dorsi* (LDM) development while neglecting other muscles. In this study, we integrated Bulk RNA Sequencing (bulk RNA-seq) and Liquid Chromatography–Mass Spectrometry (LC-MS) analyses of *Soleus* (SOL), *Gastrocnemius* (GAS), and *Psoas major* muscles (PMM) across three key stages in Duroc pigs. We identified nine critical genes (*S100A1*, *MBOAT2*, *CA3*, *GYG2*, *ACTN3*, *ENO3*, *SLC3A2*, *SLC16A10*, and *GAPDH*) and eight metabolites potentially involved in regulating both skeletal muscle development and fiber-type transformation. The heterogeneity between SOL and GAS was low at birth but increased gradually during development. In contrast, PMM exhibited higher heterogeneity than SOL and GAS from birth. Notably, expression levels of *MYH7*, *MYH1*, and *MYH4* displayed stage-specific and muscle type-dependent variations. Moreover, we observed a developmental shift from the MAPK signaling pathway (1–21 d) to the regulation of the actin cytoskeleton (21–120 d). Pairwise comparisons between the SOL, GAS, and PMM revealed that the signaling pathways were enriched in muscle fiber-type switching. Collectively, through the integration of bulk RNA-seq and LC-MS data, this study provides novel molecular breeding strategies for the genetic improvement of meat-producing animals.

Keywords: pig; skeletal muscle; transcriptome; metabolome; gene

1. Introduction

In adult mammals, skeletal muscle accounts for approximately 40% of body weight. It is formed of highly heterogeneous and elastic tissues and plays a vital role in metabolism and movement [1]. Skeletal muscle development is mainly divided into two stages: embryonic and postnatal [2]. Skeletal muscle development requires myogenesis, a complex process regulated by numerous genes, including *LMOD2* [3,4]. Although transcriptomics (RNA-seq) has been extensively employed to identify genes involved in skeletal muscle development and regeneration, most studies are restricted to single anatomical sites. This is especially true in the field of animal husbandry, where most attention has been paid to the *longissimus dorsi* (LDM) and *soleus* (SOL), whereas other muscles have been overlooked [5].

Recent studies have focused on this topic. Terry et al. [6] used the bulk RNA-seq technique to measure more than 20 types of muscles in rats and mice. The results showed that more than 50% of the animal gene expression levels varied among the muscles. Notably, there is an average 13% difference in gene expression between any two muscle types. Jin et al. [7] revealed transcriptional differences among 47 skeletal muscles in pigs; however, this study did not uncover transcriptional variability at longitudinal developmental stages. Yang et al. [8] compared skeletal muscle transcriptomes across 27 developmental stages in Chinese and Western pigs, revealing the epigenetic mechanisms underlying the differences in muscle growth between the two breeds. However, this longitudinal study focused only on the LDM and lacked a cross-sectional assessment of distinct muscles. Given the limitations of these previous studies, we propose conducting a multi-muscle, multi-stage, and multi-omics study. Muscle aging is an inevitable process. The transcriptional profiles of 11 muscles in young and old mice indicated that the mitochondria-enriched SOL exhibited greater resistance to aging than other skeletal muscles [9]. Thus, an integrated longitudinal and cross-sectional analysis of distinct skeletal muscles is crucial for understanding the muscular system.

Skeletal muscle heterogeneity arises from the distinct composition of myosin heavy chain (MYHC) isoforms [10], which define phenotypic variations [11]. Postnatal development is a critical period for fiber-type transformation in pigs, as the dominance of oxidative fibers progressively declines with the expansion of glycolytic fibers between days 1 and 28 [12]. However, research on muscle fiber-type transformation has primarily focused on LDM, and the spatiotemporal dynamics of this transition across distinct muscles remain unclear. Comparative transcriptomics has revealed substantial expression divergence between oxidative (PMM) and glycolytic muscles (*Longissimus lumborum, Triceps brachii,* and *Semimembranosus*) in cattle, which reflects distinct metabolic programs [13]. Transcriptomic profiling has revealed that *Extensor digitorum longus* (EDL) and PMM are fast-twitch muscles, while SOL is a slow-twitch muscle [14]. The SOL displays significantly enriched slow-twitch fibers and intramuscular adiposity compared to adjacent hindlimb muscles. Notably, slow-twitch fiber density is positively correlated with intramuscular lipid accumulation, suggesting functional interdependence between these traits [15]. These results indicate that skeletal muscles at different anatomical locations have distinct structures, functions, and developmental patterns. However, cross-sectional and longitudinal multi-omics dynamics governing the development of distinct muscle types remain unclear.

In this study, in order to address these key issues, we conducted in-depth research on various types of skeletal muscles. We selected rapid-growth Duroc pigs as the model organism. We performed transcriptomic (Bulk RNA-seq) and metabolomic (LC-MS) profiling of porcine SOL (slow-twitch muscle), GAS (fast-twitch muscle), and PMM (fast-twitch muscle) at three key developmental stages (1, 21, and 120 d). Integrative analysis revealed the essential genes and metabolites governing skeletal muscle development and fiber-type transitions. This study provides novel directions and insights for the genetic improvement and breeding of meat-producing animals.

2. Materials and Methods

2.1. Experimental Animals and Cells

All experimental procedures in this study were approved by the Hunan Agricultural University Animal Care and Use Committee (permit number CACAHU-20250405). The experiment was conducted on a standardized farm where groups of six pigs were housed in pens. The ambient temperature and relative humidity were maintained at 25 ± 2 °C and 60–70%. Natural light was provided for 6 h daily. Pigs had free access to water and were fed a diet meeting NRC nutritional standards. Eighteen pigs were divided into three groups. The pigs in the same group were full siblings, while those in different groups were half-siblings. Three skeletal muscles (*Soleus*, *Gastrocnemius*, and *Psoas major*) were collected from 18 Duroc pigs at three developmental stages (1 d, 21 d, and 120 d; $n = 6$ per stage), snap-frozen in liquid nitrogen, and stored at −80 °C for subsequent transcriptomic and metabolomic analyses. C2C12 myoblasts (ATCC, New York, NY, USA) in proliferative (GM-cultured) and differentiated (DM-induced) states were harvested for subsequent RNA-seq analysis. The analysis of these cells is presented only as a supporting dataset and not as the primary focus of the study.

2.2. Total RNA Extraction and Quantitative Real-Time PCR

Total RNA was isolated from cells and tissues using TRIzol reagent (Invitrogen, Carlsbad, CA, USA). Total RNA was reverse-transcribed into cDNA using a kit (Invitrogen, Carlsbad, CA, USA). RT-qPCR was performed using the SYBR Green Kit (TransGen, Beijing, China), and the relative expression of genes was calculated using the $2^{-\Delta\Delta Ct}$ method. All primer sequences are listed in Table 1.

Table 1. The primer sequences used in this study.

Species	Genes	F: Sequence (5′ to 3′)	R: Sequence (5′ to 3′)
Pig	*MYH1*	GAAGTTGCATCCCTAAAGGCAG	CGATGACTTGGCGTCAAAAGG
Pig	*MYH2*	GGAGGCTGAGGAACAATCCA	GCATCGGGACAGCCTTACTC
Pig	*MYH4*	AGTTCCGTAAGATCCAGCACG	CCTGTCACCTCTCAACAGAAAGA
Pig	*MYH7*	AGTCCCAGGTCAACAAGCTG	TTCCACCTAAAGGGCTGTTG
Pig	*MFAP4*	CGGCGTGTACCTCATCTACC	GCCGTTGAACCTCTTCTGGA
Pig	*PGM2L1*	CGTCTTTTCACGGAGTCGGA	TGAAGTTCTGCCACTGCCAA
Pig	*S100A1*	GAACTGGAGACAGCGATGGA	TTGTTACAGGCCACTGTGAGG
Mouse	*S100A1*	AGAGTGCCATGGAGACCCTC	GCTCAACTGGTCTCCCAGAA
Pig, Mouse	*GAPDH*	AGGGCATCCTGGGCTACACT	TCCACCACCCTGTTGCTGTAG

2.3. Western Blot Detection of Protein Expression

Total proteins were extracted from muscle tissues using RIPA lysis buffer (P0013B, Beyotime Biotech Inc., Shanghai, China) supplemented with protease and phosphatase inhibitors. Western blot analysis was performed with the following primary antibodies: MYH7 (1:1000, 22280-1-AP, Proteintech Group, Inc., Wuhan, China), MYH1 (1:10000, 67299-1-Ig, Proteintech Group, Inc., Wuhan, China), and Anti-GAPDH (1:10000, 10494-1-AP, Proteintech Group, Inc., Wuhan, China). Finally, protein bands were quantified using ImageJ 2 software.

2.4. Hematoxylin–Eosin (HE) Staining

Paraffin sections were deparaffinized, stained with hematoxylin and eosin, dehydrated, and mounted using an HE staining kit. Sections were then observed under a microscope, and images were captured for analysis. The muscle fiber cross-sectional area was quantified using ImageJ 2 (National institutes of Health), followed by a comparative analysis.

2.5. Immunofluorescence Assay

For immunofluorescence assays, fresh SOL, GAS, and PMM muscles were fixed in 4% paraformaldehyde and embedded in paraffin. Serial 20 µm sections were cut using a cryostat. Sections were incubated overnight at 4 °C with primary antibodies against fast-twitch (GB112130-100, Wuhan Servicebio Technology Co., Ltd., Wuhan, China) and slow-twitch (GB112131, Wuhan Servicebio Technology Co., Ltd., Wuhan, China) myosin heavy chains, followed by 1 h incubation with species-matched fluorescent secondary antibodies at room temperature. Images were acquired using a fluorescence microscope.

2.6. RNA-Seq and LC-MS Analyses

A total of 54 samples from three types of muscle tissues at three developmental stages (n = 6 per stage) were collected. Transcriptomic and metabolomic profiles were generated from these samples. RNA-seq and LC-MS analyses were performed by Shanghai Majorbio Bio-Pharm Technology Co., Ltd. (Shanghai, China). Specifically, to identify differentially expressed genes (DEGs) between the two groups, the abundance of each transcript was quantified based on the transcripts per million (TPM) method. RSEM was used to estimate transcript abundance. Subsequently, differential expression analysis was conducted using the DESeq2 and DEGseq packages. Transcripts exhibiting $|\log_2FC| \geq 1.5$ and an FDR < 0.05 (for DESeq2) or FDR < 0.001 (for DEGseq) were defined as statistically significantly differentially expressed genes (DEGs). KEGG pathway analyses were performed using the Goatools and Python SciPy (1.14) packages. For LC-MS, variance analysis was performed on the preprocessed data matrix using the R package ropls (1.6.2). Principal component analysis (PCA) and orthogonal partial least squares-discriminant analysis (OPLS-DA) were conducted, and model stability was assessed using 7-cycle cross-validation. Significantly altered metabolites were selected based on variable importance in projection (VIP) scores from the OPLS-DA model (VIP > 1 and $p < 0.05$). Functional interpretation was performed using pathway enrichment analysis in the KEGG database.

2.7. Statistical Analysis of the Data

The qRT-PCR data are presented as means ± standard error. An unpaired Student's t-test was performed between the treatment and control groups using GraphPad Prism 8, *, $p < 0.05$; **, $p < 0.01$.

3. Results

3.1. Phenotypic Assessment of SOL, GAS, and PMM in Duroc Pigs

The picture presents anatomical, developmental, and experimental diagrams of SOL, GAS, and PMM in Duroc pigs (Figure 1A). The volume and weight of SOL, GAS, and PMM gradually increased as they grew (Figure 1B). Similarly, the muscle fiber cross-sectional areas of SOL, GAS, and PMM increased significantly (Figure 1C). At 120 days of age, immunofluorescence results demonstrated that slow-twitch muscle fibers in the SOL exhibited the highest expression compared to GAS and PMM (Figure 1D). At 120 days of age, we examined the mRNA expression of *MYH7* and *MYH4* in these three muscles. Results indicated that SOL primarily expressed *MYH7*, while GAS and PMM primarily expressed *MYH4* (Figure 1E,F). Furthermore, the MYH7 protein (slow-twitch) is mainly expressed in SOL, while the MYH1 protein (fast-twitch) is mainly expressed in PMM and GAS (Figure 1G). Collectively, these results confirm the accuracy of the sampling sites, and the inconsistent molecular expression in these three muscles during development can be further analyzed.

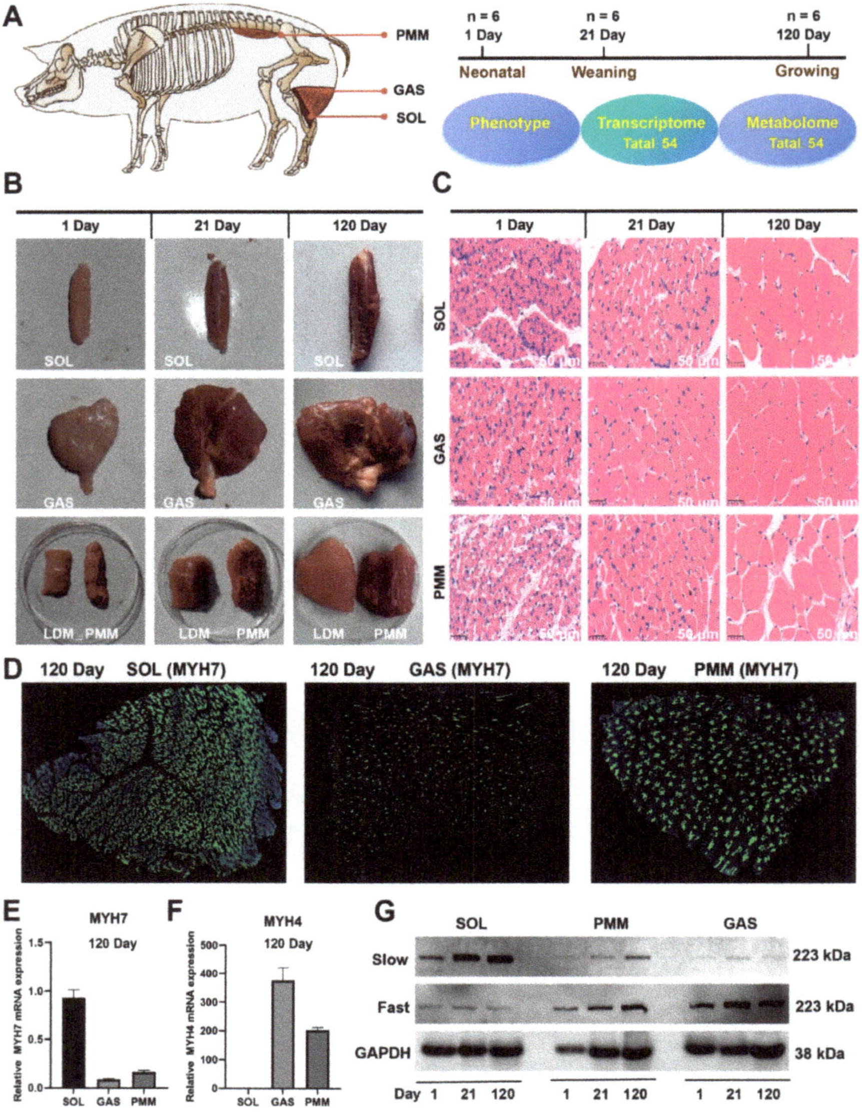

Figure 1. Phenotypic assessment of SOL, GAS, and PMM in Duroc pigs. (**A**) Experimental design diagram. (**B**) Phenotypic diagrams of SOL, GAS, LDM, and PMM are presented, with LDM as the control for PMM ($n = 6$). (**C**) HE staining of SOL, GAS, and PMM across three developmental stages ($n = 6$); Scale bar, 50 μm. (**D**) At 120 days of age, MYH7 antibody immunostaining was performed on SOL, GAS, and PMM ($n = 3$). (**E**,**F**) At 120 days of age, the mRNA expression levels of *MYH7* and *MYH4* in SOL, GAS, and PMM were detected using qRT-PCR ($n = 3$). (**G**) Protein expression levels of MYH7 and MYH1 during the development of the SOL, GAS, and PMM were analyzed by Western Blot ($n = 3$).

3.2. Identification of DEGs by Bulk RNA-Seq During Multiple Skeletal Muscle Development

Transcriptomic analyses of SOL, GAS, and PMM across three developmental stages identified 9153 genes in 54 samples (Figure 2A,B). Principal Component Analysis (PCA) revealed differences among the three muscle types across the three developmental stages and exhibited strong intra-group clustering (Figure 2C). Notably, SOL and GAS muscles had only 49 differentially expressed genes (DEGs) on day 1 after birth, and the number of DEGs increased as the muscles developed. Neonatal transcriptomes of SOL and GAS were indistinguishable by PCA, indicating transcriptional similarity (Figure 2D). PMM exhibited significant differences in gene expression from both SOL and GAS after the postnatal stage (days 1–120) (Figure 2D,F). Venn diagram analysis identified 32 consistently shared genes across the SOL, GAS, and PMM at the three developmental stages, of which 21 were key genes (Figure 2F,G). The key genes mentioned in this study are listed in Table S1. Considering that myoblasts undergo proliferation and eventually differentiate into muscle fibers, we identified 2177 DEGs in the proliferation and differentiation stages of C2C12 (Figure 2H; see Table S5 for a complete list). Finally, six shared genes (*MFAP4*, *CTSW*, *ITGB7*, *LIMK1*, *PGM2L1*, and *TYMS*) were identified in muscle tissues and C2C12. These genes may regulate critical processes involved in skeletal myogenesis (Figure 2I). Among these six genes, four were progressively upregulated during development (days 1–120), while the remaining two were downregulated (Figure 2J). qRT-PCR validation of randomly selected *MFAP4* and *PGM2L1* across five developmental stages in the LDM muscle revealed expression patterns consistent with those of the SOL, GAS, and PMM muscles, confirming the reliability of transcriptome sequencing (Figure 2K,L). In conclusion, our experiments demonstrated transcriptional similarities between SOL and GAS muscles at birth, with progressive divergence during subsequent development. Additionally, six candidate genes were identified across longitudinal multi-muscle development.

3.3. Identification of DEGs by Bulk RNA-Seq in Fast-Twitch and Slow-Twitch Muscles

Transcriptome analysis revealed muscle-specific developmental dynamics of MYHC isoforms across different skeletal muscle types. In SOL, MYHC isoform proportions follow the order *MYH7* > *MYH1* > *MYH4*, with *MYH7* expression progressively increasing during development (days 1–120). In GAS, *MYH4* proportion progressively increases during development, whereas the proportions of *MYH1* and *MYH7* decrease. In PMM, MYHC isoform proportions follow the order *MYH1* > *MYH7* > *MYH4*. During development, the expression of *MYH7 and MYH4* progressively increased, whereas that of *MYH1* decreased (Figure 3A). These results demonstrate that SOL comprises slow-twitch muscle fibers (*MYH7*), whereas GAS and PMM predominantly comprise fast-twitch muscle fibers (*MYH1*). We identified 131 DEGs between fast-twitch and slow-twitch muscles, of which 113 were annotated (Figure 3B; see Tables S2 and S3 for a complete list). Of these 113 genes, 80 were highly expressed in SOL, while the other 33 were highly expressed in GAS and PMM (Figure 3B). Notably, transcription factors (TFs) enriched in slow-twitch muscle include *ZIC1*, *ZIC4*, *ESRRB*, *ZNF385*, and *RFX2*, whereas *SOX6*, *MAF*, and *HOXD8* are predominant in fast-twitch muscle (Figure 3C). In addition, the secreted proteins (SPs) *DKK3* and *FGF9* are enriched in slow-twitch muscles. We identified nine shared genes related to muscle fiber types and development (Figure 3D), namely *S100A1*, *MBOAT2*, *CA3*, *GYG2*, *ACTN3*, *ENO3*, *SLC3A2*, *SLC16A10*, and *GAPDH* (Figure 3E). qRT-PCR validation revealed significantly higher *S100A1* expression in SOL than GAS of both pigs and mice (Figure 3F,G). The expression level of *S100A1* gradually increased with skeletal muscle development in pigs and mice (Figure 3H,I). The qRT-PCR results were consistent with those of RNA-seq.

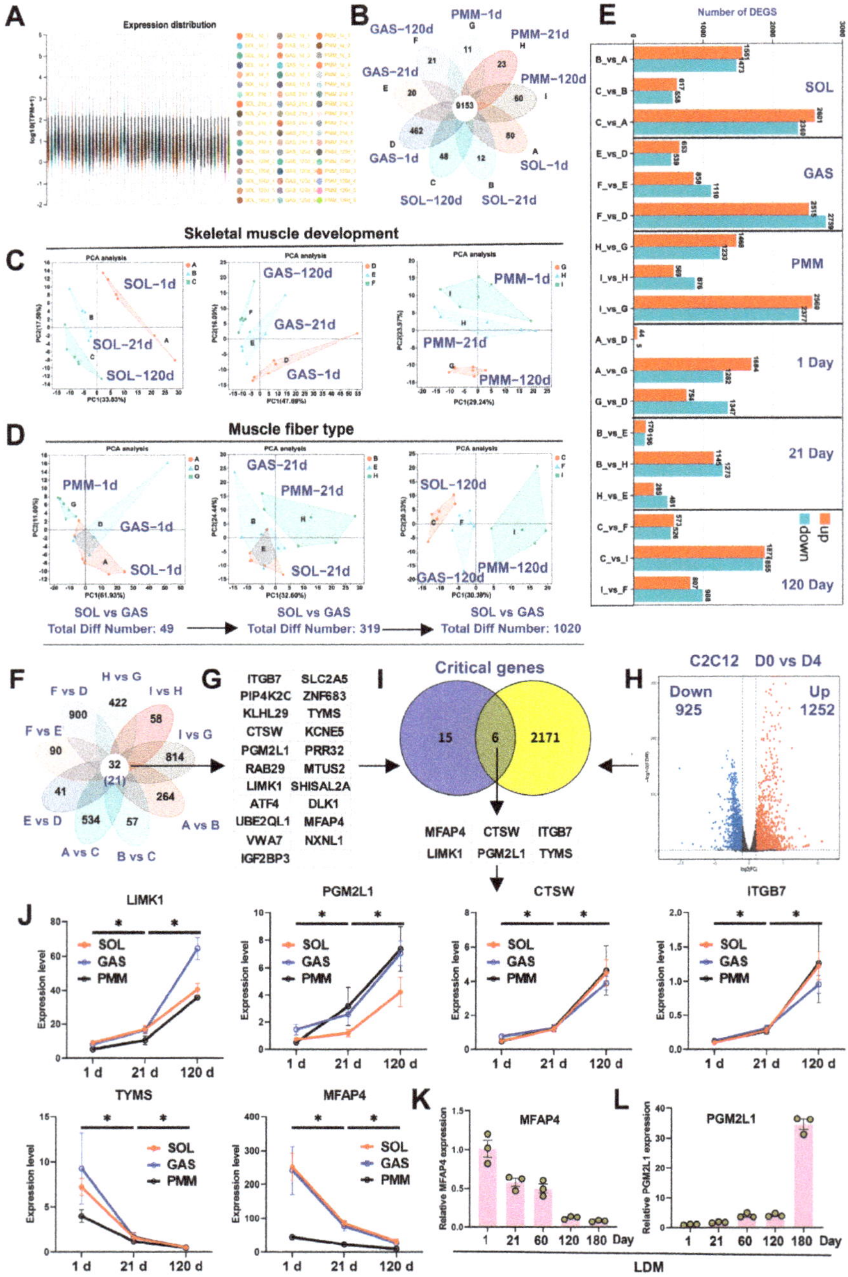

Figure 2. Identification of DEGs across multiple skeletal muscles using RNA-Seq. (**A**) Transcriptome-wide expression distribution in 54 muscle tissue samples. (**B**) Genes commonly expressed among the 54 muscle samples. (**C**,**D**) Principal Component Analysis (PCA) of skeletal muscle samples. (**E**) Number of DEGs between each pair of groups in the skeletal muscle samples. (**F**,**G**) Detected genes shared during longitudinal development in SOL, GAS, and PMM. (**H**) We performed transcriptome screening for DEGs between the proliferation and differentiation phases of C2C12 cells using data derived from our previous study [4]. (**I**) Six genes identified, which are shared between muscle development and C2C12. (**J**) Expression trends of six genes during three developmental stages in

three muscle types. (**K**,**L**) Detection of the expression levels of *MFAP4* and *PGM2L1* genes during the development of LDM using qRT-PCR. * $p < 0.05$.

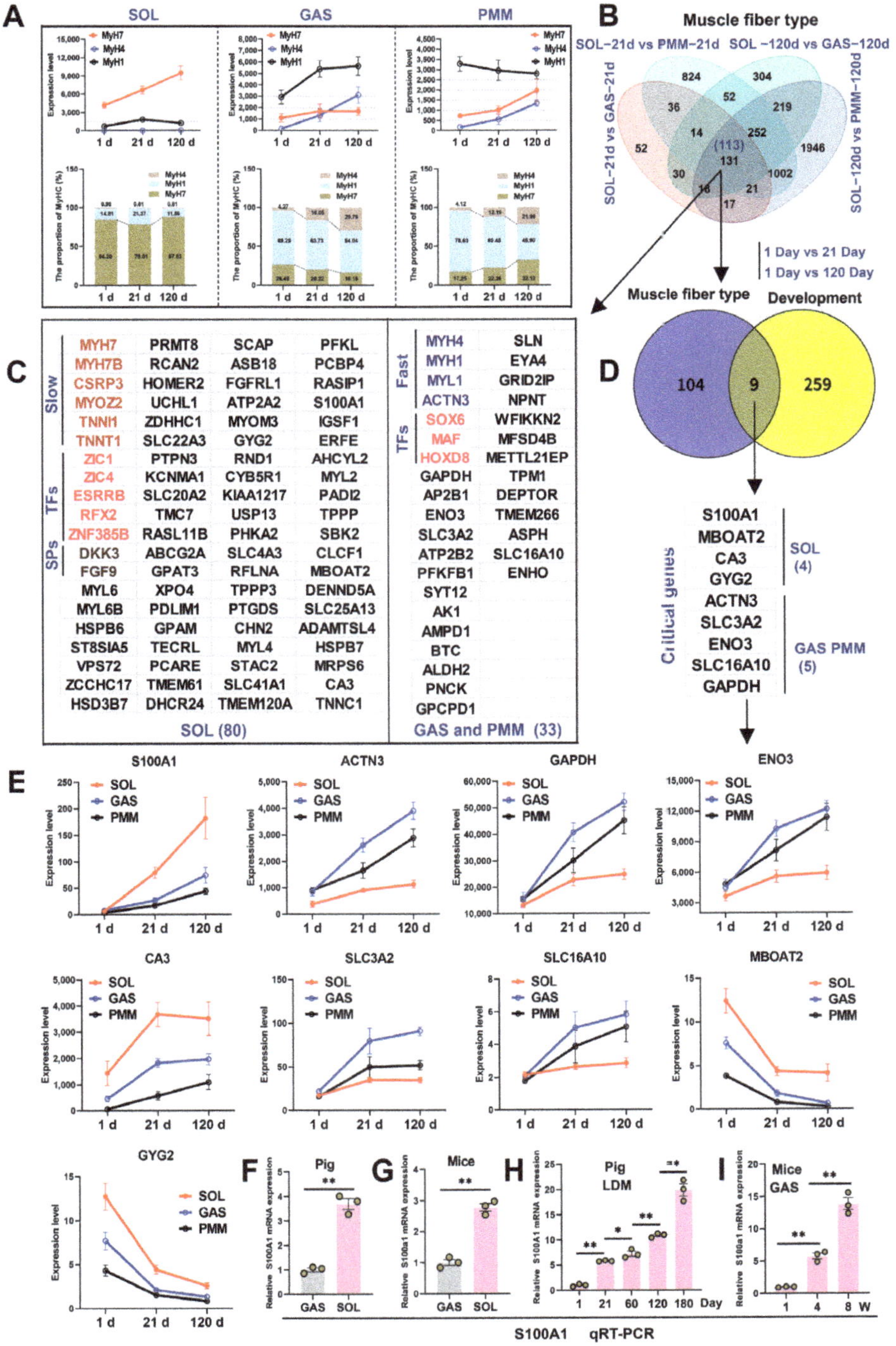

Figure 3. Identification of DEGs by bulk RNA-seq in fast-twitch and slow-twitch muscles. (**A**) Transcriptional expression levels of MYHC isoforms tracked across three developmental stages in SOL,

GAS, and PMM. (**B**) Screening of DEGs in fast-twitch versus slow-twitch muscles using Venn diagrams. (**C**) Overall, 80 genes are highly expressed in the SOL, while 33 genes are highly expressed in both the GAS and PMM. (**D**) Genes identified to be shared between skeletal muscle fibers and multiple muscle development. (**E**) Expression trends of nine genes during three developmental stages in three muscle types. (**F,G**) *S100A1* expression levels were detected by qRT-PCR in the GAS and SOL of both pigs and mice. (**H,I**) *S100A1* expression levels were detected by qRT-PCR across five developmental stages in porcine LDM and three developmental stages in mouse GAS (1, 4, and 8 weeks). * $p < 0.05$; ** $p < 0.01$.

3.4. LC-MS Reveals Critical Metabolites for Skeletal Muscle Development and Myofiber Transformation

We conducted LC-MS analysis on 54 samples, and the results showed that the reproducibility among the samples met expectations and could be used for subsequent analyses (Figure 4A). In these samples, 1574 common metabolites were identified (Figure 4B). Partial Least Squares Discriminant Analysis (PLS-DA) revealed differences among the three muscle types across three developmental stages, and strong intra-group clustering (Figure 4C). Notably, SOL and GAS muscles had only 196 differentially expressed metabolites (DMs) on day 1 after birth, and the number of DMs increased as the muscle developed (days 1–120). The PLS-DA model showed a gradual separation between SOL and GAS over time (Figure 4E). PMM exhibited significantly different metabolites compared to both SOL and GAS after the postnatal stage (Figure 4D). Venn diagram analysis identified 56 metabolites shared by SOL, GAS, and PMM across the three developmental stages. Among these, 13 metabolites were persistently upregulated and 43 were persistently downregulated (Figure 4F). We identified 92 DMs between fast-twitch and slow-twitch muscles. Among them, 69 DMs were enriched in slow-twitch muscle, and 23 DMs were enriched in fast-twitch muscle (Figure 4H). Ultimately, we identified eight crucial metabolites from the study of skeletal muscle development and muscle fiber type (Figure 4G).

Notably, these eight metabolites are enriched in slow-twitch muscle, with their abundance progressively declining during development (Figure 4I), and we have described the KEGG Pathway of Pantothenic Acid and Levan (Figure 4J). The key metabolites mentioned in this study are listed in Table S4. In conclusion, our experiments demonstrated metabolic similarities between SOL and GAS muscles at birth, with progressive divergence during subsequent development. Additionally, eight candidate DMs were identified.

3.5. Analysis of KEGG Pathway Enrichment in Different Stages of Muscle Development

The Kyoto Encyclopedia of Genes and Genomes (KEGG) is a comprehensive pathway database enabling the functional annotation of gene and metabolite regulatory networks. The pig growth period was divided into two developmental stages: birth to weaning (days 1–21) and weaning to rapid growth period (days 21–120). Given the limited number of available enriched metabolites, we prioritized the analysis of gene-enriched KEGG pathways. Notably, at the first developmental stage (days 1–21), DEGs across all three muscle types showed co-enrichment in MAPK signaling, cell cycle, and apoptosis pathways (Figure 5A–D). During the second developmental stage (days 21–120), DEGs in all three types of skeletal muscles were enriched in the regulation of the actin cytoskeleton pathway, while fast-twitch muscles (GAS and PMM) exhibited enrichment in the PI3K-Akt signaling pathway (Figure 5E–H). In conclusion, our findings demonstrate a developmental shift from the MAPK signaling pathway to actin cytoskeleton regulation pathways during muscle development (Figure 5I).

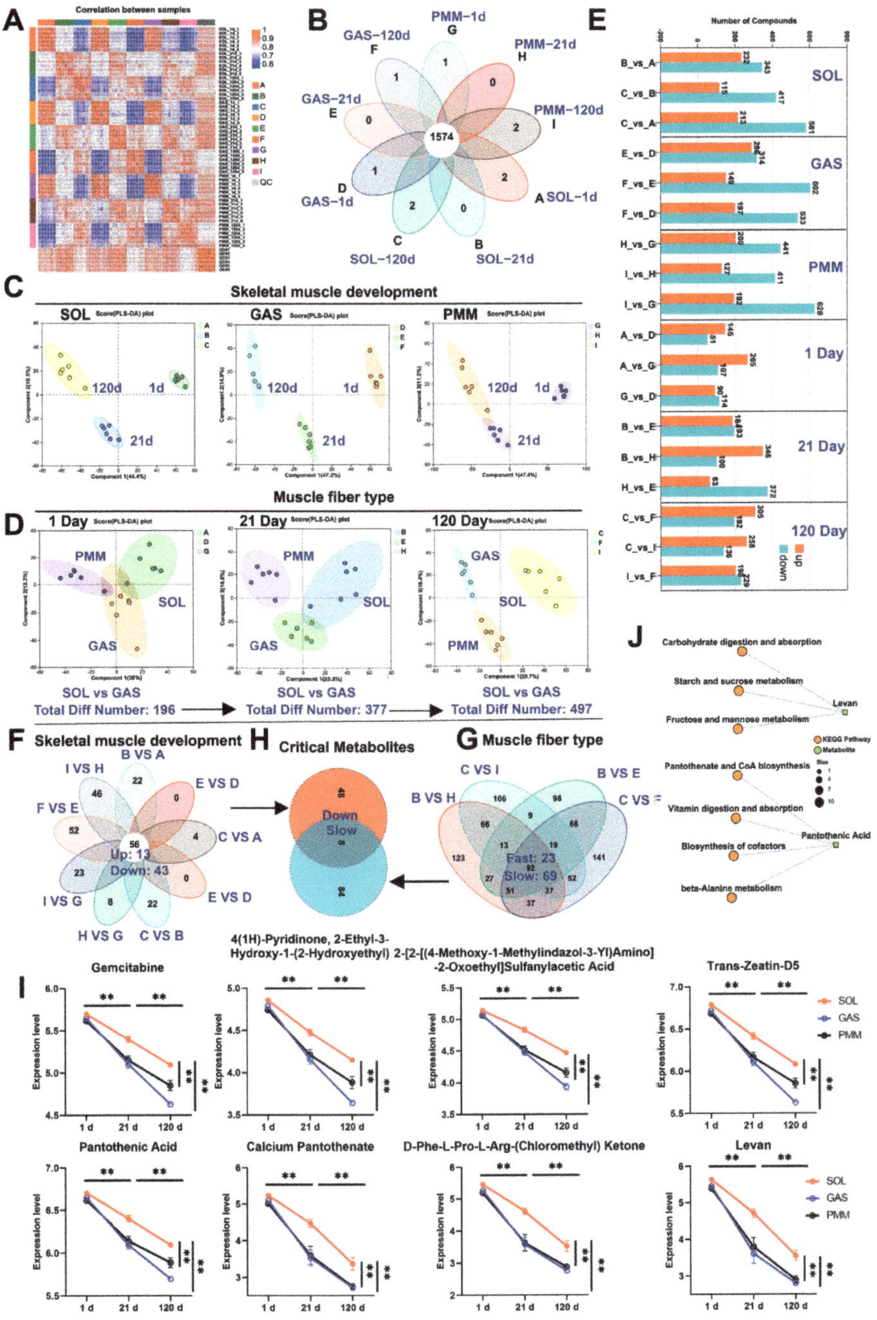

Figure 4. LC-MS analysis identified key metabolites governing skeletal muscle development and myofiber type transformation. (**A**) Inter-group correlation analysis of 54 muscle samples (nine groups, n = 6). (**B**) Venn diagram analysis of common metabolites across nine groups. (**C**,**D**) Partial Least Squares Discriminant Analysis (PLS-DA) of skeletal muscle samples. (**E**) Number of differential metabolites (DMs) between each pair of groups (n = 6) in the skeletal muscle samples. (**F**) DMs shared across longitudinal development in SOL, GAS, and PMM muscles identified using Venn diagrams. (**G**) Metabolites differentially abundant in fast-twitch and slow-twitch fibers identified. (**H**) Eight

metabolites associated with muscle development and fiber-type transformation identified. (**I**) Expression trends of eight metabolites during three developmental stages in three muscle types. (**J**) KEGG pathways of metabolites pantothenic acid and levan. ** $p < 0.01$, *** $p < 0.001$.

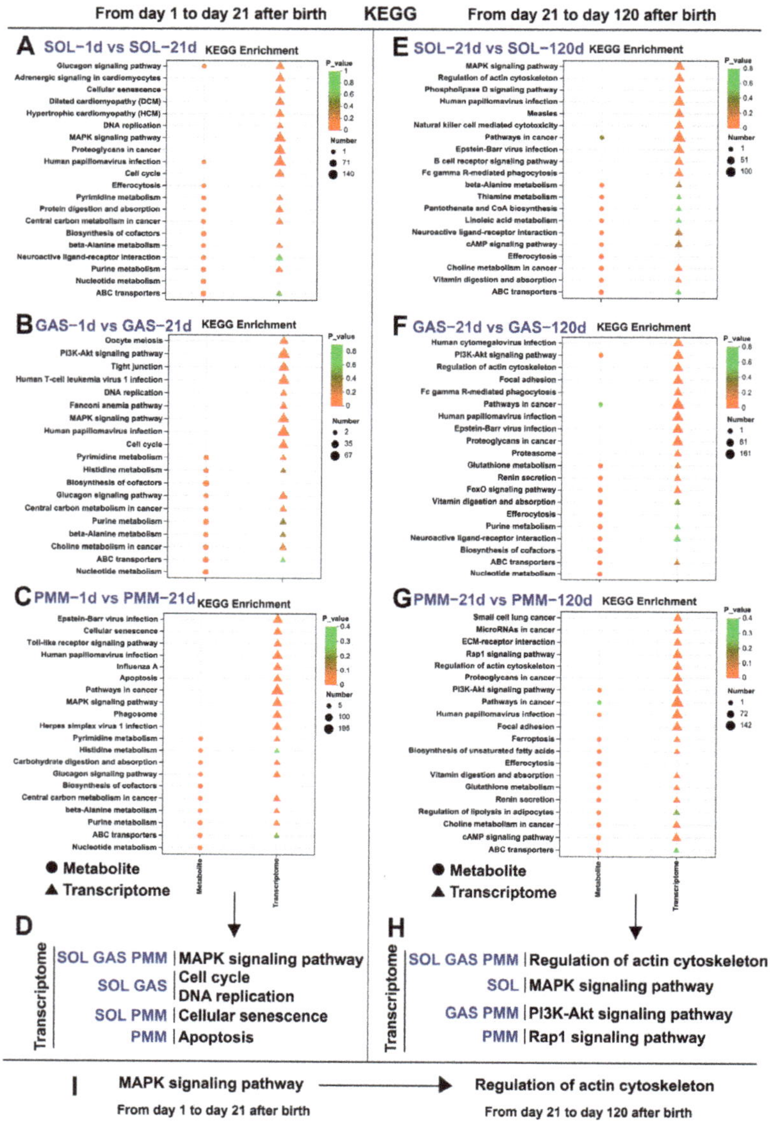

Figure 5. Integrated bulk RNA-seq and LC-MS data revealed KEGG pathways enriched in muscle development. (**A–C**) From postnatal days 1 to 21, the top ten enriched KEGG pathways in the SOL, GAS, and PMM were identified. (**D**) Important KEGG pathways are summarized. (**E–G**) From postnatal day 21 to day 120, the top ten enriched KEGG pathways in the SOL, GAS, and PMM muscles were identified. (**H**) Important KEGG pathways are summarized. (**I**) Transformation of KEGG pathways during the two developmental stages is summarized.

3.6. Analysis of KEGG Pathway Enrichment in Different Muscles

Given the limited available enriched metabolites, we prioritized the analysis of gene-enriched KEGG pathways. On postnatal day 21, DEGs in slow-twitch (SOL) and fast-twitch

(GAS and PMM) muscles were predominantly enriched in two pathways: hypertrophic cardiomyopathy (HCM) and dilated cardiomyopathy (DCM). Notably, DEGs in SOL versus GAS comparisons were significantly enriched in the glycolysis/gluconeogenesis and PPAR signaling pathways (Figure 6A–D). On postnatal day 120, DEGs in slow-twitch (SOL) and fast-twitch (GAS and PMM) were enriched in the HCM pathway (Figure 6E–H). Furthermore, DEGs between SOL and PMM were enriched in the Wnt signaling pathway (Figure 6H). We observed differences between the two fast-twitch muscles (GAS and PMM). On postnatal day 21, the DEGs were enriched in the HCM and DCM pathways. On postnatal day 21, the DEGs were enriched in the oxidative phosphorylation pathway (Figure 6D,H). In conclusion, our findings revealed differences among the three muscle types. Notably, although both PMM and GAS are fast-twitch muscles dominated by *MYH1*, distinct transcriptional profiles persist.

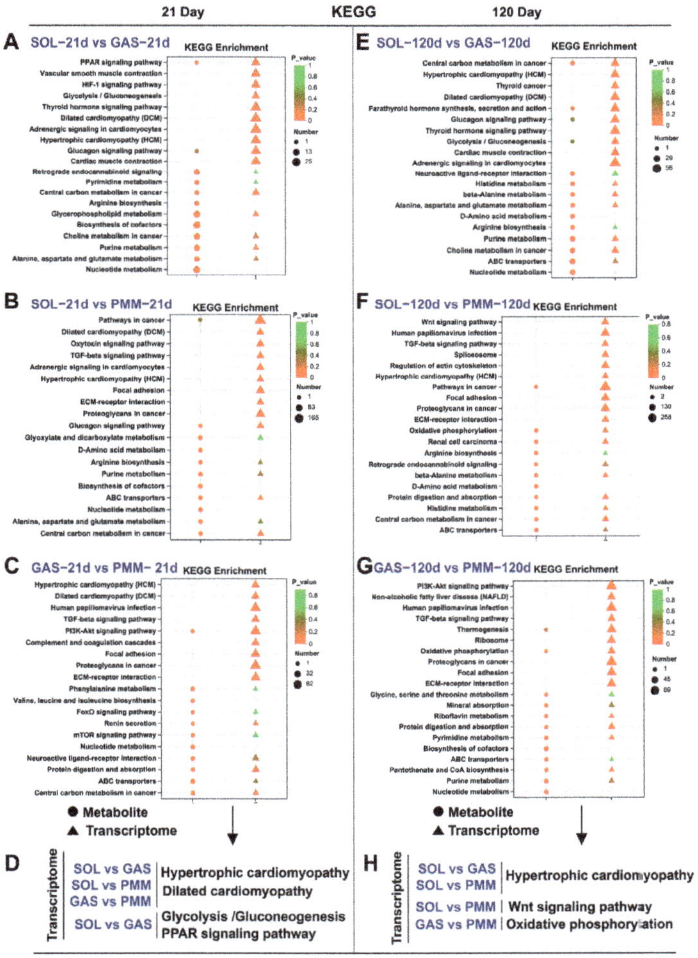

Figure 6. Integrated bulk RNA-seq and LC-MS data revealed KEGG pathways enriched in distinct muscle types. (**A–C**) KEGG signaling pathways enriched by DEGs and DMs from pairwise comparisons among the three muscle types on postnatal day 21. (**D**) The important KEGG pathways are summarized. (**E–G**) KEGG signaling pathways enriched by DEGs and DMs from pairwise comparisons among the three muscle types on postnatal day 120. (**H**) The important KEGG pathways are summarized.

4. Discussion

Skeletal muscle development is divided into two stages: embryonic and postnatal. The embryonic stage determines the number of muscle fibers, whereas the postnatal stage relies on muscle hypertrophy [16]. Most studies have focused on temporal dynamics within single skeletal muscles or spatial comparisons across anatomical regions, but have not constructed spatiotemporal developmental maps of skeletal muscles. We hypothesized that skeletal muscle heterogeneity arises from postnatal developmental trajectory. The phenotypic results of the three skeletal muscles further confirmed this assumption. HE staining showed an increase in muscle cross-sectional area, and the pigs were in a healthy and rapid growth stage, making them ideal for our subsequent sequencing requirements. Given the close relationship between MYHC isoforms and meat quality, we first focused on their changes. The results revealed different expression patterns: the SOL predominantly expressed *MYH7*, while GAS and PMM primarily expressed *MYH1*. However, this finding does not fully align with that of a previous study in mice [14]. Furthermore, *MYH7* expression was higher in PMM than in GAS. Based on the slow-twitch fiber proportion, we believe that the meat quality ranks in the following order: SOL > PMM > GAS. Among the three muscle types, we identified key differentially expressed genes that could serve as targets for gene editing to increase the slow-twitch muscle proportion and thereby improve meat quality.

Based on MYHC isoform expression, we classified the skeletal muscles into slow-twitch SOL (*MYH7*-dominant) and fast-twitch GAS and PMM (*MYH1*-dominant), which is consistent with previous classification methods [7]. In contrast to other studies [17], this study profiled longitudinal MYHC isoform expression across three muscles. *MYH7* expression in SOL continuously increased, while that of *MYH4* was barely detectable. The GAS exhibits an expression pattern similar to that of the LDM [18], with a decreased proportion of oxidative muscle and an increased proportion of glycolytic muscle. PMM is one of the tenderest porcine muscles and has exceptional commercial value. In particular, we focus on PMM. During PMM development, both *MYH7* and *MYH4* showed increased expression and proportions, whereas *MYH1* expression decreased, with *MYH7* consistently exceeding *MYH4*. This MYHC isoform expression pattern during PMM development has not been previously reported. By integrating DEGs related to both skeletal muscle development and myoblast differentiation, we identified six genes (*MFAP4*, *CTSW*, *ITGB7*, *LIMK1*, *PGM2L1*, and *TYMS*) whose functional roles remain uncharacterized. We identified five transcription factors (TFs) enriched in slow-twitch muscle and three TFs (*SOX6*, *MAF*, and *HOXD8*) enriched in fast-twitch muscles. These TFs may be key factors distinguishing different muscle types. *SOX6* and *MAF* are closely associated with glycolytic muscle [19,20], while *HOXD8* shows differential expression in the GAS and SOL of mice [21], consistent with our results. Finally, nine DEGs were identified across muscle types and developmental stages. We focus on four genes related to slow-twitch muscle: *S100A1* and *CA3* are highly expressed in SOL, consistent with reports [22,23], whereas the values for *MBOAT2* and *GYG2* are unreported. The functions of these four genes require further investigation.

The metabolic capacity of skeletal muscles varies with developmental stage and anatomical location. In addition, metabolism varies among different pig breeds. Previous studies have identified the critical intramuscular fat deposition window by comparing the developmental stages of obese and lean pigs. Obese pigs exhibit higher energy expenditure and more active protein metabolism [24]. Using conventional LC-MS, we identified DMs at different developmental stages and in different muscle fiber types. Notably, the number of DMs in SOL and GAS gradually increased with development. This might be due to their different postnatal functions: SOL mainly maintains venous return, while GAS participates in explosive movements. Ultimately, we identified eight DMs involved in both skeletal muscle development and muscle fiber-type transformation. For example, pantothenic acid serves as a coenzyme A (CoA) precursor, which is implicated in energy metabolism and fatty acid oxidation [25], while levan polysaccharide exhibits potent anti-inflammatory and antioxidant activities [26]. Thus, these metabolites may serve as key regulatory factors influencing meat quality, but require further functional validation.

From postnatal days 1 to 21, the DEGs across all three muscles exhibited enrichment in the MAPK signaling pathway. The mitogen-activated protein kinase (MAPK) signaling pathway does not act independently [27]; instead, it forms regulatory networks with other pathways (e.g., p38 MAPK and PI3K-Akt pathways) to coordinate key physiological processes, including satellite cell proliferation and differentiation, muscle fiber hypertrophy, and fiber-type transformation [28,29]. Notably, over 60% of the genes enriched in the MAPK signaling pathway showed developmental upregulation (days 1–21); for instance, the *ATF4* gene is vital for skeletal muscle development [30–32]. In addition, cell cycle-enriched genes in SOL (86.5%) and GAS (75.6%) showed predominant developmental downregulation, whereas apoptotic-enriched genes in PMM (59.2%) were upregulated. These findings demonstrate that satellite cell cycle exit initiates myofiber differentiation, which is a prerequisite for skeletal muscle development [33]. From postnatal days 21 to 120, the DEGs in SOL, GAS, and PMM exhibited significant enrichment in the regulation of the actin cytoskeleton pathway. Studies have shown that this pathway plays a crucial role in muscle cell fusion, sarcomere assembly, and differentiation of muscle fiber types [34]. Fast-twitch muscles (GAS and PMM) showed PI3K-Akt pathway enrichment (days 1–120), whereas slow-twitch SOL exhibited MAPK pathway enrichment. This pattern may be linked to differential MYHC expression trends, as the MAPK pathway mainly drives slow-twitch formation [35], while PI3K-Akt mediates fast-twitch hypertrophy. In summary, during longitudinal development, SOL, GAS, and PMM exit the cell cycle and initiate differentiation first, but subsequently diverge from each other. Despite both being fast-twitch muscles, PMM and GAS showed significant divergence in KEGG, which is worth discussing in detail. On postnatal day 21, DEGs were enriched in DCM and HCM signaling pathways, and by day 120, enrichment shifted to oxidative phosphorylation. The PI3K-Akt pathway was enriched at both time points. This may be because the PMM has more slow-twitch muscle fibers than the GAS. Early studies have defined PMM as an oxidative muscle [36]. This study demonstrated high similarity between SOL and GAS in newborns, which can be attributed to their anatomical proximity. However, whether this pattern applies to other hindlimb muscles still needs to be verified.

The limitations of this study are that it only focused on three muscles and three developmental stages, and only employed conventional bulk RNA-seq and LC-MS techniques. Future work should integrate multiple omics (e.g., single-cell RNA sequencing, spatial transcriptomics, and ATAC sequencing) to explore the molecular mechanisms of the four species, 20 muscle types, and 15 developmental stages.

5. Conclusions

This study identified six genes and eight metabolites that are key regulators of skeletal muscle development and fiber-type transformation. SOL and GAS heterogeneity were low at birth but increased during development. *MYH7*, *MYH1*, and *MYH4* expression varied according to the stage and muscle type. A developmental shift occurred from the MAPK signaling pathway (1–21 d) to actin cytoskeleton regulation (21–120 d). These results provide a theoretical basis for livestock improvement.

Supplementary Materials: The following are available online at https://www.mdpi.com/article/10.3390/ani15182715/s1, Table S1: Differentially expressed genes of SOL, GAS, and PMM during the three developmental stages. Table S2: Differences in genes between different muscle fiber types: SOL compared with GAS and PMM. Table S3: Differential genes between GAS and PMM. Table S4: Key metabolites mentioned in this study. Table S5: Differentially expressed genes between C2C12 proliferation and differentiation [4]. Abbreviations of column headers in tables: FC, fold change; Log2FC, log2 fold change; Padjust, adjusted p-value (Benjamini-Hochberg FDR); *, $p < 0.05$; **, $p < 0.01$. Significantly differentially expressed genes were defined as those with $|\text{Log2FC}| \geq 1.5$ and $p < 0.05$, and were classified as upregulated (Log2FC > 0) or downregulated (Log2FC < 0).

Author Contributions: Conceptualization, H.M. and Y.L.; software, K.W., X.L. (Xin Li), Y.L., S.L. and X.L. (Xibing Liu); validation, K.W., X.L. (Xin Li), X.L. (Xibing Liu), S.L., B.C. and Y.L.; formal analysis, K.W., L.X., Y.L. and W.C.; investigation, K.W., X.L. (Xin Li), Y.L., L.X., X.L. (Xibing Liu), J.J. and B.C.; writing—original draft preparation, K.W., X.L. (Xin Li) and Y.L.; writing—review and editing, H.M. and Y.L. All authors have read and agreed to the published version of the manuscript.

Funding: This work was supported by the Biological Breeding National Science and Technology Major Project (no. 2023ZD0404601).

Institutional Review Board Statement: All experimental procedures in this study were approved by the Hunan Agricultural University Animals Care and Use Committee (permit number CACAHU-20250405).

Informed Consent Statement: Not applicable.

Data Availability Statement: The raw Bulk RNA-seq data are available in the NCBI database (Accession: PRJNA1321496), and the raw LC-MS data are available in the MetaboLights database (Accession: MTBLS12963). All raw data are also available from the corresponding author upon request.

Acknowledgments: The authors thank Tangrenshen Group Co., Ltd. for supporting the experiments.

Conflicts of Interest: There are no financial or personal conflict of interest to report.

References

1. Frontera, W.R.; Ochala, J. Skeletal muscle: A brief review of structure and function. *Calcif. Tissue Int.* **2015**, *96*, 183–195. [CrossRef]
2. Zhang, M.; Koishi, K.; McLennan, I.S. Skeletal muscle fibre types: Detection methods and embryonic determinants. *Histol. Histopathol.* **1998**, *13*, 201–207. [CrossRef] [PubMed]
3. Grefte, S.; Kuijpers-Jagtman, A.M.; Torensma, R.; Hoff, J.W.V.D. Skeletal muscle development and regeneration. *Stem Cells Dev.* **2007**, *16*, 857–868. [CrossRef]
4. Wang, K.; Liu, C.; Yi, L.; Liufu, S.; Chen, W.; Liu, X.; Chen, B.; Xu, X.; Liu, J.; Liu, X.; et al. LMOD2 interaction with ACTC1 regulates myogenic differentiation. *BMC Genom.* **2025**, *26*, 709. [CrossRef] [PubMed]
5. Tan, B.; Hong, L.; Xiao, L.; Wu, J.; Lu, G.; Wang, S.; Liu, L.; Zheng, E.; Cai, G.; Li, Z.; et al. Rewiring of 3D chromatin topology orchestrates transcriptional reprogramming in muscle fiber-type specification and transformation. *Nat. Commun.* **2025**, *16*, 5833. [CrossRef] [PubMed]
6. Terry, E.E.; Zhang, X.; Hoffmann, C.; Hughes, L.D.; Lewis, S.A.; Li, J.; Wallace, M.J.; Riley, L.; Douglas, C.M.; Gutierrez-Montreal, M.A.; et al. Transcriptional profiling reveals extraordinary diversity among skeletal muscle tissues. *eLife* **2018**, *7*, e34613. [CrossRef]
7. Jin, L.; Tang, Q.; Hu, S.; Chen, Z.; Zhou, X.; Zeng, B.; Wang, Y.; He, M.; Li, Y.; Gui, L.; et al. A pig BodyMap transcriptome reveals diverse tissue physiologies and evolutionary dynamics of transcription. *Nat. Commun.* **2021**, *12*, 3715. [CrossRef]
8. Yang, Y.; Yan, J.; Fan, X.; Chen, J.; Wang, Z.; Liu, X.; Yi, G.; Liu, Y.; Niu, Y.; Zhang, L.; et al. The genome variation and developmental transcriptome maps reveal genetic differentiation of skeletal muscle in pigs. *PLoS Genet.* **2021**, *17*, e1009910. [CrossRef]
9. Liu, C.; Zheng, D.; Zhang, R.; Li, H.; Tong, X.; Wu, Y.; Zhang, G.; Wang, S.; Chen, H.; Ren, Z.; et al. Transcriptional Diversity in Response to Aging Across Skeletal Muscles. *Aging Cell* **2025**, *9*, e70164. [CrossRef]
10. Rahman, F.A.; Baechler, B.L.; Quadrilatero, J. Key considerations for investigating and interpreting autophagy in skeletal muscle. *Autophagy* **2024**, *20*, 2121–2132. [CrossRef]
11. Li, C.; Wang, Y.; Sun, X.; Yang, J.; Ren, Y.; Jia, J.; Yang, G.; Liao, M.; Jin, J.; Shi, X. Identification of different myofiber types in pigs muscles and construction of regulatory networks. *BMC Genom.* **2024**, *25*, 400. [CrossRef]

2. Picard, B.; Lefaucheur, L.; Berri, C.; Duclos, M.J. Muscle fibre ontogenesis in farm animal species. *Reprod. Nutr. Dev.* **2003**, *42*, 415–431. [CrossRef]
3. Wang, W.; Zhang, T.; Du, L.; Li, K.; Zhang, L.; Li, H.; Gao, X.; Xu, L.; Li, J.; Gao, H. Transcriptomic analysis reveals diverse expression patterns underlying the fiber diameter of oxidative and glycolytic skeletal muscles in steers. *Meat Sci.* **2023**, *207*, 109350. [CrossRef] [PubMed]
4. Hettige, P.; Tahir, U.; Nishikawa, K.C.; Gage, M.J. Comparative analysis of the transcriptomes of EDL, psoas, and soleus muscles from mice. *BMC Genom.* **2020**, *21*, 808. [CrossRef] [PubMed]
5. Zhao, T.; Tian, T.; Yu, H.; Cao, C.; Zhang, Z.; He, Z.; Ma, Z.; Cai, R.; Li, F.; Pang, W. Identification of porcine fast/slow myogenic exosomes and their regulatory effects on lipid accumulation in intramuscular adipocytes. *J. Anim. Sci. Biotechnol.* **2024**, *15*, 73. [CrossRef] [PubMed]
6. Bachman, J.F.; Chakkalakal, J.V. Satellite cells in the growth and maintenance of muscle. In *Current Topics in Developmental Biology*; Elsevier: Amsterdam, The Netherlands, 2024; Volume 158, pp. 1–14. [CrossRef]
7. Liufu, S.; Lan, Q.; Liu, X.; Chen, B.; Xu, X.; Ai, N.; Li, X.; Yu, Z.; Ma, H. Transcriptome Analysis Reveals the Age-Related Developmental Dynamics Pattern of the Longissimus Dorsi Muscle in Ningxiang Pigs. *Genes* **2023**, *14*, 1050. [CrossRef]
8. Tan, W.; Liu, C.; Liu, J.; Wen, S.; Chen, Y.; Ren, R.; Gao, N.; Ding, X.; He, J.; Zhang, Y. Integrative Analysis of ATAC-Seq and RNA-Seq Identifies Key Genes Affecting Muscle Development in Ningxiang Pigs. *Int. J. Mol. Sci.* **2025**, *26*, 2634. [CrossRef]
9. Zhang, Z.; Lin, S.; Luo, W.; Ren, T.; Huang, X.; Li, W.; Zhang, X. Sox6 Differentially Regulates Inherited Myogenic Abilities and Muscle Fiber Types of Satellite Cells Derived from Fast- and Slow-Type Muscles. *Int. J. Mol. Sci.* **2022**, *23*, 11327. [CrossRef]
0. Sadaki, S.; Tsuji, R.; Hayashi, T.; Watanabe, M.; Iwai, R.; Wenchao, G.; Semenova, E.A.; Sultanov, R.I.; Zhelankin, A.V.; Generozov, E.V.; et al. Large MAF transcription factors reawaken evolutionarily dormant fast-glycolytic type IIb myofibers in human skeletal muscle. *Skelet. Muscle* **2025**, *15*, 19. [CrossRef]
1. de Wilde, J.; Hulshof, M.F.; Boekschoten, M.V.; de Groot, P.; Smit, E.; Mariman, E.C. The embryonic genes Dkk3, Hoxd8, Hoxd9 and Tbx1 identify muscle types in a diet-independent and fiber-type unrelated way. *BMC Genom.* **2010**, *11*, 176. [CrossRef]
2. Wu, J.; Zhou, D.; Deng, C.; Xiong, Y.; Lei, M.; Li, F.; Jiang, S.; Zuo, B.; Zheng, R. Expression pattern and polymorphism of three microsatellite markers in the porcine CA3 gene. *Genet. Sel. Evol.* **2008**, *40*, 227–239. [CrossRef]
3. Maco, B.; Uhrík, B.; Heizmann, C.W. Distribution of the Ca^{2+}-binding S100A1 protein at different sarcomere lengths of slow and fast rat skeletal muscles. *Gen. Physiol. Biophys.* **2001**, *19*, 237–244.
4. Guo, L.; Xu, J.; Zhou, W.; Chen, S.; Shi, H.; Han, M.; Yang, Z.; Duan, Y.; Pang, W.; Yin, Y.; et al. Metabolome and RNA-seq reveal discrepant metabolism and secretory metabolism profile in skeletal muscle between obese and lean pigs at different ages. *Sci. China Life Sci.* **2025**, *68*, 1102–1117. [CrossRef] [PubMed]
5. Whitfield, J.; Harris, R.C.; Broad, E.M.; Patterson, A.K.; Ross, M.L.; Shaw, G.; Spriet, L.L.; Burke, L.M. Chronic pantothenic acid supplementation does not affect muscle coenzyme A content or cycling performance. *Appl. Physiol. Nutr. Metab.* **2020**, *46*, 280–283. [CrossRef]
6. Srikanth, R.; Siddartha, G.; Reddy, C.H.S.; Harish, B.S.; Ramaiah, M.J.; Uppuluri, K.B. Antioxidant and anti-inflammatory levan produced from Acetobacter xylinum NCIM2526 and its statistical optimization. *Carbohydr. Polym.* **2015**, *123*, 8–16. [CrossRef]
7. Endo, T. Postnatal skeletal muscle myogenesis governed by signal transduction networks: MAPKs and PI3K-Akt control multiple steps. *Biochem. Biophys. Res. Commun.* **2023**, *682*, 223–243. [CrossRef]
8. Millward, D.J. Post-natal muscle growth and protein turnover: A narrative review of current understanding. *Nutr. Res. Rev.* **2023**, *37*, 143–168. [CrossRef]
9. Xie, S.-J.; Li, J.-H.; Chen, H.-F.; Tan, Y.-Y.; Liu, S.-R.; Zhang, Y.; Xu, H.; Yang, J.-H.; Liu, S.; Zheng, L.-L.; et al. Inhibition of the JNK/MAPK signaling pathway by myogenesis-associated miRNAs is required for skeletal muscle development. *Cell Death Differ.* **2018**, *25*, 1581–1597. [CrossRef]
0. Tan, Y.-Y.; Zhang, Y.; Li, B.; Ou, Y.-W.; Xie, S.-J.; Chen, P.-P.; Mei, S.-Q.; Huang, Q.-J.; Zheng, L.-L.; Qu, L.-H. PERK Signaling Controls Myoblast Differentiation by Regulating MicroRNA Networks. *Front. Cell Dev. Biol.* **2021**, *9*, 670435. [CrossRef]
1. Memme, J.M.; Sanfrancesco, V.C.; Hood, D.A. Activating transcription factor 4 regulates mitochondrial content, morphology, and function in differentiating skeletal muscle myotubes. *Am. J. Physiol. Cell Physiol.* **2023**, *325*, C224–C242. [CrossRef]
2. Hocsak, E.; Szabo, V.; Kalman, N.; Antus, C.; Cseh, A.; Sumegi, K.; Eros, K.; Hegedus, Z.; Gallyas, F.; Sumegi, B.; et al. PARP inhibition protects mitochondria and reduces ROS production via PARP-1-ATF4-MKP-1-MAPK retrograde pathway. *Free Radic. Biol. Med.* **2017**, *108*, 770–784. [CrossRef] [PubMed]
3. Dumont, N.A.; Wang, Y.X.; Rudnicki, M.A. Intrinsic and extrinsic mechanisms regulating satellite cell function. *Development* **2015**, *142*, 1572–1581. [CrossRef]
4. Brunello, E.; Fusi, L. Regulating Striated Muscle Contraction: Through Thick and Thin. *Annu. Rev. Physiol.* **2023**, *86*, 255–275. [CrossRef]

35. Roth, R.J.; Le, A.M.; Zhang, L.; Kahn, M.; Samuel, V.T.; Shulman, G.I.; Bennett, A.M. MAPK phosphatase-1 facilitates the loss of oxidative myofibers associated with obesity in mice. *J. Clin. Investig.* **2009**, *119*, 3817–3829. [CrossRef]
36. Shen, L.; Du, J.; Xia, Y.; Tan, Z.; Fu, Y.; Yang, Q.; Li, X.; Tang, G.; Jiang, Y.; Wang, J.; et al. Genome-wide landscape of DNA methylomes and their relationship with mRNA and miRNA transcriptomes in oxidative and glycolytic skeletal muscles. *Sci. Rep.* **2016**, *6*, 32186. [CrossRef]

animals

MDPI

Article

Genomic Characterization of the Kazakh Fat-Tailed Coarse-Wool Sheep Breed Using ROH Analysis

Altynay Kozhakhmet [1,2,3], Zhanerke Akhatayeva [1,4], Kairat Dossybayev [1,2,3,*], Marina Yermekova [1,3], Tilek Kapassuly [1,2,3], Kanagat Yergali [1,2], Aibyn Torekhanov [1], Utepbergen Bissenov [5], Xianyong Lan [6] and Beibit Kulataev [1]

1 LLP "Kazakh Research Institute for Livestock and Fodder Production", Almaty 050035, Kazakhstan; altynaitg@gmail.com (A.K.); akhatayevazhanerke@163.com (Z.A.); nurii_90@mail.ru (M.Y.); tilek.kapas@mail.ru (T.K.); ergaly.qanagat@gmail.com (K.Y.); torehanov.aibyn@mail.ru (A.T.); bnar68@yandex.ru (B.K.)
2 RSE Institute of Genetics and Physiology SC MSHE RK, 93 Al-Farabi Avenue, Almaty 050060, Kazakhstan
3 Faculty of Biology and Biotechnology, Al-Farabi Kazakh National University, Almaty 050040, Kazakhstan
4 Institute of Grassland Research, Chinese Academy of Agricultural Sciences, Hohhot 010010, China
5 Department of the Biology and Fish Farming, Kh. Dosmukhamedov Atyrau University, Atyrau 060011, Kazakhstan; bisenovy@mail.ru
6 College of Animal Science and Technology, Northwest A&F University, Yangling, Xianyang 712100, China; lanxianyong79@126.com
* Correspondence: kairat1987_11@mail.ru

Simple Summary

Sheep are an important source of food and income for people in Kazakhstan. Among the local breeds, the Kazakh fat-tailed coarse-wool sheep is especially valued because it grows well on natural pastures, produces a large amount of meat, and can survive in very hot summers and cold winters. To keep this breed healthy and productive, it is important to understand its genetic background. In our study, we used 500 sheep to learn how much variety exists within the breed and to find genomic regions to underline useful traits. We discovered multiple genes that are connected to growth, fat storage, reproduction and wool production. This knowledge will help farmers and breeders improve their flocks while protecting the genetic resources that are so valuable for the future of sheep farming in Kazakhstan.

Abstract

Sheep breeding is an important sector of livestock production in the Republic of Kazakhstan. The Kazakh fat-tailed coarse-wool sheep holds a prominent position among local breeds due to its high meat productivity, resilience to extreme climatic conditions, and efficient use of pasture resources. This study focuses on the analysis of runs of homozygosity (ROH) to evaluate the genetic diversity level, inbreeding and to detect selection signatures in the Kazakh fat-tailed coarse-wool sheep breed. A total of 500 animals were genotyped using the OvineSNP50 BeadChip (Illumina, San Diego, CA, USA). As a result, a total of 41,728 ROH segments were identified, with an average length of 1.59 Mb, distributed across the entire genome. The most prominent homozygous regions were detected on chromosomes OAR10, OAR13, and OAR22, which might be associated with selection signatures. Genomic inbreeding coefficients (F_{ROH} and F_{GRM}) showed a strong positive correlation ($r = 0.58$, $p < 0.001$), supporting the effectiveness of ROH-based analysis. Several candidate genes were detected, including *MYF5*, *PRDM16*, *TGM3*, *SLC26A4* and *SMAD5* which are notably involved in muscle formation, wool traits, and fat metabolism. The findings have substantial practical value for breeding programs and for managing genetic diversity in sheep farming enterprises in the Republic of Kazakhstan.

Keywords: runs of homozygosity; sheep; inbreeding; SNP; F_{ROH}; F_{GRM}

1. Introduction

Sheep breeding is an important sector of livestock production in the Republic of Kazakhstan, providing the population with meat and wool. According to the Bureau of National Statistics under the Agency for Strategic Planning and Reforms of the Republic of Kazakhstan, the total sheep population in the country exceeded 22.6 million head in 2024 [1]. The Kazakh fat-tailed coarse-wool sheep holds a prominent position among local breeds due to its high meat productivity, resilience to extreme climatic conditions, and efficient use of pasture resources [2,3]. The Kazakh fat-tailed coarse-wool breed was recognized in 1994 [4] through complex reproductive crossbreeding of Kazakh fat-tailed sheep with Edilbay, Sarajin, Tajik, and Degeres rams. Distinctive features of the breed include a moderately sized fat tail, an optimal meat-to-bone ratio, and strong adaptation to desert and semi-desert environments. In light of the growing demand for sheep-derived products and the need for sustainable agricultural development, improving the genetic potential of meat breeds, applying modern selection methods, and monitoring genetic diversity in pasture populations have become increasingly important.

In this context, genomic evaluation tools, especially approaches based on runs of homozygosity (ROH), have gained considerable relevance. Today, ROH-based assessments are widely used to evaluate genetic diversity and inbreeding levels in livestock and to identify genomic regions that influence economically important traits [5–8]. ROHs are continuous homozygous stretches of the genome that are inherited from common ancestors in an identical-by-descent (IBD) state [9]. The number, length, and chromosomal location of ROH segments allow researchers to estimate inbreeding levels, reconstruct demographic history, and detect functionally important genomic regions associated with valuable traits [10,11]. The length of ROH segments indicates the timing of inbreeding events; long tracts (>10 Mb) point to recent consanguinity (within the past five generations), whereas medium (2–5 Mb) and short segments (<1 Mb) suggest more ancient common ancestry (up to 30 generations ago) [12–14]. The genomic inbreeding coefficient F_{ROH}, calculated from ROH segments, reflects the proportion of the genome in an autozygous state. It is determined by dividing the total length of ROHs by the total autosomal genome length and provides a quantitative measure of individual inbreeding levels without relying on pedigree records [14–16]. An alternative way to estimate inbreeding is through the genomic inbreeding coefficient F_{GRM}, based on the genomic relationship matrix (GRM). This approach evaluates the degree of inbreeding by measuring genotypic similarity between individuals and reflects the proportion of genomic variation due to shared ancestry [17].

The practical value of ROH-based evaluation has been demonstrated in various livestock species. In sheep, ROH islands have been linked to traits such as fertility, meat quality, and milk fat content [18–21]. For instance, ROH analysis using the Illumina OvineSNP50 BeadChip in African sheep breeds identified candidate genes such as *ANPEP*, *HDDC3*, *ST3GAL3*, and *GRK4* [20]. Selli et al. (2021) reported shared genomic regions of high homozygosity and heterozygosity across 17 global sheep populations [22]. Similar investigations in goats have shown the utility of F_{ROH} estimates and identification of homozygous regions associated with productive and adaptive traits in Chinese breeds [8,22,23]. In cattle, ROH exploration enables the estimation of inbreeding coefficients and the localization of regions containing genes that affect milk yield, body weight gain, and disease resistance [24,25].

The present study aimed to identify the level of inbreeding, ROH islands associated with productive traits, and functional annotation of the respective genes and biological pathways in Kazakh fat-tailed coarse-wool sheep. The results obtained may serve as a foundation for future breeding strategies aimed at improving the productivity of the Kazakh fat-tailed coarse-wool breed

2. Materials and Methods

2.1. Sample Collection

A total of 500 Kazakh fat-tailed coarse-wool sheep were sampled from six farms in the southern and southeastern regions of Kazakhstan (Figure 1), namely Medkhan (43.29° N, 76.44° E), Razakhun (43.54° N, 73.86° E), Tokan-1 (45.27° N, 79.35° E), Alakol-Agro (46.17° N, 80.95° E), Nur (50.45° N, 80.29° E), and Elkentai (50.40° N, 80.20° E). Sampling was carried out during the weaning period, from 22 February 2023 to 22 February 2025.

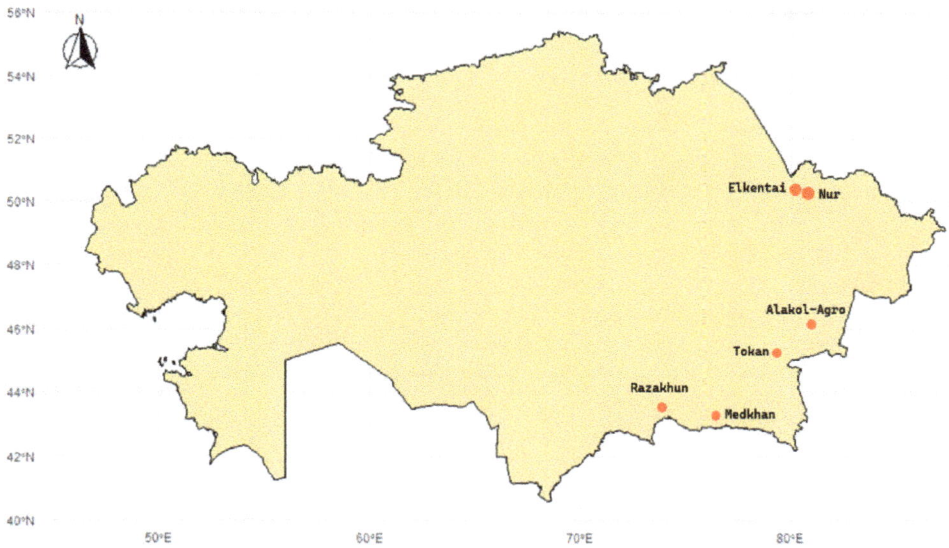

Figure 1. Map showing the geographical locations of the sampled farms in Kazakhstan.

Blood samples were collected via jugular venipuncture into EDTA-containing vacuum tubes and stored at −20 °C until transport to the laboratory.

2.2. DNA Extraction and Genotyping

DNA was extracted using the GeneJET Genomic DNA Purification Kit (Thermo Scientific, Waltham, MA, USA) following the manufacturer's protocol. Concentration was measured on a NanoDrop 2000 spectrophotometer (Thermo Scientific, Waltham, MA, USA), with most values ranging between 50 and 70 ng/μL. To evaluate purity and check for possible degradation or contamination, selected samples were run on a 1% agarose gel.

SNP genotyping was performed using the OvineSNP50 Genotyping BeadChip (Illumina, San Diego, CA, USA). Samples were prepared according to the Infinium HD Assay Ultra Protocol using a Tecan robotic liquid handling system, and scanned on an iScan platform (Illumina, San Diego, CA, USA). Raw data were processed in GenomeStudio (v2.0.). SNPs with a call rate of at least 98% were retained during quality control. Final datasets were exported in PED and MAP file formats using the PLINK Input Report Plug-in.

Only autosomal SNPs were retained by excluding those located on sex chromosomes (X and Y) using bash script. Quality control (QC) for missing data was then performed in PLINK (V1.90) by following criteria: SNPs and individual samples with a missing genotype rate >5% were removed. After quality control, a total of 49,363 SNPs and 497 individuals successfully passed and were used for subsequent analyses. The number of individuals retained from each population was as follows: Tokan (n = 71), Nur (n = 98), Alakol-Agro (n = 65), Elkentai (n = 98), Razakhun (n = 86), and Medkhan (n = 79).

2.3. ROH Detection

Runs of homozygosity were identified using PLINK (v1.90) with the following threshold parameters: minimum SNP density within a ROH was set to at least 1 SNP per 100 kb (--homozyg-density 100), the maximum gap allowed between two consecutive SNPs within a ROH was 1000 kb (--homozyg-gap 1000), the minimum ROH length was 1000 kb (1 Mb) (--homozyg-kb 1000), and the minimum number of SNPs constituting a ROH was 15 (--homozyg-snp 15). Additionally, a maximum of one heterozygous SNP per window was allowed (--homozyg-window-het 1), and a sliding window size of 15 SNPs was used to scan the genome for ROHs (--homozyg-window-snp 15) [16].

2.4. Estimation of Genomic Inbreeding and Correlation Analysis

Genomic inbreeding was evaluated using two complementary metrics. The first, genomic relationship matrix (F_{GRM}), was calculated in PLINK (v1.90) and GCTA (v1.94.1) using the --make-grm and --grm-inbreeding options, based on a genomic relationship matrix constructed from all autosomal SNPs. This coefficient reflects how much observed homozygosity deviates from what would be expected under random mating. The second metric, inbreeding coefficient based on ROH (F_{ROH}), was defined as the proportion of the autosomal genome covered by ROH segments. ROH lengths were taken directly from the PLINK output. Both F_{GRM} and F_{ROH} values were compiled into a single table using individual IDs.

To evaluate the relationship between F_{GRM} and F_{ROH}, Pearson's correlation coefficient was calculated using the cor.test function from the stats package in RStudio (v2024.12.1+563), separately for each population and for the full dataset (n = 497). The analysis was repeated using four minimum ROH length thresholds: 1, 4, 8, and 16 Mb. Results are presented as tables and scatter plots with regression lines and 95% confidence intervals in RStudio (v2024.12.1+563) using the ggplot2 package.

2.5. Gene Annotation and Functional Enrichment Analysis

Gene annotation for SNPs in the top 1% on each chromosome was performed using the BioMart platform (version: Ensembl release 114) based on the ARS-UI_Ramb_v2.0 reference genome. Subsequent annotation in UniProt database (https://www.uniprot.org, accessed on 21 August 2025) allowed accurate classification of the candidate genes as proteins. To gain deeper biological insight, we then used the DAVID databases (https://david.ncifcrf.gov, accessed on 21 August 2025) to detect the Kyoto Encyclopedia of Genes and Genomes (KEGG) pathways and Gene Ontology (GO) terms.

3. Results

3.1. ROH Analysis Results

A total of 41,728 ROH segments were identified across the 497 sheep included in the dataset, with an average segment length of 1588.5 kb. The number and mean length of ROHs varied between populations, ranging from 5656 segments in Alakol-Agro to 8374 in the Nur population. The longest average ROH length was observed in Razakhun

(1734.6 mb), while the shortest was in Alakol-Agro (1380.9 mb). A summary of ROH counts and average lengths per population is presented in Table 1.

Table 1. Number and average length of ROH segments in each population.

Population	Total Number of ROH (n)	Mean ROH Length (mb)
AlAg	5656	1.381
Elkentai	8267	1.721
Medhan	6613	1.530
Nur	8374	1.490
Razahun	6321	1.735
Tokan-1	6497	1.645
Total	41,728	1.589

Note: values rounded to one decimal point for clarity.

ROH segments were further classified by length into five categories: 1–2 Mb, 2–4 Mb, 4–8 Mb, 8–16 Mb, and >16 Mb. Table 2 shows the average number of ROHs per length category for each sheep population. Across all populations, the majority of ROH segments were short, predominantly within the 1–2 Mb range, with a gradual decrease in the number of segments as length increased. Segments longer than 16 Mb were rare or absent in most groups.

Table 2. Distribution of ROH segments by length classes across six sheep populations.

FID	ROH_1-2 Mb	ROH_2-4 Mb	ROH_4-8 Mb	ROH_8-16 Mb	ROH_>16 Mb	Total ROH
AlAg	81 ± 9.3	5.4 ± 2.6	1.5 ± 0.9	1.7 ± 1.5	NA	87 ± 10.2
Elkentai	75.7 ± 11	6 ± 2.8	2.2 ± 1.7	2.3 ± 1.9	3 ± 3.9	84.4 ± 12.2
Medhan	77 ± 12.3	5.2 ± 2.2	2.3 ± 2.2	2.4 ± 2.1	1.9 ± 1.2	83.7 ± 13.6
Nur	78.1 ± 8.7	5.9 ± 2.4	1.6 ± 0.8	1.3 ± 1.2	3 ± 4	85.4 ± 9.2
Razahun	66.2 ± 13.7	5 ± 2.8	2.3 ± 1.6	2.4 ± 1.9	3.2 ± 2.6	73.5 ± 14.9
Tokan	81.9 ± 10.3	6.8 ± 4.5	2.4 ± 3.1	1.7 ± 1.5	2.7 ± 4.3	91.5 ± 14.8

Note: NA, not available; values are presented as mean ± SD.

The chromosomal distribution and variation in ROH length are shown in Figure 2. ROH segments were unevenly distributed across the autosomes, with notable clusters on OAR10 (30–45 Mb), OAR13 (50–70 Mb), and OAR22 (20 Mb), reflecting shared homozygous regions among individuals.

To further characterize homozygosity at the SNP level, we assessed the frequency of individual SNPs occurring within ROH regions across the genome (Figure 3). ROH count and total ROH length also varied across individuals and populations, as shown in Figure 4.

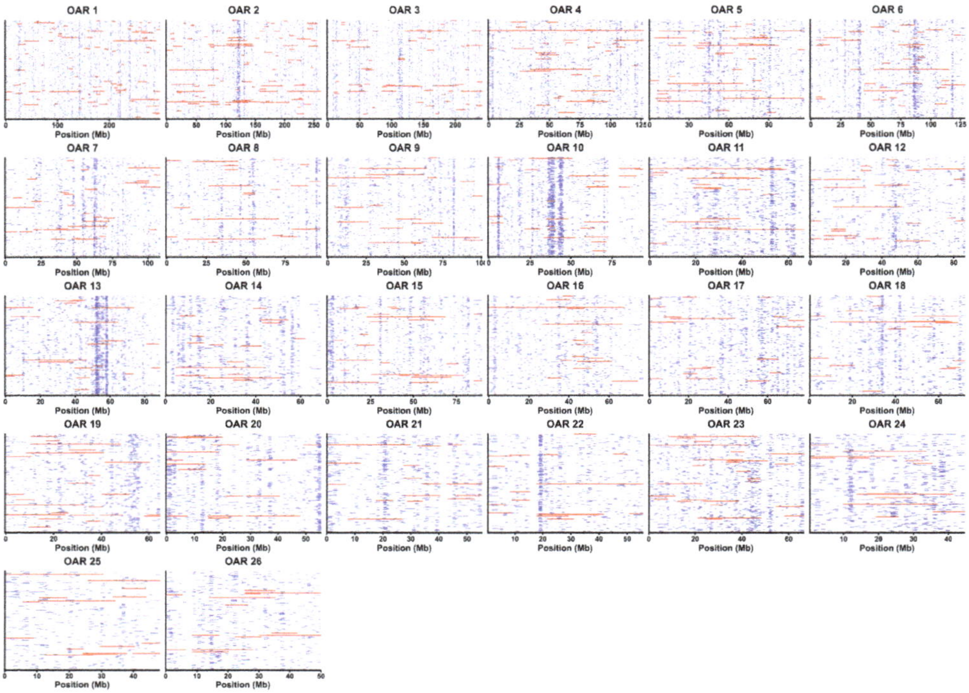

Figure 2. Chromosomal distribution of ROH segments across all autosomes. Horizontal bars represent individual ROHs. Segments shorter than 5 Mb are shown in blue; segments longer than 5 Mb are shown in red.

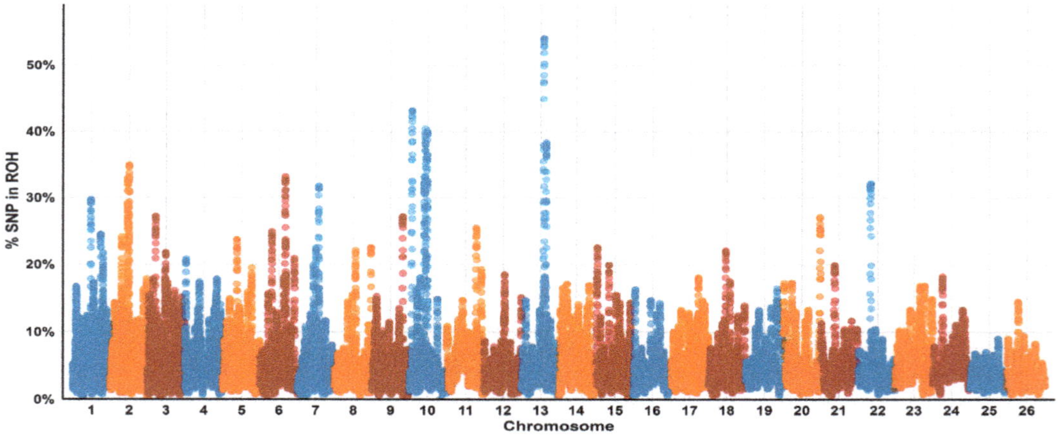

Figure 3. Manhattan plot showing SNP frequency within ROH across 26 autosomes in Kazakh fat-tailed coarse-wool sheep. The x-axis represents chromosome number and SNP position; the y-axis shows the proportion of individuals in which each SNP falls within ROH. The dashed line marks the 99.5th percentile threshold. Clusters above this level are seen on chromosomes 10 and 13, indicating regions of elevated homozygosity.

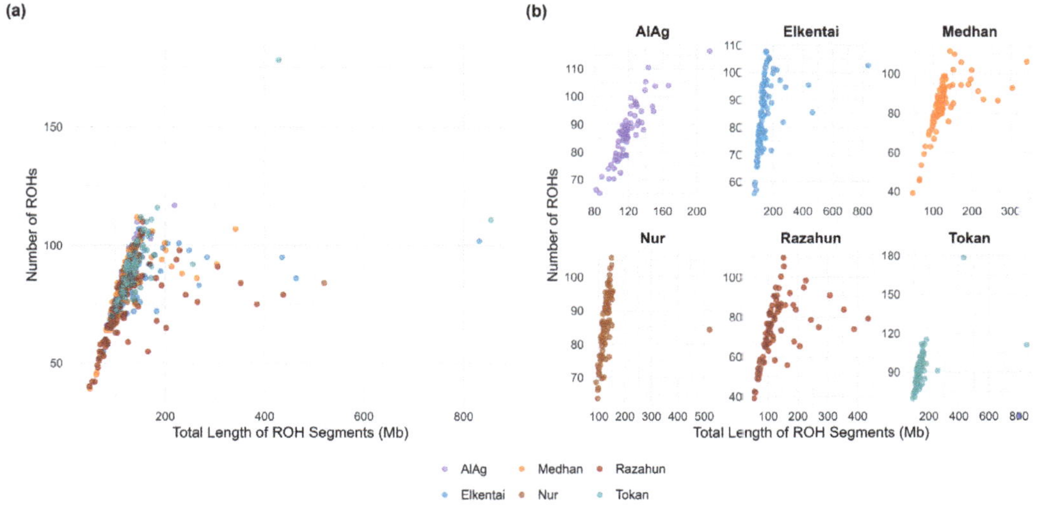

Figure 4. ROH distribution in Kazakh fat-tailed coarse-wool sheep: (**a**) Scatter plot showing the number of ROH segments (y-axis) and their total length in Mb (x-axis) for all 497 individuals. Most animals fall within 40–120 ROH segments and 100–300 Mb, although some exceed 180 segments and 800 Mb. (**b**) The same plot faceted by population (AlAg, Elkentai, Medhan, Nur, Razahun, Tokan) for within-group comparison of ROH count and total length.

3.2. Genomic Inbreeding Estimates and Correlation Between F_{ROH} and F_{GRM}

Genomic inbreeding levels were assessed using two metrics: F_{ROH}, calculated as the proportion of the autosomal genome covered by ROH segments, and F_{GRM}, derived from the genomic relationship matrix. The average values of F_{ROH} by length class and the mean F_{GRM} with 95% confidence intervals for each population are presented in Table 3. Genomic inbreeding coefficients varied among populations, with total F_{ROH} ranging from 0.046 in AlAg to 0.058 in Tokan. Mean F_{GRM} values were broadly consistent with F_{ROH}, although their wide confidence intervals reflected sampling limitations and estimation variability.

Table 3. Genomic inbreeding coefficients in sheep populations.

Population	F_{ROH} 1–4 Mb	F_{ROH} 4–8 Mb	F_{ROH} 8–16 Mb	F_{ROH} >16 Mb	Total	F_{ROH}-F_{GRM} (Mean ± CI)
AlAg	0.045	0.001	0.001	0.000	0.046	0.017 (−0.012–0.046)
Elkentai	0.043	0.003	0.003	0.007	0.056	0.017 (−0.037–0.281)
Medhan	0.042	0.002	0.002	0.003	0.049	0.006 (−0.054–0.097)
Nur	0.044	0.002	0.001	0.002	0.049	0.015 (−0.009–0.167)
Razahun	0.037	0.002	0.004	0.006	0.049	−0.007 (−0.093–0.102)
Tokan	0.047	0.004	0.002	0.005	0.058	0.025 (−0.014–0.304)

Note: Average F_{ROH} values by ROH length class and total F_{ROH} are shown for each population. Mean F_{GRM} values with 95% confidence intervals are also included.

Pearson's correlation coefficients between individual F_{ROH} and F_{GRM} values were calculated for four ROH length thresholds (>1 Mb, >4 Mb, >8 Mb, and >16 Mb). The results are shown in Table 4. Correlations were consistently high within populations, ranging from 0.58 in AlAg to 0.98 in Tokan. Across the full dataset (n = 497), the correlation between F_{ROH} and F_{GRM} was $r = 0.58$ (t = 15.91, df = 495, $p < 2.2 \times 10^{-16}$; 95% CI: 0.52–0.64), indicating a moderately strong and statistically significant association.

Table 4. Pearson correlation coefficients between F_{ROH} and F_{GRM} at different minimum ROH length thresholds (>1 Mb, >4 Mb, >8 Mb, >16 Mb) across sheep populations.

Population	*r* (>1 Mb)	*r* (>4 Mb)	*r* (>8 Mb)	*r* (>16 Mb)	*r* (Total)
AlAg	0.451	0.397	0.486	NA	0.583
Elkentai	0.233	0.698	0.706	0.894	0.959
Medhan	0.576	0.483	0.554	0.649	0.831
Nur	0.103	0.181	0.819	0.878	0.932
Razahun	0.604	0.639	0.488	0.654	0.832
Tokan	0.361	0.514	0.752	0.918	0.980

Despite differences in methodology, F_{ROH} reflecting genomic regions in a homozygous state over extended stretches and F_{GRM} capturing genome-wide allelic similarity, both metrics showed substantial agreement. Their relationship is visualized in Figure 5, where scatter plots with regression lines and correlation statistics are shown for each population.

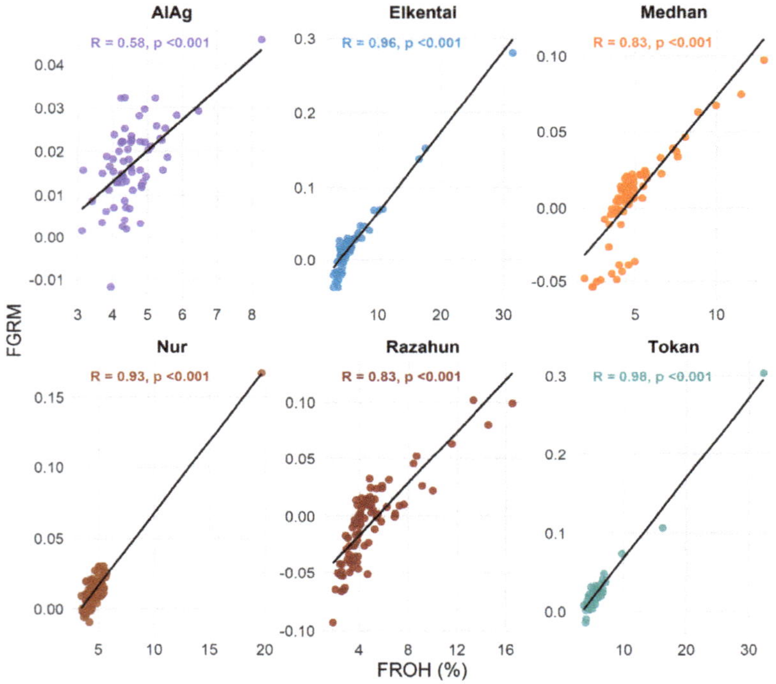

Figure 5. Correlation between F_{ROH} and F_{GRM} within each population. Each panel shows a scatter plot of individual F_{ROH} (x-axis) and F_{GRM} (y-axis) values. Regression lines and Pearson's R coefficients with *p*-values are indicated.

Moreover, Figure 6 displays violin plots illustrating the distribution of F_{ROH} and F_{GRM} across populations. F_{ROH} values were highest in Tokan, Elkentai, and Medhan, suggesting elevated levels of recent inbreeding. AlAg showed the lowest F_{ROH} levels. The distribution of F_{GRM} values was more concentrated around zero and less variable, consistent with its sensitivity to total homozygosity regardless of segment length. Together, these results indicate that both F_{ROH} and F_{GRM} are informative and complementary in capturing different aspects of genomic inbreeding.

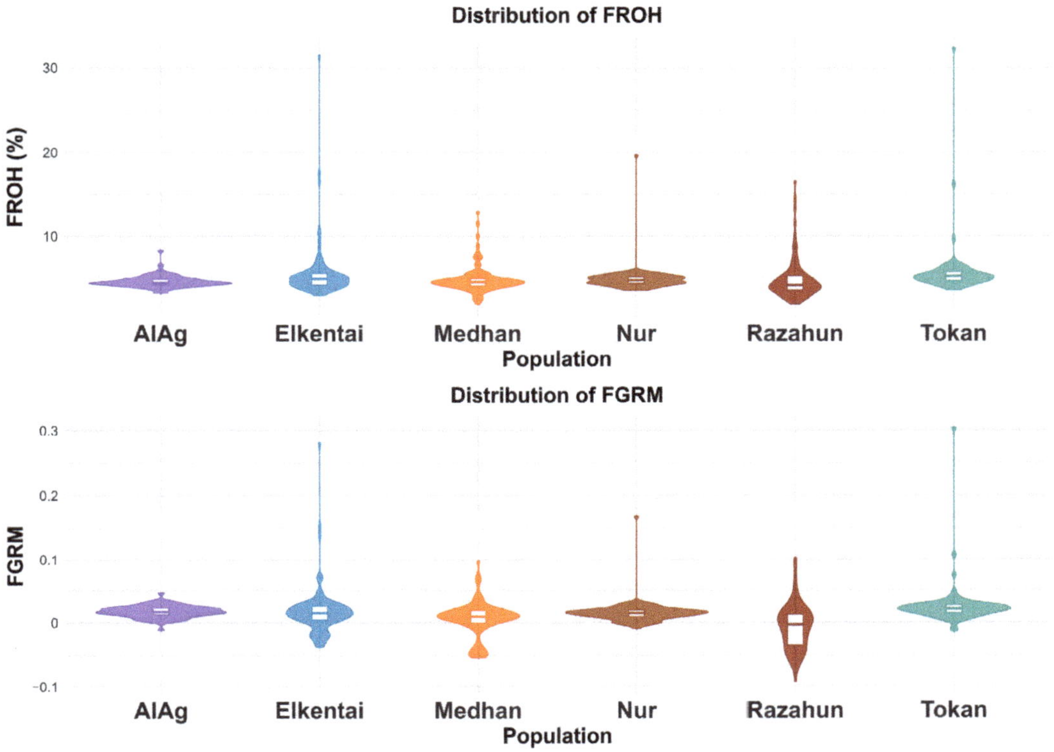

Figure 6. Distribution of F_{ROH} and F_{GRM} across populations. Violin plots show the distribution of genomic inbreeding coefficients by population. The upper panel displays F_{ROH}, the lower panel shows F_{GRM}.

3.3. Gene Annotation and Functional Clustering

The GO and KEGG analyses highlighted pathways strongly related to metabolism and growth regulation. The enrichment of transforming growth factor beta signaling (TGF-β), extracellular matrix regulation, metabolic processes, and nutrient transport reflects the complex interplay of pathways underlying sheep productive traits (Figure 7). These results suggest that genes involved in muscle development, energy metabolism, and structural tissue integrity are central to growth performance, carcass quality, and potentially fat deposition in sheep. The identification of arginine, proline, and TCA cycle-related pathways further emphasizes the importance of efficient nutrient metabolism in determining productivity traits (Tables S1 and S2).

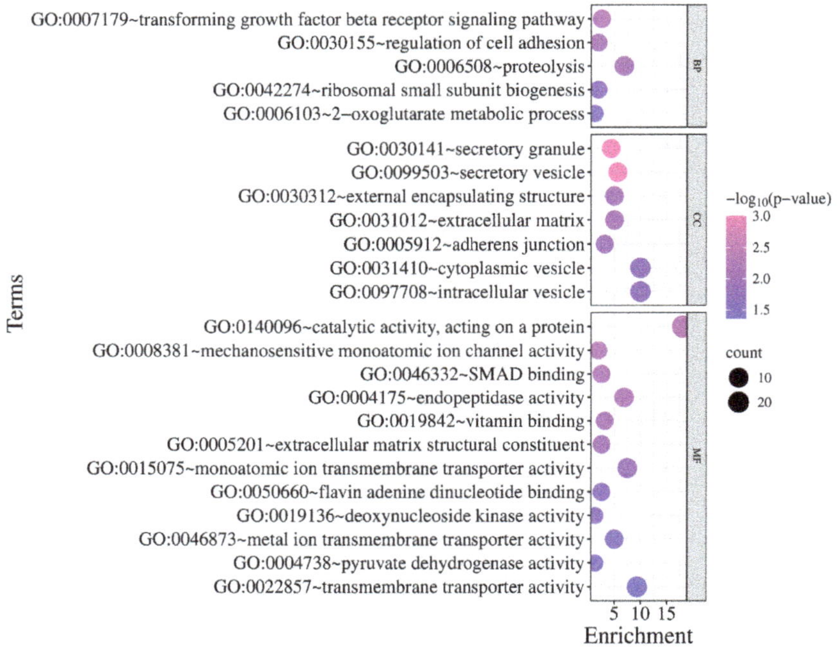

Figure 7. The GO and KEGG pathway enrichment analyses of candidate genes. BP, Biological Process; CC, Cellular Component; MF, Molecular Function. The x-axis represents the enrichment score of each GO term, while the y-axis lists the GO term names. The size of each bubble indicates the number of genes enriched in that GO term (count), and the bubble color corresponds to the statistical significance (−log10 *p*-value).

4. Discussion

Sheep play an important role in the lives of people in Kazakhstan, serving as a vital source of livestock products and cultural heritage. Regular monitoring of genomic diversity and inbreeding is essential for the sustainable development of the breed and the preservation of its adaptive traits. This is the first study to identify ROH in Kazakh fat-tailed coarse-wool sheep.

Analysis of extended runs of homozygosity revealed clusters of long ROHs on chromosomes OAR10, OAR13, and OAR22. In contrast, short and medium-sized ROHs were predominant in other genomic regions and were evenly distributed across the genome, consistent with background recombination. Genomic inbreeding coefficients (Table 3) showed clear population differences. Total F_{ROH} was lowest in AlAg (0.046) and highest in Tokan (0.058), with Elkentai (0.056) also displaying elevated values. The distribution of ROH length classes reflected demographic history: short segments (1–4 Mb) predominated in all groups, indicating the contribution of distant ancestral inbreeding, whereas the presence of long segments (>16 Mb) in Elkentai and Tokan reflected recent consanguinity.

Mean F_{GRM} values were low (from −0.007 in Razahun to 0.025 in Tokan), and the wide confidence intervals indicated estimation variability in smaller populations. The difference between F_{ROH} and F_{GRM} underlined the complementary nature of the two indicators, with F_{ROH} providing the most consistent measure of realized autozygosity.

From a breeding standpoint, the higher levels of genomic inbreeding in Tokan and Elkentai require careful monitoring to prevent further increases, while the comparatively low inbreeding in AlAg demonstrates better retention of genetic diversity. Practical management should include genomic control of mating schemes, avoidance of close-relative pairings, and the use of unrelated sires to maintain effective population size. These measures will help balance genetic improvement with the preservation of genetic diversity, supporting sustainable development of the studied sheep populations. In Iranian fat-tailed sheep, notable ROH islands were detected on chromosomes OAR2, OAR6, and OAR10 [26]. In our previous study, we also observed a high prevalence of short genomic segments in Kazakh meat–wool sheep [18]. Likewise, the majority of ROH segments in Moroccan sheep ranged between 1 and 6 Mb [27]. In addition, among the examined Chinese sheep breeds, Large-tailed Han sheep displayed the highest level of ROH-based inbreeding, while Hulun Buir sheep exhibited the lowest [28]. The contribution of long ROHs (>8 Mb), typically associated with recent inbreeding, was relatively small. However, variation in their abundance between flocks may reflect differences in demographic history and selection pressure.

The strong correlation between F_{ROH} and F_{GRM} (r = 0.58) confirms that F_{ROH} is a reliable indicator of genomic inbreeding, especially in populations with limited genetic diversity. The highest correlations were observed in Tokan, Elkental, and Nur, indicating high levels of inbreeding and reduced genetic variability. In contrast, moderate correlations in Medhan and Razahun may reflect the influence of additional factors such as gene flow or admixture. The lowest correlation was found in AlAg, suggesting greater genetic variability and weaker association between homozygosity and relatedness. Notably, different genomic inbreeding coefficients indicated low levels of inbreeding within the Hu sheep population [29]. A previous study reported that inbreeding coefficients (F_{GRM} and F_{ROH}) indicated higher levels of inbreeding and reduced genetic diversity in Iranian wild sheep and goats compared with their domestic counterparts, highlighting the impact of domestication and management practices in maintaining greater genetic variability in farmed populations [30].

An ROH study in global sheep populations has identified several genes associated with growth, body weight, and meat quality traits [31]. In this study, among the genes associated with productivity traits, particular attention is given to the transcription factors myogenic factor 5 (*MYF5*), which play a central role in myogenesis. These genes have been identified as important markers of meat performance in livestock. In Chinese Qinchuan cattle, polymorphisms in *MYF5* are associated with growth and carcass quality [32], while in goats, expression variability in both genes has been linked to muscle fiber morphology, including cross-sectional area and density [33]. In dairy cattle (e.g., Brown Swiss and Holstein), *MYF5* alleles have shown associations with feed efficiency [34], indicating the broader relevance of these markers beyond beef breeds. PR/SET domain 16 (*PRDM16*) is a key regulator of lineage determination between brown adipose tissue and skeletal muscle. Studies have shown that *PRDM16* expression in myoblasts promotes brown adipogenesis via interaction with transcription factors such as PPARγ and C/EBPβ, while its inactivation in brown fat precursors induces expression of myogenic markers (MyoD, Myogenin) and leads to muscle formation in MYF5-positive lineages [35]. These mechanisms are of practical interest in animal production, as the balance between fat and muscle affects carcass composition and meat quality. According to Lee and Hong (2024), the solute carrier family 26 member 4 (*SLC26A4*), which encodes anion transporters, is involved in regulating intracellular pH via Cl^-/HCO_3^- exchange [36]. Changes in acid–base balance may indirectly influence lipogenesis, including the expression of key lipogenic enzymes. This regulation may affect fat deposition and meat marbling.

The transglutaminase 3 (*TGM3*) gene, which encodes transglutaminase 3, was identified among genes associated with wool quality. John et al. (2012) reported that knockout of Tgm3 in mice resulted in severe hair shaft defects, including cuticle deformation and reduced tensile strength, despite an intact epidermal barrier [37]. These effects were attributed to disrupted cross-linking between trichohyalin and keratins in the absence of

TGase 3 activity. As keratin matrix stability underpins wool structure, *TGM3* might be a functionally relevant candidate for improving fleece characteristics in sheep breeding programs.

SMAD5 is a receptor-regulated SMAD (R-SMAD) activated by bone morphogenetic proteins (BMPs) within the TGF-β signaling pathway. It plays a crucial role in processes directly linked to productive traits in sheep [38]. A previous study reported *SMAD5* expression in Sertoli and Leydig cells during various developmental stages of the Tibetan sheep testis [39]. In addition, RNA-sequencing of ovine ovary tissue revealed that this gene might play a role in the actions of estrogen [40].

Collectively, the revealed genes may serve as valuable targets for marker-assisted selection to improve muscle development and wool traits in Kazakh fat-tailed coarse-wool sheep and warrant further investigation to clarify their functional roles.

5. Conclusions

This study revealed patterns of genetic diversity and inbreeding in Kazakh fat-tailed coarse-wool sheep. The most pronounced ROH were located on chromosomes OAR10, OAR13, and OAR22, potentially reflecting signatures of selection. Across populations, homozygosity was predominantly concentrated in short segments (1–4 Mb). A strong positive correlation was observed between the genomic inbreeding coefficients F_{ROH} and F_{GRM}. Furthermore, several functionally important genes (*MYF5, PRDM16, TGM3, SLC26A4, SMAD5*) were identified within these regions, which play roles in productivity traits. Taken together, these findings provide new insights into the molecular mechanisms underlying productivity, and they offer a foundation for more accurate genetic evaluation and the implementation of marker-assisted selection programs in this breed.

Supplementary Materials: The following supporting information can be downloaded at: https://www.mdpi.com/article/10.3390/ani15182714/s1. Table S1: Summary of gene annotations retrieved from UniPorts; Table S2: The KEGG-enriched terms.

Author Contributions: Conceptualization, K.D.; methodology, A.K., M.Y., K.Y. and T.K.; software, A.K., T.K. and K.D.; validation, A.K. and A.T.; formal analysis, Z.A., T.K. and K.D.; investigation, A.K., M.Y. and K.Y.; resources, U.B., A.T. and B.K.; data curation, K.D. and A.K.; visualization, A.K. and Z.A.; writing—original draft preparation, A.K.; writing—review and editing, Z.A., K.D., M.Y. and X.L.; supervision, K.D. and X.L.; project administration, K.D. and A.T.; funding acquisition, K.D., B.K. and A.T. All authors have read and agreed to the published version of the manuscript.

Funding: This research was funded by program-targeted funding for 2023–2025 of the Ministry of Education and Science of the Republic of Kazakhstan, BR21882201 "Improvement of meat productivity of fat-tailed sheep by new methods of selection, genetics and biotechnology".

Institutional Review Board Statement: The study was conducted in accordance with the Local Bioethics Committee, and the protocol was approved by the Ethics Committee of LLP «Kazakh Research Institute for Livestock and Fodder Production» (№2) on 24 July 2023.

Informed Consent Statement: Not applicable.

Data Availability Statement: The data are available upon request.

Acknowledgments: We sincerely thank the staff of all participating collective farms for their valuable assistance in collecting blood samples and recording the body measurements of the sheep.

Conflicts of Interest: The authors declare no conflicts of interest.

Abbreviations

The following abbreviations are used in this manuscript:

ROH	Runs of homozygosity
FROH	Inbreeding coefficient based on ROH
FGRM	Inbreeding coefficient based on genomic relationship matrix
MAS	Marker-assisted selection
SNP	Single-nucleotide polymorphism

References

1. Bureau of National Statistics of the Agency for Strategic Planning and Reforms of the Republic of Kazakhstan. Available online: https://stat.gov.kz/ru/industries/business-statistics/stat-forrest-village-hunt-fish/dynamic-tables/ (accessed on 15 April 2025).
2. Katasheva, A.; Iskakov, K.; Kulataev, B.; Abdramanov, A.; Sattorov, S. Improving the efficiency of the production of mutton of kazakh fat-tailed coarse-wool sheep. *Izdenister Natigeler* **2024**, *2*, 55–62. (In Russian) [CrossRef]
3. Malmakov, N.; Kulataev, B.; Iskakov, K.; Tastaganov, M. Features of effective breeding methods to improve the reproduction ability of Kazakh coarse-wooled sheep. *Izdenister Natigeler* **2023**, *3*, 5–10. (In Russian)
4. Zhumadila, K.; Zhumadilaev, N.K.; Esengaliev, K.; Burambaeva, N.B. Recommendations for Breeding Work with Kazakh Fat-Tailed Coarse-Wool Sheep: Mynbayevo, Kazakhstan. 2009, p. 28, ISBN 978-601-7226-02-2. Available online: http://rpo-kaz.kz/learn/ (accessed on 21 August 2025). (In Russian).
5. Ma, R.; Liu, J.; Ma, X.; Yang, J. Genome-wide runs of homozygosity reveal inbreeding levels and trait-associated candidate genes in diverse sheep breeds. *Genes* **2025**, *16*, 316.
6. Deng, T.X.; Ma, X.Y.; Duan, A.Q.; Lu, X.R.; Abdel-Shafy, H. Genomic insights into selection signatures and candidate genes for milk production traits in buffalo population. *Animal* **2025**, *19*, 101427. [CrossRef]
7. Ma, K.; Li, X.; Ma, S.; Zhang, M.; Wang, D.; Xu, L.; Chen, H.; Wang, X.; Qi, A.; Ren, Y.; et al. Analysis of population structure and selective signatures for milk production traits in Xinjiang Brown cattle and Chinese Simmental cattle. *Int. J. Mol. Sci.* **2025**, *26*, 2003. [CrossRef]
8. Nandolo, W.; Mészáros, G.; Banda, L.J.; Gondwe, T.N.; Lamuno, D.; Mulindwa, H.A.; Nakimbugwe, H.N.; Wurzinger, M.; Utsunomiya, Y.T.; Woodward-Greene, M.J.; et al. Timing and Extent of inbreeding in African goats. *Front. Genet.* **2019**, *10*, 537. [CrossRef]
9. Silva, G.A.A.; Harder, A.M.; Kirksey, K.B.; Mathur, S.; Willoughby, J.R. Detectability of runs of homozygosity is influenced by analysis parameters and population-specific demographic history. *PLoS Comput. Biol.* **2024**, *20*, e1012566. [PubMed]
10. Curik, I.; Ferenčaković, M.; Sölkner, J. Inbreeding and runs of homozygosity: A possible solution to an old problem. *Livest. Sci.* **2014**, *166*, 26–34. [CrossRef]
11. Meyermans, R.; Gorssen, W.; Aerts, N.; Hooyberghs, K.; Bhaskaran, B.C.; Chapard, L.; Buys, N.; Janssens, S. Genomic characterisation and diversity assessment of eight endangered Belgian sheep breeds. *Animal* **2024**, *18*, 101315. [CrossRef] [PubMed]
12. Ceballos, F.C.; Joshi, P.K.; Clark, D.W.; Ramsay, M.; Wilson, J.F. Runs of homozygosity: Windows into population history and trait architecture. *Nat. Rev. Genet.* **2018**, *19*, 220–234. [CrossRef]
13. Peripolli, E.; Munari, D.P.; Silva, M.V.G.B.; Lima, A.L.F.; Irgang, R.; Baldi, F. Runs of homozygosity: Current knowledge and applications in livestock. *Anim. Genet.* **2017**, *48*, 255–271. [CrossRef]
14. Ferenčaković, M.; Hamzić, E.; Gredler, B.; Solberg, T.R.; Klemetsdal, G.; Curik, I.; Sölkner, J. Estimates of autozygosity derived from runs of homozygosity: Empirical evidence from selected cattle populations. *J. Anim. Breed. Genet.* **2013**, *130*, 286–293. [CrossRef] [PubMed]
15. Halvoník, A.; Moravčíková, N.; Chalupková, M.; Kasarda, R. Commonly used genomic estimators of individual inbreeding in livestock. *Czech. J. Anim. Sci.* **2024**, *69*, 269–279. [CrossRef]
16. Howrigan, D.P.; Simonson, M.A.; Keller, M.C. Detecting autozygosity through runs of homozygosity: A comparison of three autozygosity detection algorithms. *BMC Genom.* **2011**, *12*, 460. [CrossRef]
17. Dadousis, C.; Ablondi, M.; Cipolat-Gotet, C.; van Kaam, J.T.; Marusi, M.; Cassandro, M.; Sabbioni, A.; Summer, A. Genomic inbreeding coefficients using imputed genotypes: Assessing different estimators in Holstein-Friesian dairy cows. *J. Dairy Sci.* **2022**, *105*, 5926–5945. [CrossRef]
18. Illa, S.K.; Mumtaz, S.; Nath, S.; Mukherjee, S.; Mukherjee, A. Characterization of runs of homozygosity revealed genomic inbreeding and patterns of selection in indigenous Sahiwal cattle. *J. Appl. Genet.* **2024**, *65*, 167–180. [CrossRef]

19. Amandykova, M.; Akhatayeva, Z.; Kozhakhmet, A.; Kapassuly, T.; Orazymbetova, Z.; Yergali, K.; Khamzin, K.; Iskakov, K.; Dossybayev, K. Distribution of runs of homozygosity and their relationship with candidate genes for productivity in Kazakh meat–wool sheep breed. *Genes* **2023**, *14*, 1988. [CrossRef] [PubMed]
20. Tao, L.; He, X.; Wang, X.; Di, R.; Chu, M. Litter size of sheep (*Ovis aries*): Inbreeding depression and homozygous regions. *Genes* **2021**, *12*, 109. [CrossRef] [PubMed]
21. Dzomba, E.F.; Chimonyo, M.; Pierneef, R.; Muchadeyi, F.C. Runs of homozygosity analysis of South African sheep breeds from various production systems investigated using OvineSNP50k data. *BMC Genom.* **2021**, *22*, 7. [CrossRef]
22. Selli, A.; Ventura, R.V.; Fonseca, P.A.S.; Buzanskas, M.E.; Andrietta, L.T.; Balieiro, J.C.C.; Brito, L.F. Detection and visualization of heterozygosity-rich regions and runs of homozygosity in worldwide sheep populations. *Animals* **2021**, *11*, 2696. [CrossRef]
23. Islam, R.; Li, Y.; Liu, X.; Berihulay, H.; Abied, A.; Gebreselassie, G.; Ma, Q.; Ma, Y. Genome-wide runs of homozygosity, effective population size, and detection of positive selection signatures in six Chinese goat breeds. *Genes* **2019**, *10*, 938. [CrossRef] [PubMed]
24. An, Z.X.; Shi, L.G.; Hou, G.Y.; Zhou, H.L.; Xun, W.J. Genetic diversity and selection signatures in Hainan black goats revealed by whole-genome sequencing data. *Animal* **2024**, *18*, 101147. [CrossRef]
25. Kim, E.S.; Cole, J.B.; Huson, H.; Wiggans, G.R.; Van Tassell, C.P.; Crooker, B.A.; Liu, G.; Da, Y.; Sonstegard, T.S. Effect of artificial selection on runs of homozygosity in U.S. Holstein cattle. *PLoS ONE* **2013**, *8*, e80813. [CrossRef]
26. Mulim, H.A.; Brito, L.F.; Pinto, L.F.B.; Ferraz, J.B.S.; Grigoletto, L.; Silva, M.R.; Pedrosa, V.B. Characterization of runs of homozygosity, heterozygosity-enriched regions, and population structure in cattle populations selected for different breeding goals. *BMC Genom.* **2022**, *23*, 209. [CrossRef]
27. Abdoli, R.; Mirhoseini, S.Z.; Ghavi Hossein-Zadeh, N.; Zamani, P.; Moradi, M.H.; Ferdosi, M.H.; Sargolzaei, M.; Gondro, C. Runs of homozygosity and cross-generational inbreeding of Iranian fat-tailed sheep. *Heredity* **2023**, *130*, 358–367. [CrossRef]
28. Kusza, S.; Badaoui, B.; Wanjala, G. Insights into the genomic homogeneity of Moroccan indigenous sheep breeds though the lens of runs of homozygosity. *Sci. Rep.* **2024**, *14*, 16515. [CrossRef]
29. Liu, J.; Shi, L.; Li, Y.; Chen, L.; Garrick, D.; Wang, L.; Zhao, F. Estimates of genomic inbreeding and identification of candidate regions that differ between Chinese indigenous sheep breeds. *J. Anim. Sci. Biotechnol.* **2021**, *12*, 95. [CrossRef]
30. Bao, J.; Xiong, J.; Huang, J.; Yang, P.; Shang, M.; Zhang, L. Genetic diversity, selection signatures, and genome-wide association study identify candidate genes related to litter size in Hu sheep. *Int. J. Mol. Sci.* **2024**, *25*, 9397. [CrossRef] [PubMed]
31. Taheri, S.; Zerehdaran, S.; Javadmanesh, A. Genetic diversity in some domestic and wild sheep and goats in Iran. *Small Rumin. Res.* **2022**, *209*, 106641. [CrossRef]
32. Zhao, C.; Raza, S.H.A.; Khan, R.; Sabek, A.; Khan, S.; Ullah, I.; Memon, S.; El-Aziz, A.H.A.; Shah, M.A.; Shijun, L.; et al. Genetic variants in MYF5 affected growth traits and beef quality traits in Chinese Qinchuan cattle. *Genomics* **2020**, *112*, 2804–2812. [CrossRef]
33. Nosrati, M.; Asadollahpour Nanaei, H.; Javanmard, A.; Esmailizadeh, A. The pattern of runs of homozygosity and genomic inbreeding in world-wide sheep populations. *Genomics* **2021**, *113*, 1407–1415. [CrossRef] [PubMed]
34. Bayrak, B.; Şen, U.; Gökçek, D.; Şirin, E. Differential Myf5 and Myf6 expression and muscle fiber traits in Angora, Hair, Honamlı, and Kilis goats. *Trop. Anim. Health Prod.* **2025**, *57*, 36. [CrossRef]
35. Fadhil, M.; Zülkadir, U. Association between polymorphismsof Myf5, MSTN and CAST genes and fattening performance in Brown Swiss and Holstein cattle breeds. *Anim. Biotechnol.* **2021**, *32*, 121–129. [CrossRef] [PubMed]
36. Lee, D.; Hong, J.H. Chloride/multiple anion exchanger SLC26A family: Systemic roles of SLC26A4 in various organs. *Int. J. Mol. Sci.* **2024**, *25*, 4190. [CrossRef] [PubMed]
37. John, S.; Thiebach, L.; Frie, C.; Mokkapati, S.; Bechtel, M.; Nischt, R.; Rosser-Davies, S.; Paulsson, M.; Smyth, N. Epidermal transglutaminase (TGase 3) is required for proper hair development, but not the formation of the epidermal barrier. *PLoS ONE* **2012**, *7*, e34252. [CrossRef]
38. Sun, R.; Li, M.; He, N.; Wen, X.; Zhang, J. Molecular characterization, expression profiles of *SMAD4*, *SMAD5* and *SMAD7* genes and lack of association with litter size in Tibetan sheep. *Animals* **2022**, *12*, 2232. [CrossRef]
39. Chen, N.N.; An, X.J.; Li, T.T.; Wang, H.H.; Xu, S.R.; Man, Y.H.; Ma, Y.J. Expression and localization of BMP6/SMAD5 in Tibetan sheep testis. *Prat. Sci.* **2022**, *39*, 2434–2441.
40. Zhong, Y.; Di, R.; Yang, Y.; Liu, Q.; Chu, M. Transcriptome analysis of neuroendocrine regulation of ovine hypothalamus-pituitary-ovary axis during ovine anestrus and the breeding season. *Genes* **2021**, *12*, 1861. [CrossRef]

animals

MDPI

Review

Wool: From Properties and Structure to Genetic Insights and Sheep Improvement Strategies

Huitong Zhou [1,2], Lingrong Bai [2], Shaobin Li [1], Jiqing Wang [1] and Jon G. H. Hickford [1,2,*]

1 International Wool Research Institute, Faculty of Animal Science and Technology, Gansu Agricultural University, Lanzhou 730070, China; huitong.zhou@lincoln.ac.nz (H.Z.); lisb@gsau.edu.cn (S.L.); wangjq@gsau.edu.cn (J.W.)

2 Gene-Marker Laboratory, Faculty of Agriculture and Life Sciences, Lincoln University, Lincoln 7647, New Zealand; lingrong.bai@lincolnuni.ac.nz

* Correspondence: jonathan.hickford@lincoln.ac.nz

Simple Summary

The wool of sheep provides a fibre with exceptional properties. This makes it ideally suited for the manufacture of performance textiles, including 'next-to-skin' applications. This review provides an overview of wool and its biology, with a focus on the molecular genetics that underpin fibre traits and how that may allow for the genetic improvement of sheep to produce better wool. It will highlight the roles of keratin and keratin-associated protein genes in producing wool fibres, how they are genetically variable, their expression patterns, and how the genetic variation may affect wool fibre traits. The review also explores how genetic tools could be used to enhance wool quality while considering possible trade-offs with other traits. The integration of phenotypic, molecular, and genetic knowledge will be essential for future advancement in wool production and the improvement of fibre quality.

Abstract

The wool of sheep consists of structurally intricate natural fibres that can be processed and manufactured into a range of products. It is prized for its insulation, moisture-buffering capability, flame resistance, and biodegradability. These features arise from its unique fibre architecture and specialised protein composition, which set it apart from most other natural and synthetic fibres. However, despite these novel characteristics, wool fibre variation hampers its uses and reduces its ability to compete with other fibres. This review summarises our current knowledge of wool fibre biology. It begins with a description of wool's functional properties and performance attributes, then explores the structural foundations of these properties, the molecular basis of fibre trait variation, and prospects for improving fibre quality using genetic approaches. Particular attention is given to the wool keratin and keratin-associated protein genes, their spatiotemporal expression patterns, and genetic polymorphism that may influence fibre characteristics. Opportunities for the genetic improvement of sheep are discussed, including the use of genetic modification and marker-assisted selection. Challenges in interpreting gene–trait associations, particularly from high-throughput omics studies, are highlighted, along with the need for functionally validated genetic markers. Potential trade-offs between wool characteristics and other production and reproductive traits are considered, emphasising the need for balanced breeding approaches. By integrating insights from structural biology, molecular genetics, and breeding strategies, this review provides a foundation for wool fibre improvement.

Keywords: wool fibre; keratin; keratin-associated protein; molecular genetics; gene polymorphism; genetic association; genetic improvement; sheep; fibre trait; breeding; wool quality

1. Introduction

Environmental concerns and the demand for sustainable materials have renewed scientific and industrial interest in natural fibres. Among these fibres, the wool of sheep stands out for its functional properties that other natural and synthetic fibres have difficulty emulating. These include its insulative properties, flame resistance, resilience, and hygroscopic properties [1]. Wool can be used to manufacture environmentally responsible fibre products like clothing, and these can also be biodegradable or recyclable.

Wool is a proteinaceous fibre that is shorn from sheep, albeit not all breeds are considered to be wool producers. In this context, sheep numbers are increasing globally [2], with the International Wool Textile Organisation (IWTO) reporting that sheep numbers rose to 1.296 billion heads in 2022, with greasy wool production rising to 1977.3 million kilogrammes, and clean wool production reaching an estimated 1051.2 million kilogrammes in that year.

Despite wool's long history of use, many of its functional properties remain under-recognised, and their biological basis is not fully understood. While much of the physical structure of wool fibres has been described [3], the molecular mechanisms and environmental factors responsible for creating variation in fibre traits are only partially understood. Accordingly, a challenge remains to improve wool quality and achieve greater fibre uniformity, which is necessary if the performance and value of wool are to be further enhanced.

This review summarises our knowledge of wool's functional properties and performance attributes and explores the structural and molecular bases of fibre traits. Particular attention is given to wool keratin and keratin-associated protein genes, their expression patterns, and genetic variation that may influence fibre characteristics. Opportunities for genetic improvement, through genetic modification and selective breeding, are also discussed. Challenges in interpreting gene–trait associations, the need for validated genetic markers, and potential trade-offs between wool characteristics and other production and reproductive traits are considered. By integrating insights from structural biology, molecular genetics, and breeding strategies, this review provides a comprehensive foundation for improving wool fibres.

2. Key Functional Properties of Wool Fibre

Wool possesses a range of functional properties that contribute to its performance across various practical applications. Although other fibres may surpass wool in individual aspects of performance, it has a combination of functional properties that other fibres have difficulty matching.

2.1. Thermal and Acoustic Insulation

Wool fibres possess a naturally crimped structure. When spun into yarn or felted, they can be used to create products containing air pockets that trap air and provide an insulative layer. These pockets can reduce heat transfer, which in clothing helps to stabilise body temperature and enhance thermal comfort. As an insulative layer, the thermal conductivity of wool typically ranges from 0.038 to 0.045 W/m·K, which is comparable to conventional insulation materials such as mineral wool and expanded polystyrene products [4]. The variation in conductivity reflects variation in the wool fibres and the way in which they are manufactured into clothing, with characteristics like the degree of medulation, mean fibre diameter, and curvature, affecting thermal conductivity.

The crimped fibre structure can also contribute to acoustic insulation. As sound waves pass through wool products, they are gradually dispersed and absorbed due to friction with air and fibre surfaces. This reduces sound intensity and reverberation. Wool-based materials can demonstrate strong sound absorption coefficients, particularly in the mid-frequency range of 500 to 2000 Hz [5], albeit this study recognises that variation in the thickness of the wool material tested affects this property. Accordingly, fibre traits like curvature and fibre diameter are likely to affect the ability to absorb sound. This makes wool effective for mitigating everyday noise in both residential and commercial environments.

2.2. Moisture Regulation

Wool is hygroscopic and has the ability to absorb and release moisture without feeling damp. It has a water retention capacity of up to approximately one-third of its weight [1]. This capability distinguishes wool from many other natural and synthetic fibres and is fundamental to its thermal 'comfort' when used in clothing.

Specifically, when water molecules bind to wool fibres, they undergo an exothermic process that produces heat (described as 'heat of sorption'). Conversely, in warmer conditions (as may occur when woollen clothing is worn during physical activity), the wool fibres release absorbed water via evaporation. This is an endothermic process that cools the body by drawing heat away. Wool's heat of sorption is high; estimated at 100–115 J/g for dry fibres [1], albeit this will vary with wool type [6].

This 'bidirectional' moisture management property (both absorption and desorption) provides wool clothing with superior thermoregulatory characteristics when compared to other fibre choices, and across a range of climatic conditions. In comparison, cotton, while also hygroscopic, absorbs less moisture and only has a moderate heat of sorption (~46 J/g) [7]. This can cause cotton clothing to have a clammy feeling when wet, and it provides far less thermal benefit. Synthetic fibres such as polyester and acrylic absorb little water, with minimal heat of sorption (about 5–7 J/g) [7]. Accordingly, many synthetic fibres rely primarily on moisture wicking, rather than absorption, and thus they offer limited thermal buffering, albeit reasonable insulative properties.

2.3. Chemical and Odour Absorption

Wool is effective at absorbing and neutralising various volatile organic compounds (VOCs) and odours. It can bind airborne pollutants such as formaldehyde, sulphur dioxide, and nitrogen oxides (NOx) through chemisorption, forming stable compounds that enable long-term pollutant removal [8]. For example, wool carpets can reduce formaldehyde levels from 300 ppm to near zero within hours [9], and with many modern building products (e.g., pressed-wood products, particle boards, plywood, paint, glue, and foam insulating products) releasing formaldehyde through a process called off-gassing, this property is helpful for removing a known human carcinogen from buildings.

In clothing, wool can reduce body odour by binding the molecules responsible for those smells. Wool garments, including socks and activewear, typically develop less odour than synthetic materials, even after prolonged wear, enhancing freshness and wearer comfort [3].

Wool's natural oleophilic properties allow it to absorb oil up to 40 times its own weight, depending on the oil's viscosity [10]. This makes wool well suited for oil spill clean-up in contaminated environments, and when blended with synthetic fibres to increase its buoyancy, it can be used to recover oil from water [3].

Wool also exhibits an affinity for absorbing metals like mercury, gold, and lead [11]. Powdered wool, with increased surface area, has increased uptake rates, making it effective for removing heavy metals from industrial effluents and contaminated water. This property is useful for filtration applications, including the treatment of mercury-contaminated effluents from the chlor-alkali industry [12]. There is a suggestion that wool is an effective filter absorbent for microplastic removal from water [13].

2.4. Fire Resistance and Safety

Wool has a high limiting oxygen index (LOI) of approximately 25.2%, albeit this will vary depending on the fibre density of the materials and the presence of moisture (given wool's hygroscopic nature). This is above the oxygen concentration in air (~21%), meaning that wool is more flame-resistant than fibres such as cotton (LOI ~18.4%), nylon (~20.1%), and polyester (~20.6%) [14]. In this context, the fibre has been used for fire protection for many years, with the London Fire Brigade describing its use in woollen tunics over the period 1866–1989, until the first Nomex (a synthetic meta-aramid fibre) suits were introduced [15].

Wool also has an ignition temperature that ranges from 570 °C to 600 °C. In comparison, cotton ignites at a much lower temperature (~255 °C). Synthetic fibres like nylon and polyester ignite between 485 and 575 °C [14], but have lower melting temperatures that can lead to structural failure in products manufactured from them. The high ignition threshold of wool reduces the likelihood of it igniting when exposed to heat sources, and if ignited, wool does not melt but instead chars. This forms a protective thermal barrier that suppresses flame spread and shields underlying materials. In contrast, if synthetic fibres melt and drip, they can increase the risk of secondary burns, and the ignition of synthetic fibres can produce carbon monoxide, dioxins, hydrogen cyanide, and nitrous oxides.

When burning, wool has a lower heat of combustion (~20.5 kJ/g, compared to 22.6 kJ/g for nylon), meaning it releases less energy compared to other fibres [14]. This limits flame intensity and reduces the potential for fire propagation, further enhancing its fire safety profile.

Taken together, these properties make wool a safer choice in contexts where fire safety is critical. Its unique combination of high ignition temperature, low heat of combustion, and non-melting behaviour provides a level of protection not commonly found for other natural or synthetic fibres. As a result, wool can be used in applications for protective clothing, mattress covers, upholstery, and interior textiles, where reducing fire risk is important.

2.5. UV Shielding

Wool fibres offer effective protection against ultraviolet (UV) radiation, making them well-suited for use in outdoor garments, coverings, and applications where UV shielding is important. A study by Gambichler et al. [16] that assessed 236 spring and summer apparel fabrics revealed that wool outperformed common fabrics such as cotton, linen, viscose, and polyester in UV protection. Over 70% of wool samples achieved a UPF rating of 50+, with even the worst-performing wool fabric reaching a UPF of 40+. Wool's high UV protection is considered an advantage and is attributed to its unique fibre structure and its ability to absorb and scatter UV radiation.

2.6. Durability, Elasticity and Resilience

Wool is valued for its ability to maintain the appearance, fit, and performance of garments. It offers excellent durability, withstanding repeated wear, bending, and handling without a major loss of performance or comfort.

Wool fibres are elastic and capable of stretching up to 30% when dry and 50% when wet, while remaining capable of returning to their original shape [17]. This attribute helps garments maintain their fit or drape, enhances comfort, and contributes to wrinkle resistance by allowing the fibre to recover its form after deformation. Wool also displays outstanding resilience to repeated bending and is capable of enduring up to 20,000 bends without breaking. This is more than natural fibres such as cotton (~3000 bends) and silk (~2000 bends) [17]. This resilience reduces the likelihood of fibre breakage under repeated stress, thereby supporting long-term performance and durability.

Wool fibres are abrasion resistant, mainly due to their tough exterior layer of scales, which helps protect them from surface damage. This makes wool suitable for high-friction applications such as may be needed for carpets, upholstery, and outerwear. It recovers shape well after compression, especially in humid conditions, which helps maintain shape and appearance over time.

Together, these properties enable wool products to remain durable, comfortable, and visually appealing even after repeated use and care. As a result, wool is a reliable choice for both everyday wear and demanding applications where longevity and consistent performance are essential.

3. Health and Environmental Benefits of Wool

Beyond its functional performance, wool offers benefits for human health and wellbeing. It also has good environmental credentials.

3.1. Health and Wellbeing

Wool possesses a variety of properties that contribute to human health, comfort, and safety. Its breathability, ability to manage moisture, and softness make it suitable for next-to-skin use. Other properties help improve indoor environmental quality and reduce health risks. Together, these attributes support wellbeing across diverse everyday and specialised environments.

3.1.1. Skin Health

Contrary to traditional views that wool irritates sensitive skin, clinical studies show that superfine Merino wool (defined as fibres finer than 18.5 μm, ideally <16.5 μm for sensitive skin), is not only well tolerated but may even benefit individuals with atopic dermatitis (AD), a common chronic inflammatory skin condition.

In a randomised controlled trial, Su et al. [18] revealed that children with mild-to-moderate AD experienced a reduction in disease severity (as measured by scoring atopic dermatitis [SCORAD] scores) after wearing superfine Merino wool garments for 3 and 6 weeks compared to cotton. Participants also reported less reliance on corticosteroids and improved comfort and quality of life. Interestingly, when participants returned to cotton clothing, their symptoms worsened, highlighting the therapeutic potential of superfine Merino wool. Spelman et al. [19] extended the evaluation period to 15 weeks and revealed that children and young adults with AD who wore superfine Merino wool base-layer garments experienced improvements across multiple validated AD severity scales, including SCORAD, eczema area, and severity index (EASI). No negative effects or symptom worsening were observed. Participants also reported enhanced comfort and garment practicality. Similarly, Fowler et al. [20] reported improvements in eczema severity (EASI scores), dermatology-related quality of life, and clinical assessments in patients with AD wearing Merino wool for 6 weeks, with no adverse skin reactions reported.

Complementing these clinical trials, Zallmann et al. [21] observed no evidence that wool causes allergic contact dermatitis. Instead, most skin reactions to wool are non-immune in nature, typically caused by mechanical irritation from coarse fibres rather than any allergenic effect. Taken together, these clinical and dermatological findings strongly support the safety and potential benefits of superfine Merino wool for individuals with sensitive skin or AD. They challenge outdated perceptions of wool intolerance and reaffirm its suitability for next-to-skin applications.

3.1.2. Sleep Quality

Emerging evidence supports wool's role in enhancing sleep quality, particularly when used in next-to-skin applications such as sleepwear. Clinical studies using polysomnography have demonstrated that Merino wool sleepwear can positively influence various aspects of sleep. In a controlled study with young healthy adults (aged 20–31 years), Shin et al. [22] revealed that at a lower ambient temperature (17 °C), wool sleepwear reduced sleep-onset latency compared to cotton. The trial participants spent more time in deep sleep (N3 stage) than at a higher temperature (22 °C). These benefits are attributed to wool's ability to buffer humidity and insulate against temperature fluctuations, helping to create a dry, thermo-neutral microenvironment. No effects were observed for woollen bedding material, suggesting that sleepwear, due to its direct skin contact, has a more pronounced influence on sleep-related thermal comfort.

Similar benefits were observed in other adults (aged 50–70 years) and individuals with poor sleep quality. Chow et al. [23] reported that under warm conditions wool sleepwear resulted in shorter sleep-onset latency, less sleep fragmentation, and reduced wakefulness compared to cotton and polyester. Wool sleepwear promoted improved sleep, particularly for those aged 65 years and older, and for poor sleepers in warm ambient conditions.

A recent review by Li et al. [24] examined studies comparing fibre types in sleepwear and bedding and reported that wool sleepwear improved sleep onset in adults in cool conditions and in older adults in warm conditions. While study designs varied in temperature conditions, fibre types, and participant groups, the review highlights wool's potential to support better sleep through its thermal and moisture buffering properties. Together, these findings suggest that wool sleepwear contributes to improved sleep quality, especially in cooler or more challenging environments, and among individuals with heightened sleep sensitivity.

3.2. *Environmental Sustainability*

In addition to its contributions to human health and wellbeing, wool also offers a range of environmental advantages that reinforce its role as a fibre that is both renewable and biodegradable. Sheep continuously grow fleece, which is typically shorn off once a year. This production relies on chemical energy derived from photosynthesis that is stored in plant biomass. Sheep convert this biomass into wool, which contributes to biogenic carbon sequestration, and it offers short- to medium-term carbon storage that helps offset emissions associated with production. In this respect, one industry report states that approximately 50% of the dry weight of wool is carbon, captured from atmospheric CO_2 through photosynthesis and stored in the fibre as sheep consume pasture [25], and that each kilogram of clean wool stores approximately 1.8 kg of CO_2 equivalents.

This is, however, a simplification of a much more complex situation wherein multiple factors affect carbon sequestration and greenhouse gas (GHG) release, including, but not limited to, the variation in farming systems, sheep breed effects, allocation of the emission footprint between meat, hides, lanolin, milk, and wool production, variation in data from different regions, and variation in assessment methodology. This is discussed in some detail by the International Wool Textile Organisation (IWTO; Brussels, Belgium), the global authority for standards in the wool textile industry, and in recent research by Wiedemann et al. [26] which highlights the challenge in undertaking a life cycle assessment (LCA) for a woollen product.

While methane from sheep is a potent GHG, there is debate around the use of different metrics for assessing climate impacts for what is considered to be a short-lived climate pollutant (unlike CO_2), and with a recent suggestion that a new metric called GWP* should be used instead of the current standard Global Warming Potential (GWP) [27].

Biodegradability is another environmental advantage of wool. Composed primarily of proteins, the fibres break down in terrestrial environments, typically decomposing within a few months in soil [28], although this depends on chemical processes applied during manufacturing, including dyeing, bleaching, stain-repellence, and various shrink-resist treatments. During this breakdown, wool releases nitrogen, sulphur, phosphorus, and carbon back into the ecosystem. In contrast, synthetic fibres can persist in the environment for centuries, with this contributing to plastic pollution and ecological harm. Unlike synthetic textiles such as polyester or acrylic, wool textiles biodegrade rapidly, and the fibre does not directly contribute to microplastic pollution in aquatic environments [29]. This makes it a safer choice for both terrestrial and aquatic environments, albeit the processing of wool fibre and manufacture of wool products can involve steps that can have negative environmental impacts.

This natural biodegradation creates a closed-loop biological cycle: plants capture solar energy and atmospheric CO_2 to produce biomass, which sheep consume and convert into wool. When wool is discarded, it biodegrades in the soil, releasing carbon, nitrogen, sulphur, and phosphorus that can resupply minerals to growing plants. This creates a biologically based circular or regenerative economy.

In addition, wool's durability and longevity further strengthen its environmental credentials. Wool garments tend to be worn longer, need washing less frequently, and maintain their performance and appearance better than many fast fashion materials. Wool garments can often be refreshed by airing rather than washing due to the fibre's natural soil repellence and odour inhibiting properties [30], and washing is typically undertaken at lower temperatures and using shorter and gentler washing cycles. These qualities reduce the need for frequent replacement and lower environmental impact.

Together, these factors position wool as one of the more environmentally responsible fibres available, aligning with values underpinning regenerative agriculture, circular material flows, and climate-conscious design. Despite these advantages, it does face competition from synthetic fibres, which currently dominate global textile production due to their low cost, versatility, uniformity, strength, elasticity and ease of large-scale manufacturing. However, this dominance comes with environmental costs, including reliance on non-renewable petroleum resources, high energy inputs, high water footprints and microplastic pollution. This highlights wool's potential as a more sustainable alternative.

4. Structural Basis of Wool's Functional Properties

The functional properties of wool fibres arise from their intricate structure, which has been described and illustrated in detail elsewhere, for example, by Caven et al. [31] and Rippon [32]. This structure confers multifunctionality that is difficult to replicate with synthetic fibres. Understanding this structure is essential to appreciating the diverse performance advantages that wool fibres offer.

4.1. The Cuticle: The Surface Structure and Protective Outer Layer

The outermost layer of wool fibres is the cuticle, a sheath of overlapping, scale-like cells that constitute approximately 10% of the fibre's mass. These scales overlap in the longitudinal direction like tiles on a roof, with their edges pointing from the fibre root to its tip [32,33] (Figure 1).

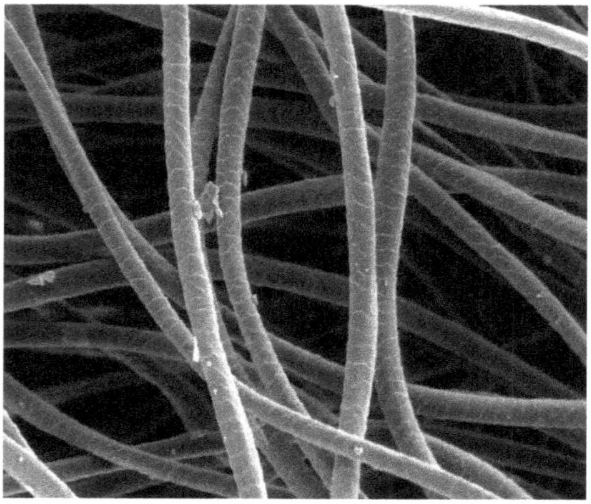

Figure 1. Electron microscopic image of scoured Merino wool fibres with a mean fibre diameter of 19.5 μm, showing the over-lapping cuticle scales.

The cuticle is composed of two major layers: the outer exocuticle and the inner endocuticle. In Merino wool, the exocuticle is approximately 0.3 μm thick, accounting for about 60% of the cuticle volume. It includes a dense outer region known as the A-layer and contains a substantial portion of the cystine present in the cuticle. The endocuticle of Merino wool, approximately 0.2 μm thick and making up about 40% of the whole cuticle, has a lower cystine content and is considered more amorphous and susceptible to chemical or enzymatic attack [32]. Surrounding each cuticle cell is the epicuticle, a thin (2–7 nm thick) proteinaceous membrane covalently bound to a monolayer of lipids, predominantly 18-methyleicosanoic acid (18-MEA; a branched-chain fatty acid). This lipid layer is approximately 0.9 nm thick and plays a key role in making wool fibres water- and dirt-resistant [32].

The cuticle structure of wool fibres varies with fibre diameter. In fine wool fibres, such as those of Merino sheep, the cuticle is typically a single cell layer thick. This single-layer arrangement primarily envelops the orthocortex, but in regions adjacent to the paracortical cells, two to three cuticle layers can be observed. In contrast, coarse wool fibres possess a thicker cuticle, which can comprise up to 15 layers of cells [23]. This increased cuticle thickness in coarse wool contributes to greater fibre rigidity and enhances resistance to external factors, but reduces fibre flexibility and softness.

These structural characteristics underpin several key functional properties of wool [32,33]. The overlapping scale arrangement creates directional friction, enabling felting and self-cleaning properties. The gaps between these scales, despite the hydrophobic surface, allow water molecules to diffuse into and out of the fibre, facilitating moisture movement. Additionally, the multi-layered cuticle and high cysteine content enhance abrasion resistance, with this contributing to the fibre's mechanical durability.

4.2. The Cortex: Internal Architecture and Contribution to Fibre Performance

Beneath the cuticle lies the cortex, which makes up approximately 90% of the fibre's volume and features a sophisticated composite architecture. The cortex is primarily composed of two types of cortical cells, orthocortical and paracortical, with intermediate mesocortical cells occasionally present [32].

In fine wool fibres, such as those of Merino sheep, these cortical cells are arranged side-by-side in semi-cylindrical arrangements that twist around each other along the fibre length [32]. In contrast, coarser wool fibres, such as those from Lincoln sheep, tend to exhibit a concentric arrangement of cortical cells, with orthocortical cells typically located centrally. Some coarse animal hairs contain only orthocortical cells, while others, particularly in very coarse fibres, contain a central medulla, an air-filled cavity that enhances the fibre's insulation properties [32]. Medullation is often observed in wool fibres of larger diameter.

While the mechanisms responsible for crimp formation remain incompletely understood, one plausible explanation is that crimp arises primarily from the bilateral arrangement of ortho- and paracortical cells, combined with differences in their physical and chemical properties. These differences, such as unequal expansion during keratinisation [34], variations in relative longitudinal cell lengths [35], and distinct responses to moisture, can generate internal stresses between the two cell types. When arranged bilaterally rather than concentrically, these stresses are asymmetrically distributed across the fibre cross-section, potentially leading to bending along the fibre axis. This process may give rise to the characteristic three-dimensional curvature observed as crimp, with the orthocortex typically oriented toward the outside of each crimp wave and the two cortical segments twisting along the fibre in phase with the crimp pattern [32].

This crimp structure plays a critical role in wool's functional performance. It introduces bulk, compressional resilience, and elasticity, which enhance comfort and durability in textile applications. The repeating waves of crimp create spaces between fibres, trapping air and contributing to thermal and acoustic insulation by reducing the transfer of heat and sound.

4.3. Microstructural Organisation Within Cortical Cells

Within cortical cells, larger structural units known as macrofibrils are aligned longitudinally along the fibre axis and can be further subdivided into finer microfibrils. In the orthocortical cells the macrofibrils are densely packed and arranged in a helical pattern, whereas in the paracortical cells, they tend to be more loosely organised and exhibit a pseudo-hexagonal packing aligned parallel to the fibre axis [32].

The microfibrils are composed of α-helical keratin intermediate filaments (KIFs), each formed by coiled-coil heterodimers of type I and type II keratin polypeptide chains. These wool keratins are orthologous to the 'hair keratins' of humans and other mammals, but because they are specifically expressed in wool fibres, they are referred to here as 'wool keratins', to distinguish them from the keratins expressed in other epithelial tissues [36]. Two keratin heterodimers align in an antiparallel arrangement to form a four-chain unit, which further assembles into larger rod-like intermediate filaments with a telescopic configuration [32]. These structures provide tensile strength and internal cohesion, enabling the fibre to withstand stretching and mechanical stress.

Surrounding the KIFs is a matrix composed of keratin-associated proteins (KAPs), including high-sulphur (HS), ultra-high sulphur (UHS), and high-glycine/tyrosine (HGT) KAPs. These proteins interact with each other and with KIFs through covalent and non-covalent mechanisms, including disulphide and isopeptide bonds, hydrogen bonding, hydrophobic interactions, and ionic forces [32]. This proteinaceous KAP matrix fills the space between KIFs, contributing to the cohesion and plasticity of the fibre and modulating its mechanical properties. Covalent cross-links enhance strength and durability, while non-covalent interactions support flexibility and energy dissipation. This collectively contributes to wool's resilience and its ability to recover after deformation.

The organisational differences between the ortho- and paracortical cells are accompanied by variations in the protein composition. A key difference lies in the relative abundance of KIFs and KAPs. Orthocortical cells contain a higher ratio of KIFs to matrix, while paracortical cells have a greater proportion of matrix [37]. Differences in KAP group profiles between these two cell types have also been reported, with HGT-KAPs more prevalent in orthocortical cells and UHS-KAPs more abundant in paracortical cells, but this pattern is not consistently observed across studies [38]. However, because only a limited subset of the KAPs were characterised at the time of these studies, the trends should be regarded as suggestive and thus be interpreted with caution.

These microstructural and biochemical differences play a vital role in defining the physical properties of the fibre. The configuration and interaction of KIFs and KAPs influences mechanical properties such as strength, extensibility, and resilience. Wool proteins contain polar and charged amino acid residues which facilitate hydrogen bonding and ionic interaction with water molecules [1]. This amphoteric nature underpins wool's high moisture regain. The amorphous regions within the matrix accommodate water and small molecules without compromising structural integrity, while the flexible protein network allows for dynamic moisture uptake and release. Together, these contribute to wool's effective buffering of temperature and humidity.

Wool proteins are rich in amino acids bearing functional groups capable of interacting with a wide range of substances. In particular, cysteine, aspartic acid, glutamic acid, serine, and threonine are present in high abundance, while tyrosine is moderately abundant (3.9 mol%) [39]. The thiol group in cysteine, the carboxyl group in aspartic and glutamic acids, the hydroxyl group in serine, threonine, and tyrosine, along with the aromatic ring in tyrosine together enable wool to bind a range of substances, including odorous molecules, VOCs, and heavy metals. This molecular composition underpins wool's capacity to sequester environmental pollutants and odours. The high nitrogen (approximately 16 wt%) and sulphur (approximately 5 wt%) content of wool proteins [40] contributes to its inherent flame retardancy by promoting char formation and suppressing melting and dripping during combustion.

Altogether, this sophisticated structural design bestows wool fibre with a range of novel performance properties. It also underpins many of wool's unique health-related and wellbeing benefits. These attributes contribute to making wool fibre a remarkable material, one that rightfully earns its title as 'nature's wonder fibre' [41].

5. Molecular Genetics of Wool Fibre Traits

Despite its functional value, wool suffers from variability in fibre traits including diameter, curvature, length, and strength. This phenotypic variation reduces fibre uniformity, complicates processing, and affects the commercial value of wool. These traits arise from intricate and hierarchical molecular processes during wool follicle development and fibre growth.

Various genes and metabolic systems have been revealed to affect wool growth. These systems can be separated into those that operate prior to, or during wool follicle activity, and those that are directly involved in producing the structural components of the fibre. Genes (and the pathways to which they contribute) of interest in the former group include, but are not limited to, the Wnt/β-catenin (the gene name *WNT* is a portmanteau of "Wingless" and "Int-1") [42,43], fibroblast growth factor (FGF) signalling [43,44], Sonic Hedgehog (Shh) [42], TGF-β/BMP signalling [42,45], Notch signalling [42,46], Hippo, PI3K-Akt, Rap1, NF-κB, and cAMP pathways [47], androgen/cortisol–Wnt crosstalk, and transcription factors as well as alternative splicing regulators [48]. These are not the focus of this review; the emphasis is instead on the keratin and KAP genes that are expressed in the wool follicle.

The expression of keratin genes (*KRTs*) and keratin-associated protein genes (*KRTAPs*) produces the proteins that form the main structural components of the fibre. A detailed understanding of the expression patterns and phenotypic effects of these genes is critical to developing effective strategies to improve wool fibre qualities.

5.1. Wool Keratin Genes

The primary structural framework of wool fibres (as described above) is built from KIFs. These KIFs are formed by type I (acidic) and type II (neutral-basic) wool keratins [49], which are encoded by a subset of *KRT* genes that are active in the wool follicle. In sheep, at least 17 *KRTs* are reported to contribute to wool fibre structure, including 10 Type I (*KRT31*, *KRT32*, *KRT33A*, *KRT33B*, and *KRT34-KRT40*, excluding *KRT37*) and 7 Type II (*KRT81-KRT87*) genes, each with distinct expression profiles within the wool follicle [50–53]. The type II genes are clustered on chromosome 3, while the type I genes are located on chromosome 11, flanking the *KRTAP* gene cluster (two on one side and the remaining eight on the other).

The expression of wool *KRTs* is regulated both spatially and temporally. Some are expressed in the cuticle (e.g., *KRT40*, *KRT82*, and *KRT84*), or the cortex (e.g., *KRT31*, *KRT33A*, *KRT33B*, *KRT34*, *KRT36*, *KRT38*, *KRT39*, *KRT81*, *KRT83*, *KRT86*, and *KRT87*), while others are expressed transiently in both fibre structures (e.g., *KRT32*, *KRT35* and *KRT85*) [53]. The initial shared expression of selected *KRTs* likely supports early filament assembly and structural continuity and precedes divergence into the compartment-specific *KRT* expression that defines the final fibre architecture. In the cortex, there is co-expression of multiple type I and type II *KRTs* [53], with this facilitating the formation of a diversity of heterodimers. Given that the cortex constitutes the majority of the fibre volume, this molecular diversity likely plays a key role in shaping fibre structure and trait variability.

Despite the compartmentalised expression of the *KRTs* in the cuticle and cortex, to date no studies have reported differential expression of *KRT* genes between the ortho- and paracortex. This is in contrast to the *KRTAPs*, which exhibit region-specific expression. The apparent absence of differential *KRT* gene expression between the cortical regions suggests that their expression may be relatively stable and uniform across the cortical structures, with this likely reflecting functional constraints related to their essential role in maintaining fibre integrity. Although subtle variation in *KRT* gene expression may exist, it may be less pronounced than the greater variability and typically lower expression levels observed for *KRTAP* genes. However, further studies are needed to determine whether this reflects true uniformity, or whether it reflects current methodological limitations in measuring *KRT* expression in the various structures of the wool fibre.

To date, genetic variation has been examined in 10 of the 17 known wool keratin genes: six type I genes (*KRT31*, *KRT32*, *KRT33A*, *KRT34*, *KRT36*, and *KRT38*) [54–58] and four type II genes (*KRT81*, *KRT83*, *KRT84*, and *KRT85*) [57,59–62]. However, most of these investigations have only focused on parts of these genes, leaving portions of them unexplored. Consequently, the full extent of their nucleotide sequence variation is poorly understood.

All the wool keratin genes investigated to date exhibit genetic variation, primarily in the form of single nucleotide polymorphisms (SNPs). These SNPs occur in both the coding and non-coding regions. They include non-synonymous SNPs that change the amino acid sequence and potentially alter protein structure, and synonymous or non-coding SNPs that may influence gene expression or mRNA processing. All of these nucleotide sequence differences could contribute to variation in wool fibre traits.

Nucleotide sequence variation in several type I and type II keratin genes has been associated with a range of wool traits, supporting their functional relevance. For example, in Southdown × Merino-cross sheep, variation in *KRT31* has been linked to greasy fleece weight (GFW), clean fleece weight (CFW), and mean staple length (MSL) [55], while variation in *KRT81* has been associated with GFW and CFW [61].

5.2. Keratin-Associated Protein Genes

The KAPs form the matrix that surrounds and cross-links the keratin filaments within the wool fibres. Unlike the larger, intron-containing *KRTs*, the *KRTAPs* are typically small, intronless and exhibit greater sequence diversity. In sheep, the *KRTAP* repertoire has expanded to contain at least 102 genes [36], more than the 89 reported in humans [63–65]. While ovine and human *KRTAPs* are typically clustered and similarly arranged, species-specific differences in clustering patterns exist. For example, humans possess two *KRTAP5* clusters on the same chromosome [63], whereas sheep appear to have one cluster. The *KRTAP10* and *KRTAP12* genes form a single cluster in humans [63], while in sheep they appear to be split into two separate clusters, albeit located on the same chromosome. The HGT-*KRTAP* cluster also differs between the two species, spanning approximately 590 kb in humans and around 890 kb in sheep (our unpublished data). This expansion in size reflects the presence of additional HGT-*KRTAP* genes in sheep [66–69].

The *KRTAP* genes are broadly categorised into three major groups (HS, UHS and HGT), based on the amino acid composition of the proteins they encode, and further divided into twenty-six families based on nucleotide sequence similarity. Of these, twelve families (KAP1-KAP3, KAP11, KAP13, KAP15-KAP16, KAP23-KAP24, KAP26-KAP28) belong to the HS group, six families (KAP4-KAP5, KAP9-KAP10, KAP12, KAP17) belong to the UHS group, and eight families (KAP6-KAP8, KAP19-KAP22, KAP36) belong to the HGT group. The number of gene members within each family varies widely, with some having one single gene, while others, such as KAP5 and KAP10, have up to 12 members.

The *KRTAPs* exhibit spatial and temporal expression patterns during wool fibre development, both complementing and diverging from the expression patterns of the *KRTs*. However, our current knowledge is based on the limited subset of *KRTAPs* identified prior to the 1990s and involves only some members of the KAP1–KAP8 families. Among these, HGT-*KRTAPs* (e.g., KAP6–KAP8 families) are expressed early in fibre development, shortly after *KRT* expression begins. They are primarily restricted to the orthocortical region. Subsequently, HS-*KRTAPs* (e.g., KAP1–KAP3 families) appear in the cortical cells in the region complementary to that expressing KAP6-KAP8, but their expression soon extends across the entire cortex [37]. In contrast, UHS-*KRTAPs* (e.g., KAP4-KAP5 families) are expressed slightly later, with KAP4 expression confined to the paracortical cells [70], and KAP5 expression occurring late in cuticle cell differentiation [58,59,71,72]. These compartment-specific expression patterns likely contribute to the asymmetric architecture of wool fibres, influencing traits such as curvature and crimp.

Due to their small size and intronless nature, the *KRTAPs* are amenable to analysis of their entire coding sequence. All of the genes investigated to date exhibit polymorphism, although the extent and nature of that polymorphism vary between genes [73]. It must also be remembered that, of the billion plus sheep globally, the *KRTAPs* have only been studies in a few individuals and breeds. The gene polymorphisms in the *KRTAPs* include SNPs and insertion/deletions (indels), with allele numbers ranging from 2 to 14 per gene [73]. Most indels are located within the coding regions, and usually within or adjacent to tandem repeat elements. They typically preserve the reading frame, but an exception is observed for *KRTAP28-1*, where the indel variation involves a variable number of dinucleotide repeats, resulting in a frameshift near the C-terminus of the encoded protein [74].

While at times considered benign, the SNPs in the flanking regions and synonymous SNPs may affect gene expression or mRNA processing, while non-synonymous SNPs can lead to changes in amino acid sequences. Indels, when present, can cause more substantial structural alterations to the encoded proteins. Collectively, these variations can influence the cross-linking potential and matrix composition, thereby potentially affecting wool fibre characteristics.

Variation in many *KRTAPs* has been associated with a variety of wool fibre traits, reinforcing their functional importance. For example, in Southdown × Merino-cross sheep, variation in *KRTAP6-1* and *KRTAP6-3* have been associated with variation in mean fibre diameter (MFD) related traits [75,76]. Variation in *KRTAP8-1* has been associated with variations in mean fibre curvature (MFC) and mean staple strength (MSS) [77], and variation in *KRTAP15-1* has been associated with wool yield and fibre diameter standard deviation (FDSD) [78].

Early studies have provided valuable insights into the expression patterns and sequence variation in some *KRTAPs*, but many ovine *KRTAPs* have yet to be characterised, and research in this area has been limited. Renewed effort to investigate *KRTAPs*, including their identification, expression profiles in the wool follicle, genetic variation, and functional roles, are essential to advancing our understanding of the genetic basis of wool fibre variation. This effort must include the study of many more sheep, and of a wider range of breeds and wool types.

5.3. Integrated Effects of Keratin and KAP Genes

The structural components of wool fibre, keratins and KAPs, are encoded by numerous genes. Their genetic diversity underpins the polygenic nature of wool traits, where fibre characteristics likely result from the combined effects of many genetic loci. In this respect, it is notable that all the wool *KRTs* and *KRTAPs* investigated to date in sheep are polymorphic, and they often possess multiple alleles [36]. The aggregated variation across both the *KRTs* and *KRTAPs* therefore likely contributes to the wide phenotypic diversity observed in fibre traits. That is, most wool traits appear to be influenced by multiple genes, each with small to moderate effects. To better understand and improve fibre traits, a comprehensive evaluation of the full catalogue of *KRTs* and *KRTAPs* is therefore essential, particularly to identify those genes with larger effects.

Interestingly, *KRT* and *KRTAP* genes located in close proximity to each other often show different trait associations. This suggests limited linkage disequilibrium and functional divergence. For example, among the type I wool *KRTs*, *KRT31* and *KRT34* are found in close proximity on chromosome 1, but variation in them shows different associations with wool traits in the same flock: *KRT31* variation is linked to GFW, CFW, and MSL [55], whereas *KRT34* variation is associated with MFD, FDSD, and MSL [54]. A similar pattern is seen among nearby type II wool *KRTs*: variation in *KRT83* is associated with MFD, FDSD, coefficient of variation in fibre diameter (CVFD), MFC, prickle factor (PF), and yield [59], while *KRT85* is linked to GFW, CFW, and PF in the same population [60].

The same phenomenon appears to occur with the *KRTAPs*. For example, *KRTAP21-1* is associated with wool yield [79], whereas the adjacent *KRTAP21-2* is linked to MSL [80]. Within the KAP13 family, *KRTAP13-2* is associated with MFD and CVFD in heterotypic hair fibres from Chinese Tan sheep, but variation in *KRTAP13-4* has no such association [81].

Taken together, these observations suggest that genes that are physically close on the chromosome may nevertheless have distinct functions. It also suggests that the multiplicity of *KRTs* and *KRTAPs* in sheep may not reflect functional redundancy, but instead they have evolved specialised or divergent roles in fibre development and structure. Accordingly, both the individual and combined effects of these genes must be considered in fibre trait studies.

Gene-by-trait associations can also vary across sheep breeds or populations, highlighting the influence of genetic background. For example, *KRT33A* is linked to fleece weight, yield, MSL, MFC, crimp frequency, and core bulk in Perendale sheep [58]; to FDSD and MSS in Merino and Merino-cross sheep [82]; and to CFW, MFD, MSL, and MSS in several Egyptian sheep breeds [83]. Similarly, *KRTAP20-1* is associated with GFW, wool yield, MFD,

FDSD, and PF in Southdown × Merino-cross sheep [84], but with MFC in the fine wool of Chinese Tan sheep [85]. These findings highlight the importance of considering breed- or population-specific genetic contexts, particularly variation in other *KRTs* and *KRTAPs*, in association studies.

The physical clustering of *KRT* and *KRTAP* genes can lead to linkage effects, where neighbouring genes may be co-inherited due to linkage disequilibrium. However, findings in sheep suggest that linkage disequilibrium within some clusters is weak, as neighbouring loci often exhibit different associations with wool traits (see discussion above). This pattern, which is somewhat unexpected given their physical proximity, might be explained by recombination or gene conversion events that disrupt linkage disequilibrium. While direct evidence of this remains limited, sequence analyses have suggested that such mechanisms may have occurred with the KAP1 gene family [86,87].

Tighter linkage may however exist in some clusters. Evidence from Longdong cashmere goats suggests that several physically adjacent genes are associated with the same traits. For example, three adjacent genes, *KRTAP22-2*, *KRTAP6-5*, and *KRTAP6-2*, have all been associated with MFD for goat fibres [88–90]. Similarly, *KRTAP28-1* and *KRTAP24-1* (also physically adjacent) have both been linked to goat fibre MFD [91,92]. These findings suggest that some gene clusters may exhibit coordinated associations with fibre traits, possibly reflecting localised linkage disequilibrium. Overall, more extensive investigation across gene clusters is needed to better understand these patterns and their implications for fibre traits.

From a breeding perspective, the polygenic and potentially linked architecture of wool traits presents both challenges and opportunities. The small effects of individual genes limit the impact of single-gene selection, making multi-loci approaches more appropriate for capturing the underlying genetic variation. Weak linkage among some clustered genes offers flexibility, allowing improvement of specific traits without unintended effects on others. Conversely, tight linkage among loci can facilitate selection when favourable alleles co-occur on the same extended haplotype but may hinder progress if favourable alleles are linked with unfavourable ones. In such cases, selection strategies must evaluate and manage these trade-offs.

Overall, the genetic architecture underlying wool traits reflects a complex interplay of multiple genes, physical clustering, recombination, and gene-specific effects. Broader and deeper genomic analyses across entire gene clusters and diverse breeds will be essential for understanding this complexity and enabling more precise and effective genetic improvement strategies.

6. Genetic Strategies for Improving Wool Quality

Wool fibres display variation in traits that affect fibre uniformity and consequently product quality. Fortunately, many important wool traits have moderate to high heritability (0.3 to 0.6+) [93,94], meaning that a substantial proportion of phenotypic variation is genetically determined and can be improved through genetic approaches, including selective breeding.

To date, this selection has progressed from simply selecting rams with desirable fibre traits for breeding, in the hope that these will be passed on to progeny, to a more precise approach using estimated breeding values (eBVs) and selection indices based upon selected and weighted eBVs for specific traits. In this respect, the most widely used scheme for wool is the Australian MERINOSELECT system [95]. This system uses what are referred to as ABVs (animal breeding values; but eBVs by definition), which are statistically derived estimates of an animal's genetic merit for specific traits, such as MFD and CFW. These ABVs are combined into selection indices such as the Fine Wool (FW) or Wool Production (WP) indexes. The system allows breeding sires (and ewes) to be rated for their wool production potential, be it for use on any given Merino sheep stud for breeding, or when selling sheep genetics.

A 2024 summary of MERINOSELECT [96] describes the genetic gains made since 2000 across three classes of wool sheep breeders (low-micron, mid-micron, and high-micron), noting that the low-micron breeders have reduced MFD and improved gastrointestinal parasite resistance; the mid-micron breeders have had 'balanced improvements across multiple traits including fat, muscle, fleece weight, fibre diameter, reproduction and wrinkle'; and the high-micron breeders have 'achieved significant gains in fat and muscle but at the expense of fleece weight, micron, wool colour, and shear force'.

In contrast to eBV-based selection, two main DNA-based strategies have been pursued to improve wool traits: genetic modification and marker-assisted selection. While genetic modification depends on understanding the molecular genetics and regulatory mechanisms underlying wool fibre traits, marker-assisted selection relies on identifying robust genetic markers linked to desirable phenotypes, even if the underlying mechanisms are not fully understood.

6.1. *Genetic Modification: High Potential, Limited Progress*

Genetic modification (also known as genetic engineering) involves the manipulation of an organism's DNA to alter specific traits. In the context of wool production, this approach holds promise for improving fibre quality by targeting genes that influence the structural and mechanical properties of the wool fibre. Unlike marker-assisted selection, which works within the limits of naturally occurring genetic variation, the approach allows for the direct modification of individual genes and enables the introduction of novel genetic changes that may not be achievable through natural selection. The most relevant targets include genes encoding wool fibre structural proteins, as well as other proteins involved in fibre development, protein cross-linking, and biochemical modifications that affect fibre properties.

6.1.1. Early Transgenic Experiments

During the 1990s and early 2000s, a series of pioneering transgenic experiments in mice and sheep explored the potential to improve fibre quality. Initial studies in transgenic mice investigated the effects of altering the expression of hair protein genes. One early experiment introduced the type II wool keratin gene *K2.10* (now *KRT83*) under control of its native promoter [97]. Transgenic mice expressing various levels of this gene displayed a range of hair phenotypes, from normal to distorted and fragile hairs, with this depending on the transgene copy number [97,98]. High levels of expression caused structural defects and hair loss, suggesting the dosage sensitivity of hair protein networks and the need for balanced expression to maintain fibre integrity.

Subsequent studies examined the effect of exchanging promoters between wool protein genes. When a *KRTAP2* gene, normally expressed late in fibre formation, was driven by the *KRT83* (formerly *K2.10*) promoter, its protein was expressed earlier than usual in the cortex. Several mouse lines exhibited variable hair phenotypes, ranging from normal to unusual, with two lines showing hair loss and one displaying a striking stubble-like coat after the first hair cycle [37,99]. Ultrastructural analyses revealed abnormal accumulation of the KAP2 protein [99], reinforcing the important roles of both protein abundance and temporal regulation of expression during fibre development.

Further experiments used the *KRT83* promoter to drive expression of a non-hair protein, SV40 T-antigen, in the mouse hair cortex. This caused severe hair deformation and fragility [100], likely due to interference with endogenous transcription factors such as AP-2, as many hair *KRT* and *KRTAP* gene promoters contain putative AP-2 binding sites [37]. Together the above findings highlight the importance of precise and appropriate transcriptional regulation of fibre structural genes to ensure normal hair development and integrity.

Transgenic sheep overexpressing *KRTs* and *KRTAPs* have been generated to study the impact of altered gene expression on wool fibre characteristics. Although these transgenes were successfully expressed in wool follicles, the resulting phenotypes were generally unfavourable. For example, overexpressing *KRT83* (formerly *K2.10*) resulted in altered fleece traits such as reduced crimp, increased lustre, and elevated lanolin content, while overexpression of *KRTAP4-2*, *KRTAP5-1*, or *KRTAP6-1* (previously *KAP4.2*, *KAP5.1*, or *KAP6.1*, respectively) reduced fibre strength [101–103]. These results suggest that even when transgenes are targeted to wool follicles, ectopic or excessive expression can disrupt the delicate balance of structural protein interactions required for normal fibre development. They also highlight the importance of finely regulated gene expression in maintaining wool quality.

Collectively, these early genetic modification experiments demonstrated that even modest disruptions in the expression level or timing of wool structural gene expression can compromise fibre quality. They suggest that the keratins and KAPs function within an interdependent and dosage-sensitive network, and any imbalance in that network could lead to weakened fibres, abnormal morphology, or other defects.

Other attempts to enhance wool production via genetic modification have focused on the overexpression of growth-regulatory genes, such as the ovine insulin-like growth factor 1 and the growth hormone genes. While some transgenic lines showed modest increases in wool production, the results were often inconsistent and sometimes accompanied by adverse health effects [104–106]. In this respect, while studies in mice involving manipulation of key signalling pathways, such as the Wnt/β-catenin, Shh, Bone Morphogenetic Protein, and EDA/EDAR pathways, have provided valuable insights into follicle development and density regulation [107–110], applying these findings to improving wool in sheep remains challenging. Although the core follicular regulatory pathways appear to be conserved across species, the biological differences between mice and sheep limit the direct translation of findings from one species to the other.

Despite early proof-of-concept experiments, progress with genetic modification has been limited due to the molecular complexity of wool fibre development and our incomplete understanding of the genetic and regulatory mechanisms involved. This highlights the need to better define the pathways governing fibre trait development and variation, and to identify critical gaps in our knowledge.

6.1.2. The Potential of CRISPR/Cas Editing for Wool Trait Improvement

Clustered Regularly Interspaced Short Palindromic Repeats (CRISPR) refers to a natural bacterial immune system that has been adapted into a powerful genome-editing technology. It functions together with a Cas nuclease (commonly Cas9), which is directed to a specific DNA target by a single-guide RNA (sgRNA). When the Cas9–sgRNA complex binds the target DNA, Cas9 introduces a double-stranded break. The cell repairs this break either by non-homologous end joining, which can introduce insertions or deletions, creating gene knockouts, or by homology-directed repair, which allows insertion or precise editing using a supplied DNA template. This makes CRISPR/Cas gene targeting specific, flexible, fast, and cost-effective.

The CRISPR/Cas editing approach has been used to engineer the loss of *FGF5* function, resulting in increased wool staple length in sheep [111]. While capable of precisely targeting genes, as with the more rudimentary transgenesis experiments described above, care is needed to ensure that the genes targeted are not pleiotropic, and that the engineering does not raise ethical or welfare concerns. In most jurisdictions, genetic engineering is regulated, and consumer acceptance of these practices is often an issue.

6.1.3. Molecular Complexity and Knowledge Gaps in Wool Genetic Modification

Although the early transgenic experiments provided valuable insights into the roles of structural and regulatory genes, they also revealed the intricate nature of the wool fibre system. Wool fibre development is governed by the coordinated action of numerous genes, including many wool *KRTs* and a diverse catalogue of *KRTAPs*. These genes are expressed in tightly regulated spatial and temporal patterns during follicle development [53], with this indicating that fibre formation relies on a finely tuned gene expression programme. It appears that the timing, location, and relative levels of expression for various genes must be coordinated to produce normal fibre structure and function.

For genetic modification to be used in future to modify wool traits, it may therefore be necessary to first identify which genes influence specific fibre characteristics and to understand the extent to which gene expression levels can be altered without disrupting the integrity of the fibre. Importantly, not only does the absolute expression level of any given protein appear to be crucial, but so too does the balance between different structural proteins. Disruptions in this balance, as seen in early transgenic experiments involving overexpression of *KRT83* or various *KRTAP* genes [101–103], have led to abnormal fibre morphology and compromised fibre integrity. These findings highlight the importance of maintaining coordinated expression.

What is more, achieving this level of precision will undoubtedly be challenging. Many *KRTs* and *KRTAPs* are clustered in the genome [73], raising the possibility of co-regulation of individual genes or the presence of shared regulatory elements. Attempts to manipulate individual genes may therefore affect adjacent genes, or broader regulatory networks, and complex genomic organisation may therefore present a barrier to targeted genetic modification.

Further complexity also likely arises from post-translational modifications. For example, differential phosphorylation of keratins and KAPs has been observed between crimped and straight wool fibre [112]. This suggests that some wool traits may be epigenetically regulated, something that may be difficult to control by genetic modification. What is more, wool traits are typically affected by a wide range of environmental factors, not least the effect of nutrition, disease, and photoperiod. Although these factors are beyond the scope of this review, they cannot be ignored in the quest to obtain more consistent wool fibre.

Taken together, these biological and technical challenges are compounded by our limited understanding of the molecular and regulatory landscape governing wool fibre development. Many *KRTAP* genes remain unidentified, and the roles of the *KRT* and *KRTAP* genes are still unclear. Available gene expression data are sparse, and mechanisms controlling gene expression, post-transcriptional processes, and the effects of environmental factors beyond the level of phenotype are largely unknown.

These conceptual challenges highlight the need for a much deeper understanding of the gene networks, expression patterns, regulatory mechanisms, and protein interactions involved in wool fibre development. Alongside these knowledge gaps, technical limitations in gene manipulation and expression control further complicate efforts to achieve successful genetic modification of wool traits. We conclude that more foundational research is required before genetic modification approaches can be effectively applied to wool trait improvement.

6.2. Genetic Selection: Practical Approaches, Ongoing Challenges

When compared to genetic modification, marker-assisted selection [113] offers a more practical and readily adoptable strategy for improving wool traits. It is a breeding method that uses DNA markers (such as SNPs, natural variation in genes, or microsatellite sequences) that are linked to, or within genes that control traits of interest. Instead of relying only on observable traits or phenotypes, breeders can test animals at the DNA level to determine whether they carry favourable alleles, which can then be selected for breeding.

Marker-assisted selection relies on capturing the benefit of naturally occurring genetic variation rather than direct manipulation of the genome. This makes it compatible with existing breeding programmes that utilise phenotypic selection and/or eBVs. Unlike genetic modification, it generally does not raise public and consumer ire. However, its success will depend on multiple factors, from the biological relevance of gene–trait associations to how genetic variation is analysed, and then how validated markers are integrated into breeding strategies.

Genomic BLUP (gBLUP) is a statistical method for genomic selection in breeding programmes that uses SNP marker data (omics data) to predict the genetic merit of individuals for complex traits [114,115]. In the simpler best linear unbiased predictor (BLUP) approach to genetic analysis, which is informed by phenotypic data, the method replaces the traditional pedigree-based relationship matrix with a genomic relationship matrix calculated from SNP genotypes. In this way, it provides more accurate eBVs by directly accounting for genetic relationships across the genome. gBLUP and SNP data from a 50 k SNP chip (approximately 50,000 genetic markers) are used by the MERINOSELECT system in Australia to evaluate Merino sheep. It produces what are known as Australian Sheep Breeding Values (ASBVs) using this evaluation approach, and enables more accurate selection of animals based on their genetic merit for various traits.

While gBLUP has shown value in livestock breeding, its application to wool traits remains challenging. Standard SNP chips may not adequately represent the genetic variation in *KRT* and *KRTAP* genes [36], and the inherently complex nature of wool traits further limits the accuracy of genomic prediction. Given these constraints, this section focuses on marker-assisted selection, which provides a practical, gene-informed approach to improving wool traits.

6.2.1. Ensuring Biological Plausibility in Gene–Trait Associations

The success of marker-assisted selection depends on using genetic markers with robust and biologically plausible associations with fibre traits. When selecting genes for association studies or interpreting such associations, it is crucial to assess whether the gene involved has a plausible biological role related to the target trait.

For example, if a gene is claimed to influence egg quality based on an observed association, we would expect it to affect a component of the egg such as the yolk, albumen, or shell, which define egg quality. If no such functional effect is observed, the association may be questionable. Similarly, genes linked to wool fibre traits should plausibly affect wool proteins (e.g., the keratins or KAPs) or follicle development processes, whether through changes in structure, abundance, assembly, or regulation. Associations lacking such biological relevance should be treated with greater caution.

Two main strategies exist for identifying gene–trait associations: the candidate gene approach and the omics-based approach. The candidate gene approach targets genes with established or suspected biological relevance to fibre traits and remains the more reliable and practical strategy for gene–trait association studies. In contrast, the omics-based approach (e.g., using SNP chips) takes a genome- or system-wide perspective and relies on large-scale statistical correlations with traits of interest rather than on prior biological knowledge of particular genes. This approach has been applied in practice, for example, in Australia, where SNP chip data are used to select sheep with superior wool genetics for breeding. While omics approaches can lead to the identification of superior pedigrees, they usually rely on markers derived from genome-wide scans rather than on genes of known or well-understood function, and thus the markers often act as anonymous indicators of superior genetics.

Although omics studies have reported many genes to be associated with wool traits, numerous candidate genes inferred from these associations lack obvious biological relevance to fibre formation or development, or they do not specifically affect wool traits without having other pleiotropic effects that may be untoward. While this does not negate their importance, it does suggest that caution is warranted when interpreting omics results, and importantly that these findings require rigorous validation before they can be used in selective breeding. Omics approaches can be valuable for describing genetic diversity and identifying selection signatures, and thus they are useful for the management and sustainable utilisation of sheep genetic resources [116].

6.2.2. Choosing the Appropriate Level of Genetic Variation

Genetic associations can be explored at distinct levels of genetic variation, ranging from SNPs to haplotypes spanning one or multiple genes. Wool *KRT* and *KRTAP* genes are often physically clustered and functionally related [73], making it appealing to analyse variation at the multi-gene haplotype level. These haplotypes may capture the combined effects of neighbouring loci, offering a more integrated genetic signal.

However, several practical limitations hinder the routine use of multi-gene haplotypes. Most of the *KRT* and *KRTAP* genes are multi-allelic [73], and as discussed earlier, even closely located genes can exhibit distinct associations with different fibre traits. This suggests that despite physical proximity, these loci may not be in strong linkage disequilibrium, leading to a large number of possible multi-gene haplotypes. Consequently, detecting associations at this level of genetic complexity will require large sample sizes to achieve adequate statistical power. Moreover, accurately detecting and phasing these complex haplotypes remains technically challenging. Existing genotyping platforms are not well-suited for this level of resolution, and while long-read sequencing holds promise, it is not yet cost-effective or practical for large-scale routine use as might be required to understand high levels of trait variation.

Given these constraints, the most feasible and biologically sound strategy is to focus on within-gene haplotypes (also referred to as alleles or variants). These haplotypes represent combinations of SNPs across the entire gene or a variable portion of the gene (if the gene is too large for full analysis) and can capture functionally relevant variation. Since each wool protein gene encodes a specific structural component of the wool fibre, analysing variation at this level is justifiable and technically achievable. Techniques such as PCR-SSCP offer reliable and cost-effective means of genotyping these haplotypes [117].

In contrast, SNP-level association analysis has limited utility for wool trait association for several reasons. First, most wool protein genes are multi-allelic [73]. Simplifying this diversity into biallelic SNPs can mask meaningful associations. Second, the high sequence similarity among related genes makes it difficult to assign SNPs to specific loci, which can compromise genotyping accuracy [36]. Third, it is often the overall structure of the protein, rather than individual amino acid residues, that determines how the protein interacts and arranges with others, thereby affecting wool fibre structure [36]. As a result, focusing on individual SNPs may fail to capture the functional complexity underlying wool fibre formation.

6.2.3. Integrating Genetic Markers into Breeding Programmes

The ultimate goal of gene–trait association research is to translate genetic insights into practical tools for improving wool quality. For marker-assisted selection to be effective, genetic markers must be robustly validated, economically viable, and effectively integrated into breeding strategies. This has been achieved with genomic selection approaches being used to improve wool traits in Australia [95], and for traits like growth, carcass and eating quality, reproduction, and health [118]. Individual genes are being used to augment sheep breeding, with DNA typing services provided by testing laboratories like the Lincoln University Gene-Marker Laboratory [119].

Prior to integration, candidate markers should undergo thorough validation across diverse breeds, populations, and production environments to ensure their reliability and consistency. This includes testing for reproducible associations with target traits and evaluating their stability under varying environmental conditions. Once validated, markers are prioritised based on several key criteria, including effect size, strength of association, linkage disequilibrium with other markers (to avoid redundancy and optimise marker panels), the technical feasibility of genotyping, and the overall cost-effectiveness.

Selected markers can then be incorporated into breeding programmes by including them in selection indices or by using them to enhance the accuracy of estimated breeding values, thereby improving the precision of selection. While integrating validated gene markers has the potential to improve selection accuracy and accelerate genetic gain, the complex genetic architecture of wool traits means that ongoing research is necessary to refine marker selection and develop effective approaches.

7. Trade-Offs and Synergies in Wool Trait Improvement

As wool marker-assisted selection programmes evolve, a key challenge is managing potential trade-offs, whether between different wool traits themselves or between wool traits and other economically important sheep production traits such as growth, meat production, and reproduction. This issue is particularly relevant in modern sheep production systems, where meat has overtaken wool as the primary value proposition, and wool is considered to be a secondary product of sheep farming [120]. Addressing these genetic conflicts is essential to achieving balanced and sustainable genetic progress.

7.1. Trade-Offs and Synergies Among Wool Traits

Of the wool traits, CFW and MFD are two of the most economically important wool traits. A common concern is whether selection to improve one may negatively affect the other. Specifically, increased CFW has sometimes been associated with increased MFD, which can reduce the value of wool [121–123]. This trade-off is partly explained by fibre physics: for a fixed number and length of fibres, finer fibres would be expected to contribute less to total fleece weight than coarser ones.

However, this relationship is not inevitable. Evidence from well-designed breeding programmes, such as the Trangie QPLU$ project, shows that simultaneous improvement of CFW and MFD is possible through selection without detriment to either trait [124]. Long-term studies, such as Ramos et al. [125], further confirm that balanced selection can reduce fibre diameter and maintain or increase fleece weight over time.

The genetic correlations underlying wool traits depend largely on pleiotropy and genetic linkage. If a gene influences multiple traits, its effects may all be favourable (creating positive synergies) or partly opposing (creating trade-offs). Similarly, different genes that are linked on the chromosome can also cause correlated responses, even if their functions are unrelated. In contrast, traits controlled by independent loci can be improved simultaneously without adverse interactions. As an example, a genetic association study revealed that specific *KRTAP1-2* variants are associated with both increased fleece weight and finer fibre diameter [126], results that are consistent with the above findings of Mortimer et al. [124] and Ramos et al. [125], and with this suggesting that favourable genetic variants can enhance both wool quantity and quality.

The underlying genetic architecture of wool traits is complex and polygenic, with numerous small-effect genes often clustered together. This complexity limits the effectiveness of traditional phenotypic selection, which may not easily separate correlated effects between traits. Marker-assisted selection offers a promising path forward by enabling breeders to identify and select for beneficial variants, while avoiding those with negative impacts. This targeted approach can enhance selection efficiency and facilitate more balanced genetic improvement across multiple wool traits of value.

7.2. Potential Effects of Wool Trait Selection on Production and Reproductive Traits

A common concern is whether selection for wool traits, particularly CFW and MFD, might adversely impact growth, carcass traits, or reproductive performance. This question is particularly relevant in both dual-purpose and wool-focused production systems, where genetic improvement must strike a balance between fleece value, meat production, and reproductive efficiency to ensure overall sheep farming profitability.

Genetically and physiologically, wool, growth, and reproductive traits are largely controlled by different genes and pathways, suggesting that they can be improved concurrently without inherent antagonism. However, all traits rely on shared metabolic resources, especially energy and nutrients. Under suboptimal nutritional or environmental conditions, competition for these limited resources may lead to physiological trade-offs, where improvement in one trait could come at the expense of another. Therefore, effective selection programmes must consider not only the genetic makeup but also the importance of providing appropriate environmental and nutritional support to fully express genetic potential across all traits.

Empirical evidence supports the feasibility of simultaneous genetic improvement of wool, growth, carcass, and reproductive traits. Mortimer et al. [127] reported generally favourable genetic correlations between wool and carcass traits in Merino sheep. While selection for increased CFW was associated with decreased carcass fat and dressing percentage, it had negligible impact on carcass muscle levels. Similarly, selection for lower MFD confirmed some unfavourable correlations with carcass weight and muscle mass, but these were minor. Mortimer et al. [128] also established moderate and favourable genetic correlations (~0.5) between fleece weight and live weight, and weak correlations between MFD and ultrasound traits such as eye muscle depth and fat depth. In Columbia and Targhee sheep, Hanford et al. [129,130] reported low to moderate positive genetic correlations between wool and growth traits, and low negative correlations between fleece weight and reproductive traits, suggesting minimal antagonism. Mortimer et al. [131] revealed low to negligible genetic correlations between wool traits and a range of meat quality and nutritional traits, indicating that selection for wool traits is unlikely to compromise meat quality.

Long-term selection studies provide further support. Ramos et al. [125] reported sustained genetic gains in both fleece weight and MFD over nearly two decades in a Merino population, with no adverse effects on ewe body condition or reproductive performance. The modest gains in reproduction were attributed to their exclusion from the selection index, rather than to unfavourable genetic correlations. This highlights the importance of including reproductive traits in selection objectives, rather than assuming inherent biological trade-offs.

A practical outcome from a dual-purpose breeding programme demonstrates the potential for co-improvement. An industry trial has identified lambs with both high-value wool traits and desirable eating quality, suggesting the concurrent genetic improvement of eating quality and wool quality is possible [132]. While based on a single trial, the results support the principle that, with appropriate selection strategies and proper management, simultaneous genetic progress across multiple economically important traits can be achieved.

Current evidence from quantitative genetic studies, long-term selection experiments, and practical breeding programmes reveals that selection for improved wool traits does not inherently compromise growth or reproductive performance. When breeding programmes are carefully designed to consider genetic correlations and are supported by appropriate management, it is both feasible and beneficial to pursue simultaneous improvement in wool, meat, and reproductive traits, thereby enhancing the overall productivity, resilience, and profitability of sheep flocks across production systems.

8. Conclusions and Future Directions

Wool is a unique fibre, valued for its exceptional functional properties, as well as its health and environmental benefits, which make it suitable as a sustainable fibre for many uses. These qualities arise from its sophisticated ordered structure, which is shaped by intricate molecular mechanisms that remain poorly understood and involve numerous genes, only a fraction of which have been identified.

Recent progress in wool genetics has improved our understanding of fibre formation and trait variation, and it opens opportunities for targeted genetic improvement, whether through genomic selection or marker-assisted selection using variation in keratin and keratin-associated protein genes. This lays a solid foundation for the improvement of wool. While genetic modification, especially precise gene-editing approaches like CRISPR/Cas, offer future potential, much greater focus is needed on identifying and characterising the genes involved in fibre formation and development. This should include research into the molecular mechanisms underlying wool trait variation. These efforts will help ensure the long-term viability, adaptability, and competitiveness of wool production in the face of evolving economic and environmental contexts.

Author Contributions: Conceptualization, H.Z. and J.G.H.H.; Literature search, H.Z., L.B., and J.G.H.H.; writing—original draft preparation, H.Z.; writing—review and editing, H.Z., S.L., J.W., and J.G.H.H.; funding acquisition, S.L. and J.G.H.H. All authors have read and agreed to the published version of the manuscript.

Funding: This work was financially supported by the Gansu Provincial Science and Technology Program (Project No. 23YFWA0001).

Institutional Review Board Statement: Not applicable.

Informed Consent Statement: Not applicable.

Acknowledgments: We would like to acknowledge the postgraduate students and researchers in our extended research group for their valuable contributions to the findings presented in this review. Their dedication and hard work have greatly enhanced our understanding and progress in this field.

Conflicts of Interest: The authors declare no conflicts of interest.

References

1. Kuffner, H.; Popescu, C. Wool fibres. In *Handbook of Natural Fibres*; Kozłowski, R.M., Mackiewicz-Talarczyk, M., Eds.; Woodhead Publisher: Cambridge, UK, 2012; Volume 1, pp. 171–195.

2. Wool Notes, Issue 3, A Summary of Wool Textile Information, Including Notes and Interesting Wool Facts. 2024. International Wool Textile Organisation 2024. Available online: https://iwto.org/sheep-wool/about-sheep/ (accessed on 18 September 2025).
3. Datta, M.; Basu, G.; Das, S. Wool, a natural biopolymer: Extraction and structure–property relationships. In *Handbook of Natural Polymers*; Sreekala, M.S., Ravindran, L., Goda, K., Thomas, S., Eds.; Elsevier: Amsterdam, The Netherlands, 2023; Volume 1, pp. 441–469.
4. Latif, E. A review of low energy thermal insulation materials for building applications. In Proceedings of the 2nd International Conference on Green Energy and Environmental Technology (ICGEET), Bangkok, Thailand, 7–8 February 2020.
5. Berardi, U.; Iannace, G. Acoustic characterization of natural fibers for sound absorption applications. *Build. Environ.* **2015**, *94*, 840–852. [CrossRef]
6. Ormondroyd, G.; Curling, S.; Mansour, E.; Hill, C. The water vapour sorption characteristics and kinetics of different wool types. *J. Text. Inst.* **2017**, *108*, 1198–1210. [CrossRef]
7. Why Wool Feels warm. My Textile Notes. Available online: https://mytextilenotes.blogspot.com/2009/10/why-wool-feels-warm.html (accessed on 15 July 2025).
8. Mansour, E. Wool Fibres for the Sorption of Volatile Organic Compounds (VOCs) from Indoor Air. Ph.D. Thesis, Bangor University, Gwynedd, UK, 2018.
9. McNeil, S. The Removal of Indoor air Contaminants by Wool Textiles. Technical Bulletin, 2015. Available online: https://www.researchgate.net/profile/Steve-Mcneil/publication/353757473_The_Removal_of_Indoor_Air_Contaminants_by_Wool_Textiles/links/610f02bf0c2bfa282a2f469c/The-Removal-of-Indoor-Air-Contaminants-by-Wool-Textiles.pdf (accessed on 5 August 2025).
10. Johnson, N.; Wood, E.; Ingham, P.; McNeil, S.; McFarlane, I. Wool as a technical fibre. *J Text. Inst.* **2003**, *94*, 26–41. [CrossRef]
11. Sid Masri, M.; Reuter, F.W.; Friedman, M. Interaction of wool with metal cations. *Text. Res. J.* **1974**, *44*, 298–300. [CrossRef]
12. Laurie, S.H.; Barraclough, A. Use of waste wool for the removal of mercury from industrial effluents, particularly those from the chlor-alkali industry. *Int. J. Environ. Stud.* **1979**, *14*, 139–149. [CrossRef]
13. Allahkarami, E.; Allahkarami, E.; Rezai, B. A brief review on utilizing natural adsorbents for microplastic removal from wastewater: A sustainable approach to environmental protection. *Results Eng.* **2025**, *27*, 106441. [CrossRef]
14. International Wool Textile Organisation. Wool & Fire. Available online: https://iwto.org/wp-content/uploads/2020/04/IWTO_Wool-Fire.pdf (accessed on 15 July 2025).
15. London Fire Brigade. Firefighters' uniforms. London Fire Brigade Museum. Available online: https://www.london-fire.gov.uk/museum/london-fire-brigade-history-and-stories/equipment-and-uniforms/firefighters-uniforms/ (accessed on 15 July 2025).
16. Gambichler, T.; Rotterdam, S.; Altmeyer, P.; Hoffmann, K. Protection against ultraviolet radiation by commercial summer clothing: Need for standardised testing and labelling. *BMC Dermatol.* **2001**, *1*, 6. [CrossRef]
17. CHARACTERISTICS OF WOOL Fact Sheet. Available online: https://www.wool.ca/images/uploads/files/care/wool-fact-sheets-charcteristics.pdf (accessed on 16 July 2025).
18. Su, J.; Dailey, R.; Zallmann, M.; Leins, E.; Taresch, L.; Donath, S.; Heah, S.; Lowe, A. Determining effects of superfine sheep wool in INfantile eczema (DESSINE): A randomized paediatric crossover study. *Brit. J. Dermatol.* **2017**, *177*, 125–133. [CrossRef] [PubMed]
19. Spelman, L.; Supranowicz, M.J.; Davidson, K.A.; Johnston, J.J.; Yau, B.; Holland, T.L. An investigator blinded, clinical trial assessing the efficacy of superfine merino wool base layer garments (SMWBG) in children with atopic dermatitis (AD) measuring SCORAD1, EASI2, POEM3 and DSA4 scores. *Biomed. J. Sci. Tech. Res.* **2018**, *7*, 5687–5693. [CrossRef]
20. Fowler, J.F.; Fowler, L.M.; Lorenz, D. Effects of merino wool on atopic dermatitis using clinical, quality of life, and physiological outcome measures. *Dermatitis* **2019**, *30*, 198–206. [CrossRef] [PubMed]
21. Zallmann, M.; Smith, P.K.; Tang, M.L.; Spelman, L.J.; Cahill, J.L.; Wortmann, G.; Katelaris, C.H.; Allen, K.J.; Su, J.C. Debunking the myth of wool allergy: Reviewing the evidence for immune and non-immune cutaneous reactions. *Acta Derm. Venereol.* **2017**, *97*, 906–915. [CrossRef]
22. Shin, M.; Halaki, M.; Swan, P.; Ireland, A.H.; Chow, C.M. The effects of fabric for sleepwear and bedding on sleep at ambient temperatures of 17 °C and 22 °C. *Nat. Sci. Sleep* **2016**, *8*, 121–131. [CrossRef] [PubMed]
23. Chow, C.M.; Shin, M.; Mahar, T.J.; Halaki, M.; Ireland, A. The impact of sleepwear fiber type on sleep quality under warm ambient conditions. *Nat. Sci. Sleep* **2019**, *11*, 167–178. [CrossRef]
24. Li, X.; Halaki, M.; Chow, C.M. How do sleepwear and bedding fibre types affect sleep quality: A systematic review. *J. Sleep Res.* **2024**, *33*, e14217. [CrossRef]
25. The Woolmark Company. (n.d.). Where Does Carbon Come From? Available online: https://www.woolmark.com/globalassets/_06-new-woolmark/_industry/research/factsheets/gd2405-where-does-carbon-come-from_122.pdf (accessed on 16 July 2025).
26. Wiedemann, S.G.; Biggs, L.; Nebel, B.; Bauch, K.; Laitala, K.; Klepp, I.G.; Swan, P.G.; Watson, K. Environmental impacts associated with the production, use, and end-of-life of a woollen garment. *Int. J. Life Cycle Assess.* **2020**, *25*, 1486–1499. [CrossRef]

27. Allen, M.R.; Shine, K.P.; Fuglestvedt, J.S.; Millar, R.J.; Cain, M.; Frame, D.J.; Macey, A.H. A solution to the misrepresentations of CO2-equivalent emissions of short-lived climate pollutants under ambitious mitigation. *npj Clim. Atmos. Sci.* **2018**, *1*, 16. [CrossRef]
28. Hodgson, A.; Leighs, S.J.; van Koten, C. Compostability of wool textiles by soil burial. *Text. Res. J.* **2023**, *93*, 3692–3702. [CrossRef]
29. Collie, S.; Brorens, P.; Hassan, M.M.; Fowler, I. Marine biodegradation behavior of wool and other textile fibers. *Water Air Soil Poll* **2024**, *235*, 283. [CrossRef]
30. Laitala, K.; Klepp, I.G. Wool wash: Technical performance and consumer habits. *Tenside Surfact. Det.* **2016**, *53*, 458–468. [CrossRef]
31. Caven, B.; Redl, B.; Bechtold, T. An investigation into the possible antibacterial properties of wool fibers. *Text. Res. J.* **2018**, *89*, 510–516. [CrossRef]
32. Rippon, J.A. The structure of wool. In *The Coloration of Wool and Other Keratin Fibres*; Lewis, D.M., Rippon, J.A., Eds.; John Wiley & Sons: Hoboken, NJ, USA, 2013; pp. 1–42.
33. Connell, D.L. Chapter 11. Wool finishes: The control of shrinkage. In *Textile Finishing*; Heywood, D., Ed.; Society of Dyers and Colourists: Hampshire, UK, 2003; pp. 372–397.
34. Mercer, E. The relation between external shape and internal structure of wool fibers. *Text. Res. J.* **1954**, *24*, 39–43. [CrossRef]
35. Harland, D.P.; Vernon, J.A.; Woods, J.L.; Nagase, S.; Itou, T.; Koike, K.; Scobie, D.A.; Grosvenor, A.J.; Dyer, J.M.; Clerens, S. Intrinsic curvature in wool fibres is determined by the relative length of orthocortical and paracortical cells. *J. Exp. Biol.* **2018**, *221*, jeb172312. [CrossRef]
36. Zhou, H.; Bai, L.; Li, S.; Li, W.; Wang, J.; Tao, J.; Hickford, J.G. Genetics of wool and cashmere fibre: Progress, challenges, and future research. *Animals* **2024**, *14*, 3228. [CrossRef]
37. Powell, B.C.; Rogers, G.E. The role of keratin proteins and their genes in the growth, structure and properties of hair. *EXS* **1997**, *78*, 59–148.
38. Plowman, J.E.; Paton, L.N.; Bryson, W.G. The differential expression of proteins in the cortical cells of wool and hair fibres. *Exp. Dermatol.* **2007**, *16*, 707–714. [CrossRef]
39. Rippon, J.A.; Evans, D.J. Improving the properties of natural fibres by chemical treatments. In *Handbook of Natural Fibres*; Kozłowski, R.M., Mackiewicz-Talarczyk, M., Eds.; Elsevier: Amsterdam, The Netherlands, 2020; pp. 245–321.
40. Popescu, C.; Höcker, H. Hair—The most sophisticated biological composite material. *Chem. Soc. Rev.* **2007**, *36*, 1282–1291. [CrossRef]
41. Leeder, J.D. *Wool: Nature's Wonder Fibre*; Australasian Textiles Publishers: Victoria, Australia, 1984.
42. Rishikaysh, P.; Dev, K.; Diaz, D.; Qureshi, W.M.; Filip, S.; Mokry, J. Signaling involved in hair follicle morphogenesis and development. *Int. J. Mol. Sci.* **2014**, *15*, 1647–1670. [CrossRef] [PubMed]
43. Zhang, R.; Li, Y.; Jia, K.; Xu, X.; Li, Y.; Zhao, Y.; Zhang, X.; Zhang, J.; Liu, G.; Deng, S.; et al. Crosstalk between androgen and Wnt/β-catenin leads to changes of wool density in FGF5-knockout sheep. *Cell Death Dis.* **2020**, *11*, 407. [CrossRef]
44. Xu, X.L.; Wu, S.J.; Qi, S.Y.; Chen, M.M.; Liu, Z.M.; Zhang, R.; Zhao, Y.; Liu, S.Q.; Zhou, W.D.; Zhang, J.L.; et al. Increasing GSH-Px activity and activating Wnt pathway promote fine wool growth in FGF5-edited sheep. *Cells* **2024**, *13*, 985. [CrossRef]
45. Lv, X.; Chen, L.; He, S.; Liu, C.; Han, B.; Liu, Z.; Yusupu, M.; Blair, H.; Kenyon, P.; Morris, S.; et al. Effect of nutritional restriction on the hair follicles development and skin transcriptome of Chinese Merino sheep. *Animals* **2020**, *10*, 1058. [CrossRef] [PubMed]
46. Xavier, S.P.; Gordon-Thomson, C.; Wynn, P.C.; McCullagh, P.; Thomson, P.C.; Tomkins, L.; Mason, R.S.; Moore, G.P. Evidence that Notch and Delta expressions have a role in dermal condensate aggregation during wool follicle initiation. *Exp. Dermatol.* **2013**, *22*, 659–662. [CrossRef]
47. Li, C.; Feng, C.; Ma, G.; Fu, S.; Chen, M.; Zhang, W.; Li, J. Time-course RNA-seq analysis reveals stage-specific and melatonin-triggered gene expression patterns during the hair follicle growth cycle in Capra hircus. *BMC Genom.* **2022**, *23*, 140. [CrossRef]
48. Yuan, X.; Meng, K.; Wang, Y.; Wang, Y.; Pan, C.; Sun, H.; Wang, J.; Li, X. Unlocking the genetic secrets of Dorper sheep: Insights into wool shedding and hair follicle development. *Front. Vet. Sci.* **2024**, *11*, 1489379. [CrossRef]
49. Fraser, R.B.; Parry, D.A. Structural hierarchy of trichocyte keratin intermediate filaments. *Adv. Exp. Med. Biol.* **2018**, *1054*, 57–70.
50. Powell, B.; Crocker, L.; Rogers, G. Hair follicle differentiation: Expression, structure and evolutionary conservation of the hair type II keratin intermediate filament gene family. *Development* **1992**, *114*, 417–433. [CrossRef] [PubMed]
51. Powell, B.C.; Crocker, L.A.; Rogers, G.E. Complete sequence of a hair-like intermediate filament type II keratin gene. *DNA Seq.* **1993**, *3*, 401–405. [CrossRef] [PubMed]
52. Wilson, B.W.; Edwards, K.J.; Sleigh, M.J.; Byrne, C.R.; Ward, K.A. Complete sequence of a type-I microfibrillar wool keratin gene. *Gene* **1988**, *73*, 21–31. [CrossRef] [PubMed]
53. Yu, Z.; Wildermoth, J.E.; Wallace, O.A.; Gordon, S.W.; Maqbool, N.J.; Maclean, P.H.; Nixon, A.J.; Pearson, A.J. Annotation of sheep keratin intermediate filament genes and their patterns of expression. *Exp. Dermatol.* **2011**, *20*, 582–588. [CrossRef]
54. Chai, W.; Zhou, H.; Gong, H.; Hickford, J.G. Variation in the ovine *KRT34* promoter region affects wool traits. *Small Rumin. Res.* **2022**, *206*, 106586. [CrossRef]

55. Chai, W.; Zhou, H.; Gong, H.; Wang, J.; Luo, Y.; Hickford, J.G. Nucleotide variation in the ovine *KRT31* promoter region and its association with variation in wool traits in Merino-cross lambs. *J. Agric. Sci.* **2019**, *157*, 182–188. [CrossRef]
56. Chen, Z.; Zhao, F.; He, Z.; Sun, H.; Xi, Q.; Yu, X.; Ding, Y.; An, Z.; Wang, J.; Liu, X. Expression localization of the *KRT32* gene and its association of genetic variation with wool traits. *Curr. Issues Mol. Biol.* **2024**, *46*, 2961–2974. [CrossRef]
57. Sulayman, A.; Tursun, M.; Sulaiman, Y.; Huang, X.; Tian, K.; Tian, Y.; Xu, X.; Fu, X.; Mamat, A.; Tulafu, H. Association analysis of polymorphisms in six keratin genes with wool traits in sheep. *Asian-Australas. J. Anim. Sci.* **2018**, *31*, 775. [CrossRef]
58. Sumner, R.; Forrest, R.; Zhou, H.; Henderson, H.; Hickford, J.G. Association of the *KRT33A* (formerly *KRT1.2*) gene with live-weight and wool characteristics in yearling Perendale sheep. *Proc. N. Z. Soc. Anim. Prod.* **2013**, *73*, 158–164.
59. Chai, W.; Zhou, H.; Forrest, R.H.; Gong, H.; Hodge, S.; Hickford, J.G. Polymorphism of *KRT83* and its association with selected wool traits in Merino-cross lambs. *Small Rumin. Res.* **2017**, *155*, 6–11. [CrossRef]
60. Chai, W.; Zhou, H.; Gong, H.; Wang, C.; Hickford, J.G. Variation in the exon 3-4 region of ovine *KRT85* and its effect on wool traits. *Animals* **2024**, *14*, 2272. [CrossRef]
61. Li, W.; Bai, L.; Zhou, H.; Zhang, Z.; Ma, Z.; Wu, G.; Luo, Y.; Tanner, J.; Hickford, J.G. Ovine *KRT81* variants and their influence on selected wool traits of commercial value. *Genes* **2024**, *15*, 681. [CrossRef] [PubMed]
62. Yu, X.; Li, S.; Zhou, H.; Zhao, F.; Hu, J.; Wang, J.; Liu, X.; Li, M.; Zhao, Z.; Hao, Z. Spatiotemporal expression and haplotypes identification of KRT84 gene and their association with wool traits in Gansu Alpine Fine-Wool sheep. *Genes* **2024**, *15*, 248. [CrossRef] [PubMed]
63. Rogers, M.A.; Schweizer, J. Human KAP genes, only the half of it? Extensive size polymorphisms in hair keratin-associated protein genes. *J. Investig. Dermatol.* **2005**, *124*, vii–ix. [CrossRef]
64. Rogers, M.A.; Winter, H.; Langbein, L.; Wollschläger, A.; Praetzel-Wunder, S.; Jave-Suarez, L.F.; Schweizer, J. Characterization of human KAP24.1, a cuticular hair keratin-associated protein with unusual amino-acid composition and repeat structure. *J. Investig. Dermatol.* **2007**, *127*, 1197–1204. [CrossRef]
65. Rogers, M.A.; Langbein, L.; Praetzel Wunder, S.; Giehl, K. Characterization and expression analysis of the hair keratin associated protein KAP26.1. *Brit. J. Dermatol.* **2008**, *159*, 725–729. [CrossRef]
66. Gong, H.; Zhou, H.; Dyer, J.; Hickford, J.G.H. The sheep KAP8-2 gene, a new KAP8 family member that is absent in humans. *SpringerPlus* **2014**, *3*, 528. [CrossRef]
67. Gong, H.; Zhou, H.; Wang, J.; Li, S.; Luo, Y.; Hickford, J.G. Characterisation of an ovine keratin associated protein (KAP) gene, which would produce a protein rich in glycine and tyrosine, but lacking in cysteine. *Genes* **2019**, *10*, 848. [CrossRef]
68. Zhou, H.; Gong, H.; Wang, J.; Dyer, J.M.; Luo, Y.; Hickford, J.G. Identification of four new gene members of the KAP6 gene family in sheep. *Sci. Rep.* **2016**, *6*, 24074. [CrossRef]
69. Zhou, H.; Li, W.; Bai, L.; Wang, J.; Luo, Y.; Li, S.; Hickford, J.G. Ovine *KRTAP36-2*: A new keratin-associated protein gene related to variation in wool yield. *Genes* **2023**, *14*, 2045. [CrossRef] [PubMed]
70. Fratini, A.; Powell, B.C.; Hynd, P.I.; Keough, R.A.; Rogers, G.E. Dietary cysteine regulates the levels of messenger-rnas encoding a family of cysteine-rich proteins of wool. *J. Investig. Dermatol.* **1994**, *102*, 178–185. [CrossRef] [PubMed]
71. Jenkins, B.J.; Powell, B.C. Differential expression of genes encoding a cysteine-rich keratin family in the hair cuticle. *J. Investig. Dermatol.* **1994**, *103*, 310–317. [CrossRef]
72. MacKinnon, P.; Powell, B.; Rogers, G. Structure and expression of genes for a class of cysteine-rich proteins of the cuticle layers of differentiating wool and hair follicles. *J. Cell Biol.* **1990**, *111*, 2587–2600. [CrossRef]
73. Zhou, H.; Gong, H.; Wang, J.; Luo, Y.; Li, S.; Tao, J.; Hickford, J.G. The complexity of the ovine and caprine keratin-associated protein genes. *Int. J. Mol. Sci.* **2021**, *22*, 12838. [CrossRef]
74. Bai, L.; Wang, J.; Zhou, H.; Gong, H.; Tao, J.; Hickford, J.G. Identification of ovine *KRTAP28-1* and its association with wool fibre diameter. *Animals* **2019**, *9*, 142. [CrossRef]
75. Zhou, H.; Gong, H.; Li, S.; Luo, Y.; Hickford, J.G.H. A 57-bp deletion in the ovine KAP6-1 gene affects wool fibre diameter. *J. Anim. Breed. Genet.* **2015**, *132*, 301–307. [CrossRef] [PubMed]
76. Li, S.; Zhou, H.; Gong, H.; Zhao, F.; Wang, J.; Luo, Y.; Hickford, J.G. Variation in the ovine KAP6-3 gene (*KRTAP6-3*) is associated with variation in mean fibre diameter-associated wool traits. *Genes* **2017**, *8*, 204. [CrossRef]
77. Gong, H.; Zhou, H.; Li, W.; Wang, J.; Li, S.; Luo, Y.; Hickford, J.G.H. Variation in ovine *KRTAP8-1* is associated with variation in wool fibre staple strength and curvature. *J. Agric. Sci.* **2019**, *157*, 550–554. [CrossRef]
78. Li, W.; Gong, H.; Zhou, H.; Wang, J.; Liu, X.; Li, S.; Luo, Y.; Hickford, J.G.H. Variation in the ovine keratin-associated protein 15-1 gene affects wool yield. *J. Agric. Sci.* **2018**, *156*, 922–928. [CrossRef]
79. Li, S.; Zhou, H.; Gong, H.; Zhao, F.; Wang, J.; Liu, X.; Hu, J.; Luo, Y.; Hickford, J.G. Identification of the ovine keratin-associated protein 21-1 gene and its association with variation in wool traits. *Animals* **2019**, *9*, 450. [CrossRef] [PubMed]
80. Li, S.; Zhou, H.; Gong, H.; Zhao, F.; Wang, J.; Liu, X.; Hu, J.; Luo, Y.; Hickford, J.G. The mean staple length of wool fibre is associated with variation in the ovine keratin-associated protein 21-2 gene. *Genes* **2020**, *11*, 148. [CrossRef] [PubMed]

81. Bai, L.; Zhou, H.; He, J.; Tao, J.; Hickford, J.G. Characterisation of three ovine *KRTAP13* family genes and their association with wool traits in Chinese Tan sheep. *Animals* **2024**, *14*, 2862. [CrossRef] [PubMed]
82. Itenge, T.; Hickford, J.; Forrest, R.; McKenzie, G.; Frampton, C. Association of variation in the ovine KAP1.1, KAP1.3 and K33 genes with wool traits. *Int. J. Sheep Wool Sci.* **2010**, *58*, 1.
83. Farag, I.; Darwish, H.; Darwish, A.; Eshak, M.; Ahmed, R. Genetic polymorphism of *KRT1.2* gene and its association with improving of some wool characteristics in Egyptian sheep. *Asian J. Sci. Res.* **2018**, *11*, 295–300. [CrossRef]
84. Gong, H.; Zhou, H.; Bai, L.; Li, W.; Li, S.; Wang, J.; Luo, Y.; Hickford, J.G. Associations between variation in the ovine high glycine-tyrosine keratin-associated protein gene *KRTAP20-1* and wool traits. *J. Anim. Sci.* **2019**, *97*, 587–595. [CrossRef]
85. Bai, L.; Zhou, H.; Tao, J.; Hickford, J.G. Effects of *KRTAP20-1* gene variation on wool traits in Chinese Tan sheep. *Genes* **2024**, *15*, 1060. [CrossRef]
86. Rogers, G.R.; Hickford, J.G.H.; Bickerstaffe, R. Polymorphism in two genes for B2 high sulfur proteins of wool. *Anim. Genet.* **1994**, *25*, 407–415. [CrossRef]
87. Zhou, H.; Visnovska, T.; Gong, H.; Schmeier, S.; Hickford, J.; Ganley, A.R. Contrasting patterns of coding and flanking region evolution in mammalian keratin associated protein-1 genes. *Mol. Phylogenet. Evol.* **2019**, *133*, 352–361. [CrossRef]
88. Chen, Z.; Cao, J.; Zhao, F.; He, Z.; Sun, H.; Wang, J.; Liu, X.; Li, S. Identification of the keratin-associated protein 22-2 gene in the Capra hircus and association of its variation with cashmere traits. *Animals* **2023**, *13*, 2806. [CrossRef]
89. Li, S.; Xi, Q.; Zhao, F.; Wang, J.; He, Z.; Hu, J.; Liu, X.; Luo, Y. A highly polymorphic caprine keratin-associated protein gene identified and its effect on cashmere traits. *J. Anim. Sci.* **2021**, *99*, skab233. [CrossRef]
90. Cao, J.; Wang, J.; Zhou, H.; Hu, J.; Liu, X.; Li, S.; Luo, Y.; Hickford, J.G. Sequence variation in caprine *kRTAP6-2* affects cashmere fiber diameter. *Animals* **2022**, *12*, 2040. [CrossRef]
91. Wang, J.; Zhou, H.; Hickford, J.G.; Zhao, M.; Gong, H.; Hao, Z.; Shen, J.; Hu, J.; Liu, X.; Li, S. Identification of caprine *KRTAP28-1* and its effect on cashmere fiber diameter. *Genes* **2020**, *11*, 121. [CrossRef]
92. Wang, J.; Zhou, H.; Luo, Y.; Zhao, M.; Gong, H.; Hao, Z.; Hu, J.; Hickford, J.G. Variation in the caprine KAP24-1 gene affects cashmere fibre diameter. *Animals* **2019**, *9*, 15. [CrossRef] [PubMed]
93. Mortimer, S.I.; Atkins, K. Genetic evaluation of production traits between and within flocks of Merino sheep. I. Hogget fleece weights, body weight and wool quality. *Crop Pasture Sci.* **1989**, *40*, 433–443. [CrossRef]
94. Safari, E.; Fogarty, N.; Gilmour, A. A review of genetic parameter estimates for wool, growth, meat and reproduction traits in sheep. *Livest. Prod. Sci.* **2005**, *92*, 271–289. [CrossRef]
95. Sheep Genetics. MERINOSELECT. Available online: https://www.sheepgenetics.org.au/resources/merinoselect/ (accessed on 4 August 2025).
96. Australian Wool Innovation. *MERINOSELECT ASBVs by Fibre Diameter and Their Genetic Trends*; Beyond the Bale. Issue 101. Australian Wool Innovation Limited: Sydney, Australia, 2024; pp. 20–21. Available online: https://www.wool.com/globalassets/wool/beyond-the-bale/issue-101-december-2024/btb-2024_december_web.pdf (accessed on 18 September 2025).
97. Powell, B.C.; Rogers, G.E. Cyclic hair-loss and regrowth in transgenic mice overexpressing an intermediate filament gene. *EMBO J.* **1990**, *9*, 1485–1493. [CrossRef] [PubMed]
98. Powell, B.C.; Nesci, A.; Rogers, G.E. Regulation of keratin gene expression in hair follicle differentiation. *Ann. N. Y. Acad. Sci.* **1991**, *642*, 1–20. [CrossRef] [PubMed]
99. Rogers, G.E.; Powell, B.C. Organization and expression of hair follicle genes. *J. Investig. Dermatol.* **1993**, *101*, 50S–55S. [CrossRef]
100. Keough, R.; Powell, B.; Rogers, G. Targeted expression of SV40 T antigen in the hair follicle of transgenic mice produces an aberrant hair phenotype. *J. Cell Sci.* **1995**, *108*, 957–966. [CrossRef]
101. Powell, B.; Walker, S.; Bawden, C.; Sivaprasad, A.; Rogers, G. Transgenic sheep and wool growth: Possibilities and current status. *Reprod. Fertil. Dev.* **1994**, *6*, 615–623. [CrossRef]
102. Bawden, C.S.; Powell, B.C.; Walker, S.K.; Rogers, G.E. Expression of a wool intermediate filament keratin transgene in sheep fibre alters structure. *Transgenic Res.* **1998**, *7*, 273–287. [CrossRef]
103. Bawden, C.; Penno, N.; Walker, S.; Hynd, P.; Rogers, G. Genetic manipulation to modify wool properties and fibre growth rates. *Proc. New Zealand Soc. Anim. Prod.* **2000**, *60*, 147–154.
104. Damak, S.; Su, H.Y.; Jay, N.P.; Bullock, D.W. Improved wool production in transgenic sheep expressing insulin-like growth factor 1. *Biotechnology* **1996**, *14*, 185–188. [CrossRef]
105. Su, H.Y.; Jay, N.; Gourley, T.; Kay, G.; Damak, S. Wool production in transgenic sheep: Results from first-generation adults and second-generation lambs. *Anim. Biotech.* **1998**, *9*, 135–147. [CrossRef]
106. Adams, N.; Briegel, J. Multiple effects of an additional growth hormone gene in adult sheep. *J. Anim. Sci.* **2005**, *83*, 1868–1874. [CrossRef]

107. Andl, T.; Reddy, S.T.; Gaddapara, T.; Millar, S.E. WNT signals are required for the initiation of hair follicle development. *Dev. Cell* **2002**, *2*, 643–653. [CrossRef] [PubMed]
108. Celso, C.L.; Prowse, D.M.; Watt, F.M. Transient activation of β-catenin signalling in adult mouse epidermis is sufficient to induce new hair follicles but continuous activation is required to maintain hair follicle tumours. *Development* **2004**, *131*, 1787–1799. [CrossRef] [PubMed]
109. Mou, C.; Jackson, B.; Schneider, P.; Overbeek, P.A.; Headon, D.J. Generation of the primary hair follicle pattern. *Proc. Natl. Acad. Sci. USA* **2006**, *103*, 9075–9080. [CrossRef]
110. Sick, S.; Reinker, S.; Timmer, J.; Schlake, T. WNT and DKK determine hair follicle spacing through a reaction-diffusion mechanism. *Science* **2006**, *314*, 1447–1450. [CrossRef] [PubMed]
111. Li, W.R.; Liu, C.X.; Zhang, X.M.; Chen, L.; Peng, X.R.; He, S.G.; Lin, J.P.; Han, B.; Wang, L.Q.; Huang, J.C.; et al. CRISPR/Cas9-mediated loss of FGF5 function increases wool staple length in sheep. *FEBS J.* **2017**, *284*, 2764–2773. [CrossRef]
112. He, D.; Chen, L.; Luo, F.; Zhou, H.; Wang, J.; Zhang, Q.; Lu, T.; Wu, S.; Hickford, J.G.; Tao, J. Differentially phosphorylated proteins in the crimped and straight wool of Chinese Tan sheep. *J. Proteomics* **2021**, *235*, 104115. [CrossRef]
113. Soller, M. Marker assisted selection—An overview. *Anim. Biotech.* **1994**, *5*, 193–207. [CrossRef]
114. Gutierrez-Reinoso, M.A.; Aponte, P.M.; Garcia-Herreros, M. Genomic analysis, progress and future perspectives in dairy cattle selection: A review. *Animals* **2021**, *11*, 599. [CrossRef]
115. Chakraborty, D.; Sharma, N.; Kour, S.; Sodhi, S.S.; Gupta, M.K.; Lee, S.J.; Son, Y.O. Applications of omics technology for livestock selection and improvement. *Front. Genet.* **2022**, *13*, 774113. [CrossRef] [PubMed]
116. Saravanan, K.A.; Panigrahi, M.; Kumar, H.; Bhushan, B.; Dutt, T.; Mishra, B.P. Genome-wide analysis of genetic diversity and selection signatures in three Indian sheep breeds. *Livest. Sci.* **2021**, *243*, 104367. [CrossRef]
117. Sunnucks, P.; Wilson, A.C.; Beheregaray, L.B.; Zenger, K.; French, J.; Taylor, A.C. SSCP is not so difficult: The application and utility of single-stranded conformation polymorphism in evolutionary biology and molecular ecology. *Mol. Ecol.* **2000**, *9*, 1699–1710. [CrossRef]
118. Sheep Genetics, University of New England, Armidale NSW. Available online: https://www.sheepgenetics.org.au/about-us/ (accessed on 18 September 2025).
119. Lincoln University Gene-Marker Laboratory, Canterbury, New Zealand. Available online: https://research.lincoln.ac.nz/testing-analytical-services/gene-marker-lab (accessed on 18 September 2025).
120. National Research Council. *Changes in the Sheep Industry in the United States: Making the Transition from Tradition*; National Academies Press: Washington, DC, USA, 2008.
121. Wuliji, T.; Dodds, K.; Land, J.; Andrews, R.; Turner, P. Selection for ultrafine Merino sheep in New Zealand: Heritability, phenotypic and genetic correlations of live weight, fleece weight and wool characteristics in yearlings. *Anim. Sci.* **2001**, *72*, 241–250. [CrossRef]
122. Safari, E.; Fogarty, N.; Gilmour, A.; Atkins, K.; Mortimer, S.; Swan, A.; Brien, F.; Greeff, J.; Van Der Werf, J. Genetic correlations among and between wool, growth and reproduction traits in Merino sheep. *J. Anim. Breed. Genet.* **2007**, *124*, 65–72. [CrossRef] [PubMed]
123. Huisman, A.; Brown, D. Genetic parameters for bodyweight, wool, and disease resistance and reproduction traits in Merino sheep. 3. Genetic relationships between ultrasound scan traits and other traits. *Anim. Prod. Sci.* **2009**, *49*, 283–288. [CrossRef]
124. Mortimer, S.; Taylor, P.; Atkins, K.; Pope, C. The Trangie QPLU $ selection lines: Responses in clean fleece weight and fibre diameter on completion of ten rounds of selection. In *Trangie QPLUS Merinos, Proceedings of the Trangie QPLUS Open Day, Trangie Agricultural Research Centre, Trangie, Australia, 11 May 2006*; NSW Department of Primary Industries: Orange, Australia, 2006; pp. 7–11.
125. Ramos, Z.; Blair, H.T.; De Barbieri, I.; Ciappesoni, G.; Montossi, F.; Kenyon, P.R. Productivity and reproductive performance of mixed-age ewes across 20 years of selection for ultrafine wool in Uruguay. *Agriculture* **2021**, *11*, 712. [CrossRef]
126. Gong, H.; Zhou, H.; Hodge, S.; Dyer, J.M.; Hickford, J.G. Association of wool traits with variation in the ovine KAP1-2 gene in Merino cross lambs. *Small Rumin. Res.* **2015**, *124*, 24–29. [CrossRef]
127. Mortimer, S.; Hatcher, S.; Fogarty, N.; van der Werf, J.; Brown, D.; Swan, A.; Jacob, R.; Geesink, G.; Hopkins, D.; Edwards, J.H. Genetic correlations between wool traits and carcass traits in Merino sheep. *J. Anim. Sci.* **2017**, *95*, 2385–2398. [CrossRef]
128. Mortimer, S.; Hatcher, S.; Fogarty, N.; Van Der Werf, J.; Brown, D.; Swan, A.; Greeff, J.; Refshauge, G.; Edwards, J.H.; Gaunt, G. Genetic parameters for wool traits, live weight, and ultrasound carcass traits in Merino sheep. *J. Anim. Sci.* **2017**, *95*, 1879–1891.
129. Hanford, K.J.; Van Vleck, L.D.; Snowder, G. Estimates of genetic parameters and genetic change for reproduction, weight, and wool characteristics of Columbia sheep. *J. Anim. Sci.* **2002**, *80*, 3086–3098. [CrossRef]
130. Hanford, K.J.; Van Vleck, L.D.; Snowder, G. Estimates of genetic parameters and genetic change for reproduction, weight, and wool characteristics of Targhee sheep. *J. Anim. Sci.* **2003**, *81*, 630–640. [CrossRef]

131. Mortimer, S.; Hatcher, S.; Fogarty, N.; Van Der Werf, J.; Brown, D.; Swan, A.; Jacob, R.; Geesink, G.; Hopkins, D.; Edwards, J.H. Genetic correlations between wool traits and meat quality traits in Merino sheep. *J. Anim. Sci.* **2017**, *95*, 4260–4273. [CrossRef]
132. Sheep Central. Research Shows Good Eating Quality and Wool is Possible. 2023. Available online: https://www.sheepcentral.com/research-shows-good-eating-quality-and-wool-is-possible/ (accessed on 23 August 2025).

animals

MDPI

Article

Metabolomic Analysis of Cauda Epididymal Fluid in Yaks and Cattle

Gan Yang [1,†], Xiaolong Yang [1,†], Dongju Liu [1], Wending Zhou [1], Anjun Zhou [1], Yan Xiong [1,2], Xianrong Xiong [1,2], Wei Fu [1,2], Jian Li [1,2], Daoliang Lan [1,2,*] and Shi Yin [1,2,*]

[1] College of Animal & Veterinary Sciences, Southwest Minzu University, Chengdu 610041, China; 18288246608@163.com (G.Y.); yxlong0123@163.com (X.Y.); qingjiu001020@163.com (D.L.); zhouwd0123@gmail.com (W.Z.); zhouanjun168@163.com (A.Z.); xiongyan0910@126.com (Y.X.); xianrongxiong@163.com (X.X.); fuwei@swun.edu.cn (W.F.); lijian@swun.cn (J.L.)

[2] Key Laboratory of Animal Science of National Ethnic Affairs Commission of China, Southwest Minzu University, Chengdu 610041, China

* Correspondence: landaoliang@163.com (D.L.); 22100122@swun.edu.cn (S.Y.)

† These authors contributed equally to this work.

Simple Summary

Yaks, native to high-altitude regions, hold significant economic and research value, yet exhibit lower reproductive capacity and reduced sperm motility compared to cattle. The epididymis, where sperm maturation occurs, features the cauda region as the primary storage site, with cauda epididymal fluid secreted by epithelial cells forming the microenvironment for sperm maturation. This study used liquid chromatography-mass spectrometry (LC-MS/MS)-based metabolomics to analyze and compare cauda epididymal fluid components between yaks and cattle. We found that under positive and negative ion detection modes, 1098 and 1297 metabolites were identified, respectively, with the most abundant categories being organic acids and derivatives, organoheterocyclic compounds, and lipids and lipid-like molecules. Compared with cattle, yak cauda epididymal fluid showed 79 significantly upregulated and 212 downregulated metabolites in the positive ion mode, and 110 upregulated and 230 downregulated metabolites in the negative ion mode. Among the differentially expressed metabolites across both species, a total of 14 metabolites positively correlated with sperm motility or functional maturation while four negatively associated with them. Our results establish a valuable metabolomic reference dataset for yak reproductive biology and provide new insights into the molecular regulation of sperm function.

Abstract

The epididymis is crucial for sperm maturation, with its caudal region storing mature sperm. Yaks show poorer sperm motility and higher abnormality rates than cattle, but the metabolic mechanisms remain unknown. This study compared cauda epididymal fluid metabolites between six yaks and six cattle using untargeted metabolomics. A total of 1098 and 1297 metabolites types annotated by the Human Metabolome Database were identified in yak and cattle cauda epididymal fluid, respectively, using positive and negative ion modes. The Yak cauda epididymal fluid exhibited distinct metabolic profiles compared with cattle. A total of 79 metabolites were upregulated and 212 were downregulated in the positive ion mode, while 110 were upregulated and 230 were downregulated in the negative ion mode. Among these, 14 metabolites were reported to promote sperm quality, function or metabolism by reducing oxidative stress, blocking premature sperm capacitation and spontaneous acrosome reaction, enhancing mitochondrial energy metabolism or facilitating flagellar motility in cattle or other species. Four were reported to impair the quality or

 https://doi.org/10.3390/ani15192861

function of sperm via increasing the DNA methylation, inhibiting spermatozoa motility, upregulating the ROS levels and diminishing sperm motility. Taken together, this study established a valuable metabolomic reference dataset for yak reproductive biology and provided new insights into the molecular regulation of sperm function.

Keywords: epididymal cauda fluid; yak; metabolomics; sperm; male fertility

1. Introduction

The maturation of spermatozoa is closely associated with the epididymis, an organ composed of highly convoluted, narrow ducts. Anatomically, the epididymis is divided into three regions: caput (head), corpus (body), and cauda (tail). The luminal microenvironment, jointly formed by epididymal epithelial cells and their secreted epididymal fluid, plays an indispensable role in sperm transport, maturation, and storage [1–6]. The rhythmic contraction of smooth muscle and the directional beating of cilia within the epididymal ductal epithelial cells facilitate spermatozoa transport toward the vas deferens. Besides, the epididymis absorbs rete testis fluid secreted by Sertoli cells, thereby preventing it from impeding sperm transit [7–9]. The luminal fluid comprises various components, including proteins, RNAs, inorganic ions, and small organic molecules [10–12]. As spermatozoa traverse the epididymis, they interact with key proteins and small molecular compounds in the luminal fluid, undergoing a series of changes including nuclear condensation, alterations in plasma membrane composition, cytoskeletal restructuring, and modifications in protein and non-coding RNA cargoes. By the time sperm exit the epididymis, the majority have acquired both progressive motility and the capacity to fertilize oocytes [13–16].

Prior to ejaculation, the cauda epididymis serves as a critical reservoir for mature spermatozoa, storing 50% to 80% of the sperm within the epididymal lumen [7,17]. Compared to other epididymal regions, the cauda fluid has higher osmotic pressure, promoting sperm dehydration, tight nuclear condensation, and genetic stability [7,17]. Its low oxygen content suppresses aerobic metabolism, reduces energy consumption, and extends survival [18,19]. The acidic environment also inhibits sperm motility [20,21]. During ejaculation, interactions between cauda fluid proteins and accessory gland secretions may influence sperm function. The unique physicochemical properties and functions of the cauda fluid directly protect the structural and functional integrity of sperm [22,23] Accordingly, an in-depth investigation of this microenvironment is crucial for elucidating the regulatory mechanisms underlying sperm maturation and fertilization. Furthermore, the epididymal fluid has emerged as a valuable source of biomarkers for fertility assessment. Several epididymal-derived molecules, such as specific proteins and metabolites, are currently under investigation for their potential clinical use in diagnosing male infertility in both humans and animals [24–27]. Identifying such biomarkers could not only enhance our understanding of sperm maturation defects but also lead to practical diagnostic tools and therapeutic strategies for improving reproductive outcomes.

Metabolomics focuses on the qualitative and quantitative analysis of all small-molecule metabolites (e.g., amino acids, organic acids, and sugars) within organisms. In recent years, metabolomics has been widely applied to study the composition of cauda epididymal fluid in single species, such as humans and mice [6,28,29]. However, cross-species comparative metabolomics of epididymal fluid has not been reported. The yak is a native livestock species mainly distributed in the Qinghai–Tibet Plateau and surrounding high-altitude regions, possessing significant scientific research and economic value. Compared with cattle, male yaks under high-altitude conditions exhibit distinct sexual maturity patterns, typically producing mature sperm at 2–3 years of age (compared to 1–1.5 years in cattle) [9,30,31]. This developmental difference may reflect metabolic adaptation strategies to extreme environments. Existing studies have reported lower sperm motility, higher abnormality rates, and poorer post-thaw recovery in the yak frozen semen [32]. We hypothesize that significant metabolic differences exist in the cauda epididymal fluid between yaks and cattle, reflecting distinct physiological adaptations to their respective environments, which may be associated with the species-specific variation in sperm maturation kinetics and functional characteristics. In this study, we applied an untargeted metabolomics approach to compare the epididymal fluid between yaks and cattle, aiming to establish a reference metabolomic dataset for future reproductive research and to elucidate the regulatory mechanisms of sperm function in high-altitude environments. The findings may provide potential targets for enhancing yak reproductive efficiency and practical applications in livestock breeding.

2. Materials and Methods

2.1. Sample Collection

Samples were obtained from a slaughterhouse in Qingbaijiang District, Chengdu, Sichuan Province. Six healthy male Maiwa yaks and six healthy male cattle (a crossbreed of Simmental cattle and Sichuan local yellow cattle), aged 3–4 years with comparable body weights (mean ± SEM: yak—385 ± 67 kg; cattle—425 ± 50 kg). The animals were raised under typical local extensive grazing conditions based on natural pasture and supplemented with concentrate during winter to keep a BCS > 2.5. In addition, general management considered veterinary clinical assessment and routine vaccination and parasite control programs for endemic diseases. The average ambient temperature was 18.6 ± 4.3 °C and relative humidity was 75 ± 11% during the sampling period. All animals were slaughtered on the same day, were selected (n = 6 for each species). Epididymal caudae were dissected under laminar flow, rinsed with a PBS, and placed in a dish containing 2 mL PBS. After making 3–4 incisions (~1 cm), the luminal fluid was expressed gently with forceps. Following 5-min incubation at ambient temperature, the fluid was aspirated and centrifuged at 1500× g for 10 min at 4 °C. The supernatant was further centrifuged at 12,000× g for 10 min at 4 °C, aliquoted, and stored at −80 °C for metabolomic analysis (n = 6 per species). All procedures were approved by the Animal Care and Ethics Committee of Southwest Minzu University (SWU-202501136).

2.2. Metabolite Extraction and Quality Control

Aliquots (100 μL) of the samples were transferred to microcentrifuge tubes and resuspended in pre-chilled 80% methanol (Shanghai Macklin Biochemical Co., Ltd. M813903-4 × 4L, Shanghai, China) via vigorous vortex mixing. The mixtures were then incubated on ice for 5 min and centrifuged at 15,000× g and 4 °C for 20 min. Then 80 μL of the supernatant was diluted with 40 μL of LC-MS grade water to achieve a final methanol concentration of 53%. The diluted samples were subsequently transferred to fresh microcentrifuge tubes and centrifuged again at 15,000× g and 4 °C for 20 min. Finally, the resulting supernatant was subjected to LC-MS/MS analysis [33,34].

To account for potential technical variability and ensure analytical robustness, all samples were randomized prior to the metabolite extraction process. Additionally, a pooled quality-controlled (QC) sample, created by combining equal aliquots from all samples, was analyzed alongside the experimental samples to monitor instrument performance and reproducibility throughout the acquisition sequence.

2.3. The UHPLC-MS/MS Analysis

UHPLC-MS/MS analyses were conducted at Novogene (Beijing, China) using a Vanquish UHPLC system (Thermo Fisher Scientific, Germering, Germany) coupled to either an Orbitrap Q Exactive™ HF or HF-X mass spectrometer (Thermo Fisher Scientific, Germering, Germany). Samples were separated on a Hypersil Gold column (with a 100 mm length, 2.1 mm internal diameter, and 1.9 μm particle size) at a flow rate of 0.2 mL/min under a 12-min linear gradient. Mobile phase A consisted of 0.1% (*v*/*v*) formic acid (Beijing Inno Chem Science & Technology Co., Ltd., 270480010, Beijing, China) in water, and mobile phase B was methanol. The gradient program was as follows: 2% B (0–1.5 min), 2–85% B (1.5–4.5 min), 85–100% B (4.5–14.5 min), 100–2% B (14.5–14.6 min), and 2% B (14.6–16.6 min, equilibration). Mass spectrometry was performed using both the positive and negative ion modes with a spray voltage of 3.5 kV, a capillary temperature of 320 °C, a sheath gas flow rate of 35 psi, aux gas flow rate of 10 L/min, an S-lens RF level of 60, and an aux gas heater temperature of 350 °C.

2.4. Data Processing and Metabolite Identification

Raw UHPLC-MS/MS data were processed using XCMS (Xtracted Chromatograms and Mass Spectra) for metabolite peak detection, alignment, and quantification. Metabolite identification was performed by matching adduct ions with a mass tolerance of 10 ppm against the Human Metabolome Database (HMDB) spectral library. Background ions identified in blank samples were excluded prior to normalization of raw quantification values using the formula: Relative peak areas = Raw quantitative value of samples/(The sum of quantitative value of samples/The sum of quantitative value of QC1). Compounds exhibiting > 30% coefficient of variation (CV) in the QC sample relative to peak areas were filtered out, yielding final annotated metabolite identities and normalized abundances. All analyses were executed on a Linux platform (CentOS, version 6.6) using custom R (version 3.4.3)/Python pipelines (version 2.7.6).

2.5. Data Analysis

Statistical analyses were performed primarily in R language. Pearson correlation coefficients between differential metabolites were computed using cor () in R, with statistical significance (p-value ≤ 0.05) determined by cor.mtest() from the R stats package (version: 3.4.3). Partial least squares-discriminant analysis (PLS-DA) was conducted in the metaX software suite (version: 2.67) [35]. All data were tested for normality prior to statistical analysis using the Shapiro-Wilk test. Data satisfying normality assumptions (p-value > 0.05) were analyzed with Student's t-test for comparisons between two groups. Univariate significance (p-value) was determined by Student's t-test.

Metabolites were annotated using the HMDB (https://hmdb.ca/metabolites accessed on 5 June 2025), KEGG (https://www.genome.jp/kegg/pathway.html accessed on 5 June 2025) and LIPID MAPS (http://www.lipidmaps.org accessed on 5 June 2025) databases. The screening of differential metabolites primarily relied on three parameters: Variable Importance in the Projection (VIP), Fold Change (FC), and p-value. The VIP was derived from the first principal component of the PLS-DA model [36], indicating the contribution of a metabolite to group separation. The FC denotes the fold change, calculated as the ratio of the mean quantified values across all biological replicates in the comparison groups. The p-value, derived from the T-test [37], represents the significance level of the difference. Metabolites were identified as differential with thresholds set at VIP > 1.0, (FC > 1.5 or FC < 0.667), and p-value ≤ 0.05 [38–40].

Volcano plots visualizing $-\log_{10}$ (p-value) vs. $\log_2$ (FC) were generated using in R to identify metabolites of interest. Hierarchical clustering heatmaps were visualized using the pheatmap R package (version: 3.4.3). Significant correlations (p-value ≤ 0.05) were plotted with the corrplot package (version: 0.95). Functional annotation of metabolites and pathway analysis used the KEGG database. Metabolic pathway enrichment was assessed using the metric, Enrichment Ratio = (x/n)/(y/N), where x = number of significant metabolites in a pathway, n = total significant metabolites, y = number of metabolites annotated to the pathway in KEGG, and N = total annotated metabolites in the dataset. Pathways that satisfied p-value ≤ 0.05 after multiple testing correction were considered significantly enriched.

The Gene Set Enrichment Analysis (GSEA) was conducted on the KEGG entries based on quantitative changes in metabolites. The analysis calculates an Enrichment Score (ES), which quantifies the degree to which a metabolite set is concentrated at the extremes of the ranked list. Statistical significance is then evaluated through the p-value of the ES, the Normalized Enrichment Score (NES) (a size-normalized version of ES), and the False Discovery Rate (FDR), derived from NES. Pathways meeting the following criteria: $|NES| > 1$, p-value ≤ 0.05, and FDR < 0.25 were considered significantly enriched [41,42]. The analyses and data processing were primarily performed using Python (version: 3.5.0), while visualizations were generated with R language (version: 3.4.3). Detailed information on the software used is provided in Table S1.

3. Result

3.1. Data Quality Control and Metabolite Identification in Cauda Epididymal Fluid from Yak and Cattle

Distinct morphological differences were documented between yak and cattle epididymides. Our findings revealed that the length and weight of the yak epididymis are both lower than those of cattle at similar body weights (Figure S1A,B). In addition, the yak epididymis exhibited a vibrant red coloration due to dense vascular distribution, whereas the cattle epididymis appeared pale pink. Notably, both the caput and cauda of the yak epididymis demonstrated prominent enlargement, while their corpus (body) segment remained comparatively short and thick. Conversely, the cattle epididymis displayed moderate expansion at the head and tail portions with an elongated, slender corpus (Figure S1C,D).

In this study, the QC samples from yak and cattle epididymal cauda fluid exhibited a Pearson correlation coefficient of 0.994 (Figure S2A,B), confirming robust data integrity. The Partial Least Squares Discrimination Analysis (PLS-DA), a supervised discriminant analysis method, was validated through permutation testing to detect overfitting. Model reliability was defined by R^2 (goodness-of-fit) exceeding Q^2 (predictive ability), with the Q^2 regression line intercepting the Y-axis below zero [43,44]. Our PLS-DA permutation results demonstrated $R^2 > Q^2$ and negative Y-axis intercepts for Q^2 regression lines under both positive and negative ion modes (Figure S2C,D), validating model reliability for subsequent analyses.

A total of 1098 kinds of metabolites were identified in the cauda epididymal fluid from yak and cattle in positive ion mode, annotated by the HMDB database, spanning fifteen major categories. Among these, lipids and lipid-like molecules (264 types of metabolites, 24.04%), organic acids and derivatives (n = 246, 22.40%), and organoheterocyclic compounds (n = 242, 22.04%), collectively accounted for the most significant proportions. Other categories included benzenoids (n = 129, 11.75%), organic oxygen compounds (n = 73, 6.65%), organic nitrogen compounds (n = 42, 3.83%), phenylpropanoids and polyketides (n = 39, 3.55%), organosulfur compounds (n = 24, 2.19%), nucleosides, nucleotides, and analogues (n = 17, 1.55%), alkaloids and derivatives (n = 12, 1.09%), hydrocarbons (n = 1, 0.09%), lignans, neolignans and related compounds (n = 1, 0.09%), hydrocarbon derivatives (n = 1, 0.09%), and homogeneous non-metal compounds (n = 1, 0.09%), with an additional six metabolites (0.55%) classified as unclassified compounds (Figure 1A; Tables S2 and S3). In contrast, a total of 1297 metabolites were identified in the cauda epididymal fluid of yak and cattle in the negative ion mode and annotated by the HMDB database, classified into 17 major categories. These metabolites were categorized as follows: organic acids and derivatives (n = 309, 23.82%), lipids and lipid-like molecules (n = 290, 22.26%), organoheterocyclic compounds (n = 244, 18.81%), benzenoids (n = 168, 12.95%), organic oxygen compounds (n = 109, 8.40%), nucleosides, nucleotides, and analogues (n = 62, 4.78%), phenylpropanoids and polyketides (n = 57, 4.39%), alkaloids and derivatives (n = 16, 1.23%), organic nitrogen compounds (n = 12, 0.93%), organosulfur compounds (n = 8, 0.62%), organohalogen compounds (n = 5, 0.39%), lignans, neolignans and related compounds (n = 4, 0.31%), homogeneous non-metal compounds (n = 2, 0.15%), organophosphorus compounds (n = 1, 0.08%), and unclassified compounds (n = 10, 0.77%) (Figure 1B; Tables S4 and S5).

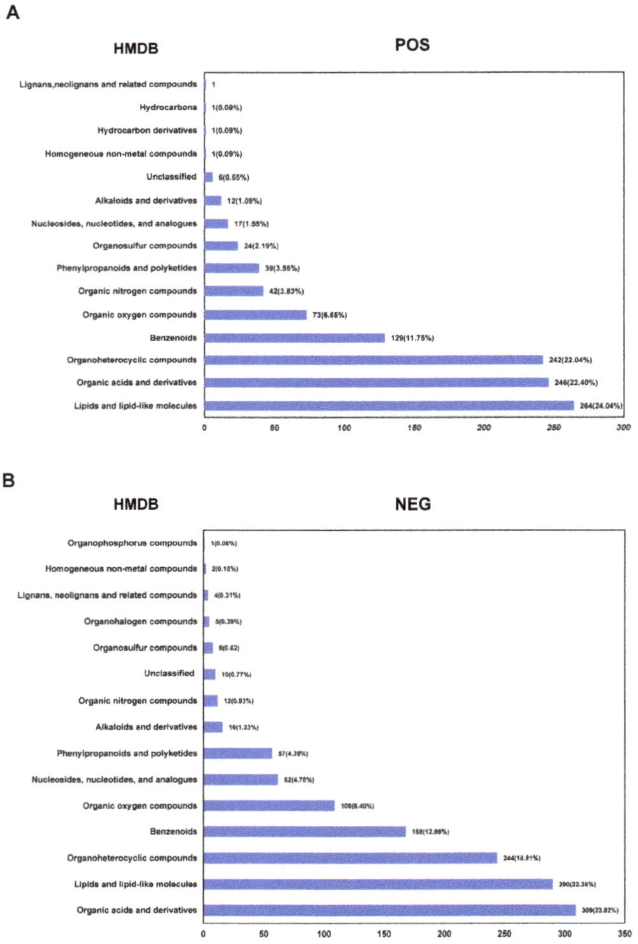

Figure 1. Statistical analysis of metabolites in epididymal tail fluid of yak and cattle based on the HMDB database. The HMDB classification and annotation statistics for metabolites under positive (**A**) and negative (**B**) ion modes.

The LIPID MAPS is a comprehensive database containing biologically relevant lipid structures and annotations [45,46]. Based on the LIPID MAPS classification and annotation of lipid metabolites in the cauda epididymal fluid of both yak and cattle, the results indicated the presence of six major lipid categories in positive ion mode, including fatty acyls, glycerolipids, glycerophospholipids, sphingolipids, sterol lipids, and polyketides. Within these, the category containing the most subclasses was fatty acyls (n = 112), and the subclass with the highest number of metabolites was fatty acids and conjugates (n = 37) (Figure 2A; Tables S6 and S7). In contrast, the lipids in the negative ion mode included seven major categories: fatty acids, glycerolipids, glycerophospholipids, sphingolipids, sterol lipids, polyketides, and prenol lipids. Among these, fatty acyls contained the most subclasses (n = 189), and the subclass with the highest number of metabolites was fatty acids and conjugates (n = 101) (Figure 2B; Tables S8 and S9).

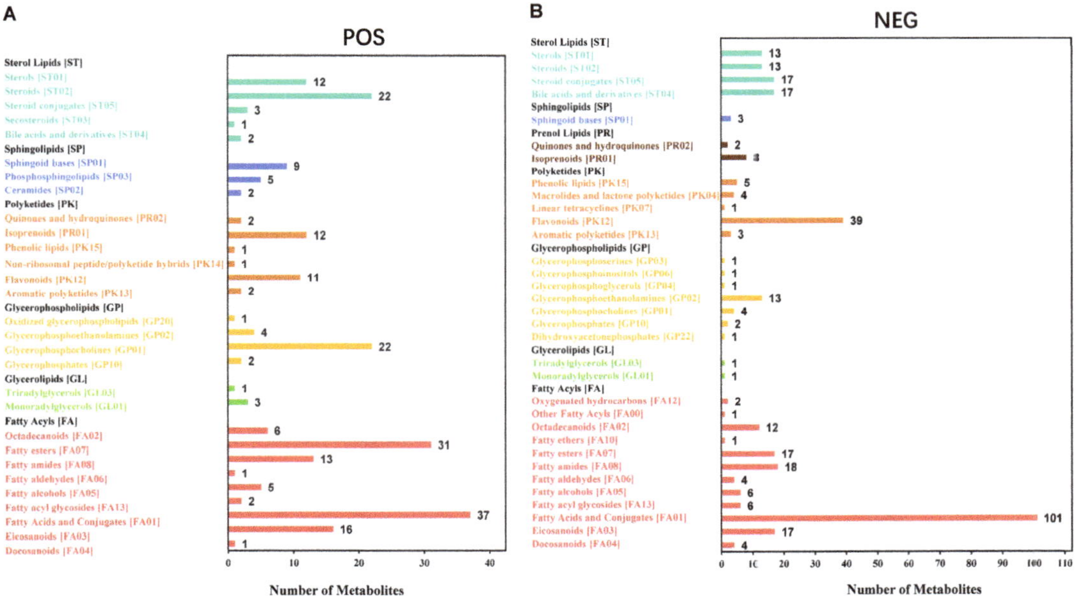

Figure 2. Statistical results of lipid metabolites in the cauda epididymis of yaks and cattle. Classification of lipid metabolites under positive (**A**) and negative (**B**) ion modes in cauda epididymis of yaks and cattle according to the LIPID MAPS database. The x-axis represents the number of metabolites. The y-axis represents the annotated LIPID MAPS lipid classification. This figure shows the number of metabolites corresponding to the main class level under the eight major lipid categories in the LIPID MAPS.

3.2. Identification of Differential Metabolites in Cauda Epididymal Fluid Between Yak and Cattle

The clustering heatmap of total differential metabolites across 12 samples is presented in Figure 3A,B. Among the 1861 metabolites identified in positive ion mode, using the screening criteria of VIP > 1.0, FC > 1.5 or FC < 0.667, and p-value $\leq$ 0.05 for significant differences, 79 metabolites were significantly upregulated, while 212 were significantly downregulated in yak cauda epididymal fluid compared with cattle (Figure 3C; Table S10). In addition, among the 2360 metabolites identified in the negative ion mode, 110 were significantly upregulated and 230 were significantly downregulated in yak cauda epididymal fluid compared with cattle (Figure 3D; Table S11).

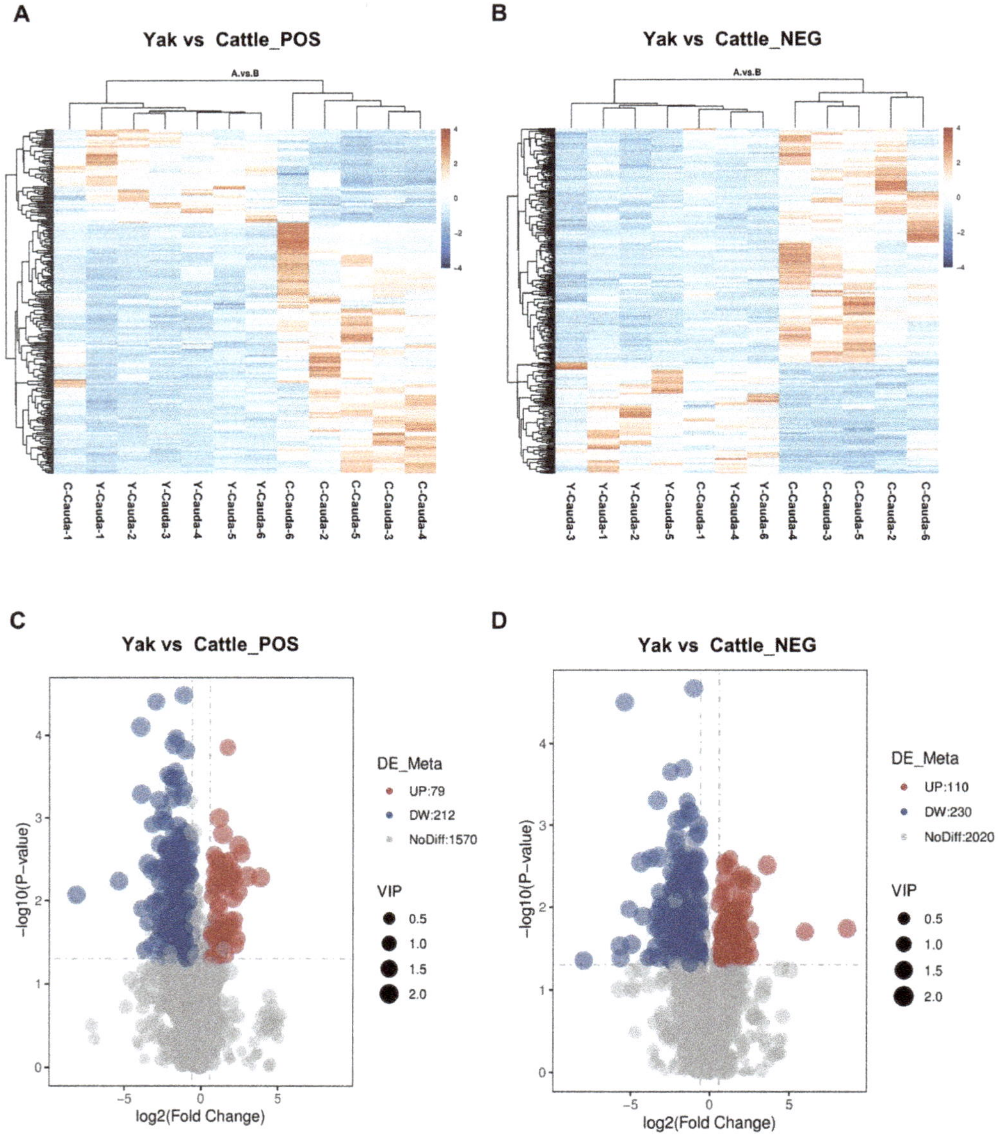

Figure 3. Analysis of differential metabolites in cauda epididymal fluid from yak and cattle. Clustering heatmap of differentially expressed metabolites in positive (**A**) and negative (**B**) ion modes. Y-Cauda-1 to Y-Cauda-6 represent six biologically independent samples of epididymal cauda fluid from yak and C-Cauda-1 to C-Cauda-6 represent six biologically independent samples of epididymal cauda fluid from cattle. The metabolites were clustered along the vertical axis, while sample groups were arranged along the horizontal axis. Shorter dendrogram branches indicate higher similarity between the connected elements (metabolites or samples). The volcano plots comparing significantly differentially expressed metabolites in positive (**C**) and negative (**D**) ion modes: The *x*-axis represents $\log_2$ (fold change) ($\log_2$FC) in metabolite expression between two groups, and the *y*-axis indicates the level of differential significance ($-\log_{10}$ (*p*-value)). Each point denotes an individual metabolite, with point size corresponding to VIP values. The metabolites showing statistically significant upregulation were highlighted in red, and those with significant downregulation were depicted in blue.

Lollipop plots were generated to visualize the 12 metabolites exhibiting the most significant fold changes in positive and negative ion modes (Figure 4A,B). In the comparison of yak vs. cattle cauda epididymal fluid in positive ion mode, the three metabolites with the most significant upregulation were 1-Methyl-corypalline, 3-(4-chlorophenyl)-7-hydroxy-2-methyl-4H-chromen-4-one, and 2-(7-methyl-2-oxo-2H-chromen-4-yl)-N-(1,3-thiazol-2-yl) acetamide. Conversely, the three metabolites with the most significant downregulation were 8-Oxoguanosine, Aspergillitine, and Gamma-glutamyl-s-allylcysteine (Figure 4A). In the negative ion mode, for the yak vs. cattle cauda epididymal fluid comparison, the three metabolites with the most significant upregulation were (-)-Epigallocatechin, Catechin gallate, and 24-Epicastasterone. The three metabolites with the most significant downregulation were 8-Oxodeoxycoformycin, N3,5′-Cycloxanthosine, and 2,8-Dihydroxyadenine (Figure 4B).

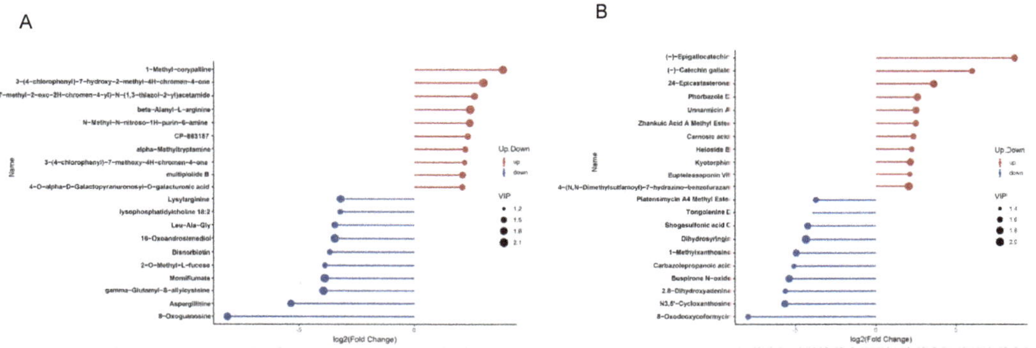

Figure 4. Lollipop plots for significantly differentially expressed metabolites in positive (**A**) and negative (**B**) ion modes. The lollipop plots visualize the top ten up/downregulated metabolites based on absolute fold-change magnitude. The *x*-axis represents $\log_2$FC values and the *y*-axis lists metabolite names. The point size corresponds to VIP values, with the red points denoting significantly upregulated metabolites and the blue points indicating significantly downregulated metabolites.

To assess the coordinated or antagonistic relationships between different metabolites, Pearson correlation coefficients were calculated among all significant differential metabolites in the cauda epididymal fluid of yak vs. cattle to evaluate the co-variation trends between metabolites. The top 20 significant differential metabolites ranked by ascending *p*-value, in positive and negative ion mode, are shown (Figure S3A,B; Tables S12 and S13). A correlation value of 1 indicates a perfect positive correlation (red), while −1 indicates a perfect negative correlation (blue). Uncolored areas represent correlations with p-values > 0.05. As shown in the figure, the vast majority of significant differentially expressed metabolites exhibited positive correlations. These results are also visualized in the chord diagrams (Figure S3C,D).

3.3. *Functional Enrichment Analyses of Differential Metabolites in Cauda Epididymal Fluid Between Yak and Cattle*

To investigate the functional implications of significant metabolic differences, the KEGG pathway enrichment analysis was conducted via the GSEA. By applying the GSEA to the KEGG entries based on quantitative metabolite changes, significantly altered metabolites in positive ion mode were enriched in three pathways: 'Biosynthesis of amino acids' (map01230), 'Metabolic pathways' (map01100), and 'ABC transporters' (map02010). Significantly altered metabolites in the negative ion mode were enriched in five pathways, including '2-Oxocarboxylic acid metabolism' (map01210), 'Tyrosine metabolism' (map00350),

Biosynthesis of amino acids (map01230), Purine metabolism (map00230), and 'Metabolic pathways' (map01100) (Figure 5A,B; Tables S14–S19).

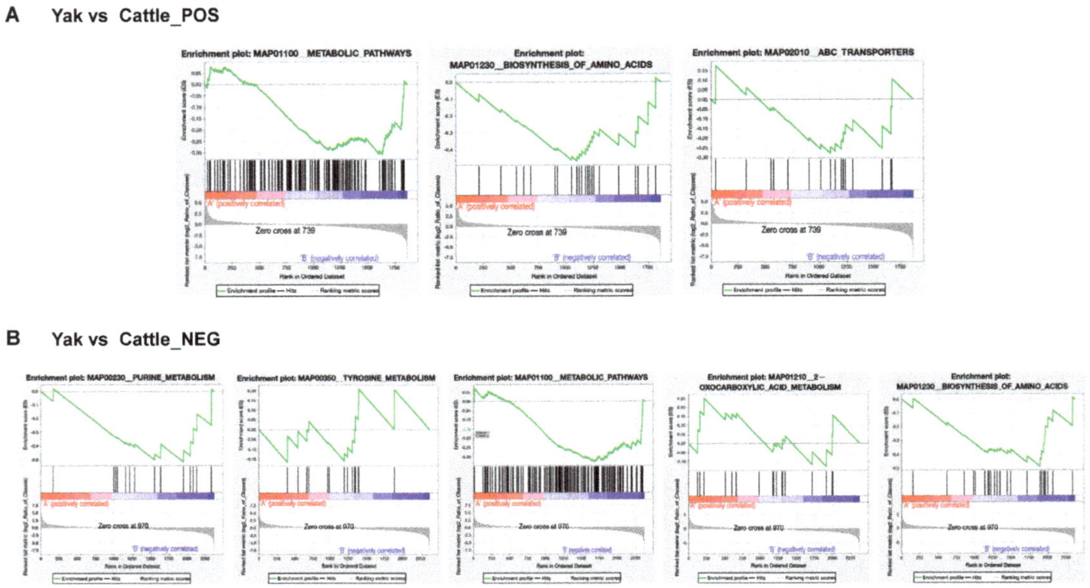

Figure 5. The Gene Set Enrichment Analysis (GSEA) of significantly altered cationic (**A**) and anionic metabolites (**B**). The upper section displays an enrichment score (ES) line graph for functional terms. The peak in this graph represents the ES value of a metabolite functional set. A higher absolute ES value indicates stronger enrichment, where a positive ES signifies enrichment of the functional set-in upregulated metabolites, while a negative ES denotes enrichment in downregulated metabolites. In the middle section, the positions of functional set members within the ranked expression list are visualized: the black vertical marks indicate members of the functional set; and the colored bars represent the ranked expression list (ordered by correlation), with red for upregulated members and blue for downregulated ones. The lower section shows the log_2FC values of all members in the expression set.

3.4. Identification of Differential Metabolites Related to Sperm Quality, Function or Metabolism

Among the differential metabolites between yak and cattle epididymal cauda fluid, six metabolites in positive ion mode and twelve metabolites in negative ion mode were associated with sperm quality, mature function or metabolism (Figure 6; Tables 1 and 2). Among these, five metabolites (Flavone, Sulforaphane, Glutathione, Danshensu, and Idebenone) were reported to enhance sperm motility through their antioxidant capacity; one metabolite, Indolelactic acid, was found to reduce sperm count by elevating oxidative stress levels; and five metabolites (Adenosine, Niflumic acid, Chlorpromazine, BIFENTHRIN, and IBMX) were linked to sperm acrosome reaction, capacitation, or chemotaxis. In addition, three metabolites (Acetylcarnitine, Taurine, and Carnosic acid) were documented to improve sperm energy metabolism; one metabolite (N6-Methyladenosine) was associated with sperm DNA methylation; and another metabolite (Glyceraldehyde 3-phosphate) was implicated in sperm flagellar movement. Additionally, two metabolites (Tranilast, Succinylproline) were clinically reported to affect sperm motility, though their mechanisms remain unclear.

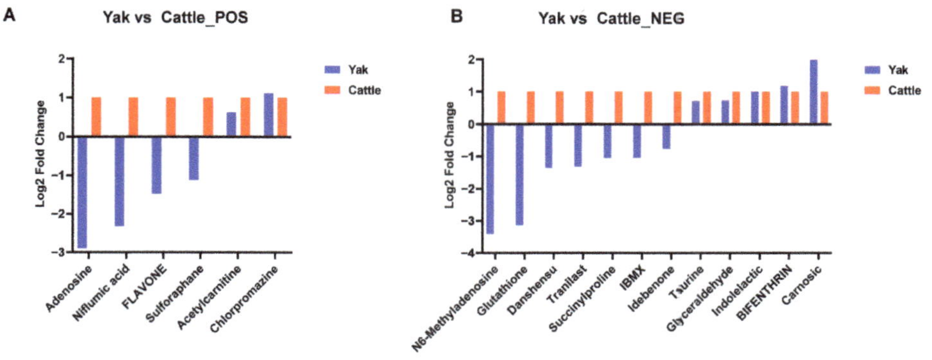

Figure 6. The fold-change of significantly differential expressed metabolites related to sperm mobility, mature or function in epididymal fluid between yak and cattle in positive (**A**) and negative (**B**) ion modes, the *x*-axis indicates the names of metabolites and the *y*-axis represents $\log_2$FC values.

Table 1. Differential metabolites related to sperm quality, function or metabolism in epididymal fluid between yak and cattle detected in positive ion mode.

Name	Formula	Classification	Log_2FC	*p*-Value	Function Related to Sperm
Adenosine	$C_{10}H_{13}N_5O_4$	Nucleosides, nucleotides, and analogues	−2.8923	0.0145	Inhibition of spontaneous acrosome reaction [47]
Niflumic acid	$C_{13}H_9F_3N_2O_2$	Benzenoids	−2.3234	0.0470	Blocking sperm acrosome reaction and chemotaxis [48,49]
FLAVONE	$C_{15}H_{10}O_2$	Phenylpropanoids and polyketides	−1.4766	0.0042	Protecting against acrosome damage, capacitation, apoptosis, and prevents lipid peroxidation during sperm cryopreservation [50]
Sulforaphane	$C_6H_{11}NOS_2$	Organosulfur compounds	−1.1168	0.0246	Reducing oxidative stress and endoplasmic reticulum stress in sperm, thus improving sperm motility and fertilization rates in frozen semen [51–54]
Acetylcarnitine	$C_9H_{17}NO_4$	Lipids and lipid-like molecules	0.6234	0.0375	Enhancing sperm energy metabolism and motility while reducing apoptosis [25,55].
Chlorpromazine	$C_{17}H_{19}ClN_2S$	Organoheterocyclic compounds	1.1107	0.0110	Inhibition of epididymal sperm capacitation and acrosomal reaction [56]

Table 2. Differential metabolites related to sperm quality, function or metabolism in epididymal fluid between yak and cattle detected in the negative ion mode.

Name	Formula	Classification	Log_2FC	*p*-Value	Function Related to Sperm
N6-Methyladenosine	$C_{11}H_{15}N_5O_4$	Nucleosides, nucleotides, and analogues	−3.3979	0.0177	Increased N6-Methyladeno-sine content is a risk factor for asthenozoospermia and affects sperm motility [57].
Glutathione	$C_{10}H_{17}N_3O_6S$	Organic acids and derivatives	−3.1306	0.0152	Increasing sperm count, motility and fertilization rate by reducing ROS levels [58–60].
Danshensu	$C_9H_{10}O_5$	Phenylpropanoids and polyketides	−1.3530	0.0481	Improving seminal quality and antioxidant capacity [61].
Tranilast	$C_{18}H_{17}NO_5$	Phenylpropanoids and polyketides	−1.3146	0.0024	Increasing sperm count [62,63].
Succinylproline	$C_9H_{13}NO_5$	Organic acids and derivatives	−1.0481	0.0248	Inhibiting spermatozoa motility [64].
IBMX	$C_{10}H_{14}N_4O_2$	Organoheterocyclic compounds	−1.0398	0.0028	Ameliorate sperm capacitation and motility and mobility [65].
Idebenone	$C_{19}H_{30}O_5$	Lipids and lipid-like molecules	−0.7619	0.0246	Improving post thaw sperm quality by mitigating oxidative stress [66–68].
Taurine	$C_2H_7NO_3S$	Organic acids and derivatives	0.7165	0.0287	Enhancing sperm mitochondrial energy metabolism [69].
Glyceraldehyde 3-phosphate	$C_3H_7O_6P$	Organic oxygen compounds	0.7332	0.0473	Facilitating sperm flagellum motility [70].
Indolelactic acid	$C_{11}H_{11}NO_3$	Organoheterocyclic compounds	1.0075	0.0377	Increasing ROS level, decreasing antioxidant activity and reducing sperm number [71].
BIFENTHRIN	$C_{23}H_{22}ClF_3O_2$	Benzenoids	1.1856	0.0299	Diminishing motility, spontaneous acrosome reaction, and capacitation in sperm [72,73].
Carnosic acid	$C_{20}H_{28}O_4$	Lipids and lipid-like molecules	2.3463	0.0296	Improving the quality and mitochondrial function of frozen-thawed sperm [74].

4. Discussion

Previous research indicates that the epididymal lumen maintains an acidic microenvironment, which is crucial for maintaining spermatozoa in a quiescent state and preventing premature capacitation [75,76]. In this study, under the positive ion mode, 'organic acids and derivatives' constituted the largest category of metabolites identified in yak epididymal fluid. This finding suggests that such substances likely play a crucial role in maintaining the low pH environment within the yak epididymal cauda fluid.

Lipid-like molecules also constitute a significant portion of the metabolome in the epididymal cauda fluid of both yaks and cattle. Under positive ion mode, 411 lipid-like metabolite features were identified, while 602 were found under negative ion mode. Database annotation revealed that fatty acids constituted the most abundant subclass at the second-level classification in both modes. This suggests that fatty acids may play a crucial role in sperm maturation within the yak epididymis. Existing research indicates that fatty acids are vital for sperm maturation in humans, pigs, and cattle. For example, unsaturated fatty acids enhance boar sperm motility after storage at 6 °C [77]. Short-chain fatty acids have been established to modulate human sperm migration via olfactory receptor 51E2 activity [78]. Furthermore, saturated fatty acids accelerate the linear motility of bull sperm through mitochondrial ATP production [79]. We propose that the altered fatty acid composition observed in yak epididymal fluid, particularly the relative deficit in certain polyunsaturated fatty acids, may contribute to differences in membrane properties and metabolic microenvironment compared with cattle. These differences warrant further investigation to determine their potential implications for sperm function adaptation in high-altitude environments.

During storage in the cauda epididymis, sperm maintain a chronic hypoxic state to minimize metabolic activity and prevent premature capacitation [17,80]. The cauda epididymal environment, rich in specific proteins and enzymes and with regulated pH, contributes to maintaining this quiescence. Comparative metabolomic profiling between yaks and cattle revealed differential abundance of certain metabolites in the cauda epididymal fluid. Five antioxidant metabolites, including Glutathione, Flavone, Sulforaphane, Danshensu, and Idebenone [50–54,58–61,66,68,81], exhibited significantly decreased levels in yak cauda epididymal fluid compared to that of cattle. Conversely, a pro-oxidant metabolite, Indolelactic acid [71], showed a marked increase in yak cauda epididymal fluid. While oxidative stress is known to damage sperm membrane structure, induce lipid peroxidation, impair energy metabolism, and accelerate DNA fragmentation and apoptosis [82,83], the functional implications of these observed metabolic differences between species require further targeted investigation. Future studies aimed at validating these findings could explore the targeted supplementation of these antioxidants or inhibitors of the identified pro-oxidants into yak frozen or fresh semen to assess their potential in enhancing sperm vitality and quality by improving the antioxidant and oxidant balance.

Under physiological conditions, sperm capacitation primarily occurs within the endometrium and oviduct; however, premature capacitation occurs when this process initiates in suboptimal microenvironments or concludes prematurely under poorly regulated in vitro conditions [82,84]. During natural fertilization, the interaction between sperm and the zona pellucida of the oocyte triggers the acrosome reaction, releasing acrosomal enzymes to dissolve the zona and enable sperm-egg binding. The spontaneous release of acrosomal enzymes prior to zona pellucida contact or other physiological inducers is classified as a spontaneous acrosome reaction [84,85]. Both processes cause sperm to prematurely lose their egg-binding capacity, compromising fertilization success. Our comparative metabolomic analysis revealed differential abundance of several metabolites in yak cauda epididymal fluid compared with cattle. Four metabolites with significantly different abundances in yak cauda epididymal fluid compared with cattle, including Adenosine, Niflumic acid, Chlorpromazine, and Bifenthrin, which have been reported in existing literature to influence premature capacitation and spontaneous acrosome reaction [47–49,56,72,73], showed significant abundance differences. Among these metabolites, Adenosine and Niflumic acid were significantly downregulated in yak cauda epididymal fluid, while Chlorpromazine and Bifenthrin were significantly upregulated. Additionally, one metabolite, IBMX, previously associated with promotion of premature capacitation [65], was significantly downregulated in yak cauda epididymal fluid. This observed disparity in the abundance of metabolites with purported opposing functions suggests a compositional shift in the yak epididymal microenvironment. The metabolic differences identified here provide a basis for future targeted research into their functional significance regarding sperm capacitation regulation in yaks. Further investigation via in vitro fertilization (IVF) assays is necessary to determine whether these metabolic changes directly compromise yak sperm fertilizing capacity.

Sperm remain in a quiescent state within the epididymis for extended periods. During this period, mitochondria minimize the ATP synthesis rates and inhibit oxidative phosphorylation activity to prevent premature energy depletion [17,80]. Our comparative metabolomic analysis revealed that the levels of three metabolites previously associated with mitochondrial energy metabolism in other systems, including Taurine, Carnosic acid, and Acetylcarnitine [26,55,69,74], were significantly upregulated in yak cauda epididymal fluid compared to that of cattle. This finding points to a species-specific difference in the metabolic composition of the epididymal microenvironment between yaks and cattle. The observed metabolic differences highlight a comparative divergence between yaks and cattle and provide a basis for future targeted research into their functional significance regarding sperm energy regulation.

N6-Methyladenosine (m6A), the most prevalent post-transcriptional modification in eukaryotic mRNA and non-coding RNA (ncRNA), regulates the RNA fate to influence gene expression and has been associated with sperm motility and energy metabolism [57]. Studies have reported differential m6A levels across the caput, corpus, and cauda epididymis of yaks, with significantly lower levels observed in the cauda compared with the corpus [86]. In the present study, we found that m6A levels in yak cauda epididymal fluid were significantly lower than those in cattle, suggesting a species-specific difference in epididymal m6A modification. This comparative metabolomic profiling highlights a distinct m6A landscape between yak and cattle, which may serve as a foundation for further targeted research on its regulatory role in sperm maturation. However, direct functional inferences regarding sperm function and fertility cannot be drawn from these untargeted findings and require subsequent validation, especially given the species-specific nature of processes such as sperm migration and fertilization in ruminants.

The Gene Set Enrichment Analysis (GSEA) effectively addresses the limitations of traditional enrichment methods in information mining, providing a more comprehensive interpretation of the regulatory roles of functional units. Its core principle involves using predefined metabolite sets to rank metabolites based on their differential expression magnitude between two sample groups [41,42]. The GSEA analysis indicated that the biosynthesis of amino acids was one of the pathways significantly enriched with differential metabolites between yak and cattle epididymal cauda fluid. Amino acid biosynthesis represents one of the core biological processes in living organisms. Beyond their fundamental role in protein synthesis, amino acids function as signaling molecules and are indispensable for diverse metabolic functions including gluconeogenesis, lipogenesis, ATP generation, and nucleotide base synthesis. Additionally, amino acids play significant roles in cellular protection against oxidative stress [87,88]. Studies have indicated that the biosynthetic capacity of certain amino acids is significantly reduced in human azoospermic patients [89]. Our research further demonstrated that, compared with cattle epididymal cauda fluid, the levels of five amino acids in yak epididymal cauda fluid–L-Saccharopine, L-Glutamine, S-Adenosylhomocysteine, L-Arogenate, and L-Valine-were significantly decreased. These findings highlight distinct metabolomic profiles between yaks and cattle, which may reflect species-specific metabolic adaptations or regulatory mechanisms. Further investigation is warranted to determine the precise roles of these specific amino acids in yak sperm metabolism and whether the observed differences are associated with specific physiological traits in yaks.

It should be noted that although epididymal fluid plays a crucial role in sperm maturation, spermatozoa are subsequently resuspended in seminal plasma and enter the female reproductive tract. Therefore, the metabolites identified in this study as potential regulators of yak sperm quality and function in the epididymis may be modified, diluted, or even eliminated during ejaculation and entry into the female reproductive tract. Future research should include further quantification of these metabolites in seminal plasma, as well as more rigorous in vitro and in vivo experiments to verify whether these metabolites indeed influence yak sperm quality or function. Additionally, it is important to emphasize that limitations such as sample size, variations in feeding management levels, and fluctuations in environmental conditions may have affected the experimental results. Subsequent studies should pay particular attention to these factors.

5. Conclusions

This study analyzed and compared the metabolomes of the epididymal cauda in yak and cattle. The findings provide foundational data and novel research perspectives for in-depth elucidation of the molecular mechanisms of sperm maturation in yaks, with potential implications for improving sperm motility and optimizing assisted reproductive technologies, such as cryopreservation, in vitro fertilization, and artificial insemination. However, given the species-specific nature of metabolic influences on sperm function, direct extrapolations regarding fertility outcomes should be avoided until further validation. Future research should prioritize in vitro validation of the specific regulatory roles of these metabolites in yak sperm functionality and explore their practical applications in assisted reproduction, thus contributing to the development of targeted strategies for enhancing yak reproductive efficiency.

Supplementary Materials: The following supporting information can be downloaded at: https://www.mdpi.com/article/10.3390/ani15192861/s1, Figure S1. Morphology analysis of yak and cattle epididymis. (A,B) The lengths (C) and weights (D) of epididymis from yak and cattle. A total of six yak epididymides and six cattle epididymides were used for data analysis, with each epididymis originating from an independent individual. * p-value < 0.05 and *** p-value < 0.001, Student's t-test. (C,D) Representative images of yak (C) and cattle (D) epididymis. The caput, corpus, and cauda of the epididymis are labeled in the figure. The two epididymides (upper and lower) in each panel originate from the same individual (n = 6). Figure S2. The result of data quality control. (A,B) Quality controlled (QC) correlation analysis of epididymal cauda fluid from yaks and cattle in positive (A) and negative (B) ion modes. The x and y axes represent the QC samples of corresponding species, with numbers indicating the Pearson correlation coefficients. (C,D) Validation plots of the Partial Least Squares Discrimination Analysis (PLS-DA) in positive (C) and negative (D) ion modes. The x-axis denotes the correlation between randomly permuted group Y and original group Y while the y-axis represents scores for R^2 (goodness-of-fit) and Q^2 (predictive ability). Figure S3. Correlation analysis of differential metabolites. (A,B) Correlation heatmap of differential metabolites in positive (A) and negative (B) ion modes. This visualization displays the top 20 differentially expressed metabolites, selected by ranking pairwise Pearson correlation coefficients between all differential metabolites in ascending order of p-value significance. Correlations range from +1 (indicating a perfect positive correlation, shown in red) to −1 (indicating a perfect negative correlation, shown in blue). Areas with no color shading represent non-significant correlations (p-value > 0.05). (C,D) Chord diagram of differentially expressed metabolites in positive (C) and negative (D) ion modes. The top 20 differentially expressed metabolites, ranked by ascending p-value, were selected based on pairwise Pearson correlation coefficients for visualization in this chord diagram. Each node represents a metabolite. The node size corresponds to the magnitude of the log-transformed fold change (FC) value for that metabolite. The node color indicates the metabolite's class. The color of the curved lines connecting nodes represents the correlation between metabolites: blue indicates a negative correlation, while red indicates a positive correlation. Table S1. The information of software used in this study. Table S2. List of the HMDB-annotated metabolites detected in the positive ion mode. Table S3. Classification statistics of the HMDB for metabolites detected in the positive ion mode. Table S4. List of the HMDB-annotated metabolites detected in the negative ion mode. Table S5. Classification statistics of the HMDB for metabolites detected in the negative ion mode. Table S6. List of the LIPID MAPS annotation results for metabolites detected in the positive ion mode. Table S7. Classification statistics of LIPID MAPS for metabolites detected in the positive ion mode. Table S8 List of LIPID MAPS annotation results for metabolites detected in the negative ion mode. Table S9 Classification statistics list of the LIPID MAPS for metabolites detected in the negative ion mode. Table S10. List of significant differential metabolites analysis in the positive ion mode. Y Cauda1 to Y Cauda6 represent six biologically independent samples of epididymal cauda fluid from yak and C Cauda1 to C Cauda6 represent six biologically independent samples of epididymal cauda fluid from cattle. The same as below. Table S11. List of significant differential metabolites detected in the negative ion mode. Table S12. List of significant differential metabolites detected in the positive ion mode. Table S13. List of significant differential metabolites detected in the negative ion mode. Table S14. The KEGG classification list of differential metabolites detected in the positive ion mode. Table S15. The KEGG classification list of differential metabolites detected in the negative ion mode. Table S16. The KEGG enrichment list of differential metabolites detected in the positive ion mode. Table S17. The KEGG enrichment list of differential metabolites detected in the negative ion mode. Table S18. The GSEA enrichment results detected in the positive ion mode. Table S19. The GSEA enrichment results detected in the negative ion mode.

Author Contributions: Conceptualization, G.Y. and X.Y.; methodology, G.Y., X.Y. and D.L. (Dongju Liu); software, W.Z.; validation, A.Z. and Y.X.; formal analysis, X.X. and W.F.; investigation, J.L.; resources, Y.X., X.X., D.L. (Daoliang Lan), and S.Y.; data curation, Y.X.; writing—original draft preparation, G.Y.; writing—review and editing, S.Y., X.Y. and J.L.; visualization, W.F. and A.Z.; supervision, D.L. (Daoliang Lan) and S.Y.; project administration, S.Y.; funding acquisition, S.Y. All authors have read and agreed to the published version of the manuscript.

Funding: This work was supported by The National Key Research and Development Program (2023YFD1300603), China Agriculture Research System of MOF and MARA (CARS-37), the Key Research and Development Program of Sichuan Provincial Science and Technology Program (2023YFN0001) and Southwest Minzu University Double World-Class Project (XM2023021).

Institutional Review Board Statement: All procedures were ap-proved by the Animal Care and Ethics Committee of Southwest Minzu University (SWU-202501136); 20 June 2025.

Informed Consent Statement: Not applicable.

Data Availability Statement: The raw data of metabolomic sequencing were deposited in the MetaboLights database (https://www.ncbi.nlm.nih.gov/bioproject, accessed on 20 July 2025) with the accession number MTBLS13067. Other data presented in this study are included in the published article and Supplementary Materials. Further inquiries can be directed to the corresponding author.

Acknowledgments: We thank Beijing Novogene Zoetry Co., Ltd. for their technical support in sequencing, as well as to the Researchers editorial team (http://www.home-for-researchers.com, accessed on 28 July 2025) for their insightful suggestions on manuscript revision. The authors have reviewed and edited the output and take full responsibility for the content of this publication.

Conflicts of Interest: The authors declare no conflicts of interest.

Abbreviations

The following abbreviations are used in this manuscript:

HMBD	Human Metabolome Database
KEGG	Kyoto Encyclopedia of Genes and Genomes
PBS	Phosphate Buffered Saline
QC	Quality Control
CV	Coefficient of variation
PLS-DA	Partial least squares-discriminant analysis
VIP	Variable Importance in the Projection
FC	Fold Change
GSEA	Gene Set Enrichment Analysis
ES	Enrichment Score
NES	Normalized Enrichment Score
FDR	False Discovery Rate
PLS-DA	Partial Least Squares Discrimination Analysis
R^2	Goodness-of-fit
Q^2	Predictive ability
ncRNA	non-coding RNA
XCMS	Xtracted Chromatograms and Mass Spectra

References

1. Lee, V.; Hinton, B.T.; Hirashima, T. Collective cell dynamics and luminal fluid flow in the epididymis: A mechanobiological perspective. *Andrology* **2024**, *12*, 939–948. [CrossRef]
2. Sosnicki, D.M.; Cohen, R.; Asano, A.; Nelson, J.L.; Mukai, C.; Comizzoli, P.; Travis, A.J. Segmental differentiation of the murine epididymis: Identification of segment-specific, GM1-enriched vesicles and regulation by luminal fluid factors. *Biol. Reprod.* **2023**, *109*, 864–877. [CrossRef]
3. Cichowska, A.W.; Wisniewski, J.; Bromke, M.A.; Olejnik, B.; Mogielnicka-Brzozowska, M. Proteome Profiling of Canine Epididymal Fluid: In Search of Protein Markers of Epididymal Sperm Motility. *Int. J. Mol. Sci.* **2023**, *24*, 14790. [CrossRef] [PubMed]
4. Weber, A.; Argenti, L.E.; de Souza, A.P.B.; Santi, L.; Beys-da-Silva, W.O.; Yates, J.R., 3rd; Bustamante-Filho, I.C. Ready for the journey: A comparative proteome profiling of porcine cauda epididymal fluid and spermatozoa. *Cell Tissue Res.* **2020**, *379*, 389–405. [CrossRef]
5. Xue, X.R.; Ma, Y.M.; Hu, N.X.; Zhao, C.C.; Jin, S.H. Experimental Comparison of Post-thaw Motility and Survival Time of Semen from Different Breeds of Bulls in Plateau Area. *China Dairy Cattle* **2017**, 17–20. [CrossRef]
6. Rosenfeld, C.S.; Javurek, A.B.; Johnson, S.A.; Lei, Z.; Sumner, L.W.; Hess, R.A. Seminal fluid metabolome and epididymal changes after antibiotic treatment in mice. *Reproduction* **2018**, *156*, 1–10. [CrossRef] [PubMed]
7. James, E.R.; Carrell, D.T.; Aston, K.I.; Jenkins, T.G.; Yeste, M.; Salas-Huetos, A. The Role of the Epididymis and the Contribution of Epididymosomes to Mammalian Reproduction. *Int. J. Mol. Sci.* **2020**, *21*, 5377. [CrossRef]
8. Sullivan, R.; Saez, F.; Girouard, J.; Frenette, G. Role of exosomes in sperm maturation during the transit along the male reproductive tract. *Blood Cells Mol. Dis.* **2005**, *35*, 1–10. [CrossRef]
9. Wang, G.; Li, Y.; Yang, Q.; Xu, S.; Ma, S.; Yan, R.; Zhang, R.; Jia, G.; Ai, D.; Yang, Q. Gene expression dynamics during the gonocyte to spermatogonia transition and spermatogenesis in the domestic yak. *J. Anim. Sci. Biotechnol.* **2019**, *10*, 64. [CrossRef]
10. Dacheux, J.L.; Belleannée, C.; Jones, R.; Labas, V.; Belghazi, M.; Guyonnet, B.; Druart, X.; Gatti, J.L.; Dacheux, F. Mammalian epididymal proteome. *Mol. Cell. Endocrinol.* **2009**, *306*, 45–50. [CrossRef]
11. Gatti, J.L.; Castella, S.; Dacheux, F.; Ecroyd, H.; Métayer, S.; Thimon, V.; Dacheux, J.L. Post-testicular sperm environment and fertility. *Anim. Reprod. Sci.* **2004**, *82–83*, 321–339. [CrossRef]
12. Belleannée, C.; Labas, V.; Teixeira-Gomes, A.P.; Gatti, J.L.; Dacheux, J.L.; Dacheux, F. Identification of luminal and secreted proteins in bull epididymis. *J. Proteom.* **2011**, *74*, 59–78. [CrossRef]

3. Jones, R.C.; Dacheux, J.L.; Nixon, B.; Ecroyd, H.W. Role of the epididymis in sperm competition. *Asian J. Androl.* **2007**, *9*, 493–499. [CrossRef]
4. Chioccarelli, T.; Manfrevola, F.; Porreca, V.; Fasano, S.; Altucci, L.; Pierantoni, R.; Cobellis, G. The Cannabinoid Receptor CB1 Stabilizes Sperm Chromatin Condensation Status During Epididymal Transit by Promoting Disulphide Bond Formation. *Int. J. Mol. Sci.* **2020**, *21*, 3117. [CrossRef]
5. Shan, S.; Xu, F.; Bleyer, M.; Becker, S.; Melbaum, T.; Wemheuer, W.; Hirschfeld, M.; Wacker, C.; Zhao, S.; Schütz, E.; et al. Association of α/β-Hydrolase D16B with Bovine Conception Rate and Sperm Plasma Membrane Lipid Composition. *Int. J. Mol. Sci.* **2020**, *21*, 627. [CrossRef]
6. Sellem, E.; Marthey, S.; Rau, A.; Jouneau, L.; Bonnet, A.; Le Danvic, C.; Guyonnet, B.; Kiefer, H.; Jammes, H.; Schibler, L. Dynamics of cattle sperm sncRNAs during maturation, from testis to ejaculated sperm. *Epigenetics Chromatin* **2021**, *14*, 24. [CrossRef]
7. Zhou, W.; De Iuliis, G.N.; Dun, M.D.; Nixon, B. Characteristics of the Epididymal Luminal Environment Responsible for Sperm Maturation and Storage. *Front. Endocrinol.* **2018**, *9*, 59. [CrossRef]
8. Koziorowska-Gilun, M.; Koziorowski, M.; Fraser, L.; Strzeżek, J. Antioxidant defence system of boar cauda epididymidal spermatozoa and reproductive tract fluids. *Reprod. Domest. Anim. = Zuchthyg.* **2011**, *46*, 527–533. [CrossRef]
9. Murdoch, R.N.; Jones, R.C.; Wade, M.; Lin, M. The ultrastructure and metabolism of ejaculated tammar wallaby sperm are impaired by swim-up procedures when compared with sperm from the cauda epididymidis. *Reprod. Fertil. Dev.* **1999**, *11*, 263–271. [CrossRef]
0. Mayorga, L.S.; Bertini, F. The origin of some acid hydrolases of the fluid of the rat cauda epididymidis. *J. Androl.* **1985**, *6*, 243–245. [CrossRef]
1. Caflisch, C.R.; DuBose, T.D., Jr. Cadmium-induced changes in luminal fluid pH in testis and epididymis of the rat in vivo. *J. Toxicol. Environ. Health* **1991**, *32*, 49–57. [CrossRef]
2. Nagdas, S.K.; Britney, T.; Simpson, D.; Salters, T.; Raychoudhury, S.S. Fibrinogen-Related Protein, Fgl2, of Hamster Cauda Epididymal Fluid: Enzymatic Characterization, and Identification of Fgl2-Binding Proteins and Ligand of Defective Hamster Sperm Organelles. *Int. J. Biochem. Physiol.* **2023**, *8*, 000228. [CrossRef]
3. Erikson, D.W.; Way, A.L.; Chapman, D.A.; Killian, G.J. Detection of osteopontin on Holstein bull spermatozoa, in cauda epididymal fluid and testis homogenates, and its potential role in bovine fertilization. *Reproduction* **2007**, *133*, 909–917. [CrossRef] [PubMed]
4. Liu, B.Y.; Zhang, B.L.; Gao, D.Y.; Li, Q.; Xu, X.Y.; Shum, W. Epididymal epithelial degeneration and lipid metabolism impairment account for male infertility in occludin knockout mice. *Front. Endocrinol.* **2022**, *13*, 1069319. [CrossRef] [PubMed]
5. Souza, A.P.B.; Lopes, T.N.; da Silva, A.F.T.; Santi, L.; Beys-da-Silva, W.O.; Yates, J.R., 3rd; Bustamante-Filho, I.C. Changes in porcine cauda epididymal fluid proteome by disrupting the HPT axis: Unveiling potential mechanisms of male infertility. *Mol. Reprod. Dev.* **2020**, *87*, 952–965. [CrossRef]
6. Khaw, S.C.; Wong, Z.Z.; Anderson, R.; Martins da Silva, S. L-carnitine and L-acetylcarnitine supplementation for idiopathic male infertility. *Reprod. Fertil.* **2020**, *1*, 67–81. [CrossRef]
7. Xu, C.; Yuan, Y.; Zhang, C.; Zhou, Y.; Yang, J.; Yi, H.; Gyawali, I.; Lu, J.; Guo, S.; Ji, Y.; et al. Smooth muscle AKG/OXGR1 signaling regulates epididymal fluid acid-base balance and sperm maturation. *Life Metab.* **2022**, *1*, 67–80. [CrossRef]
8. Zhao, X.; Nie, J.; Zhou, W.; Zeng, X.; Sun, X. The metabolomics changes in epididymal lumen fluid of CABS1 deficient male mice potentially contribute to sperm deformity. *Front. Endocrinol.* **2024**, *15*, 1432612. [CrossRef]
9. Serri, O.; Boguenet, M.; Chao de la Barca, J.M.; Bouet, P.E.; El Hachem, H.; Blanchet, O.; Reynier, P.; May-Panloup, P. A Metabolomic Profile of Seminal Fluid in Extremely Severe Oligozoopermia Suggesting an Epididymal Involvement. *Metabolites* **2022**, *12*, 1266. [CrossRef]
0. Zhao, S.; Chen, T.; Luo, X.; Chen, S.; Wang, J.; Lai, S.; Jia, X. Identification of Novel lncRNA and Differentially Expressed Genes (DEGs) of Testicular Tissues among Cattle, Yak, and Cattle-Yak Associated with Male Infertility. *Animals* **2021**, *11*, 2420. [CrossRef] [PubMed]
1. Cao, M.; Wang, X.; Guo, S.; Kang, Y.; Pei, J.; Guo, X. F1 Male Sterility in Cattle-Yak Examined through Changes in Testis Tissue and Transcriptome Profiles. *Animals* **2022**, *12*, 2711. [CrossRef]
2. Zhu, Y.; Yu, J.; Yang, Q.; Xie, Y.; Li, X.; Chen, Z.; Xiong, Y.; Fu, W.; He, H.; Yin, S.; et al. Mitochondria-targeted antioxidant MitoQ improves the quality of low temperature-preserved yak semen via alleviating oxidative stress. *Anim. Reprod. Sci.* **2025**, *273*, 107680. [CrossRef]
3. Want, E.J.; O'Maille, G.; Smith, C.A.; Brandon, T.R.; Uritboonthai, W.; Qin, C.; Trauger, S.A.; Siuzdak, G. Solvent-dependent metabolite distribution, clustering, and protein extraction for serum profiling with mass spectrometry. *Anal. Chem.* **2006**, *78*, 743–752. [CrossRef]
4. Barri, T.; Dragsted, L.O. UPLC-ESI-QTOF/MS and multivariate data analysis for blood plasma and serum metabolomics: Effect of experimental artefacts and anticoagulant. *Anal. Chim. Acta* **2013**, *768*, 118–128. [CrossRef]

35. Wen, B.; Mei, Z.; Zeng, C.; Liu, S. metaX: A flexible and comprehensive software for processing metabolomics data. *BMC Bioinform.* **2017**, *18*, 183. [CrossRef] [PubMed]
36. Kieffer, D.A.; Piccolo, B.D.; Vaziri, N.D.; Liu, S.; Lau, W.L.; Khazaeli, M.; Nazertehrani, S.; Moore, M.E.; Marco, M.L.; Martin, R.J.; et al. Resistant starch alters gut microbiome and metabolomic profiles concurrent with amelioration of chronic kidney disease in rats. *Am. J. physiology. Ren. Physiol.* **2016**, *310*, F857–F871. [CrossRef] [PubMed]
37. Rao, G.; Sui, J.; Zhang, J. Metabolomics reveals significant variations in metabolites and correlations regarding the maturation of walnuts (*Juglans regia* L.). *Biol. Open* **2016**, *5*, 829–836. [CrossRef]
38. Heischmann, S.; Quinn, K.; Cruickshank-Quinn, C.; Liang, L.P.; Reisdorph, R.; Reisdorph, N.; Patel, M. Exploratory Metabolomics Profiling in the Kainic Acid Rat Model Reveals Depletion of 25-Hydroxyvitamin D3 during Epileptogenesis. *Sci. Rep.* **2016**, *6*, 31424. [CrossRef]
39. Haspel, J.A.; Chettimada, S.; Shaik, R.S.; Chu, J.H.; Raby, B.A.; Cernadas, M.; Carey, V.; Process, V.; Hunninghake, G.M.; Ifedigbo, E.; et al. Circadian rhythm reprogramming during lung inflammation. *Nat. Commun.* **2014**, *5*, 4753. [CrossRef] [PubMed]
40. Sreekumar, A.; Poisson, L.M.; Rajendiran, T.M.; Khan, A.P.; Cao, Q.; Yu, J.; Laxman, B.; Mehra, R.; Lonigro, R.J.; Li, Y.; et al. Metabolomic profiles delineate potential role for sarcosine in prostate cancer progression. *Nature* **2009**, *457*, 910–914. [CrossRef]
41. Subramanian, A.; Tamayo, P.; Mootha, V.K.; Mukherjee, S.; Ebert, B.L.; Gillette, M.A.; Paulovich, A.; Pomeroy, S.L.; Golub, T.R.; Lander, E.S.; et al. Gene set enrichment analysis: A knowledge-based approach for interpreting genome-wide expression profiles. *Proc. Natl. Acad. Sci. USA* **2005**, *102*, 15545–15550. [CrossRef] [PubMed]
42. Benson, M.D.; Eisman, A.S.; Tahir, U.A.; Katz, D.H.; Deng, S.; Ngo, D.; Robbins, J.M.; Hofmann, A.; Shi, X.; Zheng, S.; et al. Protein-metabolite association studies identify novel proteomic determinants of metabolite levels in human plasma. *Cell Metab.* **2023**, *35*, 1646–1660.e3. [CrossRef] [PubMed]
43. Boulesteix, A.L.; Strimmer, K. Partial least squares: A versatile tool for the analysis of high-dimensional genomic data. *Brief. Bioinform.* **2007**, *8*, 32–44. [CrossRef]
44. Wang, J.B.; Pu, S.B.; Sun, Y.; Li, Z.F.; Niu, M.; Yan, X.Z.; Zhao, Y.L.; Wang, L.F.; Qin, X.M.; Ma, Z.J.; et al. Metabolomic Profiling of Autoimmune Hepatitis: The Diagnostic Utility of Nuclear Magnetic Resonance Spectroscopy. *J. Proteome Res.* **2014**, *13*, 3792–3801. [CrossRef] [PubMed]
45. Fahy, E.; Subramaniam, S.; Murphy, R.C.; Nishijima, M.; Raetz, C.R.; Shimizu, T.; Spener, F.; van Meer, G.; Wakelam, M.J.; Dennis, E.A. Update of the LIPID MAPS comprehensive classification system for lipids. *J. Lipid Res.* **2009**, *50*, S9–S14. [CrossRef]
46. Liebisch, G.; Fahy, E.; Aoki, J.; Dennis, E.A.; Durand, T.; Ejsing, C.S.; Fedorova, M.; Feussner, I.; Griffiths, W.J.; Köfeler, H.; et al. Update on LIPID MAPS classification, nomenclature, and shorthand notation for MS-derived lipid structures. *J. Lipid Res.* **2020**, *61*, 1539–1555. [CrossRef]
47. Bellezza, I.; Minelli, A. Adenosine in sperm physiology. *Mol. Asp. Med.* **2017**, *55*, 102–109. [CrossRef]
48. Espinosa, F.; de la Vega-Beltrán, J.L.; López-González, I.; Delgado, R.; Labarca, P.; Darszon, A. Mouse sperm patch-clamp recordings reveal single Cl- channels sensitive to niflumic acid, a blocker of the sperm acrosome reaction. *FEBS Lett.* **1998**, *426*, 47–51. [CrossRef]
49. Guerrero, A.; Espinal, J.; Wood, C.D.; Rendón, J.M.; Carneiro, J.; Martínez-Mekler, G.; Darszon, A. Niflumic acid disrupts marine spermatozoan chemotaxis without impairing the spatiotemporal detection of chemoattractant gradients. *J. Cell Sci.* **2013**, *126*, 1477–1487. [CrossRef]
50. Partyka, A.; Kostrzewa Susłow, E.; Dymarska, M.; Ligocka, Z.; Smalec, B.; Kalinin, J.; Meco, M.; Niżański, W. Flavone and 3-hydroxyflavone supplementation in cryopreservation medium protects canine sperm against apoptosis and lipid peroxidation. *Theriogenology* **2024**, *226*, 319–327. [CrossRef]
51. Akarsu, S.A.; Güngör, İ.H.; Cihangiroğlu, A.; Acısu, T.C.; Koca, R.H.; Türk, G.; Sönmez, M.; Gür, S. Effect of sulforaphane on long-term storage of rabbit semen. *Anim. Reprod.* **2023**, *20*, e20230001. [CrossRef]
52. Mu, Y.; Yin, T.L.; Huang, X.X.; Hu, X.; Yin, L.; Yang, J. Sulforaphane ameliorates high-fat diet-induced spermatogenic deficiency in mice. *Biol. Reprod.* **2019**, *101*, 223–234. [CrossRef]
53. Ahmed, H.; Ijaz, M.U.; Riaz, M.; Jahan, S. Sulforaphane inclusion in a freezing medium augments post-thaw motility, functional and biochemical features, and fertility potential of buffalo (*Bubalus bubalis*) spermatozoa. *Res. Vet. Sci.* **2023**, *158*, 196–202. [CrossRef]
54. Jiang, X.P.; Tang, J.Y.; Xu, Z.; Han, P.; Qin, Z.Q.; Yang, C.D.; Wang, S.Q.; Tang, M.; Wang, W.; Qin, C.; et al. Sulforaphane attenuates di-N-butylphthalate-induced reproductive damage in pubertal mice: Involvement of the Nrf2-antioxidant system. *Environ. Toxicol.* **2017**, *32*, 1908–1917. [CrossRef]
55. Anand, L.; Choudhury, A.; Sarin, S.K. Reply. *Hepatology* **2019**, *70*, 755–756. [CrossRef] [PubMed]
56. Oyovwi, M.O.; Nwangwa, E.K.; Ben-Azu, B.; Rotue, R.A.; Edesiri, T.P.; Emojevwe, V.; Igweh, J.C.; Uruaka, C.I. Prevention and reversal of chlorpromazine induced testicular dysfunction in rats by synergistic testicle-active flavonoids, taurine and coenzyme-10. *Reprod. Toxicol.* **2021**, *101*, 50–62. [CrossRef]

57. Ma, Q.; Gui, Y.; Ma, X.; Zhang, B.; Xiong, W.; Yang, S.; Cao, C.; Mo, S.; Shu, G.; Ye, J.; et al. N6-methyladenosine writer METTL16-mediated alternative splicing and translation control are essential for murine spermatogenesis. *Genome Biol.* **2024**, *25*, 193. [CrossRef]
58. Irvine, D.S. Glutathione as a treatment for male infertility. *Rev. Reprod.* **1996**, *1*, 6–12. [CrossRef] [PubMed]
59. Ishizuka, Y.; Nishimura, M.; Matsumoto, K.; Miyashita, M.; Takeo, T.; Nakagata, N.; Hosoi, Y.; Anzai, M. The influence of reduced glutathione in fertilization medium on the fertility of in vitro-matured C57BL/6 mouse oocytes. *Theriogenology* **2013**, *80*, 421–426. [CrossRef]
60. Itahashi, T.; Oikawa, T.; Numabe, T. Effects of glutathione treatments during sperm washing and in vitro fertilization on the in vitro early development of embryos of Japanese Black cattle. *J. Reprod. Dev.* **2022**, *68*, 74–78. [CrossRef] [PubMed]
61. Tian, X.; Li, D.; He, Y.; Zhang, W.; He, H.; Du, R.; Pang, W.; Yang, G.; Yu, T. Supplementation of salvianic acid A to boar semen extender to improve seminal quality and antioxidant capacity. *Anim. Sci. J. = Nihon Chikusan Gakkaiho* **2019**, *90*, 1142–1148. [CrossRef]
62. Hibi, H.; Kato, K.; Mitsui, K.; Taki, T.; Yamada, Y.; Honda, N.; Fukatsu, H.; Yamamoto, M. The treatment with tranilast, a mast cell blocker, for idiopathic oligozoospermia. *Arch. Androl.* **2001**, *47*, 107–111. [CrossRef]
63. Hibi, H.; Kato, K.; Mitsui, K.; Taki, T.; Yamada, Y.; Honda, N.; Fukatsu, H.; Yamamoto, M. Treatment of oligoasthenozoospermia with tranilast, a mast cell blocker, after long-term administration. *Arch. Androl.* **2002**, *48*, 451–459. [CrossRef]
64. Dwivedi, C.; Long, N.J. Effect of cholinergic agents on human spermatozoa motility. *Biochem. Med. Metab. Biol.* **1989**, *42*, 66–70. [CrossRef]
65. Piroozmanesh, H.; Jannatifar, R.; Ebrahimi, S.M.; Fazaeli, H.; Naserpoor, L.; Dimitriadis, E.; Nejatbakhsh, R. Cyclic adenosine monophosphate (cAMP) analog and phosphodiesterase inhibitor (IBMX) ameliorate human sperm capacitation and motility. *Rev. Int. Androl.* **2022**, *20* (Suppl. 1), S24–S30. [CrossRef] [PubMed]
66. Lone, F.A.; Naikoo, M.; Khatun, A.; Shah, R.A.; Pampori, Z.A.; Khan, H.M.; Ahanger, A.A. Idebenone improves quality of ram sperm by mitigating oxidative stress during cryopreservation. *Cryobiology* **2019**, *90*, 15–20. [CrossRef] [PubMed]
67. Eslami, M.; Jahan-Roshan, N.; Farrokhi-Ardabili, F. Influence of idebenone on ram semen quality stored at 4°C. *Reprod. Domest. Anim. = Zuchthyg.* **2019**, *54*, 486–497. [CrossRef]
68. Nikitaras, V.; Zander-Fox, D.; McPherson, N.O. Improving Sperm Oxidative Stress and Embryo Quality in Advanced Paternal Age Using Idebenone In Vitro-A Proof-of-Concept Study. *Antioxidants* **2021**, *10*, 1079. [CrossRef]
69. Li, Y.; Peng, Q.; Shang, J.; Dong, W.; Wu, S.; Guo, X.; Xie, Z.; Chen, C. The role of taurine in male reproduction: Physiology, pathology and toxicology. *Front. Endocrinol.* **2023**, *14*, 1017886. [CrossRef]
70. Elkina, Y.L.; Atroshchenko, M.M.; Bragina, E.E.; Muronetz, V.I.; Schmalhausen, E.V. Oxidation of glyceraldehyde-3-phosphate dehydrogenase decreases sperm motility. *Biochem. Biokhimiia* **2011**, *76*, 268–272. [CrossRef] [PubMed]
71. Hassan, H.A.; Isa, A.M.; El-Kholy, W.M.; Nour, S.E. Testicular disorders induced by plant growth regulators: Cellular protection with proanthocyanidins grape seeds extract. *Cytotechnology* **2013**, *65*, 851–862. [CrossRef] [PubMed]
72. Bae, J.W.; Hwang, J.M.; Yoon, M.; Kwon, W.S. Bifenthrin Diminishes Male Fertility Potential by Inducing Protein Defects in Mouse Sperm. *Toxics* **2024**, *12*, 53. [CrossRef] [PubMed]
73. Bae, J.W.; Kwon, W.S. The deleterious toxic effects of bifenthrin on male fertility. *Reprod. Toxicol.* **2021**, *101*, 74–80. [CrossRef] [PubMed]
74. Gungor, S.; Inanc, M.E.; Ozturk, C.; Korkmaz, F.; Bastan, I.; Cil, B.; Kastelic, J.P. Gallic and carnosic acids improve quality of frozen-thawed ram spermatozoa. *Andrologia* **2019**, *51*, e13393. [CrossRef]
75. Shum, W.W.; Da Silva, N.; Brown, D.; Breton, S. Regulation of luminal acidification in the male reproductive tract via cell-cell crosstalk. *J. Exp. Biol.* **2009**, *212*, 1753–1761. [CrossRef]
76. Shum, W.W.; Ruan, Y.C.; Da Silva, N.; Breton, S. Establishment of cell-cell cross talk in the epididymis: Control of luminal acidification. *J. Androl.* **2011**, *32*, 576–586. [CrossRef]
77. Jakop, U.; Svetlichnyy, V.; Schiller, J.; Schulze, M.; Schroeter, F.; Mueller, K. In vitro supplementation with unsaturated fatty acids improves boar sperm viability after storage at 6 °C. *Anim. Reprod. Sci.* **2019**, *206*, 60–68. [CrossRef]
78. Teveroni, E.; Di Nicuolo, F.; Vergani, E.; Bruno, C.; Maulucci, G.; Bianchetti, G.; Astorri, A.L.; Grande, G.; Gervasoni, J.; Santucci, L.; et al. Short-Chain Fatty Acids Modulate Sperm Migration through Olfactory Receptor 51E2 Activity. *Int. J. Mol. Sci.* **2022**, *23*, 12726. [CrossRef]
79. Islam, M.M.; Umehara, T.; Tsujita, N.; Shimada, M. Saturated fatty acids accelerate linear motility through mitochondrial ATP production in bull sperm. *Reprod. Med. Biol.* **2021**, *20*, 289–298. [CrossRef]
80. Cooper, T.G. Role of the epididymis in mediating changes in the male gamete during maturation. *Adv. Exp. Med. Biol.* **1995**, *377*, 87–101. [CrossRef]
81. Batini, C.; Buisseret-Delmas, C.; Conrath-Verrier, M. Olivo-cerebellar activity during harmaline-induced tremor. A 2-[14C] deoxyglucose study. *Neurosci. Lett.* **1979**, *12*, 241–246. [CrossRef]

82. Gyamera-Acheampong, C.; Tantibhedhyangkul, J.; Weerachatyanukul, W.; Tadros, H.; Xu, H.; van de Loo, J.W.; Pelletier, R.M.; Tanphaichitr, N.; Mbikay, M. Sperm from mice genetically deficient for the PCSK4 proteinase exhibit accelerated capacitation, precocious acrosome reaction, reduced binding to egg zona pellucida, and impaired fertilizing ability. *Biol. Reprod.* **2006**, *74*, 666–673. [CrossRef]
83. Breitbart, H.; Grinshtein, E. Mechanisms That Protect Mammalian Sperm from the Spontaneous Acrosome Reaction. *Int. J. Mol. Sci.* **2023**, *24*, 17005. [CrossRef]
84. Karanwal, S.; Pal, A.; Josan, F.; Patel, A.; Chera, J.S.; Yadav, S.; Gaur, V.; Verma, P.; Badrhan, S.; Chauhan, V.; et al. Higher abundance of DLD protein in buffalo bull spermatozoa causes elevated ROS production leading to early sperm capacitation and reduction in fertilizing ability. *J. Anim. Sci. Biotechnol.* **2024**, *15*, 126. [CrossRef] [PubMed]
85. Dahan, T.; Breitbart, H. Involvement of metabolic pathway in the sperm spontaneous acrosome reaction. *Theriogenology* **2022**, *192*, 38–44. [CrossRef] [PubMed]
86. Ding, Z.; Wang, X.; Guo, S.; Kang, Y.; Cao, M.; Hu, L.; Zhang, B.; Xiong, L.; Pei, J.; Yang, T.; et al. Characteristic analysis of N(6)-methyladenine in different parts of yak epididymis. *BMC Genom.* **2025**, *26*, 500. [CrossRef]
87. Paulusma, C.C.; Lamers, W.H.; Broer, S.; van de Graaf, S.F.J. Amino acid metabolism, transport and signalling in the liver revisited. *Biochem. Pharmacol.* **2022**, *201*, 115074. [CrossRef]
88. Katayama, S.; Mine, Y. Antioxidative activity of amino acids on tissue oxidative stress in human intestinal epithelial cell model. *J. Agric. Food Chem.* **2007**, *55*, 8458–8464. [CrossRef] [PubMed]
89. Guerra-Carvalho, B.; Carrageta, D.F.; Maurício, T.; Pereira, S.C.; Barros, A.; Carvalho, R.A.; Alves, M.G.; Domingues, P.; Oliveira, P.F. Metabolomics analysis of human spermatozoa reveals impaired metabolic pathways in asthenozoospermia. *Eur. J. Clin. Investig.* **2024**, *54*, e14289. [CrossRef]

animals

Article

Genomic Selection for Cashmere Traits in Inner Mongolian Cashmere Goats Using Random Forest, Gradient Boosting Decision Tree, Extreme Gradient Boosting and Light Gradient Boosting Machine Methods

Jiaqi Liu [1,2,3], Xiaochun Yan [1,2,3], Wenze Li [1,2,3], Shan-Hui Xue [1,2,3], Zhiying Wang [1,2,3,*] and Rui Su [1,2,3,*]

[1] College of Animal Science, Inner Mongolia Agricultural University, Hohhot 010018, China; eliauk_1371295874@163.com (J.L.)
[2] Sino-Arabian Joint Laboratory of Sheep and Goat Germplasm Innovation, Hohhot 010018, China
[3] Inner Mongolia Key Laboratory of Sheep & Goat Genetics Breeding and Reproduction, Hohhot 010018, China
* Correspondence: wzhy0321@126.com (Z.W.); suruiyu@126.com (R.S.)

Simple Summary

This study aims to perform genome selection of cashmere traits in Inner Mongolian cashmere goats using machine learning algorithms. By comparing the prediction accuracy of various machine learning algorithms, it explores the feasibility of applying different machine learning algorithms to genome selection of cashmere traits in Inner Mongolian cashmere goats, with the goal of improving the accuracy of genomic selection and enhancing breeding efficiency. Fiber length and cashmere production can enhance the economic value of cashmere goats. We analyzed cashmere trait data from 2299 cashmere goats, including fiber length, cashmere diameter, and cashmere production. We used RF, XGBoost, GBDT, and LightGBM for genome selection in Inner Mongolian cashmere goats. For fiber length, cashmere production, and cashmere diameter, LightGBM, RF, and GBDT achieved the highest selection accuracy after hyperparameter optimization. However, in the case of cashmere traits, the prediction accuracy of XGBoost was the lowest among all the models, at 0.541, 0.309, and 0.387 for fiber length, cashmere production, and cashmere diameter, respectively. For machine learning methods, hyperparameter tuning is essential, as it can improve prediction accuracy.

Abstract

In recent years, Machine Learning (ML) has garnered increasing attention for its applications in genomic prediction. ML effectively processes high-dimensional genomic data and establishes nonlinear models. Compared to traditional Genomic Selection (GS) methods, ML algorithms enhance computational efficiency and offer higher prediction accuracy. Therefore, this study strives to achieve the optimal machine learning algorithm for genome-wide selection of cashmere traits in Inner Mongolian cashmere goats. This study compared the genomic prediction accuracy of cashmere traits using four machine learning algorithms—Random Forest (RF), Extreme Gradient Boosting Tree (XGBoost), Gradient Boosting Decision Tree (GBDT), and LightGBM—based on genotype data and cashmere trait phenotypic data from 2299 Inner Mongolian cashmere goats. The results showed that after parameter optimization, LightGBM achieved the highest selection accuracy for fiber length (56.4%), RF achieved the highest selection accuracy for cashmere production (35.2%), and GBDT achieved the highest selection accuracy for cashmere diameter (40.4%), compared with GBLUP, the accuracy improved by 0.8–2.7%. Among the three traits, XGBoost exhibited the lowest prediction accuracy, at 0.541, 0.309, and 0.387. Additionally, following parameter optimization, the prediction accuracy of the four machine learning methods for cashmere fineness, cashmere yield, and fiber length improved by an average of 2.9%, 2.7%, and 3.8%, respectively. The mean squared error (MSE) and mean absolute error (MAE) for all machine learning methods also decreased, indicating that hyperparameter tuning can enhance prediction accuracy in ML algorithms.

Keywords: genomic selection; Inner Mongolian cashmere goats; machine learning

1. Introduction

Inner Mongolian cashmere goats are a superior local breed of China developed through long-term natural selection and artificial breeding, the cashmere produced is soft, white, and highly elastic, and the goats are renowned for their high cashmere production and excellent quality [1,2]. The Inner Mongolian cashmere goat population is primarily concentrated in the western regions of the Inner Mongolia Autonomous Region, categorized by origin into Albasi, Eriliangshan, and Alxa types. Among them, the cashmere from the Albasi type is renowned for its exceptional quality, earning it the nicknames 'soft gold' and 'fiber gem' [3]. In an effort to improve the breeding efficiency of fiber-related traits in Inner Mongolian cashmere goats, genomic selection has been applied.

Genomic selection (GS) (also known as genomic prediction, CP) uses genome-wide molecular markers and individual phenotypic data to construct a mixed model for genetic evaluation of individuals. Since Meuwissen et al. (2001) published their groundbreaking research on predicting breeding values using genome-wide single-nucleotide polymorphisms (SNP) [4], the concept of "genomic selection" has revolutionized the field of plant and animal breeding. To date, several approaches have been proposed to develop more efficient statistical methods for estimating genomic estimated breeding values (GEBV), such as the widely used genomic best linear unbiased prediction (GBLUP) in GS [5], single-step BLUP (ssGBLUP) [6], ridge regression, Bayesian regression [4], and emerging machine learning (ML) strategies, including support vector regression (SVR) [7], random forests (RF) [8], reproducing kernel Hilbert space regression (RKHS), and kernel ridge regression. According to Howard et al. [9], single-step genomic best linear unbiased prediction (ssGBLUP) exhibited higher accuracy compared to BLUP. Zhang et al. [10] implemented ABLUP, GBLUP, and ssGBLUP methods to perform genomic selection for early growth traits in Inner Mongolian cashmere goats. ssGBLUP exhibited higher prediction accuracy compared to GBLUP and ABLUP. Negro et al. [11] used BLUP, GBLUP, and ssGBLUP methods to perform genomic selection for milk-related traits in Saanen and Alpine goats. The results showed that compared to BLUP, GBLUP and ssGBLUP had higher prediction accuracy and correlation.

Machine learning (ML), also termed automated knowledge acquisition, is an efficient data-processing methodology and has been extensively utilized in the field of animal genetics and breeding [12–14]. Machine Learning efficiently handles high-dimensional genomic data and constructs nonlinear models, allowing for more accurate prediction of animal genetic value [15]. In ML, the three most widely used learning methods are: supervised learning, unsupervised learning, and reinforcement learning. Compared with traditional genomic selection methods, machine learning algorithms demonstrate superiority in computational efficiency and prediction accuracy. The computational efficiency of the entire process is further improved through optimization algorithms in machine learning. Cross-validation is utilized during model selection to make optimal use of sample information and enhance prediction accuracy. LIANG et al. [16] evaluated the accuracy of five genomic selection methods (SVR, KRR, RF, Adaboost. RT, and GBLUP) in Chinese Simmental beef cattle. Results indicated that the four machine learning methods improved accuracy by 12.8%, 14.9%, 5.4%, and 14.4% over the traditional GBLUP method Xiang et al. [17] compared the performance of linear genomic selection models and nonlinear machine learning models for genomic prediction of pig production traits. The results demonstrated that machine learning algorithms outperformed linear genomic selection models in terms of prediction accuracy for pig production traits. Among all machine learning algorithms, support vector machines (SVMs) exhibited the highest genomic prediction accuracy.

This study employed Random Forest (RF), XGBoost, Gradient Boosting Decision Tree (GBDT), and LightGBM to perform genomic selection for cashmere traits in Inner Mongolian cashmere goats, compared the genomic prediction accuracies of these machine learning methods for cashmere traits, investigated the feasibility of applying different genomic selection strategies to cashmere traits of Inner Mongolian cashmere goats, with the aim of improving genomic prediction accuracy and accelerating genetic improvement within the population.

2. Materials and Methods

2.1. Experimental Materials

Phenotypic Data: The phenotypic data used in this study were sourced from Inner Mongolian Cashmere Goats from the Inner Mongolia Yiwei White Cashmere Goat Co., Ltd., Ordos, China. The traits selected for analysis included fiber length, cashmere production, and fiber diameter. The goat farm employed a rotational herding management approach to ensure scientific feeding and care for the goats. In October of each year, the farm conducted synchronized breeding using artificial insemination, with detailed records of breeding data to ensure precise control of the reproduction process. Cashmere lambs were born in March, and pedigree data were recorded. The shearing season occurred annually between April and May, during which data on fiber length, cashmere production, and other traits were recorded. Fiber length was measured by shaving a 10 cm^2 patch on the goat's scapular area before shearing. Samples of cashmere were sent to the Cashmere Analysis Laboratory at Inner Mongolia Agricultural University. The samples underwent separation of fiber and cashmere, followed by cleaning with petroleum ether to remove debris such as soil and grease. Finally, the fiber diameter was measured using a Fiber Cashmere Fineness Analyzer (OFDA2000BT).

Genotypic Data: The genotypic data used in this study were derived from the GGP_Goat_70K SNP chip genotyping data of 2299 Inner Mongolian cashmere goats previously collected by the research team. Firstly, the individuals with recorded genotypes were extracted from a database of breeding information, and the outliers for each trait were further excluded. Values outside the range of the mean plus or minus 2.58 times the standard deviation were defined as outliers. Then, quality control of the genotype data was performed using PLINKv1.90 software. The quality control criteria were as follows: (1) exclude individuals with SNP missing rate of >10%; (2) exclude SNPs with individual missing rate of >10%; (3) exclude SNPs with minor allele frequency (MAF) of <5%; and (4) exclude SNPs violating the Hardy–Weinberg equilibrium (HWE) at $p < 1 \times 10^{-5}$. The missing genotypes were imputed using Beagle. Finally, a total of 50,728 SNPs in 2256 individuals remained for subsequent analyses.

2.2. Experimental Methods

2.2.1. Environment Configuration

In this study, we employed various machine learning methods, including Random Forest (RF), Gradient Boosted Decision Tree (GBDT), XGBoost, and LightGBM, to predict phenotypes from genomic data. All prediction processes were implemented using the Python programming language and executed in the Jupyter Notebook 7.2.2 environment. Below is the detailed environment configuration: Operating System: Windows 11 64-bit; Python Version: Python 3.12.4; Jupyter Notebook Version: 7.2.2 and conda 24.9.2. Key Libraries and Frameworks: XGBoost 1.3.3: for implementing the XGBoost algorithm; LightGBM 4.6.0: for implementing the LightGBM algorithm; scikit-learn 1.5.1: for implementing the RF and GBDT algorithms; NumPy 1.26.4: for numerical computations; Pandas 2.2.2: for data processing and analysis; Matplotlib 3.9.2.

2.2.2. Random Forest

Random Forest (RF) is a machine learning algorithm ensemble of numerous decision tree models, where predictions are collectively determined by multiple trees. First introduced by Breiman [18], RF comprises multiple decision trees. It estimates and fits numerous trees on diverse data subsets, leveraging their averages to enhance predictive accuracy and mitigate overfitting.

Random Forest algorithms can be expressed through mathematical formulas:

$$F(x) = \frac{1}{B}\sum_{b=1}^{B} h_b(x) \quad (1)$$

where $F(x)$ represents the predicted value of the random forest regression, $h_b(x)$ is the predicted result of the bth decision tree, and B denotes the number of decision trees in the random forest.

2.2.3. Gradient Boosting Decision Tree

Gradient Boosting Decision Tree (GBDT) is a representative algorithm in boosting methods. It serves as the foundation for modern powerful algorithms like XGBoost and LightGBM (LGBM), and is among the most stable machine learning algorithms in practical applications. Initially introduced as Gradient Boosting Machine (GBM), it combines the ideas of Bagging and Boosting, leverages their advantages, accepts various types of weak learners as input, and was subsequently renamed Gradient Boosting Decision Tree (GBDT) after weak learners were primarily defined as decision trees. GBDT incorporates the three key elements of Boosting: loss function (L), weak learner ($F(x)$), and combined ensemble results ($F(x)$).

The Gradient Boosting Decision Tree algorithm can be represented mathematically as:

$$F(x) = \sum_{m=1}^{M} \gamma_m h_m(x) \quad (2)$$

where $F(x)$ represents the final prediction of the Gradient Boosting Decision Tree, $h_m(x)$ denotes the mth decision tree, γ_m is the weight of the mth decision tree, and M represents the total number of decision trees.

2.2.4. Extreme Gradient Boosting

Extreme Gradient Boosting (XGBoost) is a new-generation boosting algorithm based on the Gradient Boosting Decision Tree (GBDT). XGBoost is an algorithm system centered on boosting trees, designed to implement various types of gradient-boosted trees. Initially introduced by Chen Tianqi in 2014 [19], XGBoost was first presented in his doctoral dissertation. XGBoost is an ensemble learning algorithm that enhances the performance and generalization ability of decision tree models through the Gradient Boosting Framework.

The Extreme Gradient Boosting algorithm can be formulated as

$$F(x) = \sum_{t=1}^{T} f(x)_t \quad (3)$$

where $F(x)$ represents the final prediction of the Extreme Gradient Boosting tree, $f_t(x)$ is the predicted value of the t-th decision tree, and T is the total number of decision trees.

2.2.5. Light Gradient Boosting Machine

Light Gradient Boosting Machine (LightGBM) is a highly efficient Gradient Boosting algorithm developed by the Microsoft Research Asia team. Initially designed for internal use at Microsoft to handle massive high-dimensional data, it was officially introduced in 2017 by Guolin Ke, Qi Meng, Thomas Finley, and others in a research paper [20]. LightGBM is a cutting-edge ensemble learning algorithm, similar to XGBoost, and both algorithms represent enhancements over GBDT. LightGBM offers superior computational efficiency and lower memory usage. Moreover, it demonstrates better resistance to overfitting when dealing with high-dimensional data, making it more suitable as a baseline model for preliminary exploratory modeling.

The Lightweight Gradient Boosting Machine algorithm can be formulated as

$$F(x) = \sum_{t=1}^{T} \omega_t f_t(x) \quad (4)$$

where $F(x)$ represents the final prediction of the Lightweight Gradient Boosting Machine, $f_t(x)$ is the predicted value of the t-th decision tree, and T is the total number of decision trees, and ω_t is the weight of the t-th decision tree.

2.2.6. GBLUP

$$y = 1\mu + Zg + e \quad (5)$$

in which $\boldsymbol{y}$ is the observed trait. $\boldsymbol{\mu}$ is the overall mean, **1** is a vector of 1 s, g is the vector of genomic breeding values, **e** is the vector of random errors, and Z is an incidence matrix allocating records to $\boldsymbol{g}$. The distributions of random effects were: $\boldsymbol{g}$ ~N(**0**, **G** $\sigma^2{}_g$) and e ~N(**0**, **I** $\sigma^2{}_e$), where **G** was the genomic relationship matrix (**G** matrix), and $\sigma^2{}_g$ and $\sigma^2{}_e$ were the additive genetic variance and the residual variance, respectively.

2.2.7. Genomic Prediction Accuracy Evaluation

In this study, we employed 10-fold cross-validation (10-fold CV) to evaluate the accuracy of genomic predictions. Specifically, the dataset was randomly partitioned into 10 subsets of approximately equal size, with nine used for model training and one for validation. This process was repeated 10 times, each time with a different subset reserved for validation, ensuring all samples were used for validation (Figure 1). This approach allowed us to comprehensively assess model performance across different data subsets, with cashmere producing a robust accuracy estimate.

The accuracy of genomic prediction was assessed by calculating the Pearson correlation coefficient between the corrected phenotypic values (yc) and the predicted values (pv). The Pearson correlation coefficient, ranging between −1 and 1, measures the linear relationship between two variables. A value closer to 1 or −1 indicates a stronger correlation, while a value near 0 suggests a weak correlation. A high Pearson correlation coefficient reflects a strong linear relationship between predicted and actual phenotypic values, indicating a more accurate model. To precisely evaluate model performance, we calculated the Pearson correlation coefficient between yc and pv for each validation set across the 10-fold cross-validation and averaged these coefficients to obtain the final accuracy metric for genomic prediction.

Figure 1. 10-fold Cross-Validation.

3. Results

3.1. Results of Descriptive Statistical Analysis of Phenotypic Data

The descriptive statistical analysis of phenotypic data is presented in Table 1, summarizing phenotypic traits (fiber length, cashmere diameter, and cashmere production) for 2256 Inner Mongolian cashmere goats (2011–2021). It includes the number of records, mean, standard deviation, coefficient of variation, and maximum and minimum values. Table 1 shows that the mean values for fiber length, cashmere diameter, and cashmere production were 18.90 cm, 15.23 µm, and 740.32 g, with coefficients of variation of 25.91%, 5.31%, and 29.07%, respectively. The cashmere diameter had a relatively low variation coefficient, while fiber length and cashmere production had higher ones, suggesting greater potential for improving these traits through selective breeding.

Table 1. Summary of Basic Statistical Summary of Traits.

Trait	Number of Record	Mean	SD	Coefficient of Variation
FL/cm [1]	9610	18.90	4.90	25.91%
CD/µm [2]	5356	15.23 [1]	0.81	5.31%
CP/g [3]	9040	740.32	215.19	29.07%

[1] FL: fiber length; [2] CD: cashmere diameter; [3] CP: cashmere production.

3.2. Optimal Hyperparameters of Machine Learning Models

This study used Bayes_opt optimization to tune the hyperparameters of RF, GBDT, XGBoost, and LightGBM, identifying the optimal hyperparameters for each machine learning model (Table 2). For RF, the maximum tree depth (max_depth) was set to 18, and the number of trees (n_estimators) to 250. For XGBoost, the number of trees (n_estimators) was 100, the learning rate (learning_rate) 0.1, and the maximum tree depth (max_depth) 5. For LightGBM, the learning rate (learning_rate) was 0.05, the maximum tree depth (max_depth) 5, and the number of leaves (num_leaves) 20. For GBDT, the number of trees (n_estimators) was 300, the learning rate (learning_rate) 0.1, and the maximum tree depth (max_depth) 5.

Table 2. Optimal hyperparameters for each machine learning algorithm.

Method	Optimal Hyperparameters
RF	max_depth = 18, n_estimators = 250
XGBoost	n_estimators = 100, learning_rate = 0.1, max_depth = 5
LightGBM	learning_rate = 0.05, max_depth = 5, num_leaves = 20
GBDT	n_estimators = 300, learning_rate = 0.1, max_depth = 5

3.3. Genomic Selection Accuracy Evaluation

Genomic prediction accuracy for cashmere diameter, cashmere production, and fiber length in Inner Mongolian cashmere goats was evaluated using GBLUP, RF, GBDT, XGBoost, and LightGBM with 10-fold cross–validation. The accuracies of these machine learning methods are presented in Table 3. After optimizing the parameters of the four machine learning models using Bayesian optimization, the prediction accuracy exceeded that of the default parameter settings. Following hyperparameter tuning, the ML methods showed an average improvement in prediction accuracy of 2.9%, 2.7%, and 3.8% for cashmere diameter, cashmere production, and fiber length, respectively. Additionally, after hyperparameter adjustment, all machine learning methods except XGBoost showed higher accuracy than GBLUP. For fiber length, compared to GBLUP, RF showed a 0.3% improvement in prediction accuracy, LightGBM showed a 2.1% improvement, and GBDT showed a 1.4% improvement. After parameter optimization, LightGBM had the highest selection accuracy at 56.4%. For cashmere production, RF had the highest selection accuracy at 35.2% after parameter optimization. Compared to GBLUP, RF showed a 2.5% improvement in prediction accuracy, and LightGBM showed a 0.9% improvement. For cashmere diameter, after parameter optimization, GBDT had the highest selection accuracy at 40.4%. Compared to GBLUP, RF showed a 0.5% improvement, LightGBM showed a 0.6% improvement, and GBDT showed a 0.8% improvement. Furthermore, we found that the XGBoost algorithm had the lowest prediction accuracy for cashmere traits, with values of 0.541, 0.309, and 0.387, all of which were lower than those of GBLUP.

Table 3. Predictive accuracy of five methods for three traits using 10-fold cross-validation.

Hyperparameters	Method	FL [1]	CP [2]	CD [3]
Tuning [4]	GBLUP	0.543 ± 0.0315	0.327 ± 0.0478	0.396 ± 0.0361
	RF	0.546 ± 0.0699	0.352 ± 0.0397	0.401 ± 0.0397
	XGB	0.541 ± 0.0475	0.309 ± 0.0794	0.387 ± 0.0325
	LightGBM	0.564 ± 0.0129	0.336 ± 0.0799	0.402 ± 0.0359
	GBDT	0.557 ± 0.0423	0.320 ± 0.0631	0.404 ± 0.0310
Default [5]	RF	0.540 ± 0.0509	0.336 ± 0.0699	0.396 ± 0.0354
	XGB	0.510 ± 0.0443	0.299 ± 0.0738	0.362 ± 0.0347
	LightGBM	0.533 ± 0.0445	0.314 ± 0.0755	0.381 ± 0.0330
	GBDT	0.475 ± 0.0478	0.262 ± 0.0678	0.340 ± 0.0388

[1] FL: fiber length; [2] CP: cashmere production; [3] CD: cashmere diameter; Tuning [4]: hyperparameter optimization; Default [5]: default hyperparameters.

This study compared the performance of RF, GBDT, XGBoost, and LightGBM in predicting fiber length, cashmere production, and cashmere diameter in Inner Mongolian cashmere goats. The Pearson correlation coefficient (PCC) measures the linear correlation between true and predicted values; a higher PCC indicates better model prediction accuracy. Root Mean Square Error (RMSE) quantifies the deviation between predicted and actual values; a lower RMSE signifies more accurate model predictions. As shown in Figure 2, LightGBM outperformed RF, GBDT, and XGBoost in predicting fiber length. It had the lowest RMSE of 4.023 and the highest PCC of 0.557 after optimization, indicating high prediction accuracy and stability. While XGBoost's RMSE decreased by 8.0% after optimization, its overall performance remained slightly inferior to LightGBM. Overall, LightGBM demonstrated the best predictive performance. The study compared the prediction performance of RF, GBDT, XGBoost, and LightGBM for cashmere production (Figure 3). RF showed the best performance after optimization, with the lowest RMSE of 0.221 and the highest PCC of 0.352. After optimization, LightGBM's RMSE decreased from 0.239 to 0.222, and its PCC increased from 0.314 to 0.336, showing significant improvement. The study also compared the prediction performance of the four models for cashmere diameter (Figure 4). GBDT performed the best after optimization, with the lowest RMSE of 0.731 and the highest PCC of 0.404, indicating high prediction accuracy and stability.

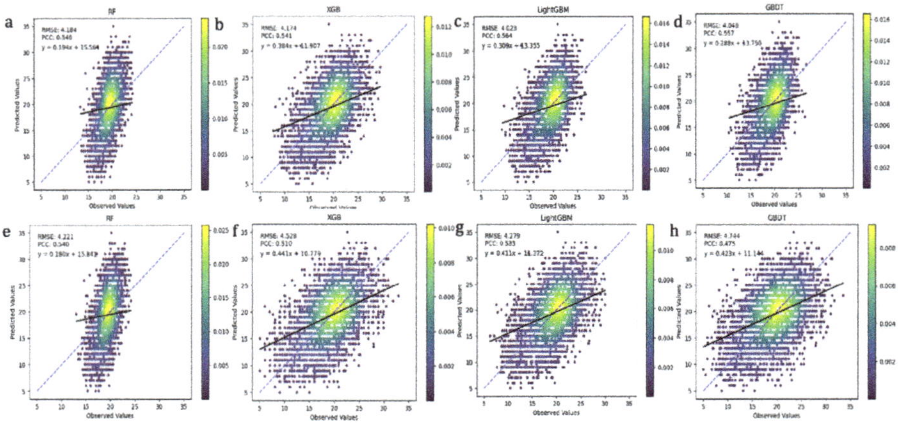

Figure 2. Impact of Four Machine Learning Methods on the Prediction Accuracy of Fiber Length Traits. The *X*-axis represents observed values, and the *Y*-axis represents predicted values. The points in the figure illustrate the relationship between model predictions and actual observations. The diagonal line indicates ideal predictions where predicted values equal observed values (Y = X line). The color bar on the right represents point density, with colors ranging from purple (low density) to yellow (high density). The blue line represents the ideal prediction scenario, where the predicted values perfectly align with the actual observed values. The black line is the linear regression line between the model predictions and the actual observations, showing the linear relationship between them. (**a–d**): Scatter plots of fiber length predictions vs. true values using RF, XGB, LightGBM, and GBDT with optimized hyperparameters. (**e–h**): Scatter plots of fiber length predictions vs. true values using RF, XGB, LightGBM, and GBDT with default hyperparameters.

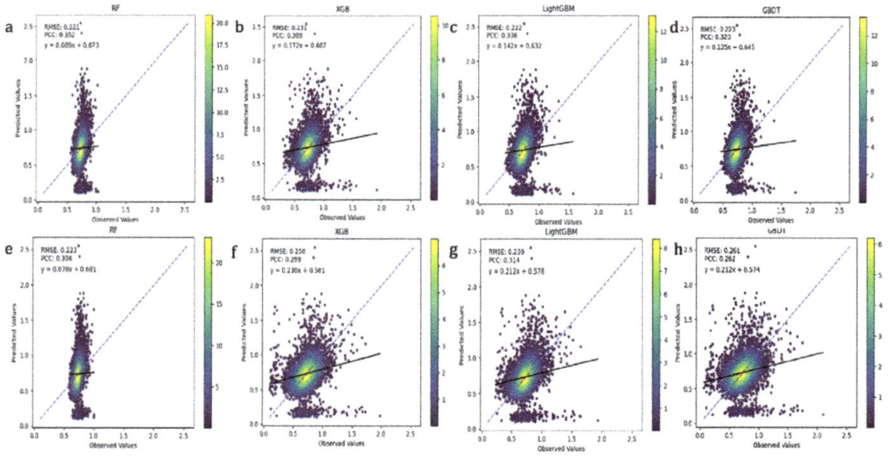

Figure 3. Impact of Four Machine Learning Algorithms on the Accuracy of Cashmere Production Traits. The blue line represents the ideal prediction scenario, where the predicted values perfectly align with the actual observed values. The black line is the linear regression line between the model predictions and the actual observations, showing the linear relationship between them. (**a–d**): Scatter plots of cashmere production predictions vs. true values using RF, XGB, LightGBM, and GBDT with optimized hyperparameters. (**e–h**): Scatter plots of cashmere production predictions vs. true values using RF, XGB, LightGBM, and GBDT with default hyperparameters.

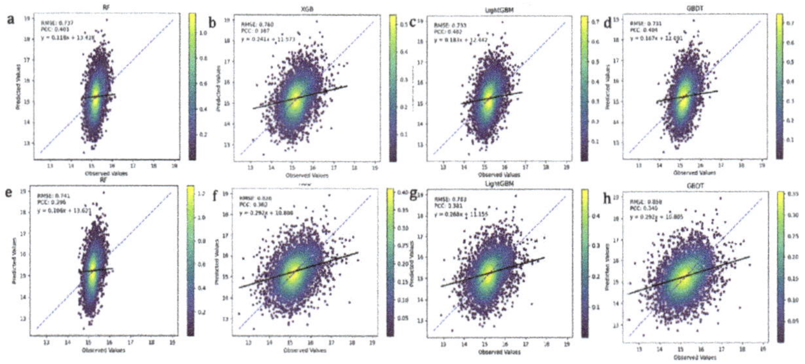

Figure 4. Impact of Four Machine Learning Algorithms on the Accuracy of Cashmere Diameter Traits. The blue line represents the ideal prediction scenario, where the predicted values perfectly align with the actual observed values. The black line is the linear regression line between the model predictions and the actual observations, showing the linear relationship between them. (**a–d**): Scatter plots of cashmere diameter predictions vs. true values using RF, XGB, LightGBM, and GBDT with optimized hyperparameters. (**e–h**): Scatter plots of cashmere diameter predictions vs. true values using RF, XGB, LightGBM, and GBDT with default hyperparameters.

3.4. Evaluation of Error Metrics for Five Method

Mean Squared Error and Mean Absolute Error are commonly used to evaluate the performance of different machine learning methods. MSE considers both prediction accuracy and bias, while MAE better reflects the actual prediction error. Smaller MSE and MAE values indicate better model accuracy, lower error, and higher reliability. As shown in Table 4, after hyperparameter tuning, all machine learning methods exhibited reduced MSE and MAE values. After optimization, LightGBM had the smallest error for fiber length (MSE = 16.359, MAE = 3.239). GBDT had the smallest error for cashmere diameter (MSE = 0.533, MAE = 0.569). The three machine learning methods showed similar errors in predicting cashmere production. Compared to GBLUP, both LightGBM and GBDT exhibited lower MSE and MAE values. Meanwhile, when XGBoost is used to predict fiber length, cashmere production, and cashmere diameter, it yields the highest MSE and MAE values, indicating that it has the lowest prediction accuracy and the largest error magnitude for cashmere traits. With default hyperparameters, the Random Forest algorithm demonstrated the smallest errors for all three cashmere traits in Inner Mongolian cashmere goats.

Table 4. Five methods' mean squared error and mean absolute error.

Hyperparameters	Method	FL [1]		CP [2]		CD [3]	
		MSE [4]/cm^2	MAE [5]/cm^2	MSE/g^2	MAE/g^2	MSE/μm^2	MAE/μm^2
Tuning	GBLUP	16.716	3.267	0.050	0.149	0.565	0.576
	RF	17.508	3.343	0.049	0.150	0.543	0.574
	XGBoost	17.419	3.340	0.053	0.154	0.578	0.591
	LightGBM	16.359	3.239	0.049	0.149	0.537	0.570
	GBDT	16.384	3.252	0.050	0.150	0.533	0.569
Default	RF	17.814	3.370	0.050	0.152	0.549	0.578
	XGBoost	20.503	3.610	0.063	0.170	0.682	0.637
	LightGBM	18.306	3.426	0.057	0.160	0.613	0.605
	GBDT	22.502	3.769	0.068	0.179	0.736	0.662

[1] FL: fiber length; [2] CP: cashmere production; [3] CD: cashmere diameter; [4] MSE: Mean Squared Error; [5] MAE: Mean Absolute Error.

3.5. Computational Efficiency of Four Machine Learning Algorithms

This study compared the average computing time per fold for different machine learning models during 10-fold cross-validation, with results shown in Table 5. Machine learning models with optimized hyperparameters generally outperformed those with default hyperparameters in terms of computing time. After hyperparameter optimization, GBDT had the lowest average computing time for all traits, at 95.51 s for cashmere production and 82.97 s for cashmere diameter. By contrast, XGB had the longest computing time for all traits. These results indicate that proper hyperparameter tuning can enhance model computational efficiency. Compared to the GBLUP method, the four machine learning algorithms demonstrated significantly higher efficiency, indicating that machine learning approaches can more effectively handle high-dimensional genomic data and construct nonlinear models.

Table 5. Average computing time to complete each fold of 10-fold CV according to different genomic prediction methods.

Hyper-Parameters	Method	FL [1]	CP [2]	CD [3]
Tuning	GBLUP	9748.25 s	19,351.12 s	62,875.9 s
	RF	131.45 s	128.14 s	139.05 s
	XGB	421.79 s	433.66 s	382.62 s
	LightGBM	199.41 s	233.12 s	222.49 s
	GBDT	107.23 s	95.51 s	82.97 s
Default	RF	189.47 s	135.45 s	189.47 s
	XGB	482.26 s	462.17 s	401.31 s
	LightGBM	247.33 s	282.89 s	235.86 s
	GBDT	200.77 s	106.18 s	86.29 s

[1] FL: fiber length; [2] CP: cashmere production; [3] CD: cashmere diameter.

4. Discussion

This study optimized the hyperparameters of RF, GBDT, XGBoost, and LightGBM using Bayes_opt, determining the best hyperparameter combinations for each machine learning method. After optimization, the average prediction accuracy for cashmere diameter, cashmere production, and fiber length improved by 2.9%, 2.7%, and 3.8%, respectively, highlighting the importance of hyperparameter tuning for ML algorithms. Liu et al. [21] compared seven ML methods (SVR, RF, GBDT, XGBoost, LightGBM, KRR, and MLP) for genomic selection accuracy in a yellow-feathered chicken population. After hyperparameter optimization, the accuracy of all seven ML methods improved, consistent with our findings. Wang et al. [22] evaluated genomic prediction accuracy in Chinese Yorkshire pigs using SVR, KRR, RF, and Adaboost.R2. The study found that RF's prediction accuracy for total number born (TNB) and number born alive (NBA) improved by 9.8% and 10.2% after hyperparameter optimization, consistent with our study. This also indicates that hyperparameter tuning is essential for improving prediction accuracy in genomic selection using machine learning algorithms. In breeding practice, a model's ability to accurately predict outcomes directly impacts the selection of superior individuals and the development of breeding strategies. Thus, hyperparameter tuning is not only a way to enhance model performance, but also a crucial step in achieving breeding objectives. Model optimization for improved prediction accuracy helps reliably identify individuals with superior cashmere traits, thereby guiding practical breeding decisions. This process ultimately improves breeding efficiency and economic outcomes by offering more precise support for breeding decisions.

Machine learning methods can flexibly capture potential nonlinear genotype-phenotype relationships and consider marker correlations and interactions. Consequently, in this study,

RF, GBDT, XGBoost, and LightGBM were used with 10–fold cross–validation to evaluate the genomic prediction accuracy of cashmere diameter, cashmere production, and fiber length in Inner Mongolian cashmere goats (Albas type). The results showed that for fiber length, the selection accuracy of LightGBM achieved the highest value (56.4%) after parameter optimization, with the lowest MSE (16.359) and MAE (3.239). GBDT showed slightly lower accuracy compared to LightGBM in this trait. For cashmere production, RF demonstrated the highest selection accuracy at 35.2% after parameter optimization. Other models showed relatively lower performance in this trait. For cashmere diameter, GBDT achieved the highest selection accuracy (40.4%) after parameter optimization, with the minimum errors (MSE = 0.533 and MAE = 0.569). However, in the case of cashmere traits, the prediction accuracy of XGBoost was the lowest among all the models, at 0.541, 0.309, and 0.387 for fiber length, cashmere production, and cashmere diameter, respectively. Meanwhile, XGBoost produces the highest MSE and MAE values when predicting cashmere traits. This reflects that XGBoost achieves the lowest accuracy and exhibits the poorest performance in predicting cashmere traits. Wang et al. [23] compared eight ML algorithms and GBLUP for genomic selection accuracy in Yorkshire pigs and found that LightGBM consistently outperformed XGBoost in prediction accuracy for raw data, which is consistent with our study. Wang et al. [24] used six traditional and five ML methods to predict three reproductive traits (litter weight, total piglets born, and number of piglets born alive). The varying prediction accuracies of different models for distinct traits mainly result from the compatibility between model characteristics and trait features. While LightGBM, Random Forest, and Gradient Boosting Decision Trees all fall under the category of ensemble learning, they each have different focuses. In this study, the RF model showed the highest prediction accuracy for cashmere production. Biologically, this might be because cashmere production is primarily controlled by the additive effects of multiple genes and some non-additive effects. Random Forest introduces randomness through random sampling and feature selection, enabling it to better handle high-dimensional data and noise. For cashmere production traits, Random Forest can effectively capture both linear and nonlinear contributions of these genes. GBDT, on the other hand, optimizes step by step, focusing on the residuals of the previous iteration in each iteration, thus better capturing complex patterns and nonlinear relationships in the data. For fiber length and cashmere diameter traits, GBDT can more effectively model these complex genetic structures. This matching between model and trait characteristics enables different models to show varying accuracies in predicting different traits. Azodi et al. [25] found that no single GP algorithm performs best across all species and traits, which is consistent with our findings. They also found that ensemble-based predictions are more stable and likely to yield better accuracy than single algorithms. Thus, for predicting cashmere traits in cashmere goats, we can consider ensemble predictions that leverage the capabilities of multiple algorithms.

Furthermore, after optimizing the hyperparameters, all machine learning methods showed reductions in both MSE and MAE. With the optimized hyperparameters, LightGBM demonstrated the smallest error for fiber length, and GBDT for cashmere diameter. The three machine learning methods showed similar errors in predicting cashmere production. When using the default hyperparameters, the Random Forest algorithm showed the smallest errors for all three cashmere traits in Inner Mongolian cashmere goats. Zhang et al. [26] assessed the genomic prediction accuracy of reproductive traits in 385 Large White pigs using GBDT, RF, LightGBM, Adaboost.R2, GBLUP, BRR, and BL. The results highlighted the critical role of hyperparameter tuning in optimizing machine learning models and showed that the LightGBM model outperformed RF in both MSE and RMSE, which is consistent with our findings. This study also explored the average computing time per fold for different machine learning models during 10-fold cross-validation. Results indicated that XGB had the longest computing time for all traits. Compared with the GBLUP method, the four machine learning algorithms showed significantly higher efficiency, indicating that machine learning approaches can more effectively handle high–dimensional genomic data and build nonlinear models. However, Chen et al. [27] found that in predicting pig growth traits and GEBV, RF had the longest training time, while XGB had the shortest, contradicting our results. These inconsistencies might arise from differences in species, data size, and computer configuration.

Compared with GBLUP, machine learning algorithms show advantages in handling complex genetic data, such as capturing nonlinear relationships and high-dimensional features. For fiber length, compared with GBLUP, RF improved prediction accuracy by 0.3%, LightGBM by 2.1%, and GBDT by 1.4%; for cashmere production, RF by 2.5%, LightGBM by 0.9%; and for cashmere diameter, RF by 0.5%, LightGBM by 0.6%, and GBDT by 0.8%. This indicates that machine learning algorithms show higher accuracy in predicting complex traits, especially those with low heritability or involving multiple gene interactions. This means machine learning algorithms can more comprehensively utilize genetic information and enhance the interpretation of genetic variations. Furthermore, machine learning algorithms can identify genetic markers more related to target traits, improving genomic prediction accuracy, and subsequently optimizing parent selection and offspring prediction. Additionally, compared with the GBLUP method, the four machine learning algorithms showed significantly higher efficiency, indicating that machine learning approaches can more effectively handle high-dimensional genomic data and build nonlinear models. The efficiency and flexibility of machine learning algorithms help reduce the cost and time of genomic selection, making breeding programs more cost-effective. Specifically, machine learning algorithms can be integrated into genomic selection programs. By customizing model selection and parameter tuning, they can achieve precise prediction and optimization for different traits and breeding objectives. This not only improves breeding efficiency but also enables early selection and reduces generation intervals.

This study has certain limitations, including a moderate sample size, potential inability to capture small–effect SNPs, and limited interpretability of machine learning models. Machine learning is an efficient data processing method. Machine learning can effectively handle high–dimensional genomic data and build nonlinear models for more accurate genomic predictions. As the reference population size increases, machine learning methods can capture more information during model training and produce more accurate predictions. This study used 70KSNP chip data for genomic predictions. It is worth considering whether using low–depth resequencing data alongside chip data could provide more training information and improve accuracy. Additionally, genome–wide association studies (GWAS) can identify genetic variants related to multiple traits. Integrating GWAS into GS models and incorporating trait–specific markers can enhance genomic predictions [28,29]. Utilizing GWAS results to enhance GS models and improve the prediction accuracy of related traits offers an alternative method to boost the biological interpretability of GS models.

5. Conclusions

In this study, we conducted genomic prediction for Inner Mongolian cashmere goats using GBLUP, RF, XGBoost, GBDT, and LightGBM, and compared the prediction accuracy, mean squared error (MSE), mean absolute error (MAE), and computing time of these methods. The results showed that for fiber length, LightGBM had the highest selection accuracy of 56.4%; for cashmere production, RF had the highest selection accuracy of 35.2%; and for cashmere diameter, GBDT had the highest selection accuracy of 40.4%. Among the three traits, XGBoost had the lowest prediction accuracy, at 0.541, 0.309, and 0.387

for fiber length, cashmere production, and cashmere diameter. Additionally, our results indicate that hyperparameter tuning is necessary when using machine learning algorithms for genome-wide selection, as it can improve prediction accuracy.

Author Contributions: Conceptualization, R.S.; methodology, J.L. and Z.W.; software, J.L.; formal analysis, J.L.; data curation, J.L. and X.Y.; writing—original draft preparation, J.L.; writing—review and editing, J.L., S.-H.X., Z.W. and R.S.; visualization, J.L.; supervision, W.L. and R.S.; project administration, R.S.; funding acquisition, R.S. All authors have read and agreed to the published version of the manuscript.

Funding: This work was financially supported by National Key Research and Development Program of China (2022YFE0113300, 2022YFD1300204), Natural Science Foundation of Inner Mongolia Autonomous Region of China (2025JQ022), Outstanding Youth Science Fund Cultivation Project of Inner Mongolia Agricultural University (BR230304), China Agriculture Research System of MOF and MARA (No. CARS-39), Special A-class Team Project for Research and Innovation Team Construction of Inner Mongolia Agricultural University (BR251201), Program for Innovative Research Team in Universities of Inner Mongolia Autonomous Region (NMGIRT2322).

Institutional Review Board Statement: The phenotype data of cashmere goats in the study were collected from the Inner Mongolia Yiwei White Cashmere Goats Company, Ordos, China. The genotype data came from the previous research results of the group. There is no ethical statement involved because the study did not include animal experiments.

Informed Consent Statement: Not applicable.

Data Availability Statement: The data that support the findings of this study are available from the corresponding authors upon reasonable request.

Acknowledgments: We are grateful to Su R for the excellent technical guidance in preparing experimental data and constructing machine learning algorithm models. We also highly appreciate the herders and workers of Inner Mongolia Yiwei White Cashmere Goats Co., Ltd. for their assistance in sample collection and phenotype determination. The authors sincerely thank the editors and anonymous reviewers for their constructive criticism and valuable suggestions that greatly improved the manuscript.

Conflicts of Interest: The authors declare no competing interests.

References

1. Han, W.; Yang, F.; Wu, Z.; Guo, F.; Zhang, J.; Hai, E.; Shang, F.; Su, R.; Wang, R.; Wang, Z.; et al. Inner Mongolian Cashmere Goat Secondary Follicle Development Regulation Research Based on mRNA-miRNA Co-analysis. *Sci. Rep.* **2020**, *10*, 4519. [CrossRef]
2. Gong, W.; Liu, J.; Mu, Q.; Chahaer, T.; Liu, J.; Ding, W.; Bou, T.; Wu, Z.; Zhao, Y. Melatonin promotes proliferation of Inner Mongolia cashmere goat hair follicle papilla cells through Wnt10b. *Genomics* **2024**, *116*, 110844. [CrossRef]
3. Li, X.; Hao, F.; Hu, X.; Wang, H.; Dai, B.; Wang, X.; Liang, H.; Cang, M.; Liu, D. Generation of Tβ4 knock-in Cashmere goat using CRISPR/Cas9. *Int. J. Biol. Sci.* **2019**, *15*, 1743–1754. [CrossRef]
4. Meuwissen, T.H.; Hayes, B.J.; Goddard, M.E. Prediction of total genetic value using genome-wide dense marker maps. *Genetics* **2002**, *157*, 1819–1829. [CrossRef]
5. Clark, S.A.; van der Werf, J. Genomic best linear unbiased prediction (gBLUP) for the estimation of genomic breeding values. *Methods Mol. Biol.* **2013**, *1019*, 321–330.
6. Forni, S.; Aguilar, I.; Misztal, I. Different genomic relationship matrices for single-step analysis using phenotypic, pedigree and genomic information. *Genet. Sel. Evol.* **2011**, *43*, 1. [CrossRef]
7. Gianola, D.; van Kaam, J.B. Reproducing kernel hilbert spaces regression methods for genomic assisted prediction of quantitative traits. *Genetics* **2008**, *178*, 2289–2303. [CrossRef] [PubMed]
8. Chen, X.; Ishwaran, H. Random forests for genomic data analysis. *Genomics* **2012**, *99*, 323–329. [CrossRef] [PubMed]
9. Howard, J.T.; Rathje, T.A.; Bruns, C.E.; Wilson-Wells, D.F.; Kachman, S.D.; Spangler, M.L. The Impact of Selective Genotyping on the Response to Selection Using Single-step Genomic Best Linear Unbiased Prediction. *J. Anim. Sci.* **2018**, *96*, 4532–4542. [CrossRef] [PubMed]

10. Zhang, T.; Gao, L.; Zhou, B.; Xu, Q.; Liu, Y.; Li, J.; Lv, Q.; Zhang, Y.; Wang, R.; Su, R.; et al. Genomic Selection for Early Growth Traits in Inner Mongolian Cashmere Goats Using ABLUP, GBLUP, and ssGBLUP Methods. *Animals* **2025**, *15*, 1733. [CrossRef]
11. Negro, A.; Cesarani, A.; Cortellari, M.; Bionda, A.; Fresi, P.; Macciotta, N.P.P.; Grande, S.; Biffani, S.; Crepaldi, P. A comparison of genetic and genomic breeding values in Saanen and Alpine goats. *Animal* **2024**, *18*, 101118. [CrossRef]
12. Alves, A.A.C.; Espigolan, R.; Bresolin, T.; Costa, R.M.; Fernandes Júnior, G.A.; Ventura, R.V.; Carvalheiro, R.; Albuquerque, L.G. Genome-enabled prediction of reproductive traits in Nellore cattle using parametric models and machine learning methods. *Anim. Genet.* **2021**, *52*, 32–46. [CrossRef] [PubMed]
13. Zhao, W.; Lai, X.S.; Liu, D.Y.; Zhang, Z.Y.; Ma, P.P.; Wang, Q.S.; Zhang, Z.; Pan, Y.C. Applications of support vector machine in genomic prediction in pig and maize populations. *Front. Genet.* **2020**, *11*, 598318. [CrossRef]
14. Montesinos López, O.A.; Montesinos López, A.; Crossa, J. Random Forest for Genomic Prediction. In *Multivariate Statistical Machine Learning Methods for Genomic Prediction*; Springer International Publishing: Cham, Switzerland, 2022; pp. 633–681.
15. Mahesh, B. Machine Learning Algorithms—A Review. *Int. J. Sci. Res.* **2020**, *9*, 381–386. [CrossRef]
16. Liang, M.; Miao, J.; Wang, X.; Chang, T.; An, B.; Duan, X.; Xu, L.; Gao, X.; Zhang, L.; Li, J.; et al. Application of ensemble learning to genomic selection in chinese simmental beef cattle. *J. Anim. Breed. Genet.* **2021**, *138*, 291–299. [CrossRef]
17. Xiang, T.; Li, T.; Li, J.; Li, X.; Wang, J. Using machine learning to realize genetic site screening and genomic prediction of productive traits in pigs. *FASEB J.* **2023**, *37*, e22961. [CrossRef] [PubMed]
18. Breiman, L. Random forests. *Mach Learn.* **2001**, *45*, 5–32. [CrossRef]
19. Chen, T.; Guestrin, C. XGBoost: A Scalable Tree Boosting System. In Proceedings of the 22nd ACM SIGKDD International Conference on Knowledge Discovery and Data Mining (KDD '16), San Francisco, CA, USA, 13–17 August 2016; Association for Computing Machinery: New York, NY, USA, 2016; pp. 785–794.
20. Ke, G.; Meng, Q.; Finley, T.; Wang, T.; Chen, W.; Ma, W.; Ye, Q.; Liu, T.-Y. Lightgbm: A highly efficient gradient boosting decision tree. *Adv. Neural Inf. Process. Syst.* **2017**, *30*, 3149–3157.
21. Liu, B.; Liu, H.; Tu, J.; Xiao, J.; Yang, J.; He, X.; Zhang, H. An investigation of machine learning methods applied to genomic prediction in yellow-feathered broilers. *Poult. Sci.* **2025**, *104*, 104489. [CrossRef]
22. Wang, X.; Shi, S.; Wang, G.; Luo, W.; Wei, X.; Qiu, A.; Luo, F.; Ding, X. Using machine learning to improve the accuracy of genomic prediction of reproduction traits in pigs. *J. Anim. Sci. Biotechnol.* **2022**, *13*, 60. [CrossRef]
23. Wang, J.; Zhang, J.; Hao, W.; Zong, W.; Liang, M.; Zhao, F.; Zhang, L.; Wang, L.; Gao, H.; Wang, L. An interpretable integrated machine learning framework for genomic selection. *Smart Agric. Technol.* **2025**, *101138*, 2772–3755. [CrossRef]
24. Wang, J.; Chai, J.; Chen, L.; Zhang, T.; Long, X.; Diao, S.; Chen, D.; Guo, Z.; Tang, G.; Wu, P. Enhancing Genomic Prediction Accuracy of Reproduction Traits in Rongchang Pigs Through Machine Learning. *Animals* **2025**, *15*, 525. [CrossRef]
25. Azodi, C.B.; Bolger, E.; McCarren, A.; Roantree, M.; de Los Campos, G.; Shiu, S.H. Benchmarking Parametric and Machine Learning Models for Genomic Prediction of Complex Traits. *G3 Genes Genomes Genet.* **2019**, *9*, 3691–3702. [CrossRef] [PubMed]
26. Zhang, H.; Bao, S.; Zhao, X.; Bai, Y.; Lv, Y.; Gao, P.; Li, F.; Zhang, W. Genome-Wide Association Study and Phenotype Prediction of Reproductive Traits in Large White Pigs. *Animals* **2024**, *14*, 3348. [CrossRef]
27. Chen, D.; Wang, S.J.; Zhao, Z.J.; Ji, X.; Shen, Q.; Yu, Y.; Cui, S.D.; Wang, J.G.; Chen, Z.Y.; Wang, J.Y.; et al. Genomic prediction of pig growth traits based on machine learning. *Yi Chuan* **2023**, *45*, 922–932. [PubMed]
28. Zhang, Y.; Zhang, M.; Ye, J.; Xu, Q.; Feng, Y.; Xu, S.; Hu, D.; Wei, X.; Hu, P.; Yang, Y. Integrating genome-wide association study into genomic selection for the prediction of agronomic traits in rice (*Oryza sativa* L.). *Mol. Breed.* **2023**, *43*, 81. [CrossRef]
29. Izquierdo, P.; Wright, E.M.; Cichy, K. GWAS-assisted and multitrait genomic prediction for improvement of seed yield and canning quality traits in a black bean breeding panel. *G3 Genes Genomes Genet.* **2025**, *15*, jkaf007. [CrossRef]

GCAT TACG GCAT genes

MDPI

Article

Analysis of Semen Proteomic Differences Among Three Genotypes of FecB Rams in Duolang Sheep

Yanlong Zhang [1,†], Zhigang Niu [2,†], Jiabao Yan [1], Yang Chen [3], Zhengfen Xue [2], Jie Xu [1], Yifan Ma [1] and Hongcai Shi [2,*]

1 College of Animal Science and Technology, Xinjiang Agricultural University, Urumqi 830052, China; 15962887647@126.com (Y.Z.)
2 Xinjiang Key Laboratory of Animal Biotechnology, Key Laboratory of Herbivorous Animal Genetics, Breeding and Reproduction, Ministry of Agriculture and Rural Affairs, Institute of Biotechnology, Xinjiang Academy of Animal Sciences, Urumqi 830011, China
3 College of Animal Science and Technology, Shihezi University, Shihezi 832061, China
* Correspondence: shc69@126.com
† These authors contributed equally to this work.

Abstract

Backgrouds: To explore the differences in semen proteins among rams of three *FecB* genotypes (++, B+, and BB) in Duolang sheep. Methods: This study employed DIA quantitative proteomics technology to identify semen proteins from four wild-type (Group A), two heterozygous (Group B), and three homozygous (Group C) rams. Results: Compared with the ++ genotype, the differentially expressed proteins (DEPs) in the semen of B+ genotype rams are significantly involved in the biological process of innate immune response and are significantly enriched in the oxidative phosphorylation pathway in KEGG analysis. From a biological perspective, the innate immune response may affect the immune health of Duolang sheep, while oxidative phosphorylation influences energy metabolism, which in turn impacts reproductive performance. Compared with the BB genotype, the DEPs in the semen of B+ genotype rams participate in biological processes such as protein phosphorylation and protein hydrolysis during cellular protein catabolism. These DEPs are also significantly enriched in pathways related to Parkinson's disease and non-alcoholic fatty liver disease in KEGG analysis. These differences may affect the cellular metabolism and physiological functions of Duolang sheep, thereby being associated with their reproductive performance. Compared with the ++ genotype, the DEPs in the semen of BB genotype rams exhibit differences in molecular function, cellular component, KEGG pathway, domain function, and subcellular localization. For instance, they are involved in threonine-type endopeptidase activity and associated with pathways like Alzheimer's disease and retrograde endocannabinoid signaling. Conclusions: These differences may have potential impacts on the physiology and reproductive performance of Duolang sheep.

Keywords: FecB; Duolang sheep; semen; proteomics

1. Introduction

The sheep *FecB* gene, also known as the *BMPR1B* (Bonemorphogeneticproteinreceptortype-1B) gene, is an acronym for the English terms fecundity and booroola [1]. The name is based on the Nomenclature Committee on Sheep and Goat Genetics because of a mutation in the gene's coding region, A764G, which can alter the trait of the number of lambs produced by Booroola Merino sheep. *FecB* manifests as three different genotypes: wild-type (++), heterozygous (B+), and pure breeds (BB). Heterozygotes (B+) can increase lambing by 0.9~1.2 lambs, while purebreds (BB) can increase lambing by 1.1~1.7 lambs [2]. Notably, the identification of the *FecB* gene and its mutations in the domestic sheep breeds Dolang and the successful breeding of three genotypes of rams and ewes do not match the performance of Booroola Merino in terms of the number of lambs produced and the distinctive phenotypes [3]. This phenomenon is accompanied by an increase in the number of abortions, weak lambs, and deaths. High-quality ram semen is the most crucial and essential component of artificial insemination to achieve optimal fertility and productive performance in rams. Sperm proteins and seminal plasma proteins are the major proteins found in semen. They are important for fertilization and semen viability and can both directly affect the molecular mechanisms and biological pathways that control sperm activity. Semen proteins can thus be employed as potential indicators for sperm dysfunction screening and semen quality assessment [4]. Proteomic analysis of sperm and seminal plasma protein profiles can reveal the intricate relationships between proteins and sperm function.

In this study, through proteomic analysis of semen from Duolang sheep rams with three FecB genotypes, this study aims to understand their impacts on reproductive performance, as well as the individual metabolism and physiological functions of rams, and to identify the potential effects on Duolang sheep and their reproduction.

2. Materials and Methods

2.1. Experimental Site and Test Animals

The experiment has passed the ethical review of the Science and Technology Ethics Committee of the Xinjiang Uygur Autonomous Region Academy of Animal Husbandry Sciences, with the ethical review number 202408012. Nine *FecB*-genotype rams, consisting of four wild-type (++) rams (Group A), two heterozygous (B+) rams (Group B), and three homozygous (BB) rams (Group C), were bred at the National Breeding Farm of Wuzheng Green Agriculture Co., Ltd. in Kashgar, Xinjiang, China. The rams were between the ages of two and three years and had a weight of roughly 120 kg. Besides the daily baseline diet, each ram was given one egg per day, supplemented with alfalfa hay (2 kg/d) and fresh carrots (1 kg/d), and taught to run 5 km two weeks before the experiment.

2.2. Equipment, Drugs, and Reagents Used

Experimental Instruments: Refrigerator, constant temperature water bath, microscope, frozen semen programmed cooling instrument, freezer, liquid nitrogen tank. Experimental Materials: Artificial vagina, semen collection cup, 5 mL vacuum test tube, 0.25 mL straw, 1 mL and 5 mL syringes, gauze, etc. Main Drugs and Reagents: Concentrated frozen semen diluent for sheep (purchased from Xinbo Biotechnology), distilled water, Vaseline, glycerol, absolute ethanol.

2.3. Semen Collection and Dilution Equilibration

Semen was collected from rams using the artificial vagina method. An estrous ewe was selected as the teaser, which was secured in a holding rack. The artificial vagina was preheated with warm water, and water was poured into it to occupy 1/3–1/2 of the volume between the outer and inner shells (i.e., when the artificial vagina was held upright, the water level reached the water filling hole). Vaseline was applied to the penis entry end for lubrication, and a semen collection cup was attached to the other end, forming a triangular shape at the Vaseline-coated end. After preparing the artificial vagina, it was placed beside the secured estrous ewe to wait. Semen collection was performed when the ram mounted the ewe and extended its genitalia. A drop of the collected semen was first placed on a glass slide, and its quality was observed with the naked eye under a microscope. Semen that showed cloudy rolling, milky white color, no abnormal odor, and a motility of over 90% was selected. The semen was diluted using the two-step dilution method: based on the semen density, it was first diluted 1–2 times with concentrated frozen semen diluent for sheep (preheated in a 37 °C water bath in advance). The diluted semen was then placed in a 4 °C refrigerator for 2 h of equilibration. After equilibration, an equal volume of freezing solution (at 4 °C) was added to dilute the original semen by 2–4 times, followed by another 2 h of equilibration in a 4 °C refrigerator.

2.4. Preparation of Straw Frozen Semen

The equilibrated semen was taken out of the refrigerator, gently shaken, and then aspirated into 0.25 mL straws, which were sealed with sealing powder. After sealing, the straws containing semen were evenly placed on a straw rack and equilibrated for approximately 5 min. After equilibration, the straw rack was quickly placed into a large container pre-filled with liquid nitrogen for fumigation. During this process, the distance between the straws and the liquid nitrogen surface was maintained at 0.8–1 cm. The container lid was closed, and fumigation was conducted for about 5 min. After fumigation on the liquid nitrogen surface, all straws were placed in a gauze bag, immersed in liquid nitrogen, and left for 10–30 min to complete the preparation of straw frozen semen. A 0.5 mL sample from each frozen semen batch was sent to Beijing Novogene Technology Co., Ltd. for Astral DIA quantitative proteomics analysis in Beijing, China.

2.5. Data Analysis

2.5.1. Protein Identification and Quantification

The DIA-NN library search program was used to search and evaluate the raw files based on the UniProt protein database. The parameters for the library search were as follows: Up to two missed cut sites were permitted, the immobilization modification was the alkylation modification of cysteine, the loss of methionine at the N-terminal end was a variable modification, and the mass variation in precursor and fragment ions was automatically detected and rectified. The DIA-NN program further limited the search results to increase the quality of the analysis findings. Only proteins with PG.Q.Value < 0.01 (FC < * [fold change, FC]) and peptides with Global. Q.Value < 0.01 were kept. Proteins with a significant difference in quantification between the two groups ($p < 0.05$, | log2FC | > * (FC > * or FC < * [fold change, FC])) were classified as differentially expressed proteins (DEP). Raw data were deconvoluted, and species libraries were searched using the DIA-NN software (Direct DIA (Basic Version)) for Blood-Plus (full platform) [5].

2.5.2. Functional Analysis of Proteins and DEP

The Interproscan software (5.72-103.0), which includes the Pfam, PRINTS, ProDom, SMART, ProSite, and PANTHER databases, was used to functionally annotate GO and IPR. COG and KEGG were used to perform functional protein family and pathway analysis of the identified proteins. Volcano plot analysis was done to identify significant DEPs, clustering heatmap analysis was employed to visualize the expression patterns, and pathway enrichment analysis using GO, IPR, and KEGG was carried out to understand the functional significance of the DEPs. The STRING DB program (http://STRING.embl.de/, accessed on 18 June 2025) was used to predict potential protein–protein interactions.

3. Results

3.1. Overview of the Number of Proteins Identified by DIA

The DIA data were imported into the DIA-NN program to achieve the simultaneous characterization and quantification of peptides. The ion-pair chromatographic peaks were retrieved, and sub-ion matching and peak area calculations were carried out. The nine samples contained 3095, 3097, 3089, 3086, 3262, 3102, 2704, 3252, and 2680 proteins, respectively (Figure 1).

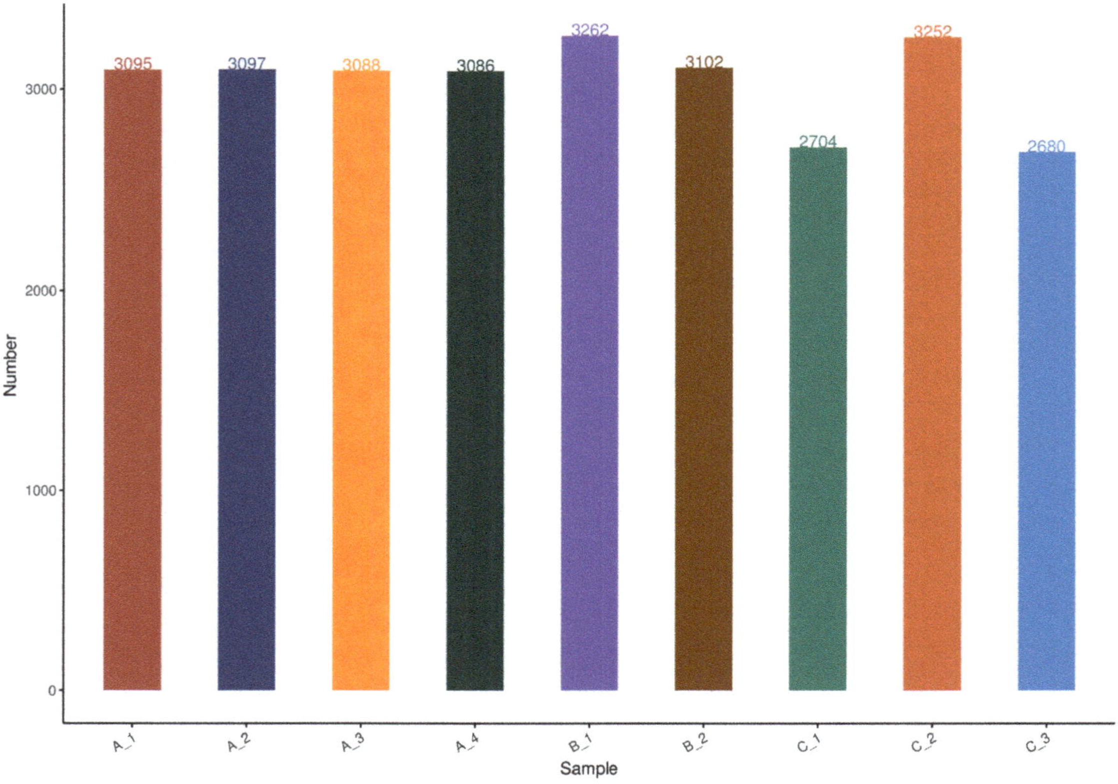

Figure 1. Overview of sample protein identification. The horizontal coordinate is the sample name, while the vertical coordinate is the number of proteins.

3.2. Functional Annotation of Proteins

GO, KEGG, KOG, Subcellular (structural domain), IPR (subcellular localization), and Tran_Factor jointly annotated 23 proteins., KEGG alone annotated 147 proteins, KOG alone annotated 1 protein, Subcellular (structural domain), IPR (subcellular localization), and Tran_Factor (transcription factor) jointly annotated 23 proteins, Subcellular (structural domain)KOG alone annotated 1 protein, and while IPR alone annotated 14 proteins, as shown in Figure 2. IPR alone, 14 proteins, and subcellular alone, 1 protein.

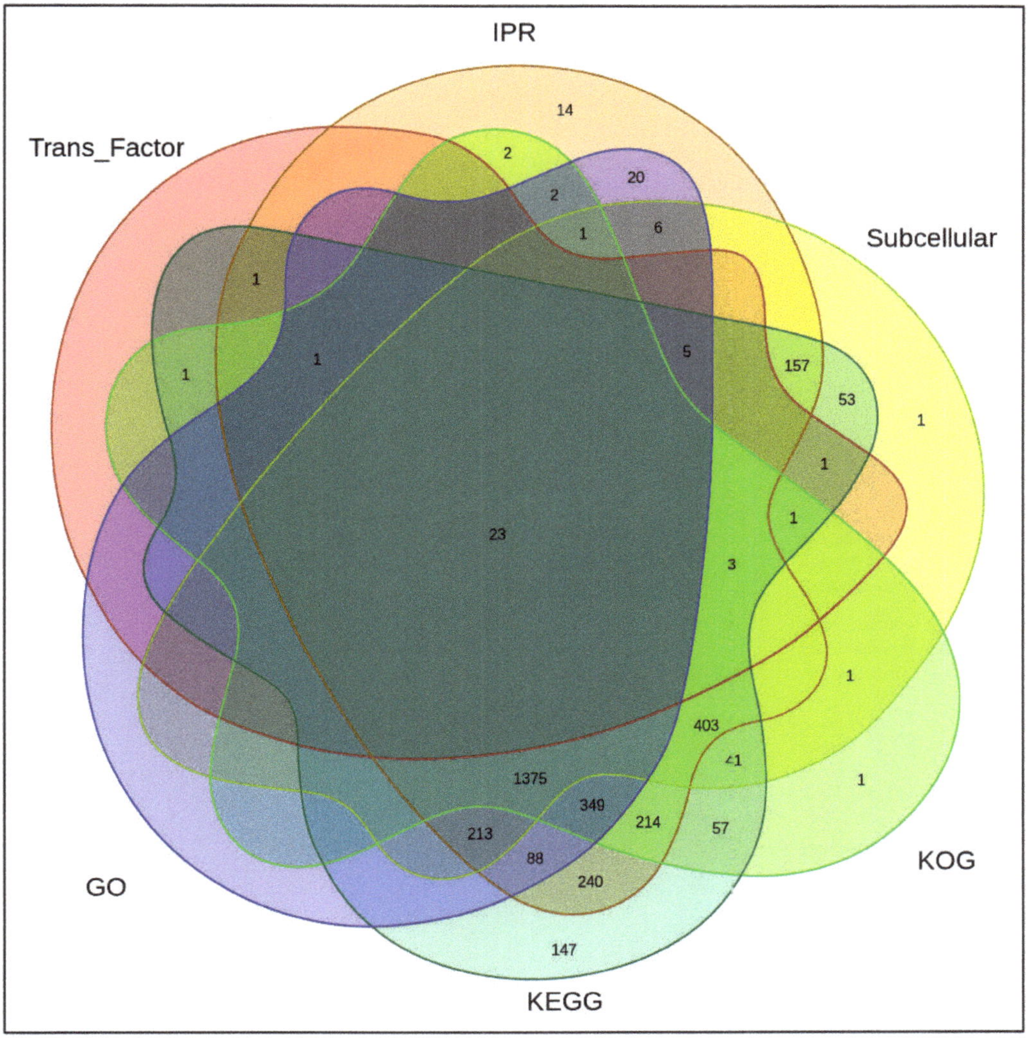

Figure 2. Functional annotation results. Each 'circle' in the figure represents the annotation results of one database. The overlapping part represents the proteins jointly annotated by multiple databases, while the non-overlapping part represents the proteins individually annotated by the corresponding databases.

3.2.1. Protein GO Functional Annotation

The GO annotation results of every protein in the samples were grouped into three primary categories: Molecular function, cellular component, and biological process (Figure 3). The molecular function primarily describes the individual functions of the gene and its product, such as the ability to bind carbohydrates or the activity of ATP to hydrolyze enzymes. The cellular component primarily describes the subcellular structure, location, and macromolecular complexes, such as nucleolus, telomere, and recognition initiation complexes. The biological process primarily describes the function of the gene and its product. Only the top 10 findings in each broad category were included in the figure (Figure 3) because of the vast number of entries in the GO annotation results. Notably, there were a lot more proteins involved in molecular processes than those involved in biological processes and cellular components. The molecular functions of proteins are essential for maintaining regular cellular life activities. Variations in protein levels impact the cell's ability to operate normally, which subsequently has an impact on the sperm flagellum's normal oscillation. Notably, the number of proteins involved in biological processes was closer to the number of proteins involved in cellular components. Changes in the composition of cellular fractions potentially affect the normal morphology and activity of the cells. In contrast, biological processes are involved in the normal motility of spermatozoa and jointly regulate sperm viability through the interactions between the various components. Sperm abnormality rates and motility are thus potentially correlated with the proteins that contribute to the cellular components of semen.

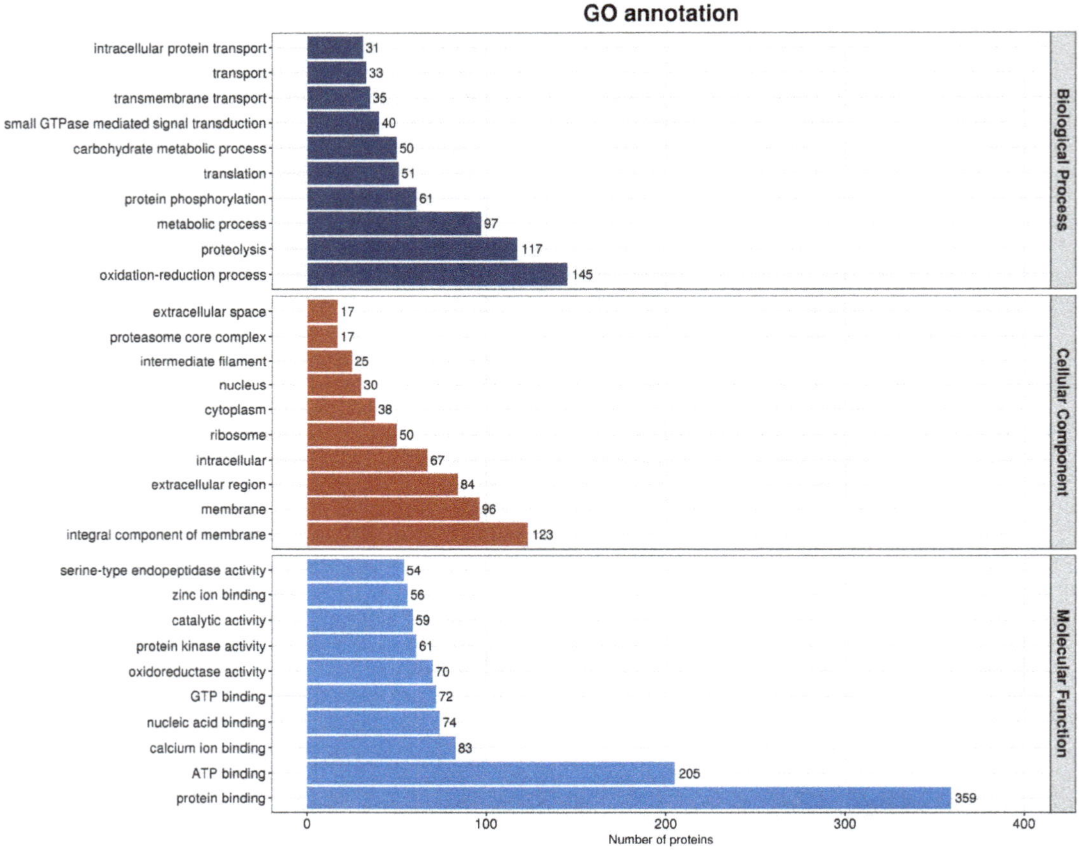

Figure 3. Histogram of all GO annotation results. The horizontal coordinates represent the number of proteins, while the vertical coordinates represent the GO entries to which the proteins are annotated.

3.2.2. Functional Annotations

Cluster of Orthologous Groups of proteins (COG) presents the classification results of all proteins (Figure 4). Notably, the frequency of posttranslational modification, protein turnover, and chaperones is significantly higher than that of other categories.

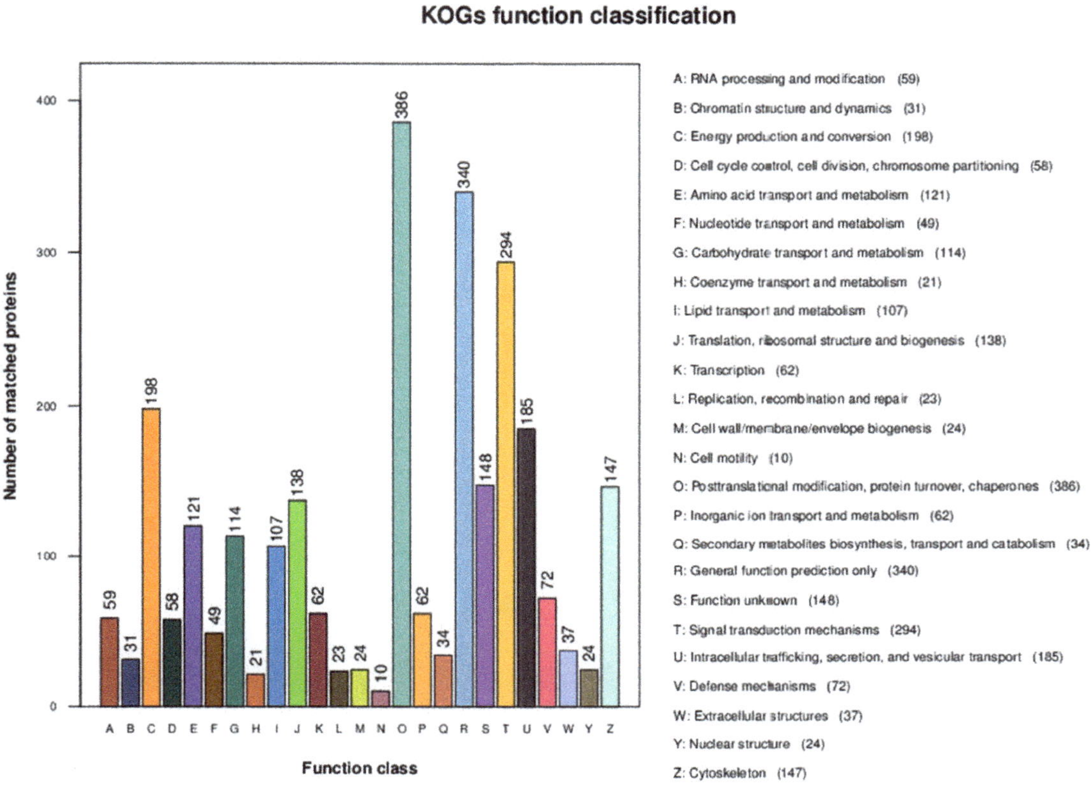

Figure 4. Histogram of all COG annotation results. The horizontal coordinate is the functional classification of the annotation, while the vertical coordinate is the number of proteins annotated to the corresponding function.

3.2.3. Functional Annotation of All KEGG

Figure 5 shows the categories of KEGG annotation findings for all proteins. Notably, 33 biological pathways were involved, with most of these pathways being engaged in five different kinds of biological activities. The proteins had a wide range of biological functions in cell physiology, environmental information processing, genetic information processing, metabolism, and tissue systems. The number of proteins involved in metabolism and tissue systems was substantially more than that involved in cell physiology, environmental information processing, and genetic information processing. There were 478 relevant proteins in global and overview maps, which was much more than the number of proteins in other pathways. These proteins revealed the architecture of the whole metabolic network, providing a more macroscopic picture of the inherent links between the main metabolic pathways in the cell rather than concentrating on the specifics of a given biochemical activity. The proteins were essential to spermatozoa's ability to sustain regular metabolic processes, giving them a strong metabolic base that ensures their regular physiological function.

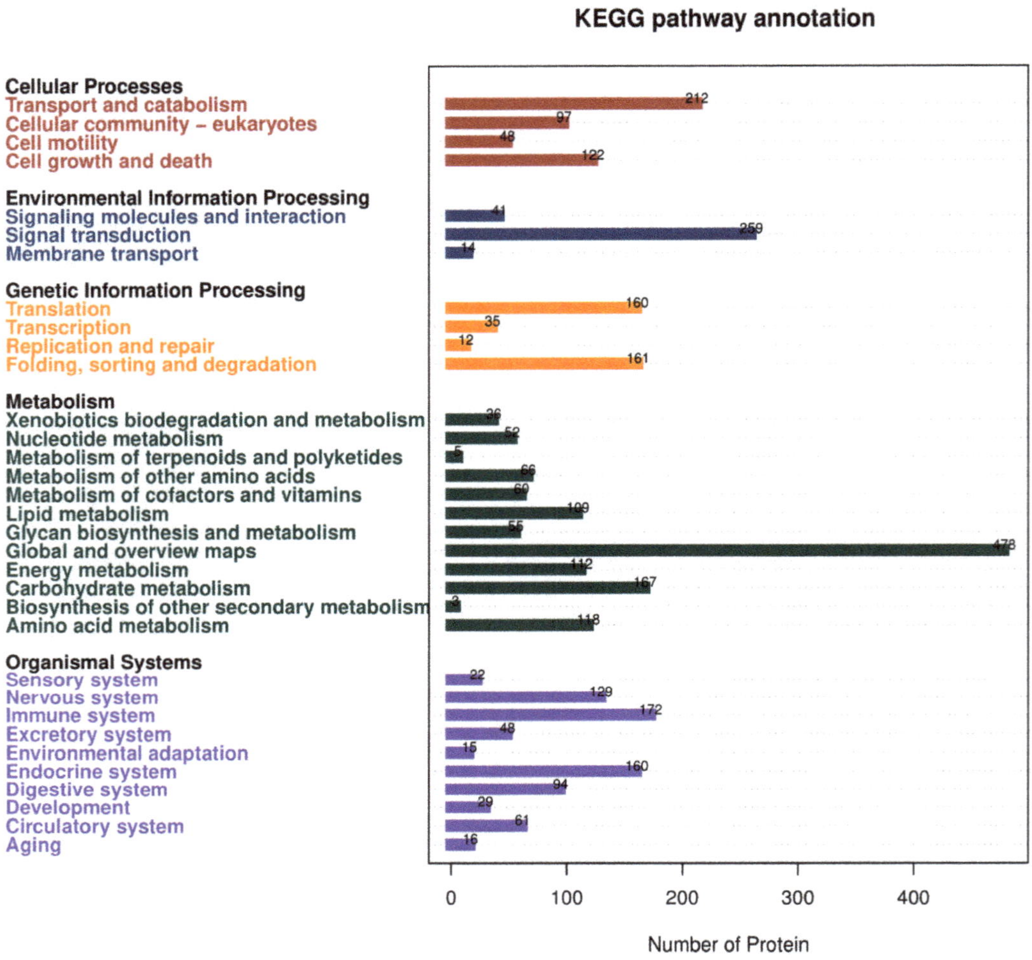

Figure 5. Histogram of KEGG annotation results. Horizontal coordinates represent the number of proteins, while the vertical coordinates represent the KEGG entries to which the proteins are annotated.

3.2.4. Annotation of All Structural Domains

The structural domains of proteins are essential to the biological functions of proteins because they are the fundamental building blocks of protein structure, function, and evolution. The structural domains of the identified proteins were thoroughly investigated to accurately understand the biological roles of proteins and their lengthy evolutionary history. Figure 6 displays the categorization of the structural domain annotation outcomes for every protein. Of note, 20 distinct structural domains were identified. There were 52 domains of protein kinase, 45 domains of RNA recognition motif, 44 domains of serine protease and trypsin, and Calcium ion. In the same line, 42 WD40 repeat proteins and 43 high-affinity structural domains unique to calcium-binding proteins were identified. These structural domain types were substantially more numerous than the others and could thus be potentially important for the execution and interactions of protein functions.

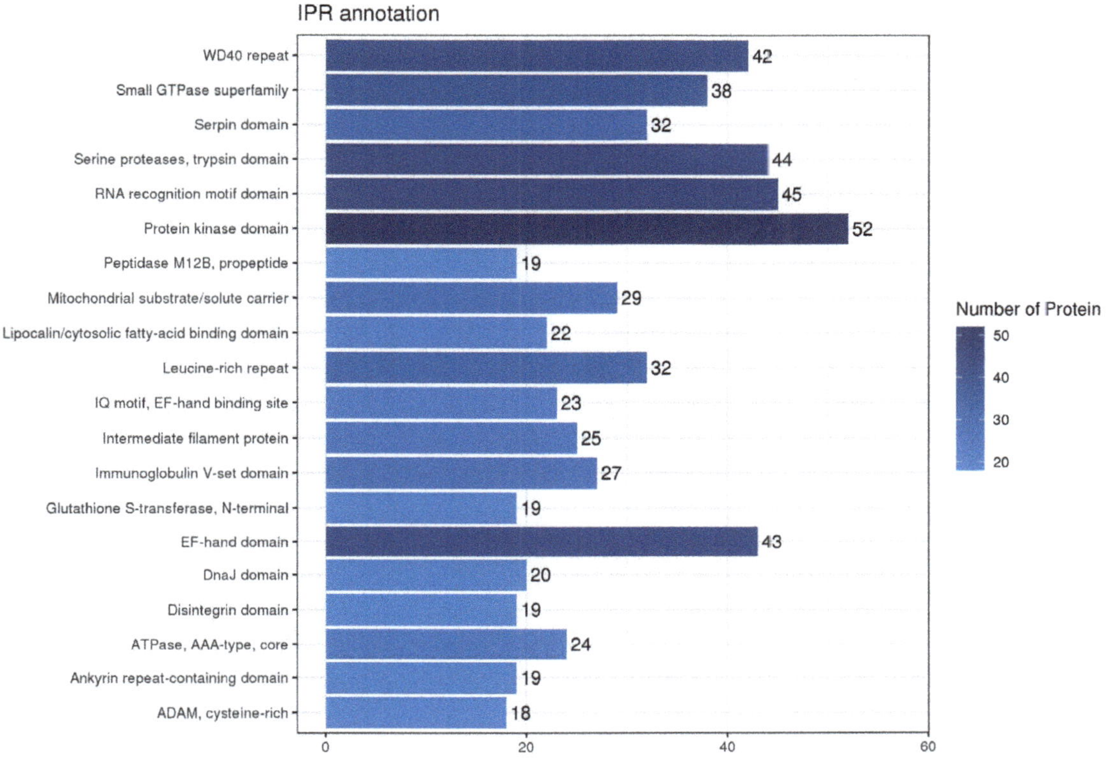

Figure 6. Histogram of structural domain annotation results. Horizontal coordinates represent the number of proteins, while the vertical coordinates represent the IPR entries to which the proteins are annotated.

3.2.5. Annotation of All Subcellular Localization Information

Cells are highly complex and well-organized structures. Numerous distinct organelles or cellular areas, including the nucleus, Golgi apparatus, endoplasmic reticulum, mitochondria, cytoplasm, and cell membrane, can be separated inside the cell based on variations in their spatial distribution features and activities. Notably, proteins are precisely delivered to certain organelles once synthesis is finished in the ribosomes because of the protein sorting signals they contain. Moreover, some proteins are localized in the cytoplasm or are released to the external environment. Noteworthy, proteins can only engage in the many functions of the cell and fulfill their assigned duties when they are carried to the appropriate location. A thorough understanding of the subcellular localization of proteins is thus crucial to deciphering the mechanisms behind an organism's activities and the control of physiological processes. Herein, 14 subcellular localization data, which included 452 cytoplasmic proteins, 451 nucleus proteins, 333 mitochondrion proteins, and 333 extracellular proteins, were analyzed based on the annotation results of subcellular localization data for all proteins (Figure 7). Of note, there were significantly more extracellular proteins than obtained in the subcellular localization data. Cytoplasmic proteins play a supportive, nutritive, connective, and protective role. In contrast. nuclear proteins are primarily involved in various proteins, DNA synthesis enzymes, RNA transcription and enzyme processing. They are also involved in various regulatory roles of protein factors, among other things, which are critical to the formation of spermatozoa. Mitochondrial proteins can generate energy and participate in cell signaling, cellular signaling and disease relevance, and apoptosis and disease relevance, which are important for sperm vitality and sperm-egg binding. Extracellular proteins have structural support, signal transduction, material transportation, and special functions that protect the structure of the sperm and provide nutrients and signals.

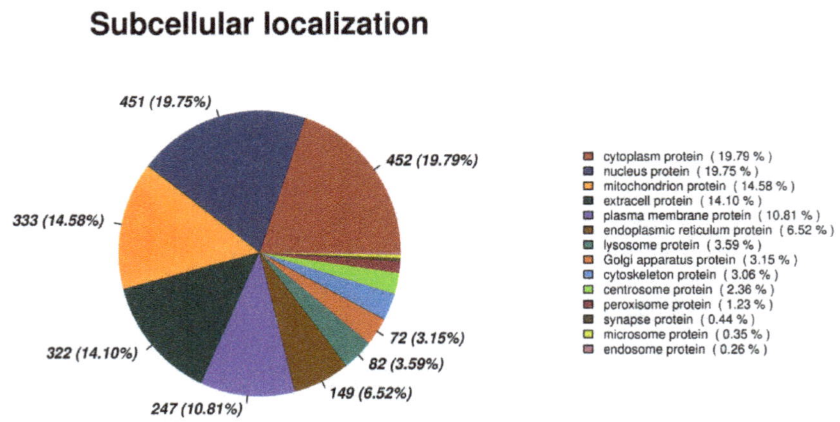

Figure 7. Pie chart of subcellular localization annotation results.

3.2.6. PCA

Principal component analysis (PCA) is a traditional statistical technique that focuses on the capacity to manipulate a collection of connected variables. The set groups were subjected to PCA typing for analysis (Figure 8). Notably, the main component analysis graph of the entire sample revealed a significant difference in Group C's proteome from that of Groups A and B. Group C also had significant differences within the group as well as overlaps between the protein components of Group C and those of Groups A and B.

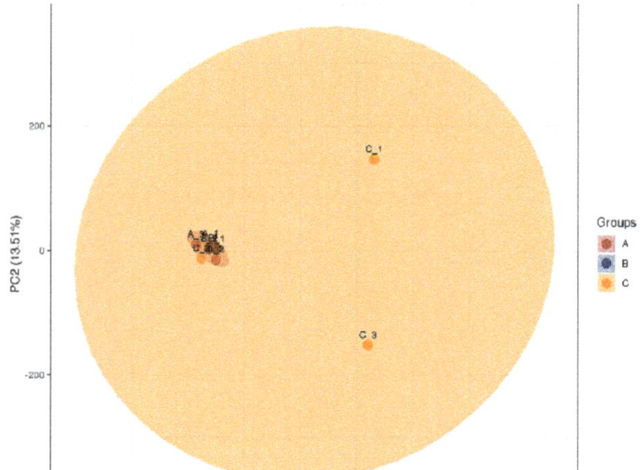

Figure 8. PCA of total samples. The horizontal and vertical coordinates PC1 and PC2 in the figure indicate the scores of the first and second ranked principal components, respectively. The scatter color indicates the experimental grouping of the sample.

The general distribution trend between the two sample groups was examined using PCA. The difference within group C was greater than that within group B. In contrast, the difference within group A is lower than that within group B (Figure 9). The samples in groups A and B were more widely distributed in the PCA, while the samples in groups B and C were more densely packed. Samples in group C were more widely distributed. PCA of groups A and C revealed that the samples in group A were densely packed while the samples in group C were scattered.

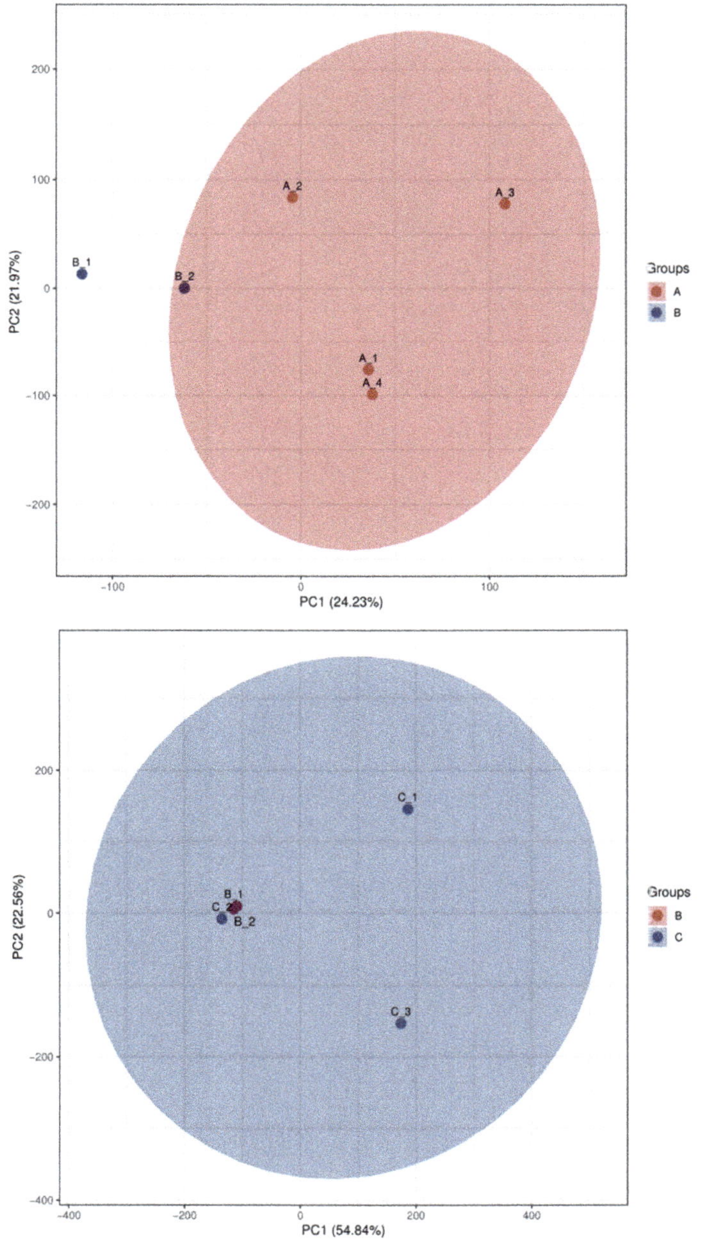

Figure 9. *Cont.*

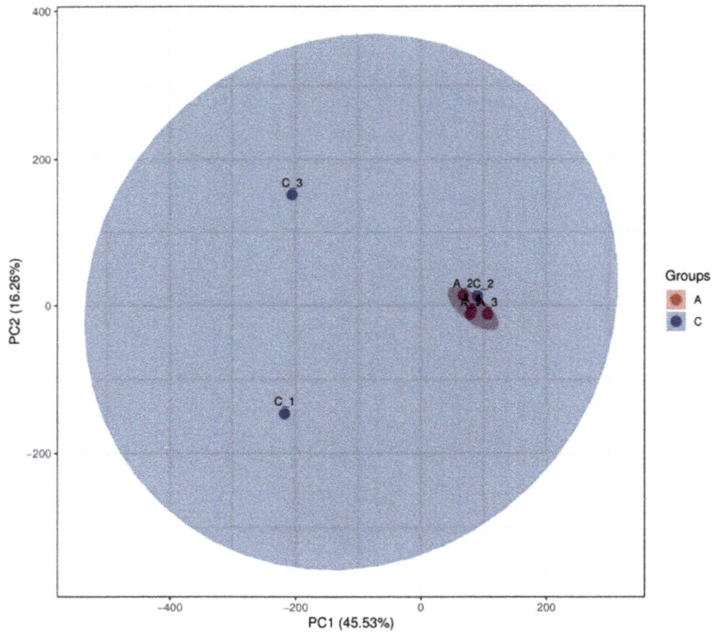

Figure 9. Comparison of PCA of pairs of samples.

3.3. *Screening for Differential Proteins*

Groups B and A yielded 2989 proteins, groups C and A yielded 2827 proteins, while groups B and C yielded 2771 proteins (Table 1). Notably, 35 proteins were down-regulated in the comparison of groups B and A, 292 proteins were down-regulated in the comparison of groups C and A, and 28 proteins were down-regulated in the comparison of groups B and C. In contrast, 116 proteins were upregulated in the comparison of groups B and C, 81 proteins were upregulated in the comparison of groups C and A, and 66 proteins were upregulated in the comparison of groups B and A. The upregulated proteins had a fold change (FC) > 1.2 and p value < 0.05, while the down-regulated proteins had a fold change (FC) < 0.83 and p value < 0.05. A volcano diagram was created after a T-test was used to analyze the significantly different proteins between the two groups (Figure 10). The multiplicity of differences for each protein was calculated as the logarithm of the base of 2. In contrast, the p value was calculated as the negative of the logarithm of the base of 10. Multiple combinations of differentially expressed proteins were compared to generate a Wayne diagram (Figure 11). The number of differentially expressed proteins shared by the BA and CA groups was 9, the CA and BC groups were 77, and the B and BC groups were 1000. There were five common differential proteins and 77 common differential proteins across the three groups and nine common differential proteins between the BC and BA groups.

Table 1. Differential protein screening.

Sample Pairs Comparing by the Former Over the Latter	The Number of Proteins Co-Identified in Both Sets of Samples	Regulated Type (Up- or Down-Regulation)	Up-Regulated Proteins (FC > 1.2 and p Value < 0.05), and Down-Regulated Proteins (FC < 0.83 and p Value < 0.05)
B.vs. A	2989	up-regulated	66
		down-regulated	35
C.vs. A	2827	up-regulated	81
		down-regulated	292
B.vs. C	2771	up-regulated	116
		down-regulated	28

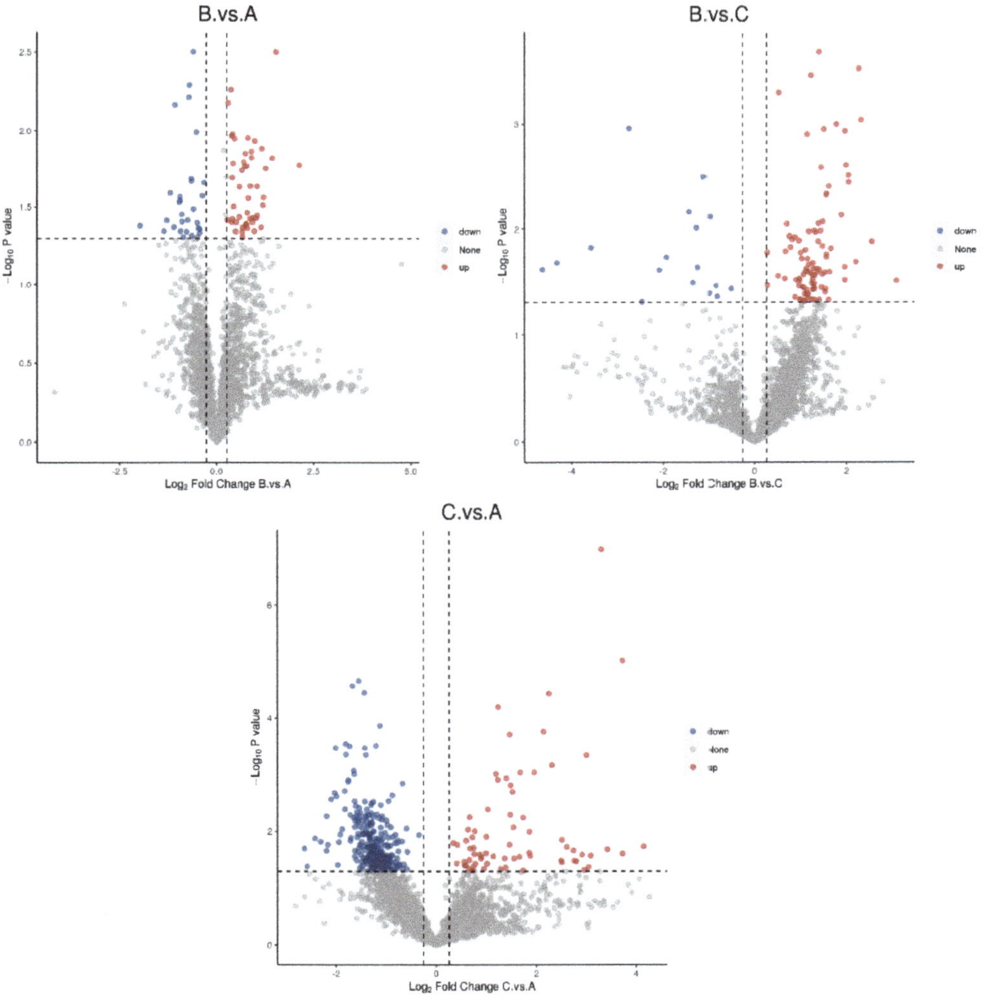

Figure 10. Volcano plots comparing the differentially expressed proteins between groups. Gray represents proteins with insignificant differences, red represents up-regulated proteins, and blue represents down-regulated proteins.

Figure 11. Wayne diagram of differential proteins. Each circle in the figure represents the differential protein results for one comparison combination. The overlap represents the differential proteins common to multiple comparison combinations, while the non-overlap represents differential proteins that only appear in one group.

3.3.1. Enrichment Analysis of Differential Proteins

Figure 12 shows the GO functional annotation analysis results of differential proteins identified by comparing two groups. The enrichment results were in three categories, and the differential proteins were expressed in all three biological functions. Group B had the most diverse types of differential proteins involved in molecular function (MF) compared to Group A, with a total of 10 molecular functions involved. The proteins were mainly enriched in RNA binding and catalytic enzyme activity, which are potentially involved in gene regulation and sperm formation. Differential proteins in biological processes (BP) were concentrated in nine processes but were mainly involved in protein transport and sulfur compound metabolism. The least differentially expressed proteins in the cellular components (CC) were mainly located in the cell membrane and endoplasmic reticulum membrane. Group B had the largest number of differentially expressed proteins involved in molecular function (MF) compared to group C, with a total of 16 molecular functions. These proteins were mainly involved in peptidase inhibitor activity, endopeptidase inhibitory activity, and calcium ion binding. These molecular functions are potentially involved in molecular functions that affect sperm oxidative stress regulate physiological metabolism, and gene expression. Differential proteins in biological processes (BP) were concentrated in 14 processes. The three main biological processes were stress response, defense response, and immune response. The least differentially expressed proteins in the cellular component (CC) were mainly located in the outer region of the cell membrane, which is a vital site for material and information exchange between cells and the external environment. There were 20 differentially expressed proteins involved in biological processes (BP) in group C compared to Group A. These proteins were mainly associated with defense response, stress response, carbohydrate metabolism, and immune response. These biological functions were potentially associated with indicators such as sperm motility and dynamic parameters. There were 11 differentially expressed proteins in molecular function (MF). They were mainly associated with enzyme inhibitory activity, DNA binding, metalloendopeptidase activity, and hydrogenase activity. The least differentially expressed proteins in cellular components (CC) were mainly located in the outer regions of the cell membrane and chromosomes.

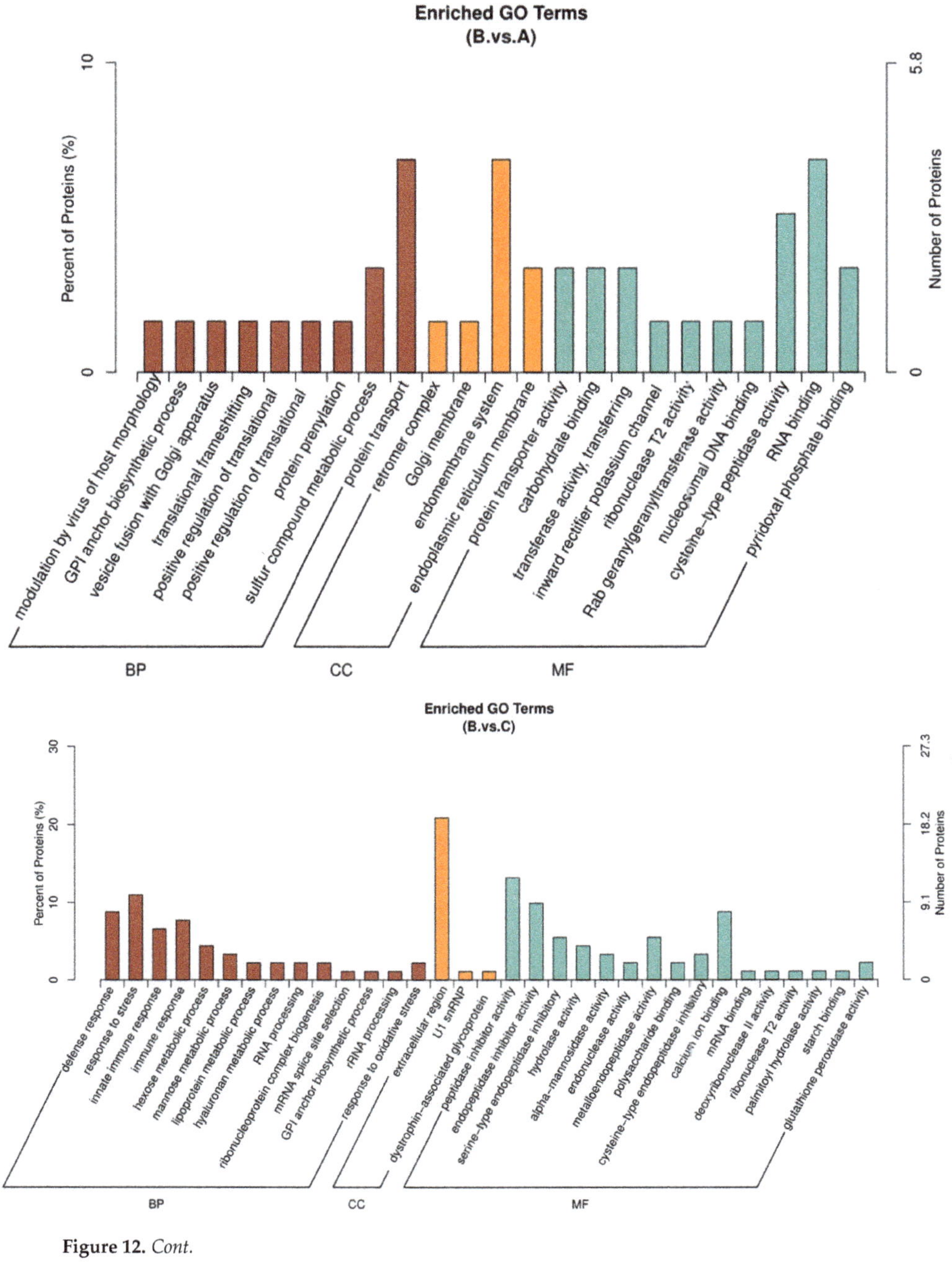

Figure 12. *Cont.*

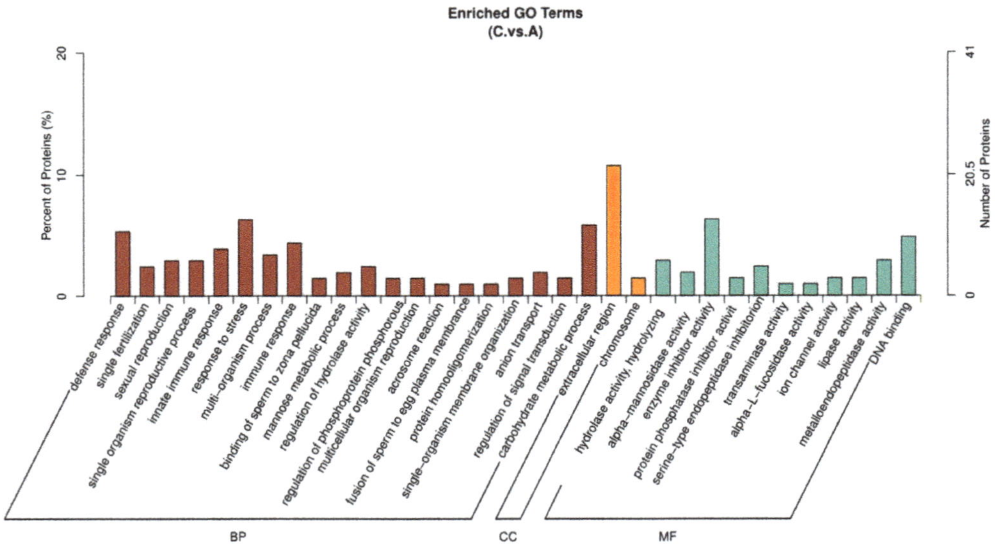

Figure 12. Histogram showing the GO enrichment results of differential proteins for the various comparison group. The enrichment results are in three categories of up to 20 species each (p value < 0.05). The vertical coordinate on the left is x/n, while the vertical coordinate on the right is the number of differential proteins annotated.

3.3.2. Comparison of KEGG Enrichment Analysis

This study used the KEGG pathway as a unit to apply hypergeometric testing for enrichment analysis of the annotation results of differential proteins. The smaller the p-value, the greater the reliability of the test and vice versa. Figure 13 is a KEGG enrichment bubble plot of the various comparison groups. Glycosylphosphatidylinositol (GPI)—anchored protein biosynthesis was the pathway with the most significant enrichment level, while endocytosis had the least significant enrichment level among the 20 pathways enriched and analyzed in the comparison of group B and A. Primary immunodeficiency, other types of O-glycan biosynthesis, and nicotine addiction were the most significantly enriched pathways, while lysosome had the least significant enrichment level among the 20 pathways enriched and analyzed in the comparison of group B and C. Glycan degradation was the pathway with the most significant enrichment level, while the regulation of the animal cytoskeleton had the least significant enrichment level among the 20 pathways enriched and analyzed in the comparison of group C and A.

Table 2 outlines the GSEA-KEGG enrichment results. Group B was significantly enriched in the OXIDATIVE PHOSPHORYLATION pathway compared to Group A. The KEGG-ID: MAP05012 Parkinson's disease and KEGG-ID: MAP04932 nonalcoholic fatty liver disease pathways were significantly enriched in group B compared to group C. The KEGG-ID: MAP05010 Alzheimer's disease and KEGG-ID: MAP04723 retrograde endocannabinoid signaling pathways were significantly enriched in Group C compared to Group A. Figure 14 summarizes the GSEA-KEGG enrichment results by displaying the top 20 | NES | entries from BP, CC, and MF. All results have been displayed in cases where the entries are less than 20.

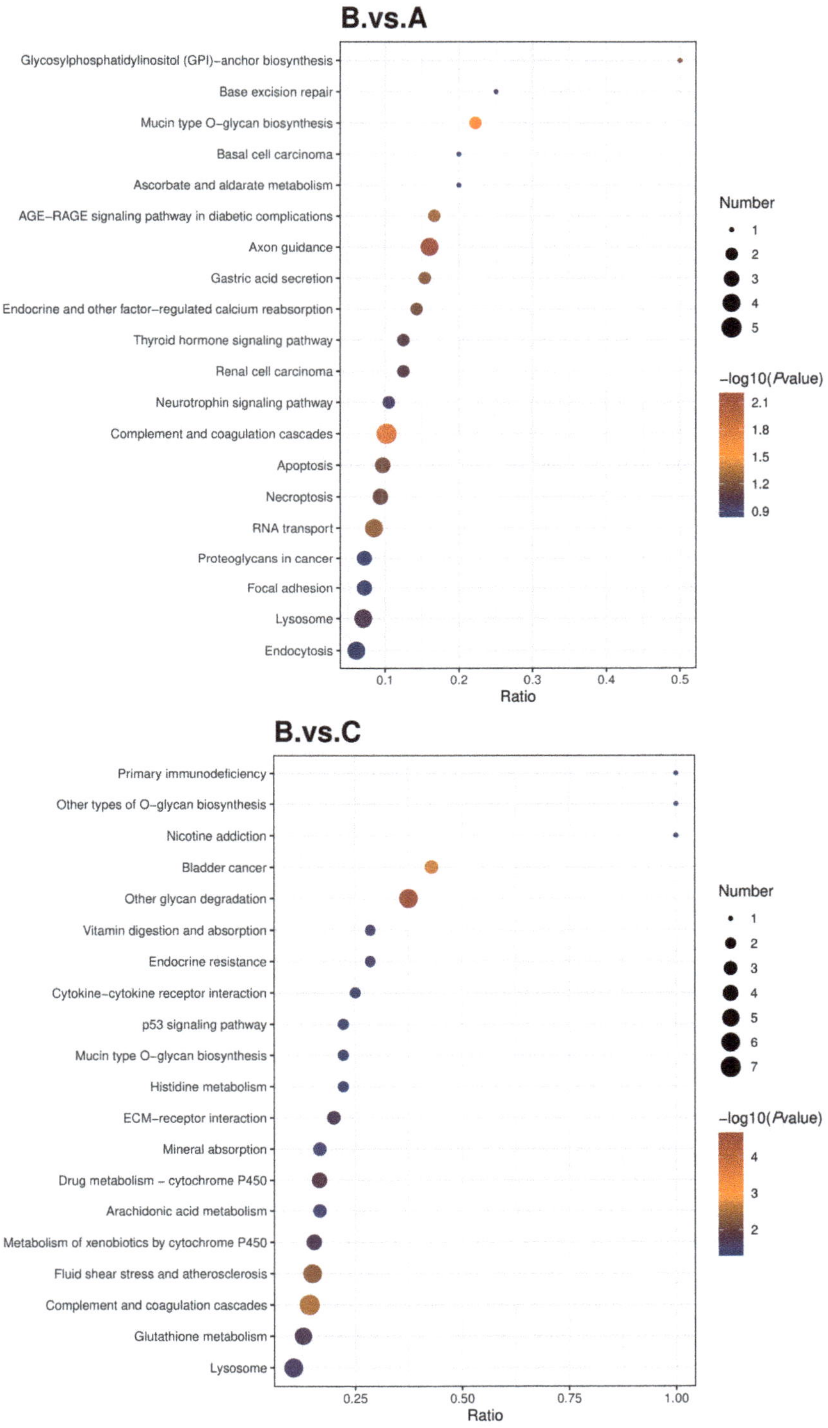

Figure 13. *Cont.*

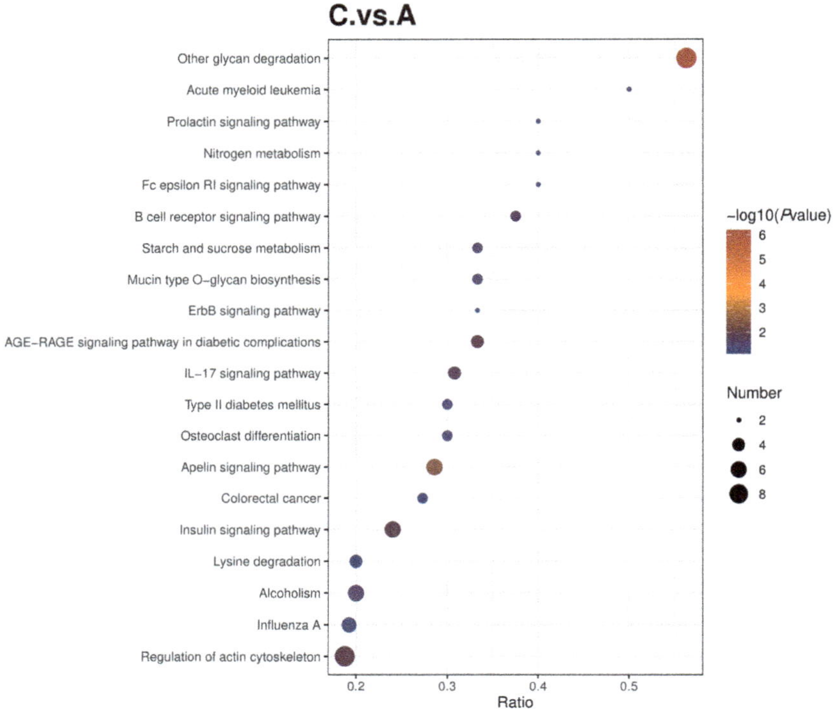

Figure 13. Bubble map showing KEGG enrichment analysis results for the different comparison groups. The horizontal coordinate in the graph is the ratio of the number of differential proteins in the corresponding pathway to the total number of proteins identified in that pathway. The larger the value, the higher the enrichment of differential proteins in that pathway and vice versa. The color of the dots represents the *p* value of the hypergeometric test. It ranges from blue to red; the redder the color, the smaller the *p* value, indicating that the test is more reliable and statistically significant. The size of the dots represents the number of differential proteins in the corresponding pathway, the larger the dot, the more differential proteins are in the pathway, and vice versa.

Table 2. GSEA-KEGG enrichment results.

KEGG_ID	KEGG_Term	*p* Value	AdjustedPv	Size	ES	NES
MAP00190	OXIDATIVE PHOS-PHORYLATION	0.0	0.019532768	97	−0.45213947	−2.0541892
MAP05012	PARKINSON'S DISEASE	0.0	0.0141317975	95	−0.45930517	−2.0012848
MAP04932	NON-ALCOHOLIC FATTY LIVER DISEASE _NAFLD_	0.0	0.045213815	70	−0.4325565	−1.8434372
MAP05010	ALZHEIMER'S DISEASE	0.0	0.07570649	98	−0.4006493	−1.7487847
MAP04723	retrograde endocannabinoid signaling	0.007633588	0.0965757	52	−0.42583093	−1.6883155

Note: (1) KEGG_ID: ID of the enriched KEGG Pathway; (2) KEGG_Term: name of the enriched KEGG Pathway; (3) *p* value: *p* value of the enrichment analysis. The *p* value is 0 when the actual *p* value is so small that it cannot be obtained accurately; (4) AdjustedPv: *p* value of the corrected (5) size: the number of proteins annotated to the functional entry; (6) ES: the enrichment score of the GSEA analysis of the entry; (7) NES: the enrichment score of the GSEA analysis of the entry after homogenization.

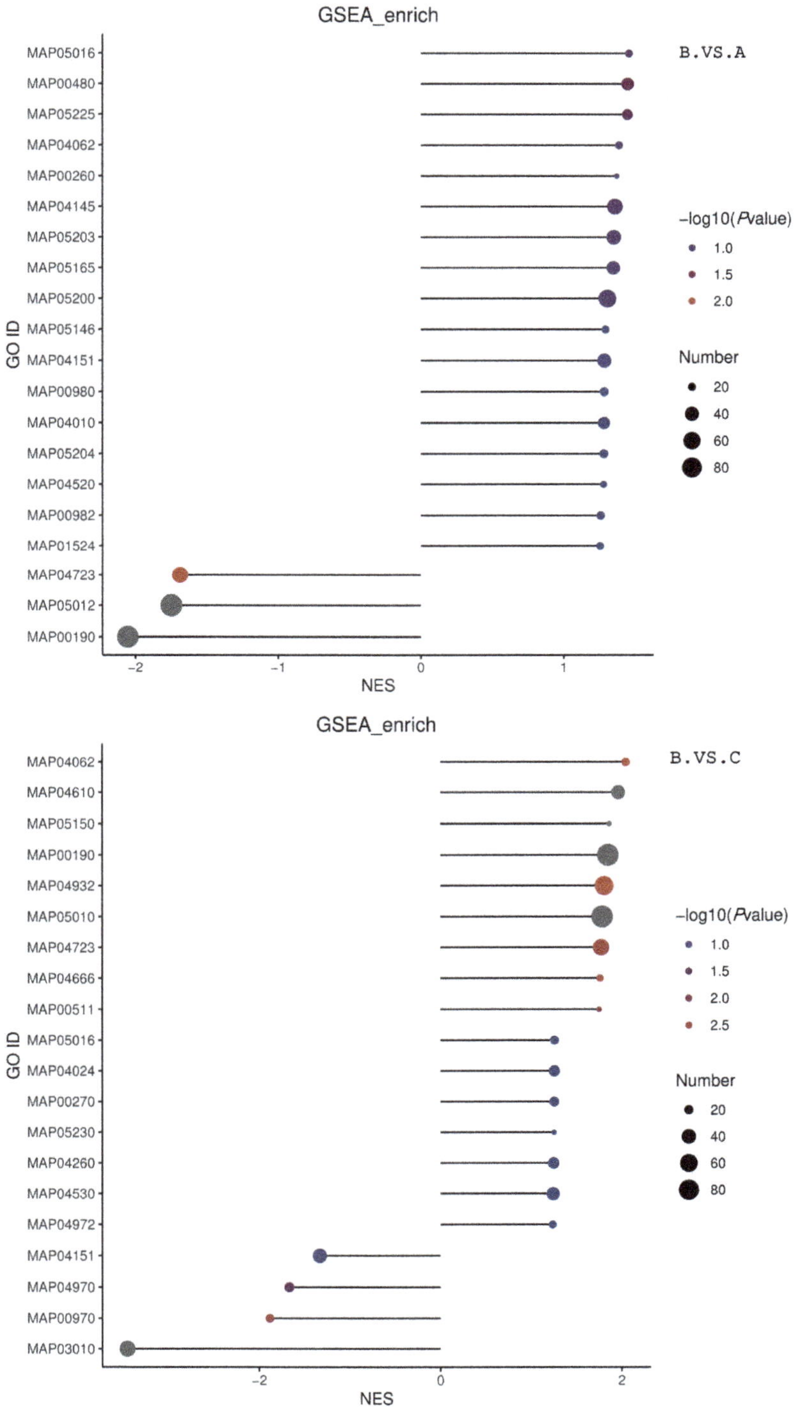

Figure 14. *Cont.*

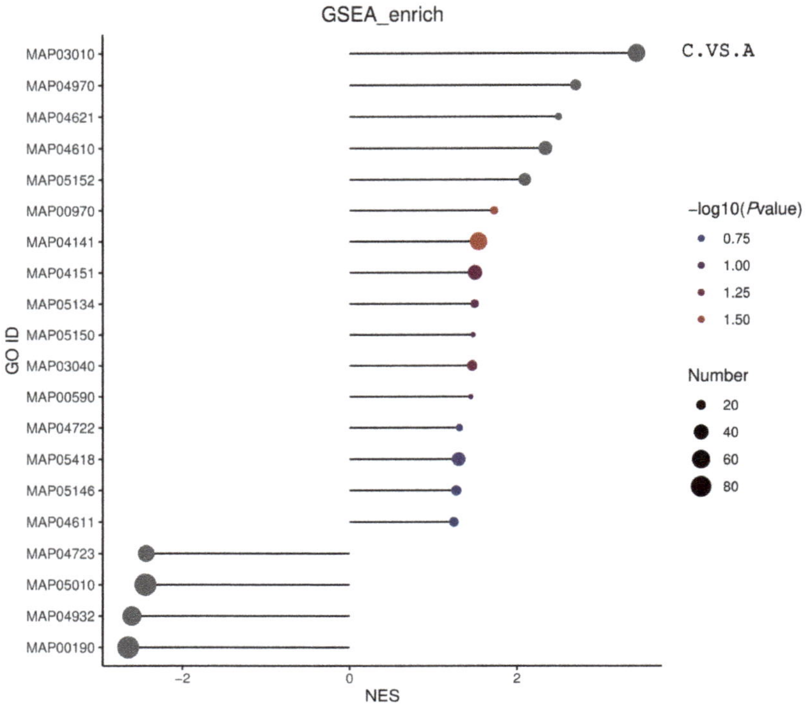

Figure 14. Matchstick plot showing the overall GSEA-KEGG ES enrichment results. The horizontal coordinate in the graph is the NES value. The larger the |NES| value, the higher the degree of differential protein enrichment in the entry and vice versa. The color of the dots represents −log10 (*p* value). The *p* value refers to the nominal *p* value in GSEA. The redder the color, the larger the −log10 (*p* value), while the smaller the *p* value, the larger the reliability of the test and, the more statistically significant the test is. The gray color denotes the pathways with a *p* value of 0. The size of the dots represents the number of differential proteins in the corresponding entry. The larger the dots, the higher the enrichment of differential proteins in the pathway and vice versa. The size of the dots represents the number of differential proteins in the corresponding entry. The larger the dots, the higher the number of differential proteins in the pathway and vice versa.

3.3.3. Structural Domain Enrichment Analysis of Comparative Samples

Figure 15 displays the findings of structural domain enrichment of differentially expressed proteins. A total of 10 structural domains were enriched in the comparison of groups B and A. Pleckstrin homology-like domains, which primarily have multiple biological functions in the cell, particularly in the regulation of cell death and tumor suppression, were not significantly enriched. In contrast, a neurotransmitter-release related protein (SNAP-25), which helps regulate the fusion of the synaptic vesicle with the cell membrane, thereby releasing neurotransmitters, was the structural domain with the highest level of enrichment. The hemopexin/matrixin (REPEAT), which protects against oxidative stress damage and plays a role in the epithelial–mesenchymal transition, was the most significantly enriched domain among 10 domains that were enriched in the comparison of groups B and C. Zona-pellucida-binding was the most significantly enriched structural domain in the comparison of groups C and A. In contrast, the enrichment level of Kunitz metazoa (Proteinase inhibitor 12, Kunitz metazoa) was not significant, while the immunoglobulin V-set structural domain, which is primarily engaged in antigen binding, diversification, and information transmission, was not considerably enriched.

Table 3 outlines the GSEA domain enrichment results. There was a significant enrichment of the PROTEIN KINASE DOMAIN protein kinase region in the comparison of groups B and A. The functions of BETA-DEFENSIN and PROTEASOME, SUBUNIT ALPHA-BETA proteasomes were significantly enriched in the comparison of groups B and C. IPR_ID: IPR001623DNAJ domain function was significantly enriched in the comparison of groups C and A. Figure 16 summarizes the GSEA domain enrichment results by displaying the top 20 | NES | entries from BP, CC, and MF. All results are displayed in cases where the number of entries is less than 20.

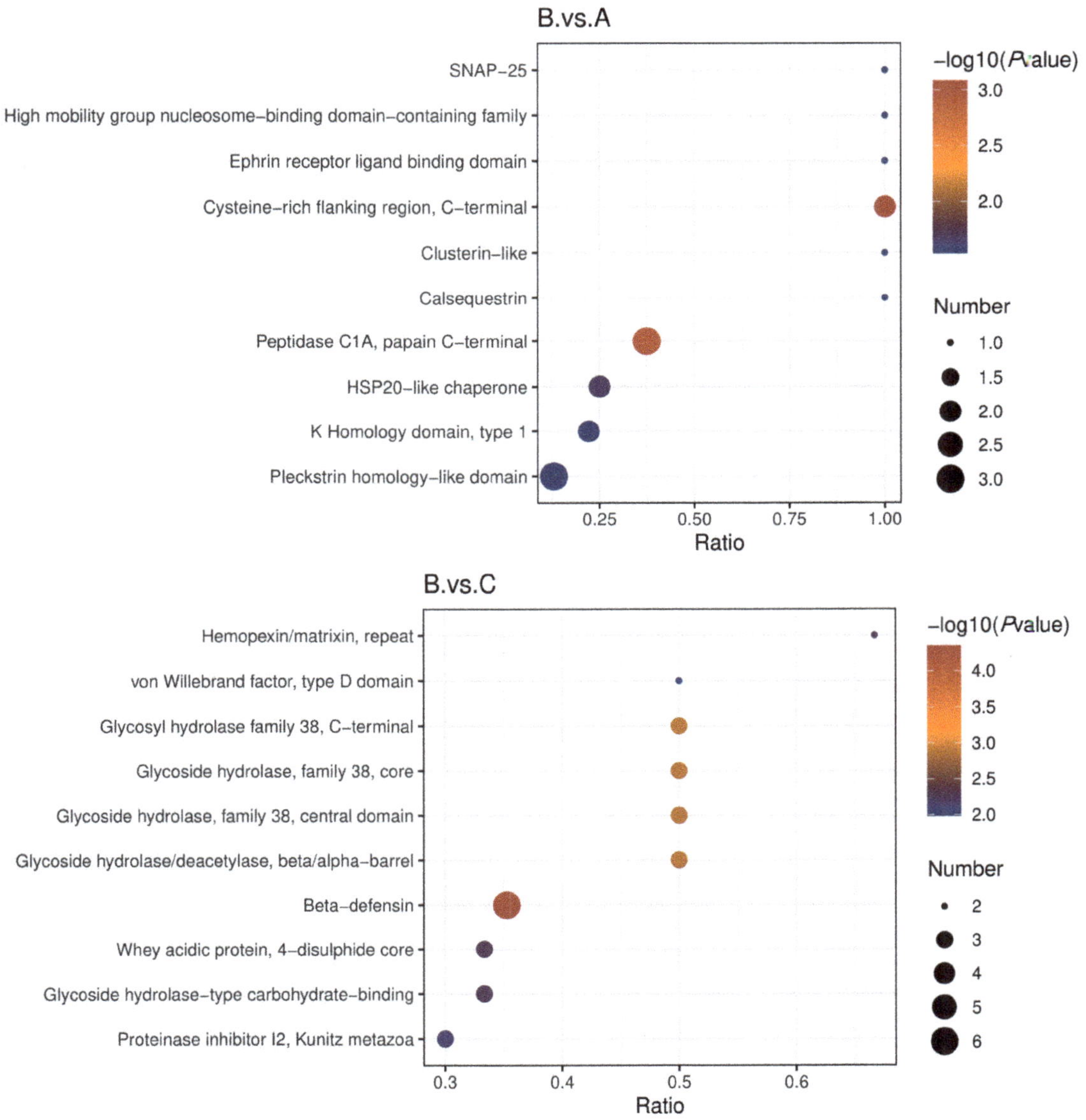

Figure 15. *Cont.*

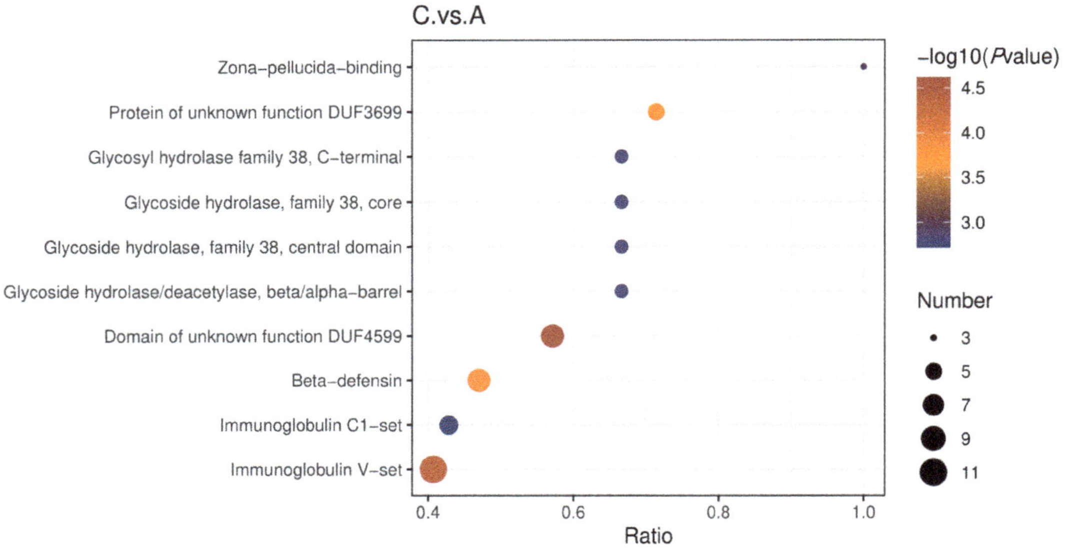

Figure 15. Enrichment bubble diagram of structural domains of the different comparison groups. The horizontal coordinate in the graph is the ratio of the number of differential proteins in the corresponding structural domain to the total number of proteins identified in that structural domain. The larger the value, the higher the enrichment of differential proteins in that structural domain and vice versa. The color of the dots represents the −log10 (p value) of the hypergeometric test. The redder the color, the larger the −log10 (p value), while the smaller the pvalue, the larger the reliability of the test and the more statistically significant it is. The size of the dots represents the number of differential proteins in the corresponding structural domains, the larger the dots. The higher the number of differential proteins in that structural domain.

Table 3. GSEA-structural domain enrichment results.

IPR_ID	IPR_Term	p Value	AdjustedPv	Size	ES	NES
IPR000719	PROTEIN KINASE DOMAIN	0.0	0.046585634	46	−0.49194428	−1.8805562
IPR025933	BETA-DEFENSIN	0.006802721	0.02746326	17	−0.63333166	−1.8570249
IPR001353	PROTEASOME, SUBUNIT ALPHA_BETA	0.0033670033	0.019854257	17	−0.6215955	−1.8289802
IPR001623	DNAJ DOMAIN	0.01650165	0.0661915	16	−0.5654077	−1.6392351

Note: (1) IPR_ID: structural domain number; (2) IPR_Term: description of structural domain function; (3) p value: p value for enrichment analysis. The p value is 0 when the actual p value is extremely small and cannot be accurately obtained; (4) AdjustedPv: corrected p value; (5) size: the function entry annotated to the number of proteins; (6) ES: enrichment score of GSEA analysis for the entry; (7) NES: enrichment score of GSEA analysis for the entry after homogenization.

3.3.4. Subcellular Localization Analysis of Differential Proteins in the Different Comparison Groups

The annotation results of subcellular localization information of proteins in different comparison groups were categorized (Figure 17). Thirteen subcellular localization items were annotated in the comparison of groups B and A. Notably, 13 extracellular protein items and 17 nucleus protein items were significantly higher than other subcellular localization information. Twenty-three extracellular proteins and eighteen nucleus proteins were significantly higher than other subcellular localization information in group B, which had ten subcellular localization information annotated overall. In group C, thirteen subcellular localization information were annotated overall, compared to fifty-two extracellular proteins and thirteen nucleus proteins in group A. Group C had a significantly higher amount of annotated subcellular localization information than group A. Among the 13 that were annotated, 52 were extracellular proteins, 43 were nucleus proteins, 34 were plasma membrane proteins, and 28 were cytoplasmic proteins.

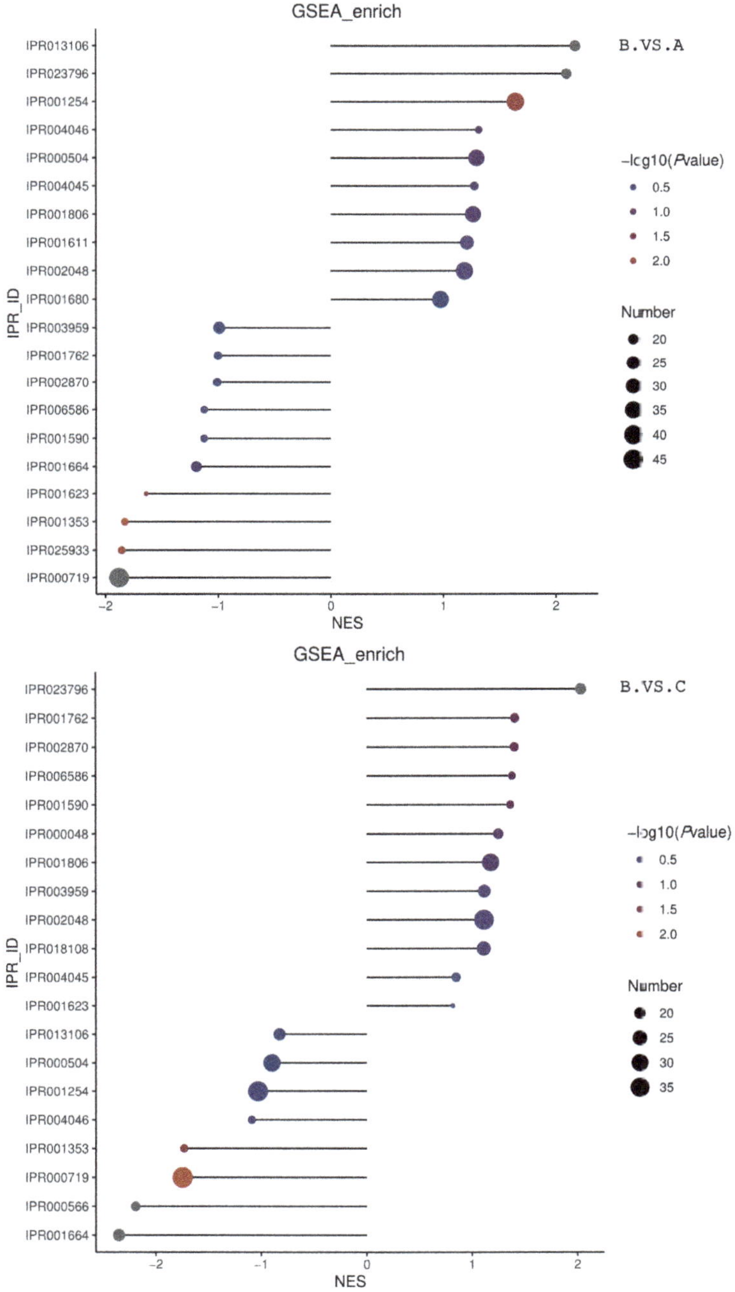

Figure 16. *Cont.*

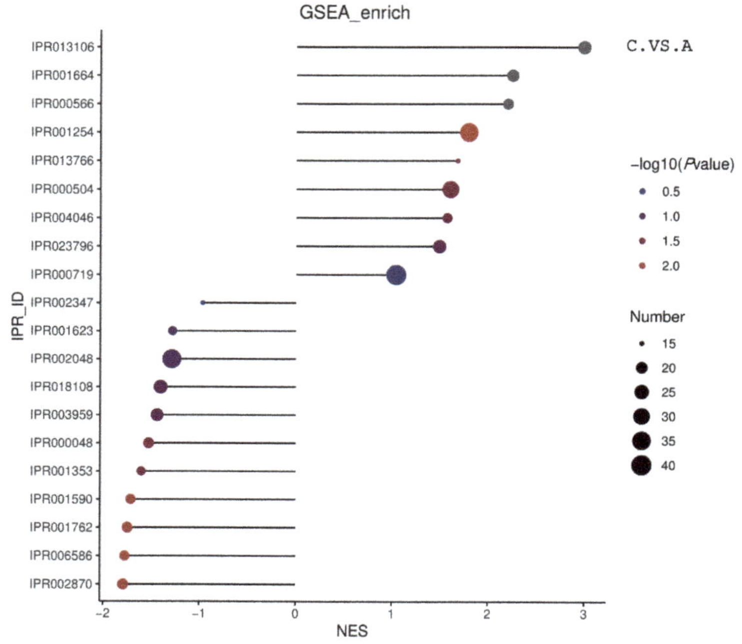

Figure 16. Matchstick diagram of the overall enrichment results of the GSEA-structural domain ES. The horizontal coordinate in the graph is the NES value. The larger the | NES | value, the higher the degree of differential protein enrichment in the entry. The color of the points represents −log10 (*p* value). *p* value refers to the nominal *p* value in GSEA. The redder the color, the larger the −log10 (*p* value), while the smaller the *p* value, the larger the reliability of the test and the more statistically significant the test is. The gray color represents the structural domains with a *p* value of 0. The size of the points represents the number of differential proteins in the corresponding entry. The larger the point, the higher the enrichment of differential proteins in that structural domain. The size of the dot represents the number of differential proteins in the corresponding entry. The larger the dot, the higher the number of differential proteins in the structural domain.

Table 4 outlines the GSEA subcellular enrichment results. The mitochondrial protein (MITOCHONDRION PROTOIN) localization information in Group B was significantly enriched compared to Group A. The localization information of extracellular and endoplasmic reticulum proteins in group C was significantly enriched compared to group A. Figure 18 summarizes the GSEA subcellular enrichment results by displaying the top 20 | NES | entries from BP, CC, and MF. All results are displayed in cases where the entries were less than 20.

Table 4. GSEA-subcellular enrichment results.

Subcell_ID	*p* Value	AdjustedPv	Size	ES	NES
MITOCHONDRION PROTEIN	0.0	0.11314847	309	−0.2983893	−1.4824772
EXTRACELL PROTEIN	0.0	0.0	266	0.6574827	2.0779767
endoplasmic reticulum productin	0.024819028	0.1322063	120	0.4547453	1.3880157

Note: (1) Subcell_ID: description of subcellular localization; (2) *p* value: *p* value for enrichment analysis. The *p* value is 0 when the actual *p* value is very small and cannot be accurately obtained; (3) AdjustedPv: corrected *p* value; (4) size: the number of proteins that this functional entry annotates to; (5) ES: the entry GSEA analysis enrichment score; (6) NES: the enrichment score of the entry after homogenization by GSEA analysis.

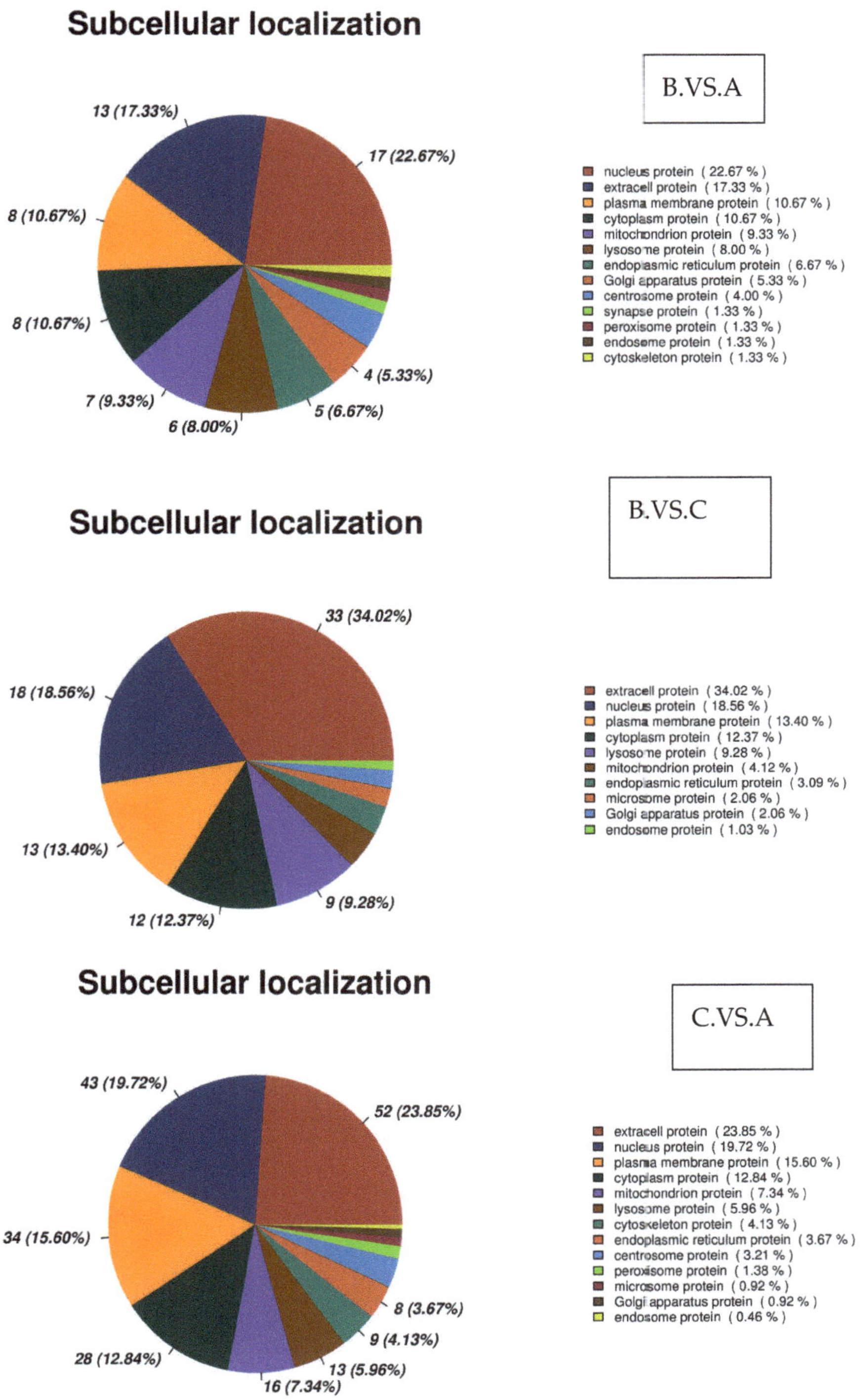

Figure 17. Analysis of subcellular localization of differential proteins.

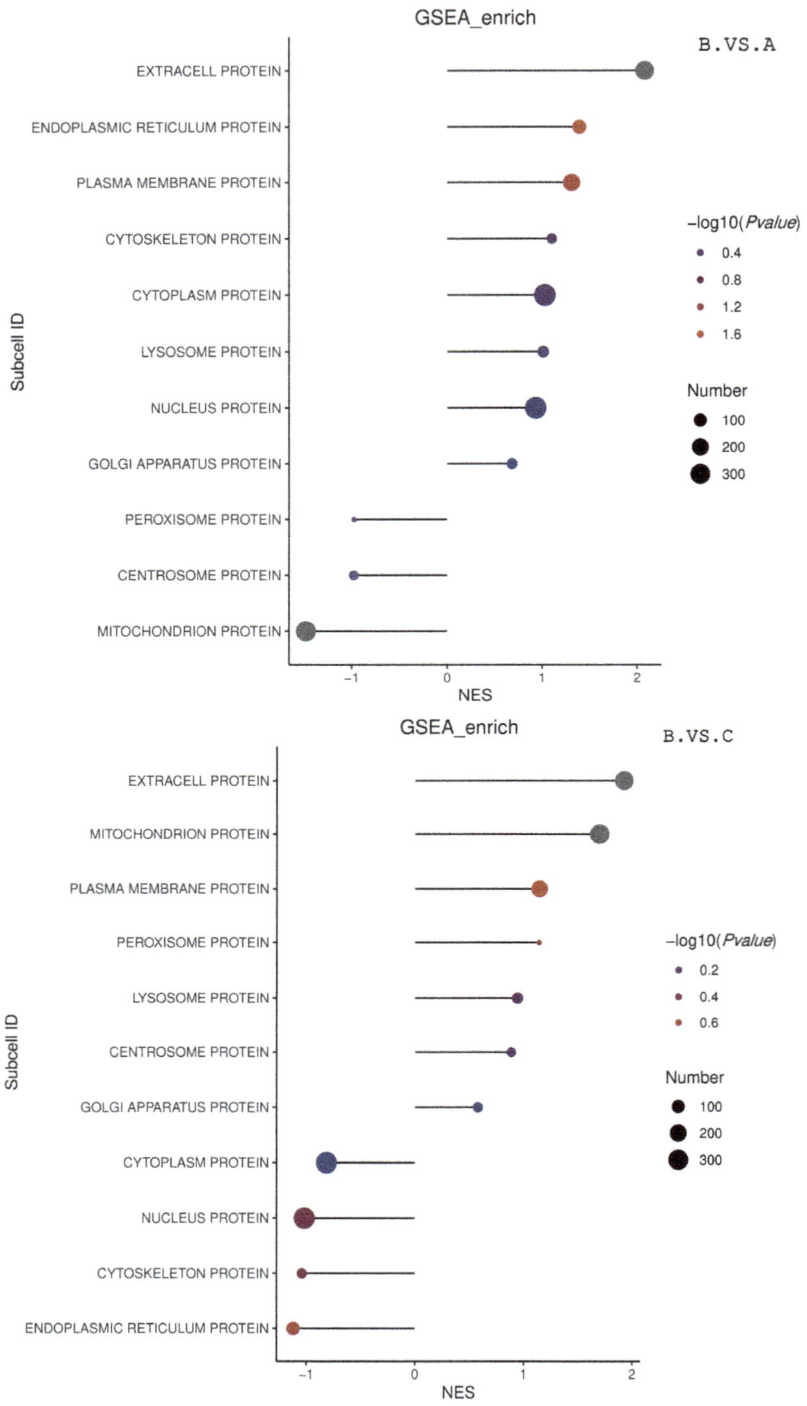

Figure 18. *Cont.*

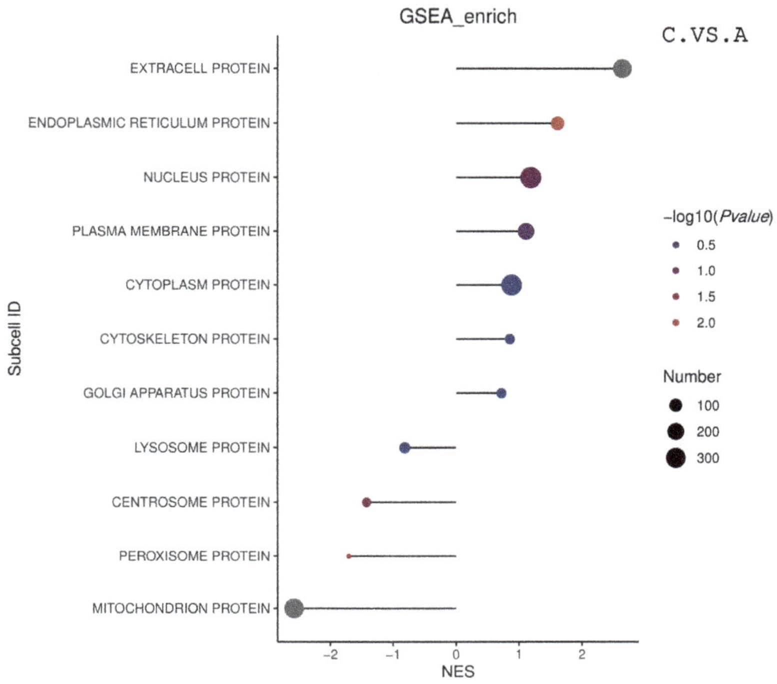

Figure 18. Matchstick plot of the overall GSEA-subcellular ES enrichment results. The horizontal coordinate in the graph is the NES value. The larger the |NES| value, the higher the degree of differential protein enrichment in the entry. The color of the dots represents −log10 (p value). The p value refers to the nominal p value in GSEA. The redder the color, the larger the −log10 (p value), while the smaller the p value, the larger the reliability of the test and the more statistically significant the test is. The gray color represents the subcellular entries with a p value of 0, the actual p value is extremely small and cannot be accurately obtained. The size of the dots represents the number of differential proteins in the corresponding entries, The larger the dots, the higher the number of differential proteins in the subcellular entries.

3.3.5. Differential Protein Interaction Analysis

The differential expression analysis results and the interaction pairings in the StringDB protein interactions database were combined to effectively create differential protein interaction networks. Figure 19 shows the protein interaction networks of the top 50 differently expressed proteins in |FC|. Each node in the network represents a protein, while its size indicates how many proteins it interacts with. The larger the node, the higher the number of proteins it interacts with. The protein's expression level in the comparison group is indicated by the node's color; orange denotes considerably high protein expression, whereas green denotes significantly low protein expression.

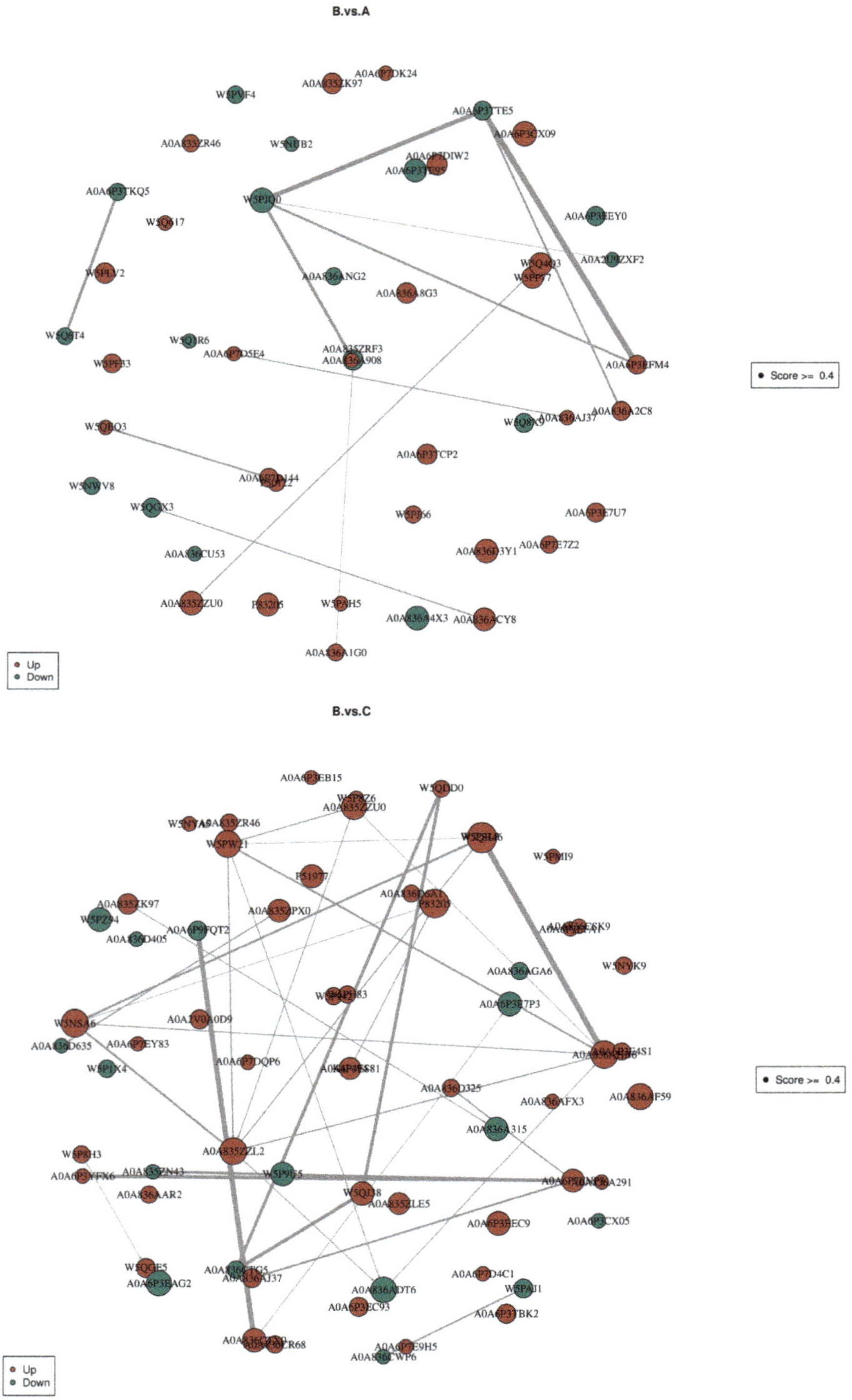

Figure 19. *Cont.*

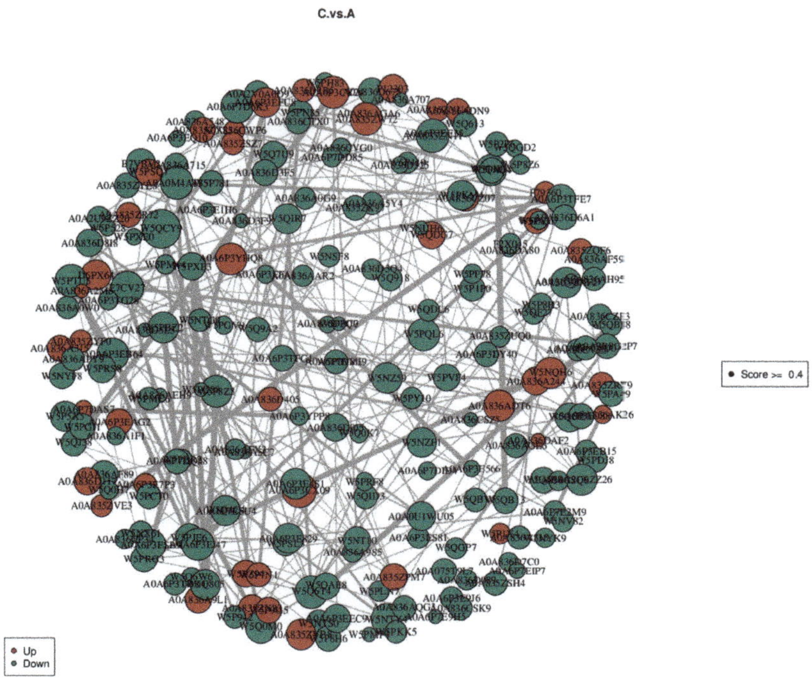

Figure 19. Protein interaction network diagram of the different comparison groups.

4. Discussion

The epididymis, seminal vesicles, prostate, and urethral bulbourethral glands are the primary sources of spermatozoa and secretions that make up an animal's semen. Semen is made up of about 10% of sperm and 90% of seminal plasma. High levels of tissue-specific proteins found in seminal plasma thus offer a wealth and extremely promising source of indicators for evaluating male fertility. Sperm and seminal plasma proteins are among the proteins found in semen. They are directly engaged in controlling sperm function in molecular processes and biological pathways and significantly affect both semen viability and the fertilization capacity degree [6]. Semen proteins are, therefore, excellent potential indicators for assessing semen quality and checking for sperm dysfunction. The advancement of proteomics technology has enabled in-depth research on cellular mechanisms unique to sperm. These kinds of investigations are thus essential for finding and creating reliable biomarkers of male fertility. Previous studies postulate that seminal plasma and sperm proteins influence the ejaculatory process and the use of sperm biotechnology in fields such as artificial insemination, semen freezing, and sex sorting.

The management of reactive oxygen species (ROS) and energy supply are the two main facets of how mitochondrial proteins affect sperm quality. Mitochondria are "energy factories" inside the cell that supply the energy required for the cell to continue its regular functions. The entire intricate process of spermatogenesis to fertilization, which includes the vital step of accessing the egg, requires an enormous amount of energy. Mitochondria are thus essential and indispensable components in this process. Numerous studies postulate that the more active mitochondria are, the more energy they can create. Highly active mitochondria would thus further boost the quantity and health of spermatozoa and give them a stable energy source for their regular physiological functions. Of note, mitochondria are the primary source of reactive oxygen species (ROS). Maintaining the normal functional activities of spermatozoa, including sperm viability, fertilization ability, and important physiological processes such as energy acquisition, acrosome response, and hyperactivation, requires a suitable quantity of ROS. Oxidative stress is caused when ROS levels rise above normal, which damages sperm cells and causes them to gradually lose their ability to function. The damage to sperm cells ultimately has a detrimental effect on fertility. Studies postulate that between 30% and 80% of male infertility cases are caused by sperm destruction brought on by ROS [7,8]. The occurrence of lipid peroxidation in sperm cell membranes is closely associated with reduced sperm viability [9]. These reports collectively demonstrate that oxidative stress plays a significant role in sperm quality decline. Spermatozoa are highly polarized cells that generate energy internally to carry out various intricate and vital physiological functions, including energy acquisition, hyperactivation, and acrosome reaction. These functions are exactly what spermatozoa need to successfully complete fertilization [10]. Halang et al. reported that sperm viability is maintained using about 80% of the energy generated by spermatozoa through glycolysis and mitochondrial oxidative phosphorylation [11]. These reports explain the relevance of the differential expression of proteins in group B compared to group A in this experiment. Besides the impact of mitochondrial proteins, post-translational modification of sperm proteins also significantly affects sperm survival. Proteins can undergo post-translational modification when modifying groups are added or removed by covalently attaching to their amino acid residues. This phenomenon alters the structure and function of proteins and eventually impacts their activity [12]. Common post-translational changes in proteins include methylation, biotinylation, alkylation, phosphorylation, and acetylation. Post-translational modifications of proteins affect numerous critical physiological processes, including sperm maturation, capacitation, motility, hyperactivation, and acquisition of fertilization potential [13]. Mature spermatozoa rely on post-translational modifications of their existing proteins to maintain their normal physiological functions [14] because they synthesize a few new proteins [15–18] and undergo minimal gene transcription and translation, except the mitochondrial genes. The most prevalent post-translational alteration in spermatozoa is protein phosphorylation, which is essential for sperm maturation, energy uptake, and energy metabolism [19,20]. Sperm proteasome was initially discovered from sea squirt (Halocynthia roretzi). Notably, sea squirt exhibits trypsin-like and chymotrypsin-like activity together with 20S core particles during fertilization [21]. Sperms are prevented from accessing the yolk when high-affinity leupeptin analogs impede proteasome action. However, sperms may resume their usual ability to fertilize once this inhibitor is eliminated. This phenomenon implies that sperm capacitation and acrosomal cytotoxicity are significantly influenced by the activity of the proteasome's β subunit [22,23]. The process of sperm capacitation potentially involves a mechanism that controls the entry of protein substrates into the proteasome's core for eventual degradation because the α subunit can control substrate entry and has been observed to undergo post-translational modification during capacitation [24]. This phenomenon also clarifies the importance of the proteins that are expressed differently in group B than in group C of this experiment.

The widespread neurodegenerative condition, Parkinson's disease (PD) [25], may have a negative impact on the male reproductive system. Patients with Parkinson's disease (PD) have several pathophysiological changes, including mitochondrial dysfunction and oxidative stress, which are the most common. Reactive oxygen species (ROS) are produced in significant quantities by oxidative stress, target lipids, proteins, and DNA in sperm cells, damaging the cell membranes and influencing sperm motility and morphology. Mitochondrial malfunction causes a decrease in sperm motility and fertilization because of inadequate sperm energy production [26]. According to neuroendocrine theory, PD patients may have disruptions in the hypothalamic–pituitary–gonadal (HPG) axis [27].

Neurotransmitter imbalances such as low dopamine levels cause the classic symptoms of Parkinson's disease and affect pituitary gonadotropin production, which in turn influences the testes' ability to produce sperm. Studies postulate that PD patients have lower testosterone levels, which directly inhibits spermatogenesis and maturation [28]. Moreover, the sperm may carry defective genes that potentially raise the offspring's chance of developing Parkinson's disease or other neurological illnesses. PD patients have lower-quality sperm, which can result in an irregular fertilization process and raise the risk of early problems in embryonic development, such as miscarriages, fetal developmental retardation, and unsuccessful embryo implantation [29]. Children borne of PD patients are at risk of neurodevelopmental abnormalities during growth and development, even in cases when conception and delivery are successful. Sheep with neurological disorders may exhibit altered sexual behavior, such as lower desire, which has an indirect impact on semen acquisition. Nonalcoholic fatty liver disease (NAFLD) is a prevalent liver illness [30] that causes metabolic problems, including insulin resistance and aberrant lipid metabolism, potentially impacting spermatogenesis negatively. Hormonal abnormalities caused by insulin resistance can impact the functioning of the HPG axis. Testicular spermatogenesis may be impacted by insulin resistance-induced changes in the insulin-like growth factor (IGF) system, including a reduction in IGF-1 levels [31]. Elevated blood levels of free fatty acids caused by abnormalities in lipid metabolism can penetrate testicular tissue and disrupt spermatogenesis and maturation. Patients with NAFLD have a diminished liver's capacity for detoxification, which causes the buildup of toxins in the body, thereby negatively impacting the reproductive system [32]. The liver produces several cytokines and inflammatory mediators that enter the bloodstream and impact the testes' milieu, causing spermatogenic cell damage, a drop in the number of viable sperm, an increase in the rate of malformation, and a decrease in viability. Sperm quality decline in males with NAFLD may have long-term consequences for the health of the offspring. Oxidative stress and metabolic disorders may damage and alter the sperm's genetic material, which may then be passed on to the offspring, increasing the risk of metabolic diseases such as obesity and diabetes mellitus. Moreover, the paternal offspring's NAFLD status may impact the embryo's early development environment. Studies postulate that certain molecular signals in the sperm can transfer metabolic disorders in the paternal offspring to the embryo, altering the embryo's gene expression and developmental trajectory. This phenomenon could potentially make the offspring more vulnerable to metabolic abnormalities as well as growth and developmental issues after birth. The aberrant metabolic environment of ewes with NAFLD may have an impact on the attachment and development of embryos. Metabolic disorders, such as hyperglycemia and hyperlipidemia, can harm the fetus and result in aberrant fetal development, including organ developmental anomalies. Alzheimer's dementia (AD) is a complicated neurodegenerative illness whose pathological process disrupts the regular functioning of the neuroendocrine system. The brain levels of neurotransmitters, including dopamine and acetylcholine, are significantly altered in AD patients [33]. Neurotransmitters are essential for cognitive function and also play a role in controlling the activity of the hypothalamic-pituitary-gonadal (HPG) axis. The rhythm and quantity of gonadotropins released by the pituitary gland, such as follicle-stimulating hormone (FSH) and luteinizing hormone (LH), are also changed when hypothalamic production of gonadotropin-releasing hormone (GnRH) is impacted [34]. The regulation of spermatogenesis in the testicles is largely dependent on LH and FSH. Excess secretion disrupts the intratesticular environment of the testicles and interferes with spermatogonial proliferation, differentiation, and spermatogenesis, thereby affecting the amount and quality of sperm [35]. Sheep with neural cell injuries may also have reduced sexual behavior and reproductive function, which impacts the amount and quality of semen produced. The development of neuroprogenitor fibril tangles because of β-amyloid (Aβ) deposition and tau protein hyperphosphorylation is a primary pathogenic characteristic of AD. These aberrantly aggregated proteins impair the neuronal function in the brain and negatively affect the testicular tissue. The function of spermatogenesis-associated proteins is affected by Aβ and aberrantly phosphorylated tau proteins interfering with the normal folding, transport, and degradation activities of proteins in the testis. Several mutations in genes linked to AD, including those for amyloid precursor protein (APP), progerin 1 (PS1), and progerin 2 (PS2), have been reported [36]. Offspring of males with AD who inherit these faulty genes are at a significantly higher risk of getting AD or other neurological illnesses. Moreover, the pathophysiology of AD is significantly influenced by epigenetic changes, such as DNA methylation and histone alterations [37]. Sperm from AD patients may have changed epigenetic markers, which can be transferred to the progeny, which potentially impacts their gene expression pattern and increases their vulnerability to neurodevelopmental and functional problems. AD patients may experience aberrant fertilization because of a decline in sperm quality [38]. Damaged sperm may not be able to give the embryo the proper genetic and epigenetic support, which increases the risk of aberrant embryonic development, including early miscarriage and unsuccessful embryo implantation [39]. The genetic material and epigenetic information of the sperm are essential for early embryonic development. The baby may experience other neurodevelopmental issues despite the embryo growing properly, leading to neurodevelopmental issues such as cognitive dysfunction and diminished learning capacity. Sheep with Alzheimer's disease may struggle to care for their offspring by neglecting to give them proper motherly attention, which might hinder the offspring's growth and development. Moreover, the poor physical state of the sick sheep affects milk supply and quality, thereby negatively impacting the offspring's growth and development. The phenomena of anomalies in ontogenetic performance, lambing %, and progeny quality in heterozygotes and purebreds have been discussed in the cited literature.

5. Conclusions

There are certain limitations due to the small sample size in each group. In this experiment, through proteomic analysis of semen from Duolang sheep rams with different FecB genotypes, the differentially expressed proteins (DEPs) in semen of the B+ genotype compared to the ++ genotype are significantly involved in the innate immune response. This indicates that these DEPs are significantly enriched in the oxidative phosphorylation pathway in the KEGG pathway, and exert effects on the immune health, energy metabolism, and reproductive performance of Duolang sheep. The DEPs in semen of the B+ genotype compared to the BB genotype play roles in protein phosphorylation and cellular protein catabolic processes. The pathways significantly enriched by these DEPs, including those related to Parkinson's disease and non-alcoholic fatty liver disease, suggest impacts on the cellular metabolism, physiological functions, and reproduction of Duolang sheep. Additionally, differences in the molecular functions and cellular components of DEPs in semen of the BB genotype compared to the ++ genotype were analyzed, so as to explore the potential effects of these differences on the physiology and reproduction of Duolang sheep.

Author Contributions: Y.Z.: Conceptualization, Data Curation, Investigation, Formal Analysis, Methodology, Resources, Valiation, Visualization, Writing—Original Draft, Writing—Review and Editing. Z.N.: Funding Acquisition, Resources, Supervision, Validation, Writing—Original Draft, Writing—Review and Editing. J.Y.: Investigation, Formal Analysis, Methodology, Resources. Y.C.: Investigation, Formal Analysis, Methodology, Resources. Z.X.: Funding Acquisition, Resources. J.X.: Investigation, Formal Analysis. Y.M.: Investigation, Formal Analysis. H.S.: Funding Acquisition,

Resources, Writing—Review and Editing. All authors have read and agreed to the published version of the manuscript.

Funding: This study was supported by Xinjiang Uygur Autonomous Region Major Science and Technology Special Project (2024A02004). The Tianshan Talent Training Program high level talents for the advancement of modern agriculture and animal husbandry (No. 2024SNGGGCC06).

Institutional Review Board Statement: The experiment has passed the ethical review of the Science and Technology Ethics Committee of the Xinjiang Uygur Autonomous Region Academy of Animal Husbandry Sciences, with the ethical review number 202408012, approved on 12 August 2024.

Informed Consent Statement: Not applicable.

Data Availability Statement: The original contributions presented in this study are included in the article. Further inquiries can be directed to the corresponding author.

Acknowledgments: First of all, I would like to give my heartfelt thanks to all the people who have ever helped me with this paper. My sincere and hearty thanks and appreciations go firstly to my supervisor, Shi Hongcai, whose suggestions and encouragement have given me much insight into these translation studies. It has been a great privilege and joy to study under his guidance and supervision. Furthermore, it is my honor to benefit from his personality and diligence, which will treasure my whole life. My gratitude to him knows no bounds. I am also extremely grateful to all my friends and classmates who have kindly provided me with assistance and companionship in the course of preparing this paper. In addition, many thanks go to my family for their unfailing love and unwavering support. Finally, I am really grateful to all those who devote much time to reading this thesis and give me much advice, which will benefit me in my later study.

Conflicts of Interest: The authors declare no conflict of interest.

Abbreviations

AD	Alzheimer's Disease
APP	Amyloid Precursor Protein
BP	Biological Process
CC	Cellular Component
COG	Cluster of Orthologous Groups
DIA	Data-Independent Acquisition
DEPs	Differentially Expressed Proteins
DTT	Dithiothreitol
FC	Fold Change
FSH	Follicle-Stimulating Hormone
GO	Gene Ontology
GnRH	Gonadotropin-Releasing Hormone
GSEA	Gene Set Enrichment Analysis
HPG	Hypothalamic-Pituitary-Gonadal
IGF	Insulin-like Growth Factor
IAM	Iodoacetamide
IPR	Inter Pro Scan
KEGG	Kyoto Encyclopedia of Genes and Genomes
LH	Luteinizing Hormone
MF	Molecular Function
NES	Normalized Enrichment Score
NAFLD	Nonalcoholic Fatty Liver Disease
PCA	Principal Component Analysis
PD	Parkinson's Disease
PS1	Presenilin 1
PS2	Presenilin 2

ROS	Reactive Oxygen Species
SDS	Sodium Dodecyl Sulfate
SNAP-25	Synaptosomal Associated Protein 25
TEAB	Triethylammonium Bicarbonate
TFA	Trifluoroacetic Acid
UHPLC	Ultra-High-Performance Liquid Chromatography

References

1. Guo, X.F.; Zhong, R.Z.; Zhang, X.S.; Zhang, J.L.; Wang, X.Y.; He, X.Y.; Chu, M.X.; Fang, Y. Effect of FecB gene mutation on superovulation in sheep. *J. Domest. Anim. Ecol.* **2024**, *45*, 43–47.
2. Guo, H.C.; Wang, X.S. Research progress of FecB gene in Small Tail Han sheep. *Livest. Poult. Ind.* **2024**, *35*, 26–28.
3. Pan, Z.Y.; Di, R.; Liu, Q.Y.; Chu, M.X. Research advances on BMPR1B as a major gene for prolificacy in sheep. *J. Domest. Anim. Ecol.* **2015**, *36*, 1–6.
4. Samanta, L.; Parida, R.; Dias, T.R.; Agarwal, A. The enigmatic seminal plasma: A proteomics insight from ejaculation to fertilization. *Reprod. Biol. Endocrinol.* **2018**, *16*, 41–68. [CrossRef]
5. Demichev, V.; Messner, C.B.; Vernardis, S.I.; Lilley, K.S.; Ralser, M. DIA-NN: Neural networks and interference correction enable deep proteome coverage in high throughput. *Nat. Methods* **2020**, *17*, 41–44. [CrossRef]
6. Iwasaki, A.; Gagnon, C. Gagnon Formation of reactive oxygen species in spermatozoa of infertile patients. *Fertil. Steril.* **1992**, *57*, 409–416. [CrossRef]
7. Yang, L. Related Studies on Proteomics of Buffalo Semen. Master's Thesis, Guangxi University, Nanning, China, 2012.
8. Ochsendorf, F.R. Infections in the male genital tract and reactive oxygen species. *Hum. Reprod. Update* **1999**, *5*, 399–420. [CrossRef]
9. Agarwal, A.; Said, T.M. Said Oxidative stress, DNA damage and apoptosis in male infertility: A clinical approach. *BJU Int.* **2005**, *95*, 503–507. [CrossRef] [PubMed]
10. Mukai, C.; Travis, A. Travis What sperm can teach us about energy production. *Reprod. Domest. Anim. Zuchthyg.* **2012**, *47* (Suppl. S4), 164–169. [CrossRef] [PubMed]
11. Li, P.F. Proteomics Comparative Analysis of Semen Quality Characteristics in Winter and Summer. Master's Thesis, Shanghai Jiao Tong University, Shanghai, China, 2019.
12. Halangk, W.; Bohnensack, R.; Kunz, W. Interdependence of mitochondrial ATP production and extramitochondrial ATP utilization in intact spermatozoa. *Biochim. Biophys. Acta* **1985**, *808*, 316–322. [CrossRef] [PubMed]
13. Agarwal, A.; Sharma, R.; Durairajanayagam, D.; Cui, Z.; Ayaz, A.; Gupta, S.; Willard, B.; Gopalan, B.; Sabanegh, E. Differential proteomic profiling of spermatozoal proteins of infertile men with unilateral or bilateral varicocele. *Urology* **2015**, *85*, 580–588. [CrossRef]
14. Wykes, S.M.; Visscher, D.W.; A Krawetz, S. Krawetz Haploid transcripts persist in mature human spermatozoa. *Mol. Hum. Reprod.* **1997**, *3*, 15–19. [CrossRef] [PubMed]
15. Baker, M.A. Proteomics of post-translational modifications of mammalian spermatozoa. *Cell Tissue Res.* **2016**, *363*, 279–287. [CrossRef]
16. Luño, V.; López-Úbeda, R.; García-Vázquez, F.A.; Gil, L.; Matás, C. Boar sperm tyrosine phosphorylation patterns in the presence of oviductal epithelial cells: In vitro, ex vivo, and in vivo models. *Reproduction* **2013**, *146*, 315–324. [CrossRef]
17. Cornwall, G.A. Role of posttranslational protein modifications in epididymal sperm maturation and extracellular quality control. *Adv. Exp. Med. Biol.* **2014**, *759*, 159–180. [PubMed]
18. Li, X.; Wang, L.; Li, Y.; Zhao, N.; Zhen, L.; Fu, J.; Yang, Q. Calcium regulates motility and protein phosphorylation by changing cAMP and ATP concentrations in boar sperm in vitro. *Anim. Reprod. Sci.* **2016**, *172*, 39–51. [CrossRef]
19. Yu, H.; Diao, H.; Wang, C.; Lin, Y.; Yu, F.; Lu, H.; Xu, W.; Li, Z.; Shi, H.; Zhao, S.; et al. Acetylproteomic analysis reveals functional implications of lysine acetylation in human spermatozoa (sperm). *Mol. Cell. Proteom. MCP* **2015**, *14*, 1009–1023. [CrossRef] [PubMed]
20. Brohi, R.D.; Huo, L.-J. Huo Posttranslational Modifications in Spermatozoa and Effects on Male Fertility and Sperm Viability. *Omics A J. Integr. Biol.* **2017**, *21*, 245–256. [CrossRef]
21. Sutovsky, P.; McCauley, T.C.; Sutovsky, M.; Day, B.N. Early degradation of paternal mitochondria in domestic pig (*Sus scrofa*) is prevented by selective proteasomal inhibitors lactacystin and MG132. *Biol. Reprod.* **2003**, *68*, 1793–1800. [CrossRef]
22. Aalberts, M.; Sostaric, E.; Wubbolts, R.; Wauben, M.W.; Nolte, E.N.; Gadella, B.M.; Stout, T.A.; Stoorvogel, W. Spermatozoa recruit prostasomes in response to capacitation induction. *Biochim. Biophys. Acta (BBA) Proteins Proteom.* **2013**, *1834*, 2326–2335. [CrossRef]
23. Sánchez, R.; Deppe, M.; Schulz, M.; Bravo, P.; Villegas, J.; Morales, P.; Risopatrón, J. Participation of the sperm proteasome during in vitro fertilisation and the acrosome reaction in cattle. *Andrologia* **2015**, *43*, 114–120. [CrossRef] [PubMed]

4. Mao, F.; Yang, Y.Y.; Su, J.; Zhao, G.P.; Li, X.N.; Wang, D.Q.; Bayinjrigala; Cao, G.F.; Wang, C.Y. Proteomic analysis of differential proteins in sheep sperm before and after capacitation. *J. Anim. Husb. Vet. Med.* **2023**, *55*, 10–16.
5. Cheng, B.; Zhang, S.S. Research progress on neuropsychiatric symptoms of Parkinson's disease. *Chin. J. Clin. Neurosci.* **2020**, *28*, 712–716.
6. Xu, G.L.; Zhou, L.Q. Research advances on male (male) reproductive damage induced by phthalates. *Chin. J. Androl.* **2024**, *30*, 66–71.
7. Mao, Z.F.; Zhan, B.Y. Relationship between hypothalamic-pituitary-gonadal axis and male aging. *Foreign Med. Sci. (Sect. Urol.)* **1993**, *5*, 196–199.
8. Wang, D.C.; Yu, S.S.; Liu, W.F.; Chen, G.J. Aggregation state of α-Synuclein protein and its role in Parkinson's disease formation. *Chin. J. Biochem. Mol. Biol.* **2006**, *8*, 615–620.
9. Tang, Z.X.; Jiang, Y.H. Research progress of antinuclear antibodies in assisted reproductive technology. *Electron. J. Pract. Gynecol. Endocrinol.* **2022**, *9*, 47–51.
0. Wang, L.F. Analysis of fatty liver prevalence and related factors in farmers of Gaobu Town, Shaoxing City. *Mod. Pract. Med.* **2011**, *23*, 1117–1119.
1. Zhang, L.; Li, G.F.; Su, L.L.; Du, L.; Zhou, D.; Cheng, X.; Lin, Z.; Qu, S. Correlation between total testosterone levels and insulin resistance in patients with acanthosis nigricans and non-acanthosis nigricans. *J. South. Med. Univ.* **2021**, *41*, 1780–1786.
2. Ji, Y.X. PINK/Parkin-Mediated Selective Autophagy Mechanism in Mitochondria of NASH Hepatocytes and Intervention Effect of Zhigan Formula. Master's Thesis, Guangxi University of Chinese Medicine, Nanning, China, 28 March 2019.
3. Huang, Q.; Liu, H.J.; Xu, P. Research progress on neurotransmitters and OSAS-related cognitive impairment. *J. Zunyi Med. Univ.* **2020**, *43*, 254–258.
4. Qi, Q.Q.; Zhang, Y. Changes in hypothalamic-pituitary-thyroid axis and hypothalamic-pituitary-gonadal axis hormones in patients with depression comorbid with obstructive sleep apnea. *China J. Pharm. Econ.* **2025**, *20*, 33–38.
5. Cheng, C.F.; Wang, B.; Zhao, T. Relationship between sex hormone levels and semen quality in patients with varicocele. *J. Henan Med. Coll.* **2023**, *35*, 152–155.
6. Zhou, Y. Clinical analysis of 68 cases of Alzheimer's disease. *Biotechnol. World* **2014**, *2*, 68.
7. Fan, T.Y. Relationship between Alzheimer's disease and epigenetics. *China New Telecommun.* **2019**, *21*, 241–243.
8. Cao, M.Y.; Zhao, Z.M.; Yan, X.; He, G.M.; Cao, J.F.; Cui, N. Retrospective analysis of factors related to fertilization failure in conventional in vitro fertilization. *J. Hebei Med. Univ.* **2019**, *40*, 309–314.
9. Ma, Y.D.; Wang, Y.C.; Yue, L.M. Research progress on endometrial glycogen metabolism in embryo implantation. *Acta Physiol. Sin.* **2022**, *74*, 255–264.

animals

MDPI

Article

Study on the Development and Formation Specifics of Longissimus Dorsi Muscles in Ziwuling Black Goats

Hailong Guo [1,†], Fuyue Shi [1,†], Lingrong Gu [1], Yanyan Wang [1], Yangyang Yue [1], Wei Huang [2], Yongqiang Yang [3], Panlong Sun [4], Wenyong Xue [5], Xiaoqiang Zhang [6], Xiaomei Zhu [7], Pengyang Shao [2], Yapeng He [2], Jianfeng Xu [1,*] and Xiu Liu [2,*]

1 Animal Husbandry and Veterinary Research Institute of Gansu Province, Pingliang 744000, China
2 Gansu Key Laboratory of Herbivorous Animal Science, College of Animal Science and Technology, Gansu Agricultural University, Lanzhou 730070, China
3 Jingning County Animal Husbandry and Veterinary Center, Pingliang 743400, China
4 Zhuanglang County Animal Husbandry and Veterinary Center, Pingliang 744600, China
5 Huating City Animal Husbandry Science and Technology Development Service Center, Huating 744100, China
6 Jingchuan County Animal Husbandry and Veterinary Center, Pingliang 744300, China
7 Chongxin County Animal Husbandry and Veterinary Center, Pingliang 744200, China
* Correspondence: xjf_fx@126.com (J.X.); liuxiu@gsau.edu.cn (X.L.)
† These authors contributed equally to this work.

Simple Summary

This study aimed to clarify the relationship between muscle development and meat quality in Ziwuling black goats. We used the longissimus dorsi muscle of 6-month-old and 12-month-old goats as samples, and analyzed muscle structure, myofiber type transformation, and molecular regulation via HE staining, fast–slow myofiber immunofluorescence double staining, and transcriptome sequencing. Results showed that compared with 12-month-old goats, 6-month-old goats had higher myofiber density, smaller myofiber diameter, and more fast-twitch myofibers (Type II); in contrast, 12-month-old goats exhibited myofiber hypertrophy (larger diameter) and an increase in slow-twitch myofibers (Type I). Transcriptome sequencing identified 387 differentially expressed genes (DEGs: 156 upregulated, 231 downregulated). GO analysis indicated DEGs were involved in processes such as skeletal muscle growth and cAMP biosynthesis, while KEGG analysis showed DEGs were enriched in arginine–proline metabolism (related to muscle maturation) and AMPK/MAPK signaling pathways (AMPK regulates fatty acid metabolism genes like *ACACB/CPT1A*). Additionally, WGCNA clustered genes into nine modules, with key modules (e.g., MEblue, MEgreen) correlating with myofiber density, diameter, and MAPK/AMPK pathways. In conclusion, we recommend focusing on protein nutrition at 6 months (to promote myofiber proliferation) and regulating energy intake at 12 months (to improve meat quality); 12 months of age is determined as the optimal slaughter age for Ziwuling black goats.

Abstract

To clarify the relationship between muscle development and meat quality in Ziwuling black goats, this study used the longissimus dorsi muscle of 6-month-old and 12-month-old goats as samples. With HE staining, fast–slow myofiber immunofluorescence double staining, and transcriptome sequencing, this study analyzed muscle structure, myofiber type transformation, and molecular regulation. Results showed that 6-month-olds had higher myofiber density and smaller diameter; 12-month-olds showed myofiber hypertrophy (larger diameter); immunofluorescence revealed more fast-twitch myofibers (Type II) at 6 months and increased slow-twitch ones (Type I) at 12 months. Transcriptome sequencing identified 387 differentially expressed genes (DEGs: 156 upregulated, 231 downregulated). GO analysis indicated that DEGs are involved in skeletal muscle growth, cAMP biosynthesis, etc.; KEGG analysis showed enrichment in arginine–proline metabolism and AMPK/MAPK signaling pathways (AMPK regulates fatty acid metabolism genes like *ACACB*/*CPT1A*; arginine–proline metabolism relates to muscle maturation). WGCNA clustered genes into nine modules (MEblue correlated with myofiber density/MAPK; MEgreen correlated negatively with diameter but positively with density, involving *PPARGC1A*/AMPK). In conclusion, protein nutrition at 6 months (promote myofiber proliferation) and regulating energy intake at 12 months (improve meat quality) are recommended, and 12 months is the optimal slaughter age.

Keywords: different ages; Ziwuling black goats; longissimus dorsi muscle; morphology; transcriptomics

1. Introduction

In the livestock and poultry breeding industry, meat quality and muscle development level are core factors that determine the industry's economic benefits and ability to meet consumer demand, and they are also key indicators for measuring the superiority of livestock and poultry germplasm resources [1]. High-quality livestock and poultry meat not only require a tender texture and rich nutritional components but also rely on the reasonable development of muscle tissue. Specifically, the morphological structure of muscle fibers (such as diameter, density, and cross-sectional area), muscle fiber types, and their dynamic transformation process directly affect meat tenderness, the accumulation of flavor substances, and nutritional value. Moreover, differences in muscle development at different growth stages are an important basis for determining the appropriate slaughter period and optimizing breeding plans [2]. For local livestock and poultry breeds, in-depth analysis of their muscle development laws and molecular regulatory mechanisms can not only provide support for the scientific protection and efficient utilization of the breeds but also fill the gaps in the research field of muscle development for specific germplasm resources, thereby promoting the sustainable development of the local characteristic breeding industry.

The Ziwuling black goat is a traditional livestock breed in Western China, mainly distributed in the producing areas around the Ziwuling Mountains. It serves as a crucial germplasm resource and an industrial pillar for the local mutton sheep industry [3]. This breed is prominently characterized by its black coat. It not only has adaptive advantages such as good roughage tolerance and strong stress resistance but also stands out for its excellent trait of tender and delicious meat. Its meat quality is significantly superior to that of other common goat breeds, making it a crucial germplasm resource and industrial pillar for the development of the mutton sheep industry in the producing areas [4]. The longissimus dorsi muscle, as the core muscle group of its body, has a developmental status that serves as a key indicator for evaluating the meat production performance and meat quality of meat-producing animals. It is directly related to the muscle fiber structure, the accumulation of nutritional components, and, ultimately, the economic value of the meat product [5]. Muscle development is a dynamic process that involves cell proliferation and tissue differentiation, with significant differences in its morphology and function across different monthly age stages [6,7]. At 6 months of age, a critical transition stage from infancy to adolescence, muscle tissue proliferates rapidly, accompanied by the initial differentiation of fiber types. At 12 months of age, the animals are close to sexual maturity, and muscle development tends to stabilize; this is an important period for the formation of meat quality traits. Indicators such as muscle fiber diameter, density, and cross-sectional area are important bases for evaluating muscle quality [8], Hematoxylin and Eosin (HE) staining technique can clearly display the morphological structure and arrangement characteristics of muscle fibers [9]. In addition, the composition and transformation of muscle fiber types are also key factors affecting meat quality. Skeletal muscle is mainly composed of fast-twitch muscle fibers (Type II) and slow-twitch muscle fibers (Type I), and the double immunofluorescence technique enables accurate identification and analysis of these fibers [10,11]. Studies have shown that muscle fiber characteristics are regulated by multiple factors such as age [12,13], diet composition [14], and rearing environment [15]. At the molecular mechanism level, transcriptomics technology has been widely applied in studies related to muscle development. For example, Pan et al. [16] identified 231 differentially expressed genes associated with meat quality and muscle development by comparing two goat breeds. Cao et al. [17] analyzed the dynamic changes in skeletal muscle development of Hu sheep from 3 days after birth to 12 months of age, identified 6865 differentially expressed genes, and suggested that genes such as ARID5B, MYOG, and ENO1 are related to muscle hypertrophy, while genes like NR1D1 and FADS1 are associated with muscle fiber type transformation. Other studies have also been conducted focusing on different breeds [18], ages [19], and developmental stages [20], gradually uncovering the molecular regulatory network of muscle development.

At present, in-depth research on the systematic developmental characteristics of the longissimus dorsi muscle during key growth and development stages is still lacking for the Ziwuling black goat, a specific local breed. Therefore, this study took the longissimus dorsi muscle of 6-month-old and 12-month-old Ziwuling black goats as the research subjects. It comprehensively used HE staining to observe tissue morphology, double immunofluorescence staining to analyze the transformation law of fast-twitch and slow-twitch muscle fiber types, and transcriptome sequencing to screen key differentially expressed genes, so as to systematically explain the developmental characteristics and molecular regulatory mechanisms of the longissimus dorsi muscle at these two important monthly age stages. The research results aim to provide a theoretical basis for the precise nutritional regulation and determination of the appropriate slaughter period of Ziwuling black goats. At the same time, the research results further enrich the basic data on muscle development of local goat germplasm resources and provide support for their scientific protection and efficient utilization.

2. Materials and Methods

2.1. Experimental Location and Time

Healthy Ziwuling black goats at 6 and 12 months of age were selected as experimental subjects from the same herdsman's flock in Huan County, Qingyang City, Gansu Province. Each age group consisted of 6 goats: the 6-month-old group (designated as B6) and the 12-month-old group (designated as B12). All goats were slaughtered to collect longissimus dorsi muscle samples, with ruminant sample collection conducted concurrently in strict accordance with the experimental protocol. Slaughter was performed via jugular vein exsanguination on the early morning of the day following sample preparation. Prior to slaughter, the experimental goats were subjected to a 12 h fasting period and a 2 h water deprivation period to standardize physiological conditions. Longissimus dorsi muscle samples were collected within 30 min post-slaughter to ensure tissue freshness. Notably, this slaughter experiment was reviewed and approved by the Animal Ethics Committee of Gansu Agricultural University, with the official approval number GAU-LC-2020-27.

2.2. Sample Collection

After slaughter, the longissimus dorsi muscle was taken. One part was fixed with 4% paraformaldehyde, which was used for histological and morphological analysis and immunofluorescence double staining; the other part was placed in 1.5 mL cryopreservation tubes (three tubes per goat) and immediately stored in a pre-prepared liquid nitrogen tank for freezing. After being transported back to the laboratory, the samples were stored at −80 °C for subsequent RNA-Seq analysis.

2.3. Histo- and Morphlolgical Analysis

A 1 cm^2 sample of longissimus dorsi muscle tissue was fixed in 4% paraformaldehyde solution. After 24 h of fixation, the tissue samples underwent a standard series of histological procedures, including dehydration, clearing, paraffin infiltration, embedding, sectioning, and staining, to prepare paraffin-embedded tissue sections. Hematoxylin–Eosin (HE) staining was performed for morphological observation: cell nuclei were stained blue–purple, while cytoplasm was stained pink. For each longissimus dorsi muscle sample, three non-adjacent HE-stained sections were selected, and digital images were captured using a microscope equipped with an imaging system.

2.4. Immunofluorescence Staining

2.4.1. Deparaffinization of Paraffin Sections to Water

Sections were sequentially immersed in environment-friendly dewaxing solution I, II, and II for 10 min each, followed by anhydrous ethanol I, II, and III for 5 min each. Finally, sections were rinsed with distilled water.

2.4.2. Antigen Retrieval

Sections were placed in EDTA (pH 8.0) retrieval buffer and treated with a microwave oven under medium power for 10 min→power off for 5 min→medium-low power for 5 min→power off for 2 min→medium-low power for 5 min (to prevent buffer evaporation and section drying). After natural cooling, sections were transferred to PBS (pH 7.4) and washed 3 times on a decolorizing shaker, 5 min each time.

2.4.3. Circle Drawing and Serum Blocking

Sections were spun dry, and a histochemical pen was used to draw a circle around the tissue. Blocking solution was added (10% donkey serum for primary antibodies of goat origin, 3% BSA for primary antibodies of other origins), and blocking was performed for 30 min.

2.4.4. Primary and Secondary Antibody Incubation

Primary antibody incubation: Two primary antibodies were mixed and prepared (rabbit-derived fast antibody, 1:300; mouse-derived slow antibody, 1:300), added dropwise to the sections, then placed flat in a humid chamber and incubated overnight at 4 °C. Secondary antibody incubation: Sections were transferred to PBS (pH 7.4) and washed 3 times on a decolorizing shaker (5 min each time). Corresponding secondary antibodies were added dropwise (CY3-labeled goat anti-rabbit IgG, 1:300; Alexa Fluor 488-labeled goat anti-mouse IgG, 1:400) and incubated at room temperature for 50 min in the dark.

2.4.5. DAPI Counterstaining and Mounting

After sections were washed 3 times with PBS (pH 7.4) on a decolorizing shaker (5 min each time), DAPI staining solution was added dropwise, and incubation was carried out at room temperature for 10 min in the dark. Sections were then washed 3 times with PBS (5 min each time), treated with auto-fluorescence quencher Solution B for 5 min, rinsed with running water for 10 min, and finally mounted with anti-fluorescence quenching mounting medium.

2.4.6. Image Acquisition

Tissue sections were loaded onto a PANNORAMIC whole-slide scanner (3DHISTECH, Budapest, Hungary). The sections moved gradually under the scanner's lens, imaging while moving to scan all tissue information on the sections into a folder containing all tissue information from the sections. The folder could be opened with CaseViewer 2.4 software for observation at any magnification from 1× to 400×. Muscle tissue was selected for imaging at 400× magnification, with the tissue filling the entire field of view as much as possible and with consistent background light for each photo. After imaging, Image-Pro Plus 6.0 (Table 1) analysis software was used to measure the diameter of muscle fibers at 5 locations under 400× magnification and calculate the average value; measure the total area of muscle fibers in 3 fields of view per section and count the number of muscle fibers in the fields of view; count the number of slow-twitch muscles, fast-twitch muscles, and total muscle fibers in 3 fields of view using immunofluorescence, and calculate the slow-twitch muscle ratio (%) = (number of slow-twitch muscles/total number of muscle fibers) × 100, and fast-twitch muscle ratio (%) = (number of fast-twitch muscles/total number of muscle fibers) × 100. (Details of the main instruments, reagents, primary antibodies, secondary antibodies, and corresponding antigen retrieval conditions used in this experiment are shown in Supplementary Tables S1–S3.)

Table 1. Summary of programs and software tools for data analysis and processing.

Software Name	Version	Purpose in This Study
Image-Pro Plus	6.0	Mainly used for image analysis, including image acquisition and morphological processing
Excel	2016	Preliminary data organization
SPSS	24.0	Used for professional statistical analysis
GraphPad Prism	9	Used for scientific data statistics and high-quality charting

2.5. Transcriptome Sequencing Analysis of Goat Longissimus Dorsi Muscle

2.5.1. RNA Extraction, Library Construction, and Sequencing

Total RNA was extracted from longissimus dorsi muscle tissue using the Trizol reagent kit (Invitrogen, Carlsbad, CA, USA). All extraction procedures were performed strictly according to the kit instructions and conducted on an ultra-clean bench. A micro-spectrophotometer (Therm Nano Drop-2000, Waltham, MA, USA) was used to detect the concentration and purity of the extracted RNA, recording the RNA concentration (ng/μL) and purity (260 nm/280 nm ratio of 1.8–2.1). The VAHTS Universal V6 RNA-seq Library Prep Kit for Illumina® (NR604-02, Vazyme, Nanjing, China) was used to construct cDNA libraries of the samples, and the VAHTSTM DNA Clean Beads kit (N411-03, Vazyme, Nanjing, China) was used to purify the products. The constructed cDNA libraries were sequenced on the Illumina NovaSeq 6000 system (San Diego, CA, USA).

2.5.2. Alignment with Reference Genome

The raw image data obtained from Illumina platform sequencing were converted into raw sequencing reads (Raw Reads) through base calling. Quality control was performed on Raw Reads to obtain high-quality sequences (Clean Reads). The Clean Data were aligned with the reference genome (Oar_rambouillet_v1.0.Ovis_aries) using HISAT to obtain Mapped Data. StringTie (v3.0.1) was used to assemble the aligned reads, and FPKM was adopted as the indicator to measure the expression levels of transcripts or genes.

2.5.3. Functional Annotation Analysis of Differentially Expressed Genes

DESeq2 (v1.49.4) was used for differential gene expression analysis, with the screening criteria of FC $\geq$ 2 and FDR < 0.01 for identifying differentially expressed genes. GO and KEGG functional enrichment analyses of the differential genes were performed using GOseq (v1.61.1).

2.5.4. Validation by Real-Time Quantitative PCR (RT-qPCR)

The reliability of the transcriptome sequencing data was verified by RT-qPCR. Eight genes were randomly selected from the screened differentially expressed genes, and their relative expression levels were determined. Information on the gene primers is shown in Table 2.

Table 2. Information on gene primers.

Gene	Primer Sequence	Primer Length	Annealing Temperature	Sequence Number	Primer Specificity
HSPB1	F: CAAGTCAGCTACCCAGTCGG R: TGTTCGGACTTTCCGGCTTC	93	60 °C	XM_018040903.1	Fairly good
LPL	F: GAGGCCTTGGAGATGTGGAC R: AATTGCACCGGTACGCCTTA	114	60 °C	NM_001285607.2	
COL1A1	F: AAATGGAGCTCCTGGTCAGATG R: AGCACCATCATTTCCTCTAGCAC	100	60 °C	XM_018064895.1	
TET2	F: GCCTAACCCACCGACTCTTC R: CTTGCTGTTTGTGCCCCATC	77	60 °C	XM_013964483.2	
SCD	F: GTGCCGTGGTATCTATGGGG R: ACAACAGCGTACCGGAGAAG	74	60 °C	NM_001285619.1	
SLC25A4	F: AGTTCACTGGTCTGGGCAAC R: TGGACCGAGACGTTGAAACC	88	60 °C	XM_018042040.1	
PPARA	F: TTCCCTCTTTGTGGCTGCTA R: GCGTCGTCAGGATGGTTGTT	135	60 °C	XM_018048905.1	
CAPN2	F: CATCCGGGTCTTTTCCGAGA R: GATGTCGTCCTCGCTGATGT	97	60 °C	XM_018060202.1	
GAPDH	F: AAGGTCGGAGTGAACGGATT R: ACGATGTCCACTTTGCCAGTA	80	60 °C	XM_005680968.3	

2.6. Data Analysis

The experimental data were initially organized using Excel 2016, then statistically analyzed via independent samples *t*-test in SPSS 24.0 software. The relative expression levels of genes were analyzed using the $2^{-\Delta\Delta CT}$ [21] method. For measurements of HE-stained sections and immunofluorescence double-stained sections, Image-Pro Plus 6.0 was used with "millimeter (mm)" as the standard unit uniformly. The results were visualized using Prism software and expressed as "mean $\pm$ standard deviation". The value of $p < 0.05$ was considered statistically significant (Table 2).

3. Results

3.1. Analysis of Morphological Characteristics of Longissimus Dorsi Muscle

By observing and analyzing HE-stained sections of the longissimus dorsi muscle from 6-month-old and 12-month-old Ziwuling black goats (Figures 1 and 2A,B), it was found that the 6-month-old group had relatively higher muscle fiber density and smaller fiber diameter, while the 12-month-old group exhibited increased muscle fiber diameter and decreased fiber density.

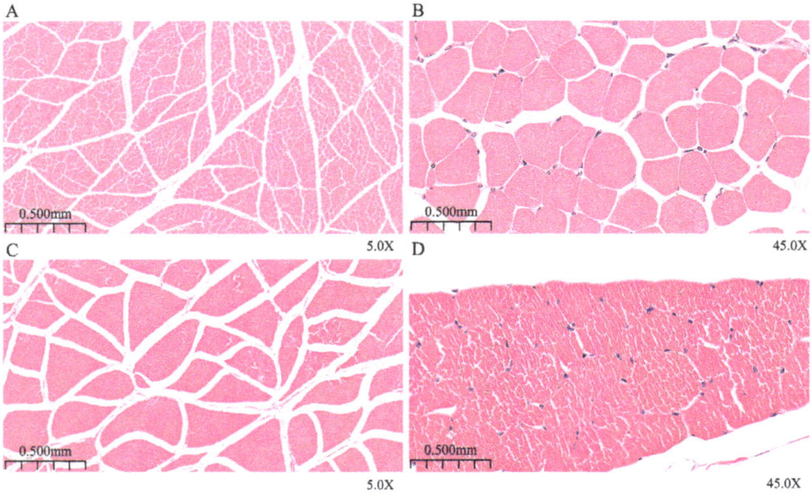

Figure 1. HE-stained histomorphological sections of the longissimus dorsi muscle. (**A**,**B**): HE-stained sections of the longissimus dorsi muscle from 6-month-old specimens; (**C**,**D**): HE-stained sections of the longissimus dorsi muscle from 12-month-old specimens.

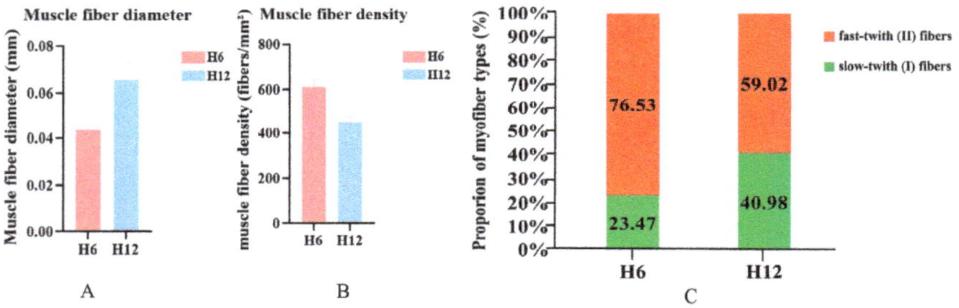

Figure 2. Measurement results of HE sections and immunofluorescence double staining. (**A**): Measurement results of muscle fiber diameter in longissimus dorsi muscle at 6 months and 12 months of age; (**B**): measurement results of muscle fiber density in longissimus dorsi muscle at 6 months and 12 months of age; (**C**): measurement results of fast-twitch muscle fibers and slow-twitch muscle fibers in longissimus dorsi muscle at 6 months and 12 months of age.

3.2. Analysis of Immunofluorescence Staining Characteristics of Longissimus Dorsi Muscle

By performing immunofluorescence staining and analysis on the longissimus dorsi muscle (Figures 2C and 3), it was found that the muscle fibers of the 6-month-old longissimus dorsi muscle were dominated by fast-twitch muscle fibers (Type II). In contrast, the proportion of slow-twitch muscle fibers (Type I) in the 12-month-old group gradually increased. Specifically, the proportion of Type I fibers was 23.47% in the 6-month-old group and increased to 40.98% in the 12-month-old group, while the proportion of Type II fibers was 76.53% in the 6-month-old group and decreased to 59.02% in the 12-month-old group.

3.3. Quality Assessment of Longissimus Dorsi Muscle RNA-Seq Data

By analyzing the base composition and quality distribution of the sequencing data (Supplementary Figure S1), it was found that the base distribution (A, T, C, G) was balanced across all 12 samples, and the Q30 values were all greater than 94.21%

(Supplementary Table S1). These results indicate that the sequencing data are of high quality and meet the requirements of subsequent experiments.

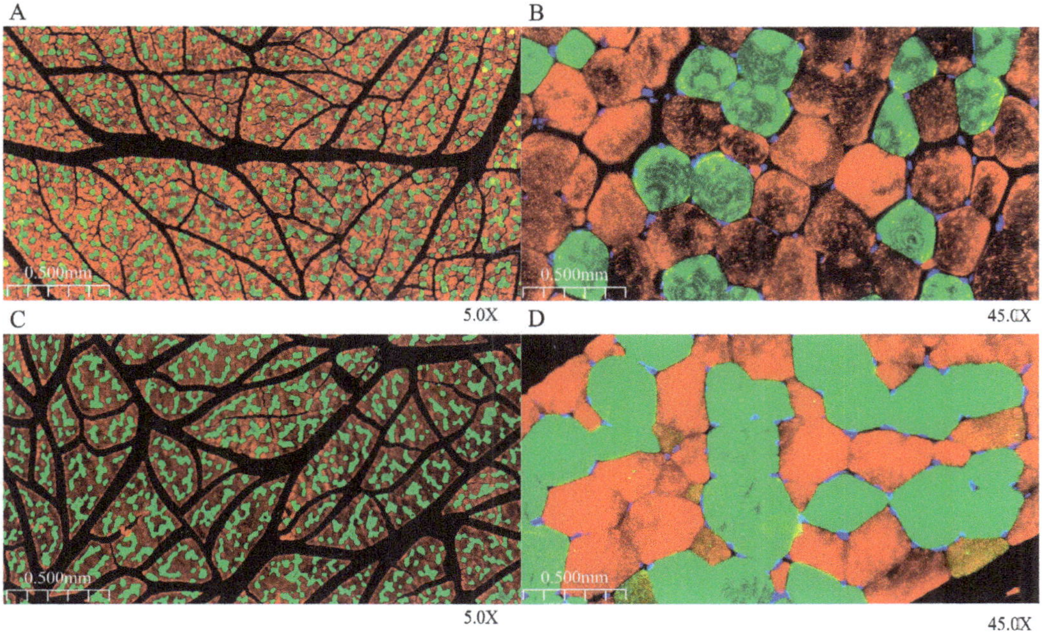

Figure 3. Immunofluorescence staining of the longissimus dorsi muscle. (**A**,**B**): immunofluorescence staining of the longissimus dorsi muscle from 6-month-old specimens; (**C**,**D**): immunofluorescence staining of the longissimus dorsi muscle from 12-month-old specimens. Note: immunofluorescence staining showing slow-twitch muscle fibers (Type I, green), fast-twitch muscle fibers (Type II, red), and myocyte nuclei (blue).

3.4. Alignment of Longissimus Dorsi Muscle RNA-Seq Data with Reference Genome

The HISAT2 software was used for rapid and accurate alignment of Clean Reads with the goat reference genome (Ovis_aries.Oar_rambouillet_v1.0) (Supplementary Table S2) to obtain the positioning information of reads on the reference genome. Subsequently, StringTie was used to assemble the aligned reads, and the transcriptome was reconstructed for subsequent analysis.

3.5. Identification of Differentially Expressed Genes

Using $|\text{Log FC}| \geq 2$ and FDR < 0.01 as the screening criteria, a total of 387 differentially expressed genes (DEGs) were identified in the B6_vs_B12 comparison group, including 156 upregulated genes and 231 downregulated genes. A heatmap was generated based on the expression levels of the screened differentially expressed genes. The results showed that the variation in the expression levels of differentially expressed genes was small within groups but large between groups (Figure 4).

Transcriptome analysis was performed on Ziwuling black goats of different ages, yielding a total of 70.17 Gb of Clean Data, with each sample generating at least 5.73 Gb of Clean Data. Principal Component Analysis (PCA) showed distinct intra-group clustering and significant inter-group differences (Figure 5 and Supplementary Table S6).

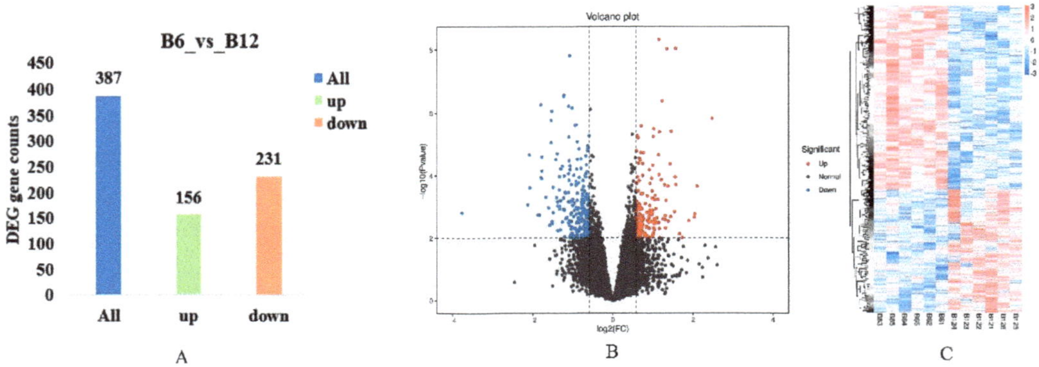

Figure 4. Analysis of differentially expressed genes in Longdong black goats. (**A**): Bar chart of differentially expressed genes; (**B**): volcano plot of differentially expressed genes; (**C**): heatmap of differentially expressed genes.

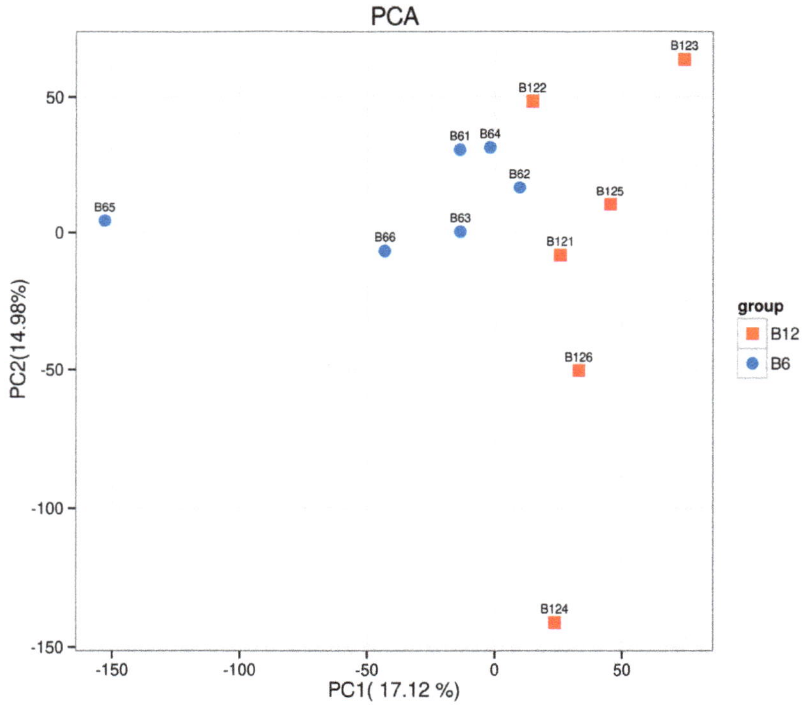

Figure 5. Transcriptome sequencing analysis of the longissimus dorsi muscle. Note: PCA showed that samples from the 6-month-old and 12-month-old groups were clearly clustered, and the differences between groups were extremely significant ($p = 0.004$) as verified by ANOVA (Supplementary Table S6).

3.6. Functional Annotation of Differentially Expressed Genes

3.6.1. GO Enrichment Analysis

To clarify the functions of DEGs in Ziwuling black goats during meat quality regulation, GO functional annotation was performed on the differentially expressed genes through biological process (GO-BP), cellular component (GO-CC), and molecular function (GO-MF). Through the analysis of the annotation results, it was found that in BP, the differentially expressed genes mainly played roles in the regulation of skeletal muscle tissue growth

(GO:0048630), cAMP biosynthetic process (GO:0006171), negative regulation of superoxide anion generation (GO:0032929), cellular amino acid metabolic process (GO:0006520), and the regulation of glucose metabolic process (GO:0010906). In CC, the differentially expressed genes were mainly involved in biological processes such as collagen-containing extracellular matrix (GO:0062023), actomyosin, actin portion (GO:0042643), and mitochondrion (GO:0005739). In MF, they mainly functioned in processes such as glutathione binding (GO:0043295), succinate dehydrogenase activity (GO:0000104), NAD+ binding (0070403), and electron transfer activity (GO:0009055) (Figure 6).

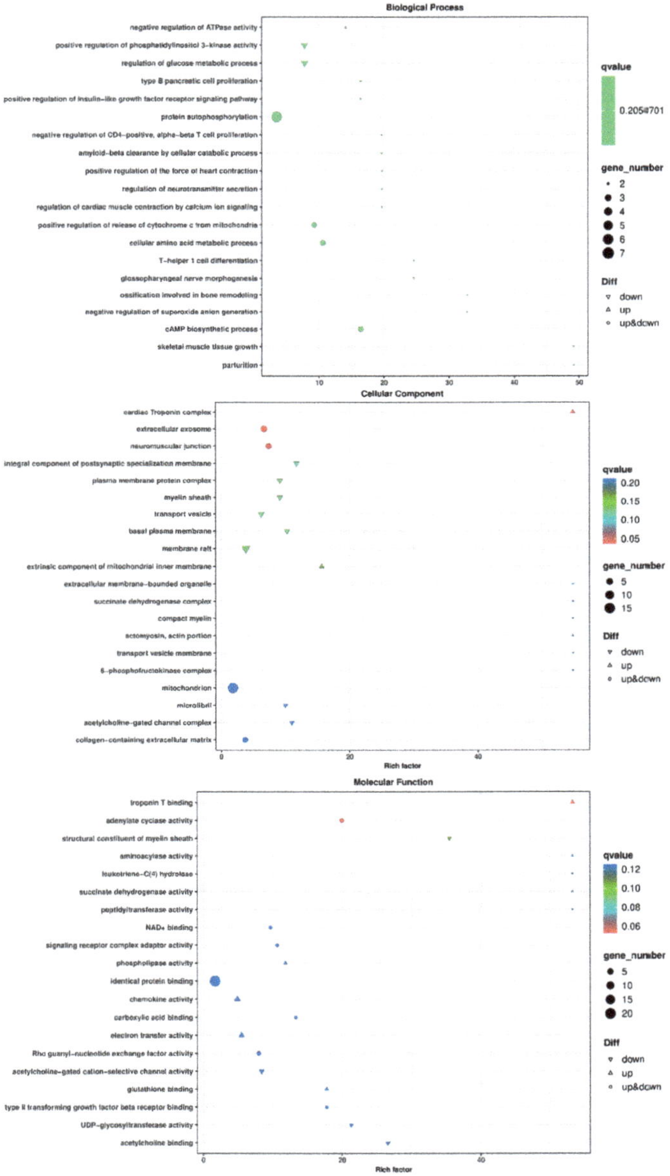

Figure 6. GO classification map of differentially expressed genes in Ziwuling black goats of different ages.

3.6.2. KEGG Signaling Pathway Enrichment Analysis

KEGG functional enrichment analysis was conducted on the DEGs of Ziwuling black goats at different ages. The results showed that these DEGs were mainly enriched in pathways including arginine and proline metabolism (ko00330), AMPK signaling pathway (ko04152), gap junction (ko04540), MAPK signaling pathway (ko04010), and Chemokine signaling pathway (ko04062). Among them, the downregulated genes were primarily enriched in pathways such as AMPK signaling pathway (ko04152) and insulin resistance (ko04911) (Figure 7 and Table 3).

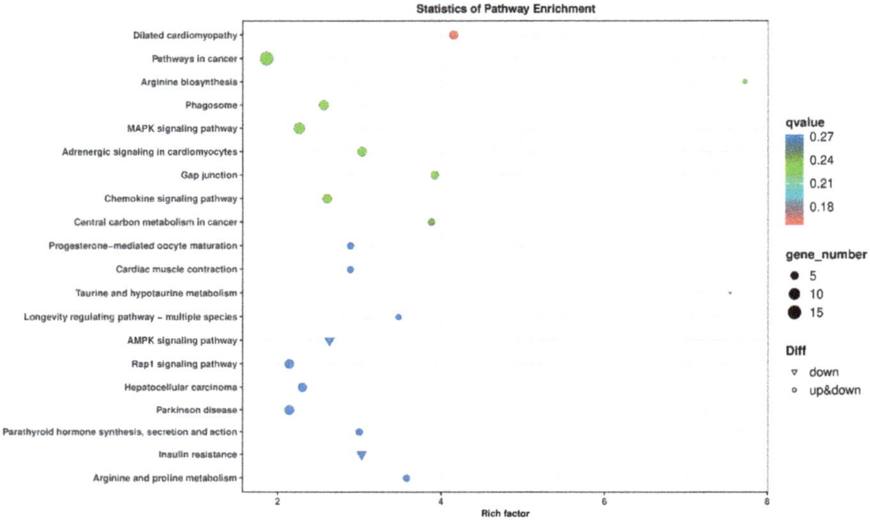

Figure 7. KEGG pathway annotation of differentially expressed genes in Ziwuling black goats of different ages.

Table 3. Partial results of KEGG pathways and differentially expressed genes enrichment.

Group	Genes	Signaling Pathways	Expression
B6_vs_B12	*ACACB* *CPT1A*	AMPK signaling pathway	Down
	ADCY6 *ADCY7*	Gap junction	Down
	NOS1 *SMOX*	Arginine and proline metabolism	Down

3.7. RT-qPCR Validation

Eight genes were randomly selected from the differentially expressed genes in the longissimus dorsi muscle of Ziwuling black goats at different monthly ages, and their expression levels were validated using the qPCR method. As shown in Figure 8, the expression trends detected by RT-qPCR were consistent with the results of RNA-Seq analysis, indicating that the RNA-Seq data had high reliability.

3.8. Weighted Gene Co-Expression Network Analysis (WGCNA)

Based on RNA transcriptome data, weighted gene co-expression network analysis (WGCNA) was performed, which revealed that the expressed genes in the rumen of

Ziwuling black goats at different months of age were clustered into nine gene modules. These gene modules exhibited varying degrees of correlation with rumen muscle fiber diameter and muscle fiber density, among which the modules with stronger correlations included MEbrown, MEblue, MEgreen, Megrey, and Mepink. Specifically, MEbrown was significantly positively correlated with muscle fiber diameter ($p < 0.05$); MEblue was significantly positively correlated with muscle fiber density ($p < 0.05$), and the co-expressed host genes (*GADD45A, FGFR3*) in these modules were associated with functions like the "MAPK signaling pathway"; MEgreen was significantly negatively correlated with muscle fiber diameter ($p < 0.05$) while being significantly positively correlated with muscle fiber density ($p < 0.05$), and the co-expressed host gene *PPARGC1A* in this module was associated with the function of the "AMPK signaling pathway" (Figure 9).

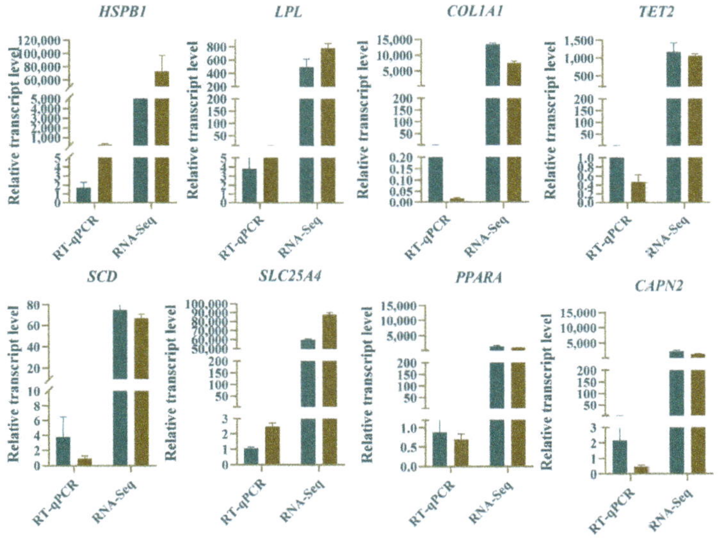

Figure 8. RT-qPCR and RNA-Seq validation. Note: For both RT-qPCR and RNA-Seq, the expression level on the left represents that of 6-month-old specimens, and the one on the right represents that of 12-month-old specimens.

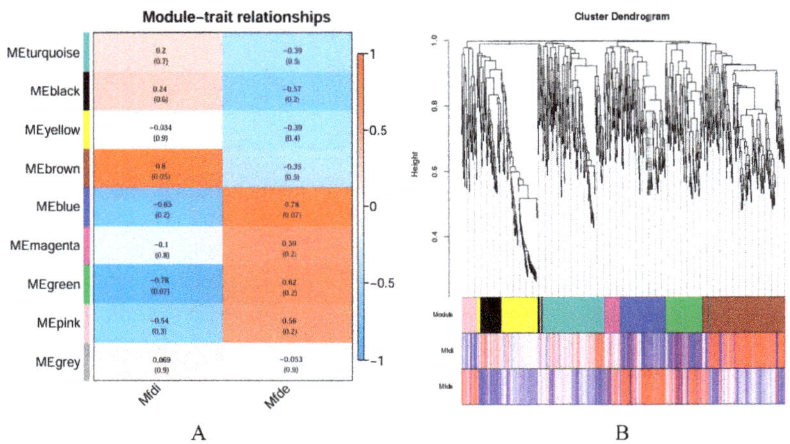

Figure 9. Weighted gene co-expression network analysis results. (**A**) Module-trait relationships; (**B**) Cluster dendrogram.

4. Discussion

Morphological parameters of muscle fibers are key indicators reflecting muscle growth status and meat quality characteristics [22]. In this study, Ziwuling black goats at 6 months of age are in the transition stage from the juvenile to the young adult period, with muscle tissue mainly undergoing rapid proliferation. This is manifested by a higher muscle fiber density and smaller diameter. This change pattern is basically consistent with the muscle development process of meat-producing animals such as pigs [23] and sheep [24], indicating that the transition of muscle fibers from "proliferation-dominated" to "hypertrophy-dominated" is a common feature of mammalian muscle growth. The composition of muscle fiber types is closely related to muscle metabolic characteristics, contractile function, and meat flavor formation [25]. In this study, the proportion of fast-twitch muscle fibers (Type II) in the muscles of 6-month-old goats is relatively high, which helps to meet the demand for rapid energy supply during the active movement of animals in the juvenile stage. In contrast, the proportion of slow-twitch muscle fibers (Type I) increases significantly at 12 months of age, reflecting a gradual shift in muscle metabolism from glycolysis-dominated to oxidative phosphorylation-dominated. Slow-twitch muscle fibers are rich in mitochondria and myoglobin, which contribute to improving muscle water-holding capacity and promoting the accumulation of flavor substances [26]. This may be the structural basis for the richer meat flavor of Ziwuling black goats at 12 months of age. Therefore, from the perspective of muscle fiber morphology and type transformation, the period of 6–12 months of age is a critical window for the formation of muscle texture characteristics in Ziwuling black goats.

Ziwuling black goat is an important local goat breed in China, and its muscle development process has breed-specific characteristics. The longissimus dorsi muscle, a typical site for muscle growth and meat production efficiency, is often used in studies on related physiological mechanisms [27]. In this study, transcriptome sequencing was performed on the longissimus dorsi muscle of Ziwuling black goats at 6 and 12 months of age, and a total of 387 DEGs were identified, including 156 upregulated genes and 231 downregulated genes. This indicates that there are significant differences in the molecular regulatory mechanisms of muscle development between the two age stages, and these genes may regulate the transformation of muscle fiber structure and metabolic function by participating in specific biological processes and signaling pathways.

GO functional enrichment analysis showed that in the biological process category, DEGs were significantly enriched in pathways such as skeletal muscle tissue growth, cAMP biosynthesis, and glucose metabolism. Among them, the expression dynamics of genes related to skeletal muscle tissue growth are directly associated with the morphological transition of muscle fiber proliferation and hypertrophy [28]. As a key second messenger, the biosynthesis process of cAMP participates in the regulation of myocyte proliferation and differentiation by regulating the PKA signaling pathway [29]. Fast-twitch muscle fibers at 6 months of age mainly rely on glycolysis, while slow-twitch muscle fibers at 12 months of age are more dependent on fatty acid oxidation. This metabolic shift may be driven by the expression regulation of nuclear receptor genes such as *PPARA*. In terms of cellular components, DEGs are significantly enriched in structures such as collagen-containing extracellular matrix and mitochondria. The upregulated expression of mitochondria-related genes provides support for the oxidative metabolic capacity of slow-twitch muscle fibers [30], which is consistent with the results of muscle fiber type transformation observed by immunofluorescence. The enrichment of extracellular matrix-related genes is closely related to the development and maturation of muscle connective tissue. In addition, the enrichment of the MAPK signaling pathway suggests that it participates in the signal transduction process of cell growth and differentiation during the rapid muscle proliferation stage at 6 months of age, further explaining the differences in muscle development between different age groups at the molecular level [31].

KEGG pathway enrichment analysis revealed the core molecular regulatory network underlying the developmental differences in the longissimus dorsi muscle between 6-month-old and 12-month-old Ziwuling black goats. These pathways synergistically participate in regulating muscle growth, metabolic transformation, and functional maturation [32]. Arginine, a conditionally essential amino acid [33], not only serves as a precursor for protein synthesis but also regulates vasodilation through the NO synthesis pathway, thereby influencing muscle blood flow and nutrient supply [34]; proline, the main component of collagen, has a metabolic level that directly affects the synthesis and remodeling of connective tissue [35,36]. In this study, the differential expressions of *NOS1* and *SMOX* genes in this pathway suggest that 12-month-old goats may enhance arginine metabolism to improve muscle blood perfusion and promote the conversion of proline to collagen. This is consistent with the observed phenotype of connective tissue accumulation, indicating that this pathway is an important basis for supporting muscle structural maturation. The AMPK signaling pathway, a core pathway for energy sensing and metabolic regulation, plays a key role in maintaining muscle energy homeostasis [37,38]. By regulating the expression of genes related to fatty acid oxidation (e.g., CPT1A) and fat synthesis (e.g., ACACB), this pathway coordinates the balance between energy supply and demand [39,40]. The downregulation of *ACACB* and *CPT1A* in this study may reflect that 6-month-old muscle relies on glycolysis and fat synthesis to meet the energy needs of rapid proliferation, while 12-month-old muscle shifts to a fatty acid oxidation-dominated metabolic pattern to support the oxidative metabolic characteristics of slow-twitch muscle fibers. This demonstrates the adaptive regulation of the AMPK pathway in response to the energy metabolism demands of different developmental stages. The enrichment of the gap junction pathway suggests that there are developmental differences in intermyocyte communication. This pathway mediates the synchronization of intercellular signals and the exchange of metabolites through connexins, thereby affecting the coordinated contraction ability of muscles [41]. The differential expression of *ADCY6* and *ADCY7* genes may be involved in intercellular signal transmission by regulating cAMP levels [42]. The function of this pathway tends to be improved in 12-month-old individuals, which helps enhance muscle contraction efficiency and aligns with the physiological characteristic of stable motor ability during sexual maturity. The MAPK signaling pathway is involved in the temporal regulation of muscle cell proliferation and differentiation [43]. Through cascade reactions, this pathway transduces extracellular signals [44] and affects processes such as cell proliferation, differentiation, and apoptosis [45]. In 6-month-old muscles, this pathway may be activated to promote satellite cell proliferation and muscle fiber formation; however, during the stable developmental stage at 12 months of age, its activity may be downregulated accordingly to avoid excessive proliferation. The enrichment of the chemokine signaling pathway suggests that immune and inflammatory responses play a potential role in muscle development [46,47]. Chemokines recruit immune cells to participate in tissue repair and regeneration [48], and their expression changes in 12-month-old muscles may be related to local immune regulation induced during muscle fiber type transformation and connective tissue remodeling [49]. The differential expression of chemokine-related genes in 12-month-old muscle may be associated with local immune responses induced by tissue remodeling (e.g., collagen cross-linking and myofiber type transformation) during muscle maturation [50], providing a microenvironmental support for the stability of muscle structure [51]. Further analysis using weighted gene co-expression network analysis (WGCNA) identified nine gene modules associated with muscle fiber diameter and density. Among these, the MEbrown module showed a significant positive correlation with muscle fiber diameter, and the co-expressed gene *GADD45A* within this module is associated with the MAPK signaling pathway. Studies have shown that *GADD45A* is involved in a variety of cellular processes and can regulate the p38MAPK pathway [52]. In goats, the relevant genes can affect myocyte differentiation by activating the p38MAPK [53]. The co-expressed gene *PPARGC1A* in the MEgreen module is associated with the AMPK signaling pathway. *PPARGC1A* plays a core role in energy metabolism and cell function regulation and may affect metabolic status as well as cell proliferation and differentiation through the AMPK pathway, thereby regulating muscle fiber morphology [54].

To sum up, the 6-month-old and 12-month-old stages are critical transition periods for the development of the longissimus dorsi muscle in Ziwuling black goats, during which significant changes occur in muscle fiber morphology, metabolic functions, and molecular regulatory mechanisms. The results of this study suggest that during the 6-month-old stage, emphasis should be placed on protein nutrition to promote muscle fiber proliferation, while during the 12-month-old stage, rational regulation of energy intake is required to optimize meat quality. Based on the characteristics of muscle fiber diameter, density, and the accumulation of related substances, it is recommended that the appropriate slaughter age for Ziwuling black goats is around 12 months old. The findings of this study not only provide a theoretical basis for further understanding of the muscle development mechanism of local goats but also lay a scientific foundation for the efficient utilization of the germplasm resources of Ziwuling black goats.

5. Conclusions

This study systematically analyzed the developmental characteristics and regulatory mechanisms of the longissimus dorsi muscle in Longdong black goats at 6 and 12 months of age. The main results are as follows: In terms of muscle morphology, the 6-month-old group showed higher muscle fiber density and smaller fiber diameter, which indicates that this stage is dominated by muscle fiber proliferation. By the 12-month-old stage, muscle fibers had undergone hypertrophic growth, with significant increases in fiber diameter and cross-sectional area; this process was accompanied by a corresponding increase in the proportion of connective tissue. Regarding muscle fiber types, fast-twitch muscle fibers (Type II) were the dominant type in the 6-month-old group. At 12 months of age, the proportion of slow-twitch muscle fibers (Type I) increased, and this change was accompanied by a metabolic shift from glycolysis to oxidative phosphorylation. Transcriptome analysis identified 387 differentially expressed genes (DEGs) in total. Functional enrichment analysis showed that these DEGs were mainly involved in biological processes such as skeletal muscle growth and glucose metabolism and were significantly enriched in pathways including arginine–proline metabolism and the AMPK signaling pathway. Collectively, these pathways drive muscle development by regulating energy homeostasis and the structural maturation of muscle fibers. Based on the above results, it is suggested that during the 6-month-old stage, protein nutrition should be strengthened to promote muscle fiber proliferation; during the 12-month-old stage, energy intake should be rationally regulated to improve meat quality.

Supplementary Materials: The following supporting information can be downloaded at https://www.mdpi.com/article/10.3390/ani15223265/s1, Figure S1: Distribution of Bases and Quality Scores. Table S1: Main instruments and related information used in this experiment. Table S2: Specific information of experimental reagents. Table S3: Primary antibodies, secondary antibodies and corresponding antigen retrieval conditions used in this experiment. Table S4: Statistics of sequencing data. Table S5: Information on alignment to goat reference genome. Table S6: Principal Component Analysis.

Author Contributions: Conceptualization, X.L.; methodology, W.H., F.S. and H.G.; formal analysis, F.S., Y.H., P.S. (Pengyang Shao) and H.G.; investigation, H.G., F.S., L.G., Y.W., Y.Y. (Yangyang Yue), W.H., Y.Y. (Yongqiang Yang), P.S. (Panlong Sun), W.X., X.Z. (Xiaoqiang Zhang) and X.Z. (Xiaomei Zhu); writing—original draft, W.H., X.L. and J.X.; funding acquisition, X.L.; resources, J.X. and X.L.; supervision, J.X. All authors have read and agreed to the published version of the manuscript.

Funding: Central Government-Guided Local Science and Technology Development Fund Project (No. 24ZYQL003); Gansu Provincial University Industry Support Program (2025CYZC-043); post-graduate Innovation Star Project of Gansu Province (2026CXZX-823); post-graduate Innovation Star Project of Gansu Province (2026CXZX-819).

Institutional Review Board Statement: All studies involving animals were carried out in accordance with the regulations for the Administration of Affairs Concerning Experimental Animals (Ministry of Science and Technology, China; revised in June 2004), and sample collection protocols were approved by the Livestock Care Committee of Gansu Agricultural University (Approval No. GAU-LC-2020-27).

Informed Consent Statement: Not applicable.

Data Availability Statement: The data presented in this study are available on request from the corresponding author.

Conflicts of Interest: The authors declare no competing interests.

References

1. Listrat, A.; Lebret, B.; Louveau, I.; Astruc, T.; Bonnet, M.; Lefaucheur, L.; Picard, B.; Bugeon, J. How muscle structure and composition influence meat and flesh quality. *Sci. World J.* **2016**, *2016*, 3182746. [CrossRef]
2. Fu, Y.; Chen, Y.; Han, X.; Tan, D.; Chen, J.; Lai, C.; Yang, X.; Shan, X.; Silva, L.; Jiang, H. Effects of muscle fiber composition on meat quality, flavor characteristics, and nutritional traits in lamb. *Foods* **2025**, *14*, 2309–2322. [CrossRef]
3. Wang, H.; Yuan, L.; Song, J.; Wang, Q.; Zhang, Y. Distribution of extracellular matrix related proteins in normal and cryptorchid ziwuling black goat testes. *Anim. Reprod.* **2022**, *19*, e20220005. [CrossRef]
4. Shen, J.; Luo, Y.; Wang, J.; Hu, J.; Liu, X.; Li, S.; Hao, Z.; Li, M.; Zhao, Z.; Zhang, Y.; et al. Integrated transcriptome analysis reveals roles of long non-coding RNAs (lncRNAs) in caprine skeletal muscle mass and meat quality. *Funct. Integr. Genom.* **2023**, *23*, 63. [CrossRef]
5. Shen, J.; Hao, Z.; Wang, J.; Hu, J.; Liu, X.; Li, S.; Ke, N.; Song, Y.; Lu, Y.; Hu, L.; et al. Comparative transcriptome profile analysis of longissimus dorsi muscle tissues from two goat breeds with different meat production performance using RNA-seq. *Front. Genet.* **2020**, *11*, 619399. [CrossRef]
6. Candek-Potokar, M.; Lefaucheur, L.; Zlender, B.; Bonneau, M. Effect of slaughter weight and/or age on histological characteristics of pig longissimus dorsi muscle as related to meat quality. *Meat Sci.* **1999**, *52*, 195–203. [CrossRef]
7. Shao, X.; Lu, X.; Sun, X.; Jiang, H.; Chen, Y. Preliminary studies on the molecular mechanism of intramuscular fat deposition in the longest dorsal muscle of sheep. *BMC Genom.* **2024**, *25*, 592. [CrossRef] [PubMed]
8. Ozawa, S.; Mitsuhashi, T.; Mitsumoto, M.; Matsumoto, S.; Itoh, N.; Itagaki, K.; Kohno, Y; Dohgo, T. The characteristics of muscle fiber types of longissimus thoracis muscle and their influences on the quantity and quality of meat from japanese black steers. *Meat Sci.* **2000**, *54*, 65–70. [CrossRef]
9. Wang, C.; Yue, F.; Kuang, S. Muscle histology characterization using h&e staining and muscle fiber type classification using immunofluorescence staining. *Bio Protoc.* **2017**, *7*, e2279.
10. Lee, S.H.; Joo, S.T.; Ryu, Y.C. Skeletal muscle fiber type and myofibrillar proteins in relation to meat quality. *Meat Sci.* **2010**, *86*, 166–170. [CrossRef] [PubMed]
11. Schiaffino, S.; Reggiani, C. Fiber types in mammalian skeletal muscles. *Physiol. Rev.* **2011**, *91*, 1447–1531. [CrossRef] [PubMed]
12. Stavaux, D.; Art, T.; McEntee, K.; Reznick, M.; Lekeux, P. Muscle fibre type and size, and muscle capillary density in young double-muscled blue belgian cattle. *Zentralbl Veterinarmed A* **1994**, *41*, 229–236. [CrossRef]
13. Siqin, Q.; Nishiumi, T.; Yamada, T.; Wang, S.; Liu, W.; Wu, R.; Borjigin, G. Relationships among muscle fiber type composition, fiber diameter and MRF gene expression in different skeletal muscles of naturally grazing wuzhumuqin sheep during postnatal development. *Anim. Sci. J.* **2017**, *88*, 2033–2043. [CrossRef]
14. Liu, W.; Gao, H.; He, J.; Yu, A.; Sun, C.; Xie, Y.; Yao, H.; Wang, H.; Duan, Y.; Hu, J.; et al. Effects of dietary allium mongolicum regel powder supplementation on the growth performance, meat quality, antioxidant capacity and muscle fibre characteristics of fattening angus calves under heat stress conditions. *Food Chem.* **2024**, *453*, 139539. [CrossRef]

15. Liu, W.; Yu, A.; Xie, Y.; Zhang, X.; Guo, B.; Xu, L.; Tao, W.; Yang, R.; Sun, C.; Hu, J.; et al. Electronic nose, flavoromics, and lipidomics reveal flavor changes in longissimus thoracis of fattening saanen goats by dietary allium mongolicum regel flavonoids. *Food Chem. X* **2025**, *29*, 102752. [CrossRef]
16. Pan, Y.; Chen, S.; Niu, S.; Bi, X.; Qiao, L.; Yang, K.; Liu, J.; Liu, W. Hybrid sequencing in different types of goat skeletal muscles reveals genes regulating muscle development and meat quality. *Animals* **2021**, *11*, 2906. [CrossRef]
17. Cao, Y.; Ai, Y.; Zhang, X.; Zhang, J.; Long, X.; Zhu, Y.; Wang, L.; Gu, Q.; Han, H. Genome-wide epigenetic dynamics during postnatal skeletal muscle growth in hu sheep. *Commun. Biol.* **2023**, *6*, 1077. [CrossRef]
18. Huo, W.; Weng, K.; Gu, T.; Zhang, Y.; Zhang, Y.; Chen, G.; Xu, Q. Effect of muscle fiber characteristics on meat quality in fast- and slow-growing ducks. *Poult. Sci.* **2021**, *100*, 101264. [CrossRef] [PubMed]
19. Xu, J.; Wen, Y.; Li, X.; Peng, W.; Zhang, Z.; Liu, X.; Yang, P.; Chen, N.; Lei, C.; Zhang, J.; et al. Bovine enhancer-regulated circSGCB acts as a ceRNA to regulate skeletal muscle development via enhancing KLF3 expression. *Int. J. Biol. Macromol.* **2024**, *261*, 129779. [CrossRef] [PubMed]
20. Lu, Y.; Zhou, J.; Li, F.; Cao, H.; Zhang, X.; Yu, D.; He, Z.; Ji, H.; Lv, K.; Wu, G.; et al. The integration of genome-wide DNA methylation and transcriptomics identifies the potential genes that regulate the development of skeletal muscles in ducks. *Int. J. Mol. Sci.* **2023**, *24*, 15476. [CrossRef]
21. Rao, X.; Huang, X.; Zhou, Z.; Lin, X. An improvement of the 2^(-delta delta CT) method for quantitative real-time polymerase chain reaction data analysis. *Biostat. Bioinf. Biomath.* **2013**, *3*, 71–85.
22. Picard, B.; Gagaoua, M. Muscle fiber properties in cattle and their relationships with meat qualities: An overview. *J. Agric. Food Chem.* **2020**, *68*, 6021–6039. [CrossRef]
23. Salomon, F.V.; Michel, G.; Gruschwitz, F. Development of fiber type composition and fiber diameter in the longissimus muscle of the domestic pig (sus scrofa domesticus). *Anat. Anz.* **1983**, *154*, 69–79.
24. Ai, Y.; Zhu, Y.; Wang, L.; Zhang, X.; Zhang, J.; Long, X.; Gu, Q.; Han, H. Dynamic changes in the global transcriptome of postnatal skeletal muscle in different sheep. *Genes* **2023**, *14*, 1298. [CrossRef]
25. Mo, M.; Zhang, Z.; Wang, X.; Shen, W.; Zhang, L.; Lin, S. Molecular mechanisms underlying the impact of muscle fiber types on meat quality in livestock and poultry. *Front. Vet. Sci.* **2023**, *10*, 1284551. [CrossRef] [PubMed]
26. Lu, Z.; Yue, Y.; Shi, H.; Zhang, J.; Liu, T.; Liu, J.; Yang, B. Effects of sheep sires on muscle fiber characteristics, fatty acid composition and volatile flavor compounds in f(1) crossbred lambs. *Foods* **2022**, *11*, 4076. [CrossRef] [PubMed]
27. Zeng, Q.; Du, Z.Q. Advances in the discovery of genetic elements underlying longissimus dorsi muscle growth and development in the pig. *Anim. Genet.* **2023**, *54*, 709–720. [CrossRef] [PubMed]
28. Arora, R.; Siddaraju, N.K.; Manjunatha, S.S.; Sudarshan, S.; Fairoze, M.N.; Kumar, A.; Chhabra, P.; Kaur, M.; Sreesujatha, R.M.; Ahlawat, S.; et al. Muscle transcriptome provides the first insight into the dynamics of gene expression with progression of age in sheep. *Sci. Rep.* **2021**, *11*, 22360. [CrossRef]
29. Huang, W.; Sha, Y.; Chen, Q.; Chen, X.; Gao, M.; Liu, X.; He, Y.; Gao, X.; Hu, J.; Wang, J.; et al. The interaction between rumen microbiota and neurotransmitters plays an important role in the adaptation of phenological changes in tibetan sheep. *BMC Vet. Res.* **2025**, *21*, 373. [CrossRef]
30. Zhang, L.; Xin, C.; Wang, S.; Zhuo, S.; Zhu, J.; Li, Z.; Liu, Y.; Yang, L.; Chen, Y. Lactate transported by *MCT1* plays an active role in promoting mitochondrial biogenesis and enhancing TCA flux in skeletal muscle. *Sci. Adv.* **2024**, *10*, 4508. [CrossRef]
31. Liu, J.; Liang, X.; Zhou, D.; Lai, L.; Xiao, L.; Liu, L.; Fu, T.; Kong, Y.; Zhou, Q.; Vega, R.B.; et al. Coupling of mitochondrial function and skeletal muscle fiber type by a mir-499/fnip1/AMPK circuit. *EMBO Mol. Med.* **2016**, *8*, 1212–1228. [CrossRef]
32. Farha, S.; Comhair, S.; Hou, Y.; Park, M.M.; Sharp, J.; Peterson, L.; Willard, B.; Zhang, R.; DiFilippo, F.P.; Neumann, D.; et al. Metabolic endophenotype associated with right ventricular glucose uptake in pulmonary hypertension. *Pulm. Circ.* **2021**, *11*, 20458940–211054325. [CrossRef]
33. Chen, M.Y.; Sun, C.Y.; Zhao, R.; Guan, X.L.; Li, M.L.; Zhang, F.; Wan, Z.H.; Feng, J.X.; Yin, M.; Lei, Q.Y.; et al. BAG2 releases SAMD4b upon sensing of arginine deficiency to promote tumor cell survival. *Mol. Cell* **2025**, *85*, 2581–2596. [CrossRef]
34. Wu, G.; Meininger, C.J.; McNeal, C.J.; Bazer, F.W.; Rhoads, J.M. Role of l-arginine in nitric oxide synthesis and health in humans. *Adv. Exp. Med. Biol.* **2021**, *1332*, 167–187. [PubMed]
35. Karna, E.; Szoka, L.; Huynh, T.; Palka, J.A. Proline-dependent regulation of collagen metabolism. *Cell Mol. Life Sci.* **2020**, *77*, 1911–1918. [CrossRef] [PubMed]
36. He, J.; Fang, B.; Shan, S.; Li, Q. Mechanical stiffness promotes skin fibrosis through piezo1-mediated arginine and proline metabolism. *Cell Death Discov.* **2023**, *9*, 354. [CrossRef]
37. Herzig, S.; Shaw, R.J. AMPK: Guardian of metabolism and mitochondrial homeostasis. *Nat. Rev. Mol. Cell Biol.* **2018**, *19*, 121–135. [CrossRef] [PubMed]
38. Wang, C.; Yan, X.; Bai, Y.; Sun, L.; Zhao, L.; Jin, Y.; Su, L. Lactobacillus improves meat quality in sunit sheep by affecting mitochondrial biogenesis through the AMPK pathway. *Front. Nutr.* **2022**, *9*, 1030485. [CrossRef]

9. Vancura, A.; Nagar, S.; Kaur, P.; Bu, P.; Bhagwat, M.; Vancurova, I. Reciprocal regulation of AMPK/SNF1 and protein acetylation. *Int. J. Mol. Sci.* **2018**, *19*, 259. [CrossRef]

0. Senatus, L.; Egana-Gorrono, L.; Lopez-Diez, R.; Bergaya, S.; Aranda, J.F.; Amengual, J.; Arivazhagan, L.; Manigrasso, M.B ; Yepuri, G.; Nimma, R.; et al. DIAPH1 mediates progression of atherosclerosis and regulates hepatic lipid metabolism in mice. *Commun. Biol.* **2023**, *6*, 280. [CrossRef]

1. Nakashima, K.; Kato, H.; Kurata, R.; Qianwen, L.; Hayakawa, T.; Okada, F.; Fujita, F.; Nakagawa, Y.; Tanemura, A.; Murota, H.; et al. Gap junction-mediated contraction of myoepithelial cells induces the peristaltic transport of sweat in human eccrine glands. *Commun. Biol.* **2023**, *6*, 1175. [CrossRef] [PubMed]

2. Simko, V.; Iuliano, F.; Sevcikova, A.; Labudova, M.; Barathova, M.; Radvak, P.; Pastorekova, S.; Pastorek, J.; Csaderova, L. Hypoxia induces cancer-associated cAMP/PKA signalling through HIF-mediated transcriptional control of adenylyl cyclases VI and VII. *Sci. Rep.* **2017**, *7*, 10121. [CrossRef]

3. Song, P.; Zhao, J.; Li, F.; Zhao, X.; Feng, J.; Su, Y.; Wang, B.; Zhao, J. Vitamin a regulates mitochondrial biogenesis and function through p38 MAPK-PGC-1alpha signaling pathway and alters the muscle fiber composition of sheep. *J. Anim. Sci. Biotechnol.* **2024**, *15*, 18. [CrossRef]

4. Martin-Vega, A.; Cobb, M.H. Navigating the ERK1/2 MAPK cascade. *Biomolecules* **2023**, *13*, 1555. [CrossRef]

5. Ma, Y.; Nicolet, J. Specificity models in MAPK cascade signaling. *FEBS Open Bio* **2023**, *13*, 1177–1192. [CrossRef]

6. Charo, I.F.; Ransohoff, R.M. The many roles of chemokines and chemokine receptors in inflammation. *N. Engl. J. Med.* **2006**, *354*, 610–621. [CrossRef]

7. Zhang, L.; Yu, M.; Deng, J.; Lv, X.; Liu, J.; Xiao, Y.; Yang, W.; Zhang, Y.; Li, C. Chemokine signaling pathway involved in CCL2 expression in patients with rheumatoid arthritis. *Yonsei Med. J.* **2015**, *56*, 1134–1142. [CrossRef] [PubMed]

8. Graca, F.A.; Stephan, A.; Minden-Birkenmaier, B.A.; Shirinifard, A.; Wang, Y.D.; Demontis, F.; Labelle, M. Platelet-derived chemokines promote skeletal muscle regeneration by guiding neutrophil recruitment to injured muscles. *Nat. Commun.* **2023**, *14*, 2900. [CrossRef] [PubMed]

9. Tidball, J.G. Regulation of muscle growth and regeneration by the immune system. *Nat. Rev. Immunol.* **2017**, *17*, 165–178. [CrossRef]

0. Puchert, M.; Koch, C.; Zieger, K.; Engele, J. Identification of *CXCL11* as part of chemokine network controlling skeletal muscle development. *Cell Tissue Res.* **2021**, *384*, 499–511. [CrossRef]

1. Sagiv, A.; Krizhanovsky, V. Immunosurveillance of senescent cells: The bright side of the senescence program. *Biogerontology* **2013**, *14*, 617–628. [CrossRef] [PubMed]

2. Li, F.H.; Han, N.; Wang, Y.; Xu, Q. Gadd45a knockdown alleviates oxidative stress through suppressing the p38 MAPK signaling pathway in the pathogenesis of preeclampsia. *Placenta* **2018**, *65*, 20–28. [CrossRef]

3. Deng, K.; Fan, Y.; Liang, Y.; Cai, Y.; Zhang, G.; Deng, M.; Wang, Z.; Lu, J.; Shi, J.; Wang, F.; et al. FTO-mediated demethylation of GADD45b promotes myogenesis through the activation of p38 MAPK pathway. *Mol. Ther. Nucleic Acids* **2021**, *26*, 34–48. [CrossRef] [PubMed]

4. Kong, S.; Cai, B.; Nie, Q. PGC-1alpha affects skeletal muscle and adipose tissue development by regulating mitochondrial biogenesis. *Mol. Genet. Genom.* **2022**, *297*, 621–633. [CrossRef] [PubMed]

animals

MDPI

Article

Study on the Function of *ID2* Gene in Granulosa Cells of Ovaries of Hetian Sheep and Its Correlation Analysis with Lambing Traits

Huiping Sun [1,2], Xinkun Wang [1,2], Lexiao Zhu [1,2], Gul Muhammad Shahbaz [1,2], Ruohuai Gu [1,2], Qiaoyan Huang [1,2], Wei Li [1,2] and Feng Xing [1,2,*]

1 College of Animal Science, Tarim University, Alar 843300, China; sunhuiping223@163.com (H.S.); wdaqqbt@163.com (X.W.); 18538512047@163.com (L.Z.); shahbazbalouch42@gmail.com (G.M.S.); a18346865515@163.com (R.G.); huangqy0316@163.com (Q.H.); 18597814328@163.com (W.L.)

2 Key Laboratory of Tarim, Animal Husbandry Science and Technology, Xinjiang Production & Construction Corps, Alar 843300, China

* Correspondence: xingfeng2001@126.com; Tel./Fax: +86-0997-4681-616

Simple Summary

Sheep are an important source of meat, wool, and income for farmers, but some breeds, such as the Hetian sheep in Xinjiang, produce only a small number of lambs each year. Improving their ability to have more lambs would benefit farmers and local communities. In this study, we focused on a gene called *ID2*, which helps control how cells grow and develop. We found that this gene is especially active in the ovaries of Hetian sheep, where it likely influences fertility. By studying the DNA of 157 sheep, we discovered several genetic differences in *ID2* that were linked to larger litter sizes. We also tested how the gene works in ovarian cells and found that it makes these cells grow faster and changes the levels of important reproductive hormones. These findings show that *ID2* plays an important role in sheep reproduction. This information can help scientists and breeders develop better strategies to select sheep with stronger fertility traits, leading to improved flock productivity and better support for the sheep industry.

Abstract

Reproductive efficiency is a key determinant of sheep productivity, yet Hetian sheep remain limited by relatively low fecundity despite their adaptability to harsh environments. The inhibitor of DNA binding 2 (ID2) gene is known to regulate cell proliferation and differentiation, but its specific role in sheep reproduction is not well understood. This study aimed to characterize the *ID2* gene in Hetian sheep and to assess its functional association with ovarian granulosa cells and litter size. The coding sequence of ovine *ID2* was cloned and analyzed using bioinformatics tools. Tissue-specific expression patterns were measured by quantitative PCR at different pubertal stages. A total of 157 ewes were genotyped to identify single-nucleotide polymorphisms (SNPs) and their association with litter size. Functional studies were performed by lentiviral overexpression of ID2 in granulosa cells, with effects evaluated using CCK-8 proliferation assays, ELISA for hormone secretion, and RT-qPCR for related gene expression. *ID2* was highly expressed in the ovary, particularly during puberty. Four SNPs (g.18202368 A>T, g.18202372 G>A, g.18202431 G>C, g.18202472 G>C) were significantly associated with increased litter size. Overexpression of ID2 promoted granulosa cell proliferation, increased progesterone, decreased estradiol, and altered expression of key genes in the TGF-β/BMP-SMAD signaling pathway. The *ID2* gene plays a crucial role in ovarian function and reproductive regulation in Hetian sheep. Its polymorphisms and functional impact on granulosa cells suggest that ID2 is a promising candidate gene for marker-assisted selection to improve reproductive efficiency in sheep.

Keywords: Hetian sheep; granulosa cells; *ID2* gene; single nucleotide polymorphism; reproductive efficiency; ovary

1. Introduction

Hetian sheep are a wool breed indigenous to the Hetian region of Xinjiang and constitute one of the region's most important livestock resources. They tolerate heat, sparse feed, and disease, and have been placed by China's Ministry of Agriculture on the National List of Protected Livestock and Poultry Genetic Resources. Ewes typically reach puberty at approximately 8–12 months and are first bred at 18–24 months, with a reported lambing percentage of about 102.52% [1]. Because fecundity remains modest, improving litter size is a priority breeding objective with clear economic benefits for local communities.

Lambing performance integrates multiple reproductive processes—endocrine regulation, folliculogenesis, ovulation, fertilization, embryo implantation, and placental/fetal development [2,3]. Both among-breed and within-breed genetic variation for these traits is substantial and generally sufficient to support steady selection responses; consequently, optimizing genetic potential for reproduction is central to enhancing flock productivity [4].

ID2 (inhibitor of DNA binding 2) belongs to the ID protein family, characterized by a helix–loop–helix (HLH) structure lacking a basic DNA-binding domain. The ID2 protein has two typical biological characteristics: inhibiting cell differentiation and promoting cell proliferation. Granulosa cells (GCs) are pivotal for follicular development: their proliferation, differentiation, and steroidogenesis determine oocyte support and follicle fate [5]. GCs' number and function are governed by networks of genes and signaling pathways, including inhibitor of DNA binding 2 (ID2) [6], bone morphogenetic protein receptor type 2 (*BMPR2*) [7], anti-Müllerian hormone (*AMH*) [8], the PI3K/AKT axis [9], and BMP/SMAD cascades [10]. Transforming growth factor-β (TGF-β) signaling coordinates intercellular communication to modulate transcription, cell-cycle progression, differentiation, adhesion, migration, apoptosis, and extracellular-matrix dynamics [11,12]. For example, Xing Du et al. showed that TGF-β signaling interacts with FSHR pathways and regulates ovarian GCs apoptosis via miR-143 [13]. As a downstream effector within TGF-β networks, ID2 has been implicated in GCs proliferation and follicular development. Across species, ID2 is involved in reproductive regulation; for instance, its induced expression in rat Sertoli cells [14]. ID2 participates in Sertoli-cell differentiation and endocrine regulation and is enriched in chicken ovarian GCs and testicular Sertoli cells [15,16]; it is upregulated in bovine follicles [17] and shows highest expression in fully differentiated porcine GCs and cumulus–oocyte complexes [18] and ID2-deficient female mice exhibit severe lactation defects with decreased mammary epithelial proliferation [19], collectively supporting a plausible role for ID2 in follicular dynamics and fecundity.

With the advent of whole-genome resequencing, selection-signal scans (e.g., F_{ST} and π ratio) have become powerful tools to detect genomic regions under differential selection between single- and multiple-lambing groups. Through within-population comparisons of Hetian sheep, 290 shared selected genomic regions were identified, and 332 candidate genes were annotated in both single-lambing (SLE) and twin-lambing (TLE) groups; *ID2* was among 13 genes significantly enriched in pathways related to lambing rate (including TGF-β signaling). Genetic studies of ovine fecundity frequently focus on SNPs in candidate genes [20,21]; for example, Zhang et al. reported that T/T and T/G genotypes at rs400827589 in Small-Tailed Han sheep were associated with greater litter size than G/G ($p < 0.05$) [22]. However, no published data directly link *ID2* to lambing number in Hetian sheep.

To address this gap, we cloned the coding sequence (CDS) of ovine *ID2* from Hetian sheep and characterized its molecular features. We then evaluated associations between *ID2* polymorphisms and lambing number in Hetian ewes and examined how *ID2* overexpression affects activity and proliferation of primary ovarian GCs.

2. Materials and Methods

2.1. Test Sample Collection

Hetian sheep ewes were maintained at the Experimental Station of Tarim University under uniform housing and feeding conditions. The feeding regime was as follows: prior to mating and during gestation all ewes were offered high-quality pasture plus a supplementary concentrate/compound feed; during lactation, the feeding allowance was increased with additional succulent forages and protein-rich feed. Drinking water, vaccination and deworming protocols were identical across the flock. Breeding management was by artificial insemination: once estrus was detected (by teaser ram method and external observation of vulvar hyperemia, swelling and mucus discharge), insemination was performed under a uniform protocol. For the cloning of the *ID2* gene, a healthy Hetian ewe in puberty was selected at the Experimental Station and humanely euthanized in accordance with institutional animal care protocols. Immediately after slaughter, the hypothalamus was dissected, snap-frozen in liquid nitrogen and transferred to −80 °C until RNA extraction. For tissue expression profiling, five healthy females of similar body weight were sampled at each of three stages: Prepuberty (approximately 90 days of age), puberty (first detected estrus), and postpuberty (10 days after the first estrus). Selection criteria for these animals were as follows: clinically healthy with no reproductive disorders, body weight within ±10% of the stage mean, no antibiotic or hormone treatments within the previous 30 days. Following humane slaughter, tissue specimens (hypothalamus, pituitary, ovary, oviduct, and uterus) were aseptically excised using sterile scissors and scalpels, minced, transferred to sterile 5 mL cryovials, snap-frozen in liquid nitrogen, and stored at −80 °C until analysis. In sheep reproductive expression profiling, rapid freezing and storage at −80 °C are correct practices to preserve RNA integrity and minimize degradation across developmental or estrous stages [23]. Puberty was defined following the standard for small ruminants. Puberty was identified by twice-daily observations at 10:00 and 16:00 using teaser rams and visual inspection of external signs. Criteria included acceptance of mounting (standing estrus) together with vulvar hyperemia and swelling, and the presence of cervical/vulvar mucus. Behavioral changes were recorded contemporaneously for each ewe.

For population genetic analyses, whole blood was collected from 157 multiparous ewes (3–5 years of age) via jugular venipuncture into EDTA-containing vacuum tubes and stored at −40 °C until DNA extraction. Parity distribution of the sampled ewes (3–5 years of age) was recorded as parity 2, $n = 62$ (39.5%); parity ≥ 3, $n = 95$ (60.5%), total $n = 157$. Of these 157 ewes, 78 delivered twins and 79 delivered singletons; all were bred by artificial insemination and the number of live lambs was recorded within 24 h after parturition.

All animal procedures were approved by the Institutional Animal Care and Use Committee of Tarim University of Science and Technology and complied with the institutional Animal Experiment Guide and applicable national regulations. This study adhered to the ethical guidelines set by the Ethics Committee of Tarim University of Science and Technology (approval number SYXK 2020-009), approved on 23 April 2020.

2.2. Ovarian Collection

Healthy Hetian ewes were selected and humanely euthanized; ovarian tissues were immediately collected post-mortem. and the ovaries located on both sides of the uterus were cut off with sterile surgical scissors, rinsed (sprayed) with 75% alcohol for disinfection, and the blood was removed. They were placed in a pre-cooled PBS buffer bottle (pre-cooled with ice packs in a foam incubator) and quickly transferred to a sterile cell culture room within 4–5 h for subsequent operations. The above ovaries were cut off from the excess connective tissue around the ovaries, and then placed in a sterile beaker, added 75% alcohol, and soaked for 30 s. The alcohol was poured out, and the PBS buffer was rinsed 3–5 times.

2.3. Primer Design

The *ID2* nucleotide sequence (accession number XM_004017705.5) was retrieved from the NCBI Nucleotide database (https://www.ncbi.nlm.nih.gov) (accessed on 3 September 2025) based on the reference genomic assembly GCF_016772045.2-RS_2023_10A. Subsequently, amplification primers and genotyping primers for the ID2 gene were designed using Primer Premier 5.0 software. Fluorescence quantitative PCR primers were designed according to the CDS of the *ID2* gene of Hetian sheep (Table 1). The *ACTB* gene was used as the reference gene for qPCR analysis, because it is a classical housekeeping gene whose expression has been reported to remain stable in different tissues and under various physiological conditions in sheep [24]. Prior to the main experiment, we confirmed that *ACTB* gene Ct values varied minimally. Target gene expression levels were therefore quantified using the $2^{-\Delta\Delta Ct}$ method with *ACTB* gene as the normalizer.

Table 1. Gene cloning and quantitative PCR primers.

Genes	Primer Sequence (5′→3′)	Amplicon Size (bp)	Annealing Temperature (°C)
ID2	F: CTTCCCTCCTCCCAATCGCT R: GCAGGCAATCACCATTCAATAAAC	1100	56
ID2-1	F: GCTGTGGAAGGAGTCGTGT R: GCGCTTTGCTGTCATTTG	207	58
QID2	F: TCCAGTGAGGTCCGTTAGG R: TGATGTCGGTGTTGAGGGT	317	58
ACTB	F: GCAGATGTGGATCAGCAAGC R: TCTCGTTTTCTGCGCAAGTT	113	58

2.4. ID2 Gene Cloning

PCR amplification was carried out with the primers listed in (Table 1). Each 25 µL reaction contained 1.0 µL of ovarian cDNA, 1.0 µL of each forward and reverse primer (10 µM), 12.5 µL 2× EasyTaq® PCR SuperMix (TransGen Biotech Co., Ltd., Beijing, China), and 9.5 µL nuclease-free water. Thermocycling conditions were as follows: initial denaturation at 94 °C for 5 min; 40 cycles of 95 °C for 30 s, 56 °C for 30 s, and 72 °C for 2 min; followed by a final extension at 72 °C for 10 min and a hold at 4 °C. Amplicons were visualized on 1.5% agarose gels to confirm expected size and integrity. purified PCR products were subjected to Sanger sequencing on an Applied Biosystems platform (Sangon Biotech (Shanghai) Co., Ltd., Shanghai, China) to identify sequence variants. Sequence chromatograms were trimmed and aligned using DNAMAN Version 6.0.

2.5. Bioinformatics Analysis

The sequencing results were assembled using DNAMAN 6.0, and homology searches were performed using BLAST on the NCBI website (https://blast.ncbi.nlm.nih.gov/Blast.

cgi) (accessed on 3 September 2025). The physicochemical properties of the ID2 protein were predicted using the ProtParam website (https://web.expasy.org/protparam) (accessed on 3 September 2025). The hydrophobicity and transmembrane structure of the protein were predicted using Prot Scale software (https://web.expasy.org/protscale/) (accessed on 3 September 2025) and TMHMM 2.0 (https://services.healthtech.dtu.dk/services/TMHMM-2.0/) (accessed on 3 September 2025). Signal P 6.0 (https://services.healthtech.dtu.dk/services/SignalP-6.0/) (accessed on 3 September 2025) and Net Phosphorylation 3.1 (https://services.healthtech.dtu.dk/services/NetPhos-3.1/) (accessed on 3 September 2025) servers were used to predict signal peptides and phosphorylation sites. PSIPRED 4.0 (https://bioinf.cs.ucl.ac.uk/psipred/) (accessed on 3 September 2025) and Phyre 2.0 (http://www.sbg.bio.ic.ac.uk/phyre2/html/page.cgi?id=index) (accessed on 3 September 2025) were used to predict the secondary and tertiary structures of ID2 protein. The amino acid sequence alignment of Hetian sheep ID2 protein with other species was constructed using MEGA 11, and a phylogenetic tree was constructed using the neighbor-joining (NJ) method.

2.6. *ID2 Gene Tissue Expression*

Total RNA was isolated from five ovine reproductive tissues (hypothalamus, pituitary, ovary, oviduct and uterus) at three physiological stages using TRIzol reagent (Invitrogen, Carlsbad, CA, USA). RNA concentration and purity were assessed using a NanoDrop 8000 spectrophotometer (NanoDrop Technologies, Wilmington, DE, USA). cDNA was synthesized using the Takara Reverse Transcription Kit (TaKaRa Bio Inc., Dalian, China) for qPCR assays. Primers were designed as listed in Table 1, and *ACTB* served as the endogenous control for normalization. Expression of *ID2* in reproductive tissues before, during, and after estrus in Hetian sheep was measured on a CFX96 Real-Time PCR Detection System (Bio-Rad Laboratories, Hercules, CA, USA). Each 15 μL RT-qPCR contained 7.5 μL SYBR Green Real-time PCR Mix, 5.5 μL nuclease-free water, 0.5 μL of each primer (10 μM), and 1.0 μL cDNA (500 ng/μL). Cycling conditions were as follows: 94 °C for 30 s; 40 cycles of 94 °C for 20 s, 58 °C for 15 s, and 72 °C for 15 s; followed by melt-curve analysis (95 °C for 15 s, 60 °C for 30 s, 95 °C for 15 s). Melt-curve parameters used the instrument's default settings. Each sample was run in triplicate.

2.7. *ID2 Genotype Analysis*

A total of 157 ewes of the Hetian sheep breed were sampled for genotyping in this study. All animals were female and belonged to a single commercial flock managed under uniform conditions. No targeted selection of close-kin (such as full-sib or half-sib) individuals was performed; background relatedness may exist as the flock is a nucleus breeding herd. All ewes included had recorded live-born lambs (i.e., sample selection excluded non-lambing individuals). The phenotype used for the SNP–litter size analysis was the number of live-born lambs per lambing (within 24 h of birth) and all ewes in this set had lambing records.

Genomic DNA was extracted from blood by the phenol–chloroform method. Genotyping primers designed for the *ID2* loci were used to detect target variants. PCR was performed on a Veriti™ 96-Well Thermal Cycler (Thermo Fisher Scientific, Waltham, MA, USA). Each 25 μL reaction contained 12.5 μL 2× EasyTaq® PCR SuperMix (TransGen Biotech, Beijing, China), 9.5 μL nuclease-free water, 1.0 μL of each primer (10 μM), and 1.0 μL genomic DNA. Amplicons of the expected size were purified and submitted to Sangon Biotech (Shanghai, China) for Sanger sequencing. Trace files were inspected in Chromas 2.6.6, and sequences were aligned to the ovine reference in DNAMAN to identify polymorphic sites.

2.8. Isolation, Culture, and Identification of Ovarian GCs

Fresh, untreated ovaries were rinsed in sterile PBS. Surface follicles were incised with a sterile scalpel, and follicular fluid was allowed to drain into pre-equilibrated collection medium. The cell suspension was centrifuged at 2000 rpm for 5 min. The supernatant was discarded, leaving a white cell pellet. The pellet was resuspended in 2 mL culture medium containing antibiotics (e.g., 3% penicillin–streptomycin), then centrifuged at 1000 rpm for 5 min; this wash was repeated three times to obtain a granulosa-cell pellet. The centrifuged GCs were inoculated into a T25 culture flask containing 7 mL of DMEM medium supplemented with 15% FBS and 1% penicillin-streptomycin. Cells were evenly distributed and incubated at 37 °C in 5% CO_2 for 24 h, after which cultures were inspected and the medium was refreshed.

GCs purity was verified by FSHR immunofluorescence. Cells were fixed in 4% paraformaldehyde for 15 min, permeabilized with 0.3% Triton X-100 for 10 min, and blocked with 20% BSA for 30 min. Cultures were incubated with anti-FSHR primary antibody for 1 h at room temperature, rinsed, and then incubated with the appropriate fluorophore-conjugated secondary antibody for 30–45 min. Nuclei were counterstained with DAPI for 10 min at 37 °C and imaged on an IX73 inverted fluorescence microscope (Olympus, Tokyo, Japan).

2.9. Lentivirus Transfection of Eggs and CCK-8 Assay

A lentiviral overexpression construct for the ovine *ID2* gene (Hanheng Biotech, Shanghai, China) was used to transduce adherent GCs. Cells were seeded in 24-well plates at 1×10^5 cells/well and transduced when cultures reached 2×10^5 cells/well, following the manufacturer's instructions. After 24 h at 37 °C, the virus-containing medium was replaced with fresh complete medium, and cultures were maintained at 37 °C for an additional 48 or 72 h.

Cell viability was evaluated using the Cell Counting Kit-8 (CCK-8) (Solarbio Science & Technology Co., Ltd., Beijing, China). GCs were seeded in 96-well plates at 2000 cells/well. At 48 and 72 h, 10 µL CCK-8 reagent was added per well and plates were incubated for 2 h at 37 °C. Absorbance was read at 450 nm on a microplate reader (Tecan Group Ltd., Männedorf, Switzerland), and viability was calculated relative to the control group. Assays were performed in triplicate, and results are reported as mean ± SD.

2.10. Enzyme-Linked Immunosorbent Assay

Progesterone (P4) and estradiol (E2) in culture supernatants were quantified by ELISA to assess the effect of ID2 overexpression. Supernatants were collected at 48 and 72 h post-transduction, clarified by brief centrifugation, and assayed with commercial ELISA kits (Solarbio Science & Technology, Beijing, China) according to the manufacturers' instructions. Concentrations were reported as ng/mL (P4) and pg/mL (E2). Each sample was run in triplicate, and the results were compared with the corresponding control group.

2.11. RT-PCR Detection of ID2-Related Gene Expression

Total RNA was extracted from *ID2*-overexpressing and control cells using TRIzol reagent (Invitrogen), cDNA was synthesized using the Takara Reverse Transcription Kit (TaKaRa Bio Inc., Dalian, China) for qPCR assays. mRNA levels of predicted ID2 interactors from the STRING database (*ID1, ID3, NFIL3,* and TCF3) and pathway-related genes (*BMPR2, RBX1,* and *SMAD7*) were quantified by RT-qPCR using the primers listed in (Table 2). Expression was normalized to *ACTB* and calculated with the $2^{-\Delta\Delta Ct}$ method. All reactions were run in triplicate.

Table 2. Primer design for ID2-related genes.

Genes	Primer Sequence (5′→3′)	Amplicon Size (bp)	Annealing Temperature (°C)
ID1	F: CCAGAACCGCAAGGTGAGCA R: CGCCGATTTCGCCATTGA	161	58
ID3	F: CTCTGGCTTTCTCCTTCCTT R: CGAGTAGCAGTGGTTCATGTC	195	58
RBX1	F: CAGCGATGGATGTGGATACC R: CTCGCTGGCTCAAAACACG	370	58
BMPR2	F: GCAGGTTCTGGTGTCTAGGG R: GTGACAGGTTGCGCTCATTC	251	58
SMAD7	F: GTGCTCAAGAAACTGAAGGAAC R: TCGCAGAGTCGGCTAAGGT	314	58
NFIL3	F: CCGAAGACTTGACGACAGG R: GCCGCTTTTCCCAATACAT	135	58
TCF3	F: AAGCCACGGGACTATTTTG R: GTGCTGAACAACAGGAACTTTA	180	58
ACTB	F: GCAGATGTGGATCAGCAAGC R: TCTCGTTTTCTGCGCAAGTT	133	58

2.12. Data Processing and Statistical Analysis

Relative mRNA expression was calculated using the $2^{-\Delta\Delta Ct}$ method. The population genetic indicators were analyzed using Excel 2019 software, and the linkage disequilibrium results of the mutation sites were statistically analyzed using the SHEsis (http://analysis.bio-x.cn) (accessed on 3 September 2025) online website. One-way ANOVA in SPSS 26.0 software was used to determine the significance, with $p < 0.01$ indicating an extremely significant difference and $p < 0.05$ indicating a significant difference.

(1) Gene frequency and genotype frequency

In a biological (diploid) population, genotype frequencies represent the proportion of individuals carrying each genotype at a locus. If we denote the genotypes as AA, AB, and BB, then one computes:

Genotype frequency: P_{AA} = AA individuals/total number; P_{AB} = AB individuals/ $totalnumber; P_{BB} = BBindividuals/totalnumber$

The allele frequencies indicate how common each allele is in the gene pool. Denoting the alleles as A and B, one calculates the following:

Allele frequency: PA = (number of AA individuals + number of AB $individuals/2)/totalnumber$

PB = (number of BB individuals + number of AB individuals/2)/ $totalnumber$

(2) Hardy–Weinberg equilibrium test: $x^2 = \sum_{i-1}^{m} \frac{(f_i - np_i)^2}{np_i}$.

(3) Population homozygosity (H_o) test: $Ho = \sum_{i-1}^{n} p_i^2$.

(4) Population heterozygosity (H_e) test: $He = 1 - \sum_{i=1}^{n} p_i^2$.

(5) Effective allele (N_e) test: $Ne = \frac{1}{\sum_{i=1}^{n} p_i^2}$.

(6) Polymorphic Information Content (PIC) Test: $PIC = 1 - \sum_{i=1}^{n} p_i^2 - \sum_{i=1}^{n-1} \sum_{j=i+1}^{n} 2p_i^2 p_j^2$.

3. Test Results and Analysis

3.1. Cloning and Nucleotide Sequence Analysis of the ID2 Gene of Hetian Sheep

Total RNA was extracted from the hypothalamic tissue of Hetian sheep, reverse-transcribed into cDNA, and the coding region of the *ID2* gene was subsequently amplified from this cDNA.A specific amplicon of 1100 bp was obtained, and the coding sequence (CDS) spanned 405 bp (Figure 1A). The sequence matched the predicted ovine *ID2* mRNA (GenBank accession XM_004005671.4). At the amino-acid level, the Hetian sheep ID2 protein was identical to those of sheep, goat, and cattle (100% identity) and highly conserved relative to horse (99.3%), gorilla (98.5%), and human (98.5%) (Figure 1B). Phylogenetic analysis grouped Hetian sheep ID2 closest to other ruminants (sheep, cattle, goat), followed by horse and gorilla, with human most distant among the species compared (Figure 1C). These findings confirm that the sequence obtained corresponds to the ovine *ID2* gene.

(A)

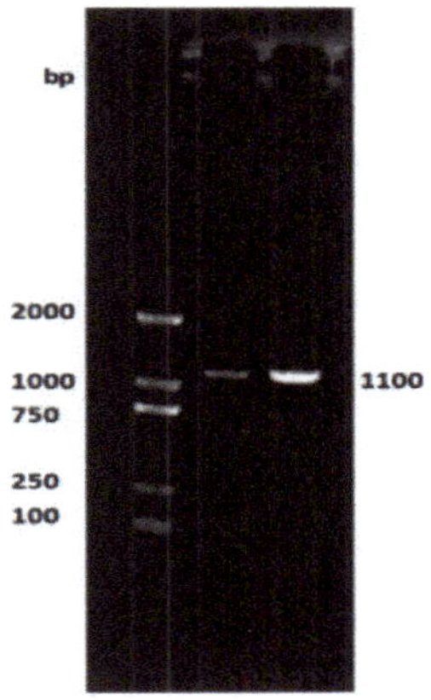

(B)

Percent Identity (upper triangle) / Divergence (lower triangle)

	1	2	3	4	5	6	7		
1		100.0	100.0	100.0	99.3	98.5	98.5	1	Hetian *sheep*
2	0.0		100.0	100.0	99.3	98.5	98.5	2	*Ovis aries*
3	0.0	0.0		100.0	99.3	98.5	98.5	3	*Bos taurus*
4	0.0	0.0	0.0		99.3	98.5	98.5	4	*Capra hircus*
5	0.8	0.8	0.8	0.8		99.3	99.3	5	*Equus caballus*
6	1.5	1.5	1.5	1.5	0.8		100.0	6	*Pan troglodytes*
7	1.5	1.5	1.5	1.5	0.8	0.0		7	*Homo sapiens*
	1	2	3	4	5	6	7		

Figure 1. *Cont.*

(**C**)

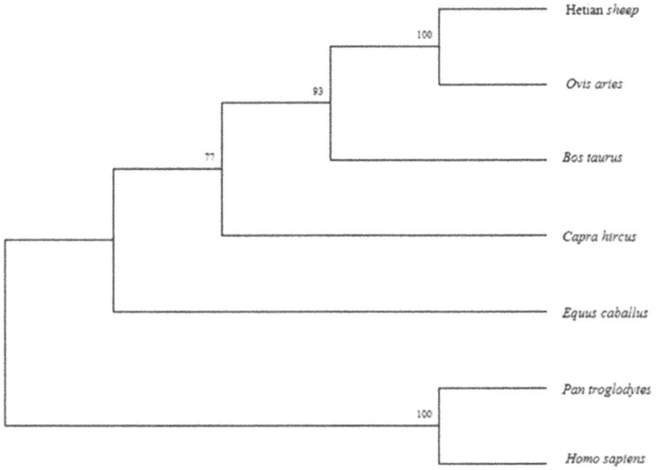

Figure 1. Cloning and nucleotide sequence analysis of Hetian sheep ID2 gene. (**A**) Detection of the coding sequence of the Hetian sheep *ID2* clone via agarose gel electrophoresis. (**B**) Nucleotide homology comparison between the sequences obtained in this study and ID2 sequences from other animal species. (**C**) Neighbor-joining tree of the sequences obtained in this study and the ID2 sequences identified in other animal species. The cloned sequences in this study are shown in a box, and the ID2 sequences from other animal species include sheep (accession number: XP_004005720.1), goat (accession number: XP_005687088.1), cattle (accession number: NP_001029403.1), horse (accession number: XP_001503661.1), orangutan (accession number: XP_034808448.1), and human (accession number: NP_002157.2).

3.2. Prediction and Analysis of the Characteristics of Hetian Sheep ID2 Protein

The physicochemical properties of the protein of the Hetian sheep *ID2* gene were interpreted, indicating that the molecular mass of the protein was 14,860.06 u, the theoretical isoelectric point pi was 7.82, and the molecular formula was $C_{648}H_{1061}N_{181}O_{203}S_7$. The total number of atoms was 2100, encoding a total of 134 amino acids, of which there were 20 serine (S) and 17 leucine (L), accounting for 14.9% and 12.7%, respectively, and no tryptophan (W) was found. The total number of negatively charged amino acid residues (Asp and Glu) was 13. The total number of positively charged amino acid (Arg + Lys) residues was 14, and the instability coefficient was 61.72 (Table 3). The hydrophobicity prediction of the ID2 protein is shown in (Figure 2A), indicating that the ID2 protein is hydrophilic. The transmembrane domain of Hetian sheep ID2 protein was predicted, and the results showed that ID2 protein did not contain a transmembrane domain (Figure 2B). By predicting the signal peptide information of the Hetian sheep ID2 protein, it was found that the ID2 protein did not have a signal peptide and was not a secretory protein (Figure 2C). Hetian sheep ID2 protein has a total of 29 phosphorylation sites, including 20 serine (S) sites, 6 threonines (T) sites, and 3 tyrosine (Y) sites (Figure 2D). The secondary structure prediction of ID2 protein showed that the protein contained 67 α-helices (50%), 5 extended chains (3.73%), 2 β-folds (1.49%), and 60 irregular coils (46.78%); α-helices and irregular coils were spread throughout the amino acid chain, with only extended chains and a few β-folds (Figure 2E).

Table 3. Amino acid composition of Hetian sheep ID2 predicted protein.

Amino Acids	Quantity	Proportion (%)	Amino Acids	Quantity	Proportion (%)
Alanine (A)	5	3.7%	Leucine (L)	17	12.7%
Arginine (R)	5	3.7%	Lysine (K)	9	6.7%
Asparagine (N)	6	4.5%	Methionine (M)	5	3.7%
Asparticacid (D)	9	6.7%	Phenylalanine (F)	2	1.5%
Cysteine (C)	2	1.5%	Proline (P)	9	6.7%
Glutamine (Q)	7	5.2%	Serine (S)	20	14.9%
Glutamate (E)	4	3.0%	Threonine (T)	6	4.5%
Glycine (G)	4	3.0%	Tryptophan (W)	0	0%
Histidine (H)	5	3.7%	Tyrosine (Y)	3	2.2%
Isoleucine (I)	9	6.7%	Valine (V)	7	5.2%

3.3. Analysis of the Expression of the ID2 Gene in Different Tissues of Hetian Sheep at Different Puberty Stages

Real-time fluorescence quantitative PCR for the expression of the *ID2* gene in five tissues of Hetian sheep, including the hypothalamus, oviduct, pituitary, ovary, and uterus, are shown in (Figure 3). The *ID2* gene was expressed in five tissues in the three periods before puberty, during puberty, and after puberty. The expression level of the *ID2* gene in the ovary was significantly higher than that in other tissues in the three periods ($p < 0.05$). The expression levels of the oviduct, pituitary, and uterus in puberty were significantly higher than those in other periods ($p < 0.05$). The expression level of the hypothalamus decreased significantly from Prepuberty to puberty ($p < 0.05$) and increased significantly from puberty to Postpuberty ($p < 0.05$).

Notably, *ID2* gene expression was highest in the ovary during puberty, significantly higher than in other tissues (e.g., hypothalamus, pituitary) at the same stage.

3.4. Descriptive Characteristics of the Genotyped Ewe Population

To provide context for the subsequent SNP–litter size association analyses, Table 4 summarizes the key descriptive statistics of the 157 genotyped ewes. It includes information on sample size, age, parity distribution, flock origin, mating method, live-born lamb counts and reproductive parameters (fertility, fecundity, prolificacy).

Table 4. Descriptive statistics and reproductive parameters of the genotyped Hetian ewe population.

Variable	Value	Notes
Sample size	157	All genotyped animals included in the SNP–litter size analysis; all selected ewes delivered live lamb(s).
Sex	Female (100%)	All ewes with lambing records.
Herd origin	Single commercial flock	Same farm and uniform management to minimize environmental heterogeneity.
Age at sampling	Mean ± SD: 3.4 ± 0.7; Range: 3.0–5.0	Age recorded at sampling.
Parity distribution	1st: 45 (28.7%); 2nd: 62 (39.5%); ≥3: 50 (31.8%)	Parity = number of previous lambings at sampling.
Mating method	artificial insemination	artificial insemination
Ewes lambed	157 (100.0%)	All sampled ewes delivered live lamb(s).
Total live-born lambs	236 (total)	Sum of live lambs born within 24 h across 157 ewes.
Lambs per ewe	1.50 ± 0.50	Calculated as 236/157 ≈ 1.50 (236/157 ≈ 1.50)
Litter size distribution	Single (1 lamb): 78 (49.7%) Twin (2 lambs): 79 (50.3%) Triplet (3 lambs): 0 (0%)—none observed	Categories based on live-born lambs within 24 h; totals sum to 157 ewes and 236 lambs.

Table 4. *Cont.*

Variable	Value	Notes
Fertility	100.0%	Fertility = proportion of mated ewes that delivered live lamb(s).
Fecundity	150.3%	Fecundity = total live lambs ÷ number of mated ewes × 100 = (236/157) × 100 ≈ 150.3%.
Prolificacy	50.3%	Prolificacy reported both as mean lambs per lambed ewe (1.50) and proportion of ewes producing multiples (50.3%).

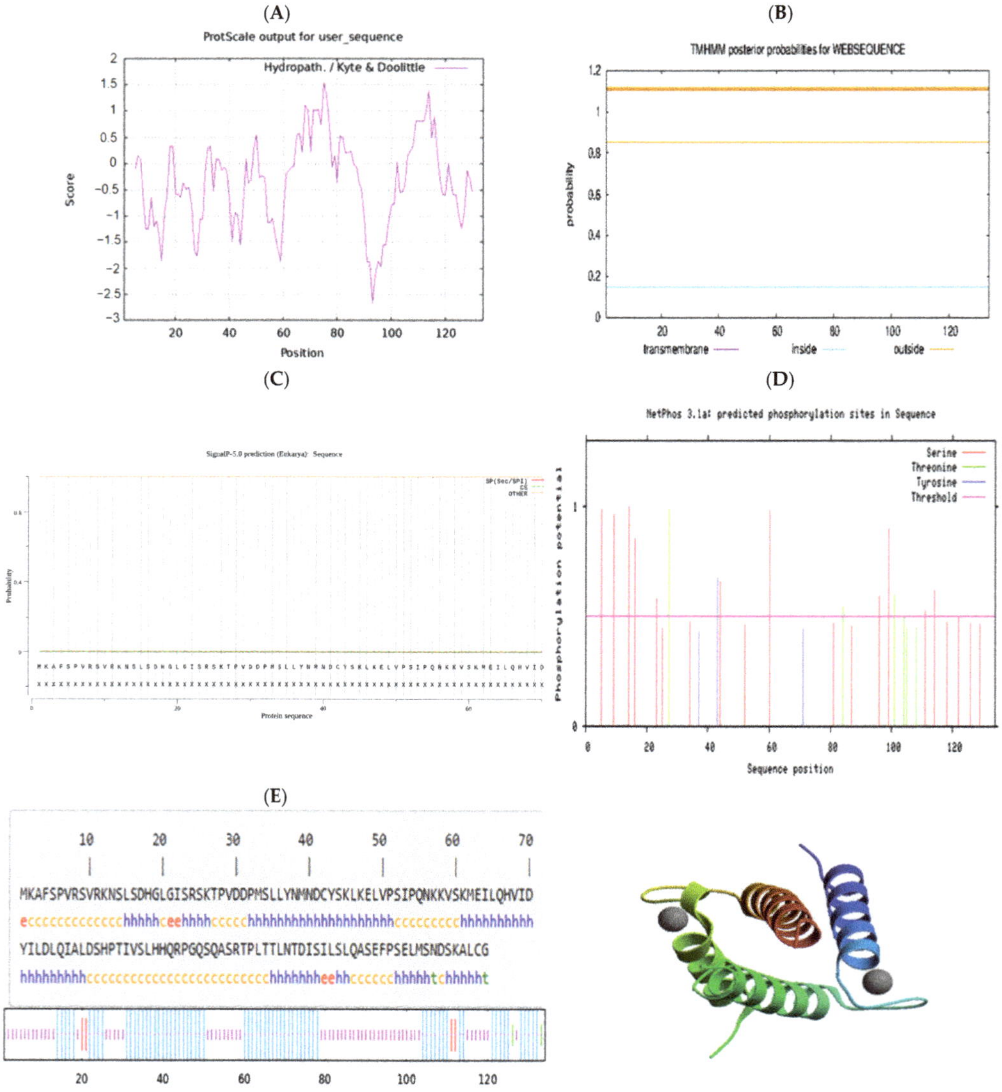

Figure 2. Characteristic analysis of sheep ID2 protein. (**A**) Hydrophilicity analysis of Hetian sheep ID2 protein. (**B**) Prediction of the transmembrane structure of Hetian sheep ID2 protein. (**C**) Prediction of the signal peptide of Hetian sheep ID2 protein. (**D**) Analysis of phosphorylation sites of Hetian sheep ID2 protein. (**E**) Prediction of secondary and tertiary structures of Hetian sheep ID2 protein.

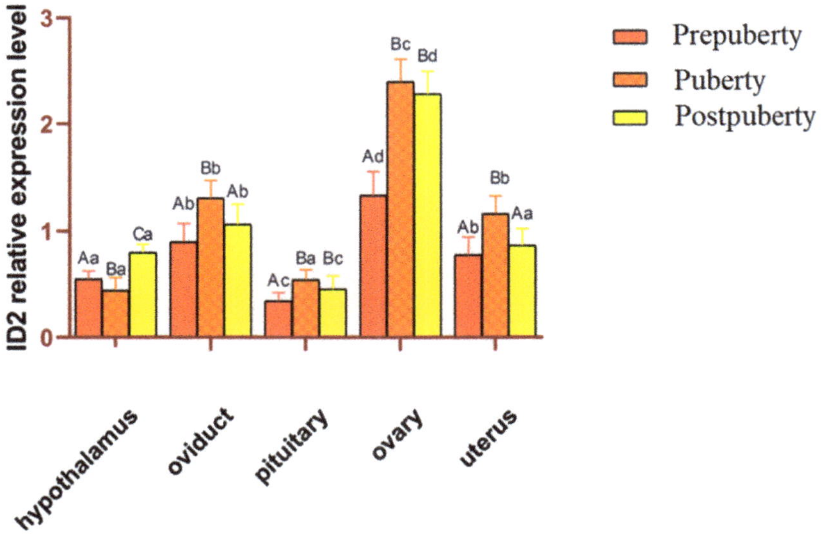

Figure 3. Relative expression levels of *ID2* gene in different tissues of Hetian sheep at different stages of puberty. (1) Values of the same tissues and different periods with different capital letter superscripts indicate significant differences ($p < 0.05$), whereas with the same capital letter or no letter superscripts, there was no significant difference ($p > 0.05$). (2) Values of the same period, different tissues with different small letter superscripts indicate a significant difference ($p < 0.05$), while with the same small letter superscripts, there was no significant difference ($p > 0.05$).

The results showed that all animals were female, from a single commercial flock under uniform management, with a mean age of 3.4 ± 0.7 years (range 3.0–5.0). All 157 ewes lambed, yielding a total of 236 live-born lambs, an average of 1.50 ± 0.50 lambs per ewe. Additional reproductive metrics include a fertility of 100.0% (157/157), fecundity of 150.3%, and a prolificacy rate (≥2 lambs) of 50.3%.

3.5. ID2 Gene PCR Amplification Results and Polymorphic Site Peak Diagram

Through PCR amplification, a specific band of 207 bp was obtained (Figure 4A). The amplified product was sent to (Sangon Biotech) for sequencing, and a total of 8 mutation sites were found (Figure 4B).

3.6. Genetic Analysis of SNP Variant Sites in ID2 Gene

A total of eight polymorphic sites were detected in the *ID2* gene in Hetian sheep, among which three sites (g.18202372 G>A, g.18202426 C>T, and g.18202472 G>C) had three genotypes, while the other five sites had two genotypes. The frequencies of the polymorphic sites are shown in Table 5. Hardy–Weinberg analysis revealed that the genotype distribution of all variant sites conformed to Hardy–Weinberg equilibrium ($p > 0.05$).

3.7. SNP and Linkage Disequilibrium of ID2 Gene

According to (Table 6), the g.18202368 A>T, g.18202372 G>A, g.18202426 C>T, g.18202451 C>T, and g.18202472 G>C sites of the *ID2* gene are moderately polymorphic (0.25 < PIC < 0.5), and the g.18202431 G>C and g.18202467 T>A sites are lowly polymorphic (PIC < 0.25). The results of linkage disequilibrium analysis of SNP mutations in the ID2 gene are shown in Figure 5. The r^2 value of the linkage between g.18202439 C>T and g.18202451 C>T is 0.54, indicating that the two mutation sites are strongly linked to each other ($D' > 0.86$, $r^2 > 0.30$), and there is no obvious linkage between the other sites.

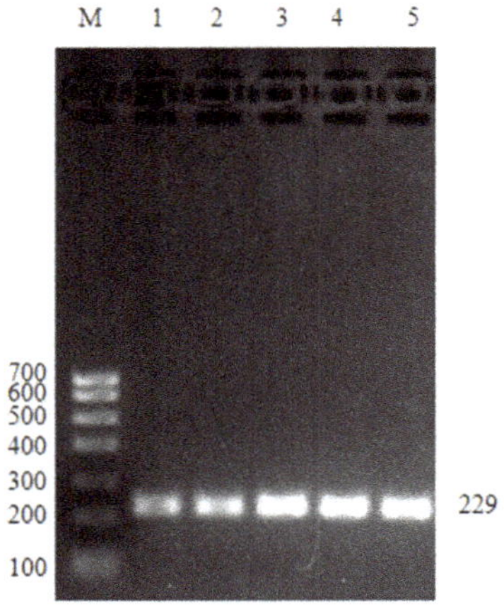

(A)

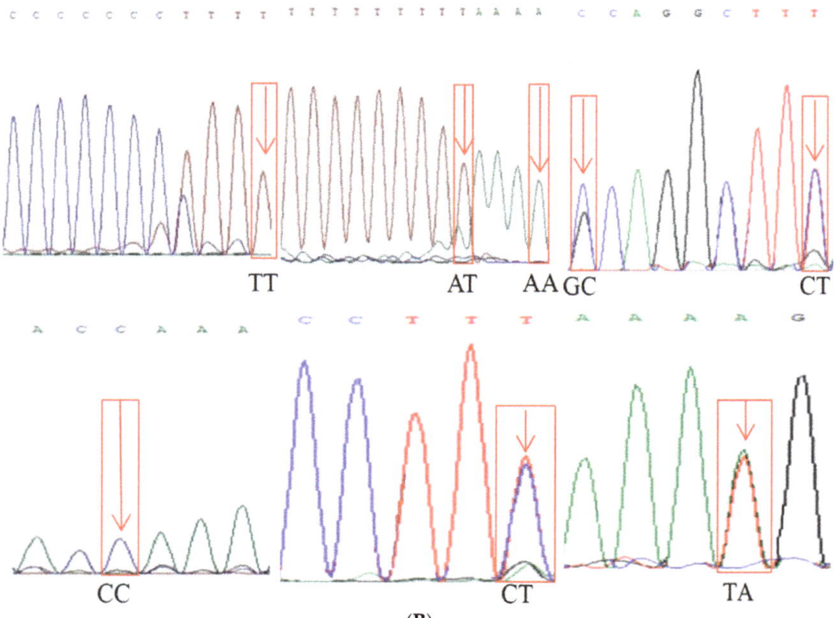

(B)

Figure 4. *ID2* gene polymorphic sites. (**A**) *ID2* gene PCR amplification target fragment. M: DNA marker; Lanes 1–5: PCR products, showing a specific band at 207 bp, indicating successful amplification of the target fragment (**B**) *ID2* gene SNP site genotype sequencing diagram. confirming the existence of genetic polymorphism in the *ID2* gene. An arrow highlights the key mutation site.

3.8. Association Analysis Between ID2 Gene Polymorphism and Litter Size

ID2 genotypes were tested for association with litter size; values are presented as mean ± SD and results are summarized in (Table 7). At g.18202368 A>T, AA individuals had a significantly higher mean litter size than AT individuals ($p < 0.05$). At g.18202372 G>A, AA individuals showed a highly significant increase in mean litter size compared with GG individuals ($p < 0.01$). At g.18202431 G>C, GC individuals had a higher mean litter size than GG individuals ($p < 0.05$). At g.18202472 G>C, CC individuals had a higher mean litter size than GG individuals ($p < 0.05$).

Table 5. Gene frequency and genotype frequency at different loci of the *ID2* gene in Hetian sheep.

Location	Genotype	Genotype Frequency (Sample Size)	Allele	Allele Frequency	X^2 (*p*-Value)
g.18202368 A>T	AA	0.96	A	0.98	0.08 ($p > 0.5$)
	AT	0.04	T	0.02	
g.18202372 G>A	GG	0.59	G	0.77	0.85 ($p > 0.5$)
	GA	0.37	A	0.23	
	AA	0.04			
g.18202426 C>T	CC	0.71	C	0.84	0.37 ($p > 0.5$)
	CT	0.26	T	0.16	
	TT	0.03			
g.18202431 G>C	GG	0.87	G	0.94	0.73 ($p > 0.5$)
	GC	0.13	C	0.06	
g.18202439 C>T	CC	0.93	C	0.96	0.21 ($p > 0.5$)
	CT	0.07	T	0.04	
g.18202451 C>T	CC	0.95	C	0.97	0.11 ($p > 0.5$)
	CT	0.05	T	0.03	
g.18202467 T>A	TT	0.96	T	0.98	0.08 ($p > 0.5$)
	TA	0.04	A	0.02	
g.18202472 G>C	GG	0.58	G	0.77	2.85 ($p > 0.5$)
	GC	0.38	C	0.23	
	CC	0.04			

Table 6. Polymorphic information content, heterozygosity, and number of effective alleles at different loci of the *ID2* gene in Hetian sheep.

Location	Genetic Homozygosity	Genetic Heterozygosity	Polymorphic Information Content	Effective Number of Alleles
g.18202368 A>T	0.96	0.43	0.42	1.04
g.18202372 G>A	0.65	0.35	0.29	1.54
g.18202426 C>T	0.73	0.27	0.23	1.37
g.18202431 G>C	0.88	0.12	0.11	0.13
g.18202439 C>T	0.93	0.68	0.65	1.07
g.18202451 C>T	0.95	0.50	0.50	1.05
g.18202467 T>A	0.96	0.04	0.04	1.04
g.18202472 G>C	0.65	0.35	0.29	1.53

Table 7. Association analysis between *ID2* gene SNP loci and lamb size in Hetian sheep.

Location	Genotype	Sample Size	Number of Lambs Litter Size
g.18202368 A>T	AA	150	1.52 ± 0.501^{a}
	AT	7	1.14 ± 0.378^{b}
g.18202372 G>A	GG	92	1.27 ± 0.447^{A}
	GA	59	1.83 ± 0.378^{B}
	AA	6	1.83 ± 0.408^{B}
g.18202426 C>T	CC	112	1.47 ± 0.501
	CT	40	1.56 ± 0.502
	TT	5	1.8 ± 0.447
g.18202431 G>C	GG	137	1.47 ± 0.501^{a}
	GC	20	1.75 ± 0.444^{b}

Table 7. *Cont.*

Location	Genotype	Sample Size	Number of Lambs Litter Size
g.18202439 C>T	CC	146	1.49 ± 0.502
	CT	11	1.73 ± 0.467
g.18202451 C>T	CC	149	1.49 ± 0.502
	CT	8	1.75 ± 0.463
g.18202467 T>A	TT	150	1.51 ± 0.501
	TA	7	1.29 ± 0.488
g.18202472 G>C	GG	92	1.41 ± 0.495 [a]
	GC	59	1.61 ± 0.492 [b]
	CC	6	1.83 ± 0.408 [b]

Note: If the sample size is less than three, no analysis will be conducted; In the same column of data, different uppercase letters on the shoulder indicate significant differences ($p < 0.01$), different lowercase letters on the shoulder indicate significant differences ($p < 0.05$), and the same or no letters on the shoulder indicate no significant differences ($p > 0.05$).

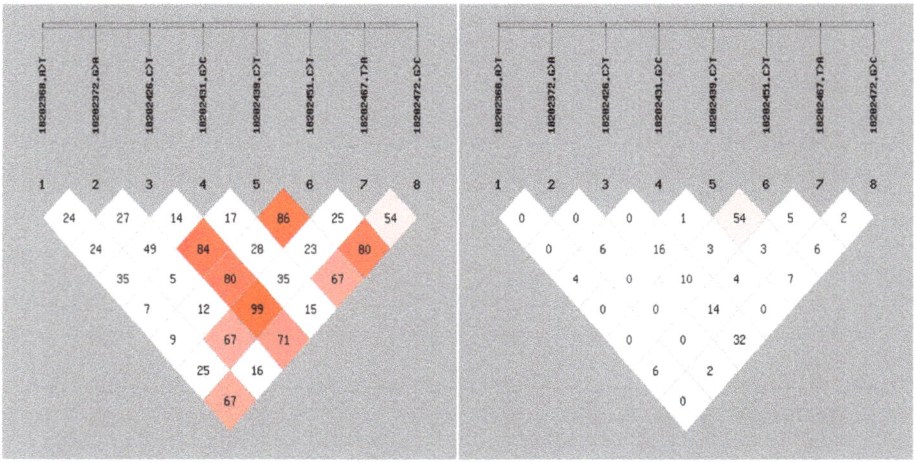

Figure 5. Linkage disequilibrium analysis of SNP mutation sites in *ID2* gene.

3.9. Identification of Ovarian GCs

The purity of the cells exceeds 90%, indicating that they are ovarian GCs. To confirm that the isolated cells were GCs, immunofluorescent staining was performed to detect FSHR expression. It was found that the GCs isolated from the ovaries of Hetian sheep specifically expressed FSHR, as shown in (Figure 6). The proportion of positive cells exceeded 95% (Figure 6C), indicating that the isolated cells were GCs.

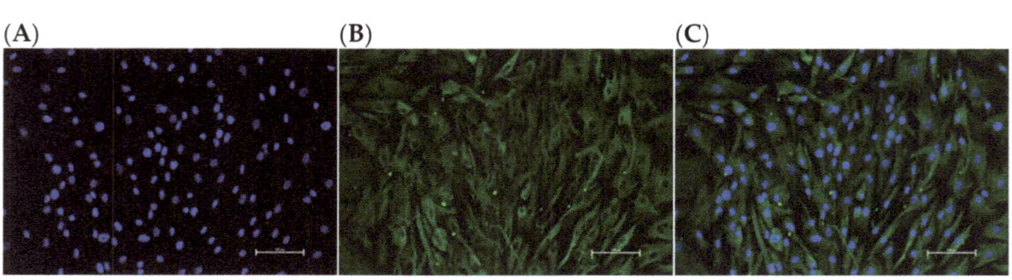

Figure 6. Identification of ovarian GCs in Hetian sheep. (**A**) DAPI-stained nuclei (blue); (**B**) FSHR-specific signal (green); (**C**) merged image. Positive cytoplasmic FSHR confirms GCs' identity. Magnification: 200×.

3.10. ID2 Overexpression Enhances Proliferation of Ovarian GCs

The *ID2* gene was measured at 48 h and 72 h. Ovarian GCs showed a large amount of green fluorescent protein under the excitation of the blue light source of the fluorescence microscope Figure 7A,B. The expression of the *ID2* gene increased with the increase in transfection time. After *ID2* overexpression, the results of qRT-PCR showed that the mRNA level of the ID2 gene transfection group at 48 h and 72 h transfection was significantly higher than that of the blank group ($p < 0.01$), and the expression level gradually increased with time Figure 7C. The CCK-8 results are shown in Figure 7D. As time increases, it can be seen that the overexpression of the *ID2* gene has a certain promoting effect on the proliferation of ovarian GCs ($p < 0.01$), which preliminarily proves that the *ID2* gene can promote the proliferation of ovarian GCs.

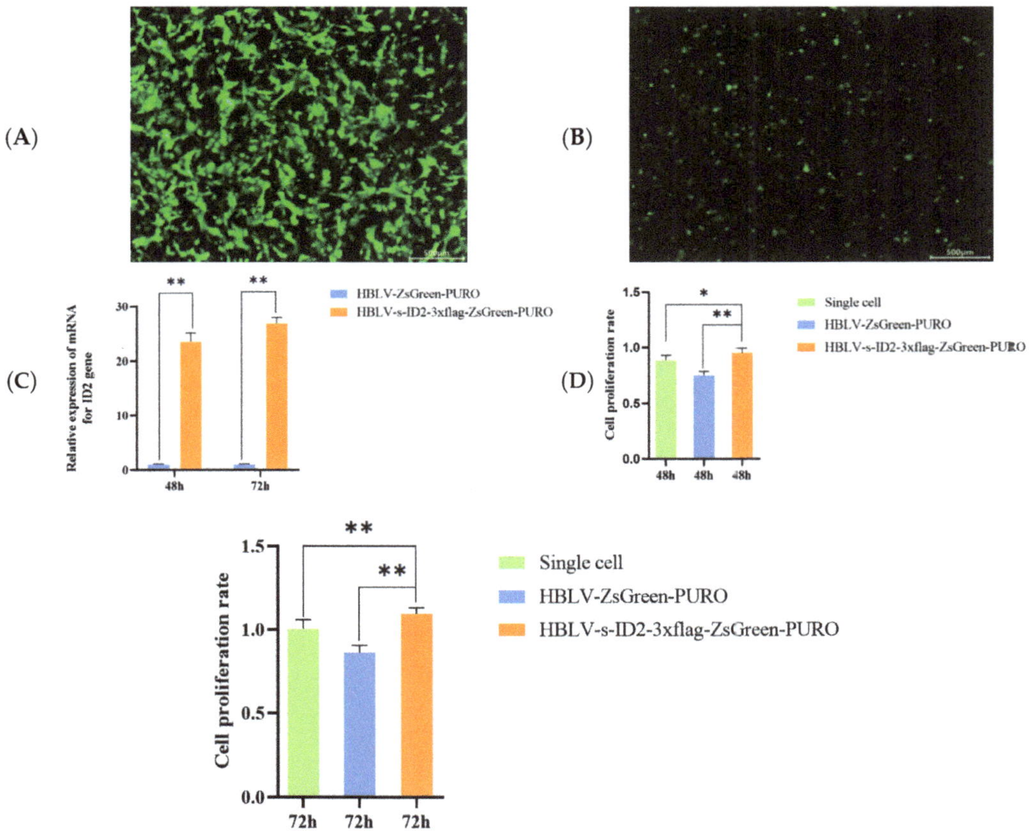

Figure 7. *ID2* is overexpressed in sheep ovarian GCs (scale bar: 500 μm). (**A**) Fluorescence image of GCs transfected with the *ID2* overexpression construct. (**B**) Fluorescence image of the negative control from empty vector transfection. (**C**) mRNA overexpression of *ID2* gene in ovarian GCs. (**D**) Effect of overexpression of *ID2* gene on proliferation of ovarian GCs. indicates extremely significant difference ($p < 0.01$), indicates significant difference ($p < 0.05$). * indicates $p < 0.05$, ** indicates $p < 0.01$.

3.11. Effects of ID2 Overexpression on P4 and E2 Hormone Production

To evaluate the effect of *ID2* on reproductive hormones, the expression levels of P4 and E2 were studied. The results are shown in (Figure 8). The effect of *ID2* overexpression on P4 concentration was significantly higher than that of the empty load group at 48 h and 72 h ($p < 0.01$); The effect of ID2 overexpression on E2 concentration was significantly lower than that of the empty load group at 48 h ($p < 0.05$), and significantly lower than that of the empty load group at 72 h ($p < 0.01$).

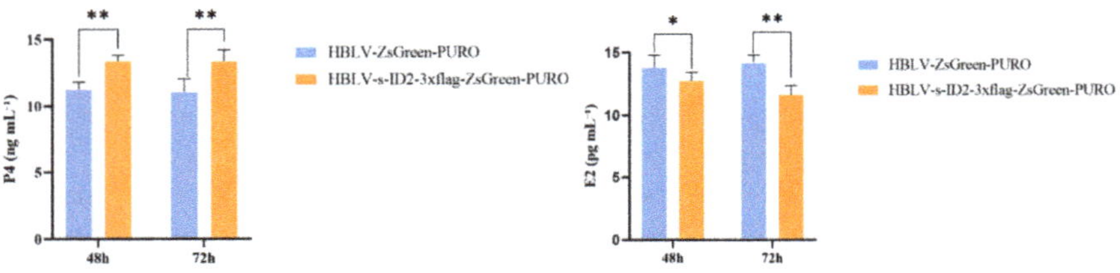

Figure 8. Effect of *ID2* overexpression on E2 and P4 concentrations. * indicates $p < 0.05$, ** indicates $p < 0.01$.

3.12. *Effects of ID2 Gene Overexpression on Reproduction-Related Genes*

ID2 gene overexpression was collected for 48 h and 72 h, and RNA was extracted and reverse transcribed to synthesize c DNA. The relative expression levels were measured by qRT-PCR using the same signaling pathway genes of *ID2* (*BMPR2*, *RBX1*, *SMAD7*) and the *ID2*-related genes (*ID1*, *ID3*, *NFIL3*, *TCF3*) found by STRING. The results are shown in (Figure 9). The relative expression levels of *BMPR2*, *RBX1*, and *SMAD7* genes in the empty group were significantly higher than those in the overexpression group ($p < 0.01$), the *BMPR2* overexpression group at 48 h was significantly lower than that at 72 h ($p < 0.01$) (Figure 9A), the *RBX1* overexpression group was significantly higher than that at 48 h ($p < 0.01$) (Figure 9B), and the *SMAD7* overexpression group had no significant changes at 48 and 72 h (Figure 9C). The relative expression level *of ID1* in the overexpression group was significantly higher than that in the empty load group ($p < 0.01$) (Figure 9D), the relative expression level of *ID3* in the overexpression group was significantly lower than that in the empty load group at 48 h ($p < 0.01$) (Figure 9E), and the relative expression level of *NFIL3* in the overexpression group was significantly higher than that in the empty load group ($p < 0.01$) (Figure 9F), and the relative expression level of *TCF3* in the overexpression group was significantly lower than that in the empty load group (Figure 9G).

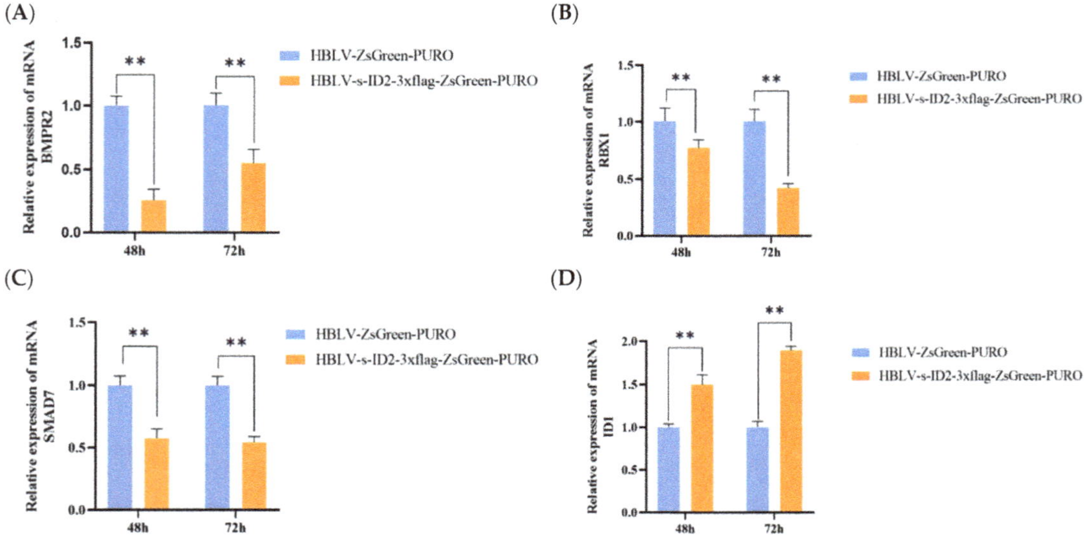

Figure 9. *Cont.*

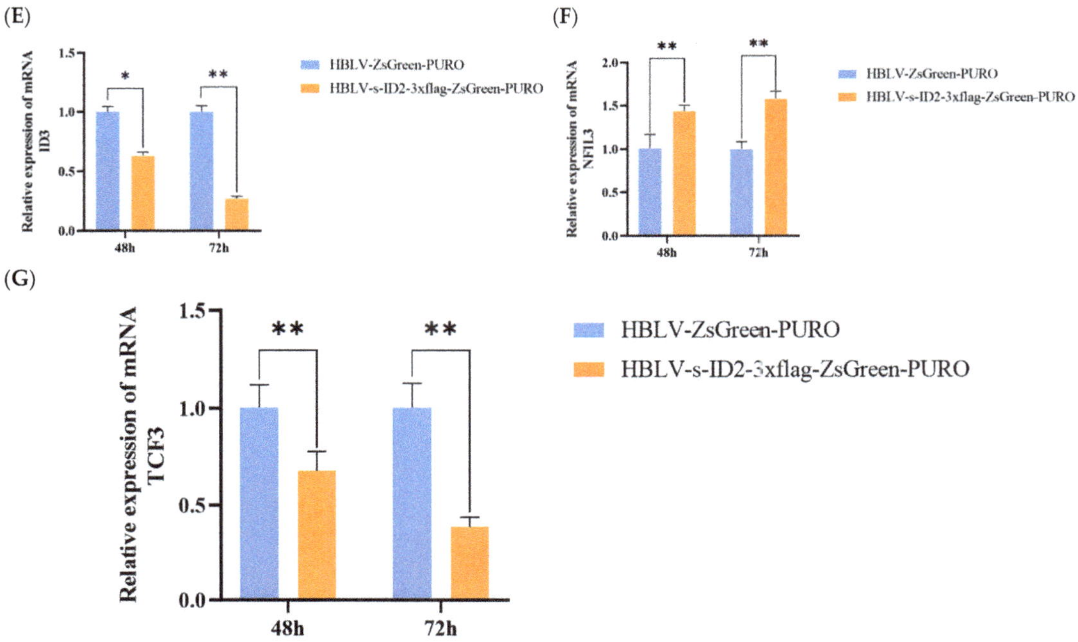

Figure 9. Effects of *ID2* overexpression on the relative mRNA expression of *BMPR2*, *RBX1*, *SMAD7*, *ID1*, *ID3*, *NFKL3*, and *TCF3* in GCs at 48 h and 72 h post-transfection. (**A**) represents the mRNA expression level of *BMPR2* at 48 h and 72 h after *ID2* overexpression. (**B**) represents the mRNA expression level of *RBX1* at 48 h and 72 h after *ID2* overexpression. (**C**) represents the mRNA expression level of *SMAD7* at 48 h and 72 h after *ID2* overexpression. (**D**) represents the mRNA expression level of *ID1* at 48 h and 72 h after *ID2* overexpression. (**E**) represents the mRNA expression level of *ID3* at 48 h and 72 h after *ID2* overexpression. (**F**) represents the mRNA expression level of *NFKL3* at 48 h and 72 h after *ID2* overexpression. (**G**) represents the mRNA expression level of *TCF3* at 48 h and 72 h after *ID2* overexpression. Bars represent mean ± SEM (n = 3 biological replicates). * indicates $p < 0.05$, ** indicates $p < 0.01$ (Independent Samples t-test).

4. Discussion

In this study, we cloned and characterized the coding sequence (CDS) of the *ID2* gene in Hetian sheep and examined its role in ovarian GCs function and reproductive performance. The high homology of ovine *ID2* with its orthologs in sheep, goats, and cattle underscores strong evolutionary conservation and supports a functionally relevant role in mammalian reproductive physiology. These findings provide new insight into molecular mechanisms underlying follicular development and fertility in sheep.

Amino-acid composition analysis showed that the Hetian sheep ID2 protein contains relatively high proportions of serine and leucine. Serine is polar, whereas leucine is hydrophobic; nonetheless, the overall bioinformatics profile predicts a hydrophilic protein, consistent with the distribution of residues and domain architecture. Prior work indicates that serine availability can promote oocyte growth and maturation, with cumulus cells supplying serine to enclosed oocytes via gap junctions [25,26]. In parallel, leucine-rich repeat signaling components such as LGR5 are expressed in the ovary and have been implicated in successful pregnancy [27]. While these observations do not imply a direct causal link between ID2 residue frequencies and ovarian signaling, they are congruent with a broader role for amino-acid–sensitive and leucine-rich pathways in reproductive biology.

RT-qPCR demonstrated tissue-specific expression of *ID2* in Hetian sheep, with the highest levels in the ovary and the lowest in the pituitary. This differential pattern is consistent with a role in ovarian development and maturation, a conclusion further supported by the effects of *ID2* on GCs viability and proliferation. *ID2* expression was also relatively high in the uterus, suggesting additional functions in female reproductive processes.

We investigated associations between *ID2* polymorphisms and litter size in 157 Hetian ewes using PCR and direct sequencing. Several single-nucleotide polymorphisms (SNPs) showed significant associations with litter size, notably g.18202368 A>T, g.18202372 G>A, g.18202431 G>C, and g.18202472 G>C. These results suggest that specific *ID2* variants are linked to fecundity in Hetian sheep. Given the moderate sample size (n = 157 ewes) in the present genotyping analysis, the possibility of sampling error cannot be excluded. Genetic association studies indicate that smaller sample sizes reduce statistical power and may increase the risk of both false negatives and less reliable effect-size estimates. Accordingly, validation of the identified polymorphisms in larger, and independent sheep populations will be necessary to determine whether these SNPs can serve as reliable adjunct markers for selection and to clarify the relationship between allele frequencies and reproductive performance.

The ovary not only supports folliculogenesis and ovulation but also regulates the estrous cycle and follicular homeostasis through steroid hormones such as E2 and P4, thereby shaping reproductive outcomes. During follicle growth, GCs are essential for maintaining the intrafollicular microenvironment by coordinating the secretion of hormones and paracrine factors. Transforming growth factor-β (TGF-β) is a pleiotropic cytokine that controls diverse cellular processes [28]. TGF-β family ligands are intimately involved in ovulation, fertilization, and the establishment and maintenance of pregnancy; even early reproductive events—including male and female germline specification—are regulated by TGF-β–related proteins [29]. Inhibitory SMADs (SMAD6/7) act as intracellular antagonists. In particular, SMAD7 serves as a critical negative regulator of TGF-β/BMP signaling by competing with receptor-regulated SMADs, thereby influencing oocyte–somatic-cell communication and GCs function [30]. Manipulating SMAD7 expression alters follicular development in mice [31], and SMAD7 has been implicated in TGF-β–induced apoptosis, affecting folliculogenesis. Moreover, SMAD7-mediated signaling can attenuate TGF-β dependence on *ID2* [32]. Because the Id gene family comprises major downstream targets of BMP/SMAD signaling, decreases in *ID2* expression may suppress TGF-β activity. Consistent with this, Juliano et al. [33] identified *ID2* as a predicted target of exosomal miRNAs regulating TGF-β/BMP components in mares; *ID2* was detected in GCs and in follicular-fluid exosomes at mid-estrus and preovulatory stages. In addition, BMPR2, a type II TGF-β receptor with serine/threonine kinase activity, has been reported to regulate germ-cell activity and to activate *ID2* [34]. BMPR2 supports ovarian development and GCs proliferation [35] and plays essential roles in follicular growth and function across mammals and birds [36]. Beyond the TGF-β superfamily, *ID2* may crosstalk with PI3K/AKT pathways that transduce gonadotropin and growth-factor signals to promote GCs proliferation, inhibit apoptosis, and support oocyte maturation [37,38]. Although prior studies have connected *ID2* to TGF-β superfamily signaling in follicular regulation, its specific functions in Hetian-sheep GCs have been scarcely explored, motivating the present work.

We constructed a lentiviral *ID2* overexpression vector and transduced primary ovine GCs. *ID2* mRNA and protein levels were markedly elevated at 48 h and 72 h relative to empty-vector controls, accompanied by increased cell proliferation, consistent with *ID2*'s mitogenic potential in follicular growth and maturation. ELISA assays showed higher P4 and lower E2 following *ID2* overexpression, indicating involvement in luteinization and corpus luteum function [39]. These hormone shifts align with evidence that steroidogenic profiles during folliculogenesis are tightly controlled by transcriptional and signaling networks, including the TGF-β superfamily [40]. CCK-8 assays further confirmed significantly greater proliferation in the *ID2* group at both time points versus controls.

To probe pathway effects, we quantified transcripts of *ID1, ID3, NFIL3, TCF3, BMPR2, RBX1*, and *SMAD7* in GCs. *ID2* overexpression significantly upregulated *ID1* and *NFIL3* and downregulated *ID3* and *TCF3*. Within the TGF-β axis, *BMPR2, RBX1*, and *SMAD7* were all reduced. Kowanetz et al. [41] reported that chronic TGF-β exposure suppresses *ID2/ID3*, whereas BMP-7 induces them in U-2 OS cells, illustrating a bidirectional responsiveness of *ID2/ID3* within the TGF-β/BMP network. Our results are consistent with *ID2* functioning not only downstream of BMP signaling but also feeding back on the TGF-β/BMP pathway. As an I-Smad, SMAD7 antagonizes TGF-β/BMP signaling by competing for receptor binding and recruiting SMURF1/2 to promote R-Smad degradation, establishing a negative feedback loop [42]. In a chondrogenesis model, ID2 overexpression enhanced BMP signaling by suppressing SMAD7, regulating postnatal cartilage formation—mechanistically echoing our observation of SMAD7 downregulation with ID2 overexpression [43]. Notably, RBX1, a core subunit of SCF-type E3 ubiquitin ligases that can target Smad proteins for degradation, was also decreased, further suggesting that ID2 influences the proteostatic control of Smad signaling components. Finally, the observed upregulation of ID1 and NFIL3 agrees with reports that NFIL3 can directly regulate ID2 expression [44], pointing to an interconnected transcriptional circuit.

In summary, *ID2* overexpression modulates GCs fate through TGF-β superfamily–associated pathways: on one hand, suppression of *BMPR2* and *SMAD7* is consistent with enhanced proliferation and luteinization; on the other, coordinated changes in downstream transcription factors—*ID1/ID3, TCF3*, and *NFIL3*—reshape gene-expression programs that support follicular growth and corpus luteum function.

5. Conclusions

In this study, we successfully cloned the *ID2* gene in Hetian sheep, characterized its protein features, and examined its expression across gonadal tissues. The results showed that *ID2* was highly expressed in ovarian tissues, especially during the pubertal stage. We identified four SNP loci in *ID2* (g.18202368, g.18202372, g.18202431, g.18202472) that are significantly associated with lambing number. Furthermore, in vitro overexpression experiments showed that *ID2* promotes the proliferation of sheep granulosa cells, enhances steroid hormone secretion, and likely acts via modulation of the TGF-β and BMP/SMAD signaling pathways. In conclusion, this study provides a new theoretical basis for understanding the proliferation of sheep GCs and the regulation of lambing performance.

Author Contributions: F.X. was responsible for the conceptualization, supervision, project administration, and funding acquisition. H.S. and X.W. designed and performed the methodology, while H.S., X.W., G.M.S., R.G. and L.Z. carried out the experimental investigation. Data curation and analysis were performed by W.L., Q.H., X.W. and H.S. The original draft of the manuscript was prepared by H.S. and L.Z., and critical review and editing were conducted by F.X. and G.M.S. All authors have read and agreed to the published version of the manuscript.

Funding: This study was funded by the National Natural Science Foundation of China (Grant/Award Number: 31960655), the President's Fund Project of Tarim University (Grant/Award Number: TDZKCX202401), and the Major Science and Technology Special Project of Xinjiang Uygur Autonomous Region (Grant/Award Number: 2024A02004-3).

Institutional Review Board Statement: For experimental animals, all protocols were performed in accordance with the "Guide to Animal Experimentation" and approved by the Use Committee under the norms of the Ethics Committee of Tarim University of Science and Technology (SYXK 2020-009).

Informed Consent Statement: Not applicable.

Data Availability Statement: The raw data supporting the conclusions of this article will be made available by the authors on request.

Acknowledgments: The authors thank the staff of the Experimental Station of Tarim University for their assistance with animal care and sample collection.

Conflicts of Interest: The authors declare no conflicts of interest.

References

1. Hasimu, A.; Maitusong, A.; Ailijiang, M. Development trends of Hetian sheep in Yutian County: A survey and analysis. *Xinjiang Anim. Husb.* **2013**, *190* (Suppl. S2), 20–21. (In Chinese)
2. Ajafar, M.H.; Kadhim, A.H.; Al-Thuwaini, T.M. The reproductive traits of sheep and their influencing factors. *Rev. Agric. Sci.* **2022**, *10*, 82–89. [CrossRef]
3. Al-Jaryan, I.L.; Al-Thuwaini, T.M.; Merzah, L.H.; Alkhammas, A.H. Reproductive physiology and advanced technologies in sheep reproduction. *Rev. Agric. Sci.* **2023**, *11*, 171–180. [CrossRef]
4. Notter, D.R. Genetic improvement of reproductive efficiency of sheep and goats. *Anim. Reprod. Sci.* **2012**, *130*, 147–151. [CrossRef]
5. Tai, Y.; Yang, X.; Han, D.; Xu, Z.; Cai, G.; Hao, J.; Zhang, B.; Deng, X. Transcriptomic diversification of granulosa cells during follicular development between White Leghorn and Silky Fowl hens. *Front. Genet.* **2022**, *13*, 965414. [CrossRef] [PubMed]
6. Hogg, K.; Etherington, S.L.; Young, J.M.; McNeilly, A.S.; Duncan, W.C. Inhibitor of differentiation (Id) genes are expressed in the steroidogenic cells of the ovine ovary and are differentially regulated by members of the transforming growth factor-beta family. *Endocrinology* **2010**, *151*, 1247–1256. [CrossRef] [PubMed]
7. Rajesh, G.; Mishra, S.R.; Paul, A.; Punetha, M.; Vidyalakshmi, G.; Narayanan, K.; Bag, S.; Bhure, S.; Chouhan, V.S.; Maurya, V.; et al. Transcriptional and translational abundance of BMP2, 4, 6, 7 and their receptors BMPR1A, 1B and BMPR2 in buffalo ovarian follicle and the role of BMP4 and BMP7 on estrogen production and survival of cultured granulosa cells. *Res. Vet. Sci.* **2018**, *118*, 371–388. [CrossRef] [PubMed]
8. Mamsen, L.S.; Bøtkjær, J.A.; Kristensen, S.G.; Pors, S.E.; Jeppesen, J.V.; Kumar, A.; Kalra, B.; Ernst, E.; Andersen, C.Y. High variability of molecular isoforms of AMH in follicular fluid and granulosa cells from human small antral follicles. *Front. Endocrinol.* **2021**, *12*, 617523. [CrossRef]
9. He, J.; Ao, C.; Li, M.; Deng, T.; Zheng, S.; Zhang, K.; Tu, C.; Ouyang, Y.; Lang, R.; Jiang, Y.; et al. Clusterin-carrying extracellular vesicles derived from human umbilical cord mesenchymal stem cells restore the ovarian function of premature ovarian failure mice through activating the PI3K/AKT pathway. *Stem Cell Res. Ther.* **2024**, *15*, 300. [CrossRef]
10. Kumar, S.; Rajput, P.K.; Bahire, S.V.; Jyotsana, B.; Kumar, V.; Kumar, D. Differential expression of BMP/SMAD signaling and ovarian-associated genes in the granulosa cells of FecB introgressed GMM sheep. *Syst. Biol. Reprod. Med.* **2020**, *66*, 185–201. [CrossRef]
11. Loos, B.; Salas-Bastos, A.; Nordin, A.; Debbache, J.; Stierli, S.; Cheng, P.F.; Rufli, S.; Wyss, C.; Levesque, M.P.; Dummer, R.; et al. TGFβ signaling sensitizes MEKi-resistant human melanoma to targeted therapy-induced apoptosis. *Cell Death Dis.* **2024**, *15*, 925. [CrossRef]
12. Meng, Z.; Li, T.; Li, J.; Ding, S.; Liu, Y.; Zhao, G.; Chen, C.; Zhao, P.; Zhou, L. LncRNAPVT1 is associated with cancer-associated fibroblast proliferation through regulating TGF-β in oral squamous cell carcinoma. *Immunol. Investig.* **2024**, *53*, 1250–1263. [CrossRef]
13. Du, X.; Zhang, L.; Li, X.; Pan, Z.; Liu, H.; Li, Q. TGF-β signaling controls FSHR signaling-reduced ovarian granulosa cell apoptosis through the SMAD4/miR-143 axis. *Cell Death Dis.* **2016**, *7*, e2476. [CrossRef]
14. Scobey, M.J.; Fix, C.A.; Walker, W.H. The Id2 transcriptional repressor is induced by follicle-stimulating hormone and cAMP. *J. Biol. Chem.* **2004**, *279*, 16064–16070. [CrossRef]
15. Chaudhary, J.; Johnson, J.; Kim, G.; Skinner, M.K. Hormonal regulation and differential actions of the helix-loop-helix transcriptional inhibitors Id1, Id2, Id3, and Id4 in Sertoli cells. *Endocrinology* **2001**, *142*, 1727–1736. [CrossRef]
16. Guirgis, F.W.; Jacob, V.; Wu, D.; Henson, M.; Daly-Crews, K.B.; Hopson, C.; Black, L.P.; DeVos, E.L.; Sulaiman, D.; Labilloy, G.; et al. DHCR7 expression predicts poor outcomes and mortality from sepsis. *Crit. Care Explor.* **2023**, *5*, e0929. [CrossRef] [PubMed]
17. Xu, X.; Jiang, H.; Wang, D.; Rehman, S.U.; Li, Z.; Song, X.; Cui, K.; Luo, X.; Yang, C.; Liu, Q. Exploration of transcriptional regulation network between buffalo oocytes and granulosa cells and its impact on different diameter follicles. *BMC Genom.* **2024**, *25*, 1004. [CrossRef]
18. Chermuła, B.; Brązert, M.; Jeseta, M.; Ożegowska, K.; Sujka-Kordowska, P.; Konwerska, A.; Bryja, A.; Kranc, W.; Jankowski, M.; Nawrocki, M.J.; et al. The unique mechanisms of cellular proliferation, migration and apoptosis are regulated through oocyte maturational development—A complete transcriptomic and histochemical study. *Int. J. Mol. Sci.* **2018**, *20*, 84. [CrossRef] [PubMed]

9. Mori, S.; Nishikawa, S.I.; Yokota, Y. Lactation defect in mice lacking the helix-loop-helix inhibitor Id2. *EMBO J.* **2000**, *19*, 5772–5781. [CrossRef]
0. He, X.; Wang, W.; Du, X.; Chu, M. Association between single-nucleotide polymorphism in PIK3CD gene and litter size in Small Tail Han sheep. *Anim. Biotechnol.* **2023**, *34*, 3337–3342. [CrossRef] [PubMed]
1. Zhu, M.; Zhang, H.; Yang, H.; Zhao, Z.; Blair, H.T.; Zhai, M.; Yu, Q.; Wu, P.; Fang, C.; Xie, M. Polymorphisms and association of GRM1, GNAQ and HCRTR1 genes with seasonal reproduction and litter size in three sheep breeds. *Reprod. Domest. Anim.* **2022**, *57*, 532–540. [CrossRef]
2. Zhang, Z.; Liu, Q.; Di, R.; Hu, W.; Wang, X.; He, X.; Ma, L.; Chu, M. Single nucleotide polymorphisms in BMP2 and BMP7 and the association with litter size in Small Tail Han sheep. *Anim. Reprod. Sci.* **2019**, *204*, 183–192. [CrossRef] [PubMed]
3. Wang, H.; Feng, X.; Muhatai, G.; Wang, L. Expression profile analysis of sheep ovary after superovulation and estrus synchronisation treatment. *Vet. Med. Sci.* **2022**, *8*, 1276–1287. [CrossRef]
4. Vasu, M.; Ahlawat, S.; Choudhary, V.; Kaur, R.; Arora, R.; Sharma, R.; Sharma, U.; Chhabra, P.; Mir, M.; Singh, M.K. Identification and validation of stable reference genes for expression profiling of target genes in diverse ovine tissues. *Gene* **2024**, *897*, 148067. [CrossRef]
5. Hsueh, A.J.; Kawamura, K.; Cheng, Y.; Fauser, B.C. Intraovarian control of early folliculogenesis. *Endocr. Rev.* **2015**, *36*, 1–24. [CrossRef]
6. Zhang, H.; McClatchie, T.; Baltz, J.M. L-Serine transport in growing and maturing mouse oocytes. *J. Cell. Physiol.* **2020**, *235*, 8585–8600. [CrossRef]
7. Sun, X.; Terakawa, J.; Clevers, H.; Barker, N.; Daikoku, T.; Dey, S.K. Ovarian LGR5 is critical for successful pregnancy. *FASEB J.* **2014**, *28*, 2380–2389. [CrossRef]
8. Deshpande, G.; Willis, E.; Chatterjee, S.; Fernandez, R.; Dias, K.; Schedl, P. BMP signaling and the maintenance of primordial germ cell identity in *Drosophila* embryos. *PLoS ONE* **2014**, *9*, e88847. [CrossRef] [PubMed]
9. Guo, Y.; Liu, M.; Mfoundou, J.D.L.; Wang, X. Expression and distribution patterns of VEGF, TGF-β1 and HIF-1α in the ovarian follicles of Tibetan sheep. *Vet. Med. Sci.* **2022**, *8*, 2223–2229. [CrossRef]
0. Quezada, M.; Wang, J.; Hoang, V.; McGee, E.A. Smad7 is a transforming growth factor-beta–inducible mediator of apoptosis in granulosa cells. *Fertil. Steril.* **2012**, *97*, 1452–1459. [CrossRef] [PubMed]
1. Gao, Y.; Wen, H.; Wang, C.; Li, Q. SMAD7 antagonizes key TGFβ superfamily signaling in mouse granulosa cells in vitro. *Reproduction* **2013**, *146*, 1–11. [CrossRef]
2. Lian, G.Y.; Wang, Q.M.; Tang, P.M.; Zhou, S.; Huang, X.R.; Lan, H.Y. Combination of Asiatic Acid and Naringenin Modulates NK Cell Anti-cancer Immunity by Rebalancing Smad3/Smad7 Signaling. *Mol. Ther.* **2018**, *26*, 2255–2266. [CrossRef]
3. da Silveira, J.C.; Carnevale, E.M.; A Winger, Q.; Bouma, G.J. Regulation of ACVR1 and ID2 by cell-secreted exosomes during follicle maturation in the mare. *Reprod. Biol. Endocrinol.* **2014**, *12*, 44. [CrossRef]
4. Yang, S.; Yuan, Q.; Niu, M.; Hou, J.; Zhu, Z.; Sun, M.; Li, Z.; He, Z. BMP4 promotes mouse iPS cell differentiation to male germ cells via Smad1/5, Gata4, Id1 and Id2. *Reproduction* **2017**, *153*, 211–220. [CrossRef]
5. Abir, R.; Ben-Haroush, A.; Melamed, N.; Felz, C.; Krissi, H.; Fisch, B. Expression of bone morphogenetic proteins 4 and 7 and their receptors IA, IB, and II in human ovaries from fetuses and adults. *Fertil. Steril.* **2008**, *89*, 1430–1440. [CrossRef]
6. Stephens, C.S.; Johnson, P.A. Bone morphogenetic protein 15 may promote follicle selection in the hen. *Gen. Comp. Endocrinol.* **2016**, *235*, 170–176. [CrossRef]
7. Kalous, J.; Aleshkina, D.; Anger, M. A role of PI3K/Akt signaling in oocyte maturation and early embryo development. *Cells* **2023**, *12*, 1830. [CrossRef] [PubMed]
8. Baddela, V.S.; Michaelis, M.; Tao, X.; Koczan, D.; Brenmoehl, J.; Vanselow, J. Comparative analysis of PI3K-AKT and MEK-ERK1/2 signaling-driven molecular changes in granulosa cells. *Reproduction* **2025**, *169*, e240317. [CrossRef]
9. Baumgarten, S.C.; Convissar, S.M.; Zamah, A.M.; Fierro, M.A.; Winston, N.J.; Scoccia, B.; Stocco, C. FSH regulates IGF-2 expression in human granulosa cells in an AKT-dependent manner. *J. Clin. Endocrinol. Metab.* **2015**, *100*, E1046–E1055. [CrossRef] [PubMed]
0. Zhu, G.; Mao, Y.; Zhou, W.; Jiang, Y. Dynamic changes in the follicular transcriptome and promoter DNA methylation pattern of steroidogenic genes in chicken follicles throughout the ovulation cycle. *PLoS ONE* **2015**, *10*, e0146028. [CrossRef] [PubMed]
1. Kowanetz, M.; Valcourt, U.; Bergström, R.; Heldin, C.H.; Moustakas, A. Id2 and Id3 define the potency of cell proliferation and differentiation responses to transforming growth factor beta and bone morphogenetic protein. *Mol. Cell. Biol.* **2004**, *24*, 4241–4254. [CrossRef] [PubMed]
2. Miyazawa, K.; Miyazono, K. Regulation of TGF-β family signaling by inhibitory Smads. *Cold Spring Harb. Perspect. Biol.* **2017**, *9*, a022095. [CrossRef] [PubMed]

43. Sakata-Goto, T.; Takahashi, K.; Kiso, H.; Huang, B.; Tsukamoto, H.; Takemoto, M.; Hayashi, T.; Sugai, M.; Nakamura, T.; Yokota, Y.; et al. Id2 controls chondrogenesis acting downstream of BMP signaling during maxillary morphogenesis. *Bone* **2012**, *50*, 69–78. [CrossRef]
44. Male, V.; Nisoli, I.; Kostrzewski, T.; Allan, D.S.; Carlyle, J.R.; Lord, G.M.; Wack, A.; Brady, H.J. The transcription factor E4bp4/Nfil3 controls commitment to the NK lineage and directly regulates Eomes and Id2 expression. *J. Exp. Med.* **2014**, *211*, 635–642. [CrossRef] [PubMed]

International Journal of
Molecular Sciences

MDPI

Article

Genome-Wide Association Studies of Growth and Carcass Traits in Charolais Cattle Based on High-Coverage Whole-Genome Resequencing

Feng Zhang [1], Chengmei Wang [2], Aishao Shangguan [1], Xiaojun Suo [1], Mengjie Chen [1], Hu Tao [1], Fan Jiang [1], Tian Xu [1], Nian Zhang [1], Zaidong Hua [1], Jin Chai [2,*] and Qi Xiong [1,*]

[1] Hubei Key Laboratory of Animal Embryo Engineering and Molecular Breeding, Institute of Animal Husbandry and Veterinary, Hubei Academy of Agricultural Sciences, Wuhan 430064, China; zhangfeng0130@163.com (F.Z.); aishao_shangguan@163.com (A.S.); suoxj@126.com (X.S.); mengjiechen2020@163.com (M.C.); taohu@hbaas.com (H.T.); fjiang_hzau@163.com (F.J.); xutian@hbaas.ac.cn (T.X.); zn991830@163.com (N.Z.); zaidonghua@163.com (Z.H.)

[2] Key Laboratory of Swine Genetics and Breeding of the Agricultural Ministry, Key Laboratory of Agricultural Animal Genetics, College of Animal Science and Technology, Huazhong Agricultural University, Wuhan 430070, China; lwchengmei@163.com

* Correspondence: chaijin@mail.hzau.edu.cn (J.C.); xiongqi777@hbaas.com (Q.X.)

Abstract

Growth and carcass traits are key economic traits in beef cattle production, and identifying their associated genetic markers is crucial for improving breeding efficiency. Charolais cattle, as a superior beef breed, exhibit excellent performance in growth rate and meat production. The aim of this study was to utilize the preferred high-coverage whole-genome resequencing (hcWGS) as a replacement for single nucleotide polymorphism (SNP) chips to identify significant SNPs and candidate genes associated with growth (body weight, body height, cross height, body length, and chest measurement across different growth stages) and carcass traits (live backfat thickness and eye muscle area at 18 months) in 240 Charolais cattle, thereby providing guidance for beef cattle breeding. Through hcWGS (approximately 13× coverage) and quality control, 4,088,633 SNPs were identified and subsequently used for genetic analyses. Through FarmCPU-based genome-wide association studies, 196 potentially significant SNPs associated with growth traits and 29 SNPs with carcass traits were identified. Annotation analyses revealed 353 candidate genes (such as RBM33, KCTD17, PTHLH, RAC2, CHD6, TRDN, WBP1L, TLL2, CH25H, and ST13) linked to growth traits and 26 candidate genes linked to carcass traits (such as CHST11, LRRK2, RIOK2, and INTS10). Additionally, three SNPs (g.86746922C>G, g.54418624G>T, and g.71085551G>A) were validated via polymerase chain reaction–restriction fragment length polymorphism (PCR-RFLP), enabling efficient marker-assisted selection. Furthermore, eight SNPs in the Acyl-CoA oxidase 1 (ACOX1) gene were found to be associated with growth and backfat thickness traits. These findings provide valuable preliminary insights into the genetic mechanisms underlying growth and carcass traits in Charolais cattle, facilitating genome-assisted breeding.

Keywords: growth traits; carcass traits; whole-genome resequencing; GWAS; SNP; candidate genes; Charolais cattle

1. Introduction

Cattle are important livestock species that have provided mankind with meat, milk, skin, labor, etc. However, due to the long reproductive cycle and low reproductive rates of beef cattle, breeding progress remains slow. By leveraging molecular breeding techniques to enable early selection of cattle, the breeding process can be effectively accelerated. Growth and carcass traits are very important in beef cattle breeding and considerably affect resource efficiency and farmers' profits. Moreover, bovine growth and carcass traits generally exhibit moderate to high heritability, making the selection of specialized breeds within breeding programs a crucial strategy for enhancing these traits in cattle populations. Charolais cattle are widely recognized as one of the most prominent large-sized beef cattle breeds globally, and they have also become one of the most popular and geographically widespread cattle breeds worldwide. They are characterized by rapid growth, high meat yield, high dressing percentage, substantial body size, and strong adaptability. Charolais beef is distinguished by its high protein content and exceptional nutritional value, making it a preferred choice for germplasm improvement programs in numerous countries worldwide [1–4].

Genome-wide association studies (GWASs) are an effective and powerful genetic tool for identifying genetic markers associated with traits of interest. Numerous GWASs have been used to investigate growth or/and carcass traits in various beef cattle breeds such as Simmental [5], Chinese Simmental [6], Piemontese [7], U.S. Red Angus [8], Sussex cattle [9], Hawai'i beef cattle [10], Montana Tropical Composite beef cattle [11], Canchim beef cattle [3,12], crossbred beef cattle [13], and multiple beef cattle breeds [14–16]. In Charolais cattle, comparatively few studies about growth or carcass traits have been reported. The growth traits of Charolais cattle in Mexico, including birth weight, weaning weight, yearling weight, preweaning average daily gain, and postweaning average daily gain, were investigated using a 77 K single nucleotide polymorphism (SNP) chip [4]. The important traits of Irish Charolais and Limousin cattle, including carcass weight, carcass conformation, carcass fat, cull-cow weight, live weight, and feed intake, were analyzed by the custom-built International Dairy and Beef SNP chip [2]. The feed efficiency and slaughter traits in Charolais cows of France were investigated by the BovineSNP50 SNP chip [17].

With ongoing advances in high-throughput sequencing technologies and declining costs, whole-genome sequencing (WGS) is becoming the preferred method for studying livestock population evolution and molecular genetic breeding [18–20]. WGS is used to sequence the genomes of different individuals of a species with known gene sequences, and further analyze the variation information in the genomes of individuals by comparing the sequenced sequences with the reference genomes of the species to obtain variability information. Compared to SNP chips, WGS captures most genomic variation and eliminates the need for sequential genetic testing, with higher sensitivity and accuracy of detection. Although GWASs have explored growth or carcass traits in Charolais cattle [2,4,17], these studies were exclusively based on SNP chip genotyping and did not fully cover relevant phenotypic traits. Therefore, it is necessary to perform a more comprehensive analysis of additional production traits in Charolais using WGS approaches.

This study aims to apply the WGS method to screen and identify significant SNPs and candidate genes associated with both growth traits at multiple growth stages and carcass traits, so as to guide beef cattle breeding. First, high-coverage whole-genome sequencing (hcWGS, ~13×) was performed to obtain high-resolution genome-wide genotypes from 240 purebred Charolais cattle. Furthermore, a GWAS was conducted to identify genomic regions influencing growth and carcass traits in this population. Growth traits included body weight (BW), body height (BH), cross height (CH), body length (BL), and chest measurement (CM), which were measured in newborns (NBs) as well as at 3, 6, 12, and 18 months of age.

Carcass traits comprised live backfat thickness (BFT) and eye muscle area (EM) at 18 months. Meanwhile, the annotated candidate genes were analyzed by Gene Ontology (GO) and Kyoto Encyclopedia of Genes and Genomes (KEGG) enrichment. Subsequently, three representative SNP variants were selected for effect analysis, and a polymerase chain reaction–restriction fragment length polymorphism (PCR-RFLP)-based genotyping method was developed. Finally, the association between Acyl-CoA oxidase 1 (ACOX1) gene polymorphism and growth traits in Charolais cattle was analyzed. This study identified multiple statistically significant SNPs and candidate genes related to the aforementioned traits in Chinese Charolais cattle. These findings provide a reference for understanding genetic mechanisms and improving the breeding of growth and carcass traits, while also establishing an important foundation for developing genomic selection technologies of Charolais.

2. Results

2.1. Phenotypic Statistics Analysis

The phenotypic values of BW, BH, CH, BL, and CM measured in newborns and at 3, 6, 12, and 18 months of age, as well as BFT and EM measured at 18 months, are shown in Table 1. The frequency distribution histograms of the phenotypic values for each individual trait are presented in Supplementary Figures S1–S6. These results indicated that all traits followed an approximately normal distribution and were thus suitable for subsequent GWASs.

Table 1. Descriptive statistics of phenotypic values for the Charolais cattle of China.

Traits [a]	Individuals	Mean ± SD [b]	Maximum	Minimum	CV [c]
NB-BW	142	39.99 ± 7.77	57.00	15.00	0.19
NB-BH	142	68.90 ± 5.49	80.00	50.00	0.08
NB-CH	142	74.46 ± 5.88	87.00	54.00	0.08
NB-BL	142	64.25 ± 6.01	78.00	50.00	0.09
NB-CM	142	73.94 ± 6.06	87.00	55.00	0.08
3-BW	139	166.92 ± 26.59	254.00	96.00	0.16
3-BH	139	96.03 ± 4.30	105.00	84.00	0.04
3-CH	139	102.71 ± 4.55	115.00	86.00	0.04
3-BL	139	101.45 ± 6.72	115.00	80.00	0.06
3-CM	139	121.81 ± 9.24	144.00	92.00	0.07
6-BW	135	203.50 ± 29.17	287.00	126.0	0.14
6-BH	135	100.71 ± 4.10	110.00	90.00	0.04
6-CH	135	108.70 ± 5.94	153.00	96.00	0.05
6-BL	135	109.34 ± 5.89	133.00	95.00	0.05
6-CM	135	131.98 ± 6.51	152.00	110.00	0.04
12-BW	115	336.92 ± 53.49	463.00	185.00	0.16
12-BH	115	114.10 ± 4.22	123.00	103.00	0.03
12-CH	115	121.89 ± 4.79	132.00	111.00	0.03
12-BL	115	127.00 ± 6.55	143.00	110.00	0.05
12-CM	115	158.13 ± 9.33	184.00	129.00	0.06
18-BW	100	457.25 ± 75.90	654.00	280.00	0.16
18-BH	100	123.27 ± 4.90	141.00	112.00	0.03
18-CH	100	132.40 ± 4.59	143.00	121.00	4.59
18-BL	100	141.53 ± 7.25	159.00	121.00	0.05

Table 1. *Cont.*

Traits [a]	Individuals	Mean ± SD [b]	Maximum	Minimum	CV [c]
18-CM	100	177.67 ± 10.59	212.00	146.00	0.06
18-BFT	190	50.72 ± 10.91	84.90	2.70	0.21
18-EM	190	76.24 ± 10.23	106.70	50.70	0.13

[a] Traits: NB-BW (kg), body weight of newborn; NB-BH (cm), body height of newborn; NB-CH (cm), cross height of newborn; NB-BL (cm), body length of newborn; NB-CM (cm), chest measurement of newborn; 3-BW (kg), body weight at the age of 3 months; 3-BH (cm), body height at the age of 3 months; 3-CH (cm), cross height at the age of 3 months; 3-BL (cm), body length at the age of 3 months; 3-CM (cm), chest measurement at the age of 3 months; 6-BW (kg), body weight at the age of 6 months; 6-BH (cm), body height at the age of 6 months; 6-CH (cm), cross height at the age of 6 months; 6-BL (cm), body length at the age of 6 months; 6-CM (cm), chest measurement at the age of 6 months; 12-BW (kg), body weight at the age of 12 months; 12-BH (cm), body height at the age of 12 months; 12-CH (cm), cross height at the age of 12 months; 12-BL (cm), body length at the age of 12 months; 12-CM (cm), chest measurement at the age of 12 months; 18-BW (kg), body weight at the age of 18 months; 18-BH (cm), body height at the age of 18 months; 18-CH (cm), cross height at the age of 18 months; 18-BL (cm), body length at the age of 18 months; 18-CM (cm), chest measurement at the age of 18 months; 18-BFT (mm), back fat thickness at the age of 18 months; 18-EM (cm^2), eye muscle area at the age of 18 months. [b] SD, standard deviation. [c] CV, coefficient of variation.

Meanwhile, correlation analyses were conducted to elucidate the relationships among phenotypic traits (BW, BH, CH, BL, and CM). The results revealed that there were significant positive correlations among all measured traits at each age interval ($p < 0.01$) (Supplementary Tables S1–S5). Notably, the correlation between BW and CM was the strongest at all developmental stages. The correlation between BH and CH was the strongest except for 6 months of age. In addition, BL had the weakest correlation with other traits.

2.2. WGS, SNPs Distribution and Principal Component Analysis (PCA)

All 240 samples were subjected to high-depth sequencing and a total of 8711.60 gigabases (Gb) of raw data was obtained after WGS. The average Q20 value, Q30 value, and GC content were 97.80%, 92.69%, and 43.13%, respectively. The average actual depth was 13.06×, which met our expectations. After generating WGS data, quality control procedures were performed, and 8505.66 Gb of filtered clean data were obtained. Clean reads were aligned to Bos taurus reference assembly ARS-UCD1.2; the average mapping rate, mismatch rate, and genome coverage were 99.56%, 0.39%, and 98.33%, respectively. These results demonstrate that the sequencing data exhibit high accuracy and were therefore suitable for subsequent analyses.

After a series of quality control procedures, including filtering and imputation, a total of 4,088,633 high-quality SNPs were retained, spanning chromosomes 1 to 29 (Supplementary Figure S7A). PCA was employed to assess the population structure of Charolais cattle. The analysis revealed moderate population stratification, with the first two principal components (PC1 and PC2) accounting for 11.89% and 10.31% of the total genetic variance, respectively. The relatively continuous distribution of samples in the PCA plot indicated limited population substructure, suggesting the feasibility of subsequent association analyses (Supplementary Figure S7B).

2.3. Genome-Wide Association Study and Candidate Gene Identification

Subsequently, a GWAS for growth and carcass traits was conducted using the FarmCPU model implemented in rMVP. By setting the threshold at a p value of 2.44×10^{-7}, 196 significant SNPs were identified for NB-BW, NB-BH, NB-CH, NB-BL, NB-CM, 3-BH, 3-CH, 3-BL, 3-CM, 6-BW, 6-BH, 6-CH, 6-BL, 6-CM, 12-CH, 12-BL, 18-CH, 18-BL, and 18-CM, and 29 significant SNPs were identified for 18-BFT (Figures 1–4). Specifically, for NB-BW, NB-BH, NB-CH, NB-BL, and NB-CM, 12, 6, 1, 6, and 4 significant SNPs were identified, respectively; for 3-BH, 3-CH, 3-BL, and 3-CM, 7, 11, 3, and 42 significant SNPs were identified, respectively; for 6-BW, 6-BH, 6-CH, 6-BL, and 6-CM, 1, 44, 12, 8, and 7 significant SNPs were identified, respectively; and 2, 1, 12, 3, 14, and 29 significant SNPs were identified for 12-CH, 12-BL, 18-CH, 18-BL, 18-CM, and 18-BFT, respectively. Additionally, candidate genes within 0.1 Mb upstream and downstream of the identified SNPs were scanned. A total of 353 genes (such as RBM33, KCTD17, PTHLH, RAC2, CHD6, TRDN, WBP1L, TLL2, CH25H, and ST13) were identified to be associated with growth traits, along with 26 genes (such as CHST11, LRRK2, RIOK2, and INTS10) tied to 18-BFT (Supplementary Table S6).

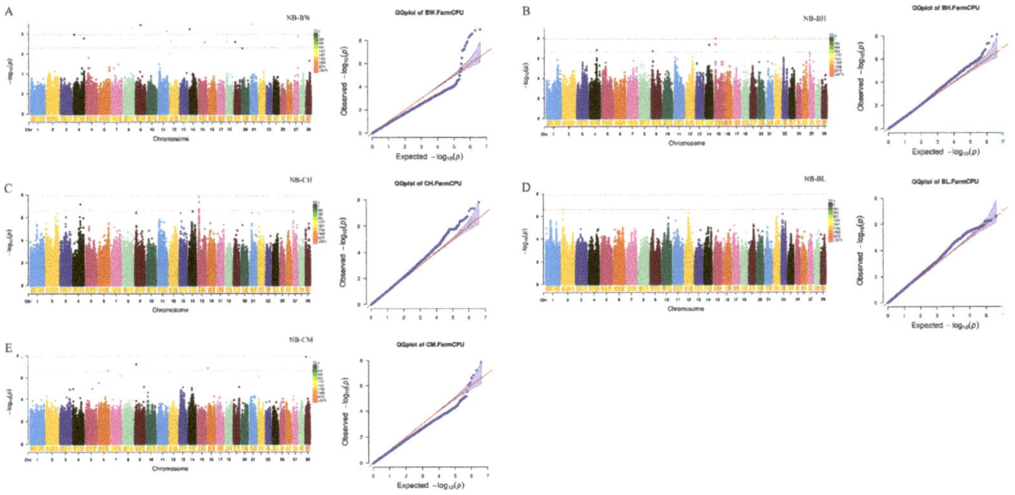

Figure 1. Manhattan and Q-Q plots for growth traits of newborn. (**A–E**) Manhattan and Q-Q plots for NB-BW (body weight of newborn), NB-BH (body height of newborn), NB-CH (cross height of newborn), NB-BL (body length of newborn), NB-CM (chest measurement of newborn), respectively. The two red dashed lines represent genome-level significance (up) and chromosomal significance (down).

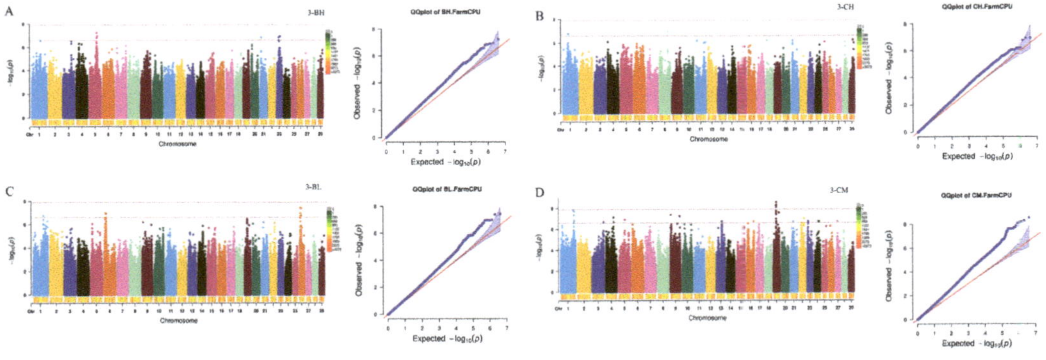

Figure 2. Manhattan and Q-Q plots for growth traits at the age of 3 months. (**A–D**) Manhattan and Q-Q plots for 3-BH (body height at the age of 3 months), 3-CH (cross height at the age of 3 months), 3-BL (body length at the age of 3 months); 3-CM (chest measurement at the age of 3 months), respectively. The two red dashed lines represent genome-level significance (up) and chromosomal significance (down).

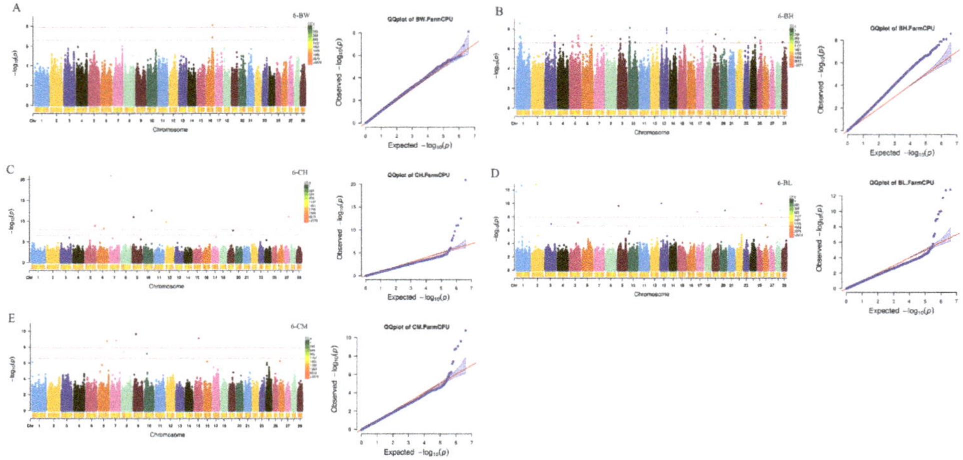

Figure 3. Manhattan and Q-Q plots for growth traits at the age of 6 months. (**A–E**) Manhattan and Q-Q plots for 6-BW (body weight at the age of 6 months), 6-BH (body height at the age of 6 months), 6-CH (cross height at the age of 6 months), 6-BL (body length at the age of 6 months), 6-CM (chest measurement at the age of 6 months), respectively. The two red dashed lines represent genome-level significance (up) and chromosomal significance (down).

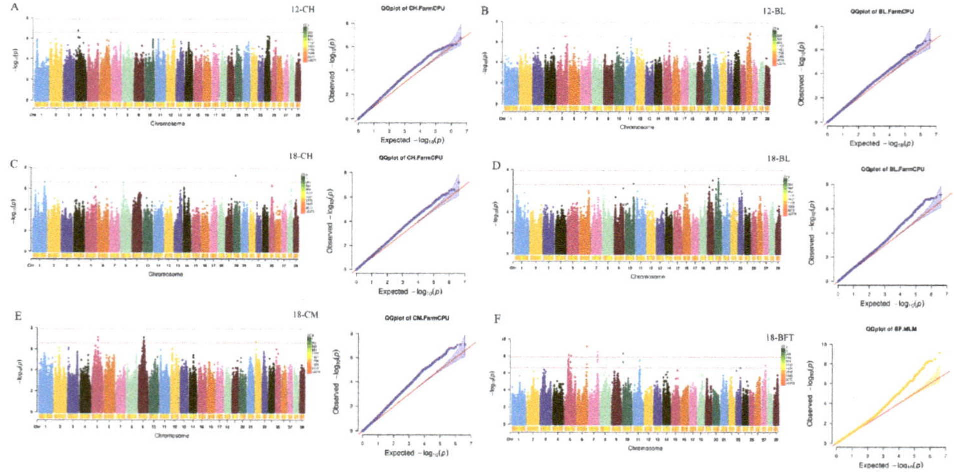

Figure 4. Manhattan and Q-Q plots for growth and back fat thickness traits at the age of 12 and 18 months. (**A–F**) Manhattan and Q-Q plots for 12-CH (cross height at the age of 12 months), 12-BL (body length at the age of 12 months), 18-CH (cross height at the age of 18 months), 18-BL (body length at the age of 18 months), 18-CM (chest measurement at the age of 18 months), 18-BFT (back fat thickness at the age of 18 months), respectively. The two red dashed lines represent genome-level significance (up) and chromosomal significance (down).

2.4. GO and KEGG Enrichment Analyses of Candidate Genes

To systematically investigate the functional characteristics of growth- and carcass-associated candidate genes, GO annotation and KEGG pathway enrichment analyses were conducted. This integrative approach enabled the identification of significantly enriched biological processes, molecular functions, cellular components, and key metabolic pathways associated with the target gene set. A total of 137, 55, and 26 candidate genes were identified for 3-CM, 6-BH, and 18-BFT, respectively. Due to the limited number of candidate genes detected for other traits, subsequent GO and KEGG enrichment analyses were performed exclusively on these three trait-associated gene sets.

GO enrichment analysis was conducted for the 137 candidate genes related to 3-CM (Figure 5A). At the Biological Process level, candidate genes were mainly enriched in bone marrow development, linoleic acid metabolic process, hepoxilin biosynthetic process, etc. At the Cellular Component level, candidate genes were mainly enriched in polysome, cytosol, lateral plasma membrane, etc. At the Molecular Function level, the enrichment primarily focused on hyaluronoglucosaminidase activity, kinase binding, ubiquitin protein ligase binding, etc. KEGG enrichment analysis revealed that these candidate genes were predominantly enriched in the Hippo signaling pathway, glycosaminoglycan degradation, and cardiac muscle contraction pathways (Figure 5B).

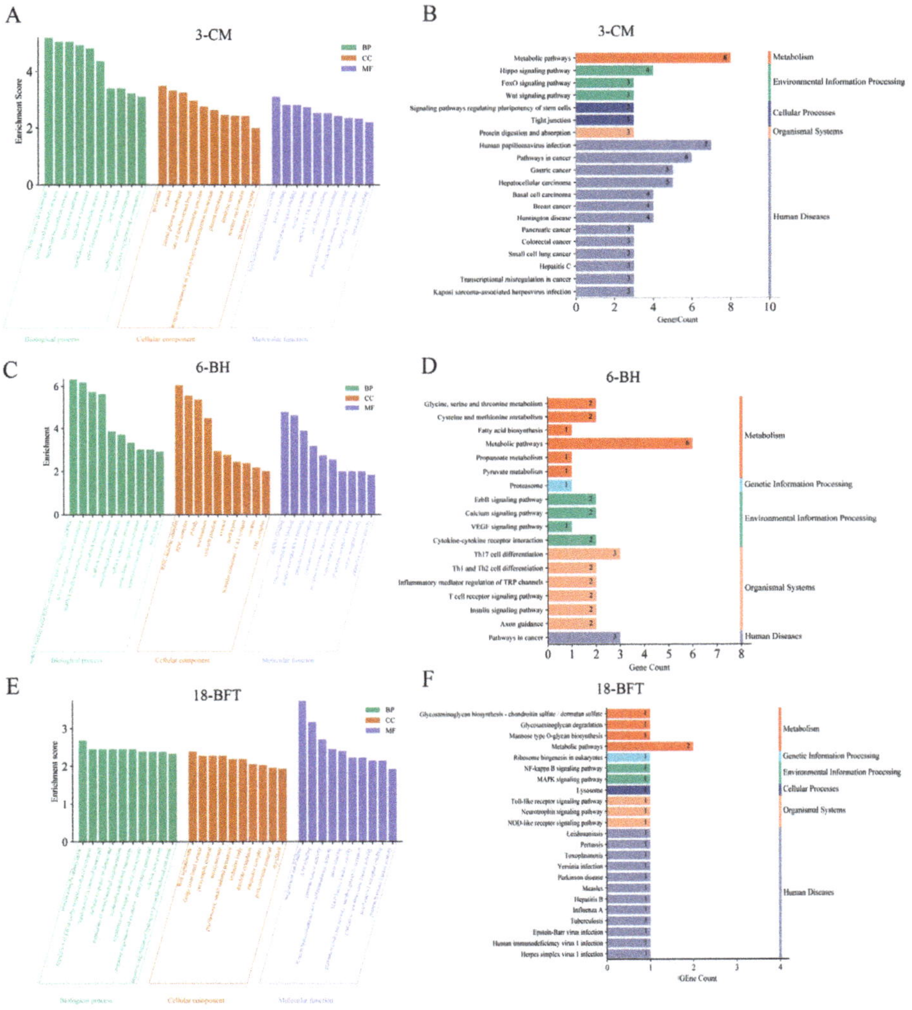

Figure 5. GO and KEGG enrichment analyses of candidate genes. GO (**A**) and KEGG (**B**) enrichment results for candidate genes associated with 3-CM (chest measurement at the age of 3 months); GO (**C**) and KEGG (**D**) enrichment results for candidate genes associated with 6-BH (body height at the age of 6 months); GO (**E**) and KEGG (**F**) enrichment results for candidate genes associated with 18-BFT (back fat thickness at the age of 18 months).

GO enrichment analysis was performed on the 55 candidate genes identified for the 6-BH trait (Figure 5C). At the Biological Process level, candidate genes were mainly enriched in miRNA loading onto RISC involved in gene silencing by miRNA, RNA secondary structure unwinding, miRNA mediated inhibition of translation, etc. At the Cellular Component level, candidate genes were mainly enriched in RISC-loading complex, RISC complex, P-body, etc. At the Molecular Function level, the enrichment primarily focused on miRNA binding, single-stranded RNA binding, double-stranded RNA binding, etc. KEGG enrichment analysis demonstrated that these candidate genes were predominantly enriched in signaling pathways, such as the VEGF signaling pathway, Insulin signaling pathway, and Glycine, Serine, and Threonine metabolism pathway (Figure 5D).

GO enrichment analysis was performed on the 26 candidate genes identified for the 18-BFT trait (Figure 5E). At the Biological Process level, candidate genes were mainly enriched in protein autophosphory, regulation of lysosomal lumen pH, etc. At the Cellular Component level, candidate genes were mainly enriched in Wnt signalosome, Golgi-associated vesicle, etc. At the Molecular Function level, the enrichment primarily focused on magnesium ion binding, ATP binding, protein kinase activity, etc. KEGG enrichment analysis indicated that these candidate genes were predominantly enriched in pathways such as the neurotrophin signaling pathway and MAPK signaling pathway (Figure 5F).

2.5. Functional SNPs Identification and PCR-RFLP Assays

In livestock production, genomic sequencing remains time-consuming and cost-prohibitive for routine implementation. To address this limitation, we developed a PCR-RFLP-based genotyping strategy to facilitate efficient marker-assisted selection in cattle breeding programs. This study identified three functionally significant SNPs, g.8674692C>G (associated with 3-BH), g.54418624G>T (associated with 3-CM), and g.71085551G>A (associated with 6-BL), each demonstrating clear genotype–phenotype relationships. At the g.8674692C>G locus, the 3-BH levels of the GG and GC genotypes were significantly higher than those of the CC genotype, and the GG genotype was significantly higher than the GC genotype, indicating that the GG genotype was the favorable allelic genotype (Figure 6A). Furthermore, the locus before and after mutation could be recognized by the restriction endonuclease *Hha* I. Following gel electrophoresis, three distinct band patterns were observed: the GG genotype displayed two bands of 341 bp and 200 bp, the GC genotype displayed three bands of 541 bp, 341 bp, and 200 bp, and the CC genotype displayed only one band of 541 bp (Figure 6B). Therefore, individuals with the GG genotype could be rapidly screened by the PCR-RFLP method. Similarly, at the g.54418624G>T locus, 3-CM levels were significantly higher in GG and TG genotypes than in TT, with no significant difference between GG and TG, indicating that the TT genotype was the unfavorable allele (Figure 6C). Furthermore, this locus could be cleaved by the restriction enzyme *Hinc* II before and after mutation. Gel electrophoresis revealed three distinct band patterns: the TT genotype displayed two bands, the TG genotype showed three bands, and the GG genotype yielded only a single band (Figure 6D). At the g.71085551G>A locus, individuals with the GG and AG genotypes exhibited significantly higher 6-BL levels compared to those with the AA genotype, with no significant difference observed between the GG and AG genotypes, suggesting that the AA genotype represents the unfavorable allele (Figure 6E). Furthermore, this locus can be digested by the restriction enzyme *Mnl* I. Gel electrophoresis analysis revealed three distinct banding patterns: the GG genotype produced two bands, the AG genotype displayed three bands, and the AA genotype yielded only a single band (Figure 6F). The established method enables precise identification of favorable alleles while facilitating culling of animals carrying disadvantageous genotypes, significantly enhancing breeding efficiency.

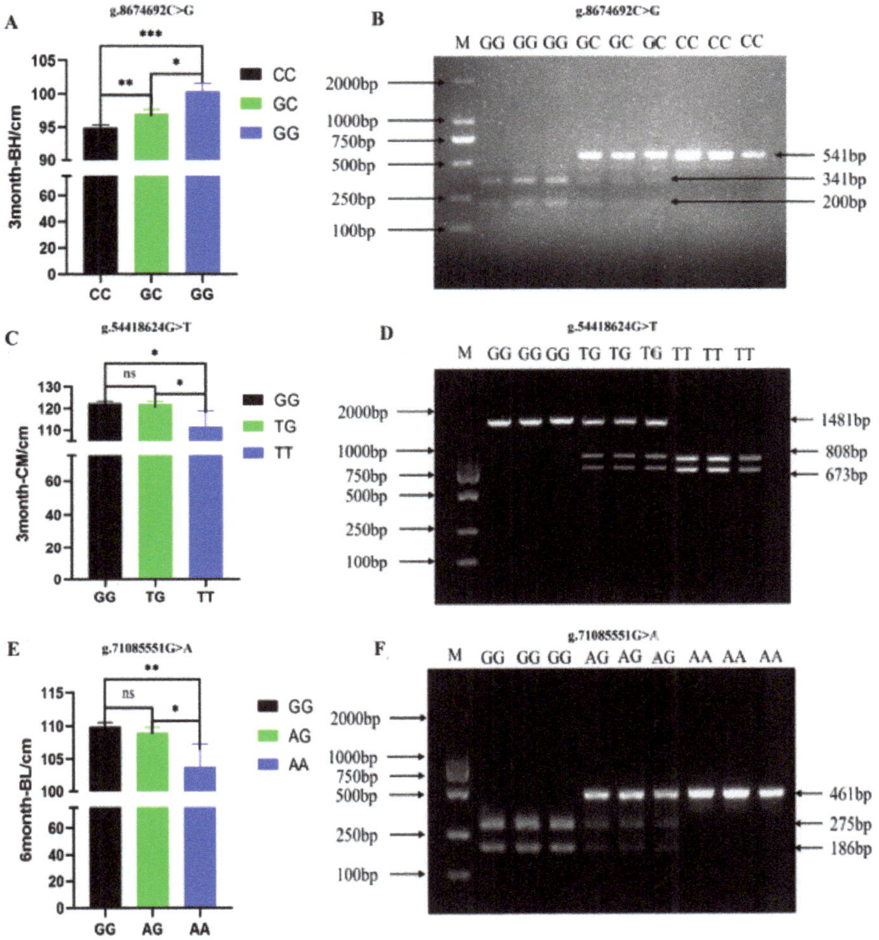

Figure 6. The results of SNP effect analysis and enzyme digestion typing The effect analysis (**A**) and enzyme digestion typing (**B**) results for g.8674692C>G associated with 3-BH (body height at the age of 3 months); the effect analysis (**C**) and enzyme digestion typing (**D**) results for g.54418624G>T associated with 3-CM (chest measurement at the age of 3 months); the effect analysis (**E**) and enzyme digestion typing (**F**) results for g.71085551G>A associated with 6-BL (body length at the age of 6 months). * $p < 0.05$; ** $p < 0.01$; *** $p < 0.001$; ns, not significant.

2.6. ACOX1 Gene Polymorphism Identification and Trait Association Analysis

In the ACOX1 gene, eight SNPs associated with growth or carcass traits were successfully identified. Specifically, three SNPs, g.55672049A>G, g.55672909C>T, and g.55673438A>G, were localized in intron 2, and the remaining five SNPs, g.55675384C>T, g.55676112A>G, g.55680392G>A, g.55681592A>G, and g.55683268C>G, were mapped to intron 3. These identified SNPs were further analyzed to determine their associations with multiple growth and carcass traits, including NB-BW, NB-BH, NB-CH, NB-BL, NB-CM, 3-BW, 3-BH, 3-CH, 3-BL, 3-CM, 6-BW, 6-BH, 6-CH, 6-BL, 6-CM, 12-BW, 12-BH, 12-CH, 12-BL, 12-CM,18-BF, and 18-EM. The traits significantly associated with the SNPs are displayed in Figure 7A–H. These results demonstrate that ACOX1 was a promising candidate gene for genetic improvement in beef cattle, with significant associations between its polymorphisms and key growth and carcass traits. The eight identified SNPs may serve as valuable molecular markers for marker-assisted selection (MAS) in the Charolais cattle breeding program. Incorporating these variants into genomic selection strategies may enhance the efficiency of breeding for superior production traits.

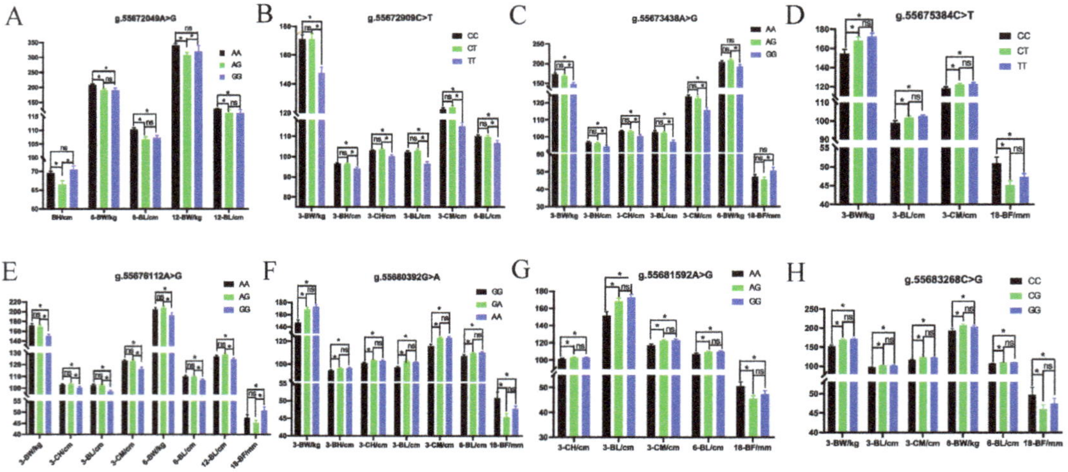

Figure 7. Association analysis of ACOX1 SNPs with growth and carcass traits in Charolais cattle. (**A**) g.55672049A>G; (**B**) g.55672909C>T; (**C**) g.55673438A>G; (**D**) g.55675384C>T; (**E**) g.55676112A>G; (**F**) g.55680392G>A; (**G**) g.55681592A>G; (**H**) g.55683268C>G. * $p < 0.05$; ns, not significant.

3. Discussion

The Charolais cattle breed, renowned for its large body size and high meat yield, serves as an ideal model for investigating genetic mechanisms underlying economically important traits in beef production. Growth and carcass traits are pivotal determinants of economic efficiency in beef production systems. The identification of genetic markers, particularly SNPs and their associated genes, provides valuable tools for marker-assisted selection programs. Such genomic approaches enable more precise breeding strategies to improve production traits and meet growing global demands for high-quality beef. In this study, high-quality SNPs were obtained from a population of 240 Charolais cattle using the hcWGC methods. GWAS was then conducted to investigate the relationship between these SNPs and growth and carcass traits, leading to the identification of 225 associated SNPs and 379 candidate genes.

3.1. Candidate Genes Identified for Growth Traits According to the GWAS Results

Growth traits crucial for livestock development include body weight, body size, and daily weight gain. It is widely acknowledged in the industry that birth, 3-month-old, 6-month-old, 12-month-old, and 18-month-old stages are critical growth nodes in beef cattle. BW, BL, BH, CH, and CM serve as key indicators for evaluating beef cattle growth status, and all of which exhibit positive correlations with production performance. Using FarmCPU-based GWAS, we identified 196 significant SNPs and 353 candidate genes related to growth traits in Charolais cattle. We observed inconsistencies in the SNPs identified for the same traits at different time points, such as in 3-CM and 6-CM. One possible explanation for this is that genetic influences on chest measurement may vary between these time points owing to biological changes as the organism grows. Additionally, variations in sample size and statistical power at each time point may have contributed to the identification of different SNPs. Furthermore, changes in genetic interactions and environmental factors over time may contribute to variations in SNPs associated with the same trait at different developmental stages. Among the 353 candidate genes, RBM33 has been identified as being significantly associated with obesity in Northern Han Chinese by targeted resequencing technology [21], and as an RNA-binding protein, RBM333 regulates m^{6}A methylation modification [22], an essential post-transcriptional regulator implicated in skeletal myogenesis [23]. KCTD17 was implicated in the cellular growth, adipogenic differentiation, and vertebrate development [24–26]. PTHLH plays an important role in chondrogenesis, brachydactyly, short stature, and developmental delay [27–29]. Similarly, RAC2 has been reported to regulate myeloid cell dysfunction, bone development, and metabolism [30,31]. In addition, CHD6 was screened to participate in porcine embryonic muscle development through comprehensive analysis of the transcriptome and miRNAome [32]. TRDN was detected to be significantly associated with adipogenesis [33,34], and it has been proposed as a candidate gene for the 6-month weight of sheep and meat quality traits of Simmental cattle [35,36]. The copy number of the WBP1L gene was significantly correlated with growth traits of Chinese cattle populations [37]. Other relevant genes, such as TLL2, CH25H, and ST13, affect muscular atrophy and adipocyte differentiation [38–40]. The identified genes may offer vital insights into selecting and breeding beef cattle with desirable growth traits. However, further validations and investigations are required to elucidate their precise functions and mechanisms of action.

3.2. Candidate Genes Identified for BFT According to the GWAS Results

Backfat thickness and eye muscle area serve as essential indicators, directly influencing meat quality and carcass yield [41]. The amount of BFT deposited on the carcass is related to the total body fat and plays a major role in beef's flavor and juiciness and dressing percentage [42]. In this study, although no SNPs associated with EM were detected, 29 significant SNPs and 26 candidate genes related to BFT were identified. Important genes include CHST11, which contains two functional intronic PPARγ binding sites and exhibits reduced intracellular lipid accumulation in mature adipocytes following its knockdown [43]; LRRK2, which regulates CPT1A expression to promote β-oxidation and thereby facilitates lipid catabolism in HepG2 cells [44], and also mediates Rab8a phosphorylation to promote lipid storage in 3T3 cells [45]; RIOK2, which as an ATPase plays a pivotal role in regulating cell growth and protein synthesis across diverse cellular contexts [46,47]; and INTS10, whose causal variant rs112861901 influences high-density lipoprotein levels in bodily metabolic processes as identified through GWAS [48]. The pathways identified by GO and KEGG enrichment analyses corresponded closely to these biological processes, including regulation of synaptic vesicle endocytosis, ATP binding, protein serine/threonine kinase activity, metabolic pathways, and the MAP signaling pathway. Thus, we speculated that these genes might affect backfat thickness by regulating the lipid metabolism. The identified backfat thickness-related candidate genes may provide an opportunity to accelerate the meat quality and carcass yield of Charolais cattle.

3.3. PCR-RFLP for Quick Identification of SNPs

PCR-RFLP is the simplest and fastest method for single-nucleotide change detection, and it has been widely applied in livestock, including sheep [49], pigs [50], and crossbred cattle [51]. In the current study, three functionally significant SNPs, g.8674692C>G, g.54418624G>T, and g.71085551G>A, were genotyped using PCR-RFLP with restriction enzymes *Hha* I, *Hinc* II, and *Mnl* I, respectively. Based on the results of genotype–phenotype association analysis, after digestion with *Hha* I, if gel electrophoresis shows only two bands (341 bp and 200 bp), it indicates that the genotype of the g.8674692C>G locus in Charolais individuals is the advantageous genotype GG, and such individuals can be prioritized as candidate breeding cattle. Similarly, after digestion with *Hinc* II, gel electrophoresis displays one band (1481 bp) or three bands (1481 bp, 808 bp, and 673 bp); this suggests that the genotype of the g.54418624G>T locus in Charolais individuals is the advantageous genotype GG or TG, and these individuals can be given priority consideration as candidate breeding cattle. After digestion with *Mnl* I, gel electrophoresis reveals two bands (275 bp and 186 bp) or three bands (461 bp, 275 bp, and 186 bp); this implies that the genotype of the g.71085551G>A locus in Charolais individuals is the advantageous genotype GG or AG, and such individuals can be focused on as candidate breeding cattle.

3.4. ACOX1 Polymorphism and Association Analysis

ACOX1 is the first and rate-limiting enzyme in the peroxisomal fatty acid β-oxidation pathway across all eukaryotes [52]. Our previous studies have demonstrated that ACOX1 promotes adipogenesis in bovine preadipocytes [53]. Additionally, prior studies have reported that polymorphisms in the ACOX1 gene were associated with the backfat thickness and meat quality grade in Chinese indigenous cattle and cultivated cattle breeds, and Korean cattle. One SNP in exon 13 of the bovine ACOX1 gene resulted in significant differences in ultrasound backfat thickness and ultrasound marbling score among different genotypes [54]. The g.224G > A SNP located in ACOX1 coding regions was significantly associated with meat quantity grade at slaughter and backfat thickness tended to be greater in Korean cattle [55]. The association between ACOX1 gene polymorphism and growth traits in Charolais cattle has never been studied. In the present study, eight polymorphic loci were identified in the Charolais cattle population, all of which are located in intronic regions. However, relatively few SNPs were associated with traits at birth or 12 months of age. Specifically, only the g.55672049A>G locus showed a significant difference in body height traits at birth, and only the g.55672049A>G and g.55676112A>G loci exhibited significant differences in body weight or body length traits at 12 months of age. In contrast, all eight loci were significantly associated with one or more traits at 3 months and 6 months of age. This suggests that molecular marker loci associated with target growth traits could be selected for early selection of the Charolais cattle population at 3 months and 6 months of age, which was consistent with previous research results [56]. SNPs located in intronic regions may exert significant impacts on gene expression and phenotypes through multiple complex mechanisms. They can directly regulate gene transcriptional efficiency by creating or disrupting binding sites for transcription factors, enhancers, or silencers; induce alternative splicing and isoform imbalance by affecting pre-mRNA splice sites or regulatory elements; and interfere with the production of functional non-coding RNAs, or exert distal regulation by altering chromatin spatial conformation and epigenetic modifications. The detailed mechanisms underlying the functional roles of SNPs in ACOX1 remain to be further elucidated.

3.5. Application of SNPs and Candidate Genes in Future Genomic Breeding of Beef Cattle

The SNPs and candidate genes identified through hcWGS in this study provide a genetic foundation for precision breeding in beef cattle. These discoveries enable MAS for key economic traits such as growth and carcass characteristics, permitting the early identification of superior individuals to accelerate genetic gain. Furthermore, they can enhance the accuracy of genomic selection (GS) models by refining the prediction of genomic estimated breeding values (GEBVs) for complex traits. The candidate genes also present direct targets for gene editing initiatives. For broad industry application, core SNPs can be incorporated into cost-effective genotyping platforms. Collectively, these resources empower a shift from traditional phenotype-based selection to genomics-informed breeding strategies, facilitating the synergistic improvement of multiple traits and promoting sustainable genetic progress in beef cattle populations.

4. Materials and Methods

4.1. Animals and Phenotype Data

A total of 240 Charolais cattle, including 69 bulls and 171 cows Charolais cattle, were obtained from the national core breeding farm in Hubei Province, China. These cattle were provided with a uniform diet and subjected to a standardized raising and management protocol, with ad libitum feeding. The experimental procedure was approved by the Animal Care and Use Committee of the Hubei Academy of Agricultural Sciences, China, and was conducted in accordance with animal welfare and ethics guidelines (No. HBAASEC-2023-011, 16 February 2023).

The measured traits of this population were as follows: BW, BH, BL, CH, and CM at five key growth stages of newborn, 3 months, 6 months, 12 months, and 18 months, as well as BFT and EM at 18 months. All weight-related traits were measured after a 12 h fasting period, all body height-related traits were measured from the highest point of the withers to the ground, body length-related traits were measured from the front edge of the shoulder to the trailing edge of the sitting bone, cross height-related traits were measured from the highest point of the hooks to the ground, and all chest measurement-related traits were measured by encircling the trailing edge of the scapula. BH, BL, CH, and CM were measured on the same flat surface, and the cattle were maintained in a natural standing posture during the measurement. Then the phenotypic distribution statistics were analyzed to determine whether the above phenotypic data conform to a normal distribution.

4.2. Whole-Genome Resequencing

Cervical venous blood was drawn from the 240 Charolais cattle, and 3–5 mL of fresh blood was used to extract genomic DNA using the DNA Extraction Kit of Genstone Biotech (TD324). DNA samples with concentration $\geq$ 40 ng/µL, quantity $\geq$1 µg, A260/A280 ratio between 1.8 and 2.0, and that passed gel electrophoresis test were used for whole-genome resequencing. After assessing the DNA quality, gene libraries were assembled, each with an average fragment size of 300 bp per sample. The sequencing was carried out by the Beijing Genomics Institute using the DNBSEQ-T7 sequencer with high-depth sequencing (~13×) to obtain the raw downstream data.

4.3. Alignment, Variant Identification and PCA

After sequencing, Fastp (version 0.19.7) [57] was used to filter the raw data using default parameters. The clean reads obtained after quality control were aligned to Bos taurus reference assembly ARS-UCD1.2 using the BWA-MEM algorithm of the Burrows-Wheeler Aligner v0.7.17- r1188 [58]. The Picard tools (http://broadinstitute.github.io/picard, accessed on 20 January 2024) were used to filter potential duplicate reads. The "Haplotype Caller", Genotype GVCFs", and "Select Variants" modules of the Genome Analysis Toolkit (GATK, version 4.2.0.0) were used to call single nucleotide polymorphisms (SNPs). The filtration of raw SNPs was conducted by using the "variant Filtration" module of GATK. Finally, quality control of the SNPs was conducted using PLINK (version 1.90) software [59] with the following parameters: –geno 0.1, –mind 0.1, –maf 0.01, and –hwe 0.000001. PCA was used to evaluate cattle population stratification based on WGS data from 240 individuals using PLINK.

4.4. Genome-Wide Association Study

GWAS was performed using R package rMVP (version 1.0.0) using the Fixed and random model Circulating Probability Unification (FarmCPU) method [27]. All parameters were set to their default values. The FarmCPU model was evaluated iteratively using the fixed-effect model (FEM) and random-effect model (REM). The FEM can be written as

$$y = Xb + Z_t u_t + S_i d_i + e$$

where y is a vector of phenotypes of the analyzed trait; b is a vector of fixed effects and covariates, including top three principal components, sex, and farm effects; u_t is a vector of the fixed effects for the tth pseudo-QTNs; X and Z_t are the corresponding incidence matrices for b and u_t; S_i is the genotype for the ith candidate SNP; d_i is the effect of the ith candidate SNP; and e is a vector of the residuals. The REM updated the pseudo-QTNs as follows:

$$y = u + e$$

where y is a vector of phenotypes of the analyzed trait; $u \sim MVN\ (0,\ 2K\sigma_u{}^2)$ with $\sigma u2$ being the unknown genetic variance, and K being the kinship matrix computed by the pseudo-QTNs; and e is a vector of the residuals.

Bonferroni multiple tests were employed to identify the genome-wide significant (0.05/number of SNPs) and suggestive (1/number of SNPs) SNPs. Significant association between SNPs and reproductive traits was visualized in Manhattan plots and QQ-plots generated using rMVP when conducting GWAS.

4.5. Candidate Gene Identification and Functional Enrichment Analysis

The candidate genes were retrieved within 0.1 Mb on either side of the significant SNPs using the gene annotation information of the beef cattle reference genome version ARS-UCD1.2 in ENSEMBL (https://asia.ensembl.org/index.html, accessed on 27 January 2024), while only protein-coding genes were retained. KOBAS-i (http://bioinfo.org/kobas/, accessed on 15 February 2024) was used to perform GO and KEGG enrichment analyses, with $p < 0.05$ considered statistically significant. Based on the enrichment results, graphing was performed using Bioinformatics (https://www.bioinformatics.com.cn/, accessed on 15 February 2024) to obtain GO enrichment and KEGG enrichment bubbles and bar graphs.

4.6. SNP Effect Analysis and Genotyping by PCR-RFLP

Based on GWAS, candidate significant SNP loci were selected for effect analysis. The Charolais cattle population was stratified into three genotypic groups for each polymorphism: homozygous reference (AA), heterozygous (Aa), and homozygous alternative (aa). Phenotypic trait values were aggregated for each genotype group, and summary statistics (mean $\pm$ standard deviation) were calculated. The effect size of each SNP variant on the target phenotype was quantified and visualized using GraphPad Prism (version 8.0).

To develop a robust PCR-RFLP genotyping assay, we first retrieved genomic sequences encompassing 1000 bp flanking regions (upstream and downstream) of each target SNP from the NCBI Genome Database (http://www.ncbi.nlm.nih.gov/, accessed on 22 February 2024). Potential restriction enzymes were systematically screened using NEBcutter 3.0 (https://nc3.neb.com/NEBcutter/, accessed on 22 February 2024) to identify those exhibiting differential digestion patterns between wild-type and variant alleles. Enzymes were selected based on the following criteria: (1) generation of distinct fragment patterns for each genotype; (2) single recognition site within the amplicon; and (3) optimal activity under standard reaction conditions. Target SNP regions were amplified in PCR reactions, and then the amplicons were digested with the selected restriction enzyme at 37 °C for 1 h. Digested fragments were separated by electrophoresis on 1.5% agarose gels. Genotypes were determined by comparing observed fragment patterns with predicted digestion profiles.

4.7. ACOX1-SNP Association Study

The association between ACOX1-SNP and phenotypes was analyzed using a linear mixed model (LMM) framework implemented by a self-developed program in SAS (version 9.4). The model included genotype (I), sire (A), age (B), and sex (C) as fixed effects, with animal individual (D) as a random effect:

$$Y_{ikjlm} = \mu + I_i + A_k + B_j + C_l + D_m + E_{ikjlm}$$

where Y_{ikjlm} represents the observed trait value, μ is the overall mean, $D_m \sim N(0, \sigma_D{}^2)$ denotes the random effect of the m-th animal individual (normally distributed with variance $\sigma_D{}^2$), and $E_{ikjlm} \sim N(0, \sigma_E{}^2)$ is the residual error.

5. Conclusions

The aim of this study was to investigate the significant SNPs and genes associated with five growth traits and two carcass traits in Charolais cattle. hdWGS was performed on 240 Charolais cattle with approximately 13× coverage. Using the FarmCPU-based GWAS, 196 and 29 potential SNPs were identified for growth and carcass traits, respectively, along with 353 and 26 candidate genes associated with these traits. In addition, to reduce costs and achieve rapid detection, a PCR-RFLP-based genotyping strategy for the g.8674692C>G, g.54418624G>T, and g.71085551G>A loci was established to facilitate early selection of breeding cattle. Lastly, eight SNPs associated with growth and backfat thickness traits were detected in the intronic region of the ACOX1 gene. These findings provide valuable preliminary insights into the genetic mechanisms underlying growth and carcass traits in Charolais cattle, and facilitate genome-assisted breeding, although further functional validation of these potential SNPs and candidate genes is required.

6. Patents

Some results reported in this manuscript have been filed for patent protection. The details are as follows:

[1] Feng Zhang, Chengmei Wang, Qi Xiong, et al., Application of an SNP molecular marker related to 3-month-old body height traits in beef cattle. China, ZL202410679360.2. Filing date: 29 May 2024; Authorization announcement date: February 7, 2025. Status: Granted.

[2] Feng Zhang, Chengmei Wang, Qi Xiong, et al., Application of an SNP molecular marker related to 6-month-old body length traits in beef cattle. China, ZL202410679460.5. Filing date: 29 May 2024; Authorization announcement date: August 29, 2025. Status: Granted.

Supplementary Materials: The following supporting information can be downloaded at: https://www.mdpi.com/article/10.3390/ijms262311411/s1.

Author Contributions: Conceptualization, J.C. and Q.X.; methodology, F.Z. and C.W.; software, C.W. and A.S.; validation, F.Z. and F.J.; formal analysis, C.W. and X.S.; investigation, M.C., H.T., and Z.H.; data curation, A.S., T.X., and N.Z.; writing—original draft preparation, F.Z. and C.W.; writing—review and editing, F.Z.; visualization, F.Z.; supervision, J.C. and Q.X.; project administration, F.Z. All authors have read and agreed to the published version of the manuscript.

Funding: This research was funded by the National Natural Science Foundation of China (32202649), Funds for Supporting High-Quality Development of Seed and Breeding Industry Projects in Hubei Province (HBZY2023B008), Hubei Provincial Science and Technology Innovation Talent Program (2025DJA031), Hubei Provincial Technology Innovation Program (2024BBB048), and the National Key Research and Development Program of China (2024YFD1300104, 2024YFD1301301).

Institutional Review Board Statement: The experimental procedure was approved by the Animal Care and Use Committee of the Hubei Academy of Agricultural Sciences, China, and was conducted in accordance with animal welfare and ethics guidelines (No. HBAASEC-2023-011, 16 February 2023).

Informed Consent Statement: Not applicable.

Data Availability Statement: The original contributions presented in the study are included in the article/Supplementary Materials, further inquiries can be directed to the corresponding author.

Conflicts of Interest: The authors declare no conflicts of interest.

Abbreviations

The following abbreviations are used in this manuscript:

GWAS	genome-wide association study
WGS	whole-genome sequencing
hcWGS	high-coverage whole-genome sequencing
NB	newborn
BW	body weight
BH	body height
CH	cross height
BL	body length
CM	chest measurement
BFT	back fat thickness
EM	eye muscle area
PCR-RFLP	polymerase chain reaction–restriction fragment length polymorphism
SNP	single nucleotide polymorphism
ACOX1	Acyl-CoA oxidase 1

References

1. de Rezende, M.P.G.; Malhado, C.H.M.; Biffani, S.; Carneiro, P.L.S.; Bozzi, R. Genetic diversity derived from pedigree information and estimation of genetic parameters for reproductive traits of Limousine and Charolais cattle raised in Italy. *Ital. J. Anim. Sci.* **2020**, *19*, 762–771. [CrossRef]
2. Keogh, K.; Carthy, T.R.; McClure, M.C.; Waters, S.M.; Kenny, D.A. Genome-wide association study of economically important traits in Charolais and Limousin beef cows. *Animal* **2021**, *15*, 100011. [CrossRef] [PubMed]
3. Santiago, G.G.; Siqueira, F.; Cardoso, F.F.; Regitano, L.C.A.; Ventura, R.; Sollero, B.P.; Souza, M.D.; Mokry, F.B.; Ferreira, A.B.R.; Torres, R.A.A. Genomewide association study for production and meat quality traits in Canchim beef cattle. *J. Anim. Sci.* **2017**, *95*, 3381–3390. [CrossRef] [PubMed]
4. Jahuey-Martínez, F.J.; Parra-Bracamonte, G.M.; Sifuentes-Rincón, A.M.; Martínez-González, J.C.; Gondro, C.; García-Pérez, C.A.; López-Bustamante, L.A. Genomewide association analysis of growth traits in Charolais beef cattle. *J. Anim. Sci.* **2016**, *94*, 4570–4582. [CrossRef]
5. Braz, C.U.; Rowan, T.N.; Schnabel, R.D.; Decker, J.E. Genome-wide association analyses identify genotype-by-environment interactions of growth traits in Simmental cattle. *Sci. Rep.* **2021**, *11*, 13335. [CrossRef]
6. Liu, Y.; Xu, L.; Wang, Z.; Xu, L.; Chen, Y.; Zhang, L.; Xu, L.; Gao, X.; Gao, H.; Zhu, B.; et al. Genomic prediction and association analysis with models including dominance effects for important traits in Chinese simmental beef cattle. *Animals* **2019**, *9*, 1055. [CrossRef]
7. Pegolo, S.; Cecchinato, A.; Savoia, S.; Di Stasio, L.; Pauciullo, A.; Brugiapaglia, A.; Bittante, G.; Albera, A. Genome-wide association and pathway analysis of carcass and meat quality traits in Piemontese young bulls. *Animal* **2020**, *14*, 243–252. [CrossRef]
8. Smith, J.L.; Wilson, M.L.; Nilson, S.M.; Rowan, T.N.; Schnabel, R.D.; Decker, J.E.; Seabury, C.M. Genome-wide association and genotype by environment interactions for growth traits in U.S. Red Angus cattle. *BMC Genom.* **2022**, *23*, 517. [CrossRef]
9. Bila, L.; Putra, W.P.B.; Malatji, D.P.; Sanarana, Y.P.; Tyasi, T.L. Genome-wide association study (GWAS) for body weights of sussex cattle (*Bos taurus*) in South Africa. *Heliyon* **2024**, *10*, e39540. [CrossRef]
10. Adhikari, M.; Kantar, M.B.; Longman, R.J.; Lee, C.N.; Oshiro, M.; Caires, K.; He, Y. Genome-wide association study for carcass weight in pasture-finished beef cattle in Hawai'i. *Front. Genet.* **2023**, *14*, 1168150. [CrossRef]

11. Grigoletto, L.; Ferraz, J.B.S.; Oliveira, H.R.; Eler, J.P.; Bussiman, F.O.; Abreu Silva, B.C.; Baldi, F.; Brito, L.F. Genetic Architecture of Carcass and Meat Quality Traits in Montana Tropical® Composite Beef Cattle. *Front. Genet.* **2020**, *11*, 123. [CrossRef] [PubMed]
12. Buzanskas, M.E.; Grossi, D.A.; Ventura, R.V.; Schenkel, F.S.; Sargolzaei, M.; Meirelles, S.L.; Mokry, F.B.; Higa, R.H.; Mudadu, M.A.; da Silva, M.V.; et al. Genome-wide association for growth traits in Canchim beef cattle. *PLoS ONE* **2014**, *9*, e94802. [CrossRef] [PubMed]
13. Snelling, W.M.; Allan, M.F.; Keele, J.W.; Kuehn, L.A.; McDaneld, T.; Smith, T.P.; Sonstegard, T.S.; Thallman, R.M.; Bennett, G.L. Genome-wide association study of growth in crossbred beef cattle. *J. Anim. Sci.* **2010**, *88*, 837–848. [CrossRef] [PubMed]
14. Wang, Y.; Zhang, F.; Mukiibi, R.; Chen, L.; Vinsky, M.; Plastow, G.; Basarab, J.; Stothard, P.; Li, C. Genetic architecture of quantitative traits in beef cattle revealed by genome wide association studies of imputed whole genome sequence variants: II: Carcass merit traits. *BMC Genom.* **2020**, *21*, 38. [CrossRef]
15. Doyle, J.L.; Berry, D.P.; Veerkamp, R.F.; Carthy, T.R.; Walsh, S.W.; Evans, R.D.; Purfield, D.C. Genomic regions associated with skeletal type traits in beef and dairy cattle are common to regions associated with carcass traits, feed intake and calving difficulty. *Front. Genet.* **2020**, *11*, 20. [CrossRef]
16. Vanvanhossou, S.F.U.; Scheper, C.; Dossa, L.H.; Yin, T.; Brügemann, K.; König, S. A multi-breed GWAS for morphometric traits in four Beninese indigenous cattle breeds reveals loci associated with conformation, carcass and adaptive traits. *BMC Genom.* **2020**, *21*, 783. [CrossRef]
17. Martin, P.; Taussat, S.; Vinet, A.; Krauss, D.; Maupetit, D.; Renand, G. Genetic parameters and genome-wide association study regarding feed efficiency and slaughter traits in Charolais cows. *J. Anim. Sci.* **2019**, *97*, 3684–3698. [CrossRef]
18. Song, X.; Yao, Z.; Zhang, Z.; Lyu, S.; Chen, N.; Qi, X.; Liu, X.; Ma, W.; Wang, W.; Lei, C.; et al. Whole-genome sequencing reveals genomic diversity and selection signatures in Xia'nan cattle. *BMC Genom.* **2024**, *25*, 559. [CrossRef]
19. Harish, A.; Lopes Pinto, F.A.; Eriksson, S.; Johansson, A.M. Genetic diversity and recent ancestry based on whole-genome sequencing of endangered Swedish cattle breeds. *BMC Genom.* **2024**, *25*, 89. [CrossRef]
20. Saleh, A.A.; Xue, L.; Zhao, Y. Screening Indels from the whole genome to identify the candidates and their association with economic traits in several goat breeds. *Funct. Integr. Genom.* **2023**, *23*, 58. [CrossRef]
21. Wu, Y.; Wang, W.; Jiang, W.; Yao, J.; Zhang, D. An investigation of obesity susceptibility genes in Northern Han Chinese by targeted resequencing. *Medicine* **2017**, *96*, e6117. [CrossRef]
22. Yu, F.; Zhu, A.C.; Liu, S.; Gao, B.; Wang, Y.; Khudaverdyan, N.; Yu, C.; Wu, Q.; Jiang, Y.; Song, J.; et al. RBM33 is a unique m(6)A RNA-binding protein that regulates ALKBH5 demethylase activity and substrate selectivity. *Mol. Cell* **2023**, *83*, 2003–2019.e2006. [CrossRef]
23. Zhang, X.; Yao, Y.; Han, J.; Yang, Y.; Chen, Y.; Tang, Z.; Gao, F. Longitudinal epitranscriptome profiling reveals the crucial role of N(6)-methyladenosine methylation in porcine prenatal skeletal muscle development. *J. Genet. Genomics* **2020**, *47*, 466–476. [CrossRef] [PubMed]
24. Shin, M.C.; Jung, Y.H.; Jeong, Y.; Oh, A.R.; Lee, S.B.; Kim, K. Kctd17-mediated Chop degradation promotes adipogenic differentiation. *Biochem. Biophys. Res. Commun.* **2023**, *653*, 126–132. [CrossRef] [PubMed]
25. Skoblov, M.; Marakhonov, A.; Marakasova, E.; Guskova, A.; Chandhoke, V.; Birerdinc, A.; Baranova, A. Protein partners of KCTD proteins provide insights about their functional roles in cell differentiation and vertebrate development. *BioEssays* **2013**, *35*, 586–596. [CrossRef] [PubMed]
26. Rizk, R.; Devost, D. KCTD Proteins Have Redundant Functions in Controlling Cellular Growth. *Int. J. Mol. Sci.* **2024**, *25*, 4993. [CrossRef]
27. Bai, M.; Yin, H.; Zhao, J.; Li, Y.; Wu, Y. miR-182-5p overexpression inhibits chondrogenesis by down-regulating PTHLH. *Cell Biol. Int.* **2019**, *43*, 222–232. [CrossRef]
28. Elli, F.M.; Mattinzoli, D.; Lucca, C.; Piu, M.; Maffini, M.A.; Costanza, J.; Fontana, L.; Santaniello, C.; Forino, C.; Milani, D.; et al. Novel pathogenetic variants in PTHLH and TRPS1 genes causing syndromic brachydactyly. *J. Bone Miner. Res.* **2022**, *37*, 465–474. [CrossRef]
29. Scheffer-Rath, M.E.A.; Veenstra-Knol, H.E.; Boot, A.M. A novel mutation in PTHLH in a family with a variable phenotype with brachydactyly, short stature, oligodontia and developmental delay. *Bone Rep.* **2023**, *19*, 101699. [CrossRef]
30. Goldberg, S.R.; Georgiou, J.; Glogauer, M.; Grynpas, M.D. A 3D scanning confocal imaging method measures pit volume and captures the role of Rac in osteoclast function. *Bone* **2012**, *51*, 145–152. [CrossRef]
31. Gu, Y.; Williams, D.A. RAC2 GTPase deficiency and myeloid cell dysfunction in human and mouse. *J. Pediatr. Hematol. Oncol.* **2002**, *24*, 791–794. [CrossRef] [PubMed]
32. Zhang, X.; Cai, S.; Chen, L.; Yuan, R.; Nie, Y.; Ding, S.; Fang, Y.; Zhu, Q.; Chen, K.; Wei, H.; et al. Integrated miRNA-mRNA transcriptomic analysis reveals epigenetic-mediated embryonic muscle growth differences between Wuzhishan and Landrace pigs1. *J. Anim. Sci.* **2019**, *97*, 1967–1978. [CrossRef] [PubMed]
33. Serão, N.V.; Veroneze, R.; Ribeiro, A.M.; Verardo, L.L.; Braccini Neto, J.; Gasparino, E.; Campos, C.F.; Lopes, P.S.; Guimarães, S.E. Candidate gene expression and intramuscular fat content in pigs. *J. Anim. Breed. Genet.* **2011**, *128*, 28–34. [CrossRef] [PubMed]

34. Alradi, M.; Askari, H.; Shaw, M.; Bhavsar, J.D.; Kingham, B.F.; Polson, S.W.; Fancher, I.S. A long-term high-fat diet induces differential gene expression changes in spatially distinct adipose tissue of male mice. *Physiol. Genom.* **2024**, *56*, 819–832. [CrossRef]
35. Xia, J.; Qi, X.; Wu, Y.; Zhu, B.; Xu, L.; Zhang, L.; Gao, X.; Chen, Y.; Li, J.; Gao, H. Genome-wide association study identifies loci and candidate genes for meat quality traits in Simmental beef cattle. *Mamm. Genome* **2016**, *27*, 246–255. [CrossRef]
36. Zhang, L.; Liu, J.; Zhao, F.; Ren, H.; Xu, L.; Lu, J.; Zhang, S.; Zhang, X.; Wei, C.; Lu, G.; et al. Genome-wide association studies for growth and meat production traits in sheep. *PLoS ONE* **2013**, *8*, e66569. [CrossRef]
37. Zhang, J.; Zhang, Z.; Liu, X.; Chai, Y.; Yang, P.; Li, J.; Huang, Y.; Li, L.; Huang, W.; Yang, G.; et al. Copy number variation of WBP1L gene revealed its association with growth traits across Chinese cattle populations. *J. Agr. Sci.* **2022**, *160*, 528–534. [CrossRef]
38. Jiang, J.; Huang, J.; Gu, J.; Cai, X.; Zhao, H.; Lu, H. Genomic analysis of a spinal muscular atrophy (SMA) discordant family identifies a novel mutation in TLL2, an activator of growth differentiation factor 8 (myostatin): A case report. *Case Rep.* **2019**, *20*, 204. [CrossRef]
39. Shen, C.; Zhou, J.; Wang, X.; Yu, X.Y.; Liang, C.; Liu, B.; Pan, X.; Zhao, Q.; Song, J.L.; Wang, J.; et al. Angiotensin-II-induced muscle wasting is mediated by 25-Hydroxycholesterol via GSK3β signaling pathway. *EBioMedicine* **2017**, *16*, 238–250. [CrossRef]
40. Wang, R.; Li, Y.; Lin, Y.; Chen, D.; Sheng, X.; Zhao, N.; Liu, W. Cloning and expression characteristic analysis of goat ST13 gene. *Sheng Wu Gong Cheng Xue Bao* **2022**, *38*, 2959–2973. [CrossRef]
41. Silva-Vignato, B.; Coutinho, L.L.; Cesar, A.S.M.; Poleti, M.D.; Regitano, L.C.A.; Balieiro, J.C.C. Comparative muscle transcriptome associated with carcass traits of Nellore cattle. *BMC Genom.* **2017**, *18*, 506. [CrossRef]
42. Pan, Y.; Sun, G.; Li, G.; Chen, S.; Liu, H.; Li, H.; Mei, C.; Yang, W.; Zan, L. Sex-specific microbiota associations with backfat thickness, eye muscle area, and rumen fermentation in Qinchuan cattle. *BMC Microbiology* **2025**, *25*, 277. [CrossRef]
43. Tasdelen, I.; Berger, R.; Kalkhoven, E. PPARγ regulates expression of carbohydrate sulfotransferase 11 (CHST11/C4ST1), a regulator of LPL cell surface binding. *PLoS ONE* **2013**, *8*, e64284. [CrossRef] [PubMed]
44. Lin, C.W.; Peng, Y.J.; Lin, Y.Y. LRRK2 Regulates CPT1A to Promote β-Oxidation in HepG2 Cells. *Molecules* **2020**, *25*, 4122. [CrossRef] [PubMed]
45. Yu, M.; Arshad, M.; Wang, W.; Zhao, D.; Xu, L.; Zhou, L. LRRK2 mediated Rab8a phosphorylation promotes lipid storage. *Lipids Health Dis.* **2018**, *17*, 34. [CrossRef] [PubMed]
46. Messling, J.E.; Agger, K.; Andersen, K.L. Targeting RIOK2 ATPase activity leads to decreased protein synthesis and cell death in acute myeloid leukemia. *Blood* **2022**, *139*, 245–255. [CrossRef]
47. Gao, Z.; Xu, C.; Fan, H. Analysis of RIOK2 Functions in Mediating the Toxic Effects of Deoxynivalenol in Porcine Intestinal Epithelial Cells. *Int. J. Mol. Sci.* **2022**, *23*, 12712. [CrossRef]
48. Hebbar, P.; Abubaker, J.A.; Abu-Farha, M.; Alsmadi, O.; Elkum, N.; Alkayal, F.; John, S.E.; Channanath, A.; Iqbal, R.; Pitkaniemi, J.; et al. Genome-wide landscape establishes novel association signals for metabolic traits in the Arab population. *Hum. Genet.* **2021**, *140*, 505–528. [CrossRef]
49. Nagdy, H.; Mahmoud, K.G.M.; Kandiel, M.M.M.; Helmy, N.A.; Ibrahim, S.S.; Nawito, M.F.; Othman, O.E. PCR-RFLP of bone morphogenetic protein 15 (BMP15/FecX) gene as a candidate for prolificacy in sheep. *Int. J. Vet. Sci. Med.* **2018**, *6*, S68–S72. [CrossRef]
50. Paek, H.J.; Li, Z.Y.; Quan, B.H.; Yin, X.J. Application of PCR-RFLP for quick identification of MSTN mutants in MSTN mutant pig breeding. *Anim. Biotechnol.* **2023**, *34*, 2231–2239. [CrossRef]
51. Parasar, P.; Bhushan, B. Characterization of BoLA class II DQA and DQB by PCR-RFLP, cloning, and sequencing reveals sequence diversity in crossbred cattle. *Anim. Biotechnol.* **2023**, *34*, 955–965. [CrossRef] [PubMed]
52. Sundaramahalingam, M.A.; Amrutha, C.; Rajeshbanu, J.; Thirukumaran, K.; Manibalan, S.; Ashokkumar, M.; Sivashanmugam, P. In silico approach for enhancing innate lipid content of *Yarrowia lipolytica*, by blocking the acyl-CoA oxidase-1 enzyme, using various analogous compounds of lipids. *J. Biomol. Struct. Dyn.* **2023**, *41*, 511–524. [CrossRef] [PubMed]
53. Zhang, F.; Xiong, Q.; Tao, H.; Liu, Y.; Zhang, N.; Li, X.F.; Suo, X.J.; Yang, Q.P.; Chen, M.X. ACOX1, regulated by C/EBPα and miR-25-3p, promotes bovine preadipocyte adipogenesis. *J. Mol. Endocrinol.* **2021**, *66*, 195–205. [CrossRef] [PubMed]
54. Jiao, Y.; Zan, L.S.; Liu, Y.F.; Wang, H.B. Molecular characterization, polymorphism of the ACOX1 gene and association with ultrasound traits in Bos taurus. *Genet. Mol. Res.* **2011**, *10*, 1948–1957. [CrossRef]
55. Lee, J.K.; Cho, Y.M.; Lee, J.H. Association of Bovine CSRP3 and ACOX1 Genes with Carcass and Meat Quality Traits. *Korean J. Agr. Sci.* **2010**, *37*, 231–238.
56. Dominguez-Castaño, P.; Marchi Maiorano, A.; Silva Lopes, J.E.; Vargas de Oliveira, M.H.; Michel Castilhos, A.; Vasconcelos Silva, J.A.I. Genetic parameters for mouth size and their influence on growth traits in pasture-raised Nelore cattle. *J. Anim. Sci.* **2023**, *101*, skad150. [CrossRef]
57. Chen, S.; Zhou, Y.; Chen, Y.; Gu, J. fastp: An ultra-fast all-in-one FASTQ preprocessor. *Bioinformatics* **2018**, *34*, i884–i890. [CrossRef]

48. Li, H.; Durbin, R. Fast and accurate long-read alignment with Burrows-Wheeler transform. *Bioinformatics* **2010**, *26*, 589–595. [CrossRef]
49. Purcell, S.; Neale, B.; Todd-Brown, K.; Thomas, L.; Ferreira, M.A.; Bender, D.; Maller, J.: Sklar, P.; de Bakker, P.I.; Daly, M.J.; et al. PLINK: A tool set for whole-genome association and population-based linkage analyses. *Am. J. Hum. Genet.* **2007**, *81*, 559–575. [CrossRef]

animals

MDPI

Article

Whole-Genome Sequencing of Dorper × Hu Hybrid Sheep for Screening Selection Signatures Associated with Litter Size

Liying Qiao [1], Ke Ma [1], Quanhong Yao [1], Siying Zhang [1], Zhixu Pang [1], Wannian Wang [1], Ke Cai [1] and Wenzhong Liu [1,2,*]

[1] Department of Animal Genetics, Breeding and Reproduction, College of Animal Science, Shanxi Agricultural University, Taigu 030801, China; liyingqiao1970@163.com (L.Q.); mk753951486250@163.com (K.M.); yqh978312@163.com (Q.Y.); zhangsy9526@163.com (S.Z.); pang_z_x@163.com (Z.P.); wannian1876@163.com (W.W.); caik1021@163.com (K.C.)

[2] Key Laboratory of Farm Animal Genetic Resources Exploration and Precision Breeding of Shanxi Province, Taigu 030801, China

* Correspondence: tglwzyc@163.com

Simple Summary

Reproductive performance is a key factor influencing sheep production efficiency. The Dorper × Hu hybrid sheep (DHS) combines the rapid growth of Dorper sheep with the high fecundity of Hu sheep and is widely used for lamb production in China. In this study, whole-genome selective sweep analysis and genome-wide association study (GWAS) were used to explore the genetic basis of high fecundity in DHS. The results revealed that DHS possesses a distinct genetic structure and high genetic diversity derived from its parental breeds. Several candidate genes related to reproduction were identified. Moreover, two SNPs—g.88680390 C>A (*SLC24A2/MLLT3*) and g.18197516 T>C (*ABCA1*)—were significantly associated with litter size. These findings provide valuable molecular markers for improving reproductive efficiency and advancing genomic breeding in meat sheep.

Abstract

Sheep are an economically important livestock species, and reproductive performance is a key trait affecting productivity. The Dorper × Hu hybrid sheep (DHS), widely bred in China, provides a valuable model for studying the genetic basis of prolificacy. This study aimed to investigate the genomic architecture and identify candidate genes associated with high litter size in DHS using whole-genome selective sweep analysis and genome-wide association study (GWAS). A total of 31 DHS individuals with complete reproductive records were sequenced and compared with publicly available genomic data from 20 Hu sheep (HUS) and 10 Dorper sheep (DPS). Population genetic structure and diversity were assessed using phylogenetic trees, principal component analysis (PCA), and ADMIXTURE analysis. To identify key genomic regions associated with litter size, we performed selective sweep analysis between the polytocous and monotocous subpopulations of DHS using multiple methods within a 50 kb sliding window framework, including F_{ST}, θ_π ratio, XP-CLR, and XP-EHH; we also conducted GWAS. DHS exhibited a distinct genetic structure with admixed ancestry and elevated genetic diversity. Genetic diversity analysis showed that DHS retained moderate levels of heterozygosity and polymorphism, comparable to or exceeding those of its parental breeds. Comparative analysis between polytocous and monotocous DHS identified reproduction-associated genes, including *MUC1*, *PLCB4*, *SIN3A*, and *ELAVL2*, enriched in pathways such as ovarian steroidogenesis, insulin secretion, and circadian entrainment. Furthermore, genome-wide association study (GWAS) identified 140 significant loci ($p < 10^{-5}$) associated with reproductive traits. From these, 10 candidate SNPs were selected for validation through single-marker association analysis

 https://doi.org/10.3390/ani15233505

in 200 DHS individuals, among which two loci—g.88680390 C>A (*SLC24A2/MLLT3*) and g.18197516 T>C (*ABCA1*)—showed significant correlations with litter size. These findings enhance our understanding of the genetic basis of prolificacy in DHS and provide valuable molecular markers for genomic selection in sheep-breeding programs.

Keywords: Dorper × Hu hybrid sheep; litter size; selective sweep analysis; whole-genome resequencing

1. Introduction

Sheep (*Ovis aries*) have served an essential source of meat, wool, and milk for thousands of years. China has a long history of sheep domestication and boasts abundant resources of indigenous breeds, which are generally categorized into three main groups: Mongolian, Kazakh, and Tibetan sheep [1]. Through long-term domestication and selective breeding, these breeds have adapted to diverse ecological regions and evolved distinct desirable traits [2], such as high reproductive performance in Small-Tailed Han and Hu sheep, superior meat quality in Tan sheep, and strong adaptability to harsh environments in Tibetan sheep. To overcome the limitations of indigenous breeds and meet the growing demand for high-quality lamb, crossbreeding local breeds with highly productive exotic breeds followed by long-term selection has become a primary genetic strategy for improving sheep production traits [3]. In China, this approach has yielded several successful composite breeds, including Luxi Black Head sheep [4] and Bamei mutton sheep [5], both of which exhibit enhanced production efficiency. Among them, the Dorper × Hu hybrid sheep (DHS), derived from crossing Dorper sheep (DPS) with Hu sheep (HUS), has emerged as one of the dominant breeds for lamb meat production in China. DHS combines the rapid growth and excellent carcass traits of Dorper sheep with the high prolificacy and non-seasonal estrus of Hu sheep [6], making it a critical genetic resource for meat sheep improvement.

In recent years, the rapid advancement of whole-genome resequencing technology, coupled with the completion of the sheep reference genome [7], has significantly accelerated research into the genetic mechanisms underlying morphological and economic traits in sheep. Whole-genome resequencing has been widely utilized to investigate sheep domestication and evolution [8], as well as to unravel the genetic basis of complex traits—such as reproductive capacity, coat color, horn type, tail type, and body size [9]. The identification of genetic variants associated with these traits deepens our understanding of sheep biology and provides a valuable foundation for genomic selection in breeding programs [10].

Among various production traits, reproductive efficiency plays a pivotal role in determining productivity and economic return in sheep production systems. It is regulated by genetic, hormone, environmental, and management factors. However, litter size is a complex trait with low to moderate heritability [11], controlled by both major genes and polygenes and influenced by environmental factors. Several major genes, such as *BMPR1B* [12], *BMP15*, and *GDF9* [13]—have been shown to significantly affect this trait. In particular, the well-characterized *FecB* mutation (*BMPR1B*: A746G/p.Q249R) markedly increases litter size, and this mutation is nearly fixed in Hu sheep [1]. The *BMP15* gene plays a vital role in follicular development by promoting granulosa cell proliferation and inhibiting follicle-stimulating hormone expression. *GDF9* is also essential for folliculogenesis, with several heterozygous mutations associated with increased litter size. Notably, candidate genes linked to prolificacy vary across breeds, reflecting the polygenic and breed-specific nature of litter size. Therefore, identifying fertility-related genes is critical for elucidating the genetic mechanisms underlying reproductive performance and laying a solid foundation for marker-assisted selection in sheep breeding.

Given that the molecular mechanisms and candidate genes regulating reproductive traits in DHS remain incompletely understood, this study integrates whole-genome resequencing data from 31 DHS individuals with publicly available genomic data from 20 HUS and 10 DPS. Through genetic diversity analysis and genetic structure analysis, we systematically clarified the phylogenetic status of DHS. Meanwhile, we compare selection signals between the polytocous subgroup (those with three consecutive multiple-lamb litters) and monotocous subgroup (those with three consecutive single-lamb litters) within the DHS population to identify candidate genes associated with high prolificacy, ultimately clarifying the key genetic variants contributing to differences in reproductive performance.

2. Materials and Methods

2.1. DNA Sample and Sequence

A total of 31 Dorper × Hu hybrid sheep (DHS), all three years of age and possessing records of three consecutive lambings, were selected from Genyuan Animal Husbandry Farm in Shanxi Province, China. This included 15 individuals that consistently delivered multiple lambs (prolific/polytocous) and 16 individuals that consistently delivered single lambs (non-prolific/monotocous) across all three parities. In Supplementary Table S1, individuals denoted with a prefix of "M" correspond to monotocous sheep (non-prolific), while those starting with "S" refer to polytocous sheep (prolific). Approximately 5 mL of jugular blood was collected from each individual into EDTA-treated tubes and stored at −20 °C until DNA extraction. To assess genetic background and admixture patterns, whole-genome resequencing data from 20 HUS and 10 DPS were obtained from the Sequence Read Archive (BioProject: PRJNA681929) for comparative analysis.

2.2. Whole-Genome Data Processing

Raw genome sequence data were quality filtered using fastp (v0.23.4) [14] with parameters "−n 1 −q 5 −u 50 −l 100" to remove reads containing an excessive number of ambiguous bases, low-quality reads, or adapter contamination. Clean reads were mapped and aligned to the *Ovis aries* reference genome (ARS-UI_Ramb_v2.0, https://www.ncbi.nlm.nih.gov/datasets/genome/GCF_016772045.1/, accessed on 1 May 2025) using BWA (v0.7.17) [15], and sequencing depth for each sample was calculated using SAMtools (v1.20) [16]. Variant calling was performed across all samples using the HaplotypeCaller module in GATK (v4.18) [17]. To reduce false positives, SNPs were filtered using the following thresholds: QD < 2.0, FS > 60.0, MQ < 40.0, SOR > 3.0, MQRankSum < −12.5, and ReadPosRankSum < −8.0. Additional stringent filters were applied to remove: (i) SNP clusters (more than one SNP within 5 bp), (ii) SNPs within 5 bp of indels, and (iii) genotypes with low quality (GQ < 20), which were labeled as "lowGQ".

2.3. Population Structure Analysis

Population genetic structure was assessed using three complementary approaches: Neighbor-Joining (NJ) tree construction, Principal Component Analysis (PCA), and ancestry inference. NJ trees were constructed in MEGA (v11.0.13) [18] using pairwise genetic distance matrices calculated by PLINK. PCA was performed using GCTA (v1.94.1) [19] based on the genomic relationship matrix calculated using the VanRaden method. ADMIXTURE (v1.30) [20] was employed to estimate individual ancestry components, with the number of ancestral populations determined to be K = 2–3.

2.4. Genetic Diversity Analysis

Genetic diversity indices including observed heterozygosity (Ho), expected heterozygosity (He), polymorphic marker ratio (P_N), minor allele frequency (MAF), and polymorphism information content (PIC) were calculated using PLINK [21]. Runs of homozygosity (ROH) were identified using the following criteria: a minimum length of 300 kb, at least 50 SNPs per ROH, and a SNP density of one per 50 kb. Up to 5 missing calls and 3 heterozygous calls were allowed per window, with a maximum gap of 300 kb between adjacent SNPs. Linkage disequilibrium (LD) decay was analyzed using PopLDdecay (v3.42) [22]. The parameters used were: MaxDist 1,000,000, MAF 0.05, Het 0.9, and Miss 0.1.

2.5. Selective Sweep Identification

Genomic selection signals were identified using VCFtools (v0.1.16) [23] for fixation index (F_{ST}) and nucleotide diversity (θ_π) ratios, XP-CLR (v1.1.2) [24] for cross-population composite likelihood ratio (XP-CLR), and selscan (v1.2.0) [25] for XP-EHH. All analyses were conducted using a 50 kb sliding window with a 10 kb step. Candidate regions were defined as the top 1% windows ranked by scores from each selection method. Genes located within these candidate regions were annotated and analyzed for functional enrichment using Gene Ontology (GO) and Kyoto Encyclopedia of Genes and Genomes (KEGG) pathways via DAVID [26].

2.6. Genome-Wide Association Study (GWAS)

A GWAS was conducted using WGS data from 31 DHS ewes (15 polytocous and 16 monotocous), all aged three years with records from three consecutive lambings. The Generalized Linear Model (GLM) implemented in GEMMA [27] was utilized to identify SNPs associated with prolificacy traits. Phenotypes were treated as a binary variable: ewes with three consecutive multiple-lamb litters were classified as polytocous (coded as 1), while those with three consecutive single-lamb litters were classified as monotocous (coded as 0). To correct for population stratification, the top ten principal components (PCs) derived from PCA were incorporated into the model.

2.7. Targeted Genotyping and Association Analysis

Based on production records, 200 DHS with three consecutive lambing events were selected for targeted SNP genotyping. Jugular blood samples (5 mL) were collected into EDTA-treated tubes for DNA extraction. Candidate nonsynonymous SNPs were identified. Genotyping was performed using the MassARRAY® platform (Agena Bioscience, San Diego, CA, USA). DNA samples were quality-controlled prior to PCR amplification of target fragments. Single-base extension reactions were carried out using specific extension primers. The resulting products were purified, spotted onto a SpectroCHIP array, and analyzed by mass spectrometry. Genotyping was achieved by distinguishing products based on molecular weight differences. All MassARRAY genotyping assays were conducted by Beijing Compass Biotechnology Co., Ltd. Detailed primer information is provided in Table S8.

For association analysis, consistent with the GLM framework used in the GWAS (Section 2.6), SNP-trait associations were also evaluated using a GLM, but adapted for the analysis of quantitative traits (litter size) with repeated measurements. The model included parity as a fixed effect to account for its contribution to litter size variation:

$$Y = \mu + W\alpha + X_s\beta_s + e$$

where Y is the litter size, μ is the population mean, X_s is the genotype vector for SNP s, β_s is the genetic effect of SNP, α is the parity effect, W is the design matrix for α, and e is the residual error.

3. Results

3.1. Genome Resequencing and Identification of Single Nucleotide Polymorphisms

Whole-genome sequencing data were generated for 31 DHS, and publicly available genomic data from 30 sheep (10 DPS and 20 HUS; PRJNA681929 [28]) were also included in the analysis. In total, the raw sequencing data amounted to 2067.57 G, of which 2023.03 G remained after quality filtering. The average Q30 was 94.26%, and the GC content was 44.82% (Supplementary Table S1). All 61 samples were aligned to the sheep reference genome, achieving an average alignment rate of 99.86% (Supplementary Table S2) and an average sequencing depth of 10.83× (Supplementary Table S3). After variant calling and filtering, an average of 12.52 million SNPs were identified per DHS individual, 12.06 million per DPS individual, and 12.39 million per HUS individual (Supplementary Table S4). Most SNPs were located in intergenic and intronic regions, while a smaller proportion were found in exonic regions, untranslated regions (UTRs), and splicing sites. The genomic distribution density of SNPs is shown in Supplementary Figure S1.

3.2. Population Genetic Structure

Based on genome data, we analyzed the genetic relationships among the three sheep populations. The phylogenetic tree, constructed using individual SNP variants, revealed that DHS, DPS, and HUS were clearly grouped into three distinct genetic clusters (Figure 1A). Furthermore, PCA further supported the genetic differentiation among the three populations. PC1 and PC2 explaining 4.55% and 3.61% of the total genetic variation, respectively, and clearly separated the individuals into three clusters corresponding to DHS, DPS, and HUS (Figure 1B). Ancestry component analysis revealed that when K = 2, DPS and HUS each formed distinct ancestral groups, while DHS exhibited a higher proportion of DPS ancestry (Figure 1C). When K = 3, each population exhibited a unique genetic signature, allowing clearer differentiation among the three groups.

3.3. Population Genetic Diversity

We systematically assessed genetic diversity, polymorphism, and genome-wide patterns among the three sheep populations to investigate how hybridization has shaped the genetic background of DHS. Nucleotide diversity (π) analysis revealed that DHS exhibited diversity levels comparable to HUS and greater than DPS (Figure 2A). ROH analysis indicated that most ROHs were concentrated in the 0–1 Mb range across all populations, but DPS had significantly more ROHs in the 1–2 Mb, 2–4 Mb, and 4–8 Mb intervals (Figure 2B). The *Ho*, *He*, and MAF of DHS were 0.250, 0.249, and 0.177, respectively—values slightly higher than those observed in DPS and HUS. The P_N in DHS reached 0.940, significantly higher than in DPS (0.668) and HUS (0.846), indicating that more polymorphic functional loci were retained due to its hybrid background (Figure 2C). The MAF distribution revealed that DHS had the lowest proportion of SNPs in the 0.4–0.5 interval (11.19%), a trend similar to those observed in DPS and HUS, indicating broadly comparable allele frequency patterns across the three populations (Figure 2D). LD analysis revealed that DHS exhibited lower LD levels across all genomic distances, particularly over short distances, with DPS showing the highest LD and HUS an intermediate degree (Figure 2E). These findings, consistent with the ROH patterns, suggest that hybridization in DHS increased genetic diversity, reduced linkage disequilibrium, and contributed to a broader genomic foundation conducive to future breeding improvement.

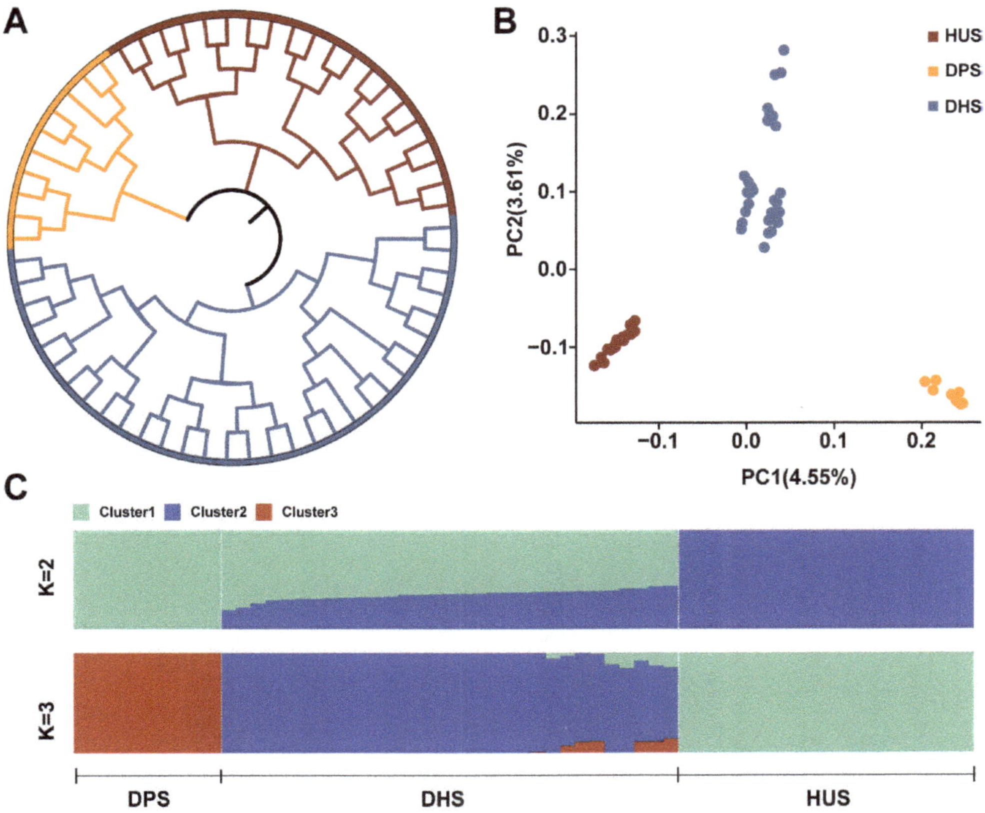

Figure 1. Population Genetics Analyses of Samples. (**A**) Phylogenetic trees of three sheep populations constructed based on the neighbor-joining method; (**B**) PCA results for three sheep populations; (**C**) Analysis of ancestral composition of three sheep populations.

3.4. Selection Signatures Between the Polytocous and Monotocous Sheep

Prior to analyzing novel selection signatures for prolificacy, we first confirmed the status of the major prolificacy gene, *FecB* (*BMPR1B* c.746G>A). The prolific B allele was fixed across all 31 DHS individuals, resulting in minimal polymorphism at this locus. Specifically, the 15 polytocous individuals comprised 14 B+ and 1 BB genotypes, while the 16 monotocous individuals consisted of 13 B+ and 3 BB genotypes. Due to this high genetic fixation and insufficient polymorphism across the two groups, the *FecB* was unable to register as a significant selective sweep region in this intraspecies comparison. This necessitates the search for other novel or quantitative loci contributing to litter size variation.

To uncover the genetic basis of prolificacy, we conducted selective sweep analysis between polytocous and monotocous DHS using four methods. Distinct selection signals were identified on multiple chromosomes, with F_{ST}, θ_π ratio, XP-CLR, and XP-EHH (Figure 3A) collectively highlighting key loci potentially associated with reproductive adaptation. In the top 1% of selected genomic regions, F_{ST}, θ_π ratio, XP-CLR, and XP-EHH, respectively, identified 470, 707, 1323, and 532 candidate genes (Figure 3B; Supplementary Table S5). *LOC101109111* and *LOC101109377* were consistently identified by all four methods, indicating their potential key roles in regulating litter size. Additional genes, such as *MUC1*, *MSH3*, *FNBP1L*, *TRIM46*, *CADM2*, and *CEP128*, were identified by at least three methods, suggesting they may be under multiple selection pressures and contribute to the genetic regulation of prolificacy. GO and KEGG analyses (Figure 3C; Supplementary Table S6) of candidate genes from selective sweep regions revealed enrichment in reproductive-related biological processes and pathways, including follicle development, hormonal regulation, ovarian steroidogenesis, insulin secretion, and circadian rhythm, suggesting their roles in regulating prolificacy and reproductive cycles.

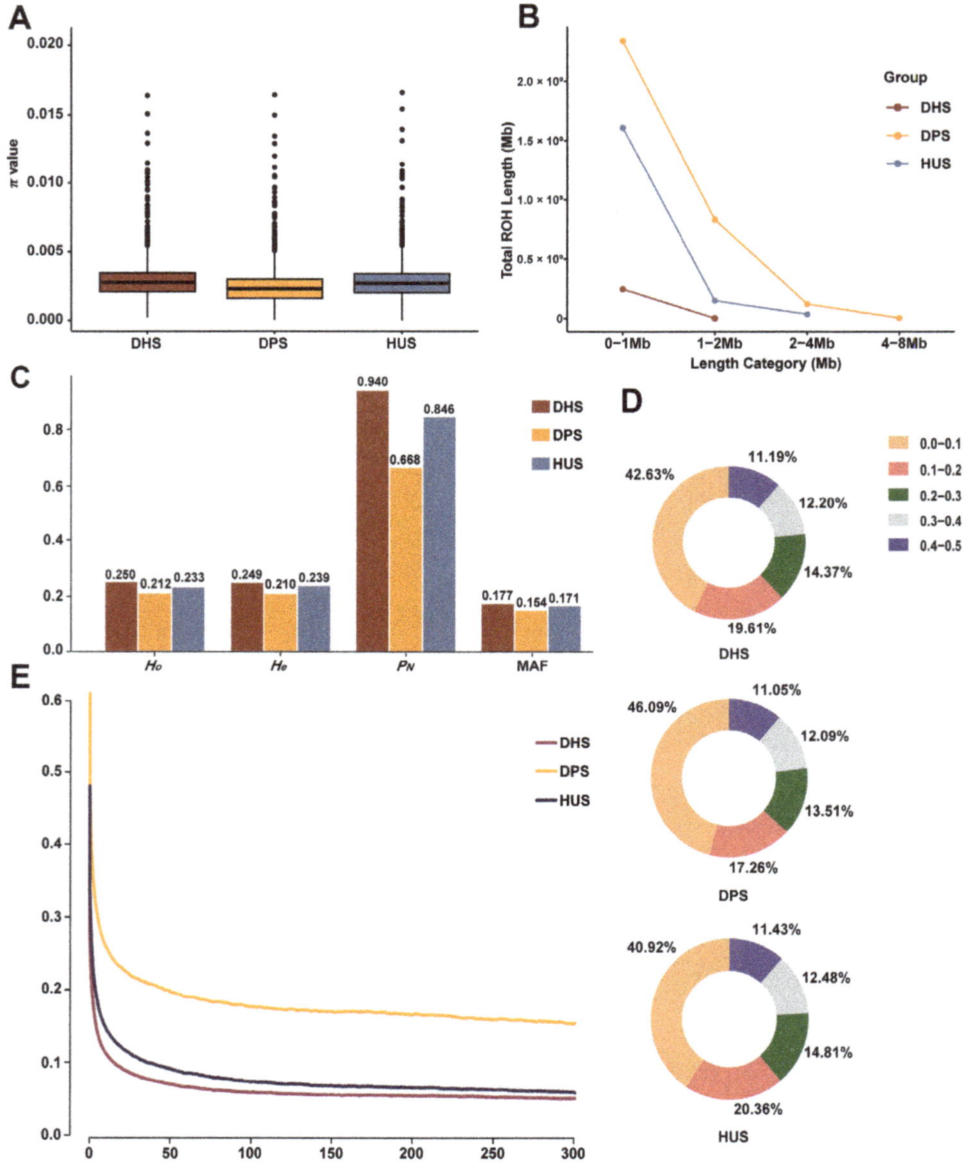

Figure 2. Genetic Diversity Statistics of Three Sheep Breeds. (**A**) Whole-genome distribution of π in three populations; (**B**) Distribution of total ROH lengths by length category for the three populations; (**C**) Statistics of genetic diversity parameters of the three populations, including *Ho*, *He*, P_N, and MAF; (**D**) Distribution of MAF in three populations; (**E**) Genome-wide average LD decay for the three breeds.

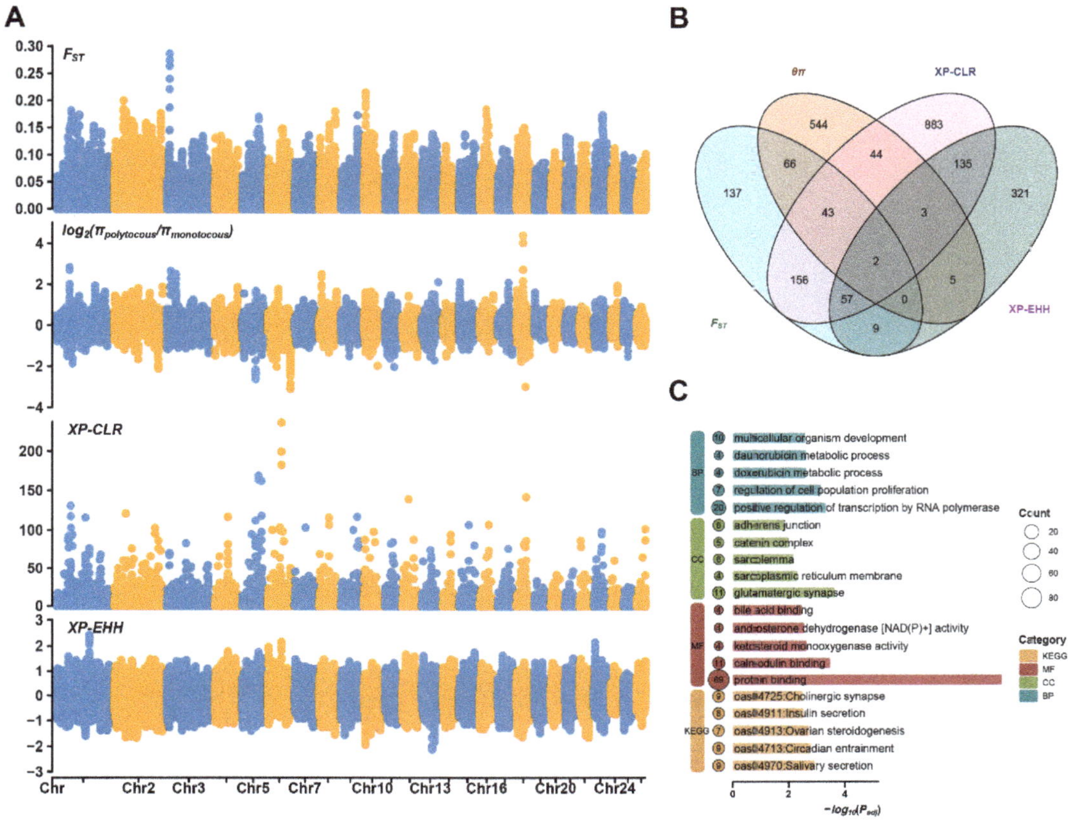

Figure 3. Selection signal analysis and functional enrichment of genes with litter size in DHS. (**A**) Manhattan plots of genome-wide selection signatures between polytocous and monotocous sheep, based on F_{ST}, $\log_2(\pi_{\text{polytocous}}/\pi_{\text{monotocous}})$, XP-CLR, XP-EHH. Different colors indicate different chromosomes. (**B**) Venn diagram of candidate genes identified by different selection methods. (**C**) GO and KEGG enrichment analysis of genes identified by at least two methods

3.5. Genome-Wide Association Analysis of Litter Size in DHS

A GWAS was performed using the GLM to identify genetic variants associated with litter size in DHS. The results are presented in Figure 4 (Manhattan and Q–Q plots). The Q–Q plot showed that all quantiles were distributed along the expected diagonal line (slope = 1), indicating that the population structure and genetic background were well controlled and the association signals were reliable. Using a threshold $p < 10^{-5}$, the analysis yielded 140 significant SNPs, which annotated 83 candidate genes highly associated with litter size (Supplementary Tables S7 and S8). Among these, 12 SNPs met a more stringent significance level of $p < 10^{-6}$. These 12 highly significant SNPs were mainly clustered on Chromosome 2 (2 loci) and Chromosome 25 (6 loci).

To validate the GWAS findings and identify potential functional mutations, targeted genotyping was performed on 200 DHS exhibiting three consecutive lambing records. The 10 candidate SNPs selected for MassARRAY® genotyping were chosen based on two primary criteria: (1) their highly significant GWAS *p*-values, which ranged from 8.77×10^{-9} to 2.94×10^{-6}, and (2) the known biological functions of their corresponding genes in reproduction and development. Specifically, genes harboring these selected SNPs included: *SLC24A2* and *MLLT3* (involved in cell signaling and gene transcription regulation); *PAX3* and *SGPP2* (related to embryonic development and lipid metabolism); *ABCA1* (affects steroid hormone biosynthesis and ovarian function); and *NRG3* and *KCNU1* (linked to neuroendocrine and sperm maturation processes). These 10 SNPs, located on Chromosomes 2, 3, 14, 23, 25, and 26, were then genotyped using the MassARRAY® platform, which showed high detection rates ranging from 81.5% to 100% (Table 1).

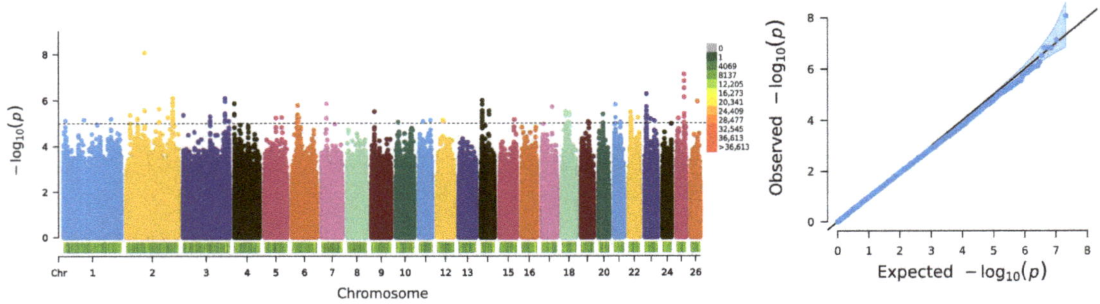

Figure 4. Manhattan Plot and QQ Plot of Genome-Wide Association Analysis for the prolificacy trait (polytocous vs. monotocous) of Dorper × Hu Hybrid Sheep. The phenotype represents the overall prolificacy status derived from the ewes' reproductive records across three consecutive parities. The blue shadow in QQ Plot indicates the confidence interval.

Table 1. MassARRAY® SNP Genotyping Site Information.

SNP	Chromosome	Position	Mutation	Gene	GWAS *p*-Value	Detection Rate
g.88680390 C>A	2	88680390	C/A	*SLC24A2, MLLT3*	8.77×10^{-9}	98.00%
g.224575852 G>A	2	224575852	G/A	*PAX3, SGPP2*	8.11×10^{-7}	92.50%
g.18197516 T>C	2	18197516	T/C	*ABCA1*	4.48×10^{-6}	100.00%
g.205268702 A>G	3	205268702	A/G	*LOC105613002*	7.96×10^{-7}	97.50%
g.205266796 T>C	3	205266796	T/C		8.55×10^{-7}	81.50%
g.6708960 G>A	14	6708960	G/A	*LOC105609364, DYNLRB2*	1.03×10^{-6}	81.50%
g.40389551 C>G	14	40389551	C/G	*ZNF536*	2.94×10^{-6}	97.00%
g.7693935 T>C	23	7693935	T/C	*DOK6*	2.98×10^{-6}	87.00%
g.36609350 G>C	25	36609350	G/C	*NRG3*	7.74×10^{-8}	99.50%
g.31340267 C>T	26	31340267	C/T	*KCNU1*	1.15×10^{-6}	99.00%

Subsequently, using the full dataset of 200 DHS ewes, the GLM was used to analyze the association between each SNP genotype and the number of offspring (Table 2). The results revealed that two sites, g.88680390 C>A and g.18197516 T>C, showed significant associations with litter size ($p < 0.05$). Specifically, ewes carrying the AA genotype at g.88680390 exhibited consistently higher average litter sizes across the first to third parities compared with other genotypes. Similarly, the TT genotype at g.18197516 was associated with larger litter sizes than the heterozygous or homozygous mutant types. These results suggest that the identified loci may have favorable allelic effects on reproductive performance in DHS and could serve as molecular markers for the selection of high-prolificacy individuals.

Table 2. Association Analysis Between SNP Mutation Sites and Litter Size in 200 DHS using Three Consecutive Lambing Records.

SNP	Genotype	Number	Detection Number (Rate)	First Parity		Second Parity		Third Parity		Overall Mean	
				Litter Size	*p*-Value	Litter Size	*p*-Value	Litter Size	*p*-Value	Litter Size	*p*-Value
g.88680390 C>A	AA	29	196 (98.00%)	1.79 ± 0.12 ^{a}	0.15	2.34 ± 0.17 ^{a}	0.86	2.59 ± 0.14 ^{a}	0.05	2.42 ± 0.11 ^{a}	0.02
	CC	75		1.72 ± 0.07 ^{a}		2.25 ± 0.09 ^{a}		2.85 ± 0.06 ^{a}		2.45 ± 0.05 ^{a}	
	CA	92		1.58 ± 0.06 ^{a}		2.26 ± 0.08 ^{a}		2.62 ± 0.07 ^{a}		2.24 ± 0.05 ^{b}	
g.224575852 G>A	AA	64	185 (92.50%)	1.72 ± 0.08 ^{a}	0.39	2.25 ± 0.11 ^{a}	0.93	2.73 ± 0.07 ^{a}	0.99	2.42 ± 0.06 ^{a}	0.16
	GG	121		1.64 ± 0.05 ^{a}		2.24 ± 0.07 ^{a}		2.74 ± 0.07 ^{a}		2.32 ± 0.05 ^{a}	
g.18197516 T>C	CC	27	200 (100.00%)	1.37 ± 0.09 ^{b}	0.01	1.85 ± 0.13 ^{b}	0.01	2.30 ± 0.17 ^{b}	0.00	1.98 ± 0.09 ^{b}	0.00
	TT	148		1.74 ± 0.05 ^{a}		2.34 ± 0.07 ^{a}		2.80 ± 0.05 ^{a}		2.44 ± 0.04 ^{a}	
	TC	25		1.56 ± 0.10 ^{ab}		2.24 ± 0.13 ^{ab}		2.64 ± 0.16 ^{ab}		2.24 ± 0.09 ^{ab}	
g.205268702 A>G	AA	46	195 (97.50%)	1.57 ± 0.08 ^{a}	0.48	2.00 ± 0.11 ^{b}	0.02	2.72 ± 0.10 ^{a}	0.97	2.23 ± 0.07 ^{a}	0.15
	AG	98		1.68 ± 0.07 ^{a}		2.32 ± 0.08 ^{ab}		2.71 ± 0.07 ^{a}		2.38 ± 0.05 ^{a}	
	GG	51		1.71 ± 0.09 ^{a}		2.43 ± 0.12 ^{a}		2.69 ± 0.09 ^{a}		2.41 ± 0.07 ^{a}	
g.205266796 T>C	CC	37	163 (81.50%)	1.57 ± 0.09 ^{a}	0.37	1.97 ± 0.12 ^{b}	0.03	2.70 ± 0.11 ^{a}	0.65	2.23 ± 0.08 ^{a}	0.08
	CT	79		1.72 ± 0.07 ^{a}		2.29 ± 0.09 ^{ab}		2.78 ± 0.08 ^{a}		2.44 ± 0.06 ^{a}	
	TT	47		1.74 ± 0.09 ^{a}		2.43 ± 0.12 ^{a}		2.68 ± 0.09 ^{a}		2.42 ± 0.07 ^{a}	
g.6708960 G>A	GG	69	188 (94.00%)	1.65 ± 0.07 ^{a}	0.76	2.13 ± 0.10 ^{a}	0.08	2.55 ± 0.08 ^{b}	0.01	2.27 ± 0.06 ^{b}	0.05
	GA	119		1.68 ± 0.06 ^{a}		2.34 ± 0.07 ^{a}		2.82 ± 0.06 ^{a}		2.42 ± 0.04 ^{a}	
g.40389551 C>G	CC	80	194 (97.00%)	1.58 ± 0.06 ^{b}	0.00	2.35 ± 0.09 ^{a}	0.38	2.66 ± 0.07 ^{a}	0.72	2.37 ± 0.06 ^{a}	0.71
	GG	35		2.00 ± 0.12 ^{a}		2.29 ± 0.13 ^{a}		2.74 ± 0.14 ^{a}		2.40 ± 0.08 ^{a}	
	GC	85		1.61 ± 0.07 ^{b}		2.18 ± 0.09 ^{a}		2.74 ± 0.07 ^{a}		2.32 ± 0.05 ^{a}	
g.7693935 T>C	CC	18	174 (87.00%)	1.72 ± 0.19 ^{a}	0.66	2.17 ± 0.20 ^{a}	0.74	2.50 ± 0.15 ^{a}	0.02	2.23 ± 0.11 ^{a}	0.41
	TT	62		1.73 ± 0.08 ^{a}		2.31 ± 0.11 ^{a}		2.63 ± 0.09 ^{a}		2.40 ± 0.06 ^{a}	
	TC	94		1.64 ± 0.06 ^{a}		2.22 ± 0.08 ^{a}		2.87 ± 0.07 ^{a}		2.39 ± 0.05 ^{a}	
g.36609350 G>C	CC	71	199 (99.50%)	1.62 ± 0.08 ^{a}	0.54	2.34 ± 0.10 ^{a}	0.61	2.59 ± 0.09 ^{a}	0.16	2.32 ± 0.06 ^{a}	0.69
	GG	35		1.63 ± 0.09 ^{a}		2.23 ± 0.12 ^{a}		2.77 ± 0.12 ^{a}		2.36 ± 0.08 ^{a}	
	GC	93		1.72 ± 0.06 ^{a}		2.22 ± 0.08 ^{a}		2.78 ± 0.06 ^{a}		2.39 ± 0.05 ^{a}	
g.31340267 C>T	CC	60	198 (99.00%)	1.63 ± 0.08 ^{a}	0.51	2.27 ± 0.11 ^{a}	0.76	2.73 ± 0.08 ^{a}	0.95	2.39 ± 0.06 ^{a}	0.72
	CT	113		1.65 ± 0.06 ^{a}		2.29 ± 0.07 ^{a}		2.70 ± 0.07 ^{a}		2.35 ± 0.05 ^{a}	
	TT	25		1.80 ± 0.12 ^{a}		2.16 ± 0.18 ^{a}		2.72 ± 0.12 ^{a}		2.30 ± 0.09 ^{a}	

Different letters (^{a} and ^{b}) indicate a significant difference between groups ($p < 0.05$). The same letters indicate no significant differences between groups ($p > 0.05$).

4. Discussion

This study attempted to identify the genetic characteristics of DHS using whole-genome data, focusing on population genetic structure and genetic diversity. The results highlight the potential and value of DHS as a composite breed with distinct genetic features for future breeding and improvement programs.

Population genetic structure analyses supported the classification of DHS as a distinct genetic cluster. These findings indicate that DHS has developed a stable genetic foundation through hybridization and has diverged from its parental populations, rather than remaining an intermediate type. These findings align with previous studies on the Purunã composite cattle breed [29] and prolific Suffolk sheep [30], further demonstrating that systematic hybrid breeding can generate novel breeds with independent genetic structures through population recombination. This provides valuable insights for meat sheep-breeding strategies.

DHS displayed greater genetic polymorphism than both HUS and DPS. Genetic diversity indicators, including nucleotide diversity (π), *Ho*, *He*, P_N, and MAF were all elevated in DHS, suggesting a greater accumulation of genetic variation during the hybridization process [31]. These findings imply that DHS harbors enhanced genetic potential for environmental adaptation and selective breeding. These observations are consistent with reports on high genetic diversity and low inbreeding in Luxi black-headed sheep [4], Bamei mutton sheep [5], and Creole goats [32]. This underscores the importance of hybrid breeding strategies in preserving genetic variability. ROH analysis showed that DHS predominantly contained short ROH segments, with relatively few long fragments. Combined with faster LD decay, this indicates a lower risk of inbreeding and a more relaxed population structure.

We identified overlapping candidate genes and enriched pathways in polytocous populations. *LOC101109111* and *LOC101109377* were consistently detected across all four selection methods. Although not extensively characterized previously, these genes may represent novel regulators of litter size. Several previously reported genes were also identified by three or more methods and appear to be under cumulative selection pressure. These include *MUC1*, which affects the uterine environment and embryo implantation and significantly influences litter size in pigs [33]; *PLCB4*, identified as a hub gene for litter size traits in sheep through meta-GWAS analysis [34]; *SIN3A*, which regulates chromatin remodeling during follicular development [35]; and *ELAVL2*, which is involved in oocyte maturation [36]. These genes are likely to play important roles in ovarian function and embryonic development. Moreover, the enrichment of pathways such as ovarian steroidogenesis [37], insulin secretion [38], Apelin signaling [39], and circadian entrainment [30] underscores the importance of coordinated endocrinological and developmental processes in prolificacy. These pathways have also been implicated in reproductive success across mammals, reinforcing the relevance of our findings.

In this study, we identified 140 significantly associated SNPs ($p < 10^{-5}$) via GWAS and selected 10 highly significant candidate loci for validation. Candidate genes in the regions harboring these SNPs and their biological functions provide critical insights into the molecular regulatory mechanisms underlying the high fecundity of DHS. Among these, the most significant association signal ($p = 8.77 \times 10^{-9}$) maps to the g.88680390 C>A locus on chromosome 2, which is located in the intergenic region between the *SLC24A2* and *MLLT3* genes. Previous studies have shown that *SLC24A2* plays an important role in the reproduction of Cameroon's native goats [40], suggesting a conserved function in sheep reproductive regulation. As a transcriptional activator, *MLLT3* can extend the transcription cycle of target genes and remodel chromatin during embryogenesis [41], and has been demonstrated to be involved in key processes of embryonic development [42]. Furthermore, significant differences in the expression levels of *MLLT3* were identified between the ovarian tissues of high- and low-fecundity Chongming white goats [43], further supporting its role in regulating reproductive traits. Another key candidate locus is g.18197516 T>C ($p = 4.48 \times 10^{-6}$), located within the *ABCA1* gene region. *ABCA1* plays a central role in transmembrane cholesterol transport, a precursor for steroid hormone synthesis. Studies have shown that *ABCA1*-deficient mice exhibit significantly reduced cholesterol ester levels in adrenal and ovarian tissues, leading to impaired reproductive capacity in homozygous female mice, characterized by fewer pregnancies and lower litter sizes [44]. Additionally, the g.36609350 G>C locus ($p = 7.74 \times 10^{-8}$) on chromosome 25 is associated with the *NRG3* gene, which has been identified as a candidate gene for litter size in sheep in multiple studies [44,45]. Although selective sweep analyses (e.g., F_{ST}, XP-CLR) and GWAS can both identify genomic regions associated with litter size variation, they capture distinct aspects of the underlying genetic architecture, resulting in incomplete overlap of detected signals. In this study, some loci (e.g., g.88680390 C>A) were identified by both methods—showing significant associations in GWAS while residing in regions with strong F_{ST} and XP-CLR signatures—whereas some GWAS-significant SNPs were outside major selective sweep regions. This discrepancy stems from the fact that selective sweep analyses focus on identifying genomic regions shaped by historical or recent directional selection within populations [46], while GWAS are more sensitive to SNPs with measurable additive genetic effects on traits [47].

To validate the GWAS findings, we performed targeted genotyping on a larger population. We confirmed that two key SNPs—g.88680390 C>A (located in *SLC24A2/MLLT3*) and g.18197516 T>C (located in *ABCA1*)—are significantly associated with increased litter size. Specifically, ewes carrying the AA genotype at g.88680390 and the TT genotype at g.18197516 exhibited higher litter sizes. Consequently, we recommend prioritizing these SNPs to enhance reproductive rates. While this study provides valuable genomic insights, it has several limitations. Although the validation utilized 200 DHS individuals, this sample size remains limited for litter size. To improve confidence in these SNPs and facilitate reliable marker-assisted selection, future work should involve genotyping in larger DHS populations and integrating functional analyses (e.g., expression Quantitative Trait Loci or eQTL analysis) to elucidate the precise causal mechanisms of these markers.

5. Conclusions

In summary, this study systematically characterized the genetic structure and genetic diversity of DHS, providing comprehensive insights for its genetic breeding research. The results confirm that DHS has formed an independent and stable genetic background through systematic hybridization, with a distinct population structure and higher genetic diversity. Additionally, this study investigated selection signatures associated with litter size in DHS, identifying multiple candidate genes and pathways that may be involved in regulating reproductive performance. Furthermore, two SNP loci—g.88680390 C>A (located in the genomic region of *SLC24A2/MLLT3*) and g.18197516 T>C (located in the genomic region of *ABCA1*)—showed a significant association with litter size, providing potential molecular markers for the improvement of sheep fecundity. These results offer important references for the development of genomic selection and breeding strategies related to sheep reproductive performance.

Supplementary Materials: The following supporting information can be downloaded at: https://www.mdpi.com/article/10.3390/ani15233505/s1, Table S1: Summary of Whole-Genome Resequencing Data Quality Before and After Filtering; Table S2: Summary of Whole-Genome Read Alignment Statistics to the Reference Genome; Table S3: Statistics of sample average sequencing depth and coverage ratio; Table S4: SNP location or type statistics; Table S5: Candidate genes associated with litter size in Dorper × Hu hybrid sheep identified by multiple selection signature methods and their overlaps; Table S6: GO and KEGG Enrichment Analysis of Candidate Genes Under Selection with litter size; Table S7: Results of Candidate SNP Loci Screening in Genome-Wide Association Analysis of Dorper × Hu Hybrid Sheep; Table S8: Site Primer Information Table. Figure S1: Heatmap of SNPs Density Distribution on Chromosomes.

Author Contributions: Conceptualization, L.Q. and W.L.; methodology, Z.P.; software, K.M. and K.C.; validation, L.Q., Z.P. and Q.Y.; formal analysis, Z.P.; investigation, W.W. and K.C.; resources, W.L.; data curation, S.Z.; writing—original draft preparation, L.Q. and Z.P.; writing—review and editing, K.M. and W.W.; visualization, Z.P. and S.Z.; supervision, Q.Y.; project administration, W.L.; funding acquisition, W.L. All authors have read and agreed to the published version of the manuscript.

Funding: This research was supported by the earmarked fund for Modern Agro-industry Technology Research System (2024CYJSTX14-06) and the Shanxi Province "Yan Yun Aries" Breeding Joint Research Project (NYGG23).

Institutional Review Board Statement: This study has been reviewed and approved by the Institutional Animal Care and Use Committee of Shanxi Agricultural University (Approval number: SXAU-EAW-2023S.NR.012023361).

Informed Consent Statement: Informed consent was obtained from all subjects involved in the study.

Data Availability Statement: The SNP variants generated in this study through variant calling using GATK are available via NutCloud at the following link: https://www.jianguoyun.com/p/DbjQdZgQ_tC_DRj9pIYGIAA (accessed on 1 July 2025).

Acknowledgments: All authors thank the Department of Genetics and Breeding, College of Animal Science, Shanxi Agricultural University for providing the necessary equipment for this study.

Conflicts of Interest: The authors declare no conflicts of interest.

References

1. Chen, K.; Zhang, Y.; Pan, Y.; Xiang, X.; Peng, C.; He, J.; Huang, G.; Wang, Z.; Zhao, P. Genomic insights into demographic history, structural variation landscape, and complex traits from 514 Hu sheep genomes. *J. Genet. Genom.* **2025**, *52*, 245–257. [CrossRef]
2. Wanjala, G.; Kusuma Astuti, P.; Bagi, Z.; Kichamu, N.; Strausz, P.; Kusza, S. A review on the potential effects of environmental and economic factors on sheep genetic diversity: Consequences of climate change. *Saudi J. Biol. Sci.* **2023**, *30*, 103505. [CrossRef]
3. Oyieng, E.; Ojango, J.M.K.; Gauly, M.; Mrode, R.; Dooso, R.; Okeyo, A.M.; Kalinda, C.; König, S. Evaluating reproduction traits in a crossbreeding program between indigenous and exotic sheep in semi-arid lands. *Animal* **2025**, *19*, 101391. [CrossRef] [PubMed]
4. Liu, Z.; Tan, X.; Wang, J.; Jin, Q.; Meng, X.; Cai, Z.; Cui, X.; Wang, K. Whole genome sequencing of Luxi Black Head sheep for screening selection signatures associated with important traits. *Anim. Biosci.* **2022**, *35*, 1340–1350. [CrossRef] [PubMed]
5. Yao, Y.; Pan, Z.; Di, R.; Liu, Q.; Hu, W.; Guo, X.; He, X.; Gan, S.; Wang, X.; Chu, M. Whole Genome Sequencing Reveals the Effects of Recent Artificial Selection on Litter Size of Bamei Mutton Sheep. *Animals* **2021**, *11*, 157. [CrossRef] [PubMed]
6. Nie, H.T.; Zhang, H.; You, J.H.; Wang, F. Determination of energy and protein requirement for maintenance and growth and evaluation for the effects of gender upon nutrient requirement in Dorper × Hu Crossbred Lambs. *Trop. Anim. Health Prod.* **2015**, *47*, 841–853. [CrossRef]
7. Cockett, N.E. The sheep genome. *Genome Dyn.* **2006**, *2*, 79–85.
8. Li, X.; Yang, J.; Shen, M.; Xie, X.-L.; Liu, G.-J.; Xu, Y.-X.; Lv, F.-H.; Yang, H.; Yang, Y.-L.; Liu, C.-B.; et al. Whole-genome resequencing of wild and domestic sheep identifies genes associated with morphological and agronomic traits. *Nat. Commun.* **2020**, *11*, 2815. [CrossRef]
9. Lagler, D.K.; Hannemann, E.; Eck, K.; Klawatsch, J.; Seichter, D.; Russ, I.; Mendel, C.; Lühken, G.; Krebs, S.; Blum, H.; et al. Fine-mapping and identification of candidate causal genes for tail length in the Merinolandschaf breed. *Commun. Biol.* **2022**, *5*, 918. [CrossRef]
10. Johnsson, M. Genomics in animal breeding from the perspectives of matrices and molecules. *Hereditas* **2023**, *160*, 20. [CrossRef]
11. Hulsman Hanna, L.L.; Taylor, J.B.; Holland, P.W.; Vonnahme, K.A.; Reynolds, L.P.; Riley, D.G. Effect of ewe birth litter size and estimation of genetic parameters on ewe reproductive life traits. *Animal* **2023**, *17*, 100900. [CrossRef]
12. Wen, Y.L.; Guo, X.F.; Ma, L.; Zhang, X.S.; Zhang, J.L.; Zhao, S.G.; Chu, M.X. The expression and mutation of BMPR1B and its association with litter size in small-tail Han sheep (*Ovis aries*). *Arch. Anim. Breed.* **2021**, *64*, 211–221. [CrossRef]
13. Zhang, Y.; Wang, H.; Li, T.; Zhang, N.; Chen, J.; Yang, H.; Peng, S.; Ma, R.; Wang, D.; Liu, Q.; et al. Association of BMP15 and GDF9 Gene Polymorphisms with Litter Size in Hu Sheep. *Genes* **2025**, *16*, 168. [CrossRef] [PubMed]

14. Chen, S.; Zhou, Y.; Chen, Y.; Gu, J. fastp: An ultra-fast all-in-one FASTQ preprocessor. *Bioinformatics* **2018**, *34*, i884–i890. [CrossRef] [PubMed]
15. Li, H.; Durbin, R. Fast and accurate short read alignment with Burrows–Wheeler transform. *Bioinformatics* **2009**, *25*, 1754–1760. [CrossRef] [PubMed]
16. Li, H.; Handsaker, B.; Wysoker, A.; Fennell, T.; Ruan, J.; Homer, N.; Marth, G.; Abecasis, G.; Durbin, R.; Genome Project Data Processing Subgroup. The Sequence Alignment/Map format and SAMtools. *Bioinformatics* **2009**, *25*, 2078–2079. [CrossRef]
17. McKenna, A.; Hanna, M.; Banks, E.; Sivachenko, A.; Cibulskis, K.; Kernytsky, A.; Garimella, K.; Altshuler, D.; Gabriel, S.; Daly, M.; et al. The Genome Analysis Toolkit: A MapReduce framework for analyzing next-generation DNA sequencing data. *Genome Res.* **2010**, *20*, 1297–1303. [CrossRef]
18. Tamura, K.; Stecher, G.; Kumar, S. MEGA11: Molecular Evolutionary Genetics Analysis Version 11. *Mol. Biol. Evol.* **2021**, *38*, 3022–3027. [CrossRef]
19. Yang, J.; Lee, S.H.; Goddard, M.E.; Visscher, P.M. GCTA: A Tool for Genome-wide Complex Trait Analysis. *Am. J. Hum. Genet.* **2011**, *88*, 76–82. [CrossRef]
20. Alexander, D.H.; Lange, K. Enhancements to the ADMIXTURE algorithm for individual ancestry estimation. *BMC Bioinform.* **2011**, *12*, 246. [CrossRef]
21. Purcell, S.; Neale, B.; Todd-Brown, K.; Thomas, L.; Ferreira, M.A.R.; Bender, D.; Maller, J.; Sklar, P.; de Bakker, P.I.W.; Daly, M.J.; et al. PLINK: A Tool Set for Whole-Genome Association and Population-Based Linkage Analyses. *Am. J. Hum. Genet.* **2007**, *81*, 559–575. [CrossRef]
22. Zhang, C.; Dong, S.-S.; Xu, J.-Y.; He, W.-M.; Yang, T.-L. PopLDdecay: A fast and effective tool for linkage disequilibrium decay analysis based on variant call format files. *Bioinformatics* **2019**, *35*, 1786–1788. [CrossRef]
23. Danecek, P.; Auton, A.; Abecasis, G.; Albers, C.A.; Banks, E.; DePristo, M.A.; Handsaker, R.E.; Lunter, G.; Marth, G.T.; Sherry, S.T.; et al. The variant call format and VCFtools. *Bioinformatics* **2011**, *27*, 2156–2158. [CrossRef]
24. Chen, H.; Patterson, N.; Reich, D. Population differentiation as a test for selective sweeps. *Genome Res.* **2010**, *20*, 393–402. [CrossRef]
25. Szpiech, Z.A. selscan 2.0: Scanning for sweeps in unphased data. *Bioinformatics* **2024**, *40*, btae006. [CrossRef] [PubMed]
26. Sherman, B.T.; Hao, M.; Qiu, J.; Jiao, X.; Baseler, M.W.; Lane, H.C.; Imamichi, T.; Chang, W. DAVID: A web server for functional enrichment analysis and functional annotation of gene lists (2021 update). *Nucleic Acids Res.* **2022**, *50*, W216–W221. [CrossRef] [PubMed]
27. Zhou, X.; Stephens, M. Genome-wide efficient mixed-model analysis for association studies. *Nat. Genet.* **2012**, *44*, 821–824. [CrossRef] [PubMed]
28. Lv, X.; Weihao, C.; Shanhe, W.; Xiukai, C.; Zehu, Y.; Tesfaye, G.; Mwacharo, J.M.; Aynalem, H.; Sun, W. Whole-genome resequencing of Dorper and Hu sheep to reveal selection signatures associated with important traits. *Anim. Biotechnol.* **2023**, *34*, 3016–3026. [CrossRef]
29. Mulim, H.A.; Brito, L.F.; Batista Pinto, L.F.; Moletta, J.L.; Da Silva, L.R.; Pedrosa, V.B. Genetic and Genomic Characterization of a New Beef Cattle Composite Breed (Purunã) Developed for Production in Pasture-Based Systems. *Front. Genet.* **2022**, *13*, 858970. [CrossRef]
30. Yang, H.; Zhu, M.; Wang, M.; Zhou, H.; Zheng, J.; Qiu, L.; Fan, W.; Yang, J.; Yu, Q.; Yang, Y.; et al. Genome-wide comparative analysis reveals selection signatures for reproduction traits in prolific Suffolk sheep. *Front. Genet.* **2024**, *15*, 1404031. [CrossRef]
31. Hedrick, P.W. Adaptive introgression in animals: Examples and comparison to new mutation and standing variation as sources of adaptive variation. *Mol. Ecol.* **2013**, *22*, 4606–4618. [CrossRef]
32. Corredor, F.-A.; Figueroa, D.; Estrada, R.; Burgos-Paz, W.; Salazar, W.; Cruz, W.; Lobato, R.; Injante, P.; Godoy, D.; Barrantes, C.; et al. Genome-wide single nucleotide polymorphisms reveal the genetic diversity and population structure of Creole goats from northern Peru. *Livest. Sci.* **2024**, *283*, 105473. [CrossRef]
33. Xiao, C.; Jinluan, F.; Aiguo, W. Effect of VNTR polymorphism of the Muc1 gene on litter size of pigs. *Mol. Biol. Rep.* **2012**, *39*, 6251–6258. [CrossRef]
34. Gholizadeh, M.; Esmaeili-Fard, S.M. Meta-analysis of genome-wide association studies for litter size in sheep. *Theriogenology* **2022**, *180*, 103–112. [CrossRef] [PubMed]
35. Luo, L.; Dang, Y.; Shi, Y.; Zhao, P.; Zhang, Y.; Zhang, K. SIN3A Regulates Porcine Early Embryonic Development by Modulating CCNB1 Expression. *Front. Cell Dev. Biol.* **2021**, *9*, 604232. [CrossRef] [PubMed]
36. Chalupnikova, K.; Petr, S.; Vadym, S.; Radislav, S.; and Svoboda, P. An oocyte-specific ELAVL2 isoform is a translational repressor ablated from meiotically competent antral oocytes. *Cell Cycle* **2014**, *13*, 1187–1200. [CrossRef] [PubMed]
37. Hadjadj, I.; Zuzana, F.; Luz, G.M.d.l.; Barbora, L.; Martin, M.; Peter, M.; Ivan, A.; Sirotkin, A.V.; Argente, M.-J. Effects of selection for litter size variability on ovarian folliculogenesis, ovarian cell proliferation, apoptosis, and production of regulatory peptides in rabbits. *Ital. J. Anim. Sci.* **2024**, *23*, 1290–1304. [CrossRef]

38. Père, M.-C.; Etienne, M. Influence of litter size on insulin sensitivity in multiparous sows1. *J. Anim. Sci.* **2019**, *97*, 874–884. [CrossRef]
39. Dobrzyn, K.; Kiezun, M.; Kopij, G.; Zarzecka, B.; Gudelska, M.; Kisielewska, K.; Zaobidna, E.; Makowczenko, K.G.; Dall'Aglio, C.; Kamiński, T.; et al. Apelin-13 modulates the endometrial transcriptome of the domestic pig during implantation. *BMC Genom.* **2024**, *25*, 501.
40. Kouam Simo, J.; Meutchieye, F.; Wouobeng, P.; Tarekegn, G.M.; Mutai, C.; Nandolo, W.; Pelle, R.; Djikeng, A.; Manjeli, Y. Genome-wide association study for the level of prolificacy in Cameroon's native goat. *J. Appl. Anim. Res.* **2024**, *52*, 2291472. [CrossRef]
41. Germano, G.; Porazzi, P.; Felix, C.A. Leukemia-associated transcription factor mllt3 is important for primitive erythroid development in zebrafish embryogenesis. *Dev. Dyn. Off. Publ. Am. Assoc. Anat.* **2022**, *251*, 1728–1740. [CrossRef] [PubMed]
42. Büttner, N.; Johnsen, S.A.; Kügler, S.; Vogel, T. Af9/Mllt3 interferes with Tbr1 expression through epigenetic modification of histone H3K79 during development of the cerebral cortex. *Proc. Natl. Acad. Sci. USA* **2010**, *107*, 7042–7047. [CrossRef]
43. Lin, Y.; Sun, L.; Dai, J.; Lv, Y.; Liao, R.; Shen, X.; Gao, J. Characterization and Comparative Analysis of Whole-Transcriptome Sequencing in High- and Low-Fecundity Chongming White Goat Ovaries during the Estrus Phase. *Animals* **2024**, *14*, 988. [CrossRef]
44. Tao, L.; He, X.; Wang, X.; Di, R.; Chu, M. Litter size of sheep (*Ovis aries*): Inbreeding depression and homozygous regions. *Genes* **2021**, *12*, 109. [CrossRef]
45. Liu, W.; Ma, S.; Lu, Q.; Tang, S.; Mamat, N.; Wang, Y.; Hong, W.; Hu, X.; Wu, C.; Fu, X. Genome-Wide Association Study Revealed Candidate Genes Associated with Litter Size, Weight, and Body Size Traits in Tianmu Polytocous Sheep (*Ovis aries*). *Biology* **2025**, *14*, 1446. [CrossRef] [PubMed]
46. Hoban, S.; Kelley, J.L.; Lotterhos, K.E.; Antolin, M.F.; Bradburd, G.; Lowry, D.B.; Poss, M.L.; Reed, L.K.; Storfer, A.; Whitlock, M.C. Finding the Genomic Basis of Local Adaptation: Pitfalls, Practical Solutions, and Future Directions. *Am. Nat.* **2016**, *188*, 379–397. [CrossRef]
47. Tam, V.; Patel, N.; Turcotte, M.; Bossé, Y.; Paré, G.; Meyre, D. Benefits and limitations of genome-wide association studies. *Nat. Rev. Genet.* **2019**, *20*, 467–484. [CrossRef] [PubMed]

International Journal of
Molecular Sciences

MDPI

Article

Spatial Multi-Omics Analysis of the Qianqiu Goat Gut Microbiome and Metabolome

Panpan Guo [1,2], Wenjuan Qin [1,3], Wencheng Song [2] and Hongquan Chen [1,*]

[1] School of Animal Science and Technology, Anhui Agricultural University, Hefei 230036, China; guopp@iim.ac.cn (P.G.); qinwenjuan0611@gmail.com (W.Q.)
[2] Hefei Institute of Physical Science, Chinese Academy of Sciences, Hefei 230031, China; wencsong@cmpt.ac.cn
[3] Key Laboratory of Anhui Local Livestock and Poultry Genetic Resources Conservation and Biobreeding of Anhui Province, Hefei 230036, China
* Correspondence: chqchq@ahau.edu.cn

Abstract

This study profiled the rumen (RM), small intestine (SI), and large intestine (LI) of 24 samples collected from eight 6-month-old Qianqiu goats (body weight 28.40 ± 1.80 kg), with the samples equally divided into three groups. A combination of methods was used, including 16S rRNA sequencing, untargeted liquid chromatography–mass spectrometry (LC-MS) metabolomics, Orthogonal Partial Least Squares Discriminant Analysis (OPLS-DA), Kyoto Encyclopedia of Genes and Genomes (KEGG) enrichment, and weighted gene co-expression network analysis-based module detection (WGCNA) with network integration. An uncommon composition of organisms dominated the SI: the hydrogenotrophic methanogens Methanobrevibacter (SI 24.51%; RM 1.92%; LI 2.19%) and Methanosphaera (SI 0.43%; RM 0.02%; LI 0.02%), together with the acetogen Acetitomaculum (SI 1.58%; RM 0.34%; LI 0.11%), were markedly more abundant compared to the RM or LI. Correlation and pathway analyses indicated that Methanobrevibacter was positively correlated with a steroid-type lipid metabolite ($r = 0.52$, $p < 0.05$) and with bile-acid-related metabolites. Acetitomaculum was positively correlated with several metabolites: 4-Hydroxyphenyl 4-hydroxybenzoate (r = 0.79, $p < 0.05$), 2-Aminoethyl dihydrogen phosphate ($r = 0.76$, $p < 0.05$), 1-Myristoyl-2-stearoyl-sn-glycero-3-phosphocholine ($r = 0.76$, $p < 0.05$), and 1,2-Dioleoyl-sn-Glycero-3-Phosphocholine ($r = 0.74$, $p < 0.05$). Together, these data define a small-intestinal microbial–metabolite module in Qianqiu goats characterized by elevated abundances of specific methanogens and acetogens in the SI. Specific positive correlations were identified between these taxa and metabolites associated with lipids and bile acids.

Keywords: integrated analysis; metabolomics; microbiome; Qianqiu goat; spatial multi-omics

1. Introduction

The four-chambered digestive system of ruminants has long been a subject of interest for researchers. This system efficiently converts low-nutrient plant material into energy and proteins, adapting surrounding life forms to the ruminant's environment [1]. It offers an insight into the study of symbiotic relationships, biological conversion efficiency, and evolutionary intelligence [2]. From the anatomical perspective, the rumen, a unique fermentative organ in ruminants, converts fiber-rich substances into volatile fatty acids through microbial action [3]. The small intestine primarily relies on enzymatic digestion and active absorption, while the large intestine plays a crucial role in maintaining electrolyte balance [4,5]. At the microbial level, differences in pH, oxygen concentration, and nutrient substrates across various gastrointestinal (GI) segments lead to significant diversification in microbial communities and metabolic pathways. Metabolites often serve as a "chemical language" between the host and microbe, regulated by the host's genetic makeup. This, in turn, influences microbial community dynamics [6,7]. For instance, the accumulation of short-chain fatty acids (SCFAs) in the rumen can modulate host energy metabolism. In the intestine, bile acid metabolism influences immune homeostasis via the Farnesoid X receptor (FXR) signaling pathway [8,9].

The Qianqiu goat is a newly recognized, Anhui-endemic local germplasm (China), distributed mainly in Tianchang City and Laian County, Anhui Province. It is white-coated, and its horns are typically counter-spiral or arched (Supplementary Figure S1). In the measured cohort, adult males averaged at 60.3 kg and females averaged at 44.4 kg in live body weight; the mean dressing percentage was 53.71% and meat yield was 48.03%. Taken together, this breed's narrow geographic range and distinctive phenotype provide a biological context for the putative metabolic adaptations observed. Data are taken from our measured cohort. To date, no public literature reports have been made for these values in Qianqiu goats.

Therefore, systematic studies on metabolic heterogeneity in the Qianqiu goat remain limited, particularly regarding the mechanisms that link multi-segment metabolic profiles to microbial functions. Current research on local goats predominantly focuses on genetic improvement and production performance, while systematic microbial analyses remain in their infancy [10,11]. Additionally, traditional culture-dependent methods are limited by the low cultivability of gut microbes (only approximately 20% can be cultured in vitro), leaving the functional potential of many uncultured microbes unexplored [12]. Because small input samples can yield high-resolution data via amplicon sequencing, this bottleneck can be overcome. However, applying this method to complex, multi-segment GI-tract analysis remains underdeveloped [13]. In addition, metabolomics has also been used to study microbial function and host genetic characteristics. Metabolomics, a branch of systems biology, uses high-throughput metabolite detection to provide a comprehensive understanding of metabolic network changes under physiological or pathological conditions [14].

There have been significant improvements in metabolite identification due to advances in technologies such as liquid chromatography–mass spectrometry (LC–MS) and nuclear magnetic resonance. This has enabled unbiased identification in complex biological samples, such as GI contents [15]. In research, metabolomics has been successfully applied to regulate rumen fermentation, mitigate methane emissions, and enhance nutrient utilization efficiency [16]. For example, studies have identified significant correlations between specific metabolic pathways and methane production in sheep under different feeding regimes [17]. Another study revealed a potential link between polyamine metabolites and intestinal barrier function in goats [18,19].

Despite some progress in the study of ruminant GI metabolism, research on the Qianqiu goat remains limited for the following reasons: (1) most current metabolomic data are based on Western commercial breeds, neglecting the unique metabolic adaptations of local breeds developed through long-term natural selection [20]; (2) many studies adopt a whole-metabolome approach, failing to analyze spatiotemporal differences in metabolic functions across different GI segments [21]; (3) the identification of differential metabolites often remains at a descriptive statistical level, lacking integrative analysis with microbiome data, which hinders in-depth exploration of the mechanisms driving metabolic heterogeneity [22].

Additionally, because the Qianqiu goat lives year-round in low-altitude, low-nutrient environments, its GI metabolic system may have evolved unique energy utilization strategies and stress-response pathways that have yet to be fully explored [23].

In this study, we combined amplicon sequencing with metabolomics to investigate microbiome composition and metabolic profiles across the rumen, small intestine, and colon of Qianqiu goats. Specifically, we aimed to address the following questions: (1) Are there statistically significant differences in the microbial community composition across different GI segments? (2) Does the microbial symbiotic network exhibit a modular structure, and are key keystone taxa associated with host metabolic phenotypes?

2. Results

2.1. Operational Taxonomic Unit (OTU) Clustering Analysis

Clustering effective bacterial tag sequences at 97% sequence identity yielded 2284 OTUs. Venn analysis (Figure 1a) revealed that 427 OTUs (18.7% of the total) were shared among all three GI segments. Additionally, 642 OTUs were shared between the rumen and small intestines, and 545 OTUs were shared between the small and large intestines. Analysis revealed 558 specific OTUs in the rumen, 234 OTUs in the small intestine, and 665 OTUs in the large intestine. Notably, the small intestine exhibited significantly fewer unique bacterial OTUs than either the rumen or the large intestine, reflecting its transitional role between the foregut and hindgut microbial communities.

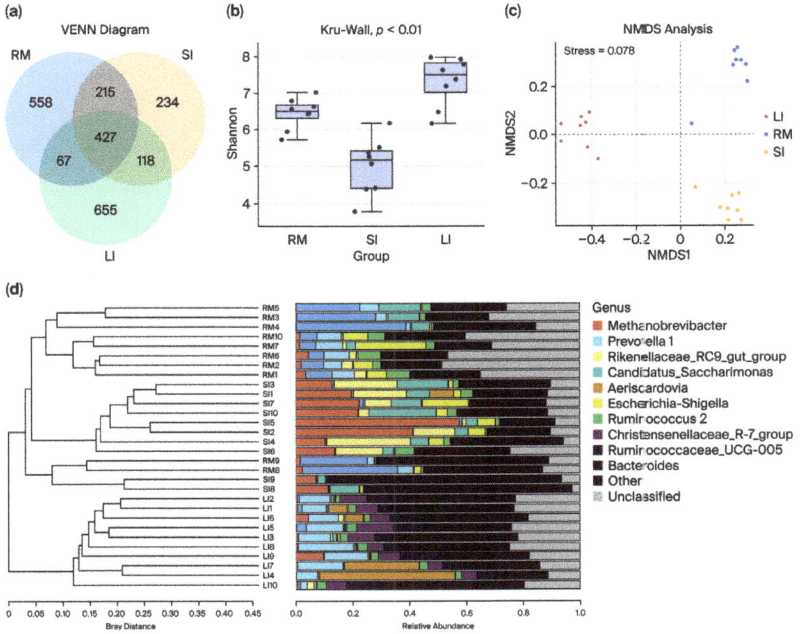

Figure 1. (**a**) Venn diagram and UpSet plot illustrating shared and unique bacterial operational taxonomic units (OTUs) among the rumen (RM), small intestine (SI), and large intestine (LI) chyme samples. (**b**) Shannon diversity index of bacterial communities in chyme samples from RM, SI, and LI. (**c**) Non-metric multidimensional scaling (NMDS) plot based on Bray–Curtis distance, showing the clustering patterns of microbial communities across the RM, SI, and LI. (**d**) Hierarchical tree illustrating the unweighted pair group method with arithmetic mean (UPGMA) clustering results based on microbial community composition in the three GI regions. In (**b**), dots indicate individual samples (Shannon diversity values) overlaid on the boxplots.

Alpha diversity analysis of luminal contents collected from the rumen, small intestine, and large intestine indicated sequencing coverage ≥ 0.99 for all samples (Figure 1b and Supplementary Figure S2), confirming that the sequencing depth was sufficient in comprehensively capturing the microbial communities. The analysis demonstrated that bacterial diversity, richness, and evenness were highest in the large intestine (colon) and lowest in the small intestine.

The NMDS ordination plots (Figure 1c,d) reveal clear clustering patterns corresponding to the rumen, small intestine, and large intestine samples. The sharp segregation of these clusters highlights marked spatial heterogeneity in microbial community composition across the GI tract of Qianqiu goats.

2.2. *Microbial Profiles of Three GI Segments*

At the phylum level (Figure 2a), *Firmicutes* dominated across all samples. In the rumen, *Firmicutes*, *Bacteroidota*, and *Actinobacteria* accounted for ~88% of the community, whereas the small intestine was characterized by *Firmicutes*, *Euryarchaeota*, and *Actinobacteria* (78.57%). The large intestine was dominated by *Firmicutes*, *Bacteroidota*, and *Proteobacteria* (89%). Along the GI tract, *Firmicutes* and *Proteobacteria* increased from the rumen to the large intestine; *Bacteroidota* decreased, then increased; and *Actinobacteria* peaked in the small intestine before declining. *Euryarchaeota* were specifically enriched in the small intestine. Statistical analysis revealed significant differences in the distribution of *Bacteroidota* and *Euryarchaeota* among the three regions. *Actinobacteria* were significantly higher in the rumen and small intestine than in the large intestine ($p < 0.05$).

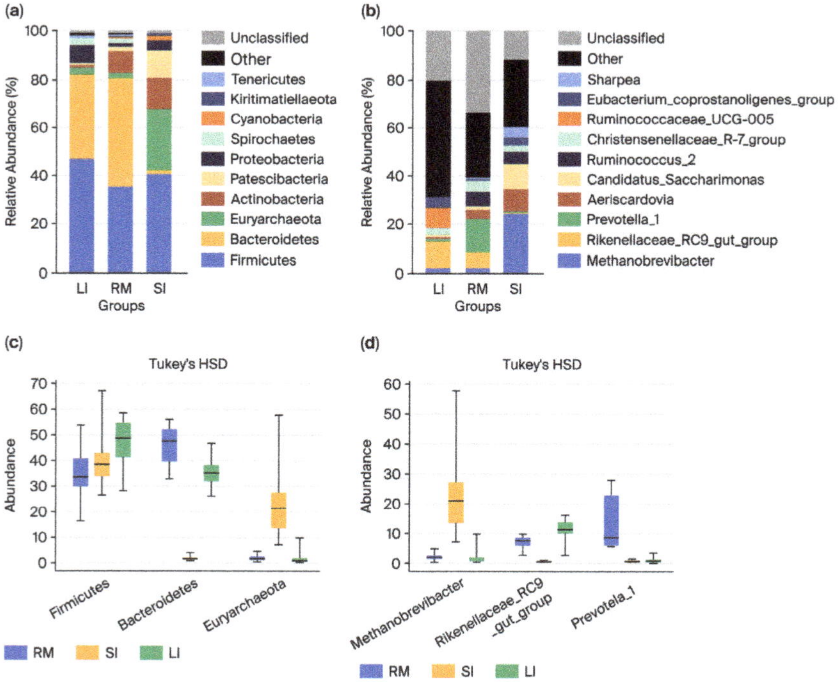

Figure 2. Microbial composition of the three gastrointestinal segments: rumen (RM), small intestine (SI), and large intestine (LI). (**a**) Phylum-level microbial composition in each segment. (**b**) Genus-level microbial composition in each segment. (**c**) Differentially abundant taxa at the phylum level between different segments. (**d**) Differentially abundant taxa at the genus level between different segments.

At the genus level (Figure 2b), the rumen was dominated by *Rikenellaceae_RC9_gut_group*, *Prevotella_1*, *Ruminococcus_2*, and *Christensenellaceae_R-7_group*. The small intestine exhibited high abundances of *Methanobrevibacter*, *Rikenellaceae_RC9_gut_group*, *Candidatus Saccharimonas*, and *Aeriscardovia*, while the large intestine was enriched in the *Rikenellaceae_RC9_gut_group*, *Ruminococcaceae_UCG_005*, *Eubacterium coprostanoligenes_group*, and *Christensenellaceae_R-7_group*. Spatial patterns revealed that *Methanobrevibacter* and *Candidatus Saccharimonas* peaked in the small intestine; *Prevotella_1*, *Ruminococcus_2*, and *Christensenellaceae_R-7_group* were highest in the rumen; and *Ruminococcaceae_UCG_005* and *Eubacterium coprostanoligenes_group* increased along the GI tract. Statistical analysis (Figure 2c,d) confirmed significant differences in the distribution of *Methanobrevibacter*, *Rikenellaceae_RC9_gut_group*, *Prevotella_1*, *Aeriscardovia*, and *Candidatus Saccharimonas among the three GI regions* (Kruskal–Wallis raw $p < 0.05$; Tukey HSD raw $p < 0.05$), reflecting region-specific microbial assemblages shaped by distinct physicochemical environments along the gastrointestinal tract of Qianqiu goats.

2.3. Analysis of Differential Genera in the GI

An in-depth analysis was conducted for the differential microbiota in the rumen, small intestine, and large intestine of Qianqiu goats, focusing on genera that exhibited significant differences in abundance between the rumen and small intestine, as well as between the large intestine and small intestine ($p < 0.05$). By selecting microbial taxa that met this criterion, 18 key genera were ultimately identified (Figure 3a; Supplementary Table S1), and a heatmap was generated to show their relative abundance in the rumen, small intestine, and large intestine (Figure 3b).

Methanobrevibacter had a relative abundance of 1.92% in the rumen, which was significantly higher at 24.51% in the small intestine, showing a notable shift in abundance. Similarly, *Candidatus_Saccharimonas* had a relative abundance of 1.63% in the rumen, which increased to 10.85% in the small intestine, displaying a significant difference. In addition, genera, such as *Prevotellaceae_UCG-003*, *Family_XIII_AD3011_group*, *Methanosphaera*, and *Acetitomaculum* also exhibited significant differences in abundance between the rumen and small intestine. For example, *Prevotellaceae_UCG-003* had a relative abundance of 0.95% in the rumen, but this was only 0.03% in the small intestine. Similarly, genera such as *Candidatus_Saccharimonas* and *Prevotellaceae_UCG-003* also showed significant differences in abundance between the large and small intestines. *Candidatus_Saccharimonas* had a relative abundance of 1.63% in the large intestine and 10.85% in the small intestine, while *Prevotellaceae_UCG-003* had a relative abundance of 0.95% in the large intestine, but its value dramatically dropped to 0.03% in the small intestine. Moreover, other genera, such as *Family_XIII_AD3011_group* and *Acetitomaculum*, displayed similar patterns of varying abundance, indicating a clear difference in their distribution between the large and small intestines. For instance, *Family_XIII_AD3011_group* had a relative abundance of 0.45% in the large intestine; conversely, in the small intestine, this value was 2.04%, showing a marked difference.

Overall, the relative abundance of several genera in the small intestine (SI) was significantly higher than in both the rumen (RM) and large intestine (LI). Additionally, certain genera, such as *Prevotellaceae_UCG-003*, which were relatively abundant in the rumen and large intestine, exhibited lower abundance in the small intestine, potentially reflecting their primary role in fiber degradation and fermentation processes [24].

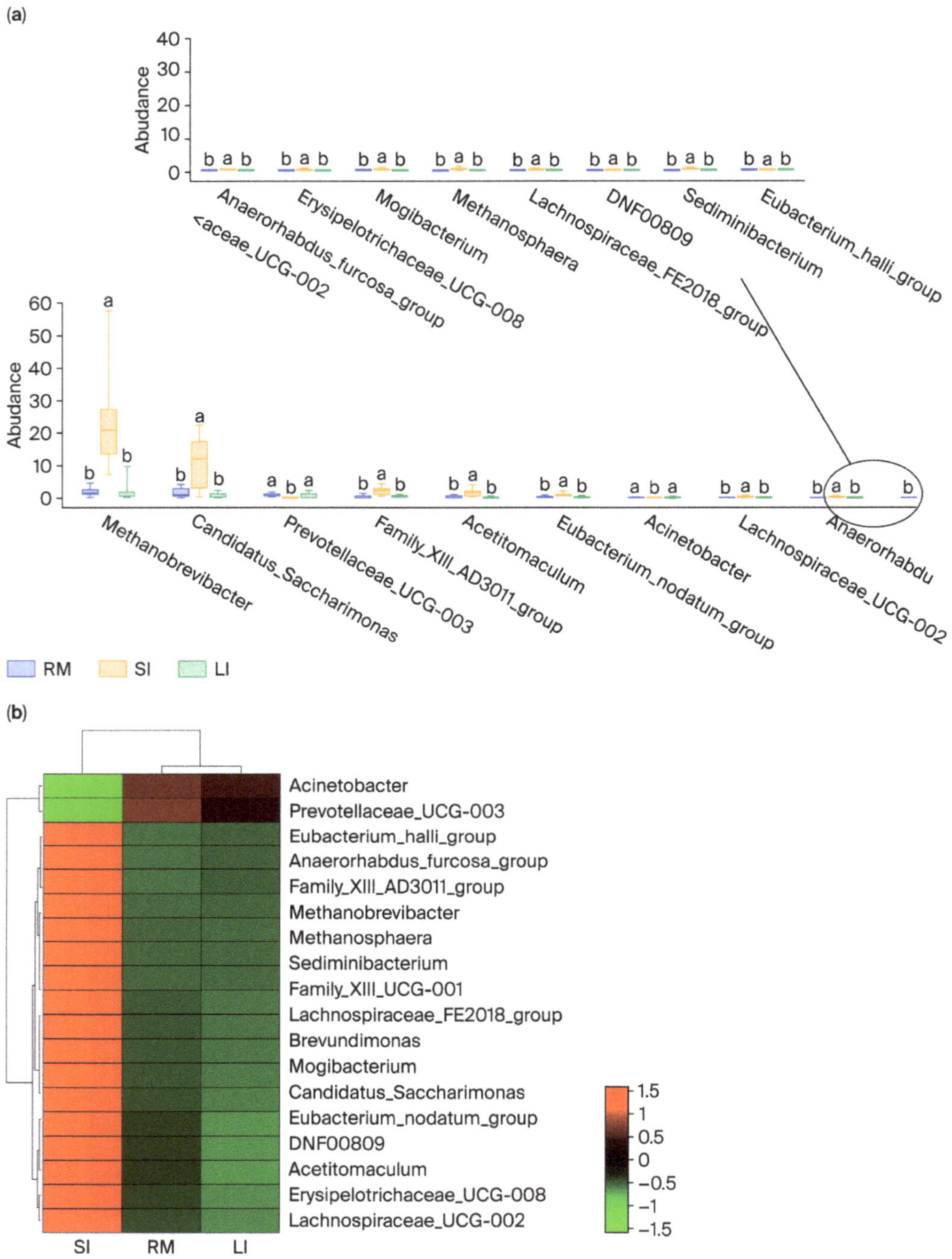

Figure 3. (**a**) Tukey's HSD analysis of key genera abundance; (**b**) heatmap of key genera abundance. Different lowercase letters above the boxplots indicate significant differences among treatments according to Tukey's HSD test ($p < 0.05$), and the circlein panel (**a**) is only used to show that the 18 differential genera are split into two coordinate axes because they could not be clearly displayed in a single axis; it has no additional statistical meaning.

2.4. Differential Metabolites in the GI Tract of Qianqiu Goats

To investigate the metabolic profiling characteristics of the GI tract in Qianqiu goats, we conducted non-targeted metabolomics using ultra-high-performance liquid chromatography–mass spectrometry on samples from the rumen, small intestine, and large intestine. The resulting Orthogonal Partial Least Squares Discriminant Analysis (OPLS-DA) model—which separates variation predictive of sample class from orthogonal (non-class-related) variation—was used to generate score plots (Figure 4a,c) and validation plots (Figure 4b,d). The analysis revealed significant differences in metabolite composition between the SI and RM groups (SI vs. RM) and the LI and SI groups (LI vs. SI), with clear group separations. Permutation tests confirmed the robustness of the models, which demonstrates excellent predictive ability and fit without the risk of overfitting.

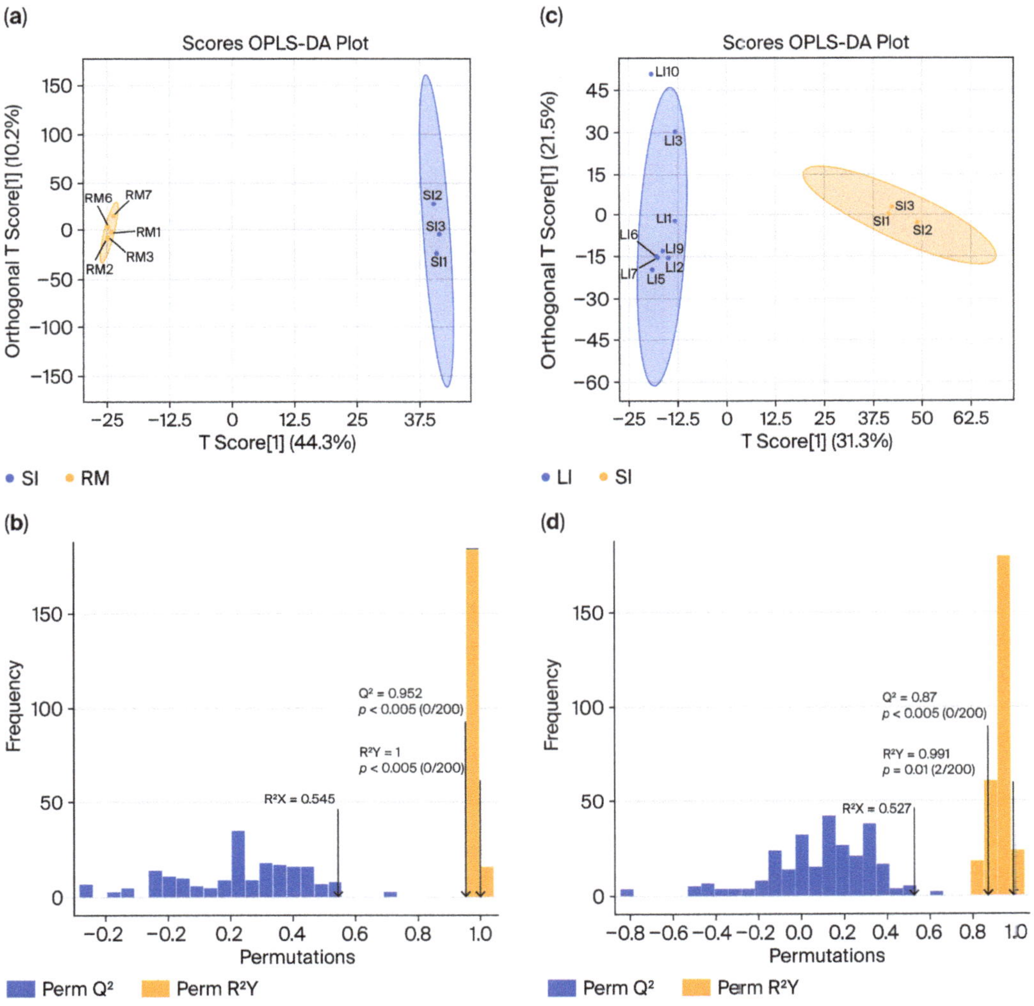

Figure 4. Orthogonal Partial Least Squares (OPLS) score plots and model validation results. (**a**) "SI vs. RM" score plot showing sample separation; (**b**) permutation/cross-validation results for the "SI vs. RM" model. (**c**) "LI vs. SI" score plot showing sample separation; (**d**) permutation/cross-validation results for the "LI vs. SI" model.

2.5. Data Quality Control, Soft-Threshold Selection, and WGCNA Module Construction

During data quality control, violin plots of metabolite intensities showed approximately symmetric, bell-shaped distributions for each sample with no aggregation of extreme outliers. This indicated that log transformation and normalization yielded comparable data across tissues (Supplementary Figure S3). Sample-level hierarchical clustering grouped all profiles strictly by tissue type, with no clear outliers identified (Figure 5a). Network construction used the topological overlap matrix (TOM) to quantify metabolite–metabolite connectedness and the dynamic tree cut algorithm to define modules; the dendrogram (top) displays metabolite clustering, and the color bar (bottom) denotes the module membership (Figure 5a). Using this procedure, several co-expression modules were identified (e.g., blue, turquoise, and brown), each representing metabolites with highly correlated expression patterns within modules and comparatively lower correlation between modules (Figure 5b). These modules provide the basis for downstream analyses of tissue-associated metabolic pathways and biomarker prioritization.

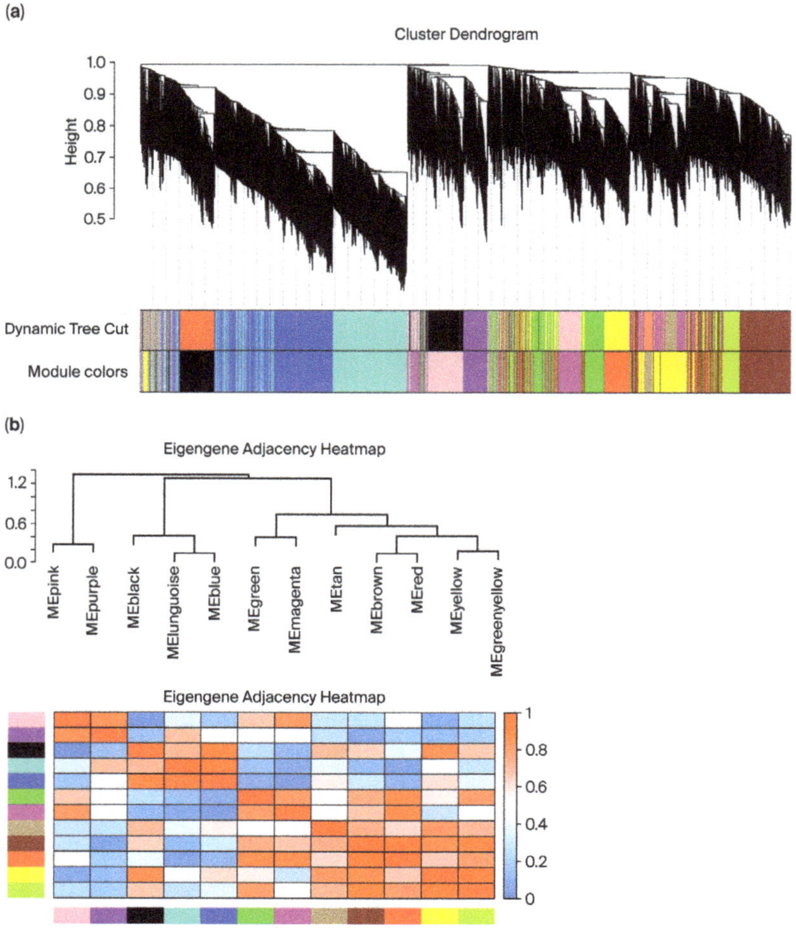

Figure 5. (**a**) Sample clustering relationship diagram; (**b**) module generation plot. In panel (**a**), the different colors under the dendrogram represent distinct co-expression modules identified by WGCNA. In panel (**b**), the color scale from blue to red indicates the eigengene adjacency (correlation) between modules, with blue representing low adjacency and red representing high adjacency.

2.6. Correlation and Expression Profiles of Tissue-Associated Modules

To examine tissue-specific metabolite patterns, WGCNA was employed to correlate MEs with tissue types numerically coded as follows: rumen = 0; small intestine = 1; and large intestine = 2. Seven metabolite modules showed significant associations ($p < 0.05$; Figure 6a). Modules with strong ME–trait correlations and internal coherence (a significant gene significance–module membership (GS–MM) correlation, Supplementary Figures S4–S8) were further analyzed across tissues (Figure 6b).

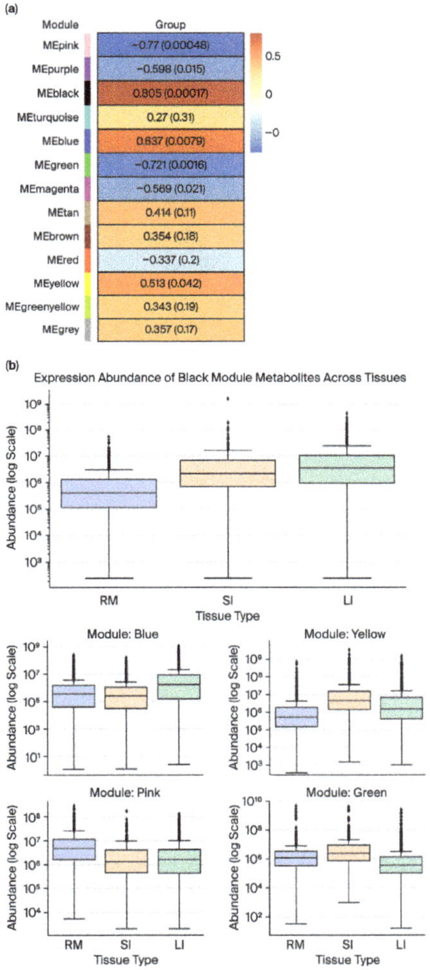

Figure 6. (**a**) Module–group correlation p-value plot (rows represent module eigengene (ME), and columns represent "group" for each intestinal segment: RM, SI, and LI), where each cell shows the correlation and p-value, (the table is color-coded by correlation). (**b**) Expression abundance of metabolite modules across gastrointestinal tissues (RM, SI, and LI) in Qianqiu goats. (Box plots show the distribution of metabolite abundances across the three intestinal segments (RM, SI, LI) for modules that are significantly associated with tissue type). In panel (**a**), the colors on the left indicate the different WGCNA modules (MEpink, MEpurple, MEblack, etc.), and the color scale in the heatmap (from blue to red) shows the strength and direction of the correlation between each module and group (blue = negative, red = positive). In panel (**b**), the colors of the box plots correspond to the different tissue types as indicated in the legend, and the dots represent the individual sample values (outliers) within each group.

The black module displayed the strongest positive correlation (r = 0.805, $p < 0.05$) and high internal consistency (GS–MM r = 0.59, $p < 0.05$). Boxplot analysis revealed significantly higher expressions in the small and large intestines compared to the rumen, with the small intestine showing the highest median. Both gut regions exhibited broad expression ranges and high-abundance outliers, which are hallmarks of hub metabolites with tissue-specific regulation.

The blue module was positively associated with distal gut tissues (r = 0.637, $p < 0.05$; GS–MM r = 0.52, $p < 0.05$), and its expression peaked in the large intestine, showing elevated medians and upper quartiles as well as extreme outliers. This highlights its functional relevance to microbial fermentation, bile acid metabolism, or ion balance in the colon.

The yellow module, although marginally correlated (r = 0.513, $p < 0.05$), demonstrated robust GS–MM alignment (r = 0.52, $p < 0.05$). Expression was highest in the small intestine, with considerable variability, suggesting small intestine-specific metabolic activity and potential as a biomarker ensemble.

In contrast, the pink and green modules were negatively correlated with tissue type (pink: r = −0.77; $p < 0.05$; GS–MM r = −0.56; $p < 0.05$; green: r = −0.721; $p < 0.05$; GS–MM r = −0.47; $p < 0.05$), indicating repressed expression in gut tissues. Expression of the pink module was highest in the rumen; however, it remained consistently low with minimal variance in the small and large intestines. This finding implies the existence of rumen-specific metabolic roles. The green module showed comparable expression in the rumen and small intestine but lower levels in the large intestine, suggesting that metabolic processes are specialized to regions and dependent on gut segments rather than subject to a uniform absorptive function.

2.7. KEGG Pathway Enrichment and Hub-Metabolite Network Analysis

Kyoto Encyclopedia of Genes and Genomes (KEGG) enrichment analysis was performed on five tissue-associated modules (black, blue, yellow, pink, and green), and the pathways enriched within each module were summarized (Figure 7a). The top 20 KEGG enrichment results are listed in Supplementary Table S2. Most significantly enriched pathways fell within the metabolism category (88.81%). Among them, glycerophospholipid metabolism (49.46%), arachidonic acid metabolism (23.10%), linoleic acid metabolism (22.74%), and choline metabolism (22.74%) were the most prominent, underscoring the central roles of lipid signaling and membrane-lipid remodeling in metabolic regulation. In addition, enrichment was observed for steroid hormone biosynthesis, fat digestion and absorption, and glycosphingolipid metabolism, further supporting links to lipid-energy metabolism and hormone-signaling regulation.

Beyond metabolism, several disease-related and environmental information-processing pathways were also represented, including insulin resistance, choline metabolism in cancer, Cushing syndrome, and the PI3K–Akt and AMPK signaling pathways. Notably, glycerophospholipid metabolism, arachidonic acid metabolism, linoleic acid metabolism, and retrograde endocannabinoid signaling showed strong enrichment in KEGG analysis ($p < 0.01$), indicating that a high proportion of the metabolites detected were mapped to these pathways relative to the background metabolite set.

We ranked metabolites in each module based on GS and visualized the tissue–module–metabolite relationships using Sankey diagrams (Supplementary Figure S9 and Figure 7b). In the rumen, core metabolites (Supplementary Table S3) from the pink and green modules included mevalonate-5-phosphate and Triglyceride (TG) (15:0/16:1/18:4). In the small intestine, core metabolites (Supplementary Table S4) from the black module included PC 14:0/18:0 and L-methionine sulfoxide. In the large intestine, core metabolites (Supplementary Table S5) from the blue module included anserine and sulfamonomethoxine.

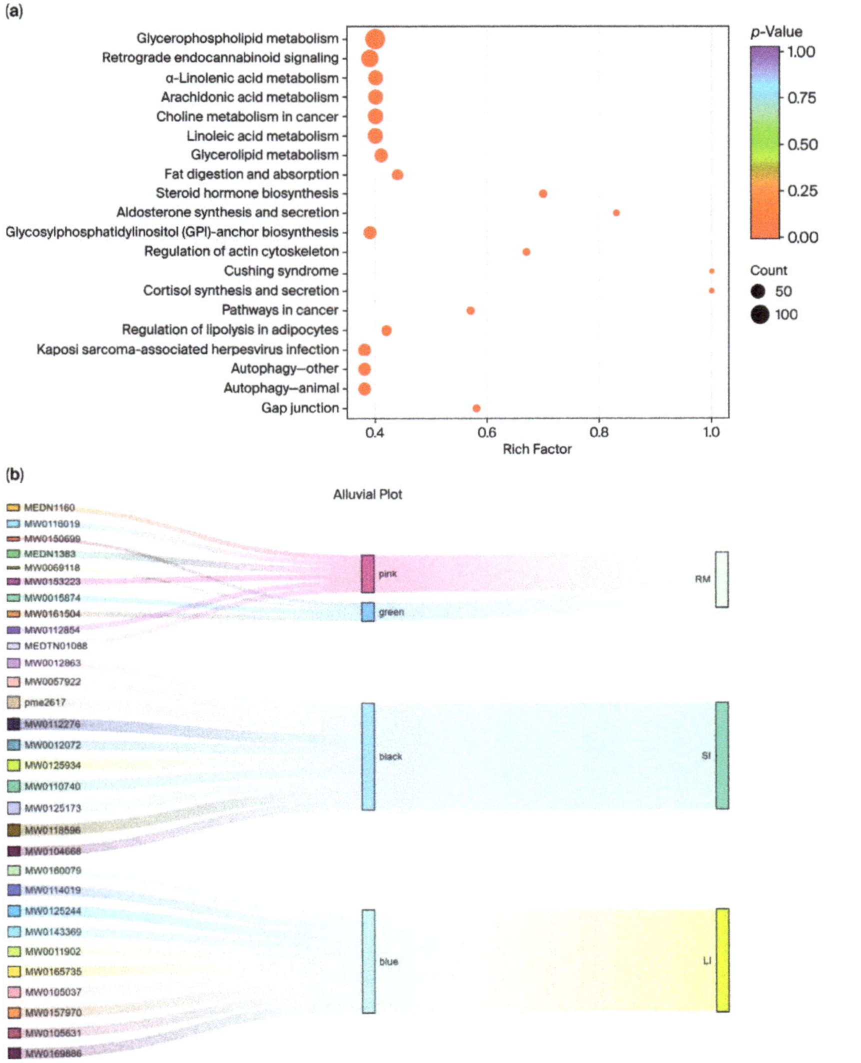

Figure 7. (**a**) KEGG pathway enrichment analysis; (**b**) Sankey visualization of the three-layer tissue–module–metabolite structure. Each ribbon links an individual metabolite (**left**, labeled by Metabolite_ID) to its WGCNA module (**middle**, colored "pink/green/black/blue") and to associated intestinal tissue (**right**, RM; SI; LI).

2.8. Integrated Microbe–Metabolite Network of Qianqiu Goats

Eighteen differential genera were further analyzed, and the relative abundances of *Methanobrevibacter*, *Acetitomaculum*, and *Methanosphaera* were significantly higher in the small intestine than in other GI segments of Qianqiu goats. This distribution pattern has been rarely observed in species [25,26]. To explore the potential role of these three genera in the differential abundance in the small intestine, a Spearman correlation analysis (a non-parametric method used to assess the strength and direction of association between two ranked variables) was conducted between the identified core metabolites and the genera (Figure 8a). The results showed that these genera were significantly correlated with several metabolites, which were enriched in specific metabolic pathways (Figure 8b).

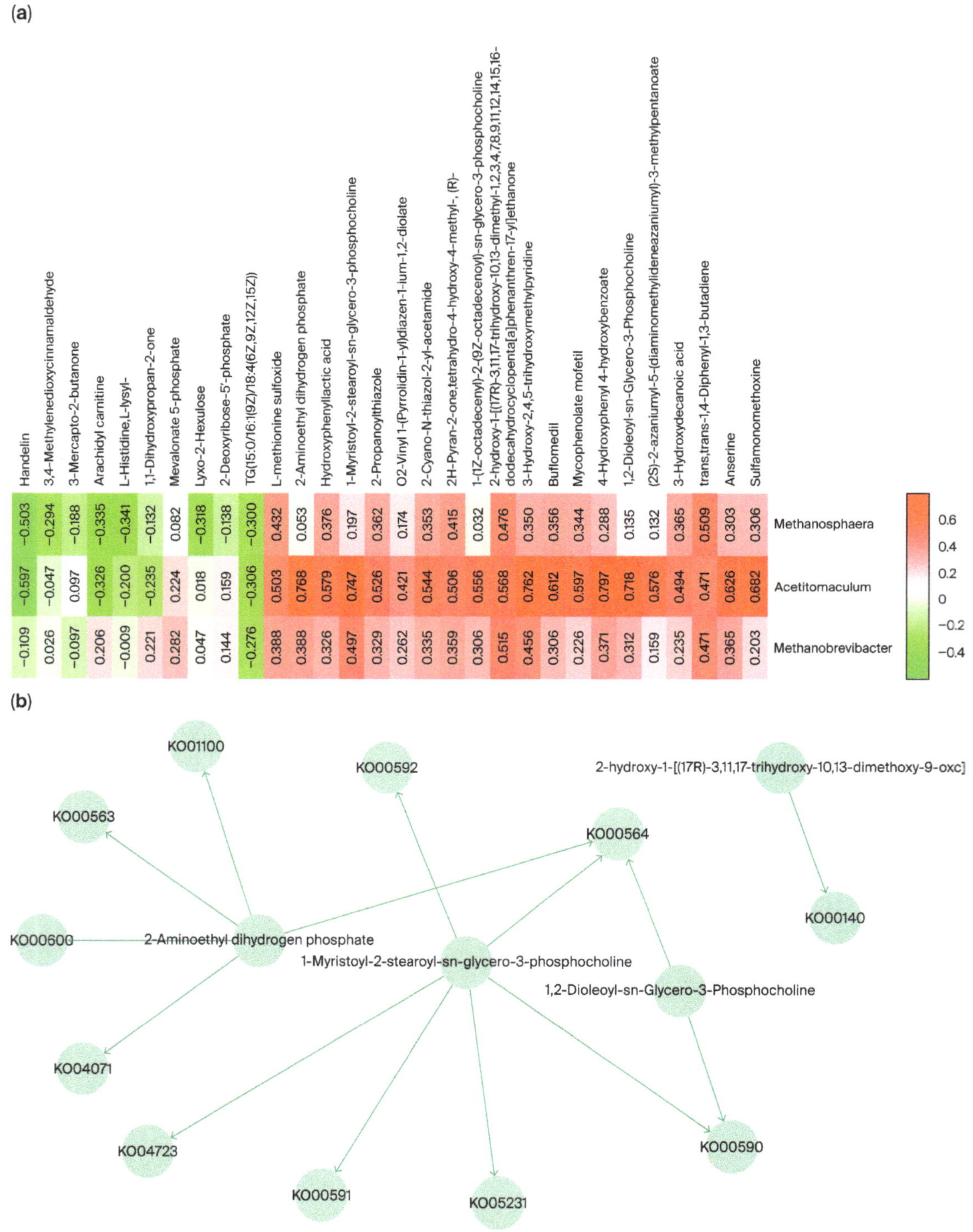

Figure 8. (**a**) Spearman correlation analysis of key metabolites and microbial abundance. (**b**) Metabolic pathway network of key metabolites associated with microbial abundance.

For *Methanobrevibacter*, the significantly correlated metabolite, 2-hydroxy-1-[(17R)-3,11,17-trihydroxy-10,13-dimethoxy-9-oxohentriacontyl], was enriched in the Ko00140 (a specific KEGG orthology (KO) identifier) steroid biosynthesis pathway. This is consistent with *Methanobrevibacter*'s documented capacity to tolerate bile-rich niches via membrane stabilization [27]. *Acetitomaculum* was significantly and positively correlated with metabolites involved in several lipid metabolism pathways, including Ko00590, Ko00591, Ko00592, Ko00564, and Ko04071, indicating its active involvement in intestinal lipid turnover [28]. For *Methanosphaera*, the significantly correlated metabolite was trans, trans-1,4-Diphenyl-1,3-butadiene, which was not enriched in any specified pathways.

3. Discussion

Recent advances in high-throughput sequencing and metagenomics have rapidly expanded research on ruminant GIT microbiota beyond the rumen [29,30]. Studies now cover multiple compartments along the entire digestive tract, including both the small and large intestines. For example, a recent large-scale study reconstructed a Goat Multi-Kingdom Microbiome Catalog (GMMC), based on 497 GIT samples from ten anatomical sites ranging from the rumen to the colon [31]. The catalog comprises 4004 bacterial MAGs, 71 archaeal MAGs, and 7204 viral genomes. This study is a comprehensive, genome-level resource covering bacteria, archaea, and viruses across the gastrointestinal tract of goats. Importantly, this study demonstrates that archaea are not confined to the rumen or large intestine; they are detectable throughout the GIT, including the small intestine (SI). In addition, previous metagenomic studies that sampled different developmental stages (e.g., goat kids) and various gut compartments (e.g., stomach, small intestine, and large intestine) have also revealed compartment-specific microbial community assembly and compositional differences along the GIT [32].

On this basis, we conducted a systematic analysis of the rumen, small intestine (SI), and large intestine (LI) of the Qianqiu goat. The results revealed marked differences in microbial community structure and composition among the three gut segments. These findings are consistent with the spatial heterogeneity along the gastrointestinal tract documented in studies such as the Goat Multi-Kingdom Microbiome Catalog (GMMC) [31]. Notably, in the small intestine (SI), pronounced enrichment of certain archaeal genera was observed, particularly Methanobrevibacter and Methanosphaera, with relative abundances up to 24.51% and 0.43%, respectively. Concurrently, bacterial genera associated with lipid and acetate metabolism, such as Acetitomaculum [25], also exhibited relatively high abundance in the SI (~1.58%).

Although previous studies have encompassed multiple GIT compartments (including the small intestine, SI), virtually no systematic, quantitative reports (e.g., relative-abundance or enrichment analyses) have been published on methanogens in the small intestine of ruminants, particularly goats or young lambs [33,34]. In prior work, even when intestinal samples (e.g., jejunum and ileum) were collected, analyses typically relied on 16S rRNA metabarcoding. Such research only documented the presence or taxonomic composition of archaea, rather than applying MAG reconstruction combined with abundance estimation to reveal archaeal enrichment in the SI [35]. The present data indicate that the small intestine may constitute a key habitat and active zone for archaea (especially certain genera). Meanwhile, the observed enrichment of bacterial populations associated with lipid and acetate metabolism within the SI suggests that this region may play a significant but previously under-appreciated role in host–microbe interactions underlying energy metabolism [36,37].

To further investigate the potential biological functions of the specific taxa enriched in the small intestine (SI), a Spearman correlation analysis was conducted between their relative abundances and the core metabolites identified in the SI. The analysis revealed that the abundance of Methanobrevibacter correlated positively and significantly with several lipid-related metabolites ($p < 0.05$), suggesting that these archaea may participate in or influence the lipid metabolic environment of the SI. This observation was consistent because a bile salt hydrolase gene (bsh) was identified in Methanobrevibacter smithii [38]. The presence of this gene has been shown to confer enhanced survival under bile stress from deconjugation of bile salts [39]. Furthermore, previous studies have provided evidence that bile acids and steroids can be transformed by microorganisms to confer ecological advantages in bile-rich environments [40]. On this basis, it is hypothesized that bile tolerance in Methanobrevibacter in the small intestine of Qianqiu goats may be enhanced by modulating lipid and steroid metabolism, thereby facilitating persistence and functional activity in this compartment.

It should be noted, however, that although BSH homologs have been widely reported across gut microbial communities [41], their distribution appears to be largely confined to bacteria [42]. Direct functional evidence of BSH activity in archaea is absent in most studies [43]. This caveat warrants caution when extrapolating bacterial-based mechanisms to archaeal taxa such as Methanobrevibacter.

Moreover, recent work has expanded the recognized repertoire of BSH functions. In addition to classical bile salt deconjugation, BSH (or BSH-like enzymes) has been shown to mediate amine N-acyltransferase activity, leading to the formation of microbial-conjugated bile acids (MCBAs), which may modulate bile acid pool composition, microbial colonization, lipid absorption, and host–microbe signaling [44–46]. Therefore, while the positive correlation between Methanobrevibacter abundance and lipid-related metabolites in the SI provides a plausible indication of lipid/steroid metabolic modulation and potential bile tolerance, functional validation, including demonstration of BSH (or analogous enzyme) expression in SI Methanobrevibacter, measurement of bile salt transformation products, and assessment of membrane lipid composition, remains necessary.

Similarly, the abundance of Acetitomaculum showed strong positive correlations with multiple metabolites associated with glycerophospholipid metabolism, ether-lipid metabolism, and other lipid metabolic pathways ($p < 0.05$). This correlation suggested that Acetitomaculum, and by extension, reductive acetogenic bacteria, may have been involved not only in acetate production [47]. Given that acetate can serve as a carbon and energy precursor (via acetyl-CoA) for lipid biosynthesis, it was hypothesized that the acetate they release might also be utilized by microbes (or the host) in the intestinal milieu for membrane lipid synthesis or remodeling [47]. This could influence microbial community structure and energy metabolism in the gut; however, direct experimental evidence is lacking. This interpretation was further supported by a recent genome-resolved metagenomic survey, which revealed that reductive acetogenic bacteria were highly enriched in the hindgut (cecum) of ruminants; in that study, acetogens were approximately 12 times more abundant in the cecum than the rumen [48]. In situ measurements in the same work showed that dissolved hydrogen and acetate levels were significantly higher in the cecum than in the rumen, consistent with enhanced reductive acetogenesis and diminished methanogenesis in the hindgut environment [48].

Based on these observations, it is reasonable to propose that Acetitomaculum (or related acetogens) is important as a "carbon/energy provider" in the hindgut: acetate produced by reductive acetogenesis might have been used as a carbon backbone or acyl-group source for membrane lipid synthesis or remodeling [47], thus contributing to the generation of metabolite signatures associated with glycerophospholipid, ether-lipid, and other lipid metabolic pathways observed in our metabolomic analysis.

Methanogens and bacteria associated with lipid/acetate metabolism were concurrently enriched in the small intestine (SI) of the Qianqiu goat. The classical view is that methanogenesis in ruminants (especially in the rumen) is driven by methanogens utilizing $H_2 + CO_2$ (or formate), which function as an H_2/electron sink to maintain fermentation balance. Thus, a hypothetical mechanism is proposed [49]. If active methanogenesis is maintained in the SI, H_2 and electrons produced by other microbes can be consumed, resulting in low H_2 partial pressure and a favorable redox balance, thus providing stable conditions for other anaerobic processes (e.g., fermentation, lipid metabolism, and intermediate metabolite production) [49]. Concurrently, under such stable conditions, lipid/acetate-metabolizing bacteria (e.g., Acetitomaculum) can utilize dietary components (or host secretions), fermentation substrates, or metabolic by-products to generate acetate or lipid/membrane-lipid intermediates [50]; these intermediates can be subsequently absorbed by the host intestinal epithelium for membrane lipid synthesis/remodeling or energy/lipid metabolism [51]. This can potentially influence epithelial membrane integrity, nutrient/lipid uptake capacity, and host energy/lipid metabolic homeostasis [52]. If this hypothesis is confirmed, the small intestine (SI) of the Qianqiu goat may be regarded not solely as a transit/absorption site, but as a distinct microbial niche and metabolic compartment: a "microbial–host metabolic compartment".

However, it is important to acknowledge several limitations of this study. Although correlation analysis was employed, it cannot establish causality, and other factors, such as the host's physiological state, secretory activity, or environmental conditions, may have contributed to the microbial and metabolic changes observed [53]. Additionally, our analysis was based on intestinal lumen contents rather than mucosa-associated microbiota; this may have excluded some important microbial interactions at the epithelial surface [54,55]. Furthermore, direct functional validation, such as measurements of methane production or metabolic enzyme activities, was not performed; thus, our conclusions were only based on correlations [56]. Finally, as this study focused solely on Qianqiu goats, the generalizability of these findings to other ruminant species or varying farming conditions remains unclear.

To address these limitations, future research should validate microbiota distribution across different gut regions [57] using metagenomics, transcriptomics, and functional analyses to elucidate specific microbial activities [58]. Additionally, further studies should assess the generalizability of this model to other ruminants and farming conditions to evaluate its broader applicability [59]. Exploring the underlying mechanisms of microbial–metabolite–host interactions and their ecological roles will help expand our understanding of gut microbiota dynamics and their implications for animal health and productivity.

In conclusion, this study introduces a new conceptual framework for understanding the spatial heterogeneity of microbiota and metabolites in the gastrointestinal tract of Qianqiu goats, particularly within the small intestine. The proposed "small intestine–methanogen/lipid/bile-adaptive micro-ecosystem" model not only challenges conventional views of ruminant gut ecology but also provides new avenues for exploring microbiota–metabolite–host interactions and optimizing feeding and nutritional strategies. While functional validation and mechanistic studies are necessary, our findings establish a foundation for future research and offer significant theoretical and practical implications for ruminant farming and feed optimization.

4. Materials and Methods

A total of 24 goats were collected from 6-month-old Qianqiu goats (Anhui Tianchang Zhou Sheep Industry Co., Ltd., Tianchang, China) (BW: 28.40 ± 1.80 kg), which were equally divided into three groups: the rumen (RM), small intestine (SI), and large intestine (LI). All animal experiments were approved by the Science and Technology Ethics Committee of Anhui Agricultural University (Protocol number: KJLL2024032).

The instruments used for sample collection included a 10 mL syringe barrel (needle-removed), 5 mL sterile sample tubes, dry ice, surgical tools, and other necessary equipment. To collect luminal contents, the barrel was carefully inserted into the rumen and the small (mid-jejunum) and large intestines (sigmoid colon), ensuring full immersion in luminal contents. The target intestinal segment was gently massaged to facilitate extraction, and a mixture of luminal contents and fluids was drawn into the syringe. The sample was transferred into labeled sterile tubes, placed temporarily on dry ice, and then processed within 30 min: parallel aliquots were stored at –20 °C and –80 °C for short-term preservation, while some were immediately immersed in liquid nitrogen for long-term storage. The exposed intestinal segment was then incised; the surrounding tissue was carefully excised, placed in sterile tubes, labeled, and stored identically.

For tissue collection, samples of gastrointestinal segments were obtained from six-month-old Qianqiu goats during slaughter. Rumen content was collected by opening the rumen directly; samples from the small and large intestines were collected via longitudinal incision, and contents were either collected directly or rinsed with saline as required. All specimens were immediately stored at approximately –80 °C to preserve integrity for subsequent analyses. Gastrointestinal tissue samples were rinsed with sterile physiological saline, transferred to labeled sterile tubes, stored at –80 °C for 24 h, and then moved to liquid nitrogen for long-term storage. All handling was performed under aseptic conditions to prevent contamination and ensure the integrity of samples used for subsequent metabolomic studies.

Genomic DNA was extracted using the HiPure Stool DNA Kit (D3141, Megagen, Guangzhou, China) following the manufacturer's instructions. The amplification steps are referenced in the literature [34].

The relative abundance of bacterial taxa was visualized using Krona [35], bar plots (ggplot2 [36]), chord plots (circlize [37]), and heatmaps (pheatmap [38]). Correlation analyses were conducted using the psych package (R package 1.0.13) [39]. Alpha diversity indices, including ACE, Chao1, and Shannon, were calculated using QIIME (v1.7.0), and non-metric multidimensional scaling (NMDS) was employed to evaluate dissimilarity in microbial community structures among the three GI segments.

For metabolomics, samples stored at −80 °C were thawed on ice and ground in liquid nitrogen. A 400 µL solution (methanol/water = 7:3, *V/V*) containing an internal standard was added to a 20 mg ground sample, and shaken at 1500 rpm for 5 min. After being placed on ice for 15 min, the sample was centrifuged at 12,000 rpm for 10 min (4 °C). In total, 300 µL of the supernatant was collected and placed in −20 °C conditions for 30 min. The sample was then centrifuged at 12,000 rpm for 3 min (4 °C). A total of 200 µL aliquots of supernatant were transferred for LC-MS analysis.

All samples were analyzed under positive ion conditions and eluted from the T3 column (Waters ACQUITY Premier HSS T3 Column 1.8 µm, 2.1 mm × 100 mm). We used 0.1% formic acid in water as solvent A and 0.1% formic acid in acetonitrile as solvent B for the following gradient: 5 to 20% for 2 min, increased to 60% for the following 3 min, before increasing to 99% for 1 min and holding for 1.5 min. Then, we returned to 5% mobile phase B for 0.1 min, which was held for 2.4 min. The analytical conditions were as follows: column temperature, 40 °C; flow rate, 0.4 mL/min; and injection volume, 4 µL. The methods alternated between full-scan MS and data-dependent MSn scans using dynamic exclusion. MS analyses were carried out with electrospray ionization in the positive ion mode and negative ion mode using full-scan analysis over *m/z* 75–1000 at 35,000 resolution. Additional MS settings are listed as follows: ion spray voltage, 3.5 KV or 3.2 KV in positive or negative modes, respectively; sheath gas (Arb), 30; aux gas, 5; ion-transfer tube temperature, 320 °C; vaporizer temperature, 300 °C; collision energy, 30, 40, 50 V; signal intensity threshold, 10^6 cps; top N vs. Top speed, 10; and exclusion duration, 3 s.

Data were log2-transformed and analyzed in R (version 4.5.2; R Foundation). Principal component analysis (PCA), hierarchical cluster analysis, and OPLS-DA (VIP > 1, $p < 0.05$) were performed with model validation via permutation testing, using the ropls package (version 1.42.0; Bioconductor 3.22). Heatmaps were generated with pheatmap (version 1.0.13). Metabolites were annotated and analyzed for pathway enrichment using the KEGG database.

To identify co-expressed metabolite clusters, WGCNA was used to construct an exploratory scale-free metabolite network (n = 10, below recommended size). Modules were identified with TOM and dynamic tree cutting (minModuleSize = 8 and mergeCutHeight = 0.25). Module eigengenes correlated with sample traits using Pearson r and raw p, implemented via the WGCNA R package (version 1.73). KEGG enrichment raw p-values were computed and used solely to rank pathways. Metabolite–metabolite correlations ($|r| \geq 0.7$) with raw p were retained for candidate prioritization.

All statistical analyses were conducted in R (v4.5.2). Alpha-diversity indices were compared among intestinal segments using one-way ANOVA followed by Tukey's HSD post hoc test; assumptions of normality and homogeneity of variance were verified using Shapiro–Wilk and Levene's tests, respectively, and data were log-transformed when necessary. Microbial-community dissimilarities (NMDSs) were assessed using Bray–Curtis distances and tested with PERMANOVA (999 permutations). Metabolomics data (log_2-transformed) were analyzed via PCA and hierarchical cluster analysis; OPLS-DA models were validated using permutation testing (200 permutations); and variables satisfying VIP > 1 and $p < 0.05$ were considered significant. Multiple-testing corrections for module–group correlations were performed by controlling the false discovery rate (FDR < 0.05). All tests were two-tailed, and significance was set at $\alpha = 0.05$.

5. Conclusions

Across the rumen and small and large intestines of Qianqiu goats, amplicon sequencing revealed a highly diverse and evenly distributed microbiota with strong functional redundancy: Prevotella dominated the rumen; Methanobrevibacter and Candidatus Saccharimonas dominated in the small intestine; and Rikenellaceae_RC9_gut_group plus Bacteroides were most significant in the large intestine. The unexpected enrichment of Methanobrevibacter and the near-absence of Prevotellaceae_UCG-003 in the small intestine indicate a unique, bile-tolerant anaerobic niche in this segment. Metabolome-wide WGCNA further identified 13 segment-specific metabolite modules; small-intestinal hubs such as 1-Myristoyl-2-stearoyl-sn-glycero-3-phosphocholine, L-methionine sulfoxide, and 2-Aminoethyl dihydrogen phosphate assemble an absorption–antioxidant–steroid metabolic network.

Integrative correlation analysis indicates that Methanobrevibacter, Methanosphaera, and Acetitomaculum of the small intestine co-occur in a hydrogen-driven syntrophic module: methanogens likely scavenge H_2 produced by Acetitomaculum, which may reduce CO_2 to acetate. Furthermore, their associations with steroid/lipid-related pathways point to possible membrane-remodeling adaptations in a bile-rich lumen. Therefore, these associations suggest a host-specific strategy in Qianqiu goats that may optimize hydrogen turnover, SCFA supply, and intestinal homeostasis, conferring a competitive advantage. However, causal verification of this hypothesis remains to be established.

Supplementary Materials: The following supporting information can be downloaded at https://www.mdpi.com/article/10.3390/ijms262411815/s1.

Author Contributions: Conceptualization, methodology, software, P.G.; validation, formal analysis, investigation, resources, W.Q.; data curation, writing—original draft preparation, P.G.; writing—review and editing, supervision, project administration, H.C.; visualization, funding acquisition, W.S. All authors have read and agreed to the published version of the manuscript.

Funding: This work was supported by The Qianqiu Goat Protection Project of Chuzhou, China (kyc23051) and Wanjiang Emerging Industry Technology Development Center 2024 Industrialization Project (WJ24CYHXM06).

Institutional Review Board Statement: The animal study protocol was approved by the Animal Ethics Committee of Anhui Agricultural University (protocol code KJLL20240321; approval date: 12 March 2024).

Informed Consent Statement: Not applicable.

Data Availability Statement: The original contributions presented in this study are included in the article/Supplementary Material. Further inquiries can be directed to the corresponding author.

Acknowledgments: The authors thank Qiang Li for their technical assistance with microbiome sequencing and bioinformatics analysis.

Conflicts of Interest: The authors declare no conflicts of interest.

References

1. Maurya, V.; Bharti, M.K. Physiology of the Digestive System of Ruminants. In *Principles of Veterinary Animal Physiology*; CRC Press: Boca Raton, FL, USA, 2025.
2. Sanjorjo, R.A.; Tseten, T.; Kang, M.-K.; Kwon, M.; Kim, S.-W. In pursuit of understanding the rumen microbiome. *Fermentation* **2023**, *9*, 114. [CrossRef]
3. Wang, L.; Zhang, G.; Li, Y.; Zhang, Y. Effects of high forage/concentrate diet on volatile fatty acid production and the microorganisms involved in VFA production in cow rumen. *Animals* **2020**, *10*, 223. [CrossRef]
4. Koepsell, H. Glucose transporters in the small intestine in health and disease. *Pflugers Arch. Eur. J. Physiol.* **2020**, *472*, 1207–1248. [CrossRef]
5. Negussie, A.B.; Dell, A.C.; Davis, B.A.; Geibel, J.P. Colonic fluid and electrolyte transport 2022: An update. *Cells* **2022**, *11*, 1712. [CrossRef]
6. McCallum, G.; Tropini, C. The gut microbiota and its biogeography. *Nat. Rev. Microbiol.* **2024**, *22*, 105–118. [CrossRef] [PubMed]
7. Nicolas, G.R.; Chang, P.V. Deciphering the chemical lexicon of host–gut microbiota interactions. *Trends Pharmacol. Sci.* **2019**, *40*, 700–711. [CrossRef]
8. Shi, J.; Su, H.; He, S.; Dai, S.; Mao, H.; Wu, D. Pan-genomic insights into rumen microbiome-mediated short-chain fatty acid production and regulation in ruminants. *Microorganisms* **2025**, *13*, 1175. [CrossRef] [PubMed]
9. Tyagi, A.; Kumar, V. The gut microbiota–bile acid axis: A crucial regulator of immune function and metabolic health. *World J. Microbiol. Biotechnol.* **2025**, *41*, 215. [CrossRef]
10. Zhang, X.; Zhong, R.; Wu, J.; Tan, Z.; Jiao, J. Dietary selection of distinct gastrointestinal microorganisms drives fiber utilization dynamics in goats. *Microbiome* **2025**, *13*, 118. [CrossRef]
11. Torres-Hernández, G.; Maldonado-Jáquez, J.A.; Granados-Rivera, L.D.; Salinas-González, H.; Castillo-Hernández, G. Status quo of genetic improvement in local goats: A review. *Arch. Anim. Breed.* **2022**, *65*, 207–221. [CrossRef] [PubMed]
12. Lagier, J.C.; Million, M.; Hugon, P.; Armougom, F.; Raoult, D. Human gut microbiota: Repertoire and variations. *Front. Cell. Infect. Microbiol.* **2012**, *2*, 136. [CrossRef]
13. Kim, M. Assessment of the gastrointestinal microbiota using 16S rRNA gene amplicon sequencing in ruminant nutrition. *Anim. Biosci.* **2023**, *36*, 364–373. [CrossRef] [PubMed]
14. Gao, P.; Huang, X.; Fang, X.-Y.; Zheng, H.; Cai, S.-L.; Sun, A.-J.; Zhao, L.; Zhang, Y. Application of metabolomics in clinical and laboratory gastrointestinal oncology. *World J. Gastrointest. Oncol.* **2021**, *13*, 536–549. [CrossRef]
15. Bingol, K. Recent advances in targeted and untargeted metabolomics by NMR and MS/NMR methods. *High-Throughput* **2018**, *7*, 9. [CrossRef] [PubMed]
16. Nunes, A.T.; Faleiros, C.A.; Poleti, M.D.; Novais, F.J.; López-Hernández, Y.; Mandal, R.; Wishart, D.S.; Fukumasu, H. Unraveling ruminant feed efficiency through metabolomics: A systematic review. *Metabolites* **2024**, *14*, 675. [CrossRef] [PubMed]

17. Ross, E.M.; Hayes, B.J.; Tucker, D.; Bond, J.; Denman, S.E.; Oddy, V.H. Genomic predictions for enteric methane production are improved by metabolome and microbiome data in sheep (*Ovis aries*). *J. Anim. Sci.* **2020**, *98*, skaa262. [CrossRef]
18. Zhang, C.; Zhen, Y.; Weng, Y.; Lin, J.; Xu, X.; Ma, J.; Zhong, Y.; Wang, M. Research progress on the microbial metabolism and transport of polyamines and their roles in animal gut homeostasis. *J. Anim. Sci. Biotechnol.* **2025**, *16*, 57. [CrossRef]
19. Goldansaz, S.A.; Guo, A.C.; Sajed, T.; Steele, M.A.; Plastow, G.S.; Wishart, D.S. Livestock metabolomics and the livestock metabolome: A systematic review. *PLoS ONE* **2017**, *12*, e0177675. [CrossRef]
20. Ma, Y.; Cai, G.; Chen, J.; Yang, X.; Hua, G.; Han, D.; Li, X.; Feng, D.; Deng, X. Combined transcriptome and metabolome analysis reveals breed-specific regulatory mechanisms in Dorper and Tan sheep. *BMC Genom.* **2024**, *25*, 70. [CrossRef]
21. Folz, J.; Culver, R.N.; Morales, J.M.; Grembi, J.; Triadafilopoulos, G.; Relman, D.A.; Huang, K.C.; Shalon, D.; Fiehn, O. Human metabolome variation along the upper intestinal tract. *Nat. Metab.* **2023**, *5*, 777–788. [CrossRef]
22. Deek, R.A.; Ma, S.; Lewis, J.; Li, H. Statistical and computational methods for integrating microbiome, host genomics, and metabolomics data. *eLife* **2024**, *13*, e88956. [CrossRef] [PubMed]
23. Qin, W.; Chen, D.; Guo, P.; Hu, L.; Zheng, X.; Cheng, J.; Chen, H. Ecogroups and maternal haplogroups reveal the ancestral origin of native Chinese goat populations based on the variation of mtDNA D-loop sequences. *Ecol Evol.* **2023**, *13*, 10382. [CrossRef] [PubMed]
24. Paradiso, R.; Borriello, G.; Censi, S.B.; Salzano, A.; Cimmino, R.; Galiero, G.; Fusco, G.; De Carlo, E.; Campanile, G. Different Non-Structural Carbohydrates/Crude Proteins (NCS/CP) Ratios in Diet Shape the Gastrointestinal Microbiota of Water Buffalo. *Vet. Sci.* **2021**, *8*, 96. [CrossRef]
25. Greening, R.C.; Leedle, J.A.Z. Enrichment and isolation of *Acetitomaculum ruminis*, gen. nov., sp. nov.: Acetogenic bacteria from the bovine rumen. *Arch. Microbiol.* **1989**, *151*, 399–406. [CrossRef]
26. St-Pierre, B.; Wright, A.D. Diversity of gut methanogens in herbivorous animals. *Animal* **2013**, *7* (Suppl. S1), 49–56. [CrossRef]
27. Bretto, E.; Urpì-Ferreruela, M.; Casanova, G.R.; González-Suárez, B. The Role of Gut Microbiota in Gastrointestinal Immune Homeostasis and Inflammation: Implications for Inflammatory Bowel Disease. *Biomedicines* **2025**, *13*, 1807. [CrossRef]
28. Jin, L.; Li, K.; Li, Z.; Huang, X.; Wang, L.; Wang, X.; Di, S.; Cui, S.; Xu, Y. Investigation into Critical Gut Microbes Influencing Intramuscular Fat Deposition in Min Pigs. *Animals* **2024**, *14*, 3123. [CrossRef]
29. Lin, L.; Lai, Z.; Zhang, J.; Zhu, W.; Mao, S. The gastrointestinal microbiome in dairy cattle is constrained by the deterministic driver of the region and the modified effect of diet. *Microbiome* **2023**, *11*, 10. [CrossRef]
30. Tardiolo, G.; La Fauci, D.; Riggio, V.; Daghio, M.; Di Salvo, E.; Zumbo, A.; Sutera, A.M. Gut microbiota of ruminants and monogastric livestock: An overview. *Animals* **2025**, *15*, 758. [CrossRef]
31. Cao, Y.; Feng, T.; Wu, Y.; Xu, Y.; Du, L.; Wang, T.; Luo, Y.; Wang, Y.; Li, Z.; Xuan, Z.; et al. The multi-kingdom microbiome of the goat gastrointestinal tract. *Microbiome* **2023**, *11*, 219. [CrossRef] [PubMed]
32. Jiao, J.; Wu, J.; Zhou, C.; He, Z.; Tan, Z.; Wang, M. Ecological niches and assembly dynamics of diverse microbial consortia in the gastrointestinal of goat kids. *ISME J.* **2024**, *18*, wrae002. [CrossRef]
33. Patra, A.K.; Park, T.; Kim, M.; Yu, Z. Rumen methanogens and mitigation of methane emission by anti-methanogenic compounds and substances. *J. Anim. Sci. Biotechnol.* **2017**, *8*, 13. [CrossRef]
34. Khairunisa, B.H.; Heryakusuma, C.; Ike, K.; Mukhopadhyay, B.; Susanti, D. Evolving understanding of rumen methanogen ecophysiology. *Front. Microbiol.* **2023**, *14*, 1296008. [CrossRef]
35. Li, C.; Geng, Y.; Wang, P.; Shi, H.; Luo, J. Comparison of microbial diversity in rumen and small intestine of Xinong Saanen dairy goats using 16S rRNA gene high-throughput sequencing. *Anim. Prod. Sci.* **2022**, *62*, 1379–1390. [CrossRef]
36. Brown, E.M.; Clardy, J.; Xavier, R.J. Gut microbiome lipid metabolism and its impact on host physiology. *Cell Host Microbe* **2023**, *31*, 173–186. [CrossRef] [PubMed]
37. Yersin, S.; Vonaesch, P. Small intestinal microbiota: From taxonomic composition to metabolism. *Trends Microbiol.* **2024**, *32*, 970–983. [CrossRef] [PubMed]
38. Heinken, A.; Ravcheev, D.A.; Baldini, F.; Heirendt, L.; Fleming, R.M.T.; Thiele, I. Systematic assessment of secondary bile acid metabolism in gut microbes reveals distinct metabolic capabilities in inflammatory bowel disease. *Microbiome* **2019**, *7*, 75. [CrossRef] [PubMed]
39. Jones, B.V.; Begley, M.; Hill, C.; Gahan, C.G.M.; Marchesi, J.R. Functional and comparative metagenomic analysis of bile salt hydrolase activity in the human gut microbiome. *Proc. Natl. Acad. Sci. USA* **2008**, *105*, 13580–13585. [CrossRef] [PubMed]
40. Tang, J.; Xu, W.; Yu, Y.; Yin, S.; Ye, B.C.; Zhou, Y. The role of the gut microbial metabolism of sterols and bile acids in human health. *Biochimie* **2025**, *230*, 43–54. [CrossRef]
41. Zhao, W.; Wang, H.; Zheng, M.; Ni, Y. Bile salt hydrolase: A key player in gut microbiota and its implications for metabolic dysfunction-associated steatotic liver disease. *Microbiome Res. Rep.* **2025**, *4*, 28. [CrossRef]
42. Song, Z.; Cai, Y.; Lao, X.; Wang, X.; Lin, X.; Cui, Y.; Kalavagunta, P.K.; Liao, J.; Jin, L.; Shang, J.; et al. Taxonomic profiling and populational patterns of bacterial bile salt hydrolase (BSH) genes based on worldwide human gut microbiome. *Microbiome* **2019**, *7*, 9. [CrossRef]

43. Dong, Z.; Yang, S.; Tang, C.; Li, D.; Kan, Y.; Yao, L. New insights into microbial bile salt hydrolases: From physiological roles to potential applications. *Front. Microbiol.* **2025**, *16*, 1513541. [CrossRef]
44. Ridlon, J.M.; Kang, D.J.; Hylemon, P.B. Bile salt biotransformations by human intestinal bacteria. *J. Lipid Res.* **2006**, *47*, 241–259. [CrossRef]
45. Rimal, B.; Collins, S.L.; Tanes, C.E.; Rocha, E.R.; Granda, M.A.; Solanki, S.; Hoque, N.J.; Gentry, E.C.; Koo, I.; Reilly, E.R.; et al. Bile salt hydrolase catalyses formation of amine-conjugated bile acids. *Nature* **2024**, *626*, 859–863. [CrossRef]
46. Garcia, C.J.; Kosek, V.; Beltrán, D.; Tomás-Barberán, F.A.; Hajslova, J. Production of new microbially conjugated bile acids by human gut microbiota. *Biomolecules* **2022**, *12*, 687. [CrossRef]
47. Debabov, V.G. Acetogens: Biochemistry, Bioenergetics, Genetics, and Biotechnological Potential. *Microbiology* **2021**, *90*, 273–297. [CrossRef]
48. Li, Q.; Huo, J.; Ni, G.; Zhang, F.; Zhang, S.; Zhang, X.; Wang, R.; Jiao, J.; Yu, Z.; Pu, X.; et al. Reductive acetogenesis is a dominant process in the ruminant hindgut. *Microbiome* **2025**, *13*, 28. [CrossRef] [PubMed]
49. Greening, C.; Geier, R.; Wang, C.; Woods, L.C.; Morales, S.E.; McDonald, M.J.; Rushton-Green, R.; Morgan, X.C.; Koike, S.; Leahy, S.C.; et al. Diverse hydrogen production and consumption pathways influence methane production in ruminants. *ISME J.* **2019**, *13*, 2617–2632. [CrossRef] [PubMed]
50. Ungerfeld, E.M. Metabolic hydrogen flows in rumen fermentation: Principles and possibilities of interventions. *Front. Microbiol.* **2020**, *11*, 589. [CrossRef] [PubMed]
51. Li, Q.; Ma, Z.; Huo, J.; Zhang, X.; Wang, R.; Zhang, S.; Jiao, J.; Dong, X.; Janssen, P.H.; Ungerfeld, E.M.; et al. Distinct microbial hydrogen and reductant disposal pathways explain interbreed variations in ruminant methane yield. *ISME J.* **2024**, *18*, 1–16. [CrossRef]
52. Golonka, R.M.; Abokor, A.A.; Ntambi, J.M.; Vijay-Kumar, M. Chapter 16—Gut microbiota interaction in host lipid metabolism. In *Lipid Signaling and Metabolism*; Academic Press: Cambridge, MA, USA, 2020; pp. 321–343.
53. Menon, R.; Ramanan, V.; Korolev, K.S. Interactions between species introduce spurious associations in microbiome studies. *PLoS Comput. Biol.* **2018**, *14*, e1005939. [CrossRef]
54. Tang, Q.; Jin, G.; Wang, G.; Liu, T.; Liu, X.; Wang, B.; Cao, H. Current Sampling Methods for Gut Microbiota: A Call for More Precise Devices. *Front. Cell. Infect. Microbiol.* **2020**, *10*, 151. [CrossRef]
55. Jiao, L.; Kourkoumpetis, T.; Hutchinson, D.; Ajami, N.J.; Hoffman, K.; White, D.L.; Graham, D.Y.; Hair, C.; Shah, R.; Kanwal, F.; et al. Spatial Characteristics of Colonic Mucosa-Associated Gut Microbiota in Humans. *Microb. Ecol.* **2021**, *83*, 811–821. [CrossRef]
56. Mutungwazi, A.; Ijoma, G.N.; Matambo, T.S. The significance of microbial community functions and symbiosis in enhancing methane production during anaerobic digestion: A review. *Symbiosis* **2021**, *83*, 1–24. [CrossRef]
57. Li, N.; Zuo, B.; Huang, S.; Zeng, B.; Han, D.; Li, T.; Liu, T.; Wu, Z.; Wei, H.; Zhao, J.; et al. Spatial heterogeneity of bacterial colonization across different gut segments following inter-species microbiota transplantation. *Microbiome* **2020**, *8*, 161. [CrossRef] [PubMed]
58. Li, X.R.; Lu, H.Y. Enhanced metagenomic strategies for elucidating the complexities of gut microbiota: A review. *Front. Microbiol.* **2025**, *16*, 1626002. [CrossRef] [PubMed]
59. Zhu, X.; Zhang, K.; Teng, X.; Xia, X.; Deng, R.; Li, J. Spatiotemporally resolved tools for analyzing gut microbiota. *Chem* **2023**, *9*, 1094–1117. [CrossRef]

genes

MDPI

Article

Integrated Transcriptomic Analysis Identifies Core Hub Genes Regulating Mammary Gland Traits (Milk Quality/Lactation) in Dairy Livestock: *Bos taurus* and *Ovis aries*

Qiang Zhang, Lulu Yang, Yunhan Li, Pengbo Gu, Riguleng Si, Shuai Li, Lin Zhu * and Wenguang Zhang *

College of Animal Science, Inner Mongolia Agricultural University, Hohhot 010018, China; 18031396471@163.com (Q.Z.)
* Correspondence: zhulinynacxhs@163.com (L.Z.); atcgnmbi@aliyun.com (W.Z.)

Abstract

Background/Objectives: Mammary gland traits (milk quality and lactation performance) are economically critical for *B. taurus* and *O. aries*, but core regulatory hub genes remain unclear due to high false positives in single-method transcriptomic analyses. This study aimed to identify robust hub genes linked to species-specific differences in mammary gland tissue via an integrated bioinformatics strategy. Methods: Raw transcriptomic data (77 *B. taurus* and 77 *O. aries* mammary gland samples) were retrieved from the European Nucleotide Archive (ENA); after quality control, differential expression gene (DEG) screening, weighted gene co-expression network analysis (WGCNA), and SHapley Additive exPlanations (SHAP)-assisted machine learning were performed, with core genes defined as the intersection of the three gene sets, and functional enrichment and protein–protein interaction (PPI) network analyses were used to prioritize hub genes. Results: A total of 13,138 high-quality genes were retained, including 6148 DEGs, 4698 WGCNA core module genes, and 500 SHAP-high-contribution genes, yielding 178 core genes that were significantly enriched in the "translation" ($p < 0.001$) pathways; hub genes were identified via PPI network analysis. Conclusions: These findings indicate that *RPS15* and *RPL7A* are core species-difference signals in mammary gland tissue of *B. taurus* and *O. aries*, providing insights into inter-species molecular differences, and this integrated strategy enhances the robustness of hub gene identification in pure bioinformatics studies.

Keywords: *B. taurus*; *O. aries*; mammary gland; WGCNA; machine learning

1. Introduction

Dairy ruminants serve as the core source of high-quality dairy products in global animal husbandry [1], with their production performance and product quality directly linking to industrial chain economic benefits and human dietary nutritional security [2]; among them, *B. taurus* and *O. aries*—the two most economically valuable dairy species with the widest breeding scale—account for over 90% of global ruminant milk production [3], acting as critical carriers for humans to obtain high-quality protein, calcium, and functional fatty acids while underpinning global food security and sustainable animal husbandry development [4]. With rising living standards and health awareness, market demand for dairy products has shifted from "quantity satisfaction" to "quality pursuit", such as with consumers preferring high-protein and high-functional fatty acid dairy products and prioritizing safety by reducing mastitis incidence to minimize drug residues [5], which has made dissecting the molecular regulatory mechanisms of mammary gland traits in *B. taurus* and *O. aries* a key direction in livestock genomics and molecular breeding [6].

As the core organ executing lactation function in dairy ruminants, mammary gland tissue determines dairy yield and quality via its complex and precise physiological functions [7]; it not only precisely regulates the whole-cycle dynamics of lactation initiation, maintenance, and termination and efficiently synthesizes and secretes nutrients like milk protein, lactose, and milk fat [8], but also constructs a robust immune defense system against pathogenic invasions such as mastitis-causing microbes [9], and the synergy of these functions forms the unique lactation phenotype of dairy animals [10,11]. The rapid development of transcriptomics has provided a powerful tool for exploring mammary gland gene regulatory networks [12]; RNA sequencing (RNA-seq), as the core method for analyzing gene expression dynamics, boasts high whole-genome coverage and accurate quantitative resolution [13], enabling systematic capture of inter-species and inter-physiological state gene expression differences in mammary tissue and providing abundant data for mining lactation and immune trait-related candidate genes [14], yet the exponential growth of transcriptomic data has brought the challenge of eliminating technical noise to screen core regulatory genes with real biological significance [11].

Traditional single analytical methods have prominent limitations: differential expression analysis is susceptible to interference from sequencing depth, batch effects, and other technical factors, leading to high false positive rates and failure to reveal inter-gene synergistic regulation [15]; weighted gene co-expression network analysis (WGCNA) can mine functionally related gene sets via module clustering and clarify gene–gene interactions but lacks quantitative evaluation of genes' "phenotypic contribution", making it hard to accurately locate core regulatory genes in complex modules [6]; additionally, conventional statistical methods struggle to quantify the regulatory weight of individual genes on species-specific phenotypes in high-dimensional transcriptomic data, restricting in-depth analysis of core mechanisms [11]. To address these flaws, multi-method integrated analysis has become the mainstream strategy in current livestock functional genomics research [16]; this strategy constructs a complete chain of "data quality control → differential screening → functional association → phenotypic verification" by complementing the advantages of different methods, effectively reducing technical biases of single approaches and improving the reliability and biological credibility of core gene screening [17]. Against this backdrop, the rise of machine learning has injected new vitality into in-depth transcriptomic data mining [18]; compared with traditional methods, machine learning models have stronger capabilities in high-dimensional data processing and complex relationship fitting [19], can capture potential species-specific molecular signals from massive gene expression data to realize accurate phenotypic prediction and gene importance ranking [20], and when combined with interpretable tools like SHapley Additive exPlanations (SHAP), can further quantify each gene's regulatory contribution to phenotypes, breaking the "black box" dilemma of traditional machine learning and providing clear directions for core gene functional verification [21].

Based on this, this study takes *B. taurus* and *O. aries* mammary gland tissue as research objects and focuses on identifying core hub genes regulating mammary gland functions via a multi-method integrated bioinformatics strategy [6]. First, the study performs strict quality control and standardization on large-scale raw mammary transcriptomic data using an integrative normalization method to eliminate low-quality data and technical noise, ensuring the reliability of subsequent analyses [22]. Then, it screens inter-species differentially expressed gene sets via differential expression analysis, mines phenotype-associated functional gene modules using WGCNA (a method widely applied in livestock trait-related gene mining) [23,24], and simultaneously employs machine learning models combined with interpretable tools to screen core genes with significant contributions to species-specific phenotypes, with core genes defined as the intersection of the three gene sets verified by the criteria of "expression difference → functional association → phenotypic contribution" [25]. On this basis, the study further clarifies the key biological pathways of core genes via functional enrichment analysis and identifies core hub genes by constructing a protein–protein interaction (PPI) network, systematically elucidating the regulatory roles of core genes in the mammary glands of the two species [20].

The implementation of this study not only aims to reveal the molecular regulatory mechanisms of mammary glands in *B. taurus* and *O. aries* and enrich the theoretical system of ruminant mammary gland biology from a systems biology perspective [26], but also strives to establish a standardized and scalable integrated analysis framework for large-scale transcriptomic data, providing methodological references for subsequent cross-species and multi-trait transcriptomic studies. Meanwhile, the core regulatory genes screened in this study will offer help for ruminant milk quality improvement (e.g., optimizing milk protein and fat composition) and disease-resistant breeding (e.g., enhancing mastitis resistance), providing important theoretical support and technical reserves for promoting the high-quality and sustainable development of the dairy industry via multi-omics-driven genetic improvement [27].

2. Materials and Methods

The screening and analysis of core hub genes in the mammary gland must be based on a scientifically standardized research design to ensure the reliability, reproducibility, and biological significance of the results. This study strictly adhered to the international experimental technical specifications for ruminant functional genomics research and bioinformatics analysis standards, innovatively constructing a multi-method integration strategy of "data quality control multi-dimensional screening functional verification" to systematically screen and analyze the core hub genes in the mammary glands of *B. taurus* and *O. aries*. To fit the actual data conditions (only 77 *B. taurus* and 77 *O. aries* mammary transcriptome datasets with no additional metadata like lactation stage or parity), this study established a streamlined yet rigorous standardized workflow: first, it implemented strict quality control on raw data via Seq2Fun (removing low-quality reads and performing batch correction) to obtain high-quality genes; second, it adopted a three-tier screening strategy combining DEG analysis, WGCNA, and SHAP-assisted machine learning, defining core genes as the intersection of the three gene sets to reduce false positives; third, it restricted functional analysis to mammary-specific pathways (e.g., milk protein synthesis) to mitigate biases from missing sample background information, ensuring the design balances scientific rigor with data constraints. The entire research process forms a closed loop, with progressive and closely linked steps.

In the data source and preprocessing stage, to ensure sample representativeness and data quality, raw data were obtained from the international public database European Nucleotide Archive (ENA), including transcriptomic data in FASTQ format from mammary gland tissues of 77 *B. taurus* and 77 *O. aries*. All mammary transcriptomic datasets used in this study (*B. taurus*: ENA PRJNA682457; *O. aries*: ENA PRJNA612351, PRJNA724691, PRJNA1064454, PRJEB71075, PRJNA301615) only provided "mammary gland tissue" as sample annotation. No additional metadata—including cell composition (epithelial/stromal/immune proportions), lactation stage, parity, or health status—was available in the original dataset records. The data was processed using Seq2Fun, referring to the mammalian database constructed in this study (including the mammals_annotation_v2.0.txt annotation file and the mammals_v2.0.fmi index file). These data covered multiple public projects to avoid the limitations of single-project data. Automated quality control was performed using Seq2Fun (v2.0.5) software: low-quality reads (sequences with >50% of bases having a Q-value < 20) and sequencing adapter sequences were removed. Clean reads were then aligned to the mammalian reference genome for gene annotation to construct an initial gene expression matrix. To further reduce technical bias and interference from low-expression genes, standardized processing was conducted: abundance filtering (excluding lowly expressed genes with TPM < 0.8 in <25% of samples), variance filtering (retaining the top 75% of genes with high variability), and batch correction using the ComBat function in the sva package. Finally, a dataset of high-quality genes was obtained, laying a high-reliability foundation for subsequent analyses.

For the screening of core transcriptomic genes, a triple independent strategy of "differential expression functional association phenotypic contribution" was adopted: (1) RNA-seq read counts were used as input (abandoning non-count metrics like TPM) and subjected to CPM-based filtering (retaining genes with CPM ≥ 1 in at least 25% of samples) and TMM (Trimmed Mean of M-values) normalization (via edgeR::calcNormFactors). Differential expression genes (DEGs) between species were then identified using the limma-voom pipeline (which models mean–variance relationships and assigns precision weights to correct RNA-seq heteroscedasticity) with the criteria of FDR-adjusted q-value (padj) < 0.01 and $|\log_2$ fold change ($\log_2$FC)$| > 1$, providing initial candidates for the core gene pool; (2) a weighted gene co-expression network analysis (WGCNA) was conducted to construct a co-expression network. The soft threshold was optimized (signed $R^2 = 0.8$, meeting scale-free topology requirements and balancing network connectivity) to build the adjacency matrix, and modules were merged with a height cut-off of 0.15, with minimum module size set to 10 and module splitting sensitivity (deepSplit) set to 3 to improve module resolution. After optimizing the soft threshold to ensure scale-free topological characteristics, gene modules were identified via module clustering. Core module genes were screened based on the criteria of "correlation coefficient $|r| > 0.7$ between module and species phenotype and $p < 0.01$", supplementing candidates from the functional association dimension; additionally, species-specific WGCNA was performed for *B. taurus* and *O. aries* separately, and module preservation analysis was conducted to evaluate cross-species module conservation; (3) four machine learning models (Random Forest, Support Vector Machine [SVM], Naive Bayes, and Decision Tree) were constructed (task: classifying mammary gland samples into *B. taurus* or *O. aries*). Among them, the SVM model adopted a radial basis function (RBF) kernel (parameter C = 1.0 determined by 5-fold cross-validation) for high-dimensional feature mapping; Random Forest was set with 100 base decision trees to mitigate overfitting, while Naive Bayes and Decision Tree served as baseline models for performance benchmarking. The dataset was divided into a training set (94 samples) and a test set (60 samples) at a ratio of 6:4. After performance evaluation, the optimal model was selected and combined with SHapley Additive exPlanations (SHAP) analysis to quantify the contribution of each gene to species classification (instead of phenotypic contribution), screening 500 high-contribution genes. Notably, SVM was prioritized for its superiority in handling high-dimensional transcriptomic data, though it has inherent interpretability limitations addressed by subsequent Kernel SHAP analysis. Kernel SHAP (adapted for non-linear SVM) was used to decode model decisions by assigning gene-level contribution scores based on game theory, with positive/negative values indicating promotion/inhibition of species classification; this addressed the "black box" limitation of traditional machine learning, though accuracy decreases for low-expression genes due to signal-to-noise constraints. Finally, the intersection of the three gene sets was obtained using a Venn diagram to identify core transcriptomic genes with triple characteristics, eliminating false positives from a single method.

In the functional annotation and protein–protein interaction (PPI) network construction section, first, the custom s2f_ids annotated by Seq2Fun were converted to standard NCBI/Ensembl gene names to avoid annotation bias. Since Seq2Fun annotates genes using "custom s2f_ids", direct use of these IDs for functional enrichment analysis may lead to annotation failure or errors due to "ID mismatch"—thus, the online platform ExpressAnalyst (a downstream analysis tool supporting Seq2Fun) was used to convert raw s2f_ids to standard Entrez IDs (based on official annotations from NCBI or Ensembl databases). A one-to-many mapping phenomenon existed during conversion: some s2f_ids correspond to multiple homologous genes or transcripts (Seq2Fun labels genes/transcripts matched by the same read as the same s2f_id), so all valid identifiers were retained. Second, Gene Ontology Biological Process (GO BP) functional enrichment analysis was performed to clarify the key biological processes involved in core genes (e.g., mammary lactation-related translation, ribosomal assembly, and immune-relevant hormone response pathways). Finally, a PPI network was constructed based on the STRING database (with an interaction confidence score ≥ 0.9) and imported into Cytoscape_v3.10.2. To avoid biases caused by relying on a single centrality index, four complementary algorithms (Degree, DMNC, EPC, and MNC) from the cytoHubba plugin of Cytoscape were used to screen for core hub genes. The core hub genes were identified by taking the intersection of the results from these four algorithms. In summary, this study ensured the scientific rigor of the research through a complete methodological chain: "data preprocessing $\rightarrow$ multi-dimensional screening $\rightarrow$ functional and network analysis". The following sections will detail the results of each step to reveal the characteristics and regulatory mechanisms of core hub genes in the mammary glands of *B. taurus* and *O. aries*.

3. Results

3.1. Transcriptomic Data Characteristics and Differential Gene Screening

The overall technical route is shown in Figure 1a. The data were processed using Seq2Fun, with reference to the mammalian database constructed in this study. The alignment rate for each sample is as follows: the average for cow samples is 97.2 ± 1.5%, and the average for sheep samples is 96.8 ± 1.7%. A preliminary 26,476-gene expression matrix was constructed; after abundance, variance filtering, and batch correction, 13,138 high-quality genes were obtained, laying a stable foundation for subsequent work.

Principal Component Analysis (PCA) results (Figure 1b) further verified transcriptomic differences between species: *B. taurus* and *O. aries* samples were completely separated into two independent clusters in the PCA plot without overlap. The first principal component (PC1) explained 42.9% of the transcriptional variation, serving as the core factor distinguishing the mammary transcriptomic characteristics of the two species. This intuitively confirmed significant species-specific differences in the gene expression patterns of mammary gland tissues between *B. taurus* and *O. aries*, providing direct evidence for the necessity of subsequent differential expression analysis.

Based on this reliable transcriptomic data, we adopted the limma-voom pipeline (with TMM normalization and CPM-based low-abundance gene filtering) for differential expression analysis, and controlled false discoveries via Benjamini–Hochberg FDR correction, resulting in 6148 reliable DEGs (Figure 1c). Among these, 3388 genes were upregulated in *B. taurus* mammary glands and 2760 genes were upregulated in *O. aries* mammary glands.

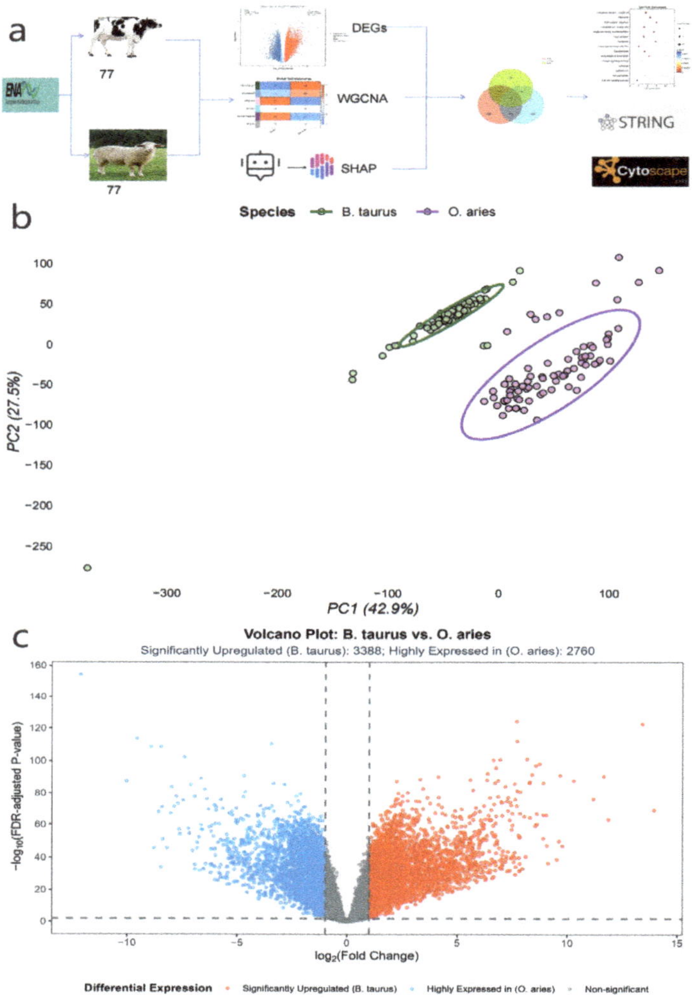

Figure 1. (**a**) Schematic diagram of the integrated methodological workflow for identifying core hub genes in mammary gland tissues of *B. taurus* and *O. aries*: The workflow included three key technical modules, namely raw transcriptomic data acquisition from the European Nucleotide Archive (ENA), multi-dimensional gene screening (differential expression analysis, weighted gene co-expression network analysis [WGCNA], and SHapley Additive exPlanations [SHAP]-assisted machine learning), and functional and interaction verification via STRING database and Cytoscape software (abbreviations: ENA, European Nucleotide Archive; WGCNA, weighted gene co-expression network analysis; SHAP, SHapley Additive exPlanations; STRING, Search Tool for the Retrieval of Interacting Genes/Proteins). (**b**) Principal Component Analysis (PCA) of mammary gland transcriptomes of *B. taurus* and *O. aries*. The scatter plot shows the clustering distribution of 77 *B. taurus* and 77 *O. aries* mammary gland samples, with the first principal component (PC1, explaining 42.9% of transcriptional variation) serving as the core factor distinguishing interspecific transcriptomic characteristics; the legend clearly differentiates the two species. (**c**) Volcano plot of differentially expressed genes (DEGs, defined as genes with false discovery rate [FDR]-adjusted p-value < 0.01 and $|\log_2$ fold change $(\log_2 FC)| > 1$) between *B. taurus* and *O. aries* mammary glands. The plot visualizes gene expression differences, with 3388 genes upregulated in *B. taurus* and 2760 genes upregulated in *O. aries*.

3.2. WGCNA Co-Expression Network Construction and Core Module Screening

From the 13,138 preprocessed high-quality genes, those without obvious co-expression trends were excluded via pre-clustering filtering, leaving 9539 genes for co-expression network construction; soft threshold screening (Figure 2a, the scale independence and mean connectivity plot) showed that when the soft-threshold power was set to 28, the network's scale-free topological fitting index (signed \(R^2\)) reached 0.8 (meeting the standard requirement for biological network topology) while maintaining a moderate average connectivity of ~120, and this parameter setting balanced the scale-free structural characteristics of real biological networks and avoided excessive sparsification. Using these optimized soft-threshold and clustering parameters (module merge cut-off = 0.15, minimum module size = 10, deepSplit = 3), the 9539 genes were clustered into 16 distinct co-expression modules (Figure 2b), and intra-module genes exhibited highly consistent expression patterns, implying potential synergistic regulatory relationships; to verify module reliability, we quantified intramodular connectivity (KIM) and performed bootstrap stability validation (100 rounds), with results showing that the KIM values of the 16 modules ranged from 0.58 to 0.87 (12 modules had KIM > 0.65, indicating strong intra-module co-expression coherence) and all modules had a bootstrap retention rate >85%, confirming the robustness of clustering results. To identify mammary function-related core modules, module–phenotype correlation analysis was conducted (Figure 2c, the module–trait relationships heatmap), and the four most prominent modules (with the strongest species–phenotype associations and extremely significant *p*-values) were screened: *darkolivegreen2* (289 genes, strongly positively correlated with **B. taurus**, r = 0.91, $p < 0.01$, predicted to regulate milk protein synthesis), *thistle4* (23 genes, positively correlated with **B. taurus**, r = 0.8, $p < 0.01$, functionally linked to bovine mammary metabolic pathways), *mediumpurple4* (4368 genes, strongly positively correlated with **O. aries**, r = 0.86, $p < 0.01$, the largest module, associated with ovine-specific mammary regulatory networks), and *lavenderblush1* (18 genes, positively correlated with **O. aries**, r = 0.76, $p < 0.01$, potentially mediating ovine mammary immune defense); these 4 core modules (totaling 4698 genes) uncovered species-specific synergistic regulatory networks in mammary gland tissues, supplemented the core gene pool from the "functional association" dimension, and provided critical "functional reliability" support for subsequent multi-dimensional core gene screening.

3.3. Machine Learning Model Performance and High-Contribution Gene Screening

This analysis was based on the 13,138 high-quality genes obtained in the previous step. Four machine learning models—Random Forest, Support Vector Machine (SVM), Naive Bayes, and Decision Tree—were constructed, and the dataset was split into a training set (94 samples) and a test set (60 samples) at a 6:4 ratio. Model performance results showed that all four models performed well in the *B. taurus-O. aries* species classification task, but significant differences existed: the SVM model exhibited the optimal performance, with accuracy, precision, recall, F1-score, and ROC_AUC all reaching 1.0, achieving error-free classification of 60 test samples (30 *B. taurus* + 30 *O. aries*) and accurately capturing species-specific molecular signals (Figure 3a). This performance was validated via nested 5-fold CV (validated accuracy: 0.993) and leave-one-project-out splits (average accuracy: 1.0), ruling out data leakage or batch effects. Model robustness was further confirmed via permutation importance (shuffled label accuracy: 0.761) and ablation experiments (gradual accuracy declines with top gene removal).The Random Forest model ranked second, with recall and ROC_AUC reaching 1.0 and an accuracy of 0.983, with only one *O. aries* sample misclassified (Figure 3b). The Naive Bayes and Decision Tree models both had an accuracy of 0.967, with slight differences in precision and recall, but maintained high classification reliability overall (Table 1) (Figure 3c,d). Considering model reliability (classification accuracy) and result interpretability (compatibility with subsequent SHAP analysis), the SVM model was ultimately selected as the optimal model. Combined with Kernel SHAP (the standard SHAP variant for non-linear SVMs), the contribution of each gene to the species classification result was quantified: by calculating the SHAP value of each gene, 500 high-contribution genes with the highest contribution rankings were screened. These genes had direct regulatory contributions to species-specific phenotypes. They provided core candidates from the "phenotypic contribution" dimension for subsequent triple-intersection analysis, further narrowing the scope of core genes.

Table 1. Performance evaluation of four machine learning models.

Model Name	Accuracy	Precision	Recall	F1_Score	ROC_AUC
Random Forest	0.983	0.968	1	0.984	1
SVM	1	1	1	1	1
Naive Bayes	0.967	1	0.933	0.966	0.997
Decision Tree	0.967	0.938	1	0.968	0.967

3.4. Core Gene Intersection, Functional Enrichment, and Hub Gene Identification

Core gene intersection analysis results showed that the intersection of DEGs (6148), WGCNA core module genes (4698), and SHAP high-contribution genes (500) was obtained using a Venn diagram, resulting in 178 core transcriptomic genes (Figure 4a). These genes were strictly screened through triple criteria, effectively eliminating false positive genes such as "differentially expressed but not functionally associated" or "module-participating but not phenotypically contributing", and thus had extremely high biological credibility, serving as the core gene set regulating mammary gland functions in *B. taurus* and *O. aries*. The s2f_ids of the 178 core intersection genes were converted to precise Entrez IDs via the aforementioned mammalian database: with the input set to "Universal (Seq2Fun orthologs)" in this database, the s2f_ids of the178 core genes (influenced by one-to-many mapping) were converted to 200 Entrez IDs (all converted entries were valid).

The results of GO BP functional enrichment analysis further clarified the biological functions of core genes (the top 20 enriched terms are shown in Figure 4b). The top-ranked enriched biological processes included translation, ribosomal small subunit assembly, ribosomal subunit assembly, response to steroid hormone, and regulation of cell population proliferation. Among these, the "translation" process was the most significantly enriched term, serving as the core link of milk protein synthesis and directly determining the efficiency and quality of protein synthesis in mammary gland tissue; "ribosomal small subunit assembly" and "ribosomal subunit assembly" are key prerequisites for the formation of functional ribosomes, which provide the structural basis for the translation process and indirectly regulate the synthesis of milk protein such as casein; "regulation of cell population proliferation" participates in the development and expansion of mammary gland tissue, laying a structural foundation for lactation by promoting the proliferation of mammary gland cells; "response to steroid hormone" is closely associated with the response of mammary gland tissue to lactation-related hormones including estrogen, which modulates the lactation activity of mammary gland cells. These findings reveal the extensive involvement of core genes in key physiological processes such as ribosome biogenesis, protein translation, cell proliferation, and hormone response, providing critical insights for elucidating their roles in mammary gland function regulation.

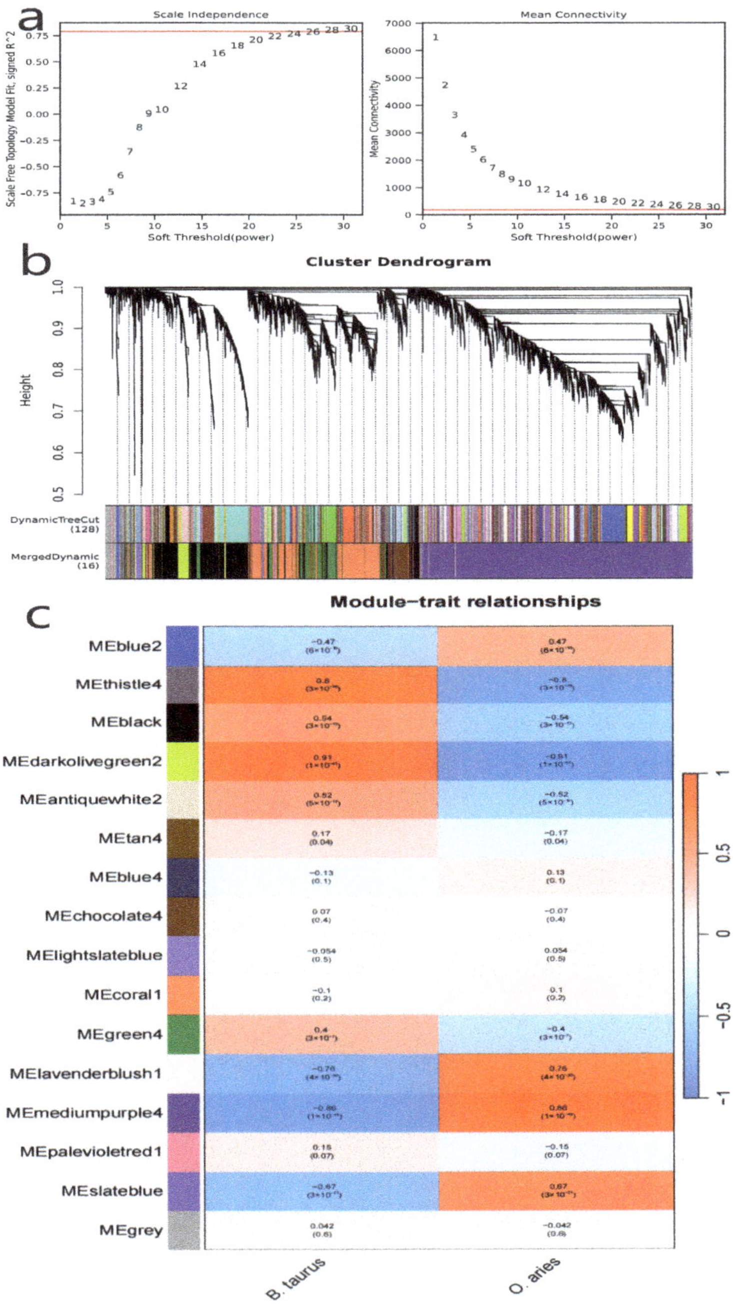

Figure 2. (**a**) Soft-threshold selection for WGCNA co-expression network construction. (**b**) Gene module clustering of *B. taurus* and *O. aries* mammary gland transcriptomes. (c) Module–phenotype correlation analysis of WGCNA co-expression network.

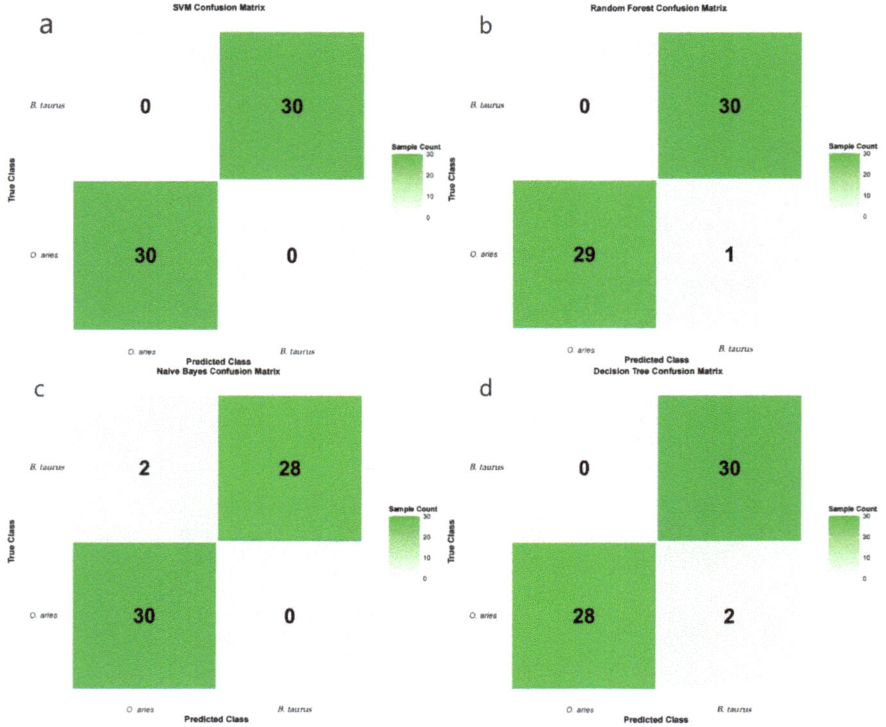

Figure 3. (**a**) SVM confusion matrix for *B. taurus* vs. *O. aries* classification. (**b**) Random Forest confusion matrix for *B. taurus* vs. *O. aries* classification. (**c**) Naive Bayes confusion matrix for *B. taurus* vs. *O. aries* classification. (**d**) Decision Tree confusion matrix for *B. taurus* vs. *O. aries* classification.

Results of the PPI network of core genes built using the STRING database showed that the network comprised 155 nodes (proteins encoded by core genes) and 277 edges (protein–protein interaction relationships), with a PPI enrichment p-value $< 1 \times 10^{-16}$—a value well below the significance threshold. This verified that the proteins encoded by core genes have highly significant non-random synergistic interactions, further proving the strong functional relevance and coordination among these core genes (Figure 4c). Additional network metrics included an average node degree of 3.57 and an average local clustering coefficient of 0.285, which characterized the structural features of the interaction network. To avoid biases from relying on a single centrality metric, this study adopted four complementary algorithms (Degree, DMNC, Edge Percolated Component [EPC], and Maximum Neighborhood Component [MNC]) via Cytoscape's cytoHubba plugin to screen core hub genes (Figure 4d, Table 2). Notably, all four metrics yielded top 10 gene lists consisting entirely of ribosomal proteins, and through cross-validation with these multi-dimensional metrics, the top 10 core hub genes identified by the Degree algorithm were specifically *RPL29*, *RPLP0*, *RPS5*, *RPL28*, *RPS9*, *RPS28*, *RPL24*, *RPSA*, *RPS15*, and *RPL7A*, all of which belong to the ribosomal protein family (with *RPS15* and *RPL7A* stably ranked in the top 10 of all four algorithms). This finding demonstrates that ribosomal protein family genes are enriched and highly connected core nodes in the mammary gland function regulatory network of *B. taurus* (cattle) and *O. aries* (sheep). By preserving ribosomal structural stability and modulating protein translation processes, these genes play an irreplaceable core role in maintaining the homeostasis of the mammary protein synthesis network and guaranteeing efficient milk protein synthesis.

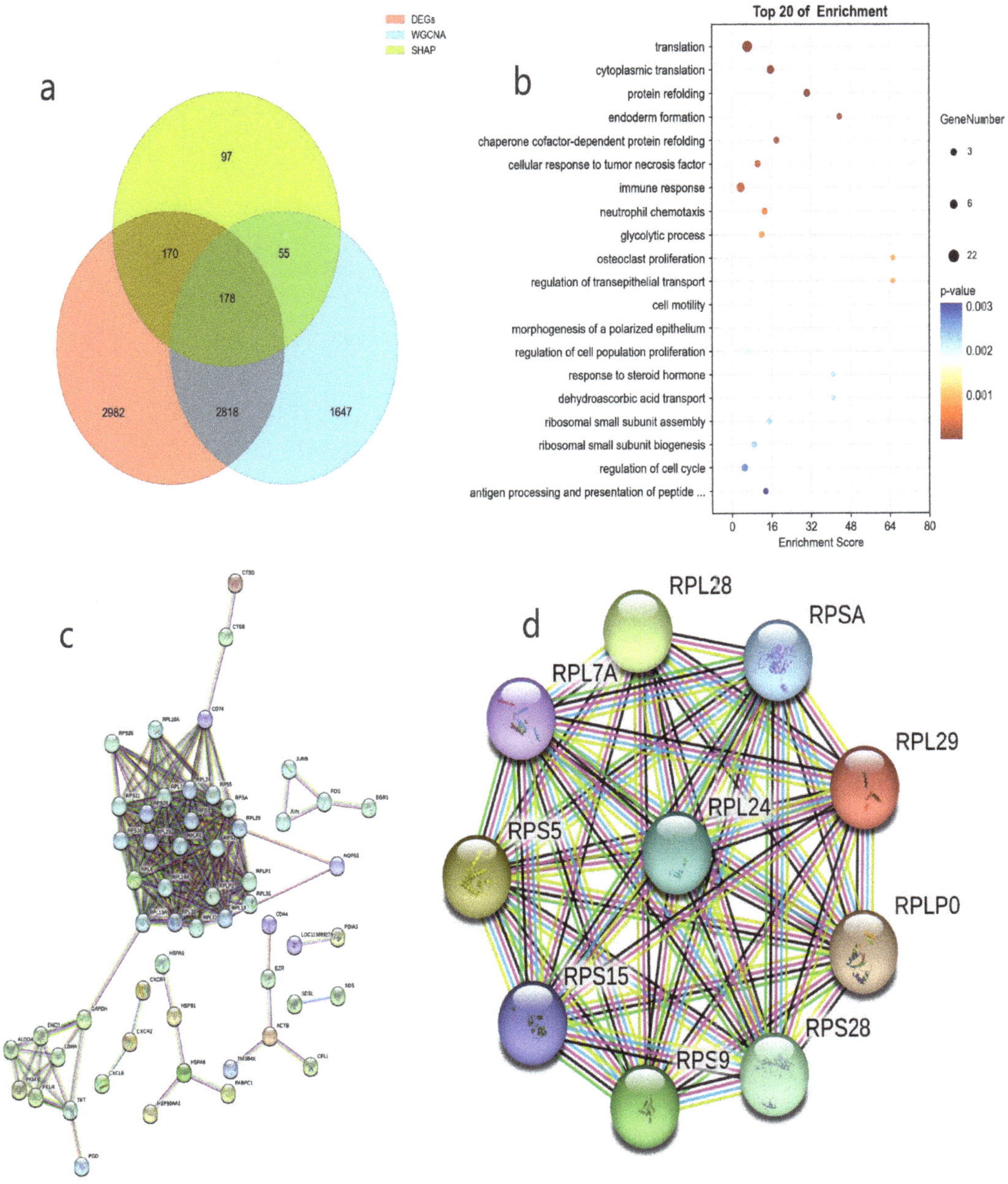

Figure 4. (**a**) Venn diagram of core gene intersection among DEGs, WGCNA modules, and SHAP high-contribution genes. (**b**) Gene Ontology Biological Process (GO BP) enrichment analysis results of core genes. (**c**) Protein–protein interaction (PPI) network of core genes. (**d**) Top 10 hub genes in PPI network identified by Degree algorithm.

Table 2. Top 10 core hub genes identified by 4 centrality metrics.

Centrality Metric	Top 10 Hub Genes (Rank, Name)	Score(s)	Proportion of Ribosomal Proteins in Top 10
Degree	1.*RPL29*; 2.*RPLP0*; 2.*RPS5*; 2.*RPL28*; 2.*RPS9*; 2.*RPS28*; 2.*RPL24*; 2.*RPSA*; 9.*RPS15*; 9.*RPL7A*	24, 23 (n = 7), 22 (n = 2)	100% (10/10)
DMNC	1.*RPLP2*; 1.*RPL12*; 3.*RPL38*; 3.*RPL13A*; 5.*RPLP1*; 6.*RPL18A*; 6.*RPL8*; 8.*RPS16*; 9.*RPS15*; 9.*RPL7A*	1.139115600700526 (n = 2); 1.1299697840265273 (n = 2); 1.1239012963372081; 1.1096551236319298 (n = 2); 1.0852465843984804; 1.0758424166431555 (n = 2)	100% (10/10)
EPC	1.*RPS9*; 2.*RPSA*; 3.*RPS5*; 4.*RPLP0*; 5.*RPL28*; 6.*RPS11*; 7.*RPL7A*; 8.*RPS15*; 9.*RPL24*; 10.*RPL29*	9.726999999999996; 9.694999999999975; 9.593999999999976; 9.564999999999976; 9.485999999999985; 9.445999999999993; 9.361999999999986; 9.332999999999998; 9.293999999999974; 9.269999999999982	100% (10/10)
MNC	1.*RPL29*; 2.*RPLP0*; 2.*RPS5*; 2.*RPL28*; 2.*RPS9*; 2.*RPS28*; 2.*RPL24*; 2.*RPSA*; 9.*RPS15*; 9.*RPL7A*	24, 23 (n = 7), 22 (n = 2)	100% (10/10)

4. Discussion

This study integrated 154 public mammary gland transcriptomic datasets (77 for *B. taurus*, 77 for *O. aries*) to systematically explore the molecular regulatory mechanisms underlying species-specific mammary gland traits in these two core dairy ruminants, which have the largest global breeding scale and highest economic value. *B. taurus* samples were derived from the PRJNA682457 project, while *O. aries* samples covered five projects (PRJNA612351, PRJNA724691, PRJNA1064454, PRJEB71075, PRJNA301615), with the full project details and corresponding accession numbers archived in Supplementary Material Table S1; the data files processed by Seq2Fun have been deposited in Supplementary Material Table S2 for further verification by the academic community (Tables S1 and S2); all samples focused exclusively on mammary gland tissue, ensuring the representativeness of species-specific analysis and data targeting. It should be noted that the asymmetric distribution of sample sources (single bovine project vs. five ovine projects) may introduce potential technical heterogeneity in the ovine cohort (e.g., batch differences across sequencing projects), which could marginally overestimate ovine-specific transcriptional effects despite batch correction via the ComBat function. A multi-method integration strategy of "differential expression analysis (DEG) + weighted gene co-expression network analysis (WGCNA) + SHAP-assisted machine learning" was adopted, forming a complete analytical system from "gene expression difference identification" to "co-expression module function annotation" and "contribution verification," which effectively avoided the inherent limitations of single methods [21]. In the data processing stage, Seq2Fun (v2.0.5) software was used for automated quality control and annotation: low-quality reads (with >50% of bases having Q-value < 20) and sequencing adapters were strictly removed to ensure raw data reliability [28]. Subsequent abundance filtering (excluding genes with TPM < 0.8 in

<25% of samples), variance filtering (retaining the top 75% of variable genes), and batch correction via the ComBat function in the sva package yielded 13,138 high-quality genes for subsequent analysis, significantly reducing technical noise interference [29,30]. For the species classification task, four machine learning models (SVM, Random Forest, Decision Tree, Naive Bayes) were systematically evaluated [31,32]. Data were split into a training set (94 samples) and a test set (60 samples) at a 6:4 stratified ratio to maintain consistent species proportions. Performance evaluation showed the SVM model was optimal, with accuracy, precision, recall, F1 score, and ROC_AUC all reaching 1.0, achieving error-free classification of 60 test samples (30 *B. taurus* + 30 *O. aries*) and accurately capturing species-specific molecular signals—consistent with the excellent performance of machine learning models in transcriptomic classification tasks in similar studies [33]. The 178 core genes, obtained through strict triple screening, were the intersection of 6148 DEGs (screening criteria: padj < 0.01, $|\log_2 FC| > 1$, identified via limma-voom with TMM normalization and FDR correction), 4698 WGCNA core genes (from 3 modules with $|r| > 0.6$ and $p < 0.01$ for species trait correlation), and 500 SHAP core genes (top contributors to species classification) [24,34]. They represent common core regulatory genes meeting "differential expression, functional association, and phenotypic contribution" in the mammary glands of both species [35,36], with functional enrichment analysis showing significant enrichment in mammary gland-related pathways such as "ribosome assembly" [37]. A PPI network constructed via the STRING database (interaction confidence threshold ≥ 0.9) contained 155 nodes and 277 edges, with a PPI enrichment p-value < 0.01, indicating significant non-random synergistic interactions among proteins encoded by core genes [38]. Identified via four complementary centrality algorithms (Degree, DMNC, EPC, and MNC, see Table 2) for hub gene screening, *RPS15* and *RPL7A* stood out as core hub genes, as they maintained stable top 10 rankings across all four metrics. Both genes are members of the ribosomal protein family, and their core regulatory position has been further validated by GO enrichment analysis (Figure 4b): these two genes show a significant enrichment feature in the milk protein-related pathway ($p < 0.001$).This result confirmed that *RPS15* and *RPL7A* play biologically meaningful roles in maintaining the stability and efficiency of the mammary gland protein synthesis network, rather than having non-specific network connectivity [39,40]. The "DEG + WGCNA + ML-SHAP" strategy was critical for improving core gene screening reliability [41,42]: it addressed flaws of single methods (e.g., high false positives in DEG, lack of phenotypic contribution quantification in WGCNA, "black box" issue in traditional machine learning) and enhanced biological credibility via three-level verification. Additionally, combining SVM with SHAP analysis further improved result interpretability, overcoming the "black box" defect of traditional machine learning and providing a scalable template for cross-species transcriptomic analysis of other livestock [43].

Despite clarifying the core regulatory network and potential hub genes of *B. taurus* and *O. aries* mammary glands, the study had four limitations: First, the lack of key metadata (cell composition, lactation stage, parity) for the public datasets prevented us from performing cell composition deconvolution or adjusting these covariates in statistical models. Moreover, the asymmetric project origin of samples (single bovine project vs. five ovine projects) may have amplified technical variation in the ovine cohort, which could lead to overestimation of ovine-specific biological effects despite batch correction. Thus, some observed cross-species transcriptomic differences may partially reflect unmeasured tissue heterogeneity, rather than inherent regulatory divergence between species. Extrapolating results to specific cell types or lactation stages is also not feasible with the current data. Second, the sample size (77 per species) might limit result generalizability, as it failed to cover the full breed diversity of the two species (e.g., Holstein/Jersey for *B. taurus*, Suffolk/Small-tailed Han for *O. aries*), potentially omitting low-expression but functionally critical genes and conflating breed-specific traits (e.g., breed-dependent milk protein synthesis efficiency) with species-level regulatory divergence. Third, relying solely on transcriptomic data made it difficult to fully reveal the regulatory chain—gene function realization requires "transcription–translation–modification–functional execution," and milk composition formation involves metabolic pathway regulation, which transcriptomic data alone cannot link to the complete "gene expression–protein function–metabolic product" chain [44,45]. Fourth, hub gene functions and species-specific expression lacked wet experimental verification, with functions only inferred via bioinformatics (missing qRT-PCR or CRISPR/Cas9 evidence), limiting their direct application in breeding [46,47]. To address these, future research could advance in three aspects: expanding sample size and breed coverage to include samples of different breeds, lactation stages (early/peak/late), and health statuses to verify core gene universality and explore breed/stage-specific regulatory genes, disentangle breed effects from species divergence, and balance project sources between species [8]; optimizing algorithm pipelines by incorporating biological priors into SVM parameter selection, using weighted WGCNA to reduce batch interference, and filtering low-expression genes before SHAP analysis to improve contribution score reliability; conducting multi-omics integration analysis by combining proteomic (detection of mammary functional protein expression) and metabolomic (detection of milk components/metabolites) data to construct a "transcriptome–proteome–metabolome" regulatory network and clarify core gene regulatory paths from "expression" to "function" to "phenotype" [16]; and strengthening hub gene functional verification and application—using qRT-PCR to confirm their expression in *B. taurus* mammary glands, performing overexpression/knockdown experiments in in vitro *B. taurus* mammary epithelial cells to verify effects on milk protein synthesis and milk fat metabolism, and exploring the potential of functionally validated genes as candidate targets for subsequent molecular marker development (pending identification of causal genetic variants linked to these genes), which could lay a preliminary foundation for supporting milk production and quality optimization in *B. taurus* breeding in the long term [27,48].

5. Conclusions

This study established a multi-method integrated strategy, including differential expression analysis, weighted gene co-expression network analysis (WGCNA), and machine learning integrated with SHapley Additive exPlanations (SHAP), to systematically identify core hub genes in the mammary glands of *B. taurus* and *O. aries*. Through rigorous data pretreatment and multi-level validation, 178 core genes were identified, which were significantly concentrated in mammary gland-related pathway assemblies. Protein–protein interaction (PPI) network analysis with four complementary centrality metrics (Table 2) revealed that hub genes *RPS15* and *RPL7A* are among the enriched and highly connected ribosomal protein genes in cross-species mammary comparisons, with prominent connectivity linked to mammary gland protein synthesis functions. These findings clarify the molecular network signatures underlying species-specific mammary gland functional profiles in *B. taurus* and *O. aries* and provide candidate molecular features that lay a foundational framework for subsequent exploration of variants linked to ruminant dairy-related pathways relevant to genetic improvement.

Supplementary Materials: The following supporting information can be downloaded at: https://www.mdpi.com/article/10.3390/genes16121483/s1, Table S1: Accession numbers of mammary gland transcriptomic datasets for *B. taurus* and *O. aries*. Table S2: Detailed information of mammary gland tissue genes processed by Seq2Fun.

Author Contributions: Conceptualization: Q.Z.; Writing—Original Draft Preparation: Q.Z.; Review and Editing: L.Y., Y.L., P.G., R.S., and S.L.; Supervision: W.Z. and L.Z. All authors have read and agreed to the published version of the manuscript.

Funding: This study was supported by the National Natural Science Foundation of China Regional Science Fund Project (Project No: 32460817) and the Youth Fund of the Natural Science Foundation of Inner Mongolia Autonomous Region (2025QN03189).

Institutional Review Board Statement: All referenced studies providing the 154 public mammary gland transcriptomic datasets (77 *B. taurus*, 77 *O. aries*, from the European Nucleotide Archive,) have confirmed compliance with ethical standards in their original manuscripts. They adhere to international "guidelines for the care and use of experimental animals". This study is an integrated transcriptomic analysis of public data, not primary research involving new animal experiments. The authors did not design/conduct/fund new animal-related work, so no additional ethical approval is required. Moreover, only studies with clear ethical statements were included, excluding those with ambiguous ethical documentation—ensuring all analyzed data and conclusions are ethically legitimate related to animal experimentation.

Informed Consent Statement: Not applicable.

Data Availability Statement: The raw mammary gland transcriptomic datasets analyzed in this study were all obtained from public databases and can be retrieved from the European Nucleotide Archive (ENA). The specific project accession numbers are as follows: datasets related to *B. taurus* were from Project PRJNA682457; datasets related to *O. aries* were from Projects PRJNA612351, PRJNA724691, PRJNA1064454, PRJNA301615, and PRJEB71075. The individual sample IDs corresponding to the aforementioned projects have been compiled in the Supplementary Materials of this article, which are available for further retrieval and verification.

Acknowledgments: We appreciate the technical assistance from members of the research team during raw data sorting and visualization.

Conflicts of Interest: The authors declare no conflicts of interest.

References

1. Cabiddu, A.; Delgadillo-Puga, C.; Decandia, M.; Molle, A.G. Extensive Ruminant Production Systems and Milk Quality with Emphasis on Unsaturated Fatty Acids, Volatile Compounds, Antioxidant Protection Degree and Phenol Content. *Animals* **2019**, *9*, 771. [CrossRef]
2. Oltjen, J.W.; Beckett, J.L. Role of ruminant livestock in sustainable agricultural systems. *J. Anim. Sci.* **1996**, *74*, 1406–1409. [CrossRef]
3. Haug, A.; Høstmark, A.T.; Harstad, O.M. *B. taurus* milk in human nutrition—A review. *Lipids Health Dis.* **2007**, *6*, 25. [CrossRef]
4. Law, F.M.; Moate, P.J.; Leaver, D.D.; Diefenbach-Jagger, H.; Grill, V.; Ho, P.W.; Martin, T.J. Parathyroid hormone-related protein in milk and its correlation with *B. taurus* milk calcium. *J. Endocrinol.* **1991**, *128*, 21–26. [CrossRef]
5. Palupi, E.; Jayanegara, A.; Ploeger, A.; Kahl, J. Comparison of nutritional quality between conventional and organic dairy products: A meta-analysis. *J. Sci. Food Agric.* **2012**, *92*, 2774–2781. [CrossRef]
6. Singh, M.; Thomson, P.C.; Sheehy, P.A.; Raadsma, H.W. Comparative transcriptome analyses reveal conserved and distinct mechanisms in ovine and *B. taurus* lactation. *Funct. Integr. Genom.* **2013**, *13*, 115–131. [CrossRef] [PubMed]
7. Beckett, L.; Xie, S.; Thimmapuram, J.; Tucker, H.A.; Donkin, S.S.; Casey, T. Mammary transcriptome reveals cell maintenance and protein turnover support milk synthesis in early-lactation cows. *Physiol. Genom.* **2020**, *52*, 435–450. [CrossRef]
8. Bionaz, M.; Loor, J.J. Gene networks driving *B. taurus* mammary protein synthesis during the lactation cycle. *Bioinform. Biol. Insights* **2011**, *5*, 83–98. [CrossRef] [PubMed]
9. Rainard, P.; Riollet, C. Mobilization of neutrophils and defense of the *B. taurus* mammary gland. *Reprod. Nutr. Dev.* **2003**, *43*, 439–457. [CrossRef]
10. Fu, Y.; Shen, N.; Wang, S.; Liu, Y.; Peng, P.; Shi, L.; Han, B.; Wang, K.; Sun, D. Integrating transcriptomic and metabolomic profiles of primiparous Holstein cows across multiple lactation periods reveals the regulatory mechanism underlying milk component traits. *J. Dairy Sci.* **2025**, *108*, 10377–10390. [CrossRef] [PubMed]

11. Fonseca, P.A.S.; Suárez-Vega, A.; Esteban-Blanco, C.; Pelayo, R.; Marina, H.; Gutiérrez-Gil, B.; Arranz, J.J. Epigenetic regulation of functional candidate genes for milk production traits in dairy sheep subjected to protein restriction in the prepubertal stage. *BMC Genom.* **2023**, *24*, 511. [CrossRef]
12. Zhou, Y.; Gong, W.; Xiao, J.; Wu, J.; Pan, L.; Li, X.; Wang, X.; Wang, W.; Hu, S.; Yu, J. Transcriptomic analysis reveals key regulators of mammogenesis and the pregnancy-lactation cycle. *Sci. China Life Sci.* **2014**, *57*, 340–355. [CrossRef]
13. Suchyta, S.P.; Sipkovsky, S.; Halgren, R.G.; Kruska, R.; Elftman, M.; Weber-Nielsen, M.; Vandehaar, M.J.; Xiao, L.; Tempelman, R.J.; Coussens, P.M. *B. taurus* mammary gene expression profiling using a cDNA microarray enhanced for mammary-specific transcripts. *Physiol. Genom.* **2003**, *16*, 8–18. [CrossRef]
14. Mumtaz, P.T.; Bhat, B.; Ibeagha-Awemu, E.M.; Taban, Q.; Wang, M.; Dar, M.A.; Bhat, S.A.; Shabir, N.; Shah, R.A.; Ganie, N.A.; et al. Mammary epithelial cell transcriptome reveals potential roles of lncRNAs in regulating milk synthesis pathways in Jersey and Kashmiri cattle. *BMC Genom.* **2022**, *23*, 176. [CrossRef]
15. Robinson, M.D.; McCarthy, D.J.; Smyth, G.K. edgeR: A Bioconductor package for differential expression analysis of digital gene expression data. *Bioinformatics* **2010**, *26*, 139–140. [CrossRef] [PubMed]
16. Sun, H.Z.; Zhou, M.; Wang, O.; Chen, Y.; Liu, J.X.; Guan, L.L. Multi-omics reveals functional genomic and metabolic mechanisms of milk production and quality in *B. taurus*. *Bioinformatics* **2020**, *36*, 2530–2537. [CrossRef] [PubMed]
17. Eklund, A.C.; Szallasi, Z. Correction of technical bias in clinical microarray data improves concordance with known biological information. *Genome Biol.* **2008**, *9*, R26. [CrossRef]
18. Hou, P.; Bao, S.; Fan, D.; Yan, C.; Su, J.; Qu, J.; Zhou, M. Machine learning-based integrative analysis of methylome and transcriptome identifies novel prognostic DNA methylation signature in uveal melanoma. *Brief. Bioinform.* **2021**, *22*, bbaa371. [CrossRef]
19. Moutsopoulos, I.; Maischak, L.; Lauzikaite, E.; Vasquez Urbina, S.A.; Williams, E.C.; Drost, H.G.; Mohorianu, I.I. noisyR: Enhancing biological signal in sequencing datasets by characterizing random technical noise. *Nucleic Acids Res.* **2021**, *49*, e83. [CrossRef]
20. Farhadian, M.; Rafat, S.A.; Hasanpur, K.; Ebrahimi, M.; Ebrahimie, E. Cross-Species Meta-Analysis of Transcriptomic Data in Combination with Supervised Machine Learning Models Identifies the Common Gene Signature of Lactation Process. *Front. Genet.* **2018**, *9*, 235. [CrossRef] [PubMed]
21. Li, S.; Guo, Y.; Huo, C.; Zhu, L.; Shi, C.; Na, R.; Gu, M.; Zhang, W. Machine Learning-Based Analysis of Differentially Expressed Genes in the Muscle Transcriptome Between Beef Cattle and Dairy Cattle. *Int. J. Mol. Sci.* **2025**, *26*, 5046. [CrossRef] [PubMed]
22. Filloux, C.; Cédric, M.; Romain, P.; Lionel, F.; Christophe, K.; Dominique, R.; Abderrahman, M.; Daniel, P. An integrative method to normalize RNA-Seq data. *BMC Bioinform.* **2014**, *15*, 188. [CrossRef]
23. Mu, T.; Hu, H.; Ma, Y.; Yang, C.; Feng, X.; Wang, Y.; Liu, J.; Yu, B.; Zhang, J.; Gu, Y. Identification of critical lncRNAs for milk fat metabolism in *B. taurus* using WGCNA and the construction of a ceRNAs network. *Anim. Genet.* **2022**, *53*, 740–760. [CrossRef]
24. Langfelder, P.; Horvath, S. WGCNA: An R package for weighted correlation network analysis. *BMC Bioinform.* **2008**, *9*, 559. [CrossRef]
25. Tang, D.; Huang, Y.; Che, Y.; Yang, C.; Pu, B.; Liu, S.; Li, H. Identification of platelet-related subtypes and diagnostic markers in pediatric Crohn's disease based on WGCNA and machine learning. *Front. Immunol.* **2024**, *15*, 1323418. [CrossRef]
26. Ferreira, A.M.; Bislev, S.L.; Bendixen, E.; Almeida, A.M. The mammary gland in domestic ruminants: A systems biology perspective. *J. Proteom.* **2013**, *94*, 110–123. [CrossRef]
27. Zheng, W.; Zhang, J.; Huang, J.; Han, B.; Liu, Y.; Sun, D. A pilot multi-omics study reveals genetic mechanisms regulating milk component traits in dairy cattle. *Commun. Biol.* **2025**, *8*, 1150. [CrossRef] [PubMed]
28. Liu, P.; Ewald, J.; Galvez, J.H.; Head, J.; Crump, D.; Bourque, G.; Basu, N.; Xia, J. Ultrafast functional profiling of RNA-seq data for nonmodel organisms. *Genome Res.* **2021**, *31*, 713–720. [CrossRef]
29. Ritchie, M.E.; Phipson, B.; Wu, D.; Hu, Y.; Law, C.W.; Shi, W.; Smyth, G.K. limma powers differential expression analyses for RNA-sequencing and microarray studies. *Nucleic Acids Res.* **2015**, *43*, e47. [CrossRef]
30. Leek, J.T.; Scharpf, R.B.; Bravo, H.C.; Simcha, D.; Langmead, B.; Johnson, W.E.; Geman, D.; Baggerly, K.; Irizarry, R.A. Tackling the widespread and critical impact of batch effects in high-throughput data. *Nat. Rev. Genet.* **2010**, *11*, 733–739. [CrossRef] [PubMed]
31. Abraham, A.; Pedregosa, F.; Eickenberg, M.; Gervais, P.; Mueller, A.; Kossaifi, J.; Gramfort, A.; Thirion, B.; Varoquaux, G. Machine learning for neuroimaging with scikit-learn. *Front. Neuroinform.* **2014**, *8*, 14. [CrossRef]
32. Mamoshina, P.; Volosnikova, M.; Ozerov, I.V.; Putin, E.; Skibina, E.; Cortese, F.; Zhavoronkov, A. Machine Learning on Human Muscle Transcriptomic Data for Biomarker Discovery and Tissue-Specific Drug Target Identification. *Front. Genet.* **2018**, *9*, 242. [CrossRef]
33. Cai, C.; Li, M.; Zhang, Y.; Meng, S.; Yang, Y.; Gao, P.; Guo, X.; Cao, G.; Li, B. Comparative Transcriptome Analyses of Longissimus thoracis Between Pig Breeds Differing in Muscle Characteristics. *Front. Genet.* **2020**, *11*, 526309. [CrossRef] [PubMed]
34. Toneyan, S.; Tang, Z.; Koo, P.K. Evaluating deep learning for predicting epigenomic profiles. *Nat. Mach. Intell.* **2022**, *4*, 1088–1100. [CrossRef]

5. Wang, B.; Men, J.; Wang, C.; Hou, X.; Zhao, F. Laminin-dependent integrin β1 signaling regulates milk protein synthesis via prolactin/STAT5 pathway in *B. taurus* mammary epithelial cells. *Biochem. Biophys. Res. Commun.* **2020**, *524*, 288–294. [CrossRef]
6. Perugini, J.; Smorlesi, A.; Acciarini, S.; Mondini, E.; Colleluori, G.; Pirazzini, C.; Kwiatkowska, K.M.; Garagnani, P.; Franceschi, C.; Zingaretti, M.C.; et al. Adipo-Epithelial Transdifferentiation in In Vitro Models of the Mammary Gland. *Cells* **2024**, *13*, 943. [CrossRef]
7. Gruden, Š.; Poklar Ulrih, N. Diverse Mechanisms of Antimicrobial Activities of Lactoferrins, Lactoferricins, and Other Lactoferrin-Derived Peptides. *Int. J. Mol. Sci.* **2021**, *22*, 11264. [CrossRef]
8. Szklarczyk, D.; Franceschini, A.; Wyder, S.; Forslund, K.; Heller, D.; Huerta-Cepas, J.; Simonovic, M.; Roth, A.; Santos, A.; Tsafou, K.P.; et al. STRING v10: Protein-protein interaction networks, integrated over the tree of life. *Nucleic Acids Res.* **2015**, *43*, D447–D452. [CrossRef] [PubMed]
9. Zhao, Q.; Zhang, Y.; Shao, S.; Sun, Y.; Lin, Z. Identification of hub genes and biological pathways in hepatocellular carcinoma by integrated bioinformatics analysis. *PeerJ* **2021**, *9*, e10594. [CrossRef] [PubMed]
0. Shen, X.; Qin, C.; Wang, Z.; Jiang, J.; Liu, Q.; Zeng, Q.; Li, L.; Li, Z. Transcriptomic and Proteomic Insights Into Buffalo Milk Fat Synthesis and the Role of IGFBP4 in BMECs. *FASEB J.* **2025**, *39*, e70971. [CrossRef]
1. Yan, S. Integrative analysis of promising molecular biomarkers and pathways for coronary artery disease using WGCNA and MetaDE methods. *Mol. Med. Rep.* **2018**, *18*, 2789–2797. [CrossRef]
2. Wang, F.; Liang, Y.; Wang, Q.W. Interpretable machine learning-driven biomarker identification and validation for Alzheimer's disease. *Sci. Rep.* **2024**, *14*, 30770. [CrossRef]
3. Guo, Y.; Li, S.; Na, R.; Guo, L.; Huo, C.; Zhu, L.; Shi, C.; Na, R.; Gu, M.; Zhang, W. Comparative Transcriptome Analysis of *B. taurus*, Porcine, and Sheep Muscle Using Interpretable Machine Learning Models. *Animals* **2024**, *14*, 2947. [CrossRef]
4. Liu, J.; Jiang, P.; Iqbal, A.; Ali, S.; Gao, Z.; Pan, Z.; Xia, L.; Yin, F.; Zhao, Z. MiR-485 targets the DTX4 gene to regulate milk fat synthesis in *B. taurus* mammary epithelial cells. *Sci. Rep.* **2021**, *11*, 7623. [CrossRef]
5. Keel, B.N.; Lindholm-Perry, A.K. Recent developments and future directions in meta-analysis of differential gene expression in livestock RNA-Seq. *Front. Genet.* **2022**, *13*, 983043. [CrossRef] [PubMed]
6. Suravajhala, P.; Kogelman, L.J.; Kadarmideen, H.N. Multi-omic data integration and analysis using systems genomics approaches: Methods and applications in animal production, health and welfare. *Genet. Sel. Evol.* **2016**, *48*, 38. [CrossRef]
7. Moghaddassi, S.; Eyestone, W.; Bishop, C.E. TALEN-mediated modification of the *B. taurus* genome for large-scale production of human serum albumin. *PLoS ONE* **2014**, *9*, e89631. [CrossRef] [PubMed]
8. Zhang, H.; Mi, S.; Brito, L.F.; Hu, L.; Wang, L.; Ma, L.; Xu, Q.; Guo, G.; Yu, Y.; Wang, Y. Genomic and transcriptomic analyses enable the identification of important genes associated with subcutaneous fat deposition in Holstein cows. *J. Genet. Genom.* **2023**, *50*, 385–397. [CrossRef] [PubMed]

genes

MDPI

Article

Single Nucleotide Polymorphisms in the Promoter Region of *MyoG* Gene Affecting Growth Traits and Transcription Factor Binding Sites in Guizhou White Goat (*Capra hircus*)

Xingchao Song, Huaixin Long, Jinzhu Meng, Yuanyuan Zhao, Zhenyang Wu and Qingming An *

Guizhou Provincial Key Laboratory for Biodiversity Conservation and Utilization in the Fanjing Mountain Region, College of Agriculture and Forestry Engineering, Tongren University, Tongren 554300, China; songxingchao_888@126.com (X.S.); 19110826607@163.com (H.L.); mjz122821@126.com (J.M.); yjy13251252247@163.com (Y.Z.); wuzhenyang0724@163.com (Z.W.)

* Correspondence: wjyaqm@gztrc.edu.cn

Abstract

Objective: Growth traits are important economic characteristics in livestock. Genetic polymorphism has great influences on the improvement of goat growth traits. As an important member of the myogenic regulatory factor (MRFs) family, *MyoG* gene polymorphisms can alter the growth characteristics in goats. In this study, we aimed to investigate the regulation mechanism of the *MyoG* gene promoter region from the perspective of single nucleotide polymorphisms (SNPs) and transcription factors. **Methods**: Genomic DNA sequencing was carried out to detect SNPs in the −1000 bp upstream to 300 bp downstream of the *MyoG* gene promoter region in 224 Guizhou White goats (*Capra hircus*), and the genetic parameters of novel SNPs were calculated. The association between SNPs and growth traits, comprising body weight, body length, body height, chest circumference and cannon circumference, were analyzed using one-way ANOVA by IBM SPSS 23.0 software according to the general linear model. Transcription factor binding sites in the promoter region of the *MyoG* gene before and after mutation were predicted using bioinformatics software programs. **Results**: Four SNPs, including g.–709C>T, g.–461G>T, g.–377G>T and g.–249G>A, were identified in the 1 246 bp promoter region of the *MyoG* gene in Guizhou White goats. Based on χ^2 test, the g.–709C>T and g.–461G>T loci were consistent with Hardy–Weinberg equilibrium, while two other SNPs were deviated from Hardy–Weinberg equilibrium in Guizhou White goats. Association analysis revealed that the body weight of those with the CT genotype at the g.–709C>T locus was greater than of those with the CC and TT genotypes in Guizhou White goats ($p < 0.05$). At the g.–461G>T locus, the body weight of individuals with the GG genotype was significantly higher than that of those with GT genotype ($p < 0.01$). The body length of individuals with the GG genotype formed by the g.–249G>A locus was significantly higher than that of those with the GA genotype ($p < 0.01$). Online software programs found that four SNPs within the promoter region of the *MyoG* gene changed some transcription factor binding sites. **Conclusions**: Mutations of the *MyoG* gene promoter region may have a significant regulatory effect on the growth traits of Guizhou White goats. The small sample size may be one of the limitations for this study; nevertheless, these findings could provide a theoretical basis for further exploring the relationship between the four SNPs studied and the growth traits in Guizhou White goats, as well as the promoter function of the *MyoG* gene.

Keywords: Guizhou White goat; *MyoG* gene; promoter region; SNPs; transcription factor

1. Introduction

Myogenic regulatory factors (MRFs) are known as vital members of the myogenesis regulatory family, which is involved in the proliferation of muscle precursor cells and in the formation of muscle fiber, as well as in postnatal muscular functions [1,2]. Previous studies have shown that the *myogenin* (*MyoG*) gene is the only member in the family of MRFs that can be expressed in animal skeletal muscle cell lines. It can activate muscle gene transcription, improve cell differentiation and is a key factor in regulating skeletal muscle growth and development. It plays a central regulatory role in the process of muscle cell generation [3]. At the same time, the *MyoG* gene interacts with other transcription regulatory factors in the MRFs family, such as *MyoD*, *Myf5* and *MRF4*, to jointly regulate animal meat quality traits and meat yield [4]. Thus, genetic variation in the *MyoG* gene may affect the growth, carcass and meat traits in livestock. In previous studies, some MRFs were confirmed to be related to meat quality traits in swine [5,6], beef cattle [7] and sheep [8,9]. In sheep, variants in *MyoD1* were reported as being associated with body traits such as thoracic girth and loin width [10]. Additionally, positive correlations between *MyoG* expression and body weight in Hu sheep were explored [11]. At present, candidate gene analysis has been achieved using relevant molecular methods [12], and the candidate gene method has been proven to be effective in identifying gene polymorphisms. If the genetic variation in quantitative traits in organisms is caused by causal mutations in candidate genes, it can be determined from this perspective that the analysis of gene physiology, biochemical function and metabolic pathways is helpful in selecting potential candidate genes for evaluation.

Guizhou White goats, with advantages such as strong adaptability, stable genetic performance, fresh flesh and having a delicious taste, make for an important germplasm resource of animal husbandry in the Guizhou province, China [13]. While being smaller in size, growing more slowly, and having lower meat production rates are contributing factors for the low economic efficiency of Guizhou White goats [14,15]. The growth characteristics of goats have high economic value, and increasing lamb yield in practical production largely depends on identifying and exploring how genetic variations in key genes that control growth rate regulate gene expression at the molecular level.

In recent years, the Genome-wide association study (GWAS) has gradually been applied to conduct research on goat body weight, body size and meat quality traits, identifying a large number of important genes and their associated SNPs for quantitative and qualitative traits, including Zhongwei goat [16], Karachai goat [17], Youzhou Dark goat [18] and Sub-Saharan Africa indigenous goat [19], enriching the scope of goat genomics research. However, there are relatively few SNPs screened in the promoter region, especially key factors related to gene transcription regulation mechanisms. Promoters are important regions for gene regulation, as they contain many sites and functional domains that bind to transcription regulatory factors. Therefore, the polymorphic sites in these regions are likely to affect the mRNA expression level of the gene to some extent, further affecting the protein function and causing changes in animal phenotypic traits.

So far, research on the relationship between the promoter region of the *MyoG* gene and growth traits is limited. The main objective of this study is to further investigate the association between novel SNPs within the promoter region of the *MyoG* gene and body size in goats. The *MyoG* gene in goats is located on chromosome 16, and there are several quantitative trait loci (QTLs) related to carcass weight [20]. Up to now, there have been no reports on the association between genetic variations in the promoter region of the *MyoG* gene in goats and growth traits. Thus, this study aims to explore the SNPs in the promoter region of the *MyoG* gene in Guizhou White goats and their effects on body measurement traits and transcription factor binding sites.

https://doi.org/10.3390/genes17010014

2. Materials and Methods

2.1. Statement of Ethics

The collection of goats samples involved in this study has been approved by the Ethics Committee of Tongren University, China (Permit number for conducting animal experiments: TRXY 2023-088; Approval date: 8 March 2023), and everything possible was done to minimize the pain of the goat as much as possible.

2.2. Animal Sources, Data Collection and DNA Extraction

Male Guizhou White (GZW) goats (n = 224 of about three years of age) were selected at random from Huazhen Animal Husbandry Co., Ltd. in Tongren City, Guizhou Province, China. According to farm records, all of the goats were GZW goats that were genetically distinct from each other and were subjected to the same environmental conditions and nutrition. According to the same standards, peripheral blood samples (5 mL) were collected from each goat through the jugular vein using a vacuum tube and frozen at −20 °C for DNA extraction. The growth traits of all selected goat individuals have been recorded by the breeding farm. Body size indicators include the following: body weight (BW), body length (BL), body height (BH), chest circumference (ChC) and cannon circumference (CaC). The blood genomic DNA of Guizhou White goats was extracted with the blood genomic DNA extraction kit (TransGen Biotech, Beijing, China), and its concentration and purity were detected by ultraviolet spectrophotometer and 0.7% agarose gel electrophoresis diluted to 80 ng/μL and stored at −20 °C.

2.3. Primer Design and PCR Amplification

Based on the flanking regions of the *MyoG* gene in *C. h.* (Accession number FJ607135) and *Bos taurus* (Accession number HM452338) as reference sequences, a pair of specific primers (Forward: 5′-CTTCCCTCCTCACCCACATT-3′, Reverse: 5′-TCCACAGACACCGACTTCCT-3′) were designed using Oligo 7.0 and Primer Premier 5.0 software to amplify the promoter region of the *MyoG* gene in Guizhou White goats. The primers were synthesized by Sangon Biotech Co., Ltd. (Shanghai, China). PCR amplification was performed in 40 μL reaction mixture, which was made up of 20 μL 2 × SanTaq Fast Mix (Sangon Biotech Co., Ltd., Shanghai, China), 1.0 μL of 80 ng/μL template DNA, 1.0 μL of 10 pmol/μL each primer and 17 μL sterilized ultrapure water. The PCR thermal cycle was programmed as follows: pre-denaturation at 94 °C for 5 min, denaturation at 94 °C for 30 s, annealing at 58.8 °C for 30 s, extension at 72 °C for 75 s and 40 cycles and a final extension at 72 °C for 10 min, stored at 4 °C. PCR products were detected by 1.2% agarose gel electrophoresis and then sent to Sangon Biotech Co., Ltd. (Shanghai, China) for sequencing. SNPs in the MyoG gene were detected by bidirectional direct sequencing.

2.4. SNPs Screening and Genotype Determination

The obtained *MyoG* gene sequences of 224 Guizhou White goats were aligned by the ClusterW Multiple alignment program in BioEdit 7.0 software and detected SNPs sites. The genotype of SNPs loci was determined using the sequencing peak mapping software Chromos 5.0, with a single peak representing a homozygous genotype and a nested peak representing a heterozygous genotype.

2.5. Genotype, Allele Frequency and Association Analysis Between SNPs Loci and Growth Traits

The genotype, allele frequency and Hardy–Weinberg equilibrium (HWE) of SNPs loci were analyzed based on the online software SHEsis (http://analysis.bio-x.cn, accessed on 8 June 2024). Polymorphism information content (*PIC*) was calculated through online software (http://www.msrcall.com/Gdicall.aspx, accessed on 18 June 2024). The mean and standard error of the different growth traits of 224 individual Guizhou White goats with different SNPs loci of the *MyoG* gene were statistically analyzed by means of IBM SPSS 23.0 software. At the same time, according to the general linear model, which is used to make a significant test on the effects of various genotypes on growth traits, one-way ANOVA was used to conduct a homogeneity of variance test and a multiple comparison test and to further analyze the correlation between SNPs locus and growth traits of Guizhou White goats. A general linear model is $y_{ij} = \mu + \alpha_i + \varepsilon_{ij}$, where y_{ij} is the individual growth trait (body weight, etc.) record, μ is population mean for each growth trait, α_i is the genotype effect and ε_{ij} is the random error. The data is displayed as mean ± standard error (SE), with $p < 0.05$ indicating statistically significant differences and $p < 0.01$ indicating extremely significant differences.

2.6. Identification of Transcription Factor Binding Sites Before and After SNPs Mutation

The possible transcription factor binding sites in the promoter region of the *MyoG* gene in Guizhou White goats were predicted using AliBaba 2.1 software (https://gene-regulation.com/pub/programs/alibaba2/index.html, accessed on 6 July 2024). Predicting changes in transcription factors before and after SNPs mutation in the promoter region of the *MyoG* gene by means of a transcription factor binding database profile (JASPAR^{2024}: https://ngdc.cncb.ac.cn/databasecommons/database/id/9171, accessed on 10 August 2024).

3. Results

3.1. Detection of MyoG Gene PCR Products

The designed primer was used to amplify the *MyoG* gene of Guizhou White goats, and the PCR products were separated by 1.2% agarose gel electrophoresis. As shown in Figure 1, the size of the amplified fragment is consistent with the target fragment. The amplified band is clear and dense, which could be used for subsequent direct sequencing analysis.

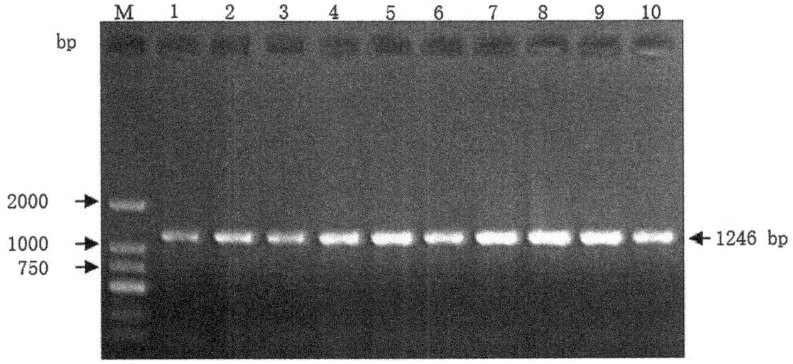

Figure 1. Agarose electrophoretic map of the *MyoG* gene specific fragment in Guizhou White goats. Lane M represents DL 2000 Marker. Lane 1 to 10 denotes target segment. The arrow indicates the 1246 bp of the *MyoG* gene fragment.

3.2. SNP Identification in the Promoter Region of the MyoG Gene

Based on the sequence structure of the bovine *MyoG* gene in the GenBank database, a total of 1246 bp nucleotide sequences of the *MyoG* gene in Guizhou White goats were obtained, including 1004 bp 5′UTR and 242 bp exon 1. The Clustal W Multiple alignment program in BioEdit 7.0 software was used to compare the sequencing results of 224 Guizhou White goats, screen for SNPs and determine the genotype of each locus based on the sequencing peak maps. The naming principle of SNPs is that the A base in the start codon

https://doi.org/10.3390/genes17010014

ATG is marked as "+1". At the same time, the obtained 1246 bp sequence was aligned with the goat genome reference sequence in the NCBI database, using BLAST (version 2.17.0) to determine the specific location of the identified SNPs in the goat chromosome. The *MyoG* gene is located on chromosome 16 of Guizhou White goats. Four SNPs were screened in the promoter region of the *MyoG* gene in Guizhou White goats, including g.–709C>T, g.–461G>T, g.–377G>T and g.–249G>A, which are located at 193,987 bp, 193,739 bp, 193,655 bp and 193,528 bp of the goat reference genome sequence NC_030823.1, and the variant IDs are rs666139870, rs636117637, rs652431992 and rs639389894, respectively. Two genotypes were formed at the g.–461G>T and g.–249G>A loci in Guizhou White goats, while three genotypes were found at the g.–709C>T and g.–377G>T loci. The sequencing results of the four SNPs are shown in Figure 2 and Supplementary Materials, Figure S1.

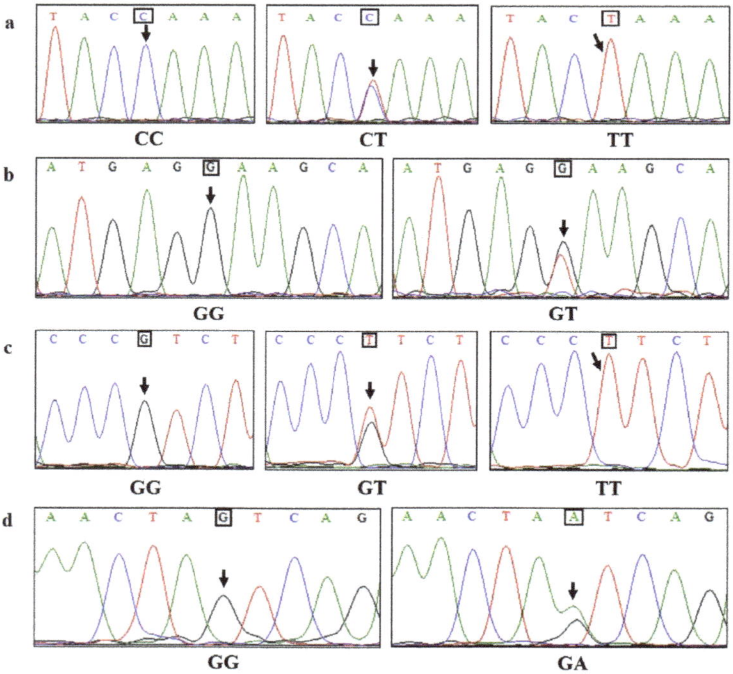

Figure 2. The sequencing chromatogram of four SNPs for the *MyoG* gene in Guizhou White goats; SNPs are indicated by arrows. (**a–d**) represent sequencing results of each genotype at g.–709C>T, g.–461G>T, g.–377G>T and g.–249G>A, respectively.

3.3. Polymorphic Parameter, Genotype and Allele Frequencies for Different SNPs

The genotype and allele frequencies of *MyoG* gene SNPs in Guizhou White goats are shown in Table 1. Three genotypes were found at the g.–709C>T locus, namely, CC, CG and GG; GG, GT and TT were identified at the g.–377G>T locus. Two genotypes were found at the g.–461G>T locus, namely, GG and GT; GG and GA were screened at the g.–249G>A locus. The χ^2-test showed that Guizhou White goats conform to the Hardy–Weinberg equilibrium at the g.–709C>T and g.–461G>T loci ($p > 0.05$), while the g.–377G>T and g.–249G>A loci significantly departed from the Hardy–Weinberg equilibrium ($p < 0.01$). At the g.–709C>T, g.–461G>T, g.–377G>T and g.–249G>A loci in the promoter region of the *MyoG* gene, the frequencies of the C, G, G and G alleles are higher than those of alleles T, T, T and A. As shown in Table 1, the population genetic parameters of four SNPs of the promoter region of the *MyoG* gene in Guizhou White goats were calculated. The *PIC*,

Ne, *Ho* and *He* of the g.–709C>T locus were 0.36, 1.92, 0.52 and 0.48 and of the g.–377G>T locus were 0.29, 1.53, 0.65 and 0.35, respectively, indicating that Guizhou White goats have moderately polymorphisms (0.25 < *PIC* < 0.5) at the g.–709C>T and g.–377G>T loci of the *MyoG* gene promoter region.

Table 1. Population genetic analysis of four SNPs of the *MyoG* gene promoter region in Guizhou White goats.

Mutation Site	Genotype Frequency [1]			Allele Frequency		χ^2	*p*-Value	*PIC*	*Ne*	*Ho*	*He*
g.–709C>T	CC 0.38 (85)	CT 0.45 (100)	TT 0.17 (39)	C 0.60	T 0.40	1.23	0.54	0.36	1.92	0.52	0.48
g.–461G>T	GG 0.86 (193)	GT 0.14 (31)	TT 0.00 (0)	G 0.93	T 0.07	1.24	0.27	0.12	1.15	0.87	0.13
g.–377G>T	GG 0.65 (145)	GT 0.26 (58)	TT 0.09 (21)	G 0.78	T 0.22	489.92	<0 01	0.29	1.53	0.65	0.35
g.–249G>A	GG 0.85 (191)	GA 0.15 (33)	AA 0.00 (0)	G 0.93	A 0.07	186.12	<0 01	0.13	1.16	0.86	0.14

[1] The number in brackets is the sample size. *PIC*: Polymorphism information content, *Ne*: number of effective alleles, *Ho*: observed heterozygosity, *He*: expected heterozygosity.

3.4. Association of MyoG Gene Promoter Region Polymorphism and Growth Traits in Guizhou White Goats

In order to investigate the association between the above mentioned SNPs and the growth traits of Guizhou White goats, a general linear model was used for association analysis by means of IBM SPSS 23.0 software. As shown in Table 2, the individuals with the CT genotype formed by the g.–709C>T locus in the promoter region of the *MyoG* gene in Guizhou White goats had significantly higher body weight than those with the CC and TT genotypes ($p < 0.05$), and the chest circumference of individuals with CT genotype was significantly higher than that of individuals with the CC and TT genotypes ($p < 0.05$). The body length, body height and circumference of individuals with the CT genotype tended to be higher than that of those with the CC and TT genotypes, but these differences are not significant ($p > 0.05$). The body weight of individuals with the GG genotype caused by the g.–461G>T locus was extremely significantly higher than those with the GT genotype ($p < 0.01$), and no significant differences were observed in body length, height, chest circumference or cannon circumference among individuals with different genotypes. ($p > 0.05$). No significant differences were observed in body weight, body length, body height, chest circumference or cannon circumference among genotypes at the g.−377G>T locus ($p > 0.05$). The body length of individuals with the GG genotype caused by the g.–249G>A locus was significantly higher than those with the GA genotype ($p < 0.01$), and no significant differences were observed in body weight, body height, chest circumference and cannon circumference among individuals with different genotypes ($p > 0.05$).

Table 2. The correlation between four SNPs in the promoter region of the *MyoG* gene and the growth traits of Guizhou White goats.

SNP	Genotype	Body Weight (kg)	Body Length (cm)	Body Height (cm)	Chest Circumference (cm)	Cannon Circumference (cm)
g.–709C>T	CC	26.84 ± 1.08^{b}	52.17 ± 1.11	58.58 ± 1.23	71.78 ± 1.31^{b}	7.30 ± 0.15
	CT	29.34 ± 1.06^{a}	52.52 ± 1.06	59.19 ± 1.22	73.58 ± 1.37^{a}	7.57 ± 0.14
	TT	27.69 ± 1.23^{b}	52.41 ± 1.39	59.09 ± 1.40	71.26 ± 1.57^{b}	7.25 ± 0.21
p-value		0.015	0.960	0.848	0.023	0.234

https://doi.org/10.3390/genes17010014

Table 2. *Cont.*

SNP	Genotype	Body Weight (kg)	Body Length (cm)	Body Height (cm)	Chest Circumference (cm)	Cannon Circumference (cm)
g.–461G>T	GG	28.74 ± 0.94 [A]	52.47 ± 0.88	58.63 ± 1.45	71.49 ± 1.17	7.46 ± 0.17
	GT	25.14 ± 1.34 [B]	51.64 ± 1.59	59.19 ± 1.22	69.43 ± 1.73	7.56 ± 0.21
p-value		0.001	0.588	0.465	0.162	0.566
g.–377G>T	GG	28.26 ± 0.99	52.28 ± 0.99	54.88 ± 1.23	71.58 ± 1.27	7.44 ± 0.19
	GT	28.14 ± 1.13	53.05 ± 1.09	55.38 ± 1.05	70.27 ± 1.67	7.51 ± 0.15
	TT	27.99 ± 2.41	52.39 ± 1.53	55.98 ± 1.42	69.43 ± 1.74	7.48 ± 0.13
p-value		0.893	0.113	0.285	0.162	0.285
g.–249G>A	GG	26.94 ± 2.24	57.39 ± 1.97 [A]	55.81 ± 1.05	70.89 ± 1.98	7.40 ± 0.12
	AA	24.78 ± 0.93	52.09 ± 1.07 [B]	59.42 ± 2.09	71.50 ± 1.29	7.52 ± 0.31
p-value		0.376	0.008	0.914	0.739	0.566

Note: In the same SNP and column, superscript in different lowercase letters indicates significant difference ($p < 0.05$), and superscript in different uppercase letters indicates extremely significant difference ($p < 0.01$). There is no significant difference in numerical values without letters or with the same letter superscript ($p > 0.05$).

3.5. SNPs in the Promoter Region of the MyoG Gene Cause Alterations in Transcription Factors

The alterations of transcription factors in the *MyoG* gene promoter region, calculated by the online software JASPAR[2024], are shown in Table 3. The g.–709C>T locus may result in the vanishing of both the C/EBPalp and serum response factor (SRF) transcription factor binding sites and in the increase of a new TATA binding protein (TBP) site. The g.–461G>T locus may result in the vanishing of the PU.1 transcription factor binding sites and in the increase of activator protein 1 (AP-1). The g.–377G>T locus may lead to the vanishing of the specific protein 1 (SP1) transcription factor binding sites and to the increase of octamer binding transcription factor 1 (OCT-1) binding sites. The g.–249G>A mutation site may result in the vanishing of the ELA1 transcription factor binding site and in the increase of ALX1 binding sites. Thus, these SNPs, especially at newly identified loci, may form new transcription factor binding sites while disrupting primitive existing binding sites and disturbing transcriptional regulation and expression of the *MyoG* gene.

Table 3. Alterations of transcription factors before and after SNPs mutation of the *MyoG* gene in Guizhou White goats.

Mutation Location	Nucleotide	Transcription Factors	Transcription Factor Binding Site Nucleotide Sequence	Transcription Factor Location
–709	C	C/EBPalp	tcttacCaaa	–715–(–706)
		SRF	tcttacCaaaaaagagaa	–715–(–698)
	T	TBP	Taaaaaagaga	–709–(–699)
–461	G	PU.1	gagGaag	–464–(–458)
	T	AP-1	tatgagTaag	–467–(–458)
–377	G	SP1	cccccaccccacccG	–391–(–377)
	T	OCT-1	Ttcttctttg	–377–(–368)
–249	G	ELA1	Gtcag	–249–(–245)
	A	ALX1	ctaAtcagatta	–252–(–241)

Note: The capital letters in the table indicate mutated bases.

4. Discussion

4.1. Correlation Between MyoG Gene SNPs and Growth Traits of Caprinae

The Guizhou White goat is an excellent local breed in the Guizhou province, with advantages such as strong adaptability, tolerance to rough feeding, tender meat and delicious

https://doi.org/10.3390/genes17010014

meat taste. However, this breed has disadvantages such as small size, low slaughter rate, low net meat rate and slow growth rate, which limit the large-scale breeding and industrial development of local goat breeds. Screening candidate genes and related molecular markers for goat growth traits at the molecular level and conducting marker-assisted selection breeding will be one of the effective means to improve goat meat production capacity and meat quality. It is also a good way to improve local goat breeds with Guizhou characteristics or cultivate new meat breeds [21]. There are few SNPs in the *MyoG* gene of goats, all of which are significantly correlated with meat quality and growth traits, indicating that the *MyoG* gene is a key candidate gene for mutton development [22]. Screening of genetic variant molecular markers of the *MyoG* gene associated with growth traits is of significant theoretical importance and has implications for the development of molecular breeding or meat production traits. At present, research on *MyoG* gene SNPs is mainly focused on pigs [23–25], cattle [26,27] and sheep [28,29], mainly involving the association between SNPs and meat quality and carcass traits. Previous studies have analyzed sheep *MyoG*, including three exons and two introns, and identified some polymorphisms. Han et al. [30] reported that there was one single nucleotide missense mutation of A109C at exon 1 of a Tibetan sheep *MyoG* gene. Statistical analysis indicated that in the Henan Oula Tibetan sheep population, individuals with the CC genotype have higher mean values for body weight, body height and body length than those with the AA genotype. Li et al. [9] analyzed the polymorphism in intron 2 of *MyoG* gene in Chinese sheep local breeds using the PCR-RFLP method. The dominant genotype was AB, and the *Eco* 72 I locus could serve as a molecular marker affecting sheep growth traits. The same SNPs of the *MyoG* gene intron 2 have also been reported in Boer goat populations, which indicates that the *MyoG* gene has a certain influence on the growth traits of goats [31]. Thus, investigating the SNPs in the promoter region of the *MyoG* gene and their associations with growth traits in Guizhou White goats represents a significant approach for discovering new molecular genetic markers used in goat breeding practices.

Previous studies have found that *PIC* value is an important parameter for marker site polymorphism in linkage analysis [32,33]. In this study, four SNPs in the promoter region of the *MyoG* gene were identified, and the relationship between SNPs and growth traits in Guizhou White goats was studied. These SNPs appear to be new and have never appeared in previous databases. In addition, their impact on the *MyoG* gene expression or their association with goat growth traits have not been explored. For the g.–709C>T and g.–377G>T loci, the *PIC* value has been obtained and were identified as moderate polymorphic in the Guizhou White goat population ($0.25 < PIC < 0.5$). This indicates that breeders can utilize significant genetic diversity throughout the entire evolutionary history of animals to select superior varieties. For the g.–461G>T and g.–249G>A loci, the *PIC* value exhibits low level polymorphism. Simultaneously, the goat population in this study was in Hardy–Weinberg equilibrium at the g.–709C>T and g.–461G>T loci. This equilibrium may be due to the stable genetic status of the Guizhou White goat population after long-term breeding, which suggests that selective breeding measures may be taken in the future to improve production traits such as growth rate, meat quality and carcass traits. Nevertheless, the g.–377G>T and g.–249G>A loci exhibited a significant deviation from the Hardy–Weinberg equilibrium, which may be due to an insufficient sample size that does not adequately represent the phenotypic distribution of these loci within these broader populations. Therefore, it is imperative that the Guizhou White goat sample size be expanded in future studies to validate this discovery.

The CT genotypes of the g.–709C>T locus appear to have a relatively higher impact on body weight and chest circumference in Guizhou White goats ($p < 0.01$). The genetic variation at this locus is directly associated with the weight and chest circumference of

https://doi.org/10.3390/genes17010014

Guizhou White goats, which may involve dominant effects or heterozygous advantages. Similarly, in Nile tilapia, heterozygosity at certain SNP loci of the *MYF5* gene is significantly associated with growth traits [34]. Therefore, heterozygous state may enhance growth potential through gene interactions. The significant effect of heterozygous genotypes suggests that these SNPs may affect growth traits through dominant or heterozygous dominance, providing a key target for genetic improvement. This result suggests that screening heterozygous genotypes in molecular marker-assisted breeding of goats has important theoretical significance and practical value, which is helpful for breeding excellent meat breeds of Guizhou White goats with rapid growth. The GG genotypes of the g.–249G>A locus appear to have a relatively higher impact on body length in Guizhou White goats ($p < 0.01$). The GG genotypes of the g.–461G>T locus appear to have a relatively higher impact on body weight in Guizhou White goats ($p < 0.01$).

Nevertheless, it must be clarified that body weight, chest circumference and body length are quantitative traits determined by many genes, including *MYOD1*, *MSTN*, *IGFs* [35–37] and so on. Therefore, the impact of single genes and mutations may be minimal. In addition, the application of *MyoG* gene polymorphism in Guizhou White goat breeding may be hindered by bias in genotype and allele distribution, even lacking some genotypes in goat populations (such as the TT genotype of the g.–461G/T locus). Therefore, their value in breeding and selection must still be validated on sufficient goat populations, preferably on different goat breeds. In addition, the next step of this study in the future should be to expand the sample size of goats and further analyze the association between SNPs haplotypes in the *MyoG* gene promoter region and goat growth traits.

The SNPs identified in the untranslated region and exons of the *MyoG* gene may have a significant impact on the expression of this gene. In Jinghai yellow chickens, individuals with the BB genotype of the *MyoG* gene, which was formed by g.36T>C mutation in exon 3, had higher bodyweight at different weeks [38]. In zebrafish embryos, the expression of the *MyoG* gene is regulated by lots of regulatory elements, especially multiple transcription factors in the promoter region [39]. Current research has only analyzed the correlation between SNPs and growth traits of Guizhou White goats. The correlation analysis between SNPs and the expression level of this gene will be the focus of our subsequent research. However, as far as we know, this study is the first to investigate the association between SNPs in the promoter region of the *MyoG* gene and growth traits in goats.

In the screening of SNPs in the exons and introns of the *MyoG* gene in goats, individuals with the AA genotype formed by the g.1823C>T mutation in intron 2 had significantly lower birth weight than individuals with the AB and BB genotypes. Selecting individuals with the B allele is expected to improve growth traits related to goat body mass [31]. In terms of other members of the myogenic regulatory factor family, myogenic differentiation factor 1 is also a gene locus significantly associated with growth traits. The Copy Number Variation loci containing the *MyoD1* gene exon in Shanbei white cashmere goats is significantly correlated with body weight, body height, hip height, chest circumference and hip circumference ($p < 0.05$), indicating that this locus is a marker-assisted selection breeding locus for goats [35]. In addition, insulin-like growth factor 1 (*IGF-1*) is an important member of the animal growth axis, mainly regulating fetal growth and development, and is a key factor for linear growth in animals after birth. Several QTLs related to growth and body shape traits were identified on goat chromosome 5, including the *IGF-1* gene [40]. Many researchers have directly sequenced different goat breeds and found a large number of SNPs in the 5′untranslated region of the goat *IGF-1* gene, some of which are significantly correlated with growth traits such as birth weight, chest width and chest depth [41,42]. However, whether the SNPs of the *MyoG* gene discovered in this study jointly affect the

https://doi.org/10.3390/genes17010014

growth traits of Guizhou White goats with the above mentioned mutation sites still needs further verification through subsequent gene interaction and related analyses.

In future research, we will expand to genome-wide association analysis (GWAS) to systematically identify more SNPs loci related to the growth traits of Guizhou White goats. In addition, it is necessary to combine functional validation experiments, such as gene editing or expression analysis, to clarify the biological mechanisms of candidate genes. Strengthening the integration of genetic resource protection and molecular breeding techniques will help achieve precise breeding for Guizhou White goats.

4.2. Effects of Transcription Factor Binding Sites Alteration in the Promoter Region on Transcriptional Regulation of MyoG Gene

The promoter region is a specific DNA sequence located upstream of a gene, which is a key and important region for RNA polymerase and transcription factors to bind and initiate gene transcription. SNPs in the promoter region may regulate gene expression by altering the binding mode of some transcription factors, which is one of the important mechanisms for understanding individual genetic differences and susceptibility to complex diseases [43,44]. SNPs in the promoter region may indeed alter transcription factor binding sites by directly disrupting or creating transcription factor binding sites and indirectly affecting binding affinity or conformation, thereby affecting gene expression [45,46].

In this study, the transcription factor prediction of the MyoG gene promoter region showed that the mutation of g.–709C>T resulted in the disappearance of the initial C/EBPalp and SRF transcription factor binding sites and in the formation of a new binding site for the TBP transcription factor. TATA binding protein (TBP) is essential for RNA polymerase III transcription. Previous studies have found that TBP plays a central role in transcriptional initiation, which was first discovered to recognize and bind to the TATA box in the gene promoter region. In addition, TBP is a key factor necessary for initiating the three major eukaryotic RNA polymerases [47,48]. Regarding the newly added TBP transcription factor binding sites, this study only used bioinformatics software for predictive analysis. Whether there is a positive over-effect still needs to be verified in the future. Similarly, the sample size should also be increased for analysis in the future.

The mutation of g.–461G>T locus resulted in the vanishing of the initial PU.1 transcription factor binding site and in the increase of a new binding site for the AP-1 transcription factor. Activation protein-1 (AP-1) is an inducible transcription factor which is composed of multiple protein complexes, including members of the fos and jun gene families. Many cells and viral genes contain AP-1 binding sites in their promoters; therefore, AP-1 has been investigated to determine whether it plays a role in regulating these genes and inducing transcription [49]. Further in-depth research is needed to determine whether the increased AP-1 transcription factor caused by the g.461G>T mutation in the promoter region of the *MyoG* gene in Guizhou White goats will affect the transcription and functional expression of this gene.

The mutation of g.–377G>T locus, led to the vanishing of the initial SP1 transcription factor binding site; meanwhile, a new binding site for the OCT-1 transcription factor was raised. OCT-1 was first found as a member of the POU transcription factor family and is widely expressed in most cells and tissues. OCT-1 plays an extremely important role in development and differentiation processes [50]. The specific role of the increased OCT-1 transcription factor in the promoter region of the *MyoG* gene in Guizhou White goats also requires further verification and exploration. The mutation of g.–249G>A locus resulted in the disappearance of the initial ELA1 transcription factor binding site and a new binding site for the ALX1 transcription factor was formed. Alx1 is a transcription factor containing homologous domains and is a highly conserved regulatory factor for bone formation in echinoderms. In sea urchins, Alx1 plays a central role in the differentiation

https://doi.org/10.3390/genes17010014

of embryonic primary mesenchymal cells (PMCs) and actively regulates the transcription of most biomineralization genes expressed in these cells [51]. The function of the Alx1 transcription factor in the development of muscle and bone tissue in vertebrates, especially goats, also needs to be further explored.

These four SNPs in the promoter region of the *MyoG* gene have altered the initial transcription factor interaction, which may, therefore, be leading to alterations in promoter activity and gene function. Alterations of the transcription factor are likely to decrease or increase the expression level of the *MyoG* gene, which indirectly affects the growth traits of Guizhou White goats. Thus, it is speculated that the alterations of transcription factors may affect the meat production capacity of Guizhou White goats, but the specific mechanism still needs to be explored further in-depth.

5. Conclusions

In the current study, the *MyoG* gene promoter region sequences of 224 Guizhou White goats were explored. Four SNPs were screened in the region −1000 bp upstream to 300 bp downstream of the goat *MyoG* gene transcription start site. The mutation within the promoter region is being reported for the first time in goats. Association analysis suggests that g.–709C>T and g.–461G>T loci in the promoter region are molecular marker sites that affect the weight index of Guizhou White goats. The g.–249G>A site is a molecular marker site that affects the body length index of Guizhou White goats. These SNPs might lead to alterations in their initial transcription factor binding sites, thus potentially changing gene expression. However, the relationship between the SNPs identified in this research and the expression level of *MyoG* gene in goats should be further studied in the future. Similarly, we will use a dual luciferase reporter gene vector to detect the promoter activity and determine the promoter active region of Guizhou White goats and verify the transcription factors of the mutation sites through EMSA or Chip methods. In short, these findings provide a theoretical basis for further exploring the regulation mechanisms of *MyoG* expression and its relationship with goat breeding.

Supplementary Materials: The following supporting information can be downloaded at: https://www.mdpi.com/article/10.3390/genes17010014/s1, Figure S1: Sanger sequencing chromatograms of different genotype for 4 mutation sites (a–d) in Guizhou White goat *MyoG* gene promoter region. Red rectangles indicate the positions of the mutation sites.

Author Contributions: X.S. and H.L. analyzed the data and drafted the manuscript. J.M. and Y.Z. prepared the blood samples for sequencing. Z.W. and H.L. performed the software and formal analysis. Q.A. initiated this study, designed the experiments and finalized the manuscript. All authors have read and agreed to the published version of the manuscript.

Funding: This research was funded by Guizhou Science and Technology Project (Qian Kehe Foundation No. ZK [2022] common 559), Doctoral Research Initiation Foundation of Tongren University (trxyDH2001), Guizhou Thousand Level Innovative Talent Project (2024-[2022]-048), Doctoral Talents project of Science and Technology Bureau of Tongren, Guizhou Province ([2020]126). Science and Technology Program of Guizhou Province (support projects MS2025-093), Doctoral Talent Program of Tongren (Tongren Scientific Research 2023-4).

Institutional Review Board Statement: All experimental procedures and sample collection in this study involving goats were approved by the Ethics Committee of Tongren University, China (Permit number for conducting animal experiments: TRXY 2023-088; Approval date: 8 March 2023), and all efforts were made to minimize the suffering of the goats.

Informed Consent Statement: Not applicable.

Data Availability Statement: The original contributions presented in the study are included in the article/Supplementary Materials. Further inquiries can be directed to the corresponding author.

https://doi.org/10.3390/genes17010014

Conflicts of Interest: The authors declare no conflicts of interest.

References

1. Kopantseva, E.E.; Belyavsky, A.V. Key regulators of skeletal myogenesis. *Mol. Biol.* **2016**, *50*, 169–192. [CrossRef]
2. Te Pas, M.F.W.; Soumillion, A.; Harders, F.L.; Verburg, F.J.; van den Bosch, T.J.; Galesloot, P.; Meuwissen, T.H. Influences of myogenin genotypes on birth weight, growth rate, carcass weight, backfat thickness, and lean weight of pigs. *J. Anim. Sci.* **1999**, *77*, 2352–2356. [CrossRef]
3. Zhan, S.; Zhai, H.; Tang, M.; Xue, Y.; Li, D.; Wang, L.; Zhong, T.; Dai, D.; Cao, J.; Guo, J.; et al. Profiling and functional analysis of mRNAs during skeletal muscle differentiation in goats. *Animals* **2022**, *12*, 1048. [CrossRef]
4. Wyszyńska-Koko, J.; Pierzchała, M.; Flisikowski, K.; Kamyczek, M.; Różycki, M.; Kurył, J. Polymorphisms in coding and regulatory regions of the porcine MYF6 and MYOG genes and expression of the MYF6 gene in m. longissimus dorsi versus productive traits in pigs. *J. Appl. Genet.* **2006**, *47*, 131–138. [CrossRef]
5. Li, M.; Liu, Q.; Xie, S.; Fu, C.; Li, J.; Tian, C.; Li, X.; Li, C. LncRNA TCONS_00323213 promotes myogenic differentiation by interacting with PKNOX2 to upregulate MyoG in porcine satellite cells. *Int. J. Mol. Sci.* **2023**, *24*, 6773. [CrossRef]
6. Verner, J.; Humpolíček, P.; Knoll, A. Impact of MYOD family genes on pork traits in Large White and Landrace pigs. *J. Anim. Breed. Genet.* **2007**, *124*, 81–85. [CrossRef]
7. Bhuiyan, M.S.A.; Kim, N.K.; Cho, Y.M.; Yoon, D.; Kim, K.S.; Jeon, J.T.; Lee, J.H. Identification of SNPs in MYOD gene family and their associations with carcass traits in cattle. *Livest. Sci.* **2009**, *126*, 292–297. [CrossRef]
8. Sousa, L.P.B., Jr.; Meira, A.N.; Azevedo, H.C.; Muniz, E.N.; Coutinho, L.L.; Mourão, G.B.; Leão, A.G.; Pedrosa, V.B.; Pinto, L.F.B. Variants in myostatin and MyoD family genes are associated with meat quality traits in Santa Inês sheep. *Anim. Biotechnol.* **2022**, *33*, 201–213. [CrossRef]
9. Li, J.Y.; Wang, X.L.; Zhang, Y.F.; Bai, J.Y.; Li, G.; Lei, Y.; Dong, Z.H.; Chen, Y.; Fan, H.D.; Wang, L.W.; et al. Polymorphism of MyoG gene intron II and its association with growth traits in sheep. *China Anim. Husb. Vet. Med.* **2022**, *49*, 1364–1373. [CrossRef]
10. Trukhachev, V.; Skripkin, V.; Telegina, E.; Yatsyk, O.; Golovanova, N.; Krivoruchko, A. Associations between newly discovered polymorphisms of the MyoD1 gene and body parameters in Stavropol breed rams. *Bulg. J. Vet. Med.* **2018**, *21*, 28–39. [CrossRef]
11. Sun, W.; Wang, P.; Ding, J.T.; Ma, Y.H.; Guan, W.J.; Chu, M.X.; Li, B.C.; Wu, W.Z.; Chen, L. Development changes of gene expression of Myostatin and Myogenin genes and their association analysis with carcass traits in Hu sheep. *Sci. Agric. Sin.* **2010**, *43*, 5129–5136. [CrossRef]
12. Mukherjee, A.; Gali, J.; Kar, I.; Datta, S.; Roy, M.; Acharya, A.P.; Patra, A.K. Candidate genes and proteins regulating bull semen quality: A review. *Trop. Anim. Health Prod.* **2023**, *55*, 212. [CrossRef]
13. Xu, H.Q. *Local Livestock and Poultry Genetic Resources in Guizhou*, 1st ed.; China Agriculture Press: Beijing, China, 2019.
14. Ruan, Y.; Dai, L.; Huang, J.; Xiao, M.; Xu, J.; An, D.; Chen, J.; Chen, X. A novel nonsynonymous SNP in the OLR1 gene associated with litter size in Guizhou white goats. *Theriogenology* **2023**, *200*, 1–10. [CrossRef]
15. Shi, S.Y.; Li, L.J.; Zhang, Z.J.; Wang, E.Y.; Wang, J.; Xu, J.W.; Liu, H.B.; Wen, Y.F.; He, H.; Lei, C.Z.; et al. Copy number variation of MYLK4 gene and its growth traits of *Capra hircus* (goat). *Anim. Biotechnol.* **2020**, *31*, 532–537. [CrossRef]
16. Han, M.; Wang, X.; Du, H.; Cao, Y.; Zhao, Z.; Niu, S.; Bao, X.; Rong, Y.; Ao, X.; Guo, F.; et al. Genome-wide association study identifies candidate genes affecting body conformation traits of Zhongwei goat. *BMC Genom.* **2025**, *26*, 37. [CrossRef]
17. Easa, A.A.; Selionova, M.; Aibazov, M.; Mamontova, T.; Sermyagin, A.; Belous, A.; Abdelmanova, A.; Deniskova, T.; Zinovieva, N. Identification of genomic regions and candidate genes associated with body weight and body conformation traits in Karachai goats. *Genes* **2022**, *13*, 1773. [CrossRef]
18. Sun, X.; Niu, Q.; Jiang, J.; Wang, G.; Zhou, P.; Li, J.; Chen, C.; Liu, L.; Xu, L.; Ren, H. Identifying candidate genes for litter size and three morphological traits in Youzhou Dark goats based on genome-wide SNP markers. *Genes* **2023**, *14*, 1183. [CrossRef]
19. Ncube, K.T.; Nephawe, K.A.; Mpofu, T.J.; Monareng, N.J.; Mofokeng, M.M.; Mtileni, B. Genomic advancements in assessing growth performance, meat quality, and carcass characteristics of goats in Sub-Saharan Africa: A systematic review. *Int. J. Mol. Sci.* **2025**, *26*, 2323. [CrossRef]
20. Wang, Z.; Lv, Q.; Li, W.; Huang, W.; Gong, G.; Yan, X.; Liu, B.; Chen, O.; Wang, N.; Zhang, Y.; et al. Chromosome-level genome assembly of the cashmere goat. *Sci. Data* **2024**, *11*, 1107. [CrossRef]
21. An, Q.; Zeng, L.; Wang, W.; Yang, J.; Meng, J.; Zhao, Y.; Song, X. Identification of FASN gene polymorphisms, expression and their relationship with body size traits in Guizhou White goat (*Capra hircus*) with different genders. *Genes* **2024**, *15*, 656. [CrossRef]
22. Zhang, Z.; Xu, F.; Zhang, Y.; Li, W.; Yin, Y.; Zhu, C.; Du, L.; Elsayed, A.K.; Li, B. Cloning and expression of MyoG gene from Hu sheep and identification of its myogenic specificity. *Mol. Biol. Rep.* **2014**, *41*, 1003–1013. [CrossRef]
23. Xue, H.L.; Zhou, Z.X. Effects of the MyoG gene on the partial growth traits in pigs. *Acta Genet. Sin.* **2006**, *33*, 992–997. [CrossRef]
24. Anton, I.; Zsolnai, A.; Komlósi, I.; Király, A.; Fésüs, L. Effect of MYOG genotypes on growth rate and production traits in Hungarian large white pigs. *Acta Vet. Hung.* **2006**, *54*, 393–397. [CrossRef]

https://doi.org/10.3390/genes17010014

25. Soumillion, A.; Erkens, J.H.F.; Lenstra, J.A.; Rettenberger, G.; Te Pas, M.F.W. Genetic variation in the porcine myogenin gene locus. *Mamm. Genome* **1997**, *8*, 564–568. [CrossRef]
26. Beever, J.E.; Fisher, S.R.; Lewin, H.A. Polymorphism identification in the ACADM, AT3, IL10, MYOG and TSHB genes of cattle. *Anim. Genet.* **1997**, *28*, 373–374. [CrossRef]
27. Wei, D.; Zhang, J.; Raza, S.H.A.; Song, Y.; Jiang, C.; Song, X.; Wu, H.; Alotaibi, M.A.; Albiheyri, R.; Al-Zahrani, M.; et al. Interaction of MyoD and MyoG with Myoz2 gene in bovine myoblast differentiation. *Res. Vet. Sci.* **2022**, *152*, 569–578. [CrossRef]
28. Sousa, L.P.B., Jr.; Meira, A.N.; Azevedo, H.C.; Muniz, E.N.; Coutinho, L.L.; Mourão, G.B.; Pedrosa, V.B.; Pinto, L.F.B. Polymorphismsin MyoD1, MyoG, MyF5, MyF6, and MSTN genes in Santa Inês sheep. *Pesqui. Agropecu. Bras.* **2019**, *54*, e01132. [CrossRef]
29. Wang, X.; Bai, J.Y.; Yang, Y.B.; Lei, X.Q.; Pang, Y.Z.; Li, H.W.; Wang, H.L.; Tang, Y.L.; Shao, J.H.; Yin, H.Y. Study on the polymorphism of MyoG exon 2 and MyoD 5′ Flanking region of Small tail han sheep. *J. Henan Agric. Sci.* **2017**, *26*, 138–141. [CrossRef]
30. Han, Y.C.; Sun, Y.G.; Chen, Z. Single nucleotide polymorphism of MyoG gene in Tibetan sheep populations and its association with body measurement traits. *J. Anim. Ecol.* **2016**, *37*, 19–23. [CrossRef]
31. Liu, Z.Z.; Gong, Y.F.; Fu, Z.X.; Zhang, C.S.; Zhang, W.X.; Song, Y.; Ma, Y.H. Genetic variation of goat MyoG gene intron II and its genetic effects on body weight. *Acta Vet. Zootech. Sin.* **2011**, *42*, 1015–1021.
32. Singh, S.P.; Kumar, R.; Kumari, P.; Kumar, S.; Mitra, A. Characterization of 5′ upstream region and investigation of TTTTA deletion in 5′ UTR of myostatin (MSTN) gene in Indian goat breeds. *Anim. Biotechnol.* **2014**, *25*, 55–68. [CrossRef]
33. Zhou, X.; Yan, Q.; Liu, L.; Chen, G.; Tang, S.; He, Z.; Tan, Z. Maternal undernutrition alters the skeletal muscle development and methylation of myogenic factors in goat offspring. *Anim. Biosci.* **2022**, *35*, 847–857. [CrossRef]
34. Wei, L.; Xiao, W.; Chen, B.; Zou, Z.; Zhu, J.; Li, D.; Yu, J.; Yang, H. Single nucleotide polymorphisms in the MRFs gene family associated with growth in *Nile tilapia*. *Mol. Biol. Rep.* **2024**, *51*, 128. [CrossRef]
35. Ren, H.; Wei, Z.; Li, X.; Wang, Q.; Chen, H.; Lan, X. Goat MyoD1: mRNA expression, InDel and CNV detection and their associations with growth traits. *Gene* **2023**, *866*, 147348. [CrossRef]
36. Na, R.; Ni, W.; E, G.; Zeng, Y.; Han, Y.; Huang, Y. SNP screening of the MSTN gene and correlation analysis between genetic polymorphisms and growth traits in Dazu black goat. *Anim. Biotechnol.* **2021**, *32*, 558–565. [CrossRef]
37. Pehlivan, E. Relationship between insulin-like growth factor-1 (IGF-1) concentrations and body trait measurements and climatic factors in prepubertal goat kids. *Arch. Anim. Breed.* **2019**, *62*, 241–248. [CrossRef]
38. Zhang, G.; Tang, Y.; Zhang, T.; Wang, J.; Wang, Y. Expression profiles and association analysis with growth traits of the MyoG and Myf5 genes in the Jinghai yellow chicken. *Mol. Biol. Rep.* **2014**, *41*, 7331–7338. [CrossRef]
39. Du, S.J.; Gao, J.; Anyangwe, V. Muscle-specific expression of myogenin in zebrafish embryos is controlled by multiple regulatory elements in the promoter. *Comp. Biochem. Physiol. Part B Biochem. Mol. Biol.* **2003**, *134*, 123–134. [CrossRef]
40. Trukhachev, V.; Belyaev, V.; Kvochko, A.; Kulichenko, A.; Kovalev, D.; Pisarenko, S.; Volynkina, A.; Selionova, M.; Aybazov, M.; Shumaenko, S.; et al. Genes expression profiles in the loin muscle of manych merino sheep with different live weight. *Bulg. J. Vet. Med.* **2016**, *19*, 19–29. [CrossRef]
41. Naicy, T.; Venkatachalapathy, R.T.; Aravindakshan, T.V.; Kurian, E. Association of a Cac8I polymorphism in the IGF1 gene with growth traits in Indian goats. *J. Genet. Eng. Biotechnol.* **2017**, *15*, 7–11. [CrossRef]
42. Song, T.; Tan, Y.; Cuomu, R.; Liu, Y.; Ba, G.; Suo, L.; Wu, Y.; Cao, X.; Zeng, X. Polymorphisms and mRNA expression levels of IGF-1, FGF5, and KAP 1.4 in Tibetan cashmere goats. *Genes* **2023**, *14*, 711. [CrossRef]
43. Vega, W.H.O.; Quirino, C.R.; Bartholazzi, A., Jr.; Rua, M.A.S.; Serapião, R.V.; Oliveira, C.S. Variants in the CYP19A1 gene can affect in vitro embryo production traits in cattle. *J. Assist. Reprod. Genet.* **2018**, *35*, 2233–2241. [CrossRef]
44. Cosenza, G.; Iannaccone, M.; Pico, B.A.; Gallo, D.; Capparelli, R.; Pauciullo, A. Molecular characterisation, genetic variability and detection of a functional polymorphism influencing the promoter activity of OXT gene in goat and sheep. *J. Dairy Res.* **2017**, *84*, 165–169. [CrossRef]
45. Mishra, C.; Kumar, S.; Panigrahi, M.; Yathish, H.M.; Chaudhary, R.; Chauhan, A.; Kumar, A.; Sonawane, A.A. Single nucleotide polymorphisms in 5′ upstream region of bovine TLR4 gene affecting expression profile and transcription factor binding sites. *Anim. Biotechnol.* **2018**, *29*, 119–128. [CrossRef]
46. Wang, P.; Li, W.; Liu, Z.; He, X.; Lan, R.; Liu, Y.; Chu, M. Analysis of the association of two SNPs in the promoter regions of the PPP2R5C and SLC39A5 genes with litter size in Yunshang black goats. *Animals* **2022**, *12*, 2801. [CrossRef]
47. Kwan, J.Z.J.; Nguyen, T.F.; Uzozie, A.C.; Budzynski, M.A.; Cui, J.; Lee, J.M.C.; Van Petegem, F.; Lange, P.F.; Teves, S.S. RNA Polymerase II transcription independent of TBP in murine embryonic stem cells. *eLife* **2023**, *12*, e83810. [CrossRef]
48. Cormack, B.P.; Struhl, K. The TATA-binding protein is required for transcription by all three nuclear RNA polymerases in yeast cells. *Cell* **1992**, *69*, 685–696. [CrossRef]
49. Foletta, V.C. Transcription factor AP-1, and the role of Fra-2. *Immunol. Cell Biol.* **1996**, *74*, 121–133. [CrossRef]

https://doi.org/10.3390/genes17010014

50. Hwang, S.S.; Kim, L.K.; Lee, G.R.; Flavell, R.A. Role of OCT-1 and partner proteins in T cell differentiation. *Biochim. Biophys. Acta (BBA)-Gene Regul. Mech.* **2016**, *1859*, 825–831. [CrossRef]
51. Guerrero-Santoro, J.; Khor, J.M.; Açıkbaş, A.H.; Jaynes, J.B.; Ettensohn, C.A. Analysis of the DNA-binding properties of Alx1, an evolutionarily conserved regulator of skeletogenesis in echinoderms. *J. Biol. Chem.* **2021**, *297*, 100901. [CrossRef]

https://doi.org/10.3390/genes17010014

animals

MDPI

Article

Regulatory Effect of circRNA3890 on the Establishment of Receptive Endometrium in Dairy Goats

Jiuzeng Cui, Linxi Li, Yonggan Su, Songrong Han, Lei Zhang * and Yuxuan Song *

College of Animal Science and Technology, Northwest A&F University, Yangling 712100, China
* Correspondence: xndkzl07@163.com (L.Z.); syx_2025@126.com (Y.S.)

Simple Summary

Mammalian embryo implantation requires the endometrium to be in a receptive state. Many non-coding RNAs, such as circRNA3890, show differential expression between the receptive endometrium and pre-receptive endometrium of dairy goats, according to our earlier research. This work experimentally showed that circRNA3890 contributes to the establishment of receptive endometrium in dairy goats by controlling the proliferation and apoptosis of dairy goat endometrial epithelial cells (gEECs) by targeting miR-26b-3p/*MDM4*. These findings will contribute to understanding circRNA and miRNA function in embryo implantation and provide a theoretical basis for further improving the embryo attachment rate.

Abstract

Receptive endometrium (RE) is essential for mammalian embryo implantation. The establishment of RE is a complex and precise dynamic process regulated by various cytokines, including non-coding RNAs (miRNAs, lncRNAs, and circRNAs). We identified candidate miR-26b-3p and circRNA3890 from our previous endometrial non-coding RNA sequencing data. CircRNA3890 adsorbs miR-26b-3p and inhibits its activity. Mouse double minute 4 (*MDM4*) is a target gene of miR-26b-3p, and circRNA3890 up-regulates the expression level of *MDM4* by inhibiting the activity of miR-26b-3p in dairy goat endometrial epithelial cells (gEECs) in vitro. circRNA3890/miR-26b-3p/*MDM4* could promote the proliferation of gEECs through the p53 signaling pathway. MiR-26b-3p could regulate the expression levels of vascular endothelial growth factor A (VEGFA) and leukemia inhibitory factor (LIF) through *MDM4* in gEECs, which contributes to the development of endometrial receptivity. Furthermore, the results showed that miR-26b-3p significantly promoted the development of RE and embryo implantation. These findings demonstrate that circRNA3890 targets and adsorbs miR-26b-3p to relieve *MDM4* inhibition and promotes EEC proliferation through the p53 signaling pathway. They reveal the regulatory effect of miR-26b-3p on receptive endometrial development and embryo implantation in vitro and in vivo.

Keywords: dairy goat; endometrial epithelial cells; endometrial receptivity

1. Introduction

The precondition for conception in mammals is successful embryo implantation [1]. Embryo implantation occurs when a healthy embryo interacts with the receptive endometrium (RE) and attaches to the endometrial lining. RE is formed by complex changes in the morphology and physiology of the endometrium [2]. Many biological processes and pathways participate in the dynamic process of RE establishment, including endometrial epithelial cells (EECs) proliferation and apoptosis [3–5]. Moreover, the establishment of RE is a necessary condition for embryo implantation. Under natural conditions, the decreased or insufficient receptivity of the mammalian endometrium is one of the reasons for the failure of healthy embryo implantation [6,7]. Abnormal endometrial receptivity leading to failed embryo implantation results in significant economic losses for the livestock industry. Therefore, studying the molecular mechanism of RE establishment for improving the conception rate of mammals is necessary. The dynamic expression and fine-tuned regulation of non-coding RNA (ncRNA) during the establishment of endometrial receptivity has been demonstrated to be a critical factor determining successful embryo implantation [4]. Exogenous supplementation or suppression of specific ncRNAs—such as through miRNA mimics or antagonists—holds promise as a novel therapeutic strategy to enhance endometrial receptivity and improve assisted reproductive technology success rates.

CircRNA is highly stable and tissue-specific in mammals. In contrast to linear RNA, circRNA does not have a 5′-terminal cap and 3′-ployA tail structure and is a closed-loop structure composed of end-to-end connections [8]. Recently, circRNA as a post-transcriptional regulator has become one of the research hotspots in many fields. Studies have shown that circRNA can function as competitive endogenous RNA (ceRNA) to adsorb miRNA and inhibit its activity, subsequently regulating the expression of the miRNA at the transcriptional or post-transcriptional level [9,10]. Furthermore, circRNA regulates the proliferation and apoptosis of many kinds of cells. For example, circCGNL1 promoted the apoptosis of pancreatic cancer cell [11], and circRNA-0013747 promoted the proliferation of mesangial cells [12]. However, it is uncertain how circRNAs affects endometrial epithelial cells and the development of RE. Our previous research revealed that circRNA3890 was down-regulated in RE, potentially participating in the establishment of receptive endometrium in dairy goats [5]. Therefore, this study selected circRNA3890 as the research subject to further elucidate its molecular mechanisms during receptive endometrial establishment.

MicroRNAs (miRNAs) are a class of endogenous non-coding RNAs that act by degrading mRNA or inhibiting protein translation [13]. They are involved in various biological functions and processes in mammals, such as cell proliferation and apoptosis [14,15]. Moreover, miRNAs play an important regulatory role during embryo implantation; for example, miR-124-3p/miR-3074-5p reduces embryo implantation [16,17], whereas let7a/b promotes embryo implantation [18]. As we previously discovered, a significant number of miRNAs, including miR-26b-3p, are differentially expressed in the receptive endometrial period and pre-receptive period in dairy goats [19]. However, the effect of miR-26b-3p and circRNA networks on gEECs and receptive endometrium is unclear.

MDM4 is a cDNA sequence that encodes p53 protein binding and inhibits the transcriptional activity of the wild-type p53 gene [20]. Bioinformatic analysis has predicted that *MDM4* is the target gene of miR-26b-3p. Moreover, studies have reported that *MDM4* promotes the proliferation of many kinds of cells [21,22]. We hypothesize that *MDM4* modulates endometrial epithelial cell proliferation and apoptosis in dairy goats, playing a key role in receptive endometrium formation.

In this experiment, we first clarified the relationship among circRNA3890, miR-26b-3p, and *MDM4* in vitro and then tested their effects on gEECs and RE marker genes. We verified the effect of miR-26b-3p on mouse endometrial receptivity and embryo value-added rate in vivo. In conclusion, we explored the effects of circRNA3890, miR-26b-3p and *MDM4* on the development of endometrial tolerance and embryo implantation. These findings will contribute to understanding circRNA and miRNA function in embryo implantation and provide a theoretical basis for further improving the embryo attachment rate.

https://doi.org/10.3390/ani16050696

2. Materials and Methods

2.1. Tissue Sample Collection

This study used Saanen dairy goats from the Longxian Dairy Goat Experimental Demonstration Base of Northwest A&F University as the research object. Twelve dairy goats with good reproductive performance were selected for concurrent estrus treatment. The day of mating was recorded as D0 of pregnancy. Our previous research indicates that the endometrium of dairy goats is in a receptive state on day 15 post-breeding (D15), whereas it is in a pre-receptive state on day 5 (D5) [5]. Endometrial samples from the anterior wall of the uterine cavity were collected on D5 (PE) and D15 (RE) of pregnancy [5,19]. All endometrial samples were washed with phosphate-buffered saline and frozen in liquid nitrogen or immediately used for the isolation and culture of gEECs.

2.2. Cell Culture

On the fifth day after breeding, uterine tissue was surgically collected from the dairy goats. The uterus was promptly placed in PBS solution containing penicillin (100 units/mL) and streptomycin (50 mg/mL) and immediately transported to the laboratory. The tissue was first rinsed with 75% ethanol, followed by PBS. In the cell culture room, endometrial tissue was excised from the uterus using a sterile scalpel, then incubated with trypsin for approximately 1 h. Digestion was terminated by adding fetal bovine serum (FBS). The cell suspension was centrifuged at 500 *g* for 10 min to pellet the cells, and the EECs were resuspended in DMEM/F12 medium containing 10% FBS. To isolate endometrial stromal cells (ESCs), the supernatant was centrifuged at 1000 *g* for 10 min. These cells were then cultured in the same cell culture system until the third passage to further purify the cells [23,24]. The HEK293T cell lines were provided by our laboratory. The gEECs or HEK293T were cultivated in Dulbecco's Modified Eagle Medium/Nutrient Mixture F-12 (DMEM/F12) (Gibco, Carlsbad, CA, USA) or DMEM (high glucose) with 10% fetal bovine serum (FBS) and penicillin/streptomycin (100 U/mL and 100 mg/mL, respectively) in a humidified atmosphere of 5% CO_2 at 37 °C.

The full length of circRNA3890 and *MDM4* coding regions was cloned and inserted into the pcDNA2.1 (+) (Geneseed Biotech, Guangzhou, China) and pcDNA3.1 (+) vectors (Promega, Madison, WI, USA). NC, NC inhibitor (NCH), MiR-26b-3p mimic, miR-26b-3p inhibitor, si-*MDM4*, and si-CircRNA3890 were synthesized by GenePharma (Shanghai, China). When the cell confluence rate reached about 50%, the small RNA was mixed with the transfection reagent following the instructions for Lipofectamine 2000 (Invitrogen, Waltham, MA, USA). The mixture was incubated at room temperature for 20 min, and subsequently added to the cell culture plate and thoroughly shaken.

2.3. Western Blot and RT-qPCR

Total RNA was extracted from tissues and cells using the Ultrapure RNA Kit (Cabio, Wuhan, China), according to the kit instructions. For mRNA analysis, complementary DNA (cDNA) was synthesized from 1.0 μg total RNA using a reverse transcription kit (TaKaRa, Dalian, China), according to the kit instructions. RT-qPCR was performed in triplicate wells using SYBR Green PCR Master Mix (TaKaRa, Dalian, China). Primer sequences used for PCR are shown in Table 1.

The total proteins were extracted from gEECs as described previously [23]. The total protein concentration was determined by BCA protein concentration assay kit (Beyotime, Shanghai, China), and the Western blot (WB) assay procedure was performed as previously described [23]. The antibodies used in this experiment are shown in Table 2.

https://doi.org/10.3390/ani16050696

Table 1. Primer sequences.

Gene	Primer Sequences (5′ → 3′)
β-actin	F: GATCTGGCACCACACCTTCT R: GGGTCATCTTCTCACGGTTG
MDM4 (qPCR)	F: GATCAGCAGGAGCAGCACATGG R: GCAGCATCTGTAGCAGCAGTAGC
MDM4 (pcDNA3.1)	F: cga*GGTACC*ATGACATCATTTTCCACCTCTACCCAGTG R: CG*CTCGAG*CTATGCTACAAAAACCTTAATAACCAACTGAATCTCCTTCT
MDM4 (Check2)	F: cgCTCGAGTATTTAATTTAATTTCTTACTGTTATTTTCTGGTAGGGAATGTTCTTGGGCA R: ATGCGGCCGCCGAGACAGAGATGTGCTACTG
circRNA3890 (qPCR)	F: AGAAGTCCTTACGGGCGAGT R: GCAGAAGCAGCTCAAGGGAC
circRNA3890 (pcD2.1-ciR)	F: GG*GGTACC*TGAAATATGCTATCTTACAGTGATGCTCCAGTGAGTGAACTGTCC R: CG*GGATCC*TCAAGAAAAAATATATTCACCTGAGTGGAAAATTTAAAGGC
circRNA3890 (Check2)	F: cgCTCGAGTGATGCTCCAGTGAGTGAACTGTCC R: ATGCGGCCGCCCTGAGTGGAAAATTTAAAGGC
MARCH6 (qPCR)	F: ACTGTCCCTTGAGCTGCTTC R: CGGGCATGCTGGTTATTGTT
VEGFA (qPCR)	F: GGGCTCTTCTCGCTCCGTAGTAG R: CCCTCTCCTCTTCCTTCTCTTCCTC
LIF (qPCR)	F: TGTAAATGCCACCTGTGCCATACG R: CATTGAGCTGTGCCAGTTGATTCTTG
U6	F: CTCGCTTCGGCAGCACA R: AACGCTTCACGAATTTGCGT
miR-26b-3p-Loop	gtcgtatccagtgcagggtccgaggtattcgcactggatacgacGAGCCAAG
miR-26b-3p-FW	gcgcgcCCTGTTCTCCATTA
Reverse Primer	GTGCAGGGTCCGAGGT

Note: The characters with underscore were restriction enzyme cutting sites of Xho I and Not I for constructing psiCHECK2. The italicized characters with underscore were restriction enzyme cutting sites of Kpn I and Xho I for constructing pcDNA3.1. The italicized characters with underscore were restriction enzyme cutting sites of Kpn I and BamH I for constructing pCD2.1-ciR. circRNA3890 (qPCR) was an outward-facing primer and divergently primed.

Table 2. Antibody information for Western blots.

Gene	Manufacturer	Product Number	Dilution Ratio
β-actin	Beyotime, Shanghai, China	AA128	1:1000
BAX	Beyotime, Shanghai, China	AB026	1:500
BCL2	Beyotime, Shanghai, China	AB112	1:500
MDM4	Proteintech, Wuhan, China	17914-1-AP	1:500
P53	Beyotime, Shanghai, China	AF0255	1:500
P21	Beyotime, Shanghai, China	AP021	1:500
CDK2	Abways, Shanghai, China	CY5020	1:1000
Cyclin E1	Abways, Shanghai, China	CY5466	1:1000
VEGFA	BBI, Shanghai, China	D160788	1:500
LIF	Boster, Wuhan, China	3A1239-2	1:500
HRP-labeled Goat Anti-Rabbit IgG (H+L)	Beyotime, Shanghai, China	A0208	1:1000
HRP-labeled Goat Anti-Mouse IgG (H+L)	Beyotime, Shanghai, China	A0216	1:1000

2.4. Luciferase Activity Assay

Using miRanda and Targetscan7.0, bioinformatic analysis of the miRNA binding sequences in *MDM4* and circRNA3890 was performed to determine the targets of miR-26b-3p. The *MDM4* 3′UTR fragment and entire circRNA3890, which each include the miR-26b-3p binding site, were cloned and put into the psiCHECK-2 vector (Promega, Madison, WI, USA) to create reporters for luciferase tests. In addition, psiCHECK2-Mut mutant plasmids with altered target sites were created using TaKaRa MutanBEST Kit (TaKaRa,

Dalian, China). The specific steps of vector construction are as described above [23,25]. All primer sequences are shown in Table 1, and all construction vectors were verified by sequencing. When the cell confluence reached roughly 60%, the wild-type (psiCHECK2-WT) or mutant (psiCHECK2-MUT) plasmids were co-transfected with the miR-26b-3p mimic into 293T cells. Each experiment was carried out three times, in triplicate. The luciferase reporter assay kit instructions (Promega, Madison, WI, USA) provided thorough instructions on detecting luciferase activity.

2.5. Proliferation Cell, Cycle, and Apoptosis Assay

The EdU incorporation assay (Ribobio, Guangzhou, China) and cell counting kit (CCK8) assay (ZETA, San Francisco, CA, USA) were used to screen for cell proliferation as previously described [23,26]. We used a cell cycle staining kit (Liankebio, Hangzhou, China) to analyze the cell cycle of different treatment groups according to the manufacturer's instructions [27]. Cell apoptosis was analyzed by V-fluorescein isothiocyanate (FITC)/propidium iodide (PI) kit (Liankebio, Hangzhou, China). The detailed procedures were conducted following the manufacturer's instructions and referring to our previous research [19]. Analyses were performed using a flow cytometer (BD Biosciences, San Diego, CA, USA).

2.6. Fluorescence In Situ Hybridization

We fixed gEECs using 4% paraformaldehyde, followed by prehybridization with 0.5% (Invitrogen, Waltham, MA, USA), and then performed fluorescence in situ hybridization as previously reported [28]. Next, we incubated cells with RNA probes in hybridization buffer for 12 h. The circRNA3890 probe was labeled with cy3. Nuclei were stained with DAPI (Vector Laboratories, Burlingame, CA, USA). Images were captured using confocal laser scanning microscopy (Leica TCS SP8, Leica Microsystems CMS GmbH, Wetzlar, Germany).

2.7. Animal Experiment

The mice used in the in vivo experiment were 8–10 weeks old ICR female mice weighing about 30–35 g and purchased from Beijing Weishanglide Biotechnology Co., Ltd. (Beijing, China) (SCXK (jing) 2016-0009). The male and female mice were caged for mating, and the vagina of the female mice was examined at 8:00 the following day. Only female mice with vaginal plugs were reserved for use, and the day plug appearance was considered the first day of pregnancy (D1). An operation was performed on day 3 of pregnancy (D3) at 9:00 to expose the left ovary and uterine horns. The uterine horn was injected with 10 μg/20 μL miR-26b-3p agomir (GenePharma, Shanghai, China) into the uterine cavity, with the contralateral uterine horn serving as the control side, where the same dose of NC was injected. On day 4 of pregnancy (D4), the uterine horns of two mice were cut longitudinally and evenly divided. One portion was scraped from the endometrial tissue for PCR detection, while the other was fixed in 2.5% glutarase fixative. Subsequently, for scanning electron microscopy detection, another operation as mentioned above was performed on an additional two mice and both sides of the uterine horns were fixed in 4% paraformaldehyde for subsequent HE and immunohistochemistry backup. On day 9 of pregnancy (D9), pictures of the uterus of three mice were taken, and the embryo value-added rate was counted.

2.8. Immunohistochemistry

Uterine tissue samples were fixed with paraformaldehyde (4%), embedded in paraffin, and sectioned. Sections were dewaxed with xylene and alcohol in turn, rehydrated, placed in citrate buffer, and cooked in a microwave oven for 10 min for antigen retrieval. Endogenous peroxidase was blocked with 3% hydrogen peroxide for 10 min at room temperature,

https://doi.org/10.3390/ani16050696

subsequently blocked with 5% BSA for 30 min, and incubated with a primary antibody (as shown in Table 2) overnight at 4 °C.

2.9. Scanning Electron Microscopy

The effects of miR-26b-3p on mouse uterine pinopodes were detected by scanning electron microscopy (SEM) [29]. Uterine tissue was fixed with glutaraldehyde (2.5%) for more than 2 h, dehydrated in non-concentrated ethanol, and subjected to critical point drying. The sample was then positioned with the observation surface (endometrial cavity) upward on the sample stage, and a vacuum film meter was used to coat the sample with conductive silver plastic. The samples were observed under JSM-6330F SEM (JEOL, Tokyo, Japan).

2.10. Statistical Analysis

All the data were processed with SPSS 22.0 (SPSS Inc., Chicago, IL, USA). The data were analyzed using *t*-test, one-way analysis of variance, and least significant difference (LSD) to determine any significant differences. A value of $p < 0.05$ was considered statistically significant. Differences with $p < 0.5$ are indicated by one asterisk and those with $p < 0.01$ by two asterisks to denote extremely significant differences.

3. Results

3.1. Identification of circRNA3890 in Dairy Goat Endometrium

To analyze the contribution of circRNA to the establishment of endometrial receptivity in dairy goats, we screened published sequencing data and found that circRNA3890 was highly expressed in PE of dairy goats [30]. CircRNA3890 is derived from Exon18 and Exon19 of membrane-associated ring-CH-type finger 6 (MARCH6) on chromosome 20, having a length of 343 bp (Figure 1a). The RT-qPCR results showed that the expression levels of circRNA3890 in PE was significantly up-regulated compared with RE (Figure 1b), which was consistent with the sequencing results. To verify the stability of circRNA3890 in gEECs of goats, we used actinomycin D (an RNA synthesis inhibitor) to treat gEECs and extracted the total RNA. The results showed that the half-life of circRNA3890 was longer than that of linear MARCH6 mRNA (Figure 1c). Furthermore, circRNA3890 was also significantly more resistant to RNase R exonuclease than linear MARCH6 (Figure 1d). The FISH experiments showed that circRNA3890 was generally expressed in the cytoplasm of gEECs (Figure 1e). These findings imply that circRNA3890 is a stable molecular marker and may play a role in the establishment of a receptive endometrium in dairy goats.

3.2. CircRNA3890 as a Competitive ceRNA Target to Suppress the Expression of miRNA-26b-3p in gEECs

Most circRNAs perform their function by adsorbing miRNAs [4,9]. In this study, we predicted the presence of a binding site for circRNA3890 and miR-26b-3p by two public programs (Targetscan 7.0 and miRanda) (Figure 2a). To verify whether circRNA3890 could bind to miR-26b-3p, a double luciferase assay was conducted. MiR-26b-3p down-regulated WT-circRNA3890 dual luciferase activity; however, there was no significant effect on MUT-circRNA3890 (Figure 2b). We successfully constructed the overexpression vector of circRNA3890 (pc3.1-circRNA3890) (Figure S1a). Two siRNAs were designed for circRNA3890 (Figure S1b), and si-circRNA3890-2 (si-circRNA3890) with high interference efficiency was selected for subsequent experiments. Furthermore, circRNA3890 decreased the expression of miR-26b-3p in gEECs (Figure 2c,d). We synthesized miR-26b-3p mimic/miR-26b-3p inhibitor and confirmed it could be used in subsequent experiments (Figure S1c). After transfection with miR-26b-3p mimic, the expression of circRNA3890 in gEECs decreased significantly, and the miR-26b-3p inhibitor significantly increased the

https://doi.org/10.3390/ani16050696

expression of circRNA3890 in gEECs (Figure 2e). In summary, circRNA3890 has a targeting relationship with miR-26b-3p and a potential negative feedback regulation relationship in gEECs.

A number of genes, including VEGFA [31] and LIF [32], have been recognized as receptive endometrial marker genes. The WB assay showed that miR-26b-3p significantly up-regulated the protein expression level of VEGFA and LIF in gEECs (Figure 2f–i). Moreover, the circRNA3890 decreased the protein levels of VEGFA and LIF in gEECs (Figure 2j–m), and miR-26b-3p attenuated the regulatory effect of the circRNA3890 on the above proteins. The above results further prove that circRNA3890 is involved in the establishment of RE by targeting adsorption of miR-26b-3p.

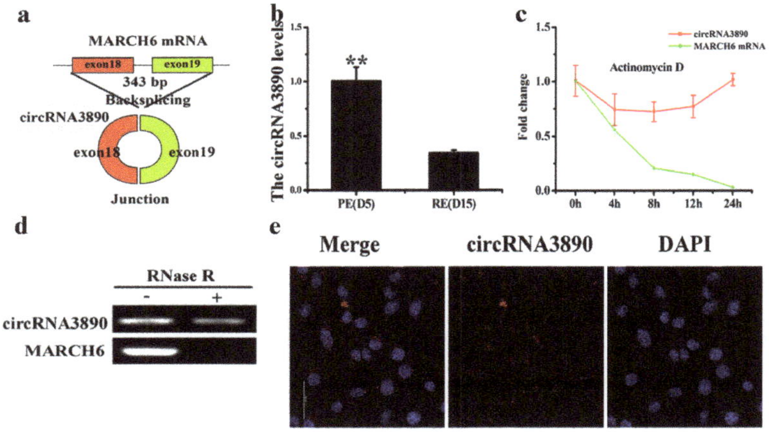

Figure 1. Expression and stability of circRNA3890 in various tissues. (**a**) Schematic for circRNA 3890 was derived from the *MARCH6* gene. (**b**) circRNA3890 levels in PE and RE. (**c**) CircRNA3890 and *MARCH6* expression levels in gEECs treated with actinomycin D (1 mg/mL). (**d**) RT-qPCR was used to ascertain the amounts of circRNA3890 and MARCH6 in gEECs that had received RNase R treatment. (**e**) Fluorescence in situ hybridization was used to detect the expression of circRNA3890 in gEECs. The scale bar represents 50 μm. Each result is shown as the mean ± SEM of three values: ** $p < 0.01$.

3.3. The Effects of circRNA3890 on gEECs

Studies have shown that apoptosis of gEECs contributes to embryo implantation [33]. To verify the function of circRNA3890 in gEECs, the proliferation effect of circRNA3890 on gEECs was detected using CCK8 and EdU. The CCK8 assay showed that circRNA3890 enhanced the cell vitality of gEECs, and the pc2.1-circRNA3890+ miR-26b-3p mimic group reduced the cell viability of gEECs compared with the pc2.1-circRNA3890 group (Figure 3a). CircRNA3890 increased the number of positive cells, and miR-26b-3p could reduce the effect of circRNA3890 on gEECs proliferation (Figure 3b,c). Overexpression of circRNA3890 decreased the protein levels of BAX and significantly increased the protein levels of BCL2 in gEECs in vitro. MiR-26b-3p could reduce the effect of circRNA3890 on these proteins in gEECs (Figure 3d,e). Furthermore, the FCM assay showed that circRNA3890 inhibited gEECs apoptosis, and that miR-26b-3p could reduce the influence of circRNA3890 on gEECs apoptosis (Figure 3f). Meanwhile, circRNA3890 increased the number of gEECs in the S phase and reduced the proportion of gEECs in the G1 phase. Moreover, miR-26b-3p could reduce the effect of circRNA3890 on the cell cycle (Figures 3g and S2). Interference with circRNA3890 was the opposite of overexpression (Figure 4). The above results show that

https://doi.org/10.3390/ani16050696

circRNA3890 promoted the proliferation of EEC and inhibited apoptosis. CircRNA3890 may participate in establishing a receptive endometrium by regulating apoptosis in gEECs.

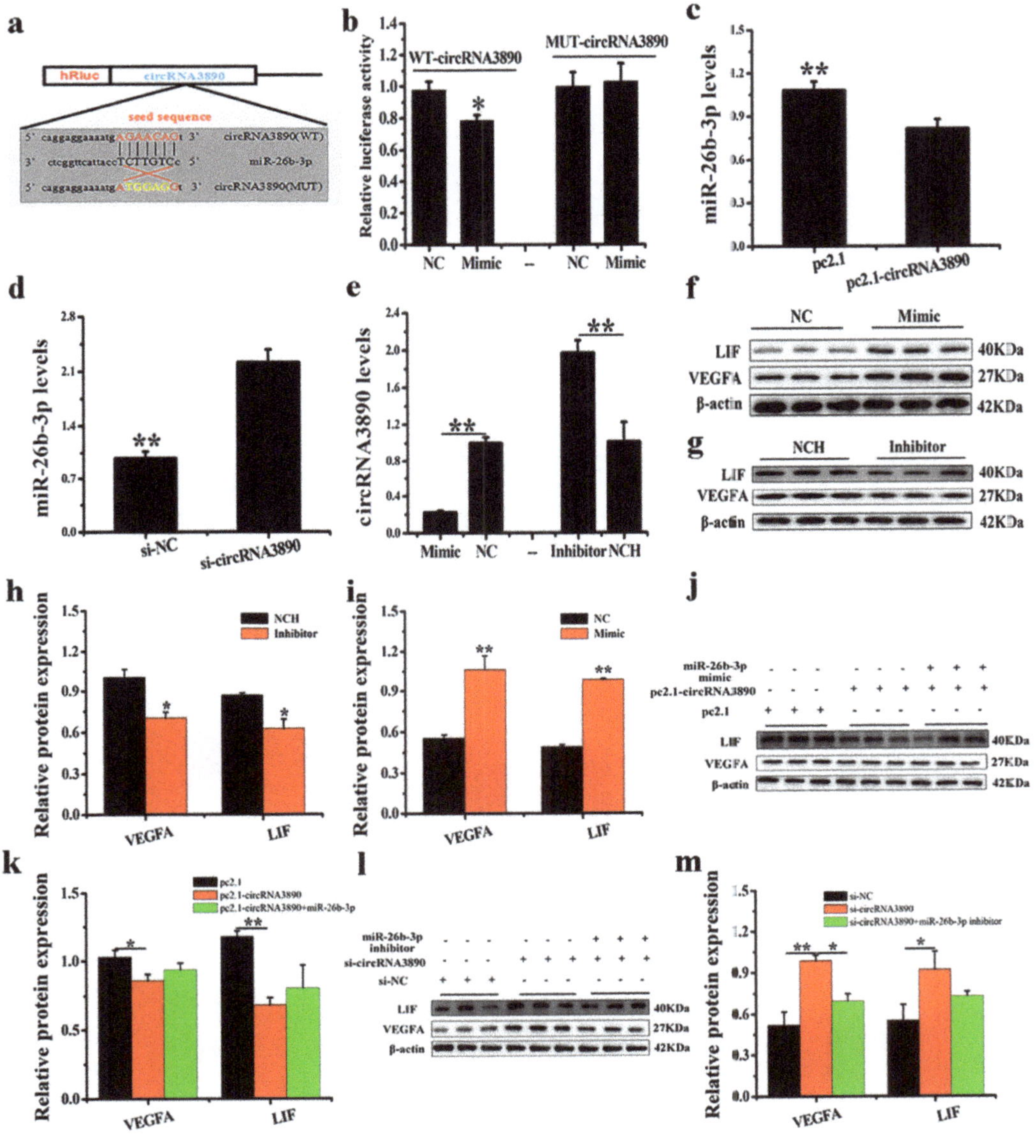

Figure 2. circRNA3890 targets miR-26b-3p and reduces its level in gEECs. (**a**) Luciferase reporter genes with binding sites (WT-circRNA3890) and binding site mutations (MUT-circRNA3890) are shown in a schematic design. The red-labeled nucleotides represent the "seed sequence" of miR-26b-3p, while the yellow ones represent mutated nucleotides. (**b**) After co-transfecting 293T cells with WT-circRNA3890, MUT-circRNA3890 or miR-26b-3p, NC (negative control), double luciferase activity was analyzed. (**c**,**d**) Regulatory effect of circRNA3890 on miR-26b-3p in gEECs was detected using qRT-PCR. (**e**) miR-26b-3p decreased the circRNA3890 levels in gEECs. (**f**–**i**) Effects of miR-26b-3p on LIF and VEGFA protein expression in gEECs. (**j**–**m**) Protein expression levels of LIF and VEGFA after transfection with pc2.1-circRNA3890/si-circRNA3890. Each result is shown as the mean ± SEM of three values: ** $p < 0.01$; * $p < 0.05$.

https://doi.org/10.3390/ani16050696

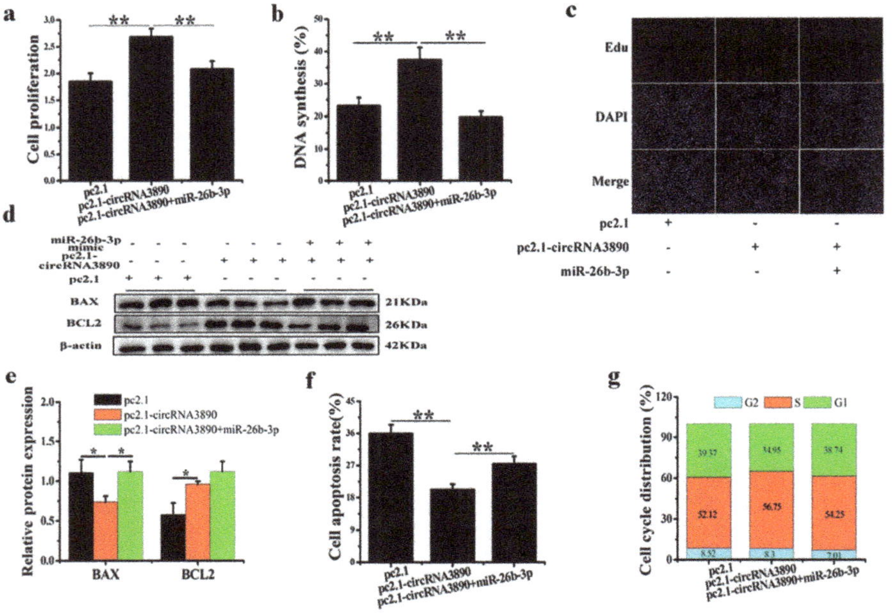

Figure 3. Regulatory effect of circRNA3890 on gEECs. (**a**–**c**) Indicators of cell proliferation were measured following treatment with CCK-8 and EdU. The scale bar represents 100 µm. (**d**,**e**) The protein expression levels of BCL2 and BAX analyzed using Western blotting after overexpression of circRNA3890 in gEECs. (**f**,**g**) FCM was used to detect apoptosis and cell phases. Each result is shown as the mean ± SEM of three values: ** $p < 0.01$; * $p < 0.05$.

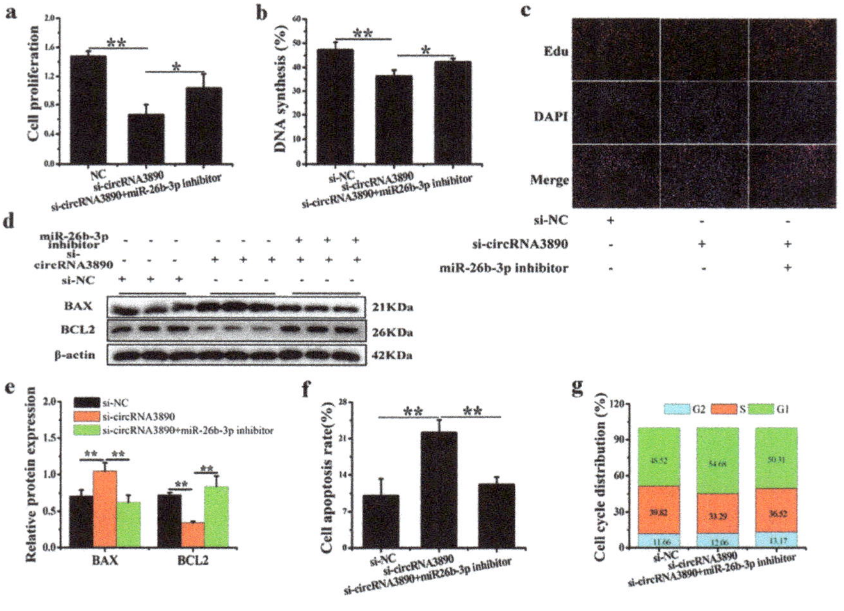

Figure 4. Interference with circRNA3890 promotes apoptosis in gEECs. (**a**) CCK-8 investigation of the effects of circRNA3890 interference on gEECs cell viability. (**b**,**c**) EdU analysis of the effects of circRNA3890 interference on gEECs cell proliferation. The scale bar represents 100 µm. (**d**,**e**) The amounts of BCL2 and BAX protein expression were determined using Western blotting following the

https://doi.org/10.3390/ani16050696

interference of circRNA3890 in gEECs. (**f**,**g**) FCM was used to examine how circRNA3890 interference affected cell phases and apoptosis. Each result is shown as the mean ± SEM of three values: ** $p < 0.01$; * $p < 0.05$.

3.4. The Effects of miR-26b-3p on gEECs

CircRNA3890 may affect gEECs through miR-26b-3p, so we examined the effect of miR-26b-3p on gEECs. The RT-qPCR results showed that the expression levels of miR-26b-3p in RE were significantly up-regulated compared with PE (Figure 5a). The EdU and CCK8 assays showed that miR-26b-3p inhibited the proliferation and decreased the cell viability of gEECs (Figure 5b–d). The FMC assay showed that miR-26b-3p promoted the apoptosis of gEECs (Figure 5e).

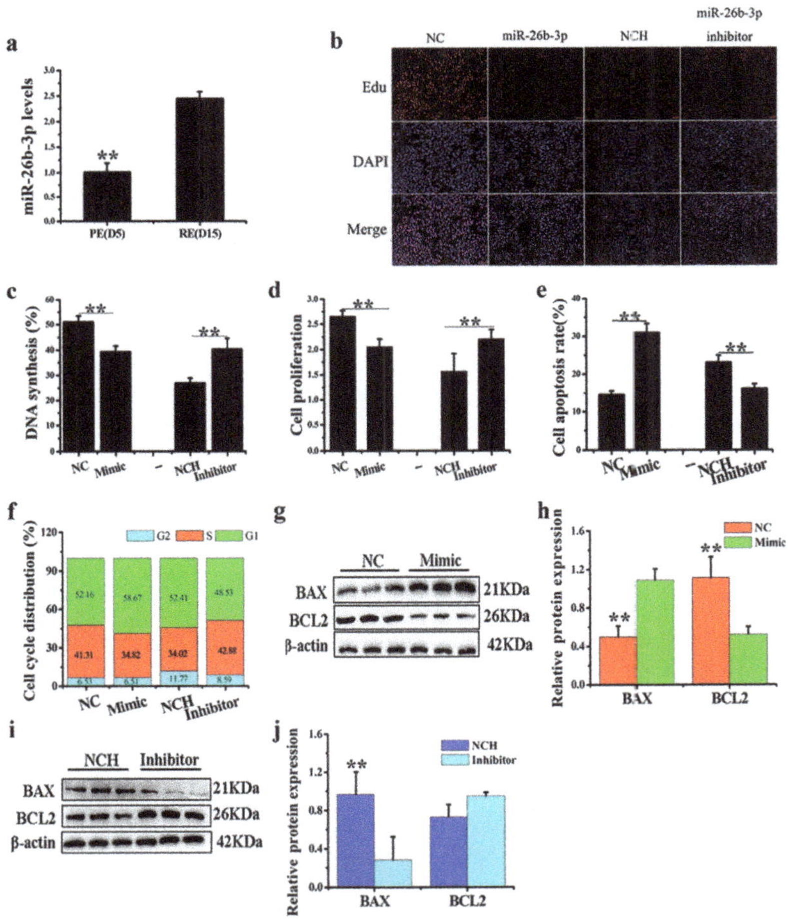

Figure 5. Regulatory effects of miR-26b-3p on gEECs in vitro. (**a**) miR-26b-3p levels in pre-receptive endometrium (PE) and receptive endometrium (RE) of dairy goats. (**b**,**c**) The influence of gEECs after transfection with miR-26b-3p mimic/inhibitor was analyzed using EdU. The scale bar represents 100 μm. (**d**) CCK-8 analysis of EEC activity after transfection with miR-26b-3p mimic/inhibitor. (**e**,**f**) Effects of apoptosis and cell cycle phases after transfection with miR-26b-3p mimic/inhibitor were analyzed by FCM. (**g**–**j**) The expression levels of BCL2 and BAX in gEECs after transfection with miR-26b-3p mimic/inhibitor were analyzed using Western blotting. Each result is shown as the mean ± SEM of three values: ** $p < 0.01$.

https://doi.org/10.3390/ani16050696

Furthermore, miR-26b-3p decreased the protein levels of BCL2 and increased the protein levels of BAX in gEECs in vitro (Figure 5f–i). The flow cytometry (FMC) analysis investigated the cell cycle; the results showed that miR-26b-3p increased the proportion of gEECs in the G1 phase while reducing the number of gEECs in the S phase (Figures 5j and S3). In conclusion, circRNA3890 regulated apoptosis in gEECs by targeting miR-26b-3p, thereby participating in the establishment of a receptive endometrium.

3.5. MiR-26b-3p Inhibits the Expression of MDM4 in gEECs

Most miRNAs are involved in post-transcriptional regulation through target genes [23,25]. We predicted that *MDM4* might be a target gene of miR-26b-3p by two public programs (Targetscan7.0 and miRanda), and the predicted binding site is located in the 3′UTR. The *MDM4* expression in PE was significantly up-regulated compared to RE (Figure 6a). We constructed wild-type (WT-*MDM4*) and mutated (MUT-*MDM4*) dual luciferase reporter vectors of the *MDM4* gene based on the predicted binding sites (Figure 6b). The double luciferase demonstrated that miR-26b-3p reduced the double luciferase activity of WT-*MDM4* but had no significant effect on MUT-*MDM4* (Figure 6c). Overexpression of miR-26b-3p significantly decreased *MDM4* mRNA and protein expression, whereas inhibition of miR-26b-3p significantly increased *MDM4* mRNA and protein expression in gEECs (Figure 6d–f). CircRNA3890 significantly promoted the protein expression level of *MDM4* (Figure 6g–i). In addition, the *MDM4* decreased the protein levels of VEGFA and LIF in gEECs (Figure 6j,k). In summary, *MDM4* was identified as the target gene of miR-26b-3p, and miR-26b-3p negatively regulates the expression of *MDM4* in gEECs.

3.6. The Effects of MDM4 on gEECs

To investigate the regulatory effect of *MDM4* on gEECs, we first constructed the *MDM4* overexpression vector (pc3.1-*MDM4*), and small interfering RNAs (siRNAs) of *MDM4* were synthesized by GenePharma (Shanghai, China). Meanwhile, the RT-qPCR and WB assays verified that the overexpression vector (pc3.1-*MDM4*) and interference (si-*MDM4*) of *MDM4* could be used in subsequent experiments (Figure S4). The CCK8 assay showed that *MDM4* enhanced the cell vitality of gEECs, and miR-26b-3p could reduce the effect of *MDM4* on cell vitality (Figure 7a,d). The EdU assay revealed that *MDM4* increased the number of positive cells and promoted the proliferation of gEECs. Moreover, miR-26b-3p could reduce the effect of *MDM4* on cell proliferation (Figure 7b,c,e,f). The WB assay showed that overexpression of *MDM4* significantly decreased the protein levels of BAX and significantly increased the protein levels of BCL2, while si-*MDM4* decreased the protein levels of BCL2 and significantly increased the protein levels of BAX, in gEECs in vitro (Figure 7g,i). Furthermore, miR-26b-3p was found to reduce the effect of *MDM4* on these proteins in gEECs in vitro (Figure 7h,j). In addition, the apoptosis results showed that *MDM4* inhibited EEC apoptosis (Figure 7k,l). The cell cycle assay showed that *MDM4* increased the number of gEECs in the S phase and reduced the proportion of gEECs in the G1 phase, while miR-26b-3p could reduce the effect of *MDM4* on the cell cycle (Figures 7m,n and S5). These results suggested that *MDM4* promoted gEECs proliferation, inhibited apoptosis, and was negatively regulated by miR-26b-3p. They also indicated that MDM4 participates in the establishment of a receptive endometrium.

https://doi.org/10.3390/ani16050696

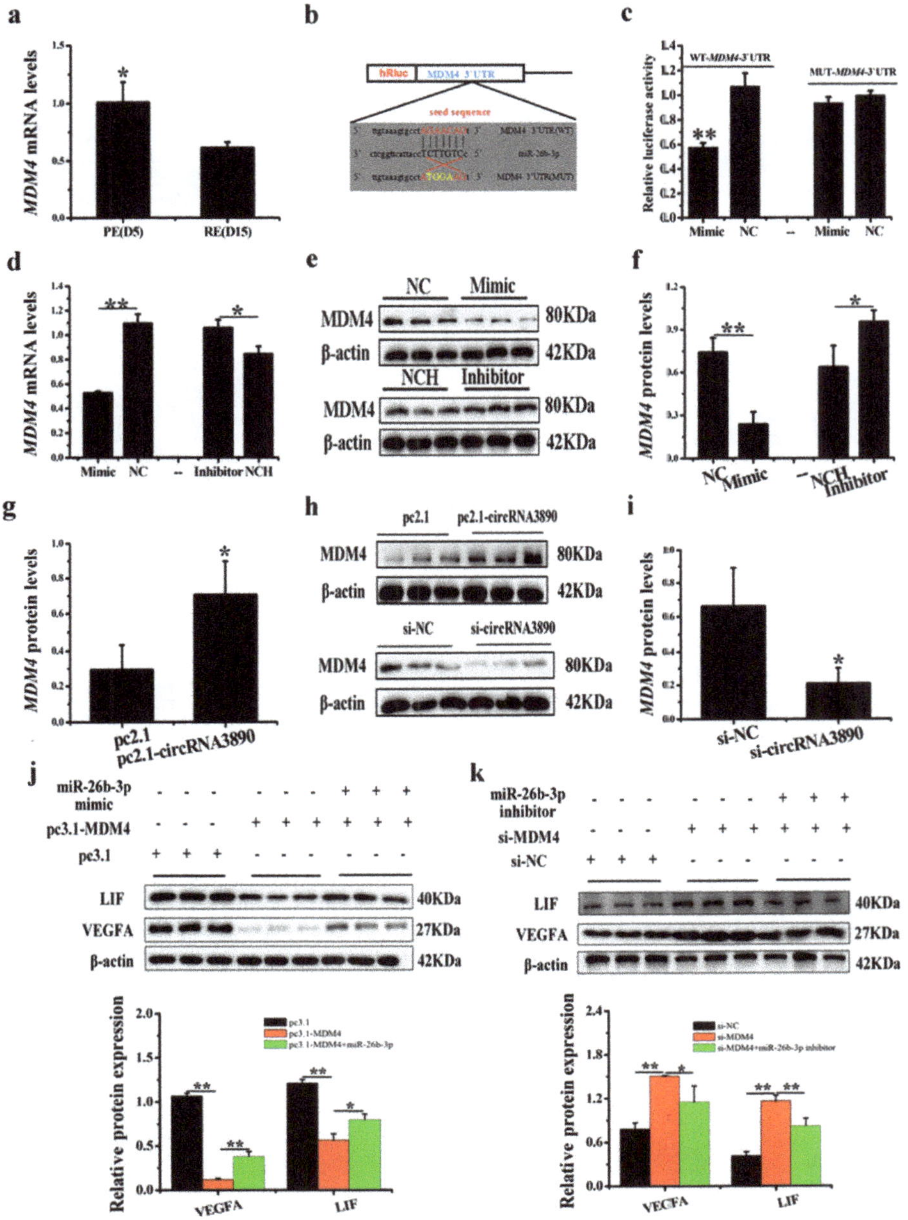

Figure 6. *MDM4* is the target of miR-26b-3p. (**a**) *MDM4* mRNA levels in PE and RE of dairy goat. (**b**) Luciferase reporter genes with binding sites (WT-*MDM4*-3′UTR) and binding site mutations (MUT-*MDM4*-3′UTR) are shown in a schematic design. The red-labeled nucleotides represent the "seed sequence" of miR-26b-3p, while the blue ones represent mutated nucleotides. (**c**) Double luciferase activity was analyzed after co-transfecting 293T cells with WT-*MDM4*-3′UTR, MUT-*MDM4*-3′UTR, or miR-26b-3p, NC (negative control). (**d–f**) The mRNA and protein expression levels of *MDM4* in gEECs after transfection with miR-26b-3p mimic/inhibitor. (**g–i**) The regulatory effect of circRNA3890 on *MDM4* protein levels was detected using Western blotting in gEECs. (**j**,**k**) Effects of *MDM4* overexpression or interference on LIF and VEGFA protein expression in gEECs. Each result is shown as the mean ± SEM of three values; ** $p < 0.01$; * $p < 0.05$.

https://doi.org/10.3390/ani16050696

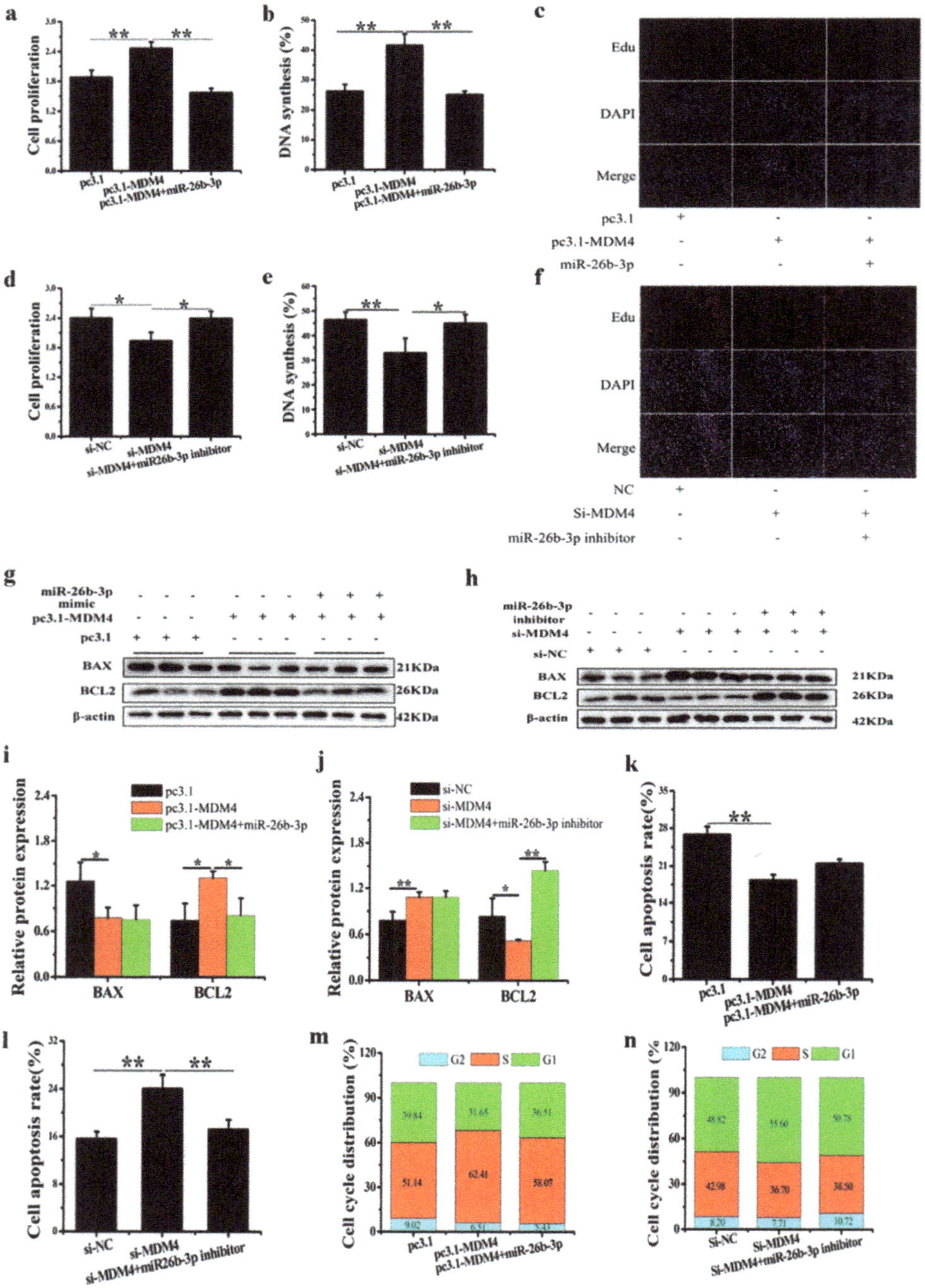

Figure 7. Regulatory effects of *MDM4* on gEECs in vitro. (**a**–**c**) The effects of *MDM4* overexpression on gEECs proliferation were analyzed using EdU and CCK-8. The scale bar represents 100 µm. (**d**–**f**) The effects of *MDM4* interference on the proliferation of gEECs were analyzed by CCK8 and EdU. The scale bar represents 100 µm. (**g**–**j**) The protein expression levels of BCL2 and BAX were analyzed using Western blotting after overexpression or interference of *MDM4* in gEECs. (**k**–**n**) The effects of overexpression or interference of *MDM4* on apoptosis and cell cycle phases were analyzed using FCM. Each result is shown as the mean ± SEM of three values: ** $p < 0.01$; * $p < 0.05$.

https://doi.org/10.3390/ani16050696

3.7. circRNA3890/miR-26b-3p/MDM4 Regulates the P53/P21 Signal Pathway in gEECs

In addition, many experiments have demonstrated that *MDM4* is a negative regulator of p53 [34,35]. Therefore, we speculated that circRNA3890/miR-26b-3p/*MDM4* may regulate the proliferation and apoptosis of gEECs through the P53/P21 signaling pathway. The WB assay showed that *MDM4* inhibited the protein levels of P53 and P21, and promoted the protein levels of CDK2 and cyclinE1 (Figure 8a–d). In addition, miR-26b-3p attenuated the regulatory effect of *MDM4* on the above proteins (Figure 8e–h). The regulation of these proteins by miR-26b-3p was opposite to that of *MDM4*. The WB assay also showed that circRNA3890 inhibited the protein levels of P53 and P21, and promoted the protein levels of CDK2 and cyclinE1 (Figure 8i–l). The above study results indicated that circRNA3890/miR-26b-3p/*MDM4* regulated the survival rate of gEECs through the P53/P21 signaling pathway.

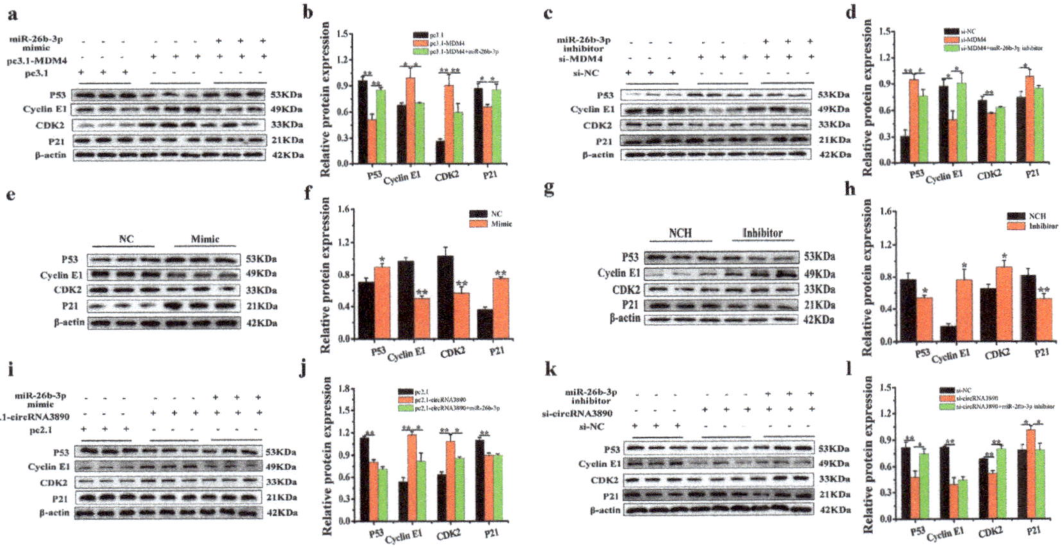

Figure 8. Effects of circRNA3890/miR-26b-3p/*MDM4* axis on key P53/P21 signaling pathway proteins. (**a–d**) Effects of *MDM4* overexpression or interference on P21, P53, CDK2, and cyclinE1 protein expression in gEECs. (**e–h**) Effects of miR-26b-3p on P21, P53, CDK2, and cyclinE1 protein expression in gEECs. (**i–l**) Protein expression levels of P21, P53, CDK2 and cyclinE1 after transfection with pc2.1-circRNA3890/si-circRNA3890. Each result is shown as the mean ± SEM of three values: ** $p < 0.01$; * $p < 0.05$.

3.8. miR-26b-3p Promotes Embryonic Engraftment in Mice

To verify the regulatory effect of miR-26b-3p on endometrial receptivity and embryo implantation in vivo, we injected miR-26b-3p-Agomir and NC-Agomir on both sides of mouse uterine horns on day 3 of gestation. The RT-qPCR assay showed that the expression of miR-26b-3p was up-regulated in the uterine horn on the side injected with miR-26b-3p-Agomir, indicating that we successfully constructed a mouse model (Figure 9a). The SEM results showed that miR-26b-3p promotes the development of the uterine pinopode process (Figure 9b). The RT-qPCR and immunohistochemical results showed that miR-26b-3p up-regulated the expression levels of VEGFA and LIF (Figure 9c,g,h). The results of the dissection test on day 9 of pregnancy showed that miR-26b-3p increased the embryo implantation rate (Figure 9d,e). The HE staining showed that miR-26b-3p had no significant effect on endometrial thickness (Figure 9f). The above results indicated that miR-26b-3p

https://doi.org/10.3390/ani16050696

promoted the establishment of mouse endometrial receptivity and increased the embryonic replication rate.

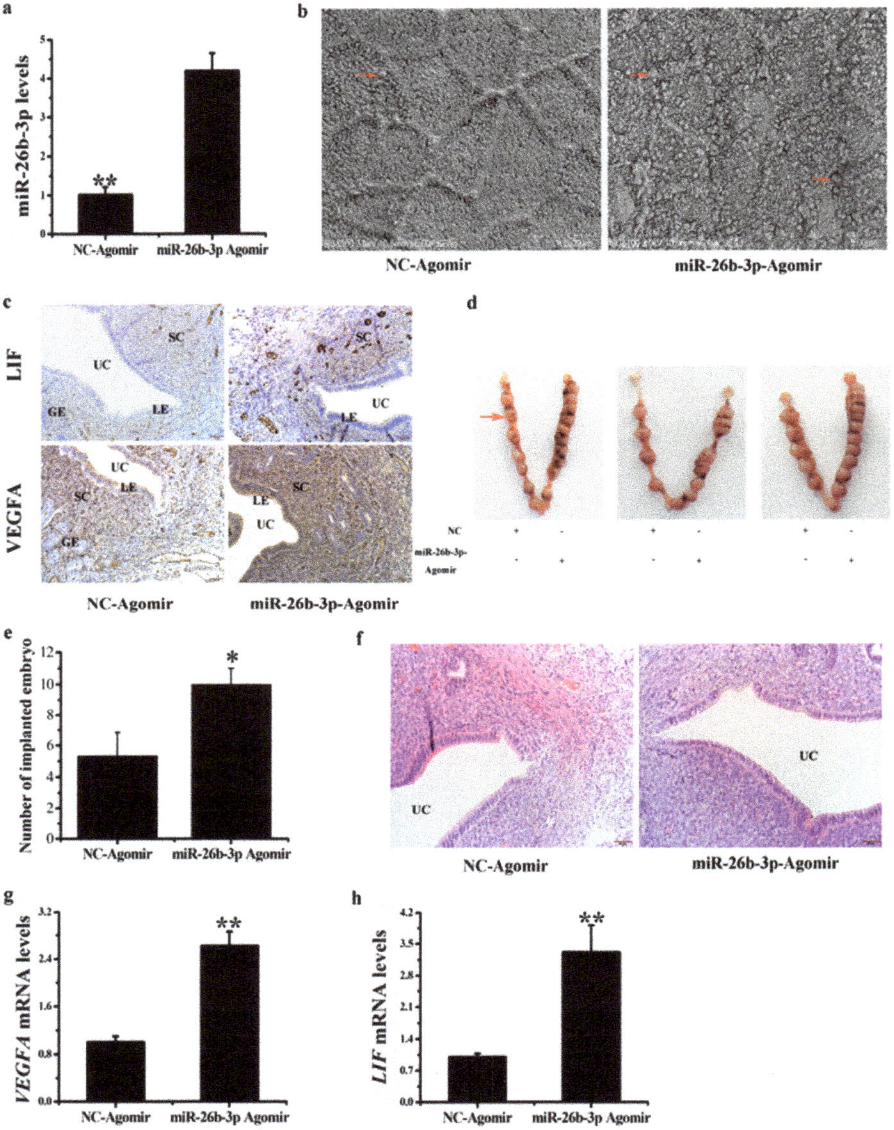

Figure 9. MiR-26b-3p promotes mouse embryo attachment. (**a**) The mRNA expression level of miR-26b-3p was detected using RT-qPCR. (**b**) The effect of miR-26b-3p agomir on endometrium pinopodes was observed using scanning electron microscopy. The scale bar represents 2 µm. The red arrows indicate annotations for pinopodes. (**c**) Immunohistochemical analysis of VEGFA and LIF expression in the endometrium. The red scale represents 50 µm. (**d**,**e**) miR-26b-3p promotes the implantation of mouse embryos. (**f**) The thickness of the endometrium was observed by HE staining. The scale bar represents 50 µm. (**g**,**h**) RT-qPCR was used to detect mRNA expression levels of receptive endometrial marker genes. UC: uterine cavity, LE: luminal epithelial cells, GE: glandular epithelial cells, SC: stromal cells. Each result is shown as the mean ± SEM of three values: ** $p < 0.01$; * $p < 0.05$.

https://doi.org/10.3390/ani16050696

4. Discussion

Recently, miRNAs and long non-coding RNAs (lncRNA) have been reported to play important regulatory roles during embryo implantation [36]. In addition, many studies have shown that miRNA and lncRNA can participate in the development of endometrial receptivity [37]. However, the role of circRNA in the establishment of endometrial receptivity remains to be further studied. Our previous study identified a circRNA 3890 associated with endometrial receptivity. CircRNA3890 acts as a ceRNA to adsorb miR-26b-3p to relieve the inhibitory effect on *MDM4*, activate the P53 signaling pathway, and regulate the establishment of receptive endometrium.

Currently, most studies on miRNAs focus on tumor molecular markers [38], and few study the role of RE. In this study, the expression level of miR-26b-3p in RE was significantly higher than that in PE, and the expression level in the uterus was relatively high. Therefore, we speculated that miR-26b-3p might be involved in the establishment of RE in dairy goats. Similar to miRNAs, miR-26b-3p can serve as a molecular therapeutic target for cancer or tumors. For example, miR-26b-3p regulates glioma cell migration and proliferation, as well as apoptosis [39]. Endometrial cells undergo proliferation and apoptosis during the estrus cycle of humans and most mammals. Studies have shown that EEC apoptosis is important for embryos to break through the barriers of the uterine surface and implant in the uterus [4,40]. The results of this experiment show that miR-26b-3p promoted apoptosis and inhibited proliferation of gEECs, increased the proportion of gEECs in the G1 phase, and reduced the number of gEECs in the S phase. These findings suggest a critical role for miR-26b-3p in the establishment of receptive endometrium.

Most miRNAs are known to participate in post-transcriptional regulation by inhibiting the expression of their target genes. For example, miRNA-143-3p regulates the differentiation of human stem cells through its target gene NFIC [41], miR-335 regulates the progression of corneal neovascularization by targeting EGFR [42]. In this study, we validated that *MDM4* is the target gene of miR-26b-3p by a dual luciferase assay. In addition, the results of this study showed that miR-26b-3p negatively regulates the expression of *MDM4* in gEECs. We speculated that miR-26b-3p might play a role in gEECs through its target gene, *MDM4*. Furthermore, *MDM4* was observed to inhibit apoptosis and promote the proliferation of gEECs, increase the number of gEECs in the S phase and reduce the proportion of gEECs in the G1 phase, while miR-26b-3p could restore the regulatory effect of *MDM4* on gEECs.

CircRNA as a new ceRNA, is mainly involved in post-transcriptional regulation as a molecular sponge for miRNA. For example, circRNA RSF1 targets miR-135b-5p to regulate the proliferation, apoptosis, and inflammation of vascular endothelial cells [43]. circ-CCNB1 acts as a ceRNA to adsorb miR-223 to regulate the proliferation and invasion of trophoblast cell, thereby affecting embryo implantation [44]. In this study, we found that circRNA3890 was significantly higher in the early endometrial receptive phase than in the receptive phase, and combined with bioinformatics tools, the analysis results showed that circRNA3890 has a targeted binding sequence to miR-26b-3p. Luciferase reporter gene and RT-qPCR analysis showed that circRNA3890 targeted down-regulation of miR-26b-3p. Our research suggests that the overexpression of circRNA3890 has a considerable impact on the suppression of miR-26b-3p expression, whereas the overexpression of linear MARCH6 mRNA has no effect. The immunofluorescence assay showed that circRNA3890 was mainly localized in the cytoplasm, which increased the possibility of binding to miRNAs. Our results indicate that miR-26b-3p promotes apoptosis in gEECs, and circRNA3890 may participate in regulating gEEC apoptosis by modulating miR-26b-3p expression. Furthermore, circRNA3890 was observed to act as a regulator of cell proliferation and apoptosis by sponge miR-26b-3p in

https://doi.org/10.3390/ani16050696

gEECs. These findings suggest that the circRNA3890–miR-26b-3p–*MDM4* axis is involved in regulating the proliferation and apoptosis of gEECs.

MDM4 is a negative regulator of tumor suppressor gene P53. In addition, the P53/P21 signaling pathway is an important signaling cascade regulator of cell survival, proliferation, apoptosis, and the cell cycle [45]. For example, HAUS6 regulates the proliferation, apoptosis, and cycle of colon cancer cells by activating the p53/p21 signaling pathway. The P53/P21 pathway is a crucial mechanism in cancer biology, playing a central role in regulating the cell cycle and apoptosis [46]. P21 is one of the most important target genes of P53 [46]. During the endometrial preparation for embryo implantation, moderate activity of the P53/P21 pathway may coordinate cell proliferation and differentiation, ensuring the endometrium reaches optimal condition during the implantation window [47]. Combining the above findings, we speculated that the circRNA3890–miR-26b-3p–*MDM4* axis regulated gEECs through the p53/p21 signaling pathway. Our findings showed that miR-26b-3p targets *MDM4* to regulate gEECs by activating the P53/P21 signaling pathway, and circRNA3890 attenuated the regulatory effect of miR-26b-3p on the p53/p21 signaling pathway. In summary, these findings further confirmed that the circRNA3890–miR-26b-3p–*MDM4* axis exerts regulatory roles in cell cycling, proliferation, and apoptosis.

p53 is known to be involved in the establishment of endometrial receptivity by regulating LIF and plays an important role in maternal reproduction [48]. Moreover, p53 regulates the protein expression levels of VEGFA [49]. We speculated that the circRNA3890–miR-26b-3p–*MDM4* axis is involved in the establishment of endometrial receptivity in dairy goats through P53. To test this hypothesis, we also examined the effect of the circRNA3890–miR-26b-3p–*MDM4* axis on the protein expression levels of the receptivity marker genes VEGFA and LIF. Our results showed that miR-26b-3p promoted the protein expression levels of VEGFA and LIF. Meanwhile, miR-26b-3p reduced the effect of *MDM4* on these proteins. circRNA3890 inhibited the regulation of miR-26b-3p on these proteins. Therefore, these findings suggest that the circRNA3890–miR-26b-3p–*MDM4* axis may be involved in regulating the establishment of endometrial receptivity in dairy goats.

Pinopodes are a good marker for evaluating endometrial receptivity [50,51]. The results of this study in vivo showed that miR-26b-3p significantly promoted the development of mouse pinopodes. The immunohistochemistry and RT-qPCR results showed that miR-26b-3p promoted the expression of endometrial receptivity marker genes such as VEGFA and LIF in mouse endometrium. In addition, miR-26b-3p promoted embryo implantation into the endometrium in mice. These findings suggest that miR-26b-3p promoted the establishment of endometrial receptivity and promoted embryo implantation in mice.

5. Conclusions

This study found that circRNA3890 sponges miR-26b-3p to relieve the inhibitory effect of the target gene *MDM4* and promoted the proliferation of gEECs through the P53/P21 signaling pathway (Figure 10). In addition, the circRNA3890–miR-26b-3p–*MDM4* axis plays an important role in the establishment of receptive endometrium, which may be a potential target for regulating embryo implantation.

https://doi.org/10.3390/ani16050696

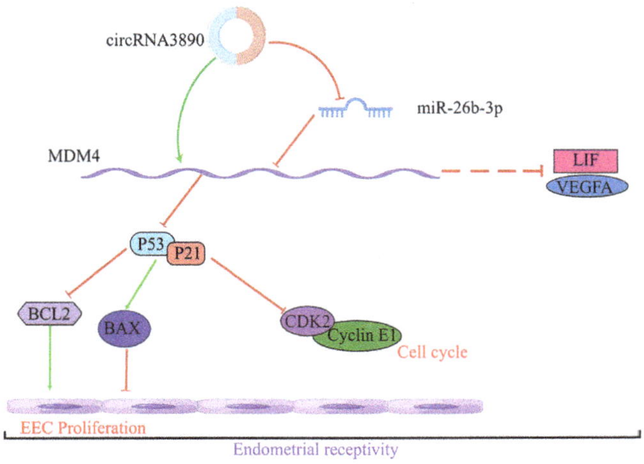

Figure 10. Model of circRNA3890 regulatory mechanism on gEECs. In this model, circRNA3890 regulates gEECs by functioning as a ceRNA for miR-26b-3p and shows a circRNA–miRNA–mRNA network, thereby exerting biological effects.

Supplementary Materials: The following supporting information can be downloaded at https://www.mdpi.com/article/10.3390/ani16050696/s1. Figure S1: Overexpression or inhibition of circRNA3890 and miR-26b-3p; Figure S2: circRNA3890 inhibited gEECs apoptosis; Figure S3: MiR-26b-3p induces gEECs apoptosis; Figure S4: Overexpression or inhibition of *MDM4* transfection efficiency in gEECs; Figure S5: *MDM4* inhibited gEECs apoptosis.

Author Contributions: J.C. conceived and designed the experiments; L.L., J.C. and S.H. performed the experiments; Y.S. (Yonggan Su) and L.L. analyzed the data; L.Z. and Y.S. (Yuxuan Song) contributed reagents/materials/analysis tools; J.C. wrote the manuscript. All authors have read and agreed to the published version of the manuscript.

Funding: This study was supported by Shaanxi Livestock and Poultry Breeding Double-chain Fusion Key Project (2022GD-TSLD-46-0202), Science and Technology Innovation 2030—Major Project Funding (2022ZD0401403), the project of central government's funds guiding local scientific and technological development in Shaanxi Province (2024ZY-CGZY-20), and basic research on free exploration in Shaanxi Province (2024ZY-JCYJ-02-06). The funders had no role in study design, data collection and analysis, decision to publish, or manuscript preparation.

Institutional Review Board Statement: All the animal use and care protocols were confirmed in accordance with the No. 5 proclamation of the Ministry of Agriculture, P. R. China. The animal protocols were approved by the Review Committee for the Use of Animal Subjects of Northwest A&F University (Approval No. DK20211015).

Informed Consent Statement: Informed consent was obtained from all subjects involved in the study.

Data Availability Statement: The data that support the findings of this study are available from the corresponding author upon reasonable request.

Acknowledgments: We are grateful to express our gratitude to Liu Xiaorui for her assistance in the SEM experiment.

Conflicts of Interest: The authors declare no conflicts of interest.

References

1. Hu, S.J.; Ren, G.; Liu, J.L.; Zhao, Z.A.; Yu, Y.S.; Su, R.W.; Ma, X.H.; Ni, H.; Lei, W.; Yang, Z.M. MicroRNA expression and regulation in mouse uterus during embryo implantation. *J. Biol. Chem.* **2008**, *283*, 23473–23484. [CrossRef] [PubMed]
2. Achache, H.; Revel, A. Endometrial receptivity markers, the journey to successful embryo implantation. *Hum. Reprod. Update* **2006**, *12*, 731–746. [CrossRef] [PubMed]
3. Gao, X.Y.; Zhang, Y.; Zhao, W.P.; Tian, E.J.; Ommati, M.M.; Wang, J.C.; Wang, H.W.; Zhou, B.H. Molybdenum interferes with MMPs/TIMPs expression to reduce the receptivity of porcine endometrial epithelial cells. *Chem. Biol. Interact.* **2025**, *405*, 111304. [CrossRef]
4. Liu, X.; Zhang, L.; Liu, Y.; Cui, J.; Che, S.; An, X.; Song, Y.; Cao, B. Circ-8073 regulates CEP55 by sponging miR-449a to promote caprine endometrial epithelial cells proliferation via the PI3K/AKT/mTOR pathway. *Biochim. Biophys. Acta Mol. Cell Res.* **2018**, *1865*, 1130–1147. [CrossRef]
5. Zhang, L.; An, X.P.; Liu, X.R.; Fu, M.Z.; Han, P.; Peng, J.Y.; Hou, J.X.; Zhou, Z.Q.; Cao, B.Y.; Song, Y.X. Characterization of the Transcriptional Complexity of the Receptive and Pre-receptive Endometria of Dairy Goats. *Sci. Rep.* **2015**, *5*, 14244. [CrossRef]
6. Margalioth, E.J.; Ben-Chetrit, A.; Gal, M.; Eldar-Geva, T. Investigation and treatment of repeated implantation failure following IVF–ET. *Hum. Reprod.* **2006**, *21*, 3036–3043. [CrossRef] [PubMed]
7. Makker, A.; Goel, M.M. Uterine leiomyomas: Effects on architectural, cellular, and molecular determinants of endometrial receptivity. *Reprod. Sci.* **2013**, *20*, 631–638. [CrossRef]
8. Thum, T.; Condorelli, G. Long noncoding RNAs and microRNAs in cardiovascular pathophysiology. *Circ. Res.* **2015**, *116*, 751–762. [CrossRef]
9. Wang, W.; Huang, L.; Lv, J.; Miao, Z.; Jin, S.; Li, S.; Cheng, Q. Silencing circRNA-ZFAND6 induces trophoblast apoptosis by activating the mitochondrial pathway through the miR-575/SOD_2 axis in unexplained recurrent spontaneous abortion. *BMC Women's Health* **2025**, *25*, 164. [CrossRef]
10. Huo, Y.; Yuan, D.; Liu, H.; Song, Y. Fusion circRNA F-circEA1 facilitates EML4-ALK1 positive lung adenocarcinoma progression through the miR-4673/SMAD4/ADAR1 axis. *Cell. Signal.* **2025**, *127*, 111571. [CrossRef]
11. Yuan, H.; Chen, C.; Li, H.; Qu, G.; Chen, L.; Liu, Y.; Zhang, Y.; Zhao, Q.; Lian, C.; Ji, A.; et al. Role of a novel circRNA-CGNL1 in regulating pancreatic cancer progression via NUDT4-HDAC4-RUNX2-GAMT-mediated apoptosis. *Mol. Cancer* **2024**, *23*, 27. [CrossRef] [PubMed]
12. Zou, H.; Deng, W.; Xu, L.; Shi, M.; Liu, L.; Gong, L.; Cui, D.; Zhang, F. CircRNA-0013747 Promotes Mesangial Cell Proliferation in Immunoglobulin a Nephropathy through Modulation of the Warburg Effect. *J. Physiol. Investig.* **2025**, *68*, 43–56. [CrossRef]
13. Ha, M.; Kim, V.N. Regulation of microRNA biogenesis. *Nat. Rev. Mol. Cell Biol.* **2014**, *15*, 509–524. [CrossRef] [PubMed]
14. Cheng, A.M.; Byrom, M.W.; Shelton, J.; Ford, L.P. Antisense inhibition of human miRNAs and indications for an involvement of miRNA in cell growth and apoptosis. *Nucleic Acids Res.* **2005**, *33*, 1290–1297. [CrossRef]
15. Yang, G.; Li, M.; Yang, Q.; Zhai, X.; Halima, J.; Hu, Q.; Lei, C.; Dang, R. Bta-miR-127 inhibits secretion, proliferation and promotes apoptosis by targeting ITGA6 in bovine Sertoli cell. *Int. J. Biol. Macromol.* **2023**, *253*, 126838. [CrossRef]
16. Yao, K.; Kang, Q.; Chen, K.; Shi, B.; Jin, X. MiR-124-3p negatively impacts embryo implantation via suppressing uterine receptivity formation and embryo development. *Reprod. Biol. Endocrinol.* **2024**, *22*, 16. [CrossRef]
17. Yang, L.; Xu, H.R.; Zhang, X.; Shi, Y.; Shi, J.X.; Chen, Q.Q.; Shen, X.R.; He, Y.P.; Tang, J.N.; Gu, W.W.; et al. Increased miR-3074-5p expression promotes M1 polarization and pyroptosis of macrophages via ERalpha/NLRP3 pathway and induces adverse pregnancy outcomes in mice. *Cell Death Discov.* **2024**, *10*, 171. [CrossRef]
18. Inyawilert, W.; Fu, T.Y.; Lin, C.T.; Tang, P.C. Let-7-mediated suppression of mucin 1 expression in the mouse uterus during embryo implantation. *J. Reprod. Dev.* **2015**, *61*, 138–144. [CrossRef]
19. Song, Y.; An, X.; Zhang, L.; Fu, M.; Peng, J.; Han, P.; Hou, J.; Zhou, Z.; Cao, B. Identification and profiling of microRNAs in goat endometrium during embryo implantation. *PLoS ONE* **2015**, *10*, e0122202. [CrossRef]
20. Shvarts, A.; Steegenga, W.T.; Riteco, N.; van Laar, T.; Dekker, P.; Bazuine, M.; van Ham, R.C.; van der Houven van Oordt, W.; Hateboer, G.; van der Eb, A.J.; et al. MDMX: A novel p53-binding protein with some functional properties of MDM2. *EMBO J.* **1996**, *15*, 5349–5357. [CrossRef] [PubMed]
21. Soliman, A.M.; Kodous, A.S.; Al-Sherif, D.A.; Ghorab, M.M. Quinazoline sulfonamide derivatives targeting MicroRNA-34a/*MDM4*/p53 apoptotic axis with radiosensitizing activity. *Future Med. Chem.* **2024**, *16*, 929–948. [CrossRef]
22. Krivec, N.; Couvreu de Deckersberg, E.; Lei, Y.; Al Delbany, D.; Regin, M.; Verhulst, S.; van Grunsven, L.A.; Sermon, K.; Spits, C. Gain of 1q confers an *MDM4*-driven growth advantage to undifferentiated and differentiating hESC while altering their differentiation capacity. *Cell Death Dis.* **2024**, *15*, 852. [CrossRef]
23. Zhang, L.; Liu, X.; Liu, J.; Ma, X.; Zhou, Z.; Song, Y.; Cao, B. miR-26a promoted endometrial epithelium cells (EECs) proliferation and induced stromal cells (ESCs) apoptosis via the PTEN-PI3K/AKT pathway in dairy goats. *J. Cell Physiol.* **2018**, *233*, 4688–4706. [CrossRef] [PubMed]

https://doi.org/10.3390/ani16050696

24. Zhang, L.; Liu, X.; Liu, J.; Zhou, Z.; Song, Y.; Cao, B.; An, X. miR-182 aids in receptive endometrium development in dairy goats by down-regulating PTN expression. *PLoS ONE* **2017**, *12*, e0179783. [CrossRef] [PubMed]
25. An, X.; Liu, X.; Zhang, L.; Liu, J.; Zhao, X.; Chen, K.; Ma, H.; Li, G.; Cao, B.; Song, Y. MiR-449a regulates caprine endometrial stromal cell apoptosis and endometrial receptivity. *Sci. Rep.* **2017**, *7*, 12248. [CrossRef]
26. Liu, X.; Zhang, L.; Yang, L.; Cui, J.; Che, S.; Liu, Y.; Han, J.; An, X.; Cao, B.; Song, Y. miR-34a/c induce caprine endometrial epithelial cell apoptosis by regulating circ-8073/CEP55 via the RAS/RAF/MEK/ERK and PI3K/AKT/mTOR pathways. *J. Cell Physiol.* **2020**, *235*, 10051–10067. [CrossRef] [PubMed]
27. Peng, S.; Song, C.; Li, H.; Cao, X.; Ma, Y.; Wang, X.; Huang, Y.; Lan, X.; Lei, C.; Chaogetu, B.; et al. Circular RNA SNX29 Sponges miR-744 to Regulate Proliferation and Differentiation of Myoblasts by Activating the Wnt5a/Ca^{2+} Signaling Pathway. *Mol. Ther. Nucleic Acids* **2019**, *16*, 481–493. [CrossRef]
28. Zhu, K.; Hu, X.; Chen, H.; Li, F.; Yin, N.; Liu, A.L.; Shan, K.; Qin, Y.W.; Huang, X.; Chang, Q.; et al. Downregulation of circRNA DMNT3B contributes to diabetic retinal vascular dysfunction through targeting miR-20b-5p and BAMBI. *EBioMedicine* **2019**, *49*, 341–353. [CrossRef]
29. Ma, L.; Zhang, M.; Cao, F.; Han, J.; Han, P.; Wu, Y.; Deng, R.; Zhang, G.; An, X.; Zhang, L.; et al. Effect of MiR-100-5p on proliferation and apoptosis of goat endometrial stromal cell in vitro and embryo implantation in vivo. *J. Cell Mol. Med.* **2022**, *26*, 2543–2556. [CrossRef]
30. Song, Y.; Zhang, L.; Liu, X.; Niu, M.; Cui, J.; Che, S.; Liu, Y.; An, X.; Cao, B. Analyses of circRNA profiling during the development from pre-receptive to receptive phases in the goat endometrium. *J. Anim. Sci. Biotechnol.* **2019**, *10*, 34. [CrossRef]
31. Miravet-Valenciano, J.A.; Rincon-Bertolin, A.; Vilella, F.; Simon, C. Understanding and improving endometrial receptivity. *Curr. Opin. Obstet. Gyn* **2015**, *27*, 187–192. [CrossRef] [PubMed]
32. Cullinan, E.B.; Abbondanzo, S.J.; Anderson, P.S.; Pollard, J.W.; Lessey, B.A.; Stewart, C.L. Leukemia inhibitory factor (LIF) and LIF receptor expression in human endometrium suggests a potential autocrine/paracrine function in regulating embryo implantation. *Proc. Natl. Acad. Sci. USA* **1996**, *93*, 3115–3120. [CrossRef]
33. Parr, E.L.; Tung, H.N.; Parr, M.B. Apoptosis as the mode of uterine epithelial cell death during embryo implantation in mice and rats. *Biol. Reprod.* **1987**, *36*, 211–225. [CrossRef] [PubMed]
34. Zhang, S.Y.; Lou, J.F.; Li, Y.F.; Zhou, F.L.; Yan, Z.Q.; Lyu, X.L.; Zhao, Y.J. Recent Progress and Clinical Development of Inhibitors that Block MDM4/p53 Protein-Protein Interactions. *J. Med. Chem.* **2021**, *64*, 10621–10640. [CrossRef]
35. Liu, Y.; Wang, X.H.; Wang, G.; Yang, Y.S.; Yuan, Y.; Ouyang, L. The past, present and future of potential small-molecule drugs targeting p53-MDM2/MDMX for cancer therapy. *Eur. J. Med. Chem.* **2019**, *176*, 92–104. [CrossRef]
36. Adu-Gyamfi, E.A.; Cheeran, E.A.; Salamah, J.; Enabulele, D.B.; Tahir, A.; Lee, B.K. Long non-coding RNAs: A summary of their roles in placenta development and pathologydagger. *Biol. Reprod.* **2024**, *110*, 431–449. [CrossRef]
37. Zhang, L.; Liu, X.; Cui, J.; Che, S.; Liu, Y.; An, X.; Cao, B.; Song, Y. LncRNA882 regulates leukemia inhibitory factor (LIF) by sponging miR-15b in the endometrial epithelium cells of dairy goat. *J. Cell Physiol.* **2019**, *234*, 4754–4767. [CrossRef]
38. Lee, Y.S.; Dutta, A. MicroRNAs in cancer. *Annu. Rev. Pathol.* **2009**, *4*, 199–227. [CrossRef] [PubMed]
39. Geng, F.; Lu, G.F.; Ji, M.H.; Kong, D.Y.; Wang, S.Y.; Tian, H.; Xie, Z.M.; Pan, M.; Gong, N.L. MicroRNA-26b-3p/ANTXR1 signaling modulates proliferation, migration, and apoptosis of glioma. *Am. J. Transl. Res.* **2019**, *11*, 7568–7578.
40. Tranguch, S.; Dey, S.K. A lifetime of deciphering complexities of embryo implantation. *Int. J. Dev. Biol.* **2014**, *58*, 79–86. [CrossRef]
41. Gao, S.; Ge, L.H.; Zhao, Y.M.; Li, P.; Li, Y.Y.; Zhao, W. Hsa-miRNA-143-3p regulates the odontogenic differentiation of human stem cells from the apical papilla by targeting NFIC. *Int. Endod. J.* **2022**, *55*, 263–274. [CrossRef]
42. Qian, J.; Yu, J.; Zhu, X.; Liang, S. MiR-335 promotes corneal neovascularization by Targeting EGFR. *BMC Ophthalmol.* **2022**, *22*, 267. [CrossRef] [PubMed]
43. Zhang, X.H.; Lu, J.Y.; Zhang, Q.H.; Luo, Q.; Liu, B. CircRNA RSF1 regulated ox-LDL induced vascular endothelial cells proliferation, apoptosis and inflammation through modulating miR-135b-5p/HDAC1 axis in atherosclerosis. *Biol. Res.* **2021**, *54*, 11. [CrossRef] [PubMed]
44. Jing, M.Y.; Xie, L.D.; Chen, X.; Zhou, Y.; Jin, M.M.; He, W.H.; Wang, D.M.; Liu, A.X. Circ-CCNB1 Modulates Trophoblast Proliferation and Invasion in Spontaneous Abortion by Regulating miR-223/SIAH1 axis. *Endocrinology* **2022**, *163*, bqac093. [CrossRef] [PubMed]
45. Li, W.; Li, K.; Wang, Z.; Fa, Z. MicroRNA-377-3p promotes cell proliferation and inhibits cell cycle arrest and cell apoptosis in hepatocellular carcinoma by affecting EGR1-mediated p53 activation. *Pathol. Res. Pract.* **2022**, *234*, 153855. [CrossRef]
46. Thangavelu, L.; Altamimi, A.S.A.; Ghaboura, N.; Babu, M.A.; Roopashree, R.; Sharma, P.; Pal, P.; Choudhary, C.; Prasad, G.V.S.; Sinha, A.; et al. Targeting the p53–p21 axis in liver cancer: Linking cellular senescence to tumor suppression and progression. *Pathol. Res. Pract.* **2024**, *263*, 155652. [CrossRef]
47. Palomares, A.R.; Castillo-Dominguez, A.A.; Ruiz-Galdon, M.; Rodriguez-Wallberg, K.A.; Reyes-Engel, A. Genetic variants in the p53 pathway influence implantation and pregnancy maintenance in IVF treatments using donor oocytes. *J. Assist. Reprod. Genet.* **2021**, *38*, 3267–3275. [CrossRef]

https://doi.org/10.3390/ani16050696

48. Hu, W.; Feng, Z.; Teresky, A.K.; Levine, A.J. p53 regulates maternal reproduction through LIF. *Nature* **2007**, *450*, 721–724. [CrossRef]
49. Gutierrez-Seijo, A.; Garcia-Martinez, E.; Barrio-Alonso, C.; Pareja-Malagon, M.; Acosta-Ocampo, A.; Fernandez-Santos, M.E.; Puig-Kroger, A.; Parra-Blanco, V.; Mercader, E.; Marquez-Rodas, I.; et al. CCL20/TNF/VEGFA Cytokine Secretory Phenotype of Tumor-Associated Macrophages Is a Negative Prognostic Factor in Cutaneous Melanoma. *Cancers* **2021**, *13*, 3943. [CrossRef]
50. Quinn, C.E.; Casper, R.F. Pinopodes: A questionable role in endometrial receptivity. *Hum. Reprod. Update* **2009**, *15*, 229–236. [CrossRef]
51. D'Ippolito, S.; Di Nicuolo, F.; Papi, M.; Castellani, R.; Palmieri, V.; Masciullo, V.; Arena, V.; Tersigni, C.; Bernabei, M.; Pontecorvi, A.; et al. Expression of Pinopodes in the Endometrium from Recurrent Pregnancy Loss Women Role of Thrombomodulin and Ezrin. *J. Clin. Med.* **2020**, *9*, 2634. [CrossRef] [PubMed]

https://doi.org/10.3390/ani16050696

MDPI AG
Grosspeteranlage 5
4052 Basel
Switzerland
Tel.: +41 61 683 77 34

MDPI Books Editorial Office
E-mail: books@mdpi.com
www.mdpi.com/books

www.ingramcontent.com/pod-product-compliance
Lightning Source LLC
LaVergne TN
LVHW070503061126
851858LV00011B/283